2nd International Conference on Emerging Trends in Manufacturing, Engines and Modelling (ICEMEM 2019)

IOP Conference Series: Materials Science and Engineering
Volume 810

Mumbai, India
23 - 24 December 2019

ISBN: 978-1-7138-4723-6
ISSN: 1757-8981

Printed from e-media with permission by:

Curran Associates, Inc.
57 Morehouse Lane
Red Hook, NY 12571

proceedings.com

Some format issues inherent in the e-media version may also appear in this print version.

Institute of Physics
Dirac House, Temple Back
Bristol BS1 6BE UK

Phone: 44 1 17 929 7481
Fax: 44 1 17 920 0979

techtracking@iop.org

Additional copies of this publication are available from:

Curran Associates, Inc.
57 Morehouse Lane
Red Hook, NY 12571 USA
Phone: 845-758-0400
Fax: 845-758-2633
Email: curran@proceedings.com
Web: www.proceedings.com

TABLE OF CONTENTS

IOP Conf. Series: Materials Science and Engineering **810** (2020) 011001 doi:10.1088/1757-899X/810/1/011001

Preface

SVKM's NMIMS, Mukesh Patel School of Technology Management and Engineering, Shirpur Campus proudly hosted the second "International Conference on Emerging trends in Manufacturing, Engines and Modelling" from 23rd -24th December, 2019. It was jointly organized by the Department of Mechanical Engineering, MPSTME Shirpur Campus and Mumbai. The inaugural function of the conference was enlightened by the august presence of **Mr. Ashutosh Pandey** - MD and CEO, Mahindra First Choice Wheels as the Chief Guest, **Dr. K.B. Patil** - Director of SVKM's NMIMS, Dhule Campus as the Guest of Honour, **Dr. R S Gaud**- Director MPTP Shirpur Campus, **Dr. Nikhlesh Kumar Sharma**- Director (Engineering) & Chairman ICEMEM 2019, **Dr. K. K. Gupta** Co-Chairman ICEMEM 2019, **Dr. Gowri Sankara Rao**- Dean(SAST) , **Dr Narayan Chandak**- Associate Dean, **Dr Basant Kumar Mohanty** Associate Dean (Research), **Dr Ashwini Deshpande** - Associate Dean (SPTM), **Shri Rahul Dande**- Chief Accountant and Administration, **Dr Rakesh Choudhari**- Head (Mechanical), **Shri Rajeev Dalal**- Deputy Registrar, All HOD's, Faculty members and more than 100 participants. The inaugural function commenced with the customary felicitation of the guests followed by the Saraswati Vandana. Dr. Nikhlesh Kumar Sharma, Director (Engineering Programs), MPSTME presented the welcome note to the guests, invited speakers, delegates and the participants. Dr. K.K. Gupta - Co-Chairman of the Conference briefed about the theme of the conference and presented the salient features of the event and expectations from them. Dr. R.S Gaud, Director, MPTP, Shirpur Campus introduced the audience about the legacy of SVKM's NMIMS and the laurels brought by the campus at various levels. His speech was followed by the address of **Dr. K.B. Patil.** He had been the Vice-Chancellor of North Maharashtra University, Jalgaon. The audience were fortunate to gain his insights on engineering education and employability in general which represented his vast experience in the field of education and academic administration. The last but the most awaited address was delivered by **Mr. Ashutosh Pandey** who had graciously accepted the invitation of the organizing committee to be the Chief Guest of the inaugural function. A civil engineering graduate with an MBA from FMS, Delhi, Mr. Pandey pointed out the need for developing multiple skills. He highlighted the current trends in the job market and discussed about the capabilities sought after by the employers. An important take-away from his address is the fact that technology environment is constantly getting disrupted and therefore everyone has to be on their toes to withstand the corresponding challenges. The address was truly captivating and thought provoking. The ceremony culminated with the vote of thanks proposed by Prof. M. Ramachandran, Convenor ICEMEM 2019 followed by the tea break. The keynote sessions on the first day were divided into two segments.

The first segment was after the tea break, which began with the kenote speech by **Dr. G. Saravana Kumar,** Professor at Department of Engineering Design, IIT Madras. A Ph. D. in Mechanical Engineering from IIT Kanpur, Dr. G. Saravana Kumar delivered a talk on Design for Additive Manufacturing and provided a lot of insights into his current research with case studies involving biomedical applications, construction of homes through additive manufacturing technology etc. The talk was knowledge filled and well received by the audience. The second invited talk in this segment was delivered by **Dr. Santosh K. Sahu**, Professor, Department of Mechanical Engineering, Indian Institute of Technology, Indore. He has received his Master's (2004) and PhD (2009) degrees from Indian Institute of Technology Bombay and Indian Institute of Technology Kharagpur India, respectively. His talk included a presentation on a research work dealing with heat transfer enhancement and pool boiling heat transfer. Quite a few queries were raised after the expert talk which was amicably resolved by the expert. The talk was followed by the lunch. The post lunch segment started with the keynote session by the international speaker, **Ts Dr. Sharizwal Ahmad Sobri**, Researcher, Faculty of Bioengineering & Technology, Universiti Malyasia Kelantan, and Malaysia. Dr. Sharizal does research in Manufacturing Systems Engineering Technology, Wood Science & Engineering, Advanced Machining - Lasers, Sequential Machining, Sustainable Manufacturing and Manufacturing/Technology Management. His research talk was on the delamination factor formula developed by him which is an important consideration in Laser drilling of micro holes. The last but not the least talk of the day was delivered by **Dr. Pradeep Dixit** from IIT Bombay. Dr. Dixit is an Assistant Professor in the Department of Mechanical Engineering at IIT Bombay. He has been Postdoctoral fellow at Georgia-Tech University from 2008-09 and a senior researcher, VTT Technical Research, Center of Finland 2009-2015. A Ph.D from reputed Nanyang Technological University, Singapore, Dr. Dixit addressed the audience on his research on machining micro-channels especially using electrochemical discharge machining. He discussed the various fundamentals of the process and subsequently pointed out the factors that affect the process including

IOP Conf. Series: Materials Science and Engineering **810** (2020) 011001 doi:10.1088/1757-899X/810/1/011001

the surface finish. The talk culminated with active question answer session and then the audience dispersed for the tea break. Around 103 manuscripts were received. Out of which, 86 papers were chosen for presentation after peer review process. These were divided into six plenary sessions. The plenary sessions I, II and III were conducted post the tea break in the three different seminar halls namely Seminar Hall- I, II and III respectively where participants presented their research work. The sessions were chaired by Dr, Tanmay Mukhopadhyay, IIT Kanpur, Dr. Vinod Pare, SGSITS Indore, Dr. Waghmare, IIT Bombay.

The second day began early with the remaining plenary sessions IV, V and VI. These were chaired by Dr. Sharizwal Ahmad Sobri, Universiti Malaysia Kelantan, Dr. Manoj Chouksey and Dr. Sudhir Tiwari from SGSITS Indore. Post tea session began with the first keynote address for the Day-Two by **Prof. Sivakumar Manickam**. Professor Sivakumar Manickam is a Chemical Engineer, specializing in process engineering of nanomaterials especially nano-pharmaceuticals. Presently he is working with University of Nottingham, Malaysia campus and his research group concentrates on the process development of cavitation based reactors towards technologically important nanomaterials. Through his research talk, he was able to captivate the audience by demonstrating the use of cavitation, an undesired mechanical engineering phenomenon, in micro-processing of materials. The talk was quite interesting and insightful. It was followed by a tea break. Second expert talk of the Day-Two was delivered by **Dr. S. Kanagraj**, from IIT Guwahati. Dr. Kanagraj received his Ph.D degree from Indian Institute of Technology Kharagpur in 2004. He did his post-doctorate at University of Aveiro, Portugal. His research is primary focussed on biomedical devices, biomaterials, assistive technology devices, prosthesis and orthosis, and materials characterization. The next talk in this series was presented by **Dr. I.A. Palani**, Associate Professor and Head, Department of Mechanical Engineering from IIT Indore. His areas of interest include smart materials and structures, laser assisted micro-manufacturing, Mechatronics systems design and soft robotics, etc. Through his talk, Dr. Palani demonstrated the working of MEMS devices and their applications. The last address before the lunch was delivered by **Dr. Gaurav Mani Khanal**. Dr. Khanal is a researcher from an Italian University. He currently works at the Department of Electronics Engineering, University of Rome Tor Vergata. His talk was based on his current project titled, 'Hardware design of memristor-based Spiking Neural Networks'. The topic of the talk was slightly different from the core theme but was still able captured the interest of the audience. It was followed by the much deserved lunch break. Post lunch segment was divided into two halls. In one hall, **Dr. Tanmoy Mukhopadhyay** who is currently working as an Assistant Professor at the Aerospace Engineering Department of IIT Kanpur presented his address on numerical modelling. Dr. Tanmoy Mukhopadhyay was a postdoctoral researcher at the Engineering Science Department of the University of Oxford, UK. He completed his PhD from the College of Engineering of Swansea University, UK. His talk was quite impressive and was mainly focussed on modelling theme of the conference. Simultaneously, Skype presentations were delivered by **Dr. Nilkantha Paudel** from Gran Sasso Science Institute, Italy. He is expert in field of data structures and software design. The topic of his talk was based on Internet Of Things and its utility in various applications including mechanical engineering design. The subsequent Skype presentation was delivered by **Dr. Yogendra Mishra**. - Professor in Nanomaterials, Mads Clausen Institute, NanoSYD, University of Southern Denmark, Denmark. He Has been the Group Leader @ Functional Nanomaterials, Institute for Materials Science, Kiel University, Germany from 2009-2011, Alexander von Humboldt Fellow@ Functional Nanomaterials, Institute for Materials Science, Kiel University, Germany from 2003-2008. His topic of the talk was on the nanotechnology and relevant applications. The session was followed by high tea. All the expert talks were quite engaging and knowledge filled. The participants were actively involved and appreciative of the quality of the expert sessions delivered.

Prepared by

Prof. M. Ramachandran
Convenor-ICEMEM 2019
Department of Mechanical Engineering
NMIMS University, Shirpur Campus
Dhule, Maharashtra, India

IOP Conf. Series: Materials Science and Engineering **810** (2020) 011002 doi:10.1088/1757-899X/810/1/011002

Peer review statement

All papers published in this volume of *IOP Conference Series: Materials Science and Engineering* have been peer reviewed through processes administered by the proceedings Editors. Reviews were conducted by expert referees to the professional and scientific standards expected of a proceedings journal published by IOP Publishing.

IOP Conf. Series: Materials Science and Engineering **810** (2020) 012001 doi:10.1088/1757-899X/810/1/012001

A Review of Hybrid polymer Bio-composites

D.V.Lohar[1], Prof. Dr.P.G.Damle[2], Prof. Dr.A.M.Nikalje[3]

1. Research scholar: Dept of Mechanical Engg. SSBT's College of Engineering, Jalgaon, India,
2. Associate professor: Dept of Mechanical Engg, SSBT's College of Engineering, Jalgaon , India
3. Associate professor: Dept of Mechanical Engg., Government College of Engineering, Aurangabad, India

corresponding author's e-mail address:drdinesh2012@rediffmail.com

Abstract. There is a large scope for polymer composites in various fields of allocations like automotive, aerospace, civil construction materials, marine applications and many more. Developing a biodegradable industry based green materials, sustainable technology and optimal processes with low ecological impact is a general societal objective but this remains a considerable challenge to achieve. Various researchers worked for development of green or bio composites which may include use of bio waste materials like coconut shell, palm fruit, egg shell, sugarcane bagasse, banana fibers, bamboo fibers, etc. There has been less progress in the application of combination or hybridization of bio waste materials as reinforcement materials for producing polymeric composites of desired characteristics. In this paper, review of hybrid polymer Bio composites is carried out in order to identify future scope to develop novel hybrid Polymer Green composite (HPGC) for industrial applications.

1. Introduction

Developing a biodegradable industry based green materials, sustainable technology and optimal processes with low ecological impact is a general societal objective, but this remains a considerable challenge to achieve. There has been less progress in the use of bio waste materials as reinforcement materials for polymeric composite. In a composite, one phase is known as the matrix while the other phase is known as the reinforcement and it is distributed within the matrix. Reinforcement is in the form of fibrous or particulate depending upon its characteristics. One of the types of composites is the Polymer-matrix composite (PMCs) which consists of polymer resin as the matrix with fibers or particulate as the reinforcement medium. To overcome the limitations of polymers, like less stiffness and less strength, and to wide up their uses in various engineering fields, different combination of fillers are often added to merge the advantages of their constituent phases. Reinforcement of polymers with hybrid (different combination of fillers) plays a vital role in the improvement of the mechanical and other properties of materials for better performance.

1.1Need of Green or Bio composites: Ecological importance have resulted interest in natural materials; reprocess characteristics and ecological safety are becoming necessity for the introduction of green or Bio materials and products. Polymer based products such as resins in thermoset plastics, are toxic and non-biodegradable. The resins and fibers used in the Bio/ green composites are biodegradable. Green composite are made from plant fibers with natural resins to produce natural composite. The term green composites also referred as bio composites and eco composites. Historical reference shows that in ancient Egypt, people used straw as the filler with the mud based matrix for materials used for walls in building homes. As reinforcement, they made mud bricks with straw and used these bricks to build walls.

1.2 Classification of Green or Bio composites- Green composites are classified on the basis of reinforcement and matrix polymer materials; they are divided into three main types as given below (a)Totally Green/Renewable composites: Matrix and reinforcement fillers are both made from renewable resources. (b)Partially or semi Green/Renewable composites: matrix is obtained from renewable resources and reinforced with a synthetic material. (c)Partially or semi Green/Renewable composites: synthetic matrix is reinforced with natural bio polymers. The classifications are based on the types of strengthening or reinforcements (a) Particulate composites (b) Fibrous composites (c) Hybrid composites

1.3 Fillers (Fibers) used in Green or Bio composites

1. Biofibres/ Natural / may be grouped in two broad categories: Wood fibers and other than wood i.e Non-wood, 2. Natural fibers such as kenaf, jute, flax, hemp, and sisal, especially as an E glass fiber substitute in the automotive industry. 3. The other fibers used are Coir (Coconut), Bamboo, Pineapple, Ramie

1.4 Filler selection: According to their origin or source, Fillers can be groped as either natural or manmade or synthetic fibers. Natural fibers are extracted from different renewable resources, namely animals, vegetable plants and minerals. Synthetic /Man-made fillers have been incorporated into research with regard to their exceptional valuable properties, Glass fibers and carbon fibers are the commonly used synthetic fillers in the hybrid composites.

2.Hybradisation in Composites:

Concept of 'hybridization 'of composites is gaining more attention by the researchers

1. Two or more fillers can be integrated into a common matrix to overcome the shortcomings of the other filler.
2. However, this concept also applies to the combination of two or more polymers and reinforced with one filler
3. Or more fillers and by adding the same filler type again, that has two or more sizes or dimensions

2.1 Parameters controlling characteristics of Hybrid composites

There are three principal parameters, which have significant effect on the properties of the resulting hybrid composite product.

1. Type of the materials used (matrix and filler), which depend mostly on the intended application. For example, it was found that bamboo/glass composites, displayed lower reinforcement than jute/glass hybrid under compressive load.
2. Preparation method which depends on the filler and the matrix under research,
3. Interaction between the fillers and the matrix. Natural fibers are treated by chemical agents in order to improve the interfacial adhesion between these components, to attain superior mechanical properties the interfacial adhesion should be strong chemical reaction or adsorption, which determines the extent of interfacial adhesion, can anchor matrix molecul es to the fiber surface.

2.2 *Factors that affect the development of polymer composites based on natural fibers*

1. Stability of thermal energy 2. Level of humidity 3.Process ability 4.Dispersion of fibers in polymer matrix 5. Adhesive fibre-matrix 6. Natural fiber surface alteration 7.Fiber aspect ratio

2.3 *Polymers from renewable resources can be classified into three major groups*:

1. Natural polymers such as starch and cellulose;
2. Natural monomers Synthetic polymers such as poly lactic acid (PLA);

3. Polymers from microbial fermentation such as polyhydroxybutyrate(PHB).

2.3 Renewable resource Bio polymer matrix types Polymers can be classified into three main groups:

1. Natural polymers including cellulose and starch;
2. Natural monomers Synthetic polymers such as PLA;
3. Microbial fermentation polymers include polyhydroxybutyrate(PHB).

3. Poly (lactic acid) (PLA)

It is the first commercial bio polymer prepared from renewable resources. Wallace Carothers developed PLA in 1932. PLA is derived from 100% renewable agricultural sources such as corn starch, potatoes, sugarcanes or others biomasses. It is recyclable, biodegradable, thermoplastic, aliphatic linear polyester s. PLA is one of the families of aliphatic polyesters usually made from α-hydroxy acids. PLA has properties which generally make possible to use it to replace with petroleum-based plastics in many application areas. It is biocompatible with non-toxic degradation products (at low concentrations), making it a natural choice for many biomedical applications. However, the demerits like high cost, water sensitivity, fast physical aging, inherent brittleness, limited impact resistance and large stiffness and slow crystallization rate confine the industrial applicability of PLA.It is now normally finds applications in a wide range of areas like apparel, food packaging, nappies and wipes,and biodegradable composites.

usual methods such as injection moulding, blow moulding, extrusion and film forming can be used to process polylactic acid (PLA), PLA has a 55-65°CTg and a melting temperature of 150-175°C.PLA's mechanical properties are compatible, or in most cases, even petrochemical polymers, s uch as polypropylene, are of better quality.PLA has therefore paid attention as a commodity polyme r capable of replacing petrochemical polymers such as polypropylene and polyethylene , particularly for single-use packaging applications.

3.1 ROM for the prediction of the Mechanical properties of hybrid polymer composites

The prediction of the mechanical behavior of hybrid materials depend on material parameters, such as: the reinforcements (fibers or particles) mechanical properties, matrix mechanical properties, distribution and dispersion of reinforcements, volume fractions of the reinforcements and test conditions. Rule of mixture (ROM) is, often applied to predict the mechanical behaviors of hybrid materials. There are several models reported in the literature, based on ROM for the prediction of the mechanical properties of hybrid composite materials. As per literature review by [10] ROM shows that the modulus of a single type polymer reinforced with reinforcement (fibers or particles) can be calculated by using Equation (1). The volume fractions of the reinforcement (fibers in most cases) and polymer matrix are also determined, as shown in Equations (2) and (3):

$E_c = E_f V_f + E_m V_m$ ---(1)

$V_m + V_f = 1$ or $V_f = 1 - V_f$ ---(2)

Hence $E_c = E_f V_f + E_m(1 - V_f)$ ----(3)

Where Ec= Modulus of Composite , Ef= Modulus of fibre and Em= Modulus of Matrix
Vf= Volume of Fibre Vm= Volume of Matrix

4. Literature review

4.1 Bamboo-glass fiber reinforced epoxy based hybrid composites:

Polymer composites of Bamboo fiber reinforced have judicious mechanical properties but by mixing of synthetic fibers or by the alkali medium treatment of fiber, it is possible to enhance their properties. Hybrid composites [2] of nine different types of (bamboo and glass fiber) have been manufactured by hand layup method for finding physical and mechanical characteristics. Fiber loading of three types of bamboo with glass fibers was 5:15 wt. %, 10: 10wt. % and 15: 5 wt. % by changing the length of fiber in each fiber loading from 0.5cm to 1.5cm respectively. Among all the composites, maximum tensile strength is 24.41MPa for fiber length of 0.5cm i.e5wt.% bamboo fiber and 15wt.% glass fiber reinforcement respectively. Whereas, 1.5cm fiber length shows maximum modulus among all the manufactured composites. The hardness of the three different samples of hybrid composites for

0.5cmfiber length varies from 13.21Hv to 15.95Hv, for 1cm fiber length 9.75Hv to 18.51Hv and for1.5cm 19.61Hv to 21.25Hv respectively.

4.2. Hybrid Glass/Carbon Fiber Reinforced Epoxy Composites:

Carbon fibers are also known as graphite fibers having diameter of about 5-10 micro meters and composed of carbon atoms. It has properties such as high resistance to chemicals, stiffness, high tensile strength, temperature tolerance and hence suitable for marines, military, aerospace applications. Epoxy resin [3] is used in the preparation the hybrid composite. The composite [3] contains resin of 70 % of its volume and 30 % of glass fiber and carbon fibers are taken in combinations. The hybrid FRP manufactured by using vacuum bags assisted wet layup process and specimens were cut from sheet of thickness 14 mm having rectangular shape. The ACI440.3R and ASTMD 3039 specifications were referred for testing procedure and the test setup consist of universal testing machine having a capacity of 500 KN was used. The maximum stiffness was observed for specimen 01 and compared to specimen 01; Specimen 02 is having low strength and stiffness. Compressive loading shows that specimen 01 has the highest strength and stiffness but the less strength and stiffness was observed in specimen 02.

4.3 Coconut shell, Walnut shells and Rice husk hybrid composites:

The coconut shell powder, walnut shells powder and Rice husk powder are used as reinforcements with bio epoxy resin Grade 3554A and Hardener 3554B to form hybrid composite specimens. The compositions [4] of fibers were 1:1 in each specimen while the resin and hardener composition was 10:1 respectively. Moisture was removed by NaOH Chemical treatment and helps to increase strength of fibers. When water absorption test was conducted then less moisture (4 g) was absorbed by hybrid composite of walnut shell and coconut shell for various application environments. The hybrid of walnut shell and coconut shell composite has more tensile strength compared to other two hybrid composites. The walnut shell and coconut shell hybrid composite has flexural strength of 14.9MPa in wet conditions and in dry condition it was 14.5 Mpa.

4.4 Coir and bagasse fiber hybrid composites

In the research work [5] an attempt has been made to produce epoxy based hybrid composites made of coir and bagasse fibres to analyze and study the effect of process parameters on thermal Analysis. Hand lay-up method was adopted for fabrication of composite panes to the size of 300mm x200mm x10mm with various weight percentage of natural fibers coconut coir(5 wt %) and bagasse (45 wt %). ASTM standard are used for testing mechanical properties such as tensile, impact strength and flexural strength.

4.5 Bamboo fiber with banana and linen fibre hybrid composites

The paper[6] is related with the comprehensive study of 2-
4 mm long bamboo fiber, banana fiber and linen fiber with random orientations of epoxy resin.
Different tests such as IZOD Impact Test and CHARPY Impact Test, FTIR (Fourier Transform Infra-Red) and to determine hardness
Rockwell test were conducted on 10 specimens of bamboo epoxy resin composite(90/10),
bamboo−banana
(90/5/5) epoxy resin composite and bamboo linen (90/5/5) epoxy resin composite .It is tested and proved that bamboo−banana epoxy resin composite indicates better results in the impact test with reference to Izod test numerical values of 4 Joules and Charpy test gives value of
5 Joules while in the FTIR test, the compatibility of fibers with polymers in bamboo−banana epoxy resin composite is the highest, while bamboo
−linen epoxy resin composite shows better results in the Rockwell toughness test with a value of 40 R HN.

4.6 Wood-Plastic hybrid composites (WPC)

Wood-plastic composites are a composite consists of wood fibers or particles and the matrix as polymer. Grounded sawdust or wood flour is used in WPCs. Other natural fibers can be used, like hemp or kenaf, jute may be also used. Injection moulding and extrusion are the manufacturing methods for WPC. The mechanical tests were performed with a Zwick Roell (Z020) testing machine.

IOP Conf. Series: Materials Science and Engineering **810** (2020) 012001 doi:10.1088/1757-899X/810/1/012001

The mechanical properties were tested according to relevant standards: According to standard EN 310 flexural strength was tested, ISO 179 refereed for impact strength, hardness according to EN-1534, and ISO 527-2 was used for tensile strength. Standard EN 323was referred for determination of density. EN 317test has been used for water absorption and swelling of thickness. One composite WF-PP was manufactured without waste filler, Other composites were manufactured, Nevertheless, part of the wood fibers (WF)
have been replaced by building waste. The building waste was mineral wool (MW), plasterboard wast e (PB) and mixed construction and demolition waste (CDW). The composite WF-PP-MW-PB consists of 24% of WF, 20% of MW and 20% of PB. The composite WF-PP-CDW contains 44% of WF and 20% of CDW, besides plastics and additives. WF-PP composite has tensile strength of 23.35Mpa, WF-PP-MW-PB has tensile strength 18.75Mpa, WF-PP-CDW has tensile strength 17.79 Mpa, Flexural modulus for WF composite was 4.44 GPa, WF-PP-MW-PB composite was 3.84 Gpa, WF-PP-CDW composite was 4.04 Gpa.

Table 1 Summery of Hybrid polymer composites

Type of Fibre	Reinforcement used	Method adopted	Testing parameters	Remark	Reference
Coir and Bagasse fiber	Epoxy resin	Hand layup	Tensile, flexural and impact strength	Semi Green composites	[1]
Bamboo, Banana and linen fibre	Polylactic Acid(PLA)	Injection moulding process	Tensile, flexural, Impact and Hardness, SEM analysis	Totally Green composites	[2]
Bamboo/ Glass fibre	Polyester resin	Hand lay-up	Flexural test, Impact test, Hardness	Semi Green composites	[3]
Bamboo/ Glass fibre	Epoxy resin	Hand lay-up	Tensile strength, flexural strength, Modulus	Semi Green composites	[4]
Glass and Carbon fibre	Epoxy resin	Wet layup moulding	Young's modulus, Tensile strength, compressive strength ,Poisson's ratio	Semi Green composites	[5]
Coconut Coir-Bagasse Fibres	Epoxy resin	Hand lay-up	Tensile, flexural, Impact and Hardness	Semi Green composites	[6
Wood-Plastic hybrid composites (WPC)	Phenolics and Epoxy	Injection Moulding, compression moulding	Tensile, flexural, Impact and Hardness, SEM analysis	Semi Green composites	[7]
Marble waste powder, carbon black,Silica	Natural rubber	Not mentioned	Tensile strength, Tear strength, Modulus, Hardness, Swelling ratio	Totally Green composites	[8]

IOP Conf. Series: Materials Science and Engineering **810** (2020) 012001 doi:10.1088/1757-899X/810/1/012001

5. Challenges and opportunities

1. Recyclability of the composites is one of the major issues for development of ecological materials. Recyclability of the composites will lead to the cost effective products as well as the corrective measure for the rise of waste materials. Green composites can replace all hazardous and waste-producing materials.
2. Life cycle analysis should be done for all newly composite materials and thus the biodegradability can be measured. This will help to select eco-friendly and acceptable materials.
3. Microfibrillar composites, their characteristics and applications created a lot of interest in research because of their innovative properties and applications.
4. Composite materials having long-term durability for continuous purposes are desirable and cost-effective.
5. Since the interface interactions exist between fiber and matrix has a significant role in property enhancement, new characterization techniques for interface will bring new opportunities.
6. Online monitoring of morphology of composites during processing is another area, which requires the attention of researchers.

6. Conclusion

Based on the above literature survey following points have emerged. Most of the researchers showed that the natural fibers vary vastly in their mechanical properties, morphology, surface composition and water uptake behavior depending on source geographical methods of extractions. The fiber as such available in nature will not have the required qualities to fulfill the demands of different industrial applications. There is a need to develop techniques and approaches to improve the properties of the natural fibers to homogenize the as-received fiber properties as required for high-end and high technology applications.

7. References

[1] Rizvi Syed Mazher Abbas, Dwivedi,Abhishek Syed Shane Raza ,AnshikaAwasthi, , Himanshu Gupta" An Investigation of Thermal Properties of Reinforced Coconut Coir-Bagasse Fibres Polymer Hybrid Composites" 2017 IJSRSET , Volume 3 ,Issue 1, Print ISSN: 2395-1990 , Online ISSN : 2394-4099

[2] Ramachandran M., Sahas Bansal, Pramod Raichurkar" Experimental study of bamboo using banana and linen fibre reinforced polymeric composites" Perspectives in Science (2016) 8, 313—316

[3] B Stanly, Jones Retnam,, M Sivapragash and P Pradeep"Effects of fibre orientation on mechanical properties of hybrid bamboo/glass fibre polymer composites" Bull. Mater. Sci., Vol. 37, No. 5, August 2014, pp. 1059–1064.

[4] Ojaswi Panda" Study on the Effect of Fiber Parameters on the Mechanical Behavior of Bamboo-Glass Fiber Reinforced Epoxy Based Hybrid Composites"B.Tech Thesis, National Institute of Technology,Rourkela-769008

[5] P. M. Bhagwat, M. Ramachandran, Pramod Raichurkar" Mechanical Properties of Hybrid Glass/Carbon Fiber Reinforced Epoxy Composites" Materials Today: Proceedings 4 (2017) 7375–7380

[6] D.Chandramohan ,A.JohnPresinKumar" Experimental data on the properties of natural fiber particle reinforced polymer composite material" Data in Brief 13 (2017)460–468

[7] Anna Keskisaari"The impact of recycled raw materials on the properties of wood-plastic composites" Phd thesis,Lappeenranta University of Technology,Lappeenranta, Finland,2017

[8] Khalil Ahmed"Hybrid composites prepared from Industrial waste: Mechanical and swelling behavior" Journal of Advanced Research, (2015) 6,pp 225–232

[9] Eustathios Petinakis, Long Yu, George Simon and Katherine Dean'Natural Fibre Bio-Composites Incorporating Poly(Lactic Acid) http://dx.doi.org/10.5772/52253

[10] M. J. Mochane" Recent progress on natural fiber hybrid composites foradvanced applications: A review eXPRESS Polymer Letters Vol.13, No.2 (2019) 159–198

[11] Nikola Španić" Bio matrices and bio composites" Conference Paper· October 2012 ,ReseachGate
[12] M Noryani,S M Sapuan, M T Mastura , M Y M Zuhri1 and E S Zainudin" Material selection criteria for natural fibre composite in automotive component: A review The Wood and Bio fiber International Conference (WOBIC 2017)
[13] M. Ramachandran, B. K. Modi" A Review on various characterisation of PLA , International Journal on Textile Engineering and Processes "Vol. 4, Issue 1 January 2018 based biodegradable composites
[14] Amrinder Singh Pannu , Sehijpal Singh and Vikas Dhawan" Amrinder Singh Pannu1 , Sehijpal Singh and Vikas Dhawan" Asian Journal of Engineering and Applied Technology ISSN 2249-068X Vol. 7 No. 2, 2018, pp.7-15
[15] M. J. Mochane" Recent progress on natural fiber hybrid composites for advanced applications: A review"
[16] V.Durga Prasada Rao" Study of hardness and flexural strength of banyan and peepal fibre reinforced hybrid composites" MATEC Web of Conferences 172, 04009 (2018) https://doi.org/10.1051/matecconf/201817204009 ICDAMS 2018

Aesthetic improvement of sheet molding compound (SMC) composite plate by using natural fiber reinforced polymer (NFRP) like jute fiber reinforced epoxy matrix composite coating

Sujay Kekare[1], B F Jogi[1], Vishal Bhise[1] and Anoop Nair[2]

Department of Mechanical Engineering, Dr. Babasaheb Ambedkar Technological University, Lonere, India.[1]
Mahindra CIE Automotive Ltd. (Composites Division), Pune, India.[2]

sujaykekare111@gmail.com, bfjogi@dbatu.ac.in, vishalbhise79@gmail.com, anoop8168@gmail.com

Abstract. In current scenario, the use of composite materials in our lives has become very common and these are found in our furniture, transportation, education, sports, building and infrastructural system. Compared with conventional materials such as metal, plastics and woods, composite materials offer specific strength, specific stiffness and tailored material properties. Composites mainly classified as metal matrix, ceramic matrix and polymer matrix composites. Polymer matrix composites are used for low temperature applications such as automotive body panels, crashworthiness, bumpers etc. Mainly carbon, glass and aramid fibers are used as reinforcement for these applications. Nowadays, natural fibers such as flax, jute, coir, cotton etc. are also used for decorative and semi structural applications in automotive and furniture industry. Polymer matrix composites are mass produced by use of sheet molding compound (SMC) compression molding technology. SMC technology has been used in manufacturing automobile parts, electrical insulation cover applications, partition panels and many semi structural applications etc. The current trend of improving the surface appearance and mechanical properties of SMC's can lead to even more applications such as exterior, structural parts and also decorative parts in automobile, furniture and other domestic applications. Automotive interior dashboard parts, door panel parts and other interiors need to be aesthetically appealing so that it feels pleasant to seat in the car. In this work, hand lay-up process is used for coating onto the SMC compression molded plate with jute fiber reinforced epoxy composite. Uncoated and coated SMC plates are compared by various characterization methods such as tensile, compression, hardness and surface roughness. Also visual appearance and gloss test were performed. Coated SMC plate has improved tensile strength, surface finish, gloss value and visual appearance. There is almost same compression strength and Rockwell hardness value obtained.

1. Introduction

Over past few decades composite materials, ceramics and plastics have been dominant emerging materials. The applications of composite material and the area in which they are being used are increasing day by day. Nowadays, the use of composite materials in our lives has become very common and these are found in our furniture, transportation, education, sports, building and infrastructural system [1]. Composite materials offers advantages such as specific strength (the strength to weight ratio), specific stiffness (the stiffness to weight ratio) and tailored material (since composites are composed of two or more "phases", they can be formulated to meet the needs of a specific application with considerable ease) over conventional materials such as metals, woods, plastics and polymers [2]. The prominent properties of fiber composites which can be utilized to manufacture complex components with lower weight and at reduced cost are high specific stiffness, specific strength, fatigue endurance, corrosion resistance, dimensional and hydro-thermal stability and desired directional properties. This is the reason why fiber reinforced composites are being used in the applications, ranging from sports goods, naval, automobiles, wind energy and bridges etc. to advanced aircraft structures. In fiber reinforced composites, polymer matrix composites (PMCs) are very popular for different types of engineering applications. They are used primarily in relatively low temperature applications. The PMCs are used in various automotive applications like crashworthiness, body panels, and bumpers and so on [3]. Despite PMCs bring advantages to many engineering applications, there are three major issues commonly criticized by the public as these are hardly to be recycled, which may cause serious environmental problems after disposal; the structures made of advanced composites may be over-strength in particular using carbon fibre reinforced polymer (CFRP) composites and there is relatively high materials cost of PMCs for domestic products. Therefore, fibres extracted from the nature have emerged in the past decades aiming at replacing traditional high strength synthetic fibres to form a new class of natural fibre reinforced polymer (NFRP) composite. The growing awareness of environmental concerns is also another element to force the engineering sectors to develop new materials from natural resources that are either reusable or renewable. Hence, natural fibres have become an attractive topic recently [4].

NFRPs have two main constituents, the polymer matrix and the organic reinforcement fiber. Polymer matrix materials used in NFRP composites can be classified into thermosets and thermoplastics. Polyurethane (PU), epoxy and polyesters are examples of thermosets and polyethylene (PE), polypropylene (PP), polystyrene (PS) and polyvinyl chloride (PVC) are examples of thermoplastics used in NFRP composites [5]. Natural fiber reinforcement materials are fibers obtained from naturally available resources such as plants, animals and minerals. Plant fibre includes jute, cotton, coir, bamboo and banana fibres which are easily available in India [6]. Basically, uses of plastics and composites in high volume vehicles are restricted to decorative or semi-structural applications. Sheet molding compound (SMC) materials are the highest performance composites in general automotive used today. Typically, SMC materials are used for grill opening panels on many car lines and closure panels (hoods, deck lids, and doors) on a few select models [7]. Sheet molding compound (SMC) is a complex composite of specialty resins (polyesters, vinyl esters, epoxy etc.) to which thickeners, inorganic fillers, fiber reinforcements, catalyst, mold release, and pigment are added. It is produced in sheet form and can be molded by compression, transfer, and thermoset injection methods. SMC enjoys a wide range of applications due to the great range in available properties. SMC can be formulated in an almost infinite number of ways affecting production cost, corrosion resistance, heat resistance, electrical properties, and flame resistance among others. The largest volume of SMC is produced by compression molding [8].

Alireza Ashori reviewed developments in the area of wood-plastic composites (WPCs) and their applications in automotive industries using plant-based fibers. The term WPCs refers to any composites that contain plant (including wood and non-wood) fibers and thermosets or thermoplastics. Currently they are very common in building, construction, furniture and automotive products. Automotive components including plant fibers are currently being used by many vehicle manufactures such as: Audi, Opel, Daimler-Chrysler, Fiat, Ford, Mercedes Benz, Peugeot, Renault, Volvo, Volkswagen and BMW [9]. Vivek Mishra et al. work describes the development and characterization of bidirectional jute fiber mat as reinforcement and epoxy resin as matrix material. Successful fabrication of the bidirectional jute fiber reinforced epoxy composite has been done by the hand lay-up technique [10]. Gon et al. discussed about Jute fiber composites. Jute fibre is a promising reinforcement for use in composites on account of its low cost, low density, high specific strength and modulus, no health risk, easy availability, renewability and much lower energy requirement for processing. In recent years, there has been an increasing interest in finding new applications for jute fibre reinforced composites that are traditionally used for making ropes, bags, hessians, sacking, mats, and carpet. To protect environment, wood can be replaced by composite materials made from natural fibres like jute, coir, sisal etc. for making home furniture, household products and building and constructions [11]. Fan-Long Jin et al. reviewed synthesis and applications of epoxy resins. Epoxy resins have been widely used for coatings, electronic materials, adhesives, and matrices for fiber-reinforced composites because of their outstanding mechanical properties, high adhesion strength, good heat resistance, and high electrical resistance [12]. There is no any research till date available on improvement of aesthetic appearance of the SMC composite plate. In this work, jute mat as a reinforcement and epoxy resin as a matrix is used. Jute-epoxy composite coating is developed on the sheet molding compound (SMC) composite plate. Purpose of this work is to improve aesthetic appearance of the SMC composite plate and comparing jute coated and uncoated samples with different characterization methods.

2. Experimentation

2.1. Materials

Jute is available in the bidirectional woven mat form in the local market. It was used 'as available' (untreated) in the raw form. Cost of this mat is 60 Rs. /meter. Epoxy resin and its hardener were purchased from Atul Ltd., Valsad, Gujrat. Polyamine hardener is suitable to cure epoxy resin at ambient temperature. Base plate upon which coating is done is made up of SMC (Sheet molding compound) compression molded composite material. It is manufactured by Mahindra CIE Automotive Ltd. (Composites Division), Pune. These plates are available in 350mm x 250mm dimension. Clear gelcoat is used as outer finishing material. It was purchased from Link Composites Pvt. Ltd., Chinchwad, Pune. It is a thixotropic, quick-curing, superior phthalic anhydride and propylene glycol based unsaturated polyester resin and its hardener is methyl ethyl ketone peroxide (MEKP). Other materials includes glass plate of size 300mm x 150mm is used as mold, weight of 7 kg, Univax Superior Wax Polish, NORTON 80 Grit abrasive paper, pneumatic cutter, brush, weighing scale, chisel, hammer and micrometre screw gauge.

2.2. Coating process on SMC composite plate

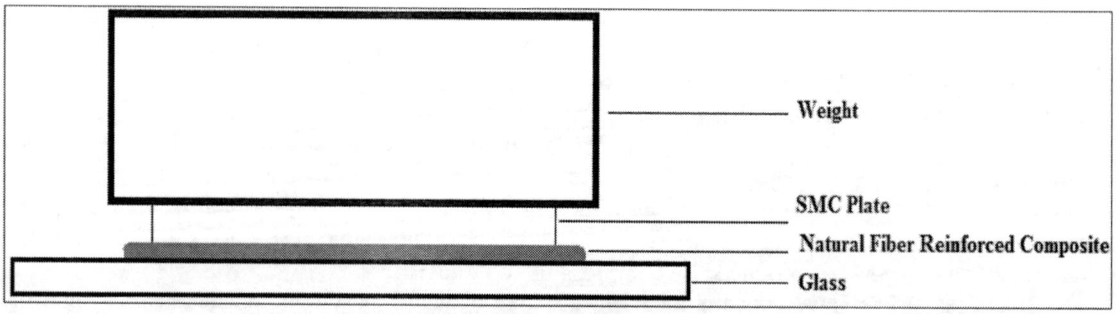

Figure 1. Schematic diagram of coating process on SMC composite plate.

Schematic diagram of coating process on SMC plate is shown in figure 1. Before coating process, surface preparation of the SMC plate is necessary to clean the surface of the plate and to get better adhesion between SMC plate and jute-epoxy composite coating. One of the surface preparation method i.e. abrasion process is done on the SMC plate. Abrasion process is done by using 80 grit sand papers. After surface preparation, wax polish was applied over the glass plate so that after completion of the process coated plate can be released from glass easily. Clear gelcoat is applied over the glass plate with brush. When clear gelcoat dries, apply properly mixed epoxy resin and hardener mixture

IOP Conf. Series: Materials Science and Engineering **810** (2020) 012002 doi:10.1088/1757-899X/810/1/012002

over it. Then, jute mat of appropriate size is kept over it and again mixture of epoxy resin and hardener is applied. Finally, SMC plate and weight were kept over the jute mat. This whole assembly is kept for curing process. Curing process completes in 24 hrs. Here in this process weight of 7 kg (approx.) is used. Sole purpose of weight is to get rid of extra resin that accumulates in the process and there won't be any air bubble trapped on the coating. The Glass is used for acquiring good surface finish on visible side of the coating. After curing completed, weight is removed. SMC plate is released with the help of chisel and hammer. Trimming of the plate is done with the help of pneumatic cutter.

2.3. Characterization

For comparison between jute coated SMC plate and uncoated SMC plate, various characterization methods are done. Mainly mechanical testing and aesthetic testing were done. Mechanical testing such as tensile, compression, hardness and surface roughness test were done. Aesthetic testing is done on the basis of visual appearance and gloss value. Universal testing machine of Star Testing System, India (Model STS 248) was used for tensile and compression tests. Rockwell hardness tester of Saroj Engineering, India (Model RAS) is used for determination of Rockwell hardness as per ASTM D 785-03 standard. Surface roughness tester of Mitutoyo (Model Surftest SJ400) was used. Gloss meter of Raj scientific company, Mumbai is used. Weighing scale is used for measuring weight of jute mat, resin, hardener etc. Tensile testing as per ASTM D 638, Compression testing as per ASTM D 695 and Rockwell Hardness testing as per ASTM D 785-03 standard were done. Standard specimens were prepared for these tests. Test plate of 300mm x 120mm size jute coated plate was prepared so that all the test specimens can be made by use of this plate. Table 1 describes approximate amount of materials required for coating of 100mm x 100mm of SMC plate. With the use of material requirement values obtained for 100mm x 100mm SMC plate coating, material required for 300mm x 120mm size SMC plate coating is calculated and depicted in the table 2.

Table 1. Material requirement for coating (100mm x 100mm plate).

Natural fiber		Polymer matrix (Resin)	Ratio	Outer coating
Type	Weight	Epoxy + Hardener	Fiber : Matrix	Clear gelcoat + Hardener
Jute	3gm	10gm+5gm	1:5	15gm+0.2ml (approx.)

Table 2. Material requirement for jute coating (300mm x 120mm plate).

Natural fiber		Polymer matrix (resin)	Ratio	Outer coating
Type	Weight	Epoxy + Hardener	Fiber : Matrix	Clear gelcoat + Hardener
Jute	11gm	36gm + 18gm=54gm	20:100	54gm + 0.8ml (approx.)

Thickness of the developed jute-epoxy composite coating on the SMC plates is determined by taking average value of three coated SMC plate samples. Three jute-epoxy coated SMC plates of 100mm x 100mm size were made by using same amount of materials. Micrometre is used for checking thickness values at different points on the plates. Thickness before coating and thickness after coating on the SMC plates were checked and thickness of the jute-epoxy composite coating is determined.

3. Results and Discussions

3.1. Aesthetic tests

3.1.1. Visual Appearance test

Visual appearance test was performed on coated and uncoated SMC plates. Figure 2 and figure 3 shows SMC plate before coating and after coating respectively. Jute fiber gives natural appearance to the SMC plate. All bidirectional jute fibers are clearly seen. By coating SMC plate with jute-epoxy composite, its aesthetic appearance has improved as seen in the figures.

Figure 2. SMC plate before coating.

Figure 3. SMC plate after coating.

3.1.2. Gloss test

Gloss values obtained are shown in the table 3 given below. SMC plates with jute-epoxy composite coating gives more values of gloss and hence it can be said that SMC plate with coating of jute-epoxy composite is glossy than uncoated SMC plate.

Table 3. Gloss test comparison.

Sample ID	Gloss value (OGU)
Uncoated SMC plate	20° - 38.1, 60° – 74.5, 85° - 92.2
Coated SMC plate	20° - 67.8, 60° – 85.3, 85° -99.7

Table 4. Tensile test results (uncoated SMC plate).

Max Load	2578.38 N
Peak Elongation	5.6 %
Break Load	101.92 N
Break Elongation	6.1 %

3.2. Mechanical tests

3.2.1. Tensile test

Uncoated SMC plate: The tensile test results show that the maximum load applied was 2578.38 N and after which the specimen failed. The maximum displacement yielded by the material was 5.6 mm. Breaking load was 101.92 N and Break Elongation was 6.1 mm. Load vs displacement graph is shown in figure 4. Tensile test results are shown in table 4.

Coated SMC plate: The tensile test results show that the maximum load applied was 6026.02 N and after which the specimen failed. The maximum displacement yielded by the material was 8.6 mm. Breaking load was 49.98 N and breaking elongation was 8.7 mm. Load vs displacement graph is shown in figure 5 and table 5 shows tensile test results for jute coated SMC composite plate.

Both the samples were found to be brittle in nature as it underwent a sudden failure. It does not show a yield point and even strain hardening have not occurred. The typical characteristic of the brittle material is to fail while the deformation is elastic and does not show any plastic deformation. Here, table 6 depicts tensile test comparison between two SMC plates. Tensile strength of jute coated samples is almost 35.15 % higher than uncoated samples. The reason could be that, there is increase in the cross sectional area after coating with jute-epoxy composite. So, it could take higher loads than without coating composite before breaking.

IOP Conf. Series: Materials Science and Engineering **810** (2020) 012002 doi:10.1088/1757-899X/810/1/012002

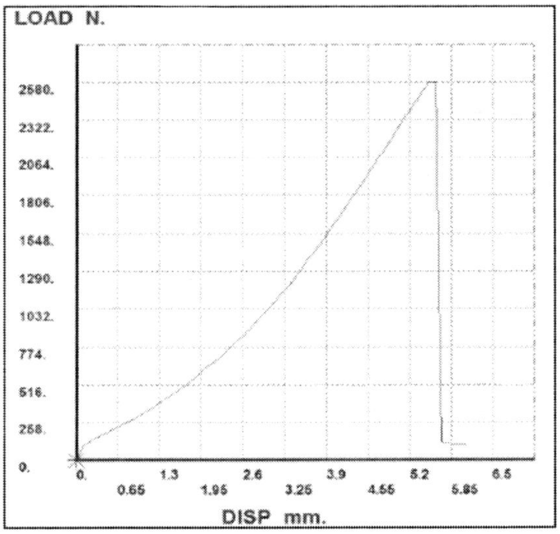

Figure 4. Load vs displacement graph (uncoated SMC plate).

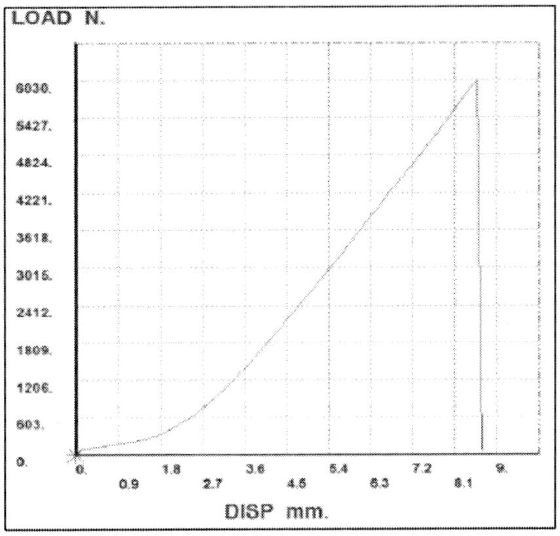

Figure 5. Load vs displacement graph (jute coated SMC plate).

Table 5. Tensile test results (coated SMC plate).

Max Load	6026.02 N
Peak Elongation	8.6%
Break Load	49.98 N
Break Elongation	8.7%

Table 6. Tensile test comparison.

Sample ID	Uncoated SMC plate	Coated SMC plate
Tensile strength (MPa)	71.622	110.44

Table 7. Compression test results uncoated SMC plate with coated SMC plate.

	Uncoated SMC Plate	Coated SMC Plate
Peak Load	5213.6 N	7536.2 N
Peak Compression	2.3 %	3.4 %

3.2.2. Compression test

Uncoated SMC plate: The compression test result show that the maximum load applied was 5213.6 N after which specimen failed. Peak compression obtained was 2.3 mm. Load vs deformation graph is shown in figure 6 and table 7 shows compression test results.

Coated SMC plate: The compression test results show that the maximum load applied was 7536.2 N and after which the specimen failed. The maximum compression yielded by the material was 3.4 mm. Load vs deformation graph is shown in figure 7 and table 7 shows compression test results. Table 8 shows compression test comparison for coated and uncoated SMC plate specimens. Compression strengths for both the samples are nearly same. The reason could be that, there is increase in the cross sectional area after coating with jute-epoxy composite. So, it could take higher loads than without coating composite before breaking.

Table 8. Compression test comparison.

Sample ID	Without coating	With jute coating
Compressive strength (MPa)	140.151	142.407

IOP Conf. Series: Materials Science and Engineering **810** (2020) 012002 doi:10.1088/1757-899X/810/1/012002

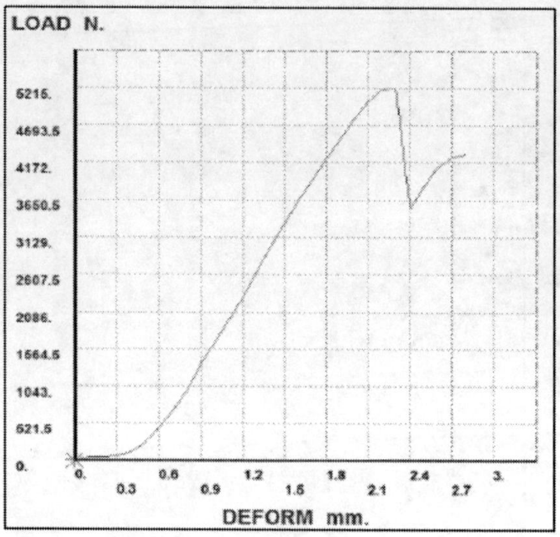

Figure 6. Load vs deformation graph (uncoated SMC plate).

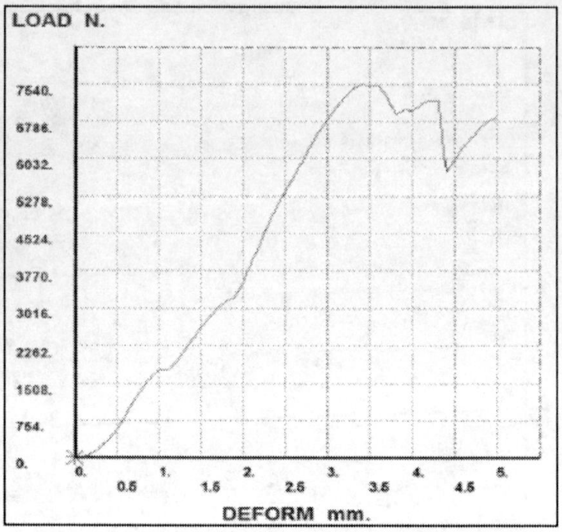

Figure 7. Load vs deformation graph (coated SMC plate).

3.2.3. Hardness test

Rockwell hardness is measured and results are shown in the following table 9 as follows. Hardness values are lower for developed jute-epoxy composite coated SMC plates. The hardness test is performed with a ball indenter with diameter ¼" and applied load of 100 kg. Rockwell hardness scale chosen was M scale. M-scale is mostly used to measure the hardness for plastics and soft materials. From the hardness values of both the samples, it was found to be almost similar. The results were found to be in close conjunction with each other. In the Rockwell test conducted the hardness value was found to be in the range 55-56 for uncoated and 50-51 for coated samples.

Table 9. Hardness test comparison.

Sample ID	Hardness (Scale 'M', 100 Kg load, ¼" Dia. ball)
Without coating	55-56
With jute coating	50-51

Table 10. Surface roughness test comparison.

Sample ID	Without coating		With jute coating	
Surface Roughness (µm)	Ra	0.30	Ra	0.13
	Rz	3.0	Rz	1.1
	Rq	0.47	Rq	0.19

3.2.4. Surface Roughness test

Roughness average (Ra), 10 point height average (Rz) and Root mean square roughness (Rq) of coated and uncoated SMC plate samples were measured in micrometre (µm) and depicted in table 10. There is decrease in all these surface roughness values for jute-epoxy composite coated SMC plate. So, it can be said that coated SMC plate has good surface finish than uncoated SMC plate.

3.3. Thickness of coating

Table 11 depicts thickness values of SMC plates before and after coating. By using values in table 11, thickness of jute coating can be obtained. Average values of coating thickness for Plate A = 0.9325, Plate B = 1.0387 and Plate C = 1.0275. Hence, thickness achieved is between 1±0.20 mm.

4. Conclusions and future scope

Sheet moulding compound (SMC) composite plates can be coated with natural fiber jute as a reinforcement material and epoxy resin can be used as a matrix material to add aesthetic appeal to them by developed method. Jute mat gives aesthetically appealing look to the SMC plates. Jute mat coating is easy to coat on the SMC plates. Jute-epoxy composite coated SMC plate specimen's gives better tensile strength values, almost same compression test values, reduced Rockwell hardness value as compared to uncoated SMC plate specimens. But jute-epoxy composite coated SMC plate specimen's gives higher values of gloss and improved surface finish values when compared with uncoated SMC plate specimens. These coated SMC plates will have attractive and aesthetically appealing look for these applications.

5. Acknowledgements

The authors would like to thank Mahindra CIE Automotive Ltd. (Composites Division), Pune for supporting in this research project.

Table 11. Thickness values before and after jute-epoxy composite coating on the SMC plates (in mm).

Pt. No.	Plate A		Plate B		Plate C	
	Before	After	Before	After	Before	After
1	3.25	4.20	3.06	4.07	3.05	4.08
2	3.19	4.15	3.13	4.11	3.17	4.31
3	3.20	4.16	3.13	4.19	3.17	4.34
4	3.31	4.25	2.97	4.06	3.16	4.15
5	3.25	4.13	3.06	4.12	3.23	4.28
6	3.41	4.28	2.98	4.06	3.25	4.16
7	3.32	4.25	3.10	4.12	3.33	4.28
8	3.20	4.17	3.09	4.10	3.35	4.33

6. References

[1] Ashby M F 1987 *London. Phil. Trans. Roy. Soc.* A **322** 393
[2] Aramide F O, Atanda P O and Olorunniwo O O 2012 *Int. J. Compos. Mater.* **2** 147
[3] Hallal A, Elmarakbi A, Shaito A and El-Hage H 2013 *Hoboken, NJ Wiley* 1
[4] Lau K T, Hung P Y, Zhu M H and Hui D 2018 *Compos. B Eng.* **136** 222
[5] Väisänen T, Haapala A, Lappalainen R and Tomppo L 2016 *Waste Manage.* **54** 62
[6] Gurunathan T, Mohanty S and Nayak S K 2015 *Compos. Part A Appl. Sci. Manuf.* **77** 1
[7] Beardmore P and Johnson C F 1986 *Compos. Sci. Technol.* **26** 251
[8] Sawallisch K 1984 *Polym.-Plast. Technol.* **23** 1
[9] Ashori A 2008 *Bioresour. Technol.* **99** 4661
[10] Mishra V and Biswas S 2013 *Procedia eng.* **51** 561
[11] Gon D, Das K, Paul P and Maity S 2012 *Int. J. Text. Sci.* **1** 84
[12] Jin F L, Li X and Park S J 2015 *J. Ind. Eng. Chem.* **29** 1

IOP Conf. Series: Materials Science and Engineering **810** (2020) 012003 doi:10.1088/1757-899X/810/1/012003

An experimental and numerical study of air flow pattern and temperature distribution of angular air distribution system in hospital operation theatre

Swati Rahate[1,a], Avinash Sarode[1,b]

[1]Mechanical Department, Lokmanya Tilak College of Engineering, Navi-Mumbai.

[a] Swati.rahate363@gmail.com, [b] avinashsarode@gmail.com

Abstract: - It is necessary to characterize the air flow pattern and temperature distribution in Operation Theatre (OT) to optimize the indoor air quality to enhance the high performance of hospital. Mostly Laminar air flow system is used in OT, which covered the surgical zone. The temperature draft observed between surgical and nonsurgical area which result in uncomfortable environment to patient and surgeon. To overcome such problem Angular Air Distribution (AAD) system is designed which covered the both area of OT. The performance of AAD system with varying inlet diffuser angle at constant inlet velocity on Scale model of OT is investigated with the help of CFD software and compared with experimental setup. Detail experimental investigation is done on a scale model of OT with the help of a smoke test and camera target method (CTM). Both numerical and experimental results showed a similar pattern of air and temperature distribution for the various design configurations.

1. Introduction

As ventilation system play important role in HVAC to keep OT environment free from bacteria. The poorly ventilated system not only affects the health of patient and surgeon from infection but also feel them uncomfortable [1]. Many national and international standards and guidelines are available for every country to design a ventilation system with environmental parameters for OT. LAF system provides unidirectional ultra clean air flow to protect the surgical zone only. If the whole OT is designed for LAF system then installation area and air circulation rate (ACH) of the ventilation system is increased which results in an increase in energy consumption and maintenance of system [2]. The traditional ventilation systems of OT provide the conditioned air to the critical zone only by maintaining laminar flow. Hospitals and other health care facilities are complex environments that require comfort and control conditioned air with flow pattern [3]. The persons present in OT such as patient and surgeon suffer from surgical site infection (SSI) and postoperative infection [4]. Indoor obstacles such as personnel, operating lamp, surgical instrument and movement of the surgeon can easily affect the unidirectional flow pattern of the ventilation system. Alternate methods are mixing, displacement, Vertical and horizontal ventilation system, which differs by the position of supply and an exhaust port in OT [2]. Ventilation and air distribution pattern have a great effect on indoor air quality which includes indoor temperature, relative humidity, air flow velocity, pressure relationship, air movement's efficiency. Therefore in this work new AAD system is designed in such way that it will distribute the air in laminar pattern in whole OT.

Many experimental and numerical analyses have been presented about ventilation system and thermal comfort in ORs. Balaras, C.A et al. [5] had presented an overview of the general indoor condition and air quality in OTs. They investigated 20 OTs in 10 different hospitals and recorded information on the type of ventilation system, temperature range, relative humidity range, etc. Chow and Yang had investigated that, thermal comfort is achieved by controlling the temperature, the humidity, and air movement. In laminar air flow supply is given at the rate of 0.46 m/s (90fpm), there is two types of LAF one is horizontal and another is vertical. In this literature velocity is measured by using hot wire anemometer and compared it with a mathematical k-e model for validation purpose [6, 7, 8]. Zang Rui et al. [9] explained that by improving airflow pattern, the particle deposition in the critical zone can be reduced but its effect becomes reduced if we increase the air change rate. He concluded that 0.25 m/s is the best velocity. Villafruela et al. [10] analyzed the effective design of the ventilation system for isolation rooms which reduce the risk of spreading infections of infected patients. Liu et al [11] investigated an alternative of horizontal flow pattern and air flow performance in an OT with a dimension of 300cm long, 296 cm wide and 240 cm high. He also evaluates the effectiveness of the horizontal unidirectional airflow to control infectious airborne particles in operation theatre. He also stated that downward unidirectional air flow has some disadvantages.

In view of above contradictory results and worldwide increases in health care costs and increasing difficulties in financing and providing all modern medical advances in OT Angular Air Distribution (AAD) System is designed whose primary goal is to provide unidirectional flow in whole OT and reduced the temperature draft to provide comfortable environment to patient and surgical staff. To test the designed AAD system, prototype scaled model of OT with 1:4 ratio is fabricated. Initially the scale model of OT is examined by numerical simulation by varying inlet angle, inlet velocity and temperature fluent software is used. The boundary conditions of numerical results are tested with grid independency test. The numerical results are validated with the help of experimental readings of scale model of OT The air flow pattern of AAD system in OT with various inlet angles are tested with the help of smoke test. The flow pattern obtained by smoke of AAD system is capture by Camera Target Method (CTM) and it is validated with CFD results. To check the human comfort and effectiveness of air distribution system, the obtained result is compared with the international standard parameter for ventilation system such as Air Diffusion Performance Index (ADPI).

2. Methodology

2.1 Geometrical similarity criteria

A reduced scale model is used to conduct a series of model experiments and investigate of ventilation system of OT. For geometrical similarities between scale model and prototype model, scale model is made with 1:4 ratios. The geometrical similarity is consider as follows

$$\frac{L_p}{L_m} = \frac{W_p}{W_m} = \frac{H_p}{H_m} = L_r \quad (1)$$

$$\frac{Ap}{Am} = Lr^2 \quad (2)$$

$$\frac{Vp}{Vm} = Lr^3 \quad (3)$$

WhereLp, Wp, Hp is length, width and height of prototype model. Lm, Wm, Hm is length, width and height of Scale model. Lr Is Geometrical ratio as explain in equation (1), (2) and (3). Table 1 shows the ratio of geometrical similarity between scale and prototype model. From table 1 $Lr = 4$.

Table 1. Dimension of Full scale and scaled model by geometrical similarity

	Geometrical Similarity				
	Length (m)	Width (m)	Height (m)	Area (m^2)	Volume (m^3)
Actual (Prototype) Model	6	6	3	36	108
Ratio	4	4	4	16	64
Scaled Model	1.5	1.5	0.75	2.25	1.6875

2.2 CAD modeling

A 3D model of OT room geometry that outline by rectangular shaped room of wooden material with 1.524 m (L) x 1.524 m (B) x 0.75 m (H) dimension as shown in figure 1. The room is equipped with 5 people and one operating bed, anaesthesia machine, surgical lamp and operating bed, which is located at middle of the OT under the area of Laminar flow. The 5 surgical staff is standing around the operating table. They were modelled in cylindrical shaped with 0.0762m diameter and 0.3048m height. Anaesthesia machine of size 0.075 m (L) X 0.0625 m (W) X 0.425 m (H) is placed head side of patient. The supply ventilation system is installed in such way that its flow will cover the critical zone or flow will wash out the critical zone. Square ceiling supply diffuser (0.5 m (l) X 0.5 m (w)) with 4 number of louvers equally spaced from edged of diffuser is located in central zone of ceiling, that guide the unidirectional and angular flow by varying angle of louvers as shown in figure 1. Four exhausts (0.00580 m2 cross section area of each) are arranged over two of opposite side wall and allow the internal air to outflow.

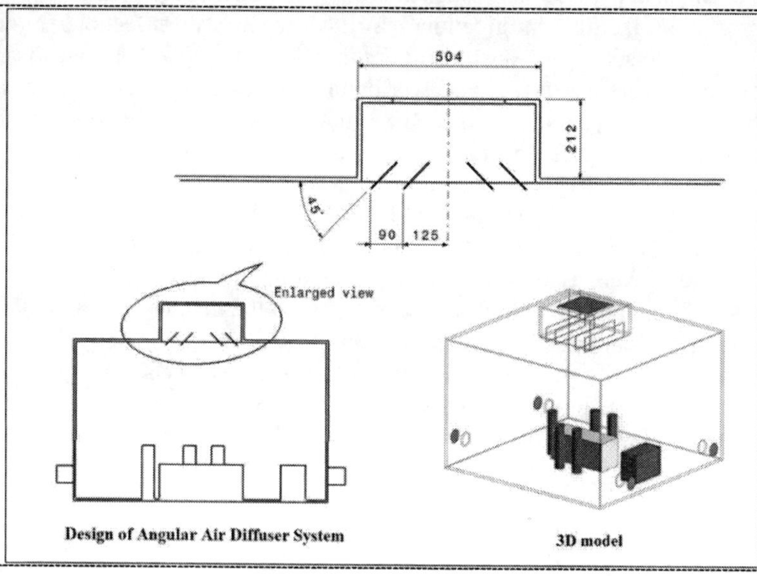

Figure 1. Solid model of OT with interior features and enlarge view of inlet-outlets

Four different additional configurations of AAD system as shown in figure 1 are to be study starting from flat to angular diffuser. For these four different ventilation systems constant air flow rate 25 ACH are assumed.

2.3 Numerical Solution

The numerical solution is carried out in commercial software which allowing solution of the governing equation by finite element approach such as CFD (Fluent). Fluent is a general purpose commercial CFD software package that solves the Navier–Stokes equations using a finite volume method and SIMPLEC algorithm. In this work, Ansys-Fluent 15.0 was used to solve the momentum, continuity and turbulence equations for fluid flow. A solid and fine tetrahedron elemental auto-mesh is used to generate mesh model. The software ANSYS mesh was used for meshing of OT geometry with many elements. The different mesh size is considered for mesh independence study. Tetrahedron cells were used with varying number of elements of 256916, 287258, 312761, 39844, and 586479. The velocity plot for different mesh size is plotted and it was approximately same for last four numbers of elements.

The criteria for converged numerical solution are taken as 1e-3 RMS with element size 586479as shown in figure 2

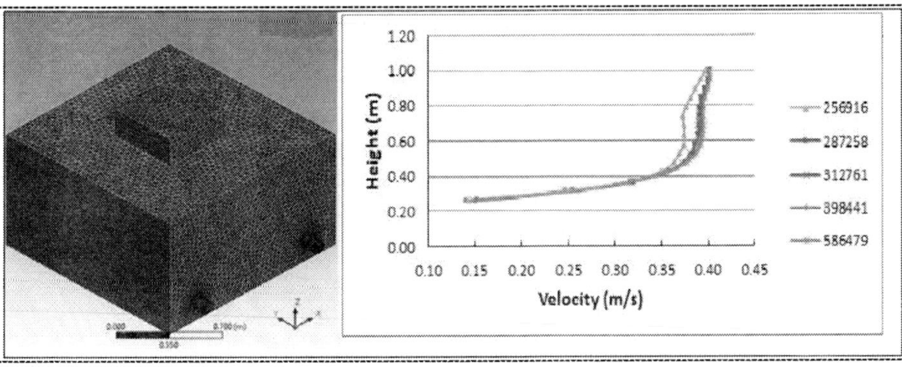

Figure 2. Mesh Independency

Airflow pattern was determined by solving a set of partial differential equations that consist of mass, momentum and energy conservations, written in equation (4), (5), and (6), below respectively:

$$\nabla (\rho \vec{v}) = S_m. \tag{4}$$

$$\nabla . (\rho \vec{v} \vec{v}) = -\nabla p + \nabla . (\bar{\bar{\tau}}) + \rho \vec{g} \tag{5}$$

$$\nabla . (\vec{v} (\rho \vec{v} + p)) = -\nabla . \left[\sum_j h_j J_j\right] + S_h. \tag{6}$$

This airflow study used the standard k–ε turbulence model equation (7) and (8) that consisted of the following Navier-Stokes equations

Turbulence Kinetic Energy (k):-

$$\frac{\partial}{\partial x_i}(\rho k u_i) = \frac{\partial}{\partial x_j}\left[\left(\mu + \frac{\mu_t}{\sigma_k}\right)\frac{\partial k}{\partial x_j}\right] + G_k + G_b - \rho\varepsilon - Y_M + S_k \tag{7}$$

Dissipation Rate of turbulence kinetic energy (ε)

$$\frac{\partial}{\partial x_i}(\rho \varepsilon u_i) = \frac{\partial}{\partial x_j}\left[\left(\mu + \frac{\mu_t}{\sigma_\varepsilon}\right)\frac{\partial \varepsilon}{\partial x_j}\right] + C_{1\varepsilon}\frac{\epsilon}{k}\left(G_k + G_{3\varepsilon}G_b\right) - C_{2\varepsilon}\rho\frac{\varepsilon^2}{k} + S_\varepsilon \tag{8}$$

Where $G_k = -\overline{u_i u_j}\frac{\partial u_j}{\partial x_i}$ to evaluate this parameter Boussinesq hypothesis is used $G_k = \mu_t S^2$ where the modulus of the mean rate of strain tensor is defined as $S \equiv \sqrt{2S_{ij}S_{ij}}$. $C_{\mu,} = 0.09$, $C_{1\varepsilon} = 1.44, C_{2\varepsilon} = 1.92, \sigma_k = 1.0, \sigma_\varepsilon = 1.3$.

Inlet velocity is taken as 0.4 m/s ,as inlet boundary condition with 5% turbulence intensity for a hydraulic diameter of OT. The exhaust port 1.3014 Pa pressure-outlet boundaries with no-slip boundary condition are combined with zero heat flux to define adiabatic walls. The walls of OT are considered as standard wall function including interior of OT. The values of physical properties of material used in numerical simulation are listed in table 2.

Table 2. Physical Properties of material used in numerical simulation

Material	ρ (kgm^{-3})	Cp (j/kg°k)	K (w/m°k)
Wood	700	2310	0.173
Air	1.225	1006	0.02435

2.4 Experimental setup

The general layout of experimental setup is shown in figure 3. It consist of three main parts such as air distribution system i.e. supply diffuser , Air Conditioning unit or Air handling unit (AHU) and testing room. Air handling unit is simple unit of vapour compression refrigeration cycle. Fan is pushing the atmospheric air over evaporating coil to get conditioned cooled air. The OT scale room model made of wooden material. This is scale by 1:4 ratio of full scale of prototype. OT room is fabricated by the plywood of 12 mm thick for three side and floor and top surface, front side of testing room is made by acrylic sheet for capturing the images of smoke test. The round supply duct of 10 cm diameter is used to supply the conditioned air from AHU outlet to inlet of test room. Four exhausts are connected to exhaust fan with the help of exhaust duct.

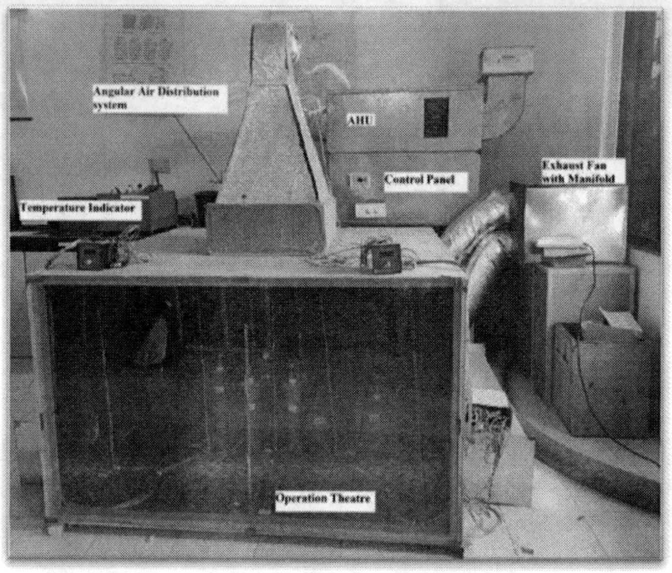

Figure 3. General layout of experimental setup

2.5 Measurement procedure and instrumentation

In order to validate the results obtained from the CFD solution, a scaled model of OT is tested by experimental measurement i.e. smoke test with help of camera target method. For clear visualization of airflow pattern inside the OT, the interior is painted as black and white coloured smoke is used with air. A smoke generator generates smoke which is accumulated in the mixing box. It is connected to inlet diffuser of OT by means of flexible ducts. The smoke entered in OT through inlet diffuser and exhausted to the atmosphere by outlet port which is placed at the bottom of the walls. Four outlets are used which is connected to exhaust through exhaust manifolds. The front side of the prototype is kept transparent so that clear images can be captured. The camera is fixed at its position using a tripod and the smoke is fed to flow inside the prototype. For smoke visualization test, smoke generator (Power 900 watt, heat up time 12 min, capacity-3 litres) with fog fluid (white cloud) is used. The blower is inbuilt in the smoke generator. The interior features of OT and the white smoke is targeted and its images are captured by high definition digital camera. Since the camera is fixed at its position, the multiple proportionate images are captured and compared on the bases of smoke distribution inside the OT. Initially, smoke visualization test is carried out to ensure no-leakage in the room. Further smoke visualization test was carried out to identify the air flow pattern with different inlet angle of AAD system and to evaluate the effectiveness of the same.

IOP Conf. Series: Materials Science and Engineering **810** (2020) 012003 doi:10.1088/1757-899X/810/1/012003

As the Human comforts is depend on the temperature draft in occupied zone. To investigate the temperature draft in whole OT temperature measurement at various points is needed. According to temperature reading the effectiveness of air distribution system is calculated. For temperature measurement fixture is design and fabricated as shown in figure 4.

As the geometry of scaled OT room is symmetrical with respective central plane, the temperature measurement is taken at on side of central plane of OT. For temperature measurement at different location, the temperature measuring fixture is placed at different location. To measure temperature draft in critical zone and beyond that zone total nine line of temperature measuring fixture is used as shown in figure 4. The temperature is measured at different plane 700 mm, 500 mm and 200 mm from the floor. The 700 mm height plane is near to ceiling, 500 mm plane at middle of OT, and 200 mm height plane near to operating bed.

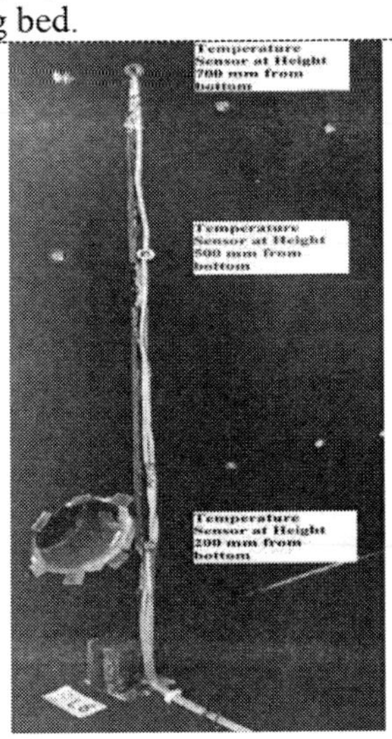

Figure 4. Temperature measuring fixture

All velocity measurement is carried out by temperature compensate Hot wire anemometer. Hot wire anemometer probe is connected to the display and storage unit. Velocity measurement was performed inside the OT, Diffuser inlet and AHU outlet. The length of probe is 3 meter long to measure the velocity at middle plane of OT. The velocity is measured at the same point of temperature sensor. To access the all 27 point of temperature sensor inside the OT, side wall of scale model is drilled with the hole to insert the probe of anemometer at different point for velocity measurement. The coordinate of velocity measuring point in OT with line fixture and it location are shown in figure 5. Initially velocity is measured at AHU outlet, then at diffuser inlet and outlet, at exhaust outlet of OT.

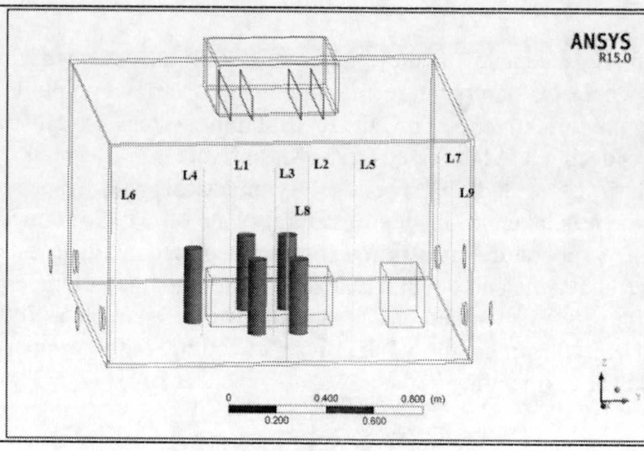

Figure 5. Line wise measurement point in OT

The widely accepted and design index that quantifies the performance of air distribution system when considering the spatial uniformity of air distribution to thermal comforts is air diffusion performance index (ADPI). Miller introduced the ADPI concept in HVAC industries for research by designing different types of air distribution system. ADPI is defined as the percentage of occupied zone falling into the acceptable velocity and temperature region determined by measuring local Effective Draft Temperature (EDT). ADPI is based on air speed and effective draft temperature (EDT) as mentioned in equation (9). ADPI method required to calculate EDT.

$$EDT = (T_P - T_R) - 8 \times (V_P - 0.15) \tag{9}$$

Where T_P is air temperature at a test point in °k, T_R is the average temperature of room or operation theatre and V_P is local air velocity in m/s. For comfort, the EDT should be within –1.7°C to +1.1°C and the air velocity should be less than 0.36 m/s. The objective of air distribution system design is to select and place the supply air diffusers in such a way that the ADPI approaches 100 %. The ADPI provides a rational way of selecting air diffusers. The ADPI as mentioned in equation (10) is defined as the percentage of measurements taken at many locations in the occupied zone of space that meets EDT criteria of–1.7°C to +1.1°C, that is

$$ADPI = (N_\theta / N) \times 100 \tag{10}$$

Where N is the total number of locations at which observations have been made, and N_θ is the number of locations at which the effective draft temperature is within –1.7°C to +1.1°C.

3. Result and discussion

3.1 Air flow pattern

CFD analysis is performed on the featured scaled geometry by using Ansys Fluent. The result of flat diffuser for air flow pattern, and turbulence plot is as shown in figure 6.

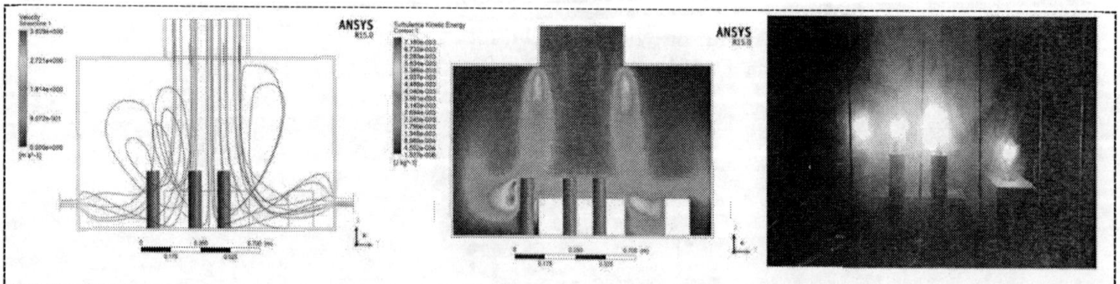

Figure 6. Distribution of air stream, turbulence energy and smoke visualization in OT with 90° louvers (Flat) angle

IOP Conf. Series: Materials Science and Engineering **810** (2020) 012003 doi:10.1088/1757-899X/810/1/012003

The air stream after striking the bed gets reflected upward causing the vortex. This vortex spread over the OT and mixes with fresh incoming primary airflow. These vortexes cause temperature draft (conada effect) and turbulence in the flow of harmful particles along with the airstream flow over the surgeons and remain in the critical zone. Turbulence kinetic energy remains low in the region from inlet diffuser to the critical zone but it increases in the obstacles offered by the surgeons. Also, the velocity is high in the same region. Hence, the primary flow carries bacteria and mixes with fresh incoming airflow.

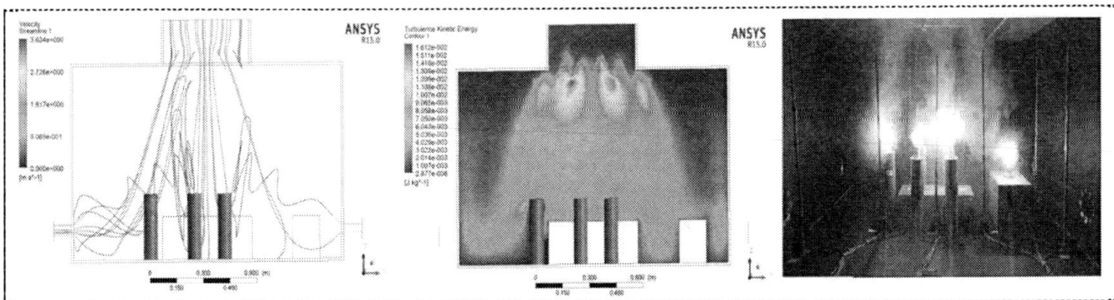

Figure 7. Distribution of air stream, turbulence energy and smoke visualization in OT with 30° louvers angle

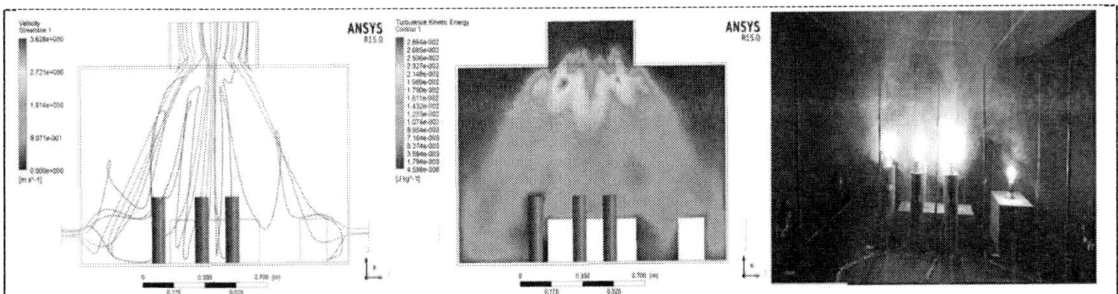

Figure 8. Distribution of air stream, turbulence energy and smoke visualization in OT with 45° louvers angle

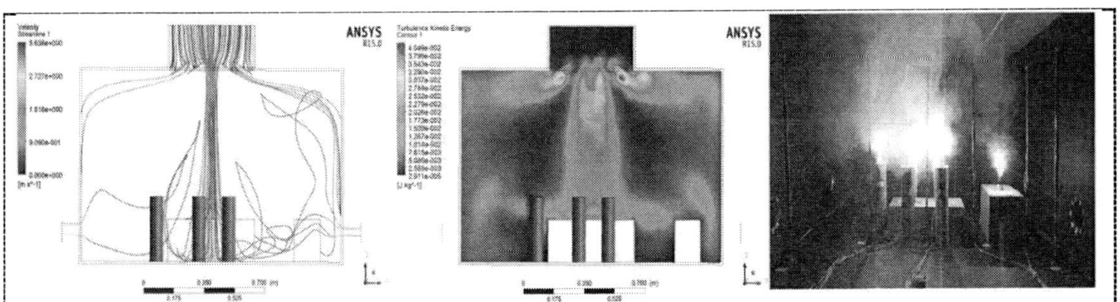

Figure 9. Distribution of air stream, turbulence energy and smoke visualization in OT with 60° louvers angle

Figure 7, figure 8 and figure 9 presents the path line exit from AAD system with 30° , 45° and 60° angle of louvers. The smoke flow from diffuser reaches to the middle of OT with minimum turbulence. The throw of AAD system seems as long and equally spread in whole OT; therefore, there is minimum vortex formation in OT.

IOP Conf. Series: Materials Science and Engineering **810** (2020) 012003 doi:10.1088/1757-899X/810/1/012003

3.2 Velocity and temperature measurement

The temperature on all measuring point i.e. 15 points inside the critical zone area is in range of 25+/-2°C in all diffuser. But in case of flat diffuser outside the critical zone the temperature is higher than critical zone. Which feel the drastic temperature draft to surgeon and patient which affect the human comforts. But in case of AAD system this temperature draft is less as compared to flat diffuser system. At the same time, time required to achieve equilibrium temperature with help of AAD system in OT is less as compared to flat diffuser system which further result in reduction in energy consumption. The temperature range and its distribution in OT with flat and AAD system is shown in figure 10 and figure 11.

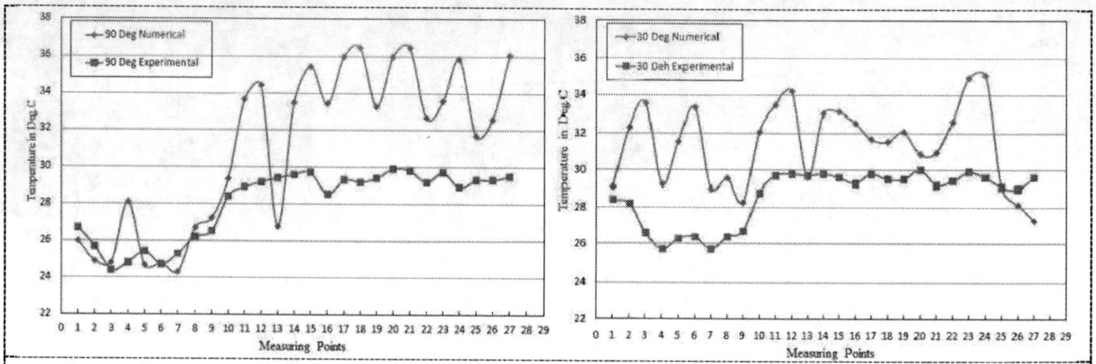

Figure 10. Distribution of Temperature in OT with 90° and 30° louvers angle.

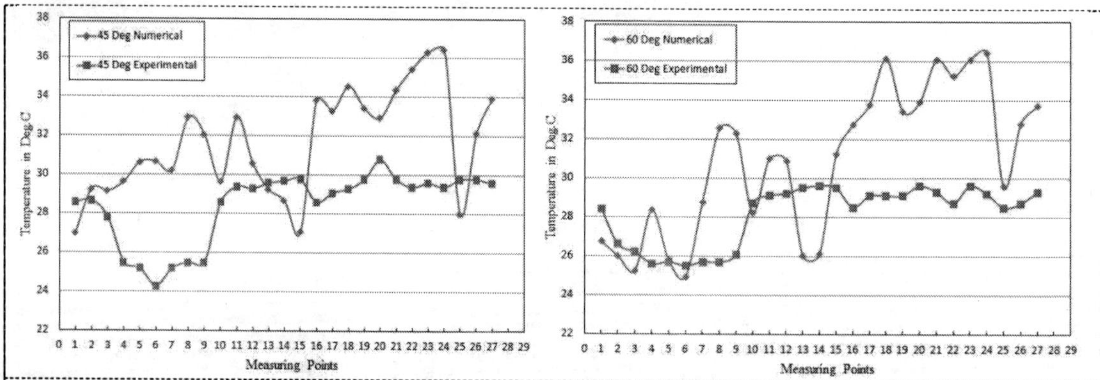

Figure 11. Distribution of Temperature in OT with 45° and 60° louvers angle.

From Figure 12 and Figure 13, it is observed that air velocity in critical zone is varying from 0.2-0.3 m/s, which is in acceptable range as mentioned in national and international standards and guidelines. From figure 20 it is observed that in case of flat diffuser air velocity in non-critical zone vary from 0 to 0.4 m/s, which results in turbulence formation in that area, but in other case i.e. 60° and 45° AAD system the velocity variation in non-critical zone is less which reduced the turbulence formation.

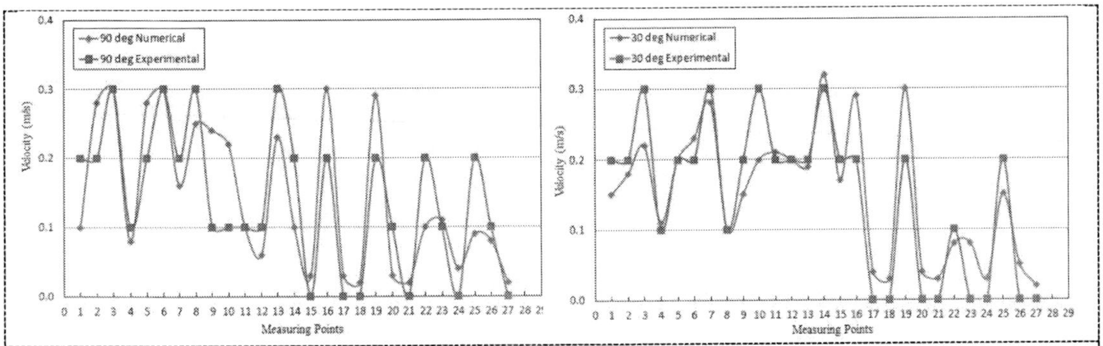

Figure 12. Distribution of velocity in OT with 90° and 30° louvers angle

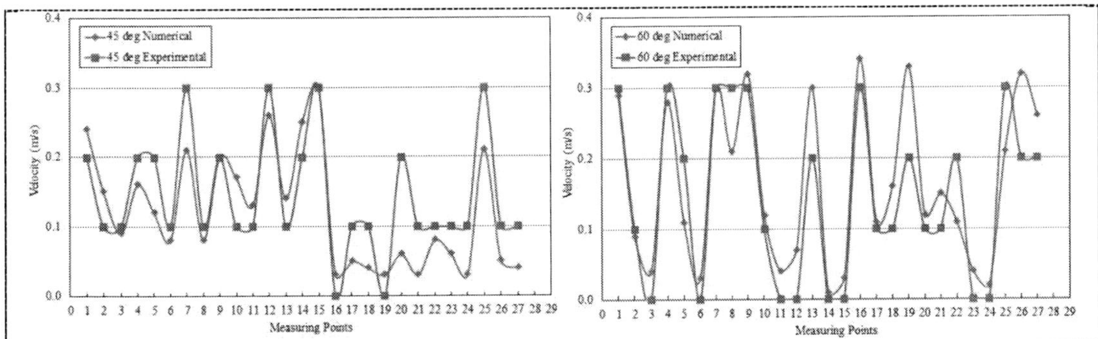

Figure 13. Distribution of velocity in OT with 45° and 60° louvers angle.

3.3 Air Diffusion Performance Index (ADPI)

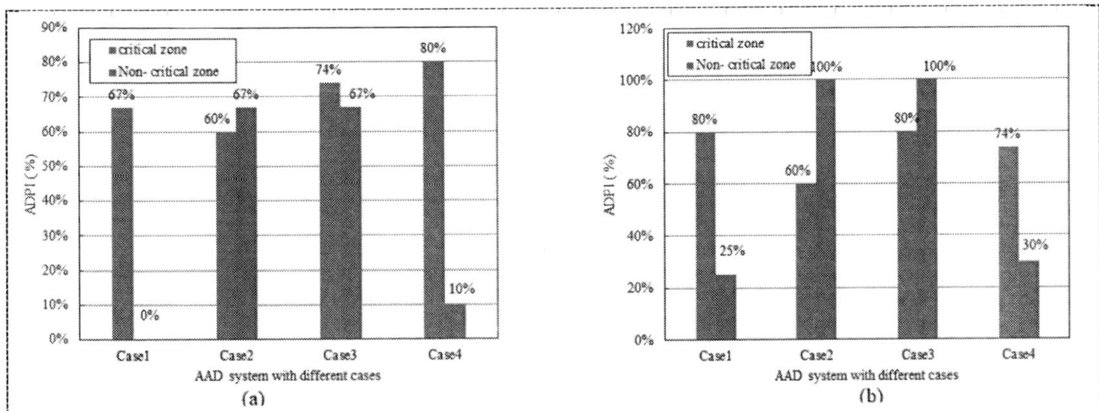

Figure 14. ADPI for AAD and Flat diffuser system in critical and non-critical zone.

ADPI model consider thermal distribution as well as the range of effective draft temperature and maximum local air speed in OT. Furthermore this work generates ADPI for all case (Case 1 -90°, Case 2 -30°, Case 3 -45°, Case 4 -60°) of designed air distribution in critical zone and non-critical zone. The measurement shows that AAD system with 45° louvers angle generate ADPI more and equal to 80% in critical and non-critical zone as compare to other cases as shown in Figure 14, which results in uniform temperature distribution with minimum temperature draft to get thermal comfort for occupant in OT.

4. Conclusion

Angular Air Distribution system can provide an alternate ventilation system for operation theatre. The results of temperature and velocity distribution in OT with AAD system is compared with Laminar air distribution system. The causes of uneven distribution of flow in laminar air distribution system are visualized by both theoretical and experimental method. The flat diffuser or laminar air distribution system provides uniform airflow at the critical zone only but badly increases the reverse flow and turbulence in critical zone when the obstruction arrives. The tendency of mixing of primary and secondary flows is higher therefore the removal of used air throughout the room is not effective. This effect may cause bacterial dispersion throughout the room and temperature draft in OT. The remedy over such problem is to increase the inlet surface area of the flat diffuser or to provide angular flow over the critical zone so that the removal of used air can be distributed evenly and exhausted quickly. It is achieved by implementation of AAD system by setting 45° louvers angle which guides the incoming air in angle and distributes uniformly throughout the OT. It is also observed that AAD system with 45° louvers angle satisfied the ADPI standard for air distribution system in critical as well as in non-critical zone which results in minimum temperature draft. The presented work methodology sets a good agreement between the numerical and experimental techniques and results.

References

[1] Ho SH, Rosario L, Rahman MM. Three-dimensional analysis for hospital operating room thermal comfort and contaminant removal. Applied Thermal Engineering. 2009 Jul 1;29(10):2080-92.
[2] Sadrizadeh S, Holmberg S, Tammelin A. A numerical investigation of vertical and horizontal laminar airflow ventilation in an operating room. Building and Environment. 2014 Dec 1;82:517-25.
[3] Woods JE, Braymen DT, Rasmussen RW, Reynolds PE, Montag GM. Ventilation requirements in hospital operating rooms—Part I: Control of airborne particles. ASHRAE transactions. 1986;92(2):396-426.
[4] Friberg B. Ultraclean laminar airflow ORs. AORN journal. 1998 Apr 1;67(4):841-51.
[5] Balaras CA, Dascalaki E, Gaglia A. HVAC and indoor thermal conditions in hospital operating rooms. Energy and Buildings. 2007 Apr 1;39(4):454-70.
[6] Chow TT, Yang XY. Performance of ventilation system in a non-standard operating room. Building and environment. 2003 Dec 1;38(12):1401-11.
[7] Chow TT, Yang XY. Ventilation performance in operating theatres against airborne infection: review of research activities and practical guidance. Journal of Hospital Infection. 2004 Feb 1;56(2):85-92.
[8] Chow TT, Yang XY. Ventilation performance in the operating theatre against airborne infection: numerical study on an ultra-clean system. Journal of Hospital Infection. 2005 Feb 1;59(2):138-47.
[9] Rui Z, Guangbei T, Jihong L. Study on biological contaminant control strategies under different ventilation models in hospital operating room. Building and environment. 2008 May 1;43(5):793-803.
[10] Villafruela JM, Castro F, San José JF, Saint-Martin J. Comparison of air change efficiency, contaminant removal effectiveness and infection risk as IAQ indices in isolation rooms. Energy and Buildings. 2013 Feb 1;57:210-9.
[11] Liu J, Wang H, Wen W. Numerical simulation on a horizontal airflow for airborne particles control in hospital operating room. Building and environment. 2009 Nov 1;44(11):2284-9.

IOP Conf. Series: Materials Science and Engineering **810** (2020) 012004 doi:10.1088/1757-899X/810/1/012004

A Review on Biological Building Blocks and its Applications in Nanotechnology

Shweta Jayachadran, Dhrriti Khetan, Kaizad Pestonjamasp, Aatur Sharma, Girish Joshi

[a] Undergraduate Student, Mechanical Engineering department, Mukesh Patel School of Technology Management and Engineering, Mumbai, India.

[b] Assistant Professor, Mechanical Engineering department, Mukesh Patel School of Technology Management and Engineering, Mumbai, India.

E-mail address: krjshweta@gmail.com, dhrritikhetan@gmail.com, kaizad07@gmail.com, aatur98@gmail.com, girish.Joshi@nmims.edu

Abstract. Biological building blocks are the basis of living organisms and its potentials are gaining more attention over a wide range of applications. major biological building blocks that exist in the universe have been taken into consideration in this paper, which are proteins, lipids, carbohydrates and nucleic acids. Due to their small size, they are such molecules which can be considered by the Nano level structures therefore have been incorporated in various applications of nanotechnology and microbiology. Taking into consideration the structural and chemical properties of these biological building blocks, further application on each individual biological building block has been covered in multiple fields, but mainly in the medical and pharmaceutical industry. The applications also concern optical imaging, precision measurement of biomolecules, Nano biomaterial synthesis and biosensors. Despite the various advantages of these applications like improved efficiency of various processes, the shared limitations of these applications stand to be difficult technique for synthesis and development due to Nano size dimension consideration. Thus, the future scope of use of biological building blocks in nanotechnology is very ambitious and promising.

1. Introduction

Biological building blocks are the basis of living organisms. They are the basic molecules that constitute to making macromolecules, cells, tissues, organs and thus organ systems. The smallest particle contributing in the making of a creature is known as a biological building block. Various biological building blocks can be combined together in multiple patterns to obtain the required results. For example, a complex bond of nucleic acid, phosphate and sugar (being the various biological building blocks) will result in the production of a DNA molecule. The wide variety of biological building blocks can be categorized into four main categories, based on their properties and functions. The four main biological building blocks which will be discussing are Lipids, Carbohydrates, Proteins and Nucleic acids. Nanotechnology is the science which pertains to the design and creation of structures and devices in the nano range, with dimensions less than 100nm [1]. With advancements in nanotechnology it is able to manipulate these naturally occurring building blocks, which can be characterised as nanoparticles, to serve specific purposes other than their own. These materials are readily available in nature, and few also show affinity for self-assembly to form structures such as nanotubes and nanowires. By exploiting these

molecules, some of the tedious processes associated with the production of nanoparticles can be eliminated. However, these molecules pose their own set of difficulties.
This has led to a wide range of possibilities for new developments in processes, from using carbohydrates in drug delivery to producing nano-electronics using lipid nanostructures. Thus, the term nano-biotechnology is coined, which is the intersection of nanotechnology and biology. This technology usually employs the bottom up approach in contrast to the top down approach of traditional methods in nanotechnology [2]. In this piece, the aim is to give a brief introduction to each of these molecules, and discuss a few applications of each of the building blocks, while also look at how these applications are impacted by the use of these building blocks. Finally, the limitations they possess in their current stage of development, and which new applications could be incorporated in these ever-growing technologies in the future are discussed.

2. Biological Building Blocks

Biological building blocks consist of Lipids, Proteins, Nucleic Acid and Carbohydrates. Lipids are classified as a large group of naturally occurring organic compounds, which are related by their insolubility in water, and solubility in non-polar solvents. Carbohydrates are the sugars, starches and fibres and are the energy source used by muscles. Proteins are formed by links of amino acids, playing an important part in formation of tissues and in defence. Nucleic acids are of two types, DNA which is responsible for storage of hereditary information and RNA which is responsible for expressing said information [3] [4].

2.1. Proteins

Proteins are one of the most vital biological building blocks. Due to its considerably large size, it is considered as a macromolecule. The structure of a protein mainly consists of multiple units of amino acids. Therefore, it can be said that protein is a polymer of structural units of amino acids. Going into the depths of the chemical composition, it is observed that an amino acid is made up of an acidic carboxyl group, amino group and an organic R group. This organic R group helps give each different amino acid its own individual identity. Various types of amino acids are present in nature. A total count of 20 has been made so far. These different types of amino acids can thus be joined together in various different patterns and orders to get the product as a variety of different proteins. The other way around to look at it is that on breaking up proteins one can release the base amino acids to the surrounding. This process of cleavage or breaking of a covalent bond by the addition of a water molecule is known as hydrolysis. Proteins perform an important role in the human body by helping in facilitating various body functions and contributing to body structures. It is a very important nutrient for the human body. Major functions of proteins can be listed as, structure and support- main components of tissues, communication and signalling, defence, regulation, catalysis.

2.2 Carbohydrates

A carbohydrate is a biomolecule consisting of carbon (C), hydrogen (H) and oxygen (O) atoms. Carbohydrates are the sugars, starches and fibers commonly found in fruits, grains, vegetables and milk products. Their main task is to provide fuel for the central nervous system and energy for working muscles. They consist of hydrogen and oxygen atom in the ratio of 2:1 (as in water) and thus with the empirical formula Cm(H2O). The carbohydrates are technically hydrates of carbon. Anthracycline antibiotic doxorubicin (DOX) has been used for the past 3 decades as a capable therapeutic agent for the treatment of a lot of human tumours, mainly breast cancer and lymphoma. The carbohydrates are technically hydrates of carbon [5].

2.3 Nucleic Acids

Nucleic acids are big and complex molecules that are present in cells and viruses. They are responsible for storage of hereditary information and expressing said information through protein synthesis. They are made up of small units called nucleotides. Nucleotides are made of 5 carbon sugar attached to a nitrogen-containing aromatic base and phosphate groups. Each nucleic acid has 4 nitrogen- containing bases out of 5 namely adenine(A), guanine(G), thymine(T), cytosine(C) and uracil(U). All nucleic acids contain A, G and C. The presence of T and U depends on the type of nucleic acid [1]. Nucleic acids are of two types namely DNA and RNA. Figure 1shows the structure, nucleotides and complimentary pairs in RNA and DNA.

DNA short for Deoxyribonucleic acid is the blueprint of all living organisms. It contains hereditary information (DNA code) and is the ideal genetic material. It is a double helix strand that has thymine(T) as the 4th nucleotide along with A, G and C. They occur in complementary pairs. A and T form a pair using 2 hydrogen bonds whereas G and C pairs [6].

Ribonucleic acid also known as RNA is responsible for decoding the DNA code and interpreting the information. It 'reads' the DNA code and formulates it by synthesizing proteins through transcription. It is single stranded and has A, G, C and uracil(U) as nucleotides. The complimentary pairings are same with the exception that thymine is replaced by Uracil [6].

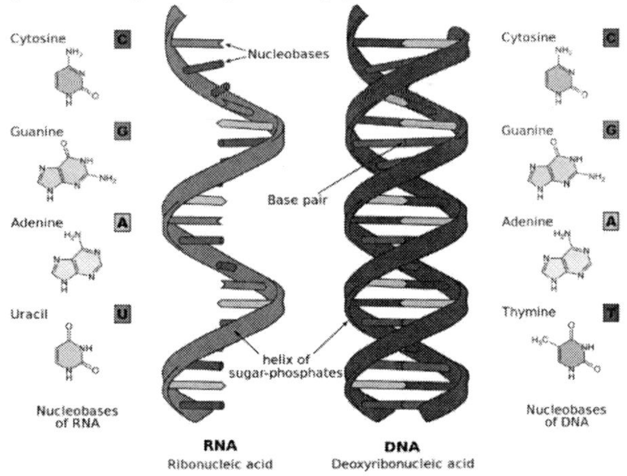

Fig. 1 Structure, nucleotides and complimentary pairs in RNA and DNA [7]

2.4 Lipids

Lipids are classified as a large group of naturally occurring organic compounds, which are related by their insolubility in water, and solubility in non-polar solvents such as alcohols and ether. Lipids are generally hydrophobic, while some are amphipathic (consisting of both hydrophobic and hydrophilic groups, with the hydrophobic group being larger). Examples of lipids include fatty acids, waxes, triglycerides and steroids. As lipids are very diverse, so are their functions. Lipids are utilized for the storage of energy, protection of internal organs, to forming structural components of cells. Cholesterol is a chain type structure of lipid as shown in figure 2, which is vital for humans, and is used for many processes including production of vitamin D and hormones a. Cholesterol molecules are attached to lipoproteins to enable their travel through the bloodstream, forming packages with lipids on the inside and protein on the outside [8-10].

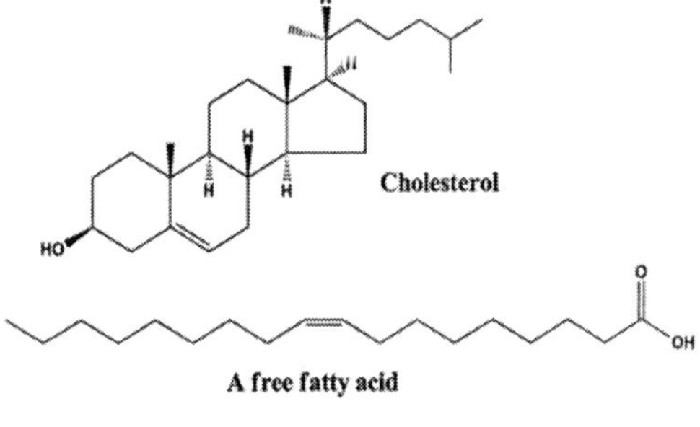

Fig. 2 Chain of Fatty acid and Cholesterol [11]

Considering the structural and chemical properties of all the biological building blocks, applications on each individual biological building block have been established, mainly in the medical and pharmaceutical industry. The applications also concern optical imaging, precision measurement of biomolecules, Nano biomaterial synthesis and biosensors. Further, more significant applications of these categories, which are listed down below.

3. Applications of Biological Building Blocks

3.1 Proteins:

3.1.1 Precision Measurement of Biomolecules and cells:

Previously, measurement of biomolecules and cells has been done using bulk assay. Bulk assay is a method of estimating the concentration of the required substance in a bulk material. This method has been proved to be quite inaccurate in various experiments as it gives no exact estimate due to bulk approach. In turn, the greatly reduced size of nanostructures acts as an advantage for this method, enabling it to probe and read the dynamics of various cells and molecules with great precision giving high resolution results. Structures like biological channels and pores are capable of regulating ion transport to an order of 0.1nm. Nano scale pores are also capable of distinguishing between various molecules on the basis of size and chemical properties. Inspired by this biological functionality, it has been hypothesized that Nano pores could unwind and unzip DNA so that individual nucleotides translocate sequentially in single file.

There are two types of Nano pores that take into consideration for this application, which are biological Nano pores and artificial Nano pores. The first one is a protein based pore known as α-haemolysin (αHL), which has good structured apertures, lacking mechanically stable lipid bilayers. The second one is an artificial Nano pore, thus it does not have a chemical structure like that or organic compounds, due to which they showcase a decrease in selectivity.

For the application, certain properties of both the Nano pores are essential. This lead to the development of a hybrid between the two Nano pores. For the synthesis of this hybrid Nano pore, a αHL is inserted into the structure of an artificial Nano pore. This hybrid leads to an overall improvement of the Nano pore properties as the two Nano pores combine to overcome each other's drawbacks. The hybrid nanoparticle thus has better selectivity, coming from the biological Nano pore, as well as mechanical stability, coming from the artificial Nano pore.

An issue that this application is yet to overcome is the problem in the synthesis of precise 1nm diameter of Nano pores, which is very difficult to implement. This application has a lot more to achieve, which involves the use of multiple Nano pores in parallel. [12]

3.1.2 Circulating Tumour Cells (CTC) Chips:

Cancer is one of the most fatal diseases that mankind has had to encounter. Till date, a definite cure of cancer has not yet been discovered. Cancer is caused due to various abnormalities of healthy cells of the human body, as a result of a number of causes. The resultant abnormal cell or cancerous cell is thus known as a malignant cell. The reason why cancer is such a fatal disease is not just because malignant cells are difficult to diminish, but also that the means to curb them from spreading to more places of the body is difficult. This is known is metastasis.

Metastasis, in malignant tumours, is caused mainly due to Circulating Tumour Cells (CTC's). CTC's are shed by metastatic malignant tumours directly into the bloodstream of the nearby blood vessels. CTC's can be considered to be extremely rare as their concentration is approximately one per billion in clinical samples, when intermixed with blood cells. This makes it even more difficult to detect these CTC'S, which is why nanotechnology is being approached as a more sophisticated technique to help ease the same.

Now, a biomarker is a molecule which helps identify a particular type of molecule among all the different ones. A well-known biomarker for CTCs is EpCAM, which stands for Epithelial cell adhesion molecule (Transmembrane glycoprotein), which is expressed by cells of epithelial origin but lacking in blood cells.

For better detection of CTC's, CTC-Chips were devised. CTC-Chips work on the principle of increased surface area as shown in figure 3. Blood is collected from a patient and is pushed through the surface of the CTC Chip pneumatically. Each post is coated with EpCAM antibodies. This thus leads to the required detection The enhanced surface area is associated with micro/nanostructures for adhesion-based capture using arrays of EpCAM-functionalized micro pillars. The application of CTC Chips was clinically accepted to help track the disease at hand, and to track its growth. This application had some drawbacks due to the practicality of the application with pertaining to its small sized dimension. When the clinical sample is poured over the CTC Chip, due to the small size of the micro pillars, it causes slow mixing. Another limitation of this application is that it cannot be used to detect whole tumours asit would take a long time, over a few years, before a tumour could be diagnosed. Lastly, the production of these nanostructures are very difficult due to the small lengths and specified production techniques [12].

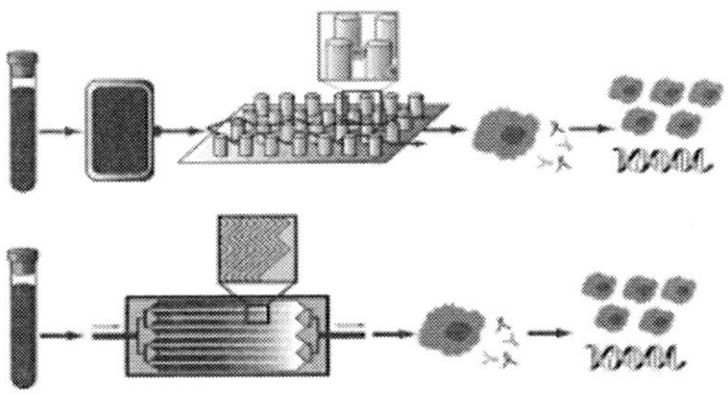

Fig. 3 Schematic of working of Circulating Tumour Cells [13]

3.1.3 Biosensors:
Biosensors are devices which are used to detect and quantify chemical and biological substances in a sample. Working principle of biosensor is given in figure 4. They do so by incorporating sensory device with a transducer. As most of the substance to be detected by a biosensor are in the range of nanometres, having biosensors on the same scale makes them more efficient systems. Also, nanoparticles possess a huge surface area to volume ratio because of their size, and thus are extremely sensitive to even small changes on surface [14].

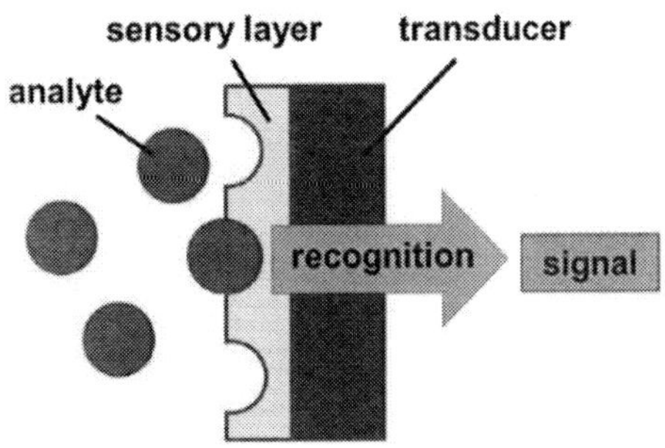

Fig. 4 Principle of a biosensor [15]

Proteins and peptides have been tested experimentally to be used as nanowires for biosensing devices, by modifying the electrodes to form a porous nanowire. The required enzyme must be immobilised on the surface of the protein nanostructure by methods such as adsorption and cross-linking the molecules. Such devices have been shown to be capable of detecting glucose and hydrogen peroxide. Although the devices created for each of the above molecules have different methodologies, the basic principle is the same as shown in figure 5.

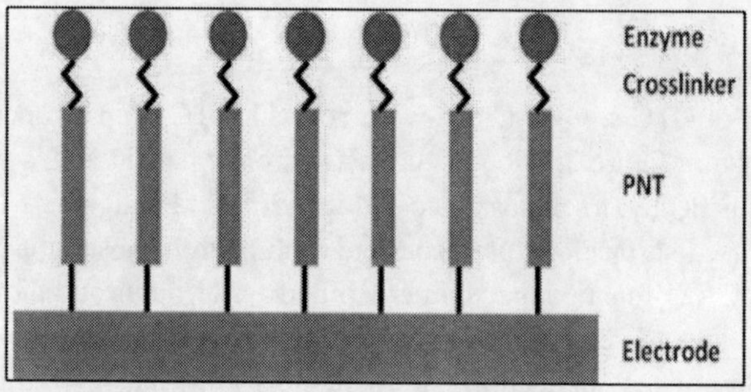

Fig. 5 Modified nanowire Biosensing device

Protein nanowires can be formed by the self-assembly of naturally occurring fibrous proteins, which makes them easily available in large quantities. As proteins are made up of amino acid chains, there is scope for modification of proteins for desired results. Another benefit of such devices is the biocompatibility of proteins and peptides. However, as this is a developing technology, a lot of further research is needed before these devices can move from experiments to reality. [16]

3.2 Carbohydrates

3.2.1 Drug Delivery:

The drug to be supplied to the affected organ is encapsulated in a nano carrier, which is an allotrope of carbon and is supplied to the diseased organ where the drug is released. Nanoparticles are very useful in delivering target specific drugs. This prevents the drug from affecting the non-diseased organs hence reducing unwanted side effects.

The nanoparticle/ nano carrier encapsulates the drug and delivers it to the pre designated target area. At 80,000 rpm and a duration of 30 minutes the encapsulation efficiency of Doxorubicin- Dextran Sulfate (DOX-DS) nano-complexes can be analysed by the help of ultracentrifugation of complexes. A free drug solution is used as a control to ensure that the free drug does not precipitate at that high a speed of rotation. Spectroscopic measurement at 480 nm gives the concentration of the remaining DOX. Encapsulation efficiency depends on the nano carrier used. Since DOX is the main drug used in the case of a tumour, the encapsulation capacity of DOX by DOX-DS nano-complexes is taken into consideration and it was observed that it is highly affected by DS/DOX ratio. On increasing the DS/DOX ratio to 0.5, the peak of 95% is achieved as shown in graph. [17]

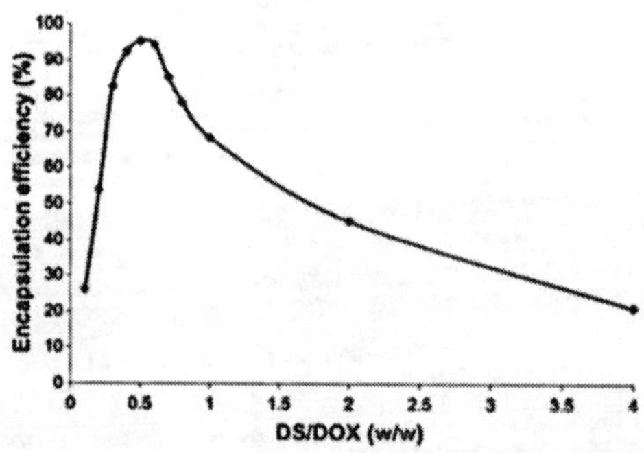

Fig. 6 Variation of values of DS/DOX with respect to encapsulation efficiency [17]

Now on approaching the target site, the drug has to be successfully untangled from the nano carrier and affect the infected organ. In-vitro drug release studies depict that 14% of DOX drugs are released from DOX-DS in the first 24 hours, followed by a slower release of 32% in total over the next 15 days. Apart from self-assembly, emulsion techniques are interesting and versatile methods for the in situ formation of carbohydrate-based nanocarriers, nanocapsules can be prepared in an inverse miniemulsion.

3.2.2 To overcome the DOX resistance:
Anthracycline antibiotic doxorubicin (DOX) has been used for the past 3 decades as a capable therapeutic agent for the treatment of a lot of human tumors, mainly breast cancer and lymphoma. Many solutions have been devised by the means of nanoparticle carriers. Drug resistance can be overcome by nano carriers by enclosing themselves in an endoscope and then allowing it to enter the cell which helps by-pass the P-glycoprotein efflux pump (drug resistant mechanism) and combating the efflux of drug from the affected resistant tumor. With the help of Nanotechnology, the binding of drugs to polymers can be successfully achieved. A positive charge is possessed by DOX and it's been observed that it forms many complexes with multiple anionic polymers such as polyacrylic acid γ-polyglutamic acid, polyaspartate, polyglutamate, block ionomers of aspartate, benzyl aspartate, benzyl glutamate, and polyethylene oxide. Polyelectrolyte complexes are formed by these polyanions with DOX.

A highly anionic carbohydrate called Dextran Sulfate (DS) has been studied for improving the efficiency of encapsulation of DOX. Janes et al used DS to improve loading of DOX into chitosan nano particle. In further studies DOX was introduced with various other polymer dextrans having different molecular weights to achieve modified cytotoxicity and improved DNA-binding characteristics. Lam et al [18] could successfully show that the DOX-DS conjugate depicted excellent antitumor activity as compared to the free removal rate of DOX from P-glycoprotein over-expressing multi-drug-resistant cells once conjugated to dextran.

3.2.3 Bionanomaterials:
Bionanomaterials are molecular materials that are made up partially or completely of biological molecules (such as proteins, enzymes, RNA, DNA, oligosaccharides, lipids, viruses, and cells etc) which results in molecular structures that have a dimension in nanoscale. Carbohydrates have been reacted with various nanoparticles in order to obtain multiple different bionanomaterials, which are then known as nano-carbohydrates. Nano-carbohydrates can be used in Carbohydrate lectin recognition, Biosensing, Screening, Cell imaging and bacterial detection. Some Nano-carbohydrates that have been synthesized are glycosylated nanoparticles of gold nanoparticles, fullerenes, quantum dots and single wall nanotubes.

Nano-carbohydrate optical imaging is an important application of glycosylated quantum dots. Quantum dots are small (< 10 nm) fluorescent semiconductor nanocrystals. They possess luminescent properties which are unique as compared with more established fluorescent proteins and organic dyes. Their fluorescence emissions are stable and are dependent on the size of the particle [19]. Cadmium selenium–zinc sulphide quantum dots find their application in visualising /capillaries which are hundreds of micrometres deep into the skin of living mice. Cadmium-based quantum dots are used to image the cerebral vasculature. Methoxy-PEGylated quantum dots are used to visualise arterioles and capillary networks in mouse hind limb skeletal muscle [20].

3.3 Nucleic Acid
3.3.1 DNA Nanotechnology:
The foundation of DNA Nanotechnology was given by Nadrian C. Seeman, an America nanotechnologist and crystallographer who has constructed the first practical Holliday junction. It was realized that DNA bunched junctions can be linked together from their sticky ends into a crystalline 3D material. Thus, assembling of crystals became more precise due to predictable interactions. This formulation started in-depth study in DNA and its uses in nanotechnology. DNA branched junctions are very geometrically flexible and instable because of branch migration due to two-fold symmetry at branch points. Seeman himself reported that Holliday junctions can be made immobile by making arms with unique

IOP Conf. Series: Materials Science and Engineering **810** (2020) 012004 doi:10.1088/1757-899X/810/1/012004

sequences/remove symmetry in the assembled molecule. To overcome this issue, Seeman and Erik Winfree developed a more rigid double-crossover (DX) motif. It has two DNA strands, double helix, linked together with two strand exchange rather than a single strand as seen in Holliday junction. These offered the stability that was lacking and hence was used in constructing DNA structures [21].

The first 3D object created was in Seeman's lab, which was a cube of DNA, also the very first DNA Object. DNA objects are just DNA constructs that can be used to position and orient molecules. The objects can hold macromolecules as guests inside them which are then manipulated to orient them as desired as in figure 7 [22].

Fig. 7 Assembling of macromolecules using DNA constructs

Information in DNA was used to form predesigned crystalline arrays. These arrays are made from DNA tiles made from DX molecules which can grow into 2D crystals that can be categorized by AFM. In the following years, many 2D and 3D discrete DNA structures were reported. It was observed that DNA could be used as a component for molecular machines. Nano mechanical devices were also developed. Various other tile motifs were developed after DX tiles. It was also found that aperiodic crystals can made from DNA [23].

One of the major discoveries was DNA origami. In 2006, Paul Rothemund proposed a method for DNA construction in which a scaffold of a single DNA strand folds into itself to form the desired pattern. Non-periodic 2D structures like a smiley or Map of America were also made. This achievement created surface area of around 100nm to work on [21]. Various molecular machines were also created. Some of them are discussed in this paper.

DNA Tweezer is used for removing drugs from inside parts. It contains three strands: A, B and C. Strand A latches onto half of strand B and half of strand C, and so it joins them all together. Strand A acts as a hinge so that the two "arms" AB and AC can move. The structure floats with its arms open wide. They can be pulled shut by adding a fourth strand of DNA (D) "programmed" to stick to both of the hanging, unpaired sections of strands B and C (other side).The closing of the tweezers was confirmed by tagging strand A at either end with light-emitting molecules that do not emit light when they are close together. To re-open the tweezers add a further strand (E) with the right sequence to pair up with strand D. Once paired up, they have no connection to the machine B-A-C, so float away. The DNA machine can be opened and closed repeatedly by cycling between strands D and E [24].

A DNA walker is a class of nucleic acid nano-machines where a nucleic acid "walker" is able to move along a nucleic acid "track". The concept of a DNA walker was first defined & named by John H. Reif in 2003. DNA walkers have functional properties such as a range of motion extending from linear to 2 and 3-dimensional, the ability to pick up and drop off molecular shipment, performing DNA-template synthesis, and increased velocity of motion DNA walkers have potential applications ranging from nano-medicine to nano robotics. Many different fuel options have been studied including DNA hybridization, hydrolysis of DNA and light. The applications of DNA walkers include nanomedicine, diagnostic sensing of biological samples, nanorobotics.

3.3.2 RNA Aptamer and Apta-sensors:
RNA, just like DNA, can be manipulated and designed. They have different base-pairings than DNA and have large number of loops that makes it easy to manipulate and make complex structures. RNA Nanotechnology can be considered as a subdivision of DNA Nanotechnology however, with the potential in this field, it can be soon regarded as its own discipline [21]. RNA forms an ideal material for various applications due to the versatility of RNA structure, low free energy in RNA annealing, structure control options and self- assembly property. It can be used to construct arrays and superstructures. RNA nanostructures are designed co-transcriptionally. Since, transcription is slow in RNA assembly, nascent RNA strand can be used to get desired shape by using hairpins and RNA-based association motifs. Self-assembly property of RNA ladders can direct arrangement of cationic gold nanoparticles. RNA's ease of conjugation and biocompatibility can be utilized in tissue engineering [25].

RNA aptamer is a family of oligonucleotides with functions similar to that of antibodies in their ability to recognize specific ligands through the formation of binding pockets. These can be used as inhibitors or in targeted therapeutic delivery. They have various advantages being used as targeting agents. They are low cost and are convenient to synthesize and modify, have rapid tissue penetration and long term stability. RNA aptamers are incorporated into RNA nanoparticles by sequence fusion. Many RNA aptamers are present for targeting cell surface. These aptamers can be used for in vitro diagnosis. Apta-sensors, aptamer based biosensors, detect small molecule targets and peptide markers. Using apta-sensors can lead to early detection and diagnosis [26].

3.4 Lipids:
3.4.1 Nanoelectronics:
Nanoelectronics is a field of nanotechnology which deals with electronics with sizes in the range of nanometres. These electronic devices can be incorporated in a wide range of applications including telecommunications and information processing. Human biological systems are capable of handling tasks such as material and information transport, and thus can be useful for the production of novel nanoelectronics. By designing electronics based on biological systems and processes, it can incorporate biological building blocks such as lipids to create nanoelectronics devices.

In this respect, lipids have been used to create electronics such as batteries and rectifiers by combing lipids bilayers with protein pores. Transistors are one of the most promising devices to be influenced by the new generation of nanoelectronics. Transistors have been produced by the combination of lipid bilayer shells with proteins, and the combination lipids and graphene. Lipid nanotechnology has proven to be useful as lipids are biologically inert, and also form nano sized containers by self-assembly. Although lipids in nanoelectronics are exciting prospects for the future, the scalability of these designs for high production rates are under consideration [27].

3.4.2 Drug delivery:
As discussed earlier, drug delivery is one such application where the induction biological building blocks provides interesting findings. Like carbohydrates, lipids can also be used for drug delivery. This is achieved with the formation of a lipid-drug conjugate, which is a drug molecule modified by lipids. Drugs have been modified with steroids, fatty acids, glycerides and phospholipids, where lipid nanoparticles can be used as a delivery system. Drugs are encapsulated in the lipophilic core of the lipid nanoparticles. Both solid and liquid nanoparticles have been used for these purposes, with solid nanoparticles showing better loading and delivery to specific organs.

Promising results have been seen in cancer treatment, where lipid-drug conjugates have been used to improve tumour targeting and to overcome drug resistance. Tumour cells have a high uptake of cholesterol, thus drugs modified with the lipid cholesterol show better targeting of tumour cells than free drugs. Lipid-drug conjugates have also been used in orally administered drugs, and have shown an improvement in performance of the drug, by increased interaction with cell membranes and increase in uptake and permeation of the drug [28].

4. Conclusion and discussions

The importance of nanotechnology in biology is unique and its amalgamation has brought about advances that were not possible before. In this paper, it has seen their various applications though not all due to space constrictions. However, there are limitations to using this technology at this date. Due to increased surface area, chemical reactivity of these particles increase immensely and hence, their behaviours in different conditions cannot be predicted. This increased reactivity produces reactive oxygen species (ROS) that can damage DNA, membranes and proteins. They are highly depended on their environment causing disintegration or aggregation leading to decrease in quantity or variations in sizes. Since nano-medicine or nano-devices will be and is currently employed worldwide, the conditions across the world are not same and hence their stability has become a topic of discussion. The major issue in this field now is accurate target delivery, thermodynamic instability and high production cost among others. The production costs can be overcome with more availability of the technology and its advances.

Researchers are still working on designing lipid molecules that have better functions than natural lipids. Few have already been incorporated in gene delivery. Profiling of proteins and their interactions has led to advances in biomarker identification, development of vaccine and profiling in immunology. Proteins have also been used in development of Nano devices which still need more research to be attain their potential. New Applications of Nano-carbohydrates are emerging due to their efficiency, some of them in bacteria detection, screening and bio sensing. Nucleic acid based nanomaterials are being used in drug delivery in cancer therapeutics. They still have major shortcomings and hence, can't be used in various fields. It has been highly believing that with the advancement of nanotechnology in the future all shortcomings seen in this paper can be overcome. This domain is still relatively new and with more research can become the one of the most prominent and impactful field.

5. References:

[1] Silva, Gabriel A. "Introduction to nanotechnology and its applications to medicine." Surgical neurology 61.3 (2004): 216-220.

[2] Oded Shoseyoy, Ilan Levy, NanoBioTechnology: BioInspired Devices and Materials of the Future, Springer Science & Business Media, 07-Feb-2008

[3] Urry, Lisa A., Michael Lee Cain, Steven Alexander Wasserman, Peter V. Minorsky, and Jane B. Reece. Campbell biology. Pearson Education, Incorporated, 2017.

[4] Kuszewski, John, Charles Schwieters, and G. Marius Clore. "Improving the accuracy of NMR structures of DNA by means of a database potential of mean force describing base– base positional interactions." Journal of the American Chemical Society 123.17 (2001): 3903-3918.

[5] Jessie Szalay; "What Are Carbohydrates?", Live Science, 2017

[6] Richard J. Roberts "Nucleic acid" Encyclopædia Britannica, Encyclopædia Britannica, inc.,2019

[7] Roland1952 "Chemical structures of nucleobases", "Nucleic Acid" Wikipedia, The Free Encyclopedia, 2001

[8] William Reusch "Natural products" – Lipid Membranes Eric Martz, Univ. Mass. Amherst cf. H Heller, M Schaefer, & K Schulten, J. Phys. Chem 2013

[9] Thomas E. Thompson "Lipid" Encyclopædia Britannica, Encyclopædia Britannica, inc. 2019

[10] https://medlineplus.gov/cholesterol.html ,United States National Library of Medicine Eoin Fahy "Common Lipids" lamps at he English Wikipedia, The free Encyclopedia 2008

[11] Wong, Ian Y., Sangeeta N. Bhatia, and Mehmet Toner. "Nanotechnology: emerging tools for biology and medicine." Genes & development 27, no. 22 (2013): 2397-2408.

[12] Su, Derrick W., and Jorge Nieva. "Biophysical technologies for understanding circulating tumor cell biology and metastasis." Translational lung cancer research 6, no. 4 (2017): 473.

[13] Nikhil Bhalla, Pawan Jolly, Nello Formisano, and Pedro Estrela, Introduction to biosensors 2016

[14] Bhalla N, Jolly P, Formisano N, Estrela P. Introduction to biosensors. Essays Biochem. 2016;60(1):1–8. doi:10.1042/EBC20150001

[15] Domigan, Laura J. "Proteins and peptides as biological nanowires: Towards biosensing devices." Protein Nanotechnology. Humana Press, Totowa, NJ, 2013. 131-152.

[16] Yousefpour, Parisa, Fatemeh Atyabi, Ebrahim Vashegani Farahani, Ramin Sakhtianchi, and Rassoul Dinarvand. "Polyanionic carbohydrate doxorubicin–dextran nanocomplex as a delivery system for anticancer drugs: in vitro analysis and evaluations." International journal of nanomedicine 6 (2011): 1487.

[17] Lam, Wing, Hingleung Chan, Mengsu Yang, Shukhan Cheng, and Wangfun Fong. "Synergism of energy starvation and dextran-conjugated doxorubicin in the killing of multidrug-resistant KB carcinoma cells." Anti-cancer drugs 10, no. 2 (1999): 171-178.

[18] Nasrollahzadeh, Mahmoud, Zahra Issaabadi, Mohaddeseh Sajjadi, S. Mohammad Sajadi, and Monireh Atarod. "Types of Nanostructures." In Interface Science and Technology, vol. 28, pp. 29-80. Elsevier, 2019.

[19] Hussain, Chaudhery Mustansar, ed. Handbook of Nanomaterials for Industrial Applications. Elsevier, 2018.

[20] Seeman, Nadrian C., and Hanadi F. Sleiman. "DNA nanotechnology." Nature Reviews Materials 3, no. 1 (2017): 1-23.

[21] Seeman, Nadrian C. "Nanomaterials based on DNA." Annual review of biochemistry 79 (2010): 65-87.

[22] Seeman, Nadrian C. "An overview of structural DNA nanotechnology." Molecular biotechnology 37, no. 3 (2007): 246.

[23] Guo, Peixuan. "The emerging field of RNA nanotechnology." Nature nanotechnology 5, no. 12 (2010): 833.

[24] Jasinski, Daniel, Farzin Haque, Daniel W. Binzel, and Peixuan Guo. "Advancement of the emerging field of RNA nanotechnology." ACS nano 11, no. 2 (2017): 1142-1164.

[25] Mashaghi, Samaneh, Tayebeh Jadidi, Gijsje Koenderink, and Alireza Mashaghi. "Lipid nanotechnology." International journal of molecular sciences 14, no. 2 (2013): 4242-4282.

[26] [28] Danielle Irby, Chengan Du, and Feng Li "Lipid-Drug Conjugate for Enhancing Drug Delivery." Molecular pharmaceutics vol. 14,5 (2017): 1325-1338

IOP Conf. Series: Materials Science and Engineering **810** (2020) 012005 doi:10.1088/1757-899X/810/1/012005

Automotive vehicle clutch engagement characteristics for different types of force trajectories

D Raghuvanshi[1], Vinod Pare[2], M Chouksey[3] and B R Rawal[4]

Mechanical Engineering Department, Shri G.S. Institute of Technology and Science, Indore-452003, India

Email: [2]vinod.pare@gmail.com, [3]machouksey@gmail.com

Abstract. Vehicle driveline of automobile, which transmit engine energy to the wheels, is complex system with nonlinear behavior. In manual transmission clutch engagement performance depends upon driver's skill and is significantly affected by trajectory of clutch force during engagement process. In this work different types of clutch force trajectories are analyzed based on engagement performance which can be incorporated in automated manual clutch system. A driveline of a vehicle is modeled using multi body dynamics software and simulation were carried out for different type of force trajectories i.e. linear, parabolic and spline, which is followed by clutch force during engagement. The results of simulation were used to analyze the quality of clutch engagement and vehicle performance by looking into clutch lock-up phase and vehicle launch. The clutch lock time and clutch lock speed for various types of trajectories were calculated and compared. These results were also compared with analytical results reported by researchers. The results would be useful in selecting appropriate type of clutch actuating force trajectories for automation of clutch system of automobiles.

1. Introduction

Numerical simulation of various components of automobiles, e.g. of suspension system, transmission system etc., helps in studying the dynamics of the vehicle [1]. A clutch system is an integral part of manual transmission in vehicle driveline. A controlled engagement operation of clutch is required for smooth driving of vehicle which largely depends upon driver's experience and skills. Accuracy in clutch engagement operation is essential to minimize driveline oscillations and vehicle body, heat generation and loss of power due to slippage, wear and enhance driving comfort. Since clutch is repetitively used for gear shift operation and vehicle starting, some variation may be expected due to human limitations [2].

One of the important factors which affect the engagement quality of clutch system is the trajectory of clutch force [2] which is defined as path followed to acquire the final clutch force during engagement phase and depends upon the skill of driver. The clutch engagement process has three phases due to stick-slip motion. Transitions between these phases introduce non-linearity in vehicle dynamics, which result in vehicle jerk [3]. In order to minimize driver intervention and make the clutch engagement or clutch locking process automatic and efficient, a control system with a predefined trajectory may be employed. This automated manual clutch will help in to improve the driving performance as well as fuel efficiency [2, 4, 5].

Various type of paths or trajectories have been discussed in literature from low order-linear type, which provide constant loading rate to high order spline type for the automation of clutch system. The engagement characteristics whether smooth or sudden during starting and end of engagement depends upon types of trajectory profile [6]. The type of path is also responsible for any overshooting of clutch force while engagement.

Literature shows that various parameters have been used for appraisal of engagement quality of clutch [7]. Gao et al. [8] suggested minimizing lockup time, friction losses and vehicle jerk as indicators for evaluation of clutch system. Chen et al. [3] proposed to minimize jerk, wear and frictional dissipation by smooth transition. Maximum engine speed, lockup time, Jerk and Shuffle were used by Tripathi [1] to evaluate engine quality.

The ride comfort will increase with selection of appropriate trajectory. This trajectory may be implemented by using embedded mechatronics system which consist of an actuator governed by control system to develop automated manual clutch. Researchers [4, 6, 14] have proposed automated manual transmission that uses electro-hydraulic and electro-mechanical system which are driven by control unit for gear shift. Gao et al. [8] conducted simulation for two-speed automotive manual transmission and proposed to use fitted trajectory for feed forward input and used a PID controller for feedback control against modelling error. Chen et al. [3] proposed model referenced adaptive control (MRAC) instead conventional control to control friction torque which reduce the effect of discontinuity in process of clutch engagement due to stick-slip.

Various researchers [4, 5, 9, 10, 11] reported mathematical model of driveline considering non linearity and carried out transient analysis. Couderc et al. [9] also compared simulation with experiment results. Centea et al. [10] used a nonlinear multi body dynamic model of drive line and clutch system to investigate the torsional vibration. Chen et al. [3] used Popov hyper-stability criterion to compensate the discontinuity in stick-slip process numerical simulation of clutch engagement. Duque et al. [12] presented mathematical model for clutch system while considering Coulomb, Viscous and Stribeck friction and found good correlation with Stribeck model in terms of energy as well as slipping time. Vasca et al. [5] discussed a model of torque transmissibility as function of friction and contact pressure.

In this work, the driveline of the vehicle is modeled using a multibody dynamic software MSC ADAMS and simulation was performed. Three different types of engagement trajectories to be followed by clutch force were used in simulation of clutch system to study and compare the dynamic response of driveline and clutch engagement performance.

2. Vehicle driveline model

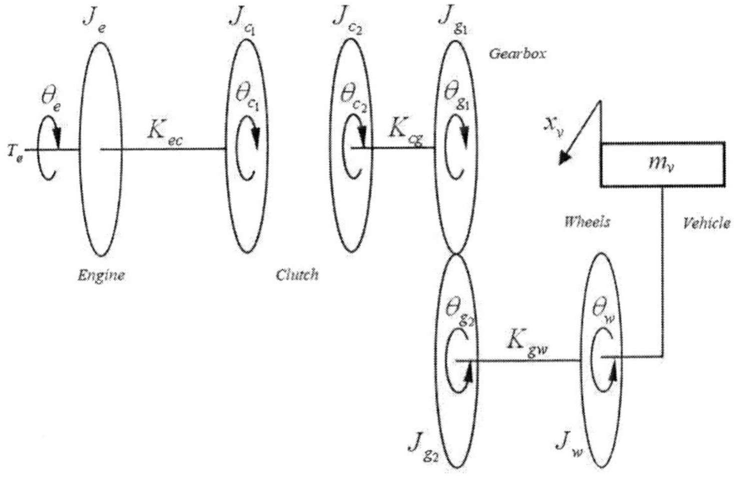

Figure1. Model of vehicle drive line [1,3,6]

The vehicle driveline makes up of various components which transmit power from engine crankshaft to wheel through clutch and transmission system [13]. Figure 1 shows schematically the dynamic model of driveline of an automotive [2, 6]. The meaning of various symbols used in model are explained in list of symbols at the end of the paper. Many works have been proposed of dynamic models comprising of simultaneously algebraic differential equations along with their solution [2, 4, 7].

3. Modeling and simulation

A multibody dynamics simulation software ADAMS was used to model and characterize the vehicle driveline and clutch engagement. The vehicle driveline has been modelled in ADAMS software as per the schematic diagram shown in figure 1. Further simulation study has been carried out on the developed model using various input parameters. The numerical model considers lumped inertia for engine, clutch, gear and wheels. The numerical values of various parameters associated with medium size light motor vehicle were used to model the driveline as used by [7] and have been listed in table 1.

Table 1: Vehicle parameters used in simulation model [7]

Parameter	Value	Unit
Moment of Inertia (Engine)	0.16	kg m^2
Moment of Inertia (Clutch Disk 1)	0.016	kg m^2
Moment of Inertia (Clutch Disk 2)	0.016	kg m^2
Moment of Inertia (Gear 1)	0.04	kg m^2
Moment of Inertia (Gear 2)	0.04	kg m^2
Moment of Inertia (Transmission and Wheels)	0.24	kg m^2
Torsional Stiffness (Engine-Clutch Shaft)	32000	N-m/rad
Torsional Stiffness (Clutch-Gear Shaft)	3200	N-m/rad
Torsional Stiffness (Counter Shaft and Axle)	16000	N-m/rad
Damping Coefficient (Crank Shaft)	100	N-m-s/rad
Damping Coefficient (Clutch-Gear Shaft)	4	N-m-s/rad
Damping Coefficient (Counter Shaft and Axle)	90	N-m-s/rad
Damping Coefficient (Support Bearings)	0.012	N-m-s/rad
Mass of Wheel	5	kg
Mass of Vehicle	1000	kg
Wheel Radius	0.3	m
Maximum Engine Torque	120	N-m
Friction Coefficient (Clutch)	0.3	
Clutch Plate ID	0.2	m
Clutch Plate OD	0.3	m
Number of Friction Surfaces	2	

Constrains were specified based on functional requirement of the system. The translational movement of clutch plate was specified according to profile to trajectory. Constant value of friction coefficient was used for friction torque function [12]. The gear ratio and loading condition were considered constant.

The linear and angular displacement and velocities at critical points were used as state variables in simulation. Figure 2 shows the isometric view of the complete driveline model in ADAMS/view. Simulation of this model has been done to calculate the parameters.

4. Simulation of vehicle driveline

Three different generic trajectories i.e. linear, parabolic and spline were selected to study their effect. These paths were specified in model by using appropriate function to see the effect on quality of engagement and the ride comfort output. The clutch release time was taken as 0.7 second, as it corresponds to small-size car. The maximum clutch force was taken as 2000 N and the coefficient of friction for the clutch is taken as 0.5. Simulation of vehicle driveline was carried out to obtain the dynamic behavior of drive line. An important element of dynamic response of driveline is the variation of flywheel shaft and clutch shaft speeds during engagement. A detailed analysis of variation of engine crank shaft and clutch shaft Speed was carried out in ADAMS. Figure 3 shows the effect of different trajectory functions (for the clutch engagement) on to the variation of angular velocities of flywheel and clutch over the time. It can be shown that the trajectory path significantly affects the flywheel and clutch angular velocities.

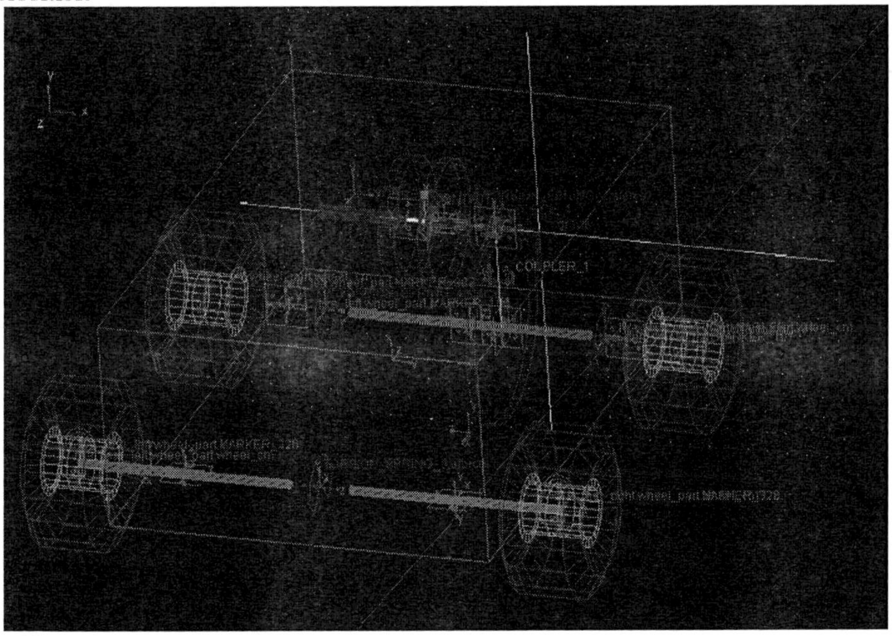

Figure 2. Complete Driveline Model in ADAMS

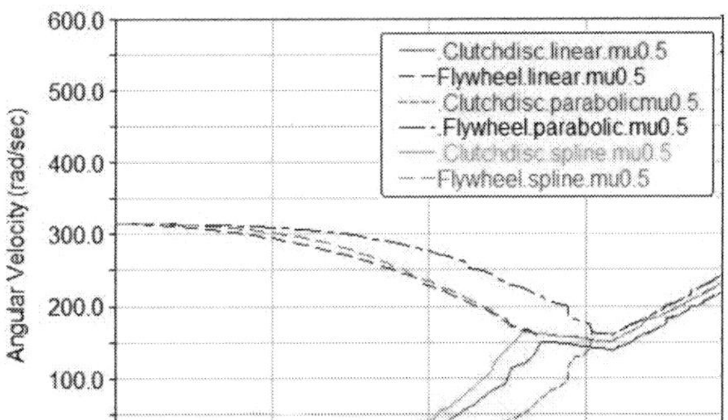

Figure 3. Effect of different trajectories on variation of flywheel and clutch angular velocity

5. Results and discussion

The simulations were conducted by considering three different type of paths for clutch force during engagement. The engagement quality of clutch system was evaluated by calculating clutch lock-up speed and clutch lock-up time which are significantly affected by nature of trajectories as shown in the figure 3.

The clutch lock-up speed is a speed corresponding to lock-up point, where engine crankshaft speed becomes equal to clutch shaft. The speed value corresponding to lock-up point depends upon engine specification and varies from engine to engine. A too low lock-up speed may result into killing of the engine [2].

During engagement process clutch shaft speed increases and engine/ flywheel shaft speed decreases due to inertial of driven parts i.e. wheels and vehicle body, as shown in figure 3. The clutch lock-up speed was noted more or less same for spline type path (~160 rad/ sec) which is around 10% higher than linear type trajectories. This profile also provide smooth starting and smooth ending of clutch engagement process and almost no oscillations were observed in driven shaft during engagement. However few oscillations can be seen just before lockup point in linear and parabolic profile, which may be attributed to sudden start and sudden end type of characteristics of the two profiles.

The lock-up time represents the delay in achieving the locking up of clutch system. The lockup time directly affects the heating and wear due to slip of friction lining. The results obtained from simulation are shown in graphical form in figure 4. The spline force trajectory showed overall better results in terms of lock-up speed and lock-up time. Similar results were also reported by [6], however they found slippage and friction work slightly high compared to linear profile but still in medium range.

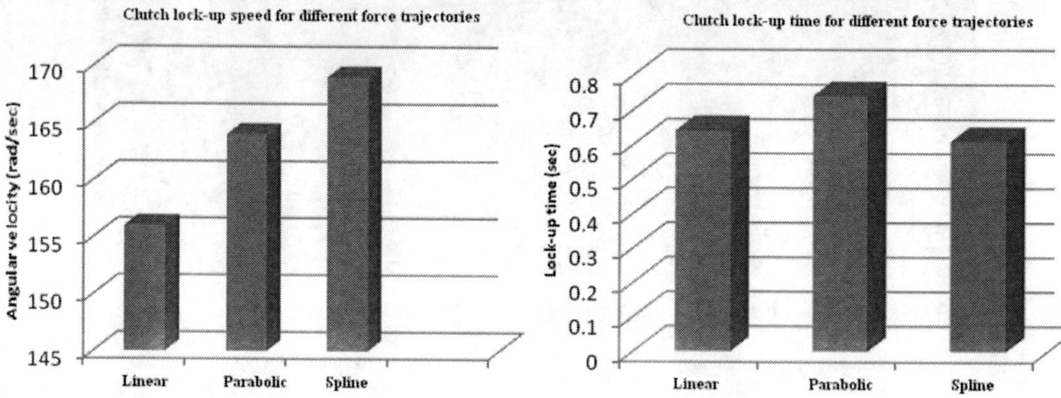

Figure 4. Clutch lock-up speed and lock-up time for different force trajectories

6. Conclusion

The trajectory followed during clutch engagement significantly affects the dynamics of driveline and driving comfort. Following points were noted from simulation of driveline for three different types of trajectories i.e. linear, parabolic and spline based on engagement quality.

Clutch lock-up speed is highest for spline force trajectory and lowest for linear force trajectory among three profiles for given input parameters

Clutch lock-up time was more or less same for linear and spline force trajectories and lower than time taken by parabolic force trajectory for given input parameters.

Spline force trajectory showed overall better results in terms of lock-up speed and lock-up time based on engagement process however other factors like jerk, slippage and frictional heating also need consideration for trajectories in automated clutch system.

7. List of symbols used

θ_e	Angular rotation of flywheel	J_{g1}	Moment of inertia of input gear
θ_{c1}	Angular rotation of clutch friction disc	J_{g2}	Moment of inertia of output gear
θ_{c2}	Angular rotation of clutch pressure plate assembly	B_{ec}	Torsional damping coefficient of crankshaft bearing
θ_{g1}	Angular rotation of input gear of gearbox	B_{cg}	Torsional damping coefficient of clutch output shaft bearing

θ_{g2}	Angular rotation of output gear of gearbox	B_{gw}	Torsional damping coefficient of transmission shaft bearing
θ_w	Angular rotation of wheel	K_{cg}	Torsional stiffness of clutch output shaft
J_{wc}	Moment of inertia of wheel about its own axis	K_{ec}	Torsional stiffness of crankshaft
J_e	Moment of inertia of flywheel and crankshaft	K_{gw}	Torsional stiffness of transmission shaft
J_{c1}	Moment of inertia of clutch friction disc	m_v	Vehicle mass
J_{c2}	Moment of inertia of clutch pressure plate	m_w	Wheel mass

8. Acknowledgement
The authors would like to gratefully acknowledge the help and guidance of Late Dr K. Tripathi for the work.

9. References
[1] Sharma SK, Pare V, Chouksey M, Rawal BR 2016. Numerical studies using full car model for combined primary and cabin suspension, Procedia Technology 23, 171-178
[2] Tripathi K 2014 Engagement Control of Automotive Clutch by Mechatronic System Using Pre-determined Force Trajectories *J. Inst. Eng. India Ser. C* **95** 109–17
[3] Chen L, Xi G and Yin C L 2011 Model referenced adaptive control to compensate slip-stick transition during clutch engagement *Int.J Automot. Technol.* **12** 913–20
[4] Lucente G, Montanari M and Rossi C 2007 Modelling of an automated manual transmission system *Mechatronics* **17** 73–91
[5] Vasca F, Iannelli L, Senatore A and Scafati MT 2008 June. Modeling torque transmissibility for automotive dry clutch engagement. In *2008 American Control Conf.* 306-311 IEEE
[6] Tripathi K 2013 A Novel Approach for Enhancement of Automobile Clutch Engagement Quality Using Mechatronics Based Automated Clutch System *J. Inst. Eng. India Ser. C* **94** 9–20
[7] Tripathi K and Agrawal MD 2008 Dynamic Modelling of Engagement of Automotive Clutch with Diaphragm Spring. *J-Inst. of Eng India Part ME*, **88(N)** 10.
[8] Gao B, Xiang Y, Chen H, Liang Q and Guo L 2015 Optimal Trajectory Planning of Motor Torque and Clutch Slip Speed for Gear Shift of a Two-Speed Electric Vehicle *Journal of Dynamic Systems, Measurement, and Control* **137** 061016
[9] Couderc Ph, Callenaere J, Der Hagopian J, Ferraris G, Kassai A, Borjesson Y, Verdillon L and Gaimard S 1998 Vehicle driveline dynamic behaviour: experimentation and simulation *Journal of Sound and Vibration* **218** 133–57
[10] Centea D, Rahnejat H and Menday M T 2001 Non-linear multi-body dynamic analysis for the study of clutch torsional vibrations (judder) *Applied Mathematical Modelling* **25** 177–92
[11] Andersson N, 2012 Driveline model calibration and validation in an automotive 4-cylinder Diesel application. *In Proceedings of ISMA2012-USD2012; ed. by P Sas, D Moens, and S Jonckheere*, Leuven, Belgium; 17-19 September, 2012.
[12] Duque EL, Barreto MA and de Toledo Fleury A 2012 Use of different friction models on the automotive clutch energy simulation during vehicle launch. In *ABCM symposium series mechatronics* **Vol 5** 1375-1389 ABCM Rio de Janeiro, Brazil
[13] Gillespie TD 1992 *Fundamentals of vehicle dynamics* **Vol. 400** Warrendale, PA: Society of automotive engineers
[14] Dolcini P, Béchart H and de Wit CC 2005 Observer-based optimal control of dry clutch engagement. *Proceedings of the 44th IEEE Conference on Decision and Control* 440-445 IEEE

IOP Conf. Series: Materials Science and Engineering **810** (2020) 012006 doi:10.1088/1757-899X/810/1/012006

Applications and challenges of arc welding methods in dissimilar metal joining

R Chaudhari[1], P K Loharkar[1] and A Ingle[2]

[1]Department of Mechanical Engineering
SVKM's NMIMS MPSTME Shirpur Campus
Dhule, India
[2]Department of Mechanical Engineering
SVKM's NMIMS MPSTME Mumbai, India

Email ID:rakesh.chaudhari@nmims.edu

Abstract. Dissimilar metal welding is a significant metal joining process with an extensive application in industrial domain. The quality of the welded joint is of critical importance in these applications and quite an issue since the base metals are different primarily due to different physical, chemical and mechanical characteristics. In this paper, the challenges encountered in dissimilar metal joining are enlisted with a brief discussion on the corresponding causes. As discussed in the literature, several methods exist to form dissimilar joints. However, arc-based fusion methods are prominently used. Therefore, an overview of these methods is presented along with the applications in which these methods have been used. The study provides a brief account of the materials that have been successfully joined using these methods and the influence of significant parameters including filler materials on the effectiveness of dissimilar joining.

1. Introduction

Dissimilar metal welding is the joining of two metals that have different chemical and mechanical properties. The difficulty to weld these metals arises due to different behavior and the existence of different alloys systems [1]. These types of joints are required in several applications such as automotive assembly lines, construction industries as well as power plants [2]. The significant criteria for the materials involved in such a diverse environment are the ability to withstand extreme working conditions. In addition, the use of dissimilar metal welded joints minimizes the overall cost in the relevant applications. These joints ensure the optimum design of welded components suitable for the specific applications. It enables the usage of appropriate materials, which subsequently enhances the damage tolerance of the components as required in several applications [3]. For instance, the designer can use high-cost material with superior properties only in case of critical requirements.

The major concern with dissimilar joints is their reliability, which is a function of the weld zone. Also, the metals to be welded consists of a mixture of materials with different compositions. Moreover, in practical applications, the material properties of the base metals are known. Hence, when these metals are joined, large magnitudes of thermal and residual stresses arise owing to unevenly distributed heat [4]. The key challenges in dissimilar joining are stated in the subsequent sub-section.

1.1. Challenges in welding dissimilar metals

The challenges posed by the process of dissimilar metal joining, which are to be accounted for in order to enable its application at the industrial scale [5, 6] are as follows.

- The formation of a transition zone between the joining dissimilar metals: The strength of welded joints depends on the solubility of joining metals. The metals with good mutual solubility form high strength joints. The metals with little or no solubility when joined pose strength related issues.
- The formation of undesirable intermetallic compounds (IMCs): IMCs increase the crack sensitivity thereby making the welded joints susceptible to corrosion. Also, the compound layer formation reduces the ductility of the welded component.
- Different coefficients of thermal expansion: The varying thermal contraction of weld metal and joining metals creates residual stresses in welds. The residual quantity and distribution of these stresses are different across the dissimilar weld joint (DMW). The wide difference in these residual stresses develops internal stresses in HAZ leading to service failure.
- Different melting points of joining metals: It causes one metal to melt and overheat before the other when subjected to the same amount of heating, thereby producing residual stresses.
- The difference of the electrochemical potential of joining metals: This may increase the susceptibility to corrosion at HAZ. Two metals of extremely different electrochemical potential scale may cause a high amount of corrosion. All these challenges are linked to the chemical, physical and metallurgical characteristics of the metals to be joined.

The welding of dissimilar metal is possible only when the joining metals are soluble with each other. There exist numerous welding methods to produce high strength dissimilar metal welds. These are classified broadly into three types as shown in figure 1. The methods based on arc welding are more commonly used to join dissimilar metals in power and processing industry. The welding methods under this category are, shielded metal arc welding, gas tungsten arc welding, gas metal arc welding and submerged arc welding which are abbreviated as SMAW, GTAW, GMAW and SAW respectively. An important point to be considered is that alloying of base metals and filler metals is critical and need special consideration while forming dissimilar metal welds [7]. The other category of methods is the low dilution methods. Electron beam welding, laser beam welding, abbreviated as EBW and LBW respectively, pulse arc welding (PAW) are three major methods under this category. While third category designated as non-fusion welding methods comprises of FSW-friction-stir welding, USM-ultrasonic welding and ERW-electric resistance welding, diffusion welding are used for special applications where less amount of alloying is observed between the joining metals [8].

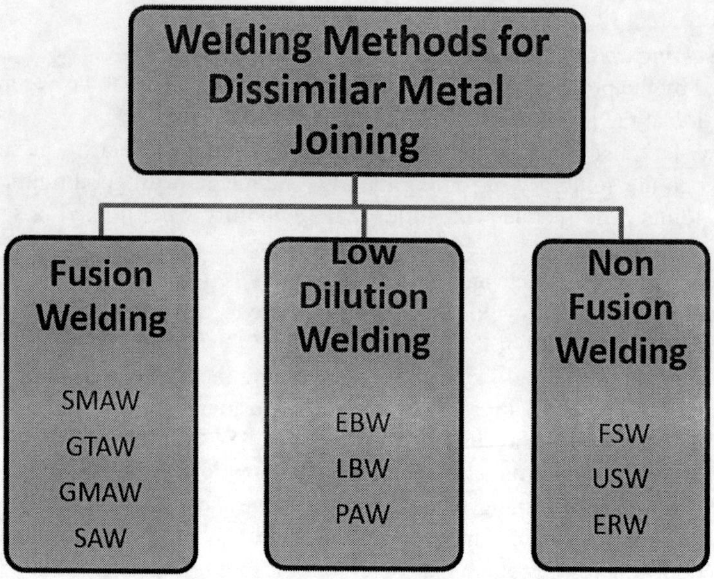

Figure 1. Classification of dissimilar welding methods.

In this paper, a discussion on arc welding methods that have been used in joining of dissimilar metals along with their types and applications is presented. The aim is to consolidate the knowledge pertaining to the domain of dissimilar joining using such methods that could serve as a reference for relevant future research.

2. Fusion welding methods for joining dissimilar metals

In fusion welding, two-metals are melted by heat source and fused together. An electric arc produces the required heat for welding or the heat is also derived from electrons or photons beams, or by the resistance offered by metals to passing of the electric current. Few fusion welding processes requires pressure to join two metals. The need and type of filler material is decided by the requirement of joining process [9]. The fusion welding methods have been successfully used for joining combinations of various metals. However, fusion welding faces many challenges due to formation of brittle IMCs at weld zone along with segregation of melting phases [10]. Development of residual stresses due to compositional difference of joining metals is another significant issue in this method. In spite of this, fusion welding is most preferred technique in joining of different types of ferrous alloys [7].

The fusion welding processes are further classified according to the source of heat generation as listed in table 1.

Table 1. Classification of fusion welding methods.

Methods of Heat Generation		
Arc	Beam	Resistance
SMAW	EBW	Spot Welding
GTAW	LBW	Seam Welding
GMAW		Flash Welding
SAW		Upset Welding

3. Arc welding methods for dissimilar metal joining

Joining of dissimilar metals has been performed extensively using arc-based fusion welding. The arc is produced by electrical means in these processes. Brief description of these methods is presented in the following sub-sections.

3.1. Gas Tungsten Arc Welding (GTAW)

GTAW is also known as tungsten inert gas- TIG. Tungsten electrode is used produce the weld. It doesn't get consumed. The arc developed between the workpiece and electrode melts the metals to be joined to form the weld pool [11]. The use of filler material depends on thickness of metals. Argon or helium or sometimes mixture of both is used to protect electrode to avoid oxidation [12]. It is most suited to weld thin sections due to lower heat inputs and can be performed on variety of metals.

3.2. Gas Metal Arc Welding (GMAW)

In GMAW, metals are joined by melting them with electric arc formed between a filler wire (electrode) and base metals. The gas is used to protect the weld from undesirable atmospheric ingress during solidification of molten weld pool. The inert gases like argon and helium are often used for protecting the arc and the molten weld pool. Hence, GMAW alternatively is called the metal-inert gas (MIG) welding process. The process is very clean due to use of inert shielding gas. It produces a uniform and slag-free weld zone [13].

3.3. Shielded Metal Arc Welding (SMAW)

A flux-coated electrode is used in this process to join the metals. The heat is produced by passing an electric current through the electrode. The molten metal is protected from atmospheric air by gaseous shield. It produces clean weld due to protection of electrode by slag. SMAW is mostly used for welding high-strength steels susceptible to hydrogen cracking [13].

3.4. Submerged Arc Welding (SAW)

In this process, base metals are joined by melting them using high energy arc produced between consumable wire electrodes and base metal. The arc is shielded by molten slag and granular flux therefore no separate shielding gas is required. It is used to joint thick sheets where long longer welds [12].

Following section discusses the state of research in the area of dissimilar metal joining using arc based welding methods causing complete fusion of the base metals.

4. Arc welding methods: Application overview

Even though there are several challenges in dissimilar metal joining, significant amount of research has been done to investigate its feasibility and effectiveness in terms of properties exhibited by the joints. In the subsequent sub-sections, a brief review of significant applications of each of the arc-based fusion welding method used in dissimilar metal joining have been presented.

4.1. Applications of GTAW

This method is used for welding of different ferrous and nonferrous metals such as steel and its alloys, aluminium and its alloys, titanium and its alloys. In the case of dissimilar metal welding using the GTAW, selection of filler metal becomes very important due to the varying base-metal composition [14].

Borrisutthekul et al. [15] and Lin et al. [16] investigated the feasibility of using GTAW to form the joint between steel and aluminium alloys. It has been reported that this welding technique can be effectively used to weld aluminium, stainless steel and titanium. However, some limitation in reliability and strength were experienced. Khalid et al. [17] experimentally investigated the dissimilar metal welds of ferrous and nonferrous metals such as steel and aluminium formed by GTAW. It was found that the accuracy and quality of weld formed has been greatly influenced by various welding parameters like power supply, welding speed and shielding gas.

Dissimilar metal welding of 5A06 aluminium alloy and AISI 321 stainless steel was carried out using GTAW [18]. The effect of filler metal on the surface of the welding groove and metallurgical characteristics of the joint was also studied. Liu et al. [19] investigated dissimilar metal weld of magnesium (Mg) alloy and copper (Cu) as lap joint prepared using GTAW welding using iron (Fe) as an interlayer. The welded joint fractured at the intersection of Fe plate and Mg alloy during tensile testing. The mechanical
properties of the joint decreased at interface between Mg and Fe due to formation of metallic oxides. Lin et al. [16] performed investigation on dissimilar metals welds of aluminium alloy and galvanized steel formed by GTAW brazing. The purpose was to study aluminium filler metal distribution over the surface of steel. The results of mechanical testing on the welded joint revealed that a zinc-coated layer has improved the dispersion of liquid aluminium filler metal on steel surface. The fusion zone near aluminium showed coarse grains due to grain coarsening. Improper brazing performed at the root of the lap joint decreased the strength of joint causing fracture at same location.

Arivazhagan et al. [20] compared properties dissimilar metal weld of SS 304 and 4140 low alloy steel formed by GTAW, EBW and friction welding techniques. The weld formed by EBW exhibited higher tensile strength relative to GTAW and FSW joints. On the other hand, ductility of weldment formed by EBW and GTAW were larger than weldment by friction welding. The GTAW weldment showed highest impact strength in comparison to other welding processes.

Bonaventur et al. [21] evaluated residual stresses in weld between ferritic steel and 316 stainless steel. The welds were made with and without the use of Inconel-82 for buttering process on ferritic steel side. The failure test carried out revealed that the failure has occurred in the HAZ of ferritic steel side. It was observed that residual stress in the welded joints was a significant factor that caused failure of welded joints. It was also investigated that buttering of Inconel-82 applied in the welded joint helped to reduce residual stresses developed in the HAZ of the ferritic steel and ultimately avoided failure at lower stress values.

Qinglei et al. [22] investigated hardness and element distribution and phase constituents in microstructures of welded joint of Mo-Cu composite and 18-8 cutlery stainless steel formed by GTAW using Cr-Ni filler wires. Higher micro hardness was measured at the weld zone due to formation of γ-Fe (Ni) phases and Fe0.54Mo0.73 compound. The microscopic observations revealed the presence of austenite as well as ferrite phases in the weld zone. In addition, increase in micro hardness was observed at the boundary of fusion zone between composite material and weld zone. The maximum micro hardness was observed at interface between fusion zone and composite as hard and brittle intermetallic compounds layer of Mo and Cu was formed.

4.2. Applications of GMAW

This method is widely used for joining of the variety of metal combinations [23]. Strong welds of dissimilar metals are obtained for metals that do not mix in solid-state but mix in the liquid state. The filler metal used has a great effect on weld zone composition. The important factors that decide the quality of dissimilar metal welds are the current and the speed of arc travel [24]. The method is widely used to form dissimilar metal joints of ferrous alloys as well as alloys of aluminium and copper.

Kumar et al. [25] investigated the weld of copper plates with iron filler using GMAW with special focus to understand the microstructural changes occurring during the process. The welded joint produced higher hardness at the weld zone. The effect of coating of aluminium and galvanized steel on dispersion of fusion metal was studied by Murakami et al. [26] in the joining of 2B50 aluminium and 1Cr18Ni9Ti steel alloy. Wang et al. [27] investigated microstructure of aluminium and magnesium DMW with Al-Si5 filler metal formed by cold metal transfer-CMT GMAW. The microstructural observations showed the formation of brittle intermetallic compound in the weld metal due to use of low heat input during the welding and addition of silicon as filler metal. Pure copper filler was used in the process of joining

Mg and Al [28]. The microstructure observations showed that weld of Mg and Al with moderate bonding strength could be successfully formed using CMT technique.

In the similar studies using this method, butt weld between 304 stainless steel and carbon steel was formed using ER309 filler metal [29] while welding between AA6082 aluminium alloy and AA6092/SiC/25p composite was performed [30].

Ul-Hamid et al. [29] investigated the butt weld between 304 stainless steel and carbon steel formed by MIG using ER309 filler metal. The metallurgical analysis revealed cracks at the intersection of weld root. The interface exhibited the formation of hard martensitic phase due to carbon separation at the temperature of welding followed by quenching. Ferrite formation was observed in the carbon steel near the weld zone primarily due to diffusion.

Lean et al. [30] studied the weldability of dissimilar metal weld of AA6082 and AA6092/SiC/25p composite using gas metal arc. The investigations were carried out on two joints formed by using ER5356 and ER4043 filler metals. Both joints showed similar tensile strength, but less than the similar metal joint of only aluminium alloys. In both joints, the fracture occurred at HAZ of AA6082 aluminium alloy. The heat treatment performed after welding increased the tensile strength of the joints and was equal to unreinforced aluminium alloy.

Taban et al. [31] analysed the mechanical and metallurgical characteristics of dissimilar metal welds of ferritic stainless steel and S355 structural steel formed by shielded metal arc (SMA) welding using AISI 309 as a filler metal. The fracture occurred in base metal during tensile testing. The HAZ exhibited low toughness of about 17 J to 30 J when measured by charpy impact test.

4.3. Applications of SMAW

Verma et al. [32] investigated inter-granular corrosion behaviour of dissimilar duplex stainless steel and 316L austenitic stainless steel formed by SMAW process. The joint was formed by using duplex 2209 electrode with variation in heat input that resulted in formation of ferrite at interface due to higher cooling rate. Also, high heat input had enhanced corrosion resistance owing to formation of large amount of austenite [26].

Comparison of electrode performance was carried out by Bala, et al. [33]. Dissimilar metal joint of duplex stainless steel (DSS) to carbon steel (CS) formed by SMAW using E2209 and E309 electrodes were investigated. It was found that the joint formed with E2209 electrodes showed better hardness and toughness than that formed with E309 electrode. On the other hand, E309 resulted in superior resistance to corrosion but it showed very high pitting susceptibility. The study by Lee et al. [34] developed DMW of Alloy 690 and SUS 304L using SMAW with Ti as a filler material. It has been inferred that Ti addition leads to increased elongation of the weld. At the same time, tensile strength remains unaltered. However, at a larger percentage of Ti content, reduced tensile strength was observed owing to reduced joint weldability.

The hardness and microstructures of super-austenitic stainless steel alloys joints was investigated. The base metals showed austenitic structures whereas weld metals showed dendritic structure. Larger hardness was observed in the weld zone due to faster cooling and thermal cycle during welding. This resulted in decrease in yield stress and the toughness is also observed in weld zone compared to base metal [35].

4.4. Applications of SAW

Labanowski [36] worked on determining the outcome of varying process parameters of SAW on the metallurgical behaviour of welded joint of Duplex 2205 and austenitic 316L steels. Microstructure observations indicated no excessive segregation of ferrite in heat affected zones (HAZ) of the welds. The lowest stress corrosion cracking resistance is observed at HAZ of duplex steel. This was due to the formation of large coarse ferrite grains and precipitation of austenitic phase. A novel application in the form of narrow-gap, SAW was evolved to study the fatigue behaviour of base metals(BM), HAZ and weld zone of the joint of advanced 9Cr and CrMoV steels [37].

The welded joint of chrome molybdenum steel ASTM A387 Gr.12 and ASTM A240 TP 304L formed by explosion welding, submerged arc welding and hot rolling were experimentally investigated by Branko et al. [38]. The as-welded (AW) as well as post heat treated (PHT) specimen were tested for measuring shear strength and impact energy. It was fond that PHT samples had significantly modified impact energy of the joint. The cladding procedure used in welding as well heat treatment had also affected the shear strength of the joint.

5. Conclusion

Dissimilar metal welding is a critical area in the field of materials engineering as well as manufacturing. The process of welding is significantly influence by the properties of the two base metals being joined. Even a slight change in composition critically affects the weld metallurgy and quality. The paper has focussed on providing an overview of arc-based methods used in dissimilar metal joining and presented an account of their applications in joining combination of materials such as aluminium-magnesium alloys, steel-steel alloys, copper-magnesium alloy, aluminium–steel alloys etc. The effect of addition of filler metals in terms of improved metallurgical properties have also been discussed along with welding parameters. The study provides a consolidated information of dissimilar metal welding formed using various types of arc welding methods that is useful for future research.

6. References

[1] Chaudhari R, Parekh R and Ingle, A 2014 International Conference on Machine Learning, Electrical and Mechanical Engineering (ICMLEME: 2014).
[2] Ghorbani S, Ghasemi R, Ebrahimi-Kahrizsangi R and Hojjati-Najafabadi A 2017 *Materials Science and Engineering: A* ***688*** 470-479
[3] Yuce C, Tutar M, Karpat F and Yavuz N 2016 *Metals* ***6*****(10)** 245
[4] Avery R E 1991 Nickel Development Institute
[5] Kou S 2003 John Wiley & Sons,Inc *New Jersey USA* 431-446
[6] Kah P, Shrestha M and Martikainen J 2014 *Applied Mechanics and Materials* **440** 269-276 Trans Tech Publications
[7] Cornu J 2013 Springer Science & Business Media
[8] Węglowski M S, Błacha S and Phillips A 2016 *Vacuum* ***130*** 72-92
[9] Lippold J C 2015 John Wiley & Sons Incorporated
[10] Mohammadi J, Behnamian Y, Mostafaei A, Izadi H, Saeid T Kokabi A H and Gerlich A P 2015 Materials Characterization **101** 189-207
[11] Jeyaprakash N, Haile A and Arunprasath M 2015 **4(2)** 11-20
[12] Balaji C, Kumar S A, Kumar S A and Sathish R 2012 International Journal of Engineering Science and Technology **4(5)** 2053
[13] Nagesh D S and Datta G L 2002 Journal of Materials Processing Technology **123(2)** 303-312.
[14] Ramkumar K D, Arivazhagan N and Narayanan S 2012 Materials & Design **40** 70-79
[15] Borrisutthekul R, Mitsomwang P, Rattanachan S and Mutoh Y 2010 Energy Research Journal **1(2)** 82-86
[16] Lin S B, Song J L, Ma G C and Yang C L 2009 Frontiers of Materials Science in China **3(1)** 78-83
[17] Hussain A K, Lateef A, Javed M and Pramesh T 2010 International journal of applied engineering research **1(3)** 518
[18] Song J L, Lin S B, Yang C L, Ma G C and Liu H 2009 Materials Science and Engineering: A **509 (1-2)** 31-40
[19] Liu L M, Wang S X and Zhu M L 2006 Science and Technology of Welding and Joining **11(5)** 523-525

[20] Arivazhagan N, Singh S, Prakash S and Reddy G M 2011 Materials & Design **32(5)** 3036-3050
[21] Bonaventur A Ayrault D, Montay G and Klosek V 2011 Materials Science Forum **681** 182–187
[22] Qinglei J, Yajiang L, Puchkov U A, Juan W and Chunzhi X 2010 International Journal of Refractory Metals and Hard Materials **28(3)** 429-433
[23] Joseph A, Harwig D, Farson D F and Richardson R 2003 Science and Technology of Welding and Joining **8(6)** 400-406
[24] Cary H B and Helzer S C 2005 Modern Welding Technology Prentice Hall New York
[25] Kumar K S, Phanikumar G, Dutta P and Chattopadhyay K 2002 Journal of materials science **37(11)** 2345-2349
[26] Murakami T, Nakata K, Tong H and Ushio M 2003 ISIJ international **43(10)** 1596-1602
[27] Wang J Feng J C and Wang Y X 2008 Materials Science and Technology **24(7)** 827-831
[28] Shang J, Wang K, Zhou Q, Zhang D, Huang J and Li G 2012 Materials & Design **34** 559-565
[29] Ul-Hamid A, Tawancy H M and Abbas N M 2005 *Engineering Failure Analysis 12*(**2**) 181-191
[30] Lean P P, Gil L and Ureña A 2003 Journal of Materials Processing Technology **143** 846-850
[31] Taban E, Deleu E, Dhooge A and Kaluc E 2008 Welding Journal **87(12)** 291-297
[32] Verma J, Taiwade R V, Khatirkar R K, Sapate S G and Gaikwad A D 2017 Transactions of the Indian Institute of Metals **70(1)** 225-237
[33] Srinivasan P B, Muthupandi V, Dietzel W and Sivan V, 2006 *Materials & design 27*(**3**) 182-191
[34] Lee H T, Jeng S L, Yen C H and Kuo T Y 2004 Journal of Nuclear Materials **335(1)** 59-69
[35] Kchaou Y, Haddar N, Hénaff G, Pelosin V and Elleuch K 2014 *Materials & Design 63* 278-285
[36] Łabanowski J 2007 Archives of Materials Science and Engineering **28(1)** 27-33
[37] Wang B, Lan H and Lei B 2019 Advances in Materials Science and Engineering
[38] Mateša B, Samardžić I and Dunđer M 2014 Tehnički vjesnik-Technical Gazette **21(5)** 1035-1040.

A review on characteristics of natural fibre composite

Chetan Prakash Chaudhari, Kiran Bhole

Sardar Patel College of Engineering, Andheri West, Mumbai 400058, India

chetanpc117@gmail.com, kiran_bhole@spce.ac.in

Abstract. Since the last we saw huge growth in innovation and research in field of natural fibre composite (NFC). Availability of Natural Fibres (like coir, abaca, banana, jute, sisal, cotton) are huge in India. It is not surprising that the natural fibre composite is attracting the researchers all around the world because of superiority of its compared to other fibre composites and their capabilities in wide range of applications. The present study aims at various applications of natural fibre composite and their mechanical characteristics. This review also provides the various aspects that affects the mechanical characteristics of natural fibre.
Keywords: Natural Fibre, Composite, Mechanical Properties, Applications

1. Introduction

1.1. Natural fibre composites

Over the closing 30 years' composite were the dominating and rising materials due to gradual growth of its applications. Composites have established their value as light weight substances; the present venture is to make composites non costly. It is obvious that just by improving the manufacturing techniques is not sufficient particularly for composites to triumph over the price barrier. Therefore, it is so much essential to have incorporated attempt in manufacturing, design and material selection. [15]
Composites are material structures that offer properties no longer obtainable from any single material and they are structures of compatible materials made by physically combining materials which are specific in characteristics and composition.
High specific strength of Fibre-reinforced composites have performed a dominant role in a various application. The manufacturing sector consider elimination of traditional fibre composites critical due to environmental troubles. By natural fibre composites we suggest a substance which is combined with fibres from natural/renewable resources. Classification of Natural fibres is done according to their place of origin that include fibres crafted from plant, animal and mineral.
Animal fibre normally contain proteins; examples animal hair, silk fibre and avian fibre. We can get animal hair fibres from furry mammals, for example hair of horse, goat and sheep. Examples of silk fibre are the fibres from silk worms. Avian fibre is a fibre from flying animals and its examples are feathers and feather fibre.
Mineral fibres can be divided into the three classes: First one is Asbestos which is unique natural mineral fibre. Second is Ceramic fibres which consist of glass fibres, Al_2O_3, SiC, and B_4C and last one is Metal fibres which includes aluminium fibres.

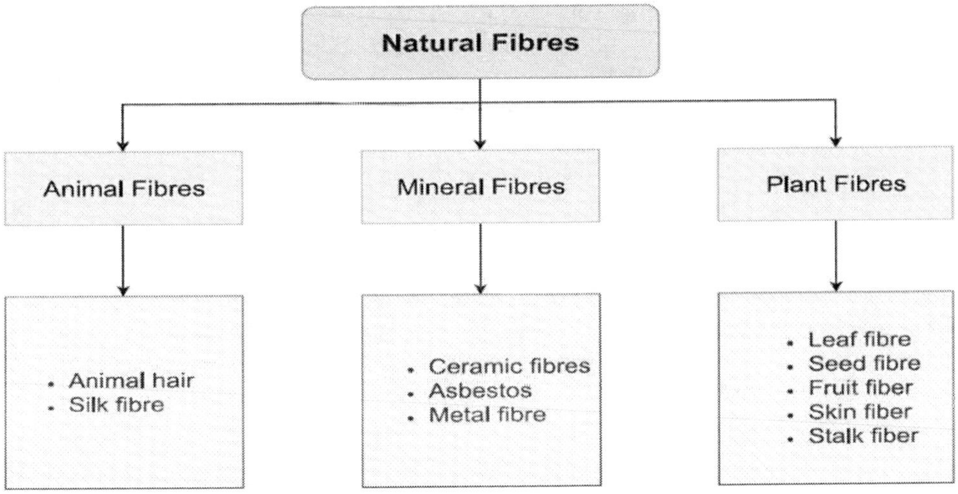

Figure 1. Classification of Natural Fibre.

Plant fibres usually contain cellulose. Cellulose based fibre serve inside making of paper as well as cloth. Seed fibres are obtained from a seed. Leaf fibre are fibres obtained by plant leaves. Skin fibres are accumulated by plant stem. Fruit fibre are gathered from plant having fruit. Stalk fibre are the fibres that are truly plant stalks.

1.2. Advantages and limitations of NFC (Natural Fibre Composite)

Figure below showcase some of the key advantages and limitations of NFC, i.e. Natural Fibre Composite.

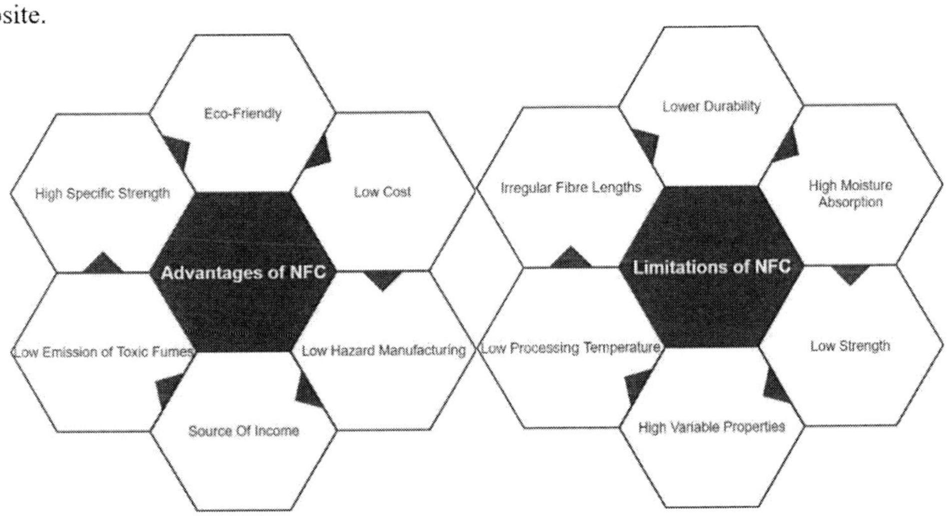

Figure 2. Advantages and Limitations of NFC.

1.3. Study of natural fibres

Abaca fibre is a form of tough natural fibres specially produced in Philippine, which uses the Musa plant leaf. Musa plant is like banana but the fruit of musa is not edible. Abaca fibre said to have a specific flexural strength, high rotting resistant and has high tensile strength as much as glass fibre. That's why bags, paper and ropes uses it as raw material. Abaca fibre application in latest years has prolonged to car enterprise, for example Mercedes-Benz. Stringent quality required by the outdoor components of street vehicles is also stated to satisfy by abaca fibre with a great resistance to elements, stone strike as well as dampness. It's widely known fact that the properties of substance are influenced by

microstructure of substance. As a result, researchers often by changing the microstructure of material, the mechanical properties of material gets improve. As structure of natural fibre is dependent on fibre growing circumstance, there is possibility to achieve natural fibres of required characteristics through deciding on fibres from one-of-a-kind positions of plant. [8]
Jute is not so unusual agricultural fibres that are cultivated almost solely in India, Thailand and Bangladesh. Jute fibre show off reasonably high mechanical properties that other fibres. 2500000 tones are a total yearly international production of jute. In Europe the flax is also the most crucial as well as demanding bast fibres. Spain, Belgium, Holland, France and UK produces nearly 80% of the total worldwide flax. Relatively flax fibres are stiffer to handle, crisper and stronger. Flax and jute fibres won brilliant attention of everyone during past decade as a material for reinforcing of composites. Out of total worldwide jute production Bangladesh and India supplied ninety percent jute. [8, 15, 16]
Sisal have a botanical name of Agave sisalana. It's an agave species native to southern Mexico but largely naturalized and cultivated in lots of different nations. Different products are made by using stiff fiber which is yield by sisal. It's once in a while called "sisal hemp", because for hundreds of years' hemp changed into main supply for fibre, and different fibre assets. [8, 15]

Table 1. Mechanical characteristics of synthetic and natural fibre.

Fibre	Length (mm)	Density (gm/cm^3)	Failure Strain (%)	Young's Modulus (GPa)	Tensile Strength (MPa)	Specific Young's Modulus ($GPa/gm\ cm^{-3}$)	Specific Tensile Strength ($MPa/\ gm\ cm^{-3}$)
E-glass	Continuous	2.5	2.5	70	2000-3000	29	800-1400
Coir	20-150	1.2	15-30	4-6	131-220	3.3-5	110-180
Cotton	10-60	1.5-1.6	3-10	5.5-13	287-800	3.7-8.4	190-530
Sisal	900	1.3-1.5	2-2.5	9.4-28	507-855	6.7-20	362-610
Hemp	5-55	1.5	1.6	58-70	550-1110	39-47	370-740
Flax	5-900	1.5	1.2-3.2	27-80	345-1830	18-53	230-1220
Jute	1.5-120	1.3-1.5	1.5-1.8	10-55	393-800	7.1-39	300-610

A good way to use Coconut as reinforcement in resin of polyester was studied by chemically modifying the coconut fibres by using alkaline treatment. Coconut fibres had been reformed for the duration of one hour with NaOH solution of 1% weight/volume. The reformed fibres were assessed by means of Fourier Transform infrared spectroscopy, thermal analysis, X-ray diffractometry and scanning electron microscopy. Compression molding technique have been used to prepare the composites of 10% weight of fibres. Fatigue and tensile tests were used to evaluate the mechanical properties. The fractured specimen's surfaces were inspected with a purpose to evaluate the fracture mechanisms. Fatigue performance of composites shows reduction in fatigue lifespan on application of higher tension. Total worldwide production of coir fibre is 250 thousand tons. Sixty percent of worldwide white coir fibre is supplied by India. [2, 16]

2. Key performance affecting factors of natural fibre composite

2.1. Matrix choice

The vital part of a fibre-composite is matrix as it offers an obstacle in opposition to detrimental environments and shields the fibre surface from mechanical abrasion. It also handovers load into fibres. As polymeric matrices can be treated at lower temperature and are lighter in weight, they most commonly utilized in Natural Fibre Composite. Thermoset as well as thermoplastic polymers along with natural fibres were utilized for matrices.
Matrix choice is constrained via temperature when natural fibres starts to degrade as so many natural fibres are thermally unsteady beyond 200^0 C. However, under few instances it's feasible to be treated at better temperature aimed at brief time frame. Because of this drawback, only those thermoplastics that become softer beneath this temperature is useable as a matrix (polyethylene (PE), polystyrene, polyvinyl chloride, polyolefin, PP and thermosets). Indeed, PE and PP are the 2 maximum usually followed

thermoplastic matrices used for natural fibre composites. The most important thermosets used are VE resins, phenol formaldehyde, epoxy resin and unsaturated polyester (UP). [4].

2.2. Fibre choice
Fibre kind is normally categorized on the basis of its foundation like mineral, animal, or plant. Cellulose is a main structural element of entire plant fibres (flax, ramie and hemp fibre are among the cellulose centered natural fibres with the best tensile strengths and specific Young's moduli). Plenty of greater stiffness and strengths are attainable with the better performing plant fibres compare to easily accessible animal fibres (Silk is an exception). Silk may have great strength, however is enormously costly and has poor stiffness. Therefore, to get a composite satisfying structural requirements the most appropriate choice is plant centered fibres. Commonly, plant based fibre can be harvested in quick durations and can properly be grownup in many nations. Furthermore, geography regarding fibre obtainability has a primary role in fibre choice. The consciousness, for example interest on flax fibre has been in Europe, while Asia have been more interested in sisal, kenaf, ramie, jute and hemp.
Generally, better performance is accomplished with types that have greater cellulose content and with alignment of cellulose micro fibrils greater with fibre direction which have a tendency to happen in bast fibres (ramie, jute, hemp and flax). The characteristics of natural fibres range significantly dependent on structure and chemical composition, which is related to storage procedures, extraction method, harvesting time, growing conditions, fibre type as well as treatment. Strength has been reported to decrease through 15% over period of 5 days afterward optimal harvest time. By hand mined flax fibres have 20% more strength than mechanically extracted fibres.

Figure 3. Coconut, Banana and Sisal Fibre

Generally, strength will increase as moisture level increase and reduces with increase in temperature. The ratio of transverse Young's modulus to longitudinal Young's modulus of jute has been predicted equal to 1/7. Fibre length, used in aspect ratio of fibre (which is l/d), is a crucial aspect that influences the mechanical characteristics of composites. Tensile load in case of short fibre composite transmitted from the matrix right into a fibre. Fibre load bearing efficiency usually increases as fibre length increases. However too lengthy fibres can also get tangled at some stage during mixing which may cause terrible fibre dispersion that may lessen the general reinforcement performance. [4]

2.3. Fibre orientation
The finest mechanical characteristics are usually being found in composites where the fibre is line up parallel to the path of applied load. Yet, obtaining alignment of natural fibre is more challenging than of continuous synthetic fibre. More or less alignment is accomplished at some stage in injection moulding. However, elongated natural fibre may be carded and positioned by hand in sheets previous to matrix to obtain greater level of fibre alignment. On the other hand, to facilitate an unbroken yarn production conventional fabric processing of fibres may be hired along with spinning. We might use wrap spinning to get aligned fibre, which can convert short into a continuous form by using an uninterrupted strand bound around irregular fibre.

IOP Conf. Series: Materials Science and Engineering **810** (2020) 012007 doi:10.1088/1757-899X/810/1/012007

Dynamic Sheet Forming (DSF) is a lately applied substitute for the textile path with its widespread infrastructure necessities to take along fibre alignment in composite. Conventionally in paper manufacturing DSF is used for aligning fibre. In DSF, small fibres are mixed in water and sprayed with the help of a nozzle on a spinning drum. Spinning drum is blanketed by wire net to eliminate the water, which brings alignment inside the rotation path and spray. This has supplied enhancements of mechanical performance in comparison to different small fibre processing techniques. [4]

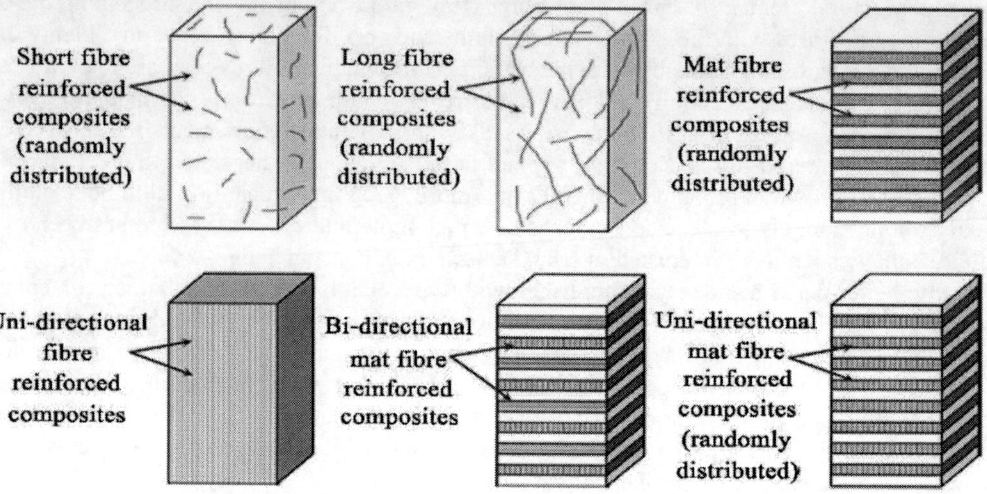

Figure 4. Different types of fibre orientation

An organized numerical examination of conical heater become carried out for assessing the orientational consequences and the fabric is studied below 35 and 50 kW/m^2 irradiances for the fire response properties of wool-PP/flax-PP composite. The examine virtually shows a huge impact of the pattern orientation on fire reaction properties of NFC. [6]

2.4. Interface strength

The Interfacial bond between matrix and fibre performs critical function in figuring out mechanical properties for composite. As stress is transmitted among fibres and matrix throughout interface, desirable interfacial bond is essential for obtaining best reinforcement. However, there is generally restricted interaction aimed at plant centred fibre composite among the matrices and hydrophilic fibres that might be usually hydrophobic causing weak interfacial bond. Weak interfacial bond restricts mechanical performance of plant centred fibre composite. Matrix and fibre need to be carried into intimate touch for occurrence of bonding; wettability is a vital originator for bonding. Interfacial defects may arise due to inadequate fibre wetting which might cause stress concentration. Wettability of fibre can be enhanced by chemical and physical treatment.

Interfacial bonding can arise by using mechanical interlocking mechanisms, chemical bonding, inter-diffusion bonding and electrostatic bonding. For rough fibre surface the mechanical interlocking take place to extra volume. Electrostatic bonding simply has substantial impact on metallic interfaces. When chemical groups within the matrix and the fibre surface can react to formulate bonds then chemical bonding occurs. Chemical bonding may be carried out by using a coupling agent which acts as passage among matrix and fibre.

Extensive studies have been done so one can acquire enhanced interfacial bonding for natural fibre composites. These studies can be broadly distributed into chemical and physical approaches. Physical treatments consist of fibre beating, electron radiation, heat treatments, ultraviolet (UV), plasma and corona while chemical treatments consist of acrylonitrile treatments, alkali, acetyl, silane, benzyl, acryl, isocyanate, titanate, zirconate and coupling agent. The most frequently used chemical treatments are maleated anhydride embedded coupling agent, alkali, acetyl and silane. However, enzyme treatment is turning into more and more famous with its specific advantage related to environmental friendliness. For enhanced interfacial bonding alkali treatment eliminates fibre constituents which uncover cellulose and rises surface roughness per area. [4]

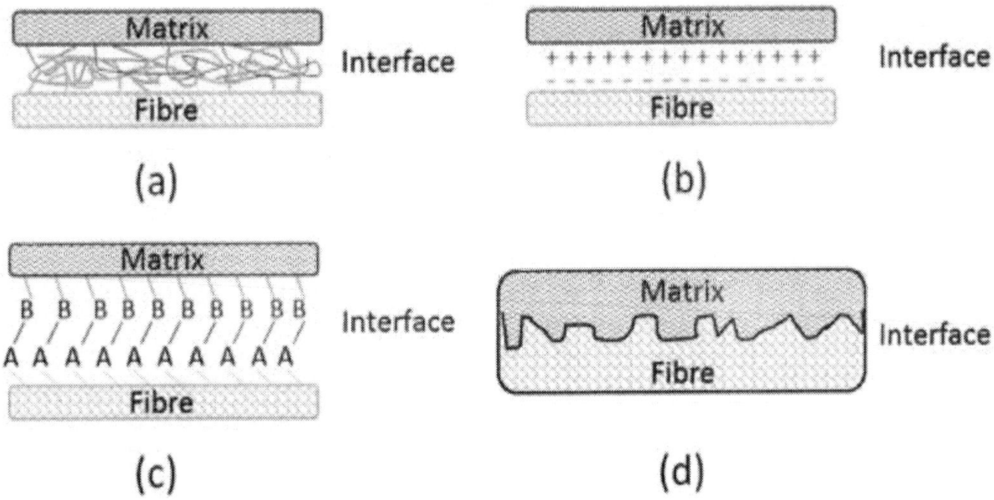

Figure 5. Matrix-fibre interface

2.5. Porosity

Porosity an is a part of natural fibre composite, acknowledged to have massive effect on mechanical characteristics of composite. Plenty of attempt has been gone to lower the porosity in synthetic based fibre composite. It rises because of addition of air in the course of processing, inadequate fibre wettability and different hollow structures inside fibre bundles (that may turn into closed in the course of high pressure processing). Porosity also rises because of less compaction abilities of fibres. In natural fibre composites porosity increase with increase in fibre content. It will increase more quickly after surpassing the geometrical compaction limit. Depend on fibre orientation and type of fibre; composites of flax/PP had been seen to have growing porosity of 4% to 8% of volume for increase in fibre content from 56% to 72% of matrix. [4]

2.6. Thermal effect on machinability

In an examine which specializes in the combined results of sample temperature and tool rake angle when slicing one directional flax fibres reinforced polypropylene composites investigates the effect of these 2 essential parameters on cutting performance of natural fibre composites and cutting contact stiffness. In this study, the chip formation and in-situ cutting forces are noted for each cutting situation. Scanning Electron Microscopy (SEM) observations are used to evaluate the state of machined surface. Under unique sample temperatures and tool rake angles orthogonal cutting of one directional flax fibres/polypropylene has been accomplished. This is done to investigation the thermal aspect of flax fibres on tribological cutting performance. The removed chip remains continuous beneath every slicing situation due to the excessive ductility prompted with the aid of PP matrix and flax fibres. Nevertheless, decreasing the temperature of sample reduces the curling of chip and increases the obvious material stiffness which result in improvement in machinability of uni-directional flax fibre / polypropylene (UDF/PP) composites. Excessive flax fibres deformations at some point of cutting interaction is induced by negative value of rake angle which degrades machinability of the UDF/PP. However, the machinability and cutting contact stiffness of UDF/PP can be improving by machining using slight positive rake angle. [1]

2.7. Flammability

For the improvement in composite products and the material selection it is exceedingly essential to have the information of flammability and thermal decomposition of the bio-centered fibres, their composites and polymers. Natural fibre composites has high flammability which is a major disadvantage for applications of engineering (ex. building insides and aerospace). The reason behind the high flammability of NFC is combustible fibres and hydrocarbon polymers. Therefore, greater innovative studies associated with the material choice, flame retardant treatments or production procedure is necessary to attain an equilibrium among flammability and overall mechanical performance of composites. [10]

2.8. Fibre configuration (Banana)

In a research of polypropylene (PP) composite, with reinforcement of continuous banana fibre by weightiness percentage ratio of 70:30 of polymer to fibre, were have been developed. The mechanical characteristics of composite materials were investigated for the outcomes of dissimilar configurations of continuous banana fibre. Raw banana fibre, banana fibre mat, and banana fibre yarn are three dissimilar configurations of banana fibre as shown in underneath table.

Table 2. Configuration of fibre and matrix (wt.%) in the polymer composites.

Composites	Fibre (%wt.)	Matrix (%wt.)	Total (%wt.)
Unreinforced PP	0	100	100
PP/Raw Banana Fibre	30	70	100
PP/Banana Yarn	30	70	100
PP/Banana Mat	30	70	100

The bar diagram in figure 6(a) shows three exceptional configurations of banana fibre with Tensile Strength for the PP/Banana fibre composites. Banana fibre reinforced natural composites gives higher tensile strength in comparison to unreinforced PP, which confirmed that the higher mechanical strength is possessed by banana fibre reinforced composites than unreinforced PP. The best enhancement of tensile strength is given by composite of PP/Banana yarn. Composite of PP/Banana yarn yields very best tensile strength of 77.74 MPa compare to 33.27 MPa of PP/Banana mat and 66.26 MPa of PP/Raw banana fibre.

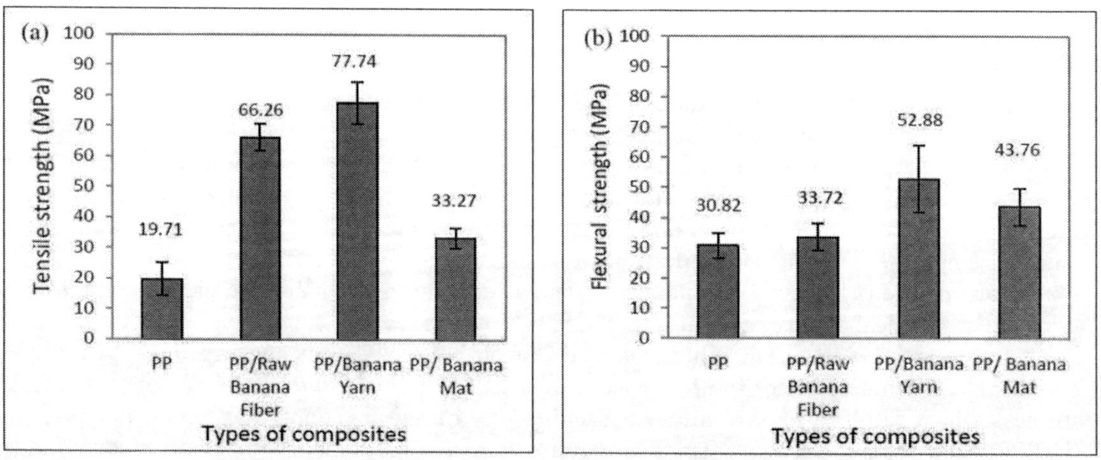

Figure 6. Bar chart of (a) tensile strength and (b) flexural strength for different types of configuration of banana fibre composites.

The bar diagram in figure 6(b) shows three exceptional configurations of banana fibre with flexural strength of the composites of PP/Banana fibres. It's miles clear that the three specific reinforced PP

composites of banana fibre gives greater flexural strength than unreinforced PP. The findings indicate compared to former 2 arrangements banana yarn arrangement gives finest results as it demonstrates the very best 52.88 MPa of flexural strength.
Therefore, PP/Banana yarn composite possesses the best flexural strength in addition to tensile strength and as a result, it could be realized that for getting the finest mechanical properties, the greatest choice of fibre configuration in PP composite is banana fibre yarn. [3]

2.9. Time at temperature
Temperature is usually considered as restricting criteria for the production of natural fibre composites (commonly not surpassing 200^0C). The impact for time at temperature of natural fibres is showcase by a study on fibres of jute weighing 880 gm^{-2}. It studies fibre mechanical properties after air heating for different times and temperatures, by means of taking batches of fibre which have been heated by using oven to fixed temperatures (step of 10^0C and 180-220^0C) for predetermined times/instances (5, 10, 15, 30, 60 and 120 minutes). Five models each at mixture of time and temperature had been tested and below ambient situations the mechanical properties for jute fibres had been measured. The study shows quick degradation at upper temperature and at low temperature prolonged heating. The statistics used by the observe is based on a restricted set of assessments and subsequently deciding most appropriate processing values of time-temperature has not been feasible yet. [7]

3. Applications of natural fibre composite
Many researchers have paid their attention on the usage of natural fibres in cementitious centered structural materials as reinforcements. Natural fibres usage in structural applications isn't cutting-edge as far hundred years ago in lots of small cities in Korea and China people already mixed straws and dust to construct partitions in rural areas., Natural fibre composites in automotive business are suitable to feature as front sheets of automobiles and they're no longer prime structural constituents. Weight and cost of cars might be partly reduced though replacing conventional aluminium and fibre glass composites with natural fibre composites in those components. [5, 16]
Even though the high strength fibre consisting of carbon fibre has more strength than the natural fibre, strength of glass fibre is comparable to natural fibre. Natural fibre is able to be carried out as very valuable composite if the suitable resin is selected. The eco-friendly hood shape design for automobile is feasible via the usage of natural fibre composite. [9]
The coconut fat collected from feedstock's can be carefully chosen as bio based Phase Change Material (PCM) due to its decent Thermal Energy Storage (TES) properties. This PCM can be included in a laboratory prepared composite matrix of cellulose fibres and natural clay. Coconut fat can be chosen as bio based PCM. [11]
Kinetic energy of air is converted into electrical power via tool named wind turbine. Amongst the wind turbine parts which includes blades of wind turbine, nacelle, tower, gear field and hub are normally prepared from carbon fibres and glass fibres for resistance to corrosion, light weight and higher strength. The most important drawbacks of these fibres (carbon and glass fibres) are the health risk, non-biodegradable, their fabrication value and availability. Consequently, the goal of investigation is to update those materials by using natural fibres. Bio-degradable substances with greater mechanical properties can be formed via natural fibre composites. Reinforced composites of natural fibre can be a capable nominee in which they are able to substitute conventional material for the manufacturing of various sections of a wind turbine. [12]
Natural fibres signify ecological option to be used both in noise manipulate engineering and constructing production compared to traditionally used synthetic materials. Due to hydrothermal characteristics of natural fibres, they are mainly engaged in the constructing industry. However, opportunity to additionally use natural fibres for acoustic will significantly rise their attraction in the market. An experimental examination of hemp-fibres diagnosed the effect of every step of the manufacturing method and shows use of natural hemp fibre can enhance the sound absorption property of sustainable thermal insulating materials. [13]

Currently, less costly housing and power conservation are a number of the topical problems in building creation. Few researches have shown the feasibility for incorporating the plant-based natural fibres in construction materials as it reduces heat transfer in structures and cost of material handling. A study found out that mud bricks with fibre reinforcement were light in weight than traditional mud brick, however fibre reinforced bricks possessed greater compressive strength than traditional mud brick. [14] So many car enterprises have done very much of research with the purpose of use of NFCs for their products. Various examines has been done in Europe by the car manufacturing enterprises to rise the applications of NFCs in automobile business, particularly in car interior for instance door-trim panels, truck linens, door linens, parcel shelves and seat backs. [15]

4. Conclusion

Much research and growth has taken place in current a long time within the mechanical performance of natural fibre composite. The properties (mechanical and physical) of natural fibres varies from fibre to fibre. This paper has reviewed the studies that focusses on composite of natural fibre and factors affecting the performance of natural fibre composite (like fibre choice, orientation, configuration, matrix choice as well as thermal parameters). Natural composites such as sisal, jute, abaca and coconut are utilized in lots of applications of engineering, due to its superior properties (which include light weight, specific strength, worthy mechanical properties, lower cost, eco-friendly as bio degradable). Overall, growth of natural fibre composite uptake maintains at a fast rate and there would seem like a very tremendous destiny beforehand for their utility.

5. References

[1] Faissal Chegdani, Behrouz Takabi, Bruce L. Tai, Mohamed El Mansori and Satish T.S. Bukkapatnam, "Thermal effects on tribological behavior in machining natural fiber composites" 2018 *46th SME North American Manufacturing Research Conference, Procedia Manufacturing* **26** 305–316

[2] Mulinari, D.R., Baptista, C.A.R.P, Souza, J. V. C., Voorwald and H.J.C, "Mechanical properties of coconut fibers reinforced polyester composites" 2011 *Procedia Engineering* **10**, 2074–2079

[3] N. Amir, Kamal Ariff Zainal Abidin and Faizzaty Binti Md Shiri, "Effects of fibre configuration on mechanical properties of banana fibre/pp/mapp natural fibre reinforced polymer composite" 2017 *Advances in Material & Processing Technologies Conference, Procedia Engineering* **184** 573 – 580

[4] K.L. Pickering, M.G. Aruan Efendy and T.M. Le, "A review of recent developments in natural fibre composites and their mechanical performance" 2016 *Composites: Part A* **83** 98–112

[5] Kin-tak Lau, Pui-yan Hung, Min-Hao Zhu and David Hui, "Properties of natural fibre composites for structural engineering applications" 2018 *Composites: Part B* **136** 222–233

[6] Swagata Dutta, Nam Kyeun Kim, Raj Das and Debes Bhattacharyya, "Effects of sample orientation on the fire reaction properties of natural fibre composites" 2019 *Composites: Part B* **157** 195–206

[7] Alexander Hart and John Summerscales, "Effect of time at temperature for natural fibres" 2017 *3rd International Conference on Natural Fibers- Advanced Materials for a Greener World, Procedia Engineering* **200** 269–275

[8] C.Elanchezhian, B.Vijaya Ramnath, G.Ramakrishnan, M.Rajendrakumar,V.Naveenkumar and M.K.Saravanakumar, "Review on mechanical properties of natural fiber composites" 2018 *Materials Today: Proceedings* **5** 1785–1790

[9] Changduk Kong, Haseung Lee and Hyunbum Park, "Design and manufacturing of automobile hood using natural composite structure" 2016 *Composites: Part B* **91** 18-26

[10] N.K. Kim, S. Dutta and D. Bhattacharyya, "A review of flammability of natural fibre reinforced polymeric composites" 2018 *Composites Science and Technology* **162** 64–78

[11] Lisa Boussaba, Amina Foufa, Said Makhlouf, Gilles Lefebvre and Laurent Royon, "Elaboration and properties of a composite bio-based PCM for an application in building envelopes" 2018

Construction and Building Materials **185** 156–165

[12] Ganesh R Kalagi, Rajashekar Patil and Narayan Nayak, "Experimental study on mechanical properties of natural fiber reinforced polymer composite materials for wind turbine blades" 2018 *Materials Today: Proceedings* **5** 2588–2596

[13] Andrea Santoni , Paolo Bonfiglio, Patrizio Fausti, Cristina Marescotti, Valentina Mazzanti, Francesco Mollica and Francesco Pompoli, "Improving the sound absorption performance of sustainable thermal insulation materials: Natural hemp fibres" 2019 *Applied Acoustics* **150** 279–289

[14] Obinna Onuaguluchi and Nemkumar Banthia, "Plant-based natural fibre reinforced cement composites: A review" 2016 *Cement and Concrete Composites* **68** 96-108

[15] Layth Mohammed, M. N. M. Ansari, Grace Pua, Mohammad Jawaid and M. Saiful Islam, "A Review on Natural Fiber Reinforced Polymer Composite and Its Applications" 2015 *International Journal of Polymer Science*

[16] D Saravana Bavan and GC Mohan Kumar, "Potential use of natural fiber composite materials in India" 2010 *Journal of Reinforced Plastics and Composites OnlineFirst* 1-4

IOP Conf. Series: Materials Science and Engineering **810** (2020) 012008 doi:10.1088/1757-899X/810/1/012008

Application of advanced optimization techniques in the selection of optimal process parameters and failure prediction of anti-frictional bearings-A review

C N Wani[1*], R D Dandagwhal[2]

[1,2]Mechanical Engineering Department,

NMIMS, MPSTME, Shirpur,

Maharashtra, 425405, India

chandrakantnwani@gmail.com

Abstract. Engineering design generally considered as goal-oriented, controlled, conclusion making process to manufacturing a product. While designing a mechanical component, designers always consider certain objective like tensile and flexural strength, deflection, weight, wear and tear, fatigue, etc. It is very difficult to accommodate all the objectives govern by multiple variables in the design of components. For this purpose, concentration has been given on the optimization of certain components of the mechanical design specially bearing. Bearings are extensively used in machine to reduce the friction and enhance rotational motion. Any critical failure of it, would not only affect the overall systems performance but also its reliability, safety, availability and cost-effectiveness. Objectives like long fatigue life i.e. higher dynamic load capacity, rating life need to be achieved collectively in the design of the rolling element bearing. The objective of this article is to highpoint influence of the advanced optimization techniques in design optimization and to determine and understand various methods of fault diagnosis of bearings to accomplish superior performance and to withstand global competitive environment.

1. Introduction

Bearing is simple, precious and of great utility member of machine system. The purpose of bearing is to provide relative motion, free movement and support to the transmission component of machine line shaft [1]. Bearings classified into two broader groups i.e. hydrodynamics bearing also known as sliding contact bearing and anti-friction or rolling element bearing. There is basic difference in two types of bearing, later is selected form the started catalogue of the manufacturer while the first is custom design. Hydrodynamic bearing design is application based and considered the parameters like higher life, rigidity, capacity and less friction. One general form of hydrostatic bearing is radial bearings finds it bid in heavy load applications like machine tools and some of the measuring devices. However, out of number of parameters, oil canal width ratio is crucial parameter to optimize the capacity [2].
State of art design of the bearing is not limited to hydrodynamic or rolling element. Advanced forms of the bearing includes the magnetic bearing, externally pressurized air bearing, thrust bearing, gas bearing and many more. Magnetic bearing is the modified form of the journal bearing in which fluid is replaced by the magnetic field. By using magnetic force, rotor is stabilized in at the center and seamlessly move without wear and lubrication [3]. Air bearing is also extended form of journal bearing in which air cushions provided to permit bearing action with minimum friction. Number of parameters like pressure of supply air, characteristics of porous pads like geometry and permeability make the optimal design of air bearing complex [4,5]. One more form of the hydrodynamic bearing known as gas bearing is also

widely used for heavy applications. Tight tolerance in the bearing clearance is the key factor for the stability of the gas bearing [6].

A wide form of bearing that used in low to medium load applications is the anti-friction or rolling element bearings. Dynamics capacity is most important objective mostly consider while selecting the rolling element from the manufacturers standard catalogue [7]. Load carrying capacity of the rolling element also depends on type of element and related geometrical parameters. With the geometrical changes in the design of the rolling element bearing the load carrying capacity of the bearing in different directions can be enhanced like radial, axial and thrust forces [8]. Allot of forms of the anti-friction bearing that widely used are deep groove ball bearing, angular contact bearing, single and double row taper roller bearing. Due to non-linear nature of the bearing it is look simple but difficult to design, any failure of the bearing may results into catastrophic breakdown in the machine or the part of the machine [9].

Every time after creating or designing a new product, designer tries to achieve the best solution for the product and therefore need to perform optimization. Though, optimization in design is not new, but it is time intense and involves systematic procedure to identify the right grouping process parameters and relevant objectives. Engineering design problems classified under several criteria's that are number, nature and permissible values of design variables, existence of constraints, number, nature and reparability of the objective function, physics of problem, and the nature of the environment **[6]**. Application of optimization is wide and cover almost every application of engineering. Metal cutting process is key factor manufacturing support system which consists of allot of input-output and in process parameters [7]. Due to complexity and uncertainty in the machining process it is difficult to perform computing manually, so, soft computing techniques are prefer to applied for machining process and optimizing them **[8]**. Relatively very less work was carried on the design optimization of the rolling element bearing. Objective of this paper is to review the application of advanced optimization techniques to rolling element bearing design. Review has been made on the basis of design optimization of the rolling element bearing along with highlighting the scope of nature inspired optimization techniques in the field of design and fault diagnosis of the rolling element bearing. The aim of present work is to highlight the applicability of the traditional and advanced optimization techniques in the selection of optimal process parameters and to predict the failure mode of the bearings.

2. Bearings

2.1. Hydrodynamic Journal bearings

Hydrodynamics journal bearings work on the hydrodynamics lubrication theory which statuses that the two contacting surfaces to be separated thin layer of viscous material while transmitting the motion. These bearings are very well known for their characteristic features like excellent damping, minimum friction due nonphysical contact and high load carrying capacity. Various equations has been adopted in direct or in modified form like Reynolds equation and different optimization techniques. Important design variables includes length of the bearing, geometrical specification of groove and radial clearance between journal and the bearing, etc.

Basic principle of the hydrodynamic journal bearing i.e separation of mating surfaces by a thin layer of viscous material has been adopted in various forms which results in the modified form of the bearing like air bearing, magnetic bearing. In case of air bearing separation between two surfaces achieved by providing air film. Pressure of supply air and characteristics of porous material are important design variables to achieve optimal performance of the air bearing [5]. In case of magnetic bearings, magnetic field used to keep the rotor in concentric position in the bearing casing. Number of accountability aspects of magnetic bearing has been considered by the researchers like reduction in actuator mass and power loss and enhancement in the external static load capability of the rotor [10]. The combination of shaft with multiple discs rigidly or flexibly connected to shaft is termed as rotor. Unbalance response amplitude and stability of rotor is critical performance criteria of the bearings. In case of high-speed operations stability of rotor and controlled vibrations necessary to avoid catastrophic failures. So, higher power-to-weight ratio is desirable [11]. Number of objective functions governed by number of parameters led the application of advanced optimization techniques in the design of the hydrodynamics journal bearings. Also, number of researchers applied traditional optimization technique to optimize the design of the above mentioned bearings. But, limited scope of traditional optimization techniques to handle number of objectives together nature inspired optimization technique replaced the prior. Table 1 shows the review of application of optimization techniques on hydrodynamics, air and magnetic bearings.

Table 1. Review of application of optimization techniques

Author, Year	Type of the bearings	Optimization technique	Design parameters	Remarks
Zhang and Talke [12], 2005	Air bearings	Combined genetic algorithm-subregion (CGS) approach	Flying height, Roll angle	Objective function shown suitable convergence in three considered cases. CGS is superior convergence rate than Genetic algorithm (GA) and less computational time than sub-region approach (SA)
Song et al. [13], 2005	Journal bearings	Enhance artificial life optimization algorithm (EALA)	the radial clearance, length to diameter ratio, average viscosity	EALA shows the superior results than artificial life algorithm (ELA). EALA results shows enhancement in journal speed and radial clearance while length to diameter ratio tends to lower.
Cavdar [2], 2006	Radial journal bearings	Computer aided optimization	Power ratio, width to diameter ratio, pocket length in axial direction and width ratio	Approach is suitable up to 3000 rev/sec. If the rotation is increased above limit than flow become turbulent.
Wang et al. [4], 2009	Air bearings	Genetic algorithm, cluster OpenMP	pressure of supply air , width of square porous pad , distance of porous pad edge to bearing edge , and permeability	Three task used to compute the performance of the algorithm which consist of change in the grid size and air bearing models. The moderate and larger size grain tasks effective in CLOMP.

			of the porous material	
Rao and Tiwari [3], 2009	Magnetic bearings	Genetic algorithm	radius of the shaft, the clearance between the shaft and the inner radius of the magnetic bearing, the inner radius of the bearing	Performance parameters and the controller of the magnetic bearing highly depends on air gaps, bias currents and length of permanent magnets. GA can effectively handle optimization of multiple objectives with number of process parameters.
Wang and Cha [5], 2010	Air bearings	hypercube-dividing method (HDM)	film thickness, air pressure in the film and porous pad	The new approached proposed termed as HDM found superior to produce Pareto optimal solution. If the minimum size of hypercube is defined properly, performance of the HDM can be enhanced.
Gorasso et al. [14], 2015	Hydrodynamic journal bearings	Simulations and Artificial Intelligence Tools	power loss, mass flow, temperature rise, maximum pressure and minimum lubricant film thickness	Optimal value of mass flow help to restricts the temperature and enhances the maximum pressure and fluid layer thickness. Although, dimensional parameters like diameter and length of bearing not much modified by algorithm.
Zhong et al. [10], 2016	Magnetic bearings	Genetic algorithm	length of the coil, width of the coil, spin speed, pole width	Two modified forms of the genetic algorithm, NSGA-II and NCGA used and obtained results were compared with Pareto optimal solution. The temperature distribution and hot spot generation in the bearing has analyzed by FEA on 2D model.
Akbarzadeh et al. [15], 2017	Thermodynamics journal bearings	MOPSO algorithm	radial clearance and	Two different case of multiple optimized solved considering bearing power loss, the minimum oil film thickness, and the maximum pad temperature as objective function. Along with the effect of design variables, effect of bearing speed,

			length-to-diameter ratio	bearing load, and inlet oil temperature On the overall performance of the journal bearing is demonstrated.
Soorajkrishna et al. [11], 2018	Rotor bearing	Genetic algorithm	internal diameter of the shaft, the external diameter of an element, length of an element, and density of the material	Single and multi-objective optimization carry out on rotor bearing. The scaled goaled programming which is modified form of previously used Genetic algorithm shown considerable improvement to achieve optimal solution. Obtained results were validated with the FEA results.
Guenat and Schiffmann [6], 2019	Gas journal bearings	Monte-Carlo method	Effect of pressure, speed, viscosity, radius, mass and centrifugal growth on manufacturing tolerances	Author proposed a new methodology of multi-objective optimization to enhance the range of design parameters. Optimized results help to enhance the manufacturing tolerance in the design of the grooved journal bearing.

2.2. Rolling element bearings

Application of optimization in design is bounded into certain space and process with single or multiple goals. Common objectives in the design of mechanical components consist of aspects of strength, deflection, weight, wear, corrosion, etc. depending on the requirements. However, design optimization of complete mechanical system is complicated with number of objectives governed by number of design variables [16]. So, application of optimization technique considering single component is always better to reduce the complexity. For this purpose, concentration has been given on the optimization of certain components of the mechanical design specially bearing. To understand the current state of application of optimization in gears and bearings, a literature review has been presented. As discussed, bearing is one of the important component of the mechanical system and early detection of faults is crucial requirement to avoid catastrophic failure of the whole system. On other hand, design of the bearing depends on the dynamics capacity of the rolling element bearing which depends on number of design variables [17].

Number of researchers concentrated on various objectives related to enhancement in efficiency of rolling element bearings like fatigue life, corrosion resistance, reduction in overheating, friction, vibration and noise. But in normal operating conditions, the mail failure that most of the rolling bearings face is contact fatigue. For the longer life of the bearings, it is necessary to increase fatigue life which is directly proportional to the dynamics load capacity. A standard rolling element bearing consists of inner ring, outer ring and the rolling element. Rolling element is sphere in case of ball bearings while cylindrical roller in case of roller bearings. However, design seems to be simple of the rolling bearing but non-linear behaviour and inner geometry only responsible for the deflection and load distribution [18]. As explained above section, fatigue life is function of the dynamic load rating. Table 2 consists of review of application of optimization techniques in rolling element bearings.

Table 2. Review of application of optimization techniques

Author, Year	Type of the bearings	Optimization technique	Remarks
Rao and Tiwari [19], 2007	Rolling element bearing-Ball bearings	Genetic Algorithm	Maximization of fatigue life considered as objective while number of parameters measured were pitch dia of the bearing, sphere i.e. rolling element diameter and there number with inner and outer race curvature. Convergence of the parameters studied and the fatigue life considerably improved.
Gupta et al. [17], 2007	Rolling element bearing-Ball bearings	Genetic Algorithm	A multi-objective optimization problem in which dynamics, static capacity and electrodynamic minimum film thickness considered as objectives. Non-dominated sorting based genetic algorithm (NSGA-II) used to obtain Pareto Optimal solution considering eight geometrical parameters.
Kumar et al. [20], 2008	Cylindrical roller bearings	Genetic Algorithm	Considered dynamic fatigue failure as objective function. Number of geometrical parameters were more in cylindrical roller bearings which makes the system non-linear and complex. Genetic algorithm is capable to deal with number of parameters. Obtained results were superior and compared with standard results available in catalogue.
Wei and Chengzu [21], 2010	High speed Angular Contact ball bearings	Non-dominated sorting based genetic algorithm (NSGA-II)	Considered two objectives i.e. frictional loss in rotation and longer serving life with five process parameters. A high speed angular ball bearing (AC 7007) optimized in which 7.5 superior results obtained than standard values. Stiffness can be consider as another objective function.
Mendi et al. [8], 2010	Deep groove ball bearing	Genetic Algorithm	Reduction in volume consider as objective of a gear box which consists of reduction of volume of gear, shaft and bearings. Genetic algorithm used to perform said optimization in which obtained results shown a superiority than analytic method results.
Dragoni [22], 2013	Radial cylindrical roller bearings	Linear optimization	Focused on maximization of static and dynamic rating life considering multiple size constraints. Dynamic load rating depends on five geometrical parameters whereas static load rating depends on two geometrical parameters. For the dimensionally unconstrained condition, roller diameter and pitch diameter helps to enhance dynamic load rating.

			After certain permissible range of pitch diameter, dynamic load rating start decreasing due to abrupt changes in curvature diameter.
Thoriya et al. [23], 2015	Rolling contact bearing	Teaching learning based optimization	A review has been carried out on the design optimization of rolling element bearing using TLBO algorithm in which single and multi-objective conditions considered. In most of design problem of rolling contact bearings, dynamic capacity considered as objective which defines the life of the bearing.
Lostado et al. [24], 2015	Double row taper roller bearing	Multiple response optimization and FEM	A new methodology implemented to optimizes the life of double row taper roller bearing considering Preload, radial load, axial load and torque as the performance parameters. Three parameters value obtained from the Finite element modeling obtained results had shown considerable improvement as compared to available standard results.
Kim et al. [25], 2016	Angular contact ball bearings	Multi-objective discrete optimization	A hybrid approach in which genetic algorithm and regression optimizer used for efficient discrete optimization. Angular contact ball bearing in the application of the shaft of grinder used for the optimization. Constraints considering film thickness, friction, manufacturing and fatigue life while six other geometrical variables selected as design variables. Optimized results of stiffness shown considerable improvement.
Shaikh and Kamble [18], 2018	Deep groove ball bearings	Jaya Algorithm	A newly developed advanced optimization technique named as Jaya Algorithm used for design optimization of deep groove ball bearing which consists of number of design variables. Considered algorithm shown superior performance over the other advanced optimization and traditional techniques.
Martinez et al. [26], 2018	Taper roller bearings	Data mining classification technique	For the proper work prediction of the taper roller bearing an optimal relation need to define considering contact stresses, fatigue spalling and pitting. A finite element models followed by design of experiments synchronized in such a way that the overall simulation cost to be reduced. All the FE model validated with good accuracy.
Dandagwhal and Kalyankar [9], 2019	Deep groove ball bearing and	Teaching learning based algorithm	Two different models of rolling contact bearing i.e. deep groove ball bearing and cylindrical roller bearing has been considered. Both the models have different geometrical specification which consists

cylindrical roller bearing	of number of variables leads the optimization process become complicated. Results obtained from optimization process validated with finite element analysis results which shown close tolerance.

3. Discussions on current limitations and future trends

Hydrodynamic bearing and rolling element bearings has basic difference in the design and the process parameters which leads to use different type of optimization approach while find out optimal values of respective parameters. Hydrodynamic bearings available in number of forms like air bearing, magnetic bearing, journal bearing, gas journal and rotor bearing. All types have same basic working principles but as the lubricant replaces by magnetic field or air or gas, we need to consider respective parameters for optimization. Power loss, mass flow, temperature rise, maximum pressure and minimum lubricant film thickness are few general objective functions optimized by various researchers. Still number of process and design parameters related to traditional form of journal bearing and of advanced or modified form need to be analyses properly. Various researchers focused on design optimization of the rolling element bearings due to quite simple structure as compared to hydrodynamic bearings. Still non-linear relations of inner and outer race with rolling element i.e. sphere or roller makes optimization problem complex. A common objective function, dynamic capacity of the bearing used by many researcher as the life of the bearing directly proportional with it. Still, various objective functions like lubrication film thickness, heat generation due to friction need to be explored along with fatigue life. Certain factors like preload and material properties of the bearing material need to be considered in the process of optimization.

Also, predictive maintenance of the bearing is prime concern to avoid the catastrophic failure of the system. Early detection of faults is also emerging area for the application of advanced optimization technique [27]. Normally envelope analysis used demodulation technique which provided very little flexibility and it is very difficult to vary the working conditions. Advanced optimization technique like genetic algorithm has strong ability to optimize the optimal resonance demodulation in which initial estimates provided by kurtogram. Li et al. [28] used the support vector machine with ant colony optimization to predict the failure of the bearings. Best and worst solution updated for the pheromone trails density. Results of optimization were validated by experimentation and superior than SVM-GA combination. In another study, Xi et al. [29] used neural network modified by glowworm swarm optimization (GSO) used to determine six different faults features. Initial weight and threshold values of neural network optimized using GSO. ANN found good scope in the fault detection in the bearings. Same principle used by Unal et al. [30] to diagnose the faults in bearing, fitness value validated by experimentation. Recent use of artificial intelligence and machine learning tool also shown the potential of technique in soft computing [31]. Kolodziejczyk et al. [32] used the AI and ML to correlate the working conditions with the mating bodies and to explain the impairment mechanisms.

4. Conclusion

From the review of studies on the design optimization of the rolling element bearings, it is seen that emphasis is on the maximizing life of the bearing. Although, literature available for design optimization for multiple objectives likes dynamic capacity, the static capacity and the elasto-hydrodynamic minimum film thickness, dynamic capacity have drawn the attention of a greater number of researchers. Relatively less work has been carried out in the area of design optimization of the rolling element bearings and further application of optimization techniques required particularly for the objectives of maximum wear life, the maximum static load rating, the minimum frictional moment and the minimum spin to the roll ratio along with maximum dynamic capacity. The generic framework for application of design optimization in rolling element bearing attempts to provide a single, unified, and systematic approach to determine optimal or near-optimal design parameters. Also, scope of optimization techniques in the fault diagnosis of the rolling element bearing had been discussed. With the continuous research in the field of optimization, no doubt the quest to pursue better and improved rolling element bearing design optimization will continue.

5. References

[1] Anderson W J 1983 *NASA Ref. Publ. 1105*
[2] Cavdar K 2006 *Ind. Lubr. Tribol.* **58** 129–34
[3] Rao J S and Tiwari R 2009 *Mechatronics* **19** 945–64
[4] Wang N, Tsai C M and Cha K C 2009 *Tribol. Int.* **42** 1180–6
[5] Wang N and Cha K C 2010 *Tribol. Int.* **43** 1631–8
[6] Guenat E and Schiffmann J 2019 *Tribol. Trans.* **62** 1041–50
[7] Chakraborty I, Kumar V, Nair S B and Tiwari R 2003 *Eng. Optim.* **35** 649–59
[8] Mendi F, Başkal T, Boran K and Boran F E 2010 *Expert Syst. Appl.* **37** 8058–64
[9] Dandagwhal R D and Kalyankar V D 2019 *Arab. J. Sci. Eng.* **44**
[10] Zhong W, Palazzolo A and Kang X 2017 *J. Vib. Acoust. Trans. ASME* **139** 1–15
[11] Soorajkrishna S, Sekhar A S and Shankar K 2019 *Proc. Inst. Mech. Eng. Part C J. Mech. Eng. Sci.* **233** 4270–87
[12] Zhang J and Talke F E 2005 *Tribol. Int.* **38** 566–73
[13] Song J D, Yang B S, Choi B G and Kim H J 2005 *Tribol. Int.* **38** 403–12
[14] Pennacchi P 2015 *Mech. Mach. Sci.* **21**
[15] Akbarzadeh P, Mikaeeli S Z and Rahimiyan M 2018 *Proc. Inst. Mech. Eng. Part J J. Eng. Tribol.* **232** 657–71
[16] Rao R V., Savsani V J and Vakharia D P 2012 *Inf. Sci. (Ny).* **183** 1–15
[17] Gupta S, Tiwari R and Nair S B 2007 *Mech. Mach. Theory* **42** 1418–43
[18] Shaikh H and Kamble A 2018 *SSRN Electron. J.* 2–4
[19] Rajeswara Rao B and Tiwari R 2007 *Mech. Mach. Theory* **42** 233–50
[20] Kumar K S, Tiwari R and Reddy R S 2008 *Int. J. Comput. Methods Eng. Sci. Mech.* **9** 321–41
[21] Yu W and Ren C 2010 *Int. Conf. Comput. Control Ind. Eng. CCIE 2010* **1** 320–4
[22] Dragoni E 2013 *Proc. Inst. Mech. Eng. Part C J. Mech. Eng. Sci.* **227** 2393–401
[23] Thoriya P J, Saradava B J and Puranik P S 2015 *J. Mater. Metall. Eng.* **5** 41–3
[24] Lostado R, Escribano García R and Fernandez Martinez R 2016 *Int. J. Mech. Mater. Des.* **12** 353–73
[25] Kim S W, Kang K, Yoon K and Choi D H 2016 Design optimization of an angular contact ball bearing for the main shaft of a grinder *Mech. Mach. Theory* **104** 287–302
[26] Fernandez Martinez R, Lostado Lorza R, Santos Delgado A A and Piedra Pullaguari N O 2018 *Adv. Eng. Softw.* **123** 13–24
[27] Zhang Y and Randall R B 2009 *Mech. Syst. Signal Process.* **23** 1509–17
[28] Li X, Zheng A, Zhang X, Li C and Zhang L 2013 *Meas. J. Int. Meas. Confed.* **46** 2726–34
[29] Xu Q, Liu Y Q, Tian D and Long Q 2014 *Adv. Mater. Res.* **860–863** 1812–5
[30] Unal M, Onat M, Demetgul M and Kucuk H 2014 *Meas. J. Int. Meas. Confed.* **58** 187–96
[31] Kankar P K, Sharma S C and Harsha S P 2011 *Expert Syst. Appl.* **38** 1876–86
[32] Kolodziejczyk T, Toscano R, Fouvry S and Morales-Espejel G 2010 *Wear* **268** 309–15

Current Research in Powder Mixed Electric Discharge Machining: A Review

Sachin S Chaudhari[1], Dr A M Nikalje[2], Dr G E Chaudhari[3]

[1]Mechanical Engineering Department, Government College of Engineering, Jalgaon, Maharashtra, India
[2] Mechanical Engineering Department, Government College of Engineering, Aurangabad, Maharashtra, India
[3]Mechanical Engineering Department, J.T.Mahajan College of Engineering, Faizpur, Maharashtra, India

corresponding author's e-mail : sachinschaudhari7@gmail.com

Abstract. Electric Discharge Machining (EDM) is a recognized and commonly used non-conventional machining method for the manufacture of complicated shapes and the processing of particularly hard materials, which are very difficult to machine by any standard machining processes. In this process, there is no direct contact between the electrode and the work-piece, this non-contact machining technique is constantly rising from the press tool and dies the process creation Process for apps for micro-level machining. In the latest decades, several researchers have focused on enhancing machining process capability such as enhanced rates of material removal with improved surface topography. Powder mixed electrical discharge machining (PMEDM) may be a newly developed technique for improving the EDM process capabilities of the EDM process in this trend. In this paper, a broad history, mechanism of PWEDM method, and literature review are presented.

1. Introduction

Electrical discharge machining (EDM) is one of the oldest and preferred non-traditional machining procedures involving the removal of material through the operation of short-length electrical discharges and highly current density between the tool electrode and the work-piece. Complex shape machining is made relatively simple also in difficult-to-machine materials. EDM technique is mostly and widely used in the press tools and dies, aerospace, automotive and surgical parts manufacturing industries. EDM has effectively applied to electrically conductive materials such as hot work dies steel, carbides, stainless steel (SS), different types of alloys, etc. Despite their physical and metallurgical characteristics. The rate of removal of materials and the quality of surface generated in machining are directly associated to the quantity of energy used to erosion material from the surface of the work-piece [1].

EDM is not a conventional technique of machining that has been commonly used. It makes molds and dies. EDM also used for machining components of the automotive, dies and press tool industries [2]. The history of electrical discharge machining (EDM) methods was found in the 1770s by an English scientist after the creation of the relaxation circuit (RC-resistance-capacity) by B.R and N.I. Lazarenko [3], keeping the distance across of the gap between the tool and the work-piece maintained by easy servo control using this method, they discovered a decrease in tonnage and made EDM more lucrative. EDM has been the ideal reducer of orbital and planetary motion methods, pulse generator, computerized numerical control (CNC) and adaptive control scheme since 1940. However, the EDM was not fully utilized until scientists of Russia studied how the erosive impact of the method could be used and be regulated machining purposes [4].

Die Sinking EDM utilizes spark ignition erosion to metal removal. The power supply produces electrical impulses between the work-piece and the tool. A inter electrode gap between the electrode and the work-piece enables flow of dielectric oil. When adequate voltage is reached to pre-set value, the dielectric oil start ionizes and regulated sparks melts and vaporizes the work-piece. Pressurized dielectric oil cools the vaporized metal and removes the material from the gap.

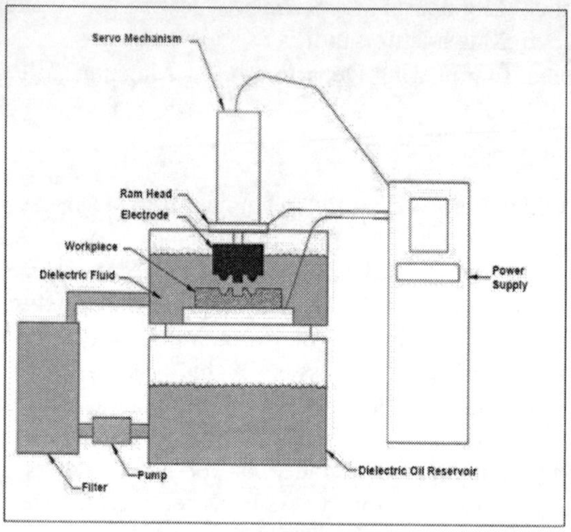

Figure 1. Die Sinking EDM

The filter mechanism cleans the suspended particles out of the dielectric oil. The oil passes through the chilling unit to remove the heat produced by the method of chisel erosion. With the used of chilling unit, a steady temperature of oil can be maintain that helps with the precision of the machine. See in figure Die Sinking EDM, like wire-cut EDM, is a method of spark erosion. Die Sinking EDM, on the other hand, generates sparks along the surface of the electrode created. The gap between the electrode and worke-piece is maintained by a servo control unit. The servo control unit stops the electrode from touching the piece of work. a short circuit would be created and no cutting would happen, when the electrode were to touch the work-piece, [5].

1.1 Powder mixed electrical discharge machining (PMEDM):

In order to overcome constraints associated with conventional EDM, the highly conductive powder particles are get mixed in the dielectric fluid as shown in fig.2, which decreases its insulating capacity and raises an inter-electrode gap. So this PMEDM is also called as a hybrid material removal method. From the point of view of industrial applications, die steel is a very important material and therefore, for the purpose of experimentation, D3 die steel with copper electrode and EDM oil as dielectric has been used. The PMEDM is very complicated in nature and is regulated by a big amount of parameters that have an effect on different reactions. The current research is an effort to explore the ability of PMEDM to increase the material extraction frequency, reduce the tool wear rate (TWR) with enhanced surface roughness (SR) of Die Steels [6].

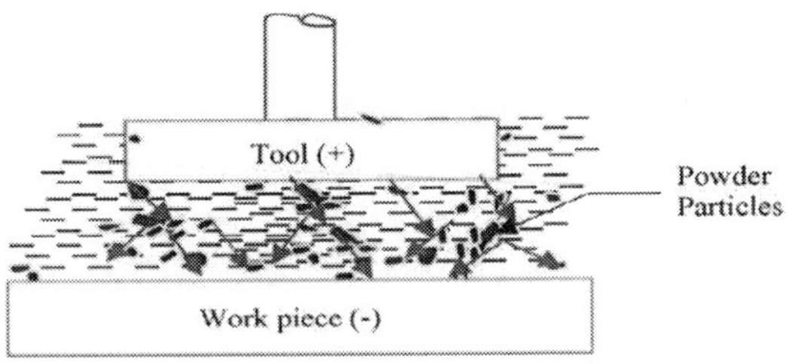

Figure 2. Powder mixed electrical discharge machining

The appropriate amount of fine abrasive powder is blended with the dielectric fluid. The process of removing the hybrid material is called Powder Mixed EDM (PMEDM) in which the minimum pulse energy is constantly working and the efficiency of the EDM method is considerably affected. The insulation resistance of the dielectric fluid gets decreases due to electrically conductive powders and increases the gap between the tool and the work-piece. The EDM method becomes much more stable and facilitates the machining process. Productivity, MRR, SR. However, most of the research was conducted to estimate the surface finish of the process. It would provide a mirror-like surface finish that becomes a challenge in the EDM. The parameters of the powder, such as size, type, and concentration, affect the dielectric performance [7].

2. Process Parameters:

Process parameters are as follows:

2.1 Electrical Parameters:

2.1.1. Peak Current:

It is essentially the most significant machining input parameter in the EDM. Peak current is the quantity of energy used in the EDM and the measure in the amperage unit. At every pulse-on-time, the current goes increases till it attain the predefined level that is expressed as the peak current. The maximum amount of amperage is to be determined according to the surface of the cut during both die sinking and wire-EDM processes. In rough operations, cavities or details with large surface zones, higher amperages are applied. Higher currents increase MRR, although surface ruggedness and tool wear are sacrificed. [8].

2.1.2. Discharge Voltage:

There is built-up of open circuit voltage between two electrodes before any current starts flowing between them. Once the current flows through the plasma channel, the open circuit voltage falls and is a crucial variable that affects the energy of the spark responsible for greater MRR, greater tool wear rate and rough surfaces.

2.1.3. Pulse-On-Time or Pulse Duration:

It is available in units of microseconds. Because every cut is made in pulses on time, the duration and cycles per second (frequency) of such pulses play a crucial role. The current can flow on a cycle basis. During this phase, dielectric ionizes and sparks occurs. It is the productive spark cycle system in which current flows and machining take place. In this on-time operation, the quantity of material extracted is directly proportional to the amount of power consumed [9].

2.1.4 Pulse-Off-Time or Pulse Interval:

It's the interval of time between two successive pulses. The supply voltage is disrupted during the switch-off period. During this period, dielectric deionizes and recovers its strength. This interval, the molten material can be solidified and separated by flushing the arc. The time to switch off must be held to a minimum. The sequence is finished when the appropriate pulse interval earlier than the start of the next cycle is allowed. The interval of the pulse affects the rate and durability of the cut. In concept, the shorter and smaller the interval, the faster the operation is performed. If the interval is too short then the excluded workpiece material will not flush off with the dielectric fluid and also not deionize the dielectric fluid. This will result in the next spark being unstable and slowed down by cutting downtimes as long and stable [10].

2.1.5. Polarity:

The electrode polarity may be positive or negative. However, the surplus material is separated from the positive side. Polarity is usually measured by experiments and involves tool materials, job materials, current densities and combinations of the pulse length. To avoid arcing, modern energy supplies insert the "swing pulse" of reverse polarity at set intervals. The normal proportion for every 15 normal pulses is 1 Swing Pulse.

2.1.6. Inter-electrode Gap:

The gap between the electrodes is essential for the stabilization of the spark and accurate flushing. The stability of the system separation and responding speed are the most significant preconditions for excellent operation; the existence of the knock-back is especially unwanted. To react to short circuits or open space circumstances, the response rate must be high. The width of the gap may not be immediately estimated but the average gap voltage can be deducted. The servo tool system maintains a fixed value of the working gap. The main purpose is to use electromechanical systems (direct current or step-by-step motors) and electro-hydraulic systems keep medium voltage gap [9].

2.2. Powder Parameters:

2.2.1 Powder types:

The added powder into the dielectric fluid can enhance MRR and reduce the wear rate of the tool and significantly enhance the surface quality of the work-piece. However, the various powders could influence the performance features of the EDM method differently. The following characteristics must exist in the powder which can be suspended in EDM dielectric fluid; It should be electrical conductive, non-magnetic, must have good suspension capabilities, good thermal conductivity, non toxic and odorless. Many investigators have used various kinds of powders such as; Aluminum (Al), Chromium (Cr), Graphite (C), Copper (Cu), Silicon (Si), Nickel (Ni), Silicon Carbide (SiC), Titanium (Ti), Tungsten (W), Bron Carbide (B_4C), Alumina (Al_2O_3), Carbon Nano-tubes (CNT), Diatomite [9].

2.2.2. Powder Concentration:

Process effectiveness and stabilization are achieved due to the proper particle concentration. The optimum concentration is determined by powder properties for increasing performance. More concentration than ideal value leads to short circuits, arcing, or unstable processes because of excessive particles like residual particles, which are generally considered by too many debris as the dominant reason for spark concentration. Excessive particles in the discharge gap also cause settling and bridging issues, which lead to surface deterioration [11].

2.2.3 Powder Size:

Powder particle size is identified as a significant powder parameter to achieve the desired PMEDM performance. The effect is more abbreviated for higher pulse-on-time values due to higher contamination and lesser deionization between work-piece and tool [12] and bigger particles resulting in greater gap growth. The additive particles ' size is a crucial factor for the quality of the working surface, Yih-Fong and Fu-chen[13] indicated. The highest surface finish is achieved by smaller particles (70-80 nm), but recast layer thickness get increased. These small particles (70-80 nm) produced, according to Tzeng and Lee [12], the lowest discharge gap, the maximum MRR, and also least tool wear rate.

3. Literature review:

On the basis of the literature review, the researchers' contribution to PMEDM studies is discussed below:

The impact of aluminum powder on the AISI D3 steel machining has been found by Jamadar et al. [14]. The process performance variables are investigated such as material removal rate (MRR), the tool wear rate (TWR) and surface roughness (SR). The machining efficiency was considerably affected by aluminum mixed with dielectric liquid. Razak et al. [15] explored that concentration of SiC micro powder in PMEDM in Hot Die steel and its particle size. A number of trials conducted by orthogonal Taguchi array with three levels and two variables.

The impact of Aluminum powder on the errosion of metal matrix composites (Al / SiC) was researched by Hussain et. al. [16]. Different parameters have been selected and test findings show that the removal rate of materials improves with a small peak 2A current. With the trending technologies addition of micro particles like MoS_2 was held in the dielectrical fluid in a microform by Prihandana, Mahardika[17] and additional ultrasonic vibration. Results indicate that ultrasonic vibration and micro-particle addition considerably boost the surface quality of the MRR.

Many scientists have come to realize about implementing nanoparticles for many applications in nano-technology with increasing speed. The list of nanoparticles used in EDM, although few, have shown that the best surface finish obtained by introducing nanoparticles in machining process as a powder mixed in EDM. Jahan, Rahman[18] studied the effect of nano particles and micro particles in WC-Co EDM and reported that the smoother and clean surface was achieved as compared to the sinking micro EDM. The highest and lowest SR values were 37 nm and 17μm, respectively.

Tan et al. [19] explored the effect of nanopowder particles on surface roughness and discharge gap ranges during the micro EDM of stainless steel mould of AISI 420 grade. SiC and Al_2O_3 nanopowders have been recorded to decrease surface roughness during the micro EDM of stainless steel. Jahan and team studied impact of graphite nanopowder on material SKH-51 grade tool steel micro-EDM [20]. Subsequently, the effects of graphite, aluminium and Al_2O_3 nanopowders on surface topography were studied [21]. The conclusion was that, smoother and mirror-like surfaces were found with semi conductive graphite nanopowder among the three types of powders.

Table 1. Current Research Literature carried out for PMEDM with different powders and the impact of varying input parameters with distinct optimization techniques is shown in the following table.

Sr No	Name of Authors Year Material: Work-piece and Electrode	Dielectric media	Parameters	Performance Measures	Findings

IOP Conf. Series: Materials Science and Engineering **810** (2020) 012009 doi:10.1088/1757-899X/810/1/012009

01	Pushpendra S. Bharti et.al. [22] 2011 Inconel 718 Copper EDM oil	Kerosene	Pulse On Time, Pulse Off Time, Shape Factor, Current, duty cycle, gap voltage, flushing pressure and lift time of tool electrode.	Surface Roughness (SR) & Material Removal Rate(MRR)	The orthogonal L27 array has used for conduct experiments The EDM process ANN model was then created through the training of the ANN with experimental data. Experimentally confirmed the prediction capacity of a trained ANN model. The trained network was finally used by regulating elitist NSGA-II to achieve the Pareto-optimal solution set. Matlab conducted a simulation. So they obtained 103 Pareto-optimal solutions and due to conflicts in response variables, they found that range between 26-55 is best to run to get an optimized solution.
02	Uttam Kumar Mohanty [23]. 2017 High carbon steel Copper Standard dielectric	Kerosene	Pulse On Time , Current and Volatage.	MRR, TWR, SR and Radial Overcut (ROC).	Experiments were conducted by L9 orthogonal array, then the simpler regression model has developed the establishment of correlations between input and output parameters also ANOVA used to see most influencing process parameter. And finally, VIKOR Index method was used for multi-objective optimization used to optimize simultaneously performance variables. They found that among process parameters under study current is the most influence on the performance of EDM.
03	B. Singaravel et.al [24]. 2019 AISI D2 Steel Cryo-treated Copper	Kerosene, Jatropha and cottonseed oils.	Types of dielectric, Current and Pulse On Time.	MMR. TWR, and SR	In order to decide the optimum parameters, a new Hybrid TAGUCHI and VIKOR method is introduced, and VIKOR index value is used for optimum parameters. They observed that MMR is increased by higher spark energy density and less tool wear is recorded due to Cryo treated electrode.
04	Muhammad Han et.al [25]. 2019 AISI D2 Steel copper	Graphene nanoplatlates mixed in kerosene	Pulse On Time, Pulse Off Time, Discharge Current, Powder concentration.	MMR,SR and White Layer Thickness(WLT)	For the design of experiments, they used a Box-Behnken design based on Response Surface Methodology (RSM). The most important factors have been identified by ANOVA. In terms of desirability, multi-objective optimization is carried out. They discovered that pulse-off time has a significant effect on WLT among input parameters. For larger discharge current and conch, greater MMR was obtained. Low discharge current and high concentrations of nanoparticles and minimum SR were obtained. Multi-objective optimization and optimal process performance with desirability function.

05	Deepak Kumar Naik et.al [26]. 2018 Nickel free Austenite stainless steel Copper	Nickel Powder Mixed with Kerosene	Pulse On Time, Peak Current and Powder concentration	MMR.SR and ROC	Taguchi L16 array used for DOE, A new approach Principal component analysis (PCA) - the technique for order preference by similarity to ideal solution (TOPSIS) used to the optimized process. Results indicate that the peak current is the most significant process parameter. The proposed technique is very suitable for getting an optimum set of input parameters.
06	Shalini Mohanty et.al [27]. 2017 AlSiCp Metal matrix composite Copper	Alumina Oxide(Al2 O3) powder	Low and High Voltage current, Pulse On Time, Pulse Off Time, Flushing Pressure	MMR.TWR,and SR	For the design of experiments, they used a Box-Behnken design based on Response Surface Methodology (RSM). Particle Swarm Optimization (PSO) and Desirability Function these two methods multi-objective optimization techniques are used and they got results as the powders improve the material removal rate and surface finish. With the use of nanopowders, it has seen that MMR increased and SR decreased.
07	Soumyakant Padhee [28]. 2011 EN31 steel	Silicon Powder	Pulse On Time, Peak current, duty cycle and Powder concentration	MMR and SR	A Face Centred Central Design (FCCD) is used to conduct the experiments. RSM is applied to do study the effect of process parameters on responses and developed predictive model. It is required to obtained optimum process parameters selection that result in higher material removal with lower surface roughness, due to conflicting nature of responses so it was necessary to explore the optimization area to obtain the set of the dominant solution with help of Non-sorted genetic algorithm (NSGA).
08	Huu-Phan Nguyen et.al [29]. 2018 SKD61,SKD11,SKT4 Graphite Copper	Titanium Powder	Pulse On Time, Pulse Off Time, polarity of electrode and nano powder concentration	MMR ,SR and Micro Hardness(HV)	Taguchi L16 array used for DOE. The results showed that MMR and HV increased and SR gets decreased with the addition of titanium powder in the dielectric. The most significant parameter found as Powder concentration. They also consider studying interactions of process parameters with each other. TOPSIS used for further multi-objective optimization to obtain optimum sets of parameters.
09	Neeraj Agarwal et.al [30]. 2018 AISI D2 steel	EDM oil	Pulse On Time pulse current, Duty cycle, and gap voltage.	MMR, and SR	To predict quadratic polynomial models, a multiple regression technique was coupled to a single objective. A modern Jaya algorithm was then applied to optimize process parameters and an optimal combination set was introduced.

10	Chinmaya P.Mohanty et. al [31]. 2017 Inconel 718 Graphite	Paraffin Oil	Pulse On Time Open circuit voltage, flushing pressure, duty factor, material of electrode, discharge current.	MRR,ROC,TWR and SR	Taguchi's orthogonal array L27 was used for experimenting, ANOVA was also conducted and discovered to be significant process parameters as an electrode material, Pulse On Time and discharge current. A hybrid method has been suggested and contrasted for acquiring the highest parameter setting for the EDM method, using the Quantum Particle Swarm Search Optimization (QPSO) and PSO algorithms.
11	G. Venkata et.al [32]. 2019 Inconel 750 Copper wire	Kerosene	Pulse On Time, Pulse Off Time, Wire feed rate, water pressure	Machining speed and SR	Taguchi's orthogonal array L18 was used for experimenting, The TOPSIS is integrated with the Gray Wolf Optimizer(GWO). Method of multi-objective optimization. Multi-objective is originally converted to a single objective by using TOPSIS, with the help of Wire EDM machining a relative close value has been calculated. TOPSIS was used as an objective function (equation) for the Grey Wolf Optimizer (GWO), and the optimal setting of the parametric was established.
12	Jun Ma et.al [33]. 2017 SiCp/Al Composite Copper wire	Kerosene	Pulse On Time, Pulse Off Time, Wire tension, and Water pressure	MMR and SR	Taguchi's orthogonal array L18 was employed in experimenting with an Optimization Model for predicting MMR as well as SR. The findings were compared with the linear regression model and the back propagation neural model and the best methodology for improving the process of EDM, using the novel hybrid Gaussian Regression (GPR) and Wolf Pack Algorithm.
13	Shankar Chakraborty et.al [34]. 2018 Alumina ceramic	Water	Water pressure,jet feed speed, abrasive mass flow rate,nozzle speed	MMR and SR	Two non-linear regression models were developed based upon the experimental data and using RSM methods, which are afterward employed for multi-objective of Abrasive Water Jet Machining process(AWJM) for that purpose Grey Wolf Optimization (GWO) technique were used and they found that optimum set of parameters.
14	Osama Lari et.al [35]. 2015 AISI 1046 steel Copper	Chromium powder in EDM oil	Concentration and Grain size of powder	MMR and SR	Taguchi L9 array has been used for conducting experiments, analysis of single and multi response stage was done with Taguchi method, also multi response stage was done with Utility concept. ANN model used for close proximity with experiment values.

15	Arvind Kumar et.al [36]. 2019 AA7050-10%B4C Composite Copper	Kerosene	Pulse-on-time, Pulse-off-time, and pulse current	MMR, SR and depth of cut (DC	A new multi criteria decision making model of AHP-The Additive Ratio Assessment (ARAS) has been used for optimization multi response. For experiments Taguchi L9 array has been utilized. ARAS is a MCDM which determines the performance of chosen alternatives also compares the scores of those chosen alternatives with the best alternative.
16	Rajkamal Santosh Shukla et.al [37].	Tzeng and Chen have created a process EDM configuration, perform experiments with copper electrode JIS SKD 61 steel Pulse-on-time, Pulse-on-time, Discharge Current and Gap voltage; use a Regression Predictive Model for performance parameters such as MMR, RA, and REWR.			Authors used same model developed by Tzeng and chen and apply Fireflies Algorithm(FA) to get optimum process parameters. The Fireflies Algorithm(FA) have distinctive characteristics compared to the many other optimization algorithms, such as., Particle Swarm Optimization (PSO), Genetic Algorithm, SA,, Artificial Bee Colony algorithm (ABC), etc. This FA algorithm possesses multi-modal characteristics, high convergence rate and few control parameters.
17	Zheng-Ming Gao et.al [38] 2019	Research Article: **An Improved Grey Wolf Optimization Algorithm with variable Weights.** An enhanced Gray Wolf Improvement optimization (GWO) variable-weight algorithm (VW-GWO) program is proposed in this paper. A hypothesis has been established that the social hierarchy of the packs even works in their view positions. And then their search technique is given variable weights. A governing equation of the dominant parameter is implemented to reduce the probability of being caught in local Optima, thus exponentially decreasing from the maximum. Finally, there are 3 types of studies to check the merits of the VW-GWO algorithmic program envisaged. The original GWO and therefore the ALO, PSO and BA algorithms are compared. The selected experiment findings indicate that under entirely distinct circumstances, the predicted VW-GWO algorithmic program operates higher than the others.				
18	Tarik A. Rashid et.al [39]. 2019	Research Article : **A multi hidden recurrent neural network with a modified grey wolf optimize** Identifying the weaknesses of university learners leads to better study and can be an early warning system for improving learners. However, current schemes do not have a successful amount of satisfaction. New and dynamic measurement of hybrid systems needed for this mechanism. A hybrid scheme is used to anticipate the results of learners (a modified recurrent neural network with the tailored gray Wolf Optimizer). This planned scheme would improve faculty training and enhance the learning experiences of the learners. Results indicate that the altered recurred neural web has the most efficient precision compared to various designs with the tailored gray Wolf Optimizer.				

4. Gaps in literature:

In the past couple of decades, research on EDM machining performance variables, such as MRR, TWR and SR, has been conducted in the area of PMEDM (Powder Mixed Electric Discharge Machining). However, information about the variation to a known work piece and electrode of process parameter for a particular powder is insufficient. Furthermore, the critical size of the powder is still not determined and its variation with other process parameters. The conclusive experimental work can achieve this. In research, researchers can find the optimal value of distinct process parameters through the use of multiple types of powders, varying workpiece and electrode material.

In recent years numerous scientists have used powders for the machining of workpiece such as silicon, silicon carbide, aluminum mixed in the dielectric. Powdered materials, such as nickel, chromium, copper, graphite etc, can be blended in a dielectric process. Few trials focused on powder parameters such as powder size and powder concentration. Consequently, further studies are needed regarding the effect of process parameters. Micro and nano-powder surface modification was not much attempted yet.

5. Conclusions:

PMEDM is a very useful type of EDM which provides improved efficiency, including compared to conventional EDM, in terms of MRR, SR and TWR, and others. By blending powders into the dielectric fluid in comparison with the conventional EDM technique, the removal speed of the material enhanced. The wear frequency for tools in PMEDM is below the normal EDM method. The removal rate is highest, depending on the increase in peak current. The use with an electrolyte of a powder mixture offers a surface finish similar to a mirror. To generate better outputs, optimum set of process parameters and powder mix must be used which can be obtained through new and improved multi-objective optimization. Following an in-depth analysis of the paper, the main focus of the research is the PMEDM of conventional materials, such as steels, alloys and so on.

References:

[1] Singh.S., Maheshwari. S. and Pandey, P.C., Some investigations into the electric discharge machining of hardened tool steel using different electrode materials 2004, *J. Mater. Process. Technol.*, **149**, 272–277.

[2] Ho K.H. and Newman S.T., State of the art electrical discharge machining (EDM), 2003, *Int. J. Mach Tool Manu*, **43**,1287–1300.

[3] Salman, Ö., Kayacan, M. C., Evolutionary programming method for modeling the EDM parameters for roughness. 2008, J. Mater. Process. Technol, **200(1**), 347–355.

[4] Schumchar B.M., Krampitz R., and Kruth J.P., Historical phases of EDM development, 2013, duel influence of ''Market Pull'' and ''Science Push'', Procedia CIRP, **6**, 5–12.

[5] Sommer C, Sommer S., Complete EDM handbook, Advance Publishing, Incorporated, 2005,137-144.

[6] Irshad S.,Khan W., and Kumar A, A literature review on process parameter optimization of powder mixed electric discharge machining (PMEDM) using design of experiment, 2016, *Int. J. Sci. Res.& Dev.* **3**, Issue 11, 370-375.

[7] Abbas N.M, Solomon D.G, and Bahari M., A review on current research trends in electrical discharge machining (EDM). 2007, Int J Mach Tools Manuf.,**47**,1214–28.

[8] Patil K., Jadhav V., Study of machining parameters in EDM, 2016, *Int. J. for Res. in App. Sci & Engg.Tech.*,**4**, Issue I, 72-78.

[9] Rajkumar H., Vishwakamra M., Performance parameters characteristics of PMEDM: A Review, 2018, *Int. J.of App. Engg. Res.* **13**, 5281-5290.

[10] Rana D. Pal A. and Tiwari P., Study of powder mixed dielectric in EDM-A Review,2015, *Int. J Engg. Sci. Adv. Res.*,**1(2)**, 69-74.

[11] Marashi H., Jafarlou D., and Sarhan Ahmed AD, Hamdi Mohd. State of the art in powder mixed dielectric for EDM applications., 2016, *Precision Engineering*, **46**,11-33.

[12] Tzeng YF, Lee CY. Effects of powder characteristics on electro discharge machining efficiency, 2001, Int. J. Adv. Manuf. Technol.,**17**,586-92.

[13] Torres A, Puertas I, and Luis CJ., Modelling of surface finish, electrode wear and material removal rate in electrical discharge machining of hard-to-machine alloys.,2015, *Precision Engineering.*,**40**, 33-45.
[14] Kavade M., Jamadar M. Effect of aluminium powder mixed EDM on machining characteristics of die steel (AISI D3), 2014, *Int. J. Mech. Prod. Engg.*,**8**, 25–28.
[15] Razak M. A., Nanimina A. M., Improving EDM efficiency with silicon carbide powder-mixed dielectric fluid, 2015, *Int. J. Mat. Mechanics and Manuf.*, **3**, 40–43.
[16] Syed K. H., G. P. Anuraag, G. Hemanth, and S. A. Subahan, "Powder-mixed EDM machining of aluminium-silicon carbide composites, 2015, *Ind. J. Sci. Tech.*, **8(S2)**, 133–137.
[17] Prihandana GS, Mahardika M, and Hamdi M, Wong Y, Mitsui K. Effect of micro-powder suspension and ultrasonic vibration of dielectric fluid in micro-EDM processes—Taguchi approach, 2009, *Int. J. Machine Tools and Manuf.*, **49**, 1035-41.
[18] Jahan M, Rahman M, and Wong Y., Modelling and experimental investigation on the effect of nanopowder-mixed dielectric in micro-electrodischarge machining of tungsten carbide. *Proc. Inst. Mech. Engg. Part B, J. Engg*, **224**,725-39.
[19] Tan PC, Yeo SH, Tan YV, Effects of nanopowder additives in micro electrical discharge machining, 2008, *Int. J. Precis. Engg. Manuf.*, **9** ,22–26.
[20] Jahan MP, Rahman M, and Wong YS., Study on the nano-powder mixed sinking and milling micro-EDM of WC-Co., 2011, *J. Mat. Process. Techno.*,**53**, 167–180
[21] Jahan M P, Rahman M, and Wong YS, Modelling and experimental investigation on the effect of nano powder-mixed dielectric in microelectro discharge machining of tungsten carbide, 2011 *Proc. Inst. Mech. Engg. Part B, J. Engg.* **224**,1725–1739.
[22] Bharti P. Maheshwari S. And Sharma C., Multi-objective optimization of electric-discharge machining process using controlled elitist NSGA-II, 2012, *J. Mech. Sci. Tec.*, **26**,1875-1883.
[23] Mohanty U. Rana J. and Sharma A., Multi-objective optimization of electro-discharge machining (EDM) parameter for sustainable machining, 2017, *Materials Today: Proceedings*, **4**, 9147–9157.
[24] Singaravel B., Deva Prasad S., and Chandra Shekar K., Mangapathi Rao K., Gowtham Reddy G., Optimization of process parameters using hybrid taguchi and VIKOR method in electrical discharge machining process, 2019, Adv. Engg. Opt.Intell.Tech. Springer, Singapore., **949**, 527-536.
[25] Muhammad H., Wasim A. and Shah A., Optimization of process parameters using graphene-based dielectric in electric discharge machining of AISI D2 steel, 2019, *Int. J.Adv. Manuf. Techno.*,**103**,3735–3749.
[26] Naik D., Khan A., Experimental investigation of the PMEDM of nickel free austenitic stainless steel: a promising coronary stent material, 2019, *Silicon*, **11**, 899-907.
[27] Mohanty S., Mishra A., Nanda B. K. and Routara B.C., Multi-objective parametric optimization of nano powder mixed electrical discharge machining of AlSiCp using response surface methodology and particle swarm optimization, 2018, *Alexandria Engg.J.*,**57**, 609–619.
[28] Padhee S., Nayak N., Panda S., Dhal P. R. and Mahapatra S., Multi- objective parametric optimization of powder mixed electro-discharge machining using response surface methodology and non-dominated sorting genetic algorithm, 2012,*Sadhana*, **37**, Part 2, 223–240.
[29] Phan H. Nguyen, Long T. Banh., Viet Bui, and Dung T. Hoang., Multi-response optimization of process parameters for powder mixed electro-discharge machining according to the surface roughness and surface micro-hardness using Taguchi-TOPSIS, 2018, *Int. J. Data and Network Sci.* **2** 109–119.
[30] Agarwal N., Shrivastava N., and Pradhan M. K., A new multi-response jaya algorithm for optimisation of EDM process parameters,2018, *Material Proceedings*, **5**, Issue 11, Part 3, 23759-23768.
[31] Mohanty C., Mahapatra S., Singh M., An intelligent approach to optimize the EDM process parameters using utility concept and QPSO algorithm, 2017 *Engg. Sci.Tech.an Int. J.* **20**, 552–562.

[32] Venkata G., Narasimhamu K. L., Multi-objective optimization in WEDM of inconel 750 alloy: application of TOPSIS embedded Grey Wolf Optimizer,2019, *Advanced Engineering Optimization Through Intelligent Techniques,* **949**, 231-240.

[33] Jun Ma., Ming W. And Jinguang Du. et.al., Integrated optimization model in wire electric discharge machining using gaussian process regression and wolf pack algorithm approach while machining SiCp/Al composite, 2018, *Adv.in Mech.Engg.*, **10(9)**, 1–17.

[34] Chakraborty S., Mitra A., Parametric optimization of abrasive water-jet machining processes using grey wolf optimizer, 2018, *Materials and Manuf.Pro.*,**33**, 1471-1482.

[35] Osama L., Mashqoor A. And Shukla M., Ann modeling and parametric optimization of powder mixed electrodischage machining by using taguchi method for AISI steel, 2016, *Int. J. Innov. Engg. Tech.*, **6,** Issue 4, 228-240.

[36] Kumar A., Sayed A. H., and Rai R., Optimization by AHP-ARAS of EDM process parameters on machining AA7050-10%B4C composite,2019, *Adv.Ind. Prod.Engg.,* Lecture notes mechanical engineering. Springer, 285-296.

[37] Shukla R., Singh D., Selection of parameters for advanced machining processes using firefly algorithm, 2017, *Engg.Sci.Tech.an Int. J.*,**20**, Issue 1, 212-221.

[38] Zheng-Ming Gao, Juan Zhaoet., An improved grey wolf optimization algorithm with variable weights, 2019, *Compu. Intell. Neuroscience*, **vol.2019**, 1-13.

[39] Rashid T.A, Abbas D.K, and Turel Y.K., A multi hidden recurrent neural network with a modified grey wolf optimizer.,2019 *PLoS ONE*, **14(3)**, 213-237.

IOP Conf. Series: Materials Science and Engineering **810** (2020) 012010 doi:10.1088/1757-899X/810/1/012010

Design and Numerical evaluation of crankshaft of diesel engine for total deformation and strain

Nishita bhise, M. Ramachandran
MPSTME, SVKM's NMIMS, Shirpur, Dhule, Maharashtra 425405, India
sweetestchandran@gmail.com*

Abstract. The *crankshaft* is an engine component that converts the linear (reciprocating) motion of the piston into rotary motion. Features of a crankshaft include the crankpin journal, throw, bearing journals, counterweights, crank gear, and a power take-off (PTO). Crankshafts are machined through a sequence of automated operations that remove material using lathes and milling machines. Other off-the line processes include inspection, test and repair. High-count cylinder engines involve higher inertias, overlapping combustion events and torsional vibrations of the flexible crankshaft which strongly complicate the diagnostics. The equivalent inertias and stiffness's of the system were calculated from drawings, modelling of the crankshaft with a CAD (computer-aided design) software and finite element method. Crankshaft is Internal cylinder pressure (or torque) estimation is an important engine parameter with significant implications for diagnostic and control applications in internal combustion (IC) engines. Many speed-based diagnostics methods employ models of the engine dynamics. In a dynamic model of a small four-cylinder diesel engine with the assumption of a rigid crankshaft is employed to estimate individual cylinder power production.

1. Introduction

The *crankshaft* is an engine component that converts the linear (reciprocating) motion of the piston into rotary motion. Features of a crankshaft include the crankpin journal, throw, bearing journals, counterweights, crank gear, and a power take-off (PTO). **[1]** Crankshafts are machined through a sequence of automated operations that remove material using lathes and milling machines. Other off-the line processes include inspection, test and repair. **[2]** High-count cylinder engines involve higher inertias, overlapping combustion events and torsional vibrations of the flexible crankshaft which strongly complicate the diagnostics. The equivalent inertias and stiffnesses of the system were calculated from drawings, modelling of the crankshaft with a CAD (computer-aided design) software and finite element method. **[3]** Crankshaft is Internal cylinder pressure (or torque) estimation is an important engine parameter with significant implications for diagnostic and control applications in internal combustion (IC) engines. Many speed-based diagnostics methods employ models of the engine dynamics. In dynamic model of a small four-cylinder diesel engine with the assumption of a rigid crankshaft is employed to estimate individual cylinder power production. In the signature analysis of crankshaft speed fluctuations is utilised for detection of misfiring cylinders. In dynamic engine models with flexible crankshafts are used to construct high-gain non-linear observers for indicated pressure or torque. The estimated indicated pressure or torque can be directly used for cylinder health diagnostics as well as controls. The additional advantage due to the presented characteristics of the crankshaft oscillations are improved capability in handling engine-to-engine variations and decreased memory/storage requirements. An inherent advantage that is not investigated in this paper is the potential for performing diagnostics on faults that are not related to cylinder health. **[4]** Crankshafts of motor vehicles generally run at high rotation, between 700 and 8000 rpm, depending on the engine type. Diesel engines of vehicles usually run below 4000 rpm. However at this rotation range, significant imbalances can arise from the engine operation. The common root causes of crankshaft failures are due to fatigue phenomenon. Most of damages take place on the main journals and crankpins close to the web fillets and lubricating holes where are present high stress levels being therefore critical zones. Poor design, deficient assembly or inadequate engine operation, shaft misalignments, wrong geometry of fillets on crankpin-webs and main journals, inadequate lubricating holes geometry, as well as abnormal vibrations and resonant frequencies generally are the most probable causes of damage in engines.

IOP Conf. Series: Materials Science and Engineering **810** (2020) 012010 doi:10.1088/1757-899X/810/1/012010

2. Design of crankshaft

The crankshaft is designed in solid works software with version 2016. The dimension of the crankshaft is shown in figure 1 and figure 2. The crankshaft is designed for three cylinder engine. Figure 3 shows the solid model of the final crankshaft design.

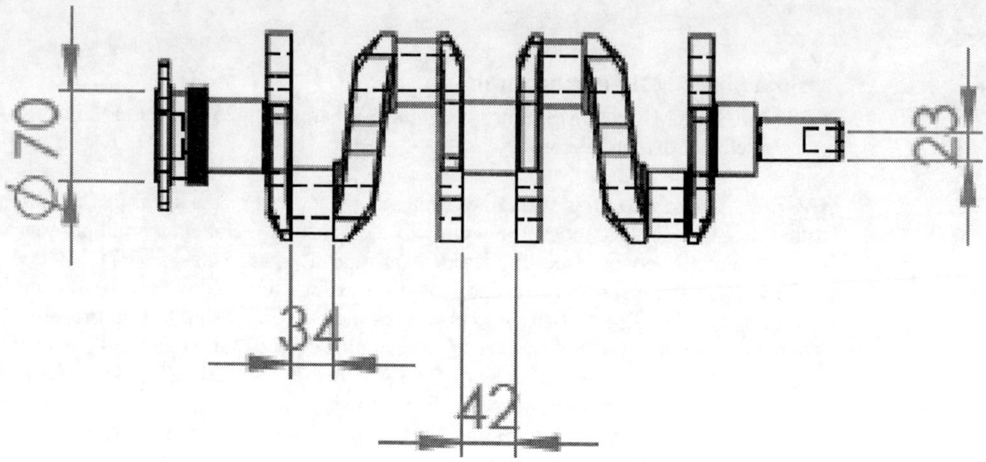

Figure 1: side view of Crankshaft

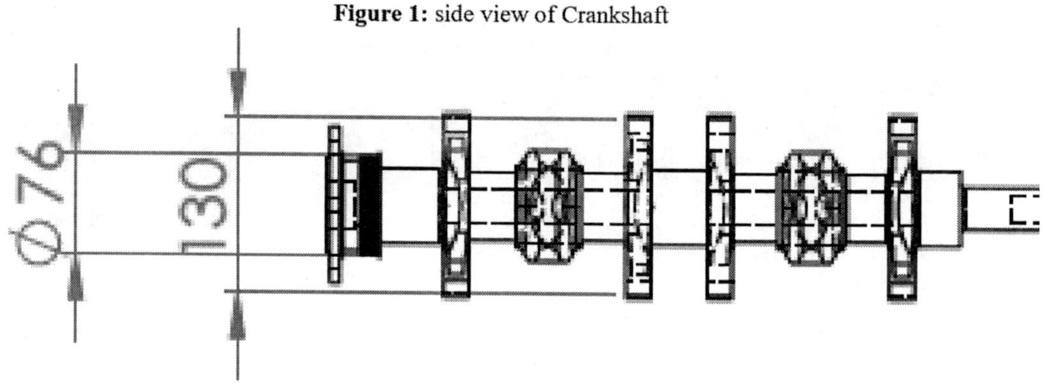

Figure 2: top view of Crankshaft

Figure 3: top view of Crankshaft

3. Ansys Analysis

The model is analysed through Ansys analysis software with version Ansys 14. For the analysis we used structural material properties because most of the either cast iron or structural steel is used as a fabrication material. Figure 4 shows the meshed model of the crankshaft.

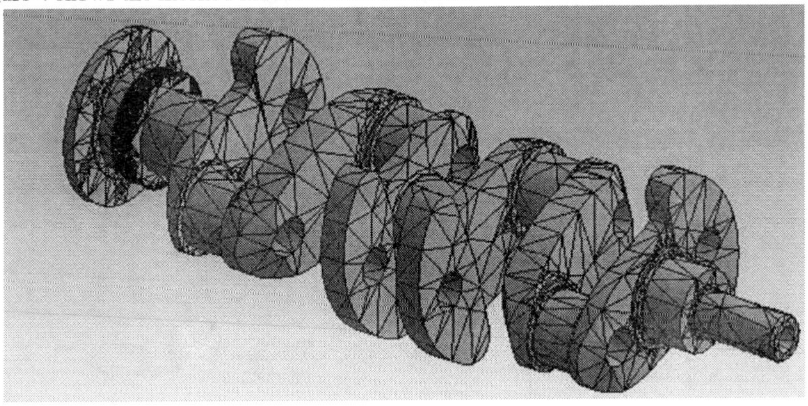

Figure 4: mesh model of Crankshaft

For analysis Bounding Box with dimensions (XxYxZ) 0.543 mx 0.135 mx 0.16 m is considered and mass of the component taken for analysis is 16.371 kg and the volume of the component is 2.0855e-003 m³. The crankshaft model is meshed with 37953 nodes and 21740 elements. The mechanical properties of structural steel which is considered for the total deformation and strain analysis are Density with value 7850 kg m^-3, Coefficient of Thermal Expansion with value 1.2e-005 C^-1, Specific Heat with value 434 J kg^-1 C^-1, Thermal Conductivity with value 60.5 W m^-1 C^-1, Resistivity with value 1.7e-007 ohm m, Compressive Yield Strength with value 2.5e+008 pa, Tensile Yield Strength with value 2.5e+008 pa, Tensile Ultimate Strength with value 4.6e+008 pa, Young's Modulus with value 2.e+011 pa, Poisson's Ratio with value 0.3, Bulk Modulus with value 1.6667e+011 Pa and Shear Modulus with value 7.6923e+010 Pa

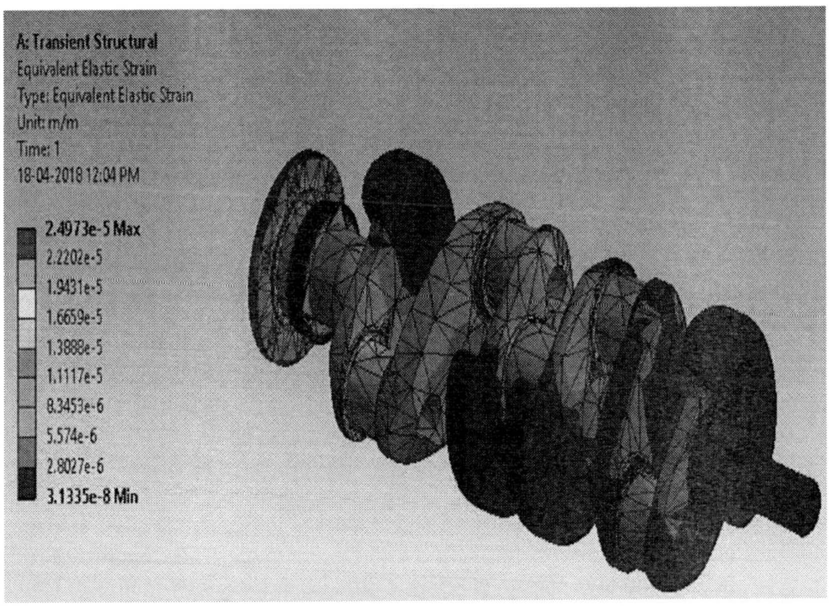

Figure 5: Equivalent strain analysis of Crankshaft

From the Figure 5 which shows Equivalent strain analysis of Crankshaft, the maximum Equivalent Elastic Strain is 2.4973e-005 m/m and the Minimum Equivalent Elastic Strain is 3.1335e-008 m/m. From the Figure 6 which shows maximum principal elastic strain of Crankshaft, the maximum principal elastic strain is 1.9219e-005 m/m and the Minimum principal elastic strain is 6.0353e-009 m/m.

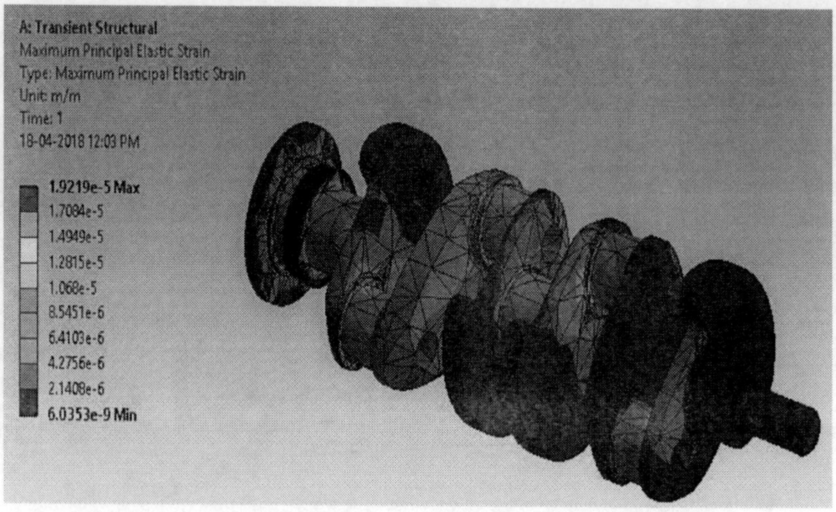

Figure 6: maximum principal elastic strain analysis of Crankshaft

From the Figure 7 which shows Total Deformation of Crankshaft, the maximum Total Deformation is 37.539 m and the Minimum Total Deformation is 5.1243e-002m.

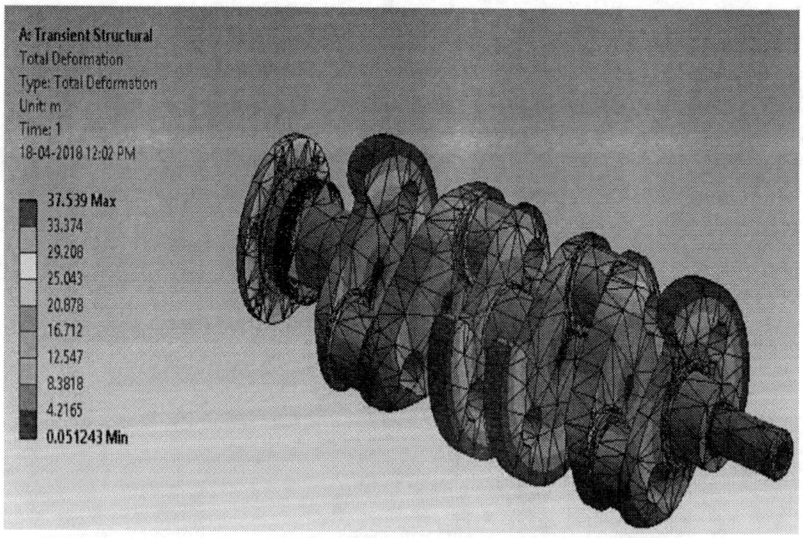

Figure 7: Total deformation analysis of Crankshaft

4. Conclusion

The crankshaft is designed in solid works software with version 2016. The dimension of the crankshaft is shown in figure 1 and figure 2. The crankshaft is designed for three cylinder engine. Figure 3 shows the solid model of the final crankshaft design. The model is analysed through Ansys analysis software with version Ansys 14. For the analysis we used structural material properties because most of the either cast iron or structural steel is used as a fabrication material. Figure 4 shows the meshed model of the crankshaft. maximum Equivalent Elastic Strain is 2.4973e-005 m/m and the Minimum Equivalent Elastic Strain is 3.1335e-008 m/m. maximum principal elastic strain is 1.9219e-005 m/m and the Minimum principal elastic strain is 6.0353e-009 m/m. maximum Total Deformation is 37.539 m and the Minimum Total Deformation is 5.1243e-002m.

References

[1]. Castro, G., A. Fernández-Vicente, and J. Cid. "Influence of the nitriding time in the wear behaviour of an AISI H13 steel during a crankshaft forging process." Wear 263, no. 7-12 (2007): 1375-1385.

[2]. T. Sunder selwyn, R Kesavan, M. Ramachandran, FMECA Analysis of Wind Turbine Using Severity and Occurrence at High Uncertain Wind in India, International Journal of Applied Engineering Research. ISSN 0973-4562 Volume 10, Number 11 (2015) pp. 10250-10253.
[3]. Joshi, Jayeshkumar J., Dipak M. Patel, and Bhandu LCIT. "Design and Failure Analysis of Single Cylinder Petrol Engine Crankshaft using ANSYS Software." International Journal of Engineering Science 10549 (2017).
[4]. Behrens, Bernd-Arno, Rouven Nickel, and Sven Müller. "Flashless precision forging of a two-cylinder-crankshaft." Production Engineering 3, no. 4-5 (2009): 381.
[5]. Farrahi, G. H., F. Hemmati, S. Abolhassani Gangaraj, and M. Sakhaei. "Failure Analysis of a Four Cylinder Diesel Engine Crankshaft Made From Nodular Cast Iron." The Journal of Engine Research 22 (2011).
[6]. M. Ramachandran, Benildus, Failure Analysis of Turbine Blade Using Computational Fluid Dynamics, International Journal of Applied Engineering Research. ISSN 0973-4562 Volume 10, Number 11 (2015) pp. 10230-10233.
[7]. Fonte, M., Bin Li, L. Reis, and M. Freitas. "Crankshaft failure analysis of a motor vehicle." Engineering failure analysis 35 (2013): 147-152.
[8]. Guagliano, M., and L. Vergani. "A simplified approach to crack growth prediction in a crankshaft." Fatigue & Fracture of Engineering Materials & Structures 17, no. 11 (1994): 1295-1306.
[9]. Ashish Mogra, M. Ramachandran and Hemant Kumar Mehta, Experimental Investigation on Droplet Formation through a Minichannel Y- Junction, IOP Conf. Series: Materials Science and Engineering 377 (2018) 012189.
[10]. Jung, Do-Hyun, Hong-Jin Kim, Young-Shik Pyoun, Alisher Gafurov, Gue-Cheol Choi, and Jong-Mo Ahn. "Reliability prediction of the fatigue life of a crankshaft." Journal of mechanical science and technology 23, no. 4 (2009): 1071-1074.
[11]. Vishal Fegade, Gajanan Jadhav, M. Ramachandran, Design, Modelling and Analysis of Tilted Human Powered Vehicle, IOP Conf. Series: Materials Science and Engineering 377 (2018) 012215.
[12]. Lee, Yung-Li, and William Morrissey. "Uncertainties of experimental crankshaft fatigue strength assessment." International Journal of Materials and Product Technology 16, no. 4-5 (2001): 379-392.
[13]. Lin, J. F., and S. J. Yuan. "Influence of internal pressure on hydroforming of double handles crankshaft." Materials Science and Engineering: A 499, no. 1-2 (2009): 208-211.
[14]. Nursoy, Mustafa, Cengiz Öner, and İbrahim Can. "Wear behavior of a crankshaft journal bearing manufactured by powder spraying." Materials & Design 29, no. 10 (2008): 2047-2051.
[15]. Paranjpe, Rohit S., and Pawan K. Goenka. "Analysis of crankshaft bearings using a mass conserving algorithm." Tribology Transactions 33, no. 3 (1990): 333-344.
[16]. Pennock, Gordon R., and John E. Beard. "Force analysis of the apex seals in the Wankel rotary compressor including the influence of fluctuations in the crankshaft speed." Mechanism and machine theory 32, no. 3 (1997): 349-361.
[17]. Kanak Kalita, M. Ramachandran, Pramod Raichurkar, S. Haldar, Free Vibration Analysis Of Laminated Composites By A Nine Node Isoparametric Plate Bending Element, Advanced Composites Letters, 25(4), 2016.
[18]. Schroeter, R. B., C. M. Bastos, and J. M. Crichigno Filho. "Simulation of the main cutting force in Crankshaft turn broaching." International Journal of Machine Tools and Manufacture 47, no. 12-13 (2007): 1884-1892.
[19]. Shu, G. Q., X. Y. Liang, and X. C. Lu. "Axial vibration of high-speed automotive engine crankshaft." International Journal of Vehicle Design 45, no. 4 (2007): 542-554.
[20]. Zhang, Chao, Husheng Zhang, and Zugan Qiu. "Fast analysis of crankshaft bearings with a database including shear thinning and viscoelastic effects." Tribology transactions 42, no. 4 (1999): 922-928.
[21]. ZHANG, Ying-jian, Wei-jun HUI, and Han DONG. "Hot forging simulation analysis and application of microalloyed steel crankshaft." Journal of Iron and Steel Research, International14, no. 5 (2007): 189-194.
[22]. Zhang, X., and S. D. Yu. "Torsional vibration of crankshaft in an engine–propeller nonlinear dynamical system." Journal of Sound and Vibration 319, no. 1-2 (2009): 491-514.
[23]. Shajit Yadav, M.Ramachandran, Analysis of Optimized Wind Turbine Failure Using Computational Fluid Dynamics, Materials Today: Proceedings, 4(2), 2017:1788–1793.

Design of three speed planetary gear hub system with gear shifting mechanism

Ankit Gupta

Assistant Professor, Mechanical Engineering Department, SVKM's NMIMS MPSTME, Shirpur Campus
ankit.gupta@nmims.edu

Abstract.

Today's 21st century has come a long way in the evolution of transmission systems. This paper aims in keen understanding and conceptual learning of Epi-cyclic gear drive system for transmission of power. The results of planetary gear transmission system yields the key characteristics such as high power density and compact design. To improve the detailed understanding of multi speed planetary gear hub systems, the work in this research will primarily guide through the utilization of these unique systems via the design of 3-speed planetary gear transmission system. Further to better predict the very potential and economical manufacturing process of planetary gears, first, the traditional manufacturing of gears is to be understood and thereby, speculating the creation of 3-D printed metal gears that will more accurately react to changing revolutionary manufacturing technologies and transmission systems , and make more effective use of these basic principles of gear system. The main objective, being, to brief the designers and manufacturers to rapidly conceptualize these basic ideas in to physically possible innovative models for the sake of revolution of the present gear technology.
Keywords: Planetary Gears, Three speed system, Transmission system, Gear Technology, Shifting mechanism.

Introduction

Although the working principle of planetary gears has been known for years, there is still a great interest towards this types of device. The mechanical Engineers are interested with the aid of this mechanism because they efficiently remedy unique troubles. Planetary gear drives are combination of three main links: a sun toothed wheel , a carrier that hosts gears and a annular gear.

In an epicyclic gear train the ring or sun gears maybe stationary, and the other rotates about own axis. The planet wheel rotate along the carrier and around the carrier. There are mainly three arrangements for a gear train. The gear train may have: One input, one output and one fixed element. The sun or the Carrier may be the input shaft , the output is the other element. In this case the mechanism is a speed retarder (or accelerator). One input, two outputs, and no fixed element. The input shaft may be the carrier, the outputs sun and annular wheels. In this case the mechanism is a differential. The output torque is distributed between the output shafts. Two inputs, one output, and no fixed element. This mechanism is a speed integrator, since the output speed is a linear mixture of the speed of the two input shafts. To obtain a nearly uniform distribution of load among the planet gears and understand the gear calculations. [1] A suitable material was selected for which material selection was done with the help of the versatility and diverse applications of Iron element. Typical properties of Iron were listed down for better evaluation of iron related mechanical models.[2] Light-weight construction of gear system and consideration of in-hand resources resulted in the development with high load ability and power density of gearbox designs. At the instant, high reliability of gears are always expected. To also Optimize the epicyclic equipment train via load evaluation, to prevent load failure from happening in the future with minimum consideration acting load between the two meshing gears.[4] This research paper aims the fulfilment of the design approach in a generalized sense keeping in mind the required necessary parameters for planetary gear system. The generalized approach thus gives the designers to brain storm with the variable design ideas which can be more flexible and advanced than the existing three speed gear hub system. It was found necessary that the paper must conclude the designer with clear and precise basic steps of planetary gear calculations and load assumptions.

Problem definition

The bicycle is great suited for a future world due to change in climate, shortage in fuel and a rising population pressure, particularly in cities. Huge problems are spent for the improvement of the bicycle, but the basic design hasn't changed for a prolonged time. The design of bicycle has been significantly improved through the use of stronger and lighter materials, but, the argument ties closely with the new and improvised need for transmission system and therefore an idea for basic three) speed internal gear hub transmission system is introduced in this research.

Material Selection

The noise reduction of gears can be achieved by selecting a correct material. A regularly unnoticed, opportunity is to apply a material that has excessive acoustic damping skills, one such material is grey cast iron. But cast iron is taken into consideration to be susceptible and brittle and it's authentic that the gray irons are pretty brittle, ductile iron isn't always. The graphite is inside the shape of small, spherical nodules in a metallic matrix ductile iron. The common material used for gears is Cast iron's matrix. Both materials which are alloyed with carbon are strong, Carbon mixed with iron to form pearlite, which consists of alternating plates of strong iron carbide and soft Ferrite. In metals the amount of pearlite is controlled by the amount of carbon. The ratio of pearlite-to-ferrite has effect on energy, hardness and machinability in all types of irons. Cast iron is a composite metallic along with of caused graphite in a metal matrix. The mechanical properties in a cast iron part is affected by the size and shape of the graphite particles.
Ductile iron= Spherical Graphite= High tensile strengths.
Gray irons= Flake graphite= low tensile strength.

The most suitable cast iron for gear manufacturing are Ductile and gray iron. Ductile iron with a ferritic matrix and 12LI4 steel has similar machining properties. The wear resistance can be improved by the increase the amount of pearlite. An absolutely pearlitic ductile iron has surest solid tensile strength (as much as 100,000 psi) with a 350 BHN hardness. Heat treating will growth power to 850,000 psi whilst quenched and tempered, and up to 970,000 psi whilst austempered.
Instead of steel in all gear applications cast iron is not an alternative material. However, all speed gears can take gain of the growth consistency in austempered Iron. Overall, cast iron may be a profitable answer for hassle created via acoustic and vibration, especially in programs where a mild decrease in power is a manageable change-off. The machining excellence of cast iron provide an environmentally pleasant opportunity to leaded steels, and its huge variety of properties allows the designer to opt the best appropriate grade for an application. Therefore, ductile iron is the most suitable type of forged iron for three-pace planetary transmission system.

IOP Conf. Series: Materials Science and Engineering **810** (2020) 012011 doi:10.1088/1757-899X/810/1/012011

Design Calculation

Light-weight production and consideration of to be had assets bring about internal hub designs with excessive power. At the same instant, durability of gear are mostly expected. Furthermore, distinctive types of teeth exist: e.g., spur and helical gears, and also double-helical gears are practicable where spur gears are considered.
All of these situations require exquisite and robust design criteria, consisting maximum load and dynamic loads under different load situations. The product design process of a gear basically includes,

1. The calculations for load.
2. To the point of structure analysis, Gear and component Design.
3. The quasi-static and dynamic behavior of the gear in actual conditions to be demonstrated.

This process does not allow for an efficient gear calculation — especially considering the antsy of the load assumptions. In this case the highly unique and, in component, standardized calculations of system factors can simplest be relevant as a long way because the accuracy of the weight assumptions allow.
Design approach for manufacturing of internal gear hub:
•The need for design and fabrication of 3-speed planetary gear transmission system generated.
•Effectively design 3- speed planetary gear system containing four planet gears and a sun gear.
•Creation of design by referencing Stormy Archer 3-speed internal gear hub.
•Assumption of load.
•The gear design calculations using design data book and the factor of safety determined using the same.
•cross verification of the components based on the design
•The CAD models developed considering different axes.
•The dynamic properties to be determined for its structure and meshing of the teeth of gears.
•The analysis performed for its structure and meshing of teeth by using
FE analysis software. Let,
S: teeth on sun gear.
P: teeth on planetary gear.
R: teeth on ring gear.
m: Module.
Ty: RPM of planetary carrier.
Tr: RPM of ring gear.
Ts: RPM of sun gear.

(R+S)Ty = R x

Tr + Ts x S Since, the sun gear is stationary. Therefore, Ts =0.
Assumed data: m = 0.65mm. P = 13.

$$R = S + 2P = 42$$

Therefore,

$$(R+S)Ty = R \times Tr$$

For first gear, ring gear will be the driver, therefore,

$$\frac{T_y}{T_r} = \frac{R}{(R+S)}$$

$$\frac{T_y}{T_r} = \frac{60}{80}$$

First gear ratio = 0.75

For third gear, planetary carrier is the driver, therefore,

$$\frac{T_y}{T_r} = \frac{80}{60}$$

Third gear ratio = 1.33
Ds = pitch circle diameter of sun gear.
Dp = pitch circle diameter of planetary gear.
Dr = pitch circle diameter of ring gear.
b = Width of the tooth.
n = RPM of rear wheel

Ds = m x S = 1 x 16 = 10.4mm
Dp = m x P = 1 x 13 = 8.45mm
Dr = m x R = 1 x 42 = 28.6mm
b = 10 x m = 10mm

$$V = 0.265 \ ^{m}/_{sec}$$

Dynamic Load Carrying capacity $P_d = 96.368 \ N$

Sut = 850
Syt = 620
Beam strength of gear $S_b = 1059.4185 \ N$

Wear Strength of gear $S_w = 1184.8519 \ N$

Now,
Power $P = 539.09 \ W$

Torque Transmitted $M_t = 8580 \ N - mm$

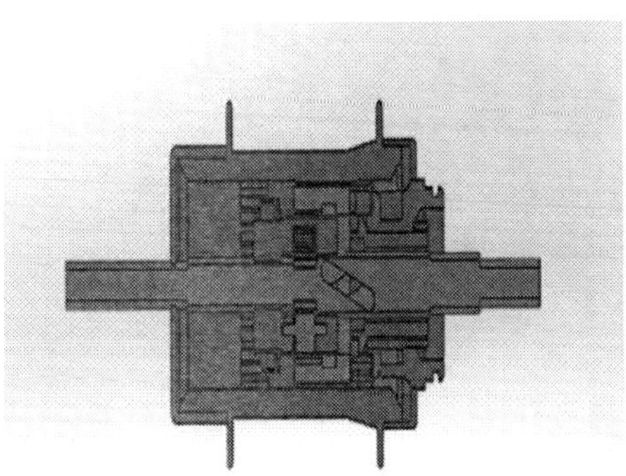

Figure 1. Section view and Isometric view of the system

Design of shifting mechanism

Despite the system has multi stage speed a mechanism, the engagement is obtained through ideally locking the rotating units via pawls with only working rotary units in action. The working units i.e. driven and driver are engaged through pawls.

1. For the First gear: The pawls on the driver unit will drive the ring gear unit and the pawls on the ring gear unit will drive the planet carrier, the pawls of the planet carrier will drive the hub shell having internal teeth.
2. Second gear: The pawls on the driver unit drives the hub shell having internal teeth. During this stage, the pawls on the ring gear unit are locked.
4. Third gear: In third gear, the complete system rotates. The pawls on the planetary carrier unit drives the ring gear unit and the pawls on the ring gear unit drives the hub having internal teeth.

Snap rings are used to restrict the sliding motion of sprocket. A sprocket is a toothed wheel upon which the chain slides. Ball bearings is used to reduce the rotational friction and guide radial and axial masses. It achieves this by using at the least races to masses. Stop ring will restrict the movement of planetary gear. Dust rings are rings provided to protect the groves from dust and contaminants.

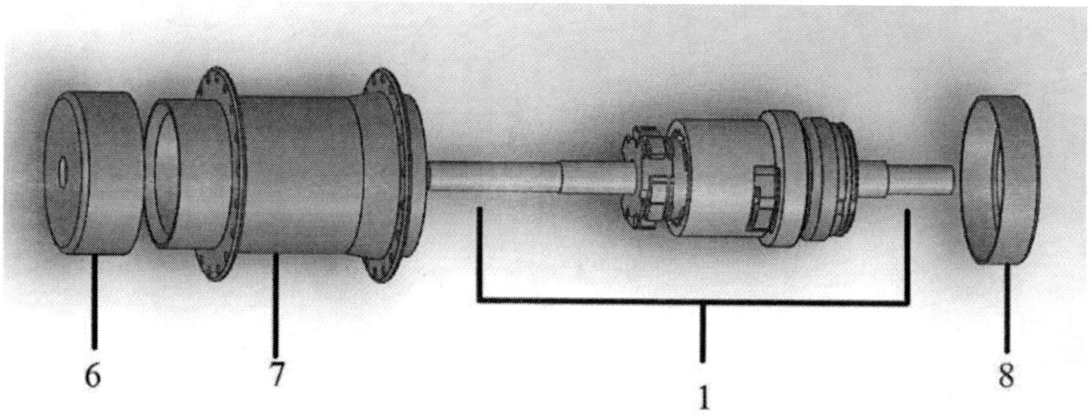

Figure 2. Exploded view of the system, 1 Internal Assembly, 6 Left hand hub cap, 7 Hub Shell, 8 Right hand hub cap

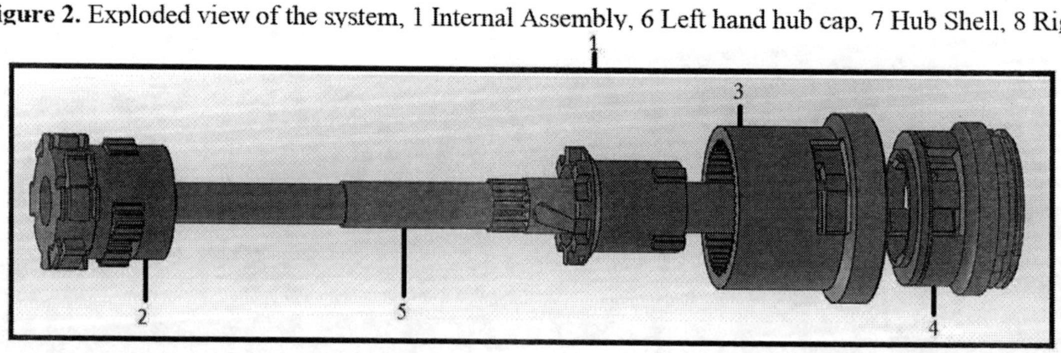

Figure 3. Exploded view of internal assembly, 2. Carrier unit, 3. Ring gear unit, 4. Driver unit, 5. Sun gear shaft unit

IOP Conf. Series: Materials Science and Engineering **810** (2020) 012011 doi:10.1088/1757-899X/810/1/012011

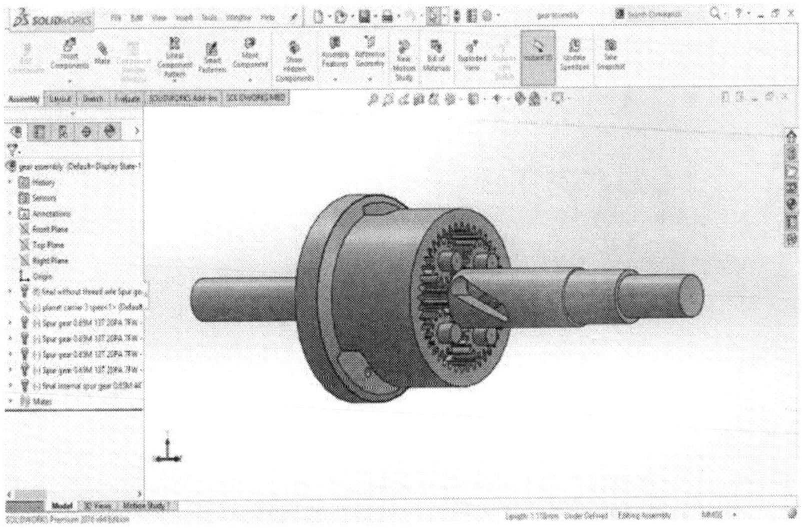

Figure 4. CAD model of final parts of assembly

Finite element analysis

The finite element analysis is capable of proving the longevity of the planetary gear. In order to reduce the modeling time, a 3-D model is created in solid modeling software. The model is transferred to ANSYS in IGES format after its construction. In ANSYS, one can click File > Import External Geometry > IGES > and check. The main focus of this analysis is:- 1) To develop and determine the bending stress using ANSYS and compare the results with conventional methods. 2) To compare the accuracy of results obtained in ANSYS by varying mesh densities.

The reaction of the gears is expressed in terms of strain, stress, displacement. The Static Structural is used as tool in the static analysis. The procedure of finite element analysis of spur gear is given below:-

- Creation of a 3D model of spur gear.
- Define the gear's material properties.
- Using finite element software meshing of the model.
- Boundary conditions for ANSYS Workbench as mentioned below

i) Fixed displacement constraint carried out on gear.

ii) Moment carried out on gear. iii) In order to arrest the displacement on x, y, z directions and rotation on x, y directions remote displacement constraint on pinion surface.

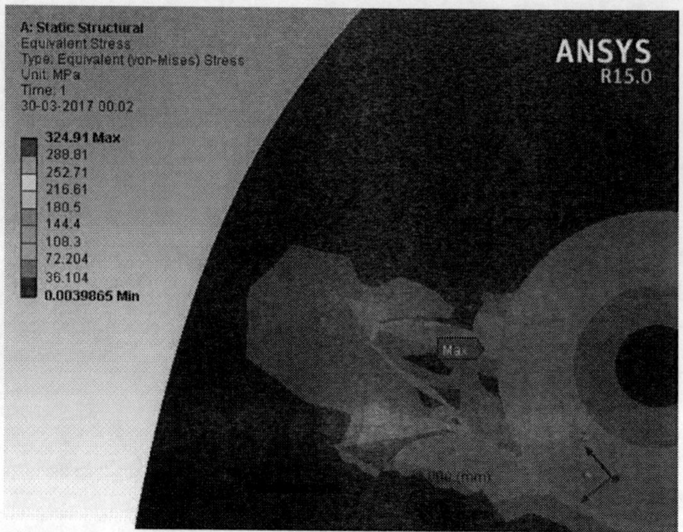

Figure 5. The CAD model of the gears used for FEA (Maximum stress – 325 MPa)

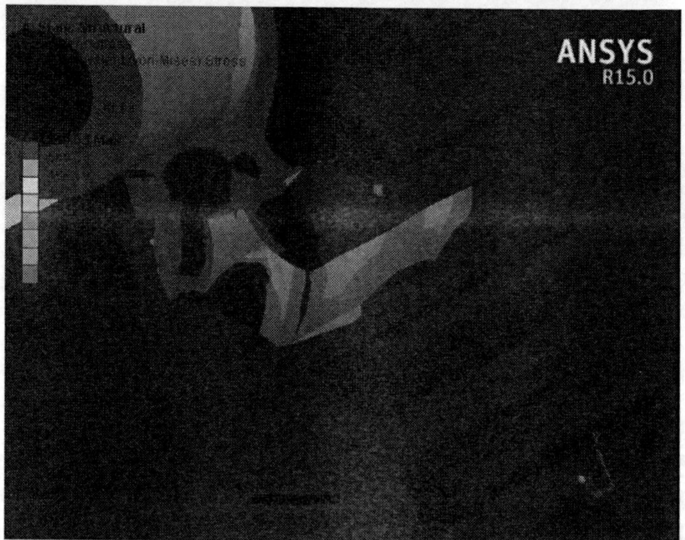

Figure 6. The CAD model of the gears used for FEA (Maximum stress – 640 MPa)

From the FEA Analysis we conclude that the Maximum stress generated is 325 MPa and 640 MPa as shown in Figure-05 and Figure-06. While the maximum ultimate stress of selected material is 340 MPa and yield strength is 840 MPa. So design is safe.

Manufacturing process

Gear production has been one of the maximum complex of the machining techniques. Objective is to cut the wide variety of operations or machines thru which a work gear desires to pass to acquire the final specifications of dimensions and teeth shape quality. The process for producing planetary gears and sun gear for internal hub, starting from raw material, are these production steps.

1. Soft- turning
2. Soft hobbing of gear teeth
3. Deburring
5. Heat treatment
6. Hard-turning
7. Grinding

4. Shaving 8.Finished gear

Three-operation clean finishing ensures easy outside diameter (Shaving, grinding and honing). However the operation can be used is two operation blank cutting and finishing. The first operation involves the soft- turning and soft-hobbing of gear teeth. The second operation involves the deburring of gears with the deburring tool. The third operation involves the shaving of gear which removes small pieces of material scrap from the hobbed gear. Further the gear is heat treated for its better strength .At last the gear undergoes, hard-turning and grinding operations.

Conclusion

Based on the studies conducted in this look at the following outcomes and conclusions may be derived. 3-Speed i.e. multi stage planetary gear system was designed and approach is suitable. The system was successful under different loading conditions. The system is compact and free from contaminants due to dust caps. Three gear ratios are successfully obtained with this system. Use of standard components in the system can be well suited and can work effectively. The grade obtained for manufactured gears in near to its set value of AGMA 13 grade quality. The gearing processes to manufacture planet and sun gears can be manufactured using automated CNC gear shapers with clean processes. Precision amongst the manufacturing of tooth profile can be improved with the improved quality standards. The complete research is economically feasible. The research and design parameters has led to efficient working of the planetary gear system.

Future scope

Future research can be carried out on light-weight design for planetary gear transmissions. Also, we can find the benefits of the usage of a carrier in first place apart from planets float in system. A logical procedure to determine gear size can be observed and worked upon in future.

References

[1] Joshi, R., & Darpe, A. K. (2018, September). Detection of Gear Tooth Crack in a Wind Turbine Planetary Gearbox. In *International Conference on Rotor Dynamics* (pp. 520-533). Springer, Cham.

[2] Chen, T. H., & Long-Chang, H. (2019). The Design and Prototype Manufacture of a Planetary Gear Reducer. *Proceedings of Engineering and Technology Innovation, 11*, 46.

[3] Chen, T. H., & Hsieh, L. C. (2018). The Design and Prototype Manufacture of Helical Gear Reducers with Two Gear Pairs. In *Applied Mechanics and Materials* (Vol. 883, pp. 37-42). Trans Tech Publications.

[4] Hsieh, L. C., & Chen, T. H. (2016). THE ENGINEERING DESIGN AND TRANSMISSION EFFICIENCY VERIFICATION OF HELICAL SPUR GEAR TRANSMISSION WITH A SINGLE GEAR PAIR. *Transactions of the Canadian Society for Mechanical Engineering, 40*(5), 981-993.

[5] Hsieh, W. H. (2013). Kinetostatic and mechanical efficiency studies on cam-controlled planetary gear trains (Part II)–Design and experiment.

[6] Wu, Y. C., & Chan, C. T. (2018). A novel bicycle rear hub transmission with a magnetic gear mechanism. *Advances in Mechanical Engineering, 10*(7), 1687814018784825.

[7] Wu, Y. C., & Chen, L. A. (2015). Design of an 8-speed internal gear hub with a rotary control mechanism for bicycles. *Tehnički vjesnik, 22*(4), 865-871.

[8] Wu, Y. C., Chen, L. A., & Chang, C. W. (2014). Prediction of efficiency of an existing 14-speed bicycle internal drive hub. *Universal Journal of Mechanical Engineering, 2*(8), 266-273.

[9] Wu, Y. C., & Sun, Z. H. (2013). Design and analysis of a novel speed-changing wheel hub with an integrated electric motor for electric bicycles. *Mathematical Problems in Engineering, 2013*.

[10] Wu, Y. C., & Lin, S. L. (2011). Conceptual design of a 16-speed bicycle drive hub. In *Applied Mechanics and Materials* (Vol. 52, pp. 279-284). Trans Tech Publications.

[11] Kim, J. (2016). Design of a compact 18-speed epicyclic transmission for a personal mobility vehicle. *International Journal of Automotive Technology*, *17*(6), 977-982.
[12] Xue, S., Entwistle, R., Mazhar, I., & Howard, I. (2016). The spur planetary gear torsional stiffness and its crack sensitivity under quasi-static conditions. *Engineering Failure Analysis*, *63*, 106-120.
[13] Gu, C., & Chen, X. (2016). A novel universal reducer integrating a planetary gear mechanism with an RCRCR spatial mechanism/Integracija planetnega gonila in prostorskega mehanizma RCRCR v novem univerzalnem reduktorju. *Strojniski Vestnik-Journal of Mechanical Engineering*, *62*(12), 730-741.
[14] Tesar, D. (2018). *U.S. Patent No. 9,862,263*. Washington, DC: U.S. Patent and Trademark Office.
[15] Xue, S. (2016). *An investigation of gear meshing behaviour of planetary gear systems for improved fault diagnosis* (Doctoral dissertation, Curtin University).
[16] Hsieh, L. C., Chen, T. H., & Tang, H. C. (2015). The design and meshing efficiency analysis of helical spur gear reducer with single gear pair for electric scooters. *Transactions of the Canadian Society for Mechanical Engineering*, *39*(3), 455-465.
[17] Cooley, C. G., & Parker, R. G. (2014). A review of planetary and epicyclic gear dynamics and vibrations research. *Applied Mechanics Reviews*, *66*(4), 040804.
[18] Lin, J., & Parker, R. G. (1999). Analytical characterization of the unique properties of planetary gear free vibration.
[19] Chen, L. A., Ren, P. W., & Wu, Y. C. (2013). Design of a Gear-Shifting Control Mechanism for 8-Speed Bicycle Drive Hub. *Smart Science*, *1*(2), 94-98.
[20] Hsieh, W. H. (2013). Kinetostatic and mechanical efficiency studies on cam-controlled planetary gear trains (Part I)– Theoretical analysis.
[21] Tsai, M. C., Huang, C. C., & Lin, B. J. (2010). Kinematic analysis of planetary gear systems using block diagrams. *Journal of Mechanical Design*, *132*(6), 065001.
[22] Tsai, M. C., Huang, C. C., & Lin, B. J. (2010). Kinematic analysis of planetary gear systems using block diagrams. *Journal of Mechanical Design*, *132*(6), 065001.
[23] Hsu, C. H. (2002). An analytic methodology for the kinematic synthesis of epicyclic gear mechanisms. *J. Mech. Des.*, *124*(3), 574-576.
[24] Nyunt, S. Z., Thanlyin, M., & Latt, A. H. 3D Modeling of Five Speed Manual Transmission System for Teaching Aid Vehicles.
[25] Uematsu, S. (1997). An application of graph theory to the kinematic analysis of planetary gear trains. *International Journal of the Japan Society for Precision Engineering*, *31*(2), 141-

Automated Gas Cutting Machine Using Open CNC

Aishwrya S. Dandgawhal[1*], D.S. S. Sudhakar[2], Bhushan T. Patil[3]
Fr. Conceicao Rodrigues College of Engineering, Bandra, India

asham1050@gmail.com, sudhakar@fragnel.edu.in, bhushantpatil@gmail.com

Abstract. This is to develop & investigate into an open CNC based gas cutting machine using freeware support to perform gas cutting operations on steel plates. A prototype for a low cost automation solution is proposed which will study the feasibility of a two axis CNC machine to improve the accuracy in implementation of gas cutting. The fundamental concept of CNC is still preserved but the operation is manipulated by a motion control system consisting of microcontrollers. This machine will provide a motion control system that provides the tool the freedom to be interpolated to interpolate various profiles. It is the independent system that allows integration of application programs, microcontrollers and drive circuits as well as G code interpreters. It is portable and user friendly hence suitable for all fabrication and production-oriented.

1. Introduction

Computer Numerical Control or CNC refers to the automation of machine tools with the use of pre-programmed sequence of the codes for machine control commands by using computers system. A CNC program is used to control, automate and monitor the movement of the machine. CNC controller works with a drive component and a combination of motors to control and move the machine axes in the requisite path thereby performing programmed motions. The Gas Cutting machines have been the necessity in manufacturing world. Consequently, there have been a lot of improvements and developments in gas cutting machine with respect to its ease in use and considering the human safety. But the traditional methods were time consuming, less accurate and inefficient. Hence, to overcome the disadvantages of traditional gas cutting machine, the automated gas cutting machine is taken into consideration. Also, a major advantage of cutting thicker plates of steel, brass and/or other electrically conductive metals like aluminum is served with better accuracy.

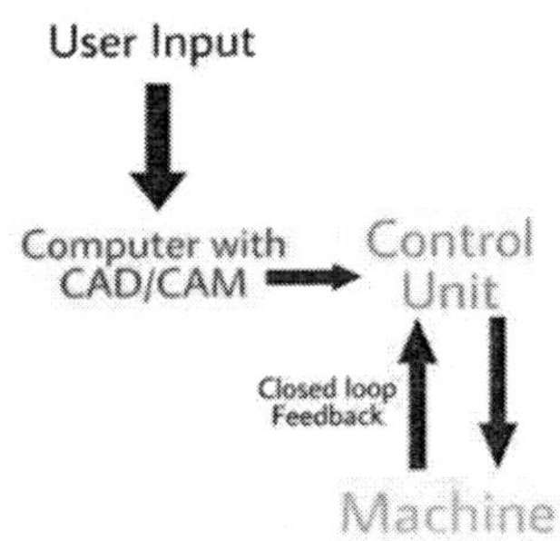

Figure. 1 indicates the basic schema involved in the prototype made

IOP Conf. Series: Materials Science and Engineering **810** (2020) 012012 doi:10.1088/1757-899X/810/1/012012

With the development of new technology open CNC architecture is gaining popularity to overcome the difficulties with proprietary controllers. The objective is to automate the gas cutting process especially in intricate contours besides supplementing with proper feed and speed to ensure the cut quality. This is an attempt to bring the freeware to create an open platform to replace propriety controllers.

Georgi M. Martinova et al. (2014) observed that it was initially difficult to control the tasks of non-traditional processes like additive manufacturing, as certain limits of classical CNC systems do not allow them to be used to solve the range of new tasks. A solution matrix for the synthesis of specialized CNC systems have been put together and a set of hardware, software devices, components and new technologies have been defined. The example of a machine selective laser sintering and a five axis water jet cutting machine would illustrate the synthesis of specialized CNC systems[1]. M. M. Goud et al. (2007) stated that flame cutting machines (which are also known as burning machines) are the machines used for cutting plates and most of these have a high end computerized or optical control with multiple cutting stations. Due to these machines, the complicated shapes are performed quickly and precisely. The procedure is rather simple, but the actual performance is a skill of science[2]. Dound Pramod et al. (2016) elaborated on stability of cutting oxygen flow, cutting speed and preheated flame is all that a flame cutter requires. In the research, experiments were performed to determine appropriate setting of nozzle size, heating oxygen pressure, cutting parameter, cutting speed and acetylene pressure to achieve the highly finished cuts for unalloyed steels upto 0.35% carbon with the help of 99.1% of oxygen which is different for different sheet thickness. One of the necessary machine tools in a workshop is the portable multipurpose electromagnetic profile machine[3]. Practically, all the metal cutting and fabrication is carried out at the shop floor where gas cutting required for different operations. The goal of Portable Multipurpose electromagnetic profile machine is to reduce cutting time of intricate curvature shapes &profiles and decrease cutting time will be a significant step towards reducing the cycle time. This will in turn improve working conditions, reduce fatigue of operator and improve accuracy at the shop floor. Xihui Yang (2017), suggested an industrial design plan for CNC machines. The design concept includes ergonomics, HMI and UED besides industrial design policies and the implementation is done through the virtual machine engines [4]. M.-Y. Yang et al. (2002) have proposed an OAC for Wire electric discharge machining. NURBS was interpolated with Taylor's series expansion [9]. Kartz, Rueven et al. (2000) have explained the available open architectural control technology by leading research groups like OMAC, OSACA, JOP, The university of Michigan etc.& Globalization of OAC standards [10]. Jorge Correa et al. (2016) proposed hardware & software for interpolation based on open source electronics. They introduced unique finite state machine (FSM) algorithms in the architecture [11]. Peng Yuhai et al. (2010) defined openness in three stages. Open protocols &communication interface of HMI, ability to replace proprietary controls for different purposes& topological structure of control equipment [12].

2. Design Features of Gas Cutting Machine

2.1. Structure

A design of XY stage gas cutting machine has been built which serves the requirements of the objective. The design model consists of a basic frame, XY axes sliders, base of sliding arrangements, lead screw, jaw couplings, motor mounting hub and holding device. The Gas Cutter (or say, in this case a tracer) is connected perpendicularly to the working table with the help of a vertical stand mounted on the Y axis. This design is robust with its arrangement. The CAD model is drawn using SolidWorks 2016 software and the 3D model of the same is generated and represented as figure 2.

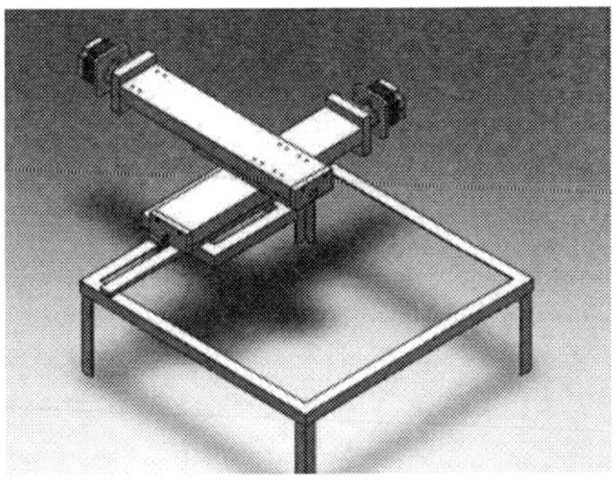

Fig. 2. CAD Model

2.2. Design Specification

The hardware basically consists of two linear actuators for two axes driven by : 2 Phase Hybrid Stepper Motor (NEMA 23) with step resolution 1.8°and torque 0.39N.m.The lead screw used is of diameter 12 mm, Length 350 mm The available working area is125X125mm.The controllers are Arduino Uno of Atmega 328. The drive circuit DM542 is selected as the motor specifications which provides the current of 4 A.
In the gas cutting process, the intricate contour to be executed is derived through a mathematical formulation like Bezier curve. These curves can be modeled by any 3D CAD modelling software or may be generated by freeware like Inkscape and the tool path is generated by Universal G code sender and through G code interpreter the motion control is executed by Arduino controllers. Inkscape is a free and open-source vector graphics editor which gives many primitive shapes besides Bezier curves for the above said purpose. Inkscape generates a file in Scalable Vector Graphics (SVG).Many other formats are compatible. Universal G code sender takes the input from Inkscape in SVG and publishes the requisite G code for GRBL/TinyG which is a G code interpreter. Then GRBL/TinyG G code interpreters will pass on the instructions to microcontrollers with the required interpolated position values, feed and speed. Arduino behaves like the brain of the system, i.e., from controlling the co-ordinates to executing the path to be traced. Once the data is retrieved from the processors, the trace begins.

2.3. Cutting Parameters

Richard, l. little (1973) has have specified all necessary components and processes which are required during gas cutting. Also, the basic features and selection of gases are determined. The oxyacetylene cutting torch has same basic components that of welding torch: an acetylene cylinder, an oxygen cylinder, regulator (oxygen and acetylene), sufficient hosing, and a torch body. A cutting head attached to the torch body is designed to have preheated orifices in the cutting tip as well as a passage for the high-pressure and pure oxygen. Cutting torches maybe of head-mixed or premixed type. In head mixed type, the most likely used type where in the oxygen and fuel (acetylene) is mixed just before entering the tip. In premixed type, the oxygen and fuel (acetylene) are mixed in the body of the torch itself. Torches can also be classified further as medium-pressure torches, in which acetylene is used in the range of 1 to 15 psi, where the acetylene is 1 psi or less. The pressure classifications are the same as those used for welding torches [5].

Plate Thickness (mm)	Nozzle Size (mm)	Acetylene Gas Pressure (bar)	Oxygen Gas Pressure (bar)
100	2.0	0.14	5.37

Table 1. Cutting Parameter for Selected Nozzle Size

Bhatia A. et al. (1998) explains the difference between cutting torch and welding torch. It is distinguished in two ways where in the cutting torch consist of a high-pressure valve which is simulated by a trigger, lever or button. In addition to acetylene and oxygen adjustment valves, when the high-pressure oxygen valve is triggered, the oxygen is released through the tip. The second difference is that the cutting torch has multiple numbers of orifice tips where in, the welding torch has just one orifice tip arrangement. Cutting torch tips have preheated orifice facility and one centrally arranged orifice directs the flow of oxygen during the cutting operation [6]. Linde A.G (2012) evaluated various fuels and the advantages and disadvantages are enumerated. Also, the high efficiency of acetylene is easy to explain: Because of the favorable molecular structure of the acetylene, the energy released during combustion, the flame propagation rate of the oxy-acetylene flame, and the high flame temperature is achieved. Also the molecular energy is being released during the breakdown of acetylene, as compared to other hydrocarbons; this is the so called formation enthalpy or energy of formation.

3. Proposed Architecture of Motion Control System

3.1. Proposed Architecture 1: System with Separate Drive Circuit Arrangement

In this system, the mechanical system (axes) are connected to the stepper motors and then to the drive circuit – DM542. The DM542 is a fully digital stepper drive developed with advanced DSP control algorithm based on the latest motion control technology. Its motor auto-identification and parameter auto-configuration feature offers quick setup to optimal modes with different motors. Compared with traditional analogue drives, DM542 can drive a stepper motor at much lower noise, lower heating, and smoother movement. Its unique features make DM542 an ideal choice for high requirement applications. The DM542 is then attached to the Arduino which is connected to the input software. The block diagram for System with Drive Circuit Arrangement is shown in Figure 3(a).

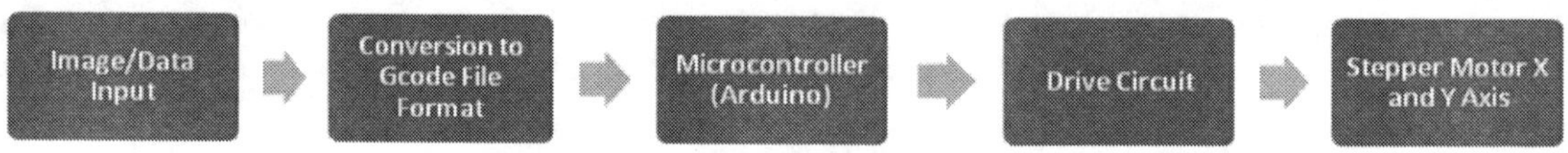

Figure. 3(a) System with Separate Drive Circuit Arrangement

3.2. Numbering Proposed Architecture 2: System with GRBL Shield Arrangement

In this system, the mechanical systems (axes) are connected to stepper motors and then to the Arduino on which the GRBL Shield is placed. The Arduino CNC Shield makes it easy to for the CNC to run for a few hours. It generally uses open (A, 1998)source firmware on Arduino to control the stepper motor. Usually, all the GRBL Shield are easily compatible with all the stepper motors and so this is one of the cheapest arrangements for the CNC machines. The block diagram for System with GRBL Shield arrangement is as shown in Figure 3(b).

CNC shield also known as GRBL shield is connected with the Arduino UNO board with proper orientation of the input polarity. The shield providesease as the basic configuration of attaching the drive circuit is already attached to it. The arrangement in proposed design 2, thus gives us a compact system.

IOP Conf. Series: Materials Science and Engineering **810** (2020) 012012 doi:10.1088/1757-899X/810/1/012012

Fig. 4. System with GRBL Shield Arrangement

4. Result and Analysis

The two-axis CNC structure developed for gas cutting was tested for geometrical and dimensional accuracy. Following are the various profile traced, the results and analysis for the same are stated below. Figure 4(a) shows a square profile of 60x60 mm, whereas figure 4(b) shows a circular profile of 60 mm diameter. A constant feed rate of 45 mm/min was followed during the trace. Figure 5 represents the actual setup of the gas cutting machine.

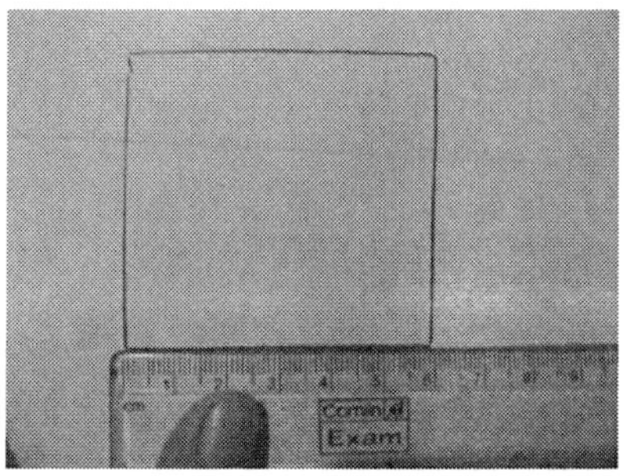

Figure. 4(a). Square Profile (60x60 mm)

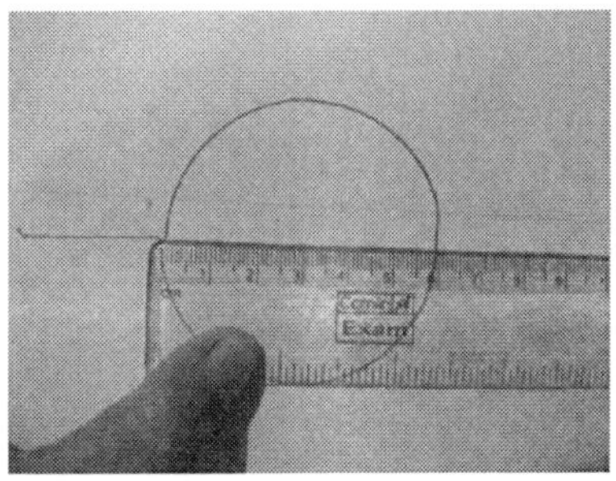

Figure. 4(b). Circular Profile (60 mm diameter)

Fig. 5. Actual Setup of Gas Cutting Machine

5. Conclusion

The XY stage CNC structure is developed for Gas Cutting. An open CNC system is implemented for linear, circular and contour interpolations. Two architectures consisting of drive circuit, microcontroller and GRBL shield is tested. The system with drivers mounted separately is found to be giving better results with a higher level of accuracy than that of GRBL shield. Various contours are tested successfully. Polygons were accurately plotted. Errors in circular interpolation are attributed due to errors in structural design. Specified feed rates are obtained with an error of ± 0.6 percentage. While experimentation, it was found that the required profile have been drawn according to the dimensions by the end-effector accurately. Maximum feed at which the machine performed accurately is about 80 mm/min, whereas, above the feed of 80 mm/min, the profile is not accurate. The setup was tested successfully with minimum errors. Thus, this gas cutting machine will satisfy its requirements within the above specified ranges.

6. References

[1] Martinov, G. M., Obuhov, A. I., Martinova, L. I., & Grigoriev, A. S. (2014). An approach to building specialized CNC systems for non-traditional processes. Procedia CIRP, 14, 511-516.

[2] M. M. Goud, P. B. (2007). Analysis of CNC Gas Cutting Machine. National Conference on Recent Developments in Manufacturing & Quality Management, (pp. 134-137). Chandigarh.

[3] Dound Pramod, P. A. (2016). Design of Profile Gas Cutting Machine. International Research Journal of Engineering and Technology , 1212-1215.

[4] Yang, X., & Cheng, K. (2017). Investigation on the industrial design approach for CNC machine tools and its implementation and application perspectives. Procedia Manufacturing, 11, 1454-1462.

[5] Richard. (1973). Welding and Welding Technology. New Delhi: TATA MCGRAW-HILL PUBLISHING COMPANY.

[6] A, B. (1998). Fundamentals of Gas Welding & Cutting, Continuing Education and Development. New York: Inc.

[7] G, L. A. (2012). Acetylene there is no better fuel gas for oxy-fuel gas processes. Hollriegelskreyth: LINDE Industrial Gases Division.

[8] Duda, J., & Pobozniak, J. (2017). The architecture of inte (Dound Pramod, 2016)lligent system for CNC machine tool programming. Procedia Manufacturing, 11, 501-508.

[9] Yang, M. Y., & Park, J. H. (2002). A study on an open architecture CNC system with a NURBS interpolator for WEDM. The International Journal of Advanced Manufacturing Technology, 19(9), 664-668.

[10] Katz, R., Min, B. K., & Pasek, Z. (2000). Open architecture control technology trends. ERC/RMS Report, 35(9).

[11] Correa, J., Toombs, N., & Ferreira, P. M. (2016, November). Implementation of an Open-Architecture Control for CNC Systems Based on Open-Source Electronics. In ASME 2016 International Mechanical Engineering Congress and Exposition (pp. V002T02A035-V002T02A035). American Society of Mechanical Engineers.

[12] Peng, Y. H., Bai, H. Q., & He, N. (2010). Research and Development of Open Numerical Control System. In Applied Mechanics and Materials (Vol. 20, pp. 254-258). Trans Tech Publications.

IOP Conf. Series: Materials Science and Engineering **810** (2020) 012013 doi:10.1088/1757-899X/810/1/012013

Analysis of Fuel from LDPE Plastic

Himanshu Satav, Pramesh Kathoke, Utsuk Shah, Dr A Deshmukh

Mukesh Patel School of Technology Management and Engineering, Mumbai, India

satavhimanshu@gmail.com, prameshkathoke@gmail.com, utsuk_shah@hotmail.com ashish.deshmukh@nmims.edu

Abstract. Plastics are perhaps the most important invention of mankind. Finding a solution to sustainably balance the usage to its environmental threat is now more important than ever. One possible solution, to make value added fuels by pyrolysis of waste plastics are discussed. Manufacturing of such fuels on a commercial scale can reduce the burgeoning plastic waste and meeting the ever-increasing fuel demand. A simple prototype to depict the pyrolysis and condensation process was constructed, and fuels samples from waste polythene bags and LDPE pellets obtained. Common physical and chemical parameters of the fuels were analysed with standard diesel as reference. The fuel obtained had some properties similar to that of diesel, and could easily be used as blends with diesel, without affecting its application.

1. Introduction

In this coming age, the biggest foreseeable revolution which is sure to come is in the field of energy. The entire world is working toward it's switch to alternate sources of energy for its daily working and functioning. It is imperative that we not only move on from our existing non-renewable sources of energy, but also deal with the mess this entire industry has created.

Consequences of rampant plastic waste mismanagement are baneful for all forms of ecology. However, waste plastics are an untapped potential source of energy, which can be effectively utilized as an efficacious source of energy in form of fuel. One feasible solution that can be taken seriously is to make value added fuels by a thermal decomposition process called pyrolysis. Manufacturing of such fuels can help tackle the problem Plastic Solidwaste Management and reduce the plastics poisoning our environment. There are limitations to common solutions like plastic recycling. Recycling is heavily dependent on sorting and segregation which is economically not as fruitful as metal recycling.

Pyrolysis is a chemical reaction which involves thermal energy breaking down larger hydrocarbons into smaller smaller hydrocarbon in anaerobic condition. In Pyrolysis, the object molecules are subjected to very high temperatures, due to which every molecule of material is broken down into a smaller molecule. Pyrolysis takes place only in the absence of oxygen. The goal of plastic pyrolysis is to break long carbon chains into smaller carbon chains. Pyrolysis is also known as thermal cracking, cracking thermolysis, depolymerisation, etc. Pyrolysis of plastic at above 400 °C results in compounds with smaller hydrocarbon chains, which resembles fuel. It also produces a colourless gas, and leaves behind a black charry residue. Allowing the products to settle, it leaves behind an oil with thick consistency. The properties of liquid product are referred to as pyro oil. It can be treated to make common fuels like petrol and diesel. Different types of plastics can be used for pyrolysis. These include common types like LDPE, HDPE, PE, PET, PS, PP, and PVC. The yield of usable pyro oil depends on the type of plastic used. LDPE and HDPE produce a higher yield of required pyro oil. [1]

IOP Conf. Series: Materials Science and Engineering **810** (2020) 012013 doi:10.1088/1757-899X/810/1/012013

2. Literature Review

Vijaykumar ,B Chanashetty et al. [1], explained the pyrolysis process with details on their apparatus, its breakdown effect on the molecular chains of LDPE molecules, explaining each and every major step involved in the pyrolysis process in detail. It also studied the effects of catalytic pyrolysis and discussed the advantages of the process.

Stella Bezergianni and Dimitrios Karonis [2], focused on the distillation of the obtained fuel. They obtained a fraction of the pyro oil, and performed catalytic hydrogenation on the same to improve its characteristics and make it very similar to standard grade diesel.

Onwudili JA, Insura N, Williams PT. [3], performed pyrolysis on Polystyrene and LDPE. It discussed the optimum temperature for pyrolysis to attain maximum yield for both the inputs. It stated that pyrolysis of LDPE above 425°C would decrease the yield. It showed aliphatic compounds were the dominant composition, but with increased temperature and residence time, the proportion of aromatic compounds in the oil increased. It also studied the effect of having a mix of LDPE and PS as input, which reduced its degradation temperature, resulting in more oil yield and lesser char.

Kumar S, Singh RK. [4], it discussed the process parameters with yield of fuel obtained. Different holding duration at different temperatures were studied and analysed and its effect on percentage yield and composition of the products obtained.

Distillation [5], basic Test in Quality Control of Automotive Fuels by Ma Mercedes del Coro Fernandez Et Al., performed fractional distillation on standard fuels petrol and diesel and studied its % v/v recovery characteristics. It has used the ISO3405 standard as the basis of the test procedure.

3. Methodology

A sample of plastic was weighed accurately and added in to the reactor inlet. The inlet is at the side of the reactor. It had a screw cap, which was made air tight by winding Teflon tape around the screw threads. The material of the reactor was considered during the design so as to not react with the fumes produced during the process. Mild Steel had the right mix of properties. It was suitable for resisting temperatures of above 500°C, and was also easy to source and procure, apart from being cost effective. It was also relatively inert to the contents of plastic inside. An LPG Stove was used as a heating source. A connecting pipe, which was a bull bar of a bumper, was used to connect the reactor to the condenser pipe. The condenser pipe was a soft copper pipe of 1inch diameter. Soft copper has the excellent thermal conduction properties, and was also easy to wind and handle. All the connections between MS to MS were sealed by TiG welding for better quality of weld and also to ensure better air sealing. The Mild Steel Pipe to copper pipe connection was sealed with the helping of brazing. The apparatus was assembled as shown in Figure 1 below. The entire apparatus was then checked for leakages.

After feeding the plastic through the inlet, it was heated in the reactor. The temperature of the reactor was closely monitored at timely intervals. White fumes start to appear at around 300°C. After the fumes stop, condensed fuel drops start dropping from the condenser pipe. The process continues till all the plastic has undergone pyrolysis. The condenser pipes were cooled by wrapping damp jute bags around them. The fuel was collected hence.

IOP Conf. Series: Materials Science and Engineering **810** (2020) 012013 doi:10.1088/1757-899X/810/1/012013

Figure 1: 3D rendering of the Apparatus used

Table 1. The fuel obtained from samples

Type of plastic	*Weight*	Temperature of Initial yield	Maximum yield	Duration of run	Yield volume
LDPE Pellets	1.8kg	330-340°C	420-470°C	100 min	1.5 L
Polythene Bags	800g	330-340°C	420-470°C	48 min	400mL

Table 1 shows the yield obtained from pyrolysis of the samples. LDPE pellets, recorded a higher yield than waste polythene bags. The fuel was sent to a chemical laboratory for detailed examination.

4. Results and Analysis

Pour Point: A sample is held in a tilted container. The lowest temperature at which this sample begins to pour is known as the pour point of that sample. Pour point is an important parameter when considering properties of fuels. A lower pour point can be a crucial factor in considering transportation through pipelines. [6]

Table 2. Pour Point of samples

Parameter	Test Method	LDPE Pellet	Polythene bags	Standard Diesel	Petrol
Pour Point	ASTM D97 - 17a	<-42°C	<-42°C	3°C to 15°C	-40 to -60°C

Table 2 shows the pour point of the samples and the test apparatus to calculate the pour point had a lower limit of -42°C.

The recorded pour points of the samples are similar to that of petrol, which freezes at around -40 to -60°C shown in figure 1. [6]

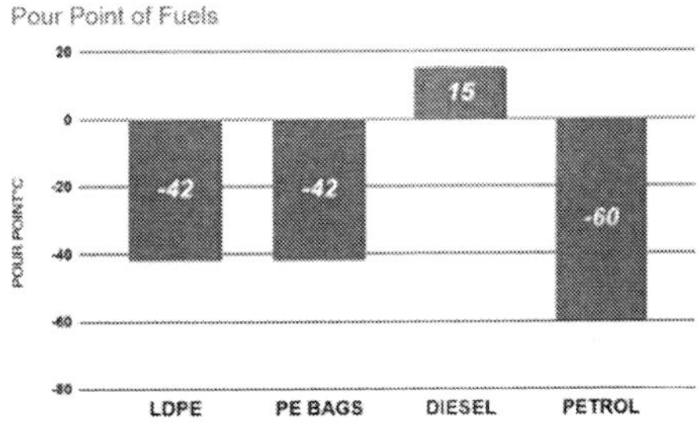

Figure 2: Pour points of samples

Flash Point: The minimum temperature at which a vapour air mixture in a closed space can catches fire, at standard atmospheric conditions is known as flash point. [7] Flash point is an important characteristic of a fuel when considering it from a safety point of view with regards to its flammability. It is an indication of the hazards while producing, storing and transporting.

Table 3. Flash Point of samples

Parameter	Test Method	LDPE Pellet	Polythene bags	Standard Diesel	Petrol
Flash Point	IP-170-2013	<-5°C	<-5°C	52-96°C	-43°C

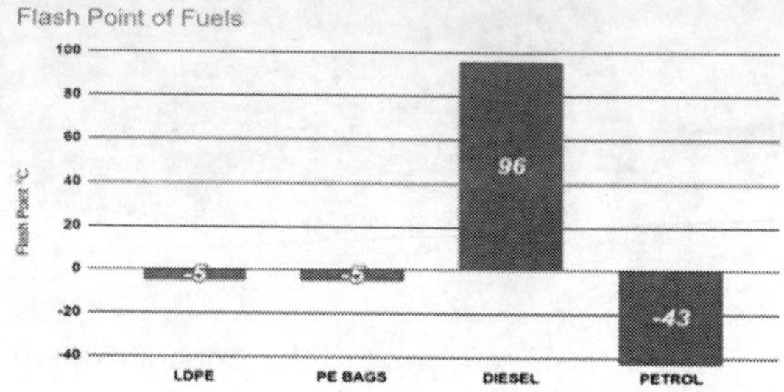

Figure 3: Flash points of samples

Distillation percent (v/v): Distillation is a process in which a mixture is separated based on their differing boiling points. It is widely used in the petroleum industry. Diesel fuels are a blend of several hydrocarbons, depending upon the source, method of distillation. Diesel fuels are comprised of hydrocarbons in the range of C12- C20, with boiling points in the range of 160°C - 380°C and petrol fuels are comprised of hydrocarbons in the range of C4-C12, with boiling points in the range of 30°C-150°C. [8]

Distillation recovery (%v/v) conveys crucial information about the composition, properties and behaviour of fuels. The distillation test measures the percentage volume of fuel recovered as the temperature increases. Figure 4 shows the percent volume recovered for LDPE Pellets. Similarly, Figure 4(a) shows the percent volume recovered for Polythene bags.

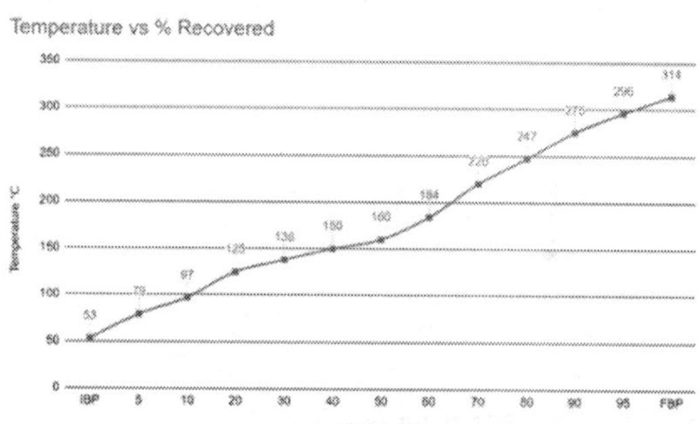

Figure 4: Recovery characteristics of LDPE pellet

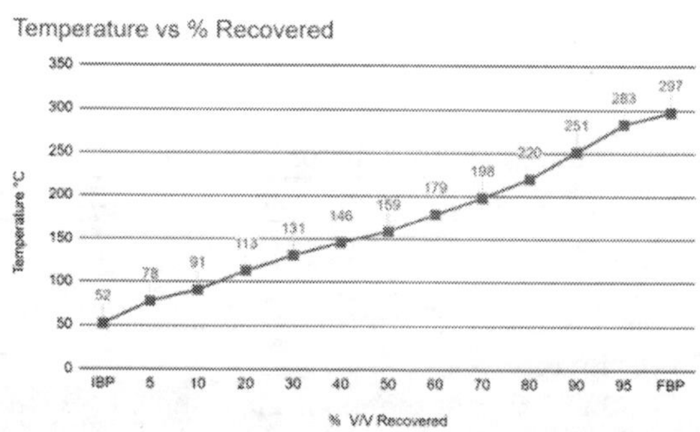

Figure 4 (a): Recovery Characteristics of polythene bags

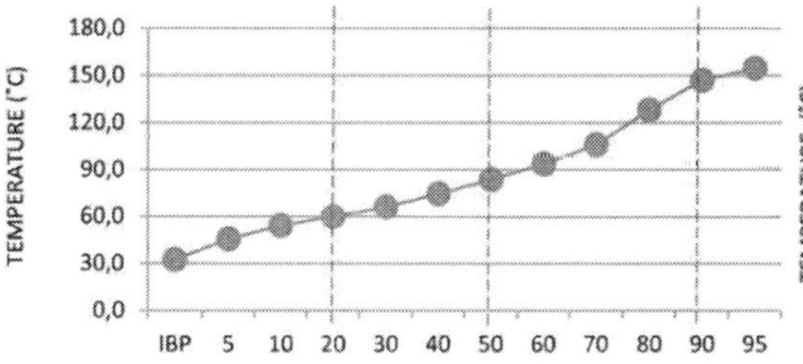

Figure 5: Recovery Characteristics of petrol [5]

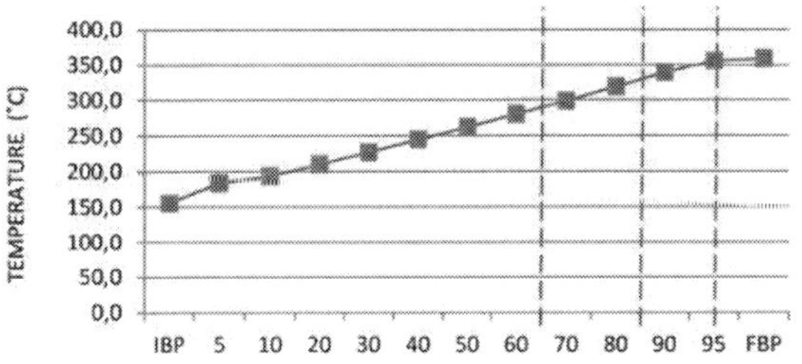

Figure 6: Recovery characteristics of diesel [5]

From analysis of the above graphs it can be firmly stated that the fuel samples i.e. from LDPE pellets and waste polythene bags, exhibit similar recovery characteristics, with FBP and IBP values being almost the same. Therefore, both the samples have almost identical composition. Figure 5 shows the percent volume recovered of standard petrol. Figure 6 shows the percent volume recovered of standard diesel.

From the above graph of petrol recovery, we can infer that petrol is recovered (95% V/V) at about 150°C, similarly we can infer that diesel starts recovery (5% V/V) at 150°C.[5]
From the graph, it can be observed that from IBP to 150°C, the fuel recovery characteristics of samples, as seen in Figure 3 and Figure 4 are similar to petrol. From 150°C to FBP, the fuel recovered follows a similar curve to that of diesel recovery characteristics. By comparing the standard fuel recovery characteristics to that of the sample fuel's recovery characteristics, it can be said that the sample fuels constitute around 40% fuel which resembles petrol. The remaining 60% fuel resembles properties similar to that of diesel.

Viscosity: Viscosity measures a fluid's resistance to flow. Kinematic viscosity is the ratio of absolute viscosity of the substance divided to its density. Kinematic viscosity quantifies the resistance of a fluid under the weight of gravity, i.e. the tendency of fluid to flow when no external force is applied. [7]

Table 4. Kinematic Viscosity of samples

Parameter	Test Method	LDPE Pellet	Polythene bags	Standard Diesel	Petrol
Kinematic Viscosity at 40 °C	ASTM D-445 17a	0.802 cSt	0.802 cSt	2 cSt to 5 cSt	0.4 cSt to 0.7 cSt

Kinematic viscosities of samples are shown in Table 4. The kinematic viscosity value of samples obtained was less than diesel, but comparable to petrol as shown in Figure 7. Hence the flowability of the samples of fuel is better than diesel. This has benefits in designing fuel injectors and making it convenient for use at lower temperatures. [9]

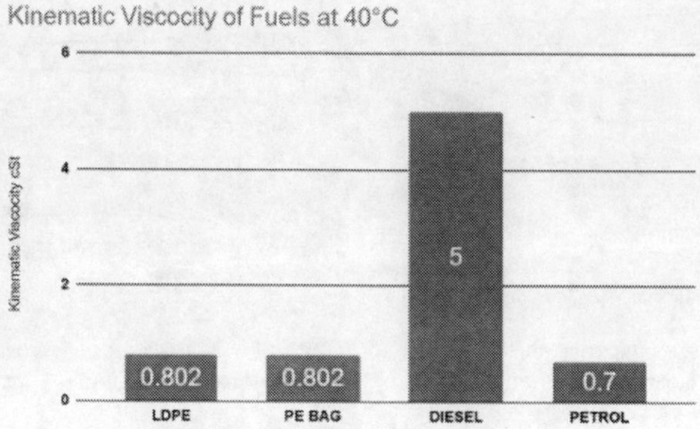

Figure 7: Flash points of samples.

Total Sulphur Content: It is an important criterion when it comes to determining quality of fuels. It is a significant pollutant. Sulphur in fuels burnt release SOx which degrades air quality. Disposed fuels in any wastewater are also another source of harmful pollutants. [11]

Table 5. Sulphur content of samples

Parameter	Test Method	LDPE Pellet	Polythene bags	Standard Diesel
Total Sulphur	ASTM D-445 17a	<17 ppm	<17 ppm	<10 ppm

Table 5 shows the sulphur content of the samples tested. The given test apparatus had a lower limit of 17 ppm. LDPE and polythene do not have sulphur as its constituent and since pyrolysis of the same takes place in an inert atmosphere, it can be said that fuels obtained have negligible sulphur content. Hence possibility of SOx pollutants from these samples is negligible. It can be said that it will pass the current Bharat Stage – VI standards for diesel.

Density: Ratio of mass of fuel to a standard volume at 15 °C.

Table 6: Density of samples.

Parameter	Test Method	LDPE Pellet	Polythene bags	Standard Diesel	Petrol
Density at 15°C	ASTM D-4052-16	0.751 g/ml	0.758 g/ml	0.82-0.86 g/ml	0.71 - 0.77 g/ml

Table 6 shows the densities of the samples, compared with densities of standard petrol and diesel. Value of density of fuels are comparable to that of petrol and marginally less than diesel as shown in Figure 8.

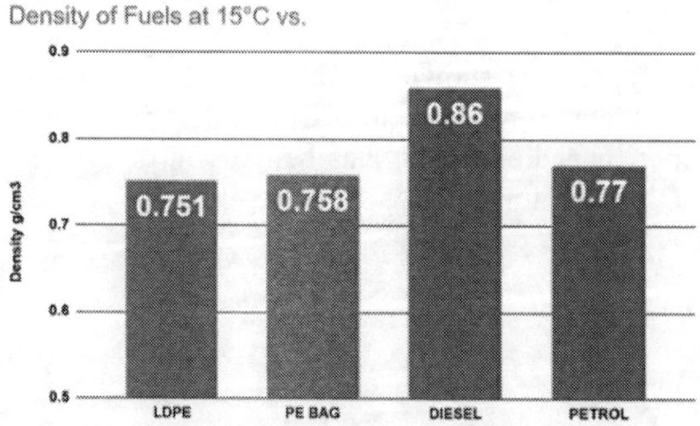

Figure 8: Density of samples

5. Conclusion

Plastic is an invention that is omnipresent. It is an invaluable resource and it needs to be accepted that it is an essential resource. Banning it only solves some of the problems plastics create. It cannot deal with the mountains of waste plastic already generated. It should not be treated completely as a bane, as it has changed human civilization. What is required is efficient management of this invaluable resource. Hence, a solid and robust plastic waste management system is the need of the hour. Pyrolysis of plastic is a solution that answers these problems effectively. Using fuel obtained from such process increases the efficiency as the Carbon atom gets used twice, once as a plastic and again as a fuel. The presence of only traces of sulphur in this fuel also makes it a great environmental positive.

LDPE Pellets and waste plastic bags were used as inputs in the pyrolysis process. Kinematic viscosity of the fuel had a value similar to that of petrol. Other parameters like pour point and flash point values were also similar to that of petrol, meaning that the petrol component of the fuel dictate some parameters of the fuel as a whole. Distillation recovery chart of the fuel shows that the fuel follows the recovery characteristics similar to that of petrol from Initial Boiling Point to temperature of 150°C, which accounts for 40% of the total sample fuel by volume. After 150°C, the recovery characteristics of the remaining fuel follow recovery characteristics similar to that of diesel, up to a Final Boiling Point which makes it suitable for it to be used in low compression diesel engines like tractors and generators.

Each figure should have a brief caption describing it and, if necessary, a key to interpret the various lines and symbols on the figure.

6. References

[1]. Vijaykumar B Chanashetty and B M Patil, (2015) 'Fuel from Plastic Waste '–International Journal on Emerging Technologies (Special Issue on NCERT 2015) Chapter 6: Pages 121-128

[2]. Stella Bezergianni & Athanasios Dimitriadis & Gian-Claudio Faussone & Dimitrios Karonis, (2017). "Alternative Diesel from Waste Plastics," Energies, MDPI, Open Access Journal, vol. 10(11), pages 1-12, October.

[3]. Onwudili JA, Insura N, Williams PT.(2009) Composition of products from the pyrolysis of polyethylene and polystyrene in a closed batch reactor: effects of temperature and residence time. J Anal Appl Pyrol 2009;Issue 86:Pages 293–303.

[4]. Kumar S, Singh RK.(2011) Recovery of hydrocarbon liquid from waste high density polyethylene by thermal pyrolysis. Braz J Chem Eng 2011;Issue 28: Pages 659–667.

[5]. Distillation: Basic Test in Qualtiy Control of Automotive Fuels by Ma Mercedes del Coro Fernandez- Feal. Luis R. Sanchez-Fernandez and Blanca Sanchez-Fernandez – Intechopen

[6]. Jumat Salimon, Nadia Salih, Emad Yousif (2012) - Improvement of pour point and oxidative stability of synthetic ester basestocks for biolubricant applications. Arabian Journal of Chemistry. Vol 5, Issue 2, Pages 193-200.

[7]. Steffen Schemme Remzi Can Samsun Ralf Peters Detlef Stolten (2017), 'Power-to-fuel as a key to sustainable transport systems – An analysis of diesel fuels produced from CO2 and renewable electricity' – Fuel, Volume 205, 1 October 2017, Pages 198-221

[8] India Fuels Diesel and Gasoline Overview BSVI, Technical Standards, Retrieved from https://www.transportpolicy.net/standard/india-fuels-diesel-and-gasoline/

[9]. Nikul Patel K. Patel, Shailesh N.Shah (2015) '11 - Biodiesel from Plant Oils' - Food, Energy, and Water The Chemistry Connection 2015, Pages 277-307

[10].Reza Sadeghbeigi (2012), 'Chapter 3 - FCC Feed Characterization' Fluid Catalytic Cracking Handbook (Third Edition)
2012, Pages 51-86

[11]. Dr. James G.Speight, (2017) 'Chapter Four - Properties of Inorganic Compound' - Environmental Inorganic Chemistry for Engineers 2017, Pages 171-229 Positioning tables

Carbon Fibre Reinforced Composite Material: Review of Properties and Processing for various Metal Matrix Materials

Rahul D. Sandhanshiv[1], D. M. Patel[2]
[1]Research Scholar: Dept. of Mechanical Engg., PSGVPM's, D.N. Patel COE, Shahada, India.
[2]Professor & Academic Dean: Dept. of Mechanical Engg., PSGVPM's, D.N. Patel COE, Shahada, India.

E-mail: rahulrds001@gmail.com

Abstract. Carbon Fibres (CF) are used as a reinforcing material from past few decades. Carbon Fibres are reinforced in Metal matrix composites (MMC), Ceramic matrix composites (CMC) & polymer matrix composites (PMC). Numerous Research work was done on Carbon Fiber Reinforced Polymer Composites (CFRP), but Carbon Fiber Reinforced Metal Matrix Composites (CFMMC) & Carbon Fiber Reinforced Ceramic Matrix Composites (CF-CMC) have wide scope to explore. This Paper reviews effect of Carbon Fibre reinforcement on Mechanical, Tribological and Morphological Properties of various Metal Matrix Materials. This paper also presents various processing techniques of Carbon Fibre Reinforced Composite Material.

1. Introduction

Composite Materials including Metal matrix composites (MMC), Ceramic matrix composites (CMC) & polymer matrix composites (PMC) are structural materials used for various Engineering applications. Composite materials combine best properties from two heterogeneous materials [1, 8]. Reinforcement and matrix are two basic constituent of composite materials. Properties of composite material depend upon selection of various types of reinforcement and matrix materials and also their manufacturing process [2-4]. From thousands of years human beings are making these combinations of reinforcement and matrix materials to obtain desired functioning properties [5]. Composite material family as MMC, CMC and PMC described very well in following Figure 1.

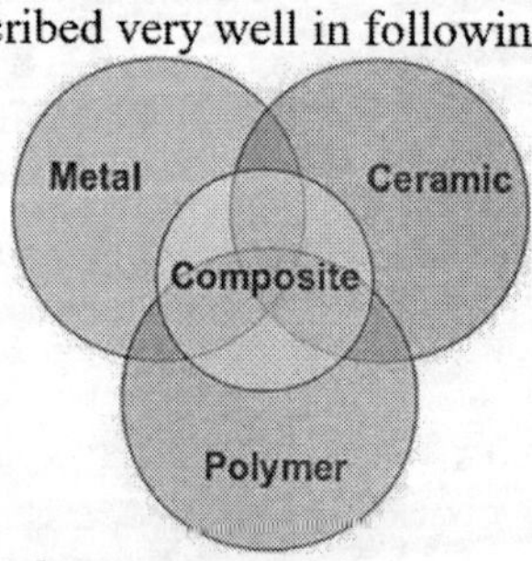

Figure 1. The Combination for composites among metals/ceramics/polymers [5].

Metal Matrix Composites are well known for high specific strength and specific stiffness. MMC were initially developed for aeronautics and space application but has limitations in automotive application due to its high cost. Now a day's research and development in MMC is going on for its application in automobiles [6]. Various Techniques are used to study chemical characterization of Fibres for reinforcement in various matrix [7]. Chemical characterization of Fibres was carried to obtain desired properties among composite material. The reinforcement in ceramic matrix improves its tensile strength and toughness. Reinforcement in ceramics forms Ceramic matrix composites (CMC). CMC has wide applications in aeronautics, space and automobiles. Cost of composite materials as compared to conventional materials is high but still has more popularity due to Lightness, High specific properties, Design and processing flexibility, Cost effectiveness, Functional superiority and Durability [8, 9].

2. Carbon Fibre: History, Processing and Characteristics

Carbon Fibres have importance as a Reinforcing material in various matrix materials due to high modulus, strength to weight ratio and resistance to chemical reaction [10-22]. Earlier carbon fibers was used in the cotton and bamboo fibers carbonization for incandescent lamp filaments. Thomas Alva Edison had used carbon fiber filaments for light bulb experiments in 1879 for the first time. When the fibers were heat-treated from 1000°C to 1500 °C and proportion of graphite in fibers vary from 0 to 100% then they are known as Carbon Fibres [11,12]. Various precursor materials are available from which carbon fibers derived. Polyacrylonitrile (PAN), viscose rayon and pitch are precursor materials for carbon fibers. Polyacrylonitrile (PAN) has advantages over other materials such as no change in original structure, better orientation and high carbon content. Carbon fibers are derived from pyrolysis of polymer fibers [13, 19]. Carbon fiber reinforced composite materials are stronger and lighter than conventional material structural parts. Carbon fiber reinforced metal matrix composites have fatigue properties better than conventional metals. Carbon Fibres used in civilian aerospace industry such as in Boeing 7E7. Composite materials contribute 58% and which results in 20% less fuel consumption [16]. Carbon fiber reinforced polymer matrix composites with the proper resins are corrosion resistant materials. Strong covalent C-C bonds in layer increases strength and stiffness otherwise weak bonding allows thermal and electrical conductivity [14]. Carbon fibers are useful as reinforcing material due to its unique properties like low density, high tensile modulus and strength, low thermal expansion coefficient, thermal stability, excellent creep resistance, chemical stability, high thermal conductivity, low electrical resistivity and decreasing cost (versus time) [15, 21].Various types of carbon fibres are used for commercial application such as High and low-modulus fibers, PAN based fibers and Pitch based fibers. High-modulus fibers are weak during shear mode and low-modulus fibers are weak during buckling mode. If modulus for PAN based fibers increases then compressive strength decreases. As PAN based carbon fibers have particle-like structure results in higher tensile and compressive strength than pitch based carbon fibers [17]. PAN based and Pitch based Carbon fibres had wide applications in automobile, aerospace, construction, medical, sports equipment, electronics etc. Figure 1 and Figure 2 shows various properties of PAN based and Pitch based Carbon fibres [16-17].

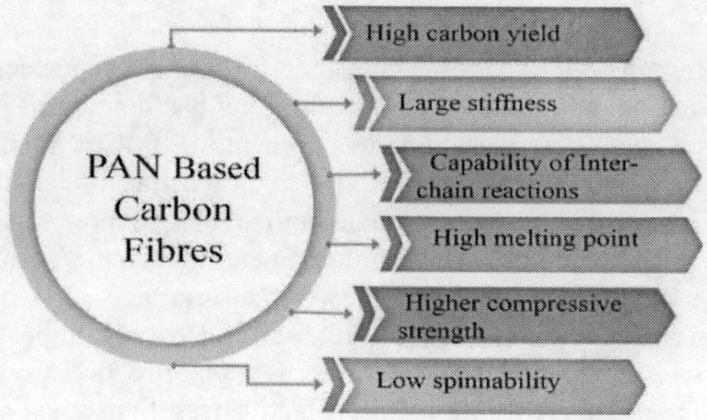

Figure 2. Properties of PAN Based Carbon Fibres.

Fibre Matrix orientation is important aspect for reinforcement in matrix. Strength of carbon fibre reinforced nickel composites with the fibres oriented at 45° or at 90° to the stress axis was low [18]. Maximum research work has been done on how to reduce processing cost of carbon fibre without changing its properties. Mechanical properties of carbon fibres are mainly depend upon precursor morphology and its condition [20, 22]. From the above it is observed that carbon fibres are attractive as reinforcement material for various matrix materials.

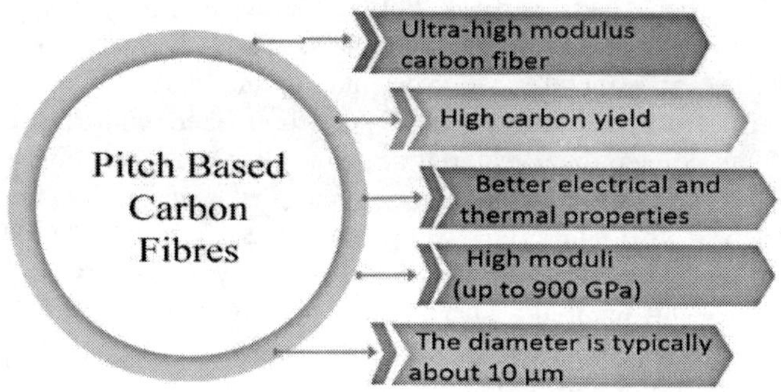

Figure 3. Properties of Pitch Based Carbon Fibres.

3. Properties and Processing of Carbon Fibre Reinforcement in various Metal Matrix Materials

Conventional metal alloys possess some properties which limits its use in aerospace and automobile application. Numerous research papers are critically reviewed to study properties and processing of carbon fibre reinforcement in aluminium and magnesium alloys. Various properties of aluminium alloy were improved with carbon fibre reinforcement [23-39]. H. Chen, A.T. Alpas [23] studied Tribological properties of carbon fibre reinforced aluminium matrix composites. Aluminium alloy A356 was selected as matrix material. Carbon fibres were coated with nickel layer. Wear tests were carried out for A356 with 4%CF, unreinforced A356 and A30 grade cast iron. For the wear test constant load and constant sliding velocity was considered. Comparison on the basis of wear behaviour of A3564 with 4%CF, unreinforced A356 and grade grey cast iron was done. It was observed that A3564 with 4%CF had high wear resistance than conventional A356 alloy and A30 grey cast iron within the certain region. ZHANG Yun-he [24] compares interface structure and thermal expansion properties of PAN M40 carbon fiber reinforced 6061Al and 5A06Al composites. Composites were processed by the squeeze-casting technology. It is observed that M40/5A06Al composite showed weaker interface reaction than M40/6061Al composite. It was found that weak interfacial reaction gives higher Coefficients of thermal expansion. M. Sánchez, J. Rams, A. Ureña [25] studied combination of centrifugal and investment casting like Centrifugal-force infiltration. Aluminium alloy composites reinforced with carbon fibre were obtained by the infiltration of preforms. In this study short carbon fibres were used in both coated and uncoated forms. Coated carbon fibres were used as reinforcement to increase wettability in aluminium alloy. This study concludes that Centrifugal-force infiltration were suitable fabrication technique, hardness of composite material increases as CF volume fraction increases and Ni-coated CF gives higher hardness than uncoated CF. H. Naji et al [26] investigated effects of aspect ratio of carbon fiber on fracture toughness of aluminium matrix composite. Aluminium matrix composite was produced using stir casting mehod. Al–8.5%Si–5%Mg selected as a metal matrix material. Three volume fractions (1, 2 and 3) and three aspect ratios (300, 500 and 800) were selected for specimen preparation. Fracture toughness was measured by three point bending test and scanning electron microscopy (SEM) was used to observe the fracture behaviour and crack deflection of composites. The results showed that the increase in CF percentage gives increased fracture toughness and also depend on volume fraction, aspect ratio of fibre. 0. PEREZ et al [27] also studied Interphases and mechanical properties of T800 carbon fibres reinforced in Al-4.5 Mg matrix composites. Al_4C_3 layer is formed with uncoated carbon fibres results in reduced strength. To avoid the same coated carbon fibres were used which gives improved strength and grain structure. Ki Hoon Jung et al [28] fabricated carbon fibre reinforced Al composite using spark plasma sintering technology. CF was reinforced in the wt% as 5, 10, 15 and 20. Mechanical testing was carried out using vickers hardness tester and Tribological test were carried out using ball on disc method. Scanning electron microscopy was used to study surface structure. It was observed that increase in CF % increases hardness and at 20wt% composite gives best wear performance.

Mechanical properties of Carbon fiber reinforced 2024 aluminium composites were improved than conventional Al 2024 alloy [29, 30]. E. Hajjari et al [29] used PAN based carbon fibre as reinforcement in Al 2024 metal matrix. Composite material was fabricated using squeeze casting method under different pressure. Mechanical properties, morphological properties and fracture surface were studied. The results showed improvement in tensile strength using nickel coating carbon fibers. As the pressure increases during the manufacturing leads to decrease in composite strength. Jacek W. Kaczmar et al [30] also studied microstructure and mechanical properties of CF-2024 aluminium alloy composite. CF-Al2024 composite material was strengthened with Al_2O_3 Saffil fibers using direct squeeze casting method. Synthesized composite showed better wear resistance at dry condition. Pull-out of fibers was rarely observed during mechanical testing. CF-Al2024 alloy composite materials reinforced with 20% of fibers exhibited highest improvement of strength. Xiong Cao et al [31] fabricated Carbon fiber reinforced AA5052 alloy composite using multi-pass friction stir processing (FSP). Microstructural, mechanical and tribological behaviour of the composites were observed. The results showed that hardness of the composites increased by 46.8% and wear volume loss was reduced by more than 70%. Mechanical and tribological behaviour of Carbon fiber reinforced Al6061 alloy composite were improved than conventional Al6061 alloy [32-35]. Al6061 alloy was used as metal matrix material due to its wide applications in aerospace, automotive and defence sectors. C.S. Ramesh et al [32] studied tribological behaviour of carbon fiber rod Al6061 alloy metal matrix composite. Carbon fiber rods of 4 mm and 6 mm diameters were used as reinforcement. Composites were fabricated using casting route. A pin on disk configuration was used to study tribological behaviour of composite. Scanning electron microscopy (SEM) studied the morphological properties. Composites exhibit higher wear resistance and lower coefficient of friction than conventional Al6061 alloy. Higher wear resistance and reduced density composite material had potential automotive applications. K.M. Sree Manu et al [33] used PAN-based bidirectional carbon fiber fabric was used as the reinforcement in Al 6061 alloy of composition Al-1.2 Mg-0.8 Si (wt%). CF-Al6061 alloy composite was fabricated by squeeze infiltration process. The infiltrated composite showed close control on the formation of aluminium carbide (Al4C3) phase results in improved properties. Self-lubrication had enhanced the wear resistance. Bi directional CF reinforcement showed increased hardness than conventional alloy. B. Bhav Singh et al [34] selected chopped carbon fibres (4–8mm length and 6μm diameter) as a reinforcement and Al6061 alloy as a metal matrix. CF-Al6061 alloy Composite material was fabricated by using stir casting method. CF coated with copper to avoid interfacial reaction and coating was done by electroless coating method. Results showed that tensile strength and yield strength of the CF-Al6061 alloy composites was found to be increasing up to 4wt% carbon fibre. Yiping Tang et al [35] studied interface and mechanical properties of short CF reinforced Al6061 alloy composite material. Different types of coating on short CF such as Ni, Cu and Al_2O_3 were done. Results showed that compared to non-coating short CF, Ni and Cu coated short CF improved wettability in Al alloy. Compounds such as Al3Ni or $CuAl_2$ are formed on short CF surface and in matrix. These compounds increase the hardness but damage the tensile strength and decreases plastic properties.

Anıl Alten et al [36] carried out mechanical characterization and fabrication of Carbon fibre reinforced Al6063 alloy composite material. Composite was processed by using squeeze casting method and Ni coating on CF was done to improve wettability. SEM, EDS and XRD methods used to determine morphological properties. Mechanical behaviour of CF-Al6063 alloy composite was carried out using three-point bending, Charpy impact and Vickers hardness tests. Results showed that increase in hardness and bending strength but impact strength reduces as CF reinforcement. 258 MPa of high strength was observed for 0.9 mm Ni coated 2wt% CF reinforced composite. The alloy Al 7075 is found to be pollution free, low energy requirements and excellent machinability. Above properties increases the tool life results in reduced production time during its fabrication process. Carbon fibre reinforced Al7075 alloy composite material showed improved properties than conventional alloy [37-39]. CF-Al7075 alloy composite was fabricated by using hot press [37], stir casting [38] and liquid metallurgy with vortex flow [39]. Madhuri Deshpande et al [37] focussed on Powder Metallurgy route that would give uniform distribution of carbon fibers in composite in the present work. Ni coated carbon fibers were reinforced in Al7075 aluminium alloy powder using hot press. Single action and double action hot press was used. Effect of volume fraction of carbon fibers on hardness was studied. Optical and Scanning electron microscopic used to study surface structure. Fabricated composites showed uniform bonding between CF and Al7075 alloy. As uncoated CF % increases hardness decreases and coated CF % increases hardness also increases. Powder Metallurgy route eliminates interfacial reaction therefore possesses a great potential to fabricate composites. Prof. S. R. Nimbalkar et al [38] reviewed fabrication of CF-Al7075 composite material using Stir casting on the basis of various parameters. Corrosion behaviour of carbon fibre reinforced Al7075 alloy composite material was studied by Nithin Kumar et al [39]. Scanning Electron Microscopy (SEM) was used to study corroded surface of composite specimen. Short carbon fibre increases then leads to more corrosion due to electrochemical between the matrix alloy and CF. Corrosion rate in NaCl solution reduces with time.

Carbon fibres were also reinforced in Al-Ni alloy and Al-Zn alloy [40, 41]. Carbon fibre reinforced Al-Ni alloy composite was fabricated by Atmospheric-pressure infiltration. This route had given reasonable mechanical properties and better control to avoid residual porosity [40]. CF-Al-Zn composite material was processed by pressure infiltration method at relatively low temperature. The tensile strength of the CF-Al-Zn composite material increased by 103% compared with that of AZ31 alloy [41]. Carbon fibre reinforced magnesium matrix composite (CF-Mg) showed better mechanical and thermal properties [42-44]. Distribution of carbon fibres in magnesium matrix had great influence on its thermal properties. Hardness and compressive strength of the composite increases by 86.6% and 42.0% as compared to conventional magnesium [42]. CF-Magnesium composite had high fatigue resistant [43]. CF-magnesium matrix composites were fabricated by using powder metallurgy method [42] and liquid-solid extrusion processed by vacuum pressure infiltration technique [43, 44].

4. Concluding Remarks

Fascinating properties of carbon Fibres makes them potential reinforcement material for various matrix materials. Carbon Fibre Reinforced polymer matrix composites (CF-PMC) had already replaced conventional material from various application. Carbon Fibre Reinforced metal matrix composites (CF-MMC) has great potential to replace existing metals and alloys from automobile application. CF-MMC has superior mechanical, tribological, morphological properties than conventional materials. CF-MMC has high strength and modulus, corrosion and wear resistance. Poor wettability of carbon fibres leads to deteriorate the composite materials. Coating is required for carbon fibres before reinforcement. Further research is required to select appropriate carbon fibre coating technique. CF-MMC has wide applications in Automobile and Aerospace industries. Potential area of research is to select suitable application in automobile that can replace by CF-MMC results in weight reduction and less fuel consumption.

References

[1] Bayraktar, E., 2015: Chapter 04108 - Section 12 Composites Materials and Technologies. Ref. Modul. Mater. Sci. Mater. Eng., 1–3.
[2] Mohammed, O., and E. Suleiman, 2017: HANDBOOK.
[3] Lynch, C. T., and J. P. Kershaw, 2018: Metal matrix composites. *Met. Matrix Compos.*, 4, 1–170.
[4] Douglas Kiser, J., K. E. David, C. Davies, R. Andrulonis, and C. Ashforth, 2018: Updating composite materials handbook-17 volume 5, ceramic matrix composites. *Ceram. Trans.*, **263**, 413–423.
[5] Haghshenas, M., 2016: Metal–Matrix Composites. *Ref. Modul. Mater. Sci. Mater. Eng.*, 0–28.
[6] Chawla, N., and K. K. Chawla, 2013: Metal matrix composites. *Met. Matrix Compos.*, **9781461495**, 1–370.
[7] Park, S. J., and M. K. Seo, 2011: Composite Characterization. 631–738 pp.
[8] Information, D. S., and M. House, 1990: composite.
[9] Wang, R.-M., S.-R. Zheng, and Y.-P. Zheng, 2011: Other mechanical properties of composite materials. *Polym. Matrix Compos. Technol.*, 455–548.
[10] W. S . Smith, 1987: Engineered Materials Handbook-Vol. **1**, p. 49.
[11] Park, S.-J., 2018: Carbon Fibers. *Fiber Technol. Fiber-Reinforced Compos.*, **210**, 123–151.
[12] Www.polycomposite.ir acceced on date 2/6/2019.
[13] Jean-Baptiste Donnet, Tong Kuan Wang and Jimmy C. M. Peng, 1998: carbon Fibres, third edition, revised and expanded, CRC press, Taylor and Francis Group.
[14] Paul J. Walsh, 2001: Carbon Fibres, *article from ASM Handbook, Volume* **21**, Composites, Copyright© ASM International.
[15] Christensen, R. M., 1994: Properties of carbon fibers. *J. Mech. Phys. Solids*, **42**, 681–695.
[16] Minus, M. L., and S. Kumar, 2005: The processing, properties, and structure of carbon fibers. *Jom*, **57**, 52–58.
[17] Chand, S., 2000: Carbon fibers for composites. *J. Mater. Sci.*, **35**, 1303–1313.
[18] Baker, A. A., 1975: Carbon fibre reinforced metals - a review of the current technology. *Mater. Sci. Eng.*, **17**, 177–208.
[19] Dorey, G., 1987: Carbon fibres and their applications. *J. Phys. D. Appl. Phys.*, **20**, 245–256.
[20] Huang, X., 2009: Fabrication and properties of carbon fibers. *Materials (Basel).*, **2**, 2369–2403.
[21] Prashanth, S., S. Km, K. Nithin, and S. Sachhidananda, 2017: Journal of Material Sciences & Engineering Fiber Reinforced Composites - A Review. **6**.
[22] Mathur, R. B., V. Gupta, O. P. Bahl, A. Tressaud, and S. Flandrois, 2000: Improvement in the mechanical properties of polyacrylonitrile (PAN)-based carbon fibers after fluorination. *Synth. Met.*, **114**, 197–200.
[23] Chen, H., and A. T. Alpas, 1996: Wear of aluminium matrix composites reinforced with nickel-

coated carbon fibres. *Wear*, **192**, 186–198.

[24] Zhang, Y. H., and G. H. Wu, 2010: Interface and thermal expansion of carbon fiber reinforced aluminum matrix composites. *Trans. Nonferrous Met. Soc. China (English Ed.*, **20**, 2148–2151.

[25] Sánchez, M., J. Rams, and A. Ureña, 2010: Fabrication of aluminium composites reinforced with carbon fibres by a centrifugal infiltration process. *Compos. Part A Appl. Sci. Manuf.*, **41**, 1605–1611.

[26] Naji, H., S. M. Zebarjad, and S. A. Sajjadi, 2008: The effects of volume percent and aspect ratio of carbon fiber on fracture toughness of reinforced aluminum matrix composites. *Mater. Sci. Eng. A*, **486**, 413–420.

[27] Perez, O., G. Patriarche, M. Lancin, and M. H. Vidal-Setif, 1993: Interphase and mechanical properties in carbon fibers/Al matrix composites. *J. Phys.*, **3**, 1693–1698.

[28] Jung, K. H., S. Nam, S. S. Kang, B. C. Ku, Y. H. Bang, and J. Y. Hwang, 2017: Tribological properties of carbon fiber-reinforced aluminum composites processed by spark plasma sintering. *Carbon Lett.*, **21**, 103–106.

[29] Hajjari, E., M. Divandari, and A. R. Mirhabibi, 2010: The effect of applied pressure on fracture surface and tensile properties of nickel coated continuous carbon fiber reinforced aluminum composites fabricated by squeeze casting. *Mater. Des.*, **31**, 2381–2386.

[30] Kaczmar, J. W., K. Naplocha, and J. Morgiel, 2014: Microstructure and strength of Al2O3 and carbon fiber reinforced 2024 aluminum alloy composites. *J. Mater. Eng. Perform.*, **23**, 2801–2808.

[31] Cao, X., Q. Shi, D. Liu, Z. Feng, Q. Liu, and G. Chen, 2018: Fabrication of in situ carbon fiber/aluminum composites via friction stir processing: Evaluation of microstructural, mechanical and tribological behaviors. *Compos. Part B Eng.*, **139**, 97–105.

[32] Ramesh, C. S., H. Adarsha, S. Pramod, and Z. Khan, 2013: Tribological characteristics of innovative Al6061-carbon fiber rod metal matrix composites. *Mater. Des.*, **50**, 597–605.

[33] Sree Manu, K. M., L. Ajay Raag, T. P. D. Rajan, B. C. Pai, V. Petley, and S. N. Verma, 2019: Self-lubricating bidirectional carbon fiber reinforced smart aluminum composites by squeeze infiltration process. *J. Mater. Sci. Technol.*.

[34] Bhav Singh, B., and M. Balasubramanian, 2009: Processing and properties of copper-coated carbon fibre reinforced aluminium alloy composites. *J. Mater. Process. Technol.*, **209**, 2104–2110.

[35] Tang, Y., L. Liu, W. Li, B. Shen, and W. Hu, 2009: Interface characteristics and mechanical properties of short carbon fibers/Al composites with different coatings. *Appl. Surf. Sci.*, **255**, 4393–4400.

[36] Alten, A., E. Erzi, Ö. Gürsoy, G. Hapçı Ağaoğlu, D. Dispinar, and G. Orhan, 2019: Production and mechanical characterization of Ni-coated carbon fibers reinforced Al-6063 alloy matrix composites. *J. Alloys Compd.*, 543–550.

[37] Gondil, R., 2016: Processing of Carbon fiber reinforced Aluminium (7075) metal matrix composite. 5–14.

[38] Rnimbalkar, P. S., and M. Satpute, 2015: Aluminiumalloy Al-7075 reinforcement and Stir casting -a Review. **3**, 38–46.

[39] Kumar, N., H. C. Chittappa, R. K. B, and S. E. Vannan, 2017: Corrosion Behaviour of Nickel Coated Short Carbon Fiber Reinforced Al Metal Matrix Composites. **12**, 659–669.

[40] Da Silva, C. C., G. M. Volpato, M. C. Fredel, and U. Tetzlaff, 2019: Low-pressure processing and microstructural evaluation of unidirectional carbon fiber-reinforced aluminum-nickel matrix composites. *J. Mater. Process. Technol.*, **269**, 10–15.

[41] Hao, X., H. Nie, Z. Ye, Y. Luo, L. Zheng, and W. Liang, 2019: Mechanical properties of a novel fiber metal laminate based on a carbon fiber reinforced Zn-Al alloy composite. *Mater. Sci. Eng. A*, **740–741**.

[42] Hou, L. G., R. Z. Wu, X. D. Wang, J. H. Zhang, M. L. Zhang, A. P. Dong, and B. D. Sun, 2017: Microstructure, mechanical properties and thermal conductivity of the short carbon fiber

reinforced magnesium matrix composites. *J. Alloys Compd.*, **695**, 2820–2826.
[43] S. Li, T. Zhang, J. Zhou, and H. Li, 2019: An analysis of the factors affecting strengthening in carbon fiber reinforced magnesium composites. *Compos. Struct.*, **209**, 328–336.
[44] Qi, L., L. Ju, J. Zhou, S. Li, T. Zhang, and W. Tian, 2017: Tensile and fatigue behavior of carbon fiber reinforced magnesium composite fabricated by liquid-solid extrusion following vacuum pressure infiltration. *J. Alloys Compd.*, **721**, 55–63.

CFD Analysis of Supersonic Combustion at Mach 2 with K-E Turbulence Model

K. K. Gupta, A.P. Singh, Shantanu Roy

[1]Department of Mechanical Engineering, SVKM'S NMIMS MPSTME Shirpur- Campus,
[2]Department of Mechanical Engineering, Sage University, Indore,
[3]Department of Mechanical Engg, Indore Institute of Science and Technology, Indore
krishnakumar.gupta@nmims.edu; hello2apsingh@gmail.com; shantanu.roy@indoreinstitute.com

Abstract. This In this paper combustion of hydrogen under unsteady and supersonic conditions has been presented. The combustor having an injector for injecting fuel parallel to the base. The main flow is also parallel to the base. Modeling of this supersonic combustion has been done using K-ε model with Finite rate chemistry combustion model. In supersonic combustion time available is very short for proper mixing. Intensity of heat release, duct geometry, injection techniques, and mixture (fuel + oxidizer) quality and flow regime are the predominating parameter which decides the flame stabilization. This numerical result demonstrates that flamelet approach is suitable in supersonic flows. Combustion characteristics for injected fuel and supersonic air flow in combustor are presented in this paper. The result shows that the mixing quality increases flame speed almost linearly. The stagnation temperature reaches up to 2420 k in the combustor.

1. Introduction

A new generation air breathing vehicles may use supersonic combustion in their propulsion system. Supersonic combustion is the essential requirement for supersonic/hypersonic flights. Generally in scramjet engines the combustor length is of the order of a meter, and the total time for mixture to be in combustor is in the order of few milliseconds. The major issues in combustion are mixing of fuel and oxidizer, flame initiation, stabilization and completion of combustion. These issues are difficult to address due to supersonic flow of air and limited combustor length. The complexity of flow pattern is very high in this combustor. At low flight speeds, the kinetic energy of the air is not sufficient for proper compression. Hence forth compression of air is required i.e turbojet engine. At higher flight speed the air flow speed is already supersonic at the entry to the combustor and will remain in supersonic condition even after diffusion. Hence forth the combustion will be taken place in supersonic flow regime. The scramjet (ramjet) which is a kind of air-breathing engine having supersonic/ hypersonic flight speeds and the combustion takes place in supersonic flow condition. Combustion efficiency in this type of combustors depends on high velocity flow, completeness in combustion and losses in pressure. [1]. Intensity of heat release, duct geometry, injection techniques, and mixture quality and flow regime are the predominating parameter which decides the flame stabilization. A stable supersonic combustion of hydrogen at mach 2 speed is performed by Aso & Hakim et al [2] resulting temperature is 2800 k and pressure is 3.5 bar. This study shows that strong bow shock is being generated due to increase in injection pressure which creates pressure losses. Different methods of fuel ignition and combustion have been reviewed [3] at low static temperatures for higher efficiency. It has been found common in experiments that inlet temperatures are higher than practical to sustain flame in low temperature hypersonic flow. But this is possible by designing the designing the combustor in a proper way. Low entry temperature is needed to compensate the intake and nozzle limitations. Scalar / scalar-velocity-turbulent frequency PDF (probability density functions) has been used in supersonic combustion by Gerlinger & Mobus et al[4]. CFD tools can be used to simulate the combustion conditions during supersonic regimes, which is having challenges of fast ignition and reduce combustor length and weight. Hydrogen is also preferred as a fuel due to its short ignition delay. Moreover high quality experimental data is not available for the evaluation of supersonic combustion parameters. Efficient mixing of fuel and air is difficult and losses in total pressure have been reported Gerlinger & Stoll et al [5] in their numerical study of mixing and combustion enhancement at Mach 2, it is due to the extremely small residence time of air in supersonic combustors. Multi step chemistry in supersonic combustion gives higher spread heat release than single step chemistry. Moreover it gives intricate details of combustion i.e. ignition distance. Kumaran & Babu [6] used a chemistry model having 37 reactions and 9 species, to investigate its effect in the supersonic combustion of hydrogen. It is required in understanding the combustion process. The front tracking method has been used for studying numerically the atomization and spray formation by Xuxk & Ohzk et al [7]. This study shows that supersonic flow in the nozzle will initiate cavitation due to liquid-vapor region formation. During numerical study Lagrangian model for turbulent combustion in supersonic flows are generally used with RANS–AMA model to simulate these supersonic co-flowing jets. GRUENIG & Mayinger [9] experimentally investigated the supersonic combustion of Kerosene at Mach 2.15 in scramjet combustor. They observed that once the combustor has ignited, the air stream temperature could be reduced below the combustor ignition level.

NUMERICAL SOLUTION: MODEL:

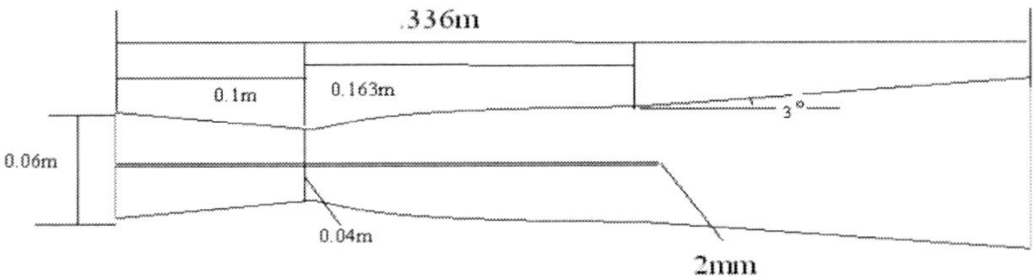

Figure 1 Supersonic combustor having co-flow jets.

GOVERNING EQUATIONS :
Navier-Stokes equations are not only useful for supersonic flight conditions and combustor geometries, but also for the determination of shock wave and shock layer. These equations are:

Continuity Equation: $\frac{\partial \rho}{\partial t}+\frac{\partial(\rho u)}{\partial x}+\frac{\partial(\rho v)}{\partial y}+\frac{\partial(\rho w)}{\partial z}=0$ (1)

X-momentum Equation: $\frac{\partial(\rho u)}{\partial t}+\frac{\partial(\rho uu)}{\partial x}+\frac{\partial(\rho vu)}{\partial y}+\frac{\partial(\rho wu)}{\partial z}=\frac{\partial \sigma_{xx}}{\partial x}+\frac{\partial \tau_{xz}}{\partial z}+\frac{\partial \tau_{yx}}{\partial y}$ (2)

Y-momentum Equation: $\frac{\partial(\rho v)}{\partial t}+\frac{\partial(\rho uv)}{\partial x}+\frac{\partial(\rho vv)}{\partial y}+\frac{\partial(\rho wv)}{\partial z}=\frac{\partial \tau_{xy}}{\partial x}+\frac{\partial \sigma_{yy}}{\partial y}+\frac{\partial \tau_{zy}}{\partial z}$ (3)

Z-momentum Equation: $\frac{\partial(\rho w)}{\partial t}+\frac{\partial(\rho uw)}{\partial x}+\frac{\partial(\rho vw)}{\partial y}+\frac{\partial(\rho ww)}{\partial z}=\frac{\partial \tau_{xz}}{\partial x}+\frac{\partial \sigma_{zz}}{\partial z}+\frac{\partial \tau_{zy}}{\partial y}$ (4)

Energy Equation: $\frac{\partial(\rho E)}{\partial t}+\frac{\partial(\rho uE)}{\partial x}+\frac{\partial(\rho vE)}{\partial y}+\frac{\partial(\rho wE)}{\partial z}=\frac{\partial(u\sigma_{xx}+w\tau_{xz}+v\tau_{xy}}{\partial x}+\frac{\partial(u\tau_{yx}+v\sigma_{yy}+w\tau_{zy}}{\partial y}+\frac{\partial(u\tau_{zx}+v\tau_{yz}+w\sigma_{zz}}{\partial z}+\frac{\partial(k\frac{\partial T}{\partial x})}{\partial x}+\frac{\partial(k\frac{\partial T}{\partial y})}{\partial y}+\frac{\partial(k\frac{\partial T}{\partial z})}{\partial z}$ (5)

For the Newtonian fluid, the normal stresses σ_{xx}, σ_{yy}, σ_{zz} are the summation of pressure and normal stress (viscous) τ_{xx}, τ_{yy}, τ_{zz}. During supersonic combustion specific energy E is

$$E=e+\frac{1}{2}(u^2+v^2+w^2) \quad (6)$$

During supersonic combustion kinetic energy of flow contributes highly to the conservation of energy as shown in the above equation.
The equation of state assuming air as a perfect gas is,

$$P=\rho RT \quad (7)$$

Where - R is the gas constant.
And for the internal energy:

$$e=C_v T \quad (8)$$

Where - C_v - specific heat for constant volume.
Assuming Prandtl number is constant (approximately 0.71 for the perfect air), the thermal conductivity may be calculated as below:

$$k = \frac{\mu C_P}{P_r} \quad (9)$$

As per the Sutherland's law viscosity μ can be calculated as follows-

$$\mu = \mu_0 \left(\frac{T}{T_0}\right)^{1.5} \frac{T_0 + 120}{T + 120} \quad (10)$$

where subscript 0 refer for standard values at sea level.
Now the generalized form of Turbulence Equations is as below:

$$(k) \frac{\partial k}{\partial t} + \frac{\partial(uk)}{\partial t} + \frac{\partial(vk)}{\partial y} + \frac{\partial(wk)}{\partial z} = \frac{\partial\left[\frac{V_T}{\sigma_k}\frac{\partial k}{\partial x}\right]}{\partial x} + \frac{\partial\left[\frac{V_T}{\sigma_k}\frac{\partial k}{\partial x}\right]}{\partial y} + \frac{\partial\left[\frac{V_T}{\sigma_k}\frac{\partial k}{\partial z}\right]}{\partial z} + (S_k = P - D) \quad (11)$$

$$(\epsilon) \frac{\partial \epsilon}{\partial t} + \frac{\partial(u\epsilon)}{\partial x} + \frac{\partial(v\epsilon)}{\partial y} + \frac{\partial(w\epsilon)}{\partial z} = \frac{\partial\left[\frac{V_T}{\sigma_k}\frac{\partial \epsilon}{\partial x}\right]}{\partial x} + \frac{\partial\left[\frac{V_T}{\sigma_k}\frac{\partial \epsilon}{\partial y}\right]}{\partial y} + \frac{\partial\left[\frac{V_T}{\sigma_k}\frac{\partial \epsilon}{\partial z}\right]}{\partial z} + (S_\epsilon = \frac{\epsilon}{k}(C_{\epsilon 1} P - C_{\epsilon 2} D)$$

where $P = 2v_T\left[\left(\frac{\partial u}{\partial x}\right)^2 + \left(\frac{\partial v}{\partial y}\right)^2 + \left(\frac{\partial w}{\partial z}\right)^2\right] + v_T\left[\left(\frac{\partial u}{\partial y} + \frac{\partial v}{\partial x}\right)^2 + \left(\frac{\partial v}{\partial z} + \frac{\partial w}{\partial y}\right)^2 + \left(\frac{\partial w}{\partial x} + \frac{\partial u}{\partial z}\right)^2\right]$ $and\ D = \epsilon$

Chemical Reaction Model:
For hydrogen-water reaction hydrogen combustion reactions module has been used.

$$2H_2 + O_2 = 2H_2O$$

Combustion Equilibrium Condition
To develop combustion equilibrium module we should know all the chemical species which are coming out from the reactions. It should also calculate the concentration of species during equilibrium condition. The main species are

$$H_2, N_2, H_2O, OH, O\ and\ NO.$$

This multi-step finite rate combustion reaction module chemical rate equations to model all reaction happening in the process. The chemical reaction rate may be calculate using the Arrhenius equation, having k as chemical reaction rate coefficient, A_p is exponential constant and $\frac{E_n}{R}$ is the activation temperature exponent.

$$k = A_p T^n e^{\left(-\frac{E_n}{RT}\right)}$$

2. Result & Discussion

Local refinement and re-coarsening of the computational grid has been as shown in figure 2. The numerical analysis has been done for both steady and unsteady cases. The computational approach is validated by comparing experimental and numerical simulation result in case of the hydrogen combustion in steady case. The K-ε model having the specialty to capture the turbulent chemical interaction. Hence the variations predicted in the temperature flow field shows the good agreement with experimental data [36].

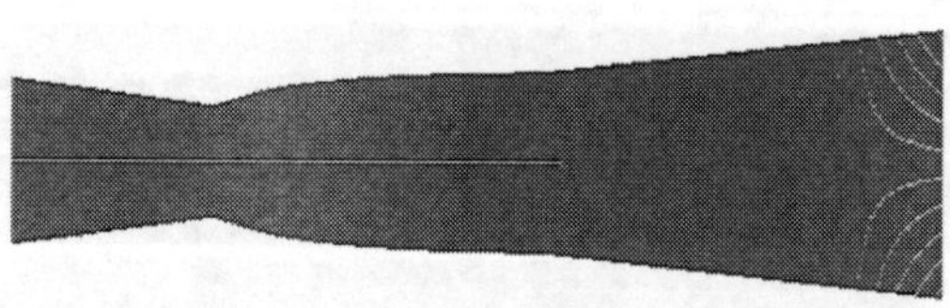

Figure 2 : Multiple adaptive grid

IOP Conf. Series: Materials Science and Engineering **810** (2020) 012015 doi:10.1088/1757-899X/810/1/012015

Table-1 Inlet flow conditions for the air and H_2 jet.

	Air	H_2
Ma	2	1
V[m/s]	830	1560
P[10^5Pa]	1	1
P[kg/m^3]	1.002	0.097
ÿO_2	0.232	0
ÿN_2	0.736	0
ÿH_2o	0.032	0
ÿH_2	0	1
K$[\frac{m2}{s2}]$	10	2400
E$\left[\frac{m2}{s3}\right]$	650	10^8

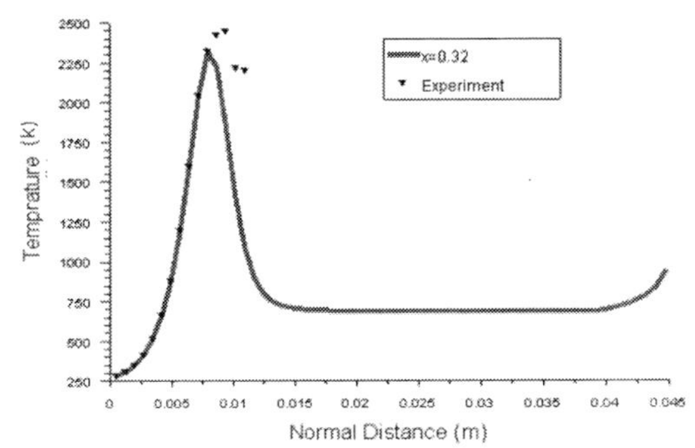

Figure 3 Comparison of computed Temperature result with experimental [deepu journal of combustion] data

Steady analysis is presented here to understand the mixing behavior. The cold flow without combustion has been investigated in this section. The validation test data is showing the capability of prediction for the code for the chemical reaction in supersonic flow field. The static pressure field shows that some spot localized within the local fuel rich zone. One can note also the presence of expansion waves followed by recompression shocks, which generated by the center-body profile and interaction between the main stream and fuel-jet. The expansion and shock waves travel downstream producing pressure fluctuation as in figure 4.it is observe that from figure 5, the deflection of path line and the accompanying density increases/decreases near the rich fuel zone. Pressure fluctuation are well reproduce in stream at x=0.205m and x=0.22m. Near the H_2 jet, figure 6, it can be seen that the pressure is more compared to the lower and upper wall of the combustion chamber.The XY plot presented for the pressure and velocity at y=o i.e. in center of the jet in figure 8 and 9. Variations of X velocity are presented at different X location in figure 9 and 11. 0A decrease in temperature is seen due to the average non-reacting hydrogen concentration is decreasing along the wall.

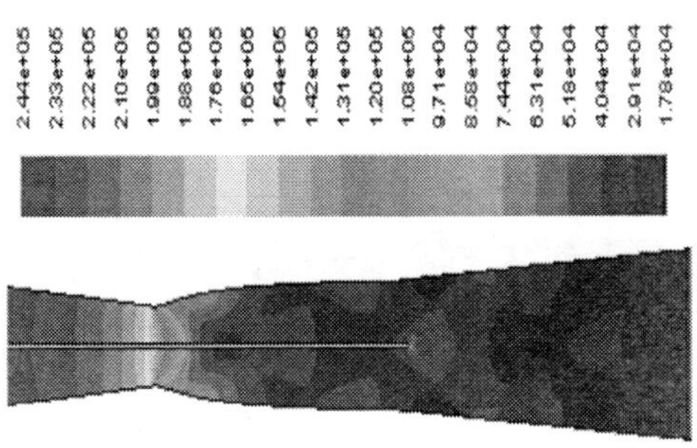

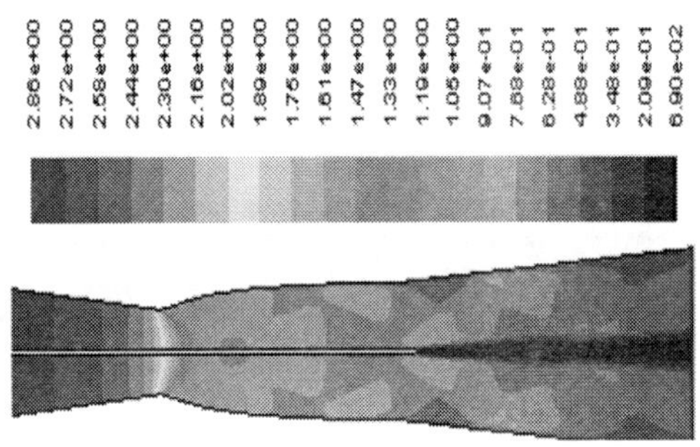

IOP Conf. Series: Materials Science and Engineering **810** (2020) 012015 doi:10.1088/1757-899X/810/1/012015

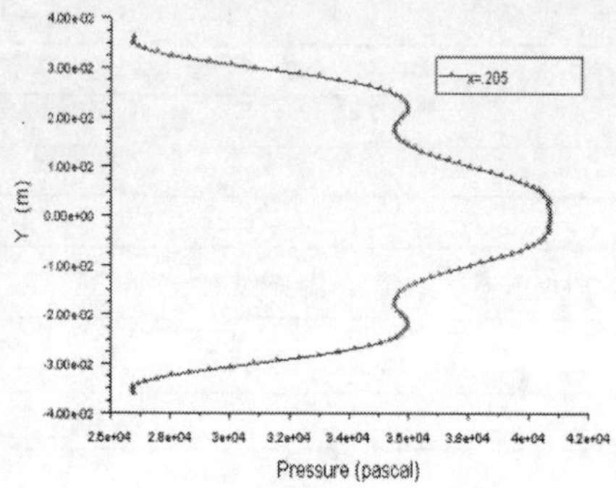

Figure 6 Contour of Static pressure

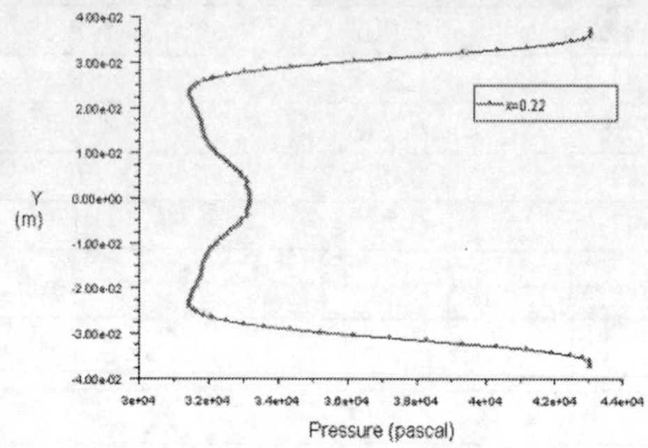

Figure 7 Contour of Static pressure

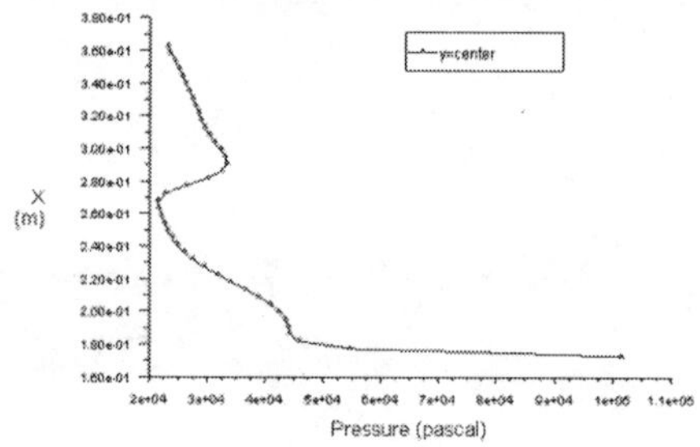

Figure 8 Contour of Static pressure

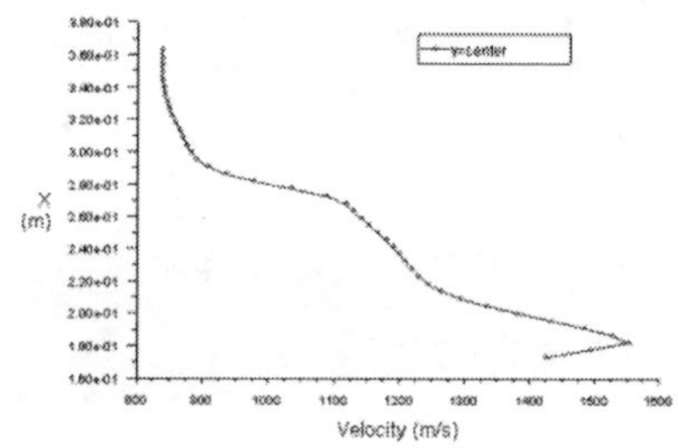

Figure 9 Contour of velocity

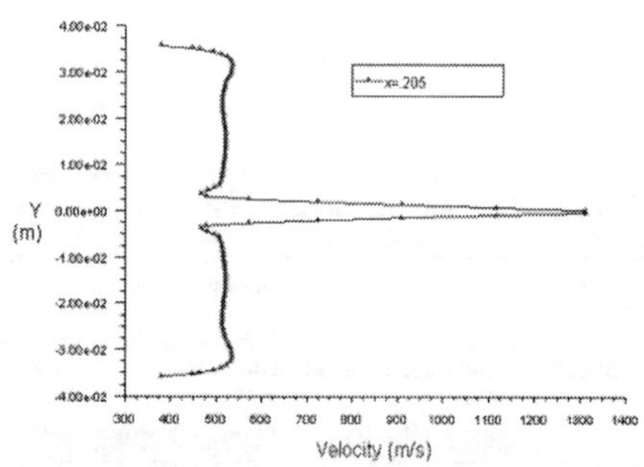

Figure 10 Contour of X velocity

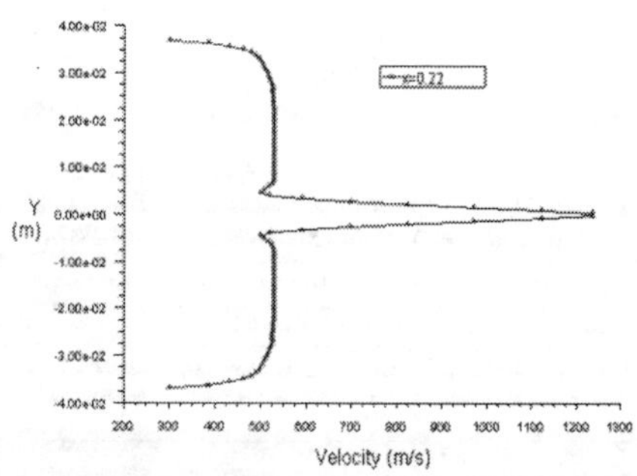

Figure 11 Contour of X velocity

IOP Conf. Series: Materials Science and Engineering **810** (2020) 012015 doi:10.1088/1757-899X/810/1/012015

3. Reactions during flow

The static pressure, density, Mach number and Stagnation temperature have shown in the computational results in figure 12, 13, 14 and 15 respectively. For this reacting flow the static pressure is different on lower and upper walls as compared to the fig 1. This rise in pressure is mainly due to the combustion is not very high because of low global equivalence ratio. Shock-wave-boundary- layer interactions play a vital role at the inlet. When these boundary layers are sufficiently strong, these shock waves impinge on the boundary layers that are sensitized by adverse pressure gradients caused by a pressure raise in the combustion chamber, which leads to flow separations and produces other adverse effects on the inlet operation. Moreover the inlet flow structure also get modified due to the new structure of shock waves induced by the local boundary-layer distortion. Some transonic spot near to fuel rich zone has been identified in this supersonic flow field. A fine balance is needed in between the flame propagation speed and the fluid velocity for the sustainability of flame. The fluid velocity is more than the flame speed in supersonic combustion region hence this flame sustainability can be solved only by the generation of recirculation region which provide sufficient residence time for the fuel-air mixing, ignition and chemical-reaction propagation. The heat releases and the flame spreads is showing upward as it moves along the wall as shown in the static temperature contour. The pressure distribution at the centre of flow and near to the upper and lower wall given in the figure 16 and 17 shows good agreement with the experimental data. The profiles of the stream wise velocity component at two different locations are presented in the figure 18 and 19. The position of the computed minimum of the velocity inside the jet tends to be shifted upward and after jet exits at x=0.2m, the maximum speed followed by the expansions waves is higher compare to x=0.32m.

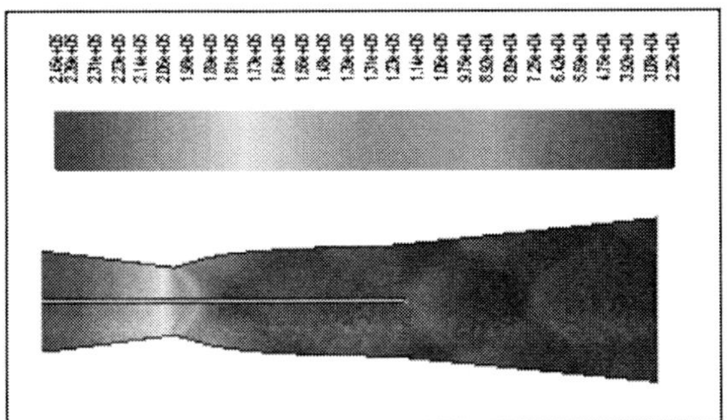

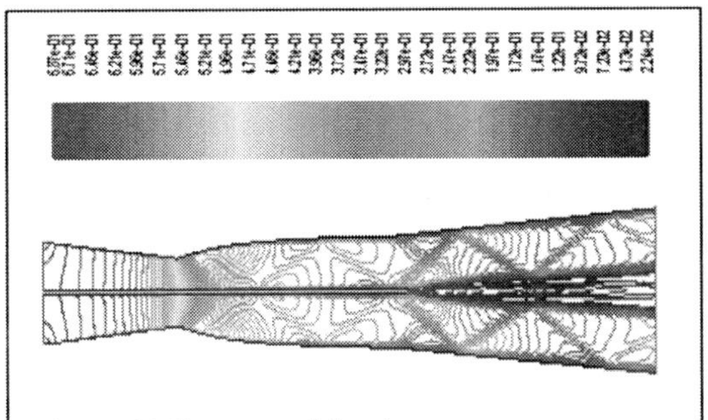

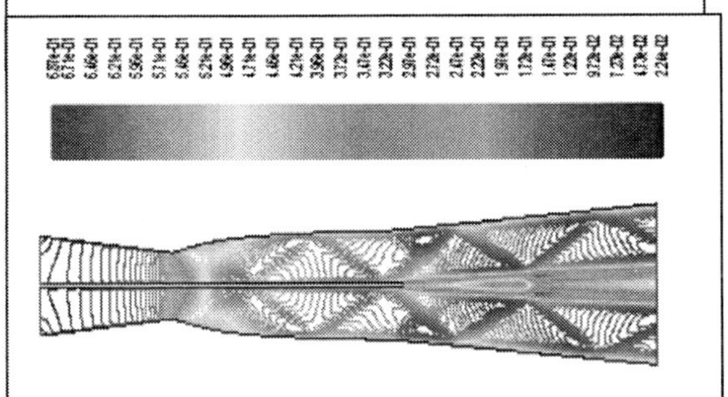

Figure 14 Contour of Mach Number

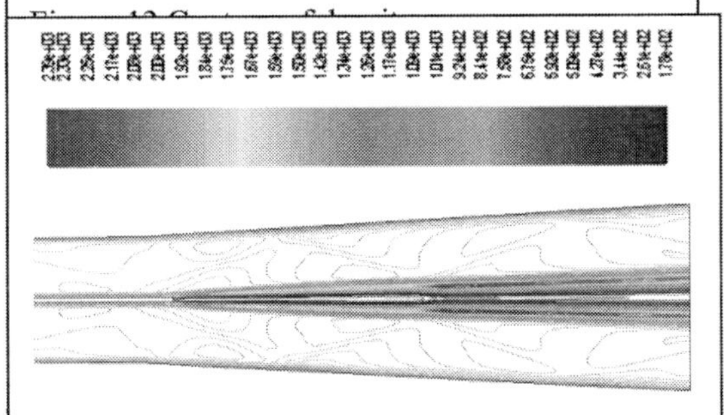

Figure 15 Contour of Static Temperature

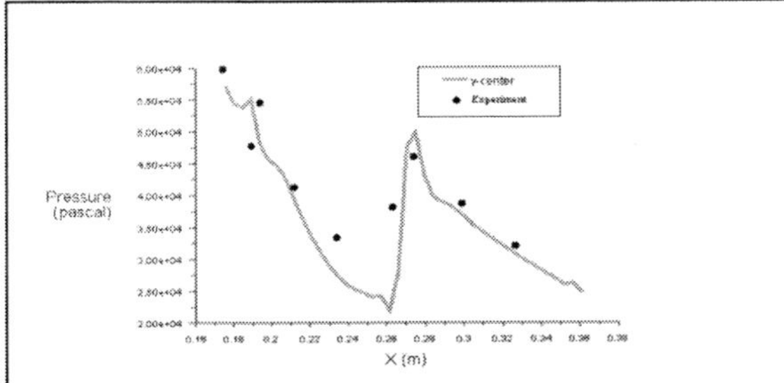

Figure 16 Pressure Distribution along the middle of the channel

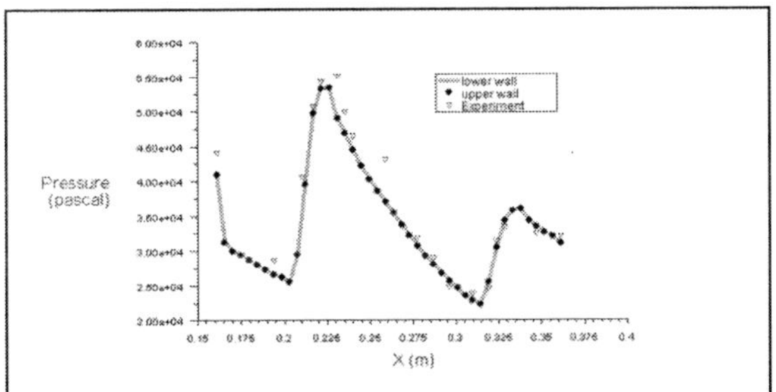

Figure 17 Pressure Distribution along the lower wall and upper wall

IOP Conf. Series: Materials Science and Engineering **810** (2020) 012015 doi:10.1088/1757-899X/810/1/012015

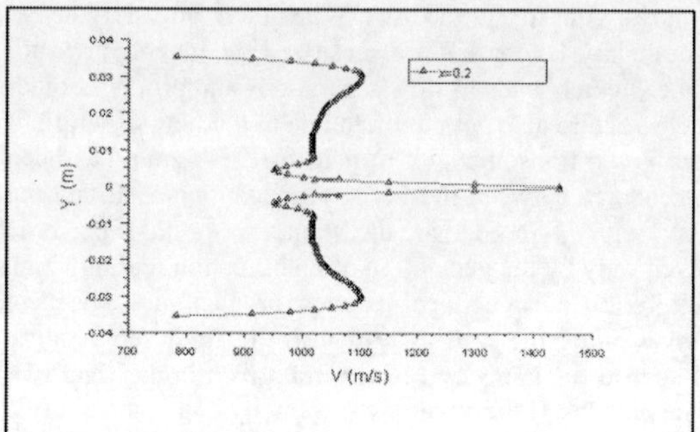

Figure18 velocity profile

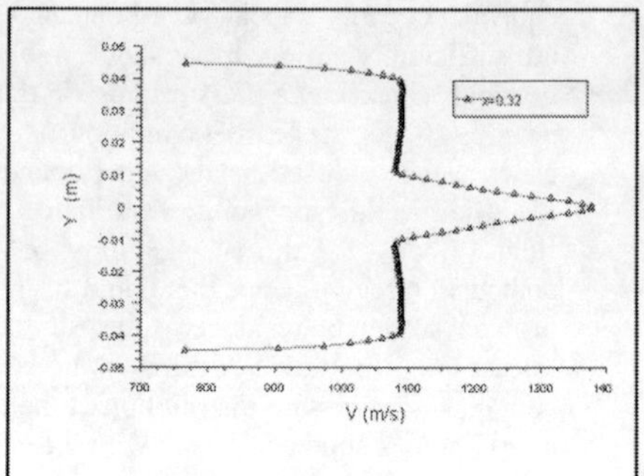

Figure 19 velocity profile

In unsteady calculations, the mixing and flame front propagation integrated over time has been shown. On each time step the fluid regime properties has been calculated. The path of the jet can be predicted by the unsteady equations and its intensification and attenuation can be predicted. Along the time line the shock dissipates completely and further temperature field and the shock wave produced by the combustion becomes study state as shown in figure 15 and figure 21. The temperature also rises sharply and within a very short time combustion takes place and it reaches up to 23.9e+03 K, it can be easily visualized that some form of shock wave has created near the flame front in the contour of temperature. It can only understand by transient numerical calculation because these waves are not present in steady state calculations. At a different time step the temperature field and also some fluid properties are presented here. At 2.0e-04 s time step the flame properties is quite different to previous one and also the flame temperature reaches up to 28.1e+03 k which is quite high. The flame front at different time step and various cross-wise x location are presented in the figure 29, 30 and 31. The temperature profile at 2.24e-03s is quite very similar to steady state calculation that means combustion is taking very minimal time. Velocity profile at a different time step presented here for various x location. From figure 32 and 33 it is clear that as the flame front increases, the velocity at different X cross-location is also increases. At a time step 2e-04 and 2.24e-03 pressure and velocity contour is depicted. In both figure24 and 25 the static pressure field shows that some spot localized due to adverse pressure gradient within the local fuel rich zone in the combustion chamber. This phenomenon leads to flow separations and produces other adverse results on the inlet operation. Fig 25 shows the profiles of numerical and experimental shock waves densities. This shocks wave and jet flows shows the increasing density as the flow behind the shock wave slows further for particular time step 24μs. The comparison shows that the shock stand–off is slightly in acceptable range.

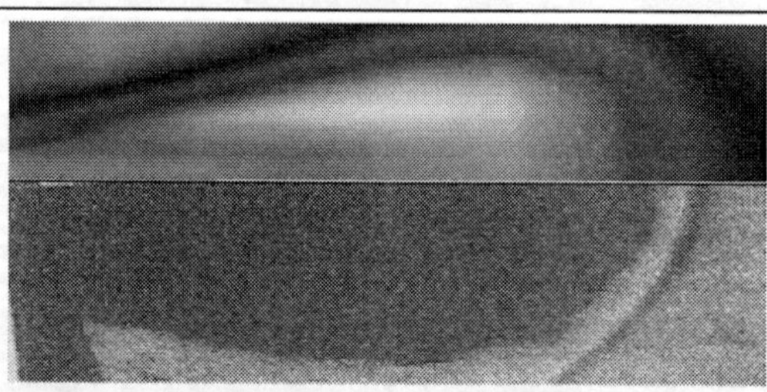

Figure 20 contour of temperature profile at 3.2e-05 s

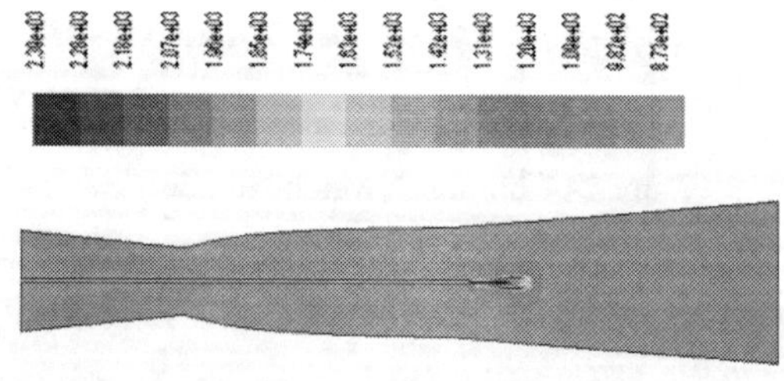

Figure 21 contour of temperature profile at 3.2e-05 s

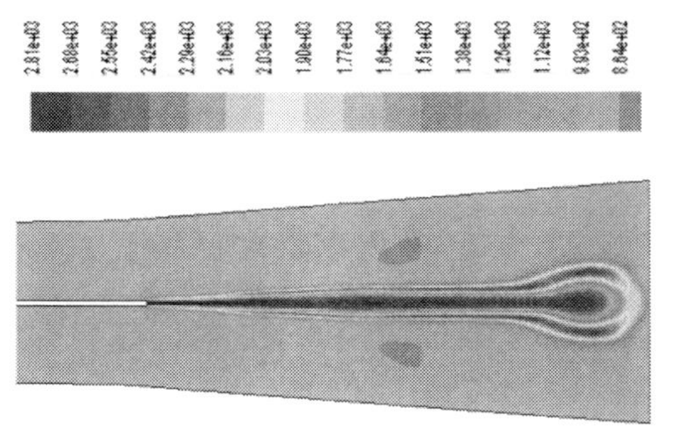

Figure 22 contour of Velocity profile at .0002s

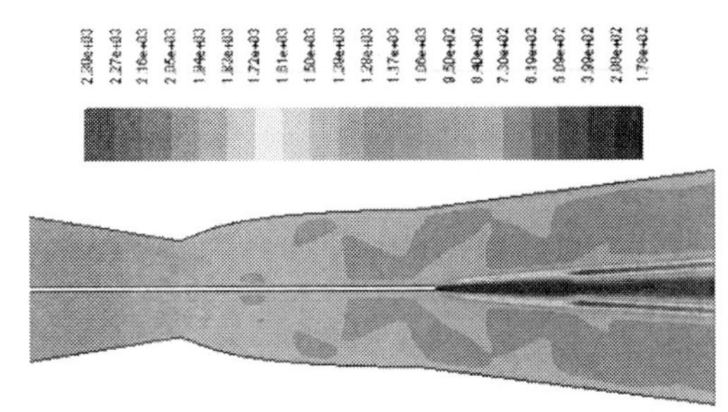

Figure 23 contour of velocity profile at 2e-04 s

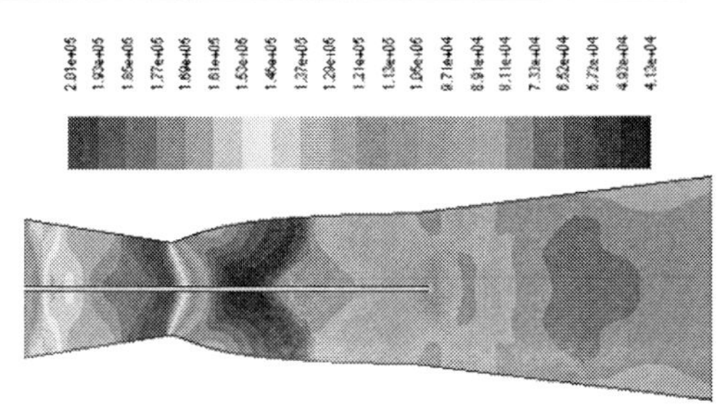

Figure 24 contour of Pressure profile at 2e-04 s

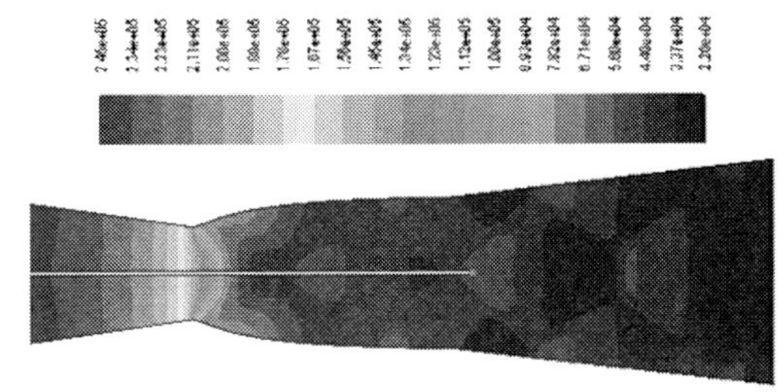

Figure 25 contour of Pressure profile at 2.24e-03s

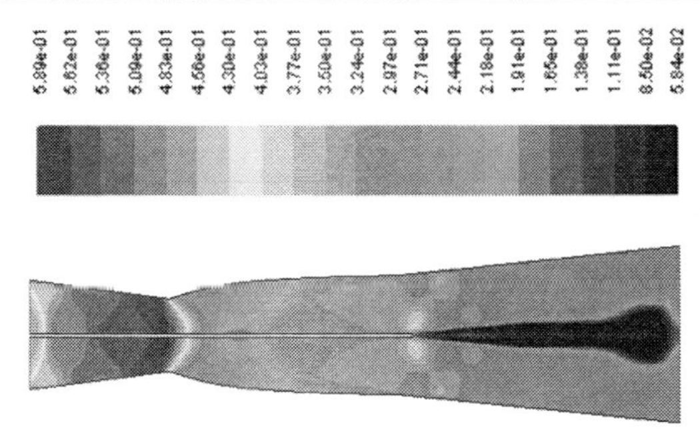

Figure 26 contour of Density profile at 2e-04 s

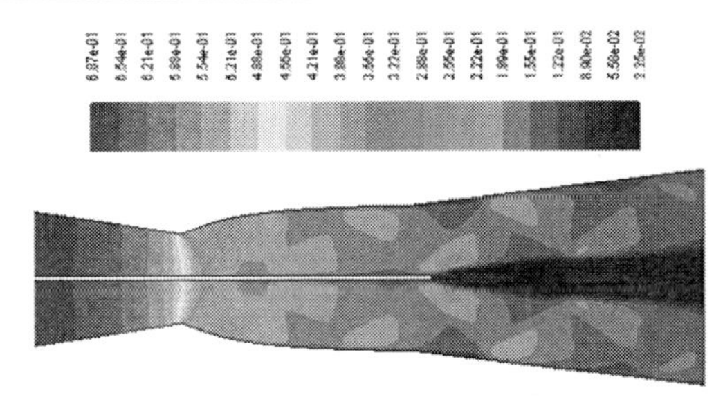

Figure 27 contour of Density profile at 2.24e-03 s

IOP Conf. Series: Materials Science and Engineering **810** (2020) 012015 doi:10.1088/1757-899X/810/1/012015

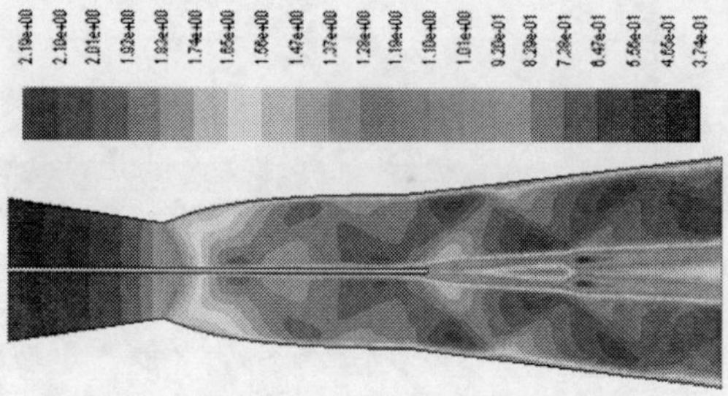

Figure 28 contour of velocity profile at 2.24e-03 s

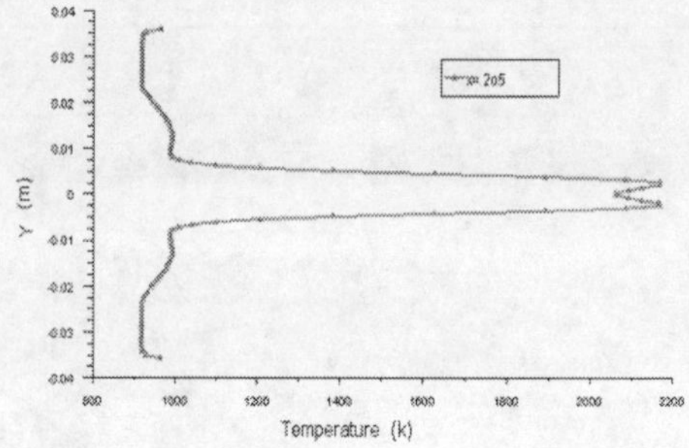

Figure 29 contour of temperature profile at 3.2e-05 s

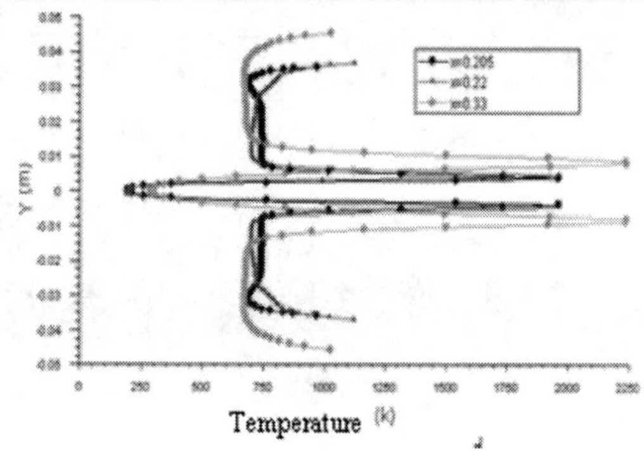

Figure 30 contour of Temperature profile at 2.224e-03 s

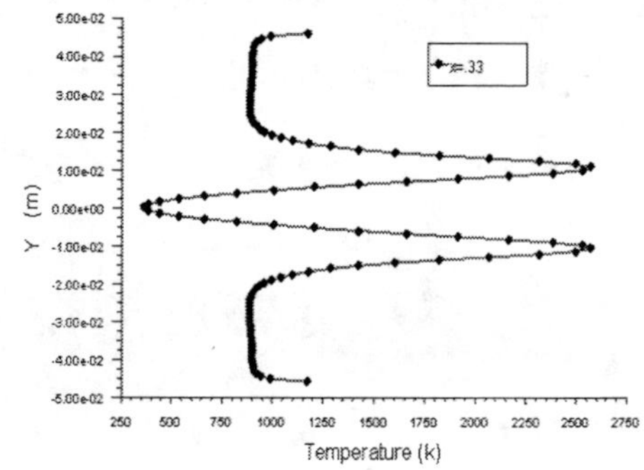

Figure 31 contour of Temperature profile at 2e-04

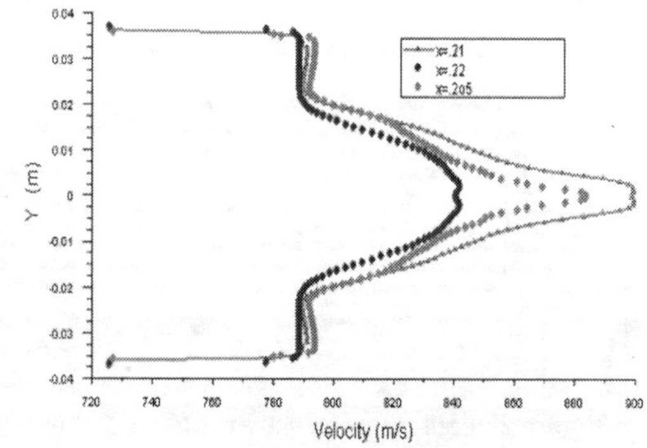

Figure 32 contour of Velocity profile at 3.2e-05 s

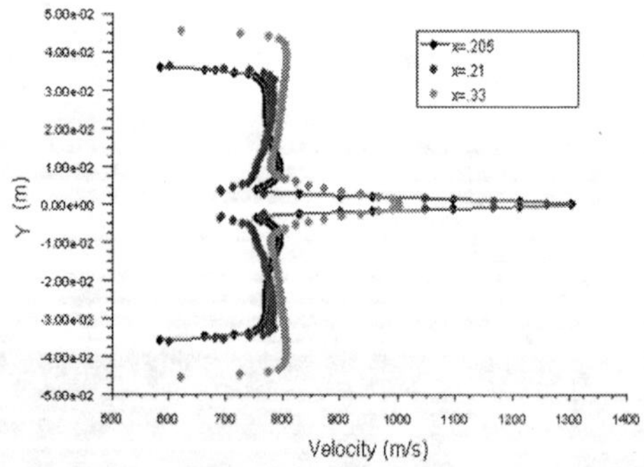

Figure 33 contour of velocity profile at 2e-04 s

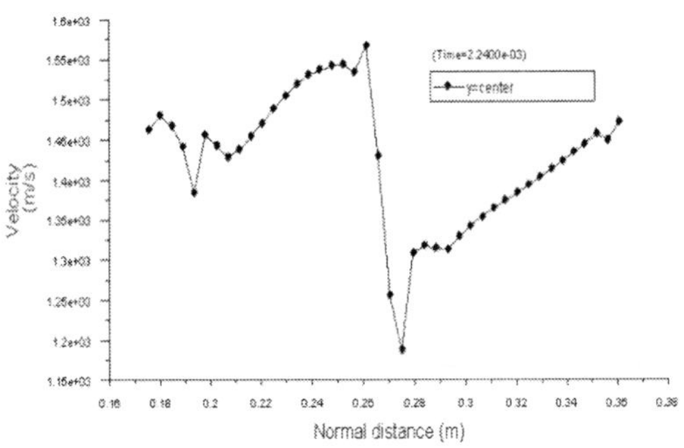

Figure 34 contour of velocity profile at 2.24e-03 s

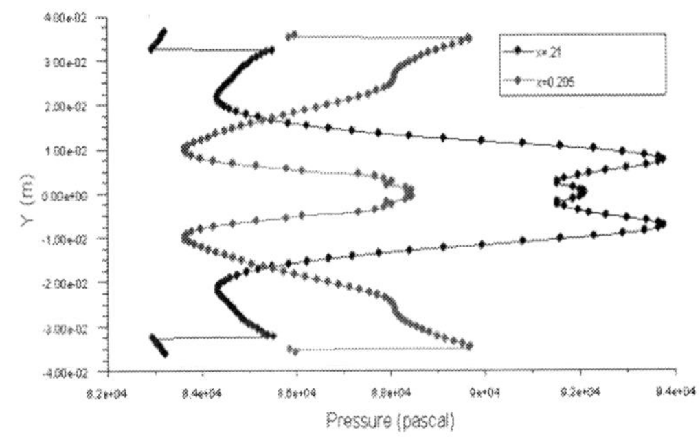

Figure 35 contour of Pressure profile at 2.24e-03 s

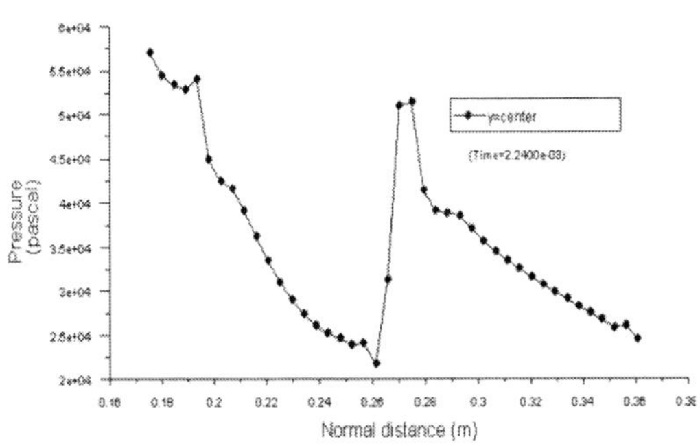

Figure 36 contour of Pressure profile at 2.24e-03 s

4. Conclusion

This computational analysis of supersonic regime combustion is very much in line with the unsteady analysis of H_2 and air combustion. The results show that the flamelet combustion approach is feasible in supersonic regime, which is being demonstrated by the results though few other parameters are still to be evaluated. Some parameters such as Mixing, penetration and combustion characteristics of injected fuel in air during combustion have been shown. The increment of jet penetration to free stream happens due to increment of jet to free stream momentum flux ratio. The results shows that better mixing of air and fuel in the combustion chamber is due to more extreme shear layers and stronger shocks, which are induced during the flow. These phenomenons will leads to losses in total pressure of the supersonic stream. Moreover results shows that the k-ε model is quite compatible to predict the fluctuation in pressure and turbulence which is reasonably isotropic. These results are also in line with the mathematical results which were using during designing of a nozzle, and experimental data is also in agreement with the computational data.

Reference

1. P.K. Tretyakov "The Problems Of Combustion At Supersonic Flow" West-East High Speed Flow Field Conference 19-22, November 2007.
2. Shigeru Aso, Arifnur Hakim, Shingo Miyamoto, Kei Inoue And Yasuhiro Tani " Fundamental Study Of Supersonic Combustion In Pure Air Flow With Use Of Shock Tunnel" , Acta Astronautica 57 (2005) 384 – 389.
3. T. Cain And C. Walton "Review Of Experiments On Ignition And Flame Holding In Supersonic Flow" Published By The America Institute Of Aeronautics And Astronautics, Rto-Tr-Avt-007-V2.
4. H. Mo¨Bus, P. Gerlinger, D. Bru¨Ggemann "Scalar And Joint Scalar- Velocity- Fequency Monte Carlo Pdf Simulation Of Supersonic Combustion" Combustion And Flame 132 (2003) 3–24.
5. Peter Gerlinger, Peter Stoll 1, Markus Kindler , Fernando Schneider C, Manfred Aigner "Numerical Investigation Of Mixing And Combustion Enhancement In Supersonic Combustors By Strut Induced Streamwise Vorticity", Aerospace Science And Technology 12 (2008) 159–168.
6. K. Kumaran, V. Babu "Investigation Of The Effect Of Chemistry Models On The Numerical Predictions Of The Supersonic Combustion Of Hydrogen", Combustion And Flame 156 (2009) 826–841.
7. Zhiliang Xuxk, Myoungnyoun Kimxk, Wonho Ohzk, James Glimmzxk, Roman Samulyakx, Xiaolin Liz And Constantine Tzanos "Atomization Of A High Speed Jet" April 14, 2005, U.S. Department Of Energy
8. Jean-Francois Izard ∗ , Arnaud Mura "Stabilization Of Non-Premixed Flames In Supersonic Reactive Flows" Combustion For Aerospace Propulsion, Cras2b:2818,
9. C. Gruenig* And F. Mayinger "Supersonic Combustion Of Kerosene/H2-Mixtures In A Model Scramjet Combustor" , Institute A For Thermodynamics, Technical University Munich, D-85747
10. Kyung Moo Kim 1, Seung Wook Baek *, Cho Young Han " Numerical Study On Supersonic Combustion With Cavity-Based Fuel Injection", International Journal Of Heat And Mass Transfer 47 (2004) 271–286 ,
11. Kenji Miki, Joey Schulz, Suresh Menon "Large-Eddy Simulation Of Equilibrium Plasma-Assisted Combustion In Supersonic Flow", Proceedings Of The Combustion Institute 32 (2009) 2413–2420, Atlanta, Ga 30332-0150,
12. P Manna, D Chakraborty "Numerical Simulation Of Transverse H2 Combustion In Supersonic Airstream In A Constant Area Duct", Vol. 86, November 2005, ARTFC.
13. G. Yu, J.G. Li, J.R. Zhao, L.J. Yue, X.Y. Chang, C.-J. Sung "An Experimental Study Of Kerosene Combustion In A Supersonic Model Combustor Using Effervescent Atomization", Proceedings Of The Combustion Institute 30 (2005) 2859–2866,
14. Shajit Yadav, M.Ramachandran, Analysis of Optimized Wind Turbine Failure Using Computational Fluid Dynamics, Materials Today: Proceedings, 4(2), 2017:1788–1793.
15. A. R. Srikrishnan , J. Kurian And V. Sriramulu "An Experimental Investigation Of Thermal Mixing And Combustion In Supersonic Flows", Combustion And Flame 107:464-474 (1996).
16. Yuan Shengxue, "On Supersonic Combustion", Voi. 42 No. 2, Science In China (Series A) February 1999.
17. Zheng Chen, Xiao Qin, Yiguang Ju *, Zhenwei Zhao, Marcos Chaos, Frederick L. Dryer, "High Temperature Ignition And Combustion Enhancement By Dimethyl Ether Addition To Methane–Air Mixtures", Proceedings Of The Combustion Institute 31 (2007) 1215–1222.
18. Y. D'angelo, "Computation Of The Flow Structure Of An Hydrogen/Air Mixture Downstream Of A Steady System Of Shock Waves", Aerospace Science And Technology, 1997, No 5, 309-327.
19. S. Javoy , V. Naudet , S. Abid , C.E. Paillard, "Elementary Reaction Kinetics Studies Of Interest In H2 Supersonic Combustion Chemistry", Experimental Thermal And Fluid Science 27 (2003) 371–377.
20. H. Nagata, M. Sasaki, T. Arai, T. Totani, And I. Kudo, "Evaluation Of Mass Transfer Coefficient And Hydrogen Concentration In Supersonic Flow By Using Catalytic Reaction", Proceedings Of The Combustion Institute, Volume 28, 2000/Pp. 713–719,
21. A. Ben-Yakar And R. K. Hanson, "Experimental Investigation Of Flame-Holding Capability Of Hydrogen Transverse Jet In Supersonic Cross-Flow", Twenty-Seventh Symposium (International) On Combustion/The Combustion Institute, 1998/Pp. 2173–2180,
22. M. Leyko , F. Nicoud , S. Moreau And T. Poinsot, "Numerical And Analytical Investigation Of The Indirect Combustion Noise In A Nozzle", Combustion For Aerospace Propulsion, Cras2b:2838.
23. Doyoung Byun And Seung Wook Baek, "Numerical Investigation Of Combustion With Non-Gray Thermal Radiation And Soot Formation Effect In A Liquid Rocket Engine", International Journal Of Heat And Mass Transfer 50 (2007) 412–422,
24. K.M. Pandey and A.P Singh, "CFD Analysis of Conical Nozzle For Mach 3 At Degrees Of Angle Of Divergence With Fluent Software" International Journal of chemical Engineering and Application ,Vol.-1, No. 2, August 2010,ISSN- 2010-0221, pp 179-185.

25. Kenichi Takita, "Numerical Simulation Of A Counterflow Diffusion Flame In Supersonic Airflow", Twenty-

Sixth Symposium (International) On Combustion/The Combustion Institute, 1996/Pp. 2877–2883.
26. J.X. Wen*, B.P. Xu And V.H.Y. Tam, "Numerical Study On Spontaneous Ignition Of Pressurized Hydrogen Release Through A Length Of Tube", Combustion And Flame 2009.
27. Wookyung Kim, Hyungrok Do, M. Godfrey Mungal And Mark A. Cappelli, "Optimal Discharge Placement In Plasma-Assisted Combustion Of A Methane Jet In Cross Flow", Combustion And Flame 153 (2008) 603–615.
28. Valeriyi.Timoshenko, Igors.Belotserkovets And Vjacheslavp.Gusinin, "Problems Of Providing Completeness Of The Methane-Containing Block-Jet Combustion In A Rocket-Ramjet Engine's Combustion Chamber", Acta Astronautica .2009.03.033.
29. B.E. Milton And K. Pianthong, "Pulsed, Supersonic Fuel Jets—A Review Of Their Characteristics And Potential For Fuel Injection", International Journal Of Heat And Fluid Flow 26 (2005) 656–671.
30. S. Zakrzewski, B.E. Milton, K. Pianthong And M. Behnia, "Supersonic Liquid Fuel Jets Injected Into Quiescent Air", International Journal Of Heat And Fluid Flow 25 (2004) 833–840.
31. A.M. Starik, N.S. Titova, L.V. Bezgin, And V.I. Kopchenov, "The Promotion Of Ignition In A Supersonic H2–Air Mixing Layer By Laser-Induced Excitation Of O2 Molecules: Numerical Study", Combustion And Flame 156 (2009) 1641–1652.
32. 27.Vishal Fegade, Gajanan Jadhav, M. Ramachandran, Design, Modelling and Analysis of Tilted Human Powered Vehicle, IOP Conf. Series: Materials Science and Engineering 377 (2018) 012215
33. P. Magre, G. Collin, O. Pin, J.M Badie, G. Olalde And M. Clement, Temperature Measurement By Cars And Intrusive Probe In An Air-Hydrogen Supersonic Combustion, International Journal Of Heat And Mass Transfer 44(2001) 4095-4105.
34. Jiyun tu, guan Heng yeoh and chaoqun liu. "Computational Fluid Dynamics" Elsevier Inc. 2008.
35. K.M. Pandey and A.P Singh, "Recent Advances in Experimental and Numerical Analysis of Combustor Flow Fields in Supersonic Flow Regime" International Journal of chemical Engineering and Application,Vol.-1, No. 2, August 2010,ISSN- 2010-0221, pp 132-137.
36. Deepu, Mukundan N., Gokhale, Sadanand. S., jayaraj, Simon, "Numerical Analysis of Supersonic combustion using unstructured Point Implict finite Volume Method", Journal of combustion Society of Japan, Vol. 48 No. 144(2006) 187-197.

Design and Analysis of Hydrogen Storage Tank with Different Materials by Ansys

S. Senthil Kumar [1], C.Bibin[1], [2]M. Ramachandran

[1]Dept. of Mech. Engg, RMK College of Engg and Technology, Chennai

[2]MPSTME, SVKM'S NMIMS University, Dhule 425405, Maharashtra, India
umailbibin@gmail.com

Abstract. Pressure vessels are used for large commercial and industrial applications such as softening, filtration and storage. It is expected that high-pressure hydrogen storage vessels will be widely used in hydrogen-fuelled vehicles. Progressive failure properties, the burst pressure and fatigue life should be taken into account in the design of composite pressure vessels. In this work, the model and analysis of hydrogen storage vessels along with complete structural and thermal analysis. Liquid hydrogen is seen as an outstanding candidate for the fuel of high altitude, long-endurance unmanned aircraft. The design of lightweight and super-insulated storage tanks for cryogenic liquid hydrogen is since long identified as crucial to enable the adoption of the liquid hydrogen. The basic structural design of the airborne cryogenic liquid hydrogen tank was completed in this paper. The problem of excessive heat leakage of the traditional support structure was solved by designing and using a new insulating support structure. The thermal performance of the designed tank was evaluated. The structure of the tank was analyzed by the combination of the film container theory and finite element numerical simulation method. The structure of the adiabatic support was analyzed by using the Hertz contact theory and numerical simulation method. A simple and effective structure analysis method for a similar container structure and point-contact support structure was provided. Bases for further structural optimization design of hydrogen tank will be provided also. The analysis will be carried out with different materials like titanium, nickel alloy and some coated powders like alumina, Titania and zirconium oxide. The results will be compared with that.

Keywords. hydrogen, storage vessels, model and analysis, liquid hydrogen, airborne, cryogenic

1. Introduction

The intemperate use of fossil fuels has led to gradually increasing drastic environmental pollution and energy crisis. Numerous research works have recently been carried out on looking for renewable resources as a replacement for conventional fossil fuels. Hydrogen has been recognized as the superior option for the future energy industry because of the characteristics of unlimited supply, zero-emission of greenhouse gases, and high energy efficiency. Hydrogen storage has become one of the predominant technical barriers limiting the widespread use of hydrogen energy. Safe, high-efficiency and economical hydrogen storage technique is a key to ensure a favourable run of hydrogen fuel cell vehicles. Among many hydrogen storage patterns including high-pressure gaseous storage, cryogenic liquid storage and chemical hydrogen storage, high-pressure gaseous storage has become the most popular technique. The basic requirements for the design of storage vessels are safety, reliability and economy. However, the composite pressure vessels may work under the high-pressure and high-temperature environment. Conventional metallic pressure vessels cannot longer be competent for the rigorous need for high strength and stiffness weight ratios. Therefore, the composite filament wound technology was introduced to improve the performance of the storage vessels. Generally, the composite materials are used for fabrication of pressure vessels by placing them in different orientations for different layers and in a common orientation within a layer. These layers are stacked in such a way to achieve high stiffness and strength. The design of the composite vessel as a fundamental research work relates the physical and mechanical properties of materials to the geometric specifications.

1.1. Liquid Hydrogen Storage

Storage and transportation of hydrogen as a liquid is another possibility. The cryogenic hydrogen is to be stored in specially insulated vessels at (-) 252.880C. The energy required to liquefy hydrogen (gas at 300oK and 1 bar pressure) is about 47 MJ / kg of hydrogen. The energy also is dependent on the size of the plant. With improved technologies and small plants involving magnetic regenerative liquefaction about half of this energy may be adequate. Thus, energy required for bulk storage and transport cryogenic liquid hydrogen gas can be about 10 to 20% lower. Like hydrogen in gaseous form, the liquid hydrogen also has the tendency to diffuse into the material of construction at high pressures and make them brittle. To check this problem of embrittlement, the storage vessels may be made of FCC (as the material of construction) with special insulation, comprising double-walled with the vacuum in between, opacifiers and multi-layer insulations.

Fig: 1.1 Hydrogen stored as a liquid

1.2. Cryogenic Storage

Liquid hydrogen is stored in cryogenic tanks at 21.2 K at ambient pressure. Because of the low critical temperature of hydrogen (33 K), the liquid form can only be stored in open systems, as there is no liquid phase existent above the critical temperature. The pressure in a closed storage system at room temperature (RT) could increase to ~10^4 bar. The simplest liquefaction cycle is the Joule-Thompson cycle (Linde cycle). The gas is first compressed and it undergoes an isenthalpic Joule-Thomson expansion, producing some liquid. The cooled gas is separated from the liquid and returned to the compressor via the heat exchanger The Joule-Thompson cycle works for gases, such as nitrogen, with an inversion temperature above RT. Hydrogen, however, warms upon expansion at RT. then cooled in a heat exchanger, before it passes through a throttle valve where For hydrogen to cool upon expansion, its temperature must be below its inversion temperature of 202 K.

1.3. Nano Structure Tube

Materials with a large specific surface area like activated or nano structured carbon and carbon nano tubes (CNTs) are possible substrates for physisorption. The main difference between CNTs and high surface area graphite is the curvature of the graphene sheets and the cavity inside the tube. In micro porous solids with capillaries, which have a width of less than a few molecular diameters, the potential fields from opposite walls overlap so that the attractive force acting upon adsorbate molecules is increased compared with that on a flat carbon surface. This phenomenon is the main motivation for the investigation of hydrogen-CNT interaction. Most work on the theoretical absorption of hydrogen in carbon nanostructures uses the Feynman (semi classical) effective potential approximation to calculate the adsorption potential or the grand canonical Monte Carlo simulation.

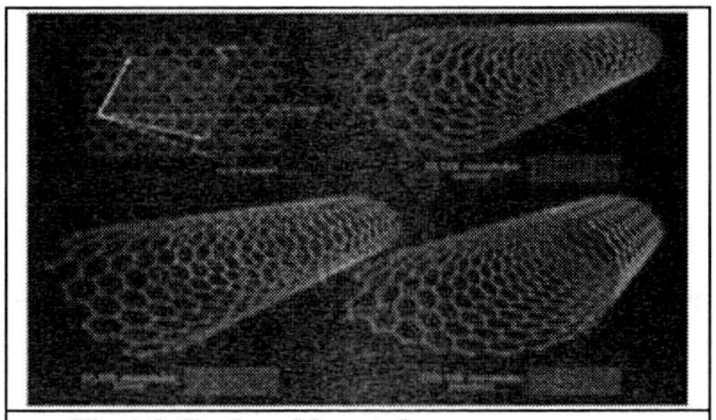

Fig 1.2 Carbon Nanotube

1.4. Workbench Ansys

The Ansys software is used to analyse a given model. The Ansys consists of more number of libraries and module for the analysis.

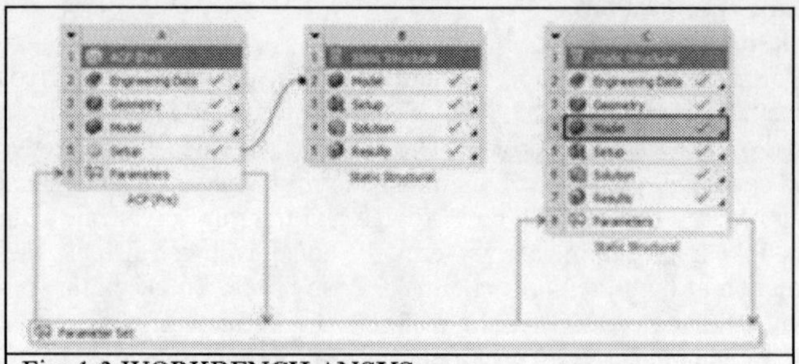

Fig: 1.3 WORKBENCH ANSYS

i. Mechanical APDL

ACP (Pre) is used to create the laminate i.e. the ply orientation and stacking different material. Loading conditions is done in static structural module and results are viewed in ACP (Post).

ii. Module

The Ansys workbench consists of several modules which help in the analysis of the multi-leaf spring.

- Engineering Data
- Geometry
- Model
- Setup
- Result.

The engineering data is used to feed the data like (young's modulus, Poisson ratio etc.) Ansys is like a black box the result depends upon what data is given. Geometry is used to create the 3D model of the object or import the model from other platforms such as creo5.0, catiav5, solid works etc. Para solid, IGES, STEP. The file is flexible when the model is being imported. The Model is used to create the mesh (hedraterdral, triangular). The nodes are classified based on the mesh type. Smaller the element size more accurate the results. In this module, the loading condition and boundary condition can also be defined based on which result converge. Setup (ACP) is used to create the composite material i.e. ply orientation, laminate direction, fibre direction. e viewed in ACP (Post) module. The module is used to create a solid model of the storage tank using different materials. In this module, the stack up the layer of the lamina is also created. Results module is used to display the output of the given input.

iii. Element Type

The tools used for simulation in Ansys 14.0 software. The software uses the element type to create a solid model. The element type which is suitable for the composite analysis is solsh190, solid185, solid186. Solsh190, solid185 uses 3D 8- node for solid structural data. Solid186 uses 3D 20-node for solid structural data which gives finer mesh thereby giving accurate results.Solsh190 is the default element type in Ansys workbench software. The Ansys software has plenty of tools and a suitable tool for creating and loading a composite is ACP (pre), ACP (post), Static Structural. The Fatigue tool is used for Flexural analysis of leaf spring.

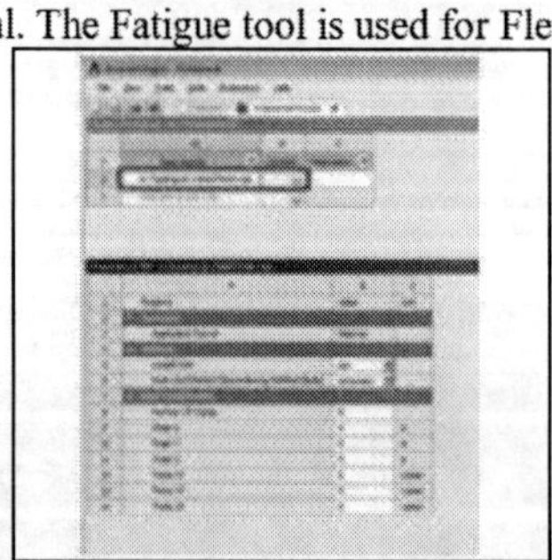

Fig 1.4 Element type ANSYS

2. MATERIALS AND METHODS

2.1. DESIGN METHODOLOGY

Modelling and analysis of 3-D models of the tank were carried out using ANSYS FEA. The methodology includes the process of analysis and this project includes the process of numerical investigation and comparison with different materials and ceramic coated powders by Ansys.

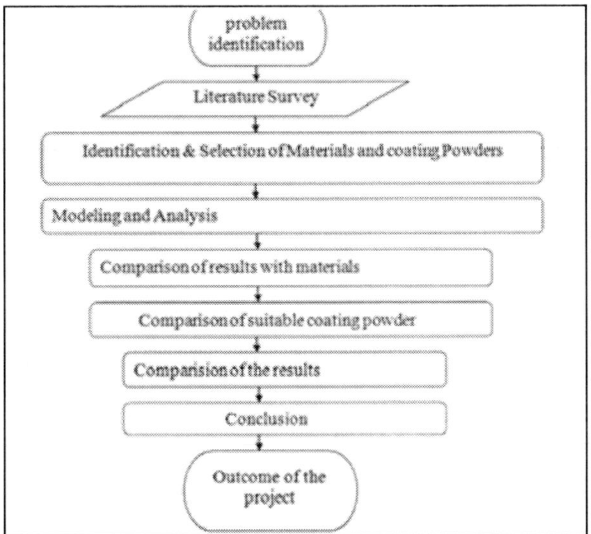

Fig 2.1 Flowchart of Methodology

2.2. MODELLING AND ANALYSIS

The analyses conducted during this contract are described in the following section. Cost analysis of emerging technologies is inherently an iterative process as the technology, designs, and performance are rapidly changing. Consequently, some of the analyses in this report are marked as preliminary to denote their interim status as of the report date, due to either on-going areas of investigation or shifting of DOE priorities. However, these preliminary analyses have significant value and are included to provide both status reporting and identification of cost drivers and manufacturing requirements.

2.3. DESIGN PARAMETERS

i. Titanium alloys

Unalloyed, commercially pure titanium has a tensile strength ranging from 275 to 590 MPa, and this strength is controlled primarily through oxygen content and iron content. The higher the oxygen and iron content, the higher the strength. Commercially alloyed titanium grades can range from a tensile strength as low as 600 MPa such as Ti-3A1-2.5V to a tensile strength as high as 1250 MPa (e.g. for the high strength alloy Ti-15Mo-5Zr-3AI).

Table 1: Mechanical properties of Titanium alloys

Mechanical properties	**Titanium**	**Nickel**	**Zirconium**
Density	4.50 g/cc	8.88 g/cc	
Hardness, Vickers	60	-	150
Tensile strength, Ultimate strength	220 MPa	317 MPa	330 MPa
Tensile strength, yield	140 MPa	59 MPa	230 MPa
Elongation at Break	54%	30%	32%
Modulus of Elasticity	116 GPa	207 GPa	94.5 GPa
Poisson's ratio	0.34	0.31	0.34
Shear Modulus	43 GPa	76 GPa	

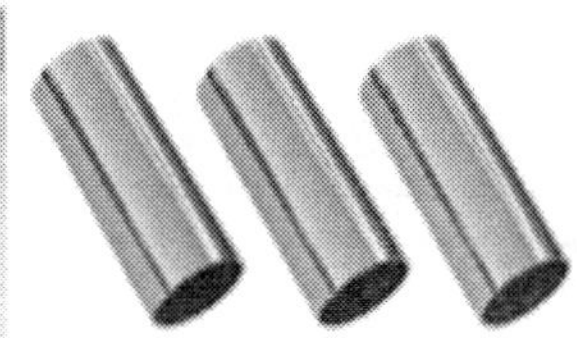

Fig:2.3 titanium **Fig 2.4 Nickel** **Fig 2.5 Zirconium**

IOP Conf. Series: Materials Science and Engineering **810** (2020) 012016 doi:10.1088/1757-899X/810/1/012016

3. RESULT AND DISCUSSIONS

3.1 *MODELLING* OF HYDROGEN STORAGE TANK

I. MODELLING OF NICKEL-ALUMINA

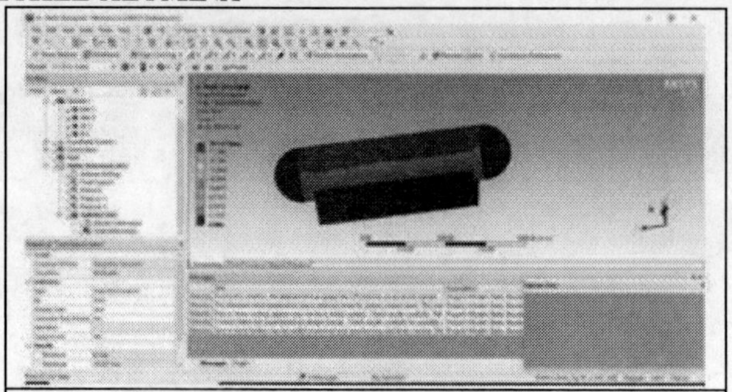

Fig:3.1 Deformation of Nickel-alumina

II. MODELLING OF TITANIUM-ALUMINA

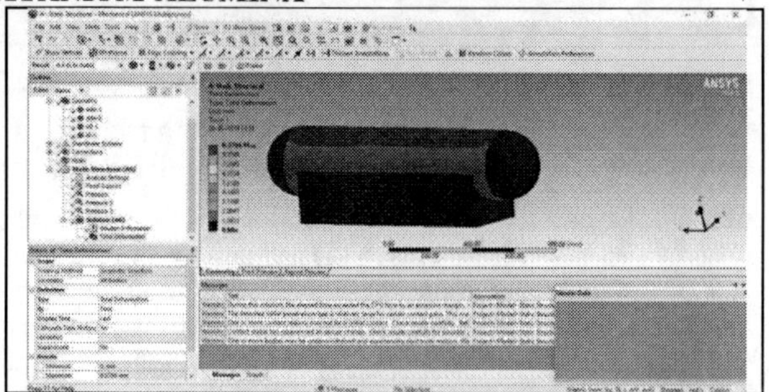

Fig:3.2 Deformation of Titanium-alumina

III. MODELLING OF TITANIUM-ZIRCONIUM

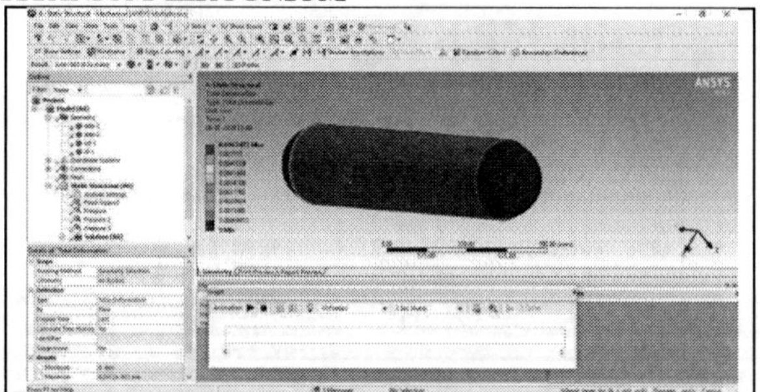

Fig:3.3 Deformation of Titanium-zirconium

IV. TOTAL DEFORMATION OF TITANIUM-ZIRCONIUM

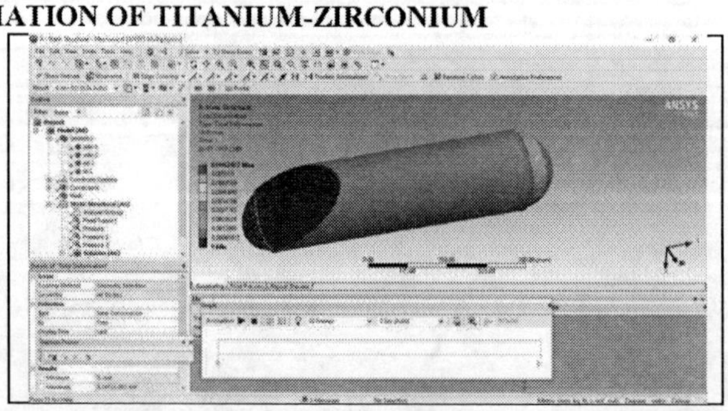

Fig 3.4 Cut section of titanium-zirconium

This shows the input data needed for the analysis and properties of the material chosen from the standard ANSYS Workbench Library. Engineering data of the material can be added when the properties are not available in the Library. This shows the solid model of the composite with fibre and matrix combined in it. The drop-off and cut off handles the global material handling of the composite. Drop-off method defines the ply's drop-off before or after the edge. The maximum equivalent strain produced in the peri-implant region was mostly within the range for bone augmentation. Under oblique loading, maximum von Mises stresses and the equivalent strain was more evident at the neck of the most distal implant on the loaded side. Under an axial load, the stress and strain were transferred to the peri-implant bone around the apex of the implant. Maximum tensile stresses that developed for either material were well below their fracture strength. The highest stresses were mainly located at the distobuccal region of the neck for the two implant materials under both loading conditions. Zirconium, Titanium and Hafnium are a group of metals with similar properties. Titanium and its alloys are increasingly used in both aerospace and chemical engineering. Zirconium and alloys based on it have important structural applications in certain nuclear reactors. These uses of Ti and Zr require that the designer has a good understanding of both their low-temperature plasticity and their high-temperature strength.

4. Conclusion

Both Zirconium and titanium are strong, long-lasting, corrosion-resistant metals that are ideal form any demanding projects. In most cases, either one would work. Titanium is more expensive, but the demand for Zirconium is increasing, which may lower the cost of titanium. So many industries, such as in dentistry, now prefer Zirconium because of its low electrical conductivity and better hypoallergenic properties.

Table 2: Comparisons of Results

MATERIAL USED	COATING USED	THICKNESS (mm)	PRESSURE (bar)	DEFORMATON (mm)
NICKEL	ALUMINA	4	600	9.38
TITANIUM	ALUMINA	4	600	19.54
TITANIUM	ZIRCONIUM	4	600	0.00625

Zirconium are strong, durable and resistant to a variety of chemicals. Titanium, however, weighs a fraction of what steel weighs. To have the same strength with less weight, titanium seems like an ideal combination. The biggest drawback to titanium is the price. The superiority of titanium over steel will cost you. But oftentimes, another similar metal often gets overlooked: Zirconium. Zirconium and titanium do share many desirable characteristics, most notably their durability and corrosion resistance. Zirconium's strength, corrosion resistance and durability make it ideal for use in pipes, fittings and heat exchangers, steel alloys, coloured glazes, bricks, ceramics, abrasives, flashbulbs, lamp filaments, artificial gemstones and some deodorants. Zirconium is also used in surgical instruments, jewellery, lab crucibles, television glass, surgical instruments and catalytic converters. Zirconium's lack of attracting particles make it a popular option for use in nuclear power plants. At one time, Zirconium was even used to treat poison ivy, before skin irritations became linked to possible side effects. This work may also help identify and develop similar hydrogen storage systems. To be a mainstream source of fuel, hydrogen must be stored safely and efficiently. Conventional high-pressure storage tanks can be dangerous and are too big and heavy for certain applications, such as hydrogen-based fuel cells in automobiles. Hydrogen-storage materials, however, incorporate hydrogen safely and compactly and temporarily hold large quantities of it that can be recovered easily under safe, controlled conditions. A hydrogen-storage material must be able to store hydrogen quickly under normal conditions that are without very high temperatures and pressures

5. References

1. Sippel M, van Foreest A. SpaceLiner rocket-powered highspeed passenger transportation concept evolving inFAST20XX. Inof the 61st International Astronautical Congress, Prague; 2010.
2. Geuskens FJJMM, Bergsma OK, Koussios S, Beukers Analysis of conformable pressure vessels: introducing theMultibubble. AIAA J 2011.
3. Haaland A. High-pressure conformable hydrogen storage for fuel cell vehicles. In: Proceedings of the U.S. DOE hydrogen program review, California; 2000.
4. Vasiliev VV. Composite pressure vessels: analysis, design and manufacturing. Bull Ridge Publishing; 2009.
5. Liang CC, ShiahSW, Jen CY, Chen HW. Optimum design of multiple intersecting spheres deep-submerged pressure hull.Ocean Eng 2004.
6. Leon GF. Advanced hull design intersecting titanium spheres.In: Proceedings of the Offshore Technology Conference, TEXAS; 1969.
7. Final report of the X-33 liquid hydrogen tank testInvestigation. Huntsville: NASA Marshall Space FlightCenter;2000.

IOP Conf. Series: Materials Science and Engineering **810** (2020) 012016 doi:10.1088/1757-899X/810/1/012016

8. M. Deshpande, K. Kalita, Ramachandran. M, Stress Mitigation in Isotropic Plates with Square Cutout using Auxiliary Holes, International Journal of Applied Engineering Research. 9(23) (2014) pp. 21975-21992
9. Aceves SM, Berry GD, Rambach GD. Insulated pressure vessels for hydrogen storage on vehicles. Int J HydrogenEnergy 1998.
10. 27.Vishal Fegade, Gajanan Jadhav, M. Ramachandran, Design, Modelling and Analysis of Tilted Human Powered Vehicle, IOP Conf. Series: Materials Science and Engineering 377 (2018) 012215.
11. Majumdar AK, Steadman TE, Maroney JL. Numerical Modeling of propellant boiloff in the cryogenic storage tank.NASA Technical Report. 2007.
12. Sippel M, Kopp A, Sinko K, Mattsson D. Advanced hypersonicCryogenic -tanks research in CHATT. In: Proceedings of the 18thAIAA International Space Planes and Hypersonic Systems and Technologies Conference, France; 2011.
13. Flu¨ gge W. Stresses in shells. Springer- Verlag BerlinHeidelberg GmbH; 1960.
14. Peter.ST.composite filament winding. ASM international materials park;2011.
15. Raithby GD, Hollands KGT. A general method of obtaining approximate solutions to laminar and turbulent free convection problems. Adv Heat Trans 1975.
16. Noda N, Hetnarski RB, Tanigawa Y. Thermal stresses. Taylor&Francis; 2003.
17. Sunder selwyn, R Kesavan, M. Ramachandran, FMECA Analysis of Wind Turbine Using Severity and Occurrence at High Uncertain Wind in India, International Journal of Applied Engineering Research. ISSN 0973-4562 Volume 10, Number 11 (2015) pp. 10250-10253.
18. Fryer DM, Harvey JF. High-pressure vessels. Chapman &Hall,1998.

Die design and optimization of cooling channel position for cold chamber high pressure die casting machine

Radhika Chavan, P S Kulkarni

Maharashtra Institute of Technology, Aurangabad, Maharashtra, India.
radhichavan18@gmail.com, prashantkulkarni83@gmail.com

Abstract. A well-designed gating system for pressure die casting has always been of great importance to achieve good quality castings. However, the cooling system employed in the die and its position has found to have a significant impact on the generation of stresses in the die. High stress concentrations contribute to the erosion of cavity surface thereby affecting the tool life and casting quality. The objective of this research work is to design a die and to analyse the effect of positioning of the cooling channels on the generation of stresses on the cavity surface. In this study, a runner gate system is designed for an ADC 12 aluminium alloy component that is to be cast on a Cold Chamber High Pressure Die Casting (HPDC) machine. Stress analysis and optimization of the cooling channel positionof the die is carried out using ANSYS Workbench. Flow simulation of the optimized design is carried using MAGMA software to understand the effect of cooling channel position on the flow characteristics. The results demonstrate the impact of cooling channel position on the stresses generated in the die and prove to be valuable in designing the die. Defect inspection and testing of porosity is carried out on the castings produced with the optimized design in operation to check if they are acceptable.

Keywords. Cooling channels, die design, flow simulation, high pressure die casting, porosity, radiography.

1. Introduction

The High Pressure Die Casting (HPDC) operation is widely employed for large scale production of non-ferrous castings that have complex shapes and requirement of high quality and dimensional accuracy. The Cold Chamber HPDC process is a pressure die casting process where the injection mechanism is not submerged in the melting furnace and rather is transferred by means of a ladle into the shot hole of the machine. The molten alloy is then pushed into the die by means of a hydraulically operated plunger. The process has advantages of minimizing the contact time between the molten alloy and components of injection mechanism thereby increasing their life. However, air entrapment due to high speed injection can result in porosity in the castings that is a common problem.

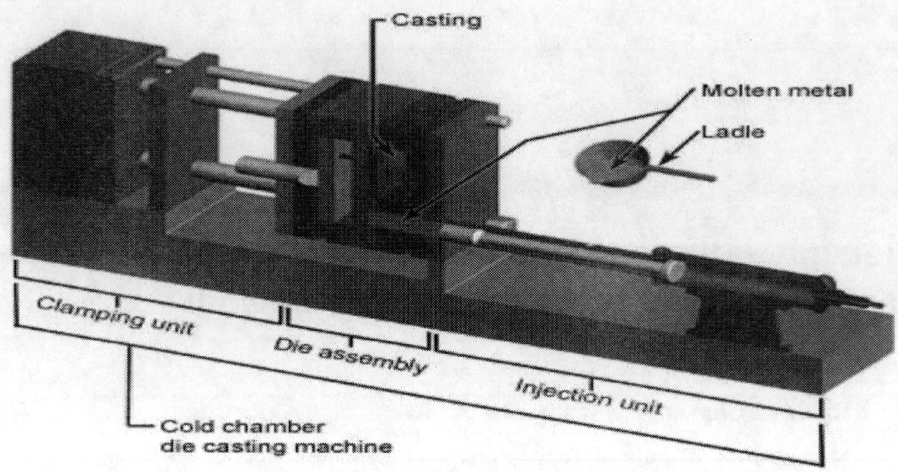

Figure 1. Cold Chamber High Pressure Die Casting machine

The cast alloy selected is ADC 12 aluminium alloy that is to be cast on 160 T Cold Chamber HPDC machine. Minimum flow temperature of the alloy is 580 °C and it enters the die at a temperature of 660°C. Low density, good corrosion resistance, high thermal and electrical conductivity, high dimensional stability and good castability are some of the characteristics of ADC 12. Aluminium die casts have often found applications in automobile products that suffer repeated loading and thermal cycling. Compared to other cast alloys, ADC 12 is found to have less casting defects. It has high production rate, good mechanical properties and ability to form intricate shapes. B. Vijaya Ramnath et al. studied the occurrence of various defects in ADC 12 such as misrun, cold shuts, shrinkage porosity that were produced with an existing flat gate system. A modified spoon-fed gating system design was proposed that overcame the drawbacks of the flat gate system. Process parameters like fill velocity, fill time and metal pressure were considered for process optimization and flow analysis using Rotork Flow 3D software [1].

In high pressure die casting operation, the most commonly employed hot work tool steels are H11 and H13. The material selected for the die is H13 which has good resistance to thermal fatigue, erosion and wear. The production cost of castings is highly dependent on life of the die which is affected by its design, heat treatment, material and various other casting parameters. The thermo-mechanical stresses generated on the die during operation, when exceeded beyond the strength of tool steel are one of the major causes of die failure [2]. Damjan Klobcar et al. analysed the influence of aluminium alloy die casting parameters, die material and die geometry on the in-service tool life. An immersion test apparatus was developed and the die casting was simulated for the special test specimens prepared of the die materials with Finite Element Method used for computation of transient stresses. Maraging steel was found to produce lower stresses and was considered superior material for die casting dies compared to other hot work tool steels [3].

A well-designed gating system for die casting has always been of great importance to achieve good quality die castings. Numerical simulation has achieved great attention and has proved to be a cost-effective way in the optimization of runner gate design. B. H. Hu et al. investigated the effect of split gate system on the filling pattern. Numerical simulation was carried out that showed the continuous gating system provided homogenous mould filling compared to the split gate system. A series of short shot experiments were carried out to validate the simulation results [4]. In die casting operation, porosity is a serious problem as it is invisible and cannot be identified visually. Kulkarni Sanjay

Kumar et. al. studied and examined the effect of die casting process parameters such as first phase speed, second phase speed, first phase length and injection pressure on porosity in SAE 380 aluminium alloy using Taguchi method. Of all the parameters, he found that higher level of injection pressure has most significant effect on porosity in castings and the injection pressure stage is more important than other parameters in the die casting operation [5].

Most of the research work is found to be made on optimization of runner gate system using various simulation tools and analysing the effects of various casting parameters on the filling operation and formation of defects. Failure of dies is a common problem but very less literature is found on analysing die failure in the design stage. Analysis of the cooling system and its effect on the generation of stresses on the tool during operation has been rarely published. This paper discusses the effect of cooling channel position on the stress generation in the die that can be optimized in the design stage itself so as to improve die life and achieve good quality castings.

2. Design of runner gate system

2.1. Component

The runner gate system is designed for an automobile component SRE Bracket. The material of SRE Bracket is ADC 12 aluminium alloy. The weight and density of the component is 265 gm and 2.7 gm/cm^3 respectively. The 3D model of the component is created using NX software and is shown as follows.

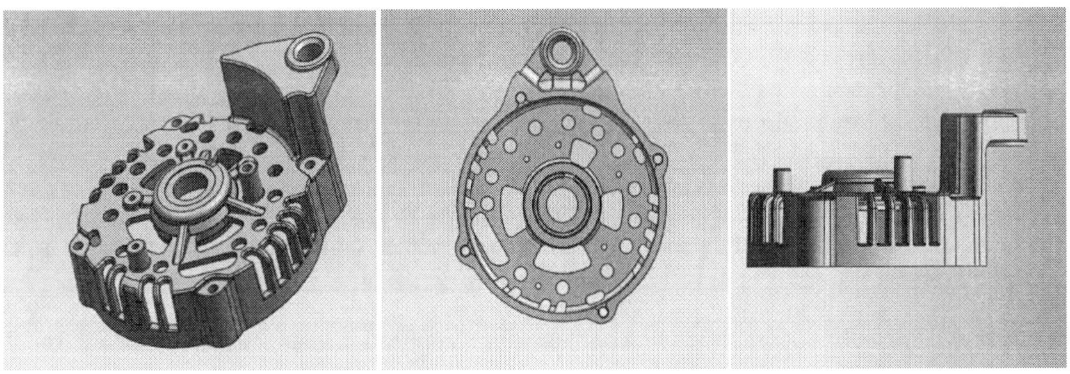

Figure 2. Model of SRE Bracket.

2.2. Cavity fill time

One of the best-known formulas for determining the die cavity fill time is that given by NADCA fill time equation by J. F. Wallace and E. A. Herman: the equation takes slightly different forms in different literature.

The equation for fill time is

$$t = K\left(\frac{\text{Ti} - \text{Tf} + S\,Z}{\text{Tf} - \text{Td}}\right) \times T \tag{1}$$

Where, t = maximum fill time in seconds

K = empirically derived constant related to thermal conductivity of die steel, sec/mm

T = characteristic thinnest average wall thickness of the casting in mm.

T_f= liquidus temperature, °C.

Ti = metal temperature at the gate, °C.

T_d= die surface temperature just before the shot, °C.

S = percent solids at the end of fills, %.

Z = solid unit conversion factor, related to the width of solidification range.

$$t = 0.0346 \left(\frac{660 - 580 + 15 \times 3.8}{580 - 180}\right) \times 3.5$$

$$t = 41 \text{ milli-sec}$$

The part of the equation between the brackets sets a relation between the consumable heat during the cavity fill time and the temperature difference between the minimum flow temperature and die cavity surface temperature. Constant K relates this to the die material thermal conductivity and constant T to the average wall thickness of the casting.

2.3. Gate velocity

The gate velocity has an influence on the casting mechanical properties and on the surface quality. High gate velocity produces higher mechanical properties and less porosity than lower gate velocity. New HPDC machines are capable of producing gate velocities up to 100 m/s, but the die erosion starts to increase already around 40 m/s. Thus, higher velocity range from 40 m/s to 100 m/s is not very practical.

Gas porosity can be reduced without raising the gate velocity by designing the gate and runner system to maintain smooth, continuous flow profiles and by designing the casting so that no backflow occurs. Backflow can occur if there are protrusions on the flow of the metal.

The gate velocity can be chosen depending on the type of application and material of casting from the following table.

Table 1. Recommended gate velocities for typical HPDC alloys

Alloy	Gate Velocity range (m/s)	Recommended gate velocities		
		Decorative parts	Engineering parts	Pressure Tight Parts
Aluminium	20 - 60	20	40	60
Zinc	30 - 50	30	40	50
Magnesium	40 - 60	40	50	60
Copper	20 - 50	20	35	50

Therefore, from the alloy type and application of SRE Bracket we have, Gate velocity, $V_g = 40$ m/s.

IOP Conf. Series: Materials Science and Engineering **810** (2020) 012017 doi:10.1088/1757-899X/810/1/012017

2.4. PQ^2 analysis

The PQ^2 Analysis is based on knowing the pumping rate capability of a die casting machine. It is used to predict what happens when a die is put on the machine. To achieve good quality castings, the gating system must be matched to the machine characteristics so as to have correct pressure and flow rate. This analysis takes into consideration both the die casting machine and the die characteristics that help to evaluate the performance of the die on the selected machine. The machine's characteristic curve describes how much pressure (P) the machine applies to the metal at a given flow rate (Q). PQ^2 analysis matches the selected gate velocity to the HPDC machine plunger hydraulic system. Using these, the runner gate system can be designed to be hydraulically and thermally efficient so that they give minimal mixing of air-metal and minimal re-melt. The value of Metal Pressure (P) can be calculated using the equation as follows

$$P = \left[\frac{\rho}{2g}\right] \times \left[\frac{Vg}{Cd}\right]^2 \tag{2}$$

Where, P = metal pressure, kg/m^2

ρ = density of the cast alloy kg/m^3

V_g = gate velocity, m/s

g = gravitational acceleration, m/s^2

C_d = coefficient of discharge

$$P = \left[\frac{2700}{2 \times 9.8}\right] \times \left[\frac{40}{0.3}\right]^2$$

$$P = 4.32 \times 10^6 \ kg/m^2$$

The metal pressure creates a breaking force F, which is proportional to the projected area of the cavity A and is calculated as,

$$F = P \times A \tag{3}$$

Where, F = breaking force, kg

P = metal pressure, kg/m^2

A = projected area of the die cavity, m^2

$$F = 4.32 \times 10^6 \times 268 \times 10^{-4}$$

$$F = 115.776 \times 10^3 \ kg$$

$$F = 115.776 \ T$$

$$F < 160 \ T$$

The breaking force is less than the machine tonnage capacity. "PQ^2 check is OK".

Therefore, the die can be operated on the selected 160 T machine.

2.5. *Gate parameters*

2.5.1 Gate area. It represents cross sectional area of the gate and can be determined as follows

$$A_g = \frac{Vpart + Vo}{Vg \times t} \quad (4)$$

Where, A_g = gate area, mm^2
V_{part} = part volume, mm^3
V_o = overflows volume, mm^3
V_g = gate velocity, m/s
t = fill time milli seconds
Overflow volume, V_o is accounted to be 10% of the part volume

$$A_g = \frac{98148.1 + 9814.81}{40 \times 41}$$

$$A_g = 65.8310 \text{ mm}^2$$

$$A_g = 66 \text{ mm}^2 \text{ (approx.)}$$

2.5.2 *Gate height* The gate height can be calculated as a fraction of the average wall thickness

$$H_g = \frac{n}{3} \sim \frac{n}{4} \quad (5)$$

Where, H_g = gate height, mm
n = average wall thickness, mm

$$H_g = \frac{3.5}{3} \sim \frac{3.5}{4}$$
$$H_g = 1.2 \text{ (approx.)}$$

The user can however interactively choose another suitable value of gate height that fulfils the gating requirements. The gate height is found to be dependent on selected gate velocity and alloy density as given by the equation
$V_g^{1.707} \times H_g \times \rho \geq J$
Where, ρ = alloy density
J = constant, 998000 for aluminium, magnesium and zinc alloys.

2.6. *Runner parameters*

2.6.1. Runner Area. The ratio of runner area to gate area varies with part design and usually ranges from 1.1 to 1.4 A_g. A larger ratio of 1.6 is used in the case of small parts in multi cavity gating system design. The type and cross section shape of the runner chosen is tangential and trapezoid respectively.

$$A_r = 1.1\ A_g \text{ to } 1.6\ A_g \quad (6)$$

The default value to calculate runner area is

IOP Conf. Series: Materials Science and Engineering **810** (2020) 012017 doi:10.1088/1757-899X/810/1/012017

$$A_r = 1.4\ A_g$$

$$A_r = 92.1634\ mm^2$$

2.6.2. Runner Height The runner height is calculated from the equation as follows and the runner width (W_r) is taken twice of the runner height (H_r)

$$H_r = \sqrt{\frac{Ar}{1.6}} \sim \sqrt{\frac{Ar}{2}} \tag{7}$$

$$H_r = 7.5896 \sim 6.7883\ mm$$

$$H_r = 7.5\ mm\ (approx.)$$

Therefore, we have the width of runner as W_r = 15 mm

The various constants and assumptions used in the calculations are as given below

Table 2. Constants and Assumptions

Constants and Assumptions	Value
Empirically derived constant, K	0.0346 sec/mm
Temperature of molten metal as it enters the die, T_i	660 °C
Min flow temperature of metal (liquidus), T_f	580 °C
Temperature of die cavity surface just before metal enters, T_d	180 °C
Percent solid fraction allowable in metal at the end of filling, S	15%
Units conversion factor, Z	3.8
Coefficient of discharge, C_d	0.3
Acceleration due to gravity, g	9.8 m/sec^2

3. Cooling system of the die

Cooling system employed in the die is line cooling which covers large surface area of the die unlike spot cooling. Cooling channels are employed in the die to maintain its temperature during operation. As internal cooling is very important, the cooling channels are placed in the high temperature regions in the vicinity of the cavity surface. They also facilitate proper solidification of the cast part after complete filing of the die cavity. If the cooling channel is placed too close to the cavity surface, the surface temperature becomes too low and low strength between the channels and the cavity surface likely results in cracking.

Hot phase of the die casting cycle is characterized by generation of high compressive stresses on the surface that get additionally increased by filling pressure of the die. Compressive stresses suppress the crack nucleation and growth which is favourable if no plastic deformation occurs. Local plastic deformations occur as a consequence of exceeding stresses beyond the yield strength of the die material which are expressed on specific locations as stress concentrators, notch effects, sharp transitions and large mass changes of the tool. High tensile stresses are produced during the cold phase that which are increased by plastic deformation of the surface material of the die. Due to cyclic loading operation of the die, low cycle fatigue occurs when the stresses exceed beyond the yield strength or ultimate tensile strength of the material at working temperature that result in nucleation and initial crack growth [6].

In die designing, the designer can intentionally challenge the distance of cooling channels from cavity surface based on some experience. The cooling channel distance from die cavity surface is usually kept twice of its diameter. However, this position does not always prove efficient in controlling the stresses developed in the die. Suitable stress analysis is carried out to optimize the cooling channel position so as to have acceptable stresses in the die. The standard cooling channel diameter taken in the research work is 8 mm and they are placed at a distance of 16 mm from the cavity surface. This positioning of cooling channels based on experience is used in the existing model.

4. Stress analysis and optimization of cooling channel position

A steady state thermo mechanical analysis of the existing model is carried out on ANSYS Workbench 14.5. The shot model of the casting with the cooling channels and the existing die assembly model used for analysis are shown in figure 3 and figure 4 respectively. The modelling of gating system and the die halves is done using NX software.

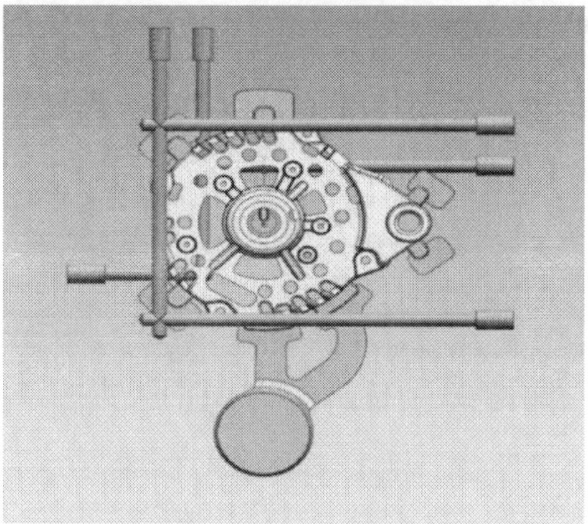

Figure 3. Shot Model with Cooling channels

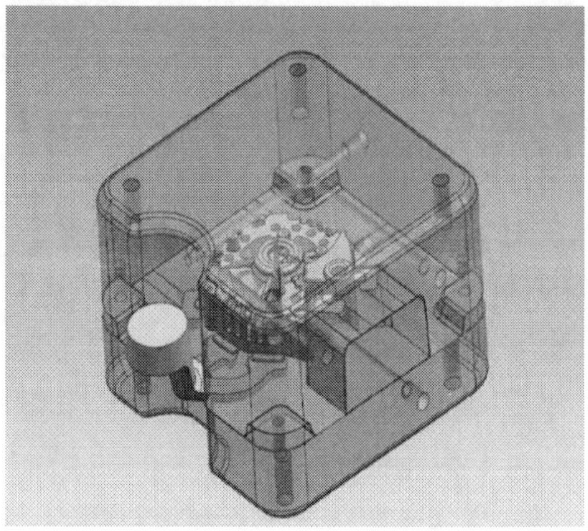

Figure 4. Existing Die assembly

4.1. Pre-processing

The step file of die assembly model created on NX software is imported in ANSYS. The assignment of materials made in analysis is as given below.

Table 3. Material Assignment

Part	Material
Component	ADC 12
Fixed side insert (FSI)	H 13
Moving side insert (MSI)	H13

IOP Conf. Series: Materials Science and Engineering **810** (2020) 012017 doi:10.1088/1757-899X/810/1/012017

4.2. Processing

The loading and boundary conditions employed for analysis are given as follows

Table4. Boundary Conditions

Condition	Part	Value
Temperature	Casting	660 °C
	FSI	180 °C
	MSI	180 °C
	Cooling channel	25 °C
Loading	FSI	Fixed
	MSI	160 T

4.3. Post-processing

4.3.1. Existing Model

The cooling channel distance in existing model is 16 mm from cavity surface. The stress values obtained with the existing position of cooling channels from the analysis as shown in figure 7 and figure 8 are found to exceed the permissible limits. The yield strength and ultimate tensile strength of the tool being 1650 MPa and 1990 MPa respectively. Thus, keeping the distance of cooling channels from cavity surface twice as the used diameter is found to be inefficient from stress point of view. The positioning of the cooling channels from the cavity surface therefore needs to be optimized so as to have stresses produced within the permissible limits.

The temperature and stress results obtained on the Moving Side Insert (MSI) and Fixed Side Insert (FSI) of the existing model of the die are shown as follows.

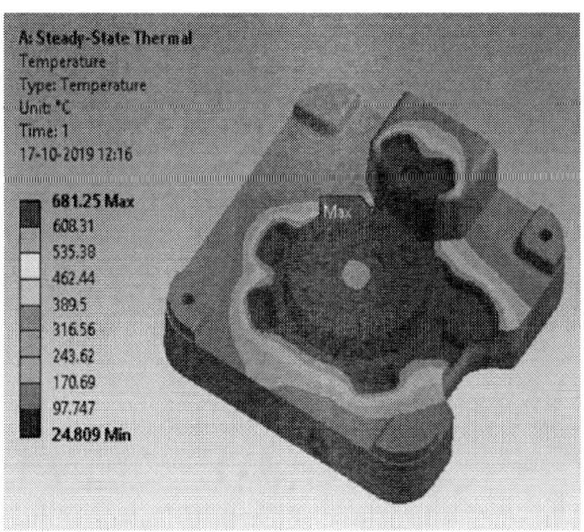

Figure 5. MSI Temperature

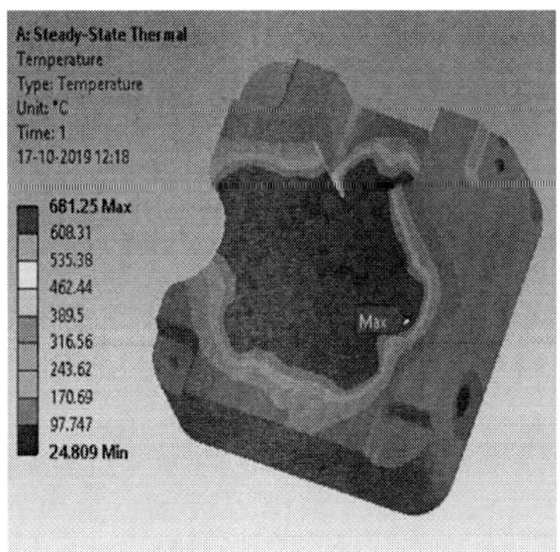

Figure 6. FSI Temperature

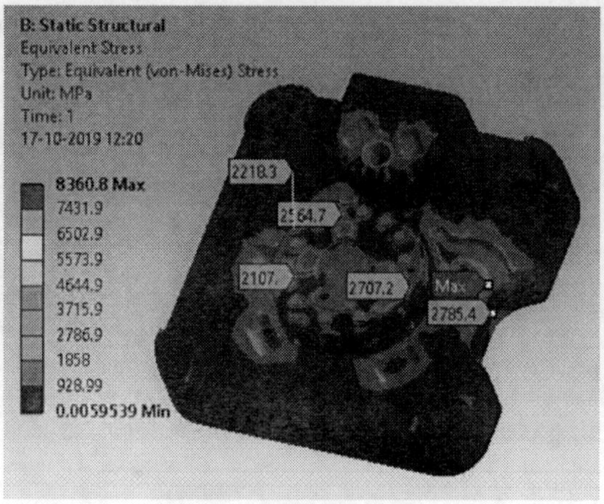

Figure 7. MSI Stress

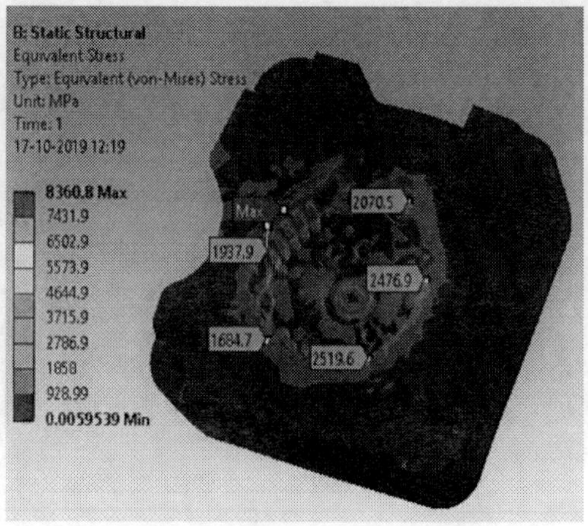

Figure 8. FSI Stress

4.3.2. Proposed model

Moving the cooling channels closer to the cavity surface were found to produce higher stresses and so decreasing the distance was not a solution. The cooling channels distance from cavity surface was then increased so that the stresses that were produced were in safe limits. After few trials, an optimum distance of cooling channels from the cavity surface was achieved that produced efficient results. The new proposed design of the die with optimized cooling channel position was found to be safe from stress point of view. The optimum cooling channel distance obtained was 20 mm from the cavity surface in FSI and MSI.

The stress and temperature results that were obtained from the proposed position were acceptable and are shown as follows.

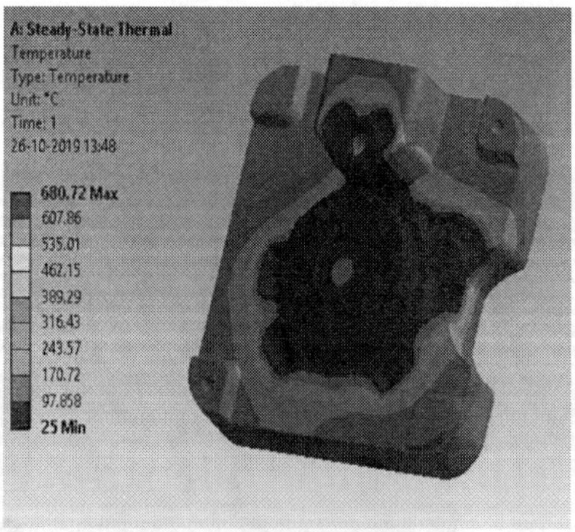

Figure 9. MSI Temperature

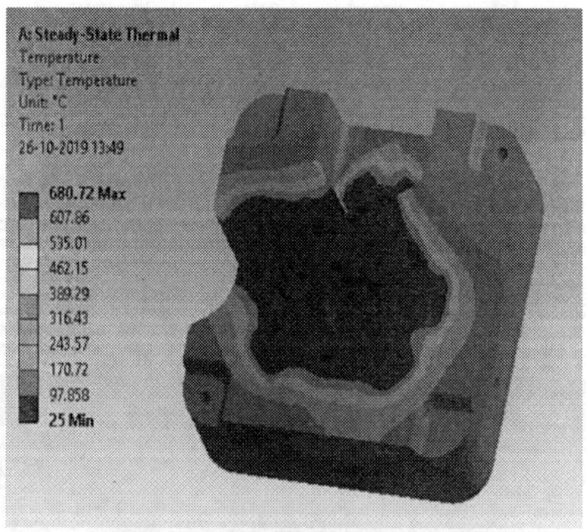

Figure 10. FSI Temperature

IOP Conf. Series: Materials Science and Engineering **810** (2020) 012017 doi:10.1088/1757-899X/810/1/012017

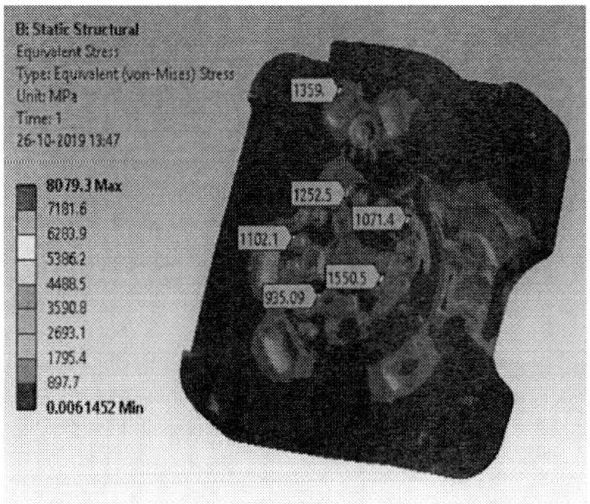

Figure11. MSI Stress

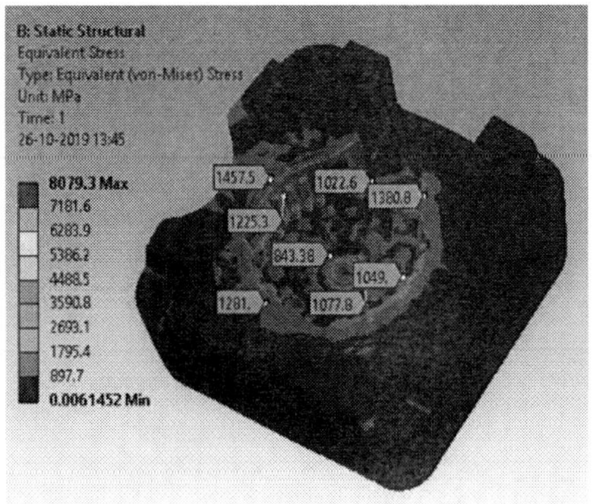

Figure 12. FSI Stress

5. Flow simulation

The proposed design of the die with optimized position of the cooling channels is validated by carrying out the flow simulation of the mould filling process. The simulation is done using the MAGMA Soft software. The input data used for simulation of the proposed design of die is as given below.

Table 5: Initial Conditions for flow simulation

Description	Value	Description	Value
Plunger diameter	50 mm	Casting weight	265 gm
Active sleeve length	285 mm	Shot weight	528 gm
Slow phase length	206 mm	Yield	52 %
Slow phase velocity	0.25 m/s	Shot projected area	268 cm^2
Second phase velocity	2.2 m/s	Molten melt temperature	660 °C
Gate area	66 mm^2	Casting pressure	800 kg/cm^2
Fill ratio	34%	Vent area considered	8 mm^2
Gate velocity	40 m/s	Machine tonnage	160 T
Cavity fill time	41 milli-sec	Density of alloy	2.7 gm/cm^3

The filling simulation of the optimized model of gating system gave metal temperature during filling in liquids range thus ensuring complete filling of the mould cavity. With the optimized position of cooling channels, the filling of the cavity is found to be good. The other results of flow simulation obtained are explained as follows.

IOP Conf. Series: Materials Science and Engineering **810** (2020) 012017 doi:10.1088/1757-899X/810/1/012017

5.1. PQ^2 analysis

PQ^2 Analysis is used to match the target gate velocity with the capabilities of High Pressure Die Casting machine's plunger hydraulic system.

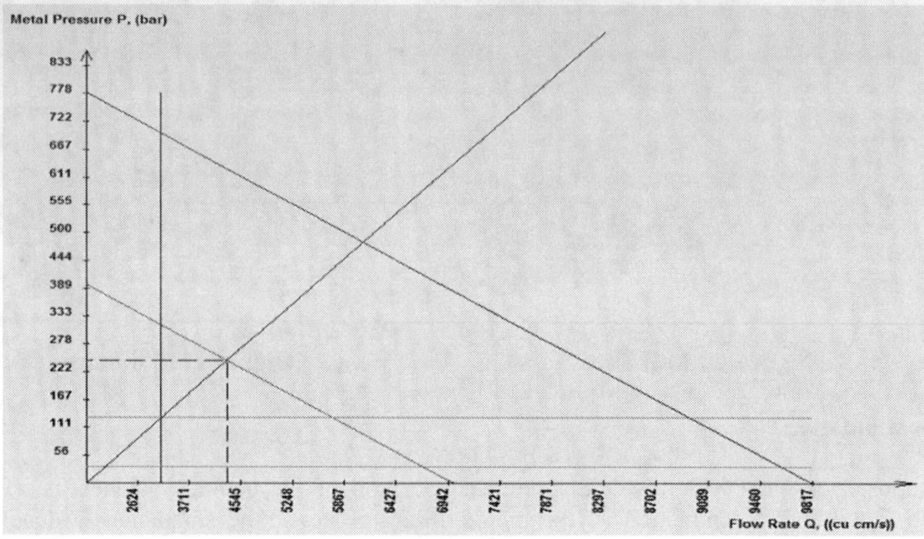

Figure 13. PQ^2 Analysis

From PQ^2 Analysis, the plunger size is found to be OK to fill the casting.

5.2. Switch over point during filling

At plunger position 206 mm, velocity changes from slow to fast. The switch over point occurs at 206 mm plunger position as shown below. The slow phase velocity being 0.25 m/s that increases to 2.2 m/s in the second phase.

Figure 14. Switch over point during filling

5.3. Air entrapment result

Maximum air entrapment is found to be 15 % which is in the acceptable range for any surface defect of casting.

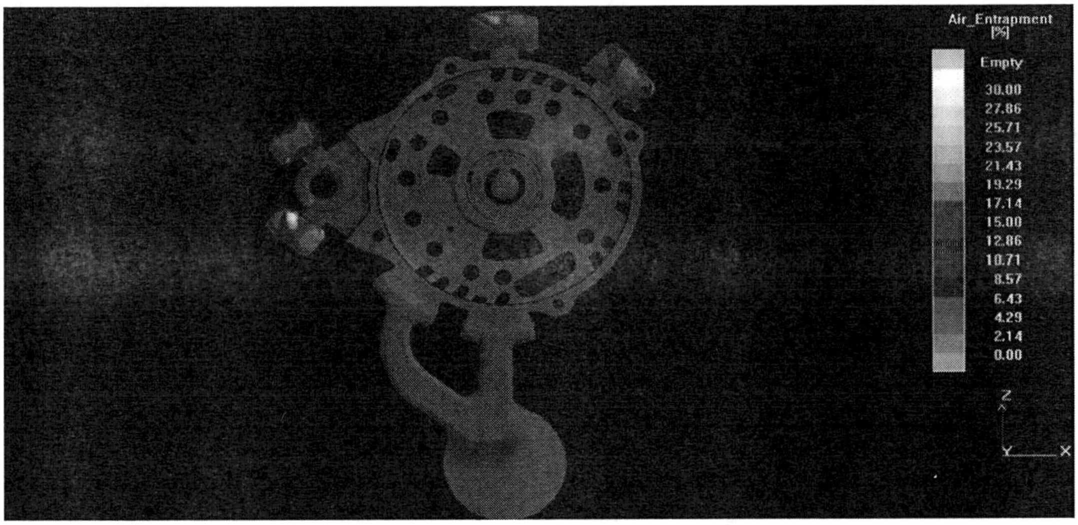

Figure 15. Air entrapment result

5.4. Hotspot_FS result

The result shows local solidification time at different places. This is maximum 12 sec which is acceptable.

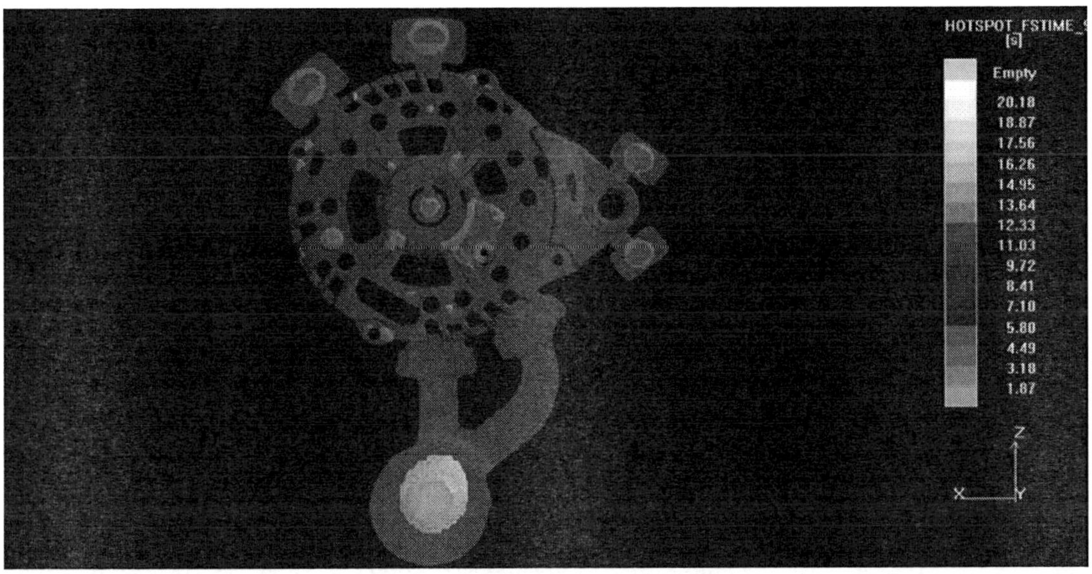

Figure 16. Hotspot_FS Result

5.5. Shrinkage porosity

The shrinkage porosity is in the acceptable range. Porosity defects however cannot be completely avoided in die castings. They can be minimized by controlling process parameters like first phase speed, second phase speed, injection pressure and first phase length in operation.

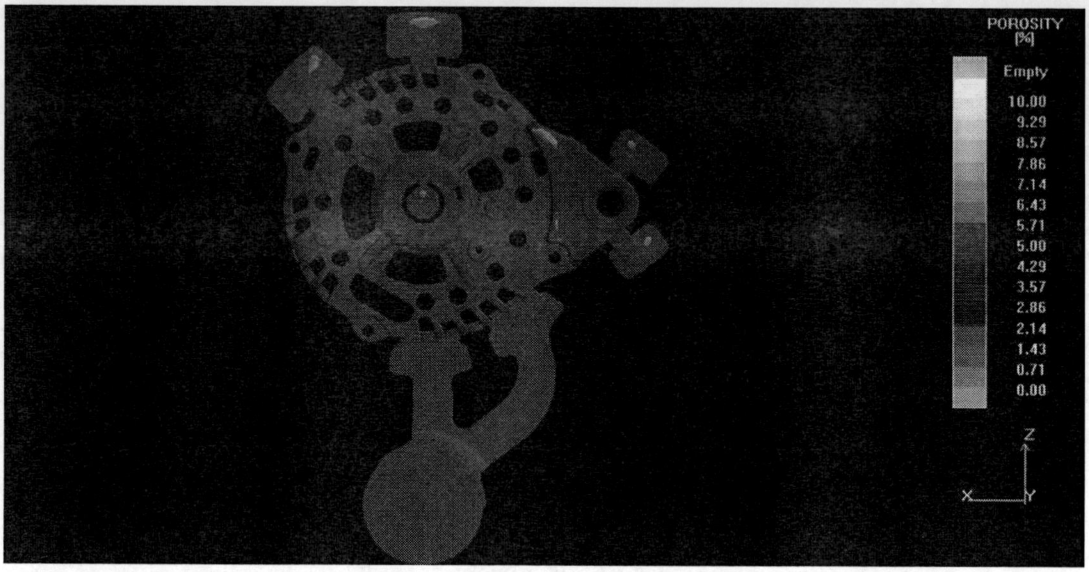

Figure 17. Shrinkage porosity result

6. Defect inspection and testing

Porosity is the most common defect found in die castings that is of vital concern. The different types of porosities that occur in castings include gas porosity, flow porosity and shrinkage porosity. Gas and shrinkage porosity often occur together and are mostly confused with each other. Gas porosity occurs as round smooth wall shaped bubbles. It has different sources like trapped air, trapped steam, vaporized lubricant vent design, etc. which cannot be completely avoided but can be minimized if proper care is taken during design and die operation. Shrinkage porosity defects are voids that appear jagged and are irregularly shaped. Increasing the injection pressure produces less difference in the void size or visibility of porosity but shrinkage porosity is found to respond quite well to metal pressure applied at the right time and place [7]. Casting defects cannot be completely avoided but they can be minimized. The defective castings are re-melted and used again. Defects less than 1% of the production rate are acceptable in high pressure die casting.

The cavity pressure should be within the range needed to minimize shrinkage porosity. For pressure to be effective it has to be present when metal solidifies as metal cannot be pressurized before the end of stroke and it can have large porosities if it solidifies before the end of the stroke. Two metal pressures are important, one without intensification called as static pressure which is applied at the end of the stroke before the intensifier comes in and the second is intensified pressure which is about 2.5 to 3 times the static pressure. The magnitude of pressure applied during operation was obtained by trial and error and is in the range of 270 to275 kg/cm^{2}so as to keep shrinkage porosity at a minimum.

IOP Conf. Series: Materials Science and Engineering **810** (2020) 012017 doi:10.1088/1757-899X/810/1/012017

The castings manufactured in operation were inspected for porosity. The numbers of rejections due to porosity from the defect inspection sheet were found to be less than 1% of the production. The defects obtained were within acceptable limits. Table 6 gives the number of defects obtained in operation for five different heat loads.

Table 6. Defect inspection sheet

Heat load	Sr. no.	Production Qty. (per day)	No. of Defects	% Defects
1	1	700	5	0.71
	2	700	6	0.85
	3	750	3	0.40
2	4	700	3	0.42
	5	700	4	0.57
	6	700	6	0.85
3	7	690	4	0.57
	8	700	4	0.57
	9	700	5	0.71
4	10	700	6	0.85
	11	670	6	0.89
	12	700	5	0.71
5	13	700	4	0.57
	14	700	5	0.71
	15	700	3	0.42

Defective casting samples were taken from different heat loads that were examined for the type and level of porosity. Five samples were taken, each from one of the five heat loads and were tested for porosity using radiography. The observations obtained from radiography are as given below.

Table 7. Radiography Observations

Sr. No	Film Size (in.)	Observations	Remarks
1	6 × 7.5	Shrinkage Cat. C	Level IV
2	6 × 7.5	Porosity Cat. A	Level II
3	6 × 7.5	Porosity Cat. A	Level II
4	6 × 7.5	Porosity Cat. A	Level III
5	6 × 7.5	Porosity Cat. A	Level II

IOP Conf. Series: Materials Science and Engineering **810** (2020) 012017 doi:10.1088/1757-899X/810/1/012017

Radiographic examination serves as a guide that enables recognition of discontinuities and their differentiation both as to type and severity level through radiographic examination. The results of radiography are compared with ASTM E505 standards. The reference radiographs as per ASTM E505 standard used for die castings are as shown in the figure below.

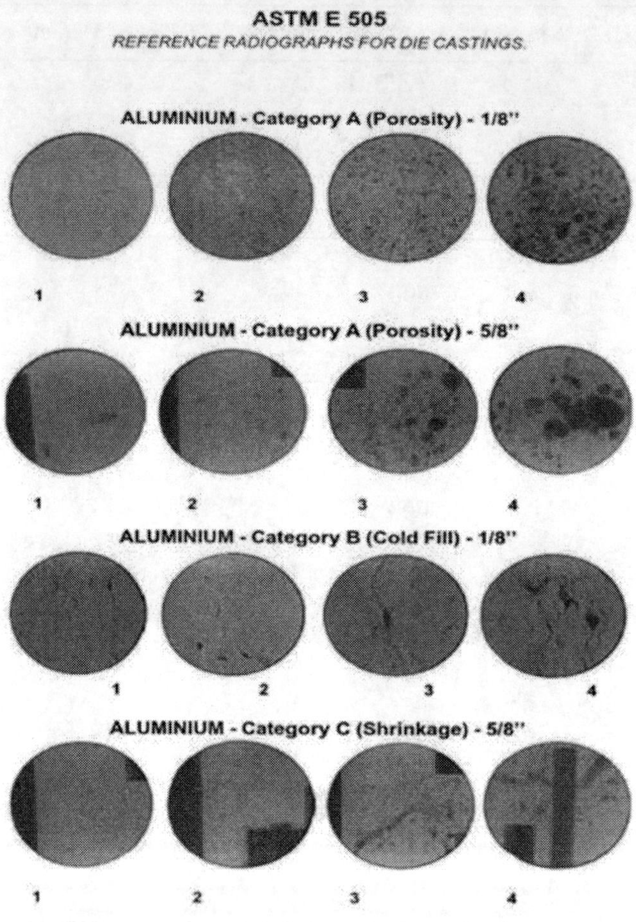

Figure 18. Reference Radiographs for Die Castings

As per ASTM E505 standards, the following descriptions are used in discontinuity identification and classification. These descriptions apply to these reference radiographs only

Category A (Porosity) – Round or elongated, smooth-edged dark spots occurring individually distributed or in clusters.

Category B (Cold Fill) – A distinct, darkened line or band of variable length and definite smooth outline, usually continuous or interconnected

Category C (Shrinkage) – Filamentary or jagged darkened areas, usually continuous or inter connected

Category D (Foreign Material) – Isolated irregular variations in film density that are either lighter or darker than surrounding areas. They may indicate the inclusion of oxide or dross or metallic compounds of different density.

Table 8. Reference Radiographs for Aluminium Die Castings

Discontinuity	Plate Thickness in. (mm)	Applicable casting Thickness in. (mm)
Category A (Porosity)	1/8 (3.2)	Up to 3/8 (9.5), incl
Category A (Porosity)	5/8 (15.9)	over 3/8 to 1 (9.5 to 25.4) incl
Category B (Cold Fill)	1/8 (3.2)	up to 3/8 (9.5), incl
Category C (Shrinkage)	5/8 (15.9)	over 3/8 to 1 (9.5 to 25.4) incl
Category D (Foreign materials)	0.200 (5.08)	up to 1 (25.4) incl

7. Results and discussion

Designers often produce a good quality part by following the best practices for designing the runner gate system, vents, overflows, etc., of the casting. The die designing is a very broad area to be studied and the process parameters have been found to have a significant influence on the formation of defects and ultimately the quality of the cast product. The positioning of the cooling channel has significant effect on the generation of stresses in the die. The existing model had cooling channels placed at a distance of 16 mm which was twice the diameter as per past experience. This model was analysed using ANSYS and resultant stresses were found to exceed permissible stress limits. A new model was proposed by optimizing the distance of cooling channels from the cavity surface so as to have permissible values of stresses. The optimized distance of cooling channels obtained was 20 mm from the cavity surface.

The proposed model with optimized cooling channel position was simulated to study its effect on the fill characteristics using MAGMA Soft software and the results obtained with the proposed design are acceptable from stress as well as filling point of view. Of the various defects that occur in castings, porosity is the most common. The defects produced in operation were inspected for porosity so as to check if they are acceptable in the optimized model. Five defective samples from five heat loads were tested using radiography to find porosity and the defect inspection sheet showed results in the acceptable range (< 1 %)

8. Conclusion

Die erosion due stress concentrations during operation is a typical die failure mode and the positioning of the cooling channels is found to significantly impact the stresses generated on the die. The project work intended to design a die and optimize the cooling channel position to ensure that the stresses developed in the die are produced within permissible limits.

The investigation made was of line cooling used in the cold chamber machine die. A coupled structural and thermal analysis of the die was carried out of the existing design using ANSYS Workbench 14.0. As the existing design resulted in stresses beyond the permissible values, a new safe design was proposed with optimized position of cooling channels. The proposed design was simulated using MAGMA Soft to understand its effect on the filling and solidification of casting and the fill results obtained were satisfactory.

A major challenge while designing a die is determining whether or not the final part has defects. A major portion of defects that occur in die casting have been found to be related with temperature fluctuations and stresses that are produced during the operation. The final cast products produced with the proposed design of the die were tested to observe the defects. The defects obtained were within the acceptable range.

Thus, this type of analysis could prove to be of value while designing the die and positioning of the cooling system. It not only helps in the prediction of stresses but also allows one to make the required modifications in the die accordingly. This type of analysis in the early design stage serves to improve the life of the die as well as the quality of the cast products.

9. References

[1] B. Vijaya Ramnath, C. Elanchezhian, Vishal Chandrashekhar, A. Arun Kumar, et.al."Analysis and optimization of gating system for commutator end bracket," Procedia Materials Science, vol 6,pp. 1312-1328, 2014.
[2] Alastair Long, David Thornhill, Cecil Armstrong and David Watson, "Predicting die life from die temperature for high pressure die castings. applied thermal engineering," vol. 44, pp.100-107, 2012.
[3] Damjan Klobcar and Janez Tusek, "Thermal stresses in aluminium alloy die castings. computational materials science,"vol. 43,pp. 1147-1154, 2008.
[4] B. H. Hu, K. K. Tong, X. P. Niu and I. Pinwill,"Design and optimization of runner and gating systems for the die casting of thin walled magnesium telecommunication parts through numerical simulation," Journal of Materials Processing Technology, vol. 105,pp. 128-133, 2000.
[5] Kulkarni Sanjay Kumar, J. K. Sawale and Sampath Rao,"Study of effect of process parameter setting on porosity levels of aluminium pressure die casting process using taguchi methodology," IOSR Journal of Mechanical and Civil Engineering,"vol. 9(4), pp. 12-17, 2013.
[6] D. Klobcar, L. Kosek, B. Kosek, J, Tusek,"Thermo fatigue cracking of die casting dies," Engineering Failure Analysis, vol. 20, pp. 43-53, 2012.
[7] William Walkington,"Gas Porosity: a guide to correcting the problems," North American Die Casting Association,publication 516, pp. 3-22, 2018.

IOP Conf. Series: Materials Science and Engineering **810** (2020) 012018 doi:10.1088/1757-899X/810/1/012018

Effect of Aluminium oxide Nano filler in Tetrafluoroethane (R-134a) Refrigerant

Dhairysheel Ambhore, Akash Tiwari, Ujval Patel, Jishnu Patil, M. Ramachandran
MPSTME, SVKM'S NMIMS University, Maharashtra, India
jishnupatil724@gmail.com

Abstract. In today's world, refrigeration plays an important role to satisfy human needs. People are doing several researches to enhance the performance of Refrigeration systems. Aiming to improve the characteristics of the old refrigerants by introducing nano particles in it, nano refrigerant is a process of combination of nano particle with the refrigerant for the sake of better refrigeration process. To study the thermal conductivity and viscosity at different volume concentration, this paper presents a review on mechanism for improvement in VCRS performance by using Nano refrigerants. We are using eco-friendly refrigerant that is R134a and nano particle Al2O3, comparing coefficient of performance (COP) of pure refrigerant and Nano refrigerant. In general, with the increment of nanoparticle concentration the viscosity is increased and instantaneous enhancement of thermo physical properties such as thermal conductivity is seen. The viscosity of Nano refrigerant has been influenced by the viscosity of the conventional refrigerant.
Keywords: Nano-refrigerants, Refrigeration cycle; Thermal conductivity, Coefficient of Performance

1. Introduction

Now days we are facing a huge problem of global warming (i.e. average temperature of earth is increasing). Government policies on energy saving have been implemented in many countries worldwide with the purpose to reduce energy consumption, consequently reducing the cost and also greenhouse gas emissions. Usage of eco-friendly refrigerant could be an effective solution to the global warming problem. Refrigeration is a human need (preservation) but it results in global warming. For refrigeration, refrigerator uses different types and composition of refrigerant. While researching we found that nano particles in refrigerant affects the efficiency and performance of refrigeration system. These results in an increase in the improvement of thermo-physical and heat transfer characteristic of refrigerant. The vapour compression cycle is a type of refrigeration system in which the vapor-compression uses a circulating liquid refrigerant as the medium which absorbs and removes heat from the space to be cooled and subsequently rejects that heat elsewhere. As we have seen above the cycle uses refrigerant. In these, we also found that nano particle in refrigerant is the good alternative and also provides a desired output. The contribution of these refrigerants represents between **1-2%** of greenhouse gas emissions. It has been reduced considerably since 1987. Tetrafluoroethane (also known as norflurane (INN), R-134a, Freon 134a, Forane 134a, Genetron 134a, Florasol 134a, Suva 134a, or HFC-134a) is a haloalkane refrigerant with thermodynamic properties similar to R 12 (dichlorodifluoromethane) but with insignificant ozone depletion potential and a somewhat lower global warming potential (1,430, compared to R-12's GWP of 10,900). It has the formula CH_2FCF_3 and a boiling point of −26.3 °C (−15.34 °F) at atmospheric pressure. R-134a cylinders are colored light blue. Attempts at phasing out its use as a refrigerant with substances that have lower global warming potential, such as HFO-1234yf, are underway. Tetrafluoroethane is a non-flammable gas used primarily as a "high- temperature" refrigerant for domestic refrigeration and automobile air conditioners. These devices began using tetrafluoroethane in the early 1990s as a replacement for the more environmentally harmful R-12 and retrofit kits are available to convert units that were originally R-12-equipped. Other uses include plastic foam blowing, as a cleaning solvent, a propellant for the delivery of pharmaceuticals (e.g. bronchodilators), wine cork removers, gas dusters, such as Dust-Off, and in air driers for removing the moisture from compressed air. Tetrafluoroethane has also been used to cool computers in some overclocking attempts. It is the refrigerant used in plumbing pipe freeze kits. The gas is often mixed with a silicone-based lubricant. It can also be used as a solvent in organic chemistry, both in liquid and supercritical fluid. Chlorodifluoromethane or difluoromonochloromethane is a colorless gas is also known as R-22. It is commonly used as a propellant and refrigerant. R-22 is often used as an alternative to the highly ozone-depleting CFC-11 and CFC-12, because of its relatively low ozone depletion potential of 0.055, among the lowest for chlorine containing halo alkanes. As an additional environmental concern, R-22 is a powerful greenhouse gas with a global warming potential equal to 1810.

2. Literature Review

Mahbubul et al.(2011) in this paper prepared Al_2O_3/R141b. Nano refrigerants by shaking mixture of the nanoparticle of 13 nm size And 99.5% purity and refrigerant continuously for 24 h at 240 rpm in An orbital incubator shaker at a constant temperature of 15 °C. Thermal conductivity increases with the increase of volume concentrations and temperatures [1]. Jiang (2013) measured the thermal conductivity of various Nano refrigerants and showed that the improvement is highly influence by nanoparticle aggregation and forming of nanoparticle clusters. They have also developed a model for Predict Nano refrigerant thermal conductivity Thermal conductivity increases with increase in its nanoparticle volume fraction [2]. Mahbubul (2011) studied the effects of various parameters on thermal conductivity of R113, R141b and R134a based Nano refrigerants dispersed with Cu, Al, Ni, CuO, Al_2O_3 and carbon nanotube. Thermal conductivity increases with increase in temperature [3]. Alawi (2012) with Nano refrigerant Al_2o_3/R141a summarized that Nano refrigerant thermal conductivity increases with increase in particle volume Fraction, (ii) Nano refrigerant thermal conductivity increases with Decrease in particle size (or increase in aspect ratio for carbon nanotube), (iii) Nano refrigerant thermal conductivity increases with increase in temperature. [4]. Mahbubul (2013) (viscosity of nanorefrigerants) with Nano refrigerant Al_2o_3/R134a of size less than 200nm states that Viscosity increases with the increase in particle volume fraction [5]. Alawi (2014) investigated the effect of volume concentration and Temperature on thermos physical properties of Al_2O_3/R134a Nano refrigerant and observed that the viscosity of Nano refrigerant increases with increase in volume concentration and a decrease in temperature [6]. BI (2011) with a Nano refrigerant of size 20nm showed that the refrigerator's performance could improve by 26.1% with 0.1% mass fraction of TiO_2 nanoparticles dispersed in R134a[7]. Subramani & Prakash (2014) reported the enhancement of freezing capacity, reduction of compressor power by 25% and enhancement of COP by 33% using Nano refrigerant.[8]. Kumar (2014) experimentally investigated the effect of Al_2O_3/R134a Nano refrigerant with PAG oil on energy consumption of refrigeration system and observed 10.3% reduction of energy consumption by using 0.2% concentration of nanoparticle [9]. Mahbubul (2013) observed 15% improvement of COP by Using Al_2O_3- R134a Nano refrigerant. They observed the effect of different properties on the COP while using Nano refrigerant. COP canenhance by 15%, 3.2% and 2.6% due to thermal conductivity, density and specific heat respectively [10]. Soliman (2015) reported 13.5% less energy consumption and 10.5% more performance by using Al_2O_3-R143a [11]. Kedziereski (2014) found out that the Nano refrigerants were significantly have a better pool boiling heat transfer. In addition, the absolute nanoparticle surface density that resides on the heat transfer surface was the main factor to the boiling heat transfer enhancement [12]. Sharif (2016) performed an experimental study to examine the relationship between the thermal conductivity and viscosity with the volume concentration of the Al_2O_3/PAG Nano lubricants specifically used in the Automotive Air Conditioning (AAC) system. The thermal conductivity was shown to be enhanced up to 4% at 1% of volume concentration. However, the viscosity increases 7.58 times higher than the PAG lubricant at 0.4% volume concentration [13]. Zawawi (2012) a year later he extended the work using SiO_2 nanoparticles dispersed into the PAG lubricant. The thermal conductivity and viscosity of the Nano lubricants reported increased with volume concentration but decreased with temperature. However, because of the Nano lubricants were specifically designed for the AAC system the author has suggested the specific concentration of SiO_2/PAG and Al_2O_3/PAG to be used. These are because the excessive viscosity of lubricant would affect the performance and power consumption of the compressor in the refrigeration system [14].

Cremaschi (2016) found nanoparticles addition in lubricant do not impede with POE characteristics of oil solubility [15]. Manoj (2014) analyzed Al_2O_3 and TiO_2 Nano lubricants and observed COP enhancement in both Nano lubricants, where Nano lubricants TiO_2/MO showed higher than Al_2O_3/MO Nano lubricants with significant energy consumption. [16]. Yusof (2015) have analyzed the energy reduction and performance enhancement by using Nano lubricants in a refrigerator system. With 0.2% volume concentration of Al_2O_3/POE Nano lubricants, the result shows 7% COP optimization with 2.1% energy consumption reduction which is highest. [17]. Kedzierski (2016) analyzed the copper oxide /POE Nano lubricants effect on pool boiling heat transfer with the refrigerant R134a. The finding shows significant enhancement in the heat transfer (up to 275%) at only 0.5% mass fraction of copper oxide /POE Nano lubricants. Additionally, they also exposed that a small increase in the thermal conductivity contributes to substantial enhancement in the heat transfer [18]. Akhavan (2013) have analyzed the effect of copper oxide/R600a/POE Nano refrigerants condensation inside a horizontal smooth tube and proved that addition of copper oxide nanoparticles enhanced the condensation heat transfer compared to lubricant mixture with baseline refrigerant. Peng (2015) stated that For normal lower boiling point refrigerants, the following two methods can be adopted: (i) two step or one step method for preparation of Nano lubricant and then mixing with refrigerant. [20]

Technical specification of test rig

Compressor Power	0.25 KW
Refrigerants	R134a, R22
Compressor	Reciprocating type, hermetically sealed

IOP Conf. Series: Materials Science and Engineering **810** (2020) 012018 doi:10.1088/1757-899X/810/1/012018

Nano Refrigerant	Al2O3
Condenser	Copper tubes, aluminium fins, forced convection air cooled
Expansion device	Capillary tubes
Evaporator	Copper coil tubing through which refrigerant is circulated, it cools water, heater is at bottom to balance the cooling
Condenser Size	9x9x2 Rows
Condenser Fan Motor	1/83 HP X 1350 RPM
Condenser Fan	8' Diameter X 4 Blade
Evaporator Size	5/16 X 30
Capillary Tube	0.050' X 10 ft.

Actual Experimental Setup

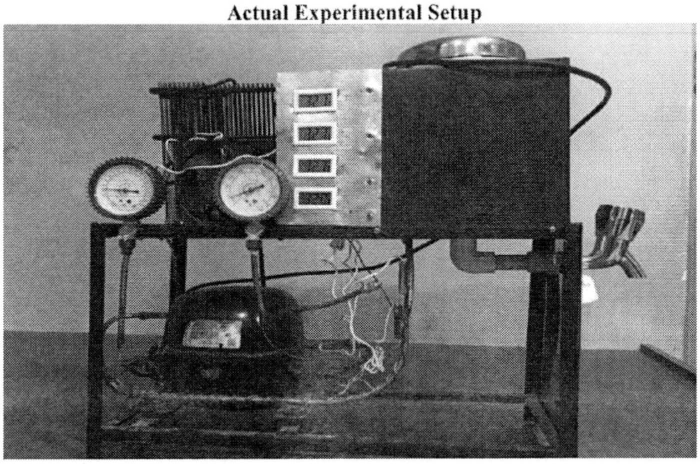

3. Preparation of Nano Refrigerant

As we have seen above, to reduce the global warming problem we have to use the eco-friendly refrigerant. So by reviewing different research paper, we decided to use refrigerant R134a, as it has zero potential to cause the depletion of ozone layer and very little greenhouse effect, it belongs to the family of hydrofluorocarbon (HFC). Its chemical name is tetrafluoroethane. It is safe for normal handling as it is non-toxic, non-flammable and non-corrosive. On other hand when nanoparticles are dispersed in refrigerant (termed as nano- refrigerant), then it directly enhances the refrigerant thermal properties and thereby performance of refrigeration system is found to be improved. The following methods are used to prepare Nano refrigerant by mixing nanoparticles in refrigerant are single step method and two step method. The single-step method is a process of combining the preparation of nanoparticles with the synthesis of Nano fluids, for which the nanoparticles are directly prepared by physical vapour deposition (PVD) technique or liquid chemical method. The two-step method for preparing Nano fluids is a process by dispersing nanoparticles into base liquids. This step-by step method isolates the preparation of the Nano fluids from the preparation of nanoparticles. As a result, collection of nanoparticles may take place in both steps, especially in the process of drying, storage, and transportation of nanoparticles. The agglomeration will not only result in the settlement and clogging of micro channels, but also decrease the thermal conductivity. Simple techniques such as ultrasonic agitation or the addition of surfactants to the fluids are often used to minimize particle aggregation and improve dispersion behaviour. Since Nano powder synthesis techniques have already been scaled up to industrial production levels by several companies, there are potential

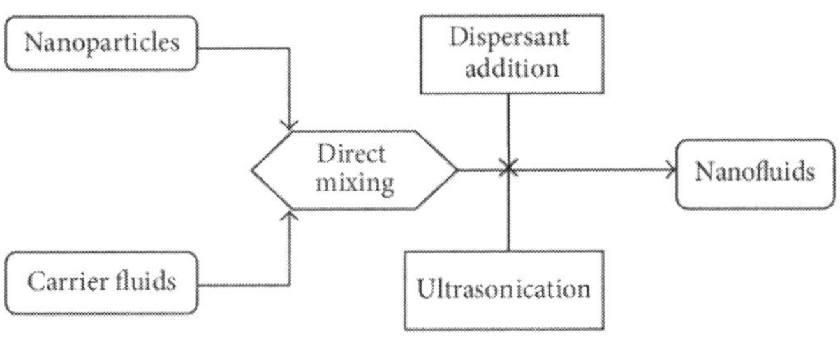

economic advantages in using two-step synthesis methods that rely on the use of such powders. But an important problem that needs to be solved is the stabilization of the suspension prepared. Two-step preparation process is extensively used in the synthesis of Nano fluids by mixing base fluids with commercially available Nano powders obtained from different mechanical, physical and chemical routes such as milling, grinding, and sol-gel and vapor phase methods. An ultrasonic vibrator or higher shear mixing device is generally used to stir Nano powders with host fluids. The most common two-step method is shown in figure,

4. Block diagram of two step method

The Nano refrigerant for this work is prepared in National Chemical Laboratory, Pune. Al_2O_3 Nanoparticles with concentration of 0.5wt%, and 0.1wt% were measured by digital weight balance. Each mass fraction of nanoparticles is mixed with R134a Resultant nano refrigerant was homogenized for 15 minutes followed by sonication for up to 4 hours using ultrasonicator.

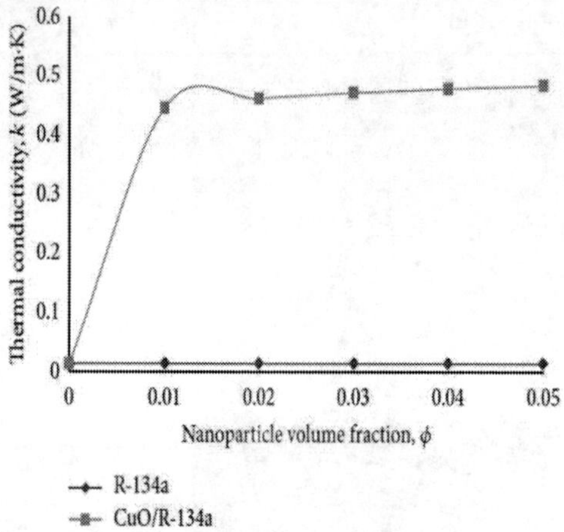

Thermal conductivity as a function of nano Particle volume concentration

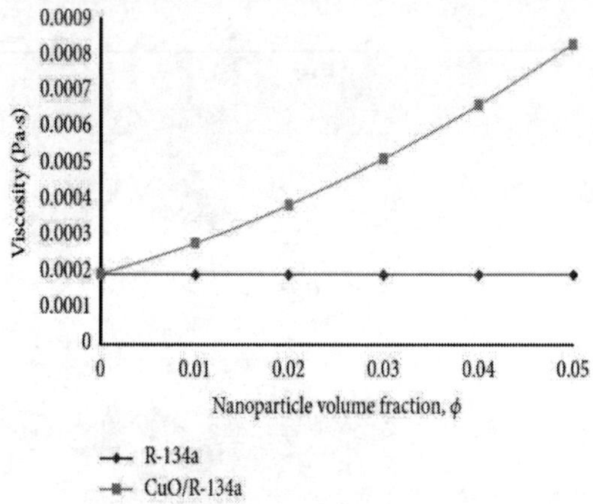

Viscosity as a function of nano particle volume concentration

Thermal conductivity of pure refrigerant i.e. R134a is 0.0139W/m·K and it gives a graph of a nano particle percentage in a refrigerant and its change in thermal conductivity with respect to it. Viscosity of pure refrigerant i.e. R134a is at 26°C is 0.0001934 Pa.s. Above graph shows the change in viscosity with respect to volume fractional of nano particle in refrigerant. It shows the linear increment after the viscosity of pure refrigerant.

Results

1) R134a

P1 (bar)	P2 (bar)	T1 °C	T2 °C	T3 °C	T4 °C
2.1	23.8	2.8	79.5	45.3	1.9

P1= Suction pressure; P2= Discharge pressure; T1= Temperature after condensation; T2= Temperature after compression; T3= Temperature after evaporation; T4= Temperature after expansion; T5= Chamber temperature; h1= Enthalpy at evaporator outlet temperature; h2= Enthalpy at condenser inlet temperature; hf3=h4= Enthalpy at condenser outlet temperature. From refrigeration table h1= 394 KJ/kg; h2= 429 KJ/kg; hf3=h4= 189 KJ/kg

$$\text{COP (theoretical)} = \frac{h1 - h4}{h2 - h1} = 5.857$$

$$\text{COP (actual)} = \frac{R.E}{W.D}$$

R.E = m x Cp x (Tf – Ti) = 2.5 x 4.187 x (26.7 – 9.4) = 210.396 kJ
Power = V x I x cos ϕ = 230 x 0.5 x 0.85 = 115 W = 0.115 kW

Therefore, WD = Power x time = 0.115 x 15 x 60 = 103.5 kJ
COP (actual) = 2.03 2) R22
From refrigeration table- h1 = 392.846 KJ/kg
h2 = 417.98 KJ/kg h4 = 162.742 KJ/kg

$$COP\ (theoretical) = \frac{h1 - h4}{h2 - h1} = 9.17$$

P1 (bar)	P2 (bar)	T1 °C	T2 °C	T3 °C	T4 °C	T5 °C
2.1	290	9.3	77.3	48.2	-13.8	8.6

$$COP\ (actual) = \frac{R.E}{W.D}$$

R.E = m x Cp x (Tf – Ti) = 2.5 x 4.187 x (26 – 8.6) = 182.13 KJ
Power = V x I x cos ϕ = 230 x 0.6 x 0.85 = 117.3 W = 0.1173 KW
W.D = Power x time = 0.1173 x 15 x 60 = 105.57 KJ
COP (actual) = 1.72

5. Conclusion

The thermal conductivities of nano refrigerants are higher than traditional refrigerants. It was also observed that increased thermal conductivity of nano refrigerants is comparable with the increased thermal conductivities of other nano refrigerants. Nanoparticle factor that can be effect to thermos-physical properties is nanoparticle size, type of nanoparticle, shape of nanoparticle size and etc. Refrigerant type need to be environmental friendly to minimized the Global Warming Potential (GWP) factor. The use of nano-refrigerants enhances the heat transfer characteristic of nano-refrigerant relative to pure refrigerant. The addition of 0.5% of Al2O3 nano particles in the base refrigerant will lead to improvement in the overall performance of the VCRs than that of pure base refrigerant. However increase in the percentage of nano particle in the base refrigerant will result in decreased system performance

References

[1]. Mahbubul IM, Saidur R, Amalina MA. Influence of particle concentration and temperature on thermal conductivity and viscosity of Al2O3/R141b nanorefrigerant. Int Commun Heat Mass Transf 2013
[2]. Vishal Fegade, Gajanan Jadhav, M. Ramachandran, Design, Modelling and Analysis of Tilted Human Powered Vehicle, IOP Conf. Series: Materials Science and Engineering 377 (2018) 012215
[3]. Alawi OA, Sidik NAC. Mathematical correlations on factors affecting the thermal conductivity and dynamic viscosity of nanorefrigerants. IntCommun Heat Mass Transf 2014;
[4]. Mahbubul IM, Fadhilah SA, Saidur R, Leong KY, Amalina SA. Thermophysical properties and heat transfer performance of Al2O3/R134ananorefrigerants. Int Heat Mass Transf 2013
[5]. Alawi OA, Sidik NAC. The effect of temperature and particles concentration on the determination of thermo and physical properties of SWCNT-nanorefrigerant. IntCommun Heat Mass Transf 2015
[6]. Bi SS, Shi L, Zhang LL. Application of nanoparticles in domestic refrigerators. Appl Therm Eng 2008
[7]. Subramani N, Prakash MJ. Experimental studies on a vapour compression system using nanorefrigerants. Int J Eng, Sci Technol 2011
[8]. Krishna Kumar Gupta, M. Ramachandran, Effect of Ethylene glycol as Phase Change Material in a Cold Storage Unit on retention of cooling, IOP Conf. Series: Materials Science and Engineering 377 (2018) 012190
[9]. Kumar DS, Elansezhian RD. Experimental study on Al2O3-R134a nanorefrigerant in refrigeration system. Int J Mod Eng Res 2012
[10]. Mahbubul IM, Saadah A, Saidur R, Khairul MA, Kamyar A. Thermal performance analysis of Al2O3/R-134a nanorefrigerant. Int J Heat Mass Transf 2015
[11]. Soliman A.M.A., Abdelrahman A.K., Taher S.H., Ookawara S. Performance enhancement of vapour compression cycle using Nano materials.In: Proceedings of the 4th international conference on renewable energy research and applications palermo, Italy; 2015
[12]. M.A. Kedzierski, R134a/Al2O3 nanolubricant mixture pool boiling ona rectangular finned surface, J. Heat Transf. 134 (12) (2012)
[13]. M.Z. Sharif, W.H. Azmi, A.A.M. Redhwan, R. Mamat, Investigation of thermal conductivity and viscosity of Al2O3/PAG nanolubricant forapplication in automotive air conditioning system, Int. J. Refrig. 70 (2016)
[14]. N.M.M. Zawawi, W.H. Azmi, A.A.M. Redhwan, M.Z. Sharif, K.V. Sharma, Thermophysical properties of Al2O3-SiO2/PAG composite nanolubricant for refrigeration system, Int. J. Refrig. 80 (2017)
[15]. M. Ramachandran, Benildus, Failure Analysis of Turbine Blade Using Computational Fluid Dynamics, International Journal of Applied Engineering Research. ISSN 0973-4562 Volume 10, Number 11 (2015) pp. 10230-10233

[16]. L. Cremaschi, T. Wong, A.A.M. Bigi, Thermodynamic and Heat Transfer Properties of Al2O3 Nanolubricants, International Refrigeration and Air Conditioning Conference, Purdue E-Pubs, Purdue, 2014

[17]. Manoj Babu, S. Nallusamy, K. Rajan, Experimental analysis on vapour compression refrigeration system using nanolubricant with HFC-134a refrigerant, Nano Hybrids, 9 2016

[18]. Akshay Ahuja, M. Ramachandran, Review on Phase Change Material as thermal energy storage for cooling, International Journal of Chem Tech Research 9(4):192-196 • May 2016

[19]. T.M. Yusof, A.M. Arshad, M.D. Suziyana, L.G. Chui, M.F. Basrawi, Experimental study of a domestic refrigerator with POE-Al2O3 nanolubricants, Int. J. Aut. Mech.Eng. 11 (2015)

[20]. M.A. Kedzierski, M. Gong, Effect of CuO Nano lubricant on R134a Pool Boiling Heat Transfer with Extensive Measurement and Analysis Details, US Department of Commerce, National Institute of Standards and Technology,USA, 2007

[21]. M.A. Akhavan-Behabadi, M.K. Sadoughi, M. Darzi, M. Fakoor-Pakdaman, Experimental study on heat transfer characteristics of R600a/POE/CuO Nano- refrigerant flow condensation, Exp. Thermal Fluid Sci. 66 2015

[22]. Peng H, Ding G, Hu H. Effect of surfactant additives on nucleate pool boiling heat transfer of refrigerant-based nanofluid,2013

[23]. Aakash Kumar Bihari, M. Ramachandran ,Kanak Kalita, Disaster management on Ballast System in marine industry, International Journal of Oceans and Oceanography, 8(2) (2014), pp. 127-135

IOP Conf. Series: Materials Science and Engineering **810** (2020) 012019 doi:10.1088/1757-899X/810/1/012019

A comparative study of EMI shielding effectiveness of metals, metal coatings and carbon-based materials

Jeevan Kittur, Bhavya Desai, Rakesh Chaudhari, Praveen Kumar Loharkar
Department of Mechanical Engineering
SVKM's NMIMS MPSTME Shirpur Campus
Dhule, India

jeevan.kittur@nmims.edu

Abstract. The increased use of electronic gadgets across almost all fields in the fast progressing 21st century has resulted in a new type of pollution. This pollution can be fatal for animals, humans as well as electronic gadgets. This new form of pollution is nothing but the electromagnetic interference or radio-frequency interference. When these harmful electromagnetic waves come in contact with the electronic devices, it causes them to malfunction thus resulting into damage of the costly equipment. Hence comes the need to protect the electronic gadgets and systems of commercial as well as defence fields from these harmful waves by providing a shielding effect. The review is a study of materials used for shielding like coated nano-fibers, nano-composites, metals, conducting polymers and carbon nanotubes (CNTs).

1. Introduction

With increase in electronic devices there is a rise in electromagnetic interference. Electrostatic discharge is one of the biggest reasons for EMI[1][2][3]. The signal radiations caused due to EMI can damage the electronic systems upto a great extent[4][5][6]. The distortions in audio or video signals on radio/ TV when an aircraft is flying at a low altitude, or the noise on microphones are the effects of these EMI. This on large exposure can cause the devices to malfunction and also affect human health. The only solution to prevent damage through these harmful radiations and protect electronic devices is to provide a shield which filters interferences[7][8]. There is a high demand of Strong and light weight Shielding Materials nowadays [9][10][11]. The materials used to manufacture these shields are being studied closely in this review.

1.1 Shielding of EMI

An electromagnetic wave has two major components- magnetic field and electric field, both perpendicular to each other. Forces are produced on the charge carriers under the influence of electric field. When the ideal surface of conductor comes in contact with electric field, the current will be induced which results in causing charge displacement in conductor that further cancels the applied electric field thereby stopping the current. In the similar way the applied magnetic field gets cancelled due to the presence of generated varying magnetic field. The conductor only responds to the static magnetic fields when it is undergoing relative motion with respect to the magnetic field [12]. The outcome is the reflection of EM radiation from conductor's surface, internal fields stay inside and external fields stay outside. Absorption and reflection occurs when an EMI wave is passed through a

shield [13][14]. The energy which is left after the phenomenon of absorption and reflection is termed as residual energy which emerges out of the shield. The ratio of this residual energy to impinging energy can be termed as Shielding Effectiveness (SE) [3].
Conductive or magnetic materials are used as barriers to reduce the electromagnetic field in space by blocking the fields this process is called electromagnetic shielding. In order to isolate the electronic device from outside world the shielding is applied to enclosures. Similarly, it is also done to keep wires isolate from environment through which the cable runs. [2][15].
Radio frequency shielding is typically the process of blocking radio frequency and electromagnetic radiation.[16] Shielding mainly reduces the coupling of electromagnetic fields, electro static fields and waves. This phenomenon is dependent on the material used, thickness of the shield, volume, and frequency of fields of interest, shape and orientation of apertures in a field to an incident EM field. [17][18][19].

$$SE = 20 \log (Et/Ei) \quad (i)$$

$$SE = 20 \log (Ht/Hi) \quad (ii)$$

where Et is electric field of transmitted wave
Ei is electric field of incident wave
Ht is magnetic field of transmitted wave
Hi is magnetic field of incident wave.

The units of Electric field and magnetic field are volts/m and amps/m respectively. Three mechanisms are responsible for attenuation of an EM wave, namely absorption (A), reflection (R) and multiple reflections (M) Fig. 1.
Thus, SE= A+R+M

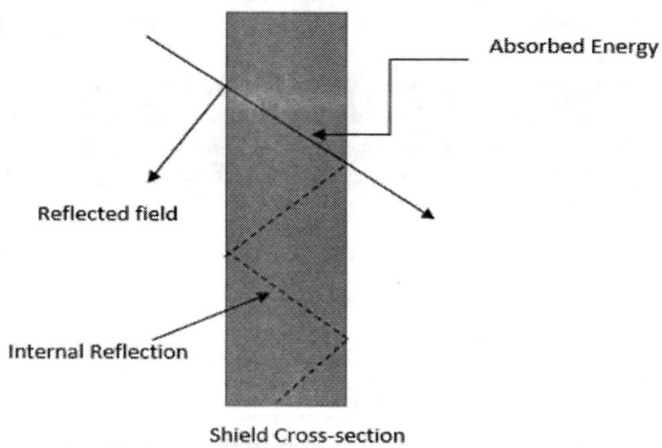

Figure 1. Attenuation of EM wave [20]

The shielding effectiveness is quantified using fours of the following methods. It's detail description can be obtained from [20].

- Open Field or Free Space Method
- Shielded Room Method
- Shielded Box Method
- Coaxial Transmission Line Method

In the subsequent sections, studies on materials used for EMI shielding are presented with a focus on their comparative effectiveness.

2. Metals used for shielding purposes

Sheet Metal, metal foam and metal screen are typically used as shielding materials. The size of irregularity or holes in the material or mesh should be lesser than wavelength of the radiation which is to be kept out. Thus these irregularities in the shield should be considered minor than the radiation wavelength [21]. Coating with metal ink or similar material inside the enclosure is another variant of shielding method. The ink has very small particulates of carrier material that is loaded with suitable metal such as copper or nickel. particles. It is then deposited on the enclosure, which produces a layer of continuous conductive material which further can be electrically connected to a chassis ground of the equipment thus providing effective shielding [22]. Radio Frequency shielding enclosures filter a range of frequency for specific conditions. Copper is usually used as RF shielding material because the absorption capacity of radio and magnetic waves of copper is considerably high if design is made properly and constructed well [23]. From the field of computers and electrical switches to the city scans and MRI scans used in hospitals the copper is used as RF Shielding material which is found satisfactorily effective [24][25].

Dou et al. [26] reported the production of shielding material by using aluminium composites i.e. aluminium 2024 along with CFA, and Aluminium 2024 along with PFA provided excellent shielding effectiveness till frequency of 100.0 MHz but declined at 600 MHz and further degraded exponentially at the frequency of 1.0 GHz in comparison with 2024 Al. Thus, it was inferred that the 2024 aluminium's shielding effectiveness was much better than CFA/2024 Al at 1.0 GHz.
EMI shielding effectiveness and electrical properties of GCu nanoparticles reinforced in PVC nano composites developed by Alghamdi et al.[5] were examined. The composites having a higher volume of GCu did show an excellent Shielding Effectiveness of 70 db in at a frequency of 20 GHz.
Yao et al. [19] concluded that electromagnetic shielding effectiveness of materials based on cement is affected significantly by adding nickel fiber. An increase in the nickel fiber content has a direct effect on increase in SE of the Nanocomposites with cement as base.
The maximum shielding effectiveness reported was 24.48 dB at 120 MHz for a composite with doped nickel fiber.
To significantly improve the shielding effectiveness of the composite, the cement matrix reinforced with nickel fiber must be treated with silica fume. Here the maximum SE achieved was 35.50 db at frequency of 30 MHz.

3. Metals used as coatings for shielding

Materials with metals as base component are used on large scale for the protection of devices from EMI shielding [27]. But the bulkiness, and their nature to corrode act as one of the main reasons to limit their use [28]. Hence to overcome the limitations, metals are used in plated form on fibres and other carbonaceous materials.
Jiang and Guo [29] investigated the electro-less plating of nickel, phosphorous and nickel, copper multilayers on polyester fabrics gave good results of EMI shielding along with corrosion free results unlike other metal conductors. The Cu-Ni coat gave better EMI shielding effectiveness than Ni-P coating. Kim and Kyong [30] reported the Emi shielding behaviours of MWCNTs filled epoxy matrix nanocomposites coated with Ni. The higher the percentage of nickel in the composite, the better the EMI shielding effectiveness. He reported the shielding effectiveness of the 17.65 % nickel composite had the best results.
Kim [31] further reported the SE results of carbon fibre reinforced composite coated with Co alloy and Ni. Since Co also has magnetic properties like Ni, he experimented with addition of Co in his next research. Hence coating with addition of cobalt can cause a massive change in conduction, i.e., absorption of EMI will be strong. Ni- Co electroplating method resulted in excellent electromagnetic interference shielding efficiency of the CF- reinforced composites. The resistivity decreased and thus the conductivity increased. The composite with 69.39% Co and 18.85 % Ni proved to be the best material.

Yuan et al. [9] reported that the low effectiveness materials are less used, copper has high shielding effectiveness value among other available metals. Also, the weight of these metals is a major concern for its application in fields of aviation and aerospace. One way to reduce the weight constraint is by coating the surfaces of light weight porous materials with metallic materials using vacuum deposition. But still these materials can cause problems such as less chemical resistance, high prone to oxidation, highly corrosive, etc.
Hence polymer composites come into picture. Yuan et al prepared Samples of CH- rGO ie. Carbon honeycomb reduced graphene oxides of varying thickness were experimented for achieving SE.

4. Carbon-based shielding material

The metal coated or metal plated are conventional materials for shielding. Stainless steel fibers, copper and nickel coated fibers are mostly used. But there are limitations in these metal based shielding materials like bulkiness[32], low corrosion resistance, and lack of structural flexibility [33][34][35][36]. Hence, it is better to design a material which can overcome a few of these disadvantages. Therefore, carbon based are given preference as a better option for shielding materials [37]. In comparison with the metals or metal based electro-magnetic shielding materials, nano-carbon based shielding materials such as graphene, CNTs (carbon nano-tubes), graphene oxide, etc. are used [20]. These materials have exceptional structural flexibility, Strength and Electrical conductivity [38][39].
Arjmand et al. [21] compared the electrical conductivity of nitrogen doped and undoped CNTs in their experiment. The presence of extra electron in Nitrogen has been a topic for research to change the electrical conductivity of Carbon. According to him the synthesis conditions are important in deciding the electrical percolation behaviour of CNTs. On experimentation, the results proved that undoped CNTs have greater electrical conductivity, improved dispersion, less percolation ratio and thus better EMI shielding. It further states that nitrogen doping hampers crystallinity and shielding behaviour of CNTs. Thus it is better to use undoped CNTs. These are carbon and its various forms are used as conductive fillers. EMI shielding materials possesses shielding as a primary function and reflection as secondary function [36].
Al-Saleh et al. [36] conducted experiments on shielding by making composites of MWCNTs and polypropylene of varying MWCNTs concentration and different thicknesses. According to the theoretical studies and experimental results of experiments conducted by him the shielding effectiveness of the material improved with the increase in MWCNTs in composite along with its thickness. Thus, shielding effectiveness of material increases with increase in the the amount of Carbon Nano tubes present in the composite along with thickness.
Zhou et al. [40] experimented and studied the synergistic effect of GF (Graphene Films) and GCF (Graphene – Carbon nanotube Films). The Electrical conductivity and thus shielding effectiveness performance of graphene films can be increased to a good extent by introducing CNTs into them. Just by addition of CNTs by 20% can improve the shielding from ~50db to ~60db [40]. The SE of samples with CNTs content from 0% to 20% gradually increased with peak SE shown by sample GCF- 20, but the SE started to decrease as soon as the CNT content increased from 20%. Thus GCF- 20 was found to be the best material for shielding with the maximum shielding effectiveness. Also, the mechanical reliability of GCF- 20 was found to be excellent as it had good flexibility and structure integrity during bending.
Verma et al. [41] prepared samples of Polyurethane based composite materials with varying amounts of CNT pillared Graphene nanostructures (GCNTs). The GCNTs were mixed in amounts of 0%, 0.5%, 1%, 2%, 3%, 5%, 10% wt in thermoplastic polyurethane matrix. The hybrid filler was tested against frequency range of 12.4 to 18 GHz. It was noted that the shielding effectiveness increased with the increase in GCNTs. On the 10% wt GCNT composite, the shielding obtained was -47db, thus blocking more than 99.99% of the incident EMI radiations.
Because of skin effect, conductive fillers of smaller diameters are preferred than those of higher diameter [42].

To enhance the EM SE of composite fibers, electrical conductive fillers of various types were added to it. These conductive fillers included continuous metal mesh and discontinuous silver coated micro balloons. Silver was used as its electrical conductivity is better than that of copper.
According to Liang et al.[10] the SE for EM in through absorption is better than the SE due to reflection. The EM of absorption was observed to be sensitive with regards to areal density. Absorption proved to be more dominant mechanism for shielding EMI.

Fletcher et al. [22] reported the nano composites formed of carbon NanoTubes with the polymer of fluorocarbon in the form of foamed elastomers were potential materials for being a Shielding element. Although no significant change in the Shielding Effectiveness was noted but a major decrease in the weight of 25%- 30% was recorded.
In the study conducted by Wu et al. [1] Graphene Carbon Fibers (GCF) and magnetic Graphene (MG) were filled in Epoxy resin based composites. The SE of the nano composite material improved largely on adding MG (Magnetic Graphene). According to the study, if the surface of Magnetic Graphene is modified, the SE will be better. This can be a major take away for future references. Another reason for better SE was the interpenetrating 3 dimensional structure of the composite.

Table 1. The SE and electrical conductivity of EP/MG /GCF composites

Sample	Content%			Shielding Effectiveness(SE_T)/db		
	GCF	MG	NMG	X-Band	Kn-Band	K-Band
GCF/EP	0.5	0	0	28.9-29.7	29.6-30.3	31.9-38.8
GCF/MG/EP	0.5	3	0	30.8-32.8	32.7-34.1	37.0-39.2
GCF/NMG/EP	0.5	0	3	31.3-33.6	33.2-37.7	36.6-41.7
GCF/MG_2/EP	0.5	6	0	31.9-34.3	34.5-38.3	41.7-41.2
GCF/MG_3/Ep	0.5	9	0	34.0-37.1	37.8-38.8	44.6-51.1

Xing et al [43] reported that a composite reinforced with ultra-thin multilayer carbon fiber of thickness less than 1 mm was prepared for Electro Magnetic Interference absorption dominated applications for frequency range of 30–1500 MHz. The Shielding Effectiveness of the UTM-CFRC increased with increase in number of shielding layers.

Table 2. The Shielding Effectiveness of the UTM-CFRC

Sample Designation	Thickness(mm)	Frequency(GHz)	SE_T(db)	A
MWCNT/PANI/PEO	3.0	0.85-4	42.0	0.15
PDMS	1.0	8-12	23.0	0.39
MCMB/MWCNT	0.6	8.2-12.4	56.0	0.16
CNFRSF	3.0	0.03-1.2	25.0	0.27
UTM-CFRC	0.81	0.03-1.5	30.1	0.53

Shielding Effectiveness of nickel-fiber reinforced cement based composite material is directly affected by compounded colloidal graphite. According to the study by Yao et al. [19] Composite with cement as base with doped 20% graphite colloid gave best SE results.The sample which gave best results of shielding effectiveness was a cement matrix with 5% vol Nickel fibre and 20% colloidal graphite. Mean

SE noted was 45.0 dB, the highest SE recorded was 51.19 dB at 1170 MHz, the least SE recorded was 35.18 dB at 1 MHz.

Li et al. [2]saw the need of requirement of materials to provide EMI shielding at higher temperatures. Hence they prepared samples of Graphene like Carbon silicon carbide composites. Better EMI shielding is delivered as graphene like carbon has higher degree of defects. The highest EMI shielding effectiveness recorded at room temperature was 36.8 db and at 600 degrees the SE was 33.8 db with the material sample being SiC100-0 of 7.6 wt % carbon mass.

Materials used by Pan et al.[7] were CNWs (Carbon Nano wires) and CNWs- CNTs(Carbon Nano Wires- Carbon Nano tubes) hybrid. CNWs have the Graphite nanosheets basal planes perpendicular to the wire axis, which is contrary to that of CNTs. 5%wt concentration proved optimal for improving the crystal structure of CNWs- CNTs hybrid. At 5.15 wt% carbon the shielding effectiveness due to absorption reaches 21.3 dB which alone accounts for 99.7% of shielding. Due to unique micro structure, CNWs did show better capability of attenuation and coefficient of absorption. These two composites are believed to give good results in harsh environments

Nasouri et al[32] reported that the method of electrospinning was used to fabricate MWCNTs/ Polyvinyl Analin composite nanofibers which are electrically conductive. The increase in the diameter and electrical shielding of composite nanofibers was directly related with the increase in amount of MWCNTs. Composite Nanofibers' reflection and absorption shielding effectiveness was optimized using Response Surface Methodology Technology.
Sample of MWCNTs with content of 7.7 wt%, 0.3cm thickness and 12Ghz frequency was put under testing. The results showed Shielding effectiveness as 8.8 db in reflection and 31.5 db in absorption.

5. Conclusion

Metal based shielding materials are used conventionally. However, there are quite a few limitations such as large weight, poor corrosion resistance. Recently carbon based conducting polymer materials have been developed as suitable alternative to counter the limitations of the metal based shielding materials. It can be inferred from the review that there are some challenges with respect to the usage of polymeric shielding materials. These are lesser shielding effectiveness and poor mechanical flexibility. In addition, further research on for using nano materials to enhance shielding effectiveness.

6. References

[1] Wu J, Ye Z, Ge H, Chen J, Liu W and Liu Z 2017 Modified carbon fiber/magnetic graphene/epoxy composites with synergistic effect for electromagnetic interference shielding over broad frequency band J. Colloid Interface Sci. 506 217–26

[2] Li Z and Wang Y 2017 Preparation of polymer-derived graphene-like carbon-silicon carbide nanocomposites as electromagnetic interference shielding material for high temperature applications J. Alloys Compd. 709 313–21

[3] Zhang C S, Ni Q Q, Fu S Y and Kurashiki K 2007 Electromagnetic interference shielding effect of nanocomposites with carbon nanotube and shape memory polymer Compos. Sci. Technol. 67 2973–80

[4] Lin S C, Ma C C M, Hsiao S T, Wang Y S, Yang C Y, Liao W H, Li S M, Wang J A, Cheng T Y, Lin C W and Yang R Bin 2016 Electromagnetic interference shielding performance of waterborne polyurethane composites filled with silver nanoparticles deposited on functionalized graphene Appl. Surf. Sci. 385 436–44

[5] Al-Ghamdi A A and El Tantawy F 2010 New electromagnetic wave shielding effectiveness at

microwave frequency of polyvinyl chloride reinforced graphite/copper nanoparticles Compos. Part A Appl. Sci. Manuf. 41 1693–701

[6] Bhadra S, Singha N K and Khastgir D 2009 Dielectric properties and EMI shielding efficiency of polyaniline and ethylene 1-octene based semi-conducting composites Curr. Appl. Phys. 9 396–403

[7] Pan H, Yin X, Xue J, Cheng L and Zhang L 2017 Microstructures and EMI shielding properties of composite ceramics reinforced with carbon nanowires and nanowires-nanotubes hybrid Ceram. Int. 43 12221–31

[8] Kim S W, Yoon Y W, Lee S J, Kim G Y, Kim Y B, Chun Y Y and Lee K S 2007 Electromagnetic shielding properties of soft magnetic powder-polymer composite films for the application to suppress noise in the radio frequency range J. Magn. Magn. Mater. 316 472–4

[9] Yuan Y, Liu L, Yang M, Zhang T, Xu F, Lin Z, Ding Y, Wang C, Li J, Yin W, Peng Q, He X and Li Y 2017 Lightweight, thermally insulating and stiff carbon honeycomb-induced graphene composite foams with a horizontal laminated structure for electromagnetic interference shielding Carbon N. Y. 123 223–32

[10] Liang J, Wang Y, Huang Y, Ma Y, Liu Z, Cai J, Zhang C, Gao H and Chen Y 2009 Electromagnetic interference shielding of graphene/epoxy composites Carbon N. Y. 47 922–5

[11] Yuen S M, Ma C C M, Chuang C Y, Yu K C, Wu S Y, Yang C C and Wei M H 2008 Effect of processing method on the shielding effectiveness of electromagnetic interference of MWCNT/PMMA composites Compos. Sci. Technol. 68 963–8

[12] Mondal S, Ganguly S, Das P, Bhawal P, Das T K, Ravindren R, Ghosh S and Das N C 2017 Effect of thermal-air ageing treatment on mechanical properties and electromagnetic interference shielding effectiveness of low-cost nano-structured carbon filled chlorinated polyethylene Mater. Sci. Eng. B Solid-State Mater. Adv. Technol. 225 140–9

[13] Gupta A and Choudhary V 2011 Electromagnetic interference shielding behavior of poly(trimethylene terephthalate)/multi-walled carbon nanotube composites Compos. Sci. Technol. 71 1563–8

[14] Wong K H, Pickering S J and Rudd C D 2010 Recycled carbon fibre reinforced polymer composite for electromagnetic interference shielding Compos. Part A Appl. Sci. Manuf. 41 693–702

[15] Lu L, Xing D, Teh K S, Liu H, Xie Y, Liu X and Tang Y 2017 Structural effects in a composite nonwoven fabric on EMI shielding Mater. Des. 120 354–62

[16] Tibbetts G G, Lake M L, Strong K L and Rice B P 2007 A review of the fabrication and properties of vapor-grown carbon nanofiber/polymer composites Compos. Sci. Technol. 67 1709–18

[17] Mu Y, Li H, Deng J and Zhou W 2017 Temperature-dependent electromagnetic shielding properties of SiCf/BN/SiC composites fabricated by chemical vapor infiltration process J. Alloys Compd. 724 633–40

[18] Singh A P, Mishra M, Hashim D P, Narayanan T N, Hahm M G, Kumar P, Dwivedi J, Kedawat G, Gupta A, Singh B P, Chandra A, Vajtai R, Dhawan S K, Ajayan P M and Gupta B K 2015 Probing the engineered sandwich network of vertically aligned carbon nanotube-reduced graphene oxide composites for high performance electromagnetic interference shielding applications Carbon N. Y. 85 79–88

[19] Yao W lin, Xiong G, Yang Y, Huang H qing and Zhou Y fen 2017 Effect of silica fume and colloidal graphite additions on the EMI shielding effectiveness of nickel fiber cement based composites Constr. Build. Mater. 150 825–32

[20] Geetha S, Kumar K K S, Rao C R K, Vijayan M and Trivedi D C 2009 EMI shielding: Methods and materials - A review J. Appl. Polym. Sci.

[21] Arjmand M and Sundararaj U 2015 Electromagnetic interference shielding of Nitrogen-doped and Undoped carbon nanotube/polyvinylidene fluoride nanocomposites: A comparative study Compos. Sci. Technol. 118 257–63

[22] Fletcher A, Gupta M C, Dudley K L and Vedeler E 2010 Elastomer foam nanocomposites for electromagnetic dissipation and shielding applications Compos. Sci. Technol. 70 953–8

[23] Das A, Hayvaci H T, Tiwari M K, Bayer I S, Erricolo D and Megaridis C M 2011 Superhydrophobic and conductive carbon nanofiber/PTFE composite coatings for EMI shielding J. Colloid Interface Sci. 353 311–5

[24] Klemperer C J von and Maharaj D 2009 Composite electromagnetic interference shielding materials for aerospace applications Compos. Struct. 91 467–72

[25] Lakshmi K, John H, Mathew K T, Joseph R and George K E 2009 Microwave absorption, reflection and EMI shielding of PU-PANI composite Acta Mater. 57 371–5

[26] Dou Z, Wu G, Huang X, Sun D and Jiang L 2007 Electromagnetic shielding effectiveness of aluminum alloy-fly ash composites Compos. Part A Appl. Sci. Manuf. 38 186–91

[27] Phang S W, Hino T, Abdullah M H and Kuramoto N 2007 Applications of polyaniline doubly doped with p-toluene sulphonic acid and dichloroacetic acid as microwave absorbing and shielding materials Mater. Chem. Phys. 104 327–35

[28] Panwar V, Kang B, Park J O, Park S and Mehra R M 2009 Study of dielectric properties of styrene-acrylonitrile graphite sheets composites in low and high frequency region Eur. Polym. J. 45 1777–84

[29] Jiang S X and Guo R H 2011 Electromagnetic shielding and corrosion resistance of electroless Ni-P/Cu-Ni multilayer plated polyester fabric Surf. Coatings Technol. 205 4274–9

[30] Kim B J, Bae K M, Lee Y S, An K H and Park S J 2014 EMI shielding behaviors of Ni-coated MWCNTs-filled epoxy matrix nanocomposites Surf. Coatings Technol. 242 125–31

[31] Kim J T, Park C W and Kim B J 2017 A study on synergetic EMI shielding behaviors of Ni-Co alloy-coated carbon fibers-reinforced composites Synth. Met. 223 212–7

[32] Nasouri K and Shoushtari A M 2017 Designing, modeling and manufacturing of lightweight carbon nanotubes/polymer composite nanofibers for electromagnetic interference shielding application Compos. Sci. Technol. 145 46–54

[33] Thomassin J M, Jérôme C, Pardoen T, Bailly C, Huynen I and Detrembleur C 2013 Polymer/carbon based composites as electromagnetic interference (EMI) shielding materials Mater. Sci. Eng. R Reports 74 211–32

[34] Varela-Rizo H, Montes de Oca G, Rodriguez-Pastor I, Monti M, Terenzi A and Martin-Gullon I 2012 Analysis of the electrical and rheological behavior of different processed CNF/PMMA nanocomposites Compos. Sci. Technol. 72 218–24

[35] Al-Saleh M H, Gelves G A and Sundararaj U 2011 Copper nanowire/polystyrene

nanocomposites: Lower percolation threshold and higher EMI shielding Compos. Part A Appl. Sci. Manuf. 42 92–7

[36] Al-Saleh M H and Sundararaj U 2009 Electromagnetic interference shielding mechanisms of CNT/polymer composites Carbon N. Y. 47 1738–46

[37] Sudha J D, Sivakala S, Prasanth R, Reena V L and Radhakrishnan Nair P 2009 Development of electromagnetic shielding materials from the conductive blends of polyaniline and polyaniline-clay nanocomposite-EVA: Preparation and properties Compos. Sci. Technol. 69 358–64

[38] Im J S, Kim J G, Lee S H and Lee Y S 2010 Effective electromagnetic interference shielding by electrospun carbon fibers involving Fe2O3/BaTiO3/MWCNT additives Mater. Chem. Phys. 124 434–8

[39] Im J S, Kim J G and Lee Y S 2009 Fluorination effects of carbon black additives for electrical properties and EMI shielding efficiency by improved dispersion and adhesion Carbon N. Y. 47 2640–7

[40] Zhou E, Xi J, Guo Y, Liu Y, Xu Z, Peng L, Gao W, Ying J, Chen Z and Gao C 2018 Synergistic effect of graphene and carbon nanotube for high-performance electromagnetic interference shielding films Carbon N. Y. 133 316–22

[41] Verma M, Chauhan S S, Dhawan S K and Choudhary V 2017 Graphene nanoplatelets/carbon nanotubes/polyurethane composites as efficient shield against electromagnetic polluting radiations Compos. Part B Eng. 120 118–27

[42] Hong X and Chung D D L 2017 Carbon nanofiber mats for electromagnetic interference shielding Carbon N. Y. 111 529–37

[43] Xing D, Lu L, Tang W, Xie Y and Tang Y 2017 An ultra-thin multilayer carbon fiber reinforced composite for absorption-dominated EMI shielding application Mater. Lett. 207 165–8

IOP Conf. Series: Materials Science and Engineering **810** (2020) 012020 doi:10.1088/1757-899X/810/1/012020

An insight of synthesis, stability and thermophysical properties of hybrid nanofluids

Rajesh Verma, Krishna Kumar Gupta

Department of Mechanical Engineering, Mukesh Patel School of Technology Management & Engineering, Shirpur 425405, India

rajesh.verma@nmims.edu, krishnakumar.gupta@nmims.edu

Abstract. Nanofluids are emerging as more suitable heat transfer fluid compare to conventionally used fluids for transferring heat in various engineering applications. Nanofluid is a fusion of very small quantity of nanoparticles and host fluid with considerable stability. The presence of nanoparticles in host fluid significantly enhance its thermal conductivity, which leads to the improvement in thermal performance. Introduction of submicron particles influence the viscus behaviour of the host fluid. In general viscosity enhancement is not desirable as it imposes a penalty in terms of increased pumping power in many engineering applications. Presence of more than one type of nanoparticles in host fluids provide opportunities to enhance thermal conductivity with control over other properties of resultant nanofluid. The nanofluid having more than one type of nanoparticles termed as hybrid nanofluid are in focus of researchers for investigation. A large variety of hybrid nanofluids are prepared and their properties investigated by various researchers. This review paper is an attempt to compile the research work related to preparation techniques, stability analysis and thermophysical properties of hybrid nanofluids. The potential applications of hybrid nanofluids and challenges associated with it are also discussed.

Key words: Hybrid nanofluid, synthetization, stability, thermophysical properties.

1. Introduction

There are industries like electronics, computer, automobile and many more, whose product's performance and reliability are very much depend on the cooling ability of the system used for the purpose. Rapid enhancement in performance and technology advancement have led to the exponential increment of the heat flux load. Water ethylene glycol, oil and many more conventionally used fluids to transfer hat need to improve significantly or replace by thermal fluids capable to transfer heat efficiently to manage ever increasing heating load. With the advancement in nanotechnology it become feasible a nd economic to get nano sized particles of metals or non-metals. In 1995 Choi [1] first time introduced the nanotechnology based thermally superior suspension containing submicron particles. He coined the term Nanofluid for this modified fluid. The characteristics of any nanofluid largely depend on the type and amount of nanoparticles used. But the researches have also shown that combination of different types of nanoparticles and their addition to the base fluid may alter the heat transfer capabilities of the resultant fluid. This class of nanofluids which is having more than one types of submicron particles in host fluid are termed as hybrid nanofluids. Hybrid nanofluids are getting more and more attention of the research community as these fluids give opportunity to manage rheological and flow properties also along with thermal performance enhancement. The present review is an effort to summaries the recent development and challenges in the field of hybrid nanofluids.

2. Synthesis of Hybrid Nanofluids

The method adopted to prepare mono/hybrid nanofluid plays an important role to get stable nanofluid with improved thermo-physical properties. Thus appropriate method of preparation is a key point in experimentation and research of mono/hybrid nanofluids. It has been observed that one step method and two step method are conventionally used to prepare nanofluid. But the two step method is most preferred as it is convenient and economic.

2.1. Two step method

The two step method comprises of two separate steps, first step is the synthesise of nanoparticles of desired material by suitable chemical or mechanical process, while the second step is the addition of already prepared nanoparticles into the host fluid to get stable nanofluid. This approach is convenient and extensively used but the nanofluids obtained by this method also have a big concern of stability as the suspended nanoparticles agglomerate under the influence of strong Vander wall and cohesive forces. Agglomeration of nanoparticles leads to instability and poor thermo-physical characteristics of nanofluid. To mitigate the agglomeration of nanoparticles different techniques like high-pressure homogenizer, addition of surface acting agents, sonication and magnetic forced stirrer are employed by the researchers. The nanofluids showed improved thermal properties but the increase in viscosity is an undesirable change in characteristics of nanofluid in most of the cases. Thus, a balance between the thermal and rheological characteristics of nanofluids is desirable for successful adoption of nanofluids in actual applications. This justifies the use of hybrid nanofluids as these fluids gives us opportunity to maintain the thermal and rheological properties both.

Various research groups prepared hybrid nanofluid on introducing different nanoparticles in to the host fluid. Asadi et. al. [2] mixed Al2O3/MWCNT nanoparticles in oil with help of magnetic stirrer and ultrasonic homogenizer. Persian and Akabari [3] procured Al2O3 (5 nm diameter) and Cu (50 nm diameter) nanoparticles. The appropriate amount of nanoparticle mixture of Al2O3 and Cu is mixed gradually in base fluid Ethylene glycol. In order to make suspension stable and agglomeration free it is subjected to ultrasonication for 7 hours. Wei et. al. [4] also procured the nanoparticles (SiC and TiO2) from outside. The SiC/TiO2 nanoparticles suspended into the base fluid diathermic oil. The volume fraction of SiC/TiO2 is determined by using the following formula

$$\varphi_v = \frac{(m_{SiC}/\rho_{SiC}) + (m_{TiO_2}/\rho_{TiO_2})}{(m_{SiC}/\rho_{SiC}) + (m_{TiO_2}/\rho_{TiO_2}) + (m_f/\rho_f)}$$

where φ_v is volume fraction of nanofluid (%), m mean mass and ρ is the density and subscript f represent base fluid. Oleic acid in appropriate amount is used as surfactant and ultra-sonication is done for 2 hours in order to make hybrid nanofluid stable.

Aparna et. al. [5] preferred to prepare Ag nanoparticles by using chemical reduction method while the Al2O3 nanoparticles are procured from outside. The appropriate ratio of Ag and Al2O3 nanoparticles mixed in distilled water. The Ag particles are also used to decorate the functionalised MWCNTs to get hybrid Ag/MWCNTs nanoparticles. These particles are more suitable to prepare stable suspension as compare to MWCNTs/base-fluid suspension [6]. Thus the two step method provide flexibility to prepare nanoparticles and their composites to enhance the desired effects in nanofluids.

2.2. One step method

As the name indicates in this method of nanofluid formation, the nanoparticles are synthesised as well as dispersed simultaneously into the base fluid. This method gives excellent stability of prepared nanofluid.

Eastman et. al. [8] used one-step procedure to disperse Cu nanoparticles into ethylene glycol. In the technique they adopted, the metallic vapour condensed directly into nanoparticles when the metallic vapour come in contact with the flowing ethylene glycol having low vapour pressure. Aureen et. al. [7] prepared CuO-Polyvinyle alcohol (PVA)/DI water hybrid nanofluid. They initially prepared the PVA-CuO nanocomposite by sol-gel method in single step, then the obtained wet precipitate is dispersed in DI water. Lo et. al. [9] prepare and dispersed CuO nanoparticles directly in to DI water by means of submerged arc nanoparticle synthesis scheme. Zhu et. al. [10] synthesis Cu nanofluid by single step chemical procedure . CuSO4·5H2O reduced with NaH2PO2·H2O in ethylene glycol(EG) in the presence of microwave irradiation to get Cu nanoparticles in ethylene glycol which produce Cu/ethylene glycol nanofluid. Botha et. al. [13] prepared nanofluid having 0.3 wt% Ag on 0.07 wt% SiO_2 in base fluid oil. AgNO3 and SiO2 introduced in to the host fluid and the mixture is heated up to 130°C and this temperature is maintained for 2 hours.

Teng et. al. [11] used plasma arc technology which is one step procedure to produce Carbon/distilled water nanofluid. Similarly Aberoumand and Jafarimoghaddam [12] and Munkhbayar et al. [14] also followed the single step method to prepare hybrid nanofluids by using Electrical Explosion Wire (E.E.W). and Pulsed-Wire Evaporation (PWE) procedure respectively. The previously prepared MWCNTs nanofluid is placed in PWE instrument. The silver nanoparticles synthesized by PWE method and suspended directly into the base fluid. Similarly Munkhbayar et al. [14] also used one step procedure pulse power evaporation (PPE) to dispersed Ag nanoparticles in the previously prepared MWCNTs/water nanofluid.

The review indicate that for preparing nanofluid in single step different techniques were adopted which provide stability and thermophysical characteristics but the high preparation cost and limitation for large scale production make this method less popular.

3. Thermo-physical Properties

Thermo-physical properties of mono/hybrid nanofluid fixed their heat transfer capabilities and overall thermal performance. Thermal conductivity, viscosity, density, and specific heat which are important thermophysical properties may either be measured using instruments or predicted by using available models given by various researchers. Thermo-physical properties rely on several factors like material, concentration, size and shape of nanoparticles, base fluid, and operating temperatures.

3.1. Thermal Conductivity

Thermal conductivity plays an important in thermal performance of any fluid hence its study for mono/hybrid nanofluid is very important. The enhancement in thermal conductivity of mono/hybrid nanofluid gives a base for the improvement in thermal performance and more energy efficient systems. The material, shape, and size of nanoparticle are important factors in deciding thermal conductivity of nanofluids.

Munkhbayar et al. [14] reported that on adding Ag nanoparticles in MWCNTs based aqueous nanofluid ensures considerable improvement in stability and thermal conductivity of resultant 'Ag/MWCNT'/water hybrid nanofluid. Maximum 14.5% rise in thermal conductivity is recorded for prepared 'Ag-MWCNT'/water hybrid nanofluid as compare to the MWCNT/water mono nanofluid. It confirms the advantage of hybrid nanofluids against mono nanofluids. The better straightness ratio increment, improved specific surface area and reduction in aggregation are the possible causes for improved thermal conductivity of prepared hybrid nanofluid.

Nine et. al. [16] synthesis and investigate the Al2O3-MWCNTs/water hybrid nanofluid for dispersion quality and thermal conductivity. They found Al2O3-MWCNTs/water hybrid nanofluid superior than Al2O3/water mono nanofluid from thermal conductivity point of view. They also conclude that the nanoparticles having cylindrical geometry as compare to spherical one contribute more to improve thermal conductivity of base fluid.

Batmunkh et al. [17] added modified "Ag" nanoparticles to the TiO2 based nanofluid to prepared three samples of Ag-TiO2 based aqueous hybrid nanofluid. Experimentally they demonstrated that the adding small quantity of modified silver nanoparticles improve thermal conductivity of TiO2 based mono nanofluid. The combination of Ag/TiO2 facilitate phonon conduction and hence improve thermal conductivity. Chen et al. [18] prepare hybrid nanofluid by suspending Ag-NPs decorated functionalized MWNT in water. The Ag-MWNT based hybrid nanofluid exhibited more thermal conductivity compare to the functionalized MWNT suspended mono nanofluid.

Ramaprabhu and Jha [20] dispersed Pd, Au, and Ag crystalline nanoparticles separately on multi walled carbon nanotubes to get metal-MWNTs composite nanoparticles. These composite nanoparticles are added to DI water and ethylene glycol separately to synthesizing hybrid nanofluids. The results confirmed that rise in temperature and concentration of metal-MWNTs in base fluid causes improvement its thermal conductivity. The thermal conductivity of hybrid nanofluid containing metal-MWCNT follow the same order which is that of constituent metal particles. The order of thermal conductivity of water based hybrid nanofluid is Ag-MWCNT/water>Au-MWCNT/water >Pd MWCNT/water and the same order is followed by EG based hybrid nanofluid also. Authors suggest that as the rise in temperature increases Brownian movement as well as thermal energy of the suspended nanoparticles which ultimately improves the thermal conductivity of the hybrid nanofluid. In their previous work [19] Cu particle-loaded MWCNTs, synthesised by chemical reduction method, suspended in two different base fluids DI water and EG in the absence of any surface active agent. Maximum thermal conductivity recorded at extremely less concentration Cu/MWCNTs which may be due to the homogeneous suspension of Cu/MWCNTs in the host fluid. Among the DI water and EG based hybrid nanofluids the DI water based experienced more improvement in thermal conductivity.

Moldoveanu et al. [21] prepare and perform experiments on water based hybrid nanofluid containing alumina and silica nanoparticles. They investigate the effect of concentration of nanoparticles (ranging from 1.0% to 3.0%) and temperature (within 20°C to 50°C) on thermal conductivity of hybrid nanofluid. A linear increase in thermal conductivity is observed along the rise in temperature. The hybrid nanofluid exhibits improvement in thermal conductivity with the growth in volume fraction of both Al2O3 as well as SiO2 particles, but the influence of increase of SiO2 is more as compare to Al2O3. A remarkable enhancement in thermal conductivity recorded with the increase in temperature at all the concentration of nanoparticle.

The nanofluids containing metallic nanoparticles have more thermal conductivity as compare to non-metallic nanoparticles. Cylindrical shaped nanoparticles contribute more as compare to spherical shaped nanoparticles in improving thermal conductivity of base fluid. On changing proportion of constituent nanoparticles only, while keeping the total concentration of nanoparticles intact, a significant variation in thermal conductivity of the hybrid nanofluid is observed.

3.2. Viscosity

Flow characteristics of working fluid is an important aspect while considering it for heat transfer applications. Penalty in terms of increase in pumping power is an undesired impact on using mono/hybrid nanofluids in thermal management devices as working fluid. Increase in viscosity is responsible for intensification of friction factor as well as pressure drop both, which ultimately rises the needed pumping power. The several research groups studied mono/hybrid nanofluid focussing on viscosity and the factors affecting it.

Bahrami et al. [23] considered the Fe-CuO/EG-water hybrid nanofluid to investigate their rheological behaviour under the influence of variation in volume fraction of nanoparticles. About 2200% growth in viscosity is recorded on changing volume fraction from 0.05% to 1.5%. Similar type of study done by Asadi and Asadi [24] considered engine oil based hybrid nanofluid containing MWCNT-ZnO nanoparticles for investigation. But they reported only 45% viscosity enhancement on increasing the solid fraction. Dalkılıç et al. [37] prepared hybrid nanofluid by suspending SiO2 along with Graphite nanoparticles in water. The viscosity of hybrid nanofluids is determined, keeping volume concentration in between 0.1% to 2%. A maximum 36.12% increase in viscosity observed for hybrid nanofluid of 2vol% at 15°C. An increase in viscosity from 0.65 to 36.32% is recorded along the growth of volumetric concentration of hybrid nanofluids. Sharma et al. [31] dispersed TiO2-CuO/C nanocomposite in base fluid ethylene glycol to prepare hybrid nanofluid. The experiments were performed to determine the viscosity and its dependency on temperature and volume fraction. About 80% enhancement in viscosity is recorded at 2% volume fraction and 313.4 K temperature as compare to ethylene glycol. Esfe et al. [36] estimate the viscosity of SAE50 based hybrid nanofluid having suspended MWCNT-CuO(30%-70%) nanoparticles at different volume fractions (0 to 1%) and temperature (25 to 50°C). The authors reported a reduction in viscosity of the fluid on adding nanoparticles from 0.0625% to 0.25% volume fraction. A maximum reduction of 15% was recorded in viscosity and then it improves as the volume fraction becomes 0.5% or more. A maximum 14% increase in viscosity observed at the volume fraction of 1%. Thus the use of hybrid nanoparticles can provide thermal fluid with enhanced heat transfer characteristics along with reduced or very low increase in pumping power as penalty.

Dardan et al. [27] compare the viscosity of pure SAE40 engine oil and hybrid nanofluid obtained on adding Al2O3-MWCNTs nanoparticles. The investigations were done at temperature ranging from 25oC to 50oC. A reduction in viscosity recorded on rising temperature. Baghbanzadeh et al. [28] prepared four types of nanoparticles SiO2, MWCNTs, SiO2 (80 wt%) + MWCNTs (20 wt%) termed as H1, and SiO2 (50 wt%) + MWCNTs (50 wt%) termed as H2. Four different types of mono/hybrid nanofluids synthesised on adding above mentioned nanoparticles separately in to the host fluid distilled water. The trend of rise in viscosity with the increase in nanoparticle concentration and fall in viscosity with the increase in temperature are same as reported by other researchers. At high concentration the nanofluid with H2 hybrid particles registered least increase in viscosity as compare to others. The addition of nanoparticles of MWCNTs affect most negatively the viscosity of base fluid. They suggested that the 20oC is the optimum temperature as it causes the least enhancement in the viscosity of base fluid. Hashemzadeh and Hormozi [34] concluded that the dynamic viscosity of hybrid nanofluids is more than the mono nanofluid after performing experiments on Al2O3/water, SiO2/water nanofluids and Al2O3-SIO2/water hybrid nanofluids. Rise in viscosity with the growth in nanoparticle concentration, and fall in viscosity with rise in temperature are the trend followed by mono/hybrid nanofluids. Afshari et al. [38] also confirmed the same trend of variation in viscosity in their investigation of MWCNTs+Al2O3/(water80%+ethylene glycol80%) hybrid nanofluid. Kumaresan et al. [26] reveal that the dependency of viscosity over temperature is a function of concentration of nanoparticles in the base fluid. Thus the intensity of change in viscosity with temperature may be controlled by regulating the concentration nanoparticles of mono/hybrid nanofluid.

Shahsavar and Bahiraei [33] dispersed $C_4H_{13}NO$ coated Fe3O4 nanoparticles and GA coated CNTs nanoparticles in to base fluid water to get hybrid nanofluid which is investigated for variation in viscosity and thermal conductivity. The hybrid nanofluid showed shear-thinning behaviour, which intensified at low values of shear rate. Akilu et al. [31] performed the rheological analysis of ethylene glycol based hybrid nanofluid containing suspended TiO2-CuO/C nanoparticles. They observed Newtonian behaviour of all samples at various shear rate and temperature. On the other hand Esfe et al. [36] reported non-Newtonian behaviour of SAE50 based hybrid nanofluid having MWCNT(30%)+CuO(70% nanoparticles.

Sundar et al. [25] investigate the impact of concentration of MWCNTs-Fe3O4 nanoparticles in base fluid water over viscosity, friction factor and pumping power. For friction factor experiments the particle concentration were 0.1% and 0.3%. The hybrid nanofluid observed 1.5 times enhancement in viscosity at 0.3% concentration and 60°C temperature and friction factor enhanced by 1.11 times as compare to the base fluid. 1.18 times increase in pumping power as penalty is also recorded at 0.3% particle loading and 22000 Reynolds number. Suresh et al. [29] focused on the friction behaviour of Al2O3 -Cu/water hybrid nanofluid. They found that the presence of 0.1% volume fraction of Al2O3-Cu in base fluid causes 16.97% average increase in friction factor and rise in pumping power also. They suggested that the agglomeration and surface adsorption are the root cause for the rise in viscosity which eventually affect the friction factor and pumping power as penalty. Allahyar et al. [30] also suggested that the particle agglomeration and clusters formation may be responsible for affecting the viscosity of hybrid nanofluid which leads to rise of pressure drop.

It has been observed that the theoretical equations may not provide the viscosity of hybrid nanofluids at all volume fraction and temperature range. Kumaresan et al. [26] experimentally investigate the discrepancies in between the results of viscosity obtained from experiments and predicted by using the well-known correlations for nanofluid viscosity. The investigations reveal that the equations are fail to predict the experimental results within desired accuracy. Even Einstein's law of viscosity fails to provide the viscosity of nanofluids having very less concentration of nanoparticles. Esfe et al. [36] collect the viscosity of SAE50 based hybrid nanofluid containing MWCNTs and CuO at various temperatures and volume fraction of nanoparticles. A mathematical correlation (equation 3) is suggested to predict the viscosity of hybrid nanofluid.

$$\frac{\mu_{nf}}{\mu_{bf}} = 0.99 + 0.49\varphi - 0.0077T + 6.81 \times 10^{-6}\dot{\gamma} - 0.00249\varphi T - 2.19 \times 10^{-7}T\dot{\gamma} - 0.19\varphi^2 + 0.00011T^2 \quad (1)$$

Shahsavar et al. [22] study the viscosity of hybrid nanofluid prepared by adding aqueous ferrofluid and water based CNT nanofluid under the influence of magnetic field. The results indicated that the rise of magnetic field strength causes the fluid more viscosity. Sahoo and Kumar [35] reported that the hybrid nanofluid having platelet shaped nanoparticles experienced maximum viscosity and the viscosity while the spherical shaped nanoparticle causes maximum rise in viscosity.

The rise in viscosity with the growth in nanoparticle concentration is the general observation in most of the research work. The appropriate ratio of constituent material of hybrid nanoparticles can help to control the viscosity change as per requirement. The temperature also plays a critical role to regulate the viscosity as with rise in temperature the nanofluid become less viscose. The effect of shape of nanoparticles, and applied magnetic field are less investigated parameter for mono/hybrid nanofluid from viscosity point of view.

3.3. Density and specific heat capacity

Thermophysical properties of mono/hybrid nanofluids are responsible for their feasibility and usefulness for any thermal device. The mono/hybrid nanofluids are focused less to investigate their density and specific heat capacity.

Density is a significant characteristic of hybrid nanofluid to investigate as it directly influences stability, friction factor, pumping power, Reynolds number along with thermal performance. The mass of a substance per unit volume can be defined as density. As the density of solid is more as compare to the liquid, hence the presence of solid nanoparticles into host liquid tend to increase the density of resultant nanofluid. The density will vary with the concentration of nanoparticles in the mono/hybrid nanofluid. The conventional mixture relations can be used to predict the density of nanofluids. For determining the density of nanofluid researchers used the equation (2), which is a modified form of the basic model.

$$\rho_{nf} = \frac{m_{np}+m_{bf}}{V_{np}+V_{bf}} = \frac{\rho_{np}V_{np}+\rho_{bf}V_{bf}}{V_{np}+V_{bf}} \quad (2)$$

Pak and Cho [39] preferred the model given by equation (3) to predict the density of nanofluid.

$$\rho_{nf} = \varphi\rho_{np} + (1-\varphi)\rho_{bf} \quad (3)$$

Several research groups[40-42] confirm that the equation (3) is capable to predict the density of nanofluid very accurately and hence it is widely used to validate the experimentally obtained density of nanofluids.

Takabi and Salehi [43] specified the viscosity of hybrid nanofluid by equation (4) which is extension of equation (3) to hybrid nanofluids.

$$\rho_{hnf} = \varphi_{np1}\rho_{np1} + \varphi_{np2}\rho_{np2} + (1-\varphi)\rho_{bf} \quad (4)$$

where $\varphi = \varphi_{np1} + \varphi_{np2}$.

Sahu and Sarkar [44] also used the equation (4) to calculate the density of hybrid nanofluids. Microencapsulated phase change material(MEPCM)-Al2O3/water hybrid nanofluid is prepared and investigated for density by Ho et al. [32]. The mixture theory formula i.e. equation (4) is used to get the predicted values of effective density for hybrid nanofluid. They validated the correctness of the equation (4) also, and the results were in accepted range.

Yarmand et al. [45] dispersed Ag decorated graphene nanoplatelets (GNP) in distilled water. The density of obtained hybrid nanofluids are investigated for different weight concentration and temperature ranging from 20 to 40°C. A minute rise in density is recorded as the nanoparticle fraction increases. About 0.09% increase in density of hybrid nanofluid is observed against the base fluid at 40°C and 0.1% fraction of nanoparticles.

The accurate and reliable knowledge of specific heat of mono/hybrid nanofluids is crucial because the analysis of energy and exergy of nanofluid depends on it. But it is comparatively less focused property of mono/hybrid nanofluids. The specific heat of nanofluid is less than the base fluid, which means for the same rise in temperature, less heat is required for nanofluid as compare to the base fluid [47]. Specific heat of nanofluid can be considered as a mixture of heat capacities of solid and liquid phases under the condition of thermal equilibrium with each other [46]. The model (equation 5) which is analogous to mixture theory is simple and widely adopted for predicting the specific heat of nanofluid [48-50].

$$c_{p,nf} = (1-\varphi)c_{p,bf} + c_{p,np} \tag{5}$$

where nf, bf, and np are abbreviations for nanofluid, base fluid, and nanoparticle respectively.

Zhou and Ni [51] studied Al2O3/water nanofluid and experimentally determine the specific heat. They observe a fall in specific heat of nanofluid on increasing Al2O3 concentration. The experimental values of specific heat are close to the values predicted from thermal equilibrium model i.e. equation (6), while the simple mixing model i.e. equation (5) fails to provide accurate values. Hence the equation based on thermal equilibrium model is more preferred to predict the specific heat of nanofluids [15, 62]

$$\rho_{nf}c_{p,nf} = (1-\varphi)\left(\rho c_p\right)_{bf} + \left(\varphi\rho c_p\right)_{np} \tag{6}$$

Shin and Banerjee [52] prepared high temperature nanofluid by doping alkaline metal chloride salt eutectics with SiO2 (1% mass concentration) nanoparticles. The specific heat capacity of nanofluid is increased by 14.5% as compare to the base fluid. The specific heat of hybrid nanofluids is given by Equation (7). The subscript of bf, hnf, p1 and p2 represent base fluid, nanofluids, hybrid nanofluids, nanoparticle type 1 and nanoparticle type 2, respectively. Madhesh et al. [53] and Mund et al. [54] used equation (7) for predicting the specific heat of Cu-TiO2/water and Al2O3-TiO2/water hybrid nanofluids, respectively.

$$c_{hnf} = \frac{(1-\varphi)(\rho c)_{bf} + (\varphi\rho c)_{p1} + (\varphi\rho c)_{p2}}{\rho_{hnf}} \tag{7}$$

The model obtained by extending conventional mixture relation is capable to predict the density up to considerable correctness. The influence of mono and composite nanoparticles on the specific heat of nanofluid are not consistent. The impact of temperature and volume fraction of hybrid nanoparticles over the density and specific heat of hybrid nanofluids need to explore further.

4. Stability

Stability is a crucial parameter for mon/hybrid nanofluids. The nanoparticles have strong tendency to agglomerate because of Vander Waal and cohesive forces, which lead to sedimentation of the particles. The hybrid nanofluid with agglomerated nanoparticles are not improved form of heat transfer fluid. The presence of large sized agglomerated particles enhance the viscosity, friction factor as well as pressure drop. Thus agglomeration and particles sedimentation make the hybrid nanofluid unsuitable for use in heat transfer application. From the literature, it is found that the techniques such as ultrasonication addition of surface active agents, and electrostatic stabilization are commonly adopted for minimizing the agglomeration.

Surfactant is a surface active agent, which function as a bridge among suspended nanoparticles and the host fluid. Surfactant creates continuity within the nanofluid. The stability of a nanofluid will depend on the nature of nanoparticles as well as host fluid both. The nanoparticles used may be of the nature of hydrophobic or hydrophilic, while the host fluid may have polar or non-polar nature. The dispersion of hydrophilic nanoparticles in to the polar host fluid and the hydrophobic nanoparticles in to the non-polar host fluid is convenient and do not required any third component in general for achieving the stability. But if the dispersion is having opposite combination of nanoparticles and host

IOP Conf. Series: Materials Science and Engineering **810** (2020) 012020 doi:10.1088/1757-899X/810/1/012020

fluid as mentioned above, then surfactant need to introduce for achieving stability of suspension [55]. The surfactant can be classified on the basis of charge on their head group. The surfactant may have negative charge, positive charge, neutral, or both positive and negative charge. Table 1 summarises the commonly used surfactant along with nanoparticles and host fluids.

Table 1. Combinations of surfactants nanoparticles and host fluid.

Reference	Surfactant	Nanoparticle	Host fluid
Zhu et al. [56]	Sodium odecylbenzene sulphonate (SDBS)	Al2O3	water
Hwang et al. [57]	Sodium dodecyl sulphate (SDS)	MWCNTs	DI water
Kim et al. [58]	Sodium dodecyl sulphate (SDS)	Al2O3, ZnO, TiO2	Water and ethylene glycol (EG)
Yu et al. [59]	Sodium dodecyl sulphate (SDS)	MWCNTs	Distilled water
Madni et al. [60]	Sodium Octonate (SOCT)	MWNTs	Dimethyl formamide (DMF)
Kim et al. [61]	Cetyl trimethyl ammonium bromide (CTAB)	MWCNTs	water
Avsec et al. [62]	Dodecyle Trimethyl Ammonium Bromide (DTAB)	MWNTs	Dimethyl formamide (DMF)
Shanbedi et al. [63], Shahsavar and Bahiraei [33]	Gum arabic (GA)	MWCNTs	water
Li et al. [64]	Polyoxyethylene (10) nonyl phenyl ether	Cu	water
Tang et al. [65]	Poly vinyl pyrrolidone (PVP)	CNTs	Dimethylacetamide
Ghadimi and Metselaar [66]	Poly vinyl pyrrolidone (PVP)	TiO2	distilled water
Choi et al. [67]	Oleic acid	Al2O3 and AlN	transformer oil
Drzazga et al. [68]	Rokanol K7	CuO	water
Utomo et al. [69]	Octyl silane	Al2O3, TiO2	water
Suresh et al. [70]	Sodium lauryl sulphate (SLS)	Al2O3–Cu	deionized water

The addition of suitable surfactant help to stabilized the nanofluid by improving the repulsive force among the nanoparticles on strengthening the electric double layer. Surfactant may also improve the charge over the surface of nanoparticles and hence may rise the zeta potential of nanofluid [57, 71]. Adoption of appropriate surfactant according to the material of nanoparticle and base fluid is an important activity as it not only decides the stability/durability of prepared mono/hybrid nanofluid but the thermophysical properties are also may affect.

Electro kinetic characteristics of suspended nanoparticles plays a crucial role to stabilize mono/hybrid nanofluids. Electric charge which may be at the outer most shell of the atom is the reason behind the electro kinetic characteristics of nanoparticles in the host fluids. Ions present on particles and in the base fluid form an electric static double layer, it helps to stabilize the suspension. At a particular pH, at which the net charge on the particle becomes zero is termed as isoelectric point (IEP). At pH_{IEP} the repulsive forces between the particles become zero, which causes the agglomeration. As the pH of suspension deviates from pH_{IEP} the force of repulsion among the nanoparticles increases and the stability of suspension improves. For example the pH_{IEP} for water based nanofluid having suspended Al2O3 and Cu nanoparticles separately are 10 and 9.6 respectively [100, 101]. Suresh et al. [70] measured the pH of water based hybrid nanofluid having Al2O3 and Cu both and reported it about 6 which is much less than the nanofluids containing Al2O3 and Cu separately. This ensures that the prepared hybrid nanofluid is stable. A model to predict the pH of hybrid nanofluid will be very helpful to estimate the stability also.

The most widely used method to stabilize the mono and hybrid nanofluid is ultrasonic vibration as it is economical and easy to use. The basic principle of this technique is to uniform disperse nanoparticles in the host fluid by using high frequency vibrations. There is no alteration in the surface characteristics of suspended nanoparticles is an important advantage of this technique. The suitable time duration needed for sonication in order to get stable nanofluid based on material, size and concentration of nanoparticles. Ultrasonic bath/vibrator along with homogenizer are commonly used devices for sonication process. But the ultrasonic horn or prob devices as compare to ultrasonic bath are more effective to reduce the size of nanoparticle [72].

Stokes law indicate that the minimum velocity needed for a spherical particle to be suspended stably in host fluid is given by equation (8) [102]. This equation indicates that smaller the size of particle more will be its velocity and hence the particle will be more stable in the suspension.

$$V = \frac{2R^2}{9\mu}\left(\rho_{np} - \rho_{bf}\right) \tag{8}$$

where V, R, ρ_{np} refers to velocity, radius and density of nanoparticle, respectively while μ, ρ_{bf} stands for viscosity and density of host fluid, respectively. Asadi et al. [72] reviewed the literatures focussing on stability of nanofluids and the factors affecting it such as sonication time and power. They observed that higher the sonication duration and power more will be the reduction in cluster size, which improves the stability of nanofluid, it is also supported by the equation 8. Thus it is significant to investigate further about the influence of sonication time duration and power to stabilize the mono/hybrid nanofluids.

5. Applications of hybrid nanofluids

The research work on mono/hybrid nanofluid revealed strong possibilities of significant improvement in thermal performance of traditional heat transfer fluids. After introducing nanofluid by Choi [1] in 1995 large number of research work has been conducted regarding application of mono and hybrid nanofluids. Heat exchanger [73-75], electronic cooling [76-79], heat pipes [80-82], engine cooling [83-85], transformer oil [67, 86], nuclear plant [87-89], solar heating [90,91], chillers and domestic refrigerators [92, 93]] are some of the application areas where mono or hybrid nanofluid may be used.

The use of mono/hybrid nanofluid not only improve the thermal performance but it may extend the life span of the device. and minimize the chance of overheating. Choi et al. [67] confirmed that the transformer oil having suspended Al2O3-AlN nanoparticles performed better as compare to pure oil. Sundar et al. [95] prepared Nanodimond-Fe3O4 nanofluid and proposed it as a new heat transfer fluid for heat exchanger application. They conduct experiments for electrical conductivity along with thermal conductivity and viscosity. Zyla et al. [96] investigate the ethylene glycol based nanofluid containing graphite and nanodiamonds. On varying ash content they reported a big impact on electrical conductivity with negligible effect on thermal conductivity. Aberoumand and Jafarimoghaddam [12] investigated dielectric strength along with thermal conductivity of transformer oil having suspended Ag-WO3 nanoparticles. A reduction in dielectric strength is recorded as compare to the pure oil. Contreras et al. [94] reviewed the research work on nanofluid application in electrical transformer. They conclude that heat transfer and dielectric properties enhancement of conventionally used transformer coolant on adding nanoparticles is reported by researchers. The thermal conductivity improvement may be associated with rise in electrical conductivity. The results related to the dielectric strength of mono/hybrid nanofluids having transformer oil as base fluid are not consistent and need to investigate further.

The performance of HVAC systems may be improved by using mono/hybrid nanofluids as primary or secondary refrigerant. Wang et al. [98] appending nanoparticle n-TiO2(r) into mineral lubricant and get new nanolubricant which is used with HFC134a based refrigeration system. The presence of nanoparticle n-Ti2(r) in general mineral oil can intensify the polarity of the oil and it helps to enhance the compatibility between the new lubricant and HFC refrigerants. This supports the returning of more lubricating oil back to compressor which is an essential condition for safe working of a refrigeration system. Ahmed and Elsaid [97] studied chilled water air conditioning system under the influence of Al2O3-TiO2/water hybrid nanofluid, as secondary working medium. The investigation reveal that the supply fluid temperature decreases more rapidly as the ratio of Al2O3 in hybrid nanofluid increases as compare to TiO2. It also has been noticed that at the same mass fraction of mono nanofluid (Al2O3/H2O) and hybrid nanofluid (0.3%Al2O3+0.1%TiO2)/H2O the evaporator and condenser pressure difference is lower in case of mono nanofluid. In HVAC system the use of nanofluid may help to ensure the presence of lubricant oil in the compressor along with thermal performance improvement.

6. Challenges associated

The researchers expresses their concern about the challenges associated with mono and hybrid nanofluid while reporting about their capabilities of improving thermal performance. The major limitation of mono or hybrid nanofluid is stability, which restricts the implementation of these nanofluids. Siddiqui et al. [99] identified that the Cu nanoparticles have poor dispersion stability with high thermal conductivity and the Al2O3 nanoparticles have good dispersion stability with poor thermal conductivity. They prepared three samples of hybrid nanofluids with different Cu and Al2O3 ratio. The most suitable ratio of Cu and Al2O3 is 0.5:0.5, at which the hybrid nanofluid have significant enhancement in thermal conductivity along with relatively better stability. The stability of mono nanofluids is achieved by various proved methods, but for hybrid nanofluids the stability becomes more complicated as the different types of nanoparticles may possess different surface charges. To get stabile hybrid nanofluids the factors such as pH value, surfactants used and the combination of nanoparticle material need to be considered.

The addition of more than one type of nanoparticles for preparing hybrid nanofluid gives us an opportunity to minimize the factors causing penalties such as viscosity, friction factor, pumping power etc. with an affordable compromise of percentage enhancement in thermal performance. The combination of any two type of nanoparticles give rise of entirely different type of characteristics to the composite nanoparticles and hence to the resultant hybrid nanofluid. The various combinations of nanoparticles such as Al2O3+MWCNTs, CU+TiO2, Cu+CNTs, Cu+MWCNTs and much more like these have been synthesized and used to prepared nanofluids. But the choice of suitable grouping of nanoparticles along with base fluid depends upon the application for which the resultant hybrid nanofluid is targeted.

The selection of proper material of constituent nanoparticles for preparing hybrid nanofluid is not properly defined. The limited studies are available on the collective effect of the factors such as sonication time, use of surfactant, along with nanoparticle combination. The high cost of nanoparticles make the mono and hybrid nanofluid less favourable for the end users. To make nanoparticles and their hybrids affordable is a challenge yet to answer.

7. Conclusion

This paper, is an attempt to get inside on the research on hybrid nanofluid related to their synthesis, important characteristics like thermal conductivity, viscosity, specific heat capacity, and density. The contradictory results are reported related to thermophysical properties of mono/hybrid nanofluids.

Getting stabile mono/hybrid nanofluid is still a major challenge that need to address for their successful implementation in industries.
Through this review following points could be concluded.

- To synthesis hybrid nanofluid the two step method is more convenient as here two or more type of nanoparticles can be combined to prepare their composite and then they can be dispersed by any of the established techniques. On the other hand one step method have benefits of stability but their high production cost and small scale production are major drawbacks.
- The metallic nanoparticles contribute more as compare to the non-metallic nanoparticles in improving thermal conductivity of host fluids. Cylindrical shaped nanoparticles are more capable as compare to spherical shaped nanoparticles to rise the thermal conductivity of host fluid. The study of thermal performance under the influence of combination of various nanoparticle materials with different shape and size could be an area to investigate.
- The hybrid nanofluid become more viscus as the concentration of nanoparticles increases on the other hand it becomes less viscus on rise the temperature. On varying the proportion of constituent nanoparticles of hybrid nanoparticle the viscosity of nanofluid may be changed, although the overall concentration of hybrid nanoparticles in the host fluid kept constant.
- The research on density and specific heat of hybrid nanofluids are less. Inconsistency is observed in the results reported regarding change in specific heat of host fluids on adding nanoparticles.
- Selection of appropriate constituent nanoparticle material of hybrid nanoparticles according to base fluid is first and important step. The introduction of surface active agents generally hinders the positive impacts of nanoparticle addition in the host fluid, hence its selection as well as concentration must be optimized to get stable suspension.
- More studies are suggested on collective influence of sonication, use of surfactant, along with nanoparticle shape, size and proper combination, on thermal behaviour of hybrid nanofluids.

References:

[1] Choi S U S 1995 in *Developments and Applications of Non-Newtonian Flows,* D. A. Singer and H. P. Wang, Eds., American Society of Mechanical Engineers, New York, **FED–231/MD-66** 99–105
[2] Asadi A, Meisam M, Rezaniakolaei A, Rosendahl L A, Afrand M, and Wongwises S 2018 *Int. J. Heat. Mass Transf.* **117** 474–486
[3] Parsian A and Akbari M 2017 *J Therm Anal Calorim.* **131** 1605–1613
[4] Wei B, Zou C, Yuan X, and Li X 2017 *Int. J. Heat. Mass Transf.* **107** 281–287
[5] Aparna Z, Michael M, Pabi S K, and Ghosh S 2019 *Powder Technology.* **343** 714–722
[6] Farbod M, and Ahangarpour A 2016 *Phys. Lett. A* **380(48)**, 4044-4048
[7] Aureen Albert A A, Harris Samuel D G, Parthasarathy V, and Kiruthiga K 2019 *Chemical Engineering Communications* **207(3)** 319-330
[8] Eastman J A, Choi S U S, Li S, Yu W, and Thompson L J 2001 *Appl. Phys. Lett.* **78** 718-720
[9] Lo C H, Tsung T T, Chen L C, Su C H and Lin H M 2005 *Journal of Nanoparticle Research* **7** 313–320
[10] Zhu H, Linb Y, and Yin Y 2004 *Journal of Colloid and Interface Science* **277** 100–103
[11] Teng T P, Cheng C M, and Feng-Yi Pai F Y 2011 *Nanoscale Research Letters* **6(1)** 293
[12] Aberoumand S and Jafarimoghaddam A 2018 *Alexandria Engineering Journal* **57 (1)** 169-174
[13] Botha S S, Ndungu P, and Bladergroen B J 2011 *Ind. Eng. Chem. Res.* **50** 3071–3077
[14] Munkhbayar B, Tanshen M R, Jeoun J, Chung H, and Jeong H 2013 *Ceramics International* **39** 6415–6425
[15] Buongiorno J, 2006 J. Heat Transfer **128(3)** 240–250
[16] Nine M J, Batmunkh M, Kim J H, Chung H S, and Jeong H M 2012 *J. Nanosci. Nanotechnol.* **12(6)** 4553–4559
[17] Batmunkh M, Tanshen M R, Nine M J, Myekhlai M, Choi H, Chung H, and Jeong H 2014 *Ind. Eng. Chem. Res.* **53** 8445–8451

[18] Chen L, Yu W, and Xie H 2012 *Powder Technology* **231** 18–20
[19] Jha N and Ramaprabhu S 2008 *J. Phys. Chem.* ***112*** 9315–9319
[20] Jha N and Ramaprabhu S 2009 *Journal of Applied Physics* **106** 084317
[21] Moldoveanu G M, Huminic G, Minea A A, Huminic A 2018 *International Journal of Heat and Mass Transfer* **127** 450–457
[22] Shahsavar A, Salimpour M R, Saghafian M, and Shafii M B 2016 *Journal of Mechanical Science and Technology* **30(2)** 809-815
[23] Bahrami M, Akbari M, Karimipour A, and Afrand M 2016 *Experimental Thermal and Fluid Science* **79** 231-273
[24] Asadi M and Asadi A 2016 *International Communications in Heat and Mass Transfer* **76** 41–45
[25] Sundar L S, Singh M K, and Sousa A C M 2014 *International Communications in Heat and Mass Transfer* **52** 73–83
[26] Kumaresan V andVelraj R 2012 *Thermochimica Acta.* **545** 180–186
[27] Dardan E, Afrand M, Isfahani A H M 2016 *Applied Thermal Engg.* **109** 524–534
[28] Baghbanzadeh M, Rashidi A, Soleimanisalim A H, and Rashtchian D 2014 *Thermochimica Acta.* **578** 53– 58
[29] Suresh S, Venkitaraj K P, Selvakumar P, and Chandrasekar M 2012 *Experimental Thermal and Fluid Science* **38** 54–60
[30] Allahyar H R, Hormozi F, and ZareNezhad B 2016 Experimental investigation on the thermal performance of a coiled heat exchanger using a new hybrid nanofluid *Experimental Thermal and Fluid Science.* **76** 324-329
[31] Akilu S, Baheta A T,and Sharma K V 2017 *Journal of Molecular Liquids.* **246** 396-405
[32] Ho C J, Huang J B, Tsai P S, and Yang Y M 2010 *International Communications in Heat and Mass Transfer* **37(5)** 490-494
[33] Shahsavar A and Bahiraei M 2017 *Powder Technology* **318** 441–450
[34] Hashemzadeh S and Hormozi F 2019 *Journal of Thermal Analysis and Calorimetry* **139** https://doi.org/10.1007/s10973-019-08626-6
[35] Sahoo R R, and Kumar V 2019 *J. Nanotechnol Material Sci.* **6(1)** 34-29
[36] Esfe M H, Dalir R, Bakhtiari R, Afrand M 2018 *International Journal of Refrigeration* **104** 123-133
[37] Dalkılıç A S, Açıkgöz Ö, Küçükyıldırım B O, and Eker A A 2018 *International Communications in Heat and Mass Transfer* **97** 30–38
[38] Afshari A, Akbari M, Toghraie D, and Yazdi M E 2018 *Journal of Thermal Analysis and Calorimetry* **132** 1001-1015
[39] Pak B C and Cho Y I 2007 *Experimental Heat Transfer: A Journal of Thermal Energy Generation, Transport, Storage, and Conversion* **11(2)** 151-170
[40] Shoghl S N, Jamali J, and Moraveji M K 2016 *Experimental Thermal and Fluid Science* **74** 339-346
[41] Vajjha R S, Das D K, and Mahagaonkar B M 2009 *Petroleum Science and Technology* **27(6)** 612-624
[42] Heyhat M M, Kowsary F, Rashidi A M, Esfehani A V, and Amrollahi A 2012 *International Communications in Heat and Mass Transfer* **39** 1272–1278
[43] Takabi B and Salehi S 2014 *Hindawi Publishing Corporation Advances in Mechanical Engineering* **6** 147059
[44] Sahu M and Sarkar J 2019 *Journal of Heat Transfer* **141** 082401
[45] Yarmand H, Gharehkhani S, Ahmadi G, Shirazi S F S, Baradaran S, Montazer E, Zubir M N M, and Alehashem M S 2015 *Energy Conversion and Management* **100** 419–428
[46] Akilu S, Sharma K V, Baheta A T, and Mamat R 2016 *Renewable and Sustainable Energy Reviews* **66** 654–678
[47] Chandrasekar M, Suresh S, and Senthilkumar T 2012 *Renewable and Sustainable Energy Reviews* **16** 3917– 3938

[48] Xuan Y and Roetzel W 2000 *International Journal of Heat and Mass Transfer* **43** 3701-3707
[49] Louis Gosselin L and da Silva A K 2004 *Applied Physics Letters* **85** 4160-4162
[50] Lee J and Mudawar I 2007 *International Journal of Heat and Mass Transfer* **50** 452–463
[51] Zhou S Q and Ni R 2008 *Appl. Phys. Lett.* **92** 093123
[52] Shin D and Banerjee D 2011 *International Journal of Heat and Mass Transfer* **54** 1064–1070
[53] Madhesh D, Parameshwaran R, and Kalaiselvam S 2014 *Experimental Thermal and Fluid Science* **52** 104–115
[54] Mund A, Pattanayak B, Jayakumar J S, Parashar K, and Parashar S K S 2019 *Advances in Fluid and Thermal Engineering* 531–540 Springer Singapore
[55] Yu H, Hermann S, Schulz S E, Gessner T, Dong Z, and Li W J 2012 *Chemical Physics* **408** 11–16
[56] Zhu D, Li X, Wang N, Wang X, Gao J, and Li H 2009 *Current Applied Physics* **9** 131–139
[57] Hwang Y, Lee J K, Lee C H, Jung Y M, Cheong S I, Lee C G, Ku B C, and Jang S P 2007 *Thermochimica Acta.* **455** 70–74
[58] Kim S H, Choi S R, and Kim D 2007 *Journal of Heat Transfer* **129** 298-307
[59] Yu J, Grossiord N, Koning C E, and Loos J 2007 *Carbon* **45** 618–623
[60] Madni I, Hwang C Y, Park S D, Choa Y H, and Kim H T 2010 *Colloids and Surfaces A: Physicochem. Eng. Aspects* **358** 101–107
[61] Kim H S, Park W, Kang M, and Jin H J 2008 *Journal of Physics and Chemistry of Solids* **69** 1209–1212
[62] Avsec J, Oblak M 2007 *International Journal of Heat and Mass Transfer* **50 (21–22)** 4331-4341
[63] Shanbedi M, Heris S Z, and Maskooki A 2015 *J Therm Anal Calorim.* **120(2)** 1193-1201
[64] Li X, Zhu D, and Wang X 2007 *Journal of Colloid and Interface Science* **310** 456–463
[65] Tang Q Y, Shafiq I, Chan Y C, Wong N B, and Cheung R 2010 *Journal of Nanoscience and Nanotechnology* **10** 4967–4974
[66] Ghadimi A and Metselaar I H 2013 *Exp. Therm. Fluid Sci.* **51** 1–9
[67] Choi C, Yoo H S, Oh j M 2008 *Current Applied Physics* **8** 710–712
[68] Drzazga M, Dzido G, Lemanowicz M, and Gierczycki A 2012 *14th European Conference on Mixing Warszawa*
[69] Utomo A T, Poth H, Robbins P T, and Pacek A W 2012 *International Journal of Heat and Mass Transfer* **55** 7772–7781
[70] Suresh S, Venkitaraj K P, Selvakumar P, and Chandrasekar M 2011 *Colloids and Surfaces A: Physicochem. Eng. Aspects* **388** 41– 48
[71] Huang J, Wang X, Long Q, Wen X, Zhou Y, and Li L 2009 *Symp Photonics Optoelectron.* **1** 1-4
[72] Asadi A, Pourfattah F, Szilágyi I M, Afrand M, Żyła G, Ahn H S, Wongwises S, Nguyen H M, Arabkoohsar A, and Mahian O 2019 *Ultrasonics – Sonochemistry* **58** 104701
[73] Arya H, Sarafraz M M, Pourmehran O, and Arjomandi M 2019 *Heat and Mass Transfer* **55(6)** 1769–1781
[74] Madhesh D and Kalaiselvam S 2014 *Applied Mechanics and Materials* **592-594** 922-926
[75] Huang D, Wu Z, and Sunden B 2016 *Experimental Thermal and Fluid Science* **72** 190–196
[76] Sohel M R, Saidur R, Khaleduzzaman S S, and Ibrahim T A 2015 *International Communications in Heat and Mass Transfer* **65** 89–93
[77] Khaleduzzaman S S, Sohel M R, Saidur R, Mahbubul I M, Shahrul I M, Akash B A, and Selvaraj J 2014 *International Communications in Heat and Mass Transfer* **57** 118–127
[78] Naphon P, Klangchart S, and Wongwises S 2009 *International Communications in Heat and Mass Transfer* **36** 834–840
[79] Lai W Y, Duculescu B, Phelan P E, and Prasher R 2006 *Thermal and Thermomechanical Proceedings 10th Intersociety Conference on Phenomena in Electronics Systems*
[80] Ghanbarpour M, Nikkam N, Khodabandeh R, and Toprak M S 2015 *nternational Communications in Heat and Mass Transfer* **67** 14–20
[81] Alawi O A, Sidik N A C, Mohammed H A, and Syahrullail S 2014 *International Communications*

in Heat and Mass Transfer **56** 50-62
[82] Han W S and Rhi S H 2011 *Thermal Science* **15** 195-206
[83] Ollivier E, Bellettre J, Tazerout M, and Roy G C *Energy Conversion and Management* **47** 879-893
[84] Sidik N A C, Yazid M, and Mamat R 2015 *International Communications in Heat and Mass Transfer* **68** 85–90
[85] Delavari V and Hashemabadi S H 2014 *Applied Thermal Engineering* **73** 378-388
[86] Sumathi S, Rajesh R, and Subburaj P 2019 *IETE Journal of Research,* 1-9
[87] Mousavizadeh S M, Ansarifar G R, and Talebi M 2015 *Nuclear Engineering and Technology* **47 (7)** 814-826
[88] Ghazanfari V, Talebi M, Khorsandi J, and Abdolahi R 2016 *Progress in Nuclear Energy* **87** 67-73
[89] Kianpour R and Ansarifar G R 2019 *Annals of Nuclear Energy* **133** 623-636
[90] Ahmadi M H, Ghazvini M, Sadeghzadeh M, Nazari M A, and Ghalandari M 2019 *Nano-Structures & Nano-Objects* **20** 100386
[91] Jin X, Lin G, Zeiny A, Jin H, Bai L, and Wen D 2019 *Renewable Energy* **141** 937-949
[92] Jatinder G, Ohunakin O S, Adelekan D S, Atiba O E, Daniel A B, Singh J, Atayero A A 2019 *Applied Thermal Engineering* **160** 114004
[93] Ahmed M S, Abdel Hady M R, and Abdallah G 2018 *Thermal Science and Engineering Progress* **5** 589-596
[94] Contreras J E, Rodriguez E A, and Taha-Tijerina J 2017 *Electric Power Systems Research* **143** 573-584
[95] Sundar L, Ramana E V, Graça M P F, Singh M K, and Sousa A C M 2016 *International Communications in Heat and Mass Transfer* **73** 62-74
[96] Żyła G, Fal J, and Estellé P 2017 *Diamond and Related Materials* **74** 81-89
[97] Ahmed M S and Elsaid A M 2019 *Applied Thermal Engineering* **163** 114398
[98] Wang R, Hao B, Xie G, and Li H 2003 *4th International Symposium on HAVC*
[99] Siddiqui F R, Tso C Y, Chan K C, Fu S C, and Chao C Y H 2019 *International Journal of Heat and Mass Transfer* **132** 1200-1216
[100] Pastoriza-Gallego M J, Casanova C, Páramo R, Barbés B, Legido J L, Piñeiro M M 2009 *Journal of Applied Physics* **106** 064301
[101] Pastoriza-Gallego M J, Casanova C, Legido J L, Pineiro M M, 2011 *Fluid Phase Equilibria* **300** 188–196
[102] Hiemenz P and Rajagopalan R, 1997 *Principles of Colloid and Surface Chemistry, Revised and Expanded.* Boca Raton: CRC Press

IOP Conf. Series: Materials Science and Engineering **810** (2020) 012021 doi:10.1088/1757-899X/810/1/012021

Active hybrid suspension system: A review

Dr. Rajnish Katarne[1], Dr. Satyendra Sharma[2]

[1]Assistant professor, Mechanical Engineering Department, NMIMS University, Mukesh Patel School of Technology Management and Engineering, Shirpur-4254053,India, [2]Associate professor, Mechanical Engineering Department, SIRT, Sage University, kailod kartal, Rau-pass road, Indore- 452020, India,

rajnish.katarne@gmail.com

Abstract. In the current research paper, researcher has tried to give a review on hybrid/active suspension system. The primary purpose of a suspension system in an automobile is to sustain vehicle's stability and maneuverability during irregular and uneven road surfaces. Owing to the fast changing technology, conventional suspension systems limit the performance necessary. However, the modern suspension systems lack the practicality of implementation. The objective of this research is to highlight the various damping systems available and thereby, taking into account the best feasible option for further research in the area. An attempt has been made by the researcher to review significant papers in this area and focus on considered parameters like performance, feasibility, capacity, safety etc.

1. Introduction

Conventional suspension systems, such as, hydraulic dampers with coil springs are outdated and lack the recent innovations and progress. In spite of much technological advancement in the automotive industries, the task of the providing comfort and safety along with meeting the design constraints has not yet been effectively solved. There is a potential need for an active hybrid suspension system with controlled damping actions that utilizes the unused energy dissipated during damping action as well as it provides better ride comfort and vehicle handling abilities to the driver. Excitations from wide range of frequencies restrict the performance of the conventional passive systems.

2. Review on Suspension

Propose a concept of active control of hybrid suspension system is discussed by Suda (1996) *et al.* In this systems researcher has talked about passive dissipate heat in suspension. By numerical simulations and basic experiments, author explained and found that the proposed hybrid control mechanism provides a satisfactory performance in reducing the vibrations as well as in energy regeneration [1]. Actively controlling the vibrations could be depicted with the help of regenerated vibration energy. The energy required for the active control of the system is provided by the damper that regenerates the vibration energy. In this Yoshihiro (1998) *et al.* proposes less energy consuming active control law using the concept of skyhook control [2]. Martins (1999) *et al.* proposed the idea of combining both the passive damper and electromagnetic actuation mechanism. The author also concluded that the use of smaller, lighter actuator results in the same performance as that of the active suspension system. They also suggested that the energy required for the system to run is powered by the suspension system which results in the active force when compared to the conventional hydraulic active system where the source provides both the passive and active forces [3]. Various predictive control techniques to magnetically control mass-spring-damper system could be use.

S. Di Cairano (2007) *et al.* developed two algorithms, one based on mechanical and other on electronic sub systems. Model Predictive Controllers (MPC) only controls the mechanical behavior of the chosen subsystem whereas, electromagnetic subsystems provides a dynamics through an inner-loop controller [4]. Z Li (2013) *et al.* presents an electromagnetic shock absorber which can also regenerate power. Their test results indicated that the stiffness of the spring is a major characteristic in vibration energy harvester. The excess bounce can be easily managed by controlling the electric load on shock absorber. Their design consists of a rack and pinion mechanism which generates power from the irregular surfaces and converts vibration energy to electrical energy [5]. The experimental study on Magneto-rheological (MR) dampers, indicating the variety range in allowable damping force under the acting magnetic field. GZ Yao (2002) *et al.* also suggested that the saturation of *MR* damper takes place as the current in the system reaches a definite value. During the *MATLAB* simulation, the response time was assumed to be very fast and the time delay was neglected. As a result, graphs of acceleration of sprung mass, suspension travel and tire deflections vs frequency were plotted and it became evident that the response was satisfactory for all three parameters around body resonance. Hence, performance characteristics of the MR dampers are very high [6].

Rashid (2007) *et al.* talked about magneto-rheological dampers and their characteristics. In this researcher conducted a deep and intense study on the controller's performance (graphically) and the simulations of a MR damper with fuzzy PID controller and a hybrid fuzzy PID controller are plotted. Adhering to the simulation results, they found that performance of the MR damper (in terms of amplitude of displacement) with a hybrid fuzzy controller reduces the disturbance considerably (90% to 92%) when compared to fuzzy controlled damper. Hence, they concluded that hybrid fuzzy controller was better in performance than the fuzzy controller [7]. The high speed ON/OFF valves, hydro-pneumatic cylinder (actuator) and sensors are used in the suspension system. Shaojun & Liu(1996) *et al.* The system was analyzed with the help of simulation to examine the dynamic performance by designing a quarter car model experiment setup. Skyhook damper control and optimal control was adopted for the simulation of the model. Results indicated that optimal regulator gave vibration isolation nearly 13% whereas the skyhook control gives about 23%. A solenoid ON/OFF valve use confirmed better performance of designed optimal regulator [8]. For the riding comfortableness and traveling safety of automobile , control strategy of automobile magneto-rheological semi-active suspension was proposed by Wang(2008) *et al.* to progress . In addition the application and research status quo of automobile damper were introduced as well as the principle of magneto-rheological effect and the composition of the magneto-rheological fluid. Working principles and models of the automobile magneto-rheological damper was analysed and the future focus was discussed after summarizing the simulation models, control method and testing technology of automobile magneto-rheological damper of automobile suspension [9].

The Eddy Current Damper (ECD) virtual model is developed by Babak (2008) et al. By using appropriate equations and a prototype was developed for testing the magnetic flux density and the damping force. The performance of the ECD is compared with the readily available dampers. However, the authors also concluded that the failure of such system is high as they rely on the use of permanent magnets which eventually gets demagnetized. Use of rare magnets can ensure the desired reliability and durability of the system [10]. The feasibility of a smart passive control system using a MR based damper system. The newly developed system uses conventional passive system (that is, without sensors, controllers or any other power sources) that includes an Electromagnetic Induction device which is capable of generating electrical energy that is further used to change the damping properties of the MR damper. After analysing the result, author was certain that it is feasible to implement a smart passive control with an MR damper reducing the complexity of the circuit and the cost of production [11]. Gysen (2010) *et al.* has proposed the idea of a tubular Permanent-Magnet actuator with a spring attached to it. The author has tried to compare the proposed model with the readily available suspension system such as hydraulic, MR, electromagnetic etc. Specific magnetizations were employed to develop the analytical model and of a quarter car model and conclusions were drawn [12]. Halbach magnetization topology to obtain the highest force generation in system. The research showed that exterior halbach magnetized topology is not capable to hold axial magnetization. The system is suitable when the voltage is low and large currents are necessary which in turn require expensive converters [13].

Hybrid suspension system that exhibited the same simulations results as that of a fully active suspension system proposed by Guido Koch 2010 et.al. An adaptive controller was designed for a quarter car model that consisted of a low bandwidth actuator and a damper. Lyapunov function approach ensured the stability of the adaptive controller, this concept proved to be beneficial as it decreased the power demand, made the system asymptotically stable and easy to implement [14]. The use of a hybrid magnet to reduce the vibration amplitudes over a wide range of frequencies. A quarter car model with single degree of freedom was analyzed with and without the use of hybrid magnet. An actuator which contained a cylindrical magnet was placed between the ends of the springs to suppress vibrations. *MATLAB* Simulink tool was used for modeling the quarter car model and the theoretical results obtained were verified experimentally using *LABVIEW* and *FAST FOURIER TRANSFORM* analyzer [15]. A hybrid energy harvester that uses two power generating mechanism that is piezoelectric and electromagnetic. The finite element and coupling mathematical models were developed and the experimental results depicted that the combination of both the energy harvesting devices had thrice the bandwidth of that a single harvesting mode device. The work presented proved to be effective in increasing the frequency bandwidth and energy output of a harvester [16]. The use of Magneto Rheological dampers in front bumpers of car to reduce the impact of collisions discussed by T.Imthiyaz Ahamed (2014) *et al.* The damper assembly included a chamber filled with MR fluid covered inside coil spring. This was intended to reduce the impact of vibrations by the combine action of the spring and MR fluid. The proposed model suggested that MR fluid accounts for absorbing 50-60 % vibrations during the event of a collision and the rest as absorbed by spring [17].

The numerous remote sensors embedded in a vehicle and harvest the energy to power. This power was to be obtained by piezoelectric conversion of vibrations transferred through uneven surfaces. Bond graph model was adapted to identify the power recovering areas when the car hit a bump at a speed of 30km/hr. The results suggested that around 0.5mW was recovered during the process. The paper suggested that increase in the number of piezoelectric harvesters resulted in more power recovery [18]. The effects of various control strategies on semi active dampers like skyhook, ground hook and hybrid on a quarter car simulation on *MATLAB was* analysed. The results obtained are further compared with the results of a commercially available passive system. It was concluded that for a distance of 0.01 meter, skyhook control proved to the most optimal choice in reference of ride comfort and stability [19]. The key purpose of a suspension system is to provide the shock absorption in automobiles. Besides carrying the weight of the vehicle it attempts to minimize/eliminate vibrations that may be induced by different sources like road surface irregularities, aerodynamics forces and non-uniformity of the tire/wheel assembly [20]. Abroon Jamal qazi (2013) discussed successful application of hybrid artificial intelligence techniques in designing a semi-active suspension system. Both passive and semi-active suspension systems have been showed in Simulink [21].

Y.M. Halde e.al.(2015) show active hybrid suspension system and focused on performance of dampers in the suspension system by using smaller electrically powered actuator increasing the performance of system. As per road conditions suspension system synchronize the performance. Few Researchers also suggested three degree of freedom vibration isolation system using active zero power controlled magnetic suspension is presented in order to isolated vibrations transmitted. With the help of experiments studies research has demonstrated the effectiveness of the isolator for dynamic vibration [22]. Zhang Jin-qiu, et.al.(2013)development of regenerative suspension is reviewed, and the energy harvesting schemes and their characteristics are remarked/summarized. In the regenerative suspensions combination of performance and energy harvesting efficiency is having scope in this type of system. The vibration generated by road roughness is controlled by robust artificial neural network control system scheme [23]. This is proposed by Ikbal Eski et.al. (2009). He designed active suspension by using neural network based control system. The performance is compared to both the proposed RNN control system and the PID controller for fortuitous in road bumpiness [24]. With the help of neural networks K N Spentzas et.al (2002) has presented a design method for non-linear hybrid suspension systems. When the roughness of road is exceeded up to certain limit then active suspension system gives best performance. Irregular action of the controller helps to minimize the energy consumed by the actuator. The target of the design is to give more comfort ride to the passenger by minimization of the vertical acceleration effects [25].

Kinetics simulation model/ working theories by applying the interdisciplinary software AMESim for hydraulic transmission electromagnetic energy-regenerative suspension are discussed by the Lin Xu (2011). Comparison of hydraulic transmission electromagnetic energy-regenerative suspension/ passive suspension, of the energy-regenerative suspension of this structure leads to improve the ride comfort /fuel economy [26]. Due to high energy consumption Active suspension has limited use in automobiles. Long Chen et.al (2016) has discussed the concept of different type of suspension in this linear motor is used in adjustable shock absorber to form the hybrid active semi active suspension (HASAS). Its energy consumption and regeneration mechanisms are revealed. And the system controller which is composed of linear quadratic regulator (LQR) controller, mode decision and switch controller, and the sliding mode control based thrust controller is developed. LQR controller is designed to maintain the suspension control objectives, while mode decision and switch controller decides the optimal damping level to tune motor thrust [27]. The modelling and force tracking control of non-linear hydraulic actuator system is used in a quarter-car active suspension system. The two loops namely inner loop and outer loop controllers are developed for the active suspension system. The function of outer loop controller is absorbing the effects of road disturbances and the inner loop controller is used to keep the actual force close to this desired force. Both the loop show significant improvement in the performance of the suspension system [28]. In hybrid electric vehicles active suspension system with simultaneous simulation of power train shown by Seyedmohsen Hosseini et.al (2008). Both the system is integrated and the exchange of power and data between two systems is simulated by the authors. Variations due to road roughness are generated and are applied to the suspension system [29].

Now days demand in on road and off road is increasing comfort and handling performance is vital. When we focus on suspensions performance, it's a trade of between various parameters like suspension travel, vibration reduction, actuator effort, road holding capability, noise and fatigue. Stratis Kanarachos et.al.(2015) proposed design of intelligent suspension system which focuses on all these permanents with the help of actuators. An algorithm continuously monitors the real time data of suspension travel and adapts accordingly [30]. A mathematical model for three axis active suspension system is proposed by Takeshi Mizuno et.al (2003). In this he shown the three motions could be treated separately in this zero-power control system is used this characteristic was confirmed experimentally [31]. Neural scheme is used with PD, PI and PID controllers to control the suspension system. This control system always works in closed loop which leads to better performance S. Yıldırım (2003) [32]. Some researchers also used semi-active control methods by using commercial magnetorheological damper in a suspension system. With the help of experimental test bed Y. SHEN et.al.(2006) study the three control methods under sinusoidal and random excitations. This experimental result verified with analytical results also. Both the methods show significance improvements in the system [33].

Use of Magneto rheological dampers proposed by Ali Fellah Jahromi,(2012) in suspension systems of passenger cars. This system provides comfortable ride by considering various parameters like roiling, vertical movement of four tyres [34]. Babak Assadsangabi, (2009) designed controller which takes consideration of ride comfort and road holding for this he suggested hybrid model reference sliding mode controller. This controller is having two stages control system [35]. By the use of MATLAB/SIMULINK Ansar Mulla et.al. (2014) analyze effect of vibration on suspension system in quarter-car model [36]. Magneto rheological damper system is presented for semi-active control system by G Z. Yao et.al. (2002) [37]. Luis C. Félix-Herrán (2008) magneto rheological damper system including the dynamics of the semi-active suspension. This leads to more comfortable ride for passengers [38]. Yechen Qin et.al. (2017) has compare/ analysis the time delay in semi-active suspension control systems. In this paper five systems has been compared and suggested increase in delay time could increase ride comfort [39]. Multi-objective microgenetic algorithm is used for semi-active suspension systems in MATLAB. This result shown it is helping to increase ride comfort [40].

3. Conclusion

In this paper, a hybrid suspension machine which mixes a passive damper with an electromagnetic actuator is proposed. The objective of the research is to reap a better performing suspension system without compromising the protection of the automobile. Literature review furnished an informational technique closer to the controlling systems for dampers. The following table 1 is the optimization of damper controlling structures on the premise of 4 parameters. If the cost parameter is considered, Electromagnetic damping device is a superb choice than Magneto-rheological or Electro-rheological systems. Although, if the performance is given a more credit score, the MR machine is a great choice and Electromagnetic damping system comes subsequent. Reviewing the desk similarly, Permanent Magnet structures show to be less green being highly-priced. Reviewing the desk similarly, Permanent Magnet systems prove to be much less efficient being steeply-priced. It may be without problems concluded that Electromagnetic Damping device may be an apt desire for further evaluation in this region.

Table 1. Optimization of damper controlling structures on the premise of 4 parameters.

Controlling system	Cost	Feasibility	Performance	Implementation
Magneto-Rheological Damper System	High	Medium	High	Low
Electro-Rheological Damper System	Very High	Less	High	Low
Electromagnetic Damping system	Medium	Medium	High	High
Permanent Magnets	Medium	Less	Low	Medium
Hydraulic Damping System	Low	More	Low	High
Eddy Current Damping system	Medium	Less	Medium	Medium

4. References

[1] Suda, Yoshihiro and Taichi Shiiba "A new hybrid suspension system with active control and energy regeneration." Vehicle System Dynamics 25, no. S1 (1996): 641-654
[2] Suda, Yoshihiro, Shigeyuki Nakadai and Kimihiko Nakano. "Hybrid suspension system with skyhook control and energy regeneration (development of self-powered active suspension)." Vehicle system dynamics 29, no. S1 (1998): 619-634

[3] Martins, Isménio, M Esteves, F Pina Da Silva and Pedro Verdelho. "Electromagnetic hybrid active-passive vehicle suspension system." In Vehicular Technology Conference, 1999 IEEE 49th, vol. 3, pp. 2273-2277. IEEE, 1999
[4] Cairano, S Di, Alberto Bemporad, Ilya V Kolmanovsky and Davor Hrovat "Model predictive control of magnetically actuated mass spring dampers for automotive applications." International Journal of Control 80, no. 11 (2007): 1701-1716
[5] Li Zhongjie, Lei Zuo, George Luhrs, Liangjun Lin and Yi-xian Qin "Electromagnetic energy-harvesting shock absorbers: design, modeling, and road tests." IEEE Transactions on Vehicular Technology 62, no. 3 (2013): 1065-1074
[6] Yao G Z, F F Yap, G Chen, W H Li and S H Yeo "MR damper and its application for semi-active control of vehicle suspension system." Mechatronics 12, no 7 (2002): 963-973
[7] Rashid M M, M A Hussain, N Abd Rahim and J S Momoh. "Development of a semi-active car suspension control system using magneto-rheological damper model." International Journal of Mechanical and Materials Engineering 2, no. 2 (2007): 93-108
[8] Shaojun Liu "An Active Suspension System Operated by High Speed Solenoid Valves and Its Control." Automotive Engineering18, no. 4 (1996)
[9] Bing, Wang, Han Bingyuan, Wang Yan and Ai Xifeng "Research Review of Automobile Magneto-rheological Damper [J]" Forest Engineering 4 (2008): 012
[10] Ebrahimi, Babak, Mir Behrad Khamesee and Farid Golnaraghi "Eddy current damper feasibility in automobile suspension: modelling, simulation and testing." Smart Materials and Structures18, no. 1 (2008): 015017
[11] Choi, Kang-Min, Hyung-Jo Jung, Heon-Jae Lee and Sang-Won Cho "Feasibility study of an MR damper-based smart passive control system employing an electromagnetic induction device." Smart Materials and Structures 16, no. 6 (2007): 2323
[12] Gysen, Bart LJ, Johannes JH Paulides, Jeroen LG Janssen and Elena A Lomonova "Active electromagnetic suspension system for improved vehicle dynamics." IEEE Transactions on Vehicular Technology 59, no. 3 (2010): 1156-1163
[13] Gysen, Bart LJ, Jeroen LG Janssen, Johannes JH Paulides and Elena A Lomonova "Design aspects of an active electromagnetic suspension system for automotive applications." IEEE transactions on industry applications 45, no. 5 (2009): 1589-1597
[14] Koch, Guido, Sebastian Spirk and Boris Lohmann "Reference model based adaptive control of a hybrid suspension system " IFAC Proceedings Volumes 43, no. 7 (2010): 312-317
[15] Easu, D and A Siddharthan "Theoretical and Experimental Analysis of a Vibration Isolation System Using Hybrid Magnet." Procedia Engineering 64 (2013): 1139-1146
[16] Shan, Xiao-biao, Shi-wei Guan, Zhang-shi Liu, Zhen-long Xu and Tao Xie. "A new energy harvester using a piezoelectric and suspension electromagnetic mechanism." Journal of Zhejiang University SCIENCE A 14, no. 12 (2013): 890-897
[17] Ahamed, T Imthiyaz, R Sundarrajan, G T Prasaath and V Raviraj. "Implementation of magneto-rheological dampers in bumpers of automobiles for reducing impacts during accidents." Procedia Engineering 97 (2014): 1220-1226
[18] Lafarge, Barbara, Christophe Delebarre, Sébastien Grondel, Octavian Curea and A Hacala. "Analysis and optimization of a piezoelectric harvester on a car damper." Physics Procedia 70 (2015): 970-973
[19] Mulla, Ansar, Sudhindra Jalwadi and Deepak Unaune. "Performance analysis of skyhook, groundhook and hybrid control strategies on semiactive suspension system." Int. J. Curr. Eng. Technol., Special Issue-3 (2014): 265-271
[20] Anil Shirahatt, Pravin Panzade and Kulkarni M M Optimal design of passenger car suspension

for ride and road holding. Journal of Brazilian Society of Mechanical Science & Engineering, 2008, 30(1): 66-76.
[21] Abroon Jamal qazi, Umar A Farooqui, Afjal Khan, M Tahir Khan, Farrukh Mazhar and Ali Fiaz AASRI Procedia 4 (2013) 160 – 166, Optimization of semi-Active suspension system using particle swarm optimization algorithm
[22] Md. Emdadul Hoque, Masaya Takasaki, Yuji Ishino, Hirohisa Suzuki and Takeshi Mizuno (2006), an active micro vibration isolator with zero power controlled magnetic suspension technology JJSME international journal 49(3) 719-726
[23] Zhang Jin-qiu, Peng Zhi-zhao, Zhang Lei and Zhang Yu, A Review on Energy-Regenerative Suspension Systems for Vehicles, Proceedings of the World Congress on Engineering 2013 Vol III, WCE 2013, July 3 - 5, 2013, London, U.K.
[24] Ikbal Eski, Sahin Yıldırım Vibration control of vehicle active suspension system using a new robust neural network control system, Simulation Modelling Practice and Theory 17 (2009) 778–793
[25] K N Spentzas and S A Kanarachos, A neural network approach to the design of a vehicle's non-linear hybrid suspension system, Proc Instn Mech Engrs (2002) Vol 216 Part B: J Engineering Manufacture,833-838
[26] Lin Xu Structure Designs and Evaluation of Performance Simulation of Hydraulic Transmission Electromagnetic Energy-Regenerative Active Suspension, 2011 SAE International, doi: 10.4271/2011-01-0760
[27] Long Chen, Dehua Shi, Ruochen Wang and Huawei Zhou Energy Conservation Analysis and Control of Hybrid Active Semiactive Suspension with Three Regulating Damping Levels,Shock and Vibration Volume 2016, Article ID 6196542,1-14
[28] Y M Sam and K Hudha, Modelling and Force Tracking Control of Hydraulic Actuator for an Active Suspension System, ICIEA 2006
[29] Seyedmohsen Hosseini, Rushed Abdolah, and Amir Khani Investigation of the Idea of Active Suspension System Application in Hybrid Electric Vehicles, Proceedings of the World Congress on Engineering 2008 Vol IIWCE 2008, July 2 - 4, 2008, London, U.K.
[30] Stratis Kanarachos and Andreas Kanarachos, Intelligent road adaptive suspension system design using an experts' based hybrid genetic algorithm, Expert Systems with Applications 42 (2015) 8232–8242
[31] Takeshi Mizuno, Masaya Takasaki, Hirohisa Suzuki and Yuji Ishino, Development of a Three-Axis Active Vibration Isolation System Using Zero-Power Magnetic Suspension, 42nd IEEE conference on decision and control maui, hawaii USA, December 2003,4493-4498
[32] S Yildirim (2003), Vibration control of suspension systems using a proposed neural network, Journal of Sound and Vibration 277 (2004) 1059–1069
[33] Y SHEN, M F Golnaraghi and G R Heppler (2006), Semi-active Vibration Control Schemes for Suspension Systems Using Magnetorheological Dampers, Journal of Vibration and Control, 12(1): 3–24, 2006
[34] Ali Fellah Jahromi, Wen Fang Xie and Rama B Bhat, Ride Control of Passenger Cars with Semiactive Suspension System Using a Linear Quadratic Regulator and Hybrid Optimization Algorithm, International Journal of Vol: 6 No: 7, 2012
[35] Babak Assadsangabi, Hybrid sliding mode control of semi-active suspension systems, Smart Materials And Structures,18 (2009) 125027 (10pp)
[36] Ansar Mulla, Sudhindra Jalwadi and Deepak Unaune, Performance Analysis of Skyhook, Groundhook and Hybrid Control Strategies on Semiactive Suspension System, International

Journal of Current Engineering and Technology ISSN 2277 - 4106, International Journal of Current Engineering and Technology, Special Issue-3, (April 2014), 265-269

[37] G Z Yao, F F Yap, G Chen, W H Li and S H Yeo, MR damper and its application for semi-active control of vehicle suspension system, Mechatronics, 12 (7), 963-973

[38] Luis C Félix-Herrán, Modeling and Control for a Semi-active Suspension with a Magnetorheological Damper including the actuator dynamics, Electronics, Robotics and Automotive Mechanics Conference 2008, 338-334,DOI 10.1109/CERMA.2008.16

[39] Yechen Qin, Feng Zhao, Zhenfeng Wang, Liang Gu and Mingming Dong, Comprehensive Analysis for Influence of Controllable Damper Time Delay on Semi-Active Suspension Control Strategies, Journal of Vibration and Acoustics, JUNE 2017, Vol. 139 / 031006-1

[40] Yi Chen, Zhong-Lai Wang, Jing Qiu and Hong-Zhong Huang, Hybrid Fuzzy Skyhook Surface Control Using Multi-Objective Microgenetic Algorithm for Semi-Active Vehicle Suspension System Ride Comfort Stability Analysis, Journal of Dynamic Systems, Measurement, and Control, JULY 2012, Vol. 134 / 041003-1

IOP Conf. Series: Materials Science and Engineering **810** (2020) 012022 doi:10.1088/1757-899X/810/1/012022

An application of AHP for parameter importance evaluation for farm machinery selection

Abhijeet U. Karmarkar

Department of Mechanical Engineering, K J Somaiya College of Engineering, Mumbai, India

abhijeet@somaiya.edu

Nandkumar R. Gilke

Department of Mechanical Engineering, K J Somaiya College of Engineering, Mumbai, India

nandkumargilke@somaiya.edu

Abstract. The farm machinery selection is a complex decision making task. It involves a variety of variables associated with farming which include social and economic aspects of the decision maker. The aim of this work is to develop decision support architecture for tractor selection process. Different parameters for selecting tractor were identified; their importance was studied by literature survey and evaluated with AHP. The study concludes that the parameter operation type receives highest importance in tractor selection process. Other parameters like, service, type of business, brand and cost are also important in tractor selection process.

Keywords: Agriculture, Farm Machinery, AHP, decision support architecture

1. Introduction

The tractor is one of the most commonly used farm machinery for various farming related applications. Thus the tractor selection is an important part of farm machinery selection and management. The farm machinery selection is the process of assessing the suitability of machinery and their possible usage in farming domain for a sustainable and profitable crop production. The farm machinery selection is also a significant process for ensuring effective utilisation of available mechanical power. This makes the decision making process most important for the farming domain. The selection process encompasses a variety of factors such as, financial, social, technical etc. The involvement of multiple parameters adds to the complexity of the decision making process and thus the problem can be modelled as a multi-criteria-decision-making (MCDM) problem. Such MCDM problem deals with evaluation of a set of alternatives while considering the set decision criteria. In recent years, researchers have introduced a large variety of tools, techniques and methods for solving different MCDM problems. Most of the MCDM approaches incorporate the pairwise comparison approach for calculation of weights for decision alternatives.

This study introduces a Decision Support Architecture (DSA) for helping decision-makers in decision making for complex MCDM problems. The DSA is a tool which can support in structured decision making. The novelty of this DSA is that the decision making process involves the implementation of the pairwise comparison method for evaluating the importance of features associated with decision making instead of the alternative. The development of DSA is the first phase of development of the Decision Support System (DSS) for supporting farmers. The alternative selection phase of the DSS is not a part of this work and will be considered separately later.

This paper is structured as follows. Section 2 describes the research method followed in this work. Section 3 provides a glance into the literature of tractor selection methodologies and related parameters. Section 4 describes the development of decision support architecture for tractor selection process. Section 5 highlights the results obtained from the study along with the analysis of obtained results. Section 6 gives the discussion on obtained results and offers the directions for future studies regarding tractor selection.

2. Materials and methods

In this study the problem of the tractor selection is under consideration. The problem typically arises due to the gap in the understanding and analysis of product requirements of the manufacturer and the buyer. The tractor manufacturer tries to develop more and more advanced versions of the product by incorporating latest trends in technology. This results in a large mix of products and the buyer has to select required product from this mix. The buyers who are typically from farming background may face difficulty in understanding farm specific requirements and thus may fail to efficiently select the product. This results in wastage of mechanical power.

In recent years, researchers have implemented different strategies for addressing the power wastage problem. A commonly used strategy is not to select tractor specifically but to select different parameters associated with tractor. These parameters include various parameters like, tractor power, implement size etc. The prediction of tractor performance is also studied by researchers. The focus of these studies was to predict how tractor will perform under the operation and soil conditions. In these types of work, objective of researchers was to support the buyer by considering technical requirements of the tractor for farming operations. Some researchers have considered other parameters or rather non-technical parameters along with technical parameters for tractor selection. These non-technical parameters include financial attributes like price, cost of maintenance etc., social parameters like, buyer's background, age, education etc. Table 1 illustrates some selected support system approaches from the literature. Most of the tabulated work as observed in table can be classified as either associated with technical or with non-technical parameters being considered for tractor selection. In only a few cases, combination of both technical and non-technical parameters is considered in a single system. Even in such systems, many other parameters associated with tractor selection are ignored. The other attribute of the support systems considered in literature is the type of MCDM approach used for addressing tractor selection problem. In most of the studies, DBMS based approach is used where a support system uses a database at background for selecting suitable tractor. In case of such system, there is a problem of empty selection, i.e. no suitable product is found which can fulfil all necessary requirements completely. The AHP based systems are advantageous in such MCDM problems as these systems can be used to find most suitable option out of available options. These systems depend on a pairwise comparison of various attributes pertaining to different options and whenever list of options or attributes is modified, the whole evaluation process for AHP model has to be conducted again. Further this method is dependent on technology advances which can cause entire pairwise comparison invalid and will require a re-evaluation of system. Table 1 highlight the technology used which can be DBMS or AHP in support system approaches from literature.

Table 1. Overview of the tractor selection and performance prediction systems from literature

Author(s)	Domain	Modeling Approach	Technology Used	Input Parameters	Output Parameters
Al-Hamed & Al- Janobi (2001)	Tractor Performance prediction	Mathematical Model	DBMS, Visual C++	Weight, Tractive Efficiency	Traction Ratio, Pull, Drawbar Power, Actual Speed
Rizwan-ul-Haq, Muhammad, Shafi (2003)	Tractor Selection	Mathematical Model	Statistical Analysis	Tractor Size, Crop, Implement	Field Time, Number of Tractors, Tractor Size
Catalán et al. (2008)	Tractor Performance Prediction	Mathematical Model	Visual Basic	Drawbar Pull, Tractor, Tires, Terrain	Dynamic Axle Load, Slip
Sahu & Raheman (2008)	Tractor Performance Predication	Mathematical Model	Visual Basic, DBMS	Tractor, Implement, Soil Condition, Operating Condition	Coefficient of Traction, Pull, Engine Torque, Slip
Mehta et al. (2011)	Tractor Selection	Mathematical Model	Visual Basic, DBMS	Field Area, Available Time, Speed	Speed, Drawbar Power
Bialobrzewski and kolator (2011)	Tractor Performance Predication	Mathematical Model	MATLAB & Simulink	Drawbar Pull, Implement	Speed, Tractive Efficiency, Slip
Patel et al. (2012)	Tractor Performance Predication	Mathematical Model	Visual Basic, DBMS	Implement, Tractor, Soil Condition, Type of Land	Draft, Drawbar Power, Slip, Tractive Efficiency,
Lotfie et al. (2013)	Tractor Selection	Mathematical Model	Visual Basic, Excel	Crop Type, Implement	Tractor Size, Power Required, Fuel Consumption
Civelek, Say (2016)	Tractor Selection	Expert System	DBMS	Tractive Efficiency, Time, Implement	Drawbar Power
Awulu et al. (2016)	Tractor Selection	Expert System	Visual Basic, DBMS	Soil Type, Operation Type, Implement	Tractor

Notes: The first column 'Domain' points out problem domain. The second column 'Modelling Approach' denotes the type of modelling approach used in study. The third column 'Technology Used' specifies the programming or conceptual tool used in the study. The fourth and fifth columns namely 'Input Parameters' and 'Output Parameters' signify the input and output parameters involved in the study.

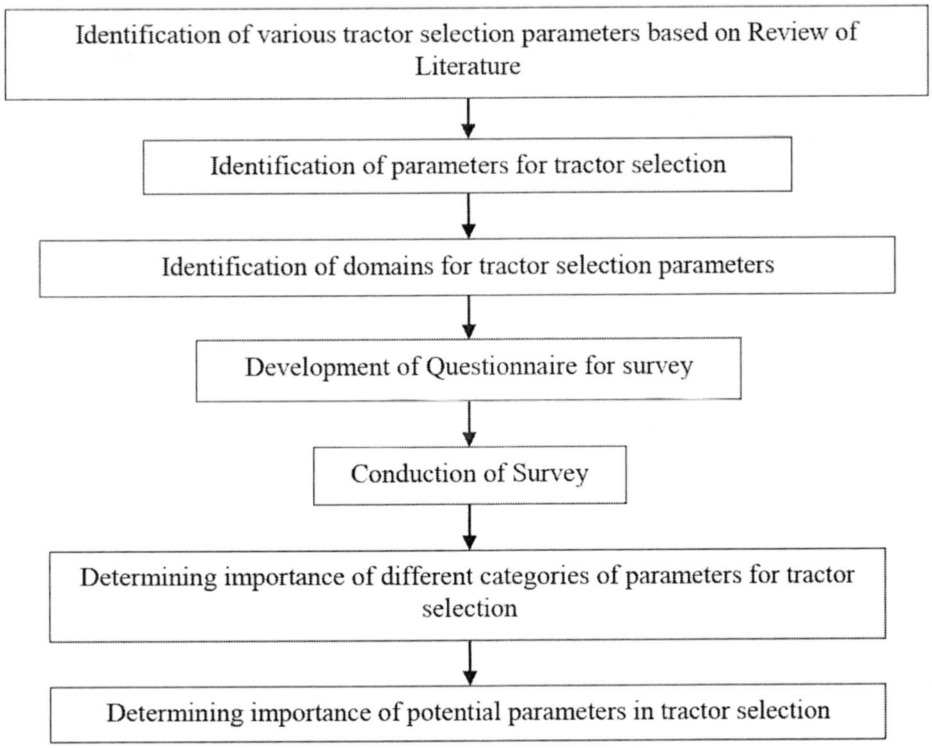

Figure 1: Methodology for determination of importance for parameters and domains

In this work, the a MCDM system is discussed which can aid in the product selection process. The MCDM system deals with problem solving through help of the DSA. The DSA deals with structuring of MCDM problem which concerns a set of criteria: (1) Parameter identification, (2) Parameter Selection, (3) Feature extraction, (4) Selection of weighing method, (5) Applying selected method, (6) Evaluation of parameter importance in decision making process.

The DSA proposed in this work is based on Analytic Hierarchy Process (AHP) methodology for evaluation of parameter importance. The goal of DSA is to provide a generalised parameter importance values for a structured decision making. The DSA is inspired by expert knowledge which is gathered through series of interviews. Total nine experts (tractor dealers representing different tractor manufacturers located in different parts of State of Maharashtra, India) participated in this research to evaluate parameter importance. The methodology for incorporating the views of these experts into the DSA design is shown in figure 1. The survey was conducted in an interview format with open ended survey questioner. The questioner was designed as an open ended questioner for effectively capturing the expert opinions. The parameter and domain importance was evaluated from expert opinion. These importance values were used for development of relative importance matrix for parameters. The developed relative importance matrix was analysed by calculating consistency ratio for the matrix and comparing the calculated value with the set threshold value.

3. Decision Support Architecture

The main aim of this study is to develop DSA for effective selection of farm machinery. The DSA is represented by means of a block diagram in figure 2. In order to develop the DSA, a prerequisite was to collect information and acquire knowledge regarding the tractor selection process. The required knowledge was acquired through a series of interview of experts. This acquired knowledge was then used for system model development. The system model comprised of various parameters and domains associated with the farm machinery selection. This parameter domain relation along with the respective weights formed the knowledge base for decision making. This knowledge base can be further used by an inference engine for decision making purpose. The design and functionality of each block is discussed in further topics.

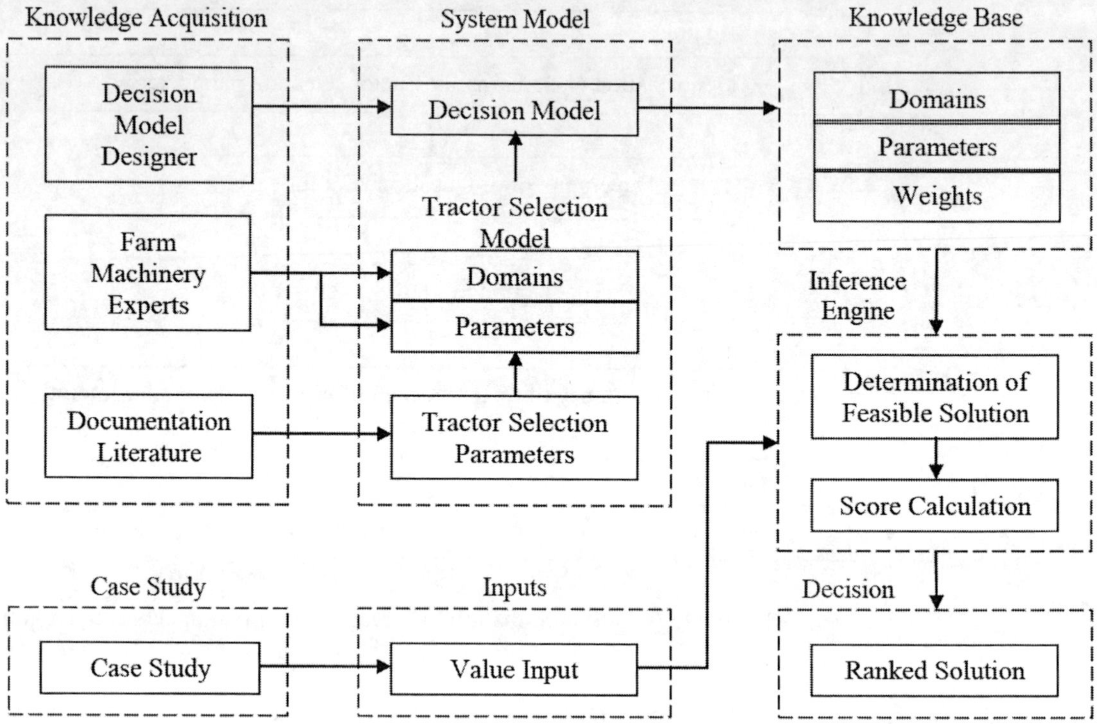

Figure 2. A model based DSA for tractor selection problem

3.1. Knowledge Acquisition

As discussed earlier, acquisition of knowledge was a prerequisite for development of DSA. The information gathered for knowledge acquisition was from available literature, system model designers and from experts working in farm machinery selection domain. After study of available literature, around 40 different parameters were identified. These parameters were largely spread over various phases of farm machinery. These parameters typically belonged to technical, financial, social, psychological and farming domains. All of these parameters were considered for developing parameter description for farm machinery selection. The inputs were also taken from farm machinery selection experts for understanding importance of identified parameters. This expert knowledge was used for system model designing.

3.2. System Model

The system model for the DSA comprised of the parameter relation model which was used for development of inference engine which are the most important aspects of any MCDM system. The domain and parameter definitions formed the base of system model. These specified parameters and domains were the products of analysis of parameters extracted from the literature. In order to facilitate

IOP Conf. Series: Materials Science and Engineering **810** (2020) 012022 doi:10.1088/1757-899X/810/1/012022

effective decision making, the parameter dependency was modelled by means of weighted domain parameter dependency. An AHP based weighted parameter matrix was used for this purpose.

The AHP is a method of decision making which incorporates the pairwise comparison of parameters with reference to input data. This AHP methodology is capable of supporting a complex decision making process while dealing with a variety of input parameters. As suggested by Saaty [18] AHP considers problem definition, decision hierarchy, pairwise comparison of parameters and assigning priorities to parameters for providing an organised way to decision making. The AHP method is based on evaluations of the available alternatives with reference to the weight or importance assigned to the criteria by the decision maker. The weight is in terms of relative importance of the criteria in decision making process. The AHP methodology is easy to implement for a decision maker in order to make a correct decisions. The AHP based problem solving approach mainly consists of four main steps, which are problem definition, formation of decision hierarchy, pairwise comparison matrix formation and priority determination [19].

Table 2. Satty's Scale for Pairwise comparison [19]

Intensity of Importance	Definition	Explanation
1	Equal importance	Two activities contribute equally to the objective
3	Weak importance of one over the other	Experience and judgment slightly favour one activity over the other
5	Essential or strong importance	Experience and judgment strongly favour one activity over the other
7	Demonstrated importance	An activity is strongly favoured and its dominance demonstrated in practice
9	Absolute importance	The evidence favouring one activity over the other is one of the highest possible validity
2-4-6-8	Intermediate values between the two adjacent judgments	When compromise is needed
Reciprocals of above nonzero numbers	If activity 1 has one of the above nonzero numbers assigned to it when compared with activity 2, then activity 2 has the reciprocal value when compared with activity 1	

The AHP methodology finds a large number of applications in decision making and decision assessment domains. In this work, a modified AHP (M-AHP) method was used for analysing the data gathered regarding the tractor selection parameters and building a maintainable and evolvable DSA. The M-AHP approach as suggested by Nefeslioglu et. al [20] is useful in compensating the expert subjectivity encountered with the parameter comparison. The M-AHP method uses basics of AHP methodology and builds on the concepts with certain modification. The steps involved in M-AHP in approach are:

- Inputs from Experts are used for assigning maximum scores for each parameter in the system.
- Based on the parameter scores, the parameter score difference matrix is prepared.

- The parameter score difference values are normalised with reference to the maximum parameter scores assigned.
- The parameter comparison matrix is then constructed with help of the conversion scale.

The parameters under considerations in this study demonstrate a correlation with few other parameters. This correlation in form of parameter inter-dependency must be considered in the decision making process. In order to incorporate this parameter dependency in DSA, multi-level implementation of M-AHP approach is performed. The parameter importance obtained from analysis of data highlights the necessity of every parameter in tractor selection process. The major challenge was to incorporate the maintainability and resolvability with DSA for tractor selection. This was achieved through a step wise implementation of approach. This stepwise approach utilised for DSA development involves, domain and parameter description, knowledge base and decision model as fundamental sub-parts.

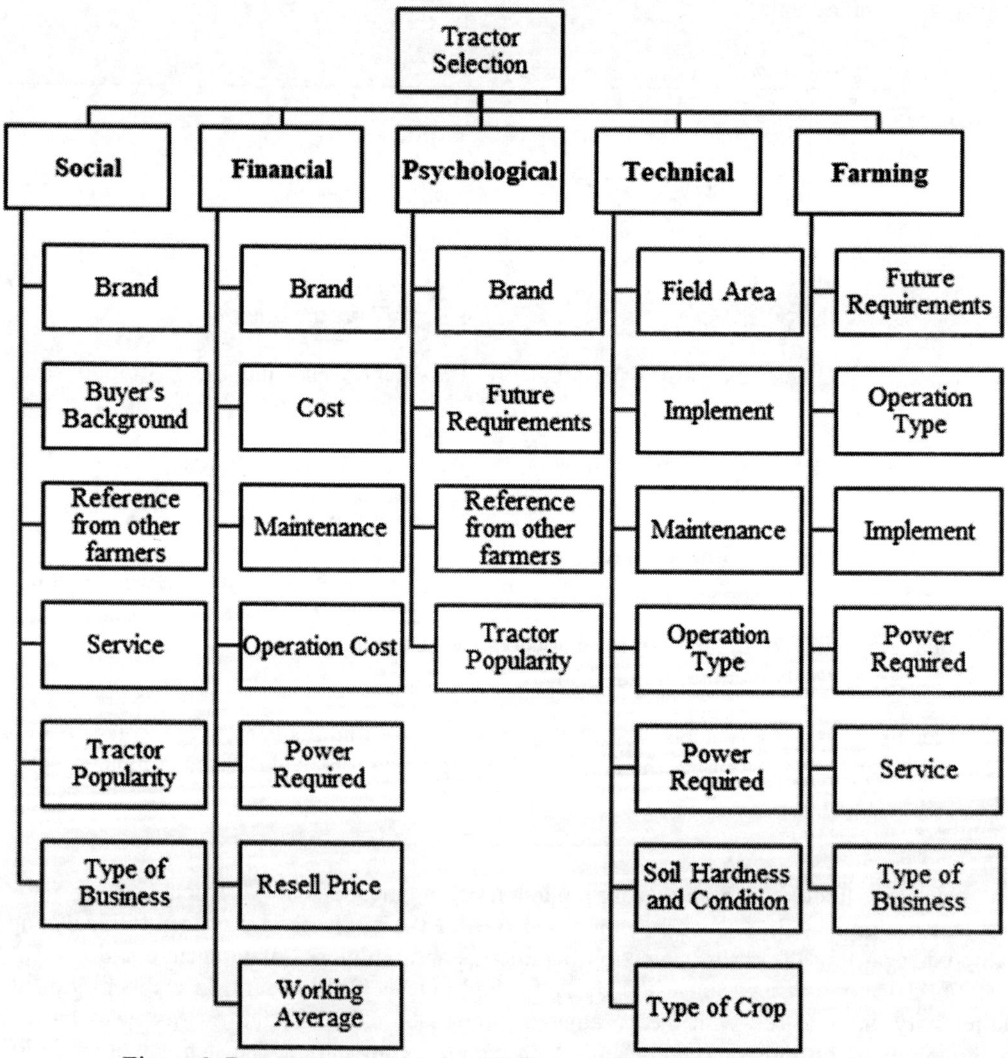

Figure 3. Parameter grouping into different tractor selection categories

3.3. Domain and Parameter Description

At the initial stage of work, various parameters associated with tractor selection were identified and studied. The similar parameters from literature were grouped together in order to formulate the parameters for tractor selection. This parameter grouping was done with reference to the similarity between the parameters. The parameters which showed similarity with other parameters were grouped into a single parameter. The parameters which exhibited an overlap of concept or meaning or definition were also grouped together. The parameters which were not pertaining to any of the above criteria were considered separately. There were 18 such individual parameters which were formulated. These 18 parameters were then classified into 5 tractor selection domains. This classification was not unique i.e. a single parameters may have been a part of multiple domains. These parameters were both technical and non-technical parameters and were quantitative or qualitative in nature. Figure 3 shows the domain wise classification of the formulated parameters.

3.4. Knowledge Base

In order to evaluate the importance of parameters, views of 9 experts were considered. These experts were tractor dealers pertaining to selected regions of state of Maharashtra. Their views on importance of each parameter in tractor selection were recorded by means of a survey. The parameter importance level was mapped by means of a predefined scale which is shown in table 3. This importance scale provided a structured data for further processing. The parameter importance was used for development of knowledge base for decision model. The knowledge base also included domain and parameter definitions along with weightages. The knowledge base was utilized for effective decision making and machinery selection.

Table 3. Importance matrix scale for survey data conversion

	Level of Importance				
	Not Important	Moderately Important	Strongly Important	Very Strongly Important	Extremely Important
Numerical Importance assigned	0	1	2	3	4

4. Result and Discussion

Table 4 illustrates the result of analysis carried out for parameter importance evaluation. The use of AHP based methodology proved effective in assessment of importance of tractor selection parameters. The analysis was carried out not only for parameter importance but also for importance of domain in tractor selection process. The domain wise importance was also used for evaluating parameter importance. The analysis of AHP model formulated was carried out by evaluating 'Consistency Ratio' of the relative importance matrix which is the heart of AHP methodology. This analysis showed consistency ratio had value within limits which proved the feasibility of relative importance matrix. The parameter 'Operation Type' was found to be of maximum importance in tractor selection process. This analysis also showed that the parameters like, 'Service', 'Type of Business', 'Brand', 'Cost', 'Buyer's Background' were also important while dealing with tractor selection. The 'Operation Cost' parameter was found to be a neglected parameter in tractor selection process. The parameters like, 'Working average', 'Resell Price', 'Field Area' were found to be of lower importance in tractor selection process.

5. Conclusion

In this work, the AHP method is implemented for development of DSA. The implementation of AHP methodology allows the evaluation of parameter importance in decision making process. The analysis of consistency ratio for the decision matrix showed that the decision matrix was acceptable. Thus this decision matrix was usable in actual decision making scenario. The analysis also showed that

IOP Conf. Series: Materials Science and Engineering **810** (2020) 012022 doi:10.1088/1757-899X/810/1/012022

parameter 'Operation Type' received maximum importance. This result was in union with the literature as many other parameters of tractor selection were directly or indirectly dependent on this parameter. Parameters like 'Service', 'Type of Business' and 'Brand' were also found to be important in tractor selection. The 'Cost' parameter was found to be less important than expected. The parameter importance evaluated in this work provides a feasible relative importance matrix for DSA. The use of step wise importance evaluation enables easy evolution and modification of DSA. The DSA developed in this study can be further used for a development of DSS for actual tractor selection.

Table 4. Parameter importance and importance wise ranking for tractor selection parameters

Potential Parameter	Parameter Ranking
Operation Type	1
Service	2
Type of Business	3
Brand	4
Cost	5
Buyer's Background	6
Maintenance	7
Implement	8
Tractor Popularity	9
Power required	10
Soil Hardness and condition	11
Type of crop	12
Reference from other farmers	13
Future requirements	14
Field Area	15
Resell Price	16
Working average	17
Operation cost	18

6. References

[1] Al-Hamed, S. A., and Al-Janobi, A. A., "A program for predicting tractor performance in Visual C++." Computers and Electronics in Agriculture, 31, (2001): 137-149.

[2] Muhammad, M., and Sabir,S., "Tractor-size selection model for multi-crop farms of Pakistan", Journal of Engineering and Applied Sciences22, (2003): 79-86.

[3] Catalán, H., Linares, P., and Méndez, V., "Tractor_PT: A traction prediction software for agricultural tractors." computers and electronics in agriculture, 60, (2008): 289-295.

[4] Sahu, R. K., and Raheman, H., "A decision support system on matching and field performance prediction of tractor-implement system." Computers and electronics in agriculture, 60, (2008): 76-86.

[5] Mehta, C. R., Singh, K., and Selvan, M. M., "A decision support system for selection of tractor–implement system used on Indian farms." Journal of Terramechanics, 48, (2011): 65-73.

[6] Kolator, B., and Białobrzewski, I., "A simulation model of 2WD tractor performance." Computers and electronics in agriculture, 76, (2011): 231-239.

[7] Patel, T., Chakravorty, A., and Karmakar, S., "Software for performance prediction and matching of tractor-implement system." Emerging Trends and Applications in Computer

Science (NCETACS), 3rd National Conference on. 2012, IEEE, 2012.
[8] Yousif,L. A.,Dahab, M. H., and El Ramlawi, H. R., "Crop-machinery management system for field operations and farm machinery selection", Journal of Agricultural Biotechology and Sustainable Development, 5, (2013): 84-90.
[9] CİVELEK, Ç., and Say, S.M., "A web-based decision support system to select proper machinery size and tractor power", Turkish Journal of Agriculture and Forestry, 40, (2016): 512-525.
[10] Awulu, J. O., Enokela, J. A., and Arubi,A., "Efficient Matching of Tillage Implements with Tractors using a Computer Program", International Journal of Agriculture Innovations and Research, 4, (2016): 1178-1182.
[11] Aybek, A.,"The Influence of Various Factors on Tractor Selection", Agricultural Mechanization in Asia, Africa, and Latin America, 37, (2006): 58-61
[12] Wu, C., and Shang, J., "Research on model of agricultural machinery selection based on principal component projection." Electronic and Mechanical Engineering and Information Technology (EMEIT), International Conference on. 2011, 7, IEEE, 2011.
[13] Miodragović, R., Tanasijević, M., Mileusnić, Z., andJovančić, P.,"Effectiveness assessment of agricultural machinery based on fuzzy sets theory", Expert Systems with Applications, 39, (2012): 8940-8946.
[14] Akinnuli, B. O., Akerele, F. O., and Nathaniel, A. B.,"Model Developed for Farm Tractors and Implements Selection for Optimum Utilization", Journal of Emerging Trends in Engineering and Applied Sciences, 5, (2014):248-255.
[15] Sivakumar, V. and Kaliyamoorthy, S., "Factors influencing the purchase of agricultural tractors: An empirical study", Journal of Business and Management, 16, (2014): 42-46.
[16] Mottaleb, K.A., Rahut, D.B., Ali, A., Gérard, B. and Erenstein, O., "Enhancing Smallholder Access to Agricultural Machinery Services: Lessons from Bangladesh",The Journal of Development Studies, 53, (2017): 1502-1517.
[17] Amini, S. and Asoodar, M.A., "Selecting the most appropriate tractor using Analytic Hierarchy Process–An Iranian case study", Information Processing in Agriculture, 3, (2016): 223-234.
[18] Saaty, T. L., "How to make a decision: the analytic hierarchy process", Interfaces, 24, (1994): 19-43.
[19] Saaty, T. L., "Decision making with the analytic hierarchy process", International journal of services sciences, 1, (2008): 83-98.
[20] Nefeslioglu, H. A., Sezer, E. A., Gokceoglu, C., Ayas, Z., "A modified analytical hierarchy process (M-AHP) approach for decision support systems in natural hazard assessments", Computers & Geosciences, 59, (2013):1-8.

IOP Conf. Series: Materials Science and Engineering **810** (2020) 012023 doi:10.1088/1757-899X/810/1/012023

Effect of fire retardancy materials in fibre reinforced composite plate for false ceilings- A Review

Shannay Rawal, M. Ramachandran
Mukesh Patel School of Technology Management and Engineering, Shirpur-4254053,India
shannay.nmims@gmail.com

Abstract. The use of fillers in composite material development and manufacturing has been in commercial use for decades. These fillers induce and/or enhance desirable characteristics in the composite matrix. Flammability is a major deterrent contributing to the reduced scope of utility of completely biodegradable composites. The ability of any material to act as a flame retardant is influenced not only by explicit factors such as the material's melting temperature and thermal degradation factor, but also by various implicit factors such as the multiple transformations taking place in the material and the compounds that are generated due to the material's exposure to elevated temperatures. The paper studies and compares the mechanism of fire retardancy of four popular fillers and aims to determine their effectiveness in introducing fire retardancy in biodegradable false ceiling materials. The results show that the properties of fire retardancy can be measured in terms of heat release rates, limiting oxygen indexes and thermogravimetric analyses. Final comments are on the testing standards used to check fire retardancy

1. Introduction

False ceiling, also known as dropped ceiling, is a secondary roof on the inside of the room. It is suspended from the main roof of the room or building. The addition of a false ceiling offers multiple advantages like temperature regulation by providing an additional layer of insulation between the room and the external surface of the building. Lights, electrical wiring and networking cables can be routed through the false ceiling, and can reconfigured with minimal investment and efforts. Gypsum ($CaSO_4 \cdot 2H_2O$) and Plaster of Paris ($CaSO_4 \cdot \frac{1}{2}H_2O$) are extensively used as primary materials in making false ceilings. Common construction materials (wood, glass, metal, fibre, composites) are also used in enhancing aesthetic appeal and for specialised applications.

Intumescent and flame retardant materials can be used in conjunction to improve retardancy in false ceilings. Intumescence is a material characteristic where exposure to heat results in the material's increased volume with decreased density, which is a method of passive fire resistance. In comparison, flame retardancy retards or prevents the propagation of fire by forming a preventive film on exposed areas or by opposing underlying chemical reactions causing the flame. The ability of any material to act as a flame retardant is influenced not only by explicit factors such as the material's melting temperature and thermal degradation factor, but also by various implicit factors such as the multiple transformations taking place in the material and the compounds that are generated due to the material's exposure to elevated temperatures. An example to support this point is ceramics. They have poor thermal conductivity and high melting temperatures, allowing them to effectively act as insulators, but their fire resistance is not notable. On the other hand, fire inhibiting materials have low conductivity in degrees of heat, but in the presence of high flame, they insulate heat effectively and deter the spread of flames.

Polymeric materials demonstrate flammability mainly due to the occurrence of thermal degradation due to exposure to heat. The mechanisms of fire-retardant systems for countering polymer flammability are yet to be comprehensively studied at the fundamental level. The shift of focus towards risk mitigation has led to recent innovations that address concerns that have been neglected in the past.

Four fire retardants were shortlisted after conducting a literature survey of the subject. Their techniques of retarding flammability are discussed below.

1. *Magnesium hydroxide* undergoes endothermic (heat absorbing) decomposition accompanied by hydration (water released as a product) in the temperature range of polymer degradation.
2. *Ammonium polyphosphate* works on the principle of intumescence. A carbon char layer having low density is formed when the surface area is exposed to heat. This carbon layer acts sacrificially to avoid further material breakdown by shielding the material from the heat source.
3. *Boric acid* acts as a dehydration catalyst, causing the water component in the composite to break down while simultaneously eliminating oxygen from the compounds at lower temperatures of heat exposure. This prevents the availability of oxygen from the composite at elevated temperatures, resulting in a faster drop in burning rate.
4. *Zinc borate* reduces the rate of polymer degradation by restricting the detrimental effect of the heat source on the exposed surface. The retardant creates a vitreous protective layer of residue that induces intumescence in the composite.

2. Flame retardancy testing

Limiting Oxygen Index tests are primarily used to check the combustibility of materials. The test indicates the least percentage of oxygen that must be present in the atmosphere to support combustion of material at a marginal level. This is achieved by placing the sample in a chamber fitted with an ignition wand and filled with a combination of oxygen and nitrogen. The critical level of oxygen is noted by lowering the percentage of oxygen in the test chamber till the burning stops. Samples with a higher critical oxygen percentage are labelled as less flammable. UL-94 testing standards for fire retardancy are to be followed for biodegradable composite materials. The naming standards followed for the classifications of materials are elaborated below. As per the official testing procedure document, "Material classified as 5VA or 5VB is subjected to a flame ignition source that is approximately FIVE Times more severe than that used in the V-0, V-1, V-2 and HB tests." Preconditioning for VB is for168 hours at 70°C.

Table 1: Flame retardancy testing

Class	Orientation	Definition	Permitted time of burn	Particle drops allowed		Plaque holes
				Flaming	Non-flaming	
HB	Horizontal	slow burning	burning rate < 76 mm/min or burning stops before 100 mm for thickness <3 mm			
V-2	Vertically upright	burning stops	30 sec	Yes	Yes	
V-1	Vertically upright	burning stops	30 sec	No	Yes	
V-0	Vertically upright	burning stops	10 sec	No	Yes	
5VB	Vertically upright	burning stops	60 sec	No	No	Yes
5VA	Vertically upright	burning stops	60 sec	No	No	No

3. Boric Acid flame retardancy materials

Boric acid inhibits the discharge of flammable gases from burning plastic materials, like cotton, wood, and paper-based merchandise. Chemical element acid conjointly releases with chemicals guaranteed water to dilute the concentration of transmutation fuel fragments and a carbon char is created that any inhibits combustion. it's used as a flame retardant in plastics, textiles, coatings, and there are different industrial merchandise that contains chemical element acid to strengthen their ability to resist exposure to flames. The glassy coating and also the char shield the substrate from chemical elements and warmth. The physical aspects of a hearth will take the shape of volatile, inflammable gases igniting and any mouldering the solid plenty via transmutation, so making additional inflammable gases till solely a carbon structure is left. The second methodology operates at a lower temperature with the carbon to provide smouldering combustion. Combustion of plastic materials will occur in 2 phases. Within the 1st section, visible flames are present; within the second section, flames are absent and are in an exceedingly vaporise state. Within the latter section, the combustion is remarked as glowing or smouldering, reckoning on whether or not or not light-weight is emitted from the burning material. chemical element acid works in an exceeding variety of the way to act as a hearth retardant, together with preventing flame combustion, promoting char formation, and suppressing smouldering, glowing and smoke. Boric acid, anon-alkaline salt can react each as a hearth retardant and smoulder drug. Chemical element acid releases water additionally to assist extinguish the hearth, however, it provides char forming price on the surface of the polysaccharide thanks to the presence of its atomic number 5 price. Every sample was kindled with a flame and burned down into the cold material. Then, minimum chemical element concentrations that might support combustion for all samples were recorded as a share.

Table 2: Boric acid fire retardant analysis results by various researchers

S. No	Author	Control	Composition	Approx. reduction in results
1.	Suharty.et.al	recycled PP	rPP/KF/acrylic acid/divinyl benzene/Al. Hyd/Mg. Hyd/ Boric Acid	55% (Burning Rate)
2.	Cavdar.et.al	HDPE/Wood Flour/Wax	HDPE/Borax impregnated Wood Flour/Wax	54% (Burning Rate)
3.	Nagieb.et.al	Wood + 10% urea formaldehyde	Wood + 10% urea formaldehyde + Paraffin wax + 5% (BA + BX)	81% (Glowing Time)

4. Magnesium Hydroxide flame retardancy materials

Magnesium hydroxide (MH) is that the environmentally most favoured, promptly obtainable, non-toxic, and low-value flame retardancys, however, their use is restricted by the requirement for too high loading level (up to quite 60%) and, the MH is stable up to three hundred C. The discharge of vapour upon decomposition of cathartic dilutes the fuel offer gift within the gas part. The comparatively high heat capacities of each cathartic and also the decomposition merchandise fashioned upon decomposition of cathartic cut back the thermal energy obtainable to degrade the plastic. The decomposition merchandise gives exaggerated insulation of the plastic from the warmth supply through char formation. The high filler content sometimes related to metal hydroxide- treated plastics act as a solid part diluting. cathartic (MH) square measure the environmentally most favoured, promptly obtainable, non-toxic, and low-value FRs, however, their use is restricted by the requirement for too high loading level (up to quite 60%). MDH undergoes energy-absorbing decomposition once water is free at 330 °C. The discharge of vapour upon decomposition of cathartic dilutes the fuel offer gift within the gas part. Inorganic flame retardants like atomic number 13 trihydroxide and cathartic square measure believed to decrease the temperature of the chemical compound surface and retard shift of the chemical compound by their energy-absorbing decomposition and water unleash. It absolutely was found that cathartic will effectively cut back the flammability (almost 50%) of fiber stuffed plastic composites. No synergistic impact was ascertained once cathartic was employed in combination with chemical element acid and atomic number 30 salt. Paper sludge/polypropylene composite decomposes at close to 300°C, however Mg (OH) two will therefore at regarding 400°C. Consequently, it will be thought that sizeable decomposition of the composite system already happens before Mg (OH) two acts as a flame retardant. (i) Flammability isn't influenced by the speed of loss of water from the mouldering filler. (ii) Atomic number 8 indices of stuffed chemical compound compounds are also associated with the structure of the parent cathartic, whereas UL94 ratings seem to be freelance of cathartic morphology. (iv) Water doesn't participate in the gas reaction resulting in increased levels of carbon monoxide gas emissions. Reduced levels of carbon monoxide gas evolution square measure ascertained. (v) Carbon volatilisation is promoted by mineral throughout exposure to atomic number 8. (vi) The mineral decomposition product seems to participate in condensed part reactions with adsorbate organic compound fragments.

Table 3: Magnesium Hydroxide fire retardant analysis results by various researchers

Author	Control specimen	Ideal Composition	Approx. reduction in results
Marosfoi.et.al	Polypropylene	PP/Organomodified clay/SEP (soft white clay)	57.7% (Burning Rate)

Jang.et.al	Paper sludge/ polypropylene (50:50 by weight)	Saytex8010 30 phr and antimony trioxide10 phr (phr= per hundred resin)	96% reduction in burning time
Chiu.et.al	Polypropylene	PP/APP/Penta-erythritol/Melamine/Mg($OH)_2$	Max HRR (86.09%) kw/m^2 Avg. HRR (82.88%)
Hornsby.et.al			Burning rate of composite with 25% of magnesium hydroxide decreases to 50% composite B and is even less than polypropylene.
Braun.et.al	High Impact Polystyrene	HIPS/$Mg(OH)_2$/Red Phosphorous	Max HRR (72.6%) kw/m2 Avg. HRR (59.6%)

5. Ammonium Phosphate flame retardancy materials

Ammonium Polyphosphate (APP) is found in the solid state at STP and has a high percentage of phosphorous (upto 30%). APP is moderately soluble in water and has a high melting point. APP is commonly available in several crystalline forms and is sold to several industrial clients for use in intumescent paints. The market names and compositions differ in crystalline dimensions, solubility in different agents and mol. wt. On contact with a heat source, the phosphorous forms residue in the form of a coating that acts to protect the surface from flames. [5]. APP displays hydrolysis and gets converted to monoammonium phosphate (orthophosphate). When APP is exposed to greater elevated temperatures or to water for longer periods, the process of hydrolysis is expedited. Short chain APP begins to decompose at temperatures above 150 °C. This allows APP to cater to a wider range of fire retardancy applications in the composite industry.

Table 4: Ammonium Phosphate fire retardant analysis results by various researchers

S. No	Author	Control specimen	Ideal Composition	Approx. reduction in results
1.	Chiu.et.al	Polypropylene	PP/APP/Penta-erythritol/Melamine (100/23/14/13)	Max HRR (84.2%) kw/m^2 Avg. HRR (81.5%)
2.	Zhao.et.a l	The composition was varied.	PVA/APP/Ni-Al LDH (85/14.7/0.3)	LOI (-29.3%)
3.	Guo.et.al	Polypropylene	EG/ WPC and M-APP/WPC	LOI (-66.52%)

6. Zinc Borate flame retardancy materials

Zinc borate (ZnB) could be an element-based mostly flame agent compatible with several chemical compound matrices. Its effective each within the solid section and within the gas section and its robust smoke suppressing action, helps to boost time of rescue just in case of fireside. ZnB is an associate chemical compound, a salt of Zn. it's a white crystalline or amorphous powder insoluble in water. Its toxicity is low. Its freezing point is 980 °C. ZnB, thanks to the delicate bond mediate filler's surface and matrix and since of their inherent incompatibility, decrease lastingness. ZnB thermal decomposition and consequently steam generation within the method causes the formation of cavities at intervals the matrix that all over again decreases lastingness. ZnB acts as a flame agent in each condensed and gas phases of the combustion method. At elevated temperatures, this compound decomposes, emotional crystallization water molecules (about 15 August 1945 by weight of water within the temperature vary of 290–450°C) that dilutes the concentration of H radicals and organic volatiles within the flame, reducing its temperature. Once decomposition, the residual salt maintained within the rotten compound can kind a vitreous-like layer that slows the permeation of volatile compounds through the surface and restricts them from getting into the gas section. ZnB decreases the utmost rate of warmth unleash (RHR) and delays the second heat unleash peak. Throughout these tests, the formation of a char layer at the surface of the flame-retarded polymers is discovered. ZnB in an associate chemical reaction at a temperature varies of 290 °C up to 450 °C releases fifteen World Trade Centre steam. Thermal decomposition and consequently steam generation within the method causes the formation of cavities at intervals the matrix that all over again decreases lastingness. ZnB addition to composites has improved composite flame retardancy property.

Table 5: Zinc Borate fire retardant analysis results by various researchers

S.No	**Author**	**Ideal composition**	**Approx reduction in result**
1.	Braun.et.all	PolyAmide 6,6/GF-Aluminium Phosphinate	Max. HRR(52.134%)
2.	Durin.et.al	20%$Mg(OH)_2$ + 20% BZn + 20% Talc	LOI (-35.7%)
3.	Ramazani.et.al	R40–ZnB–ATH30(Allumina trihydrate)-(ZnB=0.2ATH)-MAPP(Maleic Anhydride grafted polypropylene)	LOI(-37.5%)

7. Conclusion

The ability of any material to act as a flame retardant is influenced not only by explicit factors such as the material's melting temperature and thermal degradation factor, but also by various implicit factors such as the multiple transformations taking place in the material and the compounds that are generated due to the material's exposure to elevated temperatures. The paper shows the comparison between the mechanisms of fire retardancy of four conventionally used fire retardant fillers and aims to determine their effectiveness in introducing fire retardancy in biodegradable false ceiling materials. Addition of fire-retardant additives should result in an increase in limiting oxygen index (LOI), and therefore improved flame retardancy in the tested composite. The rate of water released by the retardants would increase by virtue of higher temperatures and longer periods of subjection to elevated temperatures. Due to the nature of composites, some degree of reduction in the mechanical properties would be observed due to the reduced percentage of strength inducing components. The working mechanism of intumescent additives is a chemically induced physical process that achieves fire retardancy by forming a high volume, low density buffer for protecting the area exposed to elevated temperatures.

References

1. Rothon, R. N., and P. R. Hornsby. "Flame retardant effects of magnesium hydroxide." *Polymer Degradation and Stability* 54, no. 2-3 (1996): 383-385.
2. Cavdar, Ayfer Donmez, Fatih Mengeloğlu, and Kadir Karakus. "Effect of boric acid and borax on mechanical, fire and thermal properties of wood flour filled high density polyethylene composites." *Measurement* 60 (2015): 6-12.
3. Wang, Qingwen, Jian Li, and Jerrold E. Winandy. "Chemical mechanism of fire retardance of boric acid on wood." *Wood science and technology* 38, no. 5 (2004): 375-389.
4. Marosfoi, B. B., S. Garas, B. Bodzay, F. Zubonyai, and G. Marosi. "Flame retardancy study on magnesium hydroxide associated with clays of different morphology in polypropylene matrix." *Polymers for Advanced Technologies* 19, no. 6 (2008): 693-700.
5. Shumao, Li, Ren Jie, Yuan Hua, Yu Tao, and Yuan Weizhong. "Influence of ammonium polyphosphate on the flame retardancy and mechanical properties of ramie fiber-reinforced poly (lactic acid) biocomposites." *Polymer International* 59, no. 2 (2010): 242-248.
6. Chun, Koay Seong, Salmah Husseinsyah, and Hakimah Osman. "Mechanical and thermal properties of coconut shell powder filled polylactic acid biocomposites: effects of the filler content and silane coupling agent." *Journal of Polymer Research* 19, no. 5 (2012): 9859.
7. Chun, Koay Seong, Salmah Husseinsyah, and Fatin Nasihah Azizi. "Characterization and properties of recycled polypropylene/coconut shell powder composites: Effect of sodium dodecyl sulfate modification." *Polymer-Plastics Technology and Engineering* 52, no. 3 (2013): 287-294.
8. Singh, Alok, Savita Singh, and Aditya Kumar. "Study of mechanical properties and absorption behaviour of coconut shell powder-epoxy composites." *International Journal of Materials Science and Applications* 2, no. 5 (2013): 157-161.
9. Agunsoye, J. Olumuyiwa, Talabi S. Isaac, and Sanni O. Samuel. "Study of mechanical behaviour of coconut shell reinforced polymer matrix composite." *Journal of minerals and materials characterization and Engineering* 11, no. 8 (2012): 774-779.
10. M. Ramachandran, , Sahas Bansal, Pramod Raichurkar, Scrutiny of Jute Fiber Poly-Lactic Acid (PLA) Resin Reinforced Polymeric Composite, Journal of the Textile Association, Volume 76, Issue 6, 2016, pp. 372-375.
11. Ullah, Sami, Faiz Ahmad, Azmi M. Shariff, Mohamad A Bustam, Girma Gonfa, and Qandeel F. Gillani. "Effects of ammonium polyphosphate and boric acid on the thermal degradation of an intumescent fire retardant coating." *Progress in Organic Coatings* 109 (2017): 70-82.

12. Ramazani, SA Ahmad, Arash Rahimi, Massoud Frounchi, and Saeed Radman. "Investigation of flame retardancy and physical–mechanical properties of zinc borate and aluminum hydroxide propylene composites." *Materials & Design* 29, no. 5 (2008): 1051-1056.
13. Durin-France, A., L. Ferry, J-M. Lopez Cuesta, and A. Crespy. "Magnesium hydroxide/zinc borate/talc compositions as flame-retardants in EVA copolymer." *Polymer International* 49, no. 10 (2000): 1101-1105.
14. Braun, Ulrike, Bernhard Schartel, Mario A. Fichera, and Christian Jäger. "Flame retardancy mechanisms of aluminium phosphinate in combination with melamine polyphosphate and zinc borate in glass-fibre reinforced polyamide 6, 6." *Polymer Degradation and Stability* 92, no. 8 (2007): 1528-1545.
15. Guo, Chuigen, Lin Zhou, and Jianxiong Lv. "Effects of expandable graphite and modified ammonium polyphosphate on the flame-retardant and mechanical properties of wood flour-polypropylene composites." *Polymers and Polymer Composites* 21, no. 7 (2013): 449-456.
16. Zhao, Chun-Xia, Ya Liu, De-Yi Wang, De-Long Wang, and Yu-Zhong Wang. "Synergistic effect of ammonium polyphosphate and layered double hydroxide on flame retardant properties of poly (vinyl alcohol)." *Polymer Degradation and Stability* 93, no. 7 (2008): 1323-1331.
17. Chiu, Shih-Hsuan, and Wun-Ku Wang. "Dynamic flame retardancy of polypropylene filled with ammonium polyphosphate, pentaerythritol and melamine additives." *Polymer* 39, no. 10 (1998): 1951-1955.
18. Hornsby, Peter R. "The application of magnesium hydroxide as a fire retardant and smoke-suppressing additive for polymers." *Fire and Materials* 18, no. 5 (1994): 269-276.
19. Chiu, Shih-Hsuan, and Wu-Kou Wang. "The dynamic flammability and toxicity of magnesium hydroxide filled intumescent fire retardant polypropylene." *Journal of Applied Polymer Science* 67, no. 6 (1998): 989-995.
20. P. M. Bhagwat, M. Ramachandran, Pramod Raichurkar, Mechanical Properties of Hybrid Glass/Carbon Fiber Reinforced Epoxy Composites, Materials Today: Proceedings, 4(8), 2017:1788–1793.
21. Jang, Jyongsik, and Eungki Lee. "Improvement of the flame retardancy of paper-sludge/polypropylene composite." *Polymer Testing* 20, no. 1 (2000): 7-13.
22. Marosfoi, B. B., S. Garas, B. Bodzay, F. Zubonyai, and G. Marosi. "Flame retardancy study on magnesium hydroxide associated with clays of different morphology in polypropylene matrix." *Polymers for Advanced Technologies* 19, no. 6 (2008): 693-700.
23. Nagieb, Zenat A., Mona A. Nassar, and Magda G. El-Meligy. "Effect of addition of boric acid and borax on fire-retardant and mechanical properties of urea formaldehyde saw dust composites." *International Journal of Carbohydrate Chemistry* 2011 (2011).
24. Cavdar, Ayfer Donmez, Fatih Mengeloğlu, and Kadir Karakus. "Effect of boric acid and borax on mechanical, fire and thermal properties of wood flour filled high density polyethylene composites." *Measurement* 60 (2015): 6-12.
25. Suharty, Neng Sri, Hanafi Ismail, Kuncoro Dihardjo, Muhammad Nizam, and Maulidan Firdaus. "Improvement of Inflammability and Biodegradability of Bio-composites Using Recycled Polypropylene with Kenaf Fiber Containing Mixture Fire Retardant." In *Advanced Materials Research*, vol. 950, pp. 18-23. Trans Tech Publications Ltd, 2014.

A review on microfluidic devices for separation of blood constituents

Syed Abdul Mateen[1] **Kiran S Bhole**[2]

[1,2]Department of Mechanical Engineering, Sardar Patel College of Engineering, Mumbai, India

abdulmateensyed7@gmail.com[1], kiran_bhole@spce.ac.in[2]

Abstract. This paper presents review of microfluidic device used for extraction of red blood cell, white blood cell, and plasma from the whole blood sample. . Micro fluidic based cell separation has various advantages as it reduces sample size, faster sample processing, has more sensitivity and low device cost as compared to the conventional method of blood cell separation. Basically there are two techniques for blood separation using microfluidic device, one is called active separation and another is passive separation method. The review highlights various cell separation methods. We will also take review on microfluidic based blood cell separation techniques. Separation of blood cells helps in clinical diagnosis and therapeutic research.

1. Introduction

Now a day's people are very concern about their health with increase of life cycle related disease. There is going to be huge demand for periodical blood test which we can carry out at home. Conventional methods are impossible to perform at home. Hence, development and characterization of easy to use microfluidic devices or lab on chip devices for blood test is much more needed.

Separation of blood cells is the first step towards the blood analysis, as every component provides different information. Anaemia disease is diagnosed by examine physiological changes in red blood cells levels. Various constituents of blood viz; white blood cell, plasma and red blood cell play various vital roles for human body. On v/v basis blood contains about 45%-50% of red cells, 1% for WBS-platelets and 55% is plasma which has yellow colored [1-2]. Each of these serves various important functions for body. Typically, the immune system largely dependent on white blood cell present in the blood. Pure blood has a density of 1040–1070 kg m−3 and a viscosity of 4–5 cP at 37 ° Centigrade. The ratio of WBCs against RBCs is 1: 600–900. Blood is non–Newtonian fluid. Blood cells are extensively used in clinical diagnosis. Extracting the red blood cells from a whole blood sample gives the significant information about the pathological characteristics of breast cancer cells. Now a day complete blood count (CBC) is used everywhere as it is most common blood test. This complete blood count (CBC) test shows the overall health status of patient [3].

Mainly there are two common conventional techniques for blood cell separation, one is Centrifugation and another is filtration. These methods are expensive and require lot of time for completion, skilled labour and if not used carefully might damage cells. Due to above disadvantages researchers introduces microfluidic technique for separation of blood cell. Microfluidics is a platform

IOP Conf. Series: Materials Science and Engineering **810** (2020) 012024 doi:10.1088/1757-899X/810/1/012024

that offers significant advantages. These microfluidic techniques required less volume of blood sample also it has high response time and less expensive than the conventional methods. It is therefore not surprising that the microfluidic techniques of blood cell separation have increasing interest over the last decade. There are so many microfluidic devices which have been developed for the separation of blood cells and plasma extraction. In this review, our aim is to present the current blood cell separation techniques using microfluidic channels. Various techniques have been used for blood cells separation by microfluidic system which includes bifurcation law, geometric obstructions, membrane filtration acoustic standing wave forces, filters and cross-flow filtration. Microfluidic is an ideal technique for whole blood cell separation. Due to its very small channel size, the required amount of sample volume can be minimized. Basically, the microfluidics system for blood cells separation is categorized into passive and active separation methods [3-4].Active separation methods require external source (forces) like magnetic force, dielectrophoretic force or acoustic force, or any combination of these forces to give blood cells a specific direction. While passive separation method simply uses physical properties of cell like their shape and size or their stiffness. This review studies various active and passive separation techniques based on their principle of separation.

Passive separation techniques doesn't need of cell labeling process due to which they are more cost effective and suitable for separating blood cells. Passive separation techniques has many advantages like they are simple in design, continuous operation, and they are easy to fabricate. To design effective blood cell separation micro devices, which are based on hydrodynamic principal we must need to understand the theoretical mechanism of blood flow in micro channels.

2. Mechanism of blood flow separation

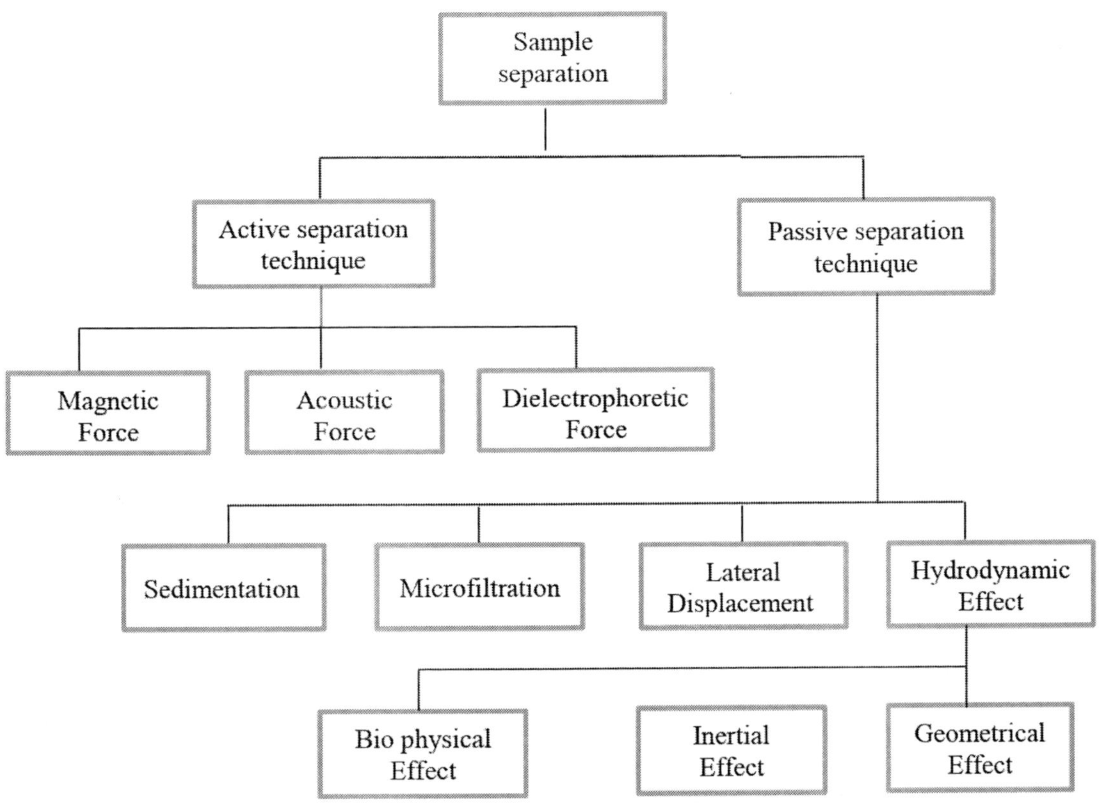

Figure 1. Classification of Blood separation techniques

There are various mechanisms for blood cells separation. They are mainly divided into two types, 1) active separation technique 2) Passive separation techniques. In active separation technique external forces are required for separation. In passive separation technique the separation takes place in designed microfluidic devised geometry by virtue of sizes of constituents of blood. These techniques are further classified into the different separation techniques. Figure1 explain in detail classification of blood constituent's separation techniques.

2.1. Hydrodynamic technique

This study presents most of the blood cells separation techniques based on microfluidic. In study it is found that most method separates only one type of cell component by extraction or rejecting. Study optimizes the design of bifurcation channel, they also added trapping units, the author introduces new techniques through which it can simultaneously separates blood components by a continuous-flow whole blood processing microfluidic device. This is working on the principle of hydrodynamic [3].Hydrodynamics filtering does not require any external force for working of a device hence hydrodynamic filtering is widely used. As compared to other separation technique, Hydrodynamic technique has simple design. There are two types of side channel in this device and both are at angle of 60 degree to the main channel (see Fig. 2). The device also contains series of white blood cells trapping units based on hydrodynamic principle. The total time required for overall process is just 20 minutes and it requires only 6μL of whole blood sample [3].

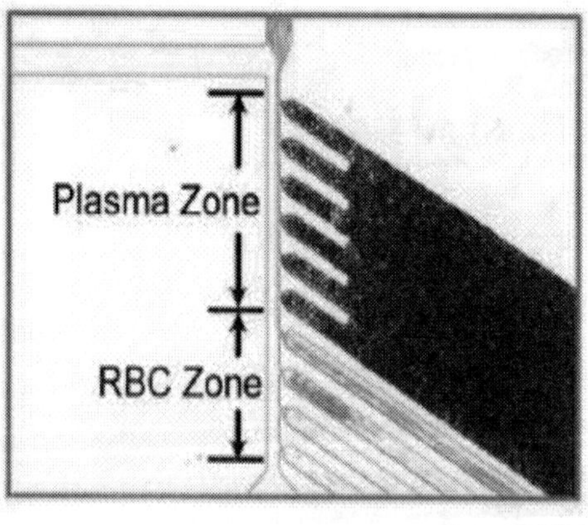

Figure 2. Entrance of the plasma zone and RBC zone [3].

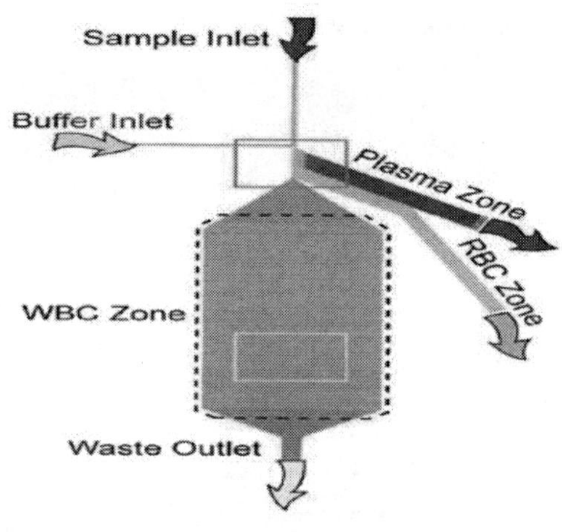

Figure 3. Microfluidic device [3].

Figure 3 shows microfluidic device which consists of two inlets one is for blood sample inlet and another is a buffer inlet. As the blood sample enters from inlet and passes through the bifurcation region i.e. the orange colour zone, plasma and red blood cells were trapped into plasma zone and red blood cell zone, respectively. Then White blood cells will flow towards the main channel and will trap in the white blood cell zone. As you can see in the figure 3 red particle are RBCs and green particle in the plasma zone are packed beads.

The author aim is to examine the microfluidic device which can simultaneously extract red blood cells, plasma and white blood cell, when only 6 μL of whole blood sample is loaded into the device in just 20 minutes. By this microfluidic device all three cells of blood can be extracted simultaneously without any interference.

Figure 4. Zones of separated Plasma, RBS and WBC [3].

In the figure 4 the green line indicates plasma zone, white and yellow lines represents RBC and WBC zone respectively [3]. The author has successfully developed the microfluidic device which can simultaneously separate plasma, red blood cell and white blood cell. This experiment has two main important features, 1) plasma and red blood cells are separately trapped in the side channels of the bifurcation region i.e. orange zone. And we can use the extracted red blood cells for blood test; also the extracted plasma has low haemolysis effect. Second important feature is that, in the design of device the author added some hydrodynamically based white blood cells trapping units in a series form to main channel. In just 20 minutes near about 1800 white blood cell can be trap in that main channel. These trapped white blood cells will be helpful for various blood tests.

2.2. Microfiltration

Microfiltration is one of the important passive separation methods; this is used in micro devices for filtration. The sample volume required for this process is very small. This technique is further classified as cross flow filtration or dead end technique. The fluid i.e. blood sample flows directly into the micro filters are considered as dead end technique. While in cross flow technique blood separation take place tangentially to flow direction. The scientists have come across with the micro filters which have a pore size of 14 μm and it contains near about 14400 holes. This technique is mainly used for plasma separation, as plasma cells are very small in size. But this technique has some disadvantages as it has low flow rate and its fabrication is complex.

2.3. Sedimentation

Sedimentation technique is based on the differences in density of cells and plasma; it is also one of the passive separation techniques. Its main function is to separate plasma from the whole blood sample. Due to the density difference blood cells sediment at the bottom and layer of plasma comes up at top from which we can easily extract plasma. It is an old technique and now days no one prefer this technique because it takes lot of time for processing and also extraction rate is very low.

3. Materials and fabrication

Selection of material and fabrication technique for the microfluidic devices is one of the major issues. Initially, glass and silicon were the materials used for fabrication. But these materials have some disadvantages like high cost, brittle nature, complex fabrication techniques. Due to these disadvantages scientist introduced alternative material which is based on polymer called polydimethylsiloxane (PDMS). The reason behind choosing this material is low cost, easy to fabricate device, elasticity and transparency. PDMS is used widely especially those who working in biomedical. Many microfluidic devices are fabricated using a standard soft lithography [14-18].

4. Techniques of blood separation

This section presents review of literature on separation of constituents of blood. Table1 presents comprehensive review of different mechanisms presented by researchers for separation of blood constituents.

Table1. Review of various mechanisms for separation of constituents of blood.

Author	Active / Passive	Mechanism	Injected sample	Sample volume	Desired target	Performance
Da-Han Kuan [3]	Passive	Hydrodynamic	Whole blood	0.3 µL/min	Plasma, RBC, WBC	Low hemolysed plasma, 1200–1800 trapped WBC
Myounggon Kim[6]	Passive	Hydrodynamic	Whole blood	0.33 µL/min	WBC	96.9% RBC purity 97.2% WBC recovery rate.
John A Davis[7]	Passive	Hydrodynamic	Whole blood	0.4 µL/min	Plasma	100% plasma recovery rate 100% cell removal rate
Ivan K. Dimov[8]	Passive	Sedimentation	Whole blood	5 µL in 10 min	Plasma	99.9–100% blood cell retention
Hye-Kyoung Seo [9]	Active Passive	Magnetic + Hydrodynamic	1000X RBC dilution using PBS	1000 µL/min	WBC, RBC	86.8% RBC separation efficiency 29.1% WBC separation efficiency
Mahdi Mohammadi et [10]	Active/ Passive	Dielectrophoretic + Hydrodynamic	Whole blood mix with1:1 heparin sodium	2 µL in 7 min	Plasma	100 nL plasma with 99% purity
Yuchao Chen[11]	Active	Acoustic	Whole blood	1000 µL/min	Platelet	>85% Platelet recovery rate >80% RBC/ WBC removal Rate

Ki-Ho Han[13]	Active	Magnetic	10X diluted whole blood using sodium hdrosulfite.	0.083 µL/min	RBC, WBC	93.5% RBC separation efficiency 97.4% WBC separation efficiency
P. Dow et[19]	Active	Acoustic	Spiked bacteria in PBS diluted whole blood to 20%Hct	10 µL/min	Bacteria	>85% RBC removal rate, And 45–60% Bacteria yield
DowSung Yang[20-21]	Passive	Hydrodynamic	Whole blood	0.16 µL/min	Plasma	100% plasma purity 15-25% plasma Extraction

4.1. Comparative analysis of passive blood separation techniques

This section presents comparison of three main passive blood separation techniques. Further section discusses their advantages and disadvantages in the Table 2. From the Table 2 we can observe that techniques based on sedimentation or microfiltration required very low amount of sample volume. But these techniques have so many challenges and disadvantages.

Table 2. Comparison of various methods of blood flow separation techniques

Method	Principle	Advantage	Disadvantage
Hydrodynamic	Biophysical effects, geometric	Continuous output and High throughput	Sample volume Counteracting yield and separation
Microfiltration	Differences in cell size and shape,	Sample volume required is very small	It has low flow rate and fabrication is complex
Sedimentation	Difference in density of cells and plasma	Volume require is 1–10 µL	Time consuming, extraction rate is very low

IOP Conf. Series: Materials Science and Engineering **810** (2020) 012024 doi:10.1088/1757-899X/810/1/012024

5. conclusion

The review has focused on various blood cells separation techniques. We mainly review different passive blood separation methods like hydrodynamic and microfiltration. Passive separation techniques have high potential to give solutions, when the difficulty arises during blood separation. Due to use of microfluidic devices the sample volume decreases by default. Among all passive blood separation techniques, the hydrodynamic technique for blood separation is reliable. We have also seen a device which only can simultaneously trap red blood cells, plasma and white blood cells by using passive hydrodynamic separation technique. Polydimethylsiloxane (PDMS) is extensively used for fabrication of many microfluidic devices. Many challenges which arise in sedimentation and filtration are solved by using hydrodynamic technique.

6. References

[1] Siddhartha Tripathi, Y V Varun Kumar, Amit Prabhakar 2015 Passive blood plasma separation at the microscale: a review of design principles and microdevices, J. Micromech. Micromechanics and Micro engineering 1

[2] Fung Y C 1981 Biomechanics-Mechanical Properties of Living Tissues, New York Springer

[3] Spell, D W, Jones, D V, Harper, W F and David Bessman, J 2004 The value of a complete blood count in predicting cancer of the colon. Cancer Detection and Prevention **28** 37

[4] Da Han Kuan1, Chia Chien Wu1, Wei Yu Su1 and Nien Tsu Huang 2018 A Microfluidic Device for Simultaneous Extraction of Plasma, Red Blood Cells, and On-Chip White Blood Cell Trapping Scientific reports 1

[5] Yuta Nakashima1, Sakiko Hata1, Yasuda 2009 Blood plasma separation and extraction from a minute amount of blood using dielectrophoretic and capillary forces, Science Direct, volume **145** 561

[6] Turgeon, M. L. Clinical Hematology 2005 Theory and Procedures ,Lippincott Williams & Wilkins

[7] Kim, M., Mo Jung, S., Lee, Jun Kang, Yang, S, 2010 A Microfluidic Device for Continuous White Blood Cell Separation and Lysis From Whole Artificial Organs **34** 996

[8] Davis, J A 2006 Deterministic hydrodynamics: Taking blood apart. Proceedings of the National Academy of Sciences **103** 14779

[9] Dimov, I K, 2011 Stand-alone self-powered integrated microfluidic blood analysis system (SIMBAS). Lab on a Chip **11** 845

[10] Seo, H K, Kim, Y H, Kim, H O and Kim, Y J, 2010 Hybrid cell sorters for on-chip cell separation by hydrodynamics and magnetophoresi. Journal of Micromechanics and Microengineering **20** 223

[11] Mohammadi,Madadi,Casals-Terré, J and Sellarès, J, 2015 Hydrodynamic and direct-current insulator-based dielectrophoresis (H-DC-iDEP) microfluidic blood plasma separation. Analytical and Bioanalytical Chemistry **407** 4733

[12] Chen, Y 2016 High-throughput acoustic separation of platelets from whole blood. Lab Chip **16** 3466

[13] Szydzik, C., Khoshmanesh, K., Mitchell, A. and Karnutsch, C, 2015 Microfluidic platform for separation and extraction of plasma from whole blood using dielectrophoresis, Biomicrofluidics **9** 064120

[14] Prasanna Gandhi and Kiran Bhole, 2011 3D Microfabrication using Bulk Lithography, ASME International Mechanical Engineering Congress and Exposition, Denver, Colorado, USA 393

[15] Kiran Bhole, Sunil Ekshinge, and Prasanna Gandhi, 2014 Fabrication of Continuously Varying Thickness Micro-cantilever using Bulk Lithography Process, ASME International Manufacturing Science and Engineering Conference, Michigan, USA, V001T01A009

[16] Kiran Bhole, Prasanna Gandhi, and T. Kundu, 2014 On the Evolution of Cured Voxel in Bulk Photopolymerization upon Gaussian Laser Exposure, Journal of Applied Physics, **116** 043105

[17] Prasanna Gandhi and Kiran Bhole, 2013 Characterization of 'Bulk Lithography' Process for Fabrication of 3D Micro-structures, ASME Journal on Nano and Micro Manufacturing, **1** 041002-1-8

[18] Prasanna Gandhi, Suhas Deshmukh, Rahul Ramtekkar, Kiran Bhole, and Alem Baraki, 2013 On-axis Linear Focused Spot Scanning Microstereolithography System: Optomechatronic Design, Analysis and Development, Journal of Advanced Manufacturing Systems, **12** 43

[19] Han, K H, and Frazier, A B, 2006 Paramagnetic capture mode magnetophoretic microseparator for high efficiency blood cell separations. Lab on a Chip **6** 265

[20] Dow, P, Kotz, K, Gruszka, S., Holder, J and Fiering, J, 2018 Acoustic separation in plastic microfluidics for rapid detection of bacteria in blood using engineered bacteriophage. Lab Chip **18** 923

[21] Yang, S. Undar, A. and Zahn, J. D, 2006 A microfluidic device for continuous, real time blood plasma separation. Lab on a Chip **6** 871

IOP Conf. Series: Materials Science and Engineering **810** (2020) 012025 doi:10.1088/1757-899X/810/1/012025

Development of CNC Micro-milled Leaf-Pattern Micro-channel Heat Sink (LP-MCHS) and Testing Using Nanofluids

S R Itankar[1], S S Mohite[2], V P Gaikwad [3]and H Parmar[4]

Email: shreyas.itankar@gmail.com

Abstract. Micro-channel heat sinks (MCHS) have attracted the attention of researchers because of their compact design and wide range of applications. MCHS is mostly used to dissipate heat where large amount of heat is generated in a confined space. In this work, leaf pattern MCHS is fabricated on pure copper block by micro-milling on CNC machining center and is tested for its thermal performance and fluid flow behaviour using pure water as well as various nanofluids such as Copper Oxide(CuO), Aluminium Oxide(Al2O3)and Silicon Oxide(SiO2). The volume concentration of nanofluids was kept constant, i.e., 0.3% volume fraction and the experiments were carried out for heat flux ranging from 65 W/cm2 to 200 W/cm2 and flow rate from 100 ml/min to 900 ml/min. The results of experiments indicate that these nanofluids, as a working fluid, enhance the heat transfer by 35% and Nusselt Number by up to 37%, however, increases the pressure drop by 18% which increases the pumping power. From the performance evaluation analysis, it was found that the SiO2 nanofluid with leaf pattern MCHS gives optimal performance.
Keywords: Micro-channel Heat Sink (MCHS), Leaf pattern micro-channel heat sink (LP-MCHS), Reynolds number (Re), Nusselt Number (Nu), Deionized water (DI), Performance Evaluation Criterion (PEC), Nanofluid (nf), Base fluid (bf), Pressure drop (ΔP).

1. Introduction

The fabrication method of micro-channel is most essential aspect in the development of microfluidic devices. Depending upon the application of micro-channel-based device, different types of fabrication methods and materials of micro-channel are preferred. The rising demand of micro-channel devices has led to the development of many other new hybrid technologies that makes micro-channel more efficient and cost-effective. Basically, micro-channels are mostly created on glass, polymeric, metallic and silicon substrate. While polymeric, glass and silicon substrate are mostly used in chemical and biomedical devices, metallic and silicon substrate are used for electronic and mechanical engineering applications. However, fabrication of these micro-channels on such substrate in large number has always been a difficult task for manufacturers due to the precision and cost required for the products.

Development in the electronic industries is associated with the increase in the heat flux. Hence, to improve the performance and life of electronic devices, it is very essential to remove heat from the confined space and maintain the efficient working temperature. By using MCHS the main goals in

[1] Department of Mechanical Engineering, Government College of Engineering, Aurangabad, India
[2] Department of Mechanical Engineering, Government College of Engineering, Aurangabad, India
[3] Department of Mechanical Engineering, Government College of Engineering, Karad
[4] Department of Mechanical Engineering, Ujjain Engineering College, Ujjain, India

electronic cooling can be achieve i.e. reduction of maximum temperature and minimization of temperature gradient on the surface of the devices. One of the ways to enhance the heat transfer is the application of additives to the working fluid. The basic idea is to improve the heat transfer for changing the properties of working fluid such as nanofluids where solid nanoparticles are added to base fluid to increase the thermal conductivity of working fluid. Nanofluids, consists of the base fluid such as water or ethylene glycol, and the nano sized particles of metal or non-metal. Group at the Argonne National Laboratory, USA first introduced the term nanofluid, about decade ago. In recent years, the nanofluids have gained much attention due to their potential advantages including higher thermal conductivity than pure base fluids, excellent stability and little increase in pressure drop.

The idea of using micro-channel heat sink (MCHS) for electronic cooling was introduced in 1981 [1]. Experimental results show that the MCHS have the capability to dissipate heat flux rate of 790 W/cm^2. Increasing the area wetted by the heat transfer fluid and decreasing the width of heat sink channels would improve the convective heat transfer for single phase flows. Nanofluids have greater potential for heat transfer enhancement because of their higher thermal conductivity [2], [3], [4], [5].

Traditional working fluids are used to dispersed nanoparticles uniformly for the preparation of Nanofluids. Traditional working fluids used for dispersion are water, alcohol, oil etc. [6]. Experimental investigation was done to measure the convective heat transfer and pressure drop for nanofluids in tube flow [7], [8]. Their result shows that heat transfer coefficient was highly enhanced, and it depends on particle volume fraction, Reynolds Number and the particle size and shape. And nanoparticles did not cause an excessive pressure drop was founded.

Together increasing the nanoparticle volume fraction and decreasing the nanoparticles diameter enhances the Nusselt number value and SiO_2 nanofluid has the highest heat transfer rate compared to CuO, Al2O3, and ZnO nanofluids which are made with water as base fluid [9]. It was also found that the heat transfer coefficient increased by increasing the nanoparticles concentration in different MCHS shapes with various nanofluids and various volume fraction when studied for the fluid flow and heat transfer characteristics [10], [11], [12].

In this work the LP-MCHS is fabricated and then it is tested for the heat transfer enhancement for its performance. Heat transfer behaviour was also tested with different nanofluids. Micro-channel and different applications of micro-channel in industries are discussed in the next section.

1.1. Micro-channel and their application

Different types of applications require different shapes, sizes, and structures of micro-channels. While most microfluidic channels involve a high form ratio, the micro-channels having a low aspect ratio also have a very great application in particle separation devices. Channels with dimensions between 1 and 999 μm are considered as micro-channels. The manufacturing method shows more importance in the performance of micro-channels. Micro-channels are mainly (except micro-grinding) manufactured by unconventional methods. Over the decade different types of micro-channels are manufactured. Commonly used microchannels cross sections includes circular, square, semi-circular, rectangular, U-shaped and Gaussian bundle micro-channel. Almost all types of micro-channels are manufactured on base material except circular cross section. Since on circular cross section, a thickness of few microns is usually left over along the length for optical and visual transparency, hence those are manufactured inside the material mostly [16].

Researchers has been attracted towards micro-channel heat sink because of its highly efficient heat exchange component and unique characteristics like compact design, operational safety, high efficiency and low volume along with its vast range of applications like chemical industry, aerospace industry, electronic communication, biomedical devices and so on. Microchannel applications are divided into the three types such as (1) biological application, (2) electronic and mechanical application, (3) chemical engineering applications.

For most microfluidic applications, the properties of the materials that matter are surface charge, molecular adsorption, electrical and thermal conductivity, machinability, optical properties, mobility of electroosmotic flows and many others. The most developed microchannel based devices are inkjet printers in microfluidics, which is possible due to the introduction of miniaturized gas chromatographs and nozzles. Some of the geometrically simplest micro-channels can be used in the manufacture of fuel cells comprising single or dual T-junctions. Different materials used to fabricate micro-channels are listed in next section.

IOP Conf. Series: Materials Science and Engineering **810** (2020) 012025 doi:10.1088/1757-899X/810/1/012025

1.2. Materials

Various types of materials have been used for different micro-channel devices. These various types of materials can be divided into three main categories according to their properties.

- Polymeric and glass substrates
- Metallic substrates
- Semiconductors, ceramics and composites

In this case copper is selected as the substrate material for engraving LP-MCHS. Copper has high thermal conductivity which is important with respect to heat transfer. Also, it is ductile, has good machinability and available everywhere at low cost. In mechanical milling process the base material should be thicker so that it can sustain the cutting forces and vibrations.

Table 1: Composition of Copper block

Cu	Zn	Pb	Sn	P	Mn	Fe
99.6	0.0885	0.005	0.005	0.01	0.01	0.06
Ni	Si	Al	S	As	Bi	Se
0.01	0.005	0.02	0.037	0.07	0.01	0.01

Fabrication method used for fabrication of micro-channel is discussed in next section. The fabrication method is selected according to the requirement of surface finish, geometrical and dimensional tolerances, precision, accuracy, surface quality, thermal and mechanical deformations etc.

2. Micro-mechanical cutting

With the advancement in high-precision machining, micro-mechanical cutting has become a key technology for the creation of micro-channels. Micro-mechanical cutting process is mainly suitable for manufacturing individual customized components rather than for large batches. It can manufacture many materials, such as Aluminium, brass, plastic, steel and polymers. It doesn't require a expensive configuration, which allows the manufacture of miniatures at an economically reasonable cost [19]. With a high level of precision machining of ultra-precise machine tools, it is possible to obtain a good surface finish and precision of shape. The high machining speed of micro-cutting is another advantage over micro-manufacturing technologies. Most commonly used mechanical cutting processes are micro-milling and micro-turning. Recently, the multi-knife micro-milling process has been studied [17] and has proved more economical and faster compared to other contemporary micro-machining processes.

Composite cutting tool was used in this process of fabricating MCHS, and the depths of the micro-channels were taken inconsistent. In addition, it was observed that the corners were not sharp but curved, which could create a vacuum during the sealing process or bonding. Micro-channels were also slotted on stainless steel using thin groove cutters and the height of the upper cutter was mainly controlled by feed rate and cutting speed. In mechanical cutting processes, main disadvantages are the generation of cracks due to mechanical stresses, wear of the cutting tool and the long processing time. The challenges come in the fabrication process are discussed in next section.

2.1 Fabrication Challenges

In above geometry, the dimensions are in micron so fabrication of such a heat sink comes under the category of microfabrication. In conventional symmetric MCHS the primary channel is running along the length of the heat sink having taper cross section. The secondary veins are branching out from primary vein having width of 500 um and secondary channels are required to fabricate at an angle to primary vein. Because of these complicated and very fine geometry it is difficult to fabricate. Again, maintaining the depth, width and surface finish is challenging which require precision and accuracy. Unavailability of the probes of CMM below 1 mm makes it difficult for dimensional inspection. So, it is necessary to do modification in traditional instruments. There are number of methods like micro-EDM, etching, laser machining, lithography, molding, micro-machining, etc. In this work the component is fabricated using CNC micro-milling process. The fabrication process and the specification of cutter and milling machine used is discussed in the next section.

2.2. Fabrication

Manufacturing of micro sized component, i.e., (1 to 999 μm) is called as microfabrication. In this work, metallic substrate is used for fabricating micro-channels. The copper is selected as metallic substrate because of its high thermal conductivity which mainly affects the heat transfer rate. Along with this benefit, coper is ductile hence shows good machinability and it is easily available at low cost. In mechanical machining process it is necessary that the substrate should be thicker component so that it should withstand the forces and vibrations generated during cutting action. First the CNC machining program is tested on aluminium block to check the correctness of the program. From the testing it is found that the corresponding carbide milling cutter can be used for the final fabrication.

Micro-milling is one of the subtractive manufacturing technologies. It is characterised by mechanical interaction of sharp tool with the workpiece material causing shearing of material along the defined path to get the specified geometry. The phenomenon of size effect comes in picture and it is defined as the specific energy generally increases with the decreasing depth of cut, this is because the effective rake angle increases as the depth of cut decreases, and the larger the rake angle, greater the specific energy.

In case of micro cutting the depth of cut is very small. The workpiece is mainly process by the cutting edge and compression. Hence it becomes dominant in the deformation of the workpiece material. This results in larger frictional force at the tool-chip interface and consequently in a greater cutting ratio.

The program was made to make one set of leaf pattern micro-channel and then the program was repeated for the 8 times to make complete set of micro-channels. Table 2 shows cutting tools Parameters used for fabrication and figure 1 shows the magnified image of cutting tool.

Table 2. Specification of cutting tool

Tool Material	Tungsten Carbide	Tungsten Carbide
Tool diameter (mm)	0.5	0.3
Cutting length (mm)	2	2
Tool length (mm)	38	38

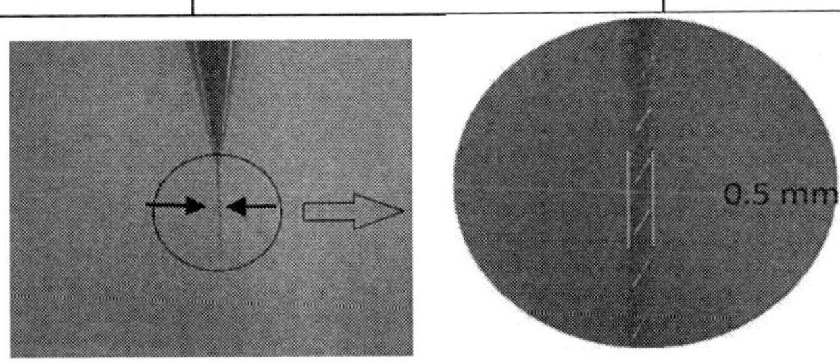

(a) (b)

Figure 1. Solid carbide end mill cutter under Zeiss Microscope at DKTE, Ichalkaranji (a) Two flute Carbide milling tool (b) Enlarged view of end mill cutter

The sequence of operation are as follows- Inlet well - Outlet well – primary veins – secondary veins.

The geometrical dimensions of LP-MCHS is shown and discussed in next section.

2.3 Geometry of micro-channel

The Figure 2 shows the geometry of the micro-channel model. This microchannel model consists of primary veins extending along the heat sink and secondary veins branching from primary veins. The primary channels are convergent or divergent. All secondary channels are convergent in nature, which helps to improve the rate of heat transfer and requires less pumping power. To ensure adequate supply of coolant by the secondary veins in a set of three main channels, the outlet of the central main channel is blocked. The central main channel is connected by main channels.

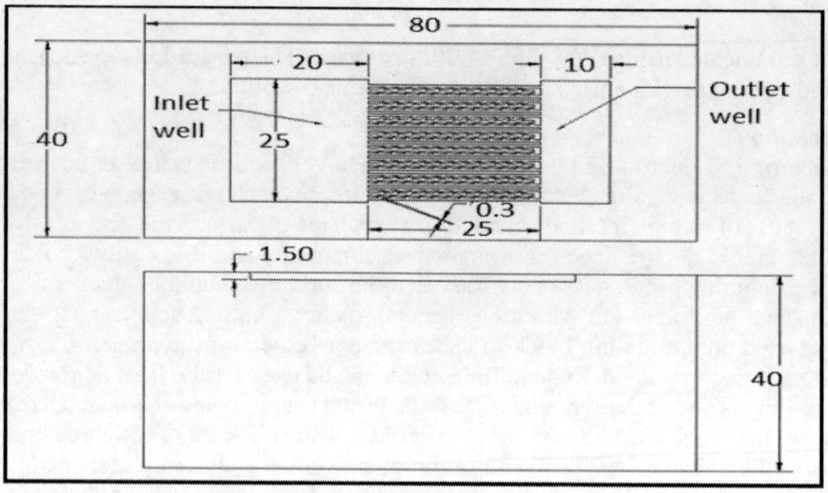

Figure 2. Geometry of Micro-channel

Table 4 presents the geometric dimensions of conventional micro-channel heat sinks and sheet patterns. The micro-channel test section comprises of 24 parallel rectangular micro-channels of width 0.5 mm, depth of 0.9 mm, length of 25 mm. The aspect ratio of micro-channel is 1.8. The fluid enters and distributes into the channels with the help of inlet well. Micro-channel is covered with transparent acrylic sheet of 6mm. The thermocouples are attached from the top of micro-channel such as one thermocouple at inlet section, and three thermocouples at outlet section. And four thermocouples attached at the bottom of the micro-channel for surface temperature measurement.

Table 3. Geometrical Dimensions of micro-channel

Characteristic	**Straight micro-channel**	**Leaf pattern micro-channel**
Material	Copper	Copper
Footprint width length (mm x mm)	25 x 25	25 x 25
Main channel width W_{ch} (μm)	500	500
Channel Depth H_{ch} (μm)	900	900
Aspect ratio	1.8	1.8

Figure 3 shows the details of one set of micro-channels. Eight such sets are clubbed together to form the required design. Fabrication of the enhanced micro-channel design cannot be done using saw blade cutter as the inclined channels of 0.3 mm width are to be cut to create the leaf venation pattern. After studying the various alternatives of machining, it was decided to fabricate the micro-channels using end mill cutter. Figure 4 shows the actual fabricated leaf pattern MCHS. The methodology used for testing thermal performance of LP-MCHS is discussed in next section.

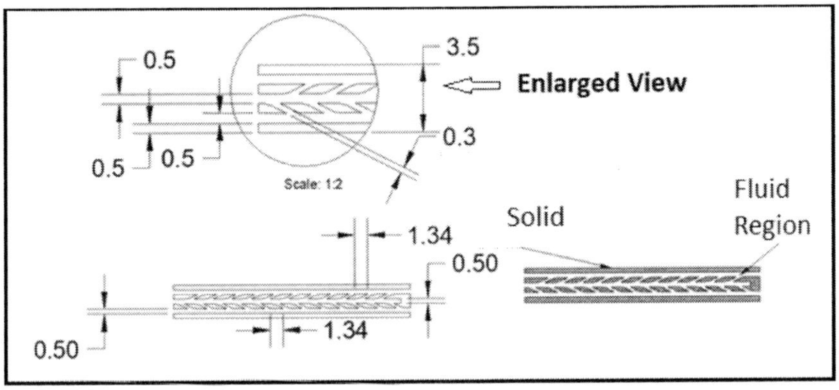

Figure 3. Details of one set of micro-channel

Figure 4. Fabricated Micro-channel

2.4 Methodology

The micro-channel are carved on the pure copper block because of its higher thermal conductivity. The fabrication of the micro-channel is carried out on 5 axis CNC machining centres. Then the leaf pattern MCHS was tested with Deionised (DI) water followed by the different nanofluids with 0.3 vol% and the particle size of 30-50 nm dispersed in DI water. The temperature readings for different heat fluxes ranging from 65 W/cm^2 to 200 W/cm^2 are taken with the varying flow rate of 100 ml/min to 900 ml/min. Surface temperature of the MCHS was taken with the K-type thermocouples. The inlet and outlet temperature were also taken with the K-type thermocouples. Flow rate was maintained and varied by the peristaltic pump. The outlet temp was also taken with the thermal camera for the cross verification. Nanofluid can enhance the heat transfer rate, but increases the pressure drop. Hence micro-channel further evaluated for performance evaluation analysis for the feasibility of use of nanofluid [14].

$$PEC = \frac{\mathrm{Nu(nf)}/Nu(bf)}{\left(\Delta P(nf)/\Delta P(bf)\right)^{\frac{1}{3}}} \qquad (1)$$

The experimental setup used for the investigation of thermal performance of LP-MCHS is discussed in following section.

3. Experimental setup

A schematic view of the experimental setup which shows main components including a peristaltic pump (Masterflex L/S, Cole-Parmer make, variable speed digital driver.), micro-channel test section, data acquisition system (DEWE-Soft Data Acquisition System), Manometer for Differential Pressure, and a collecting tank is shown in figure 5.

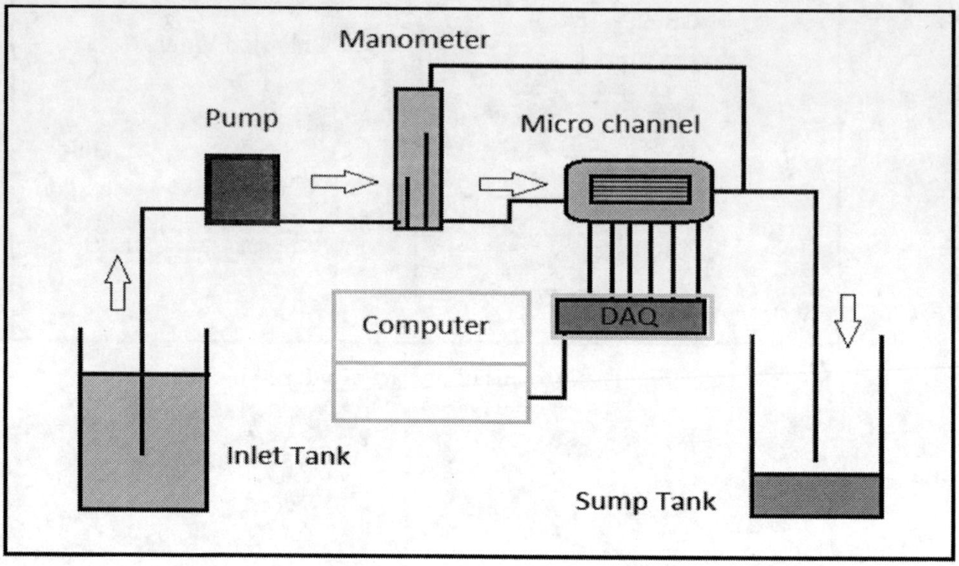

Figure 5. The schematic of micro-channel test section

The DI water and Nanofluids is pumped through the experimental by the peristaltic pump. The flow rate can be measured and controlled from the pump control panel itself. Fluid will pass through the peristaltic pump followed by micro-channel. The U tube manometer is attached to the micro-channel in parallel connection. The thermocouples are used to measure surface temperature and fluid inlet and outlet temperature. They attached to the channel connected to the DAQ system and DAQ is connected to computer. The temperature readings are recorded in DEWE-Soft X3 software. Differential pressure readings are taken manually. The results of the experimental work is discussed in the next section.

4. Results and Discussion

4.1 Dimensional Inspection

After fabrication of LP-MCHS the dimensions are measured with the help of CMM. The dimensional inspection of micro-channel is shown in table 4.

Table 4. Dimensions by CMM

Sr no	Part	Desired Dimensions (mm)	Actual Dimension (mm)	% Error
1	Hole Diameter (Left)	14	14.05	0.35
2	Hole Diameter (Right)	14	14.08	0.57
3	Depth of hole (Left)	50	50	00
4	Depth of hole (Right)	50	52	4
5	Well depth	2.5	2.38	4.8
6	Well width (Closed end)	6.5	6.34	2.46
7	Well width (open end)	6	6.05	0.83

8	Heater block diameter	14	13.95	0.36
9	Channel length	25	25	00.00
10	Channel width	0.5	0.5	00
11	Channel depth	1	0.9	10

Aspect ratio for the fabricated LP-MCHS is presented in next section.

4.2 Aspect Ratio

As aspect ratio increases, surface area available for heat transfer increases too which improves the heat transfer performance of micro-channel. so, it is desirable to produce high aspect ratio micro-channel to improve heat transfer performance. Table 6 shows the aspect of the fabricated LP-MCHS.

Table 5. Aspect Ratio

Feature	Depth (μm)	Width (μm)	Aspect Ratio
LP-MCHS	900	500	1.8

The MRR of micro end milling is discussed in next section.

4.3 MRR

Micro end milling possesses higher MRR (0.6 mm^3/min) which is higher than other fabrication processes such as EDM, chemical etching. MRR is directly related to the time required for fabrication of component. As MRR increases, the number of units produced increases. The material compatibility of micro end milling is discussed in next section.

4.4 Material Compatibility

In micro end milling process, material is removed by shearing and ploughing phenomenon. So soft and ductile materials can be effectively processed by micro end milling process, e.g. copper, aluminium, mild steel. Micro end milling is not suitable for machining of hard materials. The surface temperature variation in LP-MCHS is discussed in next section.

4.5 Surface Temp

The main objective of using micro-channels is to maintain the lower surface temperature. And from the results it is clearly observed that the leaf pattern MCHS is able to maintain the lower surface temperature for the same heat flux and flow rate when compared to conventional MCHS. The surface temperature decreases due to increase in flow rate, Reynolds number and the turbulence caused in micro-channel. Keblinski et al proposed the four possible mechanisms for heat transfer enhancement with nanofluids which are Brownian motion, ballistic phonon transport, liquid layering/ particle interference and nanoparticle clustering [18]. The heat transfer coefficient is discussed in next section.

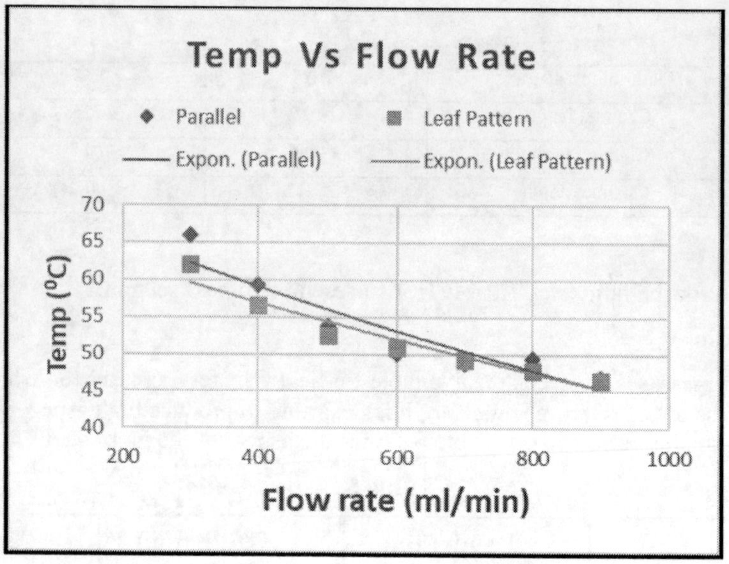

Graph 1. Variation of surface temp with the flow rate.

4.6 Heat transfer coefficient

From the measured micro-channel wall temperatures, heat absorbed by the fluid and fluid bulk temperature, the local heat transfer coefficient was calculated. Then we have calculated the Reynolds number and Nusselt number. After comparing the ratio of Re and Nu, it was found that the leaf pattern MCHS shows greater ratio which is better for the better performance of MCHS. It is shown in graph 2. Here it was observed that the leaf pattern MCHS shows average enhancement of 17% in Reynolds number and 37% in Nusselt Number. The performance evaluation analysis is discussed in next section.

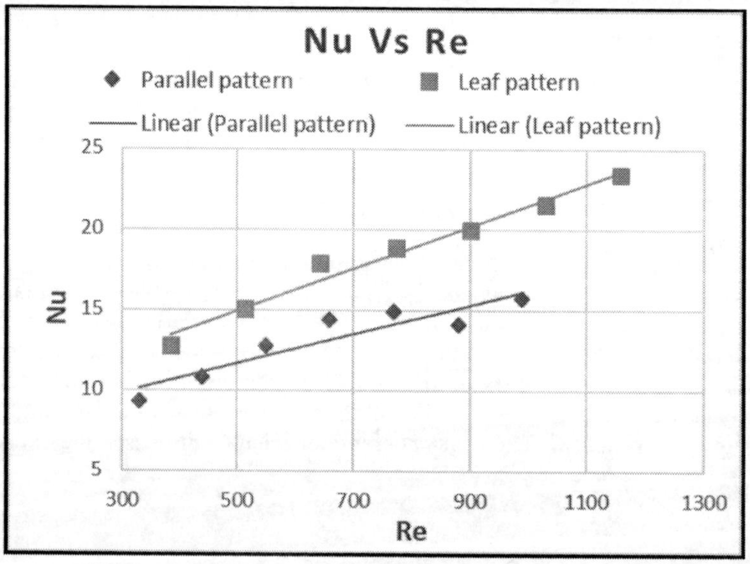

Graph 2. Nusselt No Vs Reynolds No for Leaf and Parallel MCHS for DI water.

4.7 Performance Evaluation Analysis

From the experimental result it is clearly seen that the nanofluids enhances the heat transfer but tends to increase pressure drop. Thus, to analyse that, the nanofluid performance in terms of heat transfer enhancement was studied. Hence in this study the PEC values were considered in the comparison of the heat transfer characteristics of the nanofluids and water under equal pressure drops. Specifically, the sample is good to use for the application if PEC>1, then the sample is regarded as unsuitable if PEC<1. PEC values for LP-MCHS at different heat flux is shown in graph 3.

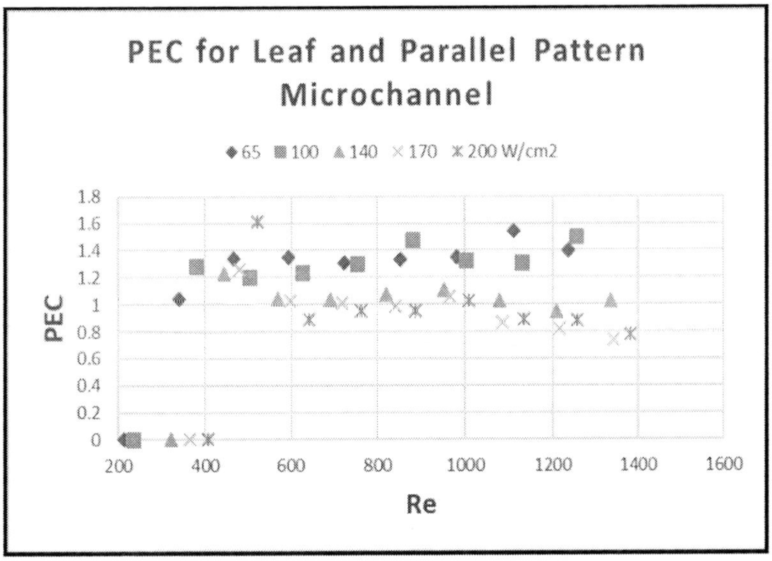

Graph 3. Reynolds No Vs PEC values

The PEC values were plotted as a function of Re. The leaf pattern MCHS achieved PEC>1 for almost all heat fluxes and flow rates. Hence, it was concluded that the leaf pattern MCHS is convenient to use and have more effectiveness than conventional MCHS. PEC values for LP-MCHS with SiO_2 nanofluid for different heat flux is shown in graph 4.

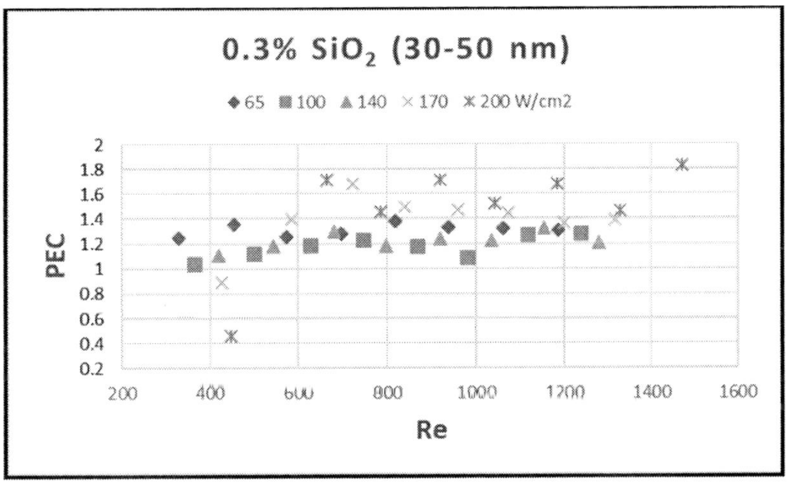

Graph 4. Reynolds No Vs PEC value for SiO2 nanofluid

The SiO2 nanofluid with leaf Pattern MCHS have achieved the PEC>1, hence the SiO_2 nanofluid was fund suitable to use and having more effectiveness. For higher heat fluxes and flow rates, the PEC value is getting increased. For Al_2O_3 and CuO nanofluids the PEC<1, hence these nanofluids were not suitable to use with leaf pattern MCHS. The pressure drop developed in micro-channel is discussed in next section.

4.8 Pressure Drop

For the same Reynolds number, the leaf pattern MCHS shows the lower pressure drop. For all heat flux and flow rates, the pressure drop was less in case of leaf pattern MCHS when compared with conventional pattern. The variation of the pressure drop was found to be increasing with the increase in Re which is shown in the graph 5. LP-MCHS shows 18% increase in pressure drop.

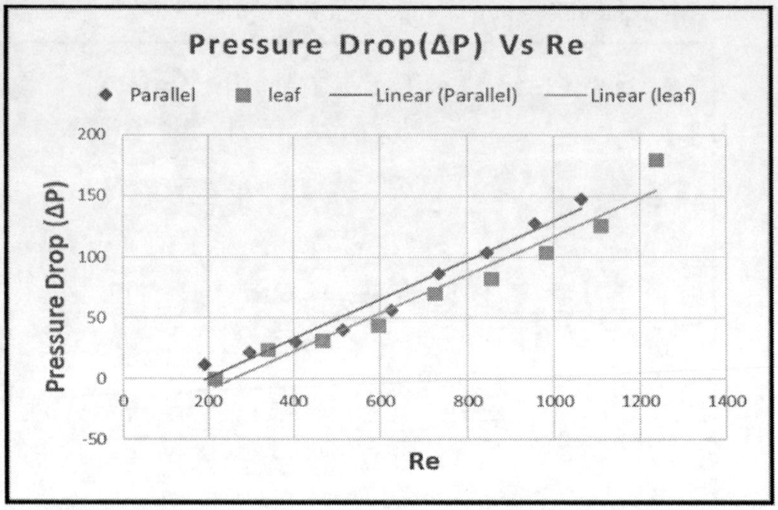

Graph 5. Pressure drop variation with Reynolds no for Leaf and Parallel MCHS

From the graph 6, it was found that the CuO and SiO_2 nanofluids shows very little variation in the pressure drop compared to the DI water. This little variation is because of the surfacecant used while preparing the nanofluid. Al_2O_3 nanofluid shows very large difference in pressure drop.

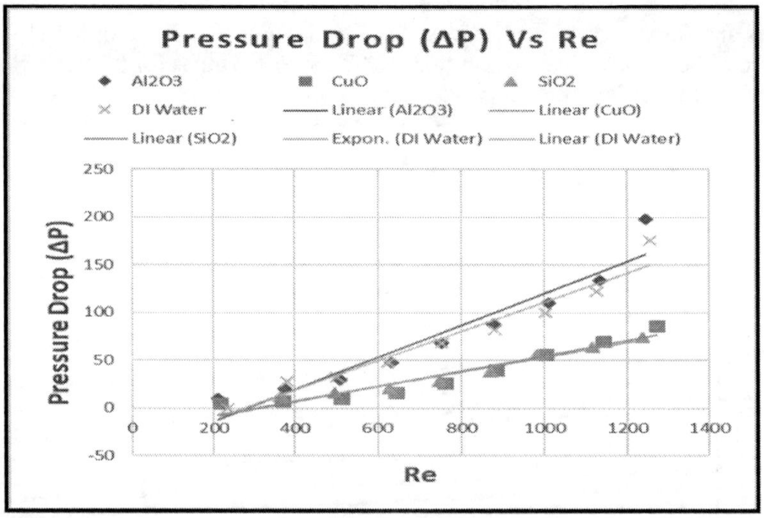

Graph 6. Pressure drop variation in LP-MCHS with nanofluids

Al_2O_3 nanofluid have larger viscosity than the CuO and SiO_2 nanofluid for the same concentration and same particle size hence the pressure drop is larger in case of Al_2O_3 nanofluid than the other nanofluids [15]. The thermal conductivity is discussed in next section.

4.9 Thermal Conductivity

The variation of thermal conductivity and flow rate is shown in graph 7 and graph 8. Thermal conductivity is seen to be increasing with the increase in flow rate. Increase in flow rate increases the contact of more nanoparticles and due to higher thermal conductivity of nanoparticles the nanofluid is able to carry out more amount of heat compared to DI water. Also increase in flow rate increases the Re, and the increase in Re indirectly increases the thermal conductivity.

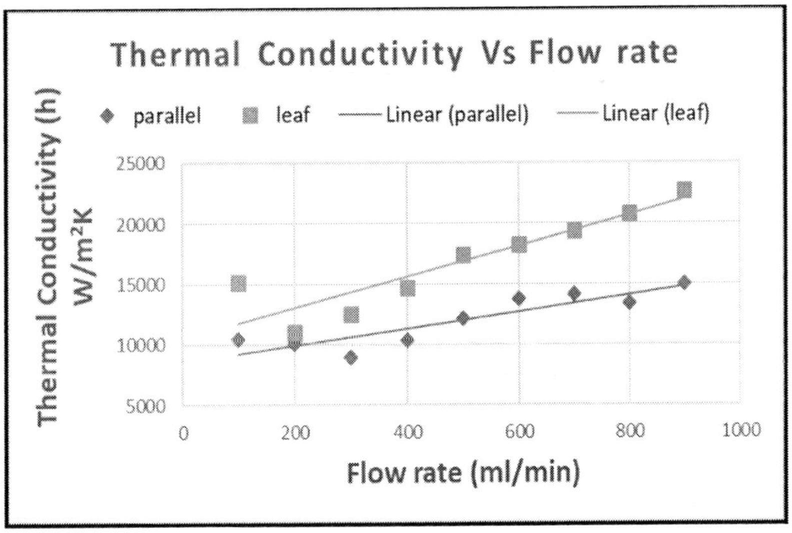

Graph 7. Variation of thermal conductivity of DI water with the variation of flow rate for Leaf and Parallel MCHS

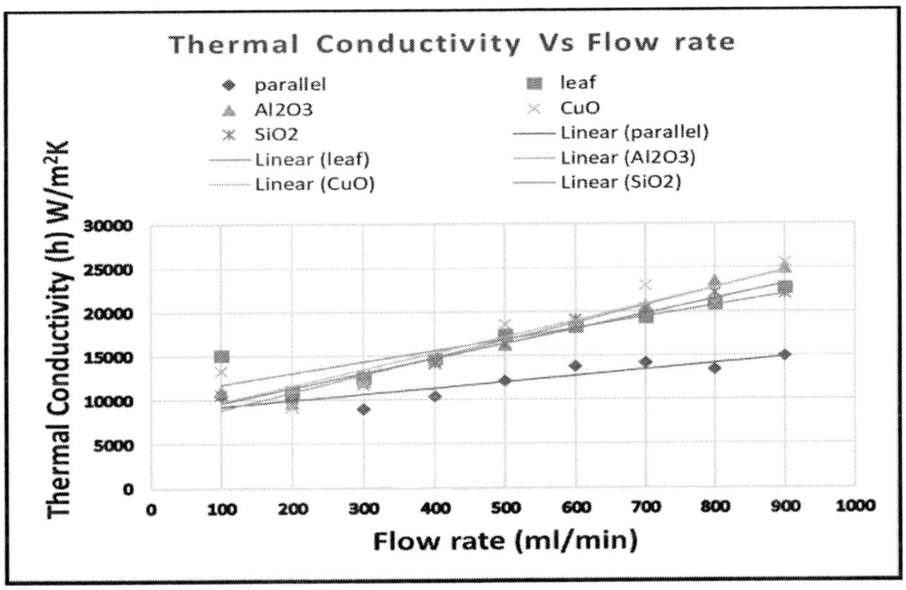

Graph 8. Variation of thermal conductivity of nanofluids with the variation in flow rate

The thermal conductivity shows significant increase above the flow rate of 500 ml/min for nanofluids. Conclusion is presented in the next section.

5. Conclusion

- In micro end milling process, material is removed by shearing and ploughing phenomenon, hence only soft and ductile materials can be processed by micro-milling. It is not suitable for hard materials.
- Micro end milling is recommended for fabrication of MCHS up to 100 μm due to limitation on the size of tool.
- Micro end milling results in good surface finish, i.e., 5 to 6 μm Rt value.

- Higher aspect ratio increases the heat transfer area. Micro end milling results in moderate aspect ratio i.e., 2 to 3.
- The production cost increases linearly with number of components, hence best suited for job production.
- Following conclusions are drawn based on the thermal experimentation results
- Leaf pattern MCHS able to maintain lower surface temperature due to the high Reynolds number, and increased Brownian motion.
- Due to the availability of higher area and the higher aspect ratio of MCHS, it shows higher heat transfer coefficient with average enhancement of 17% in Reynolds number and 37% in Nusselt number.
- For the same Reynolds number, the leaf pattern MCHS shows lower pressure drop. But for same flow rate LP-MCHS shows average 18% increase in pressure drop.
- Presence of higher thermal conductivity nanoparticles shows enhancement in thermal conductivity of nanofluids.
- SiO2 nanofluid shows lower pressure drop, higher thermal conductivity and higher PEC value, which indicates the suitability of SiO2 nanofluid for application.

6. References

[1] "D. B. Tuckerman and R. F. W. Pease. "High-Performance Heat Sinking for VLSI," IEEE Electron Devices Letters, Vol.2, No. 5, 1981, pp. 126-129"

[2] "X. W. Wang, X. F. Xu and S. U. S. Choi, "Thermal Conductivity of Nanoparticles-Fluid Mixture," Journal of Thermo-physics and Heat Transfer, Vol. 13, No. 4, 1999, pp. 474-480"

[3] "S. lee, S. U. S. Choi, S. Li and J. A. Estman, "Measuring Thermal Conductivity of Fluids Containing Oxide Nano-particles," Journal of Heat Transfer, Vol. 121, No. 2, 1999, pp. 280-289"

[4] "B-X Wang, L. P. Zhou and X. F. Peng," A Fractal Model for Predicting the Effective Thermal Conductivity of Liquid with Suspension of Nanoparticles," International Journal of Heat and Mass Transfer, Vol. 46, No. 14, 2003, pp. 2665-2672"

[5] "J. Koo and C. Kleinstreuer, " A New Thermal Conductivity Model for Nanofluids," Journal of Nanoparticle Research, Vol. 06, No. 6, 2004, pp. 577-588"

[6] "Choi, S. U. S.,1995,“ Enhancing Thermal Conductivity of Fluids With Nanoparticles," Development and Applications of Non-Newtonian Flows, ASME, New York, 1995.“

[7] "Q. li. And Y. M. Xuan,“ Convective Heat Transfer and Flow Characteristics of Cu-Water Nanofluid," Science in China Series E: Technological Sciences, Vol. 45, No. 4, 2002, pp. 408-416"

[8] "B. C. Pak and Y. I. Cho, "Hydrodynamic and Heat Transfer Study of Dispersed Fluids with Submicron Metallic Oxide Particles," Experimental Heat Transfer, Vol. 11, No. 2, 1998, pp. 151-170"

[9] "A. Abdollahi, H.A. Mohammed, "Fluid flow and heat transfer of nanofluids in micro-channel heat sink with V-type inlet/outlet arrangement," Alexandria Engineering Journal, 2017, Vlo.56, pp. 161-170"

[10] "H.A. Mohammed, P. Gunnasegaran, N. Shuaib, The impact of various nanofluid types on triangular micro-channels heat sink cooling performance, Int. Commun. Heat Mass Transf. 38 (2011) 767–773."

[11] "H.A. Mohammed, P. Gunnasegaran, N. Shuaib, Numerical simulation of heat transfer enhancement in wavy micro-channel heat sink, Int. Commun. Heat Mass Transf. 38 (2011) 63–68."

[12] "H.A. Mohammed, P. Gunnasegaran, N. Shuaib, Influence of various base nanofluids and substrate materials on heat transfer in trapezoidal micro-channel heat sinks, Int. Commun. Heat Mass Transf. 38 (2011) 194–201."

[13] "P. Keblinski, S. Phillpot, S. Choi, J. Eastman, "Mechanisms of heat flow in suspensions of nano-sized particles (nanofluids)," International Journal of Heat and Mass Transfer, 2002, Vol. 45, No. 4, pp. 855-863"

[14] "Z.F. Huang, A. Nakayama, K. Yang, C. Yang, W. Liu, "Enhancing Heat Transfer in the Core Flow by using Porous Medium insert in a tube," International Journal of Heat and Mass Transfer, Vol. 53, No. 5-6, pp. 1164-1174"

[15] "J. Bowers, H. Cao, G. Qiao, Qi Li, G. Zhang, E. Mura, Y. Ding, "Flow and Heat Transfer behaviour of nanofluids in micro-channels," Progress in natural Sciences: Materials International, 2018, Vol. 28, pp. 225-234."

[16] "Abdelgawad M, Wu C, Chien WY, et al. A fast and simple method to fabricate circular micro-channels in polydimethylsiloxane (PDMS). Lab Chip 2011; 11: 545–551."

[17] "Pan M, Zeng D and Tang Y. Feasibility investigations on multi-cutter milling process: a novel fabrication method for microreactors with multiple micro-channels. J Power Sources 2009; 192: 562–572"

[18] "Mohammad Zargartalebi, Jalel Azaiez. Heat transfer analysis of nanofluid based micro-channel heat sink, IJHMT 127(2018) 1233-1242"

[19] "Qin Yi. Micro-manufacturing engineering and technology. 1st ed. Oxford: Elsevier Inc., 2010."

[20] "Ji Zhang, Yanhua Diao, Yaohua Zhao, Yanni Zhang, Thermal-Hydraulic Performance of SiC-water and Al2O3- water nanofluids in the micro-channel, ASME 2016, Vlo. 138, 0217051-9"

[21] "Y. Yorozu, M. Hirano, K. Oka, and Y. Tagawa, "Electron spectroscopy studies on magneto-optical media and plastic substrate interface," IEEE Transl. J. Magn. Japan, vol. 2, pp. 740-741, August 1987 [Digests 9th Annual Conf. Magnetics Japan, p. 301, 1982]."

[22] "M. Young, The Technical Writer's Handbook. Mill Valley, CA: University Science, 1989."

Effect of Stratified Charge Compression Ignition on Emission and Power Output- A Review

Krishna Kumar Gupta, Ashish Mogra
Department of Mechanical Engineering, SVKM's, NMIMS, MPSTME, Shirpur, Maharashtra, India
Krishnakumar.gupta@nmims.edu

Abstract. The emissions from the vehicles causes for severe air pollution which resulted the development of stringent regulatory norms. Therefore there is a need of improvement in combustion phenomenon to decrease the NOx and PM, as they are directly dependent on combustion. To improve environmental conditions which are directly affecting human health, all automobile companies are endeavouring to satisfy strict emission standards. In this context stratified charge compression ignition (SCCI) engines seems promising and developing its application by incorporating the different types of combustion mode with the help of direct fuel injection and by ranging its capacity to full load condition. Alongside SCCI, a new concept of multi stage induction and Exhaust Gas Recirculation (EGR) can be incorporated in engine to reduce emissions, increase power output and efficiency. Various parameters have been considered to understand the combustion process for developing an engine to meet the requirements. The effect of different variables influencing the combustion process is highlighted.
Keywords: SCCI, Exhaust gas recirculation, emission.

1. Introduction

The challenges in today's world in the fields of internal combustion engine are either the rapidly decreasing crude oil or air pollution from these engines. The concern of fuel scarcity may be reduced or eliminated with the help of alternative fuel, whilst emission is controlled through modification in operating cycles of engines. Therefore there is an increasing need of developing green and environment friendly technologies. Therefore there is continuous research on different methods or techniques which can improve air quality and reduce the consumption of fuel [1-3].

In the compression ignition engines the combustion characteristics includes the spray pattern, compression ratio, swirl, type of combustion chamber etc. are important parameters to decide the volume and nature of oxidation of fuel and so the emission. Combustion phenomenon in engines is more effective when the dilution of air-fuel mixture is taken place for its maximum limit [4-6].

The combustion systems which can be preferred are Low Temperature Combustion (LTC) technology and Stratified Charge Compression Ignition (SCCI). This technology has capacity to minimize NOx, soot emissions and particulate matter. The thermal efficiency increased with compared to conventional diesel engine. Although, researchers are facing some challenges like at minimum loads conditions one can get minimum efficiency of combustion. [6-10].

2. Key Points of SCCI

To get low temperature combustion in compression ignition engine, the very necessary requirements is formation of homogeneous mixture. This may be achieved through injecting the entire fuel ratio in all cylinders. The demand of LTC is increasing day by day; this is due to longer ignition delay and higher level of exhaust gas recirculation and less content of oxygen. However, to prepare uniform mixture in cylinder large level of turbulence is required [11].

Homogeneous charge compression ignition engines and low temperature combustion has its different advantages over the conventional engines. To incorporate the same advantages in conventional combustion system fuel was mixed stratification way, this technology is called Stratified Charged Compression Ignition (SCCI). SCCI technology is also called as compound HCCI which is based on stratification mixture of fuel. This technique includes all the advantages of HCCI combustion as well as LTC. By using this technique controlling of different parameters like combustion phasing and combustion rate are directly and easily controlled [12]. It is thus, observed that some amount of in-homogeneities will always remain in SCCI engine, while applying the different methods of mixture preparation. It also shows that uniform fuel air mixture cannot be achieved. By taking the base of stratification it is found that there are three major aspects which exist and are included in the cylinder i.e. fuel composition stratification, mixture concentration stratification and temperature field stratification.

Fuel composition stratification a means controlling the homogeneous charge and it is a very important parameter that can affect the auto ignition characteristics. It was found from the published studies that the combustion efficiency increased with increasing the level of fuel stratification. However, it was also concluded that higher level of stratification generate high NOx and reduced the combustion efficiency as well as power output. Letter investigation was carried out in a single cylinder engine with two Gasoline Direction Injection (GDI) to analyses the effect of stratification. Due to fuel stratification significantly increment in indicative mean effective pressure (IMEP) and emission especially at lean limits was found. Fuel stratification was proved to decrease the rate of pressure rise as well as sensitive to the local equivalence ratio. By implementing the fuel stratification SCCI combustion can be controlled [34-38] Increasing in mixture concentration stratification, auto-ignition advanced. The auto-ignition get also advanced by increasing the amount of fuel injected later in the cylinder. Mixture stratification plays an important role to predict the emissions of CO, unburned hydro carbons and soot. The emission of NOx is also affected by mixture concentration stratification [39].

3. Elementary principle of SCCI combustion

Figure 1 shows the basic concept of outline of fuel with its fuel design and its direction for SCCI combustion.

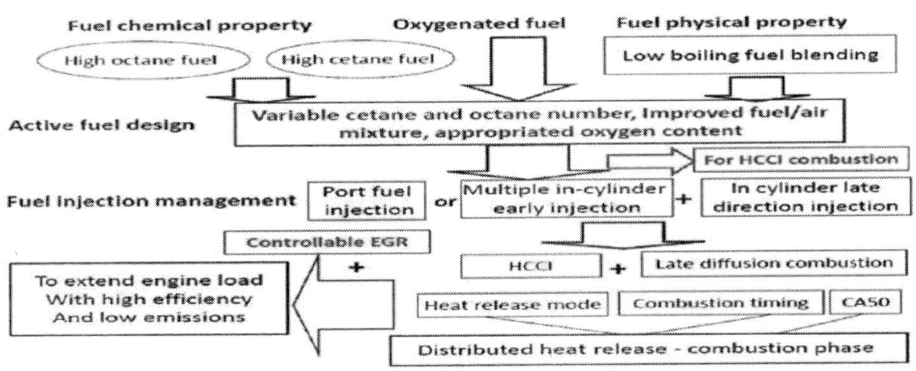

IOP Conf. Series: Materials Science and Engineering **810** (2020) 012026 doi:10.1088/1757-899X/810/1/012026

Figure 1. Elementary concept of outline of fuel with its fuel design and its direction for SCCI combustion. [12].

i. On the basis of operating parameters and selected fuels for test, the physical and chemical properties should be designed.
ii. Different methods or approach like injection of port fuel in direct injection combined with in-cylinder, injection mode, in-cylinder injection with multi hole and split injection, should be applied for selected fuel stratified concentration.
iii. The test fuels which have reverse chemical properties have to be taken that can be supplied as a port fuel injector as well as direct fuel injector.
iv. For controlling the combustion rate and peak combustion temperature, the controlled amount of exhaust gas recirculation should be used.

By following all this steps the various difficulties which the SCCI facing can be minimize.

4. Effect of engine parameters on SCCI

Engine parameters play an important role for the improvement of engine performance. To get proper SCCI combustion port fuel injection is the best method which can be used in cylinder. Specifically, for achieving the controlled SCCI combustion phasing the early injection or port fuel injection plays a significant role due to its chemistry of fuel mixture concentration [13]. This happen because of amount of heat released at low temperature is proportional to equivalence ratio. Controlling of stratification mixture can be done by changing the injection timing within the cycle. Using this technique combustion phasing can also be managed. Along with, applying different fuel injection timing combustion duration for stratified mixture can be adjusted. Consequently, the duration of combustion increased and rate of pressure rise decreased in partial fuel stratification [14]. It is observed that indicated thermal efficiency of stratified compression charge ignition engine is lower than conventional combustion engine this is due to requirements of optimized combustion parameters such as phasing control, heat release rate and design of combustion chamber. So due to this reason challenges for achieving low level of NOx emission and maximum efficiency of stratified charged combustion engine is key issues. It has been detected, by varying the proportions of premixed fuel like n-heptane or iso-octane, the stratified temperature and proper mixture formed in cylinder as shown in figure 2. Moreover, fuel mixture stratification can be changed by varying the timing of fuel injection.

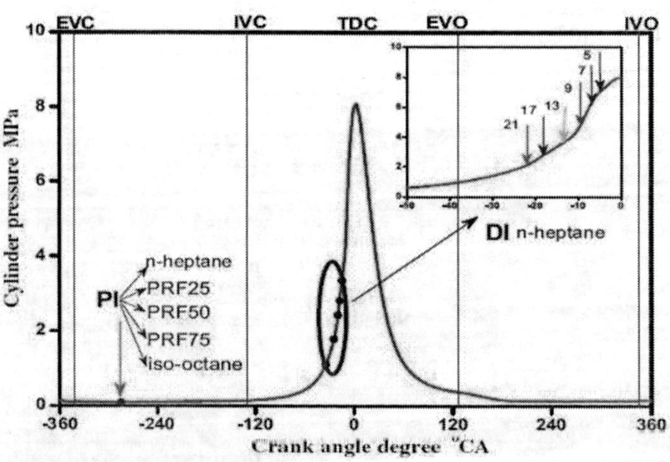

IOP Conf. Series: Materials Science and Engineering **810** (2020) 012026 doi:10.1088/1757-899X/810/1/012026

Figure 2 Stratifications of n-heptane SCCI combustion [15]

To increase the range of SCCI engine form lower load to higher load stratification of partial fuel in required. Due to stratification of partial fuel staged combustion can be achieved by decreasing the pressure rise rates.

4.1 Timing of Injection

Two stage gasoline direct injection technique is an effective method proposed by researchers for controlling the combustion rate, combustion phasing and preparing the homogeneous mixture for extending the working load of SCCI combustion [23]. By varying or changing the number of injection, width of injection pulse and dwell time for two nearer pulses, stratification of diesel fuel can be controlled using multi pulse fuel injection technology. With the help of two injections in single cycle, the fuel mixing can be improved. During first injection before TDC 80° and second injection after TDC 10°, smoke has been reduced. Using this technology thermal efficiency of SCCI engine can be reach up to the level of conventional diesel engine. Meanwhile other emission like NOx, Soot and PM also get reduced [24, 25].

Partial fuel injection method was used for achieving the homogeneous stratification mixture. In this method, injection of fuel taken place in two ways. Some amount of fuel injected in intake stroke and remaining fuel in compression stroke. It has been seen that while minimizing the rate of pressure rise and increasing the load capacity, productive potential are getting for partial fuel stratification during the SCCI combustion. It is also found that if amount of fuel stratified increased NOx emission will get increased [26]. For the closed valve injection, it has been found that maximum amount of the fuel strikes on the wall of port before entry of the cylinder so that details of injection do not matter. But during the open valve injection as the droplets enters into the cylinder the hydrocarbons (HC) emission increased. It concluded that closed valve fuel injection will be preferred and also suitable for the SCCI engines with port fuel injection system [23]. As per the stratified mixture concentration is concerned experiments showed that to increasing the load capacity of SCCI engine proper stratified mixture should be formed [21, 22].

Gasoline direct injection system showed a crucial role during the stratified combustion. Stability in combustion depends on fuel injection symmetrical or asymmetrical to the inlet port stratified mixture. Two GDI injectors also used for perfect combustion, in which one was us used for preparing homogeneous charge combustion and second was employed directly into the combustion chamber for varying the stratified charge. They found enormous advancement in indicated mean effective as well as minimum amount of emissions while keeping operating range at lean limit [29-30]. They developed the components of fuel stratification with blending of diesel fuel and iso-octane. Diesel fuel was injected directly into the combustion chamber while iso octane injected into inlet port. Research reveals by changing the fuel ratios the ignition timing can be controlled for duel fuel stratified combustion [28].

4.2 Water Injection

There has been a long history of Water injection in the field internal combustion engines [17-19]. To lighten the effect of knock as well as for lower temperature and minimize the NOx emissions in SI combustion water injection has been used. [17]. This injection was previously investigated in different combustion techniques like Homogeneous charge compression ignition (HCCI) and Premixed Charge Compression Ignition (PCCI), but now it has also been investigated in Stratified charged compression ignition (SCCI) engine [20]. Experiment analysis was performed to get depth knowledge of with or without water injection in stratified charge compression ignition

(SCCI) engine. It has been found that combustion rate is significantly increases in the range of 20 to 70 crank angle degrees before TDC for water injection as compared to without water injection [3]. Experiments were conducted on four-cylinder gasoline, SI 2.0L engine. Initially the direct water injection timing was 50 degrees before TDC. It has been found that as crank angle increased by 153 degree, water injection retarded combustion phasing to the point of misfiring [16]. It was found that by using this injection system there is incremental effect of 44% and 103% on CO emissions and unburned hydrocarbon respectively at 9.0 mg/cycle of water.

4.3 Pressure Rise Rate (PRR)

One more important parameter which needs to be considering while doing the investigation on SCCI combustion is rate of pressure rise. This is used to define the higher limit of SCCI combustion. Increase in the pressure of combustion chamber the NOx and smoke formation reduces because of dilute mixture. With the help of split injection method, the mean effective pressure decreases with no changes in NOx and smoke formations.

Figure 3 represent the relation between maximum of pressure rise rate and position of crank angle with respect to different equivalence ratio. It is observed from the figure that maximum rate of pressure rise has been increased for richer fuel air mixture and it decreased as the mixture becoming leaner. It is also observed that for every intake air temperature the rate of combustion has been very high. It can be noticed from the figure that position of crank angle for maximum rate of pressure rise shifted towards constant fuel air ratio as the inlet air temperature increased [2].

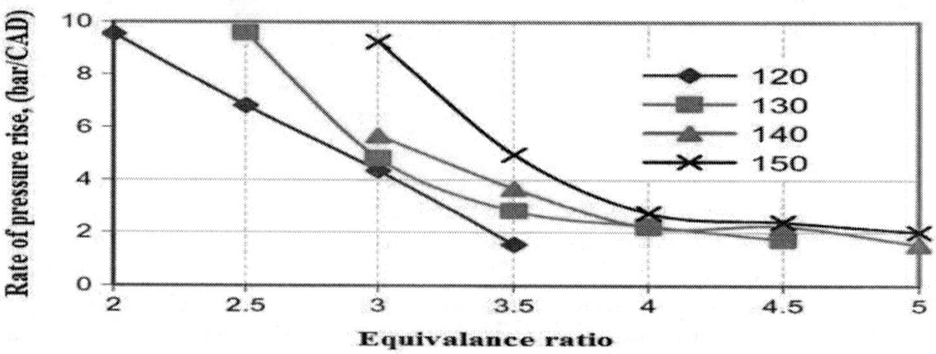

Figure 3 Maximum pressure rise rate at different equivalence ratio [2].

Experiments were performed on a single cylinder, four stroke stratification charge engine with direct injection spark ignition system. They used methanol fuel for their study. It has been observed that at maximum load and engine speed the pressure rise and break thermal efficiency decreased by 60% and 6% respectively. It has also been observed that break thermal efficiency of methanol fueled engine increased 18% at minimum load by reducing the value of compression ratio from 16 to 14. [4]. It is proved that the stratification has a capacity to change the pressure rise rate increasing or decreasing it meanwhile allowing the improvement in output of SCCI combustion. It is found that stratified fuel can also increase the combustion efficiency at low load conditions [8,9].

4.4 Effect of NOx

NOx emission directly depends on the injection timing by using different fuel quantities. In conventional combustion engine, the NOx emission is around 30-35 ppm, but using the split injection method, the NOx emission is limited to 22 ppm that is much lower than the conventional combustion engine. But overall pressure is increased, and it affects the fuel quantities. Once the fuel quantity increases, the mean effective pressure simultaneously increases. HC emission is a function of the fuel injection timing for different fuel quantities injected per stroke [19]. It has been found from that the concentration of NOx get minimum for the homogeneous charge while operating on multiple fuel injection system. This is due to there has been early injection in different form which increased the intensity of mixing until the second stages of injection starts. Which in turn reduce the formation of NOx maintain the temperature within the limit. HC emission can be controlled easily by using PFI system in SCCI engines [20].

5. Challenges of SCCI Combustion

In stratified charge compression ignition engine, as the mean effective temperature decreases, UHC and CO emissions simultaneously increases compared to conventional engines because of the lean mixture or the high exhaust gas recirculation rates. In split injection method, one portion is the induction of fuel in suction line that has challenges as diesel can't be vaporizes at normal conditions. The turbocharger has the capacity to increase the fuel pressure and temperature, but it is not sufficient to vaporize the fuel. The solution to it is to install a heat exchanger or EGR or both depending on operating conditions, which will increase the temperature of the incoming air sufficiently for the rapid vaporization of fuel. As an alternative to it, the fuel itself can be heated to the boiling point by the help of EGR.

6. Conclusion

Stratified charge compression ignition provides an effective advantage for improvement of thermal efficiency of internal combustion engines. Stratified Charge Compression Ignition (SCCI), technology can be considered as extended part of Homogeneous Charge Compression Ignition (HCCI) engine. In SCCI concept of combustion by varying different operating or design parameters of combustion like, port fuel injection, EGR, fuel management and component design, the overall performance of combustion and mixture preparation can be achieved. The SCCI concept has the advantage of both the HCCI and LTC combustion technology, by controlling can be done directly or flexibly on combustion phasing and combustion rate. Moreover, by using suitable fuel physical and chemical properties and management of design parameters like port fuel injection, split injection and EGR combustion of SCCI can be controlled. The typical feature of all such advanced combustion technique before ignition occurs is preparing of homogeneous mixture or stratified mixture. It is concluded that Hydrocarbons (HC) and Carbon oxide (CO) are less as differentiate to homogeneous charge compression ignition engine. NOx emissions are drastically reduced from conventional internal combustion engine but higher that HCCI engine. Moreover it is noted that using higher stratified charge leads to significance increment in NOx emissions.

7. References

[1] Lee, C., Chung, J., & Lee, K. (2017). Emission Characteristics for a Homogeneous Charged Compression Ignition Diesel Engine with Exhaust Gas Recirculation Using Split Injection Methodology. *Energies*, *10*(12), 2146.

[2] Maurya, R. K., & Agarwal, A. K. (2011). Experimental study of combustion and emission characteristics of ethanol fuelled port injected homogeneous charge compression ignition (HCCI) combustion engine. *Applied Energy*, *88*(4), 1169-1180.

[3] Lawler, B., Splitter, D., Szybist, J., & Kaul, B. (2017). Thermally Stratified Compression Ignition: A new advanced low temperature combustion mode with load flexibility. *Applied energy*, *189*, 122-132.

[4] Gong, C., Liu, F., Sun, J., & Wang, K. (2016). Effect of compression ratio on performance and emissions of a stratified-charge DISI (direct injection spark ignition) methanol engine. *Energy*, *96*, 166-175.

[6] Zhang, F., Liu, H. F., Yu, J., & Yao, M. (2016). Direct numerical simulation of n-heptane/air auto-ignition with thermal and charge stratifications under partially-premixed charge compression ignition (PCCI) engine related conditions. *Applied Thermal Engineering*, *104*, 516-526.

[7] Benajes, J., García, A., Monsalve-Serrano, J., & Boronat, V. (2017). Dual-fuel combustion for future clean and efficient compression ignition engines. *Applied Sciences*, *7*(1), 36.

[8] Luong, M. B., Sankaran, R., Yu, G. H., Chung, S. H., & Yoo, C. S. (2017). On the effect of injection timing on the ignition of lean PRF/air/EGR mixtures under direct dual fuel stratification conditions. *Combustion and Flame*, *183*, 309-321.

[9] Wissink, M., & Reitz, R. (2017). The role of the diffusion-limited injection in direct dual fuel stratification. *International Journal of Engine Research*, *18*(4), 351-365.

[10] Lee, J., Chu, S., Kang, J., Min, K., Jung, H., Kim, H., & Chi, Y. (2017). Operating strategy for gasoline/diesel dual-fuel premixed compression ignition in a light-duty diesel engine. *International Journal of Automotive Technology*, *18*(6), 943-950.

[11] Petersen, B., Miles, P., & Sahoo, D. (2012). Equivalence ratio distributions in a light-duty diesel engine operating under partially premixed conditions. *SAE International Journal of Engines*, *5*(2), 526-537.

[12] Liu, H., Ma, S., Zhang, Z., Zheng, Z., & Yao, M. (2015). Study of the control strategies on soot reduction under early-injection conditions on a diesel engine. *Fuel*, *139*, 472-481.

[13] Lu, X., Han, D., & Huang, Z. (2011). Fuel design and management for the control of advanced compression-ignition combustion modes. *Progress in Energy and Combustion Science*, *37*(6), 741-783.

[14] Sjöberg, M., & Dec, J. E. (2007). *EGR and intake boost for managing HCCI low-temperature heat release over wide ranges of engine speed* (No. 2007-01-0051). SAE Technical Paper.

[15] Lu, X., Shen, Y., Zhang, Y., Zhou, X., Ji, L., Yang, Z., & Huang, Z. (2011). Controlled three-stage heat release of stratified charge compression ignition (SCCI) combustion with a two-stage primary reference fuel supply. *Fuel*, *90*(5), 2026-2038.

[16] Lawler, B., Splitter, D., Szybist, J., & Kaul, B. (2017). Thermally Stratified Compression Ignition: A new advanced low temperature combustion mode with load flexibility. *Applied energy*, *189*, 122-132.
[17] Dronniou, N., & Dec, J. (2012). Investigating the development of thermal stratification from the near-wall regions to the bulk-gas in an HCCI engine with planar imaging thermometry. *SAE International Journal of Engines*, *5*(3), 1046-1074.
[18] Reitz, R. D., & Duraisamy, G. (2015). Review of high efficiency and clean reactivity controlled compression ignition (RCCI) combustion in internal combustion engines. *Progress in Energy and Combustion Science*, *46*, 12-71.
[19] Manofsky, L., Vavra, J., Assanis, D. N., & Babajimopoulos, A. (2011). *Bridging the gap between HCCI and SI: Spark-assisted compression ignition* (No. 2011-01-1179). SAE Technical Paper.
[20] Li, L., Xie, H., Chen, T., Yu, W., & Zhao, H. (2012). *Experimental Study on Spark Assisted Compression Ignition (SACI) Combustion With Positive Valve Overlap in a HCCI Gasoline Engine* (No. 2012-01-1126). SAE Technical Paper.
[21] Lawler, B., Splitter, D., Szybist, J., & Kaul, B. (2017). Thermally Stratified Compression Ignition: A new advanced low temperature combustion mode with load flexibility. *Applied energy*, *189*, 122-132.
[22] Kolodziej, C. P., Ciatti, S., Vuilleumier, D., Adhikary, B. D., & Reitz, R. D. (2014). *Extension of the lower load limit of gasoline compression ignition with 87 AKI gasoline by injection timing and pressure* (No. 2014-01-1302). SAE Technical Paper.
[23] Dempsey, A. B., Curran, S. J., & Wagner, R. M. (2016). A perspective on the range of gasoline compression ignition combustion strategies for high engine efficiency and low NOx and soot emissions: Effects of in-cylinder fuel stratification. *International Journal of Engine Research*, *17*(8), 897-917.
[24] Yang, Y., Dec, J., Dronniou, N., Sjöberg, M., & Cannella, W. (2011). Partial fuel stratification to control HCCI heat release rates: fuel composition and other factors affecting pre-ignition reactions of two-stage ignition fuels. *SAE International Journal of Engines*, *4*(1), 1903-1920.
[25] Dempsey, A., Curran, S., Storey, J., Eibl, M., Pihl, J., Prikhodko, V., ... & Parks, J. (2014). *Particulate matter characterization of reactivity controlled compression ignition (RCCI) on a light duty engine* (No. 2014-01-1596). SAE Technical Paper.
[26] Prikhodko, V. Y., Curran, S. J., Barone, T. L., Lewis, S. A., Storey, J. M., Cho, K., ... & Parks, J. E. (2010). Emission characteristics of a diesel engine operating with in-cylinder gasoline and diesel fuel blending. *SAE International Journal of Fuels and Lubricants*, *3*(2), 946-955.
[27] Lilik, G. K., & Boehman, A. L. (2011). Advanced diesel combustion of a high cetane number fuel with low hydrocarbon and carbon monoxide emissions. *Energy & Fuels*, *25*(4), 1444-1456.
[28] Kalaskar, V. B., Splitter, D. A., & Szybist, J. P. (2014). Gasoline-like fuel effects on high-load, boosted HCCI combustion employing negative valve overlap strategy. *SAE International Journal of Fuels and Lubricants*, *7*(1), 82-93.
[29] Scaringe, R., Wildman, C. B., & Cheng, W. (2010). On the high load limit of boosted gasoline HCCI engine operating in NVO mode. *SAE International Journal of Engines*, *3*(1), 35-45.
[30] Storey, J. M., Curran, S. J., Lewis, S. A., Barone, T. L., Dempsey, A. B., Moses-DeBusk, M., & Northrop, W. F. (2017). Evolution and current understanding of

physicochemical characterization of particulate matter from reactivity controlled compression ignition combustion on a multicylinder light-duty engine. *International Journal of Engine Research*, *18*(5-6), 505-519.

[31] Liang, B., Ge, Y., Tan, J., Han, X., Gao, L., Hao, L., ... & Dai, P. (2013). Comparison of PM emissions from a gasoline direct injected (GDI) vehicle and a port fuel injected (PFI) vehicle measured by electrical low pressure impactor (ELPI) with two fuels: Gasoline and M15 methanol gasoline. *Journal of Aerosol Science*, *57*, 22-31

[32] Bisetti, F., Chen, J. Y., Hawkes, E. R., & Chen, J. H. (2008). Probability density function treatment of turbulence/chemistry interactions during the ignition of a temperature-stratified mixture for application to HCCI engine modeling. *Combustion and Flame*, *155*(4), 571-584.

[33] Dec, J. E. (2009). Advanced compression-ignition engines—understanding the in-cylinder processes. *Proceedings of the combustion institute*, *32*(2), 2727-2742.

[34] Sjöberg, M., Edling, L. O., Eliassen, T., Magnusson, L., & Ångström, H. E. (2002). GDI HCCI: effects of injection timing and air swirl on fuel stratification, combustion and emissions formation. *SAE Transactions*, 407-424.

[35] Sjöberg, M., & Dec, J. E. (2006). *Smoothing HCCI heat-release rates using partial fuel stratification with two-stage ignition fuels* (No. 2006-01-0629). SAE Technical Paper.

[36] Jung, G. S., Sung, Y. H., Choi, B. C., & Lim, M. T. (2009). Effects of mixture stratification on HCCI combustion of DME in a rapid compression and expansion machine. *International journal of automotive technology*, *10*(1), 1-7.

[37] Mukhopadhyay, S., & Abraham, J. (2011). Influence of compositional stratification on autoignition in n-heptane/air mixtures. *Combustion and Flame*, *158*(6), 1064-1075.

[38] Dec, J. E., Yang, Y., & Dronniou, N. (2011). Boosted HCCI-controlling pressure-rise rates for performance improvements using partial fuel stratification with conventional gasoline. *SAE International Journal of Engines*, *4*(1), 1169-1189.

[39] Hunicz, J., Tmar, A., & Krzaczek, P. (2017). Effects of mixture stratification on combustion and emissions of boosted controlled auto-ignition engines. *Energies*, *10*(12), 2172.

Effects of laser hardening process parameters on hardness depth of Ck45 steel using Taguchi's optimization technique

Santoshkumar V. Wagh[a], Dhananjay V. Bhatt[b], Jyoti V. Menghani[c], Shivnanda S. Bhavikatti[d]

[a]*Assistant* Professor, Department of Mechanical Engineering, College of Engineering Pune, Pune-411005, India
[b]Professor, Department of Mechanical Engineering, Sardar Vallabhbhai National Institute of Technology, Surat-395007, India
[c]Associate Professor, Department of Mechanical Engineering, Sardar Vallabhbhai National Institute of Technology, Surat-395007, India
[d]Associate Professor, Department of Mechanical Engineering, College of Engineering Pune, Pune-411005, India.

Corresponding author email:waghsv@gmail.com

Abstract

Laser hardening is one of the reasonable procedures utilized to improve surface mechanical and wear resistance of target materials. This process is easy to use for improving surface properties of complex shape components with minimum time requirement. The beauty of this process is that it is possible to improve only selective surface properties without altering the remaining surface properties of bulk material. Ck45 steel has various important properties like hardness, superior wear resistance and frictional performance etc. These properties are useful in various industrial applications. The existing study is an attempt to analyse results on surface properties with the aid of process parameters of a laser hardening process on hardness depth and microstructure of laser hardened Ck45 steel the use of a design of experimental approach. From this analysis, it can be summarized that the depth of the surface layer hardness is maximized by controlling the laser beam power, laser scan speed and standoff distance, which improves the service life of Ck45 steel components hardened by laser. Comparing the experimental consequences with the effects of L9 orthogonal array approach, the proportion error was observed to be negligible with a higher co-relation coefficient (r^2) of 0.905.
Keywords: Ck45 steel, laser hardening process, hardness depth, Taguchi's design of experiment method and microstructure.

1. Introrudction

Laser surface hardening strategy is utilized to move forward laser hardened surface quality and wear resistance of components utilized for mechanical applications such as pipe, liner, crankshafts, bearing, gears and ring groves. To improve structural and mechanical properties such as hardness and wear resistance, the effects of laser hardening process on Ck45 steel cylindrical samples was investigated.[1].

Barka et. al. [2] investigated the effects of laser hardening technique parameters on case depth of laser hardened 4340 metal cylindrical specimens by using the usage of a statistical evaluation approach. The laser scanning speed has the foremost significant impacts of the process parameter on the hardness depth and the impact on the surface hardness of laser beam power. Rezagholizadeh et. al. [3] studied Ck45 steel that had a low cost and was easily available material for industrial applications, but its limitations was due to wear and corrosion resistance. In this research the focus was on the improvement of tribological properties and corrosion of Ck45 steel by Ni-B-B4C composite coating. Wagh et. al. [4,5] considered impacts of parameters of laser hardening process on Ck45 steel and cast iron materials mechanical and wear properties utilizing an orthogonal array technique. Sagaro et. al. [6] studied effects of laser hardening process parameters on U13A steel, microstructure, tribological performance and improvement in hardness depth of 260μm. Palaniradja et. al. [7] investigated the hardness and depth of case by means of optimization techniques in the laser hardening process. C. Hu. et. al. [8] studied a demonstrate utilizing the super position of Gaussian sources in two dimensions developed and utilized for the expectation of hardness depth of laser hardened medium carbon steel by utilizing CO2 laser system. Patwa el. al. [9] utilized prescient demonstrating and considered impacts laser hardening process parameters on case depth and hardness of AISI5150H steel and theoretically come to a hardness value was 63Rc and case depth of 0.5 mm. More et. al. [10] used Taguchi's method of experimentation for investigating the significant parameters used for erosion wear study on SS304 material.
The orthogonal array method was utilized to consider the effects of laser hardening process parameters on hardness depth of Ck45 steel. The process of laser hardening was completed with a laser fiber system of 400 W.

2. Experimental Details

2.1 Material utilized for the experimentation

Table 1. Elemental weight percentage of CK45 steel

Elements	C	Si	S	P	Mn
Composition weight	0.540	0.320	0.039	0.035	0.760

Now a days, Ck45 steel is used in bolts, axles, connecting rods and hydraulic clamps. Therefore, for the present experimental work, Ck45 steel is used as testing material. Table 1 shows the elemental composition in weight percent. SEM picture of the base material of CK45 steel is as shown in Fig. 1.

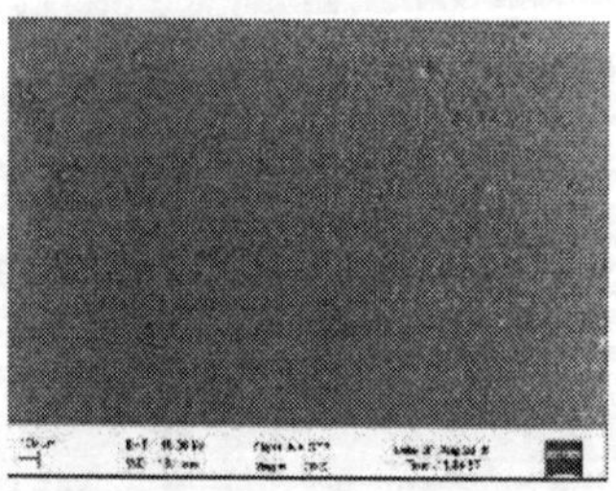

Fig. 1. SEM picture of the base material of CK45 steel

2.2 Experimental set-up and preparation of specimens

Figure 2 shows the image of fiber laser system continuous wave of 400 watt and 1070 nm laser wavelength. The experiments were conducted on the same laser system. The parameters of the laser hardening process used 210, 270 and 330 watt laser beam power, 1.0, 8.5 and 16.0 mm/s laser scanning speed and 200, 250 and 300 mm standoff distance as presented in Table 2. A laser beam diameter of 1.4

mm and argon gas used for creation of protective atmosphere during laser hardening process at a flow rate of 10 liters/min. The specimen's preparations were conducted using acetone. The specimens were cut into 20 mm x 8 mm x 8 mm by a wire electrical discharge machine.

Fig. 2. Fiber laser system (SPI 400 CW, Laser Science Pvt. Ltd. Mumbai)

Table 2.
Experimental parameters for testing laser hardening experiments and their levels.

Sr .No.	Laser parameters	Unit	Levels		
			1	2	3
1.	Laser beam power (P)	W	210	270	330
2.	Laser scanning speed (v)	mm/s	1.0	8.5	16.0
3.	Standoff distance (f)	mm	200	250	300

2.3 Metallographic and microstructure examination

To take a laser track on the sample, optimized were process parameters were used. The laser tracks were taken on the surface of the specimen and therefore the laser treated cross sectional area was polished using an emery paper of different grades using each of them as 4 % Nital to examine the cross sectional area microstructure of the laser treated samples. By using a microscope, the width, depth and microstructure were studied in the heat affected area by using a microscope made by Conation Technologies Pune. A machine was used for microhardness test at 300 gram load with 10 seconds dual time and microhardness was check on the laser hardened area of the laser hardened samples. Future Tech Corporation Japan, Vicker microhardness testing machine of model number FM700 was used to check microhardness. The microhardness was checked using heat affected zone and the base metal along the laser hardened cross sectional area.

2.4 Experimental Design

For the analysis of the influencing significant process parameters, the design of experiment technique was used and on the basis of that an output response model was developed. In the process of laser hardening, three input parameters influenced hardness depth which had out response. Taguchi's design of experiment method was more flexible to find effects on hardness depth by every input process parameter compared to other analytical methods. In this analysis study, power of laser beam power, speed of laser scanning and standoff distance were used as laser hardening process parameters to check effects on output as a hardness depth of laser treated Ck45 steel sample. Table 2 appears the levels of process parameters utilized for the experiment. Table 3 shows the process parameters. The hardness depth is that the output response and therefore the signal to noise ratio relation of the complete run. The output response values are converted to signal to noise ratio using Taguchi's L_9 experiment method. The signal to noise ratio was determined for each factor level combination. The equation for the largest was a better signal to noise ratio using a 10 log base as shown in Equation 1.

$$\text{Signal to noise ratio} = -10*\log\left(\Sigma(1/Y^2)/n\right) \quad \text{-----------------------} \quad (1)$$

where, Y is the response to the given combination of the factor level and n is the number of responses in the combination of the factor level.

3. Results and Discussion

3.1 Hardness depth analysis by Taguchi's Experimental Design

The results of the present experimental work were analyzed using MINITAB 14. Taguchi's orthogonal array L9 design was used to analyze the experimental work. Table 3 shows details of the experimental parameters and respective results of S/N ratio for all the nine sets of the experiments. The aim of the present experiment was to increase the value of hardness depth. Therefore, 'the larger the better' quality characteristic was selected and used for the present Taguchi's analysis. Table 4 shows the value of signal to noise ratio as response for the present experimental condition and the larger the better quality characteristic. Laser scanning speed was the most important significant control process parameter among all three factors that laser hardness surface as shown in Table 4, causing the hardness depth of Ck45 steel laser treated specimens. In the second and third position, I.e. laser beam power and standoff distances, the remaining two parameters were rated. The main effect plots for signal to noise ratio are graphically shown in Fig. 3. The surface hardness depth value of laser hardened sample of Ck45 steel was found maximum at 330 watt laser power, 1.0 mm/s laser scanning speed and 300 mm standoff distance.

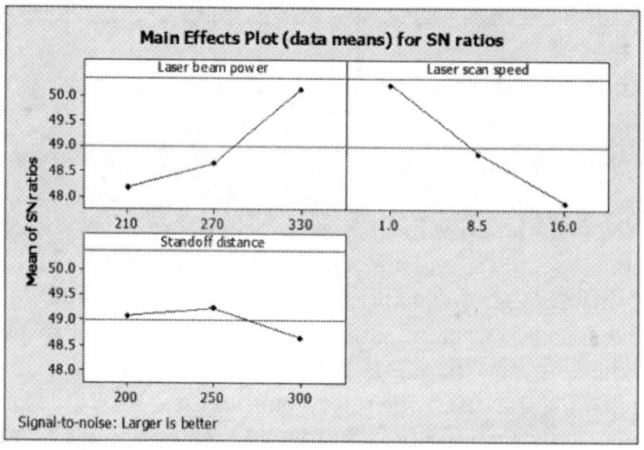

Fig. 3. Main effects plot for S/N ratios

3.2 The effect of laser power on the depth of hardness

As the power of the laser beam increases between 210 and 330 watts, the value of the signal to noise ratio and hardness depth of laser hardened sample increase too which is shown in Fig. 3. It is observed from Table 3 that increments in the hardness depth are 308.51, 325.14 and 340.25 μm respectively of laser hardened samples of Ck45 steel with variations in the laser beam power of 210, 270 and 330 watt at constant 1.0 mm/s laser scan speed and the maximum achieved hardness depth is 340.25 μm.

3.3 The effect of laser scanning speed on the depth of hardness

The effects of the process parameters as a laser scanning speed on the laser treated samples hardness depth value shows that a curve trend graph is inversely proportional as shown in Fig. 3. As the variations in the laser scanning speeds are 1.0, 8.5 and 16.0 mm/s at 330 watt constant laser beam power is observed. Table 3 shows the effects of laser scanning speed which decreases the hardness depth to 340.25, 322.32 and 300.34 μm respectively and decreases the signal to noise ratio values.

3.4 Morphology of the surface using microstructure

IOP Conf. Series: Materials Science and Engineering **810** (2020) 012027 doi:10.1088/1757-899X/810/1/012027

Figure 4 provides an overview of the nine laser hardened specimen's microstructure according to Taguchi's test sets and one untreated sample. Fig. 4(a) shows untreated sample and Fig. 4(b-j) shows nine sets of experiments as per Table 3. The effect of change in laser scanning speed on microstructure is shown in Fig. 4(b-d). The laser scanning speed varies from 1.0, 8.5 and 16.0 mm/s at 210 watt constant laser beam power. Fig. 4(b-d) presents a change in hardness width and decrease the hardness depth due to the increased laser scanning speed. Similar results can be observed in Fig. 4(e-g, h-j) at laser beam powers of 270 and 330 watt, laser scanning speed of 1.0, 8.5 and 16.0 mm/s and standoff distance of 200, 250 and 300 mm respectively. Fig. 4(e & h) presents maximum hardness depth at 270 watt, 1.0 mm/s and 250 mm and 330 watt, 1.0 mm/s and 300 mm respectively. Fig. 4(b-d, e-g and h-j) depicts constant laser beam power of 210, 270 and 330 watt and the variation in the laser scanning speed of 1.0, 8.5 and 16.0 mm/s that affects the hardness depth and it decreases that is to say that laser scan speed effects on hardness depth.

Table 3. Experimental CK45 steel laser hardness depth matrix Taguchi L9

Sr. No.	Laser Power (w)	Laser scanning speed (mm/s)	Standoff distance (mm)	Width (μm) Ck45	Hardness depth of HAZ(μm)	S/N Ratio
1.	210	1	200	1114.81	308.51	49.7854
2.	210	8.5	250	1088.34	249.10	47.9275
3.	210	16	300	1037.75	220.13	46.8536
4.	270	1	250	1094.68	325.14	50.2414
5.	270	8.5	300	1022.67	264.18	48.4380
6.	270	16	200	1063.68	231.32	47.2843
7.	330	1	300	1090.14	340.25	50.6360
8.	330	8.5	200	1093.22	322.32	50.1657
9.	330	16	250	1027.93	300.34	49.5523

Table 4. Signal to noise ratios response table larger is a better option.

Level	Laser scanning speed	Laser beam power	Standoff distance
I	50.22	48.19	49.08
II	48.84	48.65	49.24
III	47.90	50.12	48.64
Delta	2.32	1.93	0.60
Rank	1	2	3

IOP Conf. Series: Materials Science and Engineering **810** (2020) 012027 doi:10.1088/1757-899X/810/1/012027

A

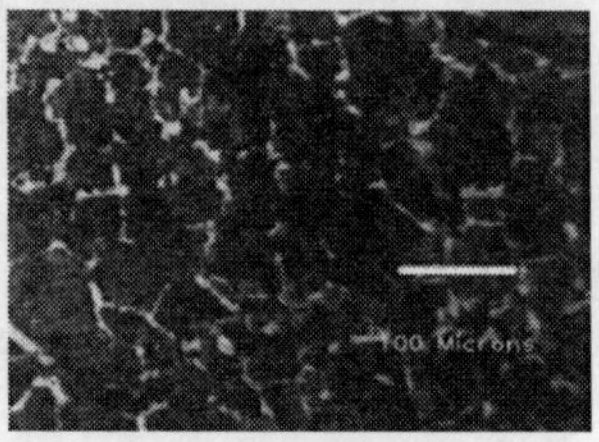

Base metal

B

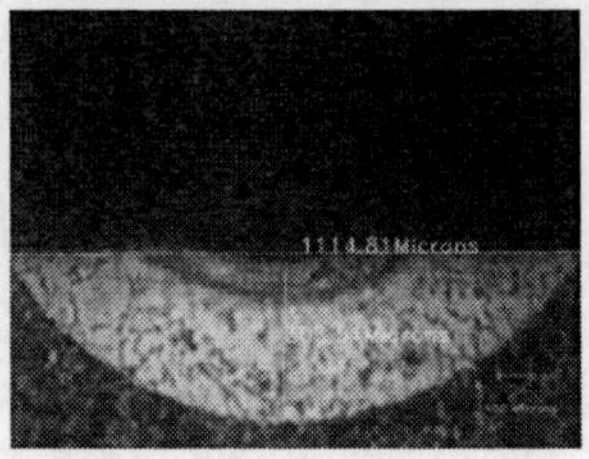

P = 210 W, v = 1 mm/s and f = 200 mm

C

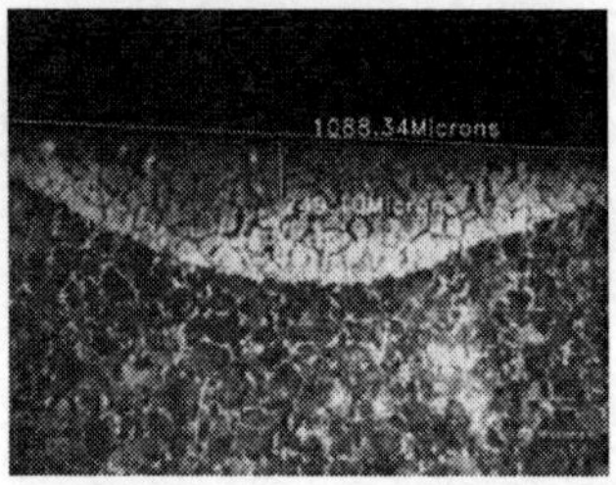

P = 210 W, v = 8.5 mm/s and f = 250 mm

D

P = 210 W, v = 16 mm/s and f = 300 mm

E

P = 270 W, v = 1 mm/s and f = 250 mm

F

P = 270 W, v = 8.5 mm/s and f = 300 mm

g

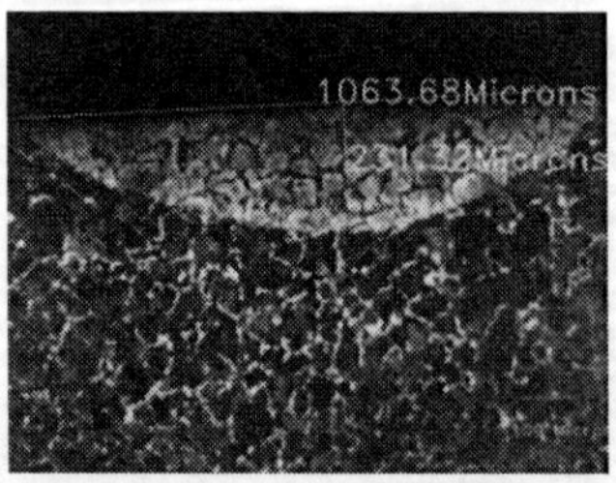

P = 270 W, v = 16 mm/s and f = 200 mm

H

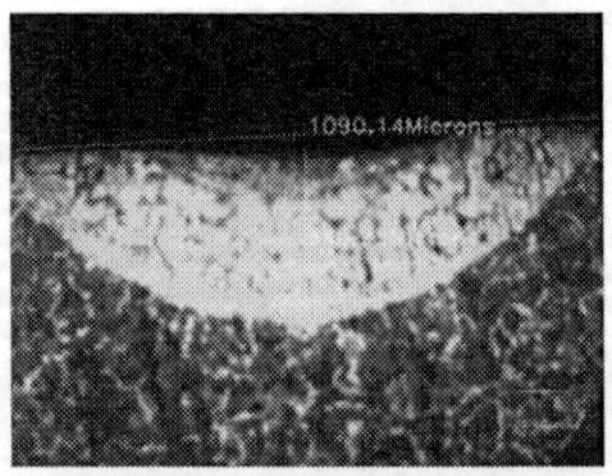

P = 330 W, v = 1 mm/s and f = 300 mm

i

J

IOP Conf. Series: Materials Science and Engineering **810** (2020) 012027 doi:10.1088/1757-899X/810/1/012027

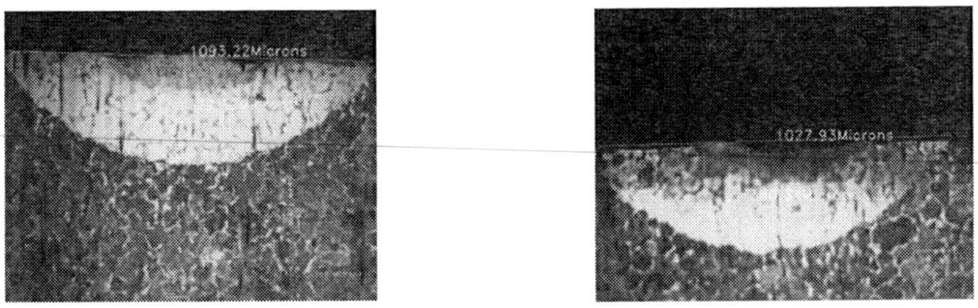

P = 330 W, v = 8.5 mm/s and f = 200 mm P = 330 W, v = 16 mm/s and f = 250 mm

Fig. 4. Samples classified as per L_9 experiments by microstructure of Ck45 steel (a) Base metal and (b-j) laser treated

Table 5. Comparison of the experimental and predictive hardness depth values.

Expt. No.	Results obtained from the experiments	Results obtained from the predictive equation	Percentage error (%)
1.	308.51	297.00	3.88
2.	249.10	253.70	1.81
3.	220.13	210.40	4.62
4.	325.14	321.59	1.10
5.	264.18	278.29	5.07
6.	231.32	253.74	8.84
7.	340.25	364.18	1.71
8.	322.32	321.63	0.21
9.	300.34	278.33	7.91

3.5 Prediction of depth of hardness using predictive equation

Due to the laser beam power, laser scan speed and standoff distance, the laser treated sample hardness depth was calculated employing a nonlinear regression formula to illustrate the relationship between the hardness depth and the combination of process control parameters. This analysis of regression was performed using Minitab 14 Software. The following regression equation from was obtained:

Hardness depth = 219 + 0.514(Laser beam power) - 4.94(Laser scan speed) -0.125(Standoff distance) - ----- (2)

In Equation 2, all the constant values are calculated by using Minitab 14 Software. The accuracy of the measured constants was verified since Equation 2 obtained a strong correlation coefficient (r^2) of 0.905. The relation between the hardness depth obtained from experimental results and the predictive formula is shown in Table 5. An affirmation test might have been performed on Ck45 steel specimen. The optimal test parameters values are used for trial I.e. laser beam power, laser scanning speed and standoff distance which acquired that maximum hardness depth. Due to the laser hardening process, the depth of hardness is obtained in the Ck45 steel product by using both nonlinear regression equation and experimental affirmation results which are shown in Table 6.

Table 6 Affirmation test comes about for hardness depth of Ck45 steel.

Laser beam power (W)	Laser scan speed (mm/s)	Standoff distance (mm)	Hardness depth in test	Hardness depth by equation	% error
330	1.0	250	333.75	352.43	5.30

A deviation of approximately 5.30 percent was found to be agreeable by comparing the experimental results and analytical results. Therefore, the derived nonlinear regression formula is shown to explain with a reasonable degree of approximation the hardness depth of the Ck45 steel material with different control factors.

4 Conclusions

The most remarkable conclusions drawn from the investigation are summarized below:

- From all the three variables, laser scanning speed is the foremost impacting noteworthy calculates of laser hardness depth of Ck45 steel, taken after by the laser beam power and standoff distance separate individually.
- The maximum hardness depth attained is at 330 watt laser beam power, 1.0 mm/s laser scan speed and 300 mm standoff distance of Ck45 steel by laser hardening process.
- Microstructure of laser hardness depth specimens of Ck45 steel highlights two different zones. The first one is the heat affected area and the second one is the base metal area. These are the main characteristics of a laser hardening process.
- The hardness layer depth range is achieved from 220.13 μm and 340.25 μm for laser treated specimens.
- The deviation in the hardness depth percentage error between predicted and experimental result was between 0.21 and 8.84 %. A higher correlation coefficient value of (r^2) of 0.905 which shows the correctness of the mathematical model used. Therefore, the model is more suitable for future study.
- From this analysis, it can be concluded that the laser surface hardness depth is maximized by monitoring the speed of laser scanning, power of laser beam and standoff distance, making progress in the gain life of the Ck45 steel specimens with laser hardened.

5. Acknowledgments

- The authors place on record their sincere thanks for the assistance provided for the experimental facility from the Department of Mechanical Engineering and Metallurgy and Material Science Department, College of Engineering Pune, Maharashtra, India.
- The authors also wish to thank the Department of Mechanical Engineering, Sardar Vallabhbhai National Institute of Technology, Surat, India for their kind support.

6. References

[1] Adel K. M., "Enhancement of dry sliding wear characteristics of CK45 steel alloy by laser surface hardening processing", Procedia Materials Science, 6 (2014) pp. 1639-1643.

[2] N. Barka, and A. El. Ouafi, "Effects of Laser Hardening Process Parameters on Case Depth of 4340 Steel Cylindrical Specimen- A Statistical Analysis", Journal of Surface Engineered Materials and Advanced Technology. 5 (2015), pp.124-135.

[3] Rezagholizadeh, M. Ghaderi, A. Heidary and S. M. Monirvaghefi, "The effect of B4C nanoparticles on the corrosion and tribological behavior of electroless Ni-B-B4C composite coatings", Surface engineering and applied electrochemistry", 51- 1 (2015), pp.18-24.

[4] S. V. Wagh, D. V. Bhatt, J. V. Menghani and Shivnanda S. Bhavikatti, "Effects of laser hardening process parameters on the mechanical and wear properties of ck45 steel using an orthogonal array", International Journal of Modern Manufacturing Technologies, X (2018), pp. 86-93.

[5] S. V. Wagh, Sudeep Ingole, D. V. Bhatt, J. V. Menghani and M. J. Rathod, [2019], "Effect of process parameters on surface properties of laser-hardened cast iron", The Minerals, Metals and Materials Society, TMS 2019, 148th, annual meeting and exhibition supplemental proceedings, pp. 732-743.

[6] Sagaro, J. S. Ceballos, A. Blanco, J. Mascarell, "Tribological behavior of line hardening of Steel U13A with Nd:YAG laser ", Wear, 225-229 (1999), pp. 575-580.

[7] Palaniradja, N. Alagumurthi and V. Soundararajan, "Hardness and case depth analysis through optimization techniques in surface hardening process", The open materials science journal, 4 (2010), pp.38-63.
[8] C. Hu and T. N. Baker, "Prediction of laser transformation hardening depth using a line source model", Acta metal, mater, 43 (1995), pp. 3563-3569.
[9] Rahul Patwa and Yung C. Shin, "Predictive modeling of laser hardening of AISI5150H steels", International Journal of Machine Tools and Manufacturing, 47 (2007), pp. 307-320.
[10] S. R. More, D. V. Bhatt, and J. V. Menghani, "Study of the Parametric Performance of Solid Particle Erosion Wear under the Slurry Pot Test Rig", International Journal of Industrial Tribology, 39-4, (2017), pp. 471-481.

IOP Conf. Series: Materials Science and Engineering **810** (2020) 012028 doi:10.1088/1757-899X/810/1/012028

Development of bench top biaxial tensile testing machine

1.Amar Jadhav 2. Dr. Priam V Pillai 3. Salim Jafri

1. Assistant Professor. Mechanical Engg. Dept., Pillai HOC College of Engineering and Technology, Rasayani, India
2. Professor, Mechanical Engg., Dept., Pillai College of Engineering, New Panvel, India
3. Assistant Professor. Mechanical Engg., Dept., Pillai College of Engineering, New Panvel, India

e-mail address : amararunjadhav@gmail.com

Abstract. Development of new biaxial testing apparatus is carried out to study and evaluate the mechanical properties of different materials. Novel materials like biological tissue, soft plastics, fibers, wires, carbon nanotubes, smart materials, textiles, ceramics, composites etc. can be tested to evaluate its mechanical properties on the developed apparatus. The Developed apparatus will be used to evaluate mechanical properties. This apparatus will be used to characterize a material's properties as a function of stain frequency, stress or a combination of these parameters. The equipment can be used to apply sinusoidal oscillating stress (load) which is applied to the specimen by actuators with the help of a servo controlled electric motors. The equipment has basic capabilities to produce static and cyclic multiaxial loading system. Generally, the test facility requires cruciform(Cross Shape) specimens which will be loaded orthogonal direction. Stress-controlled and strain controlled are two main techniques that are employed to test the specimen . As the stress-controlled analyzers are cheaper to manufacture (1) this technique will be adopted in the test rig. The test rig will be equipped with two mutually perpendicular actuators capable of incremental motions of as small as few microns. Each actuator will be equipped with a load cell to measure both tension and compression loads. The servo-electric actuators will be programmed to move independently or in a coordinated motion. Camera will be used to record and measure the strain produced in the specimen. Real world loading system can be created within the device to tests material capability in more realistic manner.

1. Introduction

During the last few decades, newer materials like smart materials, composites, ceramics etc. have become an integral part of everyday life. Their excellent mechanical performance together with their low specific weight explain why these materials, are popular for applications in various industries from aeronautics, aerospace and transportation to sports, construction and health care. Although these materials have many benefits over traditional materials such as metals, there are also some drawbacks or limitations in using them. The cost of fabrication is generally higher, repair of most materials is rather difficult and its mechanical characterization is more complex (2). The latter drawback is a result of the complexity of the material, for example in the case of composites specifically continuous fiber consisting micromechanically out of two phases (fibers and matrix) and macro-mechanically out of several lamina stacked in a certain direction. This makes the computation and testing of composite structures not straightforward (3). Uni-axial tensile tests like tension test, compression test, 4-point bend test etc. results are insufficient to predict and evaluate the mechanical behavior of composites materials subjected to biaxial stress-strain states. Many of the service application of composites, fabric soft plastic or rubber involve in multi-axial loads so these materials need to be tested in several directions. Different test like tabular inflation test, planar biaxial test ,bulge test are required to measure mechanical properties of materials which are independent of the way materials are modeled (isotropic or anisotropic) (4). A biaxial testing involves many difficulties like the design of a suitable specimen, the correct analysis of the results and the purchase or development of expensive testing equipment. The micromechanical behavior of materials which we have understood is based on uniaxial testing techniques. This deformation test gives the strain path in 180° i.e. loading and unloading. It is now broadly accepted that mechanical properties of materials under multi-axial loading conditions are different than those anticipated on the basis of test results from uniaxial tests. A planar biaxial test rig applies the stress in planar orthogonal direction with the help of actuators. This test gives the better result as compared to other budge and inflation test as the different stress can be applied in two directions. Cyclic load can also be applied to the specimen. The test can be performed on materials like biological material, smart materials, engineering textiles, membrane, carbon nano-tube, plastic, polymer, and ceramics. If we consider a membrane in a fuel cell , it is a thin sheet which is constrained in a 2D plane and not in the thickness plane. So the study of the mechanical behavior in plane becomes important to enhance the life and performance of these membranes. Meredith N. Silberstein el. al. investigated the elastic-viscoplastic behavior of Nafion (polymer electrolyte membrane) in biaxial loading condition (5) . Biaxial testing rig are extremely useful in medical and aeronautical industries.

Types of multi axial testing (6)

1. Tubular Inflation Test

This is a hydrostatic test. A specimen of tubular shape is clamped on circular jaw. The jaw must be rigid. The hydrostatic pressure created using air or water. The pressure created inside the tube generates peripheral stress. Along with this axial load can be applied at the end of the cylindrical specimen. Permeable materials like clothing fabric cannot be tested as due to permeable nature the pressure changes along the fabric so it behaves differently (7)

2. Planar Biaxial Tension test

In this test, force are applied to the cruciform shape specimen in orthogonal directions. Stress and Strain are continuously recorded during the test. Stress strain curve for materials are developed. U.S. Lindholm et. al. developed a dynamic biaxial testing machine for generating the generalized constitutive equations for materials in 1966 (8). Both the actuators are independently control.

3. Bulge Test

The material is inflated until bursting by the pressure of air or water changing the shape to sphere. The test shows bursting strength and stress strain relationship. Instant pressure and radius of curvature is measured using optical method. The bulge test has as main drawbacks that only the tension-tension quadrant can be investigated, that only thin specimens can be tested to avoid the stress gradient through the thickness and that due to the gripping, non-homogeneous stress fields could be introduced.

2. Design and Construction

IOP Conf. Series: Materials Science and Engineering **810** (2020) 012028 doi:10.1088/1757-899X/810/1/012028

Development The apparatus modeled in Solidwork 2016 SP 02 version. The 3D Model of the test instrument developed is shown below

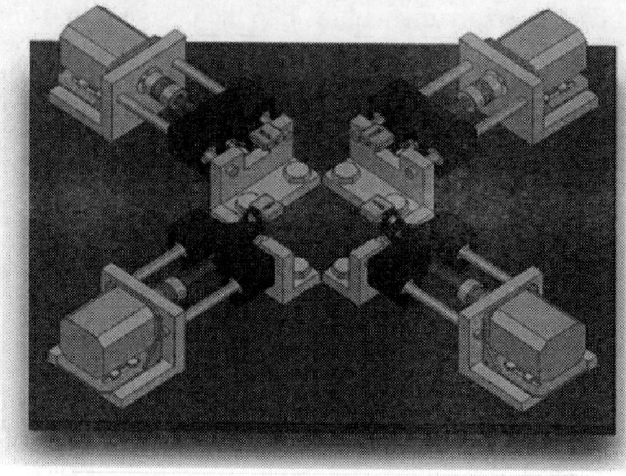

Figure 1: 3D Model of Biaxial tensile testing machine (left) and complete assembly with camera (right)

The testing equipment includes four stepper motors NEMA 17 2.2 Kg-cm. These will be used for applying loads in tension and compression mode. Motors are mounted in orthogonal direction as can be seen in the model. The Motors are attached to the head with the help of square threaded rod. Two guide rods are attached to the head with linear bearing for the desired motion. The assembly is mounted on the acrylic sheet of 10mm thick.

The components of the apparatus are manufactured using 3D printer. The moveable head is manufactured using Ultimaker 2 extended. The equipment can apply a load of 5 kg on one axis

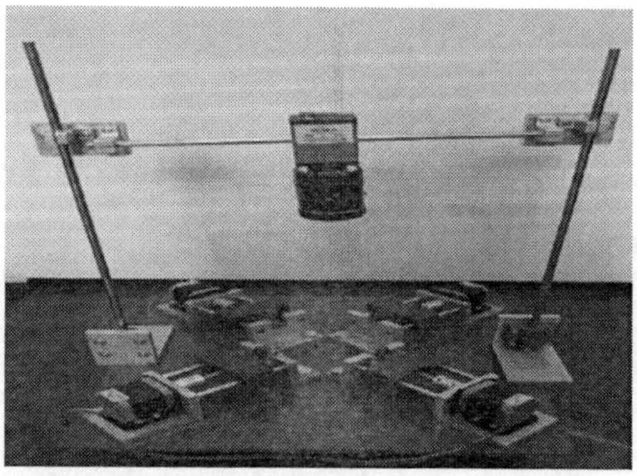

Figure 2 Biaxial tensile testing machine (left) and complete assembly with camera (right)

The cutting on acrylic sheet was done using laser cutting machine. This ensured the proper alignment of the holes with the precision of 0.1mm. The load cell of FUTEK was used for measuring stress. The NI DAQ is used to collect the data for the analysis. The Shape of the specimen to be used us cruciform.

3. Cruciform specimen design and geometry

The in-plane loading of cruciform specimens is a technique which was used for the first time in the late 60's to investigate the biaxial behavior of metals . Mönch et. al. started with the application of two perpendicular inplane loads on a rectangular plate to create a biaxial stress state in the centre of the specimen. (9) The shape of the specimen was adjusted quite quickly to a cruciform to avoid damage near the clamps and to lead failure to the center.
There are at least two principles of designing biaxial tension specimen (10):
1. Stress-strain homogeneity within the center section so that we can calculate the stress with deviations as less as possible.
2. Yield has to occur in the test section and stress concentrations to avoid fracturing in arms.

4. References

[1] *Menard, Kevin P. Dynamic mechanical analysis : a practical introduction. London : CRC Press LLC, 1999.*
[2] *kaw, Arthur K. Mechanics of Composite Materials . s.l. : CRC press book, 1997.*
[3] *Ramault, Carla. Guidelines for biaxial testing of fibre reinforced composites using a cruciform specimen. 2012.*
[4] *Development of an apparatus for biaxial testing using cruciform specimens. Makinde, A and Thibodeau, L and Neale, KW. 1992, Experimental mechanics, pp. 138--144.*
[5] *Biaxial elasticeviscoplastic behavior of Nafion membranes. Meredith N. Silberstein, Priam V. Pillai, Mary C. Boyce. 2011, Polymer, pp. 529--539.*
[6] *Biaxial Testing of Fabrics - A Comparison of Various Testing Methodologies. O.B. Ozipek, E. Bozdag, E. Sunbuloglu, A. Abdullahoglu, E. Belen, and E. Celikkanat. 2013, World Academy of Science, Engineering and Technology International Journal of Mechanical and Mechatronics Engineering.*
[7] *Experimental Methods for Measuring Fabric Mechanical Properties: A Review and Analysis. R.J. Bassetr, R. Postle, N. Pan. s.l. : Textile Research Journal, 1999.*
[8] *A Dynamic Biaxial Testing Machine. Yeakley, U. S. Lindholm and L. M. Detroit : SESA Spring Meeting, 1966.*
[9] *A method for producing a defined uniform. Monch, E., Galster, D. s.l. : Journal of Applied Physics, 1963.*
[10] *A review of planar biaxial tensile test systems for sheet metal. Hannon A, Tiernan P. s.l. : Journal of Materials Processing Technology, 2008.*

IOP Conf. Series: Materials Science and Engineering **810** (2020) 012029 doi:10.1088/1757-899X/810/1/012029

Computational Fluid Dynamics Analysis of a Class Room for Effective Utilization of Position of Air Conditioning System

Ashish Mogra 1, Pankaj Kumar Pandey 2, Krishna Kumar Gupta 3

1, 3 Department of Mechanical Engineering, SVKM's, NMIMS, MPSTME, Shirpur, Maharashtra, India
3Department of Chemical Engineering, Amity University, Jaipur, Rajasthan, India
ashish.mogra@nmims.edu

Abstract. The productivity of an individual person affected to a great extent by indoor quality of air and the condition of thermal comfort. Students as well as professors spend almost half of their day in classrooms; hence the distribution of air flow from the air-conditioning systems plays a crucial role in determining whether the students will receive the proper velocity and temperature of air upto the comfortable accepted range. In present paper a Computational Fluid Dynamic (CFD) simulation is performed on the overhead air-conditioning system of a classroom. A CFD model has been developed for position of air conditioner in class room. Originally the position of air conditioner is centrally located. A comparison has been made in between two model placed centrally and proposed diagonally position. It is found the better air flow distribution when air conditioner is placed diagonally.
Keyword: Air conditioning, CFD, boundary condition, thermal comfort, velocity distribution.

1. Introduction

Indoor comfort is now considered one of the most essential parameter as it has been scientifically proved that it affects the productivity and efficiency of an individual. Thermal comfort is related to temperature and airflow distributions in air conditioning space, which plays an important role in optimum design of air conditioning system or ventilation system. Holiday et al. [1], in their study the office room was virtually created to make it ready for flow simulation. To do that, the dimensions of the room and all furniture sets were measured accurately and then everything created was Part files which were later assembled to prepare the office room model. Yang Li [2], conducted computational fluid dynamics simulation to investigate the effect of air flow inside a class room. They selected two types of air conditioning system namely split type and centralized system. They analysed and compare the air flow of both types of system. They found better air flow distribution in centralized system as compared with the spilt air conditioning.

Sarkar and Mandal [3], performed study on CFD Modelling and Validation of Temperature and Flow Distribution in Air-conditioned Space by provided experimental study on flow pattern and velocity distribution in an enclosed space with a single inlet and outlet had been conducted. The flow pattern, velocity and temperature distributions in the above enclosed space had been investigated using three dimensional CFD simulation and the results obtained from the simulation were validated with experimental data. Rahaman and Kumar [4], did a study on CFD analysis of air conditioning of Computer Aided Design laboratory. Their result showed the air flow distribution of air conditioning system. Sun et al. [5] also provides a model of a CFD study of an indoor environment. A dynamic simulation was used to evaluate HVAC control systems based on a CFD model of a room, a mathematically modelled PID controller, and an actuation model. While the control model was beyond the scope of this current research, the CFD model was a helpful guide in gauging model inputs. Aryala and Leephakpreedaa [6], performed computational fluid dynamics study to identify the thermal comfort and consumption of energy which was affected by in split unit of air conditioning in building. They found that in split unit level of thermal comfort decreased while consumption of energy increased by 24% in some part of building. They performed computational fluid dynamics analysis of a 4 way cassette air conditioner in office room. They also made a comparison between wall mounted typical air ventilation with 4 way cassette air conditioner. The result of numerical simulation showed higher performance of 4 way cassette air conditioner with compared to wall mounted typical air ventilation [8]. This study was performed in a classroom having an overhead air conditioning system with an aim to understand air flow and temperature distribution. The positions of the existing air conditioners were also changed to compare the effects with the existing one. For that purpose Computational Fluid Dynamics (CFD) tool has been used to perform the simulation of air flow distribution and flow pattern. This simulation will help to locate or identify the various points where air distribution is insufficient.

2. Methodology

The following assumptions are made for simulation:-

1. The selected air conditioning system has been found in smooth and good operating condition.
2. There was proper arrangement of no air leaking in class room. Windows and door are assumed to be closed.
3. The temperature of atmospheric air was assumed to be constant.
4. Internal heat generation may be neglected due to less effect on properties of air.
5. All the material available in class room including tables, chair and board will be considered for simulation.
6. Atmospheric temperature and pressure take.
7. Air from the air conditioners flow vertically downwards.
8. Velocity of air is usually between the range (0.08-0.3) m/s and for the present study, air velocity is assumed to be 0.2 m/s.

   The following boundary conditions have been used in CFD simulation:

   At inlet: $V_x = V_i$, $V_y = 0$, $V_z = 0$; $T = T_i$

   At outlet: $P = P_{atm}$

   At side wall: Isothermal, $T = T_w$

   At top and bottom wall: Constant heat flux, $\frac{\partial T}{\partial y} = 0$

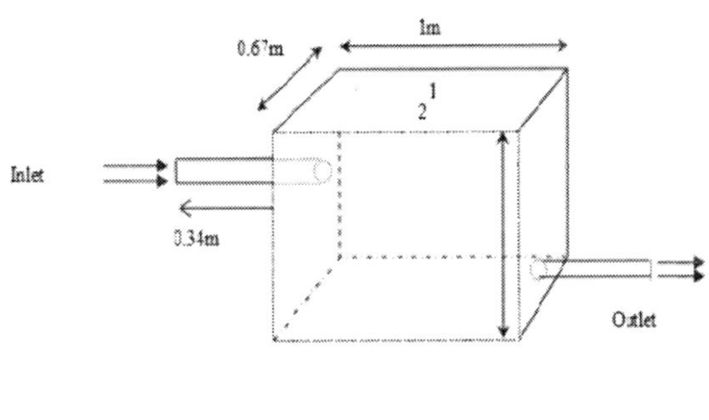

Figure 1 Boundary condition of Class room.

Condition 1: Centrally Air-Conditioner arrangement in the classroom:

The class room was virtually created to make it ready for flow simulation. To do that, the dimensions of the room and all furniture sets were measured accurately and each object was created as Part files, which were later assembled to prepare the class room model. The dimensions of the room and furniture sets are listed in Table 1. The location of all furniture sets, the air conditioning unit, and the door was kept the same as that of the original room to maintain the resemblance of the model with the class.

Table 1

ITEM	DIMENSION (cm)	THERMAL CONDUCTIVITY $(\frac{W}{m}/k)$	MATERIAL
DOOR	248x114x7	0.17-0.25	Wood
BENCH	140x78x84	0.17-0.25	Wood
WINDOW	213x233	0.90-1.5	Glass
WHITEBOARD	122x245	1.52-1.65	Ceramic (Silicon Carbide)
AC	95x95		Daikin 2 tons cassette
WALLS	303x141x35	2.07-2.94	Brick Wall
Beam	28x45	0.72-0.82	Plaster Material

The exploded and the assembly views of the office model are shown in Figures 1 and 2, respectively. Analysis under CFD Fluent will be performed on the designed model. After input of desired data, CFD models will be generated and results for different operating conditions were obtained and compared.

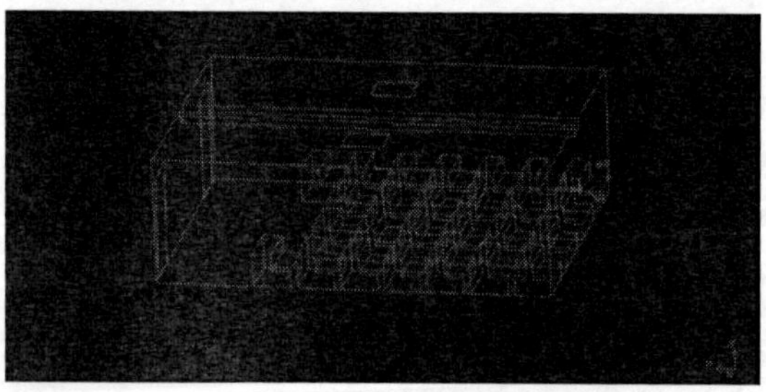

Figure 2 Line diagram of centrally located AC in class room.

Figure 3 ANSYS model of centrally located AC in class room.

Condition 2: Diagonal arrangement of air-conditioners:
All the other parameters are exactly the same as that of condition-1. Assumptions and boundary conditions are also kept similar to condition-1. Considering the center of the classroom as an origin point, the co-ordinates of the air-conditioners is as follows:

- Unit-1: (x, y, z) = (2, 3.03, 2) m
- Unit-2: (x, y, z) = (7.07, 3.03, 5.53) m

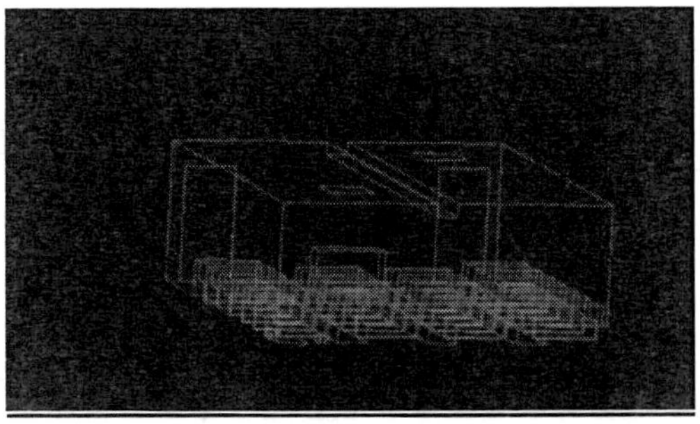

Figure 4 Line diagram of diagonally located AC in class room.

Figure 5 ANSYS model of diagonally located AC in class room.

3. Result Analysis

The air flow inside the room at two different front sections, at sections close to the window and middle of the room. Air enters the room with a high velocity and after circulating inside the rooms, it also leave the room through the vent with a relatively high velocity; but its exit velocity is lower than the entering velocity. In the sitting elevation level of the room's occupant, the high velocity zones are located in areas close to the rear glass window. The entire area in front of the window always experiences a higher air velocity. Following benefits are observed while placing air conditioners into diagonally:

- By placing one unit above the open space near door, it was possible to eliminate greater turbulence and velocity variations.
- A lower variation of velocity over the sitting area was observed.
- Variation between maximum temperature and minimum temperature was comparatively less.
- Thermal discomfort caused by the turbulence nature of air around the benches was reduced to some extent.

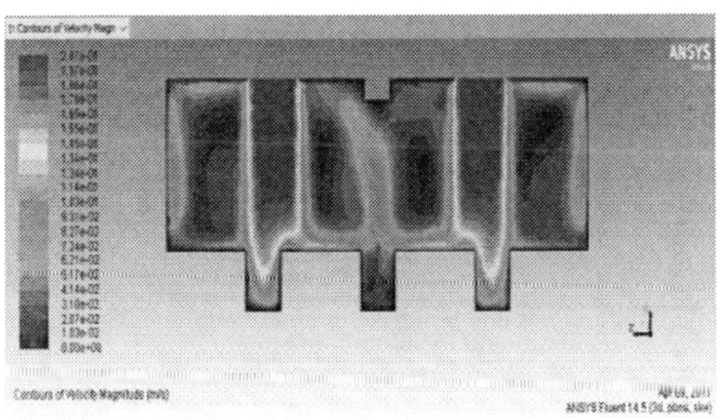

Figure 6 Velocity counter 1.5 m above the ground

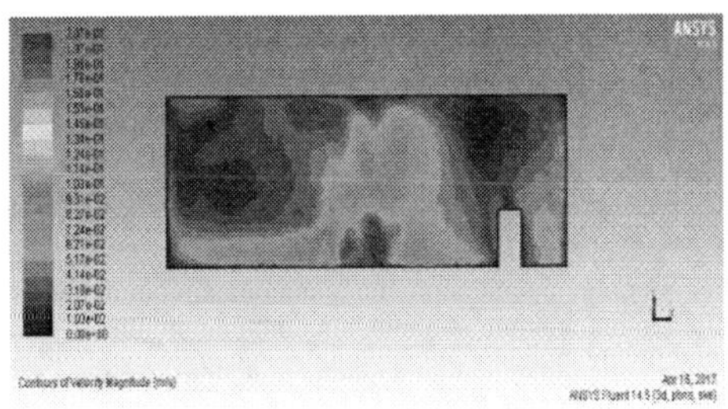

Figure 7 Velocity counter window to door.

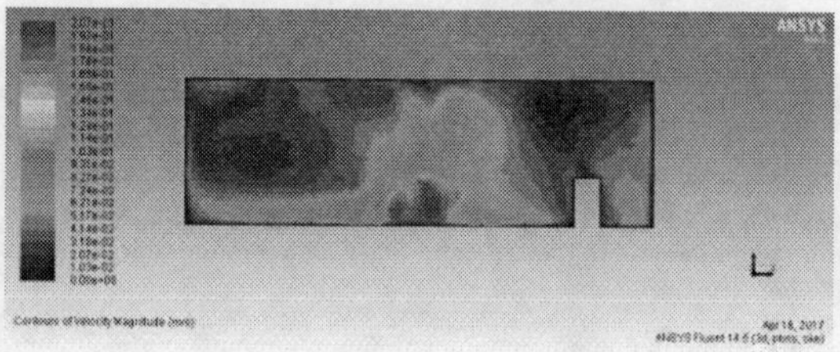

Figure 8 Velocity contour in center of classroom from door to window

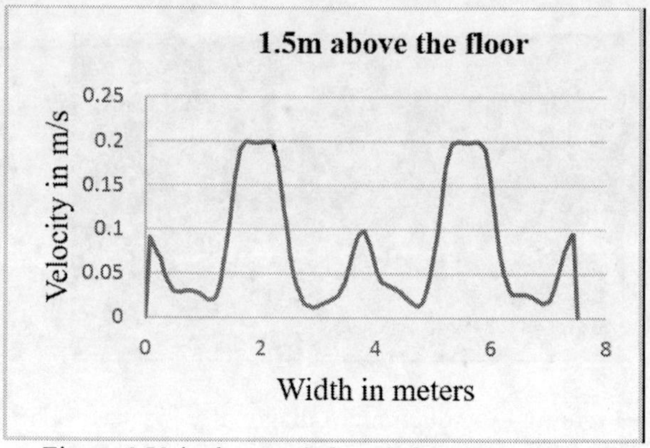

Figure 9 Velocity graph from left wall to right wall

Figure 9 shows the Velocity distribution of air in the actual classroom from a reference line drawn at 1.5meter above the ground and from left wall to right wall at the center of classroom. Velocity of air at inlet of air conditioner is considered as 0.2m/s. It has been found form the literature review that the range of velocity of air for air conditioners should be in between 0.08-0.30 m/sec depending on the fan speed. Graph has been plotted between the velocity of air and width of room.
The graph showed the variation in Velocity which ranges from 0.015m/s to 0.2m/s. This shows that velocity distribution is uneven. Maximum velocity (i.e. 0.2m/s) lies at the point just below the two air conditioners i.e. 2m and 6m from left side wall. Minimum Velocity. (i.e. 0.05m/sec) lies at the point 3m and 7m from the left side of wall because flow of air is not distributed properly. This is due to vertically downward flow of air as the benches and walls are responsible for turbulent flow of air. Graphs obtained from results are also symmetric because air conditioners are place at the same distance (2meter) from left side and right of the wall.
Figure 10 showed the comparison of simulation result between two conditions. It has been found that the distribution of air velocity is almost constant when the position of air conditioner changed from centrally to diagonally. Figure 11 showed the comparison between the flows of velocity from window to door. This is showing almost same pattern with originally placed air conditioner.

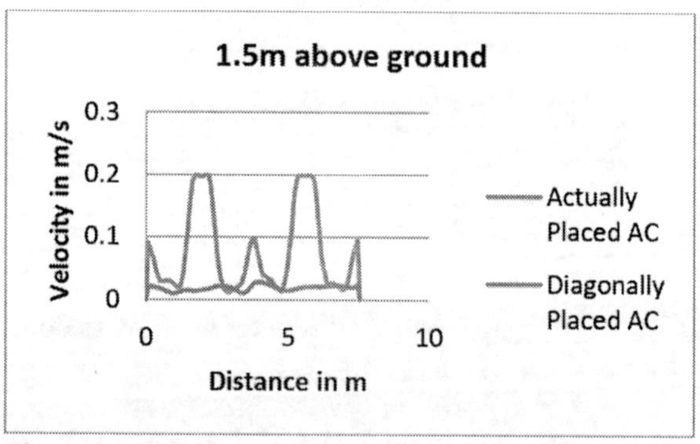

Figure 10 Velocity comparison from left wall to right wall.

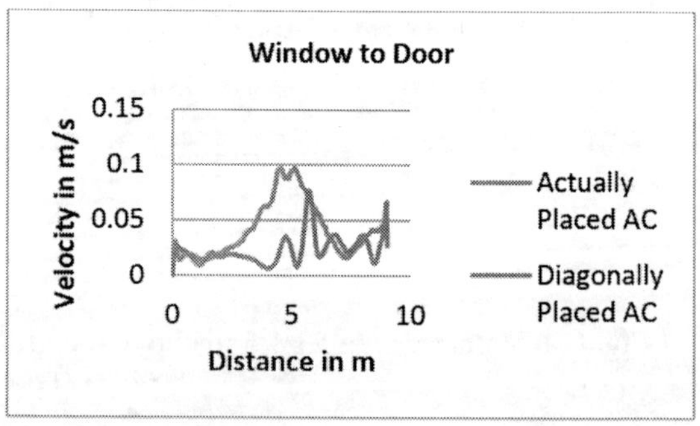

Figure 11 Velocity comparison from window to door.

4. Conclusion

Numerical simulation has been performed of a class room for determining the effect of position of air conditioner. ANYSYS fluent software has been used for simulation. Initially the air conditioner was placed centrally and then its position was changed diagonally. It was found that the thermal comfort of the peoples are affected by position of the air conditioning unit. It was found that by placing the air conditioner diagonally the velocity of flowing air is constant when measured above the 1.5 m from the ground. It has showed no effect when placed centrally and diagonally.

Reference

[1] Holiday, A., Andreou, S., & Hayder(2015). M. Air Flow Visualization in an Office Room, European International journal of science and technology,.
[2] Li, Y. (2012). Numerical simulation and analysis for indoor air quality in different ventilation. *Health*, *4*(12), 1352.
[3] Sarkar, J., & Mandal, S. (2008). CFD Modeling and Validation of Temperature and Flow Distribution in Air-Conditioned Space.
[4] K Vaseemul Rahaman, Prof. Shankar Kumar, (2015), CFD Analysis of Air Conditioning equipment of CAD Lab for Enhancing its Performance,International Journal of Recent Development in Engineering and Technology, 1(5) , pp. 68-72.
[5] Sun, Z., & Wang, S. (2010). A CFD-based test method for control of indoor environment and space ventilation. *Building and Environment*, *45*(6), 1441-1447.
[6] Aryal, P., & Leephakpreeda, T. (2015). CFD Analysis on thermal comfort and energy consumption effected by partitions in air-conditioned building. *Energy Procedia*, *79*, 183-188.
[8] Bamodu, O., Xia, L., & Tang, L. (2017). A numerical simulation of air distribution in an office room ventilated by 4-way cassette air-conditioner. *Energy Procedia*, *105*, 2506-2511.
[9] Tian, W., Sevilla, T. A., Li, D., Zuo, W., & Wetter, M. (2018). Fast and self-learning indoor airflow simulation based on in situ adaptive tabulation. *Journal of Building Performance Simulation*, *11*(1), 99-112.
[10] Yang, W., Zhu, X., & Liu, J. (2017). Annual experimental research on convective heat transfer coefficient of exterior surface of building external wall. *Energy and Buildings*, *155*, 207-214.
[11] Wang, H., & Zhai, Z. (2012). Application of coarse-grid computational fluid dynamics on indoor environment modeling: Optimizing the trade-off between grid resolution and simulation accuracy. *HVAC&R Research*, *18*(5), 915-933.
[12] Romano, F., Colombo, L. P., Gaudenzi, M., Joppolo, C. M., & Romano, L. P. (2015). Passive control of microclimate in museum display cases: A lumped parameter model and experimental tests. *Journal of Cultural Heritage*, *16*(4), 413-418.
[13] Popovici, C. G. (2017). HVAC system functionality simulation using ANSYS-Fluent. *Energy Procedia*, *112*, 360-365.

IOP Conf. Series: Materials Science and Engineering **810** (2020) 012030 doi:10.1088/1757-899X/810/1/012030

Analytical analysis on influence of dimensionless numbers in the thermal stratification of storage tank

Mainak Bhaumik

M.G.M's College of Engineering and Technology, Kamothe, New Mumbai-410209, India
mainak.bhaumik04@gmail.com

Abstract. An analytical analysis has been carried out to understand the influence of dimensionless numbers in the thermal stratification of solar domestic hot water storage (SDHW) tank. Thermal stratification is the phenomena to maintain cold water layer at the bottom of the storage tank and above the cold water maintaining of hot water layer. In between cold and hot water layer one barrier layer is considered, which is called a thermocline layer. From the thermocline layer the heat transfer exchanging starts in between hot and cold water layer. The thermocline layer thickness starts increasing, which is undesired and non-beneficial from maintaining of thermal stratification of the storage tank. This degradation of thermal stratification of the storage tank cannot be eliminated fully, but only it can be reduced down somehow. Hence this analytical study analysis is about to know the influence of dimensionless numbers like Reynold'number, Nusselt number, Prandtl number on the thermal stratification of storage Tank.

Keywords. SDHW (Solar domestic Hot water), storage tank, stratification, charging, Reynold's, Nusselt and Prandtl Number.

1. Introduction

SDHW Storage tank is used to store thermal energy, which obtained from the solar radiation and processed in the water. This thermal energy is charged into the storage tank. While being charging and storing for a particular duration is required to maintained a thermally stratified one in between cold and hot water. Expected hundred percent full stratified thermal storage tank is the desired one. But practically it is not possible to maintain a hundred percent thermally stratified storage tank due to the unavoidable interference of charging and discharging flow, continuous heat transfer in between hot and cold water, convection current etc. various experimental, analytical and CFD analysis are carried out by the various expert about the storage of thermal energy in the SDHW storage tank. The degradation of thermal stratification cannot be eliminated fully but can be reduced down and study analysis can be carried out to understand the stratification loss and impact of various factors [1-3]. The sensible heat storage, as well as latent heat storage of solar thermal energy in the water, can be done. In sensible heat storage it is not required to use phase change materials. In latent heat storage phase change materials like salt hydrates, either in the packed capsule form or by directly submerging in the water with direct interaction with water is very popularly used. Such uses of PCM in storage tank also studied and experiments also performed by the various experts and still, the studies are going on. In case of domestic storage tank PCM material is avoided or direct exposure with the water is avoided due to the human hygienic point of view. Study and analysis of storing of energy is one of the vast potential fields, especially in such case where the storing of thermal energy is from the solar light radiation. Solar energy and light radiation are freely available in the atmosphere. The main limitation of solar energy is available during day time approximately 10 to 12 hours duration from morning at 6 o'clock to the evening at 6 o'clock. Out of that effective light radiation is achieved for mostly 8 to 10 hours [4-7]. Again during mid day hours, we get much sunlight with optimum temperature. Through out the year the temperature and radiation also vary from day to day, month to month and different place to different places of different countries during different seasons. Mostly in summer season the light intensity and radiation are much more. The main factor about to perform experiment and analysis with the solar thermal energy is due to freely availability in the atmosphere. The water is cheaply, easily available in the atmosphere. So the solar energy storage tank analysis is quite cost-effective analysis. Also, the specific heat of normal water is 4.187 KJ/Kg° K at an atmospheric temperature of 20 to 25°C. Fig. 1 illustrates the schematic diagram of SDHW storage tank and system with basis minimum equipment [8-10].

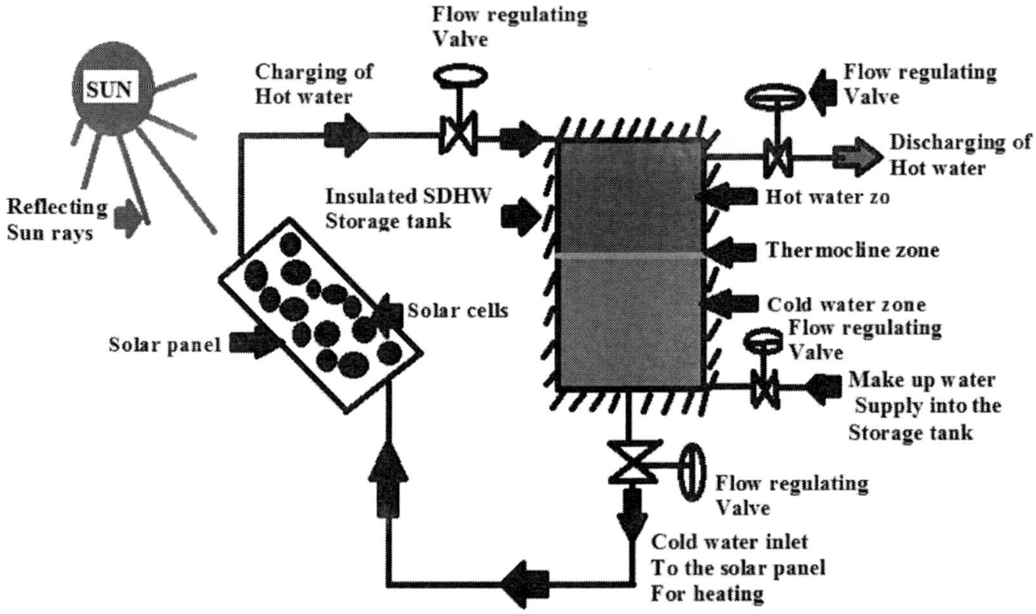

Fig. 1: Schematic diagram of solar domestic hot water (SDHW) storage system.

2. Problem definition

The selected problem under study analysis is a model storage tank of dimensional geometry of diameter (D) 500 mm, height (H) 1000 mm, thickness (t) of 18 gauge (1.27 mm) of capacity almost 200 liters which is easily available in utensil shop, to provide continues hot water supply for a nuclear family of 04 persons. The selected material of the tank is stainless steel for domestic use. Considered the tank is insulated adiabatically with an insulation thickness of 50 mm. the charging, discharging, cold water inlet and makeup water lines diametrical flow regulating valve and pipeline size (d) is 12 mm. The Fig.2 shows the schematic diagram of the Model storage tank understudy with geometrical dimensions.

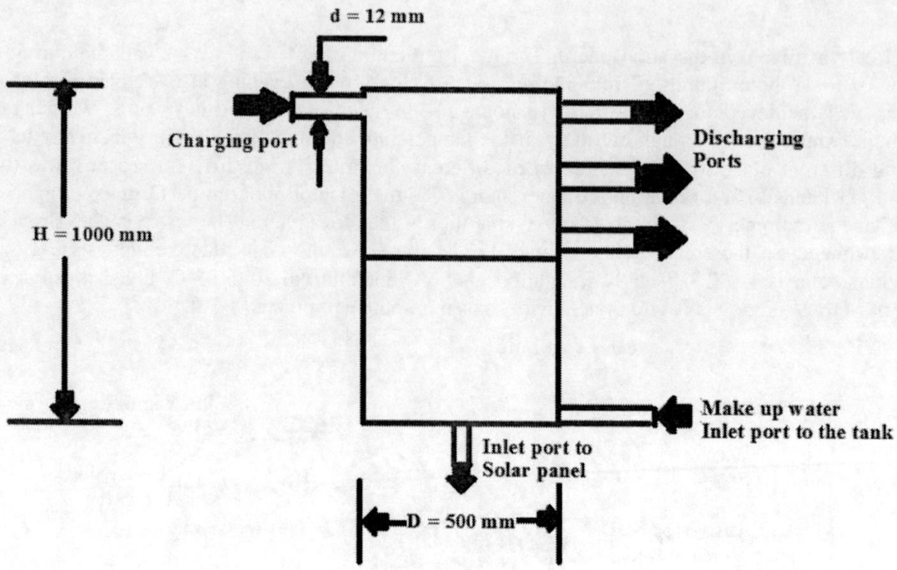

Fig. 2: The schematic geometrical diagram of the tank understudy and analysis.

3. Analytical Analysis

The analytical analysis carried out about the charging flow condition to understand the influence of flow for different values of dimensionless numbers and other parameter is temperature. The SDHW storage tank under analysis has been taken as an insulated adiabatic tank, with no heat flow through the walls of the tank. Here the minimum flow hot water temperature has been taken as 80^{0}C for analysis and up to 100^{0}C temperature has been taken under the condition for analysis and calculation. The calculations are taken on the assumption that hot water temperature is being available before to flow and store in the hot water storage tank, and once the below 80^{0}C temperature is not available, then the flow stops automatically. Other requisite parameters like absolute viscosity (μ), the specific heat of water (c_p), the thermal conductivity of water (k) has been taken out standard value from the heat transfer data book for different temperature requirement for the calculations and analysis.
To do flow analysis Reynold's Number formula has been taken under condition, which is as,

$$Re = \frac{\rho D v}{\mu} \quad \text{-----------------------------------(1)}$$

where, Re = Reynold's No.
ρ = Density of water in kg/m^3
v = Velocity of water in m/sec.
D = inlet diameter of the pipe in m.
The Nusselt No. also has been taken under consideration for analysis, which is as-

$$Nu = \frac{hl}{k} \quad \text{-----------------------------------(2)}$$

where, Nu = Nusselt No.
h = convective heat transfer coefficient in W/m^{2}K
l= D = Characteristics length or diameter of inlet flow zone in m.
k = Thermal conductivity of water in W/ m K
The Prandtl's No. which also has been utilized in the analysis is as given below,

$$Pr = \frac{c_p \mu}{k} \quad \text{-----------------------------------(3)}$$

where Pr = Prandtl No.

μ= Absolute viscosity of water in N-sec / m^2

c_p = Specific heat of the water in J / kg K

k = Thermal conductivity of water in W / m K

By utilizing the formulae, data and assuming inlet section size and other requisite parameters and calculations has been done and results in terms of graphs has been plotted and shown in the result section.

4. Results and discussions

The analytical analysis and the calculations are being presented here. For the different values of Reynold's No. starting from laminar zone Re value of 10 to turbulent region of Re value 5000 has been taken under consideration for calculation. With the assumed Re value the SDHW storage tank inlet velocity has been calculated. The same has been plotted in the graph and shown in Fig.3 for the inlet pipe size of 15 mm. There is a continuous rise in velocity with respect to increasing in the assumed Reynold's No. Also, it has been shown in Fig. 4 for the inlet pipe size of 25 mm. The other dimensionless No. which is Nusselt No. with respect to convective heat transfer flow inlet into the SDHW storage tank by the tank inlet pipeline has been calculated and plotted. Fig. 5 shows the Nusselt No. v/s convective heat transfer flow (h). The calculated values of Prandtl No. for different temperature values have been plotted in the graph. Fig. 6 shows the Prandtl No. v/s tank inlet flow temperature. It has been observed that the decreases with the increase in the temperature and hence the graph also declines.

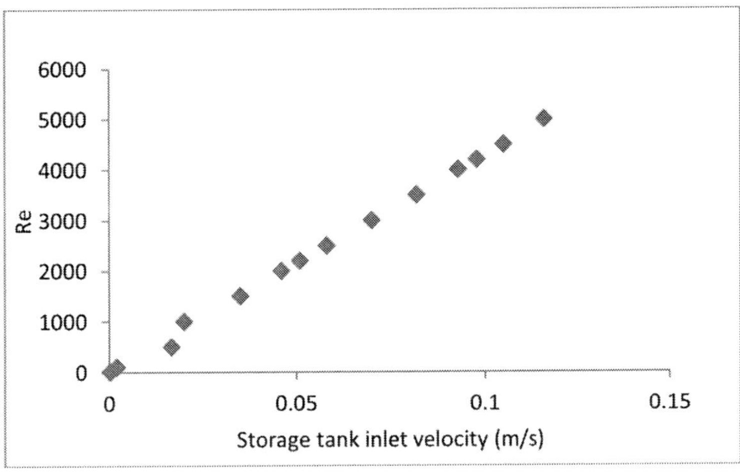

Fig. 3: Reynold's No. v/s SDHW storage tank inlet storage velocity (v in m/s) for pipe diameter of 15mm.

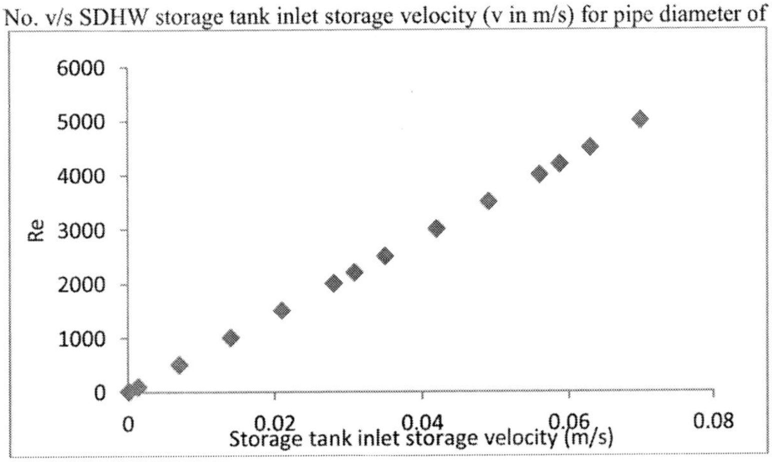

IOP Conf. Series: Materials Science and Engineering **810** (2020) 012030 doi:10.1088/1757-899X/810/1/012030

Fig. 4: Reynold's No. v/s SDHW storage tank inlet storage velocity (v in m/s) for pipe diameter of 25mm.

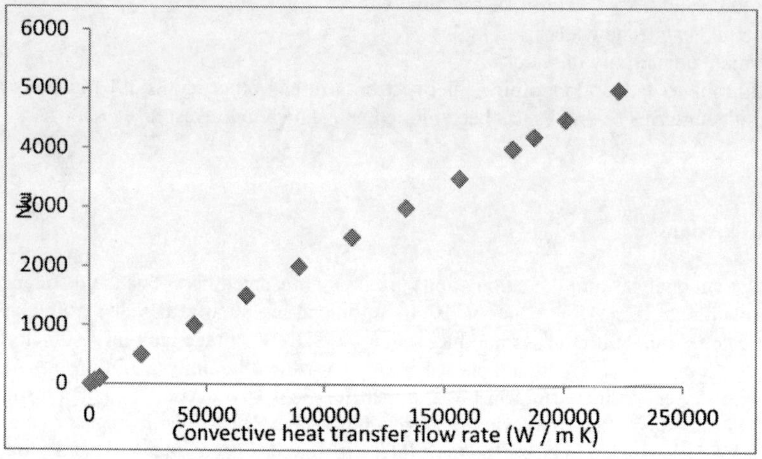

Fig. 5: Nusselt No. v/s Convective heat transfer flow rate (h in W / m K) for hot water inlet by pipe to the SDHW storage tank.

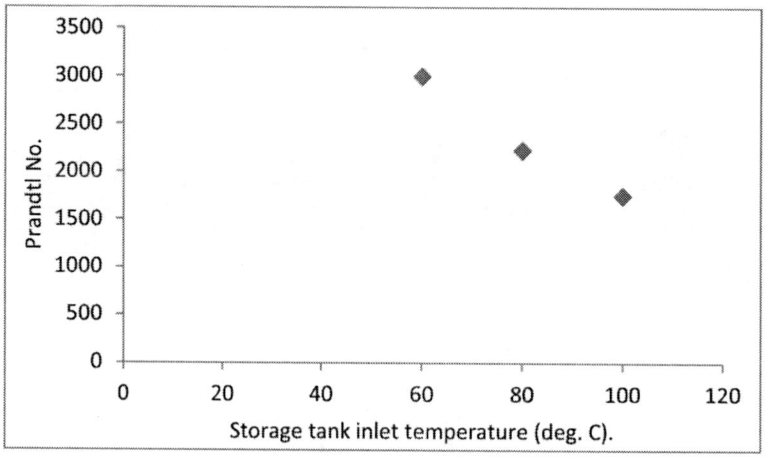

Fig. 6: Prandtl No. v/s SDHW storage tank inlet temperature (T in deg. C).

5. Conclusions

From the analysis and calculations, a conclusion can be drawn that the in case of flow and thermal analysis the dimensionless numbers play an essential role to understand the condition and influence in thermal disturbance due to influence of flow variation. With the increase in Reynold's No., the inlet velocity increases proportionately. This increase in the inlet velocity affects the thermocline zone degradation and effects to maintain a steady-state condition in the hot water temperature. With the rise in Nusselt No., the convective heat transfer flow rate also increases. Hence it is advisable to maintain a lower Nusselt No in this case of SDHW storage tank system. As in this case, the maintaining of hot water temperature inside the tank is the prime motto to utilize the stored thermal heat energy to serve the requisite purpose. The Prandtl No. achieved from different considered hot water inlet temperature of SDHW storage tank. The thermal conductivity of water increases with the increase in water temperature. So it is also advisable to maintain a higher inlet hot water temperature into the storage tank and hence lower Prandtl No. To maintain optimum thermal stratification into the SDHW storage tank, it is advisable to take lower values of dimensionless No. like (Re, Nu and Pr) for any such further calculation and analysis.

Acknowledgement

.The author is very much grateful to his heads and superiors of the MGM's Engineering institution, colleagues, students friends and family members for their continuous insisting, encouragement and support, to think scientifically in constructive, value addition way, to do calculations, research and to write passionately and to take attempt to present and publish the works.

References

[1] Li S, Zhang Y, Zhang K, Li X, Li Y and Zhang X 2014 *Energy Procedia* **48** 384-393
[2] Adolfo C, Moreira GA, Zeghmati B and Leite AP 2014 *Energy Procedia* **57** 2600-2608
[3] Bernardo L, Davidsson H and Andersson E 2016 *Energies* **9(11)** 953
[4] Abdelhak O, Mhiri H and Bournot P 2015 *In Building Simulation* **8(4)** 421-429
[5] Guo JJ, Wu JY, Wang RZ and Li, S 2011 *Applied Energy* **88(11)** 4128-4138
[6] Kenjo L, Inard C and Caccavelli D 2007 *Applied thermal engineering* **27** 1986-1995
[7] Li S, Zhang Y, Zhang K, Li X, Li Y and Zhang X 2014 *Energy Procedia* **48** 384-393
[8] Castell A, Solé C, Medrano M, Nogués M and Cabeza LF 2009 *Journal of Solar Energy Engineering* **131(2)** 024501
[9] Cruickshank CA and Harrison SJ 2009 *Journal of Solar Energy Engineering* **131(2)** 024502
[10] Walmsley M, Atkins M, Linder J and Neale J 2010 *Chemical Engineering Transactions* **21**

IOP Conf. Series: Materials Science and Engineering **810** (2020) 012031 doi:10.1088/1757-899X/810/1/012031

Effect of process parameters on additively manufactured parts using FDM process & material selection: A review

R. D. Dandagwhal[1*4], A. M. Nikalje[2,4], E. R. Deore[3,4]

[1]Mechanical Engineering Department, SVKM's NMIMS, MPSTME, Shirpur, India

[2]Department of Mechanical Engineering, Government College of Engineering, Aurangabad, India

[3]Department of Mechanical Engineering, SSVPS's BSD College of Engineering, Dhule, India

[4]Kavayitri Bahinabai Chaudhari North Maharashtra University, Jalgaon, India

E-mail: dandgwhalpes@gmail.com

Abstract. Fused deposition modelling (FDM) is second most widely used additive manufacturing process worldwide. Performance of the FDM system is highly effected by number of parameters such as environmental factors, material properties, part orientation and supports, machining parameters and working parameters. Working parameters specifically raster angle and width, layer thickness and amount of infill highly effect the mechanical characteristics of additively printed parts. Part orientation which is directly proportional with support generation, is mandatory requirement for hanging parts but it results in poor surface finish and require more post-processing. A proper selection of material is one to the key factor of success of functional prototypes produced from FDM system. Various materials categorized as biodegradable/non-biodegradable, biocompatible, water soluble/insoluble, etc. can be processed on considered system. A various commercial and open ware software's can be used to identify optimum value of working parameters for best results. In present work, a comprehensive review has been carried out to check the effect of working parameters on the performance of the FDM printed parts. Study also supported by the understanding of the various materials reinforced by FDM system and new possibilities are identified for future work.

1. Introduction

In today's well established competitive market, many companies fighting for self-stability by putting countless emphasis on product diversity and reorganization. Due to this many manufacturers led innovative products and concentrated on function markets. This results in defilement of the basing product design cycle in which traditionally product goes through several steps i.e. idea to final product. The shorter product life cycle due to rapidly changing fashions and trends required replacement to traditional manufacturing approaches [1]. The solution for these problem resolute by the use of additive manufacturing (AM) process. The first commercial additive manufacturing system developed by 3D systems in 90s. In less than three decades the technology has been evaluated tremendously and number of AM systems available in market which can process curable liquids to polymer and from polymer to metals [2].

Additive manufacturing broadly classified into three categories, liquid, powder and solid based system. The basic working principle of all systems is somewhat same i.e. based on layer by layer manufacturing, but the way of execution is different. Solid based systems based on chemical, thermal and mechanical properties of used material [3]. Extruders used in these systems are used to heat solid material to semi-liquid state and then deposit it on the surface by using some mechanism. Solidification of the deposited material results due to variation of temperature and pressure of extruder and working environment [4]. Extensively used solid based AM systems such as Laminated Object Manufacturing (LOM), Fused Deposition Modeling (FDM), Paper Lamination Technology (PLT), Multi-Jet Modeling System (MJM), Model Maker and Pattern Master, Slicing Solid Manufacturing (SSM), Melted Extrusion Modeling (MEM) are commercially accessible in market [2], [3], [5], [6].

AM systems vary as per their applications, on the basis of the process, the sources of energy, size of prototype, etc. In this work, a basic overview of solid based systems known as fused deposition modeling with detailed explanations of working principle, process parameters which effect characteristics of printed parts and the material used. Consequent section enlighten the working principle of FDM system.

2. Fused deposition modeling (FDM)

Fused deposition modeling (FDM) is second most widely used additive manufacturing technology, after stereolithography appratus. A solid plastic material in the form of a filament is unwound from a ring and delivered to an extrusion nozzle. The nozzle is heated to conrolled melting temperature of the plastic and the flow of melted plastic is controlled by on/off mechanism.

2.1 Working principle of FDM

The principle of the FDM is based on layer by layer manufacturing, thermal energy and chemical composition of the raw material. The material in the form of filament wounded on spool is melted in a especially intended head known as extruder, which deposit material on the model. As it is extruded, it is cooled rapidly due to variation of temperature and pressure and thus solidifies to form the model. The model is built in layers, like the general AM systems. Factors effecting performance of system and strength of the printed parts are environmental factors , working parameters, machine parameters, and material characteristics [1].

In fused deposition modelling system, filament of plastic feeds to extruder or nozzle through a feeding mechanism. Filament diameter or rod diameter is one of the important factor while operating the system. Extruder's heat the filament to semi liquid state and extrudes it in a thickness typically of 0.25 mm. Outlet thickness of deposited filament has direct influence on strength and aesthetics of printed part [2]. A refined automation system comprises of servo controls, sensors and precise NC system grip the extruder to deposit filament according to path generated by software [3]. In Figure 1 is shown the basics fused deposition modeling process. Platform of FDM system can be fixed on movable in x-y direction depending on the setup designed by manufacturer. The main advantage of this system is that is does not require any chemical post processing like liquid based system and other post processing are comparatively easier to do [4]. The disadvantages are poor control of z-axis, surface finish due to generation of support. FDM system able to print complicated parts but it may take days to print. This can be reduce by increasing deposition speed but it has limitation due to high viscus nature of molten plastic. Lead time can be reduced by changing the percentage of fill from fully dense to sparse mode [5]. FDM system can be operate with wide range of materials categorized as water soluble/insoluble, bio degradable/non degradable, bio-compatible/non compatible.

IOP Conf. Series: Materials Science and Engineering **810** (2020) 012031 doi:10.1088/1757-899X/810/1/012031

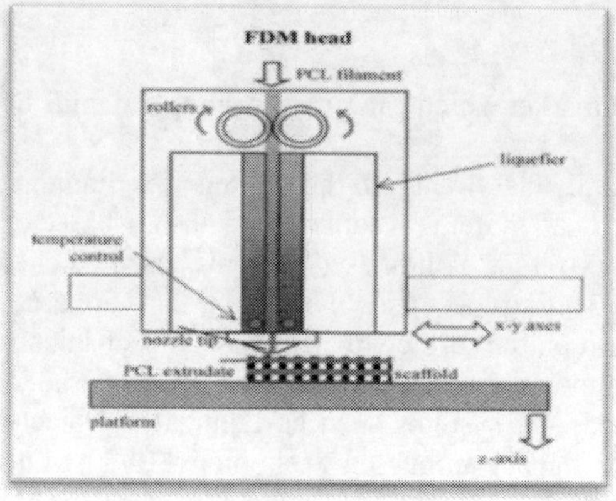

Figure 1. Working of Fused deposition modeling [6]

FDM software require a CAD model in IGES or STL format as input. STL (standard tessellation language) file is de-facto standard of all additive manufacturing systems developed by 3D systems in1983. STL file can be in binary or ASCII format. Both the formats provides the model in tessellated from i.e. triangular facets [7], [13]. Consecutive section high lighten the various process parameters that influence on the working of FDM process.

2.2 *Parameters in Fused deposition modeling*

A short working of fused deposition modeling has been explained in previous section. In various types of additive manufacturing processes, FDM finds a great scope in manufacturing and research due to wide range of raw materials, low cost set up, ease in handling of machine and many more. In order to satisfy customer needs and satisfaction, proper selection of process parameters of FDM machine is necessary. Consequent section review the importance of various process parameters used in FDM process.

2.2.1 Orientation and Support

To obtain a good result from FDM 3D printing, it is necessary to determine the optimum process parameters. Proper process parameters insures the quality of printed object, precision in dimension, reduce the wastage of material and reduces the production time and cost [14]. FDM machine is easy to handle but difficult to understand. This is due to large number of process parameters which makes it complex. Strength of 3D printed parts can be enhanced by selecting proper machine and working parameters [15]. Figure 2 shows all the factors and variables that need to be studied to optimize the performance of FDM process.

FDM parameters can be categorized as working parameters, unprocessed parameters, environmental parameters, machine parameters, part orientation, etc. Part orientation is refer to the way the part built on the platform w.r.t. x, y, and z axis. Part orientation followed by support style is another important factor need to consider while printing the object [16]. Support style is the printing style of object to prevent the part from collapsing during building process [17].

2.2.2 Working parameters

Working parameters of the FDM includes air gap, raster angle, raster width, part interior style, layer thickness, infill pattern, number of loops/contour width, etc. Layer thickness is refer to the thickness of the filament [18] placed by nozzle tip as shown in Figure 3(a). Raster width is the width of the material bead used for the raster. More the value of raster width exhibit more strength in printed part, whereas less value results in less production time and less material consumption. The value of raster width varies with the size of nozzle [19], [20].

IOP Conf. Series: Materials Science and Engineering **810** (2020) 012031 doi:10.1088/1757-899X/810/1/012031

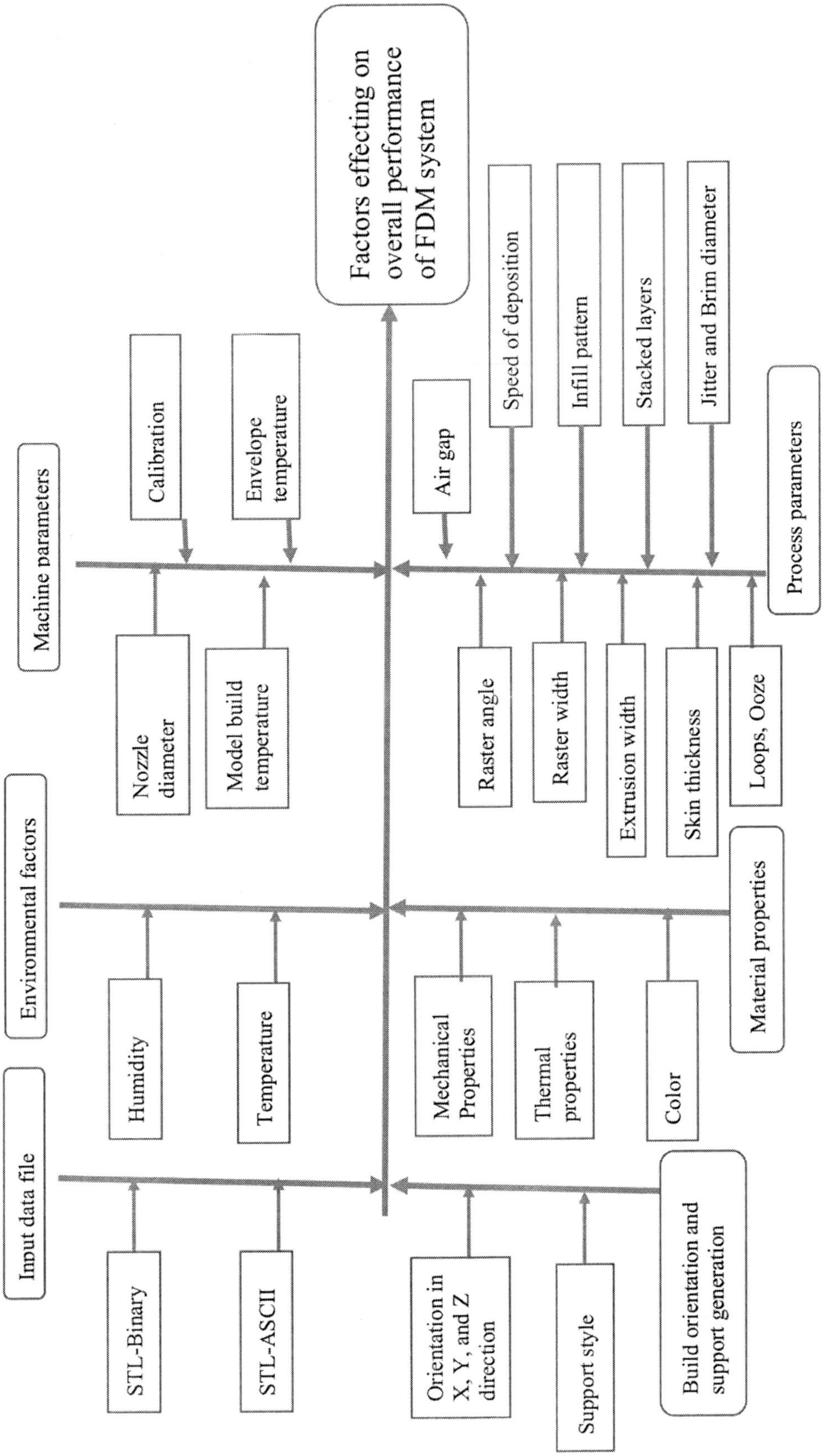

Figure 2. Process parameters in generalized FDM process

Air gap refers to the gap between adjacent raster tool paths as shown in Figure 3(b). One of the important working parameter is raster angle. It is the angle of raster pattern with respect to x axis on the bottom part layer. It is very important parameter as it directly effect on the strength of the printed part. Its value varies from 0° to 90°. Number of loops refers to the deposition of the number of filaments adjacent to skin. Infill parameters refer to the filling percentage of material in each layer of the print. For 100% fill, print pattern is always straight, however in case of fill of less than 100% pattern can be strength, circular or polygonal. Infill pattern are shown in Figure 3(c) [12]. Figure 4 shows various approaches of orientation for print [21]. Part orientation followed by support style is another important factor need to consider while printing the object.

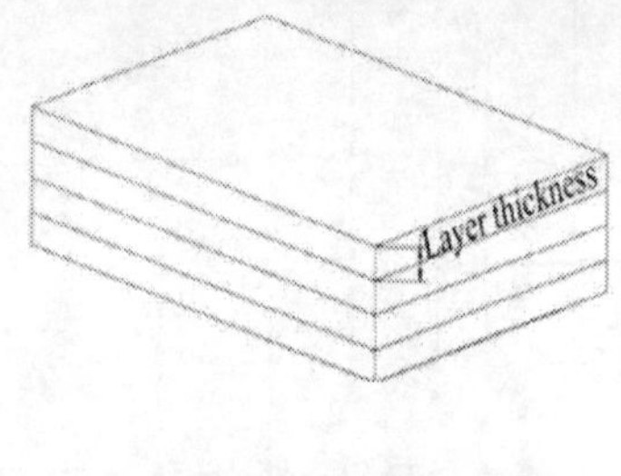

(a) Layer thickness

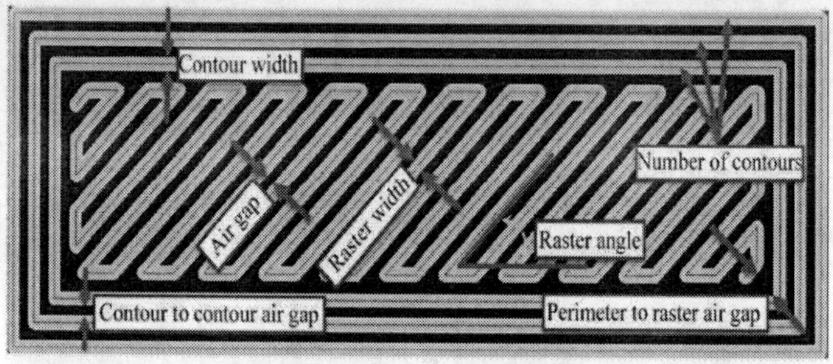

(b) Raster, raster width and gap

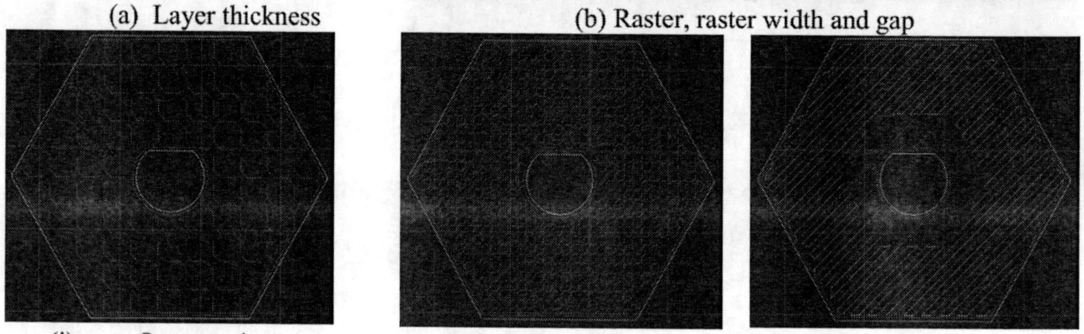

(i) Octagonal (ii) Rounded (iii) Straight

(c) Infill patterns (i, ii, iii)

Figure 3. Various parameters used in FDM (a-c)

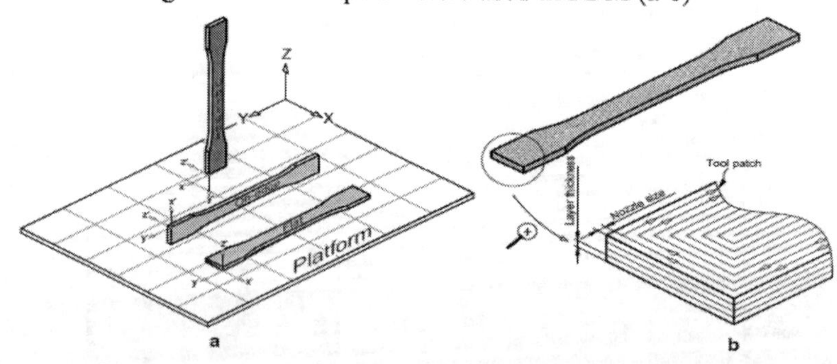

Figure 4. Part orientation and tool path generation [7,8]

2.2.3 Unprocessed parameters

Unprocessed parameters generally refer to density and the color of the material. As these two generally not effect on quality of printed part. Environmental factors need to be consider while selecting the proper material for printing. This includes temperature and humidity. As most of the materials used in FDM are of plastic category, these materials are highly effected by humidity and environmental temperature. To avoid the effect of these factor, a close working environment is required [23].

2.2.4 Machine parameters
FDM machine parameters include nozzle diameter, envelope temperature and machine calibration. Standard nozzles used in FDM machines are of diameter range 0.4 to 0.5 mm. Though 0.3 mm also available for special materials [24].

3. Literature survey

There has been wide-ranging research carried out by various researchers on the process parameter optimization of FDM process. Most of the researchers directed the research on optimizing surface roughness/finish, improvisation in mechanical properties. In the following sections, a comprehensive literature survey presented on the basis of process parameters and material used in FDM process.

3.1 Process parameter optimization
There has been wide-ranging research carried out by various researchers on the process parameter. In fused deposition modeling type of additive manufacturing process, various parameters effect on overall efficiency of printed part. Out of several parameters, process parameters has highest effect on quality of the printed part. Process or working parameters includes raster angle, raster width, air gap, nozzle diameter, temperature and many more [25]–[32]. Some of these are primary factors which have major impact on printing and the primary reason that most of the researchers concentrated on these parameters. To evaluate the effect of various parameters, it is necessary to check the effect individually while keeping other as constant. Kaveh et al. [19] considered extruded temperature, feed rate, flow rate and raster width as process parameters to enhance the dimensional accuracy and intrinsic internal cavities of the printed object. Raster angle has direct influence on the strength of the printed object. Finest raster width for each layer thickness lead to reduce air gap between rasters. Temperature is process as well as machine parameter which is important to consider for precision and internal cavity. For the considered material, 210◦C was selected as the optimum temperature.
Raster angle which decide the orientation of the filament in the printed part, responsible for the tensile strength. Orientation of the filament can be set between 0o to 90o and strength varies with the variation of the angle. So, it is necessary to decide optimum value of the raster angle for maximization of the tensile strength. Ning et al. [33] examined the outcome of the raster angle and infill on the tensile strength of Carbon fiber-reinforced plastic (CFRP) composites printed using FDM. In case of CFRP material, raster angle of 0o and 90o exhibit larger Young's modulus and yield strength than 45o whereas increase in infill speed results in reduction in strength due to poor bonding between adjusted rasters and consecutive layers.
Wenzheng et al. [34] inspected the impact of layer thickness and raster angle on the general mechanical characteristics of 3D printed polyether-ether-ketone (PEEK) objects. The average tensile strength of PEEK parts where considerably high compared to acrylonitrile butadiene styrene (ABS) parts. Compressive strength and bending strength shown considerable improvement as compared to later. Vaezi et al. [26] investigated the effect of the layer thickness and binder saturation level on the 3D printed parts and also predicted the mechanical characteristics based on average stiffness method. Results of experiment conclude the increase binder level increase tensile strength while increase in layer thickness decrease tensile strengths and increases flexural strength.
Tool path is important factor to ensure proper outcome from the system. Tool path generation approach consist of thee important sections that are identification of inclination, generation of the tool path in each sub-section and connection between discrete sub-paths. Jin et al. [35] applied optimization in tool path generation in deposition modelling. The approach is useful to generate the direction parallel to tool path based on dissimilar priorities. Sood et al. [36] used weighted principal component analysis to optimized five process parameters of FDM at three different levels.
Number of process parameters in FDM system are more which leads complication while handling multiple parameters. It is necessary to build relation between various process parameters. Lee et al. [37] optimized four working parameters that are air gap, raster angle, raster width and layer thickness, with humidity and temperature as constant with three levels. Used Taguchi's method to build relation

Table 1. Summary of the FDM process optimization

Author	Material used	Process Parameters	Remark
Kaveh et. al. [9]	HIPS	Extruder temperature, Feed rate, flow rate and raster width	Improved dimensional accuracy and reduction in intrinsic internal cavities
Vaezi et. al. [10]	ZP102 powder and Zb56 binder	Layer thickness and binder saturation level	Evaluated two layers of thickness 0.1 and 0.087 respectively, at the 90% and 125% binder saturation levels. Higher the binder level, higher the tensile and flexural strengths but results in poor dimensional accuracy and surface uniformity.
Farzad et. al. [11]	Calcium sulfate powder	Layer thickness and print orientation	Evaluation of considered parameters results in higher toughness, compressive and tensile property with 0.1125 and 0.125 mm layer thickness.
Croccolo et. al. [12]	ABS M30	Part build direction, raster angle, bead width, air gap, layer thickness and number of counters	Analytical model represented the variation in strength and stiffness properties for considered material. Experimental results show significant similarity with 4% error.
Hill and Haghi [13]	Polycarbonate	Raster orientation	Print orientation has high influence on the mechanical characteristics of the material. Failure mechanism correlated with fibre-reinforced composites model.
Ulu et. al. [14]	Material not specified	Build Orientation	Developed orthotropic material-model. Nine parameters essential to be resolute experimentally. The Young's moduli, shear moduli, and Poisson's ratios for the three principal directions where evaluated with the model. FEA simulation and experimentation shown close relevance to output of numerical model.
Espin at el. [15]	Polycarbonate	Build orientation, nozzle diameter, slice height, and diameter of the extruded filaments	Proper orientation results in improved yield strength. Tensile stresses should be perpendicular to layer's planes to avoid fragile fracture.
Ning et. al. [16]	CFRP	Raster angle and infill speed	Raster angle of 0° and 90° exhibit larger Young's modulus and yield strength than 45°
Wenzheng et. al. [17]	PEEK and ABS	Layer thickness, Raster angle	The outcome concluded that the average tensile strength of PEEK parts where considerably high compared to acrylonitrile butadiene styrene (ABS) parts.
Jin et al. [18]	ABS	Tool path generation	Tool path generation approach consist of thee important sections that are identification of inclination, generation of the tool path each sub region and connection between individual sub-paths.

IOP Conf. Series: Materials Science and Engineering **810** (2020) 012031 doi:10.1088/1757-899X/810/1/012031

Sood et al. [19]	ABS	Part orientation, road width, layer thickness, air gap, raster angle	Weighted principal component analysis to optimized five process parameters of FDM at three different levels. Improved dimensional accuracy
Lee et al. [20]	ABS	air gap, raster angle, raster width and layer thickness	Throwing distance and elasticity were considered for the optimization in FDM process.
Lee et al. [21]	ABS with plaster powder	Build direction	Nano composite deposition system (NCDS). Experimental investigation results in higher mechanical properties with optimal build direction. Also, results evaluated that 3D printer parts have low compressive strength as compared to other techniques.
Caminero et al. [22]	Carbon fiber	Ply angle, orientation	Investigated the effect of stacking sequence of composite. The distinctive behaviour of angle-ply laminate, under impact loading due to its pseudo-ductile behaviour, flexural damage testing has been assessed.
Sugavaneswaran et al. [23]	ABS, Polypropylene	Orientation, infill	New methodology based on multi materials as reinforcement. Uni-axial tensile test and fracture test shows that reinforcement results in superior mechanical characteristics.
Kim and Shin [24]	ABS	Mosture and temperature	Environmental parameters results in the decreased of the tensile strength by 67-71 per cent in hot, wet environments related to dry, room temperature conditions.

Table 2. Ultimate tensile strengths (MPa) of various materials at various raster angle used in FDM [28-30]

Raster angle	ABS [25]	Polypropylene [26]	Polycarbonate [13]	PLA [26]
90°	26	32	19	54
0 °	34	36	59.7	58

Table 3. Characteristics of various materials

Material	**References**	**Remarks**
Polylactic Acid (PLA)	[7,10,16,19,21, 27,28]	Polylactide (PLA) is widely used and biodegradable thermoplastic polymer. The unique mechanical characteristics make it superior than the other thermoplastic polymers. Various process parameters like part orientation, raster angle, raster width and infill significantly help to enhance the properties of PLA parts. Comparison of mechanical characteristics of PLA with other thermoplastic polymers has been summarized in table 2.

Acrylonitrile Butadiene Styrene (ABS)	[4,12,24,29–34]	ABS has high toughness, a high degree of moldability, and low thermal conductivity. Number of variants of ABS are available with number of different characteristics like colors, strength, thermal conductivity, etc. ABS P500 is suitable for medical and pharmaceutical applications due to its high strength. ABS-M30 is one of good replacement to virgin ABS as it is 25-75% stronger, higher durability and ideal for realistic functional test. One of the reason for wide popularity of ABS is its compatibility of combination with various materials like organic modified montmorillonite (OMMT), UHMWPE, Jute fiber, Aluminum oxide, Zinc oxide. ABS also can be added with copper and iron nano particles to enhance thermal conductivity of the base ABS.
High Impact Polystyrene (HIPS)	[9,24]	HIPS is a thermoplastic material which dissolve in d-Limonene. Due to its lightweight nature is frequently preferred as support structure material with ABS models. As it dissolve in d-Limonene, it is easy to remove support structure and post processing decidedly reduce. Also, its properties close to ABS, make it dual extruder partner with latter. For the parts which may get worn out or for the lightweight applications, HIPS is dimensionally stable and quite well-lit than ABS.
Polypropylene (PP)	[35,36]	The most cost effective 3D printing material which repel to allot of chemicals and solvents. Sometime, ethylene co-polymerising with PP to higher strength. At room temperature, PP repel to all organic solvents and fats. Whereas, sylene, tetralin and decalin i.e. non-polarity solvents used to dissolve PP. Due to rigid and robust nature of PP, it used in textile, ropes and stationary applications.
Polyether–ether–ketone (PEEK)	[17,37]	Similar to ABS P500, PEEK is medical grade material with very high temperature for printing. It frequently used in medical, pharmaceutical and food packaging applications. Also, good impact resistance, superior adhesion between layers makes is suitable for aerospace and oil industry. In 3D printed parts, PEEK shows superior mechanical properties than ABS.
Carbon fiber filaments	[16,38,39]	To increase the strength and stiffness, carbon fibre filaments reinforced into base material like PLA or ABS. number of filaments come together with carbon fibre like PLA, PETG, Nylon, ABS, and Polycarbonate. Carbon fibre reinforcement helps to enhance mechanical characteristics like tensile strength in un-axial or bi-axial direction. Whereas, it also leads clogging in extruders due to small fibres of carbon into filaments.

between various working parameters. Throwing distance and elasticity were considered as performance criteria for the optimal FDM process parameters. In other study of Kim and Shin [38], the effect of moisture and temperature on the parts prepared through FDM process and injection molding process had been investigated. Many FDM printed parts used in load-carrying application in high moisture and temperature conditions. FDM components followed Fickain diffusion theory for water absorption behaviour, irrespective of the temperature. Temperature rise resulted into increased diffusion rate however absorption rate not affected by it.

In a composite 3D printing, different stacking sequence exhibit different results. Charpy impact test and three point bending test normally conducted to check impact response and flexural behaviour of different stacking of composite. Caminero et al. [39] investigated the effect of stacking sequence of composite. Due to pseudo-ductile behaviour of angle-ply laminate, under impact loading were presented by author. Sugavaneswaran et al. [40] experimentally investigated the elastic modulus of parts printed by using polyjet 3D printers. Normally strength and stiffness of polyjet 3DP parts is typically less and have limitations to use in engineering applications. To overcome this limitations, author has explored new methodology based on multi materials as reinforcement. Addition of reinforcement result in enhancement of elastic modulus in uni-axial tensile test and fracture tests.

Lee et al. [38] deliberate the effect of build direction on mechanical characteristics of 3D printed parts. Experimentation shown that FDM parts had higher compressive strength than other AM processed parts but weaker as compare to traditional manufacturing methods. Espin at el. [20] studied the mechanical characterization of polycarbonate parts. Selection of proper orientation is important criteria when the yield strength is exceeded. If longest contour aligned to the direction of tensile force, it results in greater tensile stresses. Summary of the literature survey based on process parameters has been tabulated in Table 1.

3.2 Materials for FDM process

Mechanical properties of RP materials show the phenomenon of anisotropy. The results of anisotropy investigations are presented by various researchers [42]. Present section has been focused on the available material range in AM and their respective properties. Common materials which are being used in FDM are polyoxymethylene (POM), polyamide (PA), polypropylene (PP), nylon 66, Polyether-Ether-Ketone (PEEK), Acrylonitrile Butadiene Styrene (ABS) etc. With these materials, reinforcements like carbon fiber, glass fiber and metal fibers are used. And sometimes dry lubrication films of molybdenum di-sulphide, boron nitride, graphite flakes and poly tetra fluoro-ethylene (PTFE) are used to strengthen the base material [42]–[45]. Ultimate tensile strength of various material at various raster angle is summarized in Table 2. Table 3 is the summery of the some off the standard material that used in FDM AM system. Along with above materials, other materials like polycarbonate (PC) [16], [20], TPA and Nylon 6/6 filament also used in FDM 3D printers.

3.3 Inference from literature survey

This approach of manufacturing known as layer by layer manufacturing, has a great potential and applicability in various fields. Key highlights from the understanding of various AM systems are as follows:

1. Liquid based and powder based systems are bulky and costly as compared to solid based system due to use of laser, lens and scanning system. Photo-polymerization consumes less energy but generate fumes which is hazardous and so required a closed environment.
2. Some replacement are available to laser system which are helpful to reduce the cost of liquid based AM systems.
3. In powder based AM system, sintering process used to convert powder into solid object. With the use of high power laser system, metallic powder can be processed in SLS system to create functional prototypes.

4. Among the three classes of AM system, fused deposition modeling stands intermediate applicability. Extruder of FDM system has capability to deal with number of polymer category materials.
5. Efficiency of the FDM system depends on number of factors like environmental factors, material characteristics, machine parameters and working parameters.
6. Due large number of parameters, FDM process become complicated to handle and many researchers considered very few for optimization.
7. Various process parameters effects the overall efficiency of the FDM printed object especially raster angle, part orientation and layer thickness
8. Mechanical strength of FDM printed objects can be enhance by selecting optimum values of process parameters.
9. Wide range of materials available for FDM printers but most of these materials are polymer family which results in less strength as compare to metals. This restricts the use of FDM printed object to demonstration purpose only.
10. Above limitation can be overcome by hybridization i.e. composite structuring.
11. PLA and ABS are widely used by researchers due to good mechanical properties.
12. PEEK and ABS-P500 are biocompatible properties and hence suitable to use in biomedical applications.

4. Conclusion

FDM is widely used 3D printing technique whereas strength of printed parts is poor as compared to powder based additive manufacturing. Enhancement of mechanical characteristics is prime requirement in FDM process. Some deduction has been identified from the review are as follows:

- To improve part quality and mechanical properties FDM printed parts, it necessary to understand correlation between thermo-mechanical properties and process parameters. Thus, development of new mathematical approach is required.
- Different process parameters have significant effect on the performance of FDM system. Most of the researchers considered raster angle, part orientation and layer thickness as influencing parameter. But large number of other parameters where neglected and can be consider for future work.
- The FDM extruder able to deal with different kinds of materials such as ABS, PC-ABS, ABS-P400, ABS-P500, nylon, PLA, HIPS and PEEK. Most of researchers used ABS due to ease of availability and certain mechanical properties. Material combinations i.e. composite materials can be used to enhance mechanical properties of the fabricated parts.
- Very few work has been carried out on biocompatible materials such as ABS-P500 and PEEK.
- Many researchers focused on identification of mechanical characteristics such as tensile, compressive and flexural strength of specific materials. No work has been carried out on identification of tribological properties which are important to identify in case of fabrication of function prototypes.
- No work has been carried out on surface modification/engineering of additively fabricated parts.

5. References

[1] Chau C K, Leong K F and Lim C S 2017 **vol 91** (World Scientific Publishing Co. Pte. Ltd.)
[2] Mansour S and Hague R 2002 *Proc. Inst. Mech. Eng.* **Vol. 217 P** 453–61
[3] Kruth J P 1991 *CIRP Ann. - Manuf. Technol.* **40** 603–14
[4] Tymrak B M, Kreiger M and Pearce J M 2014 *Mater. Des.* **58** 242–6
[5] Zhang Y and Chou K 2008 *Proc. Inst. Mech. Eng. Part B J. Eng. Manuf.* **222** 959–67
[6] Mohamed O A, Masood S H and Bhowmik J L 2015 *Adv. Manuf.* **3** 42–53
[7] Chacón J M, Caminero M A, García-Plaza E and Núñez P J 2017 *Mater. Des.* **124** 143–57
[8] Zaldivar R J, Witkin D B, McLouth T, Patel D N, Schmitt K and Nokes J P 2017 *J Addit.*

Manuf. **13**
[9] Kaveh M, Badrossamay M, Foroozmehr E and Hemasian Etefagh A 2015 *J. Mater. Process. Technol.* **226** 280–6
[10] Vaezi M and Chua C K 2011 *Int. J. Adv. Manuf. Technol.* **53** 275–84
[11] Farzadi A, Solati-hashjin M, Asadi-eydivand M, Azuan N and Osman A 2014 *PLoS One* **9** 1–14
[12] Croccolo D, De Agostinis M and Olmi G 2013 *Comput. Mater. Sci.* **79** 506–18
[13] Hill N and Haghi M 2014 *Rapid Prototyp. J.* **20** 221–7
[14] Ulu E, Korkmaz E, Yay K, Ozdoganlar O B and Kara L B 2019 *J. Mech. Des.* **137** 1–9
[15] Domingo-Espin M, Puigoriol-Forcada J M, Garcia-Granada A A, Llumà J, Borros S and Reyes G 2015 *Mater. Des.* **83** 670–7
[16] Ning F, Cong W, Hu Y and Wang H 2017 *J. Compos. Mater.* **51** 451–62
[17] Wu W, Geng P, Li G, Zhao D, Zhang H and Zhao J 2015 *Materials (Basel).* **8** 5834–46
[18] Jin Y an, He Y, Fu J zhong, Gan W feng and Lin Z wei 2014 *Addit. Manuf.* **1** 32–47
[19] Sood A K, Chaturvedi V, Datta S and Mahapatra S S 2011 *J. Adv. Manuf. Syst.* **10** 241–59
[20] Lee B H, Abdullah J and Khan Z A 2005 *J. Mater. Process. Technol.* **169** 54–61
[21] Lee C S, Kim S G, Kim H J and Ahn S H 2007 *J. Mater. Process. Technol.* **187–188** 627–30
[22] Caminero M A, Rodríguez G P and Muñoz V 2016 *Compos. Struct.* **136** 345–57
[23] Sugavaneswaran M and Arumaikkannu G 2015 *Mater. Des.* **66** 29–36
[24] Hwang S, Reyes E I, Moon K sik, Rumpf R C and Kim N S 2015 *J. Electron. Mater.* **44** 771–7
[25] Rezayat H, Zhou W, Siriruk A, Penumadu D and Babu S S 2015 *Mater. Sci. Technol.* **31** 895–903
[26] Letcher T 2014 *Proc. ASME 2014 Int. Mech. Eng. Congr. Expo.* 1–8
[27] Rankouhi B, Javadpour S, Delfanian F and Letcher T 2016 *J. Fail. Anal. Prev.* **16** 467–81
[28] Chohan J S, Singh R and Boparai K S *Additive Manufacturing of Emerging Materials. Springer* 227–49.
[29] Zou R, Xia Y, Liu S, Hu P, Hou W, Hu Q and Shan C 2016 *Compos. Part B Eng.* **99** 506–13
[30] Dawoud M, Taha I and Ebeid S J 2016 *J. Manuf. Process.* **21** 39–45
[31] Weng Z, Wang J, Senthil T and Wu L 2016 *Mater. Des.* **102** 276–83
[32] Torrado A R, Shemelya C M, English J D, Lin Y, Wicker R B and Roberson D A 2015 *Addit. Manuf.* **6** 16–29
[33] Casavola C, Cazzato A, Moramarco V and Pappalettere C 2016 *Mater. Des.* **90** 453–8
[34] Klammert U, Gbureck U, Vorndran E, Rödiger J, Meyer-Marcotty P and Kübler A C 2010 *J. Cranio-Maxillofacial Surg.* **38** 565–70
[35] Hill N and Haghi M 2014 *Rapid Prototyping Journal* **20 (3)** 221-227
[36] Sodeifian G, Ghaseminejad S and Akbar A 2019 *Results Phys.* **12** 205–22
[37] El Halabi F, Rodriguez J F, Rebolledo L, Hurtós E and Doblaré M 2011 *J. Mech. Behav. Biomed. Mater.* **4** 1819–32
[38] Caminero M A, Pavlopoulou S, Lopez-Pedrosa M, Nicolaisson B G, Pinna C and Soutis C 2013 *Compos. Struct.* **95** 500–17
[39] Caminero M A, Chacón J M, García-Moreno I and Reverte J M 2018 *Polym. Test.* **68**

Experimental Investigation of Effect of Preheating of Air and Exhaust Gas Recirculation on Four Stroke Diesel Engine

Ram Thakar[1], Santosh Bhosle [2] and Subhash Lahane[3]

[1] Research Scholar, Marathwada Institute of Technology, Aurangabad
[2] Professor, Maharashtra Institute of Technology, Aurangabad, India
[3] Associate Professor, Marathwada Institute of Technology, Aurangabad,India
Address: Marathwada Institute of Technology, Aurangabad (M.S) India: PIN- 431010
Email: thakar_ram@ycmou.digitaluniversity.ac

Abstract. This paper is concerned with means for heating the inlet air passed to the cylinder by using specifically designed and manufactured heat exchanger and also use of exhaust gas recirculation to investigate its effect on performance, combustion and emission characteristics of four stroke single cylinder diesel engine. A diesel engine having a cylinder, 3.5 KW power, constant speed of 1500 RPM , water cooled, Kirloskar make, diesel engine was selected as base engine and the experiments were conducted at part load and full load conditions. The diesel engine setup was modified with (i) incorporation of heat exchanger which was designed on the basis of availability of waste heat at the engine exhaust for preheating of air and (ii) exhaust gas recirculation for waste heat recovery and reduction in the NOx. The Experiments were conducted for six setups viz. i. Basic diesel engine setup ii. Basic diesel engine setup with 6 % EGR iii. Basic diesel engine setup with 12 % EGR iv. Modified setup with heat exchanger v. Modified setup with heat exchanger and 6% EGR and vi. Modified setup with heat exchanger and 12% EGR .The experimental results obtained with modified engine setup with use of heat exchanger and EGR. The results obtained indicate thermal efficiency is decreased with heat exchanger and 12 % EGR compared to base diesel with 12% EGR. The volumetric efficiency is decreased from base diesel by 16.87 % with heat exchanger and 6% EGR at 75% load. The NOx emission is increased with increase in load for all the configurations.

1. Introduction

The depletion of fossil fuel resources at faster rate is a global phenomenon and it has emphasized the need of finding alternative sources of energy and also emphasized to improve the performance, combustion and emission characteristics of existing internal combustion engines in order to save the primary fuel. The diesel engine is concerned with many heat losses which tend to reduce its performance. The exhaust flue gas loss is the major loss that badly affect on the performance of the engine. Many researchers focusing on exhaust flue gas loss or exhaust gas loss utilization for improvement in performance, combustion and emission characteristics .Many techniques had been used to utilize exhaust gas heat potential for improving performance and reducing the emission level of diesel engine. Abd-Alla (2002) investigated on the use of exhaust gas recirculation (EGR) technique to check its potential of recovering exhaust heat from engine exhaust and reducing the NOx emissions. They suggested making optimum use of EGR by using it in airflow path rather than displacing part of inlet air. The use of EGR by this method substantially reduces NOx emissions [1]. Shi et al.(2006) made use of both internal and exhaust gas recirculation on HCCI engine and found reduction in NOx and smoke emissions due to use of homogeneous charged compression ignition system [2]. Hountals et.al (2008) verified effect of using cooled EGR gas temperature on turbocharged DI heavy duty diesel engine using multizone combustion model and shown its positive effect on brake specific fuel consumption and lowering the soot values [3]. The researchers investigated on different technologies for recovery of the exhaust heat from I.C.Engines. The techniques such as thermoelectric generation, bottoming Rankine cycle, six stroke cycle found opportunities for potential energy savings and improvement in the performance [4]. Researchers have been trying to improve thermal efficiency of diesel engines by applying different techniques. The effect of high injection pressure and cylinder air pressure have been investigated on combustion and heat release rate for DI diesel engines shown improved brake thermal efficiency [4]. Thakar et.al (2018) designed counter flow shell and tube heat exchanger and reported heating of inlet air upto 150° C [6]. Tilmann Abbe Horst et al (2013) developed dynamic model of heat exchanger for exhaust gas recovery. The dynamic heat exchanger model predicts heat recovery systems of automobile engines. This dynamic model developed was working on moving-boundary principle and is used to develop a control system for dynamic operation on test bench measurements. [7].Saiful Bari & Shekh Hossain (2013) experimented with two heat exchangers to estimate exhaust heat recovery from diesel engine where they used water as working fluid. Their experimental study found additional power of 16%. They found increased additional power from 16% to 23.7% after optimization.[8].Bumroongsri et al (2013) carried out experimentation on heat transfer coefficients of unit of waste heat recovery in the detergent manufacturing process, It resulted into reduction of heat loss by means of waste heat recovery process [9]. Hui Xie & Can Yang (2013) studied heavy duty diesel engines for waste heat recovery using Rankine cycle system to study its dynamic behavior.They defined four basic operating modes under driving cycles and established Rankine Cycle System(RCS) model. The four modes include startup mode, turbine turning mode, power mode and protection mode. The operating performances of established RCS model and actual driving cycle were discussed. The results indicate that Rankine Cycle system efficiency is as low as 3.63% which is less than half of the designed efficiency which is 7.77%.[10].Love et al (2012) carried out experimentation to improve performance of waste heat recovery systems. They tested thermoelectric devices on a bench-scale thermoelectric heat recovery apparatus and observed that simulates automotive exhaust. It is observed that thermo electric power output increases from 2 to 3.8w for higher exhaust gas flow rates and overall system efficiency decreases from 0.95% to 0.6%. There is degradation of the effectiveness of the Exhaust gas recirculation type heat exchangers over a period of time.[11] .Serrano et al (2012) introduced new solutions to the problems related to coupling between thermal engine and waste heat recovery Rankine cycle. to analyze alternative solutions for the problems related to the coupling between the WHR Rankine cycle and the thermal engine. They introduced the method of removing its turbine and adapting one of the turbochargers and trying to recover the energy by the Rankine cycle. [12].Srinivas Garimella (2012) studied simultaneoud chilled and hot water generation from the low-grade waste heat recovery steam. The high heat at 120 deg Celcius recovered from a gas stream was supplied to an absorption cycle to simultaneously produce chilled water and hot water to be used for process heating. [13]. Dae Hee Lee et al (2010) investigated experimentally on exhaust waste heat recovery of diesel engine. They employed the concept of cogeneration where electric power is generated by the generator connected to the diesel engine and the heat is recovered from exhaust gases as well as by the fin-and-tube and shell-and-tube heat exchangers. The effect of secondary combustion on efficiency and emission reduction observed. It is found total efficiency reaches a maximum 94.4% which is approximately 15-20% higher than efficiency of typical diesel engine [14] .Ismail Teke et al (2010) developed a model for determining type of heat exchanger and area for optimum waste heat recovery. Based on technical and economical parameters such as heating value of fuel, unit cost of heat exchanger etc a non-dimensional E number had been defined. This non-dimensional E numbers had been expressed in graphical forms as the functions of NTU ands ratio of heat capacities. The corresponding heat exchanger area for giving maximum heat gain can be obtained from these graphs. The efficient heat exchanger type and its area can be determined comparing or effectiveness of heat exchangers or net gains at NTU.[15].Gianfranco Caruso et al (2015) experimentally studied finned tube heat exchanger where water is used as heating medium and flowing inside the tube at different conditions. The air side heat transfer coefficients and the thermal constant conductance in the tube is studied. Petukhov's correlation was used to calculate the water heat transfer coefficient in the tube. Three tubes of different diameters 30 mm, 22 mm, and 15.6 mm were used and the fitting of the data provides a correlation that agrees very well with the experimental data. The thermal contact resistance based on Reynold's number is estimated and conformed.[16].Yakar & Karabacak (2015) investigated experimentally on perforated finned heat exchangers. They studied the thermal performance of perforated finned heat exchangers for two cases considering basic perforated finned heat exchangers and perforated finned heat exchangers with angle of Rotation 8. In order to find best angular location experiments were carried out at six different angular locations. They also compared imperforate finned heater with a perforated finned heater and observed the finned heater effectiveness is 18% higher at 608 and pressure drop is 1.16% lower than other angular positions. They concluded that the best angular position is 608. The results also shows increase in effectiveness with increase in number of transfer units.[17]. Pandiyarajan et al (2011) conducted experimental investigation on diesel engine for exhaust heat recovery using finned shell and tube heat exchanger and thermal storage system. IC engine setup with integrated shell and finned tube heat exchanger and a thermal energy storage tank used for experimentation. The designed, fabricated and tested thermal storage tank used for excess energy storage which is tested using cylindrical phase change material (PCM) capsules for sensible and latent heat storage system. The heat exchanger extract heat from the exhaust gas and excess energy is stored in thermal energy storage tank. The performance of the engine evaluated with and without heat exchanger. It is found that 10–15% of heat at higher temperature is stored the thermal storage tank.This available high temperature heat can be utilized for suitable application. The performance parameters amount of heat recovered, charging efficiency, charging rate, percentage energy saved related to heat exchanger evaluated [18].Iqbal et al (2011) considered steady, laminar, incompressible and fully developed flow subjected to constant heat flux boundary conditions in double pipe with parabolic fins and studied optimal convective heat transfer. Finite element method (FEM) is employed to compute field variables for providing function values to the optimizers.[19].Desai & Bannur (2001) conducted experimental investigation on a twin cylinder diesel engine using a shell and tube heat exchanger to recover heat from exhaust gas[20]. Lee et al (2010) carried an experimental study on the effects of secondary combustion pn efficiencies and emission reduction in the engine exhaust heat recovery system.[21].Talbi & Agnew (2002) studied turbocharged diesel engine interfaced with an absorption refrigeration unit for energy recovery from the engine exhaust gas. They estimated enhancement in the performance due to the engine exhaust heat recovery.[22]. Morcos (1998) investigated on shell and dimpled tube exchangers for waste heat recovery. They investigated exchanger heat duty, overall heat transfer coefficient, effectiveness and tube side friction factor as functions of the tube surface geometry that is plain or dimpled and for the parallel and counter flow pattern and tube Reynolds number.[23]. The present paper is the result of experimentation on single cylinder, four stoke diesel engine with the incorporation heat exchanger and exhaust gas recirculation. The counter flow shell and tube heat exchanger is designed depending on the availability of engine exhaust heat which is mounted between outlet and inlet duct of the engine. The exhaust gas recirculation at 6% EGR and 12% EGR also has been used to minimize the emission level in the engine.

Nomenclature

CO	Carbon monoxide
NOx	oxides of nitrogen
CO2	carbon dioxide
HC	hydro carbon
EGR	exhaust gas recirculation
CA	crank angle
BTE	bake thermal efficiency
VCR	variable compression ratio
WHE	without heat exchanger
HE	with heat exchanger
BSFC	brake specific fuel consumption

2. Experimental details

The present experimentation is completed on 3.5 kw capacity 4-stroke single cylinder stationary diesel engine operating at injection timing 23° BTDC and injection opening pressure of 200 bar. The engine specifications are given in Table 1 and the engine setup is shown in Figure 1. The setup is equipped with data acquisition system for measurement of combustion pressure, crank angle and other performance, combustion and emission parameters The counter flow shell and tube heat exchanger was designed on the basis of potential heat available at the engine exhaust at full load .The designed heat exchanger was placed upstream of inlet and outlet duct of the engine. The other techniques of exhaust gas recirculation were used to evaluate the performance and emission characteristics of the engine. The experimentation carried out for total number of six configurations as i. Diesel without Heat Exchanger (WHE) ii. Diesel without Heat Exchanger (WHE) and 6% EGR iii. Diesel without Heat Exchanger (WHE) and 12% EGR iv. Diesel with Heat Exchanger (HE). v. Diesel with Heat Exchanger (WHE) and 6% EGR and vi. Diesel with Heat Exchanger (WHE) and 12% EGR.

Table 1. Engine Specifications.

Title	Details
Product	VCR Engine test setup 1 cylinder, 4 stroke, Diesel (Computerized)
Product code	234
Engine	Make Kirloskar, Type 1 cylinder, 4 stroke Diesel, water cooled, power 3.5 kW at 1500 rpm, stroke 110 mm, bore 87.5 mm. 661 cc, CR 17.5
Dynamometer	Type eddy current, water cooled, with loading unit
Propeller shaft	With universal joints
Air box	M S fabricated with orifice meter and manometer
Fuel tank	Capacity 15 lit with glass fuel metering column
Calorimeter	Type Pipe in pipe
Piezo sensor	Range 5000 PSI, with low noise cable
Data acquisition device	NI USB-6210, 16-bit, 250kS/s.

IOP Conf. Series: Materials Science and Engineering **810** (2020) 012032 doi:10.1088/1757-899X/810/1/012032

Figure 1. Modified Experimental Setup with Heat Exchanger.

3. Results and Discussion

The various performance and emission parameters of the engine for the six configurations were measured and the variation of these parameters with respect to load elaborated in details in the following section.

3.1 Brake Thermal Efficiency

The performance measurement parameter of the engine is thermal efficiency. It is measure of conversion of heat released during combustion to net work output. It is observed from figure 2 the brake thermal efficiency is decreased with diesel heat exchanger as compared to base diesel without heat exchanger. Thermal efficiency is decreased from base diesel by with 2.94% with diesel without heat exchanger at part load and by 18% at full load. It may be mainly due to decrease in density of fuel at high temperature of heated air. It is also observed that thermal efficiency is also decreased with heat exchanger and 12 % EGR compared to base diesel with 12% EGR. This may be due to lower oxygen content with exhaust gas recirculation which results into lower combustion temperature. Similarly, it decreased with heat exchanger and 6% EGR compared to base diesel without heat exchanger and 6% EGR.

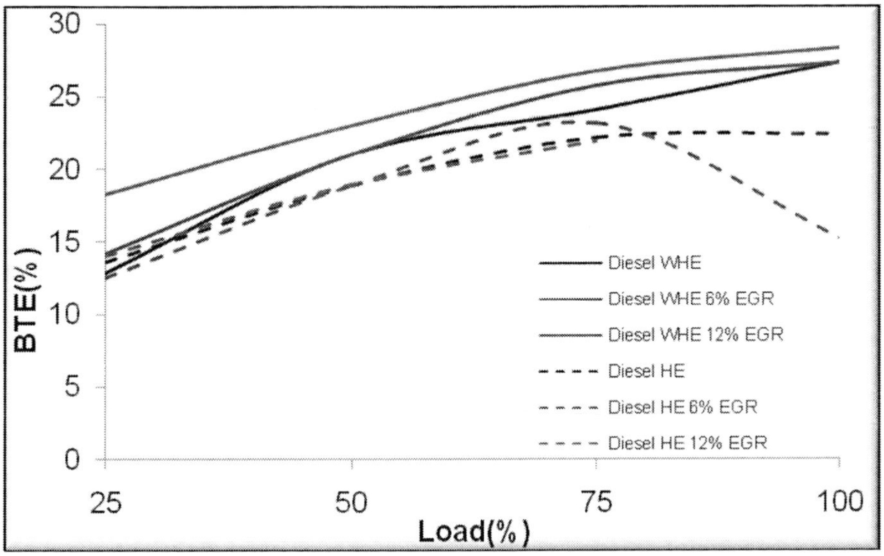

Figure 2 . Variation of brake thermal efficiency w.r.t load.

3.2 *Volumetric Efficiency*

The variation in volumetric efficiency verses load is shown in Figure 3 for base diesel and diesel with heat exchanger and exhaust gas recirculation 6% & 12% EGR for engine operating at 23° BTDC and 200 bar injection pressure. The figure 3 shows volumetric efficiency remains more or less same with respect to load for engine running at base diesel condition. The volumetric efficiency variation for normal engine lies between 74 to 55%. It is observed decreased volumetric efficiency with heat exchanger and exhaust gas recirculation as compared to engine running at base diesel with increase in load. This may be due to decrease in density of heating air in the heat exchanger which is passed inside the cylinder. The volumetric efficiency is decreased from base diesel by 16.87 % with heat exchanger and 6% EGR at 75% load. It is also observed that volumetric efficiency almost remains same and lies between 65 to 66% at all loads.

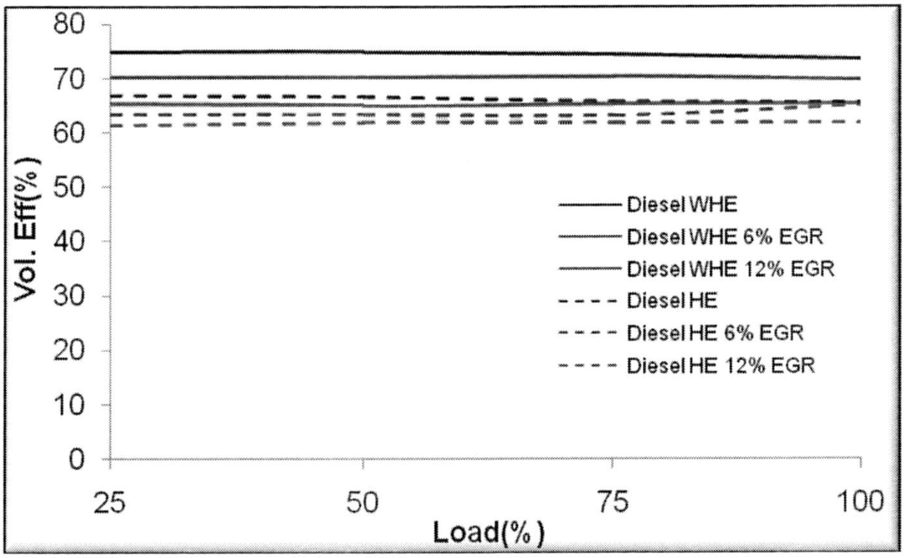

Figure 3. Variation of volumetric efficiency w.r.t load.

3.3 Brake Specific Fuel Consumption

The specific fuel consumption is derived directly from the mass of fuel consumed in the presence of oxygen and release the heat. This heat released during combustion is converted into power. It is observed from figure 4 that Brake specific fuel consumption is decreased with increase in load for base diesel without heat exchanger and similar trend of decrease in fuel consumption is also observed for diesel with heat exchanger. The brake specific fuel consumption (BSFC) is decreased from base diesel by 53% with diesel without heat exchanger at full load. This improved value of BSFC is due to preheating of air in the heat exchanger which improves the combustion in the cylinder. It is observed that the brake specific fuel consumption is increased by 51% for diesel with heat exchanger and 6% (EGR) Exhaust gas recirculation. It is also observed that the specific fuel consumption remains same for diesel with heat exchanger and with 6% EGR and 12% EGR at 50% load.

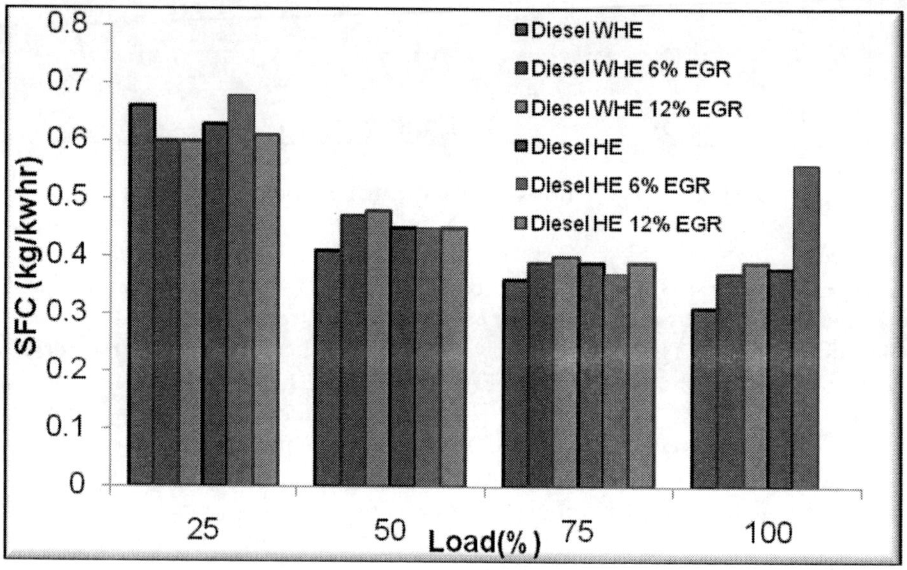

Figure 4. Comparison of Brake Specific Fuel Consumption w.r.t load.

3.4 Nitrides of Oxygen (NOx)

It is observed from figure 5 that NOx emission increases with all configurations (WHE, WHE 6% EGR, WHE12% EGR, HE, HE 6%EGR ,HE 12% EGR) as compared to base diesel. The increase in NOx is due to advancement in injection timing and availability of oxygen content in the fuel. The fixation of NOx generally takes place at higher temperature and advanced injection timings. The use of EGR and optimization of injection timing can be the solution to avoid the fixation of oxides of nitrogen as these methods tends to reduce the temperature of the charge.

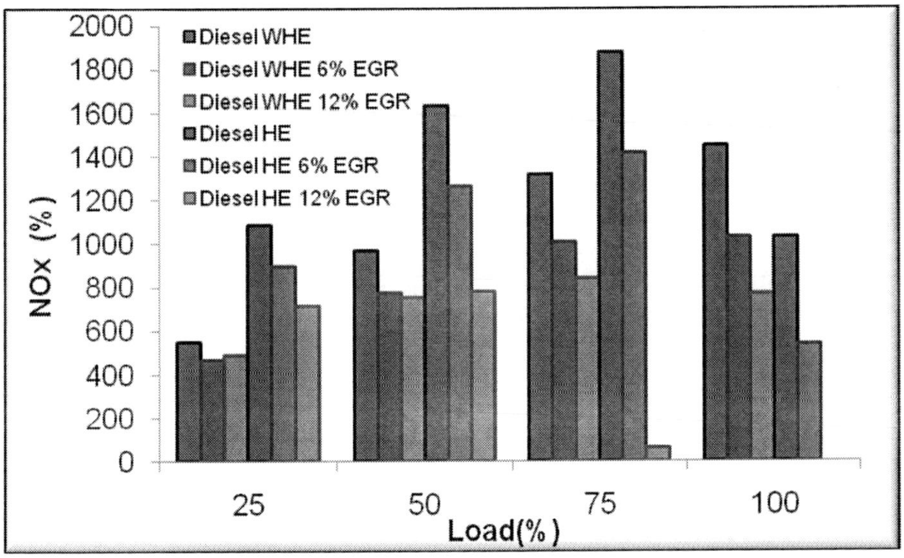

Figure 5. Comparison of NOx w.r.t load

4. Conclusions

The experimentation with use of heat exchanger for heating inlet air passed inside the cylinder does not finds fruitful results with the diesel as fuel. The brake thermal efficiency for base diesel is higher as compared to other configurations. The volumetric efficiency is decreased due to heating of air as density of heated air is decreased at higher loads.

References

[1] Abd-Alla G H 2002 Using exhaust gas recirculation in internal combustion engine: A review Energy conversion and Management **43** 1027-1042.

[2] Lei S, Yi C , Kangyao D, Haiyong P and Yuanyuan C 2006 Study of Low emission homogeneous charge compression ignition (HCCI) Engine using combined internal and external exhaust gas recirculation (EGR) Energy **31** 2665-2676

[3] Hountals D T, Mavropoulos G C and Binder K B 2008 Effect of exhaust gas recirculation (EGR) temperature for various EGR rates on heavy duty DI diesel engine performance and emissions Energy **33** 272-283

[4] Saidur M R , Muzammil W K, Hassan M H , Paria S and Hasanuzzaman M 2012 Technologies to recover exhaust heat from Internal Combustion Engines Renewable and sustainable Energy reviews **16** 5649-5659

[5] Levent Y, Tarkan S, Orkun O and Alp T E 2014 Modelling the Effect of Injection Pressure on Heat Release Parameters and Nitrogen Oxides in Direct Injection Diesel Engines Thermal Science **18**, 155-168

[6] Thakar R, Bhosle S and Lahane S 2018 Design of Heat Exchanger for waste heat recovery from exhaust gas of diesel engine Procedia Manufacturing Elsevier **20** 372-376

[7] Tilmann A H, Hermann S R, Marco S and Jürgen R 2013 Dynamic heat exchanger model for performance prediction and control system design of automotive waste heat recovery systems Applied Energy **105** 293-303

[8] Saiful B and Shekh N H 2013 Waste heat recovery from a diesel engine using shell and tube heat exchanger Applied Thermal Engineering 61 355-363

[9] Bumroongsri P, Witchayanuwat W and Kheawhom S 2013 Experimental heat transfer

coefficients of waste heat recovery unit in detergent manufacturing process Experimental Heat Transfer **26** 114-125

[10] Hui X and Can Y 2013 Dynamic behavior of Rankine cycle system for waste heat recovery of heavy duty diesel engines under driving cycle Applied Energy **112** 130-141

[11] Love N D, Szybist J P and Sluder C S 2012 Effect of heat exchanger material and fouling on thermoelectric exhaust heat recovery Applied Energy **89** 322-328

[12] Serrano J R, Dolz V, Novella R and Garcia A 2012 HD Diesel engine equipped with a bottoming Rankine cycle as a waste heat recovery system. Part 2: Evaluation of alternative solutions Applied Thermal Engineering **36**.279-287

[13] Srinivas G 2012 Low-grade waste heat recovery for simultaneous chilled and hot water generation Applied Thermal Engineering **42** 191-198

[14] Dae H L, Jun S L and Jae S P 2010 Effects of secondary combustion on efficiencies and emission reduction in the diesel engine exhaust heat recovery system Applied Energy **87** 1716-1721

[15] Ismail T, Ozden Agra S O and Hakan D 2010 Determining the best type of heat exchangers for heat recovery Applied Thermal Engineering **30** 577-583

[16] Gianfranco C, Fabio G and Antonio N 2015 An experimental study on the air-side heat transfer coefficient and the thermal contact conductance in finned tubes Heat transfer Engineering **36** 212–221

[17] Yakar, G and Karabacak R 2015 Investigation of thermal performance of perforated finned heat exchangers Experimental Heat Transfer **28** 354-365

[18] Pandiyarajan V, Chinna Pandian M, Malan E, Velraj R and Seeniraj R V 2011 Experimental investigation on heat recovery from diesel engine exhaust using finned shell and tube heat exchanger and thermal storage system Applied Energy **88**, 77-87

[19] Iqbal Z, Syed K S and Ishaq M 2011 Optimal convective heat transfer in double pipe with parabolic fins International Journal of Heat and Mass Transfer **54** 5415-5426

[20] Desai A D and Bannur Design P V 2011 Fabrication and testing of heat recovery system from diesel engine exhaust J Inst Engrs **82** 111-118

[21] Dae H L, Jun S L and Jae Suk Park 2010 Effects of secondary combustion on efficiencies and emission reduction in the diesel engine exhaust heat recovery system Applied Energy **87** 1716-1721

[22] Talbi M and Agnew B 2002 Energy recovery from diesel engine exhaust gases for Performance enhancement and air conditioning Applied Thermal Eng **22**, 693-702

[23] Morcos V H 1988 Performance of shell-and-dimpled-tube heat exchangers for waste heat recovery System CHP **8** 299-308

IOP Conf. Series: Materials Science and Engineering **810** (2020) 012033 doi:10.1088/1757-899X/810/1/012033

A Short Review on Automobile Dashboard Materials

Nitin Girdhar Shinde[1] and Dr. Dilip Mangesh Patel[2]

1 R C Patel Institute of Technology, Shirpur, Maharashtra, India.
2 D N Patel College of Engineering, Shahada, Maharashtra, India.
ngshinde23@gmail.com

Abstract. Day by day Automobile sector is going through new innovative trends in all aspects like design, appearance, performance and economy. While choosing an automobile for personal use, customer always look for aesthetics, economics and safety measures. First and far most part of an automobile that appears before user, is the dashboard. It acts as console for various components and holds the sense pleasantness of occupant space. In this paper a brief review of various types of material used in the making of dashboard is taken into account on the basis of detailed literature survey. Various types of dashboard are discussed along with their manufacturing methods. Ideal properties of a dashboard material are also taken into consideration. Testing techniques for dashboard materials are mentioned thereafter. Finally the concerns with the dashboard materials are highlighted.

1. Introduction

Automobile industry plays very important role in the economy of any country. The number of vehicles in India drastically increased by 700 times from 1951 to 2015 [1]. It is expected that globally automotive industry should work on advanced technological trends [2].Nowadays one of the main criterion for selection of personal vehicle are environmental concerns and driving comfort along with financial gains. The dashboard is one of the vital part of an automobile as it consoles many components to keep driver aware of their performance and condition for better driving comforts. It affects vehicle's structural integrity and enhances the sales appeal [3, 4]. Besides interfacing between driver and vehicle, dashboard contributes to some vital functions like aesthetics, active and passive safety, ergonomics, and object containment [5]. In early stages dashboard used to have only steering wheel and ignition switch, slowly on demand, variety of improvement in materials, product quality, appearance, and services have been seen [6].

2. Dashboard Types

As per the main characteristics and production techniques dashboards are classified as Stiff, Covered, and Foamed Dashboard. Stiff dashboard are used for the lower segment cars. Foamed dashboards have good aesthetics and touch where a soft layer is used between support and covering. Covered dashboard are intermediate solution between stiff and foamed, includes covering of an embossed material on stiff dashboard. Design wise the dashboards are classified as Monolithic and Decomposed. When the body is a single piece, it is called Monolithic and when more than one piece, it is called Decomposed. Today former is difficult to find due to availability of hybrid option having outer foamed and inner stiff part. The types of dashboards and their characteristics can be summarized as a table of figure 1 [5].

DASHBOARD TECHNOLOGICAL FAMILY									
		CHARACTERISTICS						NOTES	
TYPES	VARIANTS	Aesthetics	Touch	Freedom of style	Weight	Cost	Recycling	Problems	Segment
STIFF (injection moulding, embossed)	not painting	2	1	4	5	5	5	scratching	Lows
	painting	3.5	1	4	5	4.5	5		Lows Middle/Lows
	soft touch painting	3	2	4	5	4	5	embossing design less defined	Middle/Lows Lows
COVERED		3	3	2	3	3	5	only simple shapes	Middles
FOAMED	thermoformed covering	4	4.5	4	1	2	2/3	covering stretching	Highs Middle/Highs Middle/Lows
	slush or PU spray covering	5	5	5	1	1	2/3		Highs Middle/Highs Middle

(5 = maximum satisfaction 1 = minimum satisfaction)

Figure 1. Features of stiff, foamed and covered dashboards. [5]

3. Dashboard Making

Stiff dashboard is made by injection moulding. For Covered dashboard, the support is covered by Vacuum Thermo-forming process. In Foamed dashboard, first the support is put into the foaming mould then the support surface is joined with the mould by flaming or plasma treatment. Lastly a pre-heated material is injected and after necessary reaction time and foam formation, de-moulding can be done. Other technique is the slush molding process. At the end, finishing and trimming for air vents etc. is performed. Sometime infiltratration of support by waxing can required to remove excess foam [5].

4. Requirement of an Ideal dashboard material:

The dashboard material should be long lasting and must be resistant to wear and tear. The dashboard surface should remain scratch proof and glossy. It should be ecofriendly and hazards free. It must be sustainable to the vibration. It must be light weight for better fuel efficiency. The dashboard must be resistant to the thermal stress with proper strength, optimum stiffness, and adequate resilience. The surface of dashboard must have smooth hand-feel and pleasant visual effect along attractive pattern with proper design tolerance and no internal stress [7]. They should also be ultraviolet (UV) resistant. In addition, they must withstand higher thermal stresses with higher interior temperatures due larger and flatter windscreens [3]. The dashboard maker need to consider the exposure to sunlight and temperatures up to 130°C, also the weight and cost [8].

5. Materials used in dashboards

Initially dashboards were made of wood or metal which is now undesirable due safety, cost and appearance issues [9]. Plastics Used in Dashboards are listed in table of figure 2 [3].

Plastic	Reasons for Use	Advantages
Polypropylene (PP)	High heat resistance, superior to that of polyethylene (PE) Low cost	Helps maintain a cohesive material theme throughout the interior of the vehicle through use of the same base material in different parts like pillar covers, door panels, and IP. As these parts are in constant contact with each other, use of PP enables them to withstand temperature cycles.
Thermoplastic olefin (TPE-O or TPO)	Low gloss and density High stiffness and toughness Recyclability	Can be easily injection-molded to form various parts in IP—is used more as skin of PP foam in recent vehicles.
Acrylonitrile butadiene styrene (ABS)	High impact strength and toughness due to rubber from polybutadiene High resistance to chemical stress cracking due to acrylonitrile Easy processability	High degree of versatility as ratio of A, B, and S can be varied, enabling its use in pillars, IP, and door panels.
ABS/PC blend	Can be easily processed High heat deflection temperature Can produce many different surface types	Available in four different grades: general purpose, high flow, blow molding, and low gloss. Currently used in Jeep Cherokee and Audi A4 (B5) models.
Polyvinyl chloride (PVC) (often used with ABS)	High price/property ratio High flame retardancy	Can be blended with ABS to form sheets used for making skin of IP covers, though TPO can be an equally good alternative.
Styrene-maleic anhydride copolymers (SMA)	Excellent heat resistance compared to typical polystyrenes	Can be reinforced with glass fiber to enhance heat resistance, and with rubber to prevent plastic from becoming too brittle. Are used in IP in BMW 3 and 5 series and in Fiat Coupé, covered with polyurethane foam.
Polyurethanes	High impact strength under low temperatures High temperature stiffness, preventing heat sag	Polyether polyols have been used to form instrument and door panels via reaction injection molding (RIM). Energy-absorbing polyurethane foam, with semiclosed cell structure, has been added to pillars and door panels, enabling better protection of passengers in case of side impacts.

Figure 2. Plastics Used in Dashboards [3]

Stiff dashboards are made by Polypropylene (Polyolefin) with added talc in from of mineral along with elastomeric fillers or, compound of acrylic butadiene styrene (ABS) and Polycarbonate (PC). Polypropylene material is a cheaper and possess excellent characteristics like opacity, mouldability, and aphonicity. For Covered dashboard, initially laminated PVC (polyvinylchloride) was used, now the Thermoplastic Polyolefins (TPO) (polyolefin based material) is been used a lot. For covering, instead of using PVC and for support instead of using mixture of PC and ABS, Polypropylene (PP) is been used along with the TPO nowadays. For Foaming dashboard, polyol and isocyanate are mixed in mould at low-pressure in order to convert the mixture into a thermosetting (polyurethane) [5]. Glass fiber reinforcing in polypropylene are proven to have good resistance to impact and well balanced stiffness, and that also on relatively low cost of production of the dashboard. Polyolefin-based composites, hybrids, alloys, and blends are also are widely used in dashboards. Rubbers like ethylene-propylene and terpolymer are also used as constituents in the dashboard materials. Polyethylene with ultra-high molecular weight (UHMWPE) are also used frequently in automotive dashboards. Ethylene-vinyl acetate (EVA) copolymer with content of High vinyl acetate (HVA) are deliberately used in high noise

prone areas of an automobile dashboard. Engineering filled polypropylene (EFPP) is used to make one-piece dashboards [10].
The significant characteristics of mainly used plastic in the dashboard making can be seen from table of figure 3 [5].

Characteristics	Polyolefin base polypropylene					
	Copolymer base	Copolymer 25/30 % talc	Copolymer for dashboard	Copolymer glass reinforced	PC + ABS	ABS thermo-resistant glass reinforced
Specific weight (g/cm^3)	0.89÷0.92	1.11÷1.17	0.97÷1.00	1.10÷1.13	1.11÷1.14	1.15÷1.18
Coefficient of expansion (10^{-4}/°C)	110	50÷70	50÷60	40÷70	75÷85	40÷50
Moulding shrinkage %	1.5÷2.0	0.8÷1.1	0.8÷1.2	0.4÷0.8	0.6÷0.7	0.2÷0.3
Residual calx %	/	23÷32	10÷12	28÷32	/	15÷17
Modulus of elasticity (N/mm^2)						
a 23 °C	1000	2500	1650	4500	1900	5000
a -30°C	/	5300	/	7300	2500	5300
a 80°C	/	750	550	2700	1600	3600
Izod impact strength (KJ/m^2)						
a 23 °C	6	4÷7	13	12	45	0.65
a -30°C	/	2.5	3.5	3.5	3.5	0.4
Distorsion temperature under load (1.8 Mpa) °C	50	65	50	140	100	105
			mineral filler + anti scratching	only for non esthetic components		non aesthetical components
Opacity	+	+	+	/	-	/
Aphonity	+	+	+	+	-	-
Not scratching	+	-	+	/	++	/
Moulding	++	+	+	o	--	-
Low cost	++	++	+	++	--	-

Figure 3. Major characteristics of commonly used dashboard plastics [5].

Polyurethane is a versatile material most commonly used in the automobile industry for making dashboard. Polyurethane, due to excellent shock absorption capacity to ensure the safety of riders [11]. Polyester can be used as a thermoset resin with vinyl ester as natural fiber composites in dashboards [12]. Natural fibers such as kenaf, hemp, flax, jute, and sisal with thermoplastic and thermoset matrices can be used for making of an automobile dashboard due to less weight, cost, and CO2, and recyclability [13]. PVC blend with ABS acts similarly as PVC with plasticizer, but, the ABS doesn't evaporate so the dashboard becomes long lasting [14]. The Quality Function Deployment analysis results proposed the used of ABS, PC, styrene maleic anhydride, and PP for the car dashboard [15]. Polypropylene (PP), Acrylonitrile-butadiene-styrene (ABS), Styrene maleic anhydride (SMA), Polyphenylene Ether (PPE), and Polycarbonate (PC) are the materials commonly used in an automobile dashboard [16]. The rubber like properties of ABS along with resistant to weather and some chemicals, and good impact resistance favors its use in car dashboards [17]. ABS gives the dashboard a shiny, impervious surface and provides resilience even at low temperatures [18]. Natural-fiber composites of thermoplastics and thermosets have been used by car makers for dashboards [19]. Composite with the empty water sachet as a matrix and carbonized palm kernel shell (CPKS) particulate with iron fillings as reinforcing medium could be used to produce a car dashboard due to impact strength and low density [20]. BMW used 20% long glass fibre reinforced polypropylene (PP) to replace two-component dashboard in Borealis [21]. Nissan now uses plastic fiber from used bottles in sound insulation layers of dashboards [22]. For Transit range,

Ford are making an innovative single piece dashboard from Polypropylene Composite Materials and Alloys (PCMA) with reinforcement of glass and filling of talc. Delrin 517 acetal homopolymer from Du Pont has been used for long life of interior components like dashboard. It is ultraviolet resistant, reduces chalking of surface, and retains the gloss and colour [23]. Instead of polyethylene, different carrier materials and fillers such as boron nitride and graphite are used into the dashboard, which shown noticeable reduction in surface temperature [24]. BMW also using into dashboards, wood which is sustainably grown for its i3 electric car [25]. Composite of Polyscope styrene maleic anhydride (SMA) copolymer are also used in dashboards [26]. Composite material reinforced with glass fiber or carbon fiber / epoxy matrices can be used for dashboard making due to at least 60% weight reduction [27]. PA410 incorporated with carbon and glass-based materials and conventional epoxy resins have been checked for the use in dashboard making [28]. ABS, polycarbonate alloys, polycarbonates, polyester, and polypropylene are commonly used materials for dashboard applications [29]. Polypropylene (PP), due to its cost, is a common material for dashboard [30]. Natural fibre reinforced composites (NFRCs) has been proven as an alternative to glass and carbon fibre reinforced polymer composites in dashboards due to weight reduction and low cost [31]. Fiber composites of glass-fiber and carbon fiber performed better than an aluminum alloy in terms of environmental and economic performance in automobile interiors such as dashboard [32]. GRP composites of polyester matrix shown improved mechanical properties and with higher crush force efficiency of an automobile dashboard [33]. Renewable materials composites have been popularly used in dashboards [34]. Daimler Chrysler are producing dashboards using polypropylene and natural fibres [35]. The Super Olefin Polymer, TSOP-5 materials shown up to 18% weight reduction in dashboard [36]. Plastics like Thermoplastics viz. ABS, Polycarbonate (PC), Polymethyl methacrylate or acrylic (PMMA), Polypropylene (PP), Polyvinyl chloride (PVC), Acrylonitrile styrene acrylate (ASA), Styrene maleic anhydride (SMA), Thermoplastic Blends of ABS and Polycarbonate, Polyphenylene oxide (PPO), Thermoplastic Elastomers (TPE-U, and TPE-U) and Thermosets like Polyurethane integral-skin foams, Polyurethane semi rigid foam, and Polyurethane RIM compounds, are used in dashboards [37]. The materials like vinyls and alloys of vinyls, ABS, TPOs, TPEs (Thermoplastic elastomers), polyesters, and TPUs (Thermoplastic urethanes) are used in dashboard making. TPO has been used in injection mold dashboards [38]. As an option to fiber glass, composites with non-wood fibers like flax and hemp, as well as cellulosics have been used in automotive industries to produce dashboards [39].

6. Dashboards Materials testing:

Number of tests available to examine the functionality, reliability, and appearance of the dashboard material on the basis of dimensional stability at elevated temperatures for a certain time period. Noise tests, fatigue tests, and vertical settlements tests are carried after residual deformation. Further the tightening torques test, chemical agents and light resistance tests are performed. Lastly excessive load test, static and movement interferences checks, and aesthetics checks are conducted. By the impactor collision tests, the robustness and resilience of dashboard are determined to check the fragile breakages. Stiffness test is done to assure stiffness on the basis of dynamic stiffness performance. Ageing test also performed by various humidity conditions, carried out in customized climatic chambers [5]. Automotive dashboard are evaluated by FEM modelling with the help of specified structural codes under EEC test 781632. It simulates the impact of head of passenger with the dashboard, which can possibly occur during a front crash, where the variations in stresses are determined [40]. The general mechanical properties of automotive dashboard namely density, Young's modulus, and tensile strength should be 980-1100 Kg/m^3, 2-6 GPa, and 20-30 MPa respectively [41].

7. Dashboard Material Concerns

Painting of Polypropylene is difficult due to the absence of total adhesion, for which the surface must be coated with primers or plasma-treated. The surface obtained by blending ABS with PC can be painted only by promoter medium for adhesion. Soft touch can be get by increasing the thickness, but it can adversely affect the embossing due to crease and troughs. Materials with fare structural performance in terms of high temperature resistance and higher flexural modulus can be utilized. Material like thermoplastic polyurethane (TPU) has not been used much due to economic reasons. The recycleability of the foamed dashboards is complex as the foam which is made up of a thermosetting material, is not recyclable. If the frequency level of material is increased above 40 Hz, then the resonance caused can result in creaking [5]. The mechanical recycling costs of dashboard materials like polyurethanes are significantly higher compared to the thermal recovery costs [42]. In the dashboard injection moulding process, problems like weld line distortion, buckling deformation, uneven texture, filamentary silver, and color irregularity arises [43]. Shape errors occurs in ABS due to heat shrinkage produced with the additive manufacturing [44]. Defects like squeaks and rattles arises in frictionally incompatible materials like ABS, PC, GFPP and unfilled PP on contacting each other when used in dashboards. These materials also display sharp frictional drops due to adhesive bonds rupture [45]. A major reasons for passenger injuries in vehicles is hard contact with the dashboard. Stiffness control can be a way for reducing such injuries. This can be done by providing PP stiffener ribs between the outer and inner cover [46]. A broad study is always essential to designate and pinpoint the factors influencing the impact performance of dashboard like crucial components spacing, foams, ribbing, impact directions, crush initiators, and stiffness variations etc. It will be helpful for optimizing the dashboard structure [47]. In injection molded dashboard making, defects of weld line, buckling, texture, and color are seen commonly [48]. Problem of using of natural fibers in polymeric composites is the low compatibility between the matrix and the fibers. Nanoparticles can be used to overcome this. It removes surface contaminants and increases surface roughness which improves the surface properties of fiber and provides better bonding [49]. Microbial and bacterial infections are also one of the concerns related to dashboard material. Researches are been carried out to identify suitable antimicrobial agent [50]. The natural fiber are sensitive to moisture and temperature, which can restrict their use in application of automobile dashboard [51].

8. Conclusion

A dashboard in one of the most important part of an automobile aesthetically and functionally. Various types of materials are used for the making of a dashboard. Mostly plastics are used due to its light weight and ease in the mouldability. Polypropylene is the most commonly used material due to characteristics like opacity and aphonicity. Polyurethane is also popular material as it is good in shock absorption. ABS is used for its rubber like properties, resistant to weather and chemicals, and impact. It also gives a shiny appearance to the surface of dashboard. Natural-fiber composites are sometime used for making dashboard and so on. Along with desirable properties these materials also have some limitation like surface treatment, economy, and un-recyclability. Issues related to manufacturing like weld line distortion, buckling deformation, uneven texture, filamentary silver, and color irregularity arises also limits their versatility. Moisture and temperature sensitivity is major restricting parameter of natural fibers used in dashboards. Microbial and bacterial infections problem must also be taken into notice. On the basis of various limitations, there is always scope for researchers to identify and develop a new material for dashboard.

9. References

[1] Anumita Roychowdhury, Vivek Chattopadhaya, Gaurav Dubey, Swati Singh Sambyal, Avikal Somvanshi, Shambhavi Shukla and Tanushree Ganguly 2019, 5 June: At the Crossroads, Centre for Science and Environment, New Delhi.
[2] Mishra, S. and Malhotra, G. (2019) Is India Ready for e-Mobility? An Exploratory Study to Understand e-Vehicles Purchase Intention. Theoretical Economics Letters, 9, 376-391. https://doi.org/10.4236/tel.2019.92027
[3] Pradeep, S. A., Iyer, R. K., Kazan, H., & Pilla, S. (2017). Automotive Applications of Plastics: Past, Present, and Future. Applied Plastics Engineering Handbook, 651–673. doi:10.1016/b978-0-323-39040-8.00031-6
[4] https://www.piedmontplastics.com/applications/cluster-panels-dashboards (cit. on 21.6.2019).
[5] Morello, L., Rossini, L. R., Pia, G., & Tonoli, A. (2011). Body Interiors. The Automotive Body,

439–661.
doi:10.1007/978-94-007-0513-5_6
[6] http://164.100.133.129:81/econtent/Uploads/07%20and%2008_Instrument%20Panel%20and%20Door%20Panel%20Design%20(ver.1).pdf (cit. on 18.6.2019).
[7] Verma, D., & Senal, I. (2019). Natural fiber-reinforced polymer composites. Biomass, Biopolymer-Based Materials, and Bioenergy, 103–122.
doi:10.1016/b978-0-08-102426-3.00006-0
[8] https://www.plasticstoday.com/content/nearly-all-plasticdashboard/2786217911133 (cit. on 21.6.2019).
[9] Waterman, N. A., Neale, M., & Associates. (1981). Plastics in automobiles at SITEV '81. Materials & Design, 2(5), 242–249.
doi:10.1016/0261-3069(81)90067-4
[10] SADIKU, E. (2009). Automotive components composed of polyolefins. Polyolefin Fibres, 81–132.
doi:10.1533/9781845695552.1.81
[11] Driving With Polyurethane, a Research Article by SISOPA, the European Isocyanate Producers Association. www.isopa.org.
[12] Canan Saricam and Nazan Okur, Polyester Usage for Automotive Applications, Chapter from the book 'Polyester - Production, Characterization and Innovative Applications' http://www.intechopen.com/books/polyester-productioncharacterization-and-innovative-applications
[13] Holbery, J., & Houston, D. (2006). Natural-fiber-reinforced polymer composites in automotive applications. JOM, 58(11), 80–86.
doi:10.1007/s11837-006-0234-2
[14] https://pslc.ws/macrog/work/dash.htm (cit. on 18.6.2019).
[15] Padagannavar P (2016) Automotive Product Design and Development of Car Dashboard Using Quality Function Deployment. Adv Automob Eng 5: 136.
doi:10.4172/2167-7670.1000136
[16] http://www.plasticsconverters.eu/organisation/division/automotive (cit. on 18.6.2019).
[17] Katarína SZETEIOVÁ, 'Automotive Materials Plastics In Automotive Markets Today', Institute of Production Technologies, Machine Technologies and Materials, Faculty of Material Science and Technology in Trnava, Slovak University of Technology Bratislava
[18] https://www.quora.com/Why-are-plastics-used-in-automotive-components (cit. on 21.6.2019).
[19] B. Chehroudi, Composite Materials and Their Uses in Cars Part II: Applications, https://www.researchgate.net/publication/277303439.
[20] I.A. Samotu, M. Dauda, F.O. Anafi, and D.O. Obada (2015). Suitability of Recycled Polyethylene/Palm Kernel Shell-Iron Filings Composite for Automobile Application. Tribology in Industry, Vol. 37, No. 2 (2015) 142-153.
[21] https://www.materialstoday.com/composite-applications/news/bmw-selects-glass-reinforced-polypropylene-for/ (cit. on 21.6.2019).
[22] https://www.plasticsmakeitpossible.com/plastics-recycling/what-happens-to-recycled-plastics/use-of-recycled-plastics-in-cars-is-shifting-into-overdrive/(cit. on 21.6.2019).
[23] Reinforced polypropylene dashboard. (1988). Additives for Polymers, 18(7), 9.
doi:10.1016/0306-3747(88)90443-5
[24] https://www.compositesworld.com/blog/post/new-composite-material-lowers-temp-of-ev-dashboard (cit. on 21.6.2019).
[25] https://www.edmunds.com/car-technology/renewable-and-recycled-materials-help-make-cars-green.html (cit. on 21.6.2019).
[26] http://www.intertek.com/automotive/dashboard-quality-control-case-study/ (cit. on 21.6.2019).
[27] Sergiu SOLCAN, Calin NEAMTU, Paul BERE, Rares GHINEA, Raul ROZSOS, and Attila PAPP, Using Composite Materials For Dashboard Design of An Electric Car, Technical

University of Cluj-Napoca, Series: Applied Mathematics, Mechanics, and Engineering Vol. 61, Issue Special ,pp-215-220, September, 2018.

[28] M. Delogua, L. Zanchia, C.A. Dattiloa, and M. Pierinia, Innovative composites and hybrid materials for electric vehicles lightweight design in a sustainability perspective, Research Gate Article, September 2017.
doi: 10.1016/j.mtcomm.2017.09.012

[29] https://www.piedmontplastics.com/applications/cluster-panels-dashboards (cit. on 21.6.2019).

[30] Mohan, M., Ansari, M. N. M., & Shanks, R. A. (2016). Review on the Effects of Process Parameters on Strength, Shrinkage, and Warpage of Injection Molding Plastic Component. Polymer-Plastics Technology and Engineering, 56(1), 1-12. doi:10.1080/03602559.2015.1132466

[31] Puglia, D., Biagiotti, J., & Kenny, J. M. (2005). A Review on Natural Fibre-Based Composites—Part II. Journal of Natural Fibers, 1(3), 23–65.
doi:10.1300/j395v01n03_03

[32] P. Pec¸as, I. Ribeiro, H. Carvalho, A. Silva, H.M. Salman, and E. Henriques, Ramie and jute as natural fibers in a composite part-a life cycle engineering comparison with an aluminum part, Green Composites for Automotive Applications.
https://doi.org/10.1016/B978-0-08-102177-4.00011-2

[33] Chukwutoo Christopher Ihueze and Alfred Nwabunwanne. Enetanya, Energy Absorption and Strength Evaluation for Compressed Glass Fibre Reinforced Polyester (GRP) for Automobile Components Design in Crash Prevention Scheme, Journal of Minerals & Materials Characterization & Engineering, Vol. 11, No.1, pp.85-105 2012.
doi: 10.4236/jmmce.2012.111007

[34] Koronis, G., Silva, A., & Fontul, M. (2013). Green composites: A review of adequate materials for automotive applications. Composites Part B: Engineering, 44(1), 120–127. doi:10.1016/j.compositesb.2012.07.00

[35] Njuguna, James & Peña, I & Zhu, Huijun & Rocks, Sophie & Blázquez, María & A Desai, S. (2009). Opportunities and Environmental Health Challenges Facing Integration of Polymer NanoComposites Technologies for Automotive Applications. International Journal of Applied Polymers and Technologies. 1. 113.

[36] Matsuda, M., Nomura, T., and Iwai, H., "Material Consolidation for Automotive Interior and Exterior Parts through Development of a High Performance Material," SAE Technical Paper 982410, 1998.
https://doi.org/10.4271/982410.

[37] Mann, D., Van den Bos, J. C., & Way, A. (1999). Plastics and Reinforcements used in Automobile Construction. Automotive Plastics and Composites, 25–60.
doi:10.1016/b978-185617349-0/50005-2

[38] Hersman, M. J., Farrington, S., & Horsten, P. (1997). Material Options for Automotive Interior Trim. SAE Technical Paper Series.
doi:10.4271/970446

[39] HUDA, M. S., DRZAL, L. T., RAY, D., MOHANTY, A. K., & MISHRA, M. (2008). Natural-fiber composites in the automotive sector. Properties and Performance of Natural-Fibre Composites, 221–268.
doi:10.1533/9781845694593.2.221

[40] Molina, G. and Smeriglio, P., "Integrated CAE Simulation for Car Dashboard Design," SAE Technical Paper 940891, 1994.
https://doi.org/10.4271/940891

[41] S. M. Sapuan, N. K. Mun, A. Hambali, H. Y. Lok, A. M. Fairuz and M. R. Ishak, Prototype expert system for material selection of polymeric composite automotive dashboard, International Journal of the Physical Sciences Vol. 6(25), pp. 5988-5995, 23 October, 2011.
doi: 10.5897/IJPS11.939

[42] Driving With Polyurethane, a Research Article by SISOPA, the European Isocyanate Producers Association. www.isopa.org.
[43] Ling Suna, and XiaoMing Liub, Control Analysis of Production and Apparent Quality of Automobile Large Plastic Parts, International Workshop on Automobile, Power and Energy Engineering, Procedia Engineering 16 (2011) 438 – 443. doi: 10.1016/j.proeng.2011.08.1108
[44] Choi, Y.-H., Kim, C.-M., Jeong, H.-S. and Youn, J.-H. (2016) Influence of Bed Temperature on Heat Shrink- age Shape Error in FDM Additive Manufacturing of the ABS-Engineering Plastic. World Journal of Engineering and Techno- logy, 4, 186-192. http://dx.doi.org/10.4236/wjet.2016.43D022
[45] Eiss, N., Lee, E., & Trapp, M. (1997). Frictional Behavior of Automotive Interior Polymeric Material Pairs. SAE Technical Paper Series. doi:10.4271/972056
[46] Gokhale, A., Saravate, V., and Kshirsagar, S., "Dashboard Stiffness Control for Reducing Knee Injury in Frontal Crashes," SAE Technical Paper 2009-26-0006, 2009. https://doi.org/10.4271/2009-26-0006
[47] Malladi, A., Saifuddin, & Gadekar, G. (2011). CAE Based Dashboard Development for Head form Impact Test as per IS15223 and ECE R21 Requirements. SAE Technical Paper Series. doi:10.4271/2011-26-0014
[48] Sun, L., & Liu, X. (2011). Control Analysis of Production and Apparent Quality of Automobile Large Plastic Parts. Procedia Engineering, 16, 438–443. doi:10.1016/j.proeng.2011.08.1108
[49] Ferreira, D. P., Cruz, J., & Fangueiro, R. (2019). Surface modification of natural fibers in polymer composites. Green Composites for Automotive Applications, 3–41. doi:10.1016/b978-0-08-102177-4.00001-x
[50] J. Mathew, J. Joy, S.C. George, Potential applications of nanotechnology in transportation: A review, Journal of King Saud University - Science (2018). https://doi.org/10.1016/j.jksus.2018.03.015
[51] Witayakran, S., Smitthipong, W., Wangpradid, R., Chollakup, R., & Clouston, P. L. (2017). Natural Fiber Composites: Review of Recent Automotive Trends. Reference Module in Materials Science and Materials Engineering. doi:10.1016/b978-0-12-803581-8.04180-1.

IOP Conf. Series: Materials Science and Engineering **810** (2020) 012034 doi:10.1088/1757-899X/810/1/012034

Identification & Prioritization of Agile Manufacturing Enablers for Small and Medium Scale Industries

Suchita P Lokhande

Dept. of Mechanical Engineering, Lokmanya Tilak College of Engineering, India.
suchita.lok@gmail.com

Avinash D Sarode

Dept. of Mechanical Engineering, Lokmanya Tilak College of Engineering, India.
avinashsarode@gmail.com

Abstract. In Indian economy the small and medium scale industries are playing a significant role in manufacturing sector, global competition and dynamic demand of customer poses a big challenge in front of manufacturing industry. To mitigate this challenge the concept of agile manufacturing stood as a big relief. Agile manufacturing focuses on customer demand by giving quick response to the changes in product. Enablers are essential methodology which are significant to achieve agility. This research is divided in two parts: first part consists of Identification of Agile Enablers through extensive literature survey; second part focuses on prioritizing the identified enablers by using analytical hierarchical process (AHP). This research aims to find the solution for the problem statement given by TTC MIDC Industry Association, formerly known as Small Scale Entrepreneurs Association, Navi Mumbai, to identify the most essential enablers for implementation of Agile Manufacturing in Indian Small and Medium Scale Manufacturing industry.

1. Introduction

Rapid industrial growth gives birth to the new innovative methods and tools. The changing interest of customer is becoming very complex and dynamic with market competition. Every manufacturing industries are struggling to fulfil customer demand and try to withstand in today's competitive world. Over the time the manufacturing evolved from four major manufacturing paradigms such as Craft production, Mass production, Lean manufacturing and Agile manufacturing. The world began to observe the drastic competition in the manufacturing sector during the 1980's in order to face this, world witnessed the birth of agile manufacturing. Agile manufacturing paradigm focuses on meeting the needs of customers while manufacturing high standards of quality and controlling the overall costs involved in the production of a particular product. Building a robust framework need to identify the enablers that can help in accelerated manufacturing process Enablers act as a value-added work objects which allow estimating, obtaining visibility, getting feedback and exhibiting results. Enablers are categorized into three major type Exploration, Architecture and Infrastructure. Exploration enablers supports research prototyping and other activities needed to develop customer needs to evaluate alternatives Architecture are to build a smooth and fast development in an organization. Infrastructure enablers supports the business development and quality checking by helping in high quality testing and thus aids to faster development. In order to get the concrete solution, identified enablers need to be analyzed through various tools of statistical analysis and modelling. A framework is based on the concept of bridging the system gaps and make it dynamic and responsive, which can contribute towards the accelerated an agile manufacturing processes. Analytical Hierarchical Process is the multicriteria decision making process for a complex problem. This method has been found to be most effective and practical that can solve complex and unstructured decision. In this paper AHP is used to prioritize agile manufacturing enablers based on questionnaires.

2. Literature Review

Small-scale manufacturing industry will play a crucial role in the Indian economy in the future. It has also been observed that many units in the SME sector are plagued with technological stagnation and their quality, efficiency and productivity have declined over the years except for a few progressive units that have incrementally improved their performance. Manufacturing sector need the reorganized approach to face the challenges in the global market of 21st century in which the customer is in the center and must be satisfied in all manners. The need of customer is now highly specific and rapidly changing. Therefore, in this competitive environment the new term emerged as Agile Manufacturing. The concept of AM was originally introduced in the report entitled ''21st Century Manufacturing Enterprise Strategy'' and published by the Iacocca Institute of Lehigh University (Goldman and Nagel.,1991) as an option for managing firms in a dynamic world. (Ramesh and Devadasan.,2007) studied the literature dealing with AM criteria and derived the meaning and definitions of AM. Further, the criteria enunciated in certain papers were identified, these findings were used to design a 20 criteria agile model .(Vazquez-Bustelo and Avella.,2006) presented an initial approach to AM based on case studies on four factories in Spain. An AM conceptual model has been drawn up and several hypotheses inferred based on conclusion of case studies. Yusuf et.al (1999) identified the drivers of agility and discusses the portfolio of competitive advantages that have emerged over time because of the changing requirements of manufacturing. To further understanding of agility, authors reviewed the meaning of agility from different perspectives and suggests a comprehensive definition which can be adopted as a working definition by practitioners. Four underlining concepts of agility has emerged from the working definition and presents a representation of these concepts and their interactions. Finally, authors highlight some of the key enablers of agility and identified potential future research directions.

Gunasekaran et. al. (1998) presented a conceptual framework for the development of an agile manufacturing system and future research directions are presented in this paper. Proposed framework by author considers the customization and system integration with the help of business process redesign, legal issues, con-current engineering, computer-integrated manufacturing, cost management, total quality management and information technology on the other hand. Sanjay Kumar et. al (2013) utilized literature review approach to identify important enablers to implement lean six sigma concept. Twenty-six enablers have been identified from the literature and subsequent discussions with experts. Through this work authors contribute to identify the enablers to implement lean six sigma concept.

Aravindraj and Vinodh (2014) developed a 40-criteria agility assessment model. Besides computing agility level, the gaps across agile criteria have been identified and actions for agility improvement were subjected to implementation in the case organization. Routroy and Arjun Shankar (2015) proposed a methodology for measuring supply chain agility by capturing multiple experts' judgement regarding the Agile Manufacturing Enablers (AMEs). AMEs are defined as the parameters that have the capability to enable or enhance the present level of agility in the manufacturing system. Sarode et.al (2008) provided an extensive literature review for identification of performance measures, which in turn forms a basis for establishing a framework for performance measurements in supply chains. Ching et.al (2006) developed the absolute agility index, a unique and unprecedented attempt in agility measurement, using fuzzy logic to address the ambiguity in agility evaluation and presented the approach and a framework of a fuzzy agility evaluation. Shin, H. et.al. (2015) developed theoretical connection between strategic agility of Korean SMEs and its underlined dimensions such as technology, capability, collaborative innovation, organizational learning and internal alignment, and then proposed the conceptual model for investigation nomological network of influence among strategic agility, operational performance and firm performance. Majority authors focused on the strategic agility however it cannot influence financial performance directly or indirectly through operational responsiveness.

Potdar et.al. (2017) proposed a methodology using Interpretive Structural Modeling (ISM) integrated with Fuzzy Matriced Impacts Croises Multiplication Appliqueeaun Classment (FMICMAC) algorithm for systematic analysis of impediments of AM. To demonstrate its practicality, a case study was conducted in an Indian automobile manufacturing company. Jinturkar et.al (2014) provided a systematic approach towards the application of FAHP to supplier selection problem. Authors used FAHP to find the importance degree of each criterion as the measurable indices of the supplier. Extensive analysis of the results shows that selection of an appropriate supplier would result in improving effectiveness of supply chain.

Tseng et al. (2011) suggests a new agility development method for dealing with the interface and alignment issues among the agility drivers, capabilities and providers using the QFD relationship matrix and fuzzy logic. Eleonora Bottani et.al. (2010) has given empirical investigation performed on sample of 190 both small scale and medium scale companies, the result of analysis provided detailed investigation of the agile paradigm and outcome suggests new taxonomy for agile attributes and enablers. However, the whole findings and analysis of data was based on the email bound questionnaire, keeps the scope for the vague data also the hypothesis established is relevant to the European companies.

Ajay Gurudev et.al. (2016) proposes a set of critical success factors (CSF) for evaluating AM. He has used AHP to prioritize these factors for high performance. Author focused on only one OEM industry, which may limit the usage of the CSF for all. There is a need to carry out proposed technique in several industries for improvement of practical validity of the models developed. Soltan and Sherif (2015) presented a hierarchic framework which can be used to measure the leanness, agility, legality and overall performance of an enterprise. This enables using the multi-criteria decision-making methods especially AHP and ANP. However, his work is concentrated on decision making and lacking the capability address the issues of AM, which need the flexible and scalable manufacturing capability of the manufacturing unit. Potdar et.al.(2017)proposed a generic benchmarking framework for enhancing agility along with step by step implementation process. The proposed framework provides a systematic direction for measuring and enhancing agility on a continuous basis, said framework is conceptually developed and not empirically validated. Therefore, requires the validation in general. There is a need to investigate and study the implementation issues of this proposed benchmarking model in different manufacturing sectors.

Potdar and Routroy (2017) proposed a methodology by combining fuzzy analytic hierarchy process (FAHP) and performance value analysis (PVA) to analyze AM performance along specified time line. This proposed approach is applied to an apparel supply chain considering three-time horizons. Methodology should be tested by applying to various manufacturing units. Also, only two expert groups judgements are obtained. Therefore, there is a possibility of the judgements suffering from respondent bias. This methodology could also be modified by employing other multi-criteria decision-making techniques like fuzzy TOPSIS, etc. in the place of FAHP in order to limit the unreliability of results produced by FAHP which occurs when the number of attributes being compared becomes large. Yusuf et.al. (1999) identified the drivers of agility and discusses the portfolio of competitive advantages that have surfaced over time as a result of the changing requirements of manufacturing. To further understanding of agility, authors reviews the meaning of agility from different perspectives and suggests a comprehensive definition which can be adopted as a working definition by practitioners.

Campanelli and Fernando (2015) has conducted Systematic Literature Review (SLR) on studies published from 2002 to 2014. These studies have been identified as case studies regarding the empirical research, as solution proposals regarding the research type and as evaluation studies regarding the research validation type. However, the major problem in SLR is to handle the bias in proper way and validate the study. Sreenivasa et.al. (2012) proposed 30 criteria AM assessment tool. Using this tool, the total agility level in a pneumatic enabling products manufacturing company was assessed. After a comprehensive analysis, the drag and failure factors that prevent this company from becoming agile were identified. Allotment of marks to each 30 AM criteria need extensive research in the form of case studies and findings need to analyzed with the help of various statistical methods and tools. Pavan kumar et. al. (2016) identified and analyzed the Agile Manufacturing Barriers (AMBs) for establishing a cause and effect relationship among them. A methodology is proposed using fuzzy Decision-Making Trail and Evaluation Laboratory (DEMATEL) to capture multiple expert's qualitative judgments for mitigating the impact of the AMBs. The proposed methodology was applied to an Indian automobile manufacturing company for validation. Results are specific to Indian Automobile industry, which cannot be generalized for every manufacturing unit.

Author Soni and Kodali (2012) addressed the issue of lack of standard constructs in frameworks of lean, agile and leagile supply chain. This objective is achieved by evaluating reliability and validity of lean, agile and leagile supply chain constructs in Indian manufacturing industry. Principle component analysis is performed on these constructs to find out the pillars of each type of supply chain followed by evaluating reliability and validity of these pillars to establish the underlying constructs. Finally using the results of the study, a framework for lean, agile and leagile supply chain is proposed. As the research construct is based on the survey which need to extend to all manufacturing and service domain.

3. Problem Description

Small and Medium Scale Enterprises (SMSE) manufacturing units are struggling to cope with the growing global competition; they are influenced by intense foreign competition, rapid technological change, shorter product life-cycles and customers who are seeking for customized products and services at a greater speed. There is a need to have simple to implement and usable framework which can adopt the change and fulfil the increased demand of current market. The challenge of deploying the best practices in organization to create competitive advantage becomes more crucial as:

- The marketplace is competitive and the rate of innovation is rising, so that Agile Manufacturing must evolve and be assimilated at an ever-faster rate.
- Competitive pressures forcing industry to reduce the size of the workforce which may put impact on quality production and customer satisfaction.
- Any change in strategic direction may result in changing manufacturing process and loss of knowledge in a specific area. Employees have less time for this.
- Wastage of time and money on satisfying the customer requirements in the absence of proper agile methodology.

4. Identification of Enablers

In the process of identification of enablers number of research work is reviewed. From the extensive review of 71 research papers following enablers are identified and summarized as follows. Summary of enablers was carried out author wise on the basis of 12 set of enablers as shown in Table no.1.It shows the involvement of authors research work with reference to the enablers.

IOP Conf. Series: Materials Science and Engineering **810** (2020) 012034 doi:10.1088/1757-899X/810/1/012034

Table 1: Author-wise Literature Review of Enablers

Sr No	Author	Enablers											
		Management Responsibility (MGR)	**Manufacturing Management (MFM)**	**Workforce (WOF)**	**Technology (TECH)**	**Manufacturing strategy (MFS)**	**Knowledge Management (KNM)**	**Information Technology (INT)**	**Supply Chain Management (SCM)**	**Marketing Management (MAM)**	**Social Factors (SOF)**	**Quality Assurance& Quality Control (QAQC)**	**Customer relationship Management (CRM)**
1	Jain, S et.al. (1995)		✓	✓		✓		✓					✓
2	Cho, H.et al.(1996)		✓		✓			✓		✓			
3	Gunasekaran and Angappa (1998)		✓	✓	✓	✓		✓					
4	Y.Y.Yusuf et al. (1999)			✓	✓					✓	✓	✓	✓
5	Sharifi H. and Zhang, Z. (1999)		✓		✓					✓	✓		✓
6	Sharp et al. (1999)			✓	✓	✓		✓		✓			
7	Gunasekaran, A., (1999)			✓	✓	✓			✓				
8	Zhang, Z. and Sharifi, H. (2000)	✓	✓		✓	✓			✓	✓	✓		✓
9	Sharifi, H. and Zhang, Z (2001)		✓		✓					✓	✓		✓
10	Hormozi, A.M., (2001)			✓		✓		✓			✓		
11	Gunasekaran, A. Yusuf, Y.Y,(2002)		✓	✓			✓		✓				
12	Conboy, K. & Fitzgerald,B.,(2004)		✓		✓						✓		✓
13	Deshayes, L et al. (2005)		✓			✓	✓					✓	
14	Lin et al. (2006)			✓	✓	✓				✓	✓	✓	
15	Vázquez-Bustelo, D. and Avella (2006)			✓	✓		✓	✓	✓				
16	Lin et al. (2006)			✓	✓							✓	
17	Ye-zhuang (2006)	✓	✓	✓				✓					
18	E. Ferna and D. Va (2007)			✓	✓		✓		✓				

19.	Sherehiy B.et al (2007)				✓	✓							✓
20	Vázquez-Bustelo, D (2007)			✓	✓		✓		✓				
21	Zhang, Z. and Sharifi, H (2007)	✓		✓	✓	✓			✓	✓	✓		✓
22	Ramesh, G., S. R. Devadasan, (2007)	✓	✓	✓		✓		✓				✓	✓
23	Chandna,R., (2008)		✓	✓				✓		✓			
24	Erande, A.S. and Verma, A.K. (2008)	✓	✓	✓		✓		✓				✓	✓
25	Hasan, M.A., Shankar, R., Sarkis, J (2009)		✓	✓	✓			✓	✓				
26.	Ganguly,A. et al. (2009)				✓	✓				✓	✓		
27	Loforte Ribeiro, F. and Timóteo Fernandes, M., (2010)	✓		✓	✓				✓				
28	Abbas Toloie Eshaghy(2010)	✓	✓	✓	✓			✓	✓			✓	
29	Bottani, Eleonora, (2010)	✓			✓					✓			✓
30	Ismail, H.S et al. (2011)	✓	✓	✓								✓	
31	Y. Tseng and C. Lin, (2011)		✓		✓			✓	✓	✓	✓		✓
32	Yauch, C.A., (2011)		✓	✓			✓			✓			
33	Mashayekhi, A.N.et al.(2011)	✓	✓	✓	✓	✓		✓	✓	✓	✓		
34	Vinodh, S. and Kuttalingam, D., (2011)		✓		✓			✓					
35	Mishra, S et al. (2012)		✓		✓						✓		✓
36	Flumerfelt (2012)	✓	✓	✓	✓							✓	
37	Sreenivasa, C.G et al (2012)	✓	✓	✓	✓	✓							
38	Mohd.Asif Hasan Joseph Sarkis		✓	✓	✓	✓			✓				
39	Aravind Raj, S.et al.(2013)	✓	✓	✓	✓	✓							
40	Chang, A.Y., et al (2013)		✓		✓			✓	✓				
41	Mishra, S.(2013)	✓	✓					✓					
42	R. Dubey and A. Gunasekaran,(2014)	✓	✓	✓	✓				✓				✓
43	Avazpour, R.et al. (2014)		✓	✓	✓	✓							
44	Zhang, M et al(2013)	✓	✓		✓				✓	✓	✓		✓
45	Aravindraj, Sakthivel, and S. Vinodh, (2014)	✓	✓	✓	✓	✓							
46	Sherehiy, B(2014)	✓	✓	✓		✓							
47	Faisal, M. and Al-Esmael, B.A(2014)	✓	✓	✓									✓
48	Mishra, S (2014)	✓	✓					✓					
49	Holzner, P.,et al. (2015)		✓	✓									
50	Vinodh, S. and Aravindraj, S., (2015)	✓	✓	✓	✓	✓							

51	Jayakrishna, K et al. (2016)	✓									✓		
52	Swagatika Mishra	✓	✓					✓					
53	Stettina, C.J. and Hörz, J., (2015)	✓		✓		✓							
54	Thilak, V.M.M et al. (2015)				✓								
55	S. Routroy et al. (2015)	✓	✓	✓	✓	✓			✓				✓
56	Soltan, H. and Mostafa, S. (2015)					✓		✓	✓	✓	✓		
57	Routroy, Srikanta, and Arjun Shankar (2015)	✓	✓	✓				✓	✓	✓			✓
58	Shrikant Routroy	✓	✓	✓	✓		✓		✓				✓
59	Rahul Sindhwani			✓	✓	✓		✓					
60	Dev et al. (2016)	✓	✓	✓	✓	✓		✓				✓	✓
61	Leite, M. and Braz, V., (2016)	✓	✓	✓	✓		✓		✓				✓
62	Sindhwani, R. and Malhotra, V. (2016)	✓	✓	✓	✓			✓				✓	
63	Metaxas, I.N., (2016)	✓	✓	✓		✓	✓	✓	✓	✓	✓		✓
64	Pavan Kumar Potdar et al. (2016)		✓	✓		✓		✓	✓		✓		✓
65	Maria. do Rosário Cabrita, M.et al, (2016)		✓		✓	✓		✓			✓		✓
66	Dev, C et al. (2016)	✓		✓	✓	✓							
67	Pavan Kumar Potdar et.al, (2016)												
68	Bölükbaş, U. and Güneri, A.F., (2017)		✓				✓	✓		✓			
69	V. K. Mittal et al. (2017)	✓		✓	✓				✓				✓
70	Potdar, Pavan Kumar et al. (2017)	✓	✓		✓				✓	✓			✓
71	Potdar, Pavan Kumar et al. (2017)			✓	✓	✓				✓			✓

5. Research Methodology

Based on the problem description and literature survey, following steps were identified for research work. Proposed methodology consists of five steps as shown in Figure 1. Proposed research methodology, which includes Extensive literature survey on Agile manufacturing followed by the implementation of AM enablers by sorting of research work on the basis of enablers and categorization. First step, include the literature survey of the latest research in the Agile Manufacturing. The next step is identification of enablers from extensive literature survey from the research work on the enablers. In the third step, questionnaire is designed for pilot study to identify the sample size. A detailed questionnaire is designed for ranking the enablers, which is to be filled by the various manufacturing units in Maharashtra and Silvasa, Diu Daman U/T. Responses of the various stakeholders are collected through online form and personal interview. Use of statistical tools is proposed for the evaluation of raw information to assign the raking to enablers. Final step in the proposed research methodology is formulation of framework model to be implemented with expected outcome and future research direction.

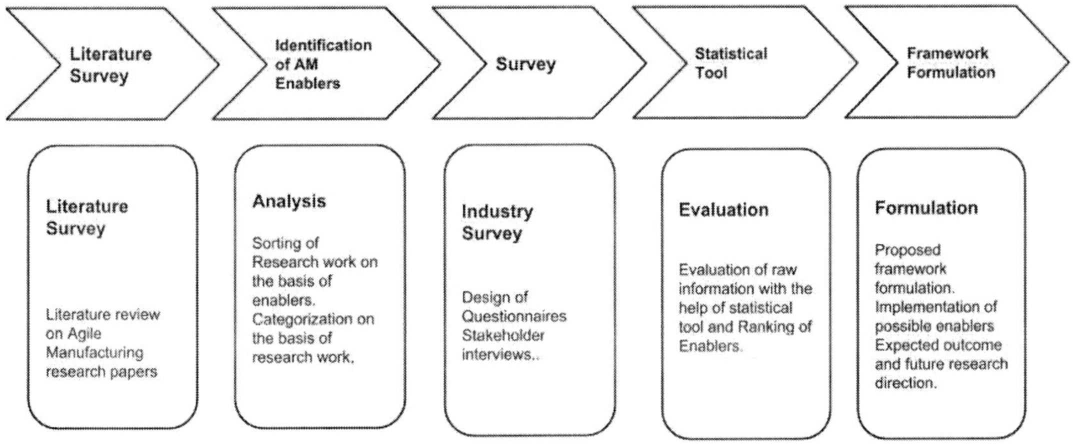

Figure 1. Proposed Research Methodology

Enablers were identified after extensive literature survey, based on number of research papers. A pilot study was designed to gather information prior to a larger study to improve the final result of research. Pilot study result was processed to get the sample size to be targeted. New questionnaire was designed to capture the input from various manufacturing units. Questionnaire were distributed in physical as well as online format (https://forms.gle/tKBEwyynk7NDd5Vf9) to capture the input from the manufacturing units outside Mumbai and Maharashtra state..

Data received through online form and physical form were analysed and processes to prioritize the agile manufacturing enablers using AHP, pairwise comparison matrix of 12 enablers are shown in table 2 and ranking of enablers are shown in table 3.

Sampling Analysis

A pilot study was conducted on 15 respondents from the personalities working in various fields. A summary of their responses and calculation thereof is attached in Appendix, sample size is calculated by following formula.

$$n = \frac{z^2 \sigma^2}{e^2}$$

Where,

n = size of the sample,

z = the value of the standard variate at a given confidence level,

σ = standard deviation of observed values,

e = acceptable error.

Design of Questionnaire

Questionnaires are designed to facilitate data collection. The two phases of data collection are discussed as Phase I: Pilot survey gives the Overview of Enablers and Phase II gives the essential Enablers. The questionnaire includes initial information like type of company, production type, industry sector, Employee strength, respondent work experience, employee size and annual turnover are summarized. Various industries from Maharashtra, Silvasa, Diu -Daman are considered for pilot survey.

IOP Conf. Series: Materials Science and Engineering **810** (2020) 012034 doi:10.1088/1757-899X/810/1/012034

In AHP pair-wise comparisons were conducted at levels, first at specific enablers and then essential level of enablers as shown in figure 2. An improved nine-point scale is used to assign relative weights to pair -wised comparisons between category of enablers and subcategory of enablers.

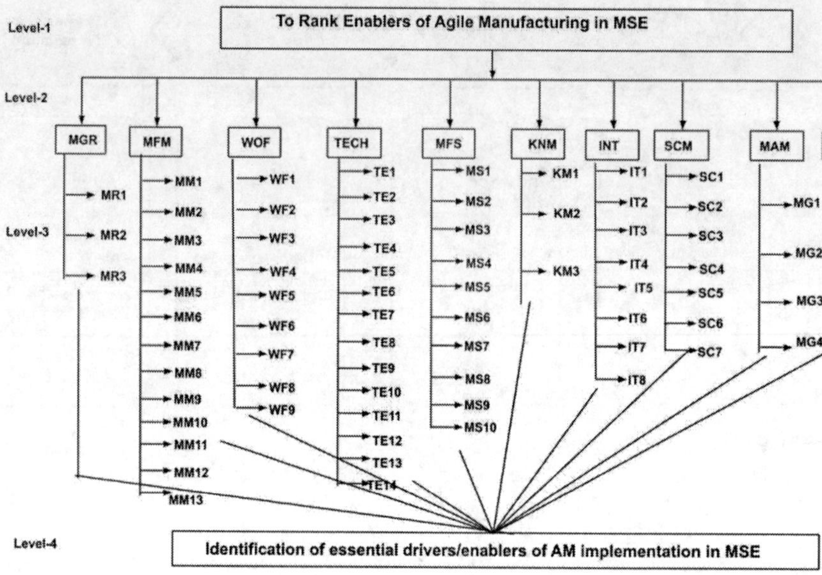

Figure 2. AHP framework for prioritizing Agile Enablers

Data Collection and Analysis

This research tries to targeted managers at middle and higher level of management from various manufacturing industries in India. Before the commencement of this research, pilot study was conducted with sample size of 15 industry representatives, academician and researchers. Obtained sample then evaluated and analyzed statically, as a result a sample size of 210 received from the population. An online questionaries' was drafted to record the input from the various industry representatives. Recorded data is analyzed by multi-criteria decision- making approach i.e. AHP to prioritize the enablers. Data obtained through various industry segments were processed with the help of statistical tool.

Table 2 Pairwise Comparison Matrix

	MGR	MFM	WOF	TECH	MFS	KNM	INT	SCM	MAM	SOF	QAQC	CRM
MGR	1	0.5	5	1	1	3	4	2	4	3	1	2
MFM	2	1	3	2	2	3	2	1	2	3	2	1
WOF	0.2	0.33	1	1	2	2	4	3	4	4	3	0.33
TECH	1	0.5	1	1	2	2	1	1	2	1	1	0.33
MFS	1	0.5	0.5	0.5	1	1	3	2	2	1	3	1
KNM	0.33	0.33	0.5	0.5	1	1	1	2	2	1	1	0.5
INT	0.25	0.5	0.25	1	0.33	1	1	0.5	3	1	1	0.5
SCM	0.5	1	0.33	1	0.5	0.5	2	1	4	1	2	0.33

IOP Conf. Series: Materials Science and Engineering **810** (2020) 012034 doi:10.1088/1757-899X/810/1/012034

MAM	0.25	0.5	0.25	0.5	0.5	0.5	0.33	0.25	1	1	1	0.33
SOF	0.33	0.33	0.25	1	1	1	1	1	1	1	1	0.2
QAQC	1	0.5	0.33	1	0.33	1	1	0.5	1	1	1	1
CRM	0.5	1	3	3	1	2	2	3	3	5	1	1
	0.14	0.13	0.10	0.075	0.08	0.055	0.048	0.066	0.033	0.046	0.057	0.13
	13.45	0.085										

Prioritizing the Agile Manufacturing Enablers.
These are the resulting weights for the criteria based on pairwise comparisons
Number of comparisons = 66
Consistency Ratio, CR = 8.6%
Principal eigen value = 13.452

After calculating all parameters with AHP final result with prioritize ranking is obtained as follows.

Table 3 Ranking of Enablers

Cat		Priority	Rank	(+)	(-)
1	MGR	15.0%	1	11.9%	11.9%
2	MFM	13.9%	2	7.3%	7.3%
3	WOF	10.9%	4	5.6%	5.6%
4	TECH	7.6%	6	3.7%	3.7%
5	MFS	8.5%	5	4.2%	4.2%
6	KNM	5.6%	9	2.3%	2.3%
7	INT	4.8%	10	2.0%	2.0%
8	SCM	6.7%	7	3.4%	3.4%
9	MAM	3.4%	12	1.4%	1.4%
10	SOF	4.6%	11	1.6%	1.6%
11	QAQC	5.7%	8	3.5%	3.5%
12	CRM	13.3%	3	7.2%	7.2%

Figure 3, shows consolidated result obtained in stacked bar format for the respective criteria. X-axis represents the various criterias and Y axis showing priority percentage.

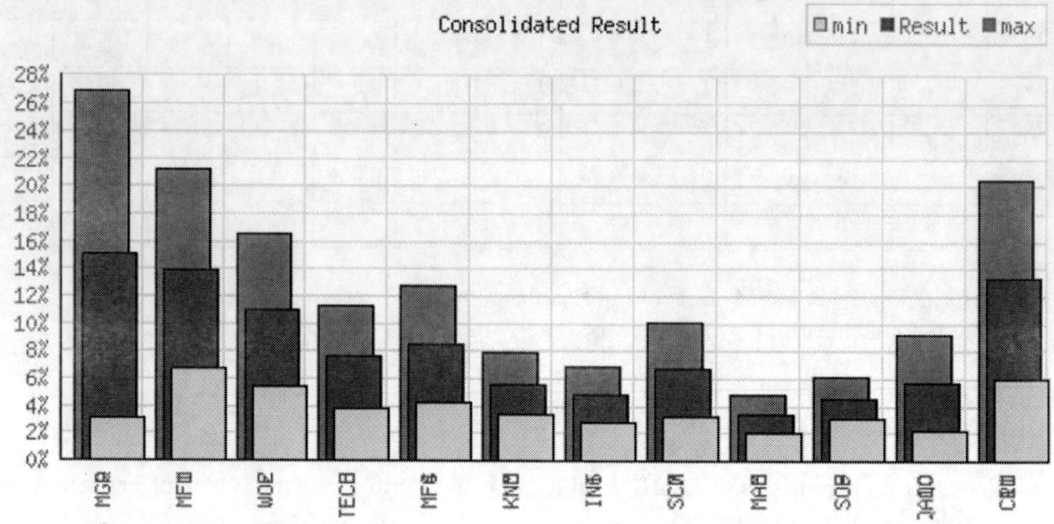

Figure 3. Consolidated Result

6. Result and Conclusion

In this study various research work based on 184 leading researchers work were studied and analyzed, as per research requirement; best 71 research papers were shortlisted for identification of enablers. The identified enablers then summarized with reference to 85 sub criteria's which were defined for 12 categories of enablers. The proposed framework will guide the ordinary skill worker to adopt the concept of Agile Manufacturing in a simpler way. It will also provide a systematic path to manage the productivity of the manufacturing units with higher level of scalability by maintaining the agility with utmost customer satisfaction. Hence adoption of this framework will lead to improvement in the customer satisfaction, overall reduction in wastage and resulted in to higher overall productivity.

References

[1] Thilak, V. M. M., S. R. Devadasan, and N. M. Sivaram. "A literature review on the progression of agile manufacturing paradigm and its scope of application in pump industry." The Scientific World Journal 2015 (2015).

[2] Soltan, Hassan, and Sherif Mostafa. "Lean and Agile Performance Framework for Manufacturing Enterprises." Procedia Manufacturing 2 (2015): 476-484.

[3] Soni, Gunjan, and Rambabu Kodali. "Evaluating reliability and validity of lean, agile and leagile supply chain constructs in Indian manufacturing industry." Production Planning & Control23, no. 10-11 (2012): 864-884.

[4] Ramesh, G., and S. R. Devadasan. "Literature review on the agile manufacturing criteria." Journal of Manufacturing Technology Management 18, no. 2 (2007): 182-201.

[5] D. Vázquez-Bustelo and L. Avella, "Agile manufacturing: Industrial case studies in Spain," Technovation, vol. 26, no. 10, pp. 1147–1161, 2006.

[6] Y. Y. Yusuf, M. Sarhadi, and A. Gunasekaran, "Agile manufacturing: the drivers, concepts and attributes," Int. J. Prod. Econ., vol. 62, no. 1, pp. 33–43, 1999.

[7] Gunasekaran, Angappa. "Agile manufacturing: enablers and an implementation framework." international journal of production research 36, no. 5 (1998): 1223-1247.

[8] Kumar, Sanjay, Naveen Kumar, Sunil Luthra, and Abid Haleem. "Enablers of Lean Six Sigma Implementation in Business Environment: A Review." growth 5 (2013): 14-16.

[9] Aravindraj, Sakthivel, and S. Vinodh. "Forty criteria-based agility assessment using scoring approach in an Indian relays manufacturing organization." Journal of Engineering, Design and Technology 12, no. 4 (2014): 507-518.
[10] Routroy, Srikanta, and Arjun Shankar. "Performance analysis of agile supply chain." International Journal of Manufacturing Technology and Management 29, no. 3-4 (2015): 180-210.
[11] Sarode, A. D., V. K. Sunnapwar, and P. M. Khodke. "A literature review for identification of performance measures for establishing a framework for performance measurement in supply chains." International Journal of Applied Management and Technology 6, no. 3 (2008).
[12] Lin, Ching-Torng, Hero Chiu, and Yi-Hong Tseng. "Agility evaluation using fuzzy logic." International Journal of Production Economics 101, no. 2 (2006): 353-368.
[13] Hojung. Shin, J. N. Lee, D. Kim, and H. Rhim, "Strategic agility of Korean small and medium enterprises and its influence on operational and firm performance," Int. J. Prod. Econ., 2015.
[14] Potdar, Pavan Kumar, Srikanta Routroy, and Astajyoti Behera. "Addressing the Agile Manufacturing Impediments using Interpretive Structural Modeling." Materials Today: Proceedings 4, no. 2 (2017): 1744-1751
[15] Jinturkar, Adesh, Sandip Deshmukh, Avinash Sarode, Vivek Sunapwar, and Prakash Khodke. "Fuzzy-AHP approach to improve effectiveness of supply chain." In Supply Chain Management Under Fuzziness, pp. 35-59. Springer Berlin Heidelberg, 2014.
[16] Tseng, Yi-Hong, and Ching-Torng Lin, "Enhancing enterprise agility by deploying agile drivers, capabilities and providers." Information Sciences 181, no. 17 (2011): 3693-3708.
[17] Bottani, Eleonora. "Profile and enablers of agile companies: An empirical investigation." International Journal of Production Economics 125, no. 2 (2010): 251-261.
[18] Dev, C. Ajay Guru, and VS Senthil Kumar. "Analysis on critical success factors for agile manufacturing evaluation in original equipment manufacturing industry-an AHP approach." Chinese Journal of Mechanical Engineering 29, no. 5 (2016): 880-888.
[19] Soltan, Hassan, and Sherif Mostafa. "Lean and agile performance framework for manufacturing enterprises." Procedia Manufacturing 2 (2015): 476-484
[20] Potdar, Pavan Kumar, Srikanta Routroy, and Astajyoti Behera. "A benchmarking approach for enhancing agility in manufacturing environment." International Journal of Productivity and Quality Management 20, no. 4 (2017): 488-512.
[21] P. K. Potdar and S. Routroy, "Performance analysis of agile manufacturing: a case study on an Indian auto component manufacturer," Meas. Bus. Excell., 2017.
[22] Yusuf, Yahaya Y., Mansoor Sarhadi, and Angappa Gunasekaran. "Agile manufacturing: The drivers, concepts and attributes." International Journal of production economics62, no. 1 (1999): 33-43.
[23] Campanelli, Amadeu Silveira, and Fernando Silva Parreiras. "Agile methods tailoring–A systematic literature review." Journal of Systems and Software 110 (2015): 85-100.
[24] Sreenivasa, C.G., Devadasan, S.R. and Murugesh, R. (2012) 'Thirty criteria-based assessment of agility in a pneumatic enabling products manufacturing company', Int. J. Services and Operations Management, Vol. 11, No. 2, pp.201–221.
[25] Pavan Kumar Potdar, Srikanta Routroy, Astajyoti Behera, "Analyzing the Agile Manufacturing Barriers using fuzzy DEMATEL", Benchmarking: An International Journal, https://doi.org/10.1108/BIJ-02-2016-0024
[26] V. K. Mittal, R. Sindhwani, V. Kalsariya, F. Salroo, K. S. Sangwan, and P. L. Singh, "Adoption of Integrated Lean-Green-Agile Strategies for Modern Manufacturing Systems," Procedia CIRP, vol. 61, pp. 463–468, 2017.
[27] Maria. do Rosário Cabrita, S. Duarte, H. Carvalho, and V. Cruz-Machado, "Integration of Lean, Agile, Resilient and Green Paradigms in a Business Model Perspective: Theoretical Foundations," IFAC-PapersOnLine, 2016.
[28] S. Routroy, P. K. Potdar, and A. Shankar, "Measurement of manufacturing agility: a case study," Meas. Bus. Excell., 2015.

[29] U. B. Ş and A. F. G. İ, "Research Article / Ara ş trma Makalesi, " A Fuzzy Multi-Criteria Decision Approach for Measuring Technology Competency Performance of SME's" vol. 8, no. 1, pp. 31–40, 2017.
[30] R. Dubey and A. Gunasekaran, "Agile manufacturing: framework and its empirical validation," Int. J. Adv. Manuf. Technol., pp. 2147–2157, 2014.
[31] H. Sharifi and Z. Zhang, Agile manufacturing in practice - Application of a methodology, vol. 21, no. 5/6. 2001.
[32] M. A. Hasan, "Interpretive structural modelling of agility enhancing management practices for agile manufacturing Joseph Sarkis Ravi Shankar," vol. 6, no. 4, pp. 361–390, 2013.
[33] C.-T. Lin, H. Chiu, and Y.-H. Tseng, "Agility evaluation using fuzzy logic," Int. J. Prod. Econ., 2006.
[34] Haq, A. Noorul, and Varma Boddu. "Analysis of agile supply chain enablers for Indian food processing industries using analytical hierarchy process." International Journal of Manufacturing Technology and Management 29, no. 1-2 (2015): 30-47.
[35] Mathiyazhagan, K., Kannan Govindan, and A. Noorul Haq. "Pressure analysis for green supply chain management implementation in Indian industries using analytic hierarchy process." International Journal of Production Research 52, no. 1 (2014): 188-202.
[36] Sarode, Avinash D., V. K. Sunnapwar, and P. M. Khodke. "Improving effectiveness of supply chain by selecting an appropriate supplier: An analytic hierarchy process approach." Journal of Advanced Manufacturing Systems 9, no. 02 (2010): 129-144.

Friction Stir Welding Processing, Materials and its Applications.

a.Rajendra shimpi b. Cheruku Sandesh Kumar c. Rajnish katarane

a.c. Department of Mechanical Engineering, SVKM's NMIMS MPSTME Shirpur Campus, Dhule, Maharashtra, India
b. ASET (Amity School of Engineering and Technology) Jaipur Campus, Rajasthan, India
Rajendra.Shimpi@nmims.edu

Abstract. Friction stir welding (FSW) is solid state welding process for joining the two similar or dissimilar materials. Friction stir welding process is alternative joining process for thin plates and pipes. In this process non consumable tool is used to join two work faces by friction. FSW process is energy efficient and environment friendly process. In this paper summery of research work done by researchers for Friction stir welding process is presented. The aftermath of this work is to elaborate the overview of numerical analysis and overview of process parameters like tool tip angle, Tool geometry; Tool and joint Materials and Joint design. FSW process can be used for joining different materials. In current research the materials used for experimentation by different researches and its applications are summarized. All the vital factors are covered here for making the process for improvement of effectiveness/ cost reduction. After refining all the process, researcher has found a scope in joining the different / alternative joining process, in different sectors.

Keywords: FSW, FSWoP, Tool tip angle, Tool Geometry.

1. Introduction

The friction stir welding (FSW) is solid state welding process for joining the wide range of materials like copper, bronze, aluminum, polymer, magnesium and steel materials which are widely used in many industrial applications. The FSW process was invented by The Welding Institute (TWI) of Cambridge, England in 1991 [1]. FSW process is simple in construction, pollution free, cost efficient and becomes very popular in automobile, aircraft, marine and aerospace industries. In FSW process the weight of the joint is minimum as compare to conventional welding processes and in this process no liquid state for weld pool, the welding joint takes place in the solid phase below the melting point of the materials. Thus the problems related to the fused material eliminated by FSW process. [2] FSW has many advantages over traditional welding techniques. FSW process has ability to weld dissimilar metals especially in automobile industry where there is need of joining steel to aluminum [3]. FSW is performed at temperature under the base metal's dissolving point. Along these lines, all issues identifying with re-cementing of welded material, similar to, fragility, split age and porosity. [4]

2. LITERATURE REVIEW

The FSW procedure includes a perplexing wonder identified with plastic deformation and material development during the welding procedure. The main elements disturbing FSW process are Welding parameters, tool geometry, numerical analysis and Joint shape. These elements show vital part in the material temperature distribution and flow configuration

Analysis of friction stir welding process by using simulation one of the complicated process. Simulation is related with "heat transfer" and "mechanics''. Also great distortions and strain proportions around the tool probe within stir area forms on the basis of simulation. Analysis of effect of several parameters like tool ''rotational speed'', ''tool design'', ''tool transverse speed'' and ''tool angle'' are easily possible by using numerical analysis. Growth of temperatures during FSW process can be predicted by using numerical modelling process. Temperature evaluation can also possible by using Computational fluid dynamics (CFD).It has observed that finite element method is best for identification and assessment of faults in FSW joints and experimental investigation also carried out by Mohammad Ali Fakih et al. He has done the detailed group of weld junctions in connotation through "CT scanning" for destruction indices (DI) [5],

Impact of FEW on residuary worries notwithstanding distortion estimation for planned airplane structure get together examination has analysed by R. W. McCune et al and he expressed that it is essential to decisively structure and mimic the FSW procedure, as the powers associated with the procedure visibly affect level of "leftover burdens.[6].Mir Zahedul H. Khandkar et al. [7], was performed FEA for guessed the residual "Thermal stresses'' inside the metal work pieces combined with FSW process. A. Murphy et al. [8], spoke to the effect of FSW process on stiffened panel for buckling potential by actualizing FEA. D.G. Richards et al. [9], used numerical investigation to break down and trial the idea of "Worldwide mechanical tensioning" to control the residual stresses created through FSW process inside the weld intersections. J.W. Yoon et al. [10], dissected the impact of FSW if there should be an occurrence of a "inherently solidified panel (ISP) structure" on its "fastening performance" by actualizing FEA. E.Feulvarch [11] was proposed a basic yet powerful moving cross section method dependent on Eulerian formalism for the numerical modelling of the FSW procedure.

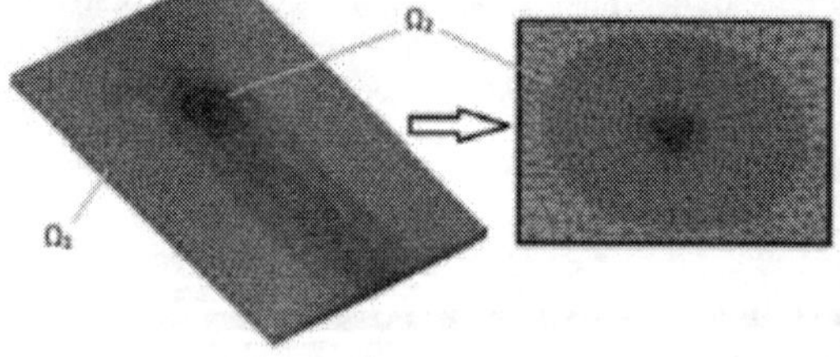

(a) Partition of the mesh

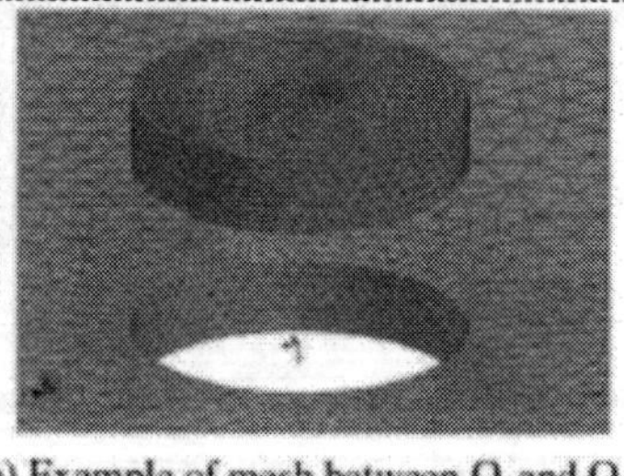

(b) Example of mesh between Ω_1 and Ω_2

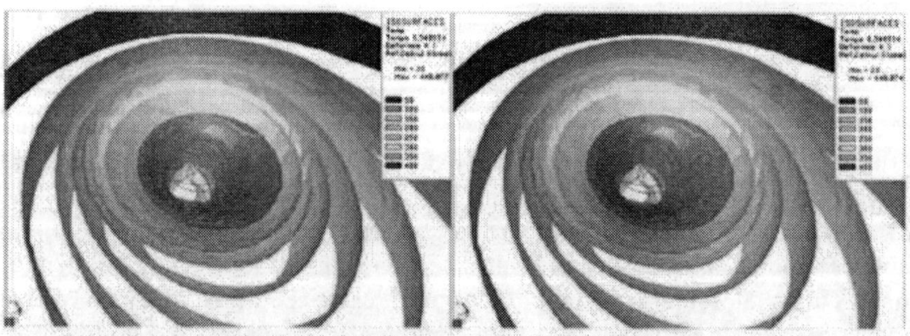

(c) Temperature distributions (°C) for $R_{\Omega2}$=12mm (left) and $R_{\Omega2}$=15mm (right)

Fig.1. Moving mesh technique for the numerical modelling of the FSW process [11]

In FSW strategy process parameters are assumes significant job. Some significant procedure parameters are ''Tool crosswise speed'', ''axial force'', ''Tool rotating speed'' and ''Tilt angle of tool''.

Shultz EF et al. [12], broke down the impact produced on the weld pattern viability because of the FSW device probe tilt point alongside the contraption quiet submission, though smearing FSW on "5083-H111 Al composite". Francesco Lambiase et al. [13], though actualizing FSW on AA5053 Al composite moved sheets and (PVC) plate, saw that "joint quality" just as effectiveness of the vitality consumed in the proper method of welding procedure improved, with the utilization of securing casing obligating lesser thermal conductivity. It continued in this way surmised, a more prominent estimation of "joint quality" having moderately littler "heating time" and synchronous decrease in "immersed energy" is conceivable adequately with a subsequent utilization of timber in place of the affixing outline material. Krishnan M et al. [14], saw that with a resulting ascend in "tool revolving speed" and "axial load", improved the components, for example, elasticity just as hardness up to a specific point, yet thus gets decreased as most extreme point is accomplished; while utilizing "Response Surface Methodology (RSM)" and ''Artificial Neural Networks (ANN)" strategy to assess the ideal joining variables of Al amalgams AA6063 and A319 implied for FSW. Nonetheless, an unfortunate impact on reactions was distinguished because of ascend in "tool traverse speed".

In FSW procedure apparatus geometry is one of significant parameter for delivering the ideal joint. It is valuable for age of warmth during procedure and it is likewise useful for material stream as the apparatus is moving consistent in transverse way. The heat produces essentially by uprightness of grinding in the focus of the tool pin shape and tool substantial, at the primary period of the tool climb. Yanni Wei et al. [15], executed friction stir segment welding for joining of Al and Ti amalgam with cutting pin has made of tungsten carbide. It has been seen that segment joints with greater quality. Shanavas Shamsudeen et al [16] researched the impact of various profiles of tool pin for joining AA 5052 H32 aluminium combination by FSW procedure utilizing reaction surface philosophy and dark social examination. The nature of the welded joint is estimated dependent on the higher estimation of rigidity and hardness of the joint got by experimentation.

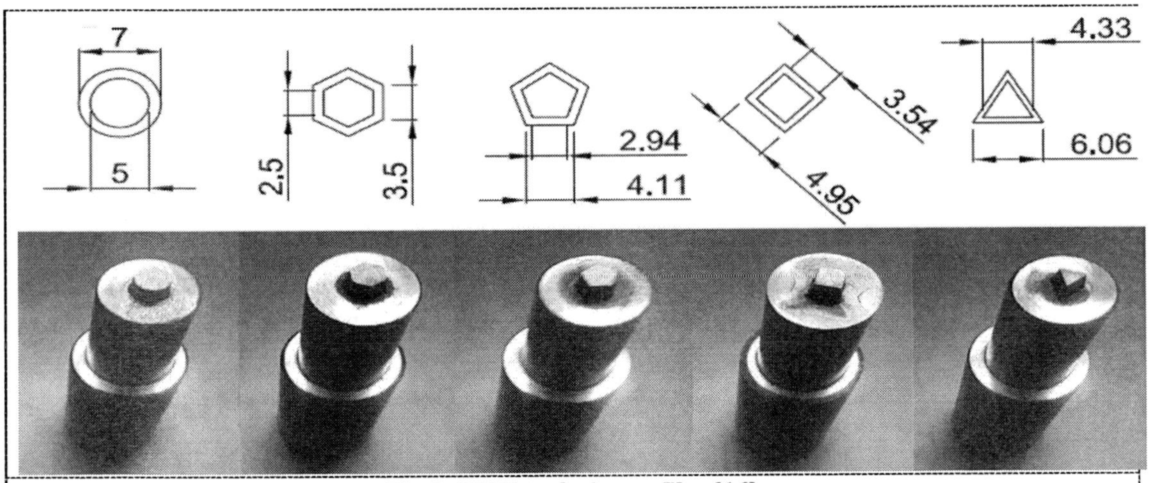

Fig.2. FSW tool pin profiles [16].

In FSW procedure Butt joint and Lap joint are the regular kinds of joints can be performed. For stay away from any partition among the connecting combined surfaces of two indistinguishably dense plates, the situation is important to secure these plates firmly towards the support plate. Likewise, affixing turns out to be unnecessarily significant over the span of beginning "tool plunge" where more prominent level of powers is being experienced. If there should be an occurrence of a segment combined, two plates attached toward the "support plate" in a lap arrangement, are joined to one another by dropping a FSW device symmetrically hooked on the base plate over the main one and navigating it lengthwise the welding course by Rajiv S et al [17]. done FSW methodology device point playing key part to joining two materials. A. Arici et al. [18] investigated those impacts from claiming device around point for FSW for joining polyethylene. Butt joint might have been produces throughout methodology what's more it might have been found that the welding parameters required noteworthy impacts around ductile properties and crack areas of the welds. The rigidity diminished with expanding device around tilt point. Kush P. Mehta et al. [19] presented the work on dissimilar material such as electrolytic hard field copper, and aluminum 6061-T651. In this experimentation effect of slope approach on the mechanical and metallurgical things were calculated. Different tests were carried out for checking the quality. It was found that fault free different copper–aluminum friction stir welding was reached by tilt angles 2°, 3°, and 4°. The supreme strength of weld was obtained at 4° tilt angle. Devaiah D. et al. [20] also examined the effect of tilt angle on mechanical properties in FSW process. it was found that 0 to 2° has effective tilt angle during FSW process. The contribution of tilt angle also evaluated for strength of welded joints and elongation of welded joint. It was found that tool tilt angle has 33% contribution to % of elongation of welded joint.

Shuai Zhang et al.[21] has created CFD prototypical in which a geometrical prototypical and a fragmented contact limit situation are planned, to examine the impacts of hardware tilt edge continuously the in-process heat move and mass exchange during FSW. Creator has discovered three promising outcome initial, a greater temperature has produced by the slope welding instrument on the propelling crosswise, which is credited to the fragmented interaction at the shoulder/work piece boundary. Second, the slope welding instrument creates a greater frictional power on the device/work piece interface, which altogether recovers the interfacial material stream speed after the device. Third, the slope welding apparatus produces a more grounded blending activity toward the substantial in the region of the fusing instrument, which is useful for the blending of materials and the development of grating mix welds.

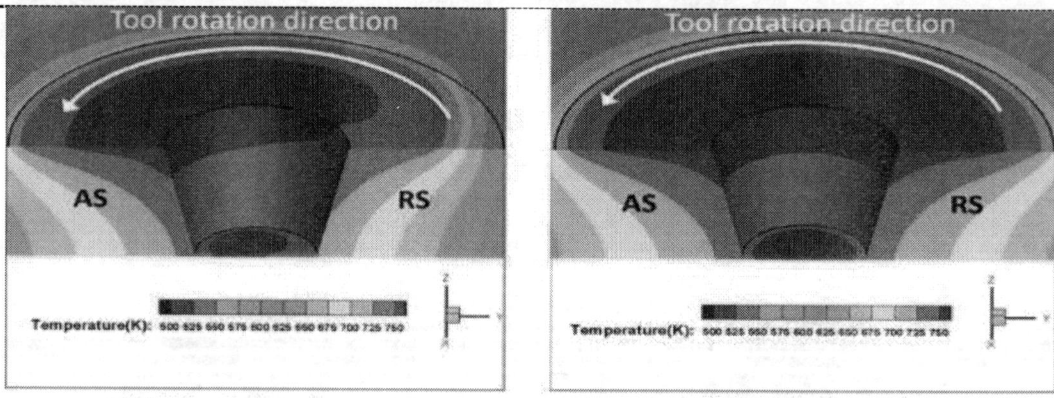

(a) 2.5° tool tilt angle (b) 0° tool tilt angle

Figure 3 Effect of Tilt angle on heat transfer during FSW process [21]

Lemos G.V.B [22] has done experimental work on nickel-chromium alloy 625 by using FSW process. The outcomes of investigation were the rotational speed badly affected on residual stresses in weld zone. Leite A.M.S. [23] were done experimental investigation on duplex stainless steel, grade UNS S82441 and polycrystalline cubic boron nitride used as a tool material. It has been found that TMAZ was the zone most susceptible to pitting. R.F. Hamade [24] were investigated the position and volume of the present defects.by using CT apparatus by using FSW technique on aluminium alloy. Yasui [25] were formulated the dissimilar metal joint by using FSW process and it was observed that Welding force and temp increases with rotating angle and fracture strength, weld strength were low as compare to base metal strength. da Cunha [26] were studied the effect FSW parameters like Rotational speed ,welding speed ,Torque, Axial force ,Heat input on GL E36 shipbuilding steel. It has been found that high gradient was occurred within a stirred zone. Wang [27] has investigated the performance of FSW joint for material used in aerospace tanks and rocket tanks. It was observed that high quality of longitudinal straight joint can be obtained by FSW. This process can further extended for joining domes to body of the tank in rocket. V. S. Gadakh [28] has performed experimental work on aluminum alloy by using FSW process and it has observed that tool rotational speed played vital role in FSW process and the author stated that Increasing the rotational speed and tilt angle of the tool at constant weld speed resulted in decrease in mechanical properties and hardness values. The author investigated the multi objective optimization of FSW process. It has been observed that tool rotation speed and welding speed shows significant influence on mechanical properties of welded joint and composite profile with conical base are improved the performance characteristics of the Friction stir welding process stated by Wakchaure [29].

Baffari D. [30] presented the effect of FSW process on composite material. It has been observed new exploratory apparatuses and opening plans were researched so as to improve the mechanical opposition of the joints and a numerical model was set up to investigation the procedure mechanics. Shujun Chen [31] investigated the effect of microstructure on the mechanical properties of friction stir welded 5A06 Al alloy. It has been found that grain orientation does not affect the tensile strength of the welded joints. At higher temperatures both the yield pressure and a definitive pliable pressure decline with temperature, while the all out extension increments and increase in hardness observed with increasing traverse speed has investigated by H. Dawson [32]. Rao et al. [33] done the experimentation for joining aluminium alloy by using different pin profile by using FSW process and predicted tool pin profiles mostly influenced the tensile properties of joint. Cylindrical profile of pin found out the most suitable profile which has no defect found during experimentation. Thermal and CFD models revealed that higher viscosity was observed in the SS side and different height levels of the weld section have different flow behaviour. Sadeghian [34] done simulation of FSW and it was observed that thermal and CFD models revealed that higher viscosity at SS side and different height levels of the weld section have different flow behavior. Bevilacqua [35] invented the effect of environmental impact on FSW process on butt joint for aluminum alloy. The outcomes given by the existence cycle appraisal examination has demonstrated that the ecological effect of rubbing mix welding is unequivocally influenced by rotational and welding speed. Ahmed, M.M [36] performed dissimilar weld joint by using FSW process. It was found that a noteworthy grain refining was happened in the NG zones of both similar and dissimilar FSW of AA7075 and AA5083. Frictions stir welding conducted by Sidhar [37] for joining of aluminum and magnesium alloy. It was resulted that dissolution of precipitates in WN and HAZ. Leitao [38] invented the utilization of a multi pass welding system for improving the attachment area in the dissimilar friction stir lap welding to reduce tool wear. Gite R.et. al. [40] was done review of FSW process. In the paper author has reported many important parameters and methodology for friction stir welding process. Javadi Y. [41] Invented optimization of residual stresses produced by friction stir welding (FSW) of 5086 aluminum plates. It has been inferred that the most critical impact on the longitudinal remaining pressure top is identified with the feed rate while the stick and shoulder breadth have no predominant impact.

Shultz, E.F et. al [42] suggested the different tool tip angle for effective FSW joining process. It has been observed that as the tool tip angle increases the gap width reduces. Leitao, C et. al [43] presented the categorization of the constitutive behavior of welds by using digital image processing techniques. Elatharasan et. al. [44] was listed the important process parameters for the friction stir welding process. It has been observed that as the tool rotational speed, welding speed and tool axial force increases UTS and YS also increased. Gibson, B.T et al. [45] has also reported the important parameters of FSW process and automation. The latest application has also discussed. Lakshminarayanan et al [46] done the experimentation on ferritic stainless steel by using FSW process and microstructure as well as mechanical characterization has presented. It has been observed that the coarse ferrite grains in the base material are changed to extremely fine grains comprising duplex structure of ferrite and martensitic because of the quick cooling rate and high strain actuated by extreme plastic twisting brought about by frictional mixing. Sun, Y et al. [47] has demonstrated Hydrogen-induced destruction and cracking in friction stir welded low carbon steel by the cathodic hydrogen charging method. It has been observed that cracking made along the thermo-mechanically affected zone boundary.

3. GAPS IN LITERATURE REVIEW

After study the existing literature review on FSW process a number of gaps have been observed

- FSW Tool probe tilt angle is also dominant parameter and very few literature is availed hence work can be extended to study the effect of tool profile by changing the tool probe angle on the mechanical properties.
- Most of the researcher investigated influence of limited process parameters for joining the Aluminium matrix composites (AMC) and Sandwich Aluminium matrix composites (SAMC).
- Very few literatures are available on the effect of post weld heat treatment on weld joint.
- AMC and SAMC sheets can be satisfactorily welded by using threaded cylindrical FSW tool. Similar joint can be produce by using another profile of tool and work can be extended to study the effect of tool profile on the mechanical properties.
- Very few literatures is available on Friction stir welding of pipes (FSWoP) as compare to friction stir welding (FSW) for plates.

All the summary of Friction stir welding important parameters, workpiece material s and tool materials are listed in table no.1.

Table 1. Experimental Process parameters. Workpiece material, Tool material and application of FSW

Author) (Year)	Input variables/parameters considered with their ranges	Response & quality characteristics	Material (s), machine, tool etc.	Material Application
Lemos et.al. [22]	- Welding speed - Axial force - Rotational speed	-Weld surface Quality -Welded depth	-Nickel-chromium alloy 625	Rigid clad pipes for transporting oil.
dos Santos Leite et.al. [23]	-Rotational speed -Welding speed - Axial force	-Microstructure -Localized corrosion resistance e	-Duplex stainless steel, grade UNS S82441	Oil , gas and Chemical transportation pipes
Hamade R.F et.al. [42]	-Feed mm/min	-Position and volume of the present defects.by using CT apparatus	-Aluminum AA6061-T6 and AA1050 -	Automotive industry
Yasui T. et.al. [25]	-Tool tilt angle	-HAZ -Microstructure -Fracture strength -weld strength	-AA6063 and S45C	High speed train, Ship & rocket Structures
da Cunha [26]	-Rotational speed (rpm) -Welding speed (mm/s) -Torque (Nm) -Axial force (kN) -Heat input (kJ/mm)	-Microstructure -UTS -YS -Elongation -Hardness	-GL E36 shipbuilding steel	Shipbuilding

Krishnan et.al [14]	- Rotational speed -Welding speed -Axial force	- UTS -YS -Elongation -Hardness	- Al-Si-Mg - AA6063 and A319 dissimilar alloys	Automobile & Shipbuilding
Devaiah D. et.al [20]	-Tool rotational speed -Welding speed -Tool tilt angle	- UTS -YS -Elongation -Hardness	- AA5083 and AA6061 Aluminum alloys	Automotive industry
Wang G. et.al [27]	-Rotation speed. - Traverse speed -Tool tilt angle -Pin Length	- UTS -YS -Elongation -Hardness	- Aluminium alloy AA2219	Aerospace and rocket tank
Kumar, T.P.et.al [43]	-Process parameters are constant. RS=1400RPM, Welding speed.60mm/min	-Temp. - Fracture zone. -weld zone. -Microstructure. - Grain size.	- Aluminium alloy AA6061 of .	Making pipes for oil or fluid transportation
Gadakh V. S. et.al [28]	-Tool Rotational Speed -Tools - Weld Speed - Tilt Angle	-Microstructure -Micro hardness	-AA2014 is a heat treatable aluminium alloy	Aircraft structure, aircraft fittings, Truck frame.
Wakchaure K. N. et.al [29]	- Tools -Tool Rotation (TR) (RPM) -Tilt Angle(TA) (in degrees) -Welding Speed (WS) (Inch/sec)	-UTS -YS -Elongation -Hardness	-AA6082-T6 aluminium alloy	Storage tanks, Marine frames, pipelines and aircraft applications
Mahdi M. et.al [44]	- Rotation speed. - Traverse speed - Tilt angle - Shoulder diameter - Pin diameter - Pin height	-UTS -YS -Elongation -Hardness	-DP700 steel	Roof for Outer structure, Body Side Outer, Floor Panel. Floor Reinforcements. Panel Inner, Rear Rails.
Baffari D. et.al [30]	-Tool rotation -Feed rate -Tilt angle -Tool force -Dwell time	-Elongation -Hardness -YS	-Aluminium alloy -AA6082-T6	Lightweight hybrid structures for automotive, aeronautics and shipbuilding
Chen S. et.al. [31]	-Rotation speed - Tilt angle -Traverse speed	-Hardness - Microstructure -Tensile test	-5A06 Aluminum alloy	?ressure vessels, ships, and automobiles, aerospace.

IOP Conf. Series: Materials Science and Engineering **810** (2020) 012035 doi:10.1088/1757-899X/810/1/012035

Dawson H. et.al. [32]	- Rotation speed. - Traverse speed	-Micro hardness -Microstructure - Temperature dependent mechanical Behavior. - Tensile testing of base material at room temp to 750 c -Weld joint test by tensile test at room temp to 500 c	- Fully ferrite MA956 ODS steel	Nuclear fission and fusion reactors
Rao M.S.S. et.al. [33]	-Rotational speed -Welding speed -tool pin	-Tensile strength -Weld joint	5mm thick IS:65032 aluminium alloys	Automobile and aircraft industries
Sadeghian B. et al [34]	-Tool rotational speed	-Material flow - Weld morphology	304 SS AA-5083	Aerospace and chemical industry.
Lambiase F.	-Plunging Load, -Dwell time	-Joining forces -Thermal analysis -Morphology of the joints -Mechanical behavior of the joints.	AA5053 sheets with polyvinyl chloride (PVC)	Automotive, aerospace, biomedical, civil constructions.
Bevilacqua . et.al. [35]	-Rotational speed -Welding speed - Tilt angle	-Metering of electrical energy absorbed -Tensile test -life cycle assessment	AA5754 aluminum alloy H13 tool steel	Shipbuilding and Vehicle bodies, Food processing, Chemical and nuclear structures.
Ahmeda M.M.Z. . et.al. [36]	-Traverse speeds -Diameter of the pin	-Tensile test - Microstructure -Hardness	Aluminum alloys AA7075-T6 and AA5083-H111	Rail cars, Vehicle bodies, Tip truck bodies, Pressure vessels.
Sidhar H. et.al. [37]	-Two thermal conditions were made by using external cooling mediums	Weldability	1424 aluminum alloy	Automotive and aerospace
Leitao C. [38]	-Rotational speed welding speed	Minimize tool wear	Aluminium and Steel	automotive parts
Evans, W. et.al. [39]	-Friction Stir Extrusion process	-Tensile test - Microstructure -Hardness	Aluminium and Steel	automotive parts

4. CONCLUSION.

In the current research researchers has found that friction stir welding is effective/cost efficient alternative method for joining the thin plates/hollow pipes. Hence same joining process can be used for joining the thin shell/tanks for many applications like automobiles, aerospace, shipbuilding etc. There is wide scope in joining thin dissimilar materials by using friction stir welding process. Friction stir process for joining of pipes (FSWOP) is also one of alternative efficient and cost effective method for joining the pipe as compare to friction stir welding of plates (FSW).There is limited research has been done in this areas. This method can be easily implemented/applied for joining the different parts/sectors

5. References

[1] W.M. Thomas, E.D. Nicholas, J.C. Needham, M.G. Murch, P. Templesmith, C.J. Dawes, G.B. *Patent Application No.9125978.8 (December 1991).*

[2] C. Dawes, W. Thomas, *TWI Bulletin 6, November/December 1995, p. 124.*

[3] M. Haghshenas, A.P. Gerlich, Joining of automotive sheet materials by friction-based welding methods: a review, *J. Eng. Sci. Technol. vol. 21 (Issue 1) (2018) 130–148.*

[4] Hofmann Douglas C, Vecchio Kenneth S. "Thermal history analysis of friction stir processed and submerged friction stir processed aluminium". *Material Science Engineering, 2007, Pages 165–75.*

[5] Mohammad Ali Fakih, Samir Mustapha, Jaafar Tarraf, Georges Ayoub, Ramsey Hamade, Detection and assessment of flaws in friction stir welded joints using ultrasonic guided waves: experimental and finite element analysis, *Mechanical Systems and Signal Processing, Volume 101,2018,Pages 516-534.*

[6] McCune RW, Murphy AA, Price MM, Butterfield JJ. The Influence of Friction Stir Welding Process Idealization on Residual Stress and Distortion Predictions for Future Airframe Assembly Simulations. *ASME. J. Manuf. Sci. Eng. 2012;134(3)*

[7] Mir Zahedul H. Khandkar, Jamil A. Khan, Anthony P. Reynolds, Michael A. Sutton, Predicting residual thermal stresses in friction stir welded metals, *Journal of Materials Processing Technology, Volume 174, Issues 1–3, 2006, Pages 195-20.*

[8] A. Murphy, W. McCune, D. Quinn, M. Price, The characterization of friction stir welding process effects on stiffened panel buckling performance, Thin-Walled Structures, *Volume 45, Issue 3, 2007, Pages 339-351*

[9] D.G. Richards, P.B. Prangnell, S.W. Williams, P.J. Withers, Global mechanical tensioning for the management of residual stresses in welds, *Materials Science and Engineering: A, Volume 489, Issues 1–2, 2008, Pages 351-36.*

[10] J.W. Yoon, G.H. Bray, R.A.F. Valente, T.E.R. Childs, "Buckling analysis for an integrally stiffened panel structure with a friction stir weld", Thin-Walled Structures,*Volume 47, Issue 12,2009,Pages 1608-1622*

[11] Feulvarch E, Roux JC, Bergheau JM. A simple and robust moving mesh technique for the finite element simulation of Friction Stir Welding. *J Comput Appl Math 2013;246:269–77.*

[12] Shultz EF, Cole EG, Smith CB, Zinn MR, Ferrier NJ, Pfefferkorn FE. "Effect of Compliance and Travel Angle on Friction Stir Welding with Gaps." *ASME. J. Manuf. Sci. Eng. 2010;132(4):041010-041010-9. doi:10.1115/1.4001581*

[13] Francesco Lambiase, Alfonso Paoletti, Valentino Grossi, Silvio Genna, Improving energy efficiency in friction assisted joining of metals and polymers, *Journal of Materials Processing Technology, Volume 250,2017,Pages 379-389*

[14] Krishnan, M.M., Maniraj, J., Deepak, R. and Anganan, K., 2018. Prediction of optimum welding parameters for FSW of aluminium alloys AA6063 and A319 using RSM and ANN. *Materials Today: Proceedings, 5(1), pp.716-723.*

[15] Yanni Wei, Jinglong Li, Jiangtao Xiong, Fu Huang, Fusheng Zhang, Syed Hamid Raza, "Joining aluminium to titanium alloy by friction stir lap welding with cutting pin", *Materials Characterization, Volume 71,2012, Pages 1-5*

[16] Shanavas Shamsudeen, John Edwin Raja Dhas ,"Optimization of Multiple Performance Characteristics of Friction Stir Welded Joint with Grey Relational Analysis", *Materials Research vol.21 no.6*

[17] Mishra, Rajiv S., and Z. Y. Ma. "Friction stir welding and processing." *Materials science and engineering: R: reports 50, no. 1-2 (2005): 1-7.*

[18] Arici, A., and S. Selale. "Effects of tool tilt angle on tensile strength and fracture locations of friction stir welding of polyethylene." *Science and technology of welding and joining 12, no. 6 (2007): 536-539.*

[19] Mehta, Kush P., and Vishvesh J. Badheka. "Effects of tilt angle on the properties of dissimilar friction stir welding copper to aluminum." *Materials and Manufacturing processes 31, no. 3 (2016): 255-263.*

[20] Devaiah, D., Kishore, K. and Laxminarayana, P., 2018. Optimal FSW process parameters for dissimilar aluminium alloys (AA5083 and AA6061) Using Taguchi Technique. *Materials Today: Proceedings, 5(2), pp.4607-4614.*

[21] Zhang, S., Shi, Q., Liu, Q., Xie, R., Zhang, G. and Chen, G., 2018. Effects of tool tilt angle on the in-process heat transfer and mass transfer during friction stir welding. *International Journal of Heat and Mass Transfer, 125, pp.32-42*

.

[22] Lemos, G.V.B., Farina, A.B., Nunes, R.M., da Cunha, P.H.C.P., Bergmann, L., dos Santos, J.F. and Reguly, A., 2019. Residual stress characterization in friction stir welds of alloy 625. *Journal of Materials Research and Technology.*

[23] Dos Santos Leite, A.M., Terada, M., Pereira, V.F., da Fonseca, E.B., de Lima, N.B. and Costa, I., 2019. On the pitting resistance of friction stir welded UNS S82441 lean duplex stainless steel. *Journal of Materials Research and Technology.*

[24] Hamade, R.F. and Baydoun, A.M.R., 2019. Nondestructive detection of defects in friction stir welded lap joints using computed tomography. *Materials & Design, 162, pp.10-23.*

[25] Yasui, T., Wu-Bian, T., Hanai, A., Mori, T., Hirosawa, K. and Fukumoto, M., 2018. Friction stir girth welding between aluminum and steel rods. *Procedia Manufacturing, 15, pp.1376-1381.*

[26] da Cunha, P.H.C.P., Lemos, G.V.B., Bergmann, L., Reguly, A., dos Santos, J.F., Marinho, R.R. and Paes, M.T.P., 2019. Effect of welding speed on friction stir welds of GL E36 shipbuilding steel. *Journal of Materials Research and Technology,8(1),pp.1041-105.*

[27] Wang, G., Zhao, Y. and Hao, Y., 2018. Friction stir welding of high-strength aerospace aluminum alloy and application in rocket tank manufacturing. *Journal of Materials Science & Technology, 34(1), pp.73-91.*

[28] Gadakh, V.S. and Kumar, A., 2018. FSW tool design using TRIZ and parameter optimization using Grey Relational Analysis.*MaterialsToday:Proceedings,5(2),pp.6655-6664.*

[29] Wakchaure, K.N., Thakur, A.G., Gadakh, V. and Kumar, A., 2018. Multi-Objective Optimization of Friction Stir Welding of Aluminium Alloy 6082-T6 Using hybrid Taguchi-Grey Relation Analysis-ANN Method. *Materials Today: Proceedings, 5(2), pp.7150-7159.*

[30] Baffari, D., Buffa, G., Campanella, D., Valvo, E.L. and Fratini, L., 2018. Experimental and numerical investigation on a new FSW based metal to composite joining technique. *Journal of Manufacturing Processes, 34, pp.758-764.*

[31]Chen, S., Li, X., Jiang, X., Yuan, T. and Hu, Y., 2018. The effect of microstructure on the mechanical properties of friction stir welded 5A06 Al Alloy. *Materials Science and Engineering: A, 735, pp.382-393.*

[32] Dawson, H., Serrano, M., Hernandez, R., Cater, S. and Jimenez-Melero, E., 2017. Mechanical properties and fracture behaviour of ODS steel friction stir welds at variable temperatures. *Materials Science and Engineering: A, 693, pp.84-92.*

[33] Rao, M.S., Kumar, B.R. and Hussain, M.M., 2017. Experimental study on the effect of welding parameters and tool pin profiles on the IS: 65032 aluminum alloy FSW joints. *Materials Today: Proceedings, 4(2), pp.1394-1404.*

[34] Sadeghian, B., Taherizadeh, A. and Atapour, M., 2018. Simulation of weld morphology during friction stir welding of aluminum-stainless steel joint. *Journal of Materials Processing Technology, 259, pp.96-108.*

[35] Bevilacqua, M., Ciarapica, F.E., D'Orazio, A., Forcellese, A. and Simoncini, M., 2017. Sustainability analysis of friction stir welding of AA5754 sheets. *Procedia CIRP, 62, pp.529-534.*

[36] Ahmed, M.M.Z., Ataya, S., Seleman, M.E.S., Ammar, H.R. and Ahmed, E., 2017. Friction stir welding of similar and dissimilar AA7075 and AA5083. *Journal of Materials Processing Technology, 242, pp.77-91.*

[37] Sidhar, H., Martinez, N.Y., Mishra, R.S. and Silvanus, J., 2016. Friction stir welding of Al–Mg–Li 1424 alloy. *Materials & Design, 106, pp.146-152.*

[38] Leitao, C., Arruti, E., Aldanondo, E. and Rodrigues, D.M., 2016. Aluminium-steel lap joining by multipass friction stir welding. *Materials & Design, 106, pp.153-160.*

[39] Evans, W.T., Gibson, B.T., Reynolds, J.T., Strauss, A.M. and Cook, G.E., 2015. Friction Stir Extrusion: A new process for joining dissimilar materials. *Manufacturing letters, 5, pp.25-28.*

[40] Gite, R.A., Loharkar, P.K. and Shimpi, R., 2019. Friction stir welding parameters and application: A review. *Materials Today: Proceedings.*

[41] Javadi, Y., Sadeghi, S. and Najafabadi, M.A., 2014. Taguchi optimization and ultrasonic measurement of residual stresses in the friction stir welding. *Materials & Design, 55, pp.27-34.*

[42] Shultz, E.F., Cole, E.G., Smith, C.B., Zinn, M.R., Ferrier, N.J. and Pfefferkorn, F.E., 2010. Effect of compliance and travel angle on friction stir welding with gaps. *Journal of Manufacturing Science and Engineering, 132(4), p.041010.*

[43] Leitão, C., Galvão, I., Leal, R.M. and Rodrigues, D.M., 2012. Determination of local constitutive properties of aluminium friction stir welds using digital image correlation. *Materials & Design, 33, pp.69-74.*

[44] Elatharasan, G. and Kumar, V.S., 2013. An experimental analysis and optimization of process parameter on friction stir welding of AA 6061-T6 aluminum alloy using RSM. *Procedia Engineering, 64, pp.1227-1234.*

[45] Gibson, B.T., Lammlein, D.H., Prater, T.J., Longhurst, W.R., Cox, C.D., Ballun, M.C., Dharmaraj, K.J., Cook, G.E. and Strauss, A.M., 2014. Friction stir welding: *Process, automation, and control. Journal of Manufacturing Processes, 16(1), pp.56-73.*

[46] Lakshminarayanan, A.K. and Balasubramanian, V., 2010. An assessment of microstructure, hardness, tensile and impact strength of friction stir welded ferritic stainless steel joints. *Materials & Design, 31(10), pp.4592-4600.*

[47] Sun, Y., Fujii, H., Imai, H. and Kondoh, K., 2015. Suppression of hydrogen-induced damage in friction stir welded low carbon steel joints. *Corrosion Science, 94, pp.88-98.*

[48] Hamade, R.F. and Baydoun, A.M.R., 2019. Nondestructive detection of defects in friction stir welded lap joints using computed tomography. *Materials & Design, 162, pp.10-23.*

[49] Kumar, T.P. and Reddy, P.P., Non-Destructive Analysis of FSW Process and Comparison with Simulation and Microstructural Analysis. *taper, 7(4), p.2.*

[50] Mahmoudiniya, M., Kokabi, A.H., Kheirandish, S. and Kestens, L.A., 2018. Microstructure and mechanical properties of friction stir welded ferrite-martensite DP700 steel. *Materials Science and Engineering: A, 737, pp.213-222.*

Comparative evaluation of machining performance of inconel 625 under dry and cryogenic cutting conditions

G B Narkhede[1*], Satish Chinchanikar[1], S S Vadgeri[2], B S Rathod[1]
[1]Vishwakarma Institute of Information Technology, Pune, India
[2]AISSMS College of Engineering, Pune-01, India

* ganesh.narkhede@viit.ac.in

Abstract. Hast alloys, wasp alloys and tool steels are the most commonly used materials in nuclear engineering and aerospace industries. However, these materials are considered to be "Difficult to machine" which attracted several researchers to work in this domain with a view to obtain optimum machining performance. The present work explored the comparative evaluation of machining performance of Inconel 625 under dry and cryogenic cutting conditions from an industrial perspective with emphasis on cutting forces, tool-chip interface temperature, surface finish and tool life. Turning experiments which were planned using the orthogonal array of L27were carried out using Tungsten carbide Al2O3 coated inserts varying the cutting speed, feed and depth of cut. Cutting forces generated during turning at varying cutting conditions were measured using three axis lathe tool dynamometer. It was found the combination of speed of 200 m/min, feed value of 0.04 mm/rev and depth of cut value of 0.2 mm gives the optimum cutting forces with superior surface finish and low tool chip interface temperature in dry cutting condition. Again for same combination of cutting parameters were tested with cryogenic coolant Liquid Nitrogen (LN2) and it was observed that cutting forces reduced up to 30%, surface roughness improved by 31.37% and temperature reduced by 71.67%.
Keywords: *Cutting forces, Cutting temperature, Cryogenic coolant, Dry turning, Inconel 625, Surface finish.*

1. Introduction

In today's competitive market and with the advancement in the technology, it becomes very much imperative to manufacture better quality products at low cost and that too in minimum time. However, still today it's a challenge to economically machine difficult-to-cut materials. Hast alloys, wasp alloys and tool steels are the most commonly used materials in nuclear engineering and aerospace industries. However, these materials are considered to be "Difficult to machine" which attracted several researchers to work in this domain with a view to obtain optimum machining performance.
Enterprises can achieve maximum revenue by manufacturing excellent quality product at lower cost by optimizing cutting parameters. The quality of surface finish is commonly used as index of product quality and it is also critical in functional behavior of the components like machine tool parts, dies etc. The optimum performance from the turning process is possible only when all the control parameters like speed, feed, depth of cut and coolant which affect the process are selected properly and with prior investigation. This in turn demands the study and analysis of the process parameters and finding out the combination of optimal cutting parameters in order to improve the performance of turning process and to reduce the cost while achieving the required geometrical tolerance.

An area of research interest in turning is the analysis of cutting forces, as minimum power consumption is a never ending process. Among the three components of cutting force namely tangential force, radial force and feed force, tangential component of force prominently influences power consumption and can be obtained using the equation as given below:
$Fc = kc \times DOC \times f$
Where, DOC = Depth of cut in mm, f is feed in mm/rev and kcis specific cutting energy coefficient in N/mm2.
From the equation and studies reported on experimental investigations of cutting forces during machining [1-5], it has been understood that cutting force is mostly influenced by the depth of cut and feed and varies with specific cutting energy coefficient for different materials. Researchers carried out comparative evaluation of cutting forces using different cutting conditions and cooling medium with a view to obtain lower values of cutting forces and hence lowering power consumption requirement in machining. However, working in thiS machining domain is constantly going on with the advancement of new work materials and new cutting edge technology [6].
Another important parameter of research interest is obtaining minimum surface roughness during machining of "difficult to machine" materials as an optimum surface finish would influence performance of mechanical parts. Industries like aerospace and nuclear engineering find wide applications of difficult to machine materials to avoid distortion caused after heat treatment. About 50 wt% of aero-engine alloys are nickel-based alloys as they exhibit higher strength to weight ratio and ability to maintain high resistance to corrosion, mechanical and thermal fatigue, mechanical and thermal shock, creep, and erosion at elevated temperatures[7]. Inconel 625 is a super alloy "difficult to machine" material which offers excellent stability during heat treatment. Inconel 625 has gained popularity for applications where a high risk of distortion or cracking exists due to increased corrosion resistance, wear, & abrasion resistance. Machinability of Inconel 625 varies from low to medium and therefore it becomes necessary to study the effects of speed, feed, depth of cut and type of coolant with the emphasis on productivity, tool life improvement and reliable performance characteristics.
Machining industries are trying for high material removal rate with improved product quality. However, the major problem to achieve this is high cutting temperature and cutting forces during machining of Inconel 625. Conventional cutting fluid fails to provide proper cooling effect [8-9]. Moreover, use of oil based coolant is more prone to environmental pollution due to chemical breakdown of cutting fluid at high temperatures and can create operator health issue due to bacterial growth in cutting fluid [9]. On the other hand, requirement of pump, storage, chilling unit, filtration system, recycling etc. adds to cost and creates problem for disposal of cutting fluid which attracted many researchers to work in the domain of sustainable manufacturing. One such alternative, which is emerging as a significantly beneficial sustainable manufacturing process, is cryogenic machining. It is reported that using cryogenic fluid over dry machining provide less cutting forces, better surface finish and improved tool life. In cryogenic cooling, the work-piece or chip changes properties of material from ductile to brittle because, the ductile chip material can become brittle when the chip temperature is lowered so it improves discontinues chips formation [10].
Yakup and Muammer [11] described the various cooling approaches in cryogenic turning. First approach was to cool the cutting point through heat conduction by liquid nitrogen (LN2)flow at the tool chip-tool interface or delivery of liquid nitrogen at the chip-tool interface through the hollow cutting tool. Directing LN2at interface increased the machining performance and does not caused significant changes in properties of the work-piece and machine tool structure. Another approach in cryogenic cooling is to cool cutting zone, particularly tool-chip interface with liquid nitrogen by using nozzle. In cryogenic jet cooling method, LN2 is sprayed at tool rake and/or the tool flank. From the available literature, it is understood that due to localized cryogenic cooling reduces the tool face temperature which in turn reduced the possibility of chips welding to the tool resulting in better surface quality.

Hong and Din [9] elaborated different cooling approaches in cryogenic machining of Ti-6Al-4V. During their study they recorded the temperature in the cutting zone by using the K-type thermocouple. Their study reported that by applying LN2 to close proximity of cutting tool, cutting temperature at the chip-tool interface can be reduced. Yusuf [12] evaluated the machining performance in cryogenic machining of Inconel 718. They observed lower values of cutting forces and better surface finish when using cryogenic cooling in comparison to dry machining. Bapat et al. [13] developed a numerical model to obtain temperature distribution.

Seong – chan Jan [14] study reported that friction coefficient between tool and work-piece depends on the cryogenic cooling approach, especially nozzle design and nozzle location. They observed lower values of friction coefficient when LN2 is delivered at the tool chip interference. Grzesik[15] conducted experiment on the influence of cryogenic cooling on process stability in turning operations and found that cryogenic machining presents a more sustainable alternative to conventional machining process via increased machining process stability and consequently increased productivity.

Dhar et al. [16] studied the influence of cryogenic cooling on tool wear, dimensional accuracy and surface finish during turning of AISI 1040 and 4340 steels. Their study concluded significant improvement in surface finish and reduction in cutting zone temperature at cryogenic cooling in comparison to dry cutting conditions. Jerold and Pradeep Kumar [17] cryogenic machining study concluded reduction in cutting temperature by 22%, cutting forces up to 38% under cryogenic cutting in comparison to dry machining. Although, a sufficient study is reported on use of cryogenic cooling during machining, very few have reported on comparative evaluation of machining performance of Inconel 625 under dry and cryogenic cutting conditions from an industrial perspective with emphasis on cutting forces, tool-chip interface temperature, surface finish and tool life. With this view, in the present study, machining performance of Inconel 625 was investigated under dry and cryogenic cutting conditions.

2. Experimentation

Experimentation work was carried out on specimen of diameter 40 mm and length 1000mm with the aim of improvement of the quality and reliability of the turning process. It was also intended to evaluate and understand the effect of the cutting forces and type of coolant on surface roughness during the turning operation.

2.1. Work-piece material

Inconel 625was used as a work-piece material which is popularly used in applications like marine applications, aerospace industries, chemical processing plant, nuclear reactors, pollution control equipment, where properties like high surface finish, high corrosion resistance and wear resistance are of prime importance. Its hardness ranges between 45-50 HRC and the chemical composition is as follows: Nickel- 58%; Chromium- 20 to 23 %; Molybdenum- 8-10 %; Iron- 5%; Aluminium- 0.4 %; Titanium- 0.4 %; Carbon- 0.1 %; Manganese- 0.5 %; Silicon- 0.5 %; Cobalt- 0.5 %.

Table 1. Physical and mechanical properties ofInconel 625

Density	0.305 Kg/m³
Melting Point range	2350-2460°C
Thermal Conductivity	74 W/m°K
Specific Heat	0.107 KJ/Kg°K
Poisson's ratio	0.312
Tensile Strength	29.2 N/mm²

2.2. Experimental Setup and Tooling

Dry and cryogenic experiments were performed varying cutting speed, feed and depth of cut as shown in Table 1. Experiments were planned using the Taguchi L27orthogonal array. Tungsten carbide inserts (Grade: TNMG 160408Al2O3 coated)were used and tool holder used was designated as WIDIA ID 4A MTJNR 2525M16.

Table 2. Experimental matrix for dry and cryogenic cutting.

Parameter	Levels		
	1	2	3
Speed (m/min)	50	125	200
Feed (mm/rev)	0.04	0.08	0.12
Depth of cut (DOC) (mm)	0.2	0.3	0.4

Kistler dynamometer 9257B was used to compute three components of cutting forces; feed force (Fx), thrust force (Fy), tangential force (Fz). It is piezo-electric dynamometer capable of-5 KN to 5 KN of direct measurement. Its high resolution enables the smallest dynamic changes in large forces to be measured. The instantaneous roughness criteria measurements (Ra) for each cutting condition were obtained by means of Mitutoyo roughness tester. Experiments were repeated four times and the result is the average of these values. Temperature at the cutting zone was recorded by using digital infrared thermometer temperature gun. The experimental set-up is shown in Figs. 1, 2 and 3.

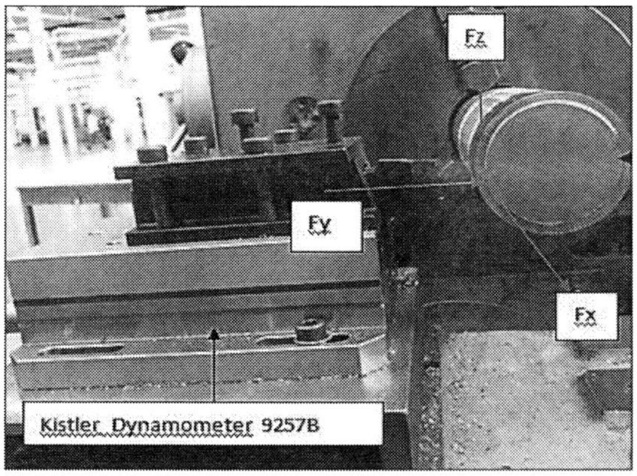

Figure 1. Cutting tool mounted on dynamometer.

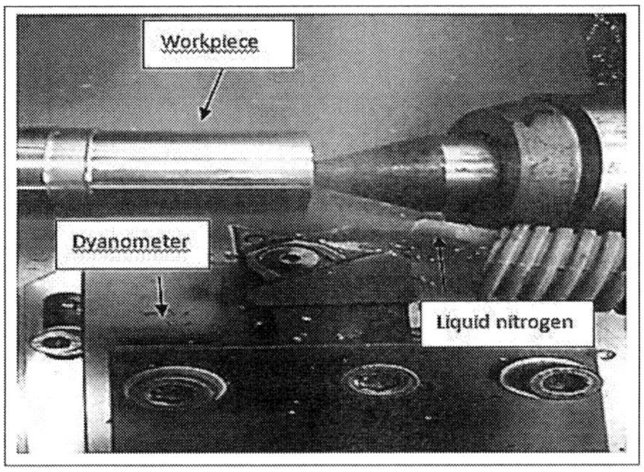

Figure 2. Cryogenic spraying and machining set-up.

Figure 3. Force measurement.

3. Result and Discussion

Dry and cryogenic experiments which were planned using the Taguchi L27orthogonal array were performed using Al_2O_3coatedtungsten carbide inserts varying cutting speed, feed and depth of cut as shown in Table 1.Dry turning experiments were performed without use of coolant resulted in higher chip-tool interface temperature. However, higher heat generated at the tool-chip interface caused annealing and softening of the work material just ahead of the tool, made it easier to shear.
However, manufacturing industries try for high material removal rate (MRR) with higher quality of machined surface. On the other hand, using higher cutting parameters to achieve higher MRR generates large amount of heat and hence, higher cutting temperature which adversely affects tool life. Many a times lead to catastrophic failure of cutting tools which significantly affects machined surface quality. Cryogenic machining is emerging as a significantly beneficial alternative and sustainable manufacturing process. With this view, to have a comparative evaluation, experiments were also performed using cryogenic cooling at the same cutting parameters which were used for dry cutting experiments.
Experiments planned using the Taguchi orthogonal array provides systematic way of conducting experiments with consideration of all the levels of the parameter and factors taken into account. The Experiments were conducted using the Taguchi L27 design of experiments (DOE) method. The factors were selected based on literature review, pilot experiments and manufacturer recommendations for dry turning of Inconel 625 using Al_2O_3 coated tungsten carbide inserts. This study involves the study of process parameters like speed, feed and depth of cut and its effect on surface roughness value. The output parameters taken into consideration are cutting force (Fx), feed force (Fy) thrust force (Fz), surface finish (Ra) and temperature (T).
Experimental results obtained during dry turning of Inconel 625 using Al_2O_3 coated tungsten carbide inserts varying cutting speed, feed and depth of cut as shown in Table 3. Underlined and highlighted parameters shown in the Table 3 observed as optimum cutting parameters for lower value of surface roughness. However, parameters are to be optimized for lower values of cutting forces and minimum temperature along with the lower surface roughness. With a view to have comparative evaluation of machining performance under dry and cryogenic cutting conditions, experiments using cryogenic coolant (LN_2) were carried out at cutting parameters which produced lower value of surface roughness under dry cutting condition. Experiments were also carried varying the cutting speed at other two levels with constant values of feed and depth of cut which produced lower value of surface roughness under dry turning. Experimental matrix and responses obtained under cryogenic turning of Inconel 625 using Al_2O_3 coated tungsten carbide inserts is shown in Table 4. Each set of turning operation was repeated four times and the measured responses are averages of four experimental observations. It can be seen lower values of surface roughness, cutting forces and chip-tool interface temperature under cryogenic turning as against higher values observed under dry cutting.

Table 3. Experimental results under dry turning.

DOC	Feed	Speed	Ra (μm)	*Fx (N)*	*Fy (N)*	*Fz (N)*	Temp (°C)
0.2	0.04	50	0.701	57.0	63.0	77.0	35
0.2	0.04	125	0.663	48.4	43.8	72.2	40
0.2	**0.04**	**200**	**0.612**	**69.5**	**70.0**	**60.1**	**60**
0.2	0.08	50	1.001	69.3	41.9	88.2	57
0.2	0.08	125	0.981	96.6	62.9	125.7	70
0.2	0.08	200	0.838	75.1	93.1	153.3	91

0.2	0.12	50	0.870	78.9	85.5	104.4	48
0.2	0.12	125	0.667	94.8	121.9	141.7	68
0.2	0.12	200	0.702	138.6	127.5	180.6	110
0.3	0.04	50	0.902	118.8	108.2	154.0	38
0.3	0.04	125	0.890	82.6	98.4	175.9	83
0.3	0.04	200	0.759	72.4	70.8	225.2	73
0.3	0.08	50	1.039	73.7	64.2	167.9	84
0.3	0.08	125	1.093	57.2	78.0	240.1	112
0.3	0.08	200	1.034	73.1	92.1	183.8	139
0.3	0.12	50	1.020	177.6	94.0	156.6	48
0.3	0.12	125	1.103	200.1	85.32	173.3	57
0.3	0.12	200	1.051	223.5	114.1	183.9	51
0.4	0.04	50	1.123	142.7	89.6	123.3	39
0.4	0.04	125	0.849	163.0	120.7	147.1	70
0.4	0.04	200	0.844	171.4	155.3	176.1	110
0.4	0.08	50	1.374	130.7	124.2	327.0	83
0.4	0.08	125	1.276	126.4	131.0	273.2	116
0.4	0.08	200	1.312	167.6	141.4	305.8	125
0.4	0.12	50	1.120	232.6	157.0	299.8	95
0.4	0.12	125	1.230	247.5	175.5	329.7	110
0.4	0.12	200	1.424	273.5	192.6	376.9	153

Underlined and highlighted parameters shown in the Table 4 observed as optimum cutting parameters for lower value of surface roughness under cryogenic cutting and can be clearly seen lower values of responses as against dry turning. Comparative evaluation of machining performance under dry turning and under cryogenic turning obtained at optimum cutting condition of depth of cut of 0.2 mm, feed value of 0.04 mm/rev and at cutting speed of 200 m/min are shown in Table 5.

Table 4. Experimental results with cryogenic coolant at optimum parameters obtained under dry cutting.

DOC	Feed	Speed	*Ra*(μm)	*Fx* (N)	*Fy*(N)	*Fz* (N)	Temp (°C)
0.2	0.04	50	0.59	51.0	54.6	67.1	13
0.2	0.04	125	0.52	44.4	38.5	59.2	16
0.2	**0.04**	**200**	**0.42**	**47.7**	**62.4**	**43.2**	**17**

It can be seen that cutting forces, surface roughness and cutting temperature reduced when turning under cryogenic cooling as against dry turning. Cryogenic cooling reduces cutting forces by up to 31% as compared to dry turning operation. Surface roughness value observed to be 0.42 μmwhich was improved by 31% when turning using cryogenic cooling. However, significant reduction around 72% in chip-tool interface temperature was observed which reduced a temperature from 60°C to a value of 17°C.

Table 5. Comparison of results for dry and cryogenic cutting.

Output Parameters	Dry Cutting	Cryogenic turning	% Change.
Surface finish (µm)	0.612	0.42	31.37
Feed force	68.30	47.70	30.16
Thrust force	70.04	62.46	10.82
Tangential force	60.01	43.20	28.01
Temperature	60	17	71.67

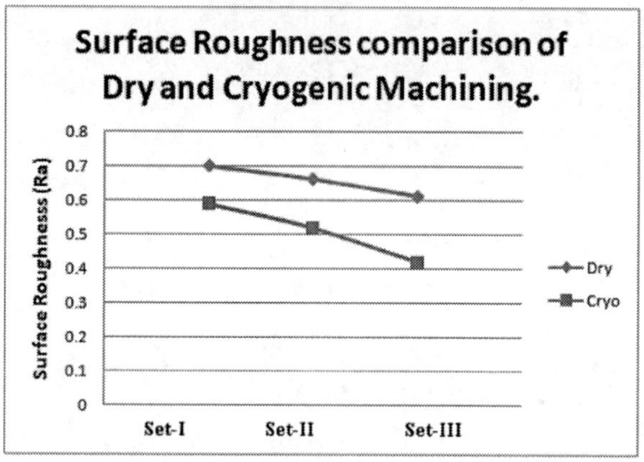

Figure 4. Surface roughness comparison.

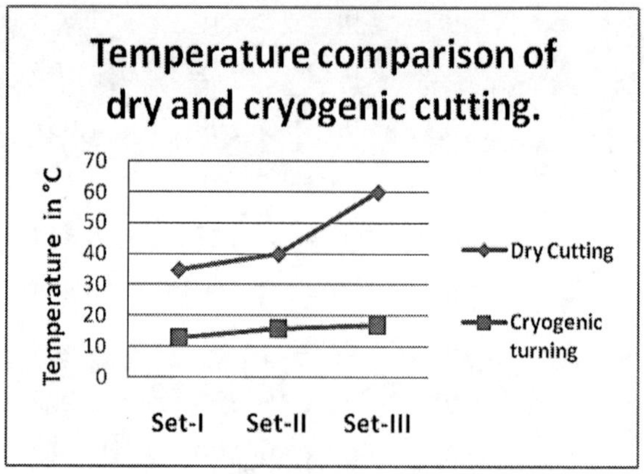

Figure 5. Temperature comparison.

4. Conclusion

Experiments were conducted using the Taguchi L27 design of experiments (DOE) method for dry turning of hard materials like Inconel 625 alloy using tungsten carbide inserts to find the combination of optimum input parameters like speed, feed and depth of cut. It was observed that depth of cut of 0.2 mm, feed value of 0.04 mm/rev and cutting speed of 200 m/min is the best combination to get lower surface roughness of 0.612 µm under dry turning and which further reduced to a value of 0.42 µm under cryogenic cooling condition. It was observed that using liquid nitrogen coolant can improve the performance of the turning process. By using liquid nitrogen coolant, cutting forces reduced up to 30%, surface finish improved by 31.37% which shows finishing operation can be replaced by turning operation with optimum parameters with cryogenic coolants. It was also observed that tool chip interface temperature reduced by 71.67% which helps to improve the tool life. In the present work, cutting parameters are optimized for lower value of surface roughness. However, parameters are required to be optimized for lower values of cutting forces and minimum temperature along with the lower surface roughness during turning of Inconel 625 alloy using tungsten carbide inserts under dry and cryogenic cooling cutting conditions.

References

[1] Satish Chinchanikar, S.K. Choudhury, "Experimental investigations to optimise and compare the machining performance of different coated carbide inserts" Proc IMechE Part B: J Engineering Manufacture, Vol. 228(9) (2014) 1104–1117.

[2] Sujit Pawar, Aniket Salve, Satish Chinchanikar, Residual Stresses during Hard Turning of AISI 52100 Steel: Numerical Modeling with Experimental Validation, Materials Today: Proceedings 4 (2), 2350-2359, 2017.

[3] Vishal Wayal, Nitin Ambhore, Satish Chinchanikar, Virendra Bhokse, Investigation on Cutting Force and Vibration Signals In Turning: Mathematical Modeling Using Response Surface Methodology, Journal of Mechanical Engineering and Automation 2015, 5(3B): 64-68

[4] Avdesh Pal, S.K. Choudhury, Satish Chinchanikar, "Machinability assessment through experimental investigation during hard and soft turning of hardened steel" Procedia of Material Science 6 (2014) 80 – 91.

[5] Pradip Netake, Satish Chinchanikar, "Hard turning under minimum quantity lubrication: Modeling of cutting force and surface roughness" Journal of Advanced Engineering Research 2 (Issue 1), 17-22, 2015.

[6] Satish Chinchanikar, S.K. Choudhury, "Machining of hardened steel: Experimental investigations, performance modeling and cooling techniques: A review" International Journal of Machine Tools and Manufacture, Volume 89, 2015, 95–109.

[7] Ezugwu EO (2005) Key improvements in the machining of difficult-to-cut aerospace super alloys. Int J Mach Tool Manuf 45(12–13):1353–1367.

[8] W. Grzesik a, K. Żak a, M. Prażmowski, "Effects of cryogenic cooling on surface layer characteristics produced by hard turning." Archives of Material science and Engineering. Volume 54, issue 1, pages 5-12.

[9] Shane Y. Hong ,Yucheng Ding, "Cooling approaches and cutting temperatures in cryogenic machining of Ti-6Al-4V". International Journal of Machine Tools & Manufacture 41 (2001) 1417–1437.

[10] Seong – Chan Jan, "Lubrication effect of liquid nitrogen in cryogenic machining friction on the tool chip interference." Journal of Mechanical Science & Technology, Vol 19, PP 936 – 944, 2005.

[11] YakupYildiz_, MuammerNalbant, "A review of cryogenic cooling in machining processes". International Journal of Machine Tools & Manufacture 48 (2008) 947–964.

[12] Yusuf Kaynak, "Evaluation of machining performance in cryogenic machining of Inconel 718 and comparison with dry and MQL machining." AdvManufTechnol (2014) 72:919–933.

[13] P.S. Bapat, P.D. Dhikale, S.M. Shinde, A.P. Kulkarni, S.S. Chinchanikar, A Numerical Model to Obtain Temperature Distribution during Hard Turning of AISI 52100 Steel, Materials Today: Proceedings 2 (4-5) (2015) 1907-1914.

[14] Seong – Chan Jan, "Lubrication effect of liquid nitrogen in cryogenic machining friction on the tool chip interference." Journal of Mechanical Science & Technology, Vol 19, PP 936 – 944, 2005.

[15] W. Grzesik a, K. Żak a, M. Prażmowski, "Effects of cryogenic cooling on surface layer characteristics produced by hard turning." Archives of Material science and Engineering. Volume 54, issue 1, pages 5-12.

[16] N. R. Dhar and Nanda S.V.Kishore, "The effects of cryogenic cooling on chips and cutting forces in turning AISI 1040 and AISI 4320 steels."Journal of Engineering Manufacture.Volume: 216 issue: 5, page(s): 713-724Issue published: May 1, 2002

[17] B. Dilip Jerold, M. Pradeep Kumar, "The Influence of Cryogenic Coolants in Machining of Ti–6Al–4V." Journal of Manufacturing Science Engineering, Volume: 135, Issue: 3. June 2013.

[18] Shivaraj S. Vadgeri, Sunil R. Patil, Sandip T.Chavan, "Static and Fatigue Analysis of Lathe spindle for MaximumCutting Force", *Materials Today: Proceedings,* ISSN 2214-7853, Volume 5, Issue 2, pp. 4438-4444, 2018.

IOP Conf. Series: Materials Science and Engineering **810** (2020) 012037 doi:10.1088/1757-899X/810/1/012037

Compression Strength Testing of Brick by FFT

Sandip Langade, Nandkumar Gilke, Kashinath Patil
K. J. Somaiya College of Engineering, Mumbai-400077, INDIA
sshankul@gmail.com, nandkumargilke@somaiya.edu, kashinath@somaiya.edu

Abstract. Brick being major construction filler material, is manufactured by various processes using different materials. Traditional method of brick manufacturing uses soil as base material, which is molded in shape and then baked to strengthen the brick. Generally the quality of core baking is tested by sound test as per IS 2077. This paper describes the manufacturing of bricks made of various compositions of bagasse ash as base material and soil as a binder. The study also describes the effect of variation in parameters such as temperature and baking time on compressive strength. The compressive strength of core for the brick is checked by sound test which is performed using FFT analyser. The effects of these parameters on compressive strength and quality of the core baking were related with each other. The mathematical model was developed. The experimental results where validated by analytical method.

1. Introduction

The coal ash from power co-generation plant is widely used for making the construction products. The brick was made with lime or cement as a binder material. Some researchers used bagasse ash for making the brick along with the cement and lime as a binder.

Table 1 Material and curing methods for making bagasse ash brick.

Sr No	Author	Material		CuringMethod with Material	Curing Period
		Filler	Binder		
1	Mangesh Madurwar and Team (2014)	Bagasse Ash	Lime	Water Cured	21 days
2	Prashant O Modani and Team (2013)	Bagasse Ash and natural sand	Cement	Water Cured	28 Days
3	R. Srinivasan and Team (2010)	25 % Bagasse Ash	Cement	Water Cured	7 and 28 days
4	Ajay Goyal and Team	Bagasse Ash	OPC	600^{0}C	5 Hrs
5	K Vidhya and Dr. S.Kandasamy	Coalash,Pond ash, Bagasse Ash,Stone Dust,Gypsum	Lime	Water Curing	21 days
6	Apurva,Kulkarni, and Team (2013)	Bagasse Ash	Lime	Water Curing	21 days

The above table 1 shows literature survey it was observed that the bagasse ash was used as filler material and lime or cement are the binding material for the brick. The water emerged process is use for curing of the brick. The researcher is using the same strength checking processes. In above case the baking process of the brick is not in used. In case of burnt bricks the sound test is a valuable test which decides the core baking of the brick. The keywords show the research papers in the topic of FFT analysis for the various rotating metal parts, gear box components, concrete beams. Issam Abu Mahafouz [7] and his team worked on crack detection and identification using vibrating signal. For this they used accelerometer to collect the vibration signal and processed by signal processor. This crack detection was for beam and machine components. Table 2 shows the study of different methods used to detect the various parameters of different parts and materials by different researchers

Table 2 Non Distractive Testing Methods, detected parameters and materials

Sr No	Author	Method of Testing	Instrument Used	Part and Material of Sample	Parameters Tested
1	Issam Abu Mahafouz	Vibration Measurement Method	FFT analyser	Beam and machine components, Material- Steel, C.I.	Crack Detection.
2	Yu chen	Vibration Measurement Method	FFT Analyser	Bearing, Material- Steel	Vibration Analysis
3	T Rameshbabu	Vibration Measurement Method	HTTP Vibration Analyser	Air Craft Engine Rotor ,Material -- Steel	Crack detection through Vibration Analyser
4	Chrudatta Choudhri	Vibration Measuring Method	FFT Analyser With Impact Ball	Cantilever Beam, Material- steel	Crack detection through Vibration Analyser
5	Jiri Borozovsky	Non Destructive Testing	Ultrasonic Pulse Testing Method	concrete, calcium silicate, and brick	Different Parameters like Crack, Porosity,

The above detailed survey of research papers from various journals was done. These survey shows regarding various materials like metals, rotating parts, concrete and beams. The researcher was not focused on the topic of sound measurement for brick through FFT Analyser. The present study focus on the FFT analysis of the bricks, to arrive at the decision whether each brick is properly baked or not from core.

2. Experimental Work–

2.1 New Composition for Bagasse Ash Brick –

The conventional brick is made up of coal powder, soil, and bagasse as a reinforcement material. The coal powder is eliminated from the conventional brick material. The function of the coal powder is performed by the bagasse ash. Table 3 illustrates the brick made of various compositions of bagasse ash and soil. B20 indicates brick with 20 per cent bagasse ash

Table3. Composition for bagasse ash brick

Brick code As	Weight of ash (in gm)	Weight of soil (in gm)	Total weight of brick (in gm)
B 20	102.03	1384	1486.03
B 40	204.07	1038	1242.07
B 60	306	692	998
B 80	412	342	745

2.2 Brick Sound Test:

After making the brick at various compositions it was baked at various temperatures with various times period. Then sound test was perform to measure the frequency of the sound with the help of FFT Analyser. This test was performed dated 11/6/2018 at Rajarambapu Institute of Technology, Islampur, Tal Walva, District Sangli, Maharashtra, India. The detailed setup fig is attached.

Fig 1 shows the setup for FFT analyser testing.

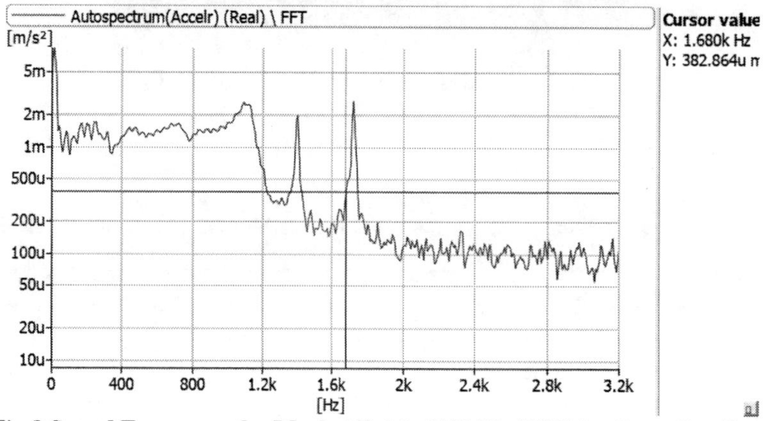

Fig 2 Sound Frequency for Market Brick 1100 Hz (FFT Analyzer Reading)

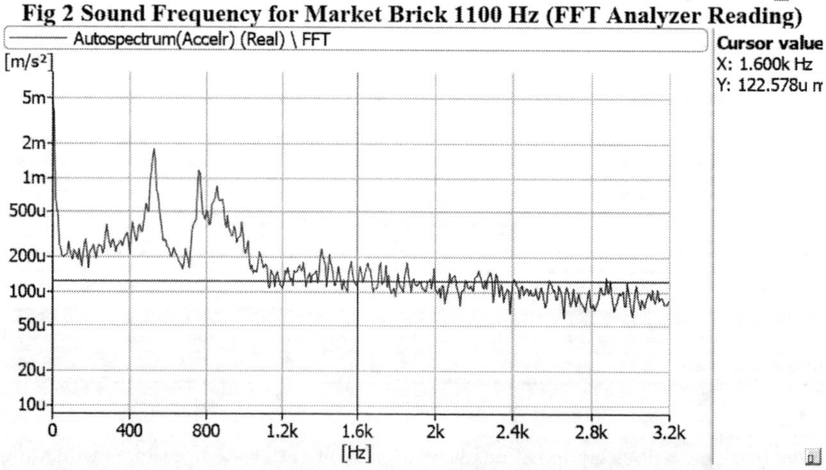

Fig 3 Sound Frequency for S1 sample Bagasse Ash Brick 528 Hz (FFT Analyzer Reading)

Fig 2 Shows the actual frequency peak curve for the market brick at FFT Analyzer screen, which has frequency of 1100 Hz. This frequency is consider as bench mark for the Bagasse Ash brick. Whereas Fig 3 shows the actual frequency curve for sample 1 made from Bagasse Ash brick which has frequency of 528 Hz.

3.3 Compressive Test-

IOP Conf. Series: Materials Science and Engineering **810** (2020) 012037 doi:10.1088/1757-899X/810/1/012037

This test is performed with the help of UTM .The compression force is applied on the brick till it break .This maximum load is recorded. After the sound test the same bricks were tested for compression strength on Universal Testing Machine. The results were compared with the standard brick. The collected sample data for compression strength and sound frequency were analysed for each sample. And experimental model is developed for sound frequency and compression strength. Based on these experimental readings the empirical formula was developed and the mathematical model was developed for compressive strength. Table 4 shows the details for Bagasse Ash Brick composition, temperature, baking time. The frequency for these brick range from 300 Hz to 738 Hz. The different sound frequency was measured for different composition, temperature and baking period. The maximum frequency for the brick was 738 Hz for 60 % ash composition brick which is baked at 600 0 C for 18 hrs. Baked brick the compressive strength for the brick was 1.13 N/mm2. The lowest frequency of the brick was 300 Hz with 0.5 N/mm2.

Table 4 Compression Strength and frequency model for Bagasse Ash Brick

Varying Parameters		Compression Strength N/mm^2				Frequency in Hz			
Temperature	**Time Hr**	**B20**	**B40**	**B60**	**B80**	**B20**	**B40**	**B60**	**B80**
Temp 600^{0}C	6	0.8	0.78	0.85	0.92	468	400	516	550
	10	0.75	0.85	0.76	0.81	476	600	446	528
	14	0.72	0.8	0.72	0.8	300	568	446	432
	18	0.75	0.75	0.8	0.7	570	528	738	324

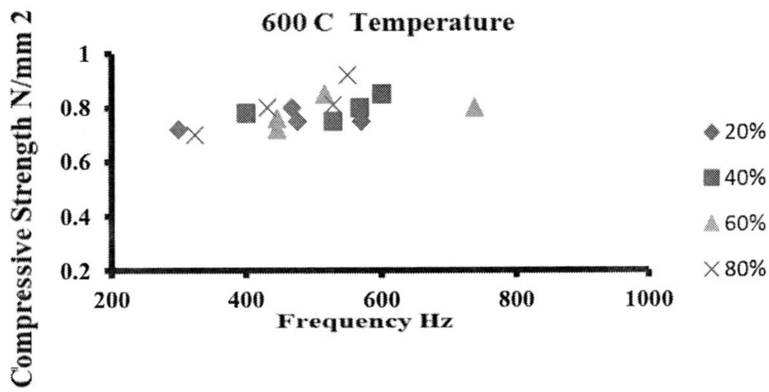

Fig 4. Compressive Strength with Frequency (Actual) for 600 ^{0}C baking temperature

Figure 3.The graph represents the variation of the frequency with change in compression strength of then brick baked at 600 0 C. It is observed that composition of the bagasse ash affects the compressive strength and frequency of the sound during the testing. The high Sound frequency for the brick indicates the core area of the brick is properly baked. The compressive strength of the brick was 0.92 N /mm2 which were the highest for this composition. The minimum frequency was 300 Hz for the brick. There was sudden drop for the free.

Table 5 Compression Strength and frequency model for 6000C Bagasse Ash Brick

Temperature	Per cent of Ash	Time Hrs	Compression Strength N/mm^2 (Experiment)	Frequency Hz	Compression Strength N/mm^2 (Analytical)	Error in %
600^{0}C	B20	6	0.8	468	0.85	-6.2
		10	0.75	476	0.81	-8
		14	0.72	300	0.79	-9.7
		18	0.75	570	0.77	-2.6
	B40	6	0.78	400	0.85	-8.9
		10	0.85	600	0.81	4.7
		14	0.8	568	0.79	1.2
		18	0.75	528	0.77	-2.6
	B60	6	0.85	516	0.85	0
		10	0.76	446	0.81	-6.5

	14	0.72	446	0.79	-9.7
	18	0.8	738	0.77	3.7
B80	6	0.92	550	0.85	7.6
	10	0.81	528	0.81	0
	14	0.8	432	0.79	1.2
	18	0.7	324	0.77	-10

Table 6 Compression Strength and frequency model for Bagasse Ash Brick

Varying Parameters		Compression Strength N/mm^2				Frequency in Hz			
Temperature	**Time Hr**	**B20**	**B40**	**B60**	**B80**	**B20**	**B40**	**B60**	**B80**
	6	0.95	0.9	0.81	0.78	544	558	506	566
	10	0.89	0.75	0.85	0.75	588	400	600	490
	14	0.72	0.72	0.8	0.76	310	410	401	480
Temp 700^{0}C	18	0.72	0.7	1.1	0.79	410	274	990	703

Table 6 shows the details for the brick composition, temperature, baking time. The frequency for these brick range from 274 Hz to 990 Hz. The different sound frequency was measured for different composition, temperature and baking period. The maximum frequency for the brick was 990 Hz for 60 % ash composition brick which is baked at 700 0 C for 18 hrs. baked brick the compressive strength for the brick was 1.1 N/mm2. The lowest frequency of the brick was 274 Hz with 0.22 N/mm2.

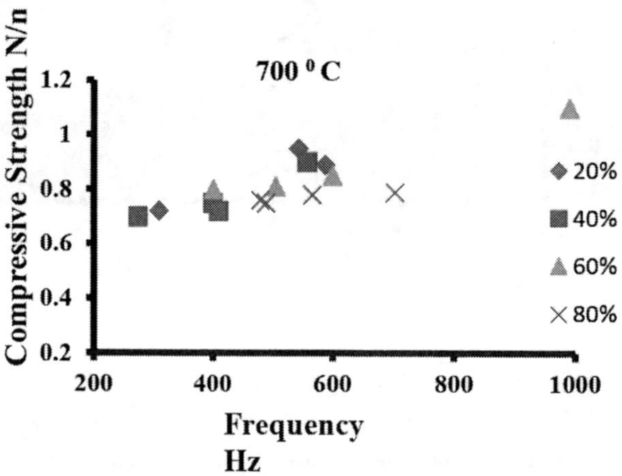

Fig 5. Compressive Strength with Frequency for 700 ⁰C baking temperature

Figure 4. The graph represents the variation of the frequency with change in compression strength. It is observed that variation in the bagasse ash composition and baking temperatures affects the frequency where the strength will remain same for two different compositions. The strength will not more than 1.1N/mm2.

Table 7 Compression Strength and frequency model for 7000 C for Bagasse Ash Brick

Temperature	**Per cent of Ash**	**Time Hrs**	**Compression Strength N/mm^2 (Experiment)**	**Frequency Hz**	**Compression Strength N/mm^2 (Analytical)**	**Error in %**
700^{0}C	B20	6	0.95	544	0.85	10.5
		10	0.89	588	0.81	8.98
		14	0.72	310	0.79	-9.72
		18	0.72	410	0.77	-6.94
	B40	6	0.9	558	0.85	5.55

	10	0.75	400	0.81	-8
	14	0.72	410	0.79	-9.72
	18	0.7	274	0.77	-10
B60	6	0.81	506	0.87	-7.40
	10	0.85	600	0.92	-8.23
	14	0.8	401	0.85	-0.62
	18	1.1	990	1.16	-5.45
B80	6	0.78	566	0.85	-8.97
	10	0.75	490	0.81	-8
	14	0.76	480	0.79	-3.94
	18	0.79	703	0.77	2.53

Table 7 shows the percentage variation in the experimental compression strength (actual) and the analytical strength for the same frequency. It is not more that 10 % plus or minus side. The experiment suggests the maximum frequency of sound for brick is concentrated between 274 to 990 Hz. The respective compression strength is between 0.5 to 1.1N/mm2. The highest frequency is 990 Hz where compression strength is 1.1N /mm2

Table 8 Compression Strength and frequency model for Bagasse Ash Brick

Varying Parameters		Compression Strength N/mm^2				Frequency in Hz			
Temperature	**Time Hr**	**B20**	**B40**	**B60**	**B80**	**B20**	**B40**	**B60**	**B80**
Temp 800^{0}C	6	0.95	0.95	0.95	0.87	500	772	620	600
	10	0.89	0.85	0.92	0.86	450	574	400	588
	14	0.85	0.75	0.88	0.82	452	474	454	454
	18	0.8	0.7	0.85	0.75	350	400	364	400

Table 8 shows the details for the brick composition, temperature, baking time. The frequency for these brick range from 350 Hz to 772 Hz. The different sound frequency was measured for different composition, temperature and baking period. The maximum frequency for the brick was 772 Hz for 40 % ash composition brick which is baked at 800 0 C for 6 hrs. baked brick the compressive strength for the brick was 0.95 N/mm2.

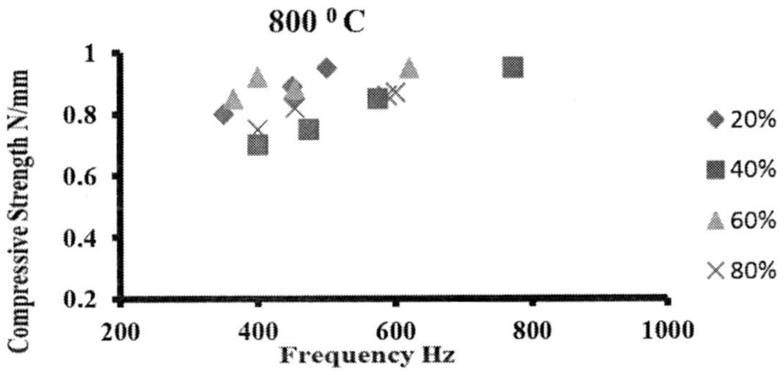

Fig 6. Compressive Strength with Frequency for 800 ^{0}C baking temperature

Figure 5. The graph represents the variation of the frequency with change in compression strength for 8000 C baking temperature. It is observed that variation in the bagasse ash composition and baking temperatures affects the frequency where the strength will remain same for two different compositions. The strength will not more than 0 .95N/mm2.

Table 9 Compression Strength and frequency model for 8000C Bagasse Ash Brick

Temperature	Per cent of Ash	Time Hrs	Compression Strength N/mm^2 (Experiment)	Frequency Hz	Compression Strength N/mm^2 (Analytical)	Error in %
800^{0}C		6	0.95	500	0.85	10.5
		10	0.89	450	0.81	8.9

B20	14	0.85	452	0.79	7.0
	18	0.8	350	0.77	3.7
	6	0.95	772	0.85	10.5
B40	10	0.85	574	0.81	4.7
	14	0.75	474	0.79	-5.3
	18	0.7	400	0.77	-10
	6	0.95	620	0.85	10.5
B60	10	0.92	400	0.81	11.9
	14	0.88	454	0.79	10.2
	18	0.85	364	0.77	9.4
	6	0.87	600	0.85	2.2
B80	10	0.86	588	0.81	5.8
	14	0.82	454	0.79	3.6
	18	0.75	400	0.77	-2.6

Table 9 shows the percentage variation in the experimental compression strength and the actual strength for the frequency. It is not more that 10 % plus or minus side. The experiment suggests the maximum frequency of sound for brick is concentrated between 350 to 772 Hz. The respective compression strength is between 0.5 to 0.95N/mm2. The highest frequency is 772 Hz where compression strength is 0.85N /mm2

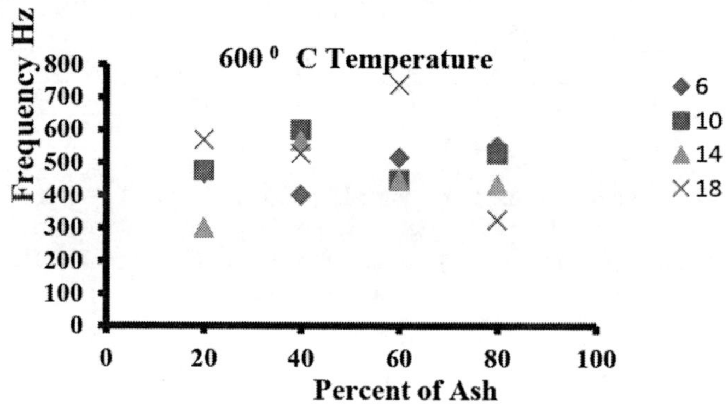

Fig 7. Percentage Composition of Bagasse Ash with Frequency for 600 ^{0}C baking temperature

Figure 6 shows the sound frequency with variation in composition of ash in brick, which are baked at temperature 600 0C. The effect of the composition and temperature represents in the graph. This will support to reduce the destructive testing of the brick. The lighter bricks are produced at higher temperature which is far less than the conventional bricks used for the Indian construction sites.

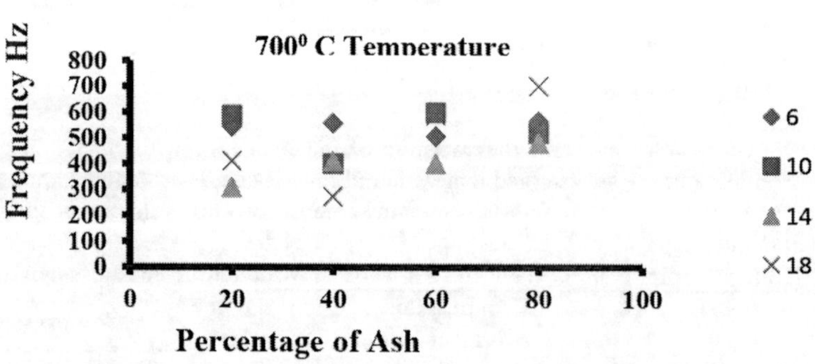

Fig 8. Percentage Composition of Bagasse Ash with Frequency for 700 ^{0}C baking temperature

Figure 7 shows the sound frequency with variation in composition of ash in brick, which are baked at temperature 700 0C. The effect of the composition and temperature represents in the graph. This will support to reduce the destructive testing of the brick. The lighter bricks are produced at higher temperature which is far less than the conventional bricks used for the Indian construction sites.

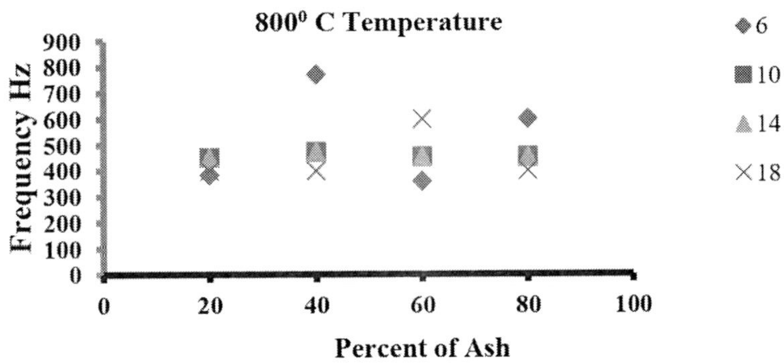

Fig 9. Percentage Composition of Bagasse Ash with Frequency for 800 ^{0}C baking temperature

Figure 8 shows the sound frequency with variation in composition of ash in brick, which are baked at temperature 8000C. The effect of the composition and temperature represents in the graph. This will support to reduce the destructive testing of the brick. The lighter bricks are produced at higher temperature which is far less than the conventional bricks used for the Indian construction sites.

Conclusion

The bricks were manufactured using various compositions of bagasse ash as a base material and soil as a binder. The percentage of bagasse ash used was ranging from 20, 40, 60 and 80 per cent. The bricks made of above composition were baked at temperature 6000C, 7000C, 8000C and baking time was varied from 6,10,14, and 18 Hrs. .The sample bricks were manufactured with above varying parameters were then tested for compression strength and sound test. The sound test was followed by compression test for above bagasse ash brick. The sound frequency of the bagasse ash brick was in the range of 400 Hz To 700 Hz. The 60 per cent composition of Bagasse Ash Brick was maximum compression strength 1.12 N/mm2.The color of the brick was red. The mathematical model and empirical formula was developed. The actual values and analytical values were compared which found to have error less than 10 per cent.

References

1.M. Mangesh V., M. Sachin A., and R. Rahul V., "Use of sugarcane bagasse ash as brick material," Current Science 107, vol. no.
6,pp.1044-1051,2014
2.Modani , Prashant O., and M. R. Vyawahare, "Utilization of bagasse ash as a partial replacement of fine aggregate in concrete,"
Procedia Engineering, vol.no. 51,pp.25-29, 2013
3. R. Srinivasan , and K. Sathiya "Experimental study on bagasse ash in concrete," International Journal for Service Learning in
Engineering, Humanitarian Engineering and Social Entrepreneurship 5, vol.no.2, pp. 60-66, 2010
4.G. Ajay, and H. Kunio, H. Ogata, and Mandula" Properties and Reactivity of Sugarcane Bagasse ash,"07006-35
5.K. Vidhya , Dr Kandasamy, Hybrid ash bricks and method of preparation"
6. Kulkarni Apurva , Samruddha Raje , and Mamta Rajgor., "Bagasse ash as an effective replacement in fly ash bricks." International Journal of Engineering Trends and Technology 4, no. 10 (2013): 4484-4489.

A Review on synthesis of AlB_2 reinforced aluminium matrix composites

Samuel Dayanand[1], Dr Satish Babu B[2]

[1]Department of Mechanical Engineering, Govt., Engineering, College, Raichur and affiliated to Visvesvaraya Technological University, Belgavi, Karnataka, India
[2] Department of Mechanical, School of Engineering, Presidency University, Bengaluru

samueldayanand@gmail.com

Abstract. In order to meet the demands of light weight high performance of materials in aerospace, automotive, marine and other devices, Aluminium matrix composites (AMCs) are developed. The combinations of matrix, reinforcement and processing route helps in proper bonding and uniform distribution in the matrix can be customized by the AMC's. The properties of AMCs are considered to be the best alternatives for making light weight and high strength applications. AMCs can be prepared by different methods with various reinforcements which are usually metallic compounds of oxygen, carbon, boron, nitrogen and their combinations. Due to its very good stiffness, hardness and wear withstand, high temperature applications aluminium diborides (AlB_2) holds attention in metal matrix composite. The paper tries to explore the capabilities of combing aluminium diboride reinforced AMCs with various methods.

1. Introduction

Nowadays, in the upcoming fields of engineering aluminium and aluminium alloy based composites are gaining importance [1]. There are two or more physically and chemically distinct phases in the Metal-matrix composites (MMCs) which are well spread to make available, the properties which cannot be obtained with either of the single phases. The reinforcement may be metallic or ceramic in MMCs but generally a metal or an alloy is used as a matrix phase. Particle reinforced, short fiber and continues fiber are the main three types of MMCs [2]. Usually aluminium, magnesium, copper and other their alloys are used as a matrix .Fibers, whiskers and particulate of various metal oxides, nitrides and carbides are forms commonly used in the reinforcement [3-4]. Chiefly due to the ease of processing, lower fabrication costs and isotropic properties particle reinforced MMCs are a better option over fiber reinforced materials [5-8] .The form, size and the shape of the reinforcement controls the characteristics of the composites. The overall properties are nevertheless affected by the quality of interface in between the bulk material (matrix) and the dispersoid (reinforcement) [9-12]. Nami et al. [13] reported that composites with Al matrix are gaining prominence due to their properties such as improved strength, stiffness, hardness, coefficient of thermal expansion, improved wear and corrosion resistances, performance at higher temperature and the increasing need of light weight and genuine materials. Sharifitabar et al. [14] found the improvement of these composites is driven by the development of these materials has been driven by structural and non structural applications in automobile and aerospace industries. Various manufacturing methods with different ceramic reinforcement have been used to fulfill these demands. Alidokht et al. [15] mentioned AMC's are potential candidates in applications such as engines, aircraft skins, landing gear struts, sea vehicles, turbine engines components, space structures etc. Michael Rajan et al. [16] and Kumar et al. [17] reported due to its unique combinations of properties AMC's have replaced conventional material in a number of applications. SiC [18], B_4C [19], Al_2O_3, TiC, B_4C, TiO_2, SiO_2, BN, AIN, fly ash, AlB_2, TiB_2 and ZrB_2 [20] particles are the different dispersoids of ceramic material commonly utilized in Al composites [21-39]. Tee et al. [40] found new processing techniques are being used to produce high performance in situ composites. Kumar et al. [41] and Gotham et al.[42] were investigated that these insitu composites to improve the interfacial capability and reduce the reinforcement size and exothermic reaction between aluminum and the ceramic compounds. Dumitru et al. [43] found the information on the in situ AlB_2 particle composites is very limited.

2.Synthesis of AlB_2 Reinforced Aluminium Matrix Composites

The two broadly classified categories for processing AlB_2 MMC's on the basis of state of matrix and reinforcement are solid-state processing and liquid state processing. The other method of classifying to process Al-AlB_2 MMC's are based on exsitu process and insitu process. The former process includes adding separately produced dispersoid materials in the form of particulate to the aluminium or Al alloy melt. Exsitu process has many restrictions like weak bonding and varying distributions of the particulate, weak isometric characteristics, reaction at interface, thermodynamic instability and lower adhesions of matrix with the dispersoids. The latter technique includes formation of dispersoids within the matrix through insitu reactions between matrix and reinforcing material such as halide salts. C.S. Ramesh et al. [44] revealed that the insitu method has several advantages such as uniform distribution of dispersoid particulates, clear interface, refined grains and improved high temperature stability. Different methods are discussed in the following sections.

2.1Solid - liquid reaction Processes

2.1.1Exothermic Reaction Process

This process also termed as XD^{TM} technology was developed in 1980's in USA [45-47]. During the process the ceramic phase forming elements say X and Y are heated along with a metallic element A above its fusion temperature but below the temperature needed to form the ceramic reinforcement material of XY phase. The nature of chemical reaction between X and Y is exothermic and as a result sub microscopic hardening particles are formed in the solvent phase as shown in 'figure 1'.

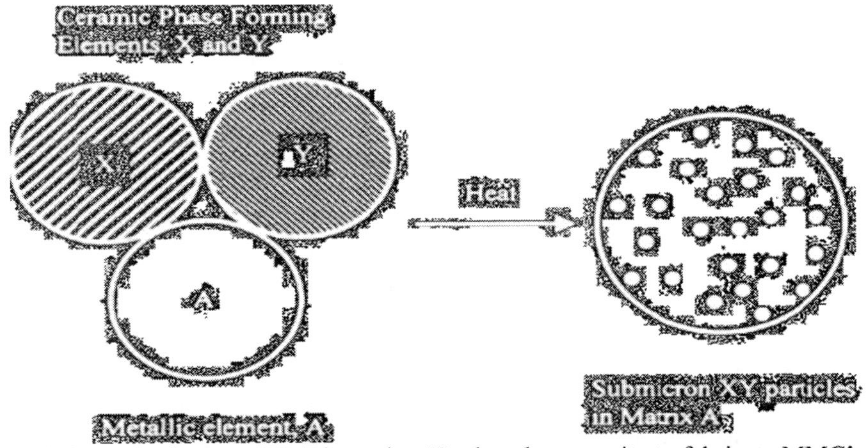

Figure 1. Schematic diagram XD Dispersion Hardened composite to fabricate MMC's [48].

IOP Conf. Series: Materials Science and Engineering **810** (2020) 012038 doi:10.1088/1757-899X/810/1/012038

Modovan P et al. [48] observed that composites with high volume fractions of ceramic reinforcement are produced. Thus required quantity of reinforcement can be obtained by adding this material and again melting with a matrix or base material. Composites with different base material like Copper, aluminum, iron, titanium etc can be produced with the ultra-fine metallic compounds of oxygen, boron, nitrogen, carbon etc. as reinforcement with the XD^{TM} technology. By using this procedure Al-AlB_2 MMC's have been produced by the various authors [49-51, 77].

2.1.2.Reactive Hot Processing (RHP)

Sirtl et.al.,[52] reported RHP is the other process used for preparing Al-AlB_2 insitu composites to produce dense aluminum composite, during the process high temperature compaction is as well as production of dispersoid particulates by insitu reaction occurs in a single step with a theoretical density of about 65% of Al. Stoichiometrical quantity of boron and Al in powder form is mixed together to form green bars by cold compaction. Combustion reaction takes place when these bars are heated to a temperature above 850^{0}C and held at this temperatures for 10 minutes in vacuum and hot pressed at 600^{0}C. Various investigators [53-56, 77] used the RHP technique to produce Al-AlB_2 composite by using B_2O_3 powder as a reinforcement.

2.1.3 Lanxide TM Process

The lanxide process is developed by the Lanxide Corporation. Earlier this process was used to process Al-AlB_2 MMC's by oxidizing the melt of Al in preforms by boron compounds using insitu technique. This process involves controlled oxidation molten al alloy to form metal ceramic mixture to form Al-AlB_2 composites is as shown in 'figure 2' [57-58, 77].

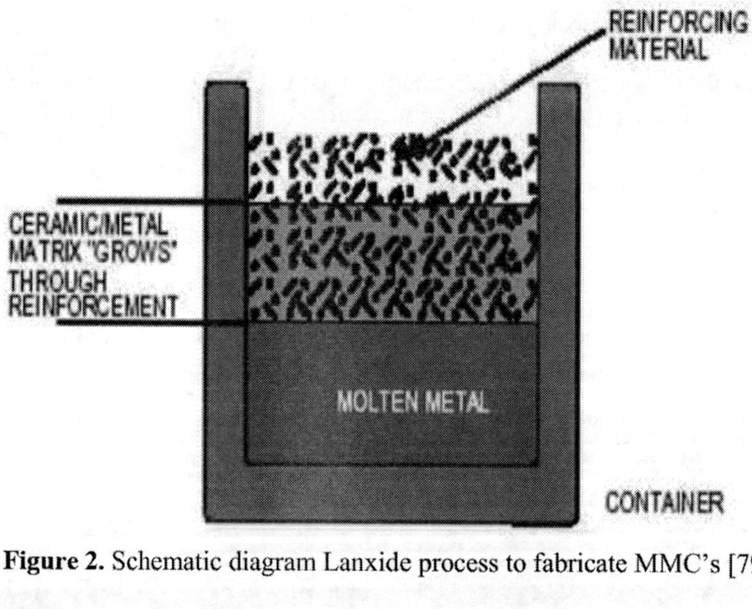

Figure 2. Schematic diagram Lanxide process to fabricate MMC's [79]

2.1.4 Stir Casting

During the process reinforcement or the material which helps in the formation of reinforcement i.e, reinforcing material is added to the melt of Al matrix with continues stirring for uniform distribution of reinforcement throughout the matrix. Al_2O_3 particles were added to Al melt with constant stirring of molten Al by S. Ray in late 1960's is as shown in 'figure 3'. The stirring of the melt using the stirrer mechanically is the most distinct aspect of this method. After the removal of the slag the melt of the matrix along with the particulate reinforcement is cast into the die. The in situ AMC with AlB_2 reinforcement particles are developed by this process using KBF_4 halide salt. High porosity in the composite is the major disadvantage of this process. In this process there is a uniform distribution of AlB_2 particles and formation clear and proper bonding Al-AlB_2 composites can be achieved. Due to the BF_3 gas there is a formation AlB_{12} compound Wang et.al [59] [77].

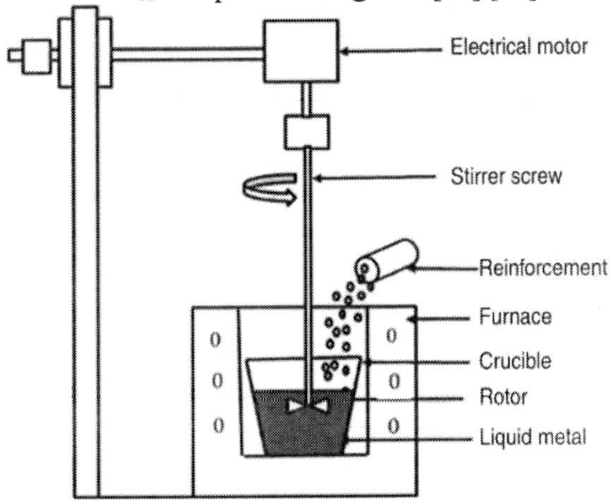

Figure 3. Schematic diagram Stir casting process to fabricate MMC's [59-60, 78]

2.1.5 Squeeze Casting

In this process to ensure optimal filling, highly porous reinforcing fibers or particle of proper shape and size are prepared. Later on the entire volume of the preform is filled with the melt of the matrix at high pressure. Fine grains are obtained as the transfer of phase takes place rapidly during the process due to directional solidification as shown in 'figure 4'. Ghamoshi et al. [60] and Tjong et al. [77] found a strong bonding between the matrix and the reinforcement is provided by this method. High productivity can be achieved because of faster pressing work. The composites prepared by this method have high density and are homogenous in nature. The cost of the tool which withstand higher temperature and pressure is high and this forms the major disadvantage of the squeeze casting process.

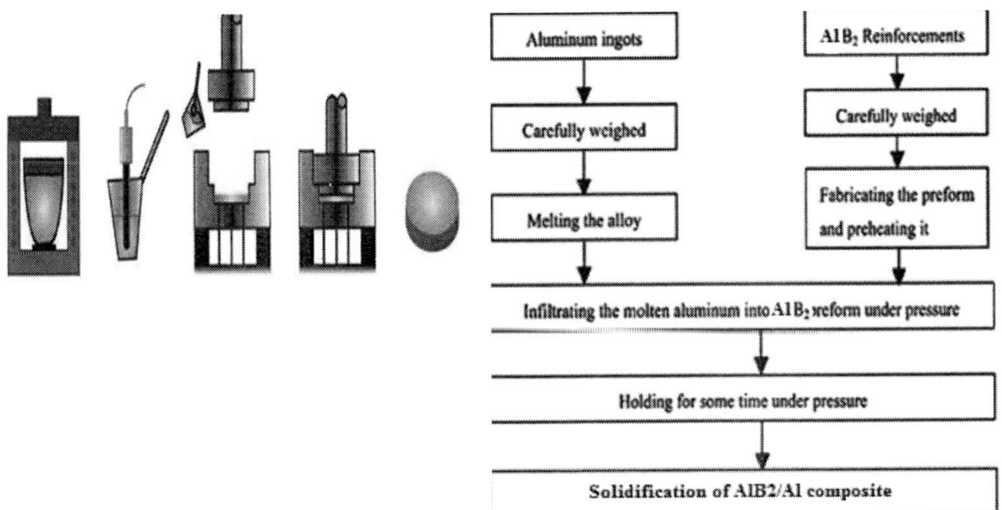

Figure 4. Shows Schematic diagram Squeeze casting process and fabricating process [60][78]

2.1.6 Flux Assisted Synthesis (FAS)

London Scandinavian Metallurgical Company developed this process and it is also known as mixed salt reaction. It is the commonly used technique to fabricate Al-AlB_2 type MMC's. Dragus Marco et.al [61], Tjong et al. [77] investigated during the process halide salts such as KBF_4 is added to the Al melt as per the required weight fraction of reinforcement. Continues stirring is carried out until the reaction

IOP Conf. Series: Materials Science and Engineering **810** (2020) 012038 doi:10.1088/1757-899X/810/1/012038

is complete, the nature of chemical reaction between the halide salt and Al is exothermic, and the sequence of reaction is as follows.

$$12KBF_4+13Al = AlB_{12}+12KAlF_4 = AlB_{12}+12KF+12AlF_3 \quad (1)$$

$$2KBF_4+3Al = AlB_2+2KAlF_4 = AlB_2+2KF+2AlF_3 \quad (2)$$

The slag being less dense floats up and is to be removed. The remaining melt of composite is poured into the die and casted into the desired shapes as shown in 'figure 5'. Ramazan Kayikci et al. [62] found this technique is economical and simple but its application is limited due to the low weight fraction of the dispersoid particles. The viscosity of molten metal Ferit et al. [63] decides quantity of AlB_2 that can be dispersed in it [64-65, 77,48].

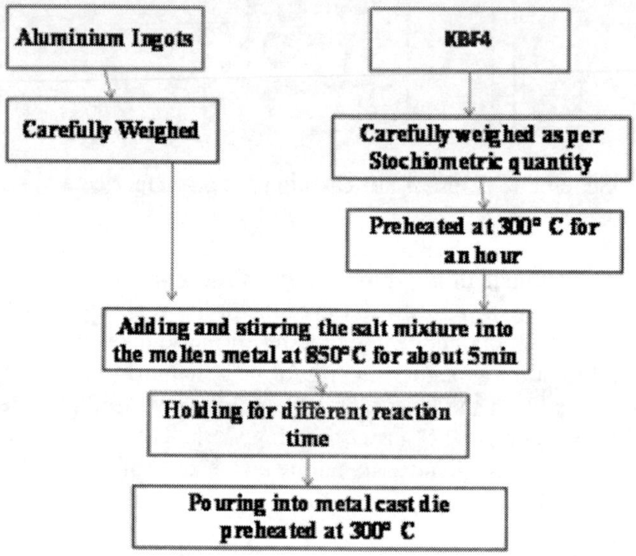

Figure 5. Shows flux assisted casting process to fabricate MMC's [48][78]

2.1.7 Centrifugal Casting:

Sakip et al. [66], N.Radhika and Omar Savas et al. [77-78] mentioned Centrifugal Casting can be used to fabricate composites with higher weight fractions upto 10% of dispersoid particulates. The thin walled cylinders are cast by centrifugal casting technique as shown in 'figure 6'.

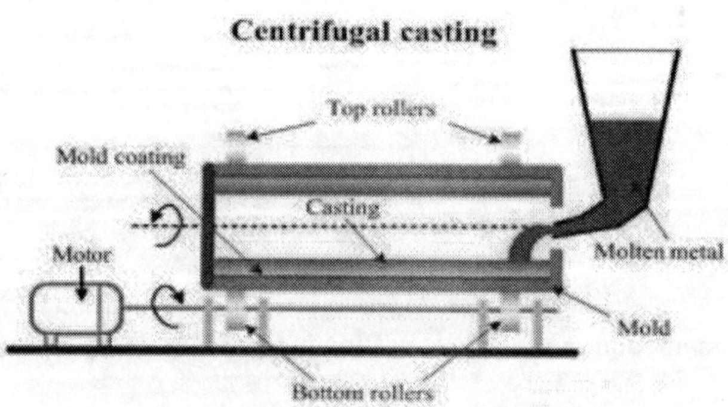

Figure 6. Schematic diagram Centrifugal casting process to fabricate MMC's [78][80]

This method involves pouring the melt into the die rotating at a speed in the range of 300-3000 rpm

IOP Conf. Series: Materials Science and Engineering **810** (2020) 012038 doi:10.1088/1757-899X/810/1/012038

about its own axis of rotation. During the rotation the melt moves along the inner periphery of the cylindrical mould and solidifies on cooling. Cylindrical castings with the finer grains can be achieved by this method.

2.1.8 Master alloy technique

Boppanna et al. [67] reported metal matrix composites were prepared using master alloys for the insitu process. Ramesh et al. [68] found in order to process AMC's with the required weight fraction of the reinforcement; reinforcement is added to the melt of matrix according to stoichiometric ratios Karantis et al. [69]. This method has been used by several investigators Wang et al. [70] to process Al-AlB_2 MMC's. Benjamin [71] and Tjong [77] reported the advantage of this method is lower amount of reaction between the matrix the reinforcement and the temperature at which the formation of reinforcement occurs is comparatively low.

2.2 Solid- state Process

2.2.1 Mechanical Alloying

Duygu Agaogullari et al. [72] found this technique is a solid state processing technique which includes repetitive cold drawing and fracture of particulates. By this technique, found alloy materials are ground to fine powder, simultaneously compressed and sintered to form a composite and properties of composites are affected by the time of mechanical alloying, quantity and the size of AlB_2 particulates and also compared to the micron scale AlB_2 particles the MMC's with AlB_2 size less than a micron shows improved hardness and wear resistances and further increases with the increasing content of AlB_2[73-75]. Balci et al. [76] observed in regard to different mechanical alloying time, for certain limit the density and hardness values of the sintered samples increases with time. Uniform distribution of ultra-fine AlB2 particles can be achieved within the matrix.

2.2.2 *Hot Isostatic Pressing*

Tjong and Mingming et al. [77] reported during the process blended powder are sintered at high temperature and pressure. As the pressurizing medium argon gas is usually used microstructure of composites fabricated using this method is decided by the ratio of the particles size of Al and ceramic material. Both the densification of MMCs and better uniform distribution of ceramic particles are favored by the Relative Particle Size as shown in 'figure7'. With the increasing AlB_2 volume content, tensile properties like modulus of elasticity and fracture strength of Al-AlB_2 composite improves[78-89].

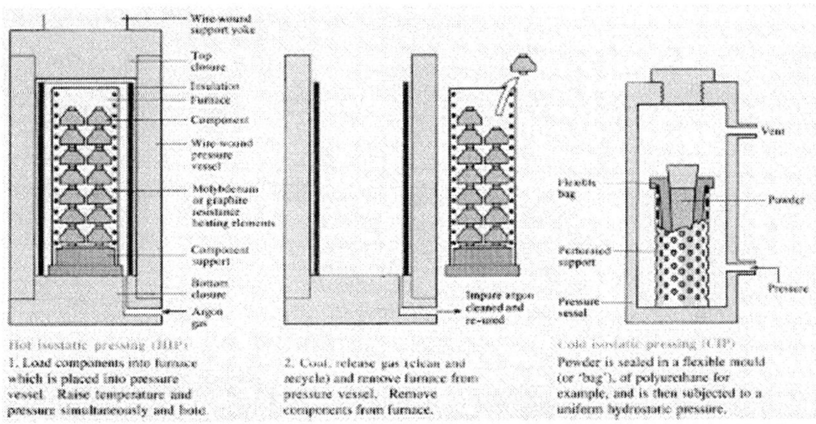

Figure 7. Schematic diagram Hot isostatic process to fabricate MMC's [81-88]

3. Conclusions:

A review on synthesis of AlB_2 reinforced aluminium metal matrix composites come to conclusion that,

- Various fabrication methods are there to synthesis AlB_2 particle have been dealt with in the present review paper. Different methods are categorized as solid state processing and liquid state processing techniques. Among them the latter is largely used to fabricate Al-AlB_2 MMC's. Stir casting route is one of the versatile, cheap and easy method to fabricate Al-AlB_2 composites, but there may be chances of cluster, agglomeration and porosity of the particles in the matrix, to overcome them proper mixing, stirring speed plays important role during casting process.
- The FAS reaction mechanism is not well understood, even though reaction mechanism of insitu formation of AlB_2 particles are reviewed and discussed.
- Insitu process can be widely used to prepare Al-AlB_2 composites owing to its simplicity and economic reasons. The advantage of this process include improved composite stability at higher temperatures, fine particulates, clean interface and an essential uniform distribution of reinforcement in matrix.
- According to the literature survey, work done in the field of Al-AlB_2 MMC's is comparatively low and might get more attraction in future.
- As the mechanism by which the insitu formation of Al-AlB_2 composite using halide salts is not completely understood, further studies may be conducted to understand the reaction mechanisms.

References:

[1]. Vijee Kumar, Kempaiah U N and Satish Babu Bopanna 2017 *J. NanoSci. Nanoengg. Appl.* **7** 1
[2]. Veeresh Kumar G. B, Rao C.S.P and Selvaraj N 2011 *J. Miner. Mat. Charact.* **10** 59.
[3]. Basavarajappa S, Chandramohan G, Subramanian R and Chandrasekar A 2006 *J.Mater.Sci.* **24** 357
[4]. Seyed Reihani S M 2006 *Mater. Des.* **27** 216
[5]. Onat A 2010 *J. Alloys Compd.* **489** 1
[6]. Samuel Dayanand and Satish Babu B 2019 IJERT **5** 2278
[7]. Wang N, Wang Z and Weatherly G C 1992 *Metall. Trans. A.* **23** 2413
[8]. Samuel Dayanand and SatishBabu B 2019 *IJIERT.* 2394
[9]. Samuel Dayanand, Satish Babu B and Auradi V 2018 *Mater. Today Proc.* **5** 10
[10]. Boppana S.B and Chennakeshavalu K 2009 *J. Minerls and Mater. Charact. and Engg. 07* 563
[11]. Seyed Reihani S.M 2006 *Mater. Des.* **27** 216
[12]. Onat A 2010 *J. Alloys Compd.* **489** 1
[13]. Nami H, Adgi H, Sharifitabar M and Shamabadi H 2011 *Mater. Des.* **32** 976
[14]. Sharifitabar M, Sarani A, Khorshahian S and Sharfiee Afarani M 2011 *J. Mater. Des.* **32** 4164
[15]. Alidokht S.A, Abdollah-Zadeh A, Soleymani S and Assadi H 2011 *J. Mater. Des.* **32** 2727
[16]. Michael Rajan H.B, Ramabalan S, Dinaharan S and Vijay S.J. 2013 *J. Mater. Des.* **44** 438
[17]. Kumar N, Gautam R.K and Mohan S 2015 *J. Mater. Des.* **80** 129
[18]. Lloyd D.J 1994 *J. Int. Mater. Rev.* **39** 1
[19]. Hassan A.M, Alrashdan A, Hayajneh M T and Mayyas A T 2009 *J.Tribol.Int.* **42** 1230
[20]. Rao R.N and Das S 2011 *J. Tribol. Int.* **45** 454
[21]. Samuel Dayanand, Satsih Babu B, Hemanth Joel and Aravinda T 2019 *J Bio Tribo Corros* **5** 60
[22]. Mandal D and Viswanathan S 2013 *J. Mater. Charact.* **86** 12
[23]. Mindivan H 2010 *Mater. Lett.* **64** 405
[24]. Baradeswaran A and Elaya Perumal A 2013 *J. Composites B.* **54** 146
[25]. Chou S. N, Huang J. L , Lii D.F and Lu H.H 2006 *J. Alloys Compd.* **419** 98
[26]. Wang H, Li G, Zhao Y and Chen G 2010 *J Mater. Sci. Eng. A.* **527** 2881
[27]. Kerti I 2005 *Mater. Lett.* **59** 3795
[28]. Kumar A, Mahapatra M .M and Jha P.K 2013 *J. Wear.* **306** 170
[29]. Akhlaghi F and Zare Bidaki A 2009 *Wear.* **266** 37
[30]. Chaudhury S.K and Panigrahi S.C 2007 *J. Mater. Process. Technol.* **182** 343
[31]. Hemanth J 2000 *J. Alloys Compd.* **296** 193
[32]. Arik H 2008 *J. Mater. Des.* **29** 1856

[33]. Ramesh C .S and Keshavamurthy R 2011 *Mater. Des.* **32** 1833
[34]. Moghaddas M.A and Kashani Bozorg S.F 2013 *J. Mater. Sci. Eng. A.* **559** 187
[35]. Lee K.B, Sim H.S, Heo S.W, Yoo H.R, Cho S.Y and Kwon H 2002 *J. Composites A.* **33** 709
[36]. Wang J, Yi D, Su X, Yin F and Li H 2009 *J. Mater. Des.* **30** 78
[37]. Selvam J.D.R, Smart D.R and Dinaharan I 2013 *J. Mater. Des.* **49** 28
[38]. Yi H, Ma N, Li X, Zhang Y and Wang H 2006 *J. Mater. Sci. Eng.* **419** 12
[39]. Samuel Dayanand and Satish Babu B 2016 *IJERT.* **5** 2278
[40]. Kumar N, Gautam R.K and Mohan S 2015 *Tribol. Ind.* **37** 244
[41]. Kumar S, Chakraborty M, Subramanya Sarma V and Murty B S 2008 *Wear.* **265** 134
[42]. Gautam G and Mohan A 2015 *J. Alloys Compd.* **649** 174
[43]. Dumitru Valentin Draught, Petru Moldovan, Mihai Butu and Emilia Usurelu 2011 U.P.B,Sci.,Bull.,Series
[44]. Ramesh C.S, Pramod S and Keshavamurthy R 2011 *J. Mater. Sci. Eng. A.* **528** 4125
[45]. Brupbacher J M, Christodoulou L and Nagle D C 1988 US Patent 4,751,048
[46]. Brupbacher J M, Christodoulou L and Nagle D C 1989 US Patent 4,836,982
[47]. Tjong S.C and Ma Z.Y 2000 *Mater. Sci. Eng. R.* **29** 49
[48]. Modovan P, Draught D.V, Butu M, Besea L. M and Preda E.A 2016 *Composites Part B.* **97** 9
[49]. Kuruvilla A.K, Prasad K.S, Bhanuprasad V.V and Mahajan Y. R 1990 *Scr. Metall. Mater.* **24** 873
[50]. Joao Batista Fogagnolo, Maria Helena Robert, Elisa M, Ruiz-Navas and Torralba J M 2004*Article* in *J.Mater.Sci.* **39** 1
[51]. Linlin Yuan, Jingtao Han, Jing Liu and Zhengyi Jiang 2016 *Tribology International. 98* 41
[52]. Sirtl E and Woerner L.M 1972 *Journal of Crystal Growth.* **16** 215
[53]. Duygu Agaogulları, Hasan Gokc¸e, Ismail Duman, Lutfi Ovec M and Oglu 2012 *J. Euro. Cera. Soc.* **32** 1457
[54]. Glorimar Ramos, Roberto Manzano and Oscar Marcelo Suarez 2012 *Sci. Eng. Compos. Mater* **19** 67
[55]. Kayikci R, Savas O, Koksal S and Demitru 2014 *ACTAPHYCICA POLONIC A.* **125** 2
[56]. Omer Savas, Ramazan Kayikci 2013 *Materials and Design.* **51** 641
[57]. Qing-Liang WANG, Hong-Sheng Zhao, Zheng-guage Li, Li Shen and Jiu-Zhou Zhao 2013 *Trans, Nonferrous Met. Soc. China.* **23** 294
[58]. Sakip Koksal, Ferit Ficici, Ramazan Kayikci and Omar Savas 2006 *Material and Design.* 124
[59]. Samuel Dayanand and Satish Babu B 2019 *IJRTE* **8** 2
[60]. Ghamoshi R and Vikrov A 2000 *Mater. Process. Technol.* **101** 1
[61]. Ramazan Kayikci and Omer Savas 2015 *J.Compo.Mater.* **49** 16
[62]. Ferit Ficici 2014 *The Minerals, Metals & Materials Society JOM,* **66** 5
[63]. Dragus Marco 2014 *U.P.B.Sci.Bull.,Series.,***76** 1454
[64]. Kayikcia R, Savas B, Koksala S and Demira A 2013 *Apmas.* **24**
[65]. Karantzails A E, Lekatou A, Gerogties M, Poulas V and Mavros H (2011) *J.Mat.Engg. and perform.* **20** 2
[66]. Sakip Koksal, Ferit Ficici, Ramazan Kayikci and Omer Savas 2014 *J.Compo.Mater.* **15** 48
[67]. Boppanna S and Chennakeshavalu K 2009 *J. Minerls. Mater. Character. Engg.* **5** 63
[68]. Kubota M 2004 Proceedings of the 9th International Conference on Aluminium Alloys Muddle Institute of Materials Engineering Australasia Ltd
[69]. KarantisA.E, Lekkatou A, Georgatis M, Poulas V and Mavros H 2011 *J. mat. .Engg. Perfor. ASM International,* **20** 198
[70]. Elecicek H, Savas O, Aydin Z, Ozdemir O K and Kayikci R 2016 *Acta Physia polonica A.* **129** 4
[71]. Benjamin J.S 1970 *Metall. Trans.* **1** 2943
[72]. Duygu Agaogulları, Hasan Gokc¸e, Ismail Duman, M, Lutfi Ovec and Oglu 2012 *J. Euro. Cera. Soc.* **32** 1457

IOP Conf. Series: Materials Science and Engineering **810** (2020) 012038 doi:10.1088/1757-899X/810/1/012038

[73]. Orban R, Cora I and Dodony I 2012 *ECCM15-15th European conference of composite materials*, venice, Italy, 24-28
[74]. Mingming Yang, Zhaozhu Zhang, Junya Yuan, Fang Guo and Xuehu Men 2017 *Compo. Sci.Technol.* **143** 75.
[75]. Z.H. Melgarejoa Z.H, P.J. Restoa P.J, Stoneb D.S and Suárezc O.M 2010 *Mater. Character.* **61** 135
[76]. Balci O, Agaogullari D, Hasan G, Ismail D and Lutfi Ovecoglu M 2014 *J. Alloys Compd. 8* 578
[77]. Tjong S.C and Ma Z.Y 2000 *Mater. Sci. Eng. R.* **29** 49
[78]. Omar savas, Ramzan Kaykaci, Ferit Ficici and Sakip Koksal 2013 *periodicals of engineering and natural science.* **1** 2
[79]. Radhika N and Arunagiri K S 2016 *Tribology in industry. 38* 3
[80]. Tjong S.C and Tam K.F 2006 *Mater. Chem. Phys.* **97** 91
[81]. Benjamin J.S and Volin T.E 1974 *Metall. Trans.* **5** 1929
[82]. Karantis A.E, Lekkatou A, Georgatis M, Poulas V and Mavros H 2011 *J. mat. .Engg. Perfor. ASM International.* **20** 198
[83]. Yucel Birol 2009 *J. Alloys Compd.* **481** 195
[84]. N. Radhika and R. Raghu 2016 *Particulate Science and Technology* ISSN: 0272-6351
[85]. Gilman P.S and Benjamin J.S 1983 *Mater. Sci.* **13** 279
[86]. Frommeyer G, Beer S and Von Oldenburg K 1994 *Z. Metallkd.* **85** 372
[87]. Balci O, Agaogullari D, Hasan G, Ismail D and Lutfi Ovecoglu M 2014 *J. Alloys Compd.* **8** 578
[88]. Narendra Kumar, Gourav Goutham, Rakesh kumar Giuham, Anitha Mohan and Sunil Mohan 2015 *J. Inst. Eng. India. Ser. D.* **7** 91
[89]. Froyan L and Verlinden B Advanced level 1, *TALAT Lecture 1402*, University of Leuven, Belgium

IOP Conf. Series: Materials Science and Engineering **810** (2020) 012039 doi:10.1088/1757-899X/810/1/012039

Experimental study on enhancement of defect detection and defect size estimation in deep groove ball bearing

Priyank Patel[1]*, V N Patel[2]

*corresponding author , [1] PG student , Mechatronics department, G H Patel College of Engineering and Technology, Vallabh Vidyanagar, India, Email : priyankpatel9099@gmail.com

[2] Professor, Mechatronics department, G H Patel College of Engineering and Technology, Vallabh Vidyanagar, India, Email: vinodgcet1@gmail.com, ORCID : 0000-0002-2068-8557

Abstract. Rolling elements bearings are widely used in rotating equipment and machines to support load and to reduce friction. The presence of micron sized defects on the mating surfaces of the bearing components can lead to failure through passage of time. The large size defects on bearing elements can be detected/identified by time domain and frequency domain analysis of its vibration signals. However, it becomes difficult to detect local bearing defects at their initial stage either due to their smaller size or presence of noise. In the present experimental study, detection of local defects like crack and pits on bearing races have been carried out. Vibrations generated by healthy bearing and bearing having circular or rectangular defect on either race of bearing have been analyzed using MATLAB software. Signal to noise ratio of vibration signal has been enhanced through self- adaptive noise cancellation. Moreover, width of rectangular defects and diameter of circular defects have been estimated through 'db5' wavelet. The estimated defect sizes have been compared and validated through measured actual crack width or pit diameter. Accuracy of results proves that wavelet analysis of time domain signal can be used with confidence to estimate the width of fatigue crack and pit of the bearing races.

Keywords: Rolling Element Bearing, Local Defects, Vibration analysis, Wavelet, Adaptive Filtering

1. Introduction: -

The performance of any rotating machine is very much dependent on the dynamic behavior of rolling bearings. As rolling element bearing plays a crucial role in the rotating machine it becomes very important to monitor the health of the bearing to avoid breakdown and accidental losses.

Most of the time bearing defects may emerge due to the excessive transferred load on the bearing, corrosion, wear or due to the improper installation of bearings. Broadly researchers have classified bearing defects as local defects and distributed defects. Localized defects are comprising of cracks, pits, spall whereas distributed defects are comprising of surface roughness, waviness, misalignment, off-size rolling element. Detection of these defects is very much necessary for condition monitoring of the bearing as well as quality inspection.

Different types of condition monitoring techniques have been used by the researchers. Different vibration and acoustic analysis methods, such as vibration measurement in time domain, vibration measurement in frequency domain, vibration measurement in combined time-frequency domain, sound measurement and acoustic emission have been used to detect the defects in rolling element bearings. Initially time domain approach has been used to detect the defect detection with limited success ratio. Root mean square (RMS) value and crest factor have been used to measure the local defects on the rolling elements of bearings. Statistical parameters like kurtosis value and probability density have been used. Defect free bearing gives a gaussian distribution. Due to the relative increase in the number of high levels of acceleration defective bearing have non-gaussian distribution. Frequency domain analysis has been used widely as it gives results very efficiently. Whenever the rolling element strikes on the defects it produces pluses of very short duration. These pulses excite the natural frequencies of rolling element bearing which can be calculated experimentally and theoretically.

Condition monitoring technique such as Proximity transducer technique has been used as this technique uses non-contact type transducer. The main advantage of non-contact type transducer is it eliminates the noise of the housing structure. Due to this reason the signal to noise is improved in this condition monitoring technique. The main disadvantage of this condition monitoring is the difficulty in the installation/mounting of probe at proper location. Signal transmission error can be avoided with proper mounting of accelerometer during vibration capture. Condition monitoring technique such as acoustic noise response has been used to diagnosis the bearing health in the modes of sound pressure measurement and sound intensity measurement. The diagnosis of bearing defects becomes difficult as it also contains surrounding noise other than bearing defect sound. Literature review shows that vibration measurement in time domain and frequency domain gives more precise and better results than other condition monitoring techniques.

Researcher [1-2] have reviewed the various methods which were used for the purpose of health diagnosis and detection of bearing defects. The purpose of condition monitoring, sound measurement, the shock pule method, and acoustic emission techniques have been used. For the purpose of vibration and wear debris analysis some techniques are used namely as shock pulse, ferrography, spectrography, oil analysis, spike energy and chip detection. To diagnose the bearings defects most of the researchers have used vibration measurement techniques. Many techniques have been applied to measure the vibration response of the bearings which is having local defects. But the problem about these signal processing techniques is they depend on the result of each other and works as complimentary for each other. Time domain and frequency domain are used to study and analyze the vibration signals.

Time domain analysis technique like root mean square (RMS), crest factor, kurtosis, spectral kurtosis, and probability density have been used by researchers [3-7]. The signal processing techniques like kurtosis, spectral kurtosis, kurtogram and band kurtosis are the extensively used signal processing techniques to determine the impulsiveness of the signal. The frequency domain is the most widely used and accepted approach to find bearing defects and to perform the higher level of signal processing techniques. Bearing elements have a characteristic rotational frequency which lies in a specific range, due to this phenomenon whenever rotating elements pass through defect on bearing elements, the vibrational energy at that characteristic defect frequency increases.

To remove the external vibration of the signal using signal processing techniques are discussed by researchers [8-9]. Some of them are: envelope analysis, adaptive noise cancellation, self-adaptive noise cancellation, high-frequency resonance technique (HFRT), wavelet analysis, stochastic resonance, cyclostationary analysis, spectral subtraction and blind source separation. It is difficult to retrieve important peaks at defect frequencies in presence of noise [9].

The bearing defects have been detected by researchers of references [10-15] with the help of adaptive noise cancellation technique (ANC). Adaptive noise cancellation requires primary signal and reference signal. Primary signal includes bearing signal and noise. Reference signal is related to the noise signal. The signal to noise ratio has increased marginally in each of the experiments performed by researchers. As a result of ANC signal to noise ratio (SNR) increased and time domain techniques like kurtosis, power spectrum, RMS values have been determined. The self-adaptive noise cancellation technique (SANC) is capable to enhance SNR with only primary signal. Khemili and Mnaouar [10] have used self-tuning adaptive filters also known as self-adaptive noise cancellation (SANC) for their theoretical and experimental study for bearing defect detection. They have concluded that SANC filters have provided better SNR than adaptive noise cancellation for simulated data as well as experimental data.

Defect size estimation has always remained challenging task for researchers. The literature survey discloses that very few researchers [16-25] have estimated the size of the faults in the rolling element bearing. Most of the researchers have used wavelet analysis for defect size estimation due to the simplicity of wavelet analysis. Khanam and Tandon [16] have estimated the bearing outer race defect size using Symlet wavelet. Symlet wavelet has identified entry and exit events of rolling elements at outer defect.

The acoustic emission burst duration has been found out by the Kumar [20]. Authors have concluded from the experimental study that the duration of burst increases with width of outer race defect. Experiments have been performed to estimate the size of outer race defects [20]. The bearing vibrational signals have been decomposed by Barszcz [23] into deterministic and non-deterministic components to detect the fault on the outer race of the ball bearing. Adaptive filter has been used to remove the noise from the vibrational signal.

Literature review reveals that a lot of research work has been carried out in the field of defect detection of rolling element bearings in absence of external vibrations. Most of researchers have estimated defect size of outer race. However, it is difficult to estimate the bearing defect size in presence of noise at the initial/beginning stage i.e. when the size of the fatigue crack or pit is minute in size. Hence, the main objective of the present experimental study is to estimate the width of fatigue crack and diameter of pit on bearing races. Moreover, self-adaptive noise cancellation technique has been adopted to improve signal to noise ratio of bearing vibration signal. Vibration signals of deep groove ball bearings having crack or pits (on either race) of varying size have been captured. Statistical parameter like kurtosis has been computed from time domain signal, while defect characteristic frequencies have been identified in frequency domain. Defects size has been estimated through 'db5'wavelet analysis of time domain signal. Estimated defect width and defect diameter has been validated with actually measured defect size. Paper has been organized as follow: section 2 discusses experimental set up, vibration responses of bearings have been presented in section 3, self-adaptive signal processing technique has been discussed in section 4, while defect size estimation and signal analysis has been carried out in section 5 and section 6, respectively.

2. Experimental details: -

Vibrations generated by healthy and defective deep groove ball bearings (SKF BB1 B420206) have been captured by accelerometer. A photographic view of experimental setup used for vibration study is shown in Figure1. A shaft is supported on two deep groove ball bearings. It receives power from motor through V-belt and pulley, attached on right hand end of the shaft. While, the test bearing (SKF BB1B420206) was mounted on cantilever portion of left-hand end of shaft. The radial load of 10 kg was hanged at the bottom of specially designed test bearing housing and accelerometer (603C01 NI) was mounted on top of this housing to capture vibration data. Data acquisition card (9234 NI) was used to transfer captured data to the computer through LAB-VIEW software. The vibration data were captured at the sampling rate of 1k. The test bearing specifications are provided in the Table 1. The collected vibration signal of the test bearing has been analyzed in time domain and frequency domain using MATLAB software.

Figure 1 Experimental setup.

Table 1. Test bearing (SKF BB1B420206) specification.

Bore of bearing, mm	30	Ball diameter (d_b), mm	10.4
Inner race diameter (d_i), mm	34.52	Diametric clearance (P_d), μm	10
Outer race diameter (d_o), mm	54.92	Grooves radii, mm	5
Pitch diameter (D_P), mm	44.72	Number of balls (N_b)	8

Vibration signals of defect free test bearings (considered as healthy bearing) and test bearings having circular defect (considered as pit) or rectangular defect (considered as fatigue crack) on its either race have been captured for analysis. Artificial defects were created through electro discharge machining. Figure 2 shows the photographic view and microscopic images of circular and rectangular defects on inner race and outer race of the bearing.

IOP Conf. Series: Materials Science and Engineering **810** (2020) 012039 doi:10.1088/1757-899X/810/1/012039

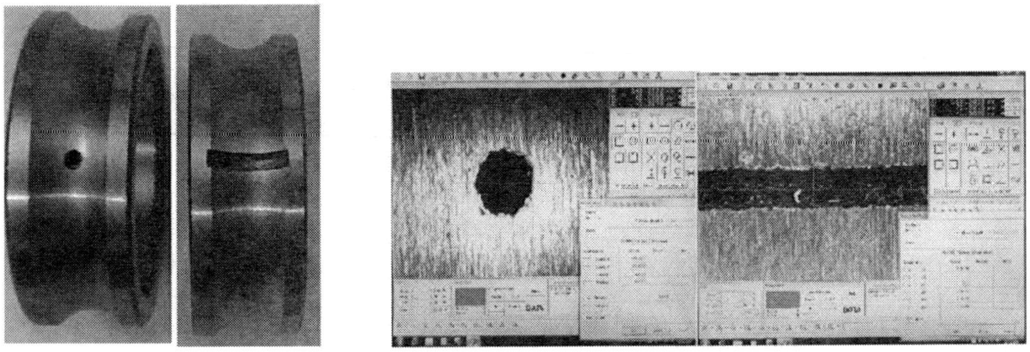

(a) Photographic view of inner race (b) Microscopic image of inner race

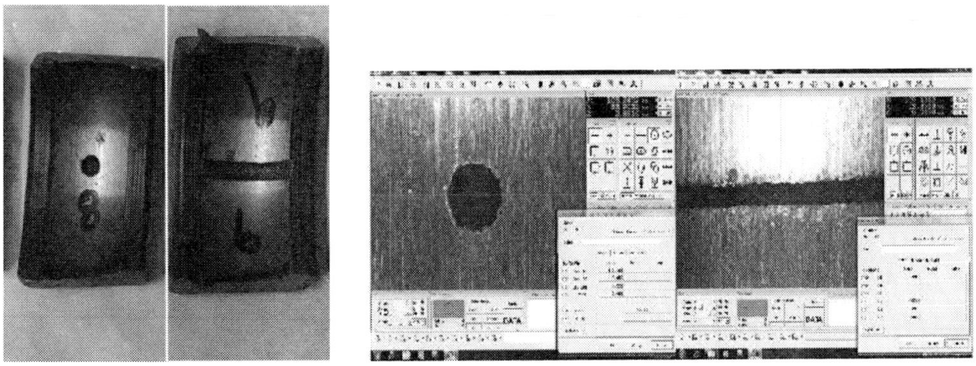

(c) Photographic view of outer race (d) Microscopic image of outer race

Figure 2 Circular and rectangular defects on bearing races.

It is worth to mention here that four artificial circular defects of different diameter on inner race and three artificial circular defects of different diameter on outer race were created. Four artificial rectangular defects of different width on inner race and three artificial rectangular defects of different width on outer race were created. Hence vibrations generated by these fourteen healthy bearings before defects creation and after defect creation were studied.

3. Vibration response of deep groove ball bearing: -

The vibration responses of all test bearings before and after defect generation have been captured and analyzed. It is worth to mention here that time domain and frequency domain analysis have been performed for healthy and defective bearings using MATLAB software. Statistical parameters like Kurtosis and Root Mean Square (RMS) value have been computed from time domain data through equation 1 and equation 2 respectively.

$$kurtosis = \frac{1}{n}\sum_{i-1}^{n}\left(\frac{x_i - \mu}{\sigma}\right)^4 \qquad (1)$$

$$\text{Root mean square} = \sqrt{\frac{1}{n}\sum_{i=1}^{n} x_i^2} \tag{2}$$

Kurtosis gives insights into the shape of the distribution of the data. Skewness is a measure of the symmetry in a distribution. A symmetrical dataset gives zero skewness. Skewness measures the relative size of the two tails. Kurtosis also measures the amount of probability in the tails. The value is many times compared with the kurtosis of the normal distribution, which is equal to 3. If the kurtosis is greater than 3 for given dataset, then the dataset has heavier tails than a normal distribution. If the kurtosis is less than 3, then the dataset has lighter tails than a normal distribution. Generally, root mean square value of defective bearing increases with the increase in the size of the defect.

3.1 Vibration response of healthy bearing: -

The vibration response of one of the healthy bearing has been presented here. Vibration data in form of acceleration captured by accelerometer has been presented in Fig.3 in form of time domain and frequency domain. As shown in Figure3(a) vibration signal has been captured for 2.5 millisecond for 4096 samples. A computed kurtosis value is 2.8246 and RMS value is 0.3887 for this healthy bearing. The shaft was rotating at 1600 RPM. The rotation speed was measured by digital tachometer. The peaks at shaft rotational frequency (F_s = 26.66 Hz) and its harmonics (2 × F_s = 53.32 Hz, 3 × F_s = 79.98 Hz) can be noticed in frequency domain of signal (refer Figure3(b)). Additional dominant frequency peaks observed around 100 Hz are due to natural frequency of the system, which were observed in all captured vibration signal irrespective of shaft rotational speed.

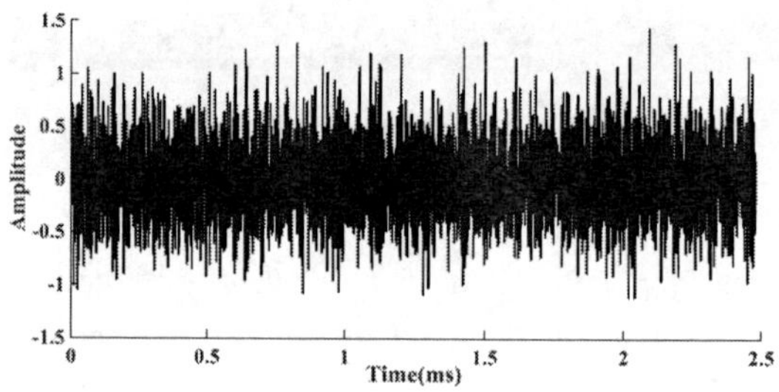

(a) Time domain

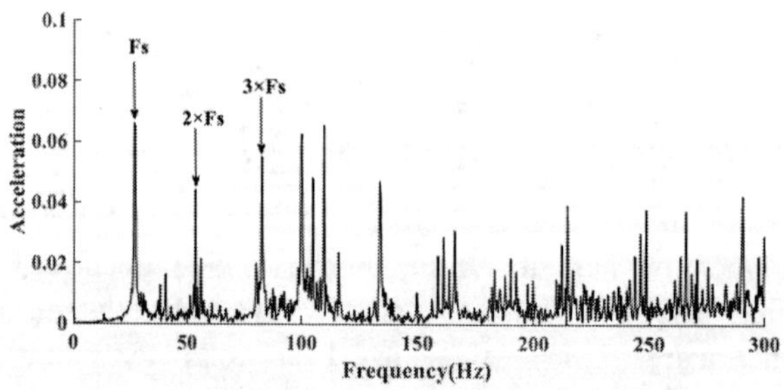

(b) Frequency domain

Figure 3 Vibration response of healthy bearing at a shaft speed of 1600 RPM.

IOP Conf. Series: Materials Science and Engineering **810** (2020) 012039 doi:10.1088/1757-899X/810/1/012039

Same methodology has been adopted for all healthy bearings (i.e. before artificial defect creation). The frequency peaks at shaft rotation speed along with their harmonics were observed in vibration acceleration spectra of all bearings. The kurtosis value was around 3 for all cases, while the maximum RMS value was 0.4774 for healthy bearings.

3.2 Vibration response of bearings having defects on its inner race: -

As discussed in section 2, rectangular and circular artificial defects were created on inner race of healthy bearings to capture vibration responses of defective bearings. In first stage rectangular slot of 0.77 mm width on one bearing and circular blind hole of 0.77 mm diameter on another bearing were created. Vibration data in form of acceleration were captured and analyzed for both bearings separately. To study the effect of defect size and shape of inner race defect on vibration response, circular hole and rectangular slot were enlarged upto 2.07 mm diameter and 1.91 mm width respectively and vibration signals were captured.

Figure 4 shows the vibration response of bearing having circular defect of 0.77 mm diameter on its inner race. The RMS value and kurtosis value computed from Figure4(a) is 0.3320 and 3.9139 respectively. Figure 4(b) shows the frequency response of defective bearing. Along with shaft rotational frequency F_s =26.63 Hz, frequency peaks at ball pass frequency of inner race (BPFI) at 131.1 Hz and its harmonics at 2×BPFI=263.1 Hz. The computed value of BPFI =131.29 Hz and its harmonics at 2×BPFI=262.58 Hz. This can be validated with Figure4(b). The inner race of bearing rotates at shaft rotation, hence side bands at BPFI±F_s are also expected. In Fig.4(b) sidebands at BPFI+F_s =157.8 Hz and BPFI- F_s = 104.9 Hz for first harmonic and 2× (BPFI+Fs) =289.3 Hz and 2× (BPFI- $F_{s)}$ =236.5 Hz for second harmonic are clearly observed.

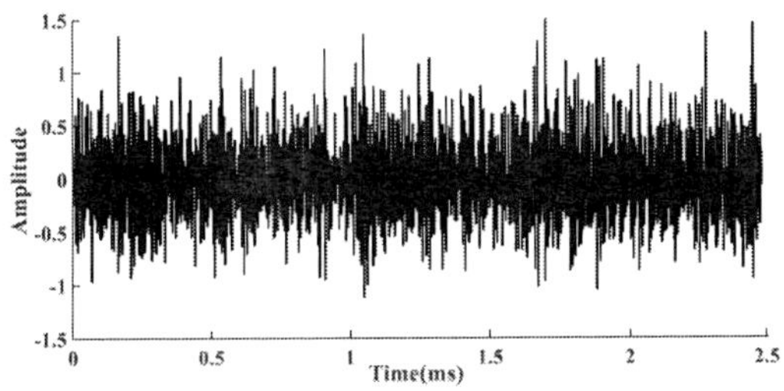

(a) Time domain

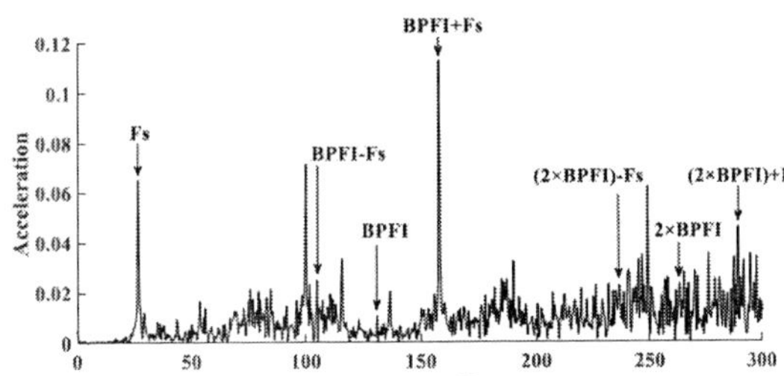

(b) Frequency domain

Figure 4 Vibration response of bearing having circular defect of 0.77 mm diameter on its inner race at shaft speed of 1600 RPM.

Figure 5 shows the vibration response of bearing having rectangular defect of 0.77 mm width on its inner race. The RMS value and kurtosis value computed from Figure5(a) is 0.2560 and 3.1116 respectively. Figure 5(b) shows the frequency response of defective bearing. Along with shaft rotational frequency F_s =26.63 Hz, frequency peaks at ball pass frequency of inner race (BPFI) at 131.6 Hz and its harmonics at 2 × BPFI = 261.5 Hz. The computed value of BPFI =131.29 Hz and its harmonics at 2×BPFI=262.58 Hz. This can be validated with Figure5(b). The inner race of bearing rotates at shaft rotation, hence side bands at BPFI±F_s are also expected. In Fig.5(b) sidebands at BPFI+F_s =158.2 Hz and BPFI- F_s = 104.5 Hz for first harmonic and 2×(BPFI+F_s) =289.7 Hz and 2× (BPFI- F_s) =236.1 Hz for second harmonic are clearly observed.

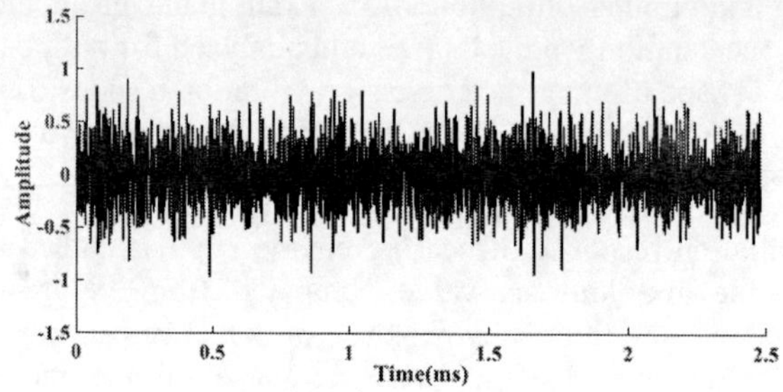

(a) Time domain

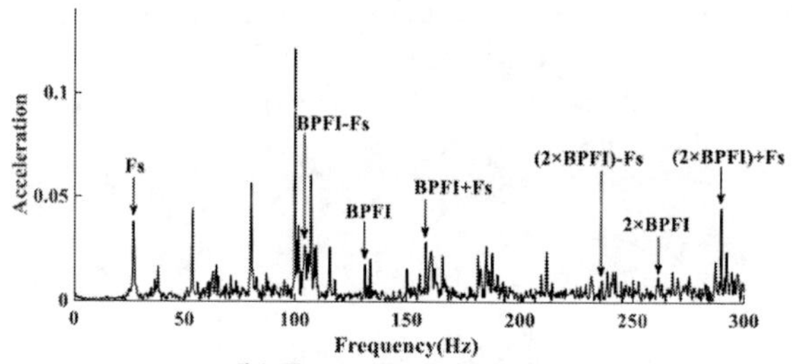

(b) Frequency domain

Figure 5 Vibration response of bearing having rectangular defect of 0.77 mm width on its inner race at shaft speed of 1600 RPM.

Same methodology has been adopted to study vibration responses of bearings having inner race defect of 1.03 mm, 1.1 mm and 1.53 mm defect diameter and 1.03 mm, 1.44 mm and 1.91mm defect width. The kurtosis value reached to 4 or more for all defective bearings and RMS value reached to 0.5094, determined from their respective time domain signals. Moreover, the expected inner race defect frequency along with side bands at BPFI ± F_s were observed in their frequency spectra.

Raw signal of Inner race Rectangular defect -1 in which shaft rotating frequency and harmonics of the defective frequency is shown by its relative names as BPFI and its harmonics 2×BPFI. The sidebands for BPFI are also shown as BPFI-F_s and BPFI+F_s. Unfortunately, defect frequency and its harmonics are not clearly visible due to the noise in the vibrational signal.

3.3 Vibration response of bearings having defects on its outer race: -

As discussed earlier, rectangular and circular artificial defects were created on outer race of healthy bearings to capture vibration responses of defective bearings. In first stage rectangular slot of 0.77 mm width on one bearing and circular blind hole of 0.77 mm diameter on another bearing were created. Vibration data in form of acceleration were captured and analyzed for both bearings separately. To study

IOP Conf. Series: Materials Science and Engineering **810** (2020) 012039 doi:10.1088/1757-899X/810/1/012039

the effect of defect size and shape of outer race defect on vibration response, circular hole and rectangular slot were created on different bearings from 0.499 mm,1.50 mm and 1.87 mm diameter and 0.72 mm, 1.49 mm and 2.16 mm width respectively and vibration signals were captured.

Figure 6 shows the vibration response of bearing having circular defect of 0.77 mm diameter on its outer race. The RMS value and kurtosis value computed from Figure6(a) is 0.4032 and 3.1089 respectively. Figure 6(b) shows the frequency response of defective bearing. Along with shaft rotational frequency F_s =26.63 Hz, frequency peaks at ball pass frequency of outer race (BPFO) at 83.53 Hz and its harmonics at 2×BPFO=166.7 Hz. The computed value of BPFO =81.74 Hz and its harmonics at 2×BPFO=163.48 Hz. This can be validated with Fig.6(b).

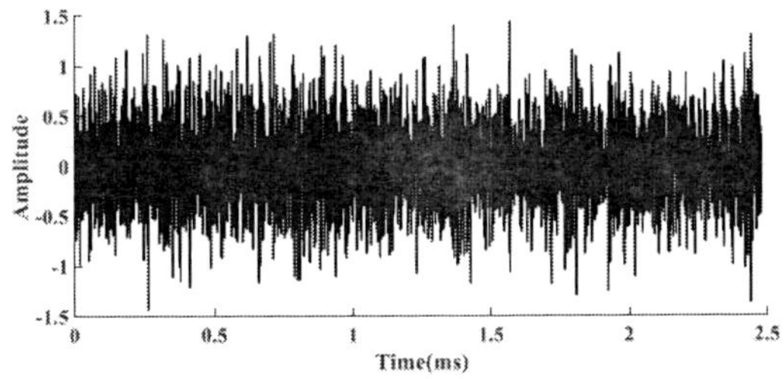

(a) Time domain

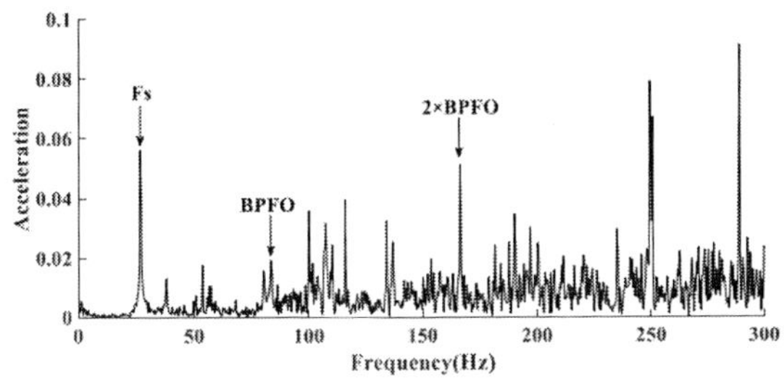

(b) Frequency domain

Figure 6 Vibration response of bearing having circular defect of 0.72 mm diameter on its outer race at shaft speed of 1600 RPM .

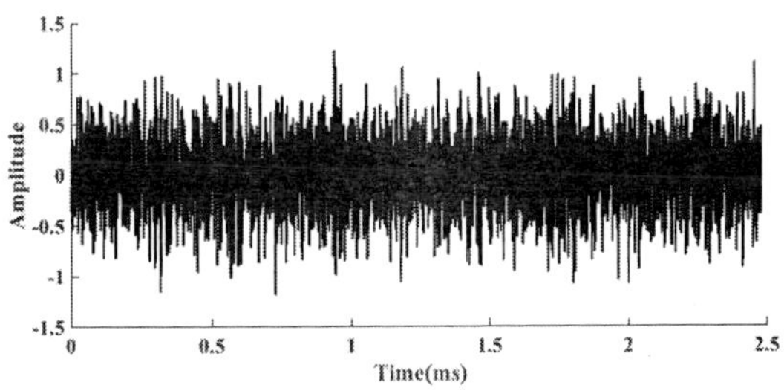

(a) Time domain

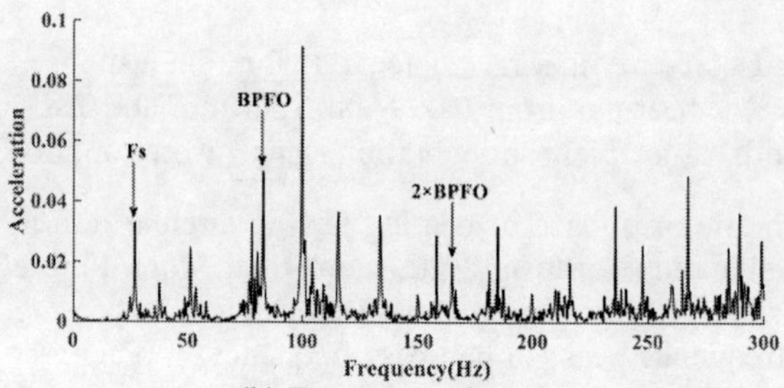

(b) Frequency domain

Figure 7 Vibration response of bearing having rectangular defect of 0.499 mm width on its outer race at shaft speed of 1600 RPM.

Figure 7 shows the vibration response of bearing having rectangular defect of 0.77 mm diameter on its outer race. The RMS value and kurtosis value computed from Figure7(a) is 0.3517 and 2.9360 respectively. Figure 7(b) shows the frequency response of defective bearing. Along with shaft rotational frequency F_s =26.63 Hz, frequency peaks at ball pass frequency of outer race (BPFO) at 82.73 Hz and its harmonics at 2 × BPFO =165 Hz. The computed value of BPFO =81.74 Hz and its harmonics at 2×BPFO=163.48 Hz. This can be validated with Figure7(b).

Same methodology has been adopted to study vibration responses of bearings having outer race defect of 0.72 mm, 1.49 mm and 2.16 mm defect diameter and 0.499 mm, 1.50 mm and 1.87 mm defect width. The kurtosis value and RMS value determined from their respective time domain signals were increased as compared to healthy bearings. The maximum RMS value reach to 0.9699 for defect width of 1.87 mm.

4. Noise cancellation in vibration signal: -

It has been observed that when defect size is small, the characteristic defect frequencies submerge in noise and its identification in frequency spectra become difficult. In this situation the adaptive noise cancellation techniques are useful. Adaptive filters have the ability to adjust their impulse response to filter out the correlated signal from the input signal. The prior information about signal and noise characteristics do not require in case of adaptive filters. The principle of adaptive noise cancellation is presented in Fig.8. The primary input signal S (vibration response of defective bearing) is corrupted by the noise $v_o(n)$ (unwanted vibration response). The reference input consists of noise $v_1(n)$ is uncorrelated with the primary signal S but correlated with the noise $v_o(n)$.

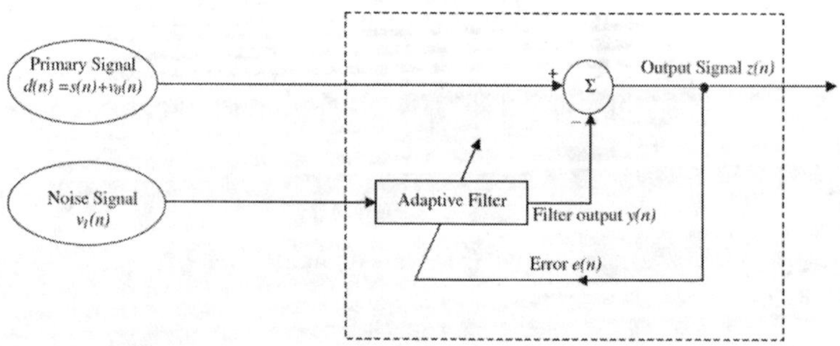

Figure 8 Basic concept of adaptive noise cancellation [9]

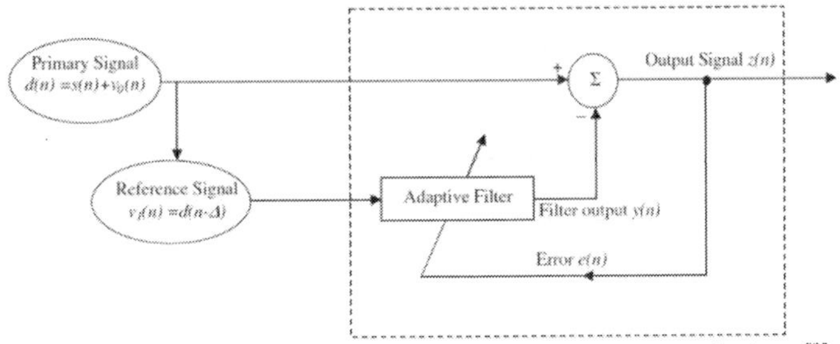

Figure 9 Basic Concept of self-adaptive noise cancellation. [9]

The principle of self-adaptive noise cancellation (SANC) is presented in Figure9. The use of self - adaptive noise cancellation signal processing technique amplifies the important frequencies in the vibration spectra. Self-adaptive noise cancellation technique does not require any reference signal. A delay version of the primary signal is used in the self-adaptive noise cancellation as the reference signal. The adaptive filter has the ability to adjust the parameter of its own and it can minimize the average power of the residual interference present in the output signal which has the error. The adaptive noise cancellation technique is highly depending on the filter length, step size and length of the signal.

4.1 Noise cancellation in defective bearing vibration signal: -

The time domain vibration signal of bearing having circular defect (signal shown in Fig.4(a) has been used as an input signal for SANC filter. The reference signal has been provided with delay of 4 samples of input signal. The filter length was 1024 and step size=0.036. Filtered vibration response of this signal is shown in Fig.11. The shaft rotating frequency along with the ball pass frequency of inner race and its sidebands are now visible clearly in Figure10. The noise in

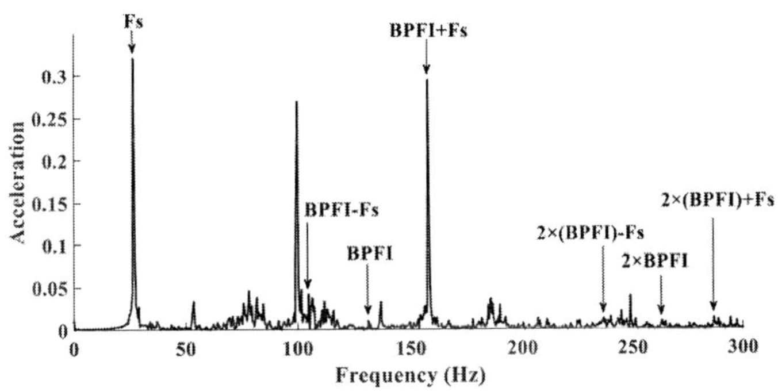

Figure 10 Vibration response of bearing having inner race circular defect after noise cancellation at a speed of 1600 RPM

vibration response has suppressed after use of self-adaptive noise cancellation. The time domain vibration signal of bearing having rectangular defect on inner race (signal shown in Fig.5(a) has been used as an input signal for SANC filter and its vibration spectra has been presented in Figure11. Selected filter length was 1024 and step size = 0.0135.

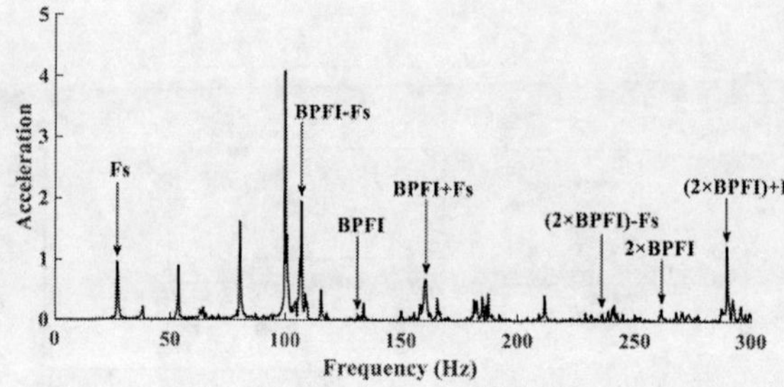

Figure 11 Vibration response of bearing having Inner race rectangular defect after noise cancellation at a speed of 1600 RPM.

The noise present in the vibration response (time domain signal in Figure 6(a)) has suppressed with the help of self-adaptive noise cancellation. Figure 12 shows the vibration response of bearing having outer race circular defect with the use of self-adaptive noise cancellation.

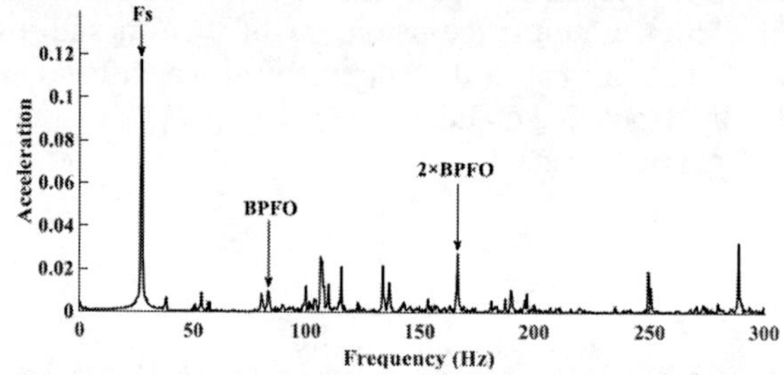

Figure 12 Vibration response of bearing having outer race circular defect after noise cancellation at a speed of 1600 RPM

Figure 13 shows the vibration response of bearing having outer race rectangular defect (for time domain signal in Figure7(a)) after the use of self-adaptive noise cancellation. The use of self-adaptive noise cancellation signal processing technique has suppressed the noise from the signal.

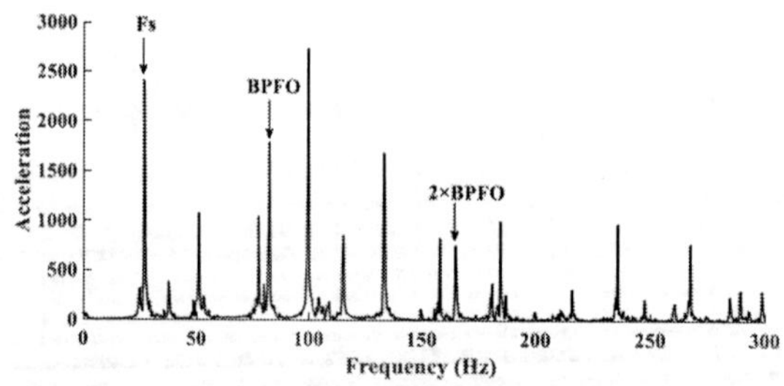

Figure 13 Vibration response of bearing having outer race rectangular defect after noise cancellation at a speed of 1600 RPM.

5. Defect size estimation: -

Wavelet analysis has gained lot of popularity among the researchers for the condition monitoring of the machine and in diagnosis of bearing faults. Wavelet analysis possesses flexibility and effectiveness in the computation and implementation. A wavelet is a waveform of effectively limited duration that has an average value of zero. Wavelet has the ability to perform local analysis of the large signal. Wavelet analysis consists of breaking up a signal into shifted and scaled version of the mother wavelet.

Approximation and details are two main part of the wavelet analysis where approximation indicates the high-scale and details indicates the low scale component. Here, high scale refers to the low frequency component whereas low scale refers to the high frequency component. The original signal passes through two given filters and emerges as two signals. Two filters are namely as low pass filter and high pass filter. The signal has been decomposed in lower resolution and reconstruct it without the loss of important information.

Wavelet analysis has the capability to expose the aspects of data of the signal like trends, breakdown points, self-similarity and discontinuities in higher derivatives. Wavelet analysis is used in many different applications which are Detecting Discontinuities and Breakdown Points, Detecting Long-Term Evolution, Detecting Self-Similarity, Identifying Pure Frequencies, Suppressing Signals, De-Noising Signals, De-Noising Images, Compressing Images and Fast Multiplication of Large Matrices.

It has been noticed that when ball passes through defect zone, the dynamic characteristic of bearing elements changes due to change in stiffness [26]. The change in stiffness results in changes in deflection pattern. The impulses generated due to impact correspond to ball position in defect zone can be identified using wavelet analysis of time domain signal.

5.1 Defect size estimation of inner race defect: -

Figure 14(a) presents the entry and exit of rolling element (ball) in defect zone of inner race. The entry point of defect zone has been marked with A (refer Figure14(a)) and it's corresponding time when ball just touches point A has been shown with A′ in zoomed view of time domain signal (refer Figure14(b)). The exit point of defect zone has been marked with B in Figure14 (a) and it's corresponding time when ball just touches point B has been shown with B′ in Figure14(b). The time difference corresponds to B′ and A′ provide information related to time spent by ball in defect zone. Inner race defects have been measured with the help of signal processing technique called wavelet analysis. Daubechies wavelet as a mother wavelet is used. Approximation of the signal is taken up to 3rd level. The same phenomenon has been used for circular defect and rectangular defect. Figure 14(b) shows time domain response at different entry and exit events of the ball when it strikes to the inner race circular defect.

Figure 15 shows the different entry and exit events of the ball when it strikes to the inner race rectangular defect. The time duration of ball at defect can be determined from these entry and exit events.

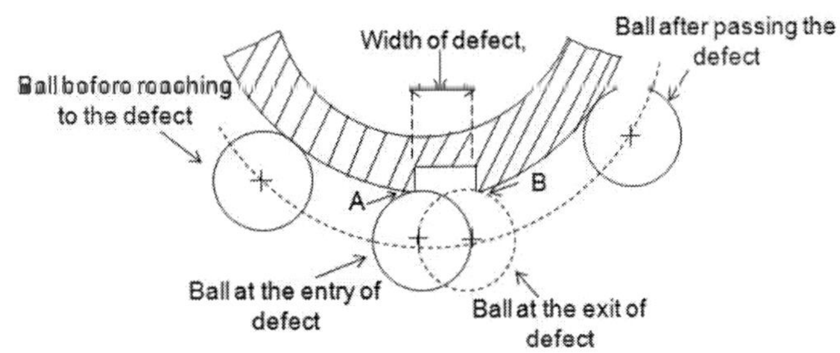

(a) Entry and exit event

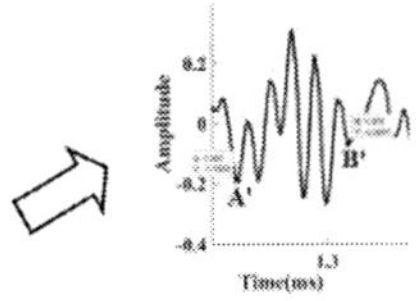

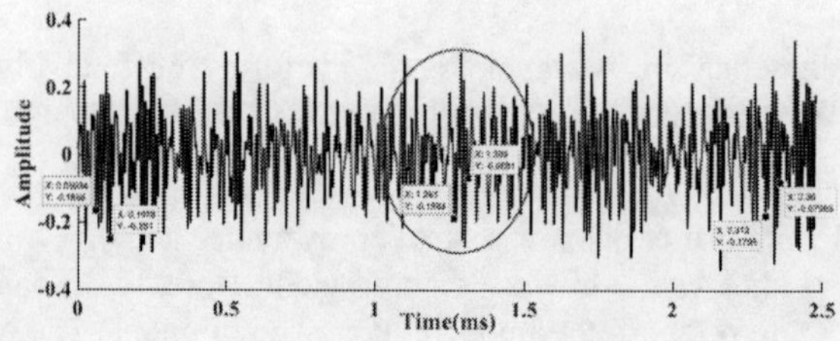

(b) Corresponding time domain signal and enlarged view of encircled part

Figure 14 Ball position at inner race defect zone .

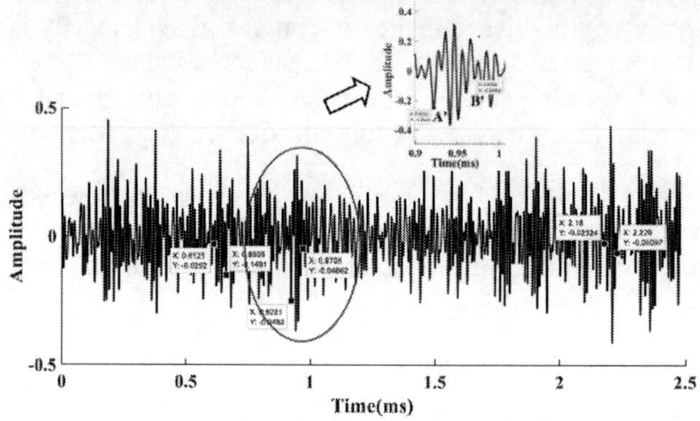

Figure 15 Approximation by decomposition of signal at 3rd level for entry and exit of ball from inner race defect zone.

5.2 Defect size estimation of outer race defect: -

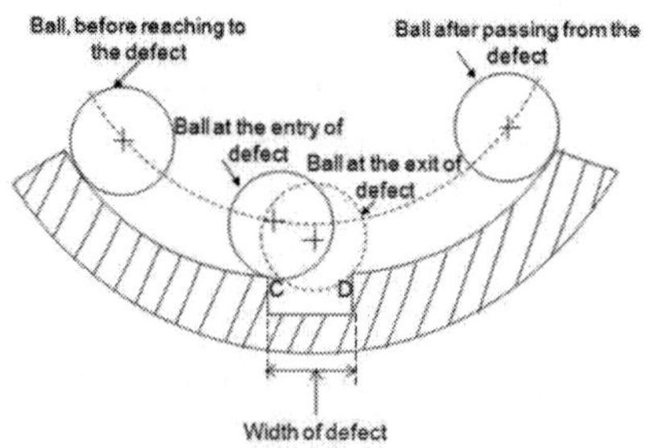

(a) Entry and exit event

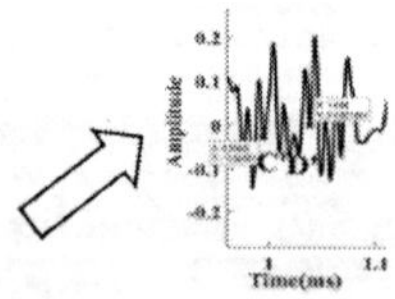

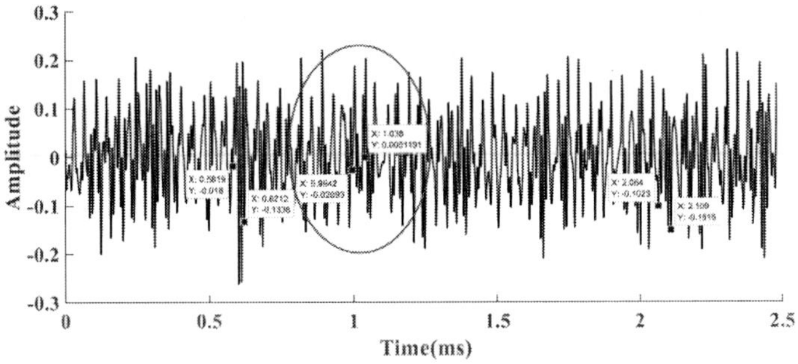

(b) Corresponding time domain signal and enlarged view of encircled part

Figure 16 Ball position at outer race defect zone.

Figure 16(a) presents the entry and exit of rolling element (ball) in defect zone of outer race. The entry point of defect zone has been marked with C (refer Figure16(a)) and it's corresponding time when ball just touches point C has been shown with C′ in zoomed view of time domain signal (refer Figure16(b)). The exit point of defect zone has been marked with D in Figure16 (a) and it's corresponding time when ball just touches point D has been shown with D′ in Figure16(b). The time difference corresponds to D′ and C′ provide information related to time spent by ball in defect zone. Outer race defects have been measured with the help of signal processing technique called wavelet analysis. Daubechies wavelet as a mother wavelet is used. Approximation of the signal is taken up to 3rd level. The same phenomenon has been used for circular defect and rectangular defect for outer race defects.

Figure 17 shows the different entry and exit events of the ball when it strikes to the outer race rectangular defect. All the defect sizes for inner race and outer race have been estimated with the help of the same signal processing technique.

6. Result and discussion

Figure 3 shows the vibration response of healthy bearing in time domain and frequency domain. The kurtosis values and root mean square (RMS) values for all the healthy bearings have been computed and tabulated in Table 3. It can be observed that the healthy bearing has kurtosis value around 3. The kurtosis value above 3 indicates manufacturing defect in bearing. However, the kurtosis value of bearing having defect on either race (refer time domain signal of Figure4 – Figure7) are around 4, this is indication of presence of local defect on bearing races. The peaks at defect frequency BPFI and 2×BPFI along with its side bands BPFI-Fs and BPFI+Fs in the case of inner race defective bearings can be observed in frequency domain. It is worth to mention here that the side bands appeared due to rotation of inner race defect at shaft rotation speed. The frequency peaks at BPFO and 2×BPFO in frequency domain of Figure6 and Figure7 indicates defect on outer race of bearing. In case of frequency domain of healthy bearing such characteristic defect frequency peaks are missing.

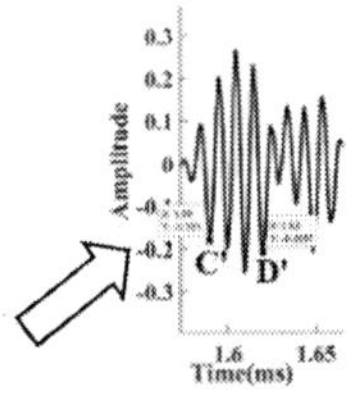

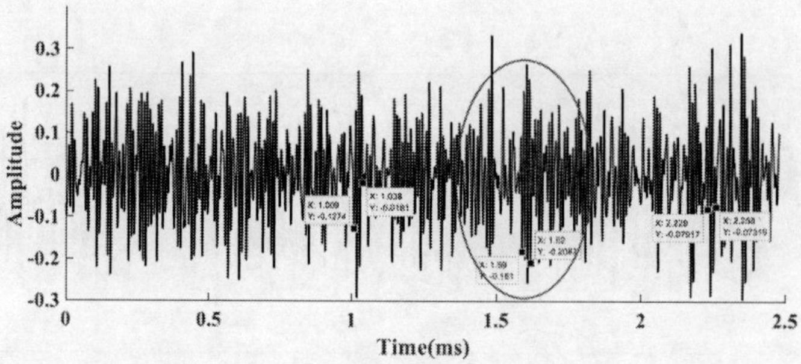

Figure 17 Approximation by decomposition of signal at 3rd level for entry and exit of ball from outer race defect zone.

The vibration response of defective bearings after the use of self-adaptive noise cancellation signal processing technique have been presented in Figure10 –Figure13. The enhancement of amplitude at defect frequency as compared to amplitude of defect frequency without use of SANC (refer frequency domain signals in presence of race defects) reveals increase of SNR. in the important peaks are massive and observed by comparing the figure without the use of self-adaptive noise cancellation and with the use of self-adaptive noise cancellation signal processing technique.

The defect size has been estimated for both rectangular defect and circular defect (on either bearing race) through wavelet decomposition of time domain signal of Figure14 – Figure17. The average time duration (Δt) i.e. time spent by rolling element in defect zone for three events has been measured. The time duration 'Δt' has been computed by taking time difference between time of exit of ball from defect and time of entry of ball in defect for same event. The estimation of defect size also depends on shaft rotational speed and bearing parameters like pitch diameter of bearing, number of balls and ball diameter. The defect size has been estimated through Eq. 3 provided by Khanam et al. [16].

$$\text{Defect size} = \frac{D_p \times \omega_s}{4}\left(1 - \frac{d_b^2}{D_p^2}\cos^2\alpha\right) \times \Delta t \qquad (3)$$

Where, ω_s is the shaft rotational speed of the ball, D_p is the pitch diameter of the bearing, d_b is the ball diameter, α is the contact angle equal to zero for deep groove ball bearing.

The estimated defect width (for rectangular defect) and defect diameter (for circular defect) through Eq. 3 and actual measured defect size have been tabulated in Table 2. It is worth to mention here that the actual defect size has been measured by optical microscope. The error between estimated value and measured value provide accuracy of defect size estimation.

Table 2 : Comparison of measured defect size and estimated defect size.

Defective race	Type of defect	Defect size (dia. or width) measured using optical microscope (mm)	Defect size (dia. or width) estimated through signal processing	% Error

			technique (mm)	
Inner race	Circular defect	0.77	0.81	5.19
		1.1	1.09	0.90
		1.53	1.57	2.61
		2.07	2.08	0.48
	Rectangular defect	0.77	0.78	1.29
		1.03	1.05	1.94
		1.44	1.39	3.47
		1.91	2.04	6.8
Outer race	Circular defect	0.72	0.71	1.38
		1.49	1.48	0.7
		2.16	2.15	0.46
	Rectangular defect	0.499	0.490	1.80
		1.50	1.50	0.00
		1.87	1.85	1.06

7. Conclusion: -

The vibrations generated by healthy and defective deep groove ball bearings have been captured experimentally and vibrations signals have been analyzed in time domain and frequency domain. It has been observed that the statistical parameters like kurtosis and RMS value increases in presence of local defects on either race of bearing. The peaks at characteristic defect frequency along with their harmonics have been observed in presence of local defect on either race. Thus, the frequency domain analysis of vibration signal can provide information regarding health of bearing. The amplitudes of defect frequency are more in case of rectangular defect as compared to circular defect amplitudes. However, it becomes difficult to identify the defect frequency peaks at the initial stage of defect and in presence of noise. In such cases, signal to noise ratio improved after use of SANC that result in enhancement of amplitude of defect frequencies. The width of rectangular defect (fatigue crack) and diameter of circular defect (pits) have been estimated accurately with maximum error of 6.8%. Authors of this paper believe that the present experimental study will provide guidelines to researchers and practicing engineers in bearing defect detection and defect size estimation.

8. References: -

[1] Tandon N, and Choudhury A 1999 A review of vibration and acoustic measurement methods for the detection of defects in rolling element bearings*Tribology international* **32** 469-480.

[2] Patel V N, Tandon N, Pandey RK 2009 Review of defect detection techniques of ball bearings in the presence of external noise *Proceedings of Ninth International Conference on Vibration Problems, IIT Kharagpur, India, January*.

[3] Tandon N, Nakra B C 1993 Detection of defects in rolling element bearings by vibration monitoring *Indian Journal of Mechanical Engineering Division* **73** 271-282.

[4] Tandon N 1994 A comparison of some vibration parameters for the condition monitoring of rolling element bearings *Measurement***12** 285-289.

[5] Antoni, J, Randall R B 2006 The spectral kurtosis: application to the vibratory surveillance and diagnostics of rotating machines *Mechanical systems and signal processing* **20**308-331.

[6] Yanxue W, Liang M 2011 An adaptive SK technique and its application for fault detection of rolling element bearings *Mechanical Systems and Signal Processing* **25** 1750-1764.

[7] Sawalhi N, Randall R B, Endo H 2007 The enhancement of fault detection and diagnosis in rolling element bearings using minimum entropy deconvolution combined with spectral kurtosis *Mechanical Systems and Signal Processing* **21** 2616-2633.

[8] Patel V N, Tandon N, Pandey R K 2012 Defect detection in deep groove ball bearing in presence of external vibration using envelope analysis and Duffing oscillator *Measurement***45** 960-970.

[9] Patel V N, Tandon N, Pandey R K 2012 Improving defect detection of rolling element bearings in the presence of external vibrations using adaptive noise cancellation and multiscale morphology *Proceedings of the Institution of Mechanical Engineers, Part J: Journal of Engineering Tribology* **226** 150-162.

[10] Khemili I, Chouchane M 2005 Detection of rolling element bearing defects by adaptive filtering *European Journal of Mechanics-A/Solids* **24** 293-303.

[11] Chaturvedi, G K, Thomas D W 1981 Adaptive noise cancelling and condition monitoring*Journal of Sound and Vibration* **76**391-405.

[12] Widrow, Bernard, et al. 1975 Adaptive noise cancelling: Principles and applications *Proceedings of the IEEE* **63** 1692-1716.

[13] Chaturvedi G K, Thomas DW 1982 Bearing fault detection using adaptive noise cancelling *Journal of Mechanical Design* **104** 280-289.

[14] Zia Ur R M, Ahamed S, Reddy D V 2011 Efficient sign based normalized adaptive filtering techniques for cancelation of artifacts in ECG signals: Application to wireless biotelemetry *Signal processing* **91** 225-239.

[15] Antoni J, Randall R B 2004 Unsupervised noise cancellation for vibration signals: part I—evaluation of adaptive algorithms *Mechanical Systems and Signal Processing* **18** 89-101.

[16] Khanam S, Tandon N, Dutt J K 2014 Fault size estimation in the outer race of ball bearing using discrete wavelet transform of the vibration signal *Procedia Technology* **14**12-19.

[17] Sawalhi N, Randall R B 2011 Vibration response of spalled rolling element bearings: Observations, simulations and signal processing techniques to track the spall size *Mechanical Systems and Signal Processing* **25** 846-870.

[18] Jena D P, Panigrahi S N 2014 Precise measurement of defect width in tapered roller bearing using vibration signal *Measurement* **55** 39-50.

[19] Purushotham V, Narayanan S, Prasad A N 2005 Multi-fault diagnosis of rolling bearing elements using wavelet analysis and hidden Markov model based fault recognition *NDT & E International* **38**654-664.

[20] Ghamd A I, Abdullah M, David Mba 2006 A comparative experimental study on the use of acoustic emission and vibration analysis for bearing defect identification and estimation of defect size *Mechanical systems and signal processing* **20** 1537-1571.

[21] Kumar R, Singh M 2013 Outer race defect width measurement in taper roller bearing using discrete wavelet transform of vibration signal *Measurement* **46** 537-545.

[22] Ahmadi M, Alireza, Howard C Q 2016 A defect size estimation method based on operational speed and path of rolling elements in defective bearings *Journal of Sound and Vibration 385* 138-148.

[23] Barszcz T 2009 Decomposition of vibration signals into deterministic and nondeterministic components and its capabilities of fault detection and identification *International Journal of Applied Mathematics and Computer Science* **19** 327-335

[24] Adewusi S A, Al-Bedoor B O 2001 Wavelet analysis of vibration signals of an overhang rotor with a propagating transverse crack *Journal of sound and vibration* **246** 777-793.

[25] Prabhakar S, Mohanty A R, Sekhar A S 2002 Application of discrete wavelet transform for detection of ball bearing race faults*Tribology International* **35** 793-800.

[26] Patel V N, Tandon N, Pandey R K 2010 A dynamic model for vibration studies of deep groove ball bearings considering single and multiple defects in races *Journal of Tribology***132** 041101.

IOP Conf. Series: Materials Science and Engineering **810** (2020) 012040 doi:10.1088/1757-899X/810/1/012040

Finite Element Analysis of Automotive Drum Brake for Prediction of Wear of Friction Liner

Dinesh Shinde[1, 3*], Mukesh Bulsara[2, 3] and K N Mistry[3]
[1]SVKM's NMIMS MPSTME Shirpur, Dist-Dhule, MH, India.
[2]G H Patel College of Engineering & Technology, V.V. Nagar, Anand, Guj, India.

[3]Gujarat Technological University, Ahmedabad, Guj, India

[*]dinesh.shinde61@gmail.com

Abstract. Brake is a crucial part of an automobile, whose performance evaluation is always, been an important task industry people are looking for. The evaluation of performance of the brake experimentally is very complex and un-economical for each and every time. This led to development of a model using soft-computing to accurately assess the brake every time. This paper presents development of such a model for drum brake using FEA and experimentally validating it for its feasibility and reliability. An automotive drum brake is chosen for study and a FEA model is developed, deformation characteristics of the brake liner surface is drawn from it and presented in terms of mean nodal height at twenty different points on its surface. The brake liner with available material (composition) is subjected to experimentation using a test set up and its surface wear is measured using a height gauge and a holding fixture. Percentage of wear is evaluated by considering initial height (new surface) and final height (weared surface). The developed FEA model is remarkably close to the exact conditions, which can be confirmed through the experimental results.

1. Introduction

Brake is a vital element of any automobile, in order to control it. It consists of rotating disc or drum lined with composite friction material and mechanism to apply the brake. Continuous sliding between the friction lining materials and rotating drum provide necessary effect of frictional resistance. Enormous research is carried out since several decades, to understand the behavior of brakes experimentally. However, it is difficult to interpret the tribological performance of each and every friction lining materials in their real operating conditions. Wear rate and coefficient of friction were the significant concerns analyzed [1]. Apart from this two performance parameters, noise and vibrational study of the brakes were also the concerns taken up by various researchers. In order to retard the high speed brake drums or discs, larger frictional effect should be provided, because of which a screeching noise is generated eventually during the braking. This noise had been called as squealing noise in studies of brakes [2-10].

Finite element analysis (FEA) is an analytical tool to solve real life problems with complex geometry, loadings and materials. The complications in the study of the friction lining materials in actual circumstances are exploratory, thus have motivated to move towards analysis using finite element analysis (FEA) to predict the wear rate and squealing noise. Wear of brake rotor disc or drum is also the point of interest for many researchers [11-23].The work presented here is an analysis of a drum brake with the help of FEA and its experimental validation to forecast the wear of friction lining materials.

2. Modelling and Analysis Methodology

This work aims at analysing automotive drum brake using finite element analysis, for nodal deformation of the friction liner surface in contact with drum and validating the outcomes using a mean height measurement of the nodes respective to the surface of the friction liner after experimentation with similar performance parameters. A standard size our wheel automotive drum brake is modeled in software package Solidworks 17.0 available at CAD/CAE lab (Mechanical Engineering Department, MPSTME SVKM's NMIMS Shirpur, Dist-Dhule, MS), with shoe and friction liner shown in figure 1. Considering the symmetrical geometry of the drum brake, a half model is taken.

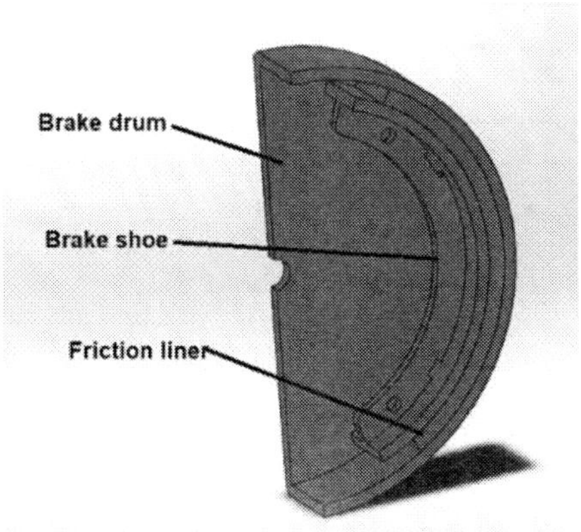

Figure 1: Half model of drum brake assembly.

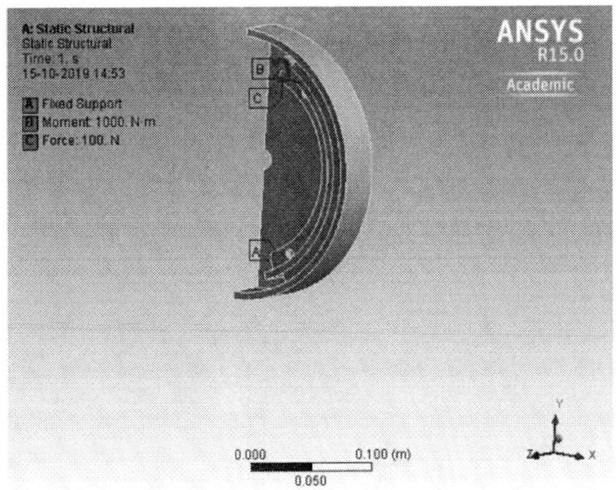

Figure 2: Constraints and loadings considered.

A static structural analysis is planned in an Ansys 15.0 software package. Assembly of brake drum with constraints and loading is shown in figure 2. Brake shoe in an automotive drum brake is mounted in a position, such that its lower filleted corner (shown as A in figure 2) is hinged using retainers, that's why it is supported as fixed. Also its upper filleted corner is placed at the hydraulic wheel cylinder, which applies the braking force, so a point load directed towards positive X-direction, is considered in analysis. The rotational effect of the drum is simulated by the moment applied about its shaft on which it is to be mounted.

The material for drum is Gray Cast Iron, for brake shoe is structural steel and friction lining material (liner) is a composite material. Respective material properties are taken from Ansys material library, whereas for friction lining material materials properties are defined in accordance with literature [24-25].

In order to endorse the results of finite element analysis, an experimental set up (figure 3) is used. The operating parameters for the experimentation are referred to FEA. The wear of friction lining materials surface is then measured as a mean height of nodes with spacing between the measurement array as 0.7 mm, using the height measuring device as shown in figure 4. An influential tribological property i.e. wear is then determined in percentage relative to the new surface.

IOP Conf. Series: Materials Science and Engineering **810** (2020) 012040 doi:10.1088/1757-899X/810/1/012040

Figure 3: Experimental Set up with, 1. Eddy current dynamometer, 2. Drum brake, 3. Inertia mass, 4. Load cell, 5. DC motor with speed regulation, 6. Control Panel, 7. Brake lever, 8. Differential gearbox, 9. Thermocouple for atmospheric temp., 10. Thermocouple for inside temperature, 11. Supporting frame.

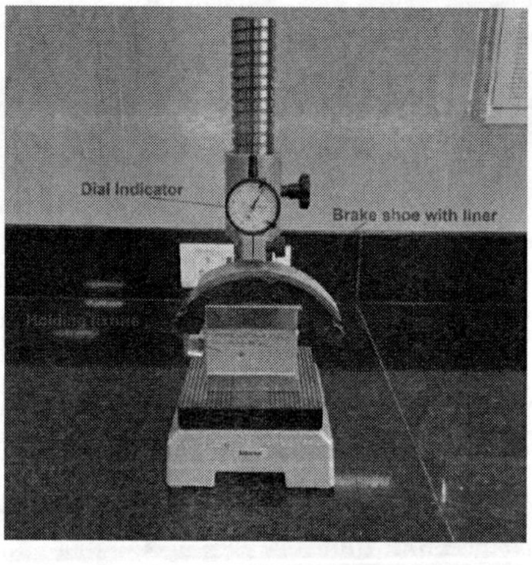

Figure 4: Nodal height measurement

3. Results and Discussions

Figure 5 shows the FEA results for total deformation along the transverse direction of friction liner, in meters. Colour representation reveals that at upper part of it deformation is quite large which goes on decreasing till the bottom part. This exactly depicted by the practical materials wear of the friction liner pad surface. This deformation is then converted into the height measured from line joining top corner and bottom corner of the shoe. After experimentation on the said set up with the same performance parameters as that of FEA, the liner got worn and its mean height at different nodal location are measured and plotted as shown in figure 6a. The resulting heights also confirm that the nodal height is quite smaller at the point where the brake shoe is hinged than at other points,. Considering new and worn out surface heights an experimental mean height distribution is calculated and plotted with FEA outcomes for comparison. The height distribution (variation) is relatively higher from left side till the midpoint of the liner surface, than right part.

IOP Conf. Series: Materials Science and Engineering **810** (2020) 012040 doi:10.1088/1757-899X/810/1/012040

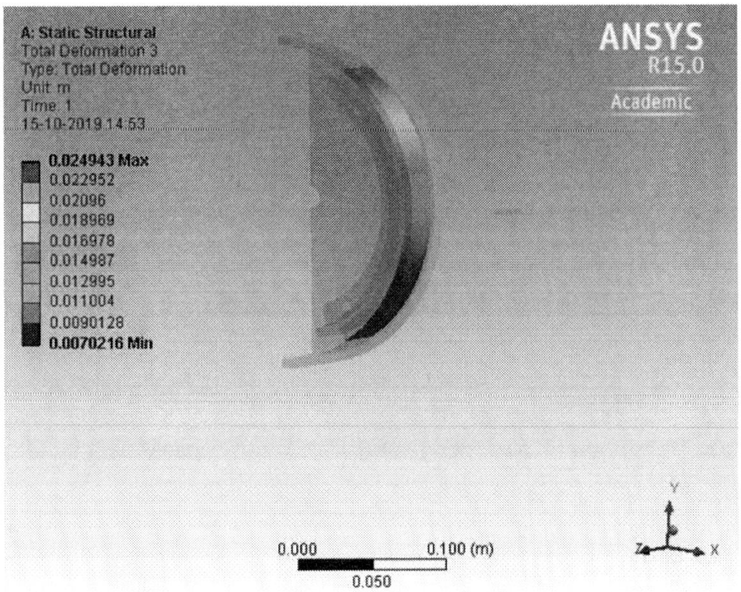

Figure 5: Resulting deformation of liner surface.

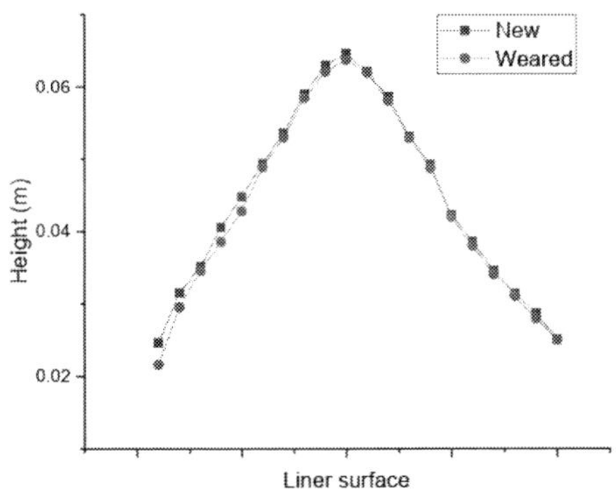

Figure 6a: Height distribution of pad surface measured from top surface of the holding fixture.

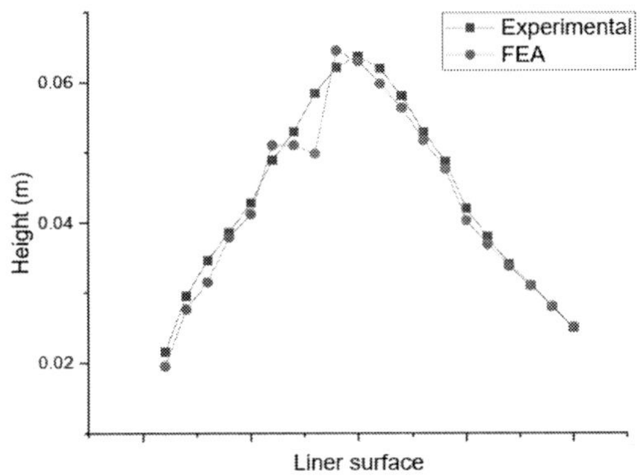

Figure 6b: Comparison of resulting height distribution of pad surface.

Figure 7 represents the comparison of percentage of wear of the liner and error found in experimental and FEA analysis. Wear of the liner goes on decreasing from the left corner (resembling to the top corner in FEA) to the right. It is found to be remarkably large at beginning part of the liner (12.5 & 20.73% respectively), the point from where the braking force is applied (0.45 &0.41% respectively). Whereas, the wear is very small at the point where is hinged (right part). The percentage errors are obtained to be conforming to the application of load and support reactions, which is in the acceptable range, revealing that the FEA model is feasible.

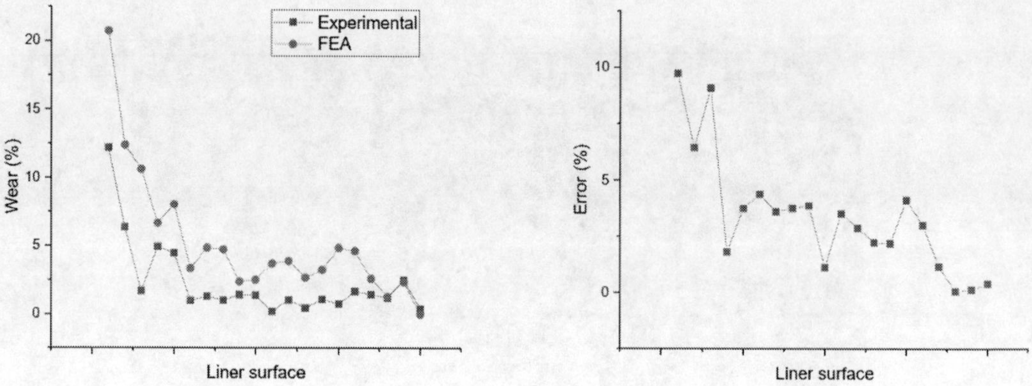

(a) (b)

Figure 7: (a) Comparison of Wear and (b) Error in Wear measurement by the two methods.

4. Conclusions

After the FEA and Experimental Analysis of the drum brake for determination and prediction of surface wear, following conclusions are made:

- An automotive critical part like drum brake can be effectively modelled and analysed using FEA, which ultimately reduces the experimentation cost and time.
- The results obtained from FEA model are quite close values to experimental percentage of wear, which endorsed the sustainability of the model.
- Measurement of surface wear of the drum brake liner can be done by determining nodal mean height distribution by using a height gauge, analog or digital.
- Wear is obtained to be large at the surface of the liner nearby to the point where braking force is applied than the area where the shoe is hinged.

5. References

[1] Dinesh Subhash Shinde, KN Mistry, Mukesh Bulsara 2019 *Proceedings of the Institution of Mechanical Engineers, Part J: Journal of Engineering Tribology*. https://doi.org/10.1177/1350650119877775.

[2] Mario Triches Junior, Samir N.Y. Gerges, Roberto Jordan 2008 *Applied Acoustics* **69** pp. 147–162.

[3] Ali Belhocine, Nouby Mahdi Ghazaly 2015 *Latin American Journal of Solids and Structures* **12** 1432-1447.

[4] Dan Yang, Zhen Yu, Leilei Zhang and Wentao Cheng 2017 *Advances in Computer Science Research* **75** 608-612.

[5] N. Coudeyras, J.-J. Sinou, S. Nacivet 2009 *Journal of Sound and Vibration* **319** 1175–1199.

[6] O. Fazio, S. Nacivet, J-.J. Sinou 2015 *Applied Acoustics* **91** 12–24.

[7] Abd Rahim Abu Bakar and Huajiang Ouyang 2006 *International Journal of Vehicle Noise and Vibration* **2(2)** 143-152.

[8] Jaeyoung Kang 2009 *International Journal of Mechanical Sciences* **51** 284–294.

[9] A. Renault, F. Massa n, B. Lallemand, T. Tison 2016 *Journal of Sound and Vibration*http://dx.doi.org/10.1016/j.jsv.2015.12.049i.

[10] H. A. Sherif 2004 *Wear* **257** 687–695.

[11] S. Oberst, J.C.S.Lai 2011 *Journal of Sound and Vibration* **330** 955–975.

[12] D. Pavelescu and M. Musat 1974 *Wear* **27** 91-97.

[13] Abd Rahim AbuBakar, Huajiang Ouyang 2008 *Wear* **264** 1069–1076.

IOP Conf. Series: Materials Science and Engineering **810** (2020) 012040 doi:10.1088/1757-899X/810/1/012040

[14] Z. Barecki and S. F. Scieszka 1988 *Wear* **127** 283 – 305.

[15] S. K. Rhee 1974 *Wear* **29** 391-393.

[16] Anders Soderberg and Soren Andersson 2009 *Wear* **267** 2243–2251.

[17] M.M. Shahzamanian, B.B. Sahari, M. Bayat, F. Mustapha, Z.N. Ismarrubie 2010 *Composite Structures* **92** 1591–1602.

[18] Priit Podra and Soren Andersson 1999 *Tribology International* **32** 71–81.

[19] A. A. Yevtushenko and P. Grzes 2010. *Numerical Heat Transfer Part A* **58** 207–226.

[20] A.A. Yevtushenko, P. Grzes 2015 *International Communication in Heat Mass Transfer* http://dx.doi.org/10.1016/j.icheatmasstransfer.2015.01.005.

[21] Dinesh Shinde and Kanak Kalita 2015 *ARPN Journal of Engineering and Applied Sciences* **10(5)** 2227-2232.

[22] A. Belhocine and O. I. Abdullah 2014 *International Journal of Advanced Design and Manufacturing Technology* **7(4)** 27-42.

[23] S.A.M Da Silva, DVV Kallon 2019. *Procedia Manufacturing* **35** 181–186,.

[24] Blau, P J 2001. *Report prepared for U.S. Department of Energy* 27.

[25] J. Brecht and A. Elven kemper, J. Betten, U. Navrath and J. B. Multhoff 2003 *SAE International*, pp. 1-10.

IOP Conf. Series: Materials Science and Engineering **810** (2020) 012041 doi:10.1088/1757-899X/810/1/012041

Influence of nozzle hole diameter on combustion and emission characteristics of diesel engine under pilot injection mode

Hiren Dave[1*], Bharatkumar Sutaria[2] and Brijesh Patel[3]
[1,2]Mechanical Engineering Department, National Institute of Technology, Surat-395007, India.
[3]Gdhyana Sanshodhana Nagari Foundation, Vadodara-390008, India.

E-mail address: davehiren.87@gmail.com

Abstract. Present work was divided in two different parts. In first part, an experimental study was carried out to analyze the effect of dwell on diesel engine combustion and emission characteristics. Experimental data showed that shorter dwell results in improved specific fuel consumption (SFC) but with increased smoke emissions penalty. To overcome the disadvantage of higher smoke emissions associated with shorter dwell, an effort was made to reduce smoke emissions under pilot injection mode by changing nozzle hole diameter (NHD) while keeping rest of the injection parameters unchanged. In second part, this was done numerically by using commercial *Converge* 3-D computational fluid dynamics (CFD) tool. In addition to baseline NHD (0.120 mm), two other NHDs i.e. 0.08 mm and 0.160 mm were considered. Numerical results showed that for given injection parameters, lower NHD resulted in faster evaporation of injected fuel as well as greater homogeneity of the mixture prepared due to better air entrainment inside fuel spray and hence led to overall faster combustion process. This faster combustion process subsequently reduced the time available for spray interactions. Therefore, in spite of getting further shortening of dwell in case of 0.08 mm NHD compared to other two cases, soot emissions were reduced by 87.2% when NHD was reduced from 0.160 mm to 0.08 mm. However, nitric oxide (NO) emission was found to increase by around 43% when NHD was reduced from 0.160 mm to 0.08 mm.

1. Introduction

Use of diesel engines is extensively increasing in heavy duty sector due to their larger power output and low carbon dioxide (CO_2) emissions. However, diesel engines suffer from greater soot and nitrogen oxides (NO_x) emissions [1]. There is an urgent necessity of reducing these emissions from diesel engines due to implementation of more and more stringent emission norms. Achieving these stringent emission norms while maintaining higher fuel economy; a prime demand of customers especially in developing countries like India has become a difficult task for diesel engine manufacturers.

Modern commercial diesel engines use pilot injection mode in order to control the formation of NO_x emissions inside engine cylinder. Dwell between pilot and main injection is one of the most significant parameter which greatly affects combustion process and hence emissions. Shorter dwell results in improved specific fuel consumption (SFC) as main fuel gets smoothly linked with pilot combusted fuel [2]. A study conducted by Huang et al. [3] reported that shorter dwell can reduce energy loss in diesel engines in terms of friction and also leads to improvement in thermal efficiency. SFC improvement with shorter dwell was also claimed by several other studies [4-7]. Shorter dwell also resulted in lesser NO_x emissions due to shorter delay period and subsequent lesser intense pre-mixed combustion of main fuel [8]. Unfortunately, soot emissions were increased with shorter dwell due to longer and more intense diffusion combustion of main fuel [9]. In diesel engines, mostly both pilot as well as main injections are performed by a single fuel injector. So, fuel is injected in the same region of combustion chamber by both these injections, especially in diesel engine without intake swirl. Therefore, liquid components of main fuel interact with still burning pilot flame which reduces the mixing time of main fuel with air and combustion takes place under locally rich conditions leading to augmented soot formation. Shorter dwell between two injections further aggravates this phenomenon and further augments the soot formation [10]. From the flame luminosity images, Cung et al. [11] confirmed that spray-combustion interactions were more aggressive with shorter dwell time compared to longer dwell time, which were highly responsible for accelerating soot formation. Increased smoke/soot emissions with shorter dwell were claimed by several other studies too [2, 5, 7, 12-14]

From above literature survey, it can be found that shorter dwell offers the advantage of improved fuel consumption but at the same time, shorter dwell affects soot/smoke emissions negatively. Few earlier studies suggested some solutions to reduce smoke/soot emissions under pilot injection mode. Binde et al. [10] suggested to use separate fuel injector for injecting pilot and main injections so that both these injections get spatially separated and soot formation due to spray interactions can be significantly reduced. Sahoo et al. [15] recommended to reduce pilot quantity for reducing soot formation under pilot injection mode. However, too small pilot quantity produces over-lean mixture which becomes difficult to ignite [16]. Park et al. [17] suggested to use higher fuel injection pressure to reduce smoke emissions under pilot injection mode. However, upper values of injection pressure are governed by the limitations of injection system. Another method is to increase swirl rate (SR), which was proposed by Li et al. [18], in which authors reported simultaneous reduction of soot KL factor (a measure of soot particle concentration, where K is absorption co-efficient which is proportional to soot volume fraction and L is the path length of measurement) and SFC. However, generation of swirling intake flow in diesel engine involves complex designing and higher cost of the intake system (intake ports, deflector etc.). In this work, an effort has been made to reduce smoke emissions under pilot injection mode by changing the nozzle hole diameter (NHD) of the fuel injector. Some earlier studies have investigated effect of NHD on diesel combustion process. Vijay kumar et al. [19] investigated effect of variation of NHD on diesel engine combustion characteristics using pure diesel as well as biodiesel fuel (B20). Authors reported that better fuel atomization with smaller NHD improved air-fuel mixing for both the test fuels which led to improvement in SFC and smoke emissions compared to greater NHD case. Experimental results obtained by Bergstrand and Denbratt [20] showed that smaller NHD significantly reduces the delay period. Moreover, improved air-fuel mixing with smaller NHD also resulted in shorter combustion duration which reduced heat loss and improved thermal efficiency along with reduction in soot and hydrocarbon (HC) emissions. Pang et al. [21] numerically investigated effect of NHD using Eulerian Stochastic Field (ESF) model. Results of the study demonstrated that fuel spray with larger NHD produces fuel-richer pre-mixed core region which augmented soot formation compared to smaller NHD case. Zhang et al. [22] investigated effect of NHD on flame lift-off length and soot formation process. Authors reported reduction in soot formation with lower NHD. Optical images confirmed that reduction in soot formation was mainly attributed to increased lift-off length in case of lower NHD.

From above literature survey, it can be found that most of the earlier studies have investigated effect of NHD on diesel combustion characteristics under conventional single injection mode and studies related to effect of NHD on combustion characteristics of diesel engine under pilot injection mode are scarce. Based on literature study related to effect of NHD on diesel combustion and emission characteristics, it can be realized that smaller NHD results in improved air-fuel mixing and reduces soot emissions and at the same time, it also results in longer injection duration due to reduced fuel flow rate because of smaller flow area [23]. So, for a given pilot quantity and given pilot as well as main injection timings, smaller NHD results in shorter dwell due to longer injection duration of pilot fuel. Therefore, use of fuel injector with smaller NHD should result in greater smoke emissions due to more aggressive spray interactions. However, it is also possible that smaller NHD may lead to lower smoke emissions due to improved fuel-air mixing under pilot injection mode. Hence, it would be interesting to analyse the effect of NHD on diesel combustion characteristics under pilot injection mode. Moreover, it is also noticed from literature studied that when effect of NHD is analysed on combustion characteristics of diesel engine, changes of spray angle and spray cone angle are inevitable which are not taken into account by most of the earlier studies. However, spray angle as well as spray cone angle have significant impact on overall combustion process and subsequent engine out emissions [24]. Hence, when effect of NHD is analysed on diesel combustion and emission characteristics, it represents combined effects of NHD, spray cone angle and spray angle. Therefore, spray angle and spray cone angle needs to be maintained constant with changing NHD in order to investigate the sole effect of NHD on diesel combustion characteristics. Considering these backdrops, present work is aimed to investigate the sole effect of NHD on combustion and emission characteristics of diesel engine under pilot injection mode. Initially, an experimental study was conducted in order to explore the effect of dwell on combustion, performance and emissions characteristics of the test engine. Then after, effect of NHD on combustion and emission characteristics of the test engine was numerically investigated using *Converge* 3-D computational fluid dynamics (CFD) tool. Base nozzle has NHD of 0.120 mm. Two other NHDs under consideration are 0.08 mm and 0.160 mm respectively. Variation of NHD was conducted while maintaining spray angle and spray cone angle constant. Moreover, total mass of injected fuel per cycle, pilot as well as main injection timings, distribution of fuel in pilot and main injection, co-efficient of discharge (C_d) and injection pressure were also maintained constant with changing NHD.

2. Experimental setup and operating conditions

All the experiments for present work were conducted on Kirlosker TV1 single cylinder diesel engine. Table 1 represents the basic engine specifications. A pressure transducer (HSM111A22, PCB) was used to measure the in-cylinder pressure. Combustion pressure was measured for 50 consecutive cycles at 1° crank angle (CA) interval and average value of these 50 cycles was used as pressure data. *"ICEngineSoft";* which is a Lab-view based software, was used to conduct performance and combustion analysis. Software calculates indicated mean effective pressure (IMEP), brake mean effective pressure (BMEP), heat release rate (HRR), brake power (BP) etc. from the obtained pressure data. An electronic fuel injection system was used in order to have flexibility in terms of variation of injection timing, number of injections per cycle and distribution of fuel in each injection. A much more detailed description of test engine, injection system, emission measurement devices, fuel properties and uncertainty of measuring instruments as well as performance parameters can be found in authors' earlier work [25].

Table 1. Test engine specifications.

Compression ratio	17.5
Displacement (cc)	661.45
Stroke (mm) × Bore (mm)	110 × 87.5
Length of connecting rod (mm)	234
Max. power (kW@rpm)	5.2@1500

2.1. Experimental operating conditions

In present work, the test engine was operated at rated speed (1500 rpm) and at 4.6 bar BMEP (75% load) for all the test runs. Main injection timing was fixed at -20° after top dead centre (ATDC) throughout the study. This is because retardation of main injection timing beyond -20° ATDC was resulting in reduction of IMEP (indicating power loss) as shown in Figure 1 which cannot be afforded. Moreover, this is the injection timing which resulted in least brake specific fuel consumption (BSFC) as indicated in Figure 1. Based on this combination, main injection timing was selected as -20° ATDC. Start of pilot injection (SOPI) was changed to vary the dwell between two injections. Injection pressure was maintained at 350 bar. The reason why injection pressure was kept limited to 350 bar only is discussed in Section 3.2. Figure 2 shows a representative line diagram of injection profile under pilot injection mode. As the aim of present work is to lower the smoke emissions under pilot injection mode, pilot quantity was fixed at 40% of total injected fuel per cycle. This is because at given pilot and main injection timings, soot emissions increase with increasing pilot quantity due to more severe spray interactions [10]. Dwell was varied from 8° to 16° in steps of 4° which are represented as run nos. 1, 2 and 3 respectively in Table 2.

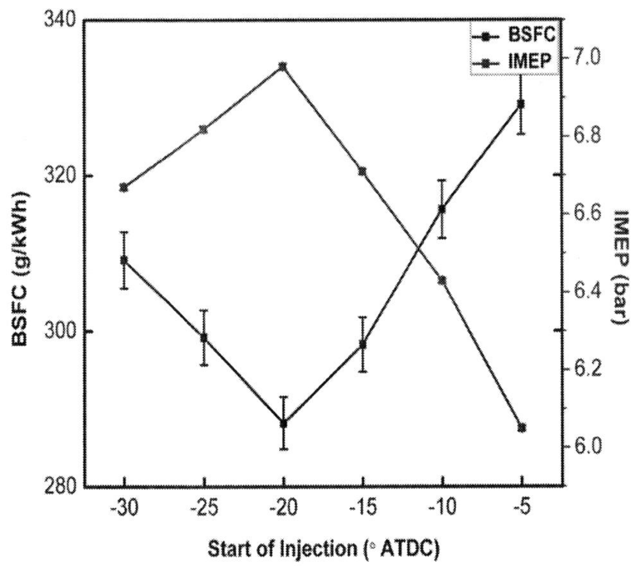

Figure 1. Effect of variation of injection timing on BSFC and IMEP.

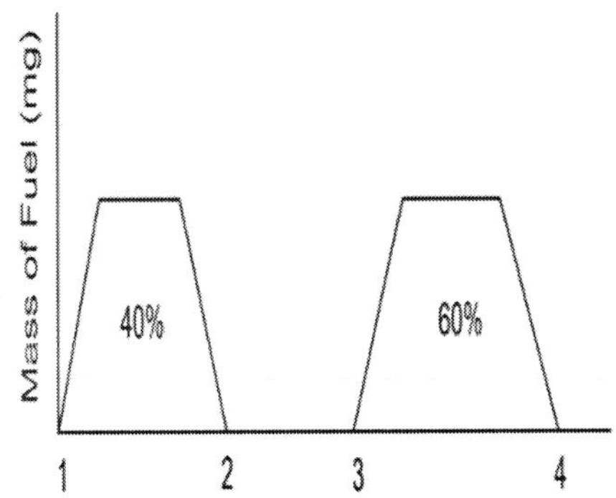

Figure 2. Representative line diagram of injection profile under pilot injection mode.

Table 2: Experimental operating conditions

Run No.	Pilot quantity (%)	SOPI (° ATDC)	Main injection timing (° ATDC)	Dwell (° CA)
1	40	-35	-20	8
2	40	-39	-20	12
3	40	-43	-20	16

3. Experimental results and discussion

3.1. Effect of dwell on combustion characteristics

In present subsection, effect of dwell on two combustion characteristics i.e. net heat release rate (NHRR) and in-cylinder pressure was analyzed.

Effect of dwell on NHRR traces has been portrayed in Figure 3. Almost all the pilot fuel gets combusted in pre-mixed mode only which led to significant rise in HRR as inferred from Figure 3. This is because temperature and pressure inside engine cylinder are not sufficient to promote auto-ignition of injected pilot fuel at the time of SOPI. Hence, injected pilot fuel passed through a long delay period and burnt instantaneously. Moreover, larger the dwell, higher the pre-mixed peak of HRR of pilot fuel as can be noticed from Figure 3. This is because SOPI is advanced in order to attain larger dwell (refer Table 2). So, in case of larger dwell, pilot fuel is injected in lower ambient in-cylinder conditions at the end of injection. Due to this, delay period of injected pilot fuel increased the further and subsequent higher pre-mixed peak of HRR for pilot fuel was obtained. Lower and earlier pre-mixed peak of main fuel with shorter dwell can be observed from Figure 3. This is because high temperature zone produced by combustion of pilot fuel shifts nearer to main injected fuel and promotes auto-ignition of main fuel more effectively in case of shorter dwell as compared to longer dwell. Therefore, main fuel passed through a shorter delay and subsequent lower and earlier pre-mixed peak of main fuel is attained. This shorter delay period of main injected fuel with smaller dwell resulted in poor mixing of main fuel with air and larger amount of main fuel got combusted in diffusion mode. This is the reason for getting more heightened as well as earlier diffusion combustion peak of main injected fuel with shorter dwell. Additionally, diffusion combustion duration of main fuel also increases with shorter dwell as expected.

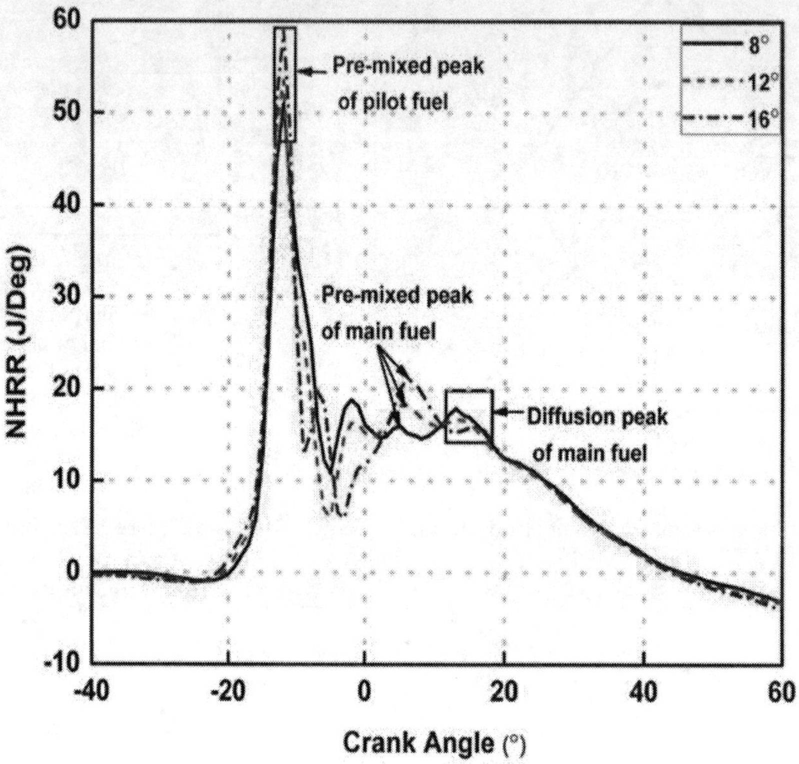

Figure 3. Effect of dwell on NHRR traces.

The variation of in-cylinder pressure traces with varying dwell is illustrated in Figure 4. Shorter the dwell, higher the peak pressure as noticed from Figure 4. Moreover, peak pressure continues to shift near top dead center (TDC) as dwell is reduced. This is because of greater impact of pilot combustion on main fuel combustion with shorter dwell which caused faster auto-ignition of main fuel. Increased pressure before TDC is representative of greater negative work, while increased pressure after TDC induces positive work on piston. In case of 8° dwell, rise of positive work and negative work (calculated from pressure-volume diagram) were found to be 6.32% and 3.08% respectively compared to 16° dwell case.

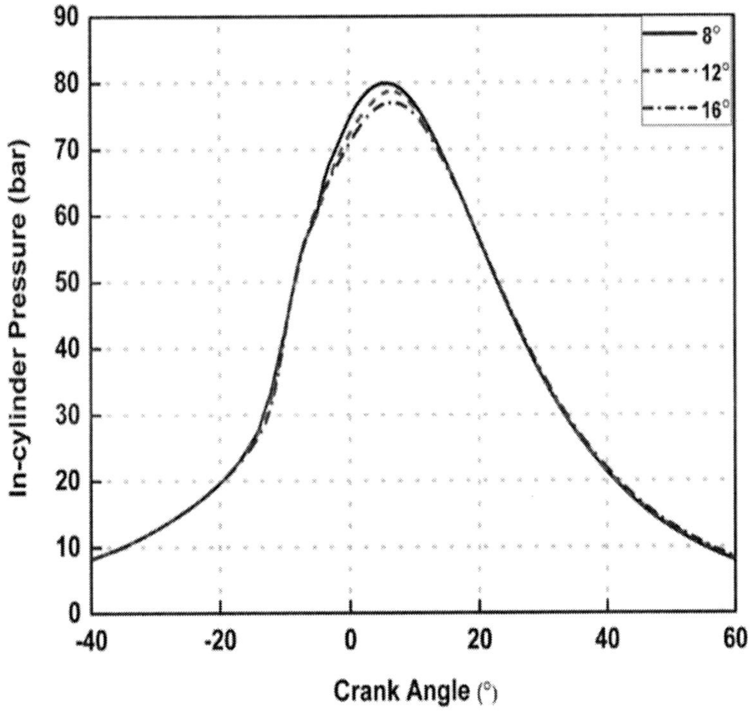

Figure 4. Effect of dwell on in-cylinder pressure traces.

3.2. Effect of dwell on emissions and BSFC

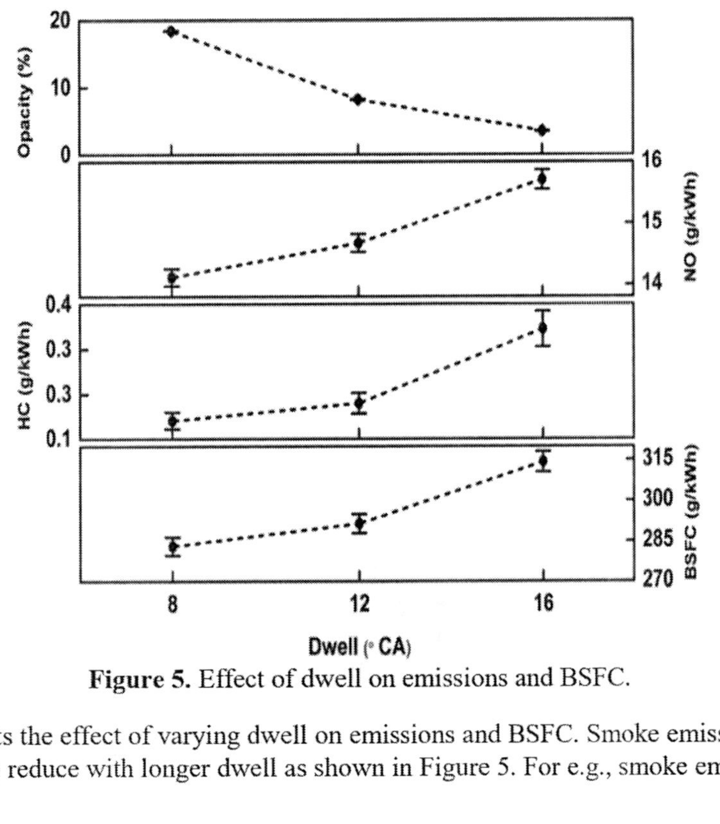

Figure 5. Effect of dwell on emissions and BSFC.

Figure 5 represents the effect of varying dwell on emissions and BSFC. Smoke emissions (represented as smoke opacity) reduce with longer dwell as shown in Figure 5. For e.g., smoke emissions reduce by

IOP Conf. Series: Materials Science and Engineering **810** (2020) 012041 doi:10.1088/1757-899X/810/1/012041

around 80% when dwell was increased from 8° to 16°. This is because severity of interactions between still burning pilot flame and liquid components of main fuel (spray interactions) reduced with longer dwell. Shorter and less severe diffusion combustion of main fuel (refer Figure 3) also supports this trend of reduced smoke emissions with longer dwell. Smoke emissions approached quite near to zero (~0% opacity) in case of 16° dwell. This is the reason why fuel injection pressure was kept limited to 350 bar in spite of much higher injection pressure offered by the injection system (up to 1200 bar). Due to the same reason, dwell was not increased beyond 16°. Nitric oxide (NO) emission was increased with increasing dwell. When dwell was increased from 8° to 16°, NO emission rose by around 11%. This is because of larger amount of heat released in pre-mixed combustion of pilot fuel as discussed earlier. A much reported *NO-smoke trade-off* can be noticed from Figure 5. HC emissions were increased with longer dwell as represented in Figure 5. This is mainly due to advanced SOPI associated with longer dwell which caused greater wall wetting as well as loss of fuel in crevice region and squish regions [26]. BSFC was reduced by 7.4% when dwell was reduced from 12° to 8°. This is due to greater impact of pilot combustion on main fuel combustion event which accelerated the combustion of main fuel.

4. Numerical simulation results and discussion

As can be noticed from experimental results, shorter dwell offers the advantage of reduced fuel consumption, but results in greater smoke emissions due to more severe spray interactions. Severity of these interactions can be reduced if dwell can be increased for a given pilot quantity as well as for given pilot and main injection timings. This is done by increasing NHD in present work. Larger NHD results in greater fuel flow rate (per °CA) and subsequent lesser fuel injection duration. Therefore, for a given pilot quantity as well as for given pilot and main injection timings, if fuel injection is performed with an injector having larger NHD, it results in shorter pilot as well as main injection duration and therefore dwell period increases compared to an injector with smaller NHD. This larger dwell might help to achieve lesser smoke emissions for a given pilot quantity as well as for given pilot and main injection timings.

With this background, effect of NHD on combustion and emission characteristics of diesel engine under pilot injection mode was numerically investigated in present work using commercial *Converge* 3-D CFD tool.

A 60° sector (1/6th sector of engine geometry), as shown in Figure 6 was created for the test engine as the fuel injector used for present work contains 6 nozzle holes as well as due to symmetric combustion chamber. Pilot injection mode with 12° dwell (run 2 in Table 2) was validated against respective experimental results. Closed cycle simulations were performed for the selected pilot injection mode with 1.4 mm base grid size. Grids were further refined in the area surrounding spray, piston and cylinder head by using *Fixed Embedding* technique available in *Converge* with embedding scale of 3. *Adaptive mesh refinement (AMR)* technique; a unique feature offered by *Converge,* was used to automatically refine the grids based on gradients of velocity and temperature. Maximum number of AMR cells was restricted to 10 lacs. Experimentally obtained combustion pressure and temperature at the start of simulation (CA of intake valve closing) were given as inputs. Basic models used in present work to simulate various combustion phenomena and emissions are listed in Table 3. A much more detailed description of different models as well as chemical mechanism used for simulation can be found in work carried out by Hiren et al. [25].

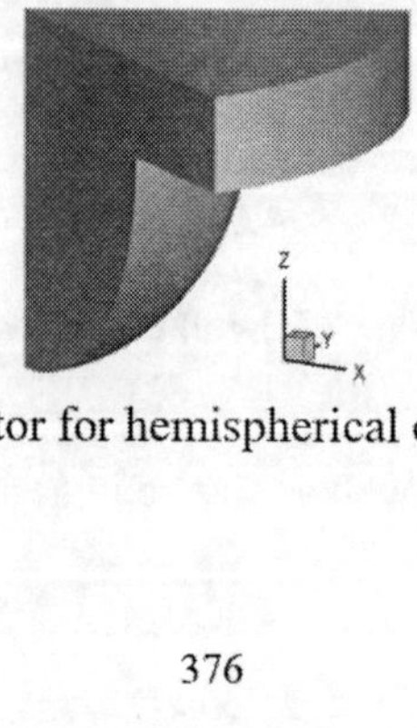

Figure 6. A 60° sector for hemispherical combustion chamber.

Table 3: Details of models used for simulating combustion process and emissions.

Process/Emission	Simulation model
Combustion process	SAGE detailed chemistry model [27]
Spray break-up	Kelvin Holmheltz-Rayleigh Taylor (KH-RT) [28]
Turbulence	Re-Normalization Group (RNG) k-ε model [29]
Spray droplet collision	Schmidit and Rutland [30]
Wall heat transfer	Han and Reitz [31]
NO_x	Extended Zeldovich [32]
Soot formation	Hiroyasu soot model [33]
Soot oxidation	Nagle-Stickland-Constable (NSC) model [34]

4.1. Code validation

Figure 7 represents comparison of experimental and simulated in-cylinder pressure traces for selected pilot injection mode (run no. 2 in Table 2). An acceptable match between experimental and simulated pressure traces can be noticed from Figure 7. Table 4 represents comparison of experimentally and numerically obtained values of NO emission for selected case. A good comparability between experimentally and numerically obtained values of NO emission can be observed from Table 4. Hence, it can be said that combustion process is modelled successfully.

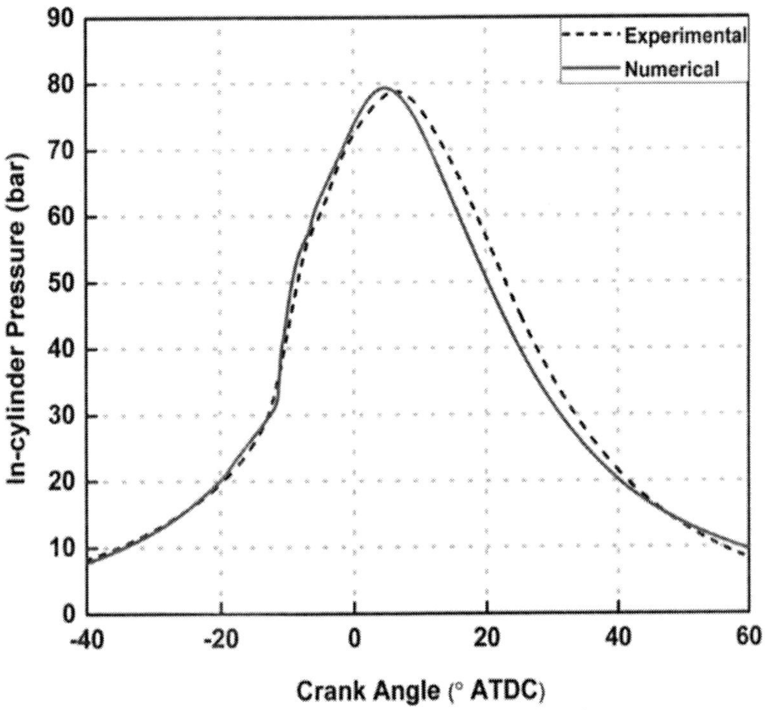

Figure 7. Comparison of experimental and simulated pressure traces.

Table 4. Comparison of experimental and simulated NO emission data.

NO (g/kWh) Experimental	NO (g/kWh) Numerical	Deviation (%)
14.68	15.33	4.24

4.2 Numerical methodology

Base injector (which is used for experimental work presented in Section 3) has NHD of 0.120 mm. Two other NHDs under consideration are 0.08 mm and 0.160 mm respectively. NHD was varied in *Converge* from *spray modelling* tab. Rest of the injection parameters like spray angle, spray cone angle, SOPI, main injection timing, total mass of injected fuel per cycle, distribution of fuel in pilot and main injection, C_d as well as injection pressure were maintained constant with changing NHD. Variation of pilot as well as main injection duration with varying NHD was calculated in *Converge* by giving injection pressure, fuel mass and C_d as inputs. Based on calculated pilot injection duration, dwell was calculated (since pilot as well as main injection timings were held constant).

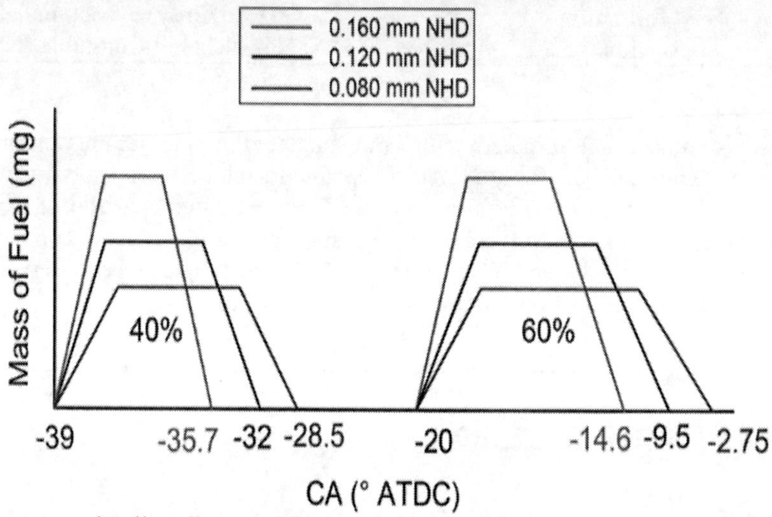

Figure 8. Representative line diagram of injection profiles under pilot injection mode with varying NHD.

Representative line diagrams of injection profile under pilot injection mode with varying NHD are represented in Figure 8. For selected pilot injection mode, change in dwell period with varying NHD is represented in Table 5. Dwell period increases with increasing NHD as inferred from Figure 8 as well as from Table 5. This is mainly attributed to shorter pilot injection duration in case of larger NHD due to increased fuel flow rate. Table 5 shows the variation of dwell with varying NHD for selected pilot injection mode.

Table 5. Variation of dwell with varying NHD for selected pilot injection mode.

Sr. No.	NHD (mm)	Pilot quantity (%)	SOPI (° ATDC)	Main injection timing (° ATDC)	Dwell (° CA)
1	0.08	40	-39	-20	8.5
2	0.120 (Baseline)	40	-39	-20	12
3	0.160	40	-39	-20	15.7

4.3. Numerical simulation results for combustion and performance characteristics

Effect of NHD on start of combustion (SOC), end of combustion (EOC) and combustion duration is tabulated in Table 6. In *Converge*, CA locations where 10% and 90% of the total injected fuel is burned are considered as SOC and EOC respectively. Based on these data, combustion duration was calculated.

Table 6. Effect of varying NHD on combustion parameters.

Sr. No.	NHD (mm)	SOC (° ATDC)	EOC (° ATDC)	Combustion duration (° CA)
1	0.08	-11.28	14.32	25.60
2	0.120	-11.29	17.24	28.53
3	0.160	-10.60	24.19	34.79

SOC is somewhat delayed in case of 0.160 mm NHD compared to other two cases as inferred from Table 6. This is due to longer delay period in case of 0.160 mm NHD due to larger diameter fuel droplets exiting from nozzle holes compared to other two cases. This can be clearly seen in Figure 9 which represents spray distribution inside combustion chamber for three different NHDs under consideration at -29° ATDC. At this CA, pilot fuel spray has sufficiently spread inside combustion chamber after striking on the piston bowl edge even for longest injection duration case (0.08 mm NHD). Larger spread area of spray inside combustion chamber at similar CA for 0.160 mm NHD case compared to other two NHDs also proves shorter injection duration with increased NHD. Moreover, density of larger diameter fuel droplets inside spray as well as on the edge of piston bowl is highest in case of 0.160 mm NHD and lowest in case of 0.080 mm NHD. In other words, larger the NHD, larger will be the fuel droplet diameter and vice-versa. These larger diameter fuel droplets in case of 0.160 mm NHD has smaller surface to volume ratio which results in poor mixing and vaporization of fuel droplets causing elongated physical delay period. However, surface area of spray droplets increased due to finer atomization in case of smaller NHD. Due to this, air entrainment inside fuel spray improved significantly and led to better air-fuel mixing in case of smaller NHD. Although, there is almost no difference in SOC location for 0.08 mm and 0.120 mm NHD cases. This might be due to longer injection duration in case of 0.08 mm NHD compared to 0.120 mm NHD case which surpassed the benefit of improved atomization. However, EOC is much earlier and combustion duration is much shorter in case of 0.08 mm compared to baseline case. Detailed explanation of this trend is given in subsequent discussion.

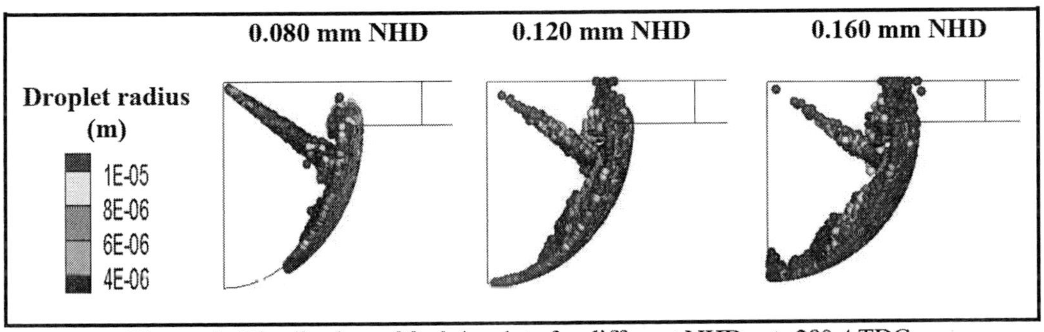

Figure 9. Distribution of fuel droplets for different NHDs at -29° ATDC.

Simulated mean gas temperature (MGT) traces with changing NHD are represented in Figure 10. Rate of rise in temperature in earlier stages of combustion as well as peak temperature are highest in case of 0.08 mm NHD as inferred from Figure 10. This is mainly attributed to faster progress of combustion process due to improved atomization in case of 0.08 mm NHD. While having similar SOC, rate of temperature rise after initiation of combustion is much more rapid in case of 0.08 mm NHD compared to baseline case as can be seen from Fig 9. This proves that even though there is no noticeable difference in SOC, progress of combustion process is much more rapid in case of 0.08 mm NHD compared to baseline case. Peak temperature is lowest in case of 0.160 mm NHD as expected but a close observation from Figure 10 reveals that temperature in the later stage of expansion process is larger in case of 0.160 mm NHD compared to baseline case. This is clearly shown in Figure 10 (A).

This might be due to longer combustion duration associated with 0.160 mm NHD case (refer Table 6) which persisted higher temperature for longer time in expansion stroke compared to baseline case.

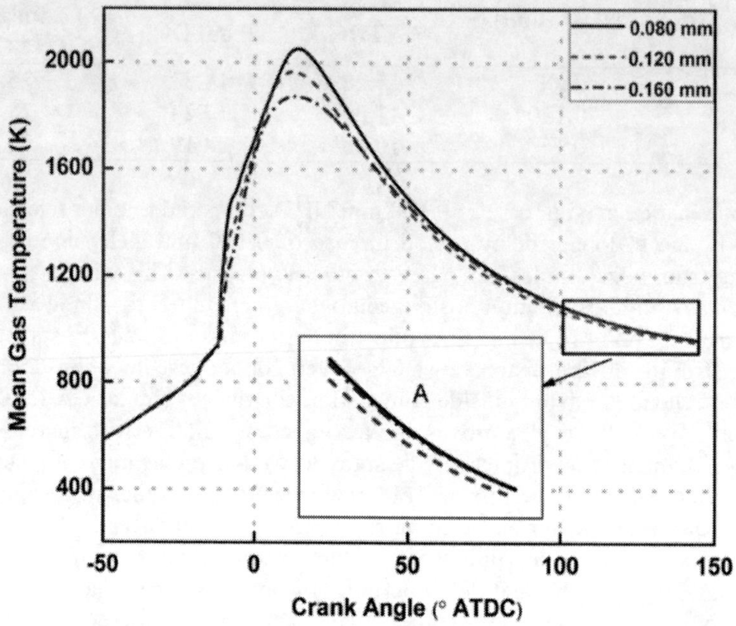

Figure 10. Effect of NHD on mean gas temperature traces.

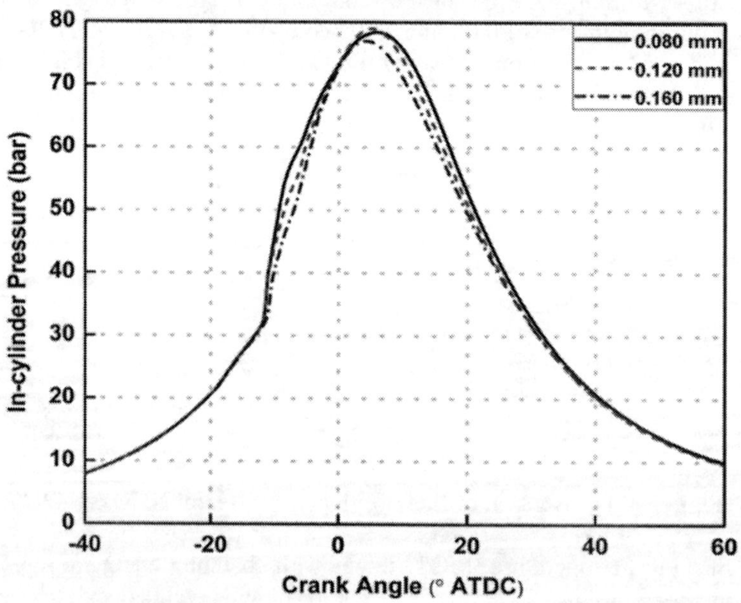

Figure 11. Effect of NHD on in-cylinder pressure traces.

In-cylinder pressure traces with varying NHD are represented in Figure 11. Baseline NHD obtained highest peak in-cylinder pressure compared to other two as can be noticed from Figure 11. Difference in peak pressure between baseline NHD and 0.080 mm NHD is quite negligible (0.72%), but noteworthy reduction in peak pressure (2.48%) is noticed with 0.160 mm NHD as compared to 0.120 mm NHD. Almost similar peak pressure in case of 0.08 mm NHD and 0.120 mm NHD is mainly attributed to longer injection duration in case of 0.08 mm NHD compared to 0.120 mm NHD case which surpassed the benefit of shortened delay period due to finer atomization. Lowest peak pressure in case of 0.160 mm NHD is mainly due to deteriorated fuel evaporation rate (due to larger diameter of fuel droplets) which lowered the rate of pressure rise during initial stages of combustion and led to overall slow progress of combustion process compared to other two cases. Moreover, peak pressure continued to shift away from top dead centre (TDC) with decreasing NHD as inferred from Figure 11. This is due to faster progress of combustion process after initiation of combustion with decreasing NHD as explained earlier. This faster progress of combustion process with lower NHD produced larger force on piston in shorter time so that pressure drop in expansion stroke due to downward motion of piston was partly compensated and pressure continued to increase in expansion stroke for longer duration. This delayed occurrence of peak pressure persisted somewhat higher pressure in expansion stroke in case of lowest NHD (0.08 mm) compared to baseline case as well as largest NHD (0.160 mm) case.

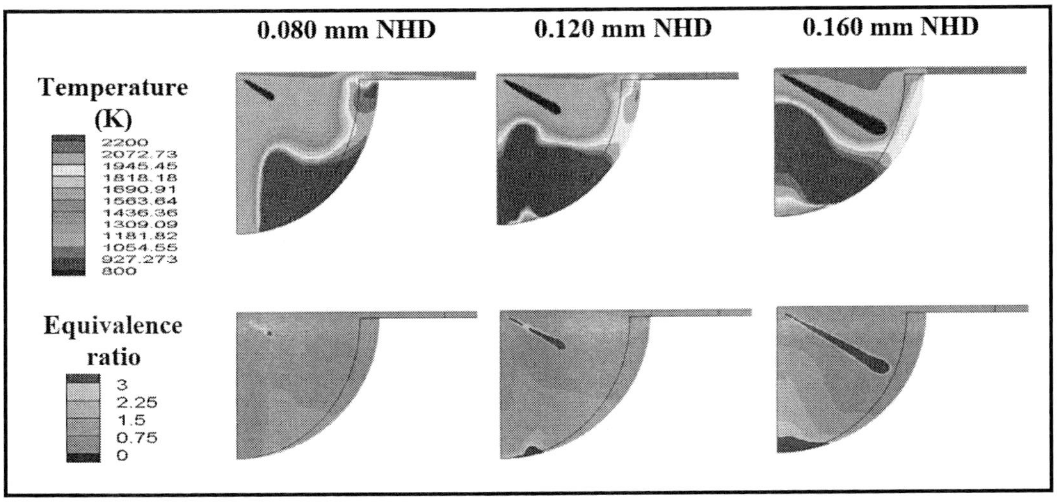

(a)

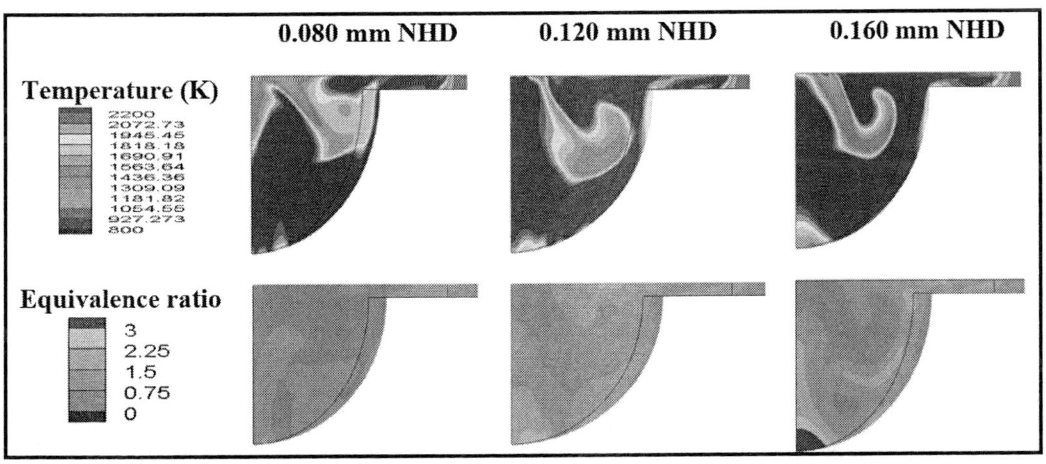

(b)

Figure 12. Effect of NHD on in-cylinder temperature and equivalence ratio distribution at (a) -8° ATDC and (b) 13° ATDC.

Two distinguished regions, i.e. a high temperature region (HTR) and a low temperature region (LTR) are clearly visible in Figure 12(a). Presence of HTR and LTR is attributed to pilot fuel combustion and evaporation of main fuel spray respectively. Greater HTR can be noticed from Figure 12(a) in case of lowest NHD compared to other two cases. This trend is further supported by Figure 10 which represents simulated MGT for all three NHD cases. Most of the HTR in case of 0.08 mm NHD is gathered around piston circumference and top of the piston, whereas HTR in other two cases is much more staggered. HTRs in case of 0.120 mm and 0.160 mm NHD are losing their contact from piston circumference and pointing towards center of combustion chamber and results in stretching of HTR. This effect is more pronounced in case of 0.160 mm NHD. This stretching of HTRs in cases of 0.120 mm NHD and 0.160 mm NHD led to reduced temperature whereas HTR in case of 0.08 mm stays on the piston circumference and maintains higher temperature. Moreover, LTR continues to decrease with decreasing NHD as can be seen in Figure 12(a). This indicates faster rate of main fuel evaporation with reducing NHD. Evaporation rate is much faster in case of 0.08 mm NHD due to finer droplets exiting from nozzle which produced smallest LTR among all the three cases. Slowest evaporation rate associated with largest NHD (0.160 mm) produced largest LTR among all the considered cases. Moreover, as stated earlier, total mass of injected fuel per cycle as well as distribution of fuel in pilot and main injection are maintained constant with varying NHDs. Therefore, greater HTR in case of 0.08 mm NHD (Figure 12(a)) compared to other two cases proves that greater pilot fuel mass has already been combusted. Additionally, main fuel evaporates at faster rate in case of 0.08 mm NHD compared to other two cases (evident by presence of smallest LTR). The combined effect of both these factors (presence of greater HTR and faster evaporation of main fuel) led to faster combustion of main fuel in case of 0.08 mm NHD compared to other two cases. This is the main reason for getting earlier EOC as well as shorter combustion duration in case of 0.08 mm NHD compared to 0.120 mm NHD case in spite of having almost similar SOC (refer Table 6). Due to this combined effect, interactions between still burning pilot flame and liquid components of main fuel continued for shorter duration and hence became less severe in case of 0.08 mm NHD compared to other two cases. This hypothesis is further supported by presence of least rich regions in case of 0.08 mm NHD compared to other two cases in Figure 12. HTR is smallest and LTR is widest in case of 0.160 mm NHD. This smallest HTR indicates that least pilot fuel has been combusted by this time compared to other two cases and widest LTR depicts that evaporation of main fuel is also slowest. Due to both these factors, combustion of main fuel continued for much longer duration which resulted in longest combustion duration in case of 0.160 mm NHD (refer Table 4) compared to other two cases. This will facilitate more time for interactions between still burning pilot flame and liquid components of main fuel and therefore these spray interactions became more severe. Due to this, combustion of main fuel took place under more locally rich condition which is evident by presence of highest rich regions in Figure 12 in case of 0.160 mm NHD compared to other two NHDs under consideration. These results are totally different from what was expected. This is because in case of 0.08 mm NHD, dwell between pilot and main injection is shortest and dwell is largest in case of 0.160 mm NHD (refer Figure 8 and Table 5). Moreover, all the injection parameters are also held constant with changing NHD as discussed earlier. Therefore, most aggressive spray interactions as well as largest rich regions were expected in case of 0.08 mm NHD (due to shorter dwell) and least aggressive spray interactions as well as smallest rich regions were expected in case of 0.160 mm NHD (due to longer dwell) based on results reported by previous studies as well as based on experimental results obtained in present work which showed greater smoke emissions with shorter dwell (refer Figure 5). The prime reason for this obtained trend is the greater homogeneity of mixture prepared in a shorter time due to improved atomization of injected fuel droplets which significantly reduced their physical delay period in case of smaller NHD. Greater homogeneity of the charge is further evident by much more consistent distribution of equivalence ratio as well as presence of least rich regions in case of 0.08 mm NHD compared to other two cases in Figure 12. This highly homogeneous mixture in case of 0.08 mm NHD compared to other two cases caused faster combustion of pilot as well as main injected fuel and drastically reduced the time available for spray interactions compared to other two cases.

Overall faster combustion process is further evident by faster growth of HTR in Figure 12 as well as least rich regions near and inside the squish regions in Figure 12(b) in case of 0.08 mm NHD.

4.4. Numerical simulation results for emission characteristics

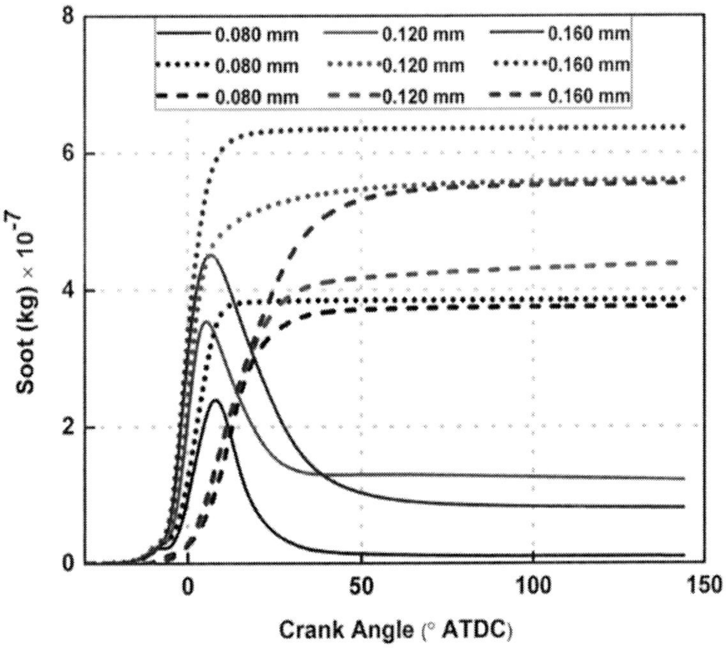

Figure 13. Computed soot emissions for different NHDs.

Computed net soot emissions (represented by full lines) along with soot formation (represented by dotted lines) and soot oxidation process (represented by dashed lines) are shown in Figure 13. First lower peak of soot formation for all three cases under consideration indicates that almost all the pilot fuel burnt in pre-mixed mode only and prime source of soot emissions is the main injected fuel. Soot formation is highest in case of largest NHD and lowest in case of smallest NHD as inferred from Figure 13. This is because spray interactions continued for longest duration in case of 0.160 mm NHD and for shortest duration in case of 0.08 mm NHD as discussed earlier. This facilitated longest time for soot formation in case of 0.160 mm NHD and least time for soot formation in case of 0.08 mm NHD respectively. Therefore, severity of spray interactions increased drastically in case of 0.160 mm compared to other two cases. This resulted in much higher rate of soot formation in case of 0.160 mm NHD compared to 0.08 mm NHD and baseline NHD cases which can be clearly visualized from Figure 13. Highest and lowest fuel rich regions with of 0.160 mm and 0.08 mm NHDs respectively in Figure 12 also supports this trend of soot emissions. Baseline NHD produced soot emissions which are in between 0.08 mm and 0.160 mm NHD cases. One important thing to be noticed from Figure 13 is that inspite of highest soot formation, net exhausted soot emissions are somewhat lower in case of 0.160 mm NHD compared to baseline NHD case. This is mainly attributed to greater temperature persisted in expansion stroke in case of 0.160 mm NHD compared to baseline NHD case as seen in Figure 10 which helped to enhance the oxidation rate of formed soot in case of 0.160 mm compared to baseline NHD case. This can be seen in Figure 13 also that rate of soot oxidation as well as duration of soot oxidation are higher in case of 0.160 mm NHD compared to baseline NHD case due to which net exhausted soot emissions are lower. When NHD is reduced from 0.160 mm to 0.08 mm, soot emissions reduced by 87.2%. Therefore, it can be said that under pilot injection mode, smaller NHD results in lesser soot formation in spite of further shortening of dwell period for a given pilot mass as well as given pilot and main injection timings. Moreover, with smaller NHD, SOPI can be further retarded (or dwell can be further reduced) to get additional improvment in BSFC with still lower or similar smoke emissions compared to larger NHD case.

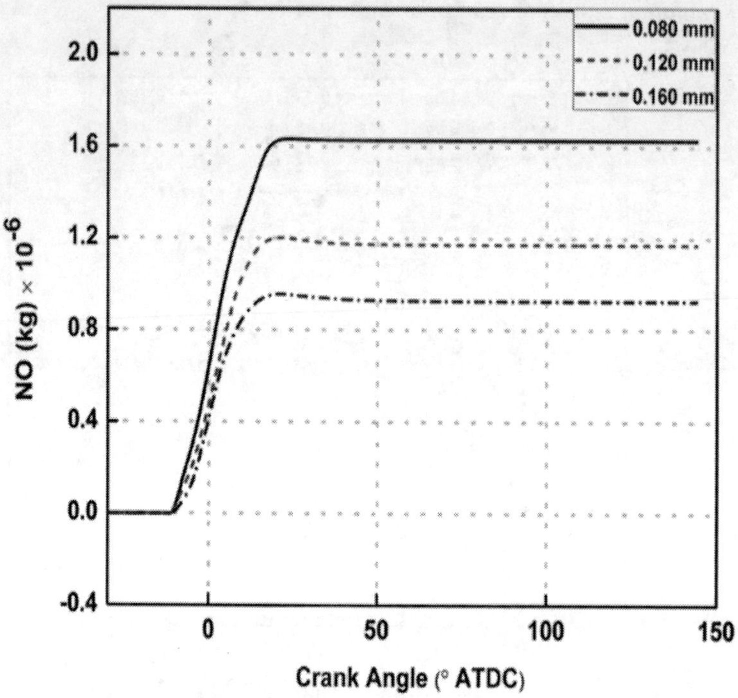

Figure 14. Computed NO emission for different NHDs.

Figure 14 represents computed NO emission for three different NHDs under consideration. NO emission is increased continuously with reduction in NHD as inferred from Figure 14. When NHD is reduced from 0.160 mm to 0.080 mm, NO emission was increased by around 43%. Highest NO emission is exhausted in case of 0.08 mm NHD. This is primarily due to faster combustion process associated with 0.08 mm NHD compared to other two cases which produced highest in-cylinder temperature as indicated in Figure 10 and augmented formation of NO emission.

Figure 15 shows variation of computed HC emissions for varying NHDs. Two different peaks of HC emissions i.e. first one for pilot fuel and second one for main fuel can be seen for all the three NHDs under consideration. HC emissions from pilot fuel are mainly due to wall wetting caused by pilot spray. Pilot injection duration is longest in case of 0.08 mm NHD and shortest in case of 0.160 mm NHD as discussed earlier. Therefore, in case of smaller NHD, pilot injected fuel has greater chance of wall wetting and greater pilot fuel mass sneak into crevice volume and squish regions, which led to higher HC emissions from injected pilot fuel. This is the reason for getting highest and lowest first peak of HC emissions in case of 0.08 mm NHD and 0.160 mm NHD respectively. A second peak of HC emissions is lowest for 0.08 mm NHD and increases continuously with increasing NHD as inferred from Figure 15. Improved mixing due to finer atomization resulted in lesser HC emissions formation from main fuel in case of smaller NHD compared to larger NHD case. Moreover, HC emissions formation from main fuel continued for longer duration in case of smaller NHD which is evident from delayed second peak of HC emissions in case of smaller NHD. This is mainly attributed to longer main injection duration associated with smaller NHD. Overall least HC emissions exhausted in case of 0.08 mm NHD (as seen from Figure 15(A)) is mainly attributed to least HC emissions formation. In spite of overall greater HC emissions formation in case of 0.160 mm NHD compared to baseline case, exhausted HC emissions are larger in case of baseline NHD compared to

0.160 mm NHD as inferred from Figure 15(A). This mainly due to higher temperature persisting in the later stage of expansion process in case of 0.160 mm NHD compared to baseline case (refer Figure 10) which facilitated oxidation of formed HC emissions for longer duration in expansion stroke.

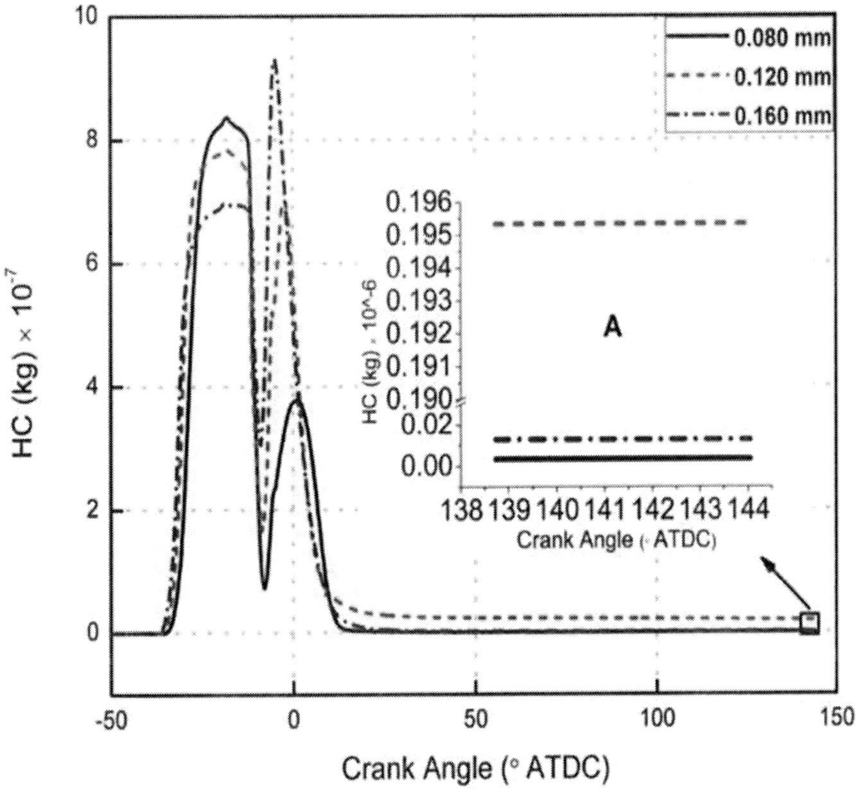

Figure 15. Computed HC emissions for different NHDs.

5. Conclusions

In first part of present work, an experimental study was conducted to analyze the effect of dwell on combustion, performance and emission characteristics of diesel engine. Experimental results showed that shorter dwell results in improvement of BSFC but with increasing smoke emissions penalty. In second part of work presented, effect of variation of NHD on combustion and emission parameters of diesel engine was investigated numerically with an aim to mitigate the disadvantage of increased smoke emissions associated with shorter dwell. All the injection parameters were kept unchanged with varying NHD so that sole effect of NHD on combustion and emission characteristics can be investigated. In addition to baseline NHD (0.120 mm), two other NHDs, i.e. 0.08 mm and 0.160 mm were considered. Numerical results showed that fuel injection with smaller NHD resulted in faster evaporation rate of injected fuel droplets and hence overall faster combustion process compared to that with larger NHD case. Due to this, spray interactions continued for shorter duration and hence became less severe in case of smaller NHD compared to larger NHD case. Because of this, least NHD (0.08 mm) resulted in lowest soot formation in spite of having shortest dwell among all the three NHDs under consideration. However, smaller NHD affected NO emission negatively.

Therefore, for a given pilot quantity as well as for given pilot and main injection timings, soot/smoke emissions can be reduced by lowering NHD. Moreover, in case of smaller NHD, dwell can be further reduced in order to achieve additional reduction in BSFC with smoke/soot emissions being still lower or equivalent to larger NHD case.

Acknowledgement

Authors sincerely acknowledge the financial support provided by SVNIT-Surat towards this work under institute annual plan head.

6. References

[1] Brijesh P and Sreedhara S 2013 *Int. J. Auto. Technol.* **14** 195
[2] Lee J, Hong K, Choi S, Yu S, Choi H and Min K 2013 *J. Mech. Sci. Technol.* **27** 1135
[3] Huang H, Huang R, Guo X, Pan M, Teng W, Chen Y and Li Z 2019 *Appl. Energy* **250** 185
[4] Yun H, Choi K and Lee CS 2016 *Energy Convers. Manag.* 111:79
[5] Park S, Kim HJ, Shin DH and Lee JT 2018 *Appl. Therm. Eng.* **140** 422
[6] Liu B, Cheng X, Liu J and Pu H 2018 *Appl. Therm. Eng.* **141** 90
[7] Huang H, Zhu Z, Zhu J, Lv D, Pan Y, Wei H and Teng W 2019 *Appl. Energy* **249** 377
[8] Okude K, Mori K, Shiino S, Yamada K and Matsumoto Y 2007 *SAE Technical Paper* 2007-01-4178
[9] Huang H, Wang Q, Shi C, Liu Q and Zhou C 2016 *Appl. Energy* **179** 1194
[10] Binde A, Busch S, Velji A, Wagner U. 2012 *SAE Int J Engines* **5** 1242
[11] Cung K, Moiz A, Johnson J, Lee SY, Kweon CB and Montanaro A 2015 *Proc Combus. Inst.* **35** 3061
[12] Jeon J and Park S 2015 *Appl. Energy* **160** 581
[13] Zheng Z, Yue L, Liu H, Zhu Y, Zhong X and Yao M 2015 *Energy Convers Manag* **90** 1
[14] Herfatmanesh MR, Lu P, Attar MA and Zhao H 2013 *Fuel* **109** 137
[15] Sahoo D, Miles PC, Trost J, Leipertz A 2013 *SAE Int J Engines* **6** 1716
[16] Qi D, Leick M, Liu Y and Lee CF 2011 *Fuel* **90** 1884
[17] Park C, Kook S and Bae C 2004 *SAE Technical Paper* 2004-01-0127
[18] Li X, Zhou H, Zhao LM, Su L, Xu H and Liu F 2016 *Energy Convers. Manag.* **129** 180
[19] Vijay Kumar M, Veeresh Babu A and Ravi Kumar P 2018 *Renewable Energy* **119** 388
[20] Bergstrand P and Denbratt I 2001 *SAE Technical Paper 2001-01-2010*
[21] Pang KM, Jangi M, Bai XS, Schramm J and Walther JH 2017 *Energy Proc.* **142** 1028
[22] Zhang W, Tian JP and Nishida K 2011 *SAE Technical Paper 2011-01-1813*
[23] Du W, Lou J and Liu F 2017 *SAE Technical Paper 2017-01-2300*
[24] Ganesh V, Deshpande S, Sreedhara S 2015 *Int. J. Eng. Res.* **17** 469
[25] Hiren D, Bharatkumar S and Brijesh P 2019 *Clean Technol. Environ Policy* **21** 905
[26] Brijesh P and Sreedhara S *Clean Technol Environ Policy* 2016 **18** 2325
[27] Senecal PK, Pomraning E, Richards KJ, Briggs TE, Choi CY, McDavid RM, Patterson MA 2003 *SAE Technical Paper 2003-01-1043*
[28] Beale JC, Reitz RD 1999 *Atomiz. sprays* **9** 623
[29] Han Z and Reitz RD 1995 *Combust. Sci. Technol.* **106** 267
[30] Schmidt DP and Rutland CJ 2000 *J Comput. Phys.* **164** 62
[31] Han Z and Reitz RD 1997 *Int. J. Heat Mass Trans.* **40** 613
[32] Zeldovich YB, Barenblatt GI, Librovich VB and Makhviladze GM 1986 *Consultants Bureau*, New York
[33] Hiroyasu H and Kadota T 1976 *SAE Technical Paper* 760129
[34] Nagle J and Stickland RF 1962 *Proc. Fifth Carbon Conf.* **1** 154

IOP Conf. Series: Materials Science and Engineering **810** (2020) 012042 doi:10.1088/1757-899X/810/1/012042

Investigation of a Magnet falling through a copper tube

[1]Sahil Jain, [2]Prita Rathi, [1]Ramachandran M*

[1]MPSTME, SVKM's NMIMS, Shirpur, Dhule, Maharashtra 425405, India,
[2]Choithram International, Indore, Madhya Pradesh 452001, India

mr.sahiljain2018@gmail.com

Abstract. Electromagnetic induction one of the most important concept around which most of the devices are developed around. My investigation explores the same concepts wherein a magnet falling through a copper tube depicts the same concept using a theoretical and experimental model. To be precise the research question is to investigation how the thickness of the copper tube affect the time taken for the magnet to fall through it. The investigation was done by dropping a magnet inside the copper tubes of different thickness and the time taken by the magnet to pass through the pipe of one meter length was noted. Moreover a theoretical model is devised using some concepts and equations. Using the theory the actual value of the gradient i.e. $\frac{1}{v}$/w was found to be 120.58 m-2 s, in the investigation, the value of the gradient was found to be 116.89 m-2 s, which is only deferring by 3.16 % form the actual value. Therefore the investigation is accurate and reliable. Also the relationship between the thickness of the copper tube and the time taken by the magnet to pass through it is found to be linear, whereas the thickness increases the time taken by the magnet also increases.

1. Introduction

Magnet has always amazed us through their attention-grabbing properties and lead to many great inventions and discoveries. Once while exploring more about magnets I came across a very interesting observation that magnet inside a copper pipe take much more time than it actually does to come down the same length in air. On further researching, moving magnet inside the copper pipe connects many important concepts in physic which have heavily effected the society. Knowing that today's world works on energy, power, current, electricity and electricity is now one of a necessity in human lives. Almost every devices and machines around us needs current to run, we often say that "The world has become Digital" and to have this digital world running we need electric current to power it. If we talk about the sources of energy then we have various number of energies but to convert those into electrical energy we need generators. Generators work on the concept of electromagnetic induction, and magnet inside a copper pipe could be a very interesting model for the students to understand the concept of electromagnetic induction and electromagnetic breaking. Furthermore the model could be used in building an emergency exit from tall buildings where a man can wear the magnet suit and then can come down slowly and safely through the copper tunnel. Hence it is worthy and significant to investigate a magnet falling through a copper pipe. In this research paper I will be investigating the time taken by the magnet to pass through it and how the thickness of the tube affects the time.
Direct Current (DC) generators work on the principal of Faraday's law [1] of electromagnetic induction. First law of faraday's electromagnetic induction states that, whenever there is changing magnetic flux

IOP Conf. Series: Materials Science and Engineering **810** (2020) 012042 doi:10.1088/1757-899X/810/1/012042

around a conductor there is emf induced in it. "Magnetic flux is the measurement of total magnetic field passing thorough a given area. Magnetic flux is also given by the following equation; Φ=BAcosθ, where "Φ" is the magnetic flux, "B" in the magnetic field strength, "A" is area and "θ" is the angle between the normal of the surface and the magnetic field vector."[1] As per Faraday's statement i.e. "change in magnetic flux induces current". So as Magnetic flux is equal to BAcosθ, which means that change in any of these factors near a conductor would induce current. Second law of Faraday states that the induced emf is equal to the rate of change of magnetic flux, which means greater the change in flux linkages the induced emf will be more.

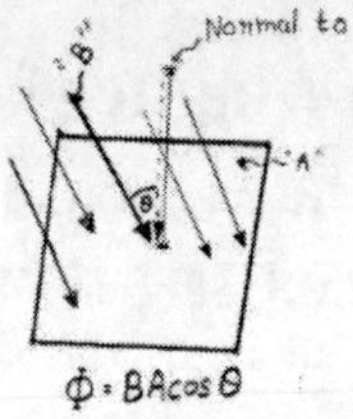

Figure 1. Faraday's Law

This principle can be applied to a falling magnet through a copper tube, when a magnet is dropped inside a copper tube it moves far slower inside the copper tube than it actually does outside under free fall. Understanding what actually happens, we will starting from the first action where the magnet is taken to a certain height where there is some gravitational potential energy (mgh) being stored in the magnet. Now when you drop the magnet in then the gravitational potential energy starts to convert into kinetic energy which helps the magnet come down. The reason for the magnet slowing down is somewhere connected to the principal we discussed above. Now when the magnet is moving inside the copper tube, which means that flux is changing (because of the moving magnet) near the conductor (i.e. the copper tube) and as per Lenz's law this induces emf in the conductor which opposes the cause i.e. the moving magnet. Lenz's law2 states that, when an emf in generated by a change of magnetic flux as per the Faraday's induction Law, the polarity of the induced emf is such that it then produces a current that is magnetic field opposes the change which is producing it. As said above that due to the falling magnet there is a change of flux which induces current, now this induced current produces an another magnetic field by turning the copper pipe into a temporary magnet. The orientation of the new magnetic field is opposite with respect to the magnet. Now these two opposing fields result in repulsion of the magnet which make the magnet travel slower. The phenomena is similar to a magnet inside a solenoid. Here the copper pipe could be referred to the solenoid as there are winding over the pipe.

Now the question arises of why does the magnet not stop at a particular point? This is because the downward force (due to the kinetic energy) is more that the repulsive for (due to Lenz law).

Figure 2. Faraday's Law of Induction

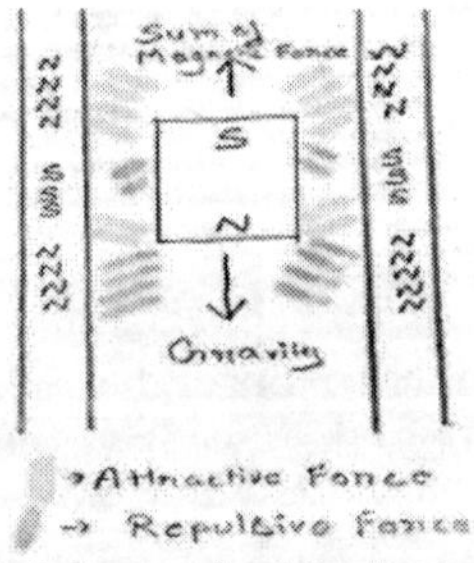

Figure 3. Magnet inside the copper tube and its pole reaction

2. Theory

IOP Conf. Series: Materials Science and Engineering **810** (2020) 012042 doi:10.1088/1757-899X/810/1/012042

As per the earlier discussion that it is convenient to imagine the pipe as a uniform subdivided rings parallel arranged, so now let the width of the unit ring be "l". When the magnetic flux changes around the ring due to the falling action of the magnet, there is electromotive force induces inside the ring which produces its own magnetic field. According to Lenz's law these fields will oppose and the repulsion caused will make the magnetic move slower. Now there are various factors which will affect the magnitude of the repulsion (upward force) taking place which affects the time taken by the magnet to pass through the pipe. So the repulsion taking place directly depends on the current induced inside the pipe which decides the magnitude of another magnetic field created to oppose the falling magnet. In the copper pipe as the magnet moves it attains constant velocity, the reason for this is, electromagnetic force being an increasing function of the velocity which will decelerate the falling magnet. The acceleration becomes zero at a point and the magnet then tends to move in a constant velocity which is also called the terminal velocity. So as the velocity "v" is constant the kinetic energy of the falling magnet inside a copper pipe remains constant and the gravitational potential energy then transforms into Ohmic heating of copper pipe, where the gravitational potential energy dissipates through eddy currents induced in the copper pipe or the unit ring. Also the rate of loss of gravitational potential energy will be equal to the rate of dissipation by the Ohmic resistance.

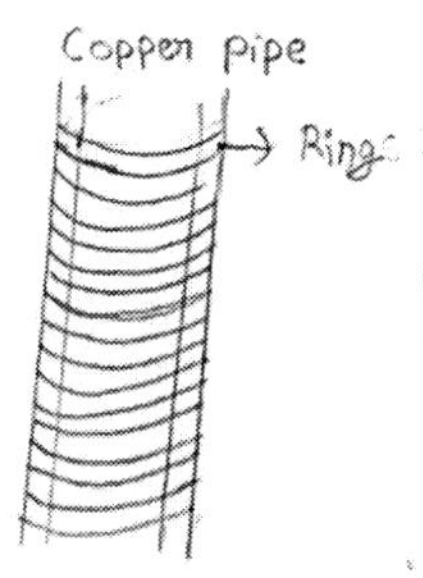

Figure 4.

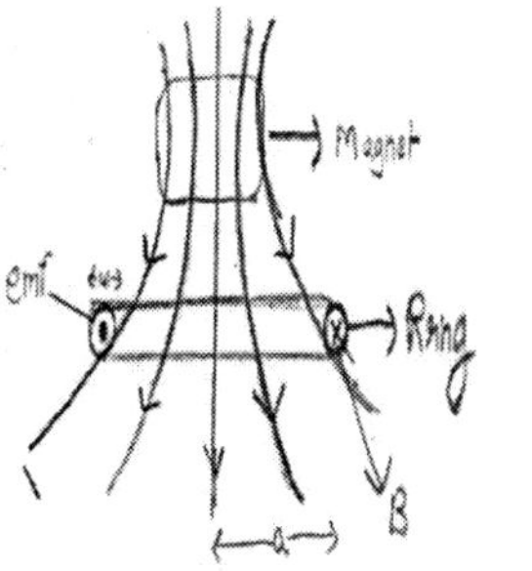

Figure 5.

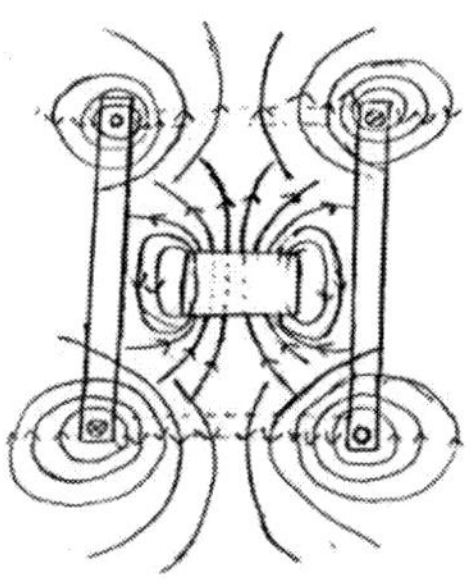
Figure 6.

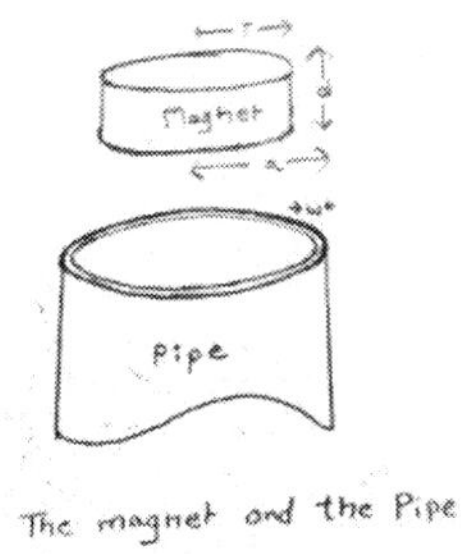

Figure 7.

Figures above show the magnet flux and its arrangement while falling inside the copper tube.

Now Considering;
"Y" being the coordinate along the pipe length, I(Y) current induced in the ring at some point Y, R the resistance of the ring we get;

$$mgv = \sum_{Y} I(Y)^2 R \qquad \text{(equation1)}$$

Considering no losses, where all the variation in electric current in the given ring at point Y is resulting into the change in flux due to magnetic motion. Here we can ignore the self-induction effect. Now let's calculate the distribution of current I(Y) in each ring. To calculate this we first need to study the rate of change of magnetic flux through a ring when the magnet passes through the pipe. Before moving forward lets note that, magnetic field of the magnet inside the pipe is somewhat similar to the one in the vacuum, this is because the magnetic permeability of copper is very close to the magnetic permeability of vacuum.

Assuming that magnetization $M = M\hat{Y}$ is uniform for the magnet, where the magnetic charge density is zero inside the magnet. Moreover at the ends of the magnet (top and bottom) the magnetic surface density will be uniform $\sigma m = M$ and $-\sigma m = -M$.
The flux created by the magnet can be assumed as fields of two disks of the same radius "r" with a distance "d". Now making another assumption/approximation, we will replace the charged disks by point monopoles of the same total charge i.e. $q_m = \prod r^2 \sigma m$.
Flux through a ring created by passing two monopoles;

$$\Phi(Y)=\frac{\mu_0 q_m}{2}\left[\frac{Y+d}{\sqrt{(Y+d)^2+a^2}}-\frac{Y}{\sqrt{Y^2+a^2}}\right] \quad \text{(equation2)}$$

Where,

μ_0 Refers to the permissibility of vacuum

Y refers to the distance from the closest monopole

Due to the fall of magnet through the ring, there is an electromotive force as per Faraday's law;

$$E(Y)=-\frac{d\Phi(Y)}{dt} \quad \text{(equation3)}$$

Equation of electric current;

$$I(Y)=\frac{\mu_0 q_m a^2 v}{2R}\left[\frac{1}{(Y^2+a^2)^{\frac{3}{2}}}-\frac{1}{\left[(Y+d)^2+a^2\right]^{\frac{3}{2}}}\right] \quad \text{(equation4)}$$

To find he rate of the ohmic dissipation, we can now calculate the sum value of the "(equation1)". For finding the power dissipated we can put continuum limit;

The limits are set to be infinity assuming that most of the energy dissipation takes place near the magnet with no losses.

$$P=\frac{\mu_0^2 q_m^2 a^4 v^2}{4R}\int_{-\infty}^{\infty}\frac{dY}{l}\left[\frac{1}{(Y^2+a^2)^{\frac{3}{2}}}-\frac{1}{\left[(Y+d)^2+a^2\right]^{\frac{3}{2}}}\right]^2 \quad \text{(equation5)}$$

Knowing that resistance equals;

$$R=\frac{\rho L}{A}$$

Where p is the electrical resistivity of the conductor, A is the area of the conductor and L is the length. So the resistance of the ring;

$$R=\frac{2\pi a\rho}{wl}$$

Where "w" is the thickness of the ring. And "ρ" is the electrical resistivity.

Writing the equation 5 again with replacing R we get;

$$P=\frac{\mu_0^2 q_m^2 v^2 w}{8\pi\rho a^2}\, f\left(\frac{d}{a}\right) \quad \text{(equation6)}$$

Here $f(x)$ is a scaling function, which is defined as;

$$f(x)=\int_{-\infty}^{\infty}dz\left[\frac{1}{(z^2+1)^{\frac{3}{2}}}-\frac{1}{\left[(z+x)^2+1\right]^{\frac{3}{2}}}\right]^2 \quad \text{(equation7)}$$

Now by substituting equation 6 into the equation 1, we get the terminal velocity of the magnet falling through the pipe;

$$v=\frac{8\pi mg\rho a^2}{\mu_0^2 q_m^2 w\, f\left(\frac{d}{a}\right)} \quad \text{(equation8)}$$

Where "v" is the terminal velocity
To simply the equation further, we will plot the function f(x);

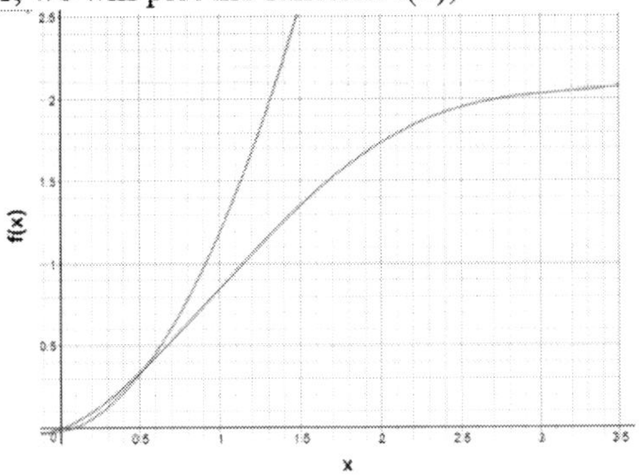

Figure 8. The function f(x) plotted for the equation.

For the small values of x, the dotted curve is an assumption being made. So the function f(x) could be now defined as;

$$f(x)=\frac{45\pi}{128}x^2 \quad \text{(equation9)}$$

Where now f(x) is the function for the parabola.

Further we can reduce the equation for terminal velocity "v" as;

$$v=\frac{1024}{45}\frac{mg\rho a^4}{\mu_0^2 p^2 w} \quad \text{(equation10)}$$

Where p= qmd which is the dipole moment of the falling magnet, "a" is the inner radius, "g" is the gravitational constant, "m" is the mass of the magnet.

From the equation above we can identify the relationship between terminal velocity and thickness (w) is inverse. As rest of the things in the equation are the constants so we can write the relation as;

$$v\alpha\frac{1}{w}$$

Since the magnet is made to fall thru constant distance in the pipe we can say that the relation of time and thickness as;

$$t\ \alpha\ w$$

Is the relationship that will be explored thru my investigation experimentally.

3. The Experiment

So as per the formula that time taken by the magnet to fall thought the copper tube could be affected by the thickness of the copper tube. This experiment will help me investigating the relationship between the thickness of the tube and the time taken by the magnet to fall through it.

Changing the thickness of the copper pipe

For this investigation, I have to wind the copper wire over the copper tube to change the thickness of the copper tube. As there are only commercial thickness of the copper tubes available in the market that too in bulk, so I have to vary the thickness of the tube by winding thin layers of thin copper wire over the pipe. The inner diameter of the copper pipe was 0.5 inch (1.27×10^{-2}m) and the thickness of the copper pipe was 20 SWG (0.914 mm).

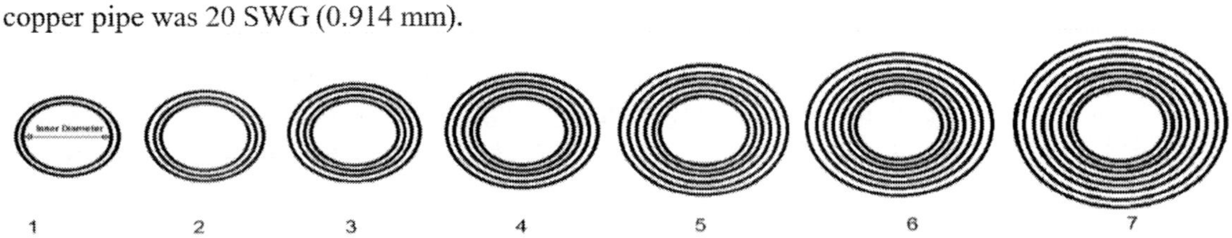

Figure 9. Different thickness of copper tube in layers

IOP Conf. Series: Materials Science and Engineering **810** (2020) 012042 doi:10.1088/1757-899X/810/1/012042

Moreover the diameter of the copper wire used for winding around the copper pipe was of 20 SWG (9.14×10^{-4}m). A total of 1090 turn were winded across the pipe evenly in single layer to increase the thickness of the pipe by 9.14×10^{-4} m. All the winding was done in a transformer industry with the help of a winding machine, which means the winding was of high quality and less uncertainty.

The variables

Table 1. Independent variable: I will be varying the thickness of the copper tube by winding different number of layers over the copper pipe. Also for being sure about the trend and for getting a better conclusion I will also vary the thickness of the magnet (Which was varies by varying the number of identical magnets) falling through the pipe. Following thicknesses were considered;

Thickness w	1	2	3	4	5	6	7
w / ($m\times10^{-3}$)	0.914	1.828	2.742	3.656	4.570	5.484	6.398

Table 2. The time calculated for the thickness of magnets;

Number of magnets	1	2	3	4	5
Thickness of magnet / ($m\times10^{-3}$)	4	8	12	16	20

Dependent variable: The time taken by the magnet to fall through the copper pipe which will be measure using a stopwatch controlled manually.

Controlled variables: The variables which were kept constant for the investigation;

- Inner diameter of the copper pipe - $(1.27\pm 0.01)\times10^{-2}$ m (0.79%)
- Length of the pipe (1.00 ± 0.01) m (1%)
- Copper wire of diameter $(9.14\ 00 \pm 0.01)\times10^{-4}$ m(0.0012%)
- Resistivity of the copper wire 2.63×10^{-2} Ohm per meter
- Material of the magnet "Neodymium"
- Diameter of the magnet $(10.00 \pm 0.01)\times10^{-3}$ m (0.005%)
- Width per magnet $(4.00 \pm 0.01)\times10^{-3}$ m (0.0125%)
- Mass of the magnet $(2.40 \pm 0.01)\ 10^{-3}$ kg (0.42%)
- Turns per layer 1090 ± 10 (0.92%)

Apparatus:

- Copper pipe, Copper wire, Neodymium magnets, Stopwatch, Winding machine and Stand with clamps.

Methodology:

To begin with the investigation, I first took the readings for the pipes own thickness which is 0.914 mm (thickness 1). For recording the time for one magnet (4 mm thickness) to fall through one meter pipe I use a manual stopwatch. I first fixed the copper pipe vertically perpendicular to the surface using the claps on to the stand. Then I took the magnet to the upper opening of the pipe ensuring that the magnet is at the center of the pipe and do not touches the wall of the pipe. Then I gently dropped the magnet into the pipe ensuring that there is no initial force applied by me to the magnet. At the time of releasing the magnet into the pipe I simultaneously stated the stopwatch. Prior to this ensure that the stopwatch is at zero seconds and there is no zero error. After taking 5 repetitive readings for the magnet I then increased the thickness of the magnet by adding one more magnet and did the same process as done for the previous one. Then same was done for all the thickness of the remaining magnets. Now after this to move on to the second thickness 1.828 mm I winded a layer of copper wire on diameter 0.914 mm over the pipe. While winding ensure that there are no gaps in between the winding and it is preferred to take the winding slow. Next 1090 complete turns were taken which ensured the whole pipe is uniformly covered with the copper wire. After winding the wire over the copper tube then the readings for the second thickness was taken in the same way as done for the previous one. The same process was repeated for all the thickness i.e. 0.914, 1.828, 2.742, 3.656, 4.570, 5.484 and 6.39 and the time taken was recorded for all the magnet thicknesses at different thickness of the pipe. A total of 5 readings were taken per case to increase the accuracy and decrease the uncertainty in the result.

IOP Conf. Series: Materials Science and Engineering **810** (2020) 012042 doi:10.1088/1757-899X/810/1/012042

4. Data collection and analysis:

Table 3. Raw data

The table shows the time recorded for the magnet to pass through the copper pipe of different thicknesses, also the table shows how the different number (width) of magnets affect the time in different thicknesses. Raw data for one magnet (4mm thickness) when dropped in different thicknesses of copper pipe.

Thickness w / (m ± 0.001) × 10^{-3}	Time Taken t / (s ±0.05)				
	For 1 magnet				
	t_1	t_2	t_3	t_4	t_5
0.914	1.78	2.28	1.78	1.97	2.10
1.828	2.44	2.05	2.49	2.15	2.43
2.742	2.17	2.23	2.06	2.05	2.17
3.656	2.69	2.36	2.42	1.97	2.42
4.570	2.70	2.49	2.50	2.62	2.69
5.484	2.57	2.50	2.54	2.50	2.68
6.398	2.89	2.62	2.76	2.63	2.56

Table 4. Processed data table

Thickness w / (m ± 0.001) × 10^{-3}	Time Taken t /s									
	For 1 magnet		For 2 magnets		For 3 magnets		For 4 magnets		For 5 magnets	
	t_{avg}	Δt_{avg}	t_{avg}	Δt_{avg}	t_{avg}	Δt_{avg}	t_{avg}	Δt_{avg}	t_{avg}	Δt_{avg}
0.914	1.98	0.25	4.18	0.03	4.58	0.19	4.23	0.30	3.53	0.13
1.828	2.31	0.22	4.27	0.27	4.62	0.28	4.10	0.18	3.82	0.10
2.742	2.14	0.09	4.35	0.19	4.71	0.51	4.21	0.09	3.82	0.03
3.656	2.37	0.36	4.44	0.18	4.87	0.19	4.38	0.21	3.97	0.09
4.570	2.60	0.11	4.56	0.20	5.08	0.23	4.69	0.27	4.18	0.16
5.484	2.56	0.09	4.63	0.13	4.80	1.00	4.78	0.29	3.99	0.07
6.398	2.69	0.17	4.65	0.10	5.13	0.17	4.79	0.10	4.27	0.09

Graph 1 – Average Time vs. Thickness of the pipe

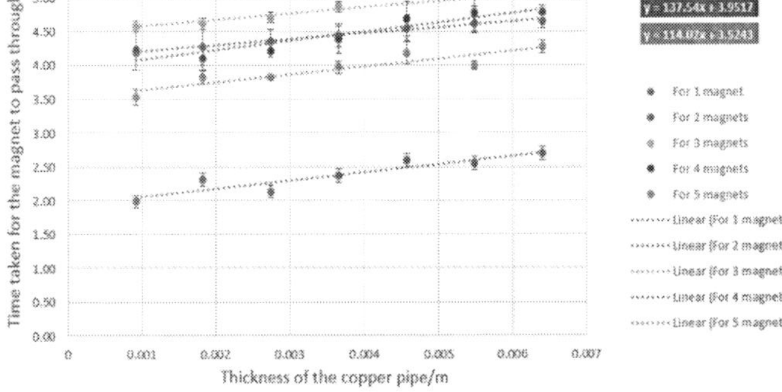

IOP Conf. Series: Materials Science and Engineering **810** (2020) 012042 doi:10.1088/1757-899X/810/1/012042

The graph above shows that how change in the thicknesses of the copper pipe leads to change in the time taken for the magnet to fall through. Now in order to verify our theory and the formula which we derived for the relation between the thickness "w" and the velocity "V" we will try putting all the values in to the equation.

Calculating the theoretical and experimental gradient; According to our theory the gradient of the line (coefficient of x from the equations of line) is equal to;

From the equation 10;

$$v = \frac{1024}{45}\frac{mg\rho a^4}{\mu_0^2 p^2}\frac{1}{w}$$

$$\frac{1}{v} = \frac{45}{1024}\frac{\mu_0^2 p^2}{mg\rho a^4}w$$

And so,

$$\frac{\left(\frac{1}{v}\right)}{w} = gradient$$

And,

$$\frac{45}{1024}\frac{\mu_0^2 p^2}{mg\rho a^4} = gradient \quad (equation 11)$$

To move forward first lets draw the graph of 1/v (inverse of velocity) vs. w (thickness);

Table 5. 1/v (inverse of velocity) and w (thickness);

w / (m ± 0.001) × 10^{-3}	$\frac{1}{v}$ / (s m^{-1})	Δ tavg /s
0.914	1.982	0.250
1.828	2.312	0.220
2.742	2.136	0.090
3.656	2.372	0.360
4.570	2.600	0.105
5.484	2.558	0.090
6.398	2.692	0.165

Graph 2:

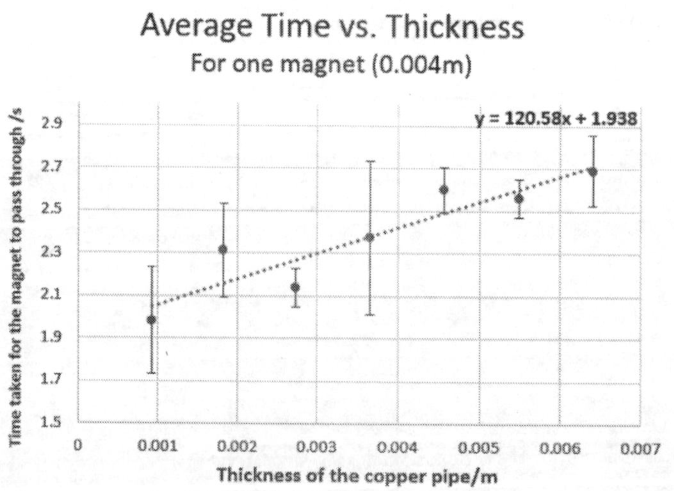

IOP Conf. Series: Materials Science and Engineering **810** (2020) 012042 doi:10.1088/1757-899X/810/1/012042

Now the gradient of the graph which is **120.58** should be equal to;

$$= \frac{45}{1024} \frac{\mu_0^2 p^2}{mg\rho a^4}$$

Where,
a is the inner radius = (6.0×10^{-3}) m , **g** is a gravitational constant = 9.8 m/s^2, **m** is the mass of magnet = (2.4×10^{-3}) kg, **μ**$_o$ is permissibility of vacuum = (1.257×10^{-6}) Henry per meter[2], **ρ** is electrical resistivity of the copper = (2.63×10^{-2}) Ω/m , **P** is the Dipole moment of the magnet

Calculating dipole moment of the magnet

To calculate the dipole moment of the magnet we will use the following equation;

$$p = \frac{2\pi r^3 B}{\mu_0} \quad (equation12)$$ [3]

Where,
P is the Dipole moment, **r** is the distance between the poles = (4.0×10^{-3}) m, **μ**$_o$ is the permissibility of vacuum = (1.257×10^{-6}) Henry per meter, **B** is the magnet flux density in Tesla.
Now we see that **r** and **μ**$_o$ are constant. To move further now we will have to find out the value of B (magnet flux density).

Calculating magnet flux density;
For finding the value of B we will use the *equation13*;

$$B = \frac{B_r}{2}\left(\frac{D+z}{\sqrt{R^2+(D+z)^2}} - \frac{z}{\sqrt{R^2+z^2}}\right) \quad (equation13)$$ [4]

Where,

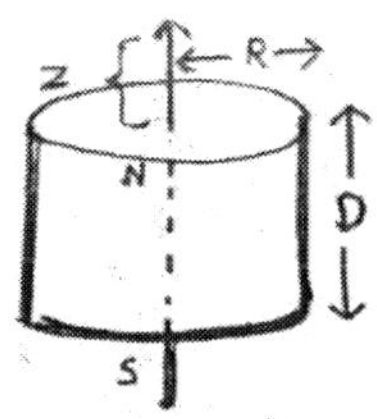

Figure 9.
Magnet parameters.

Br is the Remanence field (depends on the grade of the magnet) in Tesla and for the magnet I used Br is1.47 T [5]
R is the Radius of the magnet = (5.0×10^{-3}) m
D is the Thickness of the magnet = (4.0×10^{-3}) m
Z is the distance from the pole faced on the symmetrical axis = 0
We see that these all are again some constants and so now by putting the value of these constants in the equation 13 we will get the value of B as;
B= 0.46 T (rounded to 2 sf)
Putting the value of **B** in the previous equation 12 for calculating the dipole moment **p**;
p= 36.74 Nm/T

Now putting the value of **p** in our equation 11 we get the theoretical gradient as;
⇨ **116.89 m^{-2} s**

5. Conclusion

The aim of the essay was to investigate the relation between the thickness of the copper pipe and the time taken by the magnet to fall through it. From the graph it can be seen that as the thickness of the copper pipe increases the time taken by the magnet also increases. From the gradient of the graph it could be seen that for every 1 m increase in the thickness of the copper pipe the time increase by 120.58 seconds, which means for every 1 mm increase in thickness the time increases by 0.1206 s. The constant above in the equation of the line signifies the y-intercept, which shows the time taken by the magnet when the thickness of the copper pipe is 0 (no pipe just air) is 1.938 s. According to the equations of motion (s= ut+1/2at2) the time taken for a magnet to fall through distance of 1 m in air should be; t = 0.412 s. The difference between the theoretical and the experimental value could be considered as a systematic error in my experiment which is equal to 1.526 s. Furthermore the theoretical value I calculated and the experimental value I received for the gradient are pretty close, where the theoretical value is; 116.89 m-2 s and the experimental value obtained from the graph is; 120.58 m-2 s. Consequently the experimental derivation from the actual value is;

$$\frac{120.58-116.89}{116.89}=0.0316$$

= 3.16%

So the deviation is only 3.16% from the theoretical value, which is almost inconsequential. This shows the accuracy of my investigation and also how reliable my results are. I can now conclude that thickness of the copper pipe and the time taken by the magnet to fall through it has a linear relationship. The thicker the pipe the more the time magnets takes to pass through it. Furthermore the deviation of 3.16% of the experimental value from the theoretical value can be due to the errors instigated while performing the experiment including random and systematic errors both. To further validate the data, I have also performed the experiment with more number of magnets and the gradient of the results is almost in a range of 90-120 (refer graph 1) which is close to the predicted theoretical value.

6. Evaluation and scope of improvement:

Knowing that there is no such experiment that does not have errors and uncertainties. Below is the list of the errors which might have affected my readings and maybe be responsible for the uncertainties. List also mentions the scope of improvement to obtain a better result.

1. Uncertainty in measuring the time taken by the magnet to pass the copper tube. As the work was done manually through a stopwatch there are chances of the readings being imperfect, as human reflex time plays a major consult here. I may have turned on or off the stopwatch late or early which might have majorly affected my readings. This error could be counted as a major limitation in my methodology as a random error and to avoid this error I have taken multiple trials the error could be significantly reduced if sensors are used to record the time.
2. Difference between the material of the pipe and the material of the winding of the copper wire. Though both pipe and winded wire was both of the same metal but the electrical resistivity of both of them could differ a bit as there manufacturing is done accordingly to give them certain strength. This error could be counted as a systematic error because this applies to all thicknesses, and to avoid this error we can directly use copper pipe of different thicknesses but this wasn't easily available in the market.
3. Error while dropping the magnet into the copper pipe, there might be some initial force being applied on the magnet which could have accelerated the magnet for some distance and so the magnetic might have taken longer time to achieve terminal velocity which also reduces the distance travelled by the magnet in its terminal velocity. To avoid this error form my mythology I could have used a mechanical dropper instead.
4. Length of the pipe could be a limitation leading to disparity in the outcome, as the length of the pipe in my experiment was only 1 meters which is pretty much less. As the magnet takes some time to achieve its terminal velocity hence considering the length of the copper pipe to 1 meters in the calculation would be inappropriate. To get away with this error we cloud have used a pipe of a longer distance and also considering the time from some specific pion in the pipe and not the very top.
5. While winding the wire manually over the machine I might have missed some turns or might have increased the number of turns per layer by mistake, this could have affected the thickness of the copper pipe by disturbing the uniformities of it. to avoid this error again I could have directly use pipes of different thicknesses or instead of using copper wire to increase the thickness is could have used copper foil.
6. Air spacing between the winded coil, this is an error which can't be eliminated bur could have been reduced by using a wire if smaller diameter (thickness).
7. Use of insulation while winding, In order to increase the thickness of the pipe I had to wind one layer of copper wire over other and to do that in between I had to use and thin insulation paper. Moreover the copper

wire I used was also enamel insulated. This factor might have minutely affected the calculated thickness as the insulated material thickness gets added. Again to eliminate this error we could have used copper foil or directly use copper pipe of different thicknesses.

8. While performing the experiment carefully I heard a little noise, on investing I found that the noise was of the magnet being rubbed against the copper pipe wall which created noise. Due to this error the time taken be the magnet could have been affected due to the friction created between the pipe and the magnet, to eliminate the error I eliminated the trials in which I heard this noise and repeated that reading again.
9. Number of readings taken were less in my experiment and I only took 5 readings per trial, this again could have led to more uncertainty in time taken by the magnet to fall through the pipe. The error could be reduced by taking more number of reading.

As a further study, varying more number and grades of magnets and also varying the lengths of the pipe will help to reconfirm the theoretical model and equation predicted in my research paper.

7. References

1. Nave, R., *Faraday's Law*, Hyperphysics.phy-astr.gsu.edu, [Accessed on: 14 Nov. 2017], www.hyperphysics.phy-astr.gsu.edu/hbase/electric/farlaw.html.
2. Marco Sampaolo, *Magnetic permeability | physics.*, Encyclopedia Britannica, *Oct 25, 2016,* [Accessed on: 25 Nov. 2017], www.britannica.com/science/magnetic-permeability.
3. Physics.stackexchange.com, *Finding the magnetic moment of a permanent magnet to calculate the torque when exposed to a magnetic field*, Physics.stackexchange.com, June 11, 2015, [Accessed on: 20 Nov. 2017], www.physics.stackexchange.com/questions/189004/finding-the-magnetic-moment-of-a-permanent-magnet-to-calculate-the-torque-when-e.
4. Supermagnete.de, *Calculate magnetic flux density with formula - supermagnete.*, [Accessed on: 23 Nov. 2017], www.supermagnete.de/eng/faq/How-do-you-calculate-the-magnetic-flux-density.
5. Supermagnete.de, *Physical Magnet Data - supermagnete.*, [Accessed on: 23 Nov. 2017], www.supermagnete.de/eng/data_table.php.
6. Kanak Kalita, M. Ramachandran, Pramod Raichurkar, S. Haldar, Free Vibration Analysis Of Laminated Composites By A Nine Node Isoparametric Plate Bending Element, Advanced Composites Letters, 25(4), 2016
7. Education.com, *Lenz's Law: Magnet Through a Copper Tube | Science Project, Education.com.*, [Accessed on: 11 Nov. 2017], www.education.com/science-fair/article/copper-magnetic-affect-falling-magnet/.
8. A-levelphysicstutor.com, *Electromagnetic Induction 1, Fields & Effects - from A-level Physics Tutor.*, A-levelphysicstutor.com, 2016, [Accessed on: 16 Nov. 2017], www.a-levelphysicstutor.com/field-electro-mag-ind.php.
9. Buphy.bu.edu, *Faraday's law and Lenz's law*, Buphy.bu.edu, [Accessed on: 12 Nov. 2017], www.buphy.bu.edu/py106/notes/FaradaysLaw.html.
10. 27.Vishal Fegade, Gajanan Jadhav, M. Ramachandran, Design, Modelling and Analysis of Tilted Human Powered Vehicle, IOP Conf. Series: Materials Science and Engineering 377 (2018) 012215
11. Electrical4u.com, *Lenz's Law of Electromagnetic Induction.*, Electrical4u.com, [Accessed on: 15 Nov. 2017], www.electrical4u.com/lenz-law-of-electromagnetic-induction/.
12. BBC, *BBC - Bang Goes the Theory - Hands-on science: Magnetic Copper.*, Bbc.co.uk, 2014, [Accessed on: 7 Nov. 2017], www.bbc.co.uk/bang/handson/magneticcopper.shtml.
13. Waynesthisandthat.com, *Lenz's Law Magnet Falling Through a Copper Tube or Pipe.*, Waynesthisandthat.com, [Accessed on: 17 Nov. 2017], www.waynesthisandthat.com/Lenz%27s%20Law%20Magnet%20Falling%20Through%20a%20Copper%20Tube%20or%20Pipe.html.
14. Explorabox, *Magnet in a Copper Pipe.*, Explorabox, [Accessed on: 13 Nov. 2017], www.explorabox.org/activities/electricity-and-magnetism-2/eddy-currents/.
15. Coolmagnetman.com, *Copper Pipes Experiment.*, Coolmagnetman.com, [Accessed on: 20 Nov. 2017], www.coolmagnetman.com/magpipes.htm.

16. Simonandisa, *12.1 - Lenz's law*, Slideshare.net, May 3, 2012, [Accessed on: 17 Nov. 2017], www.slideshare.net/simonandisa/121-lenzs-law.
17. Yan Levin, Fernando L. da Silveira, Felipe B. Rizzato, Electromagnetic braking: a simple quantitative model, arxiv.org, 6 Sep. 2006, Instituto de F´ısica, Brazil, [Accessed on: 20 Nov. 2017], web, www.arxiv.org/pdf/physics/0603270.pdf
18. John A. Pelesko, Michael Cesky, Sharon Huertas, Lenz's Law and Dimensional Analysis, math.udel.edu, 6 Feb. 2004, University of Delaware, Newark, [Accessed on: 19 Nov. 2017], web, www.math.udel.edu/~pelesko/PBLOG/Pelesko_Cesky_Huertas.pdf
19. G Donoso, C L Ladera, P Mart´ın, Magnet fall inside a conductive pipe: motion and the role of the pipe wall thickness, msc.univ-paris-diderot.fr, 27 May. 2009, Universidad Simon Bol ´ ´ıvar, Venezuela, [Accessed on: on: 21 Nov. 2017], web, www.msc.univ-paris-diderot.fr/~phyexp/uploads/LaimantParesseux/Tube-Aimant2.pdf
20. Guillermo Donoso, Celso L. Ladera, Damped fall of magnets inside a conducting pipe, fisica.unlp.edu.ar, 23 Sep. 2010 , Universidad Simón Bolívar, Venezuela, [Accessed on: 21 Nov. 2017], web, www2.fisica.unlp.edu.ar/materias/FEIII/OLD/2012/AJP000193.pdf
21. Homer, David, and Mike Bowen-Jones. *Physics: Course Companion*. Oxford University Press, 2014, Print.

IOP Conf. Series: Materials Science and Engineering **810** (2020) 012043 doi:10.1088/1757-899X/810/1/012043

Mechanical, Thermal and Morphological Characterization of Basalt Fibre Composite- A Review

Mehul Nautiyal, Aditya Shrinet, Ramachandran M*
MPSTME NMIMS Shirpur Campus.
sweetestchandran@gmail.com

Abstract. In fibre reinforced polymers as well as fibre reinforced mortar, basalt fibre has been made an important replacement to glass, polymeric fibre and steel after its exploration. Now a day's BF (Basalt Fibre) has shown a rise in its attention as a replacement for glass and CF (Carbon Fibre). Basalt is a type of rock mainly dark coloured volcanic rock, which is formed after cooling lava from the volcano. In the production of fibre, basalt rocks having SiO2 (Silicon Dioxide) content of 46% is mainly used. Although basalt rock is found having various proportion of materials, only particular type of basalt rocks is used. The recently develop BFRP (Basalt Fibre Reinforced Polymer) manufactured from naturally occurring volcanic rooks shows tremendous advantages for example its recyclable, eco-friendly or does not harm the eco-system, consumes less energy in productions and as compared other types of FRP, has high performance to cost ratio. Therefore, pertaining to these advantages basalt fibre reinforced fibre has become a more reliable material to use in civil works or buildings various infrastructure. As steel material comes in contact to the corrosion, BFRP is preferred over steel as it has resistance to chemical corrosion as well as it is light weight also.

1. Introduction.

In fibre reinforced polymers as well as fibre reinforced mortar, basalt fibre has been made an important replacement to glass, polymeric fibre and steel after its exploration. 2. Now a day's BF (Basalt Fibre) has shown a rise in its attention as a replacement for glass and CF (Carbon Fibre). Basalt is a type of rock mainly dark coloured volcanic rock, which is formed after cooling lava from the volcano. In the production of fibre, basalt rocks having SiO2 (Silicon Dioxide) content of 46% is mainly used. Although basalt rock is found having various proportion of materials, only particular type of basalt rocks is used. 3. The recently develop BFRP (Basalt Fibre Reinforced Polymer) manufactured from naturally occurring volcanic rooks shows tremendous advantages for example its recyclable, eco-friendly or does not harm the eco-system, consumes less energy in productions and as compared other types of FRP, has high performance to cost ratio. Therefore, pertaining to these advantages basalt fibre reinforced fibre has become a more reliable material to use in civil works or buildings various infrastructure. As steel material comes in contact to the corrosion, BFRP is preferred over steel as it has resistance to chemical corrosion as well as it is light weight also. 4. In the era of application of eco-friendly material, there has been rise in the uses of basalt fibre and is also considered as a future material because of its high mechanical properties such as high strength and low cost of application as compared to the glass fibre. Although the chemical composition of glass and basalt fibre are nearly similar, the qualities of basalt fibre overcome the properties of glass fibres. As compared to glass fibre, basalt fibre offers high chemical as well as thermal stability, high tensile strength, and good sound and electrical insulation, high compression strength. Although the impact property of basalt fibre is very poor because of its brittle nature.5. Basalt fibre is an inorganic fibre having high strength, high operating temperature range, good modulus, less expensive, eco-friendly and better strain to failure ratio the CF (Carbon Fibre). It has been told by various researchers that basalt fibre has better compressive as well as tensile properties over glass fibres. Pertaining to these reasons, basalt fibres has attracted many people attention as a brand new reinforced material for hybrid and composite laminates.

2. Tensile test

Addition of fibre leads to an effect named bridge action, and therefore to a greater firmness of the fibre reinforced mortar and greater ductile behaviour in contrast to the simple mortar. Furthermore, the fibre reinforced mortar seems to be highly strain rate sensitive, as the tensile power Dynamic Increase Factor (DIF) risen up to 5.1, for a high strain rate of about 102 s_1.The acquired tensile qualities are explained in terms of

Dynamic Increase Factor commonly known as DIF, quantifying the rise in tensile failure stress at the measured strain rate values.

Paper no	Composition	Tensile strength (MPA)
2	BF 200g/ mm, 2.33% epoxy	2044
2	BF 280g/mm, 4.13% epoxy	2277
4	Basalt fibres/BFRP Composites 2.8% epoxy	2100
4	Basalt fibres/BFRP Composites 2.7%epoxy	2130
4	(53.1, 5.3%) 64.6 (9.9, 15.4%)	1001
4	1908 (85.9, 4.5%) 77.1 (2.5, 3.2%	1908
4	(66.3, 9.6%) (1.7, 2.7%) (29.3, 16.3%)	63.5 927.7 53
4	(67.7, 7.3%) (3.8, 5.7%)	66 18179
6	Ares = 157.98 mm2/mm for CFRP	2100
6	107.15 mm2/mm for BFRP.	3900
11	Gap's content (% wt.)	212

In reference to the acquired outcome the greatest tensile strength found in basalt fibre is 18179 MPa whilst the lowest is of 0.2 weight % Anti Peroxide, 5 weight % PP-g-MA i.e. 17.89 Mpa.

3. Flexural test

A quasi-static tensile strength characterization of the mortars was carried out as per the European Standard EN1015-11, by using an Instron 5566 testing machine, having a 5 KN of load cell, three-point bending flexural tests had been conducted on specimens of dimensions 40X40X160 mm. By using an Instron 8501 which was equipped with 50 KN of load cell, compression tests had also been directed on two fragments of specimen which failed after the flexural test. The loading rate in compression tests as well as in flexural tests had been 0.6 mm/min. After 7 to 28 days of setting or curing, the tests were directed and all the computations were performed in the sets of three.

Paper no.	Composition	Flexural Strength
6	EPOXY+B6C14 EPOXY+B10C10 EPOXY+B14C6	0.4 kN 0.3 kN 0.25 kN
12	BF/EPOXY at 0.4%MWCNT'S BF/EPOXY at 0.2%MWCNT'S BF/EPOXY at 0.3%MWCNT'S BF/EPOXY at 0.5%MWCNT'S	995.2 MPa 839.3 MPa 918.2 MPa 981.7 MPa
16	PLA + 10%basalt fibre PLA + 15%basalt fibre PLA + 30%basalt fibre PLA + 40%basalt fibre	120 MPa 140 MPa 180 MPa 182 MPa
22	Composite 750d Composite 650d Composite 650c	300 MPa 500 MPa 800MPa
23	Reinforced glass matrix (4H) Un reinforced glass matrix(2H)	96 MPa 90 MPa
26	epoxidised soybean oil and epoxy resin acrylated epoxidised soybean oil with flax fibre mat	129 MPa 27 MPa

Un-reinforced glass matrix (2H) has the lowest value of flexural test which is 90 MPa and BF/EPOXY at 0.5 % Multi-walled carbon nanotubes has the greatest value of flexure test which is 981.9MPa and hence we can find many values of basalt fibres of other types.

4. Scanning Electron Microscopy

A Scanning Electron Microscope (SEM) defined as an electron microscope which uses a focused beam of electronto scan surface of a sample and produce images. In the sample the electrons interacts with the atoms, giving rise to various signals that accommodate data regarding the composition of sample as well assurface topography. The beam comprising of electron is scanned in a raster scan pattern, and the location of the beam is merged with the intensity of the detected signal to give rise to an image. In the most random

Scanning electron microscope(SEM) mode, electron beams excites the atoms which then emit secondary electrons are detected using an Everhart-Thornley detector.The signal intensity and the no. of secondary electrons can be detected depending on the specimen topography. Scanning Electron Microscope (SEM) can attain resolution better than 10^{-9} meter or 1 nanometre. Specimens are examine dunder high vacuum in standard SEM, or in the presence of low vacuum in variable pressure or environmental SEM, and at an elevated or a wide range of cryogenic temperature with specially designed instruments. [1] A strong adhesion between matrix and basalt fibre, with no evident phase disconnection, was observed in B1 mortar. This proves that the post-peak behaviour of the fibre reinforced mortars, resulting in an evident increase in toughness, is due to the strong adhesion between fibres and binder matrix. Moreover, the absence of failed fibres allows concluding that, when fracture occurs, basalt fibres undergo a pull-out phenomenon, because the anchorage failure stress is achieved before the tensile strength of the fibre is reached. Given the better mechanical properties exhibited by B1 samples, both in terms of flexural and compression strength, B1 formulation was selected for the dynamic direct tensile characterization. [2] Epoxy-CF200 laminates show very limited energy dissipation over the whole range of stresses, while SDC values of Epoxy-BF200 sample are 5–10% higher than that showed by the corresponding GF composites. [3] H1 (i.e., stacking three carbon fibre layer at the compressive side) showing the best flexural strength and modulus from among the stacking sequence arrangements. Fatigue stress is effectively distributed in carbon fibres, thus it can withstand more stress leading to higher flexural strength. The flexural strength follows the trend: CFRP > H1 > H2 > H3 > H4 > H5 > H6 > H7 > BFRP. [4] A novel method of composite fabrication was adopted by sandwiching the TiO2 coated mats between layers of fine glass powder and subsequently hot-pressing in vacuum. Five layers of TiO2 coated basalt fibre mats were sandwiched between 6 layers of glass powder inside a graphite die with a diameter of 38 mm. The weight of the layers was calculated in order to fix the volume fraction of fibres. Orientation of the fibre mats inside the graphite die for hot-pressing at 15 vol%. The basalts fibre mats were carefully orientated in order to maintain the fibres parallel in all layers. [5] The fracture surfaces of NHL-based mortars. NHL_CH3 specimens are characterized by extensive pull-out (Fig. 4a) if compared to the other matrices (see, for instance, Fig. 4c). In addition, there are matrix particles attached to the fibres for MC and MCC composites (Fig. 4b and c), which indicates a sufficient level of adhesion between the fibre and matrix. This can be compared to the relatively smooth surface of the chopped fibres. [6] ML-506 epoxy resin based on epoxy biphenyl F with hardener HA-11, both supplied by Mokarrar Engineering Materials Co., Iran, was used as matrix material. The resin-hardener ratio was 100:15 by weight, as recommended by the manufacturer. This resin system was chosen owning to its low viscosity. (1450 centipoise) and long gel time (60 min) at 25_C. Basalt roving fibre with the trade name of BCF13-150- KV12 was purchased from Kamenny Vek Co., Russia. MWCNTs were provided by Cheap Tubes Inc., USA. The average aspect ratio (length divided to the outer diameter) of the MWCNTs was _1330

5. Impact Test

A standardized high strain rate test which is used to measure the amount of energy absorbed during its rapture is known as the Charpy impact test. It is also determined as the Charpy V-notch test. The energy which is absorbed acts as an instrument to study temperature dependent ductile brittle transition and also the measure of the material's toughness at the notch. It has many advantages, such as the results can be acquired in economic cost and in a very less time as well as it is very easy to prepare for the test and conduct it. Because of these advantages it is widely used in many big organisations. In 1900, S B Russell an American and Georges Charpy a French developed this test. The name of the test was known as the Charpy test because of the efforts made by Charpy and his technical contributions. During World War II the Charpy test was conducted to know about the problems of fracture in ships. Now a days Charpy's Impact Test is used for testing materials in many organisations, for example to calculate the effect of storm on the materials used for building big or small bridges and other high rise or massive structures.

Paper no	Composition	Impact strength (kJ/m^2)
5.	BFRE Code 100	587
	BFRE Code 130	862
11	Basalt epoxy control	95
	b- 0.1	110
	b- 0.2	105
16	Basalt 0%wt	2.5
	Basalt 10%wt	4.9
	Basalt 30%wt	9.2
	Basalt 40%wt	9.5

By the readings obtained from the test we can note that basalt of 0% weight is having the least impact strength that is 2.5 kJ/m^2 whereas BFRE Code 130 is having the greatest impact strength that is 862 kJ/m^2.

6. Differential Scanning Calorimetry

Differential Scanning Calorimeter (DSC) is generally described as a technique which comprises of the measurement of change in the quantity of heat which is required for the rise in temperature of a sample and reference. While the experiment is being conducted, constant temperature is managed for the sample and reference. Basically, temperature program is designed for Differential Scanning Calorimeter (DSC) analysis such that the temperature gradually rises of the sample holder with respect to time and the heat capacity is well known for the reference sample over the range of temperatures to be scanned. In 1962 by M. J. O'Neill and E. S. Watson the DSC technique was developed and in 1963, at Pittsburgh Conference of Analytical Chemistry and Applied Spectroscopethe technique was commercially introduced. At around 1964 by Peter L. Privalov and D. R. Monaselidze in IOP (Institute of Physics) in Tbilisi, Georgia, developed the first adiabatic differential scanning calorimeter that can be utilised in biochemistry. The instrument was termed as DSC (Differential Scanning Calorimeter) which calculates a very accurate value of heat capacity and also directly measures energy.

Sr. no.	Composite	Value (degree Celsius)
2	Glass transition 1 Glass transition 2	79-81 102-104
16	5% wt. 40% wt.	99 100
36	5% wt. 40% wt.	96 110
59	PL PL25-t	60 168
60	PLA 5%wt PLA	125 109

By the readings obtained in the test we can conclude that the impact test value of PL25-t is the highest of all that is 168°C whereas the least impact value is of PL which is 60°C.

7. Conclusion

Basalt fibre reinforced polymeric composite is preferred over steel as it has resistance to chemical corrosion as well as it is light weight. Apart from that the Basalt fibre reinforced polymeric composite also possess significant mechanical, morphological and thermal properties when compare with other synthetic and natural fibre reinforced polymeric composite. From the literature it is evidenced that the basalt fibre percentage in the Basalt fibre reinforced polymeric composite is significantly affects the mechanical, morphological and thermal properties of the composite plate.

Reference

[1]. Lopresto V, Leone C, De Iorio I. Mechanical characterisation of basalt fibre reinforced plastic. Compos Part B Eng 2011; 42(4):717-23.

[2]. Bernasconi, F. Cosmi, P. Hine, Analysis of _bre orientation distribution in short pre reinforced polymers: A comparison between optical and tomographic methods, Composites Science and Technology 72 (16) (2012).

[3]. Karuppuchamy, S. "Thermo-mechanical properties of coconut shell powder reinforced plastic composites." International Journal of ChemTech Research Vol.8, No.6 (2015): 852-857.

[4]. Kanak Kalita, U Ragavendran, M. Ramachandran, Akash Bhoi, Weighted sum multi-objective optimization of skew composite laminates, Structural Engineering and Mechanics, Vol. 69, No. 1 (2019) 21-31.

[5]. Sefiu Adekunle Bello, Isiaka Ayobi Raheem, Nasir Kolawole Raji, Study of tensile properties, fractography and morphology of aluminium (1xxx)/coconut shell micro particle composites, Journal of King Saud University - Engineering Sciences, Vol29, Issue 3, July 2017, Pages 269-277

[6]. Liyanage, Chinthani D., and Mevan Pieris. "A Physico-Chemical Analysis of Coconut Shell Powder." Procedia Chemistry 16 (2015): 222-228.

[7]. Agarwal, Rakshit, M. Ramachandran, and Stanly Jones Retnam. "Tensile Properties of Reinforced Plastic Material Composites with Natural Fiber and Filler Material." ARPN Journal of Engineering and Applied Sciences 10, no. 5 (2015): 2217-2220.

[8]. Siti Nurbazilah Ab Jabal, Yew Been Seok, Wee Fwen Hoon, carbon composition, surface porosities and dielectric properties of coconut shell powder and coconut shell activated carbon composites, ARPN Journal of Engineering and Applied Sciences, VOL. 11, NO. 6, MARCH 2016, 3832- 3838.

[9]. Shyamkumar Shah, Akshaykumar Patil, M. Ramachandran, Kanak Kalita, Effect of coal ash as a filler on mechanical properties of glass fiber reinforced material, International Journal of Applied Engineering Research, 9(22) (2014) pp. 14269-14277.

[10]. Salleh, Z., M. M. Islam, M. Y. M. Yusop, and MA Mun'aim M. Idrus. "Mechanical properties of activated carbon (AC) coconut shell reinforced polypropylene composites encapsulated with epoxy resin." APCBEE Procedia 9 (2014): 92-96.
[11]. D. Bino prince raja, B. Stanly Jones Retnam, M. Ramachandran, Analysis of mechanical properties of glass and carbon fiber reinforced polymer material, International Journal of Applied Engineering Research. ISSN 0973-4562 Volume 10, Number 11 (2015) pp. 10387-10391.
[12]. Gojny, F. H., M. H. G. Wichmann, U. Köpke, B. Fiedler, and K. Schulte. "Carbon nanotube-reinforced epoxy-composites: enhanced stiffness and fracture toughness at low nanotube content." Composites science and technology 64, no. 15 (2004): 2363-2371.
[13]. Alex. S, Stanly Johns Retnam, M. Ramachandran, A review on Biodegradability of Hybrid Bamboo/Glass fiber polymer composites, International Journal of Applied Engineering Research. ISSN 0973-4562 Volume 10, Number 11 (2015) pp. 10565-10569.
[14]. EN 1015-11 methods of test for mortar for masonary – Part 11: determination of flexural and compressive strength of hardened mortar; 1999.
[15]. Bortz DR, Heras EG, Martin-Gullon I. Impressive fatigue life and fracture toughness improvements in graphene oxide/epoxy composites, Macromolecules 2012; 45: 238- 245.
[16]. Ganguli S, Aglan H, Dennig P, et al. Effect of loading and surface modification of MWCNTs on the fracture behavior of epoxy nanocomposites. J Reinf Plast Compos 2006; 25: 175–188.
[17]. Messler RW (2004) Joining of Materials and Structures: From Pragmatic Process to Enabling Technology. Amsterdam: Elsevier Science B.V., Chapter 15, pp.697–741.
[18]. Dehkordi MT, Nosraty H, Shokrieh MM, et al. Low velocity impact properties of intra-ply hybrid composites based on basalt and nylon woven fabrics. Mater Des 2010; 31: 3835–3844.
[19]. S.L. Gao, E. Mäder, R. Plonka, Nanostructured coatings of glass fibers: improvement of alkali resistance and mechanical properties, Acta Mater. 55 (2007) 1043–1052.
[20]. Kim, Jin Ah, Dong GiSeong, Tae Jin Kang, and Jae RyounYoun. "Effects of surface modification on rheological and mechanical properties of CNT/epoxy composites." Carbon 44, no. 10 (2006): 1898-1905.
[21]. Bisanda, E. T. N., and Martin P. Ansell. "The effect of silane treatment on the mechanical and physical properties of sisal-epoxy composites." Composites Science and Technology 41, no. 2 (1991): 165-178.
[22]. Joshi, Satish V., L. T. Drzal, A. K. Mohanty, and S. Arora. Are natural fiber composites environmentally superior to glass fiber reinforced composites. Composites Part A: Applied science and manufacturing 35, no. 3 (2004): 371-376.
[23]. Sgriccia, N., M. C. Hawley, and M. Misra. "Characterization of natural fiber surfaces and natural fiber composites." Composites Part A: Applied Science and Manufacturing 39, no. 10 (2008): 1632-1637.
[24]. M. Ramachandran, , Sahas Bansal, Pramod Raichurkar, Scrutiny of Jute Fiber Poly-Lactic Acid (PLA) Resin Reinforced Polymeric Composite, Journal of the Textile Association, Volume 76, Issue 6, 2016, pp. 372-375.
[25]. Herrera-Franco, PJp, and A. Valadez-Gonzalez. "A study of the mechanical properties of short natural-fiber reinforced composites." Composites Part B: Engineering 36, no. 8 (2005): 597-608.
[26]. Ramanan. G, Rajesh Prabha. N, Diju Samuel. G, Jai Aultrin. K. S, M Ramachandran, Prediction of Machining Characteristics of Hybrid Composites Using Response Surface Methodology Approach, International Journal of Engineering & Technology, 2018, 7(3.1), 162-165.
[27]. Bachtiar, Dandi, S. M. Sapuan, and M. M. Hamdan. "The effect of alkaline treatment on tensile properties of sugar palm fibre reinforced epoxy composites." Materials & Design 29, no. 7 (2008): 1285-1290.
[28]. Sahas Bansal, M. Ramachandran, Pramod Raichurkar, Comparative Analysis of Bamboo using Jute and Coir Fibre Reinforced Polymeric Composites, Materials Today: Proceedings,4(2), 2017, 3182–3187.
[29]. ACI Committee. Guide test methods for fiber-reinforced polymers (FRP) for reinforcing or strengthening concrete structures. Farmington Hills, Mich: ACI 440.3R-04. American Concrete Institute. 40.
[30]. ACI Committee 440. Guide for the design and construction of externally bonded FRP-system for strengthening concrete structures. Detroit: ACI 440.2R-08. American Concrete Institute; 2008. 245
[31]. Raichenko, *Some Patterns in Forming and Sintering of Metal Powders*, Author's Abstract of ScD Thesis, Inst. Probl. Materialoved. AN USSR, Kiev (1967).
[32]. Asprone, Domenico, Ezio Cadoni, Fabio Iucolano, and Andrea Prota. "Analysis of the strain-rate behaviour of a basalt fibre reinforced natural hydraulic mortar." *Cement and Concrete Composites* 53 (2014): 52-58.
[33]. Yousif, B. F., A. Shalwan, C. W. Chin, and K. C. Ming. "Flexural properties of treated and untreated kenaf/epoxy composites." Materials & Design 40 (2012): 378-385.
[34]. Mohanty, A. K., ManjusriMisra, and Lawrence T. Drzal. "Surface modifications of natural fibers and performance of the resulting biocomposites: an overview." Composite Interfaces 8, no. 5 (2001): 313-343.
[35]. Gassan, Jochen. "A study of fibre and interface parameters affecting the fatigue behaviour of natural fibre composites." *Composites part A: applied science and manufacturing* 33, no. 3 (2002): 369-374.

IOP Conf. Series: Materials Science and Engineering **810** (2020) 012043 doi:10.1088/1757-899X/810/1/012043

[36]. P. M. Bhagwat, M. Ramachandran, Pramod Raichurkar, Mechanical Properties of Hybrid Glass/Carbon Fiber Reinforced Epoxy Composites, Materials Today: Proceedings, 4(8), 2017:1788–1793.
[37]. Park JM, Shin WG and Yoon DJ. A study of interfacial aspects of epoxy-based composites reinforced with dual basalt and SiC fibres by means of the fragmentation and acoustic emission techniques. Compos Sci Technol 1999; 59: 355–370.
[38]. Carmisciano S, De Rosa IM, Sarasini F, Tamburrano A and Valente M. Basalt woven fiber reinforced vinylester composites: flexural and electrical properties. Mater Des 2011; 32: 337–342.
[39]. Landucci G, Rossi F, Nicolella C and Zanelli S. Design and testing of innovative materials for passive fire protection. Fire Safety J 2009; 44: 1103–1109.
[40]. ACI 440.3R-04. Guide test methods for fiber-reinforced polymers (FRPs) for reinforcing or strengthening concrete structures. PA: American Concrete Institute 440 Committee, 2004.

IOP Conf. Series: Materials Science and Engineering **810** (2020) 012044 doi:10.1088/1757-899X/810/1/012044

Investigations on the Machining Performance using Solid Lubricant Mixed with Varying Proportions in Vegetable Oil during Hard Turning

Satish Chinchanikar[1], H.M.Bawangaonwala[2], Shubham Bokade[2], Shubham Garode[2]
Vishwakarma Institute of Information Technology, Pune – 411048, (India)
satish.chinchanikar@viit.ac.in, hakimuddin.218m0057@viit.ac.in, shubham.218m0078@viit.ac.in, shubham.218m0049@viit.ac.in

Abstract: The solid lubricant assisted machining is gaining more attention from last decade as use of cutting fluids becoming more and more prevalent around the world due to health and environmental concern. However, very few studies reported on use of solid lubricants during machining. With this view, in the present work, hard turning experiments with boric powder as a solid lubricant mixed with varying proportions in vegetable oil (refined sunflower oil) were performed using coated carbide tool. Machining performance in terms of surface roughness and cutting forces (tangential, feed and radial components of force) were investigated at different cutting conditions under dry and with three varying proportions of boric powder namely, 1%, 5% and 10% mixed with vegetable oil. It has been observed lower values of surface roughness and cutting forces when using solid lubricant assisted hard turning (SLHT) as compared to dry hard turning. Lowest values of surface roughness were observed when using 10% proportion of boric powder mixed with vegetable oil. However, cutting forces although observed to decrease with increasing percentage of proportion of boric powder in lubricating oil, no significant reduction; especially in tangential and feed components of forces, were observed when using 5% and 10% proportion of boric powder mixed with refined sunflower oil. SLHT experiments were also performed under mist lubrication (Mist-SLHT) with a view to assess the feasibility of variants of solid lubrication system in hard turning.
Keywords: hard turning, surface roughness, cutting forces, solid lubricant, vegetable oil, MQL

1. Introduction

In high precision works dimensional accuracy and geometrical tolerances are of prime importance. However, accuracy and properties of the machined component significantly get affected due to higher temperatures generated during machining. Frictional behaviour at the flank and face of the tool mainly determines the amount of heat generated during machining. Moreover, amount of heat generated in turn frictional behaviour by the tool depends on the cutting conditions used during machining. Cutting conditions includes cutting parameters namely cutting speed, feed, depth of cut, work and tool material properties and cooling medium used and it's type of application (flood cooling, minimum quantity of lubrication, cryogenic cooling etc.). High heat generated at the interface due to friction between the tool and the work-piece leads to poor machined surface quality and significantly affects the tool wear.

It is reported that approximately 320,000 tons of lubricant (neat oil, water-soluble fluids, and gases) were used per year in the European Union which clearly indicate importance of lubricant in machining. Water-soluble fluid is most commonly used in machining. However, this typically mineral-based oils known to be harmful to the worker and to the environment as they exhibits toxic product when degraded in soil. Moreover, with becoming environment consciousness enhanced laws and regulations green cutting will become important tendency of machining. Dry machining can eliminate the use of lubricant and consequently exterminate the bad effect of using mineral-based lubricant. However, dry machining limits the cutting parameters to be used during machining due to high amount of heat generated which significantly affects the quality of surface finish especially when machining at higher cutting speed and feed rate [1]. Therefore, attempts have been made by the researchers to use vegetable oil which is non-toxic and biodegradable as against the mineral oil. Most of the vegetable oil has triglycerides in its composition which enhances lubricity of oil. Moreover, viscosity of vegetable oil remains stable at high temperature due to their high viscosity index [2].

Hard turning is mostly sighted in the literature under dry condition as high temperature involved in the process makes chip deformation and shearing of the material easier. Moreover, at higher cutting speed heat is carried away by the fast flowing chip makes the use of cutting fluid redundant. However, dry cutting creates problems of chip transportation and significantly affects tool life due to increase in chip-tool and work piece-tool friction [3-5]. On the other hand, machining under wet condition (use of abundant cutting fluid) facilitates chip transportation, lowers friction and cools the work piece. However, it increases the cost of cutting and in hard turning it does not allow the work piece to soften, which is good for the process. Therefore, most of researchers performed machining with Minimum Quantity Lubrication (MQL) which represents a compromise between the advantages and disadvantages of completely dry cutting and cutting with abundant soluble oil [6-7].

Attempts have been also made by researchers using vegetable oil, solid lubricants and solid lubricants mixed in varying proportion with mineral-based oil and vegetable oils during machining. Lawal et al. [8] evaluated the performance of vegetable oil as against mineral oil during turning and observed that the vegetable oil surpasses conventional oil-in-water performance within the domain of the parameters selected in their study. Babur et al. [9] experimental investigation concluded that vegetable oil is effective in producing better surface quality and less tool wear in comparison to synthetic cutting fluids during turning. It is reported that performance of vegetable oil can be improved by utilizing powder particles such as MoS_2, graphite, boric acid etc. Sodavadia and Makwana [10] observed better performance in terms of surface finish, coefficient of friction, wear and cutting force using powder particle aided lubricant during machining. It is reported that better machining performance observed when using solid lubricants is because of lattice layered structure of powder particle, their relatively high load carrying capacity and having low steady state coefficient of friction [11].

The solid lubricant assisted machining is gaining more attention from last decade as powder particle such as boric acid, graphite, $MoS_2$are environmentally friendly and do not pose threat to operational workers. According to the Environmental Protection Agency, which is desirable in creating a greener lubrication in machining application [12].Toenshoff [13] investigated the tribological aspects of turning with ceramic tools. According to their study, cooling effect reduces the thermal load of the cutting edge, and thus increases tool life as compared to dry cutting. Further, their study reports that surface roughness of the work piece can be reduced either by chemical interaction between the work piece surface and extreme-pressure additive of the coolant or by mineral oil application. Avila and Abrao [14] investigated the effect of cutting fluid on the machining of hardened AISI 4340 steel. Their study investigated the performance of three type of cutting fluids and machining performance using cutting fluids was compared with dry machining when using mixed alumina inserts. Their experimental investigation showed that the application of a cutting fluid based on an emulsion without mineral oil resulted in longer tool life compared to dry cutting and the use of cutting fluid is responsible for reducing the scatter in the surface finish at high cutting speeds. However, Inspite of all beneficiaries these fluids produce harmful environmental results and to the human health issues. Varadaranjan et al. [15] investigated the turning with minimal fluid application and its application with dry and wet turning. Results showed that overall performance during minimum cutting fluid application is found to be superior to that dry turning and conventional wet turning on the basis of cutting force, tool life, surface finish, cutting ratio, cutting temperature, and tool-chip contact length.

Although, sufficient work have been reported on use of cutting fluids and their comparative evaluation when applied under wet, minimum quantity lubrication during machining, very few studies reported on use of solid lubricant mixed with varying proportions in vegetable oil during hard turning. Moreover, contradictory findings reported on use of cutting fluid during hard turning. With this view, in the present work solid lubricant assisted hard turning (hereafter referred as "SLHT") experiments with solid lubricant mixed with varying proportions in vegetable oil were performed to find the feasibility of solid lubricant assisted hard turning. SLHT experiments were also performed under mist lubrication (hereafter referred as "Mist-SLHT") with a view to assess the feasibility of variants of solid lubrication system in hard turning.

2. Experimentation

2.1 Experimental Set-upand Measuring Instruments

Hard turning (SLHT and Mist-SLHT) experiments were performed on hardened AISI 52100 steel (60-62 HRC) on a CNC lathe. Experiments were performed using PVD-coated nano-laminated Ti-Si-N, Ti-Al-N carbide inserts. All the inserts have identical geometry designated by ISO as CNMG 120408 with integral chip breaker geometry as MF2. A right hand side tool holder designated by ISO as PCLNR 2525M12 was used. A schematic of the experimental setup with actual photograph of the setup is shown in Figure 1. Machining experiments were performed at eight different combinations of cutting conditions with values of cutting speeds;'*V*' = 100, 150 and 200 m/min, feed '*f*' = 0.1 and 0.3 mm/rev and depth of cut '*d*' = 0.1 and 0.3 mm.

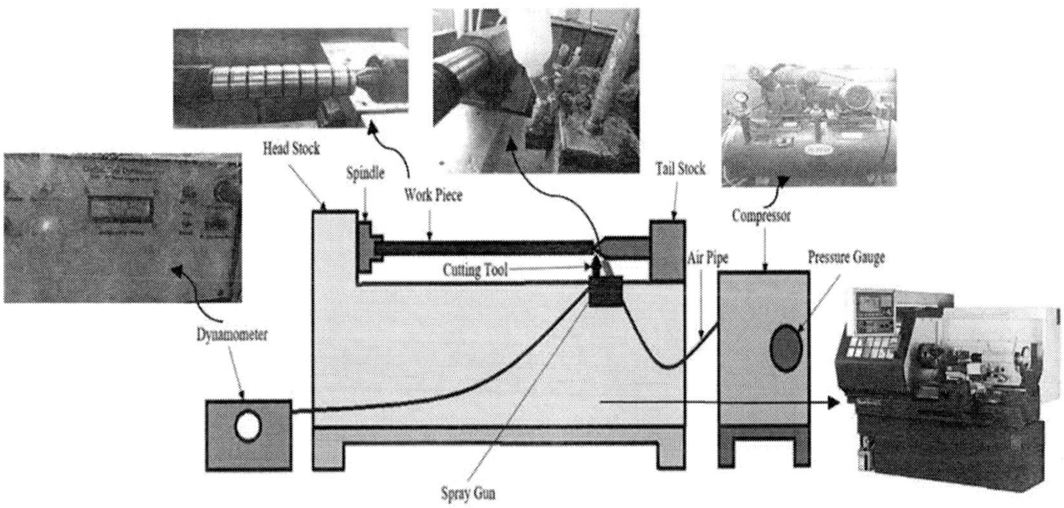

Fig. 1: Experimental set-up for solid lubricant assisted hard turning (SLHT) and Mist-SLHT

For the coolant injection along with compressed air a spray-painting gun was used. Nozzle was fixed by means of a magnetic stand in such a way that the cutting fluid or mist formed is injected at the interface of the tool and work material. Nozzle has two inlets, one for oil and other one for air. Air inlet of nozzle is connected to compressor via flexible pipe and oil inlet of nozzle is connected to the oil container. Air and oil are injected simultaneously and gets mixed just outside at the exit of the nozzle. Compressed air pressure was set at a pressure of 2 bar. Various instruments were used for measuring the responses. Surface roughness (Ra), often shortened to roughness, is a measure of the texture of a surface. It is quantified by the vertical deviations of a real surface from its ideal form. Surftronic DUO surface roughness tester (Taylor and Hobson make) was used to measure the surface roughness. The surface tester has error of 0.7μm. Cutting forces were measured by a strain gauge type lathe tool dynamometer. This strain gauge type dynamometer which was mounted on the cross slide of the CNC lathe along with cutting tool holder is used to measure the three components of a cutting force coming on the tool tip during turning. The sensor is designed in such a way that it can be rigidly mounted on the tool post, and the cutting tool can be fixed to the sensor directly. This feature will help to measure the forces accurately. The sensor is made of single element with three different wheatstones strain gauge bridge. Provision is made to fix 1/2" size tool bit at the front side of the sensor. Forces in X-Y-Z directions namely feed force (F_f), radial force (F_r) and tangential force (F_t) is seen individually and simultaneously in three digital Indicators.

2.2 Materials

The physical properties of refined sunflower oil are shown in the Table 1. The Oil contains 59% polyunsaturated fat which is an advantageous because vegetable oil that are high in unsaturated fatty acid are susceptible to oxidation. Therefore, palm kernel oil that is high in saturated fatty acid are more stable in terms of oxidation. Although, the process of oxidation is influenced by many factors, nonetheless the fatty acid content can help to predict Sunflower oil oxidation stability.

Table 1. The physical properties of refined sunflower oil.

Properties	Refined Sunflower Oil
Iodine Value	135
Dynamic Viscosity At 25°C	0.04914kg/(M*S)
Density	918.8Kg/M^3
Flash Point	232°C

The performance of lubricant is also highly influenced by the additional additives. Additives that are used in this formulation are emulsifier, anti-corrosion agent, antioxidant agent, biocide and boric acid powder. The physical properties of boric acid powder used for this investigation are shown from Table 2. The type of additives in this formulation is cautiously selected, and it is ensured that the materials are not hazardous to the operators and the environment.

Table 2: The physical properties of boric acid powder

Properties:	Values
Density, G/Cm^3	1.44
Molecular Mass, G/Mol	61.83
Melting Point, °C	160
Boiling Point, °C	300
Purity, %	99.999
Particle Average Size, Mm	< 8
Chemical Formula	H_3BO_3

3. Results and Discussion

With the view to assess feasibility of solid lubricant assisted hard turning (SLHT), hard turning experiments were performed at different cutting conditions using solid lubricant (SL) mixed in vegetable oil (VO) as a cooling medium. Boric acid was used as solid lubricant which was mixed in different proportions namely 1% by weight, 5% by weight and 10% by weight in refined sunflower oil (vegetable oil). Experiments were also performed using mist of sunflower oil and compressed air (Air; at a 2 bar pressure) and mist formed by sunflower oil carrying 5% by weight boric acid and compressed air (pressure 2 bar). A comparative evaluation of five different cooling mediums namely VO+1%SL,

VO+5%SL, VO+10%SL, VO+Air and VO+SL+Air was carried out in terms cutting forces and surface roughness at different cutting conditions. Surface roughness and three components of cutting force namely tangential force (F_t), feed force (F_f) and radial force (F_r) were measured at each cutting condition and under five different cooling mediums.

During the machining process, high temperature generated in the cutting zone is the cause for poor surface finish. Temperature can be minimized with the help of lubricant. Solid lubricant such as boric acid is one of the options to reduce the temperature at the interface between work piece and cutting tool as boric acid powder has high thermal conductivity and low coefficient of friction. Further, adding boric acid to oil lubricant might improve machining performance. High thermal conductivity of the boric acid powder mixed with lubricant oil might change heat transfer and thus aids in reducing temperature in the cutting zone. Experimental results obtained for three components of cutting force and surface roughness when using five different cooling mediums and at different cutting conditions are given below. It can be seen decrease in cutting forces with increase in cutting speed, feed and depth of cut values. Radial component of force can be seen as highest in magnitude followed by tangential force and feed force. Surface roughness also can be seen decreasing with increase in cutting speed and increased with higher values of feed and depth of cut. However, it can be seen that surface roughness gets affected significantly with feed followed by cutting speed and depth of cut. Further, it can be seen that lower values of cutting forces and surface roughness with increase in proportions of boric acid by weight in refined sunflower oil. However, significant improvement in surface roughness cannot be seen when proportion of boric acid by weight increased from 5% to 10% in refined sunflower oil. This can be justified as almost similar values of cutting forces or to some extent higher cutting forces were observed at 10% of boric acid proportion by weight as compared to 5% proportion of boric acid by weight with vegetable oil. This clearly shows that lower values of surface roughness and cutting forces could be obtained by having 5% of boric acid in proportion by weight in refined sunflower oil.

Table 3: SLHT: Cutting forces and surface roughness when used refined sunflower oil with varying % of boric acid

V (m/min)	f (mm/rev)	d (mm)	Refined sunflower oil with 1% by weight boric acid				Refined sunflower oil with 5% by weight boric acid				Refined sunflower oil with 10% by weight boric acid			
			F_t	F_f	F_r	R_a	F_t	F_f	F_r	R_a	F_t	F_f	F_r	R_a
100	0.1	0.1	120	88	161	0.41	80	64	129	0.31	70	52	132	0.31
100	0.3	0.1	140	97	212	0.62	130	81	206	0.56	117	71	192	0.54
150	0.1	0.3	123	87	172	0.54	88	73	145	0.46	85	81	167	0.62
150	0.3	0.3	170	108	224	0.91	130	55	191	0.69	110	64	177	0.71
150	0.1	0.1	90	71	145	0.33	65	59	117	0.28	60	55	128	0.38
150	0.3	0.1	123	88	173	0.56	98	68	161	0.48	87	56	174	0.58
200	0.1	0.1	70	61	133	0.81	60	55	109	0.58	55	53	119	0.61
200	0.3	0.1	98	82	197	0.72	78	64	182	0.62	70	61	203	0.7[illegible]

V: Cutting speed, f: Feed, d = Depth of cut, F_t = Tangential force, F_f = Feed Force, F_r = Radial force and R_a = Surface roughness

Cutting force values shown are in Newton and surface roughness in μm

Table 4: Mist-SLHT: Cutting forces and surface roughness (refined sunflower oil with 5 % of boricacid and compressed air)

V (m/ min)	f (mm/ rev)	d (mm)	MQL (Mist of refined sunflower oil with air at 2 bar)				MQL (Mist of refined sunflower with 5% by weight boricacid with air at 2 bar)			
			F_t	F_f	F_r	R_a	F_t	F_f	F_r	R_a
100	0.1	0.1	110	68	130	0.31	78	55	104	0.29
100	0.3	0.1	127	82	158	0.42	102	63	147	0.36
150	0.1	0.3	96	69	154	0.42	63	72	97	0.39
150	0.3	0.3	133	88	171	0.85	113	61	155	0.78
150	0.1	0.1	72	48	115	0.26	48	51	87	0.24
150	0.3	0.1	96	69	143	0.42	78	60	133	0.36
200	0.1	0.1	55	41	89	0.53	44	39	69	0.40
200	0.3	0.1	78	63	113	0.69	66	59	98	0.48

V: Cutting speed, f: Feed, d = Depth of cut, F_t = Tangential force, F_f = Feed Force, F_r = Radial force and R_a = Surface roughness
Cutting force values shown are in Newton and surface roughness in µm

Further, hard turning experiments (mist-SLHT)were carried out using mist of refined sunflower oil with compressed air and mist obtained with sunflower oil carrying 5% in proportion by weight boric acid and compressed air at the same cutting conditions that were used in solid assisted hard turning as shown in Table 3. Experimental results for cutting forces and surface roughness obtained under mist lubrication are shown in Table 4. Referring to Table 3, 4 and from Fig. 2 it can be seen that the feed rate and cutting speed are the most significant factors on the surface roughness. The lowest surface roughness obtained at lower feed value can be justified as at higher feed rate, larger cross sectional area is being removed which consequently increases the friction at the chip-tool interface and hence, increase in cutting forces. It can be seen that mist lubrication obtained using sunflower oil carrying 5% in proportion by weight boric acid and compressed air produced lower values of surface roughness of 0.24 µm at cutting speed of 150 m/min and lower feed and depth of cut values of 0.1 mm/rev and 0.1 mm respectively. However, at the same cutting condition, surface roughness value produced can be seen as 0.28 µm when using 5% proportion of boric acid in refined sunflower oil under solid lubricant assisted hard turning. Although, significant improvement in lowering surface roughness is not observed by using mist lubrication with solid lubricant, but, encouraging reduction in cutting forces can be seen when using mist lubrication. In turn, saving in cutting power could be achieved due to reduction in specific cutting energy when using solid lubricant assisted vegetable oil mist lubrication.

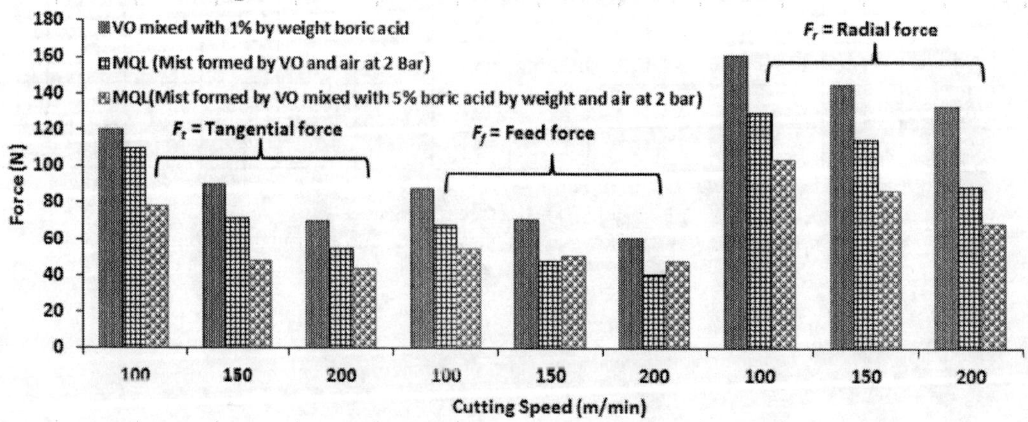

Fig. 2 Cutting forces under Vegetable oil (VO) mixed with 1% by weight of boric acid, MQL (VO+Air) and MQL (VO mixed with 5% by weight boric acid +Air)

Improvement in machining performance with solid lubricant assisted vegetable oil mist lubrication can be justified as boric acid used as solid lubricant has high thermal conductivity and low coefficient of friction aids in reducing the temperature at the interface between work piece and cutting tool. Further, adding boric acid with vegetable oil alters heat transfer due to higher thermal conductivity of solid lubricant which helped in removing heat rapidly from the cutting zone. Moreover, the micro particle size of boric acid which has higher surface area increases the heat transfer rapidly from the cutting zone. In this investigation, it has been observed that use of boric acid in vegetable oil; to be specific 5% of boric acid by weight in refined sunflower oil and that too under mist condition, i.e. Solid lubricant assisted vegetable oil mist lubrication(mist-SLHT) is most applicable cooling medium during hard turning.

4. Conclusion

In the present work, solid lubricant assisted hard turning (SLHT) experiments were performed using boric acid mixed in varying proportions mixed with refined sunflower oil. SLHT experiments were also performed under mist lubrication (mist-SLHT) with a view to assess the feasibility of variants of solid lubrication system in hard turning. Experimental results obtained during SLHT showed lower values of cutting forces and surface roughness with increase in proportions of boric acid by weight in refined sunflower oil. However, significant improvement was not seen at 10% of boric acid as compared to 5% of boric acid mixed with vegetable oil. This study suggest to use 5% of boric acid in proportion by weight in refined sunflower oil during SLHT to obtain lower values of surface roughness and cutting forces. Further, hard turning experiments performed under mist-SLHT indicated that mist lubrication obtained using sunflower oil carrying 5% of boric acid and compressed air produced lower values of surface roughness of 0.24 μm at cutting speed of 150 m/min and at lower feed and depth of cut values of 0.1 mm/rev and 0.1 mm respectively. Significant improvement in lowering surface roughness was not seen when using mist-SLHT as compared to SLHT especially at higher cutting speed of 200 m/min. However, cutting forces were observed to be lower under mist-SLHT at all the cutting conditions used in the present study. Improvement in machining performance with SLHT and mist-SLHT is seen due to higher thermal conductivity and lower coefficient of frictionof boric acid used as solid lubricant. In this investigation, it has been observed that use of boric acid (5% of boric acid by weight) mixed in refined sunflower oil under mist condition(mist-SLHT) is most feasible as a cooling medium in hard turning and saving in cutting power could be achieved due to reduction in specific cutting energy when using mist-SLHT.

5. References

[1] Satish Chinchanikar, S.K. Choudhury, "Machining of hardened steel: Experimental investigations, performance modeling and cooling techniques: A review" International Journal of Machine Tools and Manufacture,"vol. 89, pp. 95–109, 2015.

[2] Satish Chinchanikar, A.V. Salve, P. Netke, A. More, S. Kendre and Ramesh Kumar, "Comparative evaluations of surface roughness during hard turning under dry and with water-based and vegetable oil-based cutting fluids," Procedia of Material Science, vol. 5, pp. 1966-1975, 2014.

[3] Satish Chinchanikar, S.K. Choudhury, "Experimental investigations to optimise and compare the machining performance of different coated carbide inserts," Proc IMechE Part B: J Engineering Manufacture, vol. 228(9), 1104–1117, 2014.

[4] Avdesh Pal, S.K. Choudhury, Satish Chinchanikar, "Machinability assessment through experimental investigation during hard and soft turning of hardened steel" Procedia of Material Science 6 (2014) 80 – 91.

[5] Vishal Wayal, Nitin Ambhore, Satish Chinchanikar, Virendra Bhokse, "Investigation on Cutting Force and Vibration Signals In Turning: Mathematical Modeling Using Response Surface Methodology," Journal of Mechanical Engineering and Automation, vol. 5(3B), pp. 64-68, 2015.

[6] Pradip Netake, Satish Chinchanikar, "Hard turning under minimum quantity lubrication: Modeling of cutting force and surface roughness," Journal of Advanced Engineering Research, vol. 2 (Issue 1), pp. 17-22, 2015.
[7] Satish Chinchanikar, S.K. Choudhury, "Hard turning using HiPIMS-coated carbide tools: Wear behavior under dry and minimum quantity lubrication (MQL)," Measurement, vol. 55, pp. 536- 548, 2014.
[8] S.A. Lawal, I.A. Choudhury, Y. Nukman, "Evaluation of vegetable and mineral oil-in-water emulsion cutting fluids in turning AISI 4340 steel with coated carbide tools," J. Clean. Prod.,vol. 66, pp. 610–618, 2014.
[9] B. Ozcelik, E. Kuram, M. Huseyin Cetin, E. Demirbas, "Experimental investigations of vegetable based cutting fluids with extreme pressure during turning of AISI 304L", Tribol. Int.,vol. 44, pp. 1864–1871, 2011.
[10] K.P. Sodavadia, A.H. Makwana, "Experimental Investigation on the Performance of Coconut Oil Based Nano Fluid as Lubricants during Turning of AISI 304 Austenitic Stainless Steel," J. Adv. Mech. Eng.,vol. 4, 55–60, 2014.
[11] P. Deshmukh, M. Lovell, W.G. Sawyer, A. Mobley, "On the friction and wear performance of boric acid lubricant combinations in extended duration operations," Wear, vol. 260, pp. 1295–1304, 2006.
[12] P. Vamsi Krishna, R.R. Srikant, D. Nageswara Rao, "Experimental investigation on the performance of Nano boric acid suspensions in SAE-40 and coconut oil during turning of AISI 1040 steel," Int. J. Mach. Tools Manuf., vol. 50, pp. 911–916, 2010.
[13] H. Toenshoff, "Tribological aspects of hard turning with ceramic tools,"vol. 5, 1995.
[14] R.Avila, A.Abrao, "The effect of cutting fluids on the machining of hardened AISI 4340 steel," J.Material processing tech, vol. 119, pp. 21-26, 2001.
[15]A.Varadaranjan, P. Phillip, B.Ramamoorthy, "Investigation on hard turning with minimal cutting fluid Application and its comparison with dry and wet turning," Int. J. Mach tools Manuf., vol. 42, pp. 193-200, 2002.

IOP Conf. Series: Materials Science and Engineering **810** (2020) 012045 doi:10.1088/1757-899X/810/1/012045

Investigation of material removal rate on Ti-6Al-4V super alloy

Kaustab Sharma[1], Ashish Goyal[2], Ashok Kumar Sharma[3]

[1-3] Department of Mechanical Engineering, Manipal University Jaipur, Dehmi Kalan, Jaipur, Rajasthan, India, 303007

ashish.goyal@jaipur.manipal.edu

Abstract. The machining of hard material is a challenging task by the conventional machining process. The present experimental study on machining of titanium alloy by the Wire cur electric discharge machining process (WEDM). Experimental has been performed by using the pulse on-time, pulse off-time, peak current, servo voltage and wire tension as input parameters and effect of these input parameters identified on the material removal rate. The design of experimental methodology has used to perform the experiment. Total 27 experiments were performed to identify the optimum material removal rate. The analysis of variance has also applied to process parameters to investigate the significant and non-significant parameters. The experimental results show that the peak current and wire tension has the most significant parameters to influence the material removal rate.

1. Introduction

Machining of Ti-6Al-4V alloy is very tough task by traditional machining methods viz. turning, milling and grinding as of its high hardness and high melting temperature. [1-3]. Wire EDM process is the most suitable process to machine these hard materials such as titanium alloy, shape memory alloy, Inconel alloy etc. Wire electro-discharge machining process was found to be a very influential electro-thermal technique for machining Ti-6Al-4V alloy with any intricate shape. [4-6].

Manjaiah et. al found that the machined surface hardness increases with respect to the base material for longer pulse on time due to the recast layer thickness. It was concluded that the pulse on time and servo voltage have significant on material removal rate and surface roughness [7]. Goyal et. al and Goyal proposed that cryogenically treated tool electrode, current and pulse on time are the major factors that affect the material removal rate and in turn affects surface roughness also using a cryogenically treated tool electrode results in significant reduction in tool wear rate. [8-9]

Soni et. al showed that formation of micro cracks, micro globules and micro voids occur at high pulse on time [10]. Rasheed et. al found capacitance as the most influencing factor on MRR, TWR and surface roughness [11]. Kale et. al reviewed that cryogenic cooling and machining techniques improve the machining of super alloys in terms of surface integrity and tool life [12]. Feng et. al analysed that on adding Mo to NiTi there is a reduction in MRR, EWR, surface roughness, and recast layer thickness because of increased melting temperature and thermal conductivity of the material [13]. Kumar et. al reported WEDM as an adequate process to machine high strength temperature resistant materials with very little dimensional deviation, also the dimensional deviation is affected by pulse on time, pulse off time, peak current, and servo voltage [14]. Metal removal rate and surface finish were optimized by Scott et al. by explicit by design of experiment approach. The optimum machining condition has been obtaining for the MRR and SR response. Albeit many studies have been carried out in the past and reported in literatures on WEDM of Titanium alloy, but very few scholarly works have been carried out in regards of machining performance. This research focuses on WEDM machining of Ti-6Al-4V for the fabrication of slots for the industrial applications. The figure 1 presents the cause and effect diagram of the WEDM parameters. The figure shoes the important WEDM parameters that are being utilized to maintain during the machining to get the better response from the machine.

2. Experimental methodology:

The Elektra Maxicut 734, Electronica wire EDM machine has used for the experiment purpose. The figure 2 presents WEDM machine which is used for the experimental work. The zinc coated brass wire used as a tool electrode. the diameter of wire was 0.25mm. Ti-6Al-4V alloy has used as a specimen. The dimension of plates was 176mm x 55mm x 2.5mm. The peak current, pulse on time, pulse off time, wire tension and voltage has used as machining parameters and the influence of these machining parameters has been found on material removal rate. The material MRR has calculated as the ratio of the difference of weight of the workpiece before and after machining to the machining time and density of the material. The figure 3 shows the wire edm machining for the fabrication of slots. Figure 4 shows the fabricated WEDM slots.

Therefore, the MRR for the WEDM operation is calculated as

Material removal rate = $(Wb - Wa)/\ \rho * t \quad mm^3/s$ (1)

Whereas, Wb = Weight of workpiece before machining.

Wa = Weight of workpiece after machining.

t = Machining time in seconds

ρ = Density of titanium grade-5 = $4.42*10^{-3}$ g/mm^3

The influence of various operation parameters like pulse-on-time, pulse-off-time, current, wire feed rate, wire tension was studied on WEDM on the response i.e. material removal rate. Table 1 presents the selected control factors and their levels fir the machining purpose.

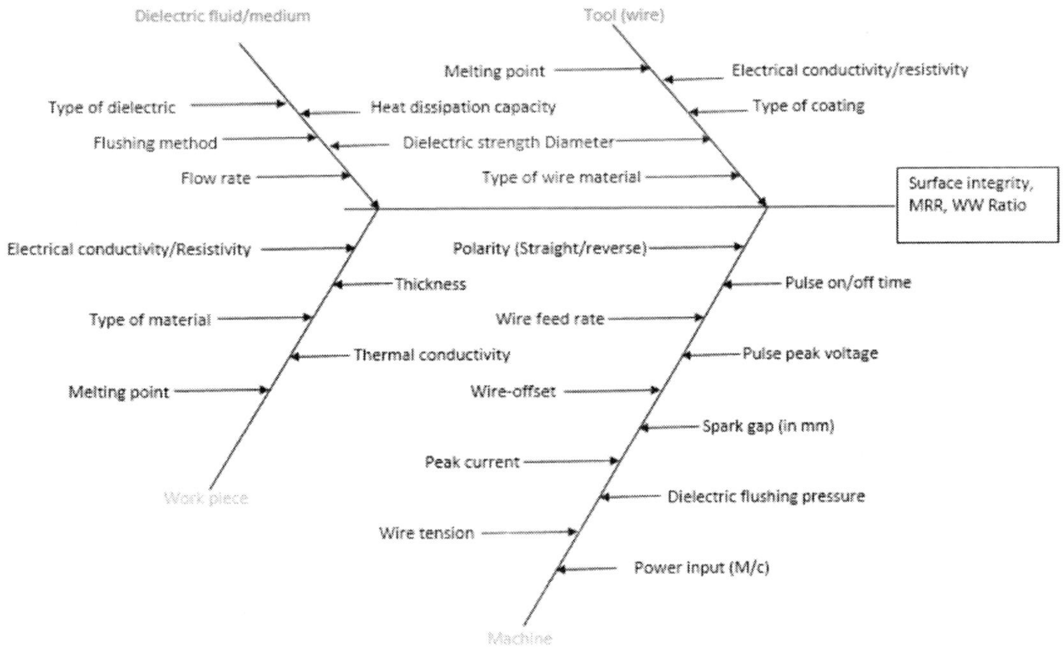

Figure 1: cause and effect diagram of WEDM process parameters

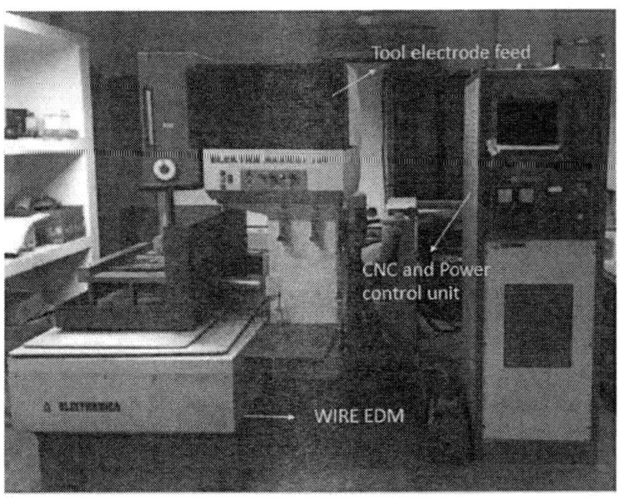

Figure 2: WEDM machine set up

IOP Conf. Series: Materials Science and Engineering **810** (2020) 012045 doi:10.1088/1757-899X/810/1/012045

Figure 3: Fabrication of slots

3. Results and discussion:

The experimental result for material removal rate of Ti-6Al-4V with different process parameters is presented. Based on results, the Signal to Noise (S/N) ratio plot for MRR has presented in figure 5. It is inferred that, the MRR increases with increase in pulse on time (T_{on}). Increase in Ton results into high discharge energy that increases the MRR. it is observed that MRR initially decreases with increase in peak current and then starts to rise. Increase in settings of peak current viz. 5 & 15 A respectively. This may be attributed to the availability of higher discharge energy, at high current levels 5 and 15 A.

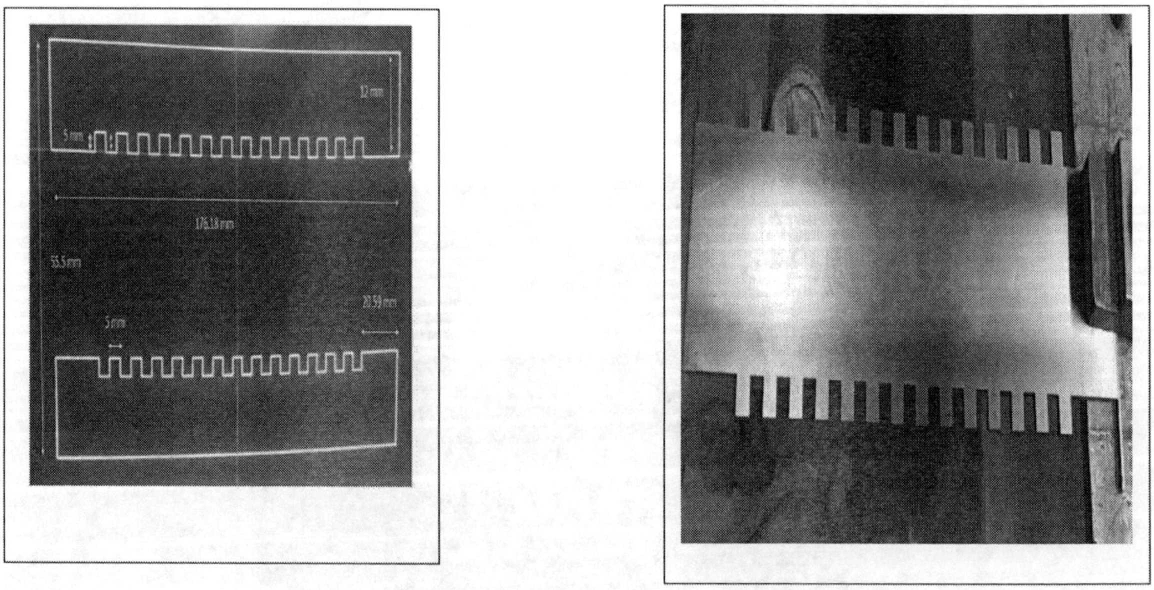

Figure 4: fabricated slots by WEDM process

Table 1: Control factor and their levels

S.No.	**Control factor**	**Level 1**	**Level 2**	**Level 3**	**Units**
1	Peak current	5	10	15	A
2.	Pulse on time	115	145	165	μs
3.	Pulse off tome	60	70	80	μs
4.	Wire tension	7	8	9	N
5.	Voltage	3-	50	70	V

Higher discharge energy rapidly overheats the work material which promotes larger melting and vaporization of the work material to generate large craters. This observation is in consonance to the past observations made by the researchers (Mohan et al. 2004). It is observed that MRR increases with increase in current and then after it start decreasing. This may be due to the presence of enough the machining debris in the di-electric fluid, which modifies the plasma channel, enlarging and widening the plasma channel, which reduces the MRR. The voltage doesn't have any significant effect on MRR. As shown in table 3, among all the process parameters considered, peak current is the most significant in terms of MRR because it has the largest impact and has been ranked one, while wire tension is the second most important process parameter due to its dominant control over the spark discharge energy leading to an increase in MRR. The graph of the main effects of one noise factor and five control factors based on the S/N ratio for the MRR is shown in figure 6.1. The optimal parametric settings are A3, B2, C1, D3 and E3.

Table 2: L27 Orthogonal array for the input parameters and MRR

S.no.	**Ton**	**Toff**	**Voltage**	**Ip**	**Wt**	**MRR**
1.	115	60	30	5	7	0.1289
2.	115	60	30	5	8	0.1302
3.	115	60	30	5	9	0.1268
4.	115	70	50	10	7	0.134
5.	115	70	50	10	8	0.1364
6.	115	70	50	10	9	0.1421
7.	115	80	70	15	7	0.1446
8.	115	80	70	15	8	0.1493
9.	115	80	70	15	9	0.1527
10.	145	60	50	15	7	0.1772
11.	145	60	50	15	8	0.1611

12.	145	60	50	15	9	0.1869
13.	145	70	70	5	7	0.1803
14.	145	70	70	5	8	0.1698
15.	145	70	70	5	9	0.1712
16.	145	80	30	10	7	0.1844
17.	145	80	30	10	8	0.1731
18.	145	80	30	10	9	0.1573
19.	160	60	70	10	7	0.1923
20.	160	60	70	10	8	0.2100
21.	160	60	70	10	9	0.228
22.	160	70	30	15	7	0.3424
23.	160	70	30	15	8	0.3689
24.	160	70	30	15	9	0.3976
25.	160	80	50	5	7	0.2753
26.	160	80	50	5	8	0.2963
27.	160	80	50	5	9	0.3211

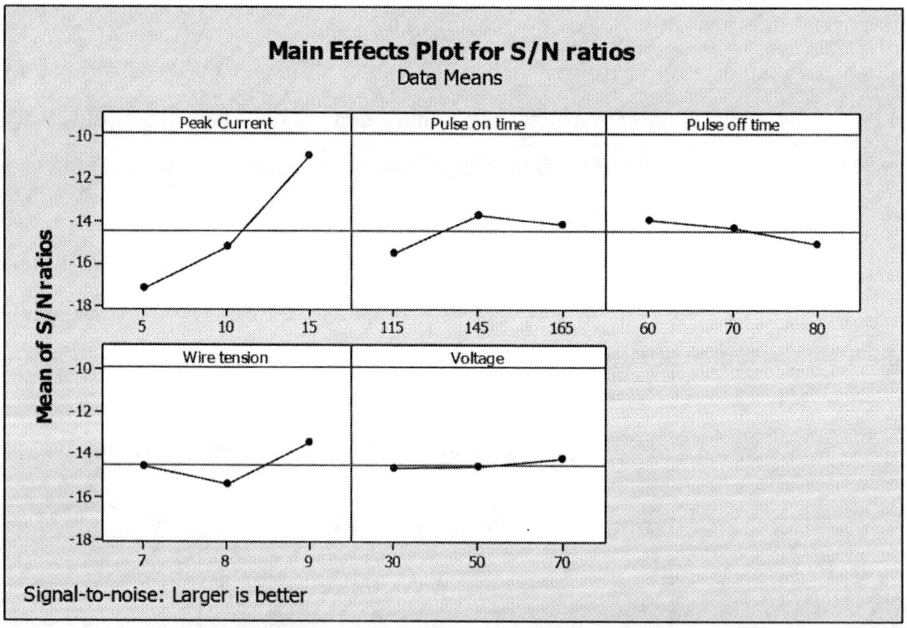

Figure 5: S/N plot for material removal rate

Table 3. Response table for material removal rate (Means)

Level	Peak current	Pulse on time	Pulse off time	Wire tension	Voltage
1	0.1383	0.1713	0.2233	0.2000	0.1955
2.	0.1735	0.2270	0.2034	0.1731	0.1995
3.	0.2924	0.2060	0.1776	0.2312	0.0138
Delta	0.1541	0.0557	0.0457	0.0581	0.0138
Rank	1	3	4	2	5

Table 4 shows the results of analysis of variance based on the S/N ratio for MRR. The S/N ratio calculated for MRR from the observed experimental data has been used to analyse ANOVA and F-test, in order to determine the effects of each parameter on the MRR. It is found that peak current, pulse on time, pulse off time and wire tension have a large impact on the MRR at 95% confidence interval level since p-value is less than 0.05. Figure 6 shows the interaction plot for S/N ratio of the MRR between the process parameters and the material removal rate. It can be clearly observed that, since the lines are not parallel to each other, there is some interaction between the input parameters and the material removal rate. The R-Seq (adj) value is 96.81% which shoes the present model have lie significant for the wire EDM process parameters.

Table 4: Analysis of Variance for MRR

Source	DF	Seq SS	Adj SS	Adj MS	F	P
Peak current	2	0.117397	0.117397	0.058699	298.01	0.000
Pulse on time	2	0.014246	0.014246	0.007123	36.16	0.000
Pulse off time	2	0.009455	0.009455	0.004727	24.00	0.000
Wire tension	2	0.015229	0.015229	0.007615	38.66	0.000
Voltage	2	0.000910	0.000910	0.000455	2.31	0.131
Error	16	0.000910	0.000910	0.000455		
Total	26	0.160389				

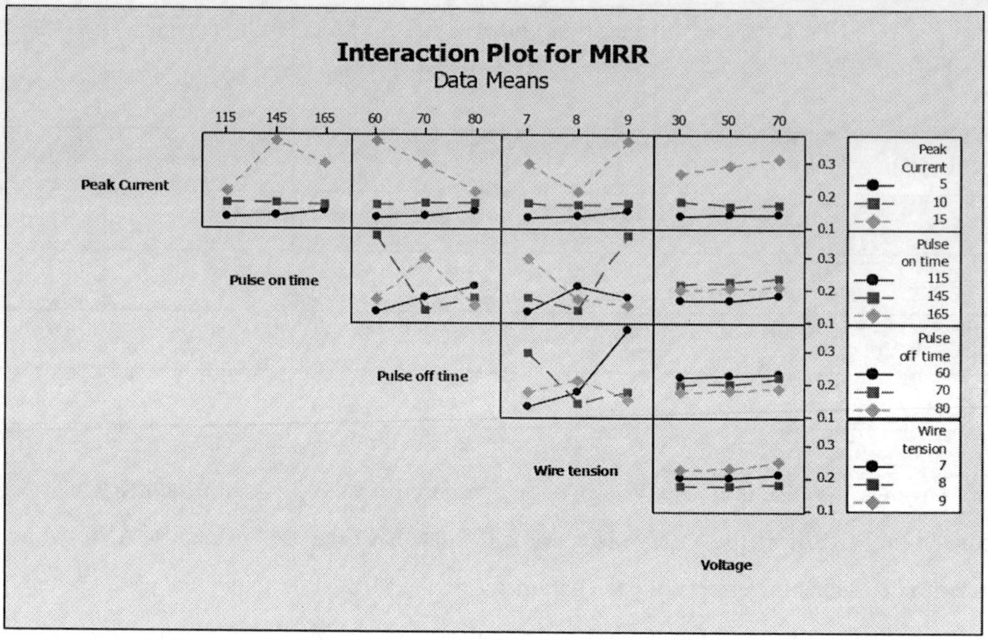

Figure 6: Interaction plot for the MRR

4. Conclusion:

- The peak current parameter has most significant effect on the material removal rate followed by the wire tension. The voltage is found the insignificant parameters for the present study.
- The material removal rate (MRR) directly increases with increase in pulse on time (Ton) and peak current (IP) while decreases with increase in pulse off time (Toff) and servo voltage (SV).
- Through this investigation, we have quantified MRR as a function of various parameters involved in the material removal process. These results will be of significant help in the future to determine the material removal parameters for most efficient results.
- Super alloys cannot normally be machined properly. With the use of Wire EDM process, super alloys which have not been used extensively can now be investigated.

Reference:

- Lin, H. C., K. M. Lin, and I. S. Cheng 2000. "The wire electro-discharge machining characteristics of TiNi shape memory alloys." High Temperature Material Processes: An International Quarterly of High-Technology Plasma Processes 4, no. 4.
- Kuriakose, Shajan, Kamal Mohan, and M. S. Shunmugam 2003. "Data mining applied to wire-EDM process." Journal of Materials Processing Technology 142, no. 1 182-189.
- Liao, Y. S., J. T. Huang, and Y. H. Chen 2004. "A study to achieve a fine surface finish in Wire-EDM." Journal of Materials Processing Technology 149, no. 1-3 165-171.

- Lok, Y. K., and T. C. Lee 1997. "Processing of advanced ceramics using the wire-cut EDM process." Journal of materials processing technology 63, no. 1-3 839-843.
- Peng, W. Y., and Y. S. Liao 2003. "Study of electrical discharge machining technology for slicing silicon ingots." Journal of Materials Processing Technology 140, no. 1-3 274-279.
- Yan, Biing Hwa, Hsien Chung Tsai, Fuang Yuan Huang, and Long Chorng Lee 2005. "Examination of wire electrical discharge machining of Al2O3p/6061Al composites." International Journal of Machine Tools and Manufacture 45, no. 3 251-259.
- Manjaiah, M., Narendranath, S., Basavarajappa, S., & Gaitonde, V. N. 2016. Influence of process parameters on material removal rate and surface roughness in WED-machining of Ti50Ni40Cu10 shape memory alloy. International Journal of Machining and Machinability of Materials, 18(1-2), 36-53.
- Goyal, A. 2017. Investigation of material removal rate and surface roughness during wire electrical discharge machining (WEDM) of Inconel 625 super alloy by cryogenic treated tool electrode. Journal of King Saud University-Science, 29(4), 528-535.
- Goyal, A., Pandey, A., & Sharma, P. 2017. Machinability of Inconel 625 Aerospace Material Using Cryogenically Treated WEDM. In Solid State Phenomena (Vol. 266, pp. 38-42). Trans Tech Publications.
- Soni, H., Sannayellappa, N., & Rangarasaiah, R. M. 2017. An experimental study of influence of wire electro discharge machining parameters on surface integrity of TiNiCo shape memory alloy. Journal of Materials Research, 32(16), 3100-3108.
- Rasheed, M. S., Al-Ahmari, A. M., El-Tamimi, A. M., & Abidi, M. H. 2012. Analysis of influence of micro-EDM parameters on MRR, TWR and Ra in machining Ni-Ti shape memory alloy. International Journal of Recent Technology and Engineering, 1(4), 32-37.
- Kale, A., & Khanna, N. 2017. A Review on Cryogenic Machining of Super alloys Used in Aerospace Industry. Procedia Manufacturing, 7, 191-197.
- S. F. Hsieh, M. H. Lin, S. L. Chen, S. F. Ou, T. S. Huang, and X. Q. Zhou, 2016. "Surface modification and machining of TiNi/TiNb-based alloys by electrical discharge machining," Int. J. Adv. Manuf. Technol., vol. 86, no. 5–8, pp. 1475–1485.
- Kumar, A., Kumar, V., & Kumar, J. 2013. Effect of machining parameters on dimensional deviation in wire electro discharge machining process using pure titanium. Journal of Engineering and Technology, 3(2), 105-112.
- D. Scott, S Boyna, K P Rajurkar, 1991 Analysis and optimization of parameter combinations in WEDM, Int. J. Prod. Res. 29. 2189–2207.

Improving the efficiency of the Lay-Up process in the manufacture of Solar Panels through time-study

Neepa M. Patel[1] and Vatsal Rathore[2]

Mechanical Engineering Department, MPSTME, NMIMS University, Mumbai – 400056, India

[1]neepa.patel@nmims.edu

[2]vatsal.rathore46@nmims.edu.in

Abstract. Lay-Up is a manual process conducted during the manufacture of solar panels where the multiple layers are set together by hand. Since this process is done manually, there is no standard time period for it. Additionally, errors in process can increase production time needlessly. This research paper employs time-study to establish a standard operation time for Lay-Up aimed at singling out steps that cause loss of valuable time, efforts and resources. The time study is carried out by making a note of all the operations involved in Lay-Up and breaking them down into individual steps. The measurements are made using a digital stop watch and time sheet.

1. Introduction

Time study is a direct and continuous observation of a task, using a timekeeping device to record the time taken to accomplish a task. It is a core tool which is used by the managers in the industrial sector to enhance the performance or the operational efficiency. This is done by breaking down the work into simpler units and setting the execution benchmarks.

This research was carried out at the central factory of one of India's leading manufacturer of solar panels. It was aimed at reducing production time and hence, increasing productivity in general. The timekeeping device used here is the digital stopwatch in mobile phones, while timesheet is used to record the data and maintain it.

2. Procedure for Manufacturing Solar Panels

The general flow of the manufacturing process for solar panels is provided below in Fig 1.

The first step is Tabbing and Stringing which takes cells and metal ribbons, and stacks them into strings (Fig 2). The strings are then laid onto EVA-covered tempered solar glass and layered with EVA (ethyl-vinyl acetate) sheets and a polymer backsheet as part of the **Lay-up** process. The appearance of the layers in a panel are as shown in Fig 3. The research was particularly carried out to improve the efficiency of the Lay-Up process.

The steps following the Lay-Up process are multiple testing stages, lamination & trimming and the attachment of a junction box alongwith an aluminium frame.

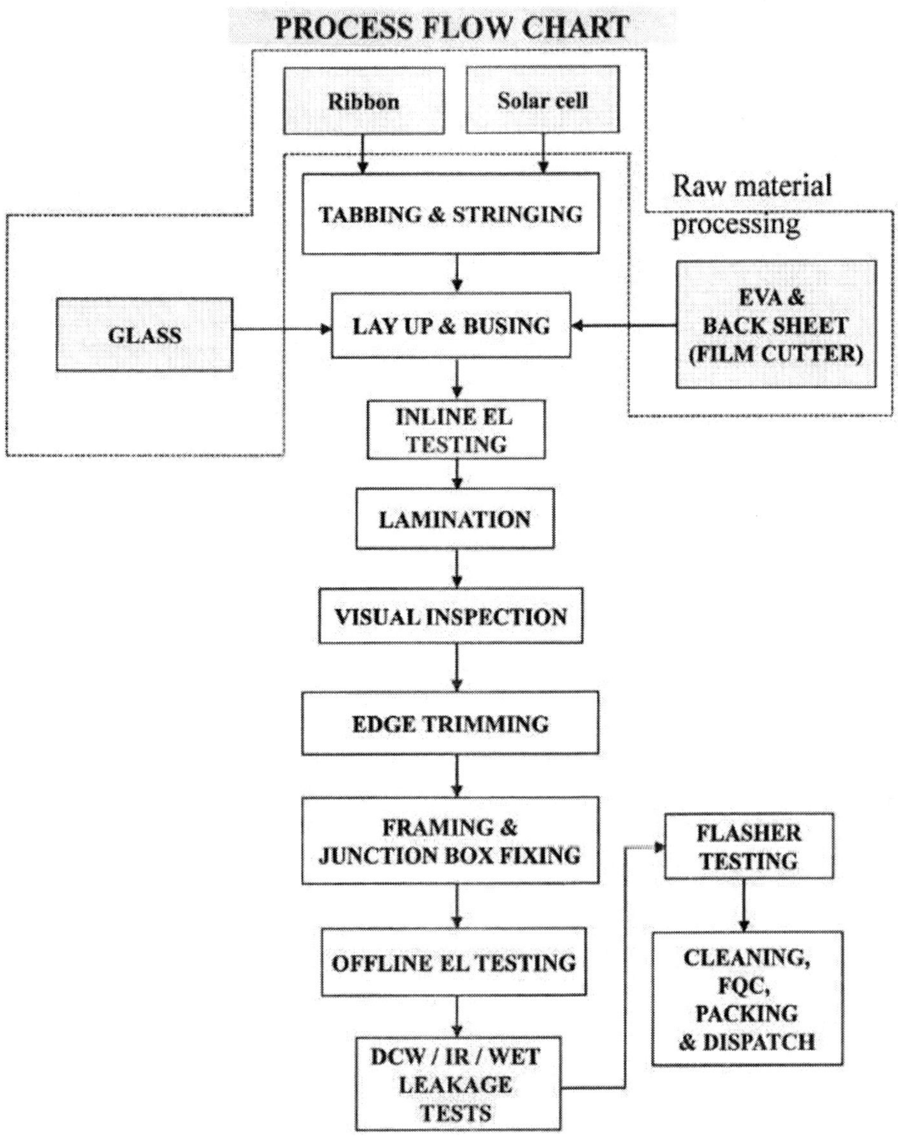

IOP Conf. Series: Materials Science and Engineering **810** (2020) 012046 doi:10.1088/1757-899X/810/1/012046

Figure 1: Flowchart of manufacture of solar panels from stringing to testing.

Figure 2: Cells laid out in strings and connected via metal ribbons and bus-bars

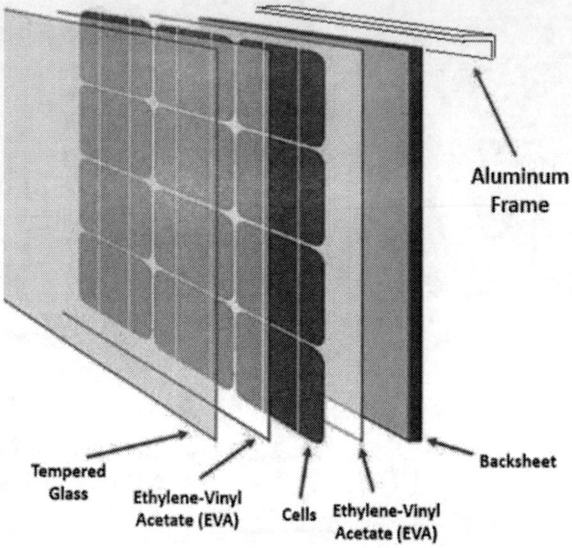

Figure 3: Various layers of a solar panel

3. Time Study Procedure for Layup

- Observation of the Lay-up process until familiarity was felt with the process.
- The workers would load the EVA-covered glass onto the Lay-Up table; the strings were already arranged on it during the Tabbing & Stringing process. This was the first step of the **Lay-Up** process.
- The workers would tear up small pieces of adhesive tape and apply them on the cells to hold them in place. The time recording for this step was divided into two sub-steps – tearing of tape, and its deposition on the cells.
- Placements of a second layer of EVA sheet, polymer back-sheet and soldering strips were recorded in a similar way.
- The soldering operation of bus-bars with the string ribbons was recorded by breaking the operation into picking up the soldering iron, soldering, and placing the iron back.
- Finally, allowances of 10 seconds and 5 seconds were given for the top-edge and bottom-edge operations respectively.

IOP Conf. Series: Materials Science and Engineering **810** (2020) 012046 doi:10.1088/1757-899X/810/1/012046

4. Data Recording

Two workers are required in the lay-up process- one works at the top-edge while the other work at the bottom edge. The entire activity is recorded by the time study method in the table below.

Activity ↓	**Observation** ⟶	**I (s)**	**II (s)**	**III (s)**	**IV (s)**	**Average Time Taken (s)**
			Top Edge			
Collection of cells-glass arrangement		1.55	3.00	1.89	2.00	2.11
Tape	Collection	2.19	6.38	9.27	4.39	5.55
	Deposition	10.10	7.13	9.94	8.42	8.89
Back-sheet	Collection	3.05	3.37	4.12	4.39	3.73
	Deposition	3.24	3.24	2.73	2.10	2.82
Bus Bars	Collection	3.31	4.79	3.18	3.38	3.66
	Deposition	5.07	7.13	6.10	5.56	5.96
Soldering-1	Pick Up	1.30	1.51	2.00	1.45	1.56
	Application	15.27	10.47	10.33	16.65	13.18
	Put Down	2.24	3.60	3.00	1.96	2.7
Probes 1&2	Collection	1.8	2.14	2.01	2.21	2.04
	Deposition	2.81	1.92	2.02	3.21	2.49
EVA strip	Collection	1.5	1.20	1.71	1.53	1.48
	Deposition	3.35	2.33	3.71	4.00	3.35
Back-sheet strip	Collection	1.72	1.73	1.14	1.57	1.54
	Deposition	3.97	4.61	4.54	3.88	4.25
Soldering-2	Pick Up	1.42	2.36	2.78	2.64	2.3
	Application	7.32	10.40	8.72	10.08	9.13
	Put Down	1.94	2.64	1.32	1.16	1.76
Probes 3&4	Collection	1.57	2.55	1.27	2.21	1.9
	Deposition	2.65	2.36	2.89	2.71	2.63
EVA	Collection	2.65	2.68	2.21	2.92	2.61
	Deposition	2.72	2.68	3.82	3.31	3.13
Back-sheet	Collection	1.11	1.10	1.89	1.00	3.05
	Deposition	1.70	1.09	2.65	2.16	1.9
Tape	Collection	3.68	3.47	2.48	3.43	3.26
	Deposition	4.42	3.44	3.11	2.64	3.40
Allowance		10	10	10	10	10
Aggregate Top Edge Production		100.1	109.32	110.83	110.96	**109.38**

Bottom Edge						
Collection and Inspection		3.03	6.24	8.05	5.74	5.77
Tapes	Collection	2.17	1.71	1.65	2.04	1.89
	Deposition	5.41	5.42	5.71	6.90	5.86
Back-sheet	Collection	2.24	3.15	1.65	2.05	2.27
	Deposition	4.86	3.27	3.81	3.00	3.73
Bus bars	Collection	2.25	3	3.89	3.4	3.14
	Deposition	6.15	7.80	8.84	10.18	8.24
Soldering	Pick Up	2.76	1.52	2.39	2.88	2.39
	Application	18.20	20.74	17.96	24.09	20.25
	Put Down	3.09	1.68	2.51	1.5	2.19
Allowance		5	5	5	5	5
Aggregate Bottom Edge Production		55.16	59.53	61.46	66.78	**60.73**

Allowance

Since no manual process can be performed identically every time, an allowance is made to account for discrepancies that might creep in. Allowances in time sheets are provided according to scope of work and the steps involved.

An allowance represents time lost due to personal factors, shift adjustments, improper equipment and fatigue.

It includes:

- **Transfer Time:** Lay-up occurs on three in-line tables and when the next in line finishes up their module, the one on the previous table is transferred onto their table, so that no table stands idle.
- **Fatigue:** Since the work is manual, the workers are bound to get tired and get hampers the production time.
- **Inefficiency:** Due to sloppy work, slipping of equipments or materials, time is lost which is allowed for here.
- **Ineffective activities:** Some steps in an operation can be performed more efficiently but non-willingness to change methods, increases productive times.

5. Observations & Calculations

1. The lay-up procedure involves the operator to make a slit in the final EVA sheet for the probes to pass through.
2. The back-sheet that's placed on it has a slit made already during its cutting.
3. Between the picking up of the cutter, slitting and putting it down, roughly 1.5 seconds pass.

Going by the popular adage, "Time is Money", saving time saves money. Reduced production time means time saved which leads to more production in the saved time and hence, less wastage of money and generation of revenue.

IOP Conf. Series: Materials Science and Engineering **810** (2020) 012046 doi:10.1088/1757-899X/810/1/012046

As observed during the lay-up, placement of EVA sheet takes up around 1.5 seconds from the picking up of the cutter to putting it back down. This happens for every single module that is manufactured in either of the plants.

Presented here is a rough account of how by saving a mere 1.5 seconds of a process saves the company so much in the longer run. While other processes too contribute to wasted time, this process makes the maximum contribution.

If the slit on EVA is made prior to production in the same way as the back-sheet, a lot of effort can be saved in total. It can be explained according to the following calculation:

1) Total time lost per module: **1.5** seconds
2) Total time taken for Lay-Up of 1 module: **109.38** seconds (=**110 seconds**)
3) Total number of production hours in a day: **24** (Source: Production Dept.)
4) Modules manufactured in a day = **1200** (Source: Production Dept.)
5) Average Time taken to manufacture one module = (24 x 60) / 1200
   = **1.2** minutes
   = **72** seconds
6) Total time lost every day = 1200 x 1.5 = **1800** seconds
   = **30** minutes
7) Total additional modules that can be manufactured in lost time = 1800 /72
   = **25**
8) Increased growth = 25/1200 = **2%** (approx.)
9) While 2% growth appears less, the total income is remarkable.
10) Approximately selling price of one module = **₹12000**
11) At the same rate, total additional revenue generated in year = ₹ 12000x25x365
    = **₹ 10, 95, 00, 000**

Hence, by just saving **1.5** seconds in a **single** process, the effort, time and revenue that can be saved = **₹ 10, 95, 00, 000**

6. Conclusion

By using Time Study Technique, 1.5 seconds saved in a single process, one can save **Rs. 10, 95, 00,000 annually.** The following are some recommendations that can lead to increased productivity and human efficiency:

- Ear-muffs for workers.
- Vibration padding to reduce noise pollution due to resonance as well as frictional wearing.
- Switching the glass loading and un-loading outlets in Stringer to reduce transportation time.
- Placement of stools with adjustable height at lay-up station to reduce fatigue in employees.
- Since the position of back-sheet strips and bus-bars is the same irrespective of position of strings, placing them **during** glass loading before Lay-up saves much time and reduces effort.

IOP Conf. Series: Materials Science and Engineering **810** (2020) 012046 doi:10.1088/1757-899X/810/1/012046

References:

[1] Naik Mihir, Agola Jaydeep, Rathod Hiren; Time and Motion Study: A Case Study on a Commercial Building at Kadodara. *IJARESM*.
[2] Jigar H Balar , SNPIT & RC.; Mr. Hiren A. Rathod, SNPIT & RC.; Mr. Rushabh Shah, SNPIT & RC: Effective Time And Motion Study on Construction Project: A Case Study of Surat City. IJSRD
[3] Finkler, Steven A., A Comparison of work-sampling and time-and-motion techniques for studies in health services research, Health Service Research Journal, Chicago. Dec 1993. Vol. 28, Issue 5.
[4] John W. Twidell and Anthony D. Weir, Renewable Energy Sources: Taylor and Francis, 2006.
[5] G.D. Rai, Non-Conventional Energy Resources: Standard Publisher, 2009.
[6] Marvin E. Mundel and David L. Danner, Motion and Time Study – Improving Productivity, 7th Edition: Pearson, 1994.
[7] International Labour Office, Introduction to Work Study, 4th Edition, 1992.

IOP Conf. Series: Materials Science and Engineering **810** (2020) 012047 doi:10.1088/1757-899X/810/1/012047

Design and Development of Masticatory Robot

1. Binsu Babu 2. Dr. Priam V Pillai

1. Mechanical Engg. Dept., Pillai College of Engineering, New Panvel, India
2. Mechanical Engg. Dept., Pillai College of Engineering, New Panvel, India

e-mail address : binsu18@student.mes.ac.in

Abstract. Design and development of a masticatory robot to measure the forces acting on teeth during mastication. Piezoelectric film bed will be used to measure the forces acting on tooth during mastication. On top of the piezoelectric film bed the lower jaw tooth will be mounted. The lower jaw will be having the upward and downward motion whereas the teeth on the upper jaw will remain fixed. Load cells will be placed on the upper jaw. Load cells would help in recording the forces acting on teeth during mastication of food. During mastication of food, due to chewing of food, the piezoelectric film bed will be deformed which will produce an electric charge that can be mathematically modelled into force by using Labview software.

1. Introduction

Human tooth is subjected to different kinds of forces throughout mastication. In dental analysis, these forces has been recorded as a variable to assess the effectuality of varied dental procedures like prostheses, orthodontic treatment, to review the results of deformities and pathologies on the masticatory system like malocclusion, temporomandibular disorders [1]. Correct determination of forces acting on teeth would help in occlusal treatments like occlusal adjustment, restorative treatment, orthodontics and orthognathic surgery [2]. It has been observed that some people who have dental implants have been reported to be failing within months or a few years after the dental implant has been done. Measurement of forces and moments acting on the implants during chewing of different textures of food will help in reducing the failure rate of dental implants. It will also help in selection of proper materials for dental implants

In piezoelectric films (PEF) the charge produced varies linearly with load and is independent of loading frequency, due to this PEF are used as force measurement devices. Recently, manufacturing of piezoelectric material in the form of thin foil sheets have been possible [1]. Using these foils as the active element, an occlusal force transducer can be fabricated in less than 2 mm thickness. It helps to measure occlusal forces with lesser jaw separation than in bite force devices using strain gauge transducers. These foils can be cut into any desired shape and size [1]. Devices using piezoelectric transducers have been used for measuring bite force in several studies [1]. In 1977 R.M.Mansour design and developed piezoelectric transducers (quartz crystals, 1.5 mm thick) for in vivo measurement of biting forces in humans [3]. Osteogenic response is accelerated by inducement of piezoelectric charges by the application of pulses of force to teeth. Klempler et al. in 1979 designed and developed pulsating force induced transducer to determine orthodontic tooth movement of maxillary molars. The tooth movement was converted into electrical signals with the help of a strain

IOP Conf. Series: Materials Science and Engineering **810** (2020) 012047 doi:10.1088/1757-899X/810/1/012047

gauge connected to the sensor [4]. In 1991 Douglas et al. developed and configured a silver inked piezoelectric film transducer for dental occlusal analysis [5]. Occlusion Analysis using piezoelectric film transducer was conducted in 2004 by Richter et al. An invivo occlusion analysis was carried out by placing piezoelectric force transducer foils of 33μm thick between pairs of teeth in centric occlusion and patients were instructed to bite as hard as possible for 1 second [6]. On the basis of biomechanical conditions T. Kawaguchi et al. in 2007 developed a piezoelectric force transducer which was inserted into a tooth crown for invivo analysis and measurement of load applied on the teeth[7].

Mastication process involves breakdown of foods due to chewing. To analyze the food dynamics and assess the masticatory efficiency the entire mastication process must be made continuously. Mathematical models have been developed based on the biomechanics of the masticatory system to simulate the human chewing process. The human chewing process has been simulated by developing mathematical models. Development of mathematical models for assessing complex masticatory efficiency is very hard. Development of a masticatory robot that can reproduce the actual mastication process and also help in collecting the detailed information pertaining to the forces applied during mastication and dynamics of food breakdown is thus essential. Temporalis, masseter and pterygoid are the three major muscles responsible for mastication. Keiser et al in 2005 developed a kinematic model of the masticatory robot. The major muscles of mastication (temporalis, masseter and pterygoid) are represented as linear actuators. The upper and lower jaw used in the model is made entirely of aluminum, and the model is maintained at equilibrium using spring-loaded cylindrical joints. Two types of jaw movements such as clenching and grinding movements were performed by the robotic model [8]. In February 2008 Xu et al. designed and developed a linear actuator which represented the temporalis muscle. But as the mechanism of the linear actuator was quite bulky it could not be used as a a linear actuator for the other two muscles of mastication. To overcome this issue in May 2008 Xu et al. developed a compact 6-RSS parallel mechanism mastication robot.[9-10]. In 2018 S.-J. et al. designed and developed a masticatory robot that could directly measure the interaction forces between each tooth and food[11]. The aim of the present study is to use piezoelectric film for measuring the forces and develop an cost effective masticatory robot.

2. Proposed Design

Development of masticatory robot in PCE Panvel research lab capable of measuring forces and moments acting on teeth during mastication of food. A CAD model of the masticatory robot will help in defining proper mechanism to the robot. Once the CAD model is ready the next plan is to develop a prototype of the masticatory robot to help in analyzing any difficulties that may lie during actual development of the masticatory robot. 3D printer would be used for manufacturing the majority of components of the prototype robot. The control system and motor control will be designed on virtual instrument software like LabVIEW. Once the prototype design is fully functional, manufacturing of the actual masticatory robot will be done. The figure below shows the proposed design of the mastication robot.

The teeth on the upper jaw and lower jaw are modelled as cylinders with diameter and height equivalent to the molar teeth. There are a total of six teeth on the upper and lower jaw. The lower jaw is fastened on a sliding base which is coupled with the stepper motor to allow the upward and downward motion of the lower jaw. A piezoelectric film bed is mounted beneath the lower jaw. The upper jaw remains fixed and is connected with load cells. All the different components like force sensors, motors, data acquisition system etc. will be interfaced in LabView to collect various data and the collected data will be calibrated for the desired output.

IOP Conf. Series: Materials Science and Engineering **810** (2020) 012047 doi:10.1088/1757-899X/810/1/012047

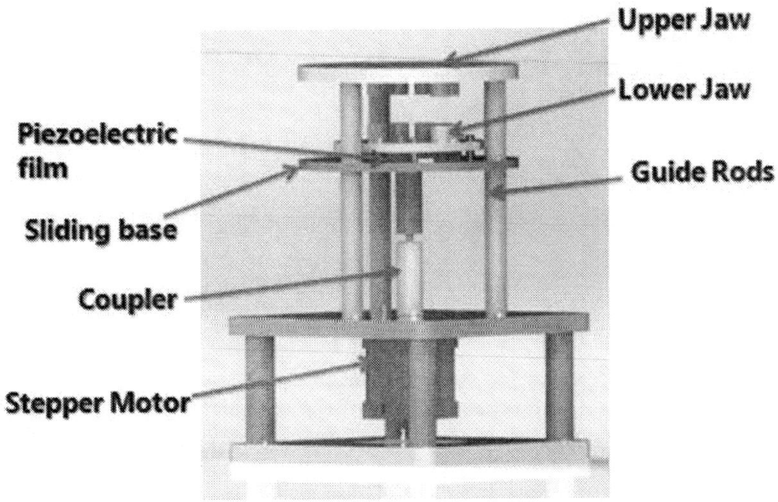

Figure 1: Proposed design of the masticatory robot.

3. References

[1] "Bite Force Recording Devices - A Review," vol. 11, no. 9, pp. 1–5, 2017.
[2] R. L. Sakaguchi, B. S. Wenande, R. Delong, and G. C. Andersonb, "Piezolectric Film Transducer for Dental Occlusal Analysis," vol. 10, pp. 145–151, 1992.
[3] R. M. Mansour, "Piezoelectric transducers for oral force monitoring," *Journal of medical engineering & technology*, vol. 1, pp. 95-97, 1977.
[4] E. Shapiro, F. W. Roeber, and L. S. Klempner, "Orthodontic movement using pulsating force-induced piezoelectricity," *American journal of orthodontics*, vol. 76, pp. 59-66, 1979.
[5] R. L. Sakaguchi, B. S. Wenande, R. Delong, G. C. Andersonb and W. H. Douglas, "Piezolectric Film Transducer for Dental Occlusal Analysis," vol. 10, pp. 145–151, 1992.
[6] Kurt Rottner and Ernst-Jürgen Richter, "Effect of occlusal morphology on the accuracy of bite force measurements using thin film transducers.," *International Journal of Prosthodontics*, vol. 17, 2004.
[7] T. Kawaguchi, T. Kawata, T. Kuriyagawa, and K. Sasaki, "In vivo 3-dimensional measurement of the force exerted on a tooth during clenching," *Journal of biomechanics*, vol. 40, pp. 244-251, 2007.
[8] WeiLiang Xu, J. Bronlund, and Jules Kieser, "A robotic model of human masticatory system for reproducing chewing behaviours," *IEEE Robotics and Automation Magazine*, vol. 12, pp. 90-98, 2005.
[9] W. L. Xu, J.-S. Pap, and J. Bronlund, "Design of a biologically inspired parallel robot for foods chewing," *IEEE Transactions on Industrial Electronics*, vol. 55, pp. 832-841, 2008.
[10] W. L. Xu et al., "Kinematics and experiments of a life-sized masticatory robot for characterizing food texture," *IEEE Transactions on Industrial Electronics*, vol. 55, pp. 2121-2132, 2008.
[11] Seung-Ju Lee, Bum-Keun Kim, Yong-Gi Chun, and Dong-June Park, "Design of mastication robot with life-sized linear actuator of human muscle and load cells for measuring force distribution on teeth," *Mechatronics*, vol. 51, pp. 127-136, 2018.

Metallurgical investigations on corrosion behavior of simple and heat treated duplex stainless steel 2205 exposed to corrosive media

A Y Chaudhari and D D Deshmukh

MET's Bhujbal Knowledge City, Institute of Engineering, Nashik, Maharashtra, India 422010.

camol1983@gmail.com, dhirgajanan@gmail.com

Abstract: Duplex stainless steel 2205 has proven to be the most acceptable material for marine and saline applications because of its superior corrosion resistance with good erosion and wear resistance. However, the disadvantage of stainless steel is its sensitivity to local corrosion in a solution containing chloride, which limits their use in seawater systems. The effect of heat treatment on corrosion properties of stainless steels can be studied in marine or chloride environments and investigation of changes in microstructure and corrosive property. This study presents the consequence of heat treatment on the microstructure and corrosion resistance and systematic approach to examine the corrosion behaviour of plain DSS 2205. Precipitation of secondary parts in DSS 2205 led to an extreme reduction in its corrosion resistance and another factor contributing to changes in ferrite and austenite ratios where preferential corrosion occurred in the DSS 2205 ferrite phase. The austenite phase shows better corrosion resistance than the ferrite phase due to maximum Ni content than ferrite. Cr is the most active element in the passive film to develop the resistance of DSS 2205 to localized corrosion. The heat treatment about 1080°C to DSS 2205 improves the corrosion resistance of steel comparatively.

1. Introduction

The degradation of materials due to chemical reactions with their environment is term as corrosion [1]. Corrosion of a material malfunction causes a gradual deterioration of its properties until it fails to meet its objectives and necessities to be replaced. Moreover, corrosion causes aesthetic problems and increasing costs of repair and rebuilding [2]. In order to succeed in corrosion protection it is important to know the fundamentals of corrosion reactions and what are their main causes. The most naturally occurring state of a material is its lowest energy states and for metals this is usually an oxide. This means that naturally all metals exist as oxides and extensive processing is required to convert the metal oxides of pure metals. Therefore, corrosion of a material can be simply defined as the driving force towards its natural, lowest energy state. If not protected properly, any metal will oxidize, resulting in rust formation or paint defects on the surface of the material.

DSS 2205 exhibits higher resistance to stress corrosion cracking and it has higher strength than austenitic stainless steel [3]. Taking advantages of these positive factors, DSS 2205 is widely used in the marine applications, tubing for heat exchanger, oil and gas, petrochemical, pulp and paper, and pollution control industries. It is well known that the DSS 2205 exhibit good weld ability, but the melting and solidification associated with fusion welding processes destroys the favourable duplex microstructure of this stainless steel. The heat treatment process is often used for the refinement of grain structure of the steels to improve the corrosion and wear resistance properties of the base material e.g. vessels, valve cone, spindle and valve seat, valves collar and pipes used in marine and other application where the material exposes to severe atmospheric conditions to improve the mechanical and tribological properties of the key element like pressure vessels, valve seat and sleeve which is made up of DSS 2205 or Inconel for chemical, petrochemical, oil/gas, marine and nuclear industries, for economic reasons. DSS has equal phase balance of about identical quantities of ferrite and austenite. It has an assorted microstructure consisting of ferrite (BCC) and austenite (FCC) phases. The corrosion properties of stainless steels can be studied in marine or chloride environments with the effect of heat treatment and investigation of changes in microstructure and corrosive property.

2. DSSs 2205

DSS 2205 is an intermediate grade between ferritic and austenitic stainless steels. DSS 2205 is more resistant to stress corrosion than austenitic types, but worse than ferritic types. They have better toughness than ferritic steels but are not as good as austenite [4]. On the other hand, the strength of DSS 2205 is greater than that of austenite. As a result of extensive research and development recently, a number of new forged alloys have been developed and relevant technical data obtained to verify the treatment of dual-phase steels as other types of steel [5]. DSSs 2205 have a structure of austenite and ferrite phases. Chromium and nickel are the principle alloying elements, the chromium is between 18-28% and the nickel is between 4.5-8%. Copper, molybdenum, tungsten, nitrogen, manganese, and silicon may be added to control structural balance and to provide certain corrosion resistance [6]. Molybdenum is typically added to also enhance crevice corrosion and pitting corrosion resistance [7]. The amount of the two phases is controlled by the composition of the alloy and its heat treatment. The ratio of the phases plays a significant role in defining their properties.

The proportion of α-ferrite and γ-austenite varies with temperature and at high temperatures the structure becomes more ferrite, less Ni and more Cr equivalents. The relationship between both phases is usually about 50% (by volume) although it can vary between 40% and 60% and even between 30% and 70%.The duplex grades are not hardened with heat treatment but are stronger than either of these two grades in the annealed condition and have approximately double the yield strength of the common austenitic grades [8]. Like the ferritic, they are ferro-magnetic, but have the good formability and weldability of the austenitics (although larger forces are required in forming due to their higher strength), and have lower thermal expansion and higher heat conductivity than austenitic grades. The duplex grades are known to be resistant to chloride SCC, pitting and crevice corrosion. In cast duplex, a structure of austenite is lands in a ferrite matrix can be observed. For wrought alloys, the morphology of the microstructure has laths of austenite in a ferrite matrix. The presence of ferrite with austenite not only provides better inter-granular corrosion resistance but also improves SCC resistance compared to fully austenitic stainless steels.

Also, the presence of austenite in the structure could enhance the mechanical properties and improve the localized corrosion resistance due to the availability of austenite stabilizing elements (N and Ni). Moreover, DSSs 2205 become increasingly ferritic as they are heated to high temperatures. However, the presence of ferrite in austenite may cause complex metallurgical reactions that include the formation of a variety of secondary phases, which have unpleasant effects on both corrosion resistance and mechanical properties. The precipitation is usually caused by decomposition of the ferrite phase, which contains large amounts of Cr and Mo, and has low solubility of N and C. Since toughness and corrosion resistance are the criteria of interest, the composition of DSS 2205 and their processing, including fabrication practice, must ensure sufficient formation of austenite and prevention of inter-metallic compounds. Figure 1 shows the precipitates in DSS and it is evident that most of these precipitates concern ferrite and ferrite promoting elements such as Cr, Mo and W. The Figure also illustrates that almost all these reactions take place over the temperature range of 300° - 1000°C [9].

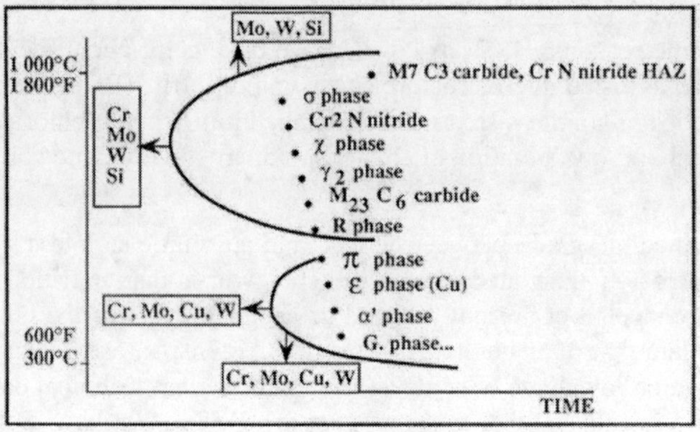

Figure 1. Possible precipitates in DSS [9]

The higher composition of alloying elements in DSS 2205 grades makes them vulnerable to creation of intermetallic and intermediate phases with inevitable form after prolonged service at high temperatures or due to incorrect heat treatment. In addition, these inter-metallic phases could be formed in DSS 2205 during isothermal ageing or quenching. The unwanted creation of secondary phases decreases the overall corrosion resistance and makes the alloy more susceptible to localized corrosion attack. Moreover, such precipitations may also affect the tribo-mechanical properties, mainly the toughness and corrosion resistance. Due to this, service temperatures are usually restricted to be less than about 300°C [10].

3. Literature review

Considering the applications of DSS and the said objectives of research a detailed literature review is carried out to understand the applicable testing methods and systematic approach for investigations. The literature studied for the investigation is summarized as follows. Mesquita et al. [11] studied the corrosion and metallurgical investigation of two super martensitic stainless steels for oil and gas environments in that, the corrosion properties of two super martensitic stainless steels were studied in chloride and H2S environments. Comparison between 1.4542 and 1.4418 SS grades was made considering the effects of microstructure, different steps of the heat treatment on the pitting and sulfide stress cracking SSC properties. Mundhenk et al. [12] in their investigation of corrosion and scaling as interrelated phenomena in an operating geothermal power plant the study refers to an experimental research and contributes to a better understanding of corrosion and scaling in an operating geothermal power plant (Soultz-sous-Forets, France).

An in situ physicochemical monitoring program has been performed in order to characterise the processed brine. In situ and laboratory corrosion experiments were performed using conventional and candidate metals. Similar or some modified studies were also reported in the literature which includes, Pohrelyuk et al. [13] investigated the resistance offered against corrosion in case of Ti–6Al–4V alloy with nitride coatings in Ringer's solution. Gastaldi & Bertolini [14] investigated the effect of temperature on the corrosion behavior of low-nickel DSS bars in concrete.

Hua et al. [15] investigated the hot corrosion behavior of TC11 titanium alloy treated by laser shock processing. Alvarez et al. [16] studied the Corrosion behavior of corrugated lean DSS in simulated concrete pore solutions. Machnikova et al. [17] studied the Corrosion inhibition of carbon steel in hydrochloric acid by furan derivatives. Tan et al. [18] had investigated the pitting corrosion resistance of 2304 DSS after autogenously PAW and subsequent short-time post-weld heat treatment at different temperatures, determined by critical pitting temperature in 1.0 M NaCl solution, etc.

IOP Conf. Series: Materials Science and Engineering **810** (2020) 012048 doi:10.1088/1757-899X/810/1/012048

4. Experimental procedure

4.1 Material and heat treatment:

In the present work, plates DSS 2205, 18 mm thick are used. The chemical % compositions of the DSS 2205 material is indicated in Table 1. The heat treatment is initially carried out at 880°C and 1080°C in the muffle furnace, followed by quenching with water in accordance with the ASTM A182. This is necessary to eliminate second phase of treatment, balancing the phase fractions and release the residual stresses during the manufacturing practice. Specimens are heat treated at 880°C and 1080°C and then water cleaned.

Table 1. Chemical composition of 2205 DSS alloys

Alloy	Cr	Ni	Mo	C	N	Mn	Fe
SS 2205	22.37 %	5.48%	3.49%	0.021%	0.20%	1.370%	Balance

4.2 Optical microscopy:

Specimen for the micro structural examination is prepared by following conventional metallographic procedures. The steps for preparation involves mountings and polishing the specimen using 120, 220, 320, 600, 1000, 1200, 2000 and 2500 grit silicon carbide paper.

These specimens are then cleaned with distilled water and then polished with 0.9, 0.3 μm diamond paste to obtain a surface like a mirror finish. After polishing, the specimens are cleaned with ethanol media for 10 min. Light microscopy is used to reveal the changes in ferrite and austenite phases and detect the presence of third phases. Following Figure 2 shows the microstructure of DSS 2205 (i) simple DSS (ii) 880°C (iii) 1080°C, where specimens annealed at different conditions, the dark part represents ferrite while the light representing austenite.

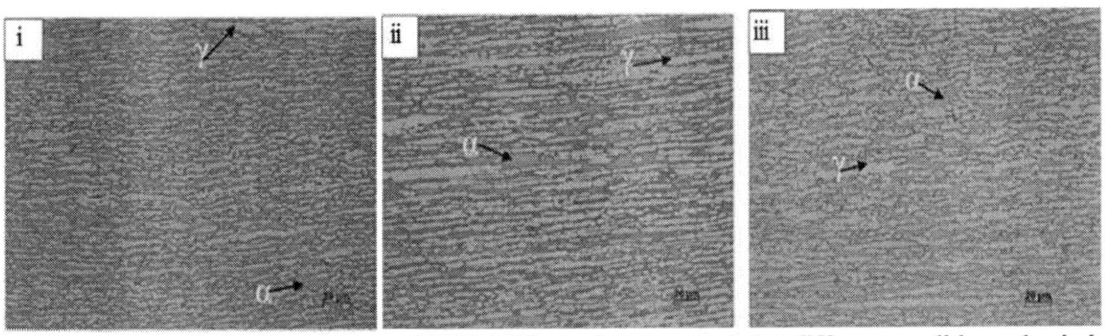

Figure 2. Optical microstructure of DSS 2205 specimens annealed at different conditions, the dark part represents ferrite while the light representing austenite (i) Simple DSS (ii) 880°C (iii) 1080°C

4.3 Scanning electron microscopy (SEM):

SEM is applied to recognize microstructural structures related to austenite, ferrite and intermediate stages. Also, specimens are checked by energy dispersive spectroscopy (EDS) to analyze the chemical composition of specimen. Figure 3 shows SEM images of 2205, indicating dark part associated with ferrite while the light representing austenite (i) Simple DSS (ii) 880°C (iii) 1080°C

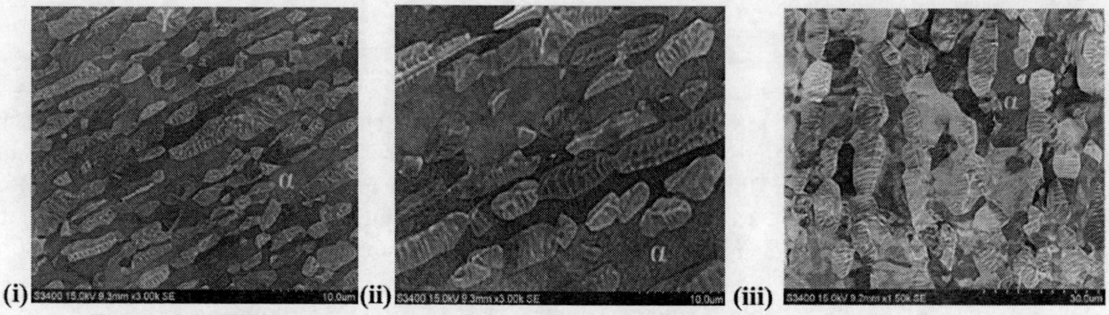

Figure 3. SEM images of 2205, indicating dark part associated with ferrite while the light representing austenite (i) Simple DSS (ii) 880°C (iii) 1080 °C

4.4 Vickers micro-hardness test:
Vickers micro hardness test is carried out using micro hardness machine (Made Mitutoyo available with Nashik Engg. Cluster), with an attached microscope, at 10-400X total magnification according to ASTM E 384 – 99 and 1000 gram load. The microhardness of simple DSS is 242 HV1.0, for specimen heat treated with 880°C microhardness is 254 HV1.0 and for the specimen 1080 °C microhardness is 258 HV1.0.

4.5 Electrochemical corrosion tests:
The specimens required for the potentiodynamic polarization test are prepared as per the requirements, these are prepared in discs shape with an area of 1 cm2 and a height of 2 mm. Three specimens are prepared for the test as mentioned earlier sections which consist of simple DSS and two heat treated DSS. The polarization tests are carried out by the potentiostat made by Gamry®, the cell contains an electrolyte which contains artificial seawater which is prepared by addition of 5 % NaCl to the 95 % of distilled water. A conventional metallographic procedure is adopted for the preparation of specimen and before being tested, all specimens are ground with SiC paper up to 3000 grit and polished.

After polishing, the specimens are cleaned with ethyl alcohol and dried in air. As per the requirements, all specimens are immersed in the electrolytes and the cell is left to stabilize the open circuit potential (OCP) before the actual conduction of tests on each specimens. The potential sweep rate is 1 mV / s and the sweep rate is -300 to 200 mV with respect to a saturated Calomel electrode. Corrosion tests are performed with artificial seawater (5% NaCl). The first is potentiodynamic polarization using a three-electrode electrochemical cell, with a platinum counter-electrode and a saturated calomel reference electrode (SCE) and the working electrode (DSS). Figure 4 shows potentiodynamic polarization curves of 2205 DSS specimen and heat treated at 880°c in artificial sea water and Figure 5 Potentiodynamic polarization curves of 2205 DSS specimen and heat treated at 1080°c in artificial sea water. The current density and corrosion rate of specimens are shown in Table 2.

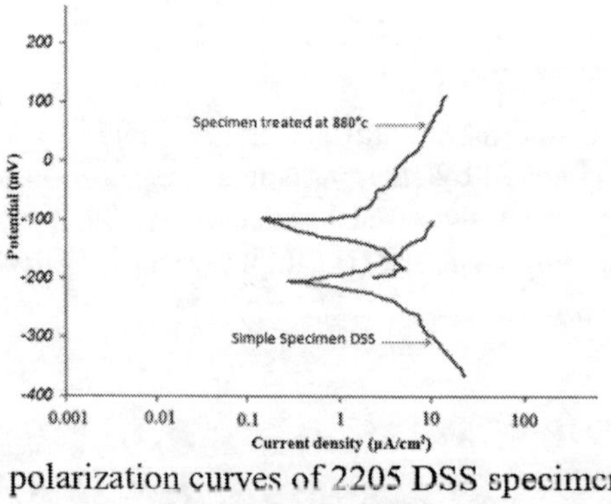

Figure 4. Potentiodynamic polarization curves of 2205 DSS specimen and heat treated at 880°c in artificial sea water

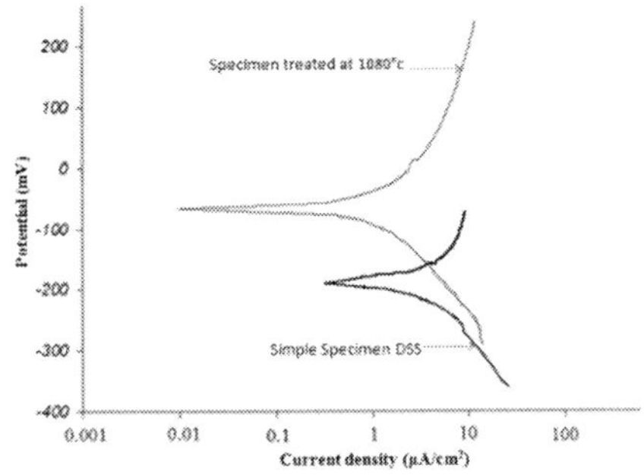

Figure 5. Potentiodynamic polarization curves of 2205 DSS specimen and heat treated at 1080°c in artificial sea water

Table 2. Current density and corrosion rate for 2205 DSS alloys specimens

Specimens	Current density (μA/c m2)	Corrosion rate (mpy)
Simple DSS	1.96	0.872
880°C	0.98	0.416
1080°C	0.67	0.318

5. Results, discussion and analysis

The microstructure of DSS 2205, as in Figure 2 shows that the optical microstructure of the simple DSS specimen also SEM image Figure 3(i) contains only ferrite and austenite with no secondary phases. In those microstructural images ferrite phases appears darker than austenite phases, after heat treatment, the evolution of secondary phases influences the corrosion and mechanical properties of the DSS 2205. The temperature of the treatment shows effects on the microstructure, secondary phases also on the volume fraction of these all phases. Moreover, the effects of these change in temperature and heat treatment of DSS 2205 can be observed in Figure 2 (i-iii) and Figure 3 (i-iii).When the DSS 2205 is the heat treated the decomposition of thermodynamic balance occurs causing the material to endeavor a more stable thermodynamic state by precipitation and formation of secondary phases. The sigma phase is preferentially nucleated at α / γ or α / α interfaces and grows through the adjacent ferrite, since the ferrite phase is unstable at elevated temperatures. Indeed, diffusion rates of alloying elements such as Cr and Mo are 100 times faster than diffusion rate values in austenite [19]. The simple DSS specimen treated contained ferrite and austenite only but secondary phases did not appear, as shown in Figure 2 and Figure 3 (i).The substantial consequences of heat treatment on microstructural morphology and distribution of phases is observed in the present investigation. Likewise, to justify the effects of the heat treatment on the constituent phase's morphology, specimens are characterized by the SEM. SEM micrographs of the specimen are shown in Figure 3 (i-iii), from these figures it can be clearly observed that the darker region represents the ferrite phase and the brighter region represents the austenite phase with differences in microstructure with the temperature of treatment. The heat treatment has an effect on the volume fraction of ferrite and austenite and on the size of its grains. The volume fraction of the ferrite phase increased with increasing aging temperature and the aging time directly affected the volumetric concentrations of the ferrite and austenite phases. The other effect of the heat treatment that appeared with the increase in the aging temperature is the increase in the size of the ferrite grains. The EDS analysis of the specimen is also carried out to analyze the % contribution of alloying elements in the material. Table 3 shows the analysis of specimens with % of composition of ferrite and austenite. The chemical composition of each phase does not vary much with the composition of the base alloys where it is shown that the ferrite phase is rich in chromium and molybdenum with regard to nickel, while the austenite phase enriches in nickel and content chromium and molybdenum lower than the ferrite phase. Moreover, Figure 6-8 shows the % distribution of alloying element in simple DSS 2205, 880°C DSS 2205 and 1080°C DSS 2205. Where, the % composition of Ni and Cr in the ferrite and austenite can be seen clearly. Ferrite phases are richer in Cr and Mo with veneration to Ni, while austenite phase enhanced in Ni with Cr and Mo lower than the ferrite.

Table 3. EDS analysis of specimens with % of composition of ferrite and austenite

Specimen details	Phases Identified	Cr Avg. %	Ni Avg. %	Mn Avg. %	Mo Avg. %	Fe Avg. %
Simple DSS	Ferrite	22.83	3.25	0.82	5.21	Balance
	Austenite	15.26	5.51	0.6	3.22	Balance
880 (°C)	Ferrite	21.17	3.42	0.68	5.65	Balance
	Austenite	19.07	5.49	0.66	3.28	Balance
1080 (°C)	Ferrite	20.17	3.64	0.6	5.33	Balance
	Austenite	19.64	5.53	0.5	3.78	Balance

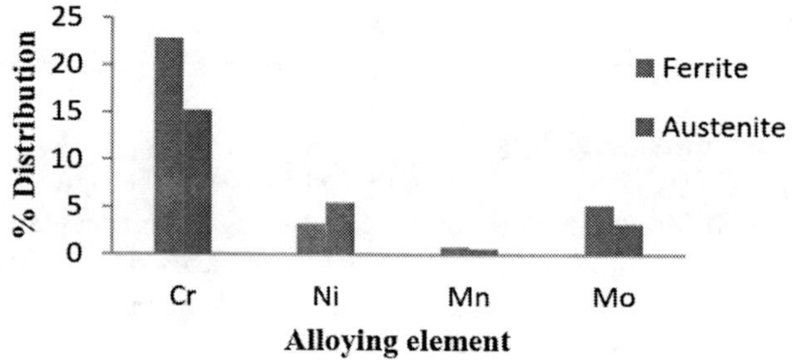

Figure 6. % Distribution of alloying element in simple DSS 2205

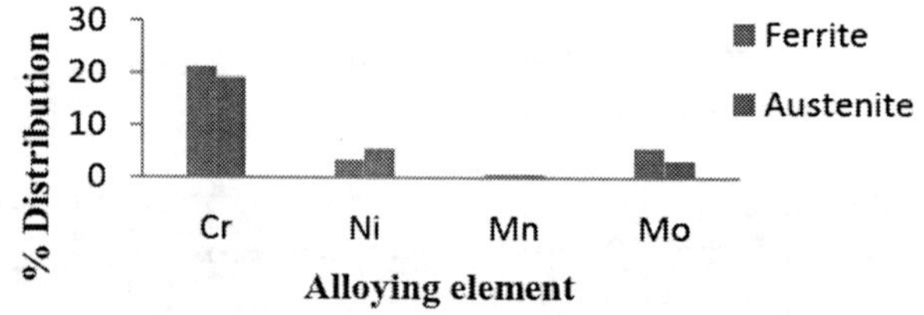

Figure 7. % Distribution of alloying element in heat treated at 880°C DSS 2205

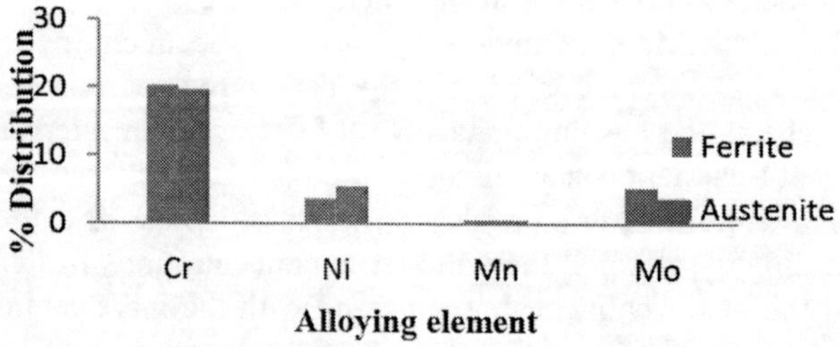

Figure 8. % Distribution of alloying element in heat treated at 1080°C DSS 2205

The ferrite / austenite ratio in which corrosion occurs preferentially in the ferrite phase of DSS affects the corrosion resistance of the steel. The austenitic phase has superior corrosion resistance compared to ferrite phase for a variety of reasons, including higher nickel content than ferrite, and in general, Cr improves the resistance of DSS to local corrosion in the existence of ferrite phase. Among the most effective elements maximum amount of Cr plays vital role in improving resistance of steel against corrosion.

However, the main alloying element in the austenitic phase is Ni, whereas in the ferrite phase it is Cr. The ferrite phase is active electrochemically than austenite phase, whereas the probability of Cr to act as an anode is more. Though, the corrosion prospective of Ni is higher than that of Cr which concludes that the main responsible element for the improvising the corrosion resistance is % of Ni in the phase. Also in austenite, higher solubility of N than ferrite is observed due to maximum availability of Mn % which enhances the corrosion resistance of austenite. Also this can be evidences from Figure 6 and Figure 7, which shows potentiodynamic polarization curves of specimens for simple 2205 DSS, heat treated at 1080°C and 880°C in artificial sea water. The results show that the best ferrite/austenite ratio for the studied specimens is obtained after treated the specimen at 1080°C. This specimen shows minimum current density compared to other specimen as shown in Table 2 and lower corrosion rate showing highest corrosion resistance.

6. Conclusions

In the present investigations an attempt has been made to investigate the behaviour of DSS 2205 under the action of corrosive media. Also, the effect of heat treatment on the microstructure and corrosion resistance has been evaluated. From the investigation it can be clearly seen that simple DSS can be protected from corrosion if treated well. Looking towards the various industrial applications of DSS 2205 this research is important for the development in materials and methodologies applied for the various steels. It is envisaged from the current analysis that the heat treatment changes the morphology of the material as well as microstructure of the steels. Not only the microstructure but also the chemical composition of the steel in phases such as austenite and ferrite changes due to heat treatment which ultimately improves the corrosion resistance of the steel. Following are the major conclusions drawn from present research.

1. The change in microstructure, due to heat treatment and secondary phase precipitation, shows a significant part in the corrosion performance of the DSS 2205.
2. Precipitation of secondary parts in DSS led to a drastic reduction in its corrosion resistance.
3. Another factor which is responsible for the change in the ferrite/austenite ratio where the corrosion occurred preferentially in the ferrite phase of DSS 2205.
4. The austenite phase shows better corrosion resistance than the ferrite phase due to maximum Ni content than ferrite.
5. Cr is the most active element in the passive film to develop the resistance of DSS 2205 to localized corrosion.
6. The heat treatment about 1080°C to DSS 2205 improves the corrosion resistance of steel comparatively.

Acknowledgement

The author would wish to convey his sincere gratitude to his parents, colleagues, friends for their kind cooperation, motivation and support during the study. This study is supported by Research Scheme under B.C.U.D. (Board of College and University Department), Savitribai Phule Pune University, Pune, India.

References

[1] Eker, B and Yuksel, E 2005 *Trakia J. Sci.* **3** 1

[2] Mele M and Siviero E 1991 *Advanced Problems in Bridge Construction* Springer Vienna **121**

[3] Hanninen H, Romu J, Ilola R, Tervo J and Laitinen A 2001 *J Mater Process Tech*. **117** 424

[4] Alvarez-Armas I and Degallaix-Moreuil S 2013 *John Wiley & Sons*

[5] Speer J, De Moor E, Findley K, Matlock D, De Cooman B and Edmonds D 2011 *Metall. Mater. Trans A* **42** 3591

[6] Al-Rabie M 2011 (Doctoral dissertation, The University of Manchester (United Kingdom)

[7] Tullmin A and Robinson A. 1988 *Corrosion*, **44** 664

[8] Sahu J, Krupp U, Ghosh R and Christ H 2009 *Mat. Sci. Eng. A* **508** 1

[9] Gunn R 1997 *Elsevier*

[10] Davis J 1994 *ASM International*

[11] Mesquita T, Chauveau E, Mantel M, Bouvier N and Koschel D 2014 *Corros. Sci.* **81** 152

[12] Mundhenk N, Huttenloch P, Sanjuan B, Kohl T, Steger H and Zorn R 2013 *Corros. Sci.* **70** 17

[13] Pohrelyuk I, Fedirko V, Tkachuk O and Proskurnyak R 2013 *Corros. Sci.* **66** 392

[14] Gastaldi M and Bertolini L 2014 *Cement Concrete Res.* **56** 52

[15] Hua Y, Bai Y, Ye Y, Xue Q, Liu H, Chen R and Chen K 2013 *Appl. Surf. Sci.* **283** 775

[16] Alvarez S, Bautista A and Velasco F 2011 *Corros. Sci.* **53** 1748

[17] Machnikova E, Whitmire K and Hackerman N 2008 *Electrochim. Acta.* **53** 6024

[18] Tan H, Wang Z, Jiang Y, Han D, Hong J, Chen L and Li J 2011 *Corros. Sci.* **53** 2191

IOP Conf. Series: Materials Science and Engineering **810** (2020) 012049 doi:10.1088/1757-899X/810/1/012049

Mathematical Modelling of Material Removal Rate During Electric Discharge Machining Using Buckingham's Pi-Theorem

Vaibhav S Gaikwad, Vijaykumar S Jatti, Satish S Chinchanikar, Nikhil S Kakuste, Keshav N Nandurkar

[1] Assistant Professor, Department of Production Engineering, K.K.Wagh Institute of Engineering Education and Research Centre, Nashik, India
[2] Associate Professor, Department of Mechanical Engineering, D.Y. Patil College of Engineering, India
[3] Professor, Department of Mechanical Engineering, Vishwakarma Institure of Information Technology, India
[4] U.G. Student, Department of Production Engineering,K.K.Wagh Institute of Engineering Education and Research Centre, Nashik, India
[5]Professor, Department of Production Engineering,K.K.Wagh Institute of Engineering Education and Research Centre, Nashik, India

Email Address : vsgaikwad113@gmail.com

Abstract. In the present work an attempt has been made to model material removal rate (MRR) by using Buckingham's Pi-theorem during electric discharge machining (EDM) of shape memory NiTi alloy. Machining of Ni-Ti alloyis mostly attempted by non-conventional methods of machining, and, especially by EDM, as this advance material imposes machinability issues when cut by conventional processes. However, EDM process has low material removal rate.In present investigation EDM of NiTi alloy is carried out using electrolytic copper toolon die sink type of electric discharge machine. The experimental-based mathematical model to predict MRR is developed varying process parameters such as gap current, pulse on time, pulse off time, voltage, workpiece electrical conductivity and tool electrode electrical conductivity during the experiments. The model developed using dimensional analysis and Buckingham pie theorempredicts the value of MRR close to the experimental observations with the accuracy of 91 % shows that the developed model could be used to predict MRR during EDM of NiTi alloy within the domain of the parameters selected in the present study.

Keywords:EDM, Ni-Ti alloy, MRR, Mathematical Modeling

1. Introduction

In industrial sector, it is very much imperative to select correct manufacturing or machining processes while machining difficult-to-machine materials and/or advanced materials. One should select the machining process

depending upon the properties and characteristics of workpiece material, tool material, dimensional accuracy and surface finish requirement. Ni-Ti alloy is one of the advanced materials which possess properties such as biocompatibility, high wear resistance and high strength. Due to these significant properties it is used in the applications such as aerospace and medical field. It is difficult to machine such alloys using convention machining; hence these can be machined using non-conventional machining processes such as electric discharge machining (EDM). In EDM material is removed by the erosive action between tool and workpiece.
Shabgard et al. [1] investigated the distribution of temperature on the EDM machined surface using ABAQUS software. During their study dimensions of discharge crater was calculated using distribution of temperature profile. From mathematical and numerical analysis carried out by them it is seen that with increase in pulse on time and gap current flushing efficiency increases. However, there are optima for pulse on time and gap currentand with further increase beyond optima point flushing efficiency observed to decrease. Also the thickness of recast layer increases for the tool and workpiece with the increase in gap current. Liu et al. [2] performed the thermal modelling of EDM processes. In their work, finite element model was developed using ABAQUS for simulating massive random discharge. Their study observed larger temperature gradient in subsurface at depth of 20 μm whereas less temperature gradient at deep subsurface.
Jithin et al. [3] investigated the flushing efficiency by finite element modelling of single spark in EDM. Their investigation concluded that crater radius and crater depth increases with rise in pulse on time and gap current. They also developed a model to predict crater aspect ratio and claimed that developed model could be used to predict crater ratio as error observed by them between predicted results and experimental results in between 9-13%. Dhupal et al. [4] investigated the machining of aluminium silicon metal matrix composite using EDM processes. In their work, particle swarm optimization technique was used to multi-objective optimization of performance measures. As a part of their work, they developed experimental-based mathematical (quadratic regression model)model to predict the values of performance measures during EDM. Their study concluded that higher values of material removal rate can be obtained by using higher values of gap current, voltage and pulse on time during EDM.
Kishan et al. [5] investigated the machining EN8 steel using EDM. During their work, they used copper and brass materials, respectively, for electrode. Their investigation observed higher MRR while machining using copper electrode as that of brass electrode. Further, their optimization study also concluded better performance measures using copper electrode as against brass electrode.Mohapatra et al. [6] performed 2-D modeling of wire used in EDM using ANSYS fluent. In their work, structural and thermal analysis was performed for the EDM wire.Their study observed maximum temperature of wire towards the contact region of the workpiece. From the literature available on EDM, it can be seen sufficient investigations have been carried out by researchers with a view to make the process more efficient for industrial applications. However, few researchers reported on electric discharge machining of Ni-Ti alloy which is considered to be advanced material having high wear resistance and high strength. This material finds application in aerospace and medical field. With this view, in the present work, electric discharge machining of Ni-Ti alloy was carried with a view to develop a mathematical model to predict material removal rate (MRR) during EDMvarying process parameters such as gap current, pulse on time, pulse off time, voltage, workpiece electrical conductivity and tool electrode electrical conductivity during the experiments.

2.Materials and Method

In present investigation *NiTi* alloy was used as workpiece material and electrolytic copper as tool electrode. Samples of *NiTi* were cut into diameter of 20 mm with 20 mm in length and tool electrode used was having diameter of 3 mm with length of 90 mm. Both the workpiece and tool were cryo-treated at -185^{0}C. Experiments were carried out varying the processes parameters namely gap current, pulse on time, pulse off time, electrical conductivity of workpiece and tool electrode materials. Drilling experiments were carried out on *NiTi* shape memory alloy on die sink type of electric discharge machine. The machine has facility of varying gap current, pulse on time,pulse off time and voltage as input parameters. Digital weighing balance;Model GR 300 with accuracy 0.0001 gm was used to measure the weight of workpiece before and after machining. MRR and tool wear rate (TWR) were calculated by the volume of material removed.Gap voltage of 55 V and pressure of flushing fluid at a value of 0.5 kg/cm^2 were kept constant throughout the experiments. Figure 1 shows the experimental set upfor cryo-treated EDM process.

Fig. 1. Experimental set up

Table 1 shows various EDM processes parameters and their levels used during cryo-treated EDM process drilling of NiTi alloy as a workpiece material and electrolytic copper as tool electrode.The processes parameters namely workpiecematerial electrical conductivity and tool electrode material electrical conductivitywere variedat two different levelsand the parameters namely gap current, pulse on time and pulse off time were varied at three different levels. Parameters were selected based on the literature review and pilot experiments performed during the study.

Table 1 shows various EDM processes parameters

Factors		Levels				
		1	2	3	4	5
A)	Workpiece electrical conductivity (S/m)	3267	4219	-	-	-
B)	Tool electrical conductivity(S/m)	10000	26316	-	-	-
C)	Gap current (A)	8	10	12	14	16
D)	Pulse on time (μs)	13	26	38	51	63
E)	Pulse off time (μs)	5	6	7	8	9

3.Mathematical Modeling

This section discusses the mathematical modeling of MRR using Buckingham's Pi theorem. This theorem is formalized by Rayleigh's method or dimensional analysis. This theorem says that "if there are n variables (might be independent and dependent variables) in a physical phenomenon and if these variable contains m fundamental dimensions (L, M, T, θ, I), then the variables are arranged into (n-m) dimensionless terms called as Pi -terms"
According to this theorem,
MRR=f (Ig ,τon ,τoff ,V σ, K, Cp, ρ, Hm)
Where, σ- Electrical conductivity of workpiece, K- Thermal conductivity of workpiece, Cp- Specific heat,Hm- Melting point,ρ - Density,Ig- Gap current,τon– Pulse on time,,τoff- Pulse off time,V- Voltage.
Therefore,Number of variables (n) =10
Number of fundamental dimensions (m) =5
Hence number of π terms= n-m=10-5=5

$$\Pi_1=(\sigma)^{a1}(K)^{b1}(C_p)^{c1}(\rho)^{d1}(H_m)^{e1}\,MRR \quad (1)$$

$$\Pi_2=(\sigma)^{a2}(K)^{b2}(C_p)^{c2}(\rho)^{d2}(H_m)^{e2}\,I_g \quad (2)$$

$$\Pi_3=(\sigma)^{a3}(K)^{b3}(C_p)^{c3}(\rho)^{d3}(H_m)^{e3}\tau_{on} \quad (3)$$

$$\Pi_4=(\sigma)^{a4}(K)^{b4}(C_p)^{c4}(\rho)^{d4}(H_m)^{e4}\,\tau_{off} \quad (4)$$

$$\Pi_5=(\sigma)^{a5}(K)^{b5}(C_p)^{c5}(\rho)^{d5}(H_m)^{e5}\,V \quad (5)$$

Table 2 Dimensions of EDM process parameters

σ	K	C_p	ρ	H_m	MRR

M	-1	1	0	1	0	0
L	-3	1	2	-3	0	3
T	3	-3	-2	0	0	-1
Θ	0	-1	-1	0	1	0
I	2	0	0	0	0	0
	a	b	c	d	e	

By equating the dimensions of equation 1, 2, 3, 4, 5 and from Table 2 we can write,
For Π_1 terms,

$$-1a_1+1b_1+0c_1+1d_1+0e_1+0=0 \quad (6)$$
$$-3a_1+1b_1+2c_1-3d_1+0e_1+3=0 \quad (7)$$
$$3a_1-3b_1-2c_1+0d_1+0e_1-1=0 \quad (8)$$
$$0a_1-1b_1-1c_1+0d_1+1e_1+0=0 \quad (9)$$
$$2a_1+0b_1+0c_1+0d_1+0e_1+0=0 \quad (10)$$

By solving equations 6, 7, 8, 9, and 10 simultaneously, we will get,
$a_1=0$; $b_1=-2$; $c_1=2.5$; $d_1=2$; $e_1=0.5$
Similarly equating the dimensions and solving the equations for Π_2, Π_3, Π_4, Π_5 we will get,
For Π_2, $a_2=-0.5$; $b_2=-1.5$; $c_2=1.5$; $d_2=1$; $e_2=0$
For Π_3, $a_3=0$; $b_3=-1$; $c_3=2$; $d_3=1$; $e_3=1$
For Π_4, $a_4=0$; $b_4=-1$; $c_4=2$; $d_4=1$; $e_4=1$
For Π_5, $a_5=0.5$; $b_5=-3.5$; $c_5=3$; $d_5=2$; $e_5=-0.5$
Hence we can write,
$\Pi_1=(\sigma)^0 (K)^{-2} (C_p)^{2.5} (\rho)^2 (H_m)^{0.5}$ MRR
$\Pi_2=(\sigma)^{-0.5} (K)^{-1.5} (C_p)^{1.5} (\rho)^1 (H_m)^0 I_g$
$\Pi_3= (\sigma)^0 (K)^{-1} (C_p)^2 (\rho)^1 (H_m)^1 \tau_{on}$
$\Pi_4=(\sigma)^0 (K)^{-1} (C_p)^2 (\rho)^1 (H_m)^1 \tau_{off}$
$\Pi_5=(\sigma)^{0.5} (K)^{-3.5} (C_p)^3 (\rho)^2 (H_m)^{-0.5} V$

Therefore we can write by Buckingham's Pi-Theorem,
$\Pi_1=K\, \Pi_2^{a2}\, \Pi_3^{a3}\, \Pi_4^{a4}\, \Pi_5^{a5}$(A)
Taking $\log_{10}$ on both side of equation (A),
$\log \Pi_1=\log K+a_2\log \Pi_2+a_3\log \Pi_3+a_4\log \Pi_4+a_5\log \Pi_5$
Let $Y= \log \Pi_1, X_1= \log \Pi_2, X_2= \log \Pi_3, X_3= \log \Pi_4, X_4= \log \Pi_5$
Log $K=a$; $a_2=b$; $a_3=c$; $a_4=d$; $a_5=e$
Therefore Equation (A) becomes
$$Y=a+bX_1+cX_2+dX_3+eX_4 \quad (B)$$
Equation B is called as linear regression it can be solved as follows,
$$\sum Y= na+b\sum X_1+c\sum X_2+d\sum X_3+e\sum X_4 \quad (11)$$
$$\sum YX_1=a\sum X_1+b\sum X_1^2+d\sum X_1X_3+e\sum X_1X_4 \quad (12)$$
$$\sum YX_2=a\sum X_2+b\sum X_1X_2+c\sum X_2^2+d\sum X_2X_3+e\sum X_2X_4 \quad (13)$$
$$\sum YX_3=a\sum X_3+b\sum X_1X_3+c\sum X_2X_3+d\sum X_3^2+e\sum X_3X_4 \quad (14)$$
$$\sum YX_4=a\sum X_4+b\sum X_1X_4+c\sum X_2X_4+d\sum X_3X_4+e\sum X_4^2 \quad (15)$$

Now by calculating all values and considering n=30 we can solve equation 11,12,13,14,15 and get,
$a=-2.59341$; $b=1.34666$; $c=-0.0102867$; $d=-0.812246$; $e=-2.12128$
Therefore,
$\log K=a$, $a_2=b$, $a_3=c$, $a_4=d$ and $a_5=e$
Hence equation (A) becomes

$$\Pi_1=2.55*10^{-3} (\Pi_2)^{1.34666} (\Pi_3)^{-0.0102867} (\Pi_4)^{-0.012246} (\Pi_5)^{-2.12128}$$

But, $\Pi_1=(\sigma)^0 (K)^{-2} (C_p)^{2.5} (\rho)^2 (H_m)^{0.5}$ MRR
Hence we will get,

$$MRR = \left(\frac{2.558*10^{-3}*K^2}{C_p^{2.5}*\rho^2*H_m^{0.5}}\right) * \left(\frac{\sigma^{0.5}*C_p^{1.5}*\rho*I_g}{K^{1.6}}\right)^{1.3466} * \left(\frac{C_p^2*\rho*H_m*\tau_{on}}{K}\right)^{-0.010267} * \left(\frac{C_p^2*\rho*H_m*\tau_{off}}{K}\right)^{-0.812246} * \left(\frac{\sigma^{0.5}*C_p^3*\rho^2*V}{K^{3.5}*H_m^{0.5}}\right)^{-2.1228}$$

Where, σ-electrical conductivity of workpiece (S/mm), K-thermal conductivity of workpiece (0.01 j/min*mm*k), C_p-specific heat (0.32 j/gk),H_m-melting point (1583 k), ρ – density (0.00645 gm/mm^3), I_g-gap current A,τ_{on}–pulse on time (min.),τ_{off}-pulse off time (min),V-voltage (V)

Table 3 shows the values of predicted and experimental material removal rate observed during EDM of NiTi alloy using electrolytic copper tool on die sink type of electric discharge machine. From Table 3, it can be seen that the experimental-based mathematical model developed to predict MRR varying process parameters such as gap current, pulse on time, pulse off time, voltage, workpiece electrical conductivity and tool electrode electrical conductivity using dimensional analysis and Buckingham pie theorem predicts the MRR value close to the experimental observations with theco-relation coefficient of 0.9099 and percentage error of 8.802 %. (Accuracy of 91%) shows that the developed model could be used to predict MRR during EDM of NiTi alloy within the domain of the parameters selected in the present study.

Table 3 Predicted and experimental MRR

I_g	V	τ_{on}	τ_{off}	σ	**MRR (Experimental) (mm^3/min)**	**MRR (Predicted) (mm^3/min)**	**% Error**
8	55	6.333E-07	1.1667E-07	3.267	1.565	1.819	13.951
10	55	6.333E-07	1.1667E-07	3.267	2.321	2.456	5.509
12	55	6.333E-07	1.1667E-07	3.267	2.741	3.140	12.701
14	55	6.333E-07	1.1667E-07	3.267	3.723	3.864	3.665
16	55	6.333E-07	1.1667E-07	3.267	4.056	4.626	12.317
8	55	6.333E-07	1.1667E-07	4.219	1.872	2.008	6.773
10	55	6.333E-07	1.1667E-07	4.219	2.440	2.712	10.024
12	55	6.333E-07	1.1667E-07	4.219	3.321	3.467	4.210
14	55	6.333E-07	1.1667E-07	4.219	3.867	4.267	9.370
16	55	6.333E-07	1.1667E-07	4.219	4.671	5.107	8.553
12	55	2.167E-07	1.1667E-07	3.267	3.051	3.175	3.902
12	55	0.000026	1.1667E-07	3.267	3.381	3.022	11.859
12	55	6.333E-07	1.1667E-07	3.267	3.061	3.140	2.517
12	55	8.5E-07	1.1667E-07	3.267	2.963	3.131	5.352
12	55	0.000063	1.1667E 07	3.267	2.812	2.995	6.107
12	55	2.167E-07	1.1667E-07	4.219	3.107	3.505	11.366
12	55	0.000026	1.1667E-07	4.219	3.602	3.337	7.953
12	55	6.333E-07	1.1667E-07	4.219	3.488	3.467	0.618
12	55	8.5E-07	1.1667E-07	4.219	3.154	3.456	8.759
12	55	1.05E-06	1.1667E-07	4.219	2.957	3.449	14.277
12	55	6.333E-07	8.3333E-08	3.267	4.573	4.127	10.809
12	55	6.333E-07	0.0000001	3.267	3.100	3.559	12.893
12	55	6.333E-07	1.1667E-07	3.267	3.569	3.140	13.661
12	55	6.333E-07	1.3333E-07	3.267	2.340	2.817	16.941
12	55	6.333E-07	0.00000015	3.267	2.721	2.560	6.293

12	55	6.333E-07	8.3333E-08	4.219	5.137	4.557	12.731
12	55	6.333E-07	0.0000001	4.219	4.127	3.929	5.029
12	55	6.333E-07	1.1667E-07	4.219	4.001	3.467	15.404
12	55	6.333E-07	1.3333E-07	4.219	2.988	3.111	3.941
12	55	6.333E-07	0.00000015	4.219	3.013	2.827	6.587
			Percentage error (%)			8.802	
			R^2			0.90997957	

4.Conclusion

In the present study attempt has been made to develop a mathematical model to predict material removal rate (MRR) during EDM (drilling) of Ni-Ti alloy using Buckingham's Pi-theoremand dimensional analysis. In present investigation, both the workpiece and electrolytic copper as tool material were cryo-treated at -185^{0}C. Experiments were carried out on die sink type of electric discharge machinevarying the processes parameters namely gap current, pulse on time, pulse off time, electrical conductivity of workpiece and tool electrode materials. It was observed that MRR increases with rise in current, pulse on time. In the present work, mathematical model developed using dimensional analysis and Buckingham pie theorem predicts the MRR value close to the experimental observations with the co-relation coefficient of 0.9099 and percentage error of 8.802 %. (Accuracy of 91%) shows that the developed model could be used to predict MRR during EDM of NiTi alloy within the domain of the parameters selected in the present study.

References

[1] M Shabgard, R Ahmadi, M Seyedzavvar , S Oliaei, "Mathematical and numerical modeling of the effect of input-parameters on the flushing efficiency of plasma channel in EDM process"Int. J. of Mach. Tools & Manu. Vol. 65,pp.79–87,2013.

[2] J. Liu, Y. Guo, "Thermal Modeling of EDM with Progression ofMassive Random Electrical Discharges" Proc. Manu., vol.5,pp.495-507,2016.

[3] S. Jithin, A. Raut, U. Bhandarkar, S. Joshi, "FE Modeling for Single Spark in EDM Considering Plasma Flushing Efficiency",Proc. Manu., vol.26,pp.617-628,2018.

[4] D. Dhupala, S. Naika, S. Dasb, "Modeling and Optimization of Al-SiC MMC through EDM Processes using Copper and Brass Electrode", Mater. Today Proc.,Vol.5,pp.11295-11303,2018.

[5] B. Kishana, B. Premkumara, S. Gajanana, K. Buchaiahc, M. Gaffara, "Development of Mathematical Model for Metal Removal Rate on EDM using Copper and Brass Electrodes",Mater. Today Proc.,Vol.5,pp.4345-4352,2018.

[6] K. Mohapatra, V. Shaibua, S.K. Sahoo, "Modeling and Analysis of Wire EDM in a Gear Cutting Process fora 2D Model", Mater. Today Proc.,Vol.5,pp.4793-4802,2018.

IOP Conf. Series: Materials Science and Engineering **810** (2020) 012050 doi:10.1088/1757-899X/810/1/012050

Methods for Low Engine Speed Torque Improvement of Natural Gas Engine for Commercial Vehicles: Experimental Observations with Turbocharger and Supercharger

P. J. Suple[1], C. R. Sonawane[1], S. S. Thipse[2], J. P. Mohite, N. B. Chougule[3]

[1]Symbiosis Institute of Technology, Mechanical Engineering Dept., Pune, India
[2]Automotive Research Association of India, Pune, India
[3]Tata Motors, Pune, India

crsonawane@gmail.com

Abstract. Diesel engines have been powering a range of commercial vehicles for a many years. Considering air pollution, there is a thrust on use of natural gas (NG). Thereafter, natural gas engines for commercial vehicles have been subject of development, particularly to meet drivability demands and emissions requirements.

Bus used for intra-city mass transportation of passengers is probably the most common form of natural gas commercial vehicle. Considering typical city applications, such vehicles is characterized by low speeds, frequent gear changes, start-stops, traffic conditions etc. For better drivability, they need higher traction at low engine speeds. This study captures few means of torque enhancement and motive is to integrate a selective ones as not much research is available mentioning enhancements specifically at low engine speeds.

Turbocharging of natural gas engine is complicated due to high exhaust temperatures. As most turbocharger manufacturers cater to requirements of diesel engine, turbochargers for natural gas are simply not available. Many compromises are thus to be made. Under this study, four different turbochargers and one supercharger shall be simulated for experimenting and optimizing to enhance torque at low engine speeds. A virtual model of reference, naturally aspirated engine is built in appropriate software and its output is verified against test bed performance, to establish model faithfulness. Next, simulation runs with different turbochargers and superchargers are carried out. Various parameters are recorded and compared. Findings are recorded and it is noted that there is room for enhancement based on different hardware capabilities.

Keywords: Low speed torque improvement, natural gas engine, turbocharging, supercharging

1. Introduction

One usually finds different kinds of vehicles in cities that serve different purposes. A closer look at energy expenditure reveals that these vehicles are consume more energy as compared to mass transport public vehicles [1].Considering conservation and environmental policies, there has been a motivation to use alternate fuels as compared to conventional ones [2]. Vehicles powered by natural gas shall be popular as an attempt to reduce emissions [3]. As the market demand surges, OEM's shall be increasingly depending on platform technology, a minor changes shall suffice market requirements [4]. As predicted, share of light and small commercial vehicles shall continue for upcoming years [5]. The design of such vehicles shall include advanced powertrain and transmission options, with more focus on alternate fuels [6]. Use of natural gas an alternate fuel shall help to curb rising pollution levels and its performance shall improve the outlook of observers towards green mobility [7]. From viewpoint of economics, design feasibility, combustion control and optimization etc. natural gas is a preferred choice [8]. In spite of limited infrastructure for re-fulling, natural gas shall continue to be a popular choice for different vehicles. This shall prove to be important step towards emission control and environment conservation. Newer methods such as dual fuel engines, more efficient storage systems etc. shall find its way in mass transport vehicles [9]. Bio-Methane, obtained by decomposing of organic materials can be substituted as a transport fuel. Before this could be done, a primary treatment consisting of increasing methane content and storage pressure is to be completed for practical usage [10].

The sale of commercial vehicles in India increased by about 22% in 2010-2011 and compounded growth rate of about 9.3%. These include different vehicle configurations. As operating cost remains key selection factor, vehicle manufacturers shall be keenly offering natural gas powered vehicles, and it is estimated that these shall constitute about 22% of total vehicles. Compulsion to use natural gas for urban mass transport and rising environmental concerns, sale of natural gas powered vehicles is set to increase [11]. Implementing new safety standards such as Universal Bus Specifications, dictates eight different aspects of vehicle design that must be catered considering passenger safety and comfort. [12]. Evolution of different technologies on vehicles now highlights manufacturer's commitment to develop prime movers on alternate fuels, most of their popular vehicles shall be available with natural gas options [13].

2. Engine modifications for low speed torque improvement

Engine chosen for purpose of study consists of six cylinders natural gas, is naturally aspirated and powers a bus for city applications. Critical specifications can be seen in table 1.

Table 1. Specifications of base engine

Parameters	Specifications
Bore x Stroke	97mm x 128 mm
No. of Cylinder	6 Cylinder, Inline
Firing order	1st-5th-3rd-6th-2nd-4th
Rated speed	2500RPM
Swept volume	5.7L
Aspiration	Natural
Compression ratio	12.5:1 +/-1

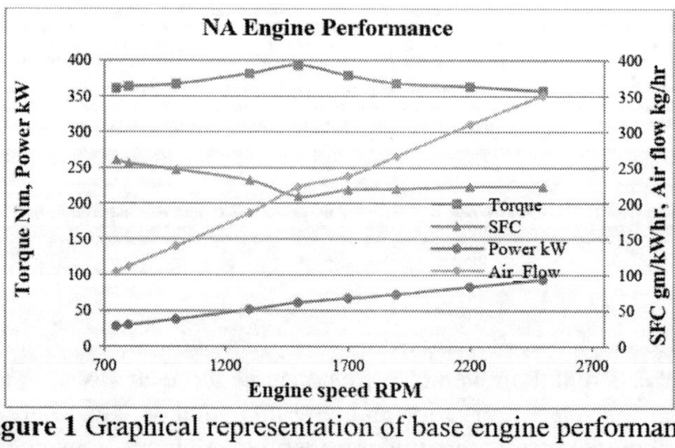

Figure 1 Graphical representation of base engine performance

Increasing torque output at low engine speeds needs changes to basic hardware. Since the engine under consideration is derived from a diesel engine, there are practical design constrains. Engine output at different speeds can be seen in the above figure 1. Other important engine operating parameters are also recorded simultaneously. Figure 2 represents engine test bed layout where interconnections action between different engine testing devices are shown in simple block diagram.

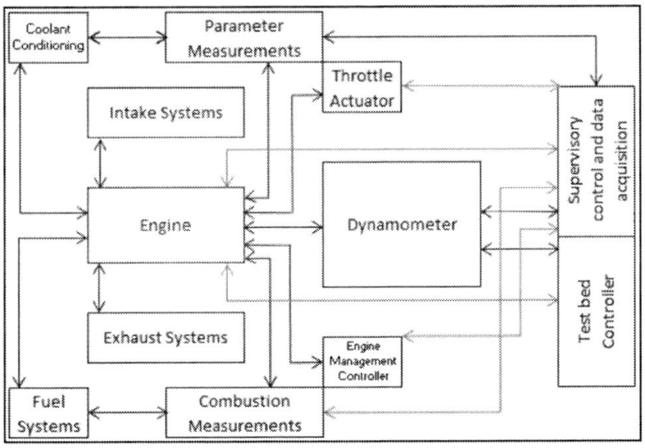

Figure 2 Layout of engine test bed

3. Different means of enhancing engine output

There are many means of increasing torque output of an engine. They can be categorized in manner in which they affect engine operation. Table 2 lists these approaches. It gives an insight of different findings about particular approach has on engine performance.

Table 2. Various approaches to engine output enhancements

Sr. No	Parameters	Effect
1.	Use of Turbocharger [14-22]	a. Provides considerable enhancement in torque and power, can extract waste exhaust energy, enhancing overall efficiency b. Lower the inertia of wheels, higher the output c. Combining supercharger and turbocharger delivers good response d. Two-stage turbocharger provides more boost delivering more engine power e. Use of VGT delivers enhanced transient response and avoids turbo lag f. Use of sequential turbocharger delivers better response and faster acceleration g. Using electrical turbocharger deliver about 20% more boost pressure
2.	Modifications to base engine [27-28, 30-32, 36-44, 47-48]	a. Reducing engine size for similar outputs helps enhance efficiency b. Leads to better mean effective pressures
3.	Use of Variable Valve Timing [34, 45-46]	a. Speed dependent higher engine output by about 15-20% b. 10% better volumetric efficiency

4.	Use of Variable valve timing and higher compression pressure [50]	a. High volumetric eff. with lesser valve overlap, irrespective of compression ratio b. Valve timing affects specific fuel consumption as well as compression ratio c. Better efficiency with higher compression ratio and valve overlaps
5.	Electronic Wastegate controls [49]	a. Better power output and low speed engine output b. Precise control over mass flow rates c. Higher in-cylinder pressures than pneumatic wastegate
6.	Swirl or Tumble air motion [33]	a. More swirl, more pressure drop, lesser volumetric efficiency b. Better mass burn fraction with better swirl, better flame radius c. Swirl delivers better flame growth rate than tumble
7.	Direct Fuel Injection [23-26, 29]	a. Higher engine efficiency by about 4-5%. b. Enhanced volumetric efficiency
8.	HCNG with high compression ratio [51, 52]	a. Better torque specific fuel consumption, partial limitations b. Better heat release, high cylinder pressures
9.	Cylinder pressure [35]	a. Diesel engines have higher cylinder pressure than CNG engine b. Use of high compression ratio and better volumetric efficiency are main reasons

4. Simulation and engine testing

In order to evaluate effect of various approaches on engine low speed torque, it is planned to first simulate a model of engine and then validate it against engine testing data available.

As a part of virtual testing, one dimensional model of engine under consideration is built in software GT ISE. Once the model converged, it was capable to predict engine performance based on changes in valve overlaps, different valve timings, different compression ratios etc. Schematic virtual representation of engine under consideration can be seen here, in figure 3. Actual components of engine and their arrangements are numerically represented. Geometric attributes and limiting conditions are applied as inputs to the model. Simulation runs are then conducted at wide open throttle to observe various parameters at different engine speeds.

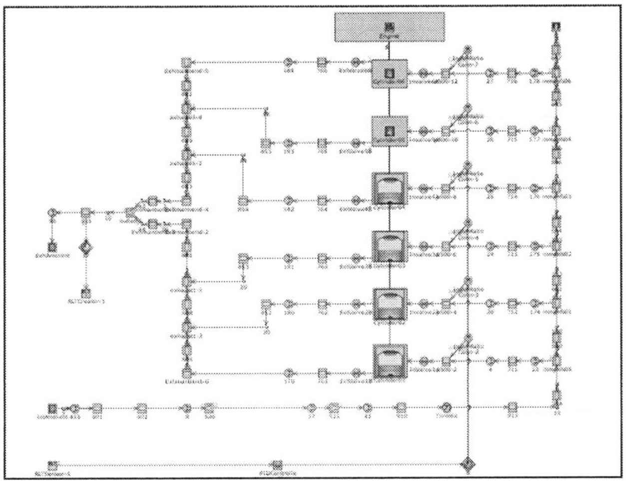

Figure 3 Base natural gas engine as modelled in software

Closeness of simulated and measured values is desired to assure resemblance of simulated model to actual engine. Thus, 1D simulation of the engine under consideration was carried out using the various dimensions and parameters from base engine specifications. The output of virtual model was then compared with data available from engine testing.

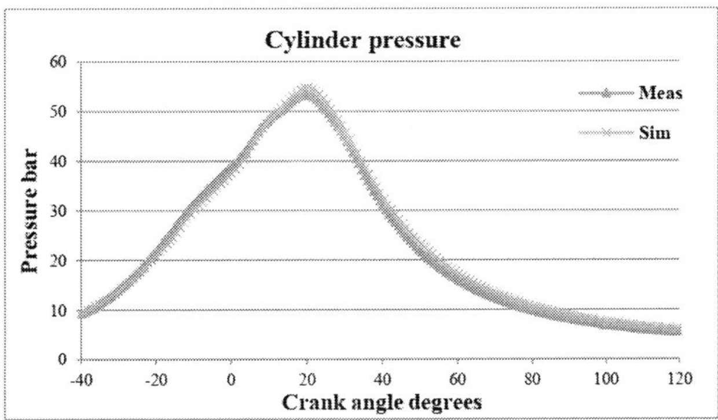

Figure 4 Comparing simulated and actual engine combustion pressure

Comparison of different engine parameters indicated that simulation is acceptable and the model is appropriately validated and faithful representation of actual engine. In figure 4, the combustion pressure trace of actual engine and simulated model show a close resemblance. Similarly, table 3 highlights measured as well as simulated values for critical engine performance parameters for engine speeds corresponding to maximum power and maximum torque. It can thus be said that as these values match closely and model is a true representation of engine under consideration, based on which other virtual trials can be carried out.

Table 3. Test data and simulation output comparison

Parameters	**Test Data**	**Sim. data**	**% Diff**
2500 RPM @ Max Power			
Torque	357.11	356.76	0.10
Power	93.44	93.401	0.04
Air flow (kG/hr)	355.65	348.29	2.07

Fuel flow (kG/hr)	22.09	21.53	2.54
BMEP	7.91	7.89	0.25
Brake thermal eff.	30.5	30.65	-0.49
1500 RPM @ Max Torque			
Torque	392.96	389.26	0.94
Power	61.69	61.11	0.94
Air flow	221.9	216.03	2.65
Fuel flow	13.38	13.13	1.87
BMEP	8.71	8.43	3.21
Brake thermal eff.	33.9	33.05	2.51

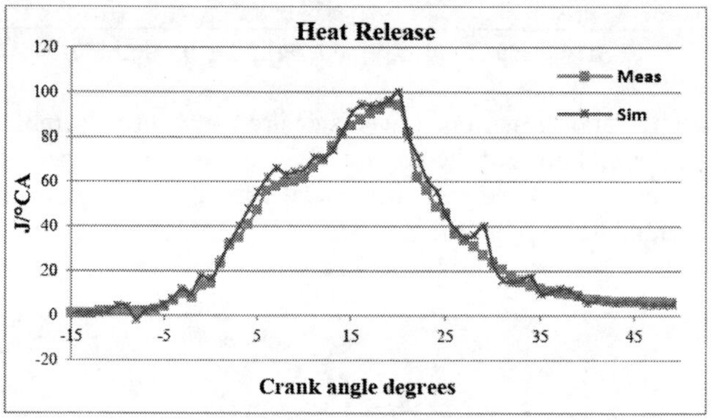

Figure 5 Heat release rate

Figure 5 indicates heat release rate for engine speed of 2500RPM, corresponding to maximum power. It can be observed that the values follow each other closely. The measured values are captured on a high speed data acquisition system and most of the noise in signal is filtered out. This validated model will now serve as base onto which different technologies such as different compression ratios, valve timings etc. shall be iterated to verify the extent by which engine output is affected and to optimize it for better low speed torque. Actual prototype assembly and respective testing shall then be carried out. It may require further optimization, which shall be determined after extensive analysis of simulated data.

Figure 6 shows comparison between simulated and measured engine torque and power. One can see that the closeness of different parameters assures model is well adapted to represent the engine configuration and any outputs based on other iterations can be taken as valid for comparing gain or loss of output.

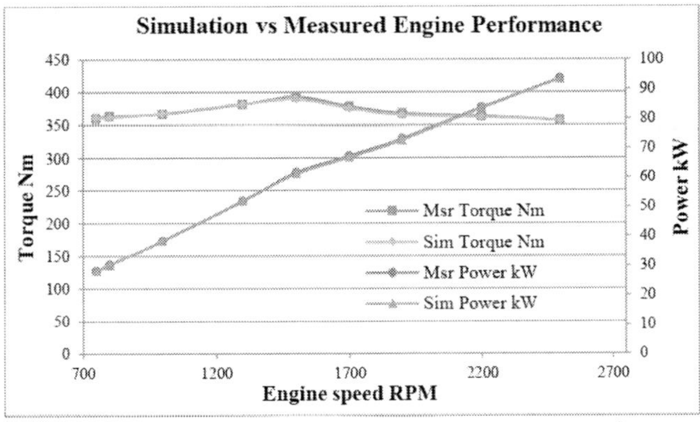

Figure 6 Simulated and measured engine torque and power

5. Turbocharging

Turbocharging: - For effectively utilizing energy of exhaust gases that would have otherwise been wasted a turbocharger is one of the convenient device. However, to do so the exhaust gases have to be moved from cylinders to the turbine, where the energy of exhaust gases is converted into mechanical work, required to drive the compressor.

However, like any other machine, the exhaust gas turbine is able to convert only a part of exhaust energy into mechanical work. The quantum of conversion depends on differential pressure across the turbine and turbine efficiency. As differential pressure increases across turbine, as the turbine efficiency increases, as the transmission efficiency increases the engine efficiency also increases.

Selection of proper turbocharger and effective design of exhaust system can help to realize higher engine performance. Adequate studies have been carried out on exhaust design and turbocharging, while various factors affecting the transient response of engine have also been examined. These studies included in this paper focuses on enhancing the output of engine at low speeds, for a typical city mass transport vehicle application.

In the scope of work mentioned herewith in this paper, a six cylinder engine is considered for experimentation purposes. The subject engine is naturally aspirated as of now and shall be turbocharged with objective on enhancing torque at low engine speeds. For this, four different turbocharger configurations and one supercharger are available. Focus shall be on speeds lesser than maximum torque speeds. One dimensional simulations, optimizations and experimental verification are engaged to carry out required optimization.

Based on the simulations and other experimental verifications carried out for optimizing output of engine at low engine speeds, the six cylinder engine under consideration shows different options and extent to which each of these affect low speed torque. The summarized results show the potential for improving low speed engine torque with different options.

One dimensional modelling of engine under consideration allows to study most points of discontinuity under the model with suitable variation to fluid flow coefficient and engine parameters. Fairly faithful representation of simulated model with actual engine is essential for optimization of various parameters on virtual model. This will ensure its co-relation with actual engine. To evaluate the extent of variations in output based on simulated inputs, a sensitivity study is also conducted.

The numerical effect of turbocharging and its various factors on engine low speed torque is obtained from one dimensional virtual engine model. To have a realistic representation of engine under consideration, appropriate simulation software is used. The output of one dimensional model is then compared with engine performance data as available from test beds that includes combustion pressure traces and other critical parameters.

The engine under consideration here is a six cylinder inline engine operating on stoichiometric air fuel ratio. In turbocharged configuration, three cylinders form one bank and rest three the other bank.

IOP Conf. Series: Materials Science and Engineering **810** (2020) 012050 doi:10.1088/1757-899X/810/1/012050

The engine will be shall be modelled accordingly, to include intercooler as well. Pipe connections, other junctions, restrictions, sequential injectors etc. are logically placed onto the model.

As the sole purpose of modelling the engine in virtual environment it to observe the effect of turbochargers and its allied systems, the pressure measurements are of prime importance. Correspondingly, the engine testing includes pressure measurements in exhaust pipes, as well. Similarly, the combustion pressure sensor is also included on the actual engine test bed testing and serves as an indication of closeness of virtual model to actual engine testing. Different temperature measurements are also included in engine measurements from test bed engine performance capturing activities.

The engine performance in virtual conditions is captured in the one dimensional model, as described. The transfer of heat from fuel to cylinder walls is based on Woschni formula. The combustion is simulated by referencing the heat release with respect to crank angles, described in terms of rotational degrees of crank shaft. The heat release patterns are captured with respect to crank angle domain. Other aspects of combustion as required for study are recorded using appropriate data acquisition systems.

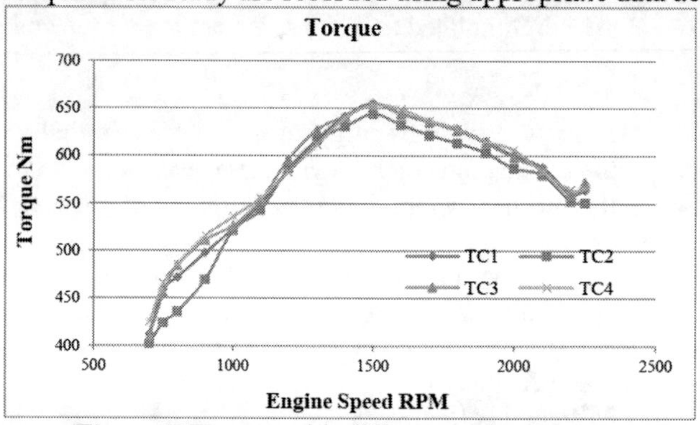

Figure 7 Torque with different turbochargers

The figure 7 shows torque output for different turbochargers assembled virtually on base engine. Variation in torque at low engine speeds, below maximum torque of 1500 RPM is to be observed. The difference in engine torque is due to different turbochargers that exhibit different characteristics. Objective is to identify these and provide a directions so that one can chose turbocharger appropriately that shall provide enhanced low speed torque.

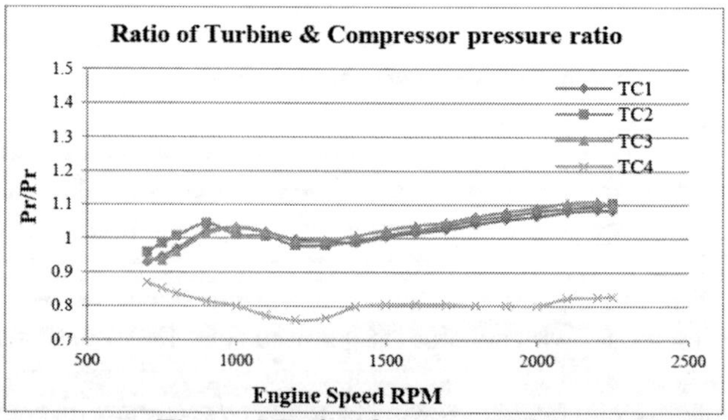

Figure 8 Ratio of pressure ratios

Figure 8 shows ratio of turbine pressure ration to the compressor pressure ratio, a dimension less parameter that helps in understanding of engine torque at low speeds. Based on the observations, it can be seen that lower this dimensionless number of ratio of turbine pressure ratio to the compressor pressure ratio, the engine output is better. Thus for better low speed torque, it is important that the compressor pressure ratio be as high as possible.

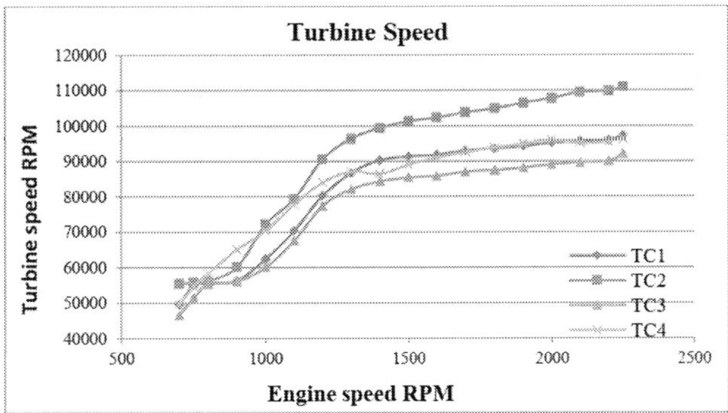

Figure 9 Ratio of pressure ratios

In the figure 9 it is seen that lower the turbocharger speed at which the desired pressure ratio can be met, better are chances of delivering higher engine torque. Thus, the speed at which pressure ratio can be achieved, in relation to intake conditions plays an important role in turbocharger selection. This parameter also affects the temperature of exhaust gases exiting the turbine.

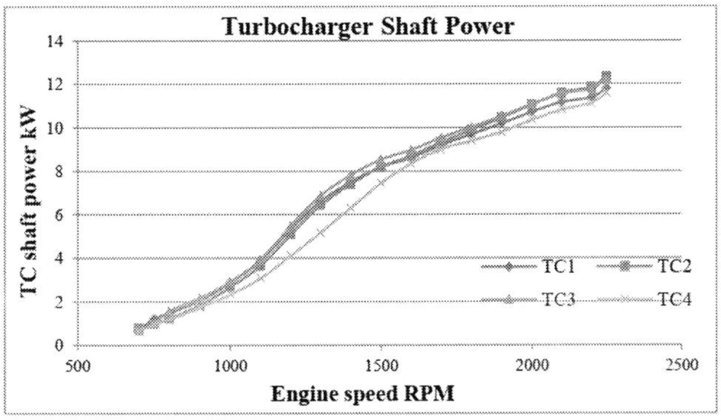

Figure 10 TC shaft power

Figure 10 refers to average power used to drive the turbocharger. This power is extracted from exhaust gases, and represents useful work done that would have been otherwise wasted. Principally this is a turbine characteristic of the turbocharger assembly. For enhancing output at low engine speeds, higher shaft power is a preferred. This assures better work extraction from the exhaust gases to spin the turbine, thus moving it into higher pressure ratio operation zone of the air compressor.

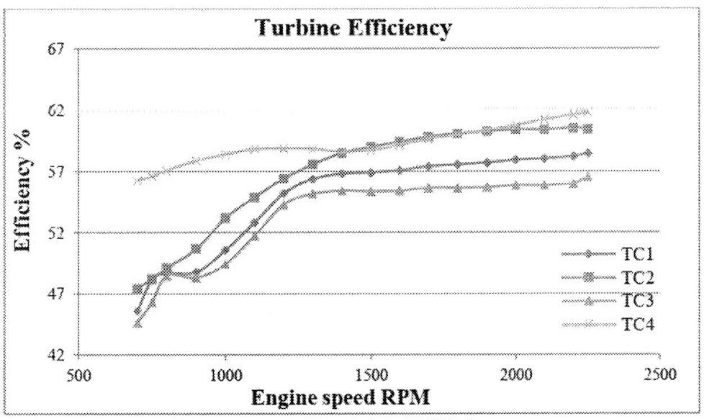

Figure 11 Turbine efficiency

Figure 11 here refers to efficiency of turbine and it is needless to say that higher the efficiency of turbine, higher will be the pressure ratio, leading to higher pressures on the compressor side. As more energy is converted into useful work, the turbine with higher efficiency shall provide higher boost and higher engine torque. This it is important the selection of turbine should by such that efficiency should be as high as possible in the desired engine operating region to have higher engine output.

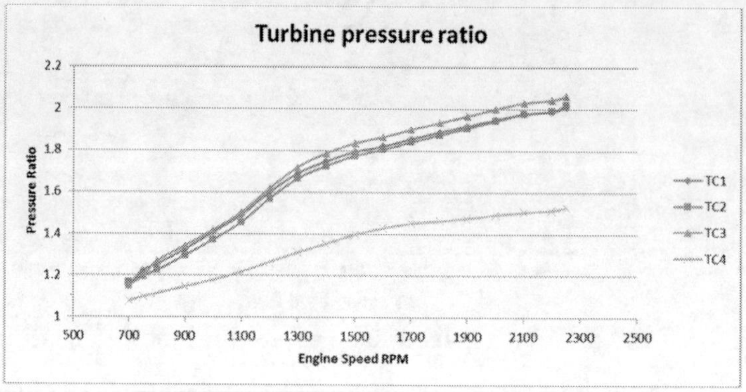

Figure 12 Turbine pressure ratio

Figure 12 refers to pressure ratio of turbine and it can be observed that as the pressure ratio of the turbine increases the engine torque also increases. However, it is important to note that efficiency of turbocharger also plays an important role. Lower pressure ratio with higher efficiency can compensate for lower output pressure, but maintains desired boost and thus the engine output.

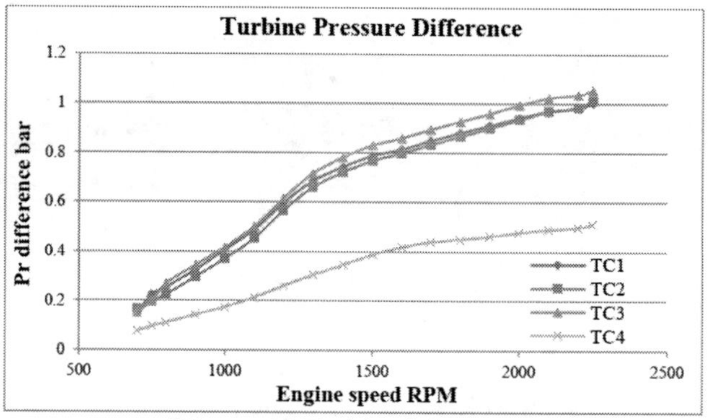

Figure 13 Turbine inlet pressure

Figure 13 refers to average pressure difference between turbine inlet and outlet pressures. Under ideal conditions, the inlet pressures are similar in each case, the variation observed in graphs is due to variation in output pressure that represents effective use of turbine. It is observed that higher the pressure difference, better is the turbine able to deliver higher boost, leading to higher engine output. However, other factors too are to be taken into consideration such as average efficiency and turbine speeds are also to be taken into consideration.

IOP Conf. Series: Materials Science and Engineering **810** (2020) 012050 doi:10.1088/1757-899X/810/1/012050

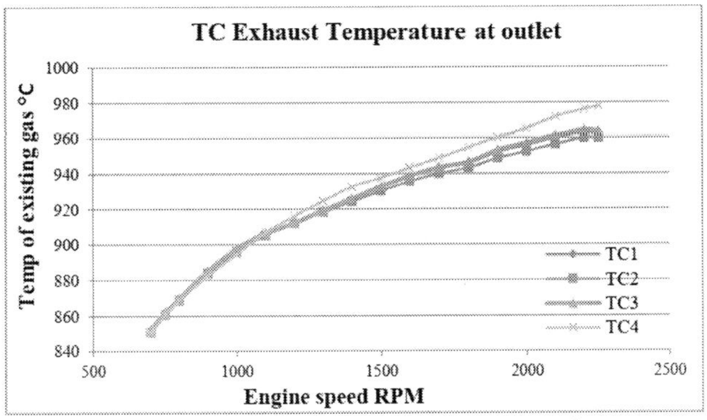

Figure 14 Exhaust temperature at outlet

From the figure 14 it is seen that higher exhaust temperatures exists only at high engine speed and load regions. In zones of lower engine speeds, the area of our interest, the exhaust temperatures for different turbochargers are comparable.

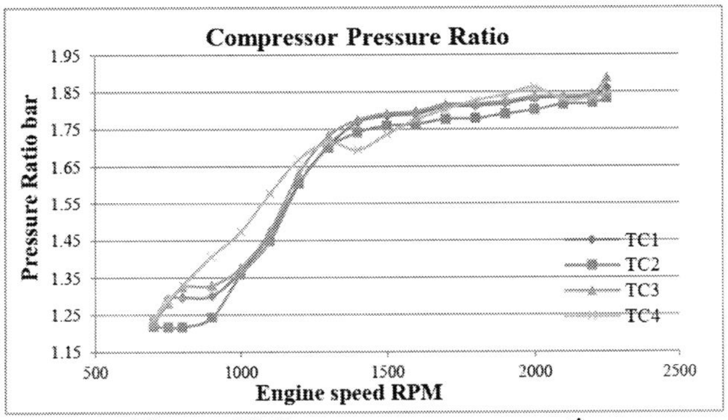

Figure 15 Compressor pressure ratio

The figure 15 above represents pressure ratio of the compressor and it can be seen that higher the pressure ratio, higher is the boost pressure and thus, higher the engine output torque in respective zone. It is to be noted that the inlet conditions to compressor are more or less the same and the pressure ratio is governed by the output pressures of the compressor.

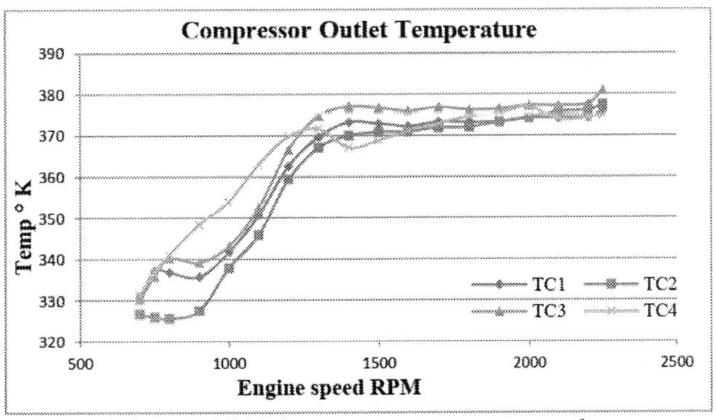

Figure 16 Exhaust temperature at outlet

As the pressure of the air at compressor outlet increases, so does its temperature. This reduces the effective mass flow rate of air as density reduces due to elevated temperatures. In most of the engine fitted with a kind of compressor arrangement to provide air at higher pressures, an intercooler is commonly found device that can add to efficiency of the system without much loss. Thus, lower the temperature of air at outlet of compressor, better will be the mass flow rate. Refer figure 16.

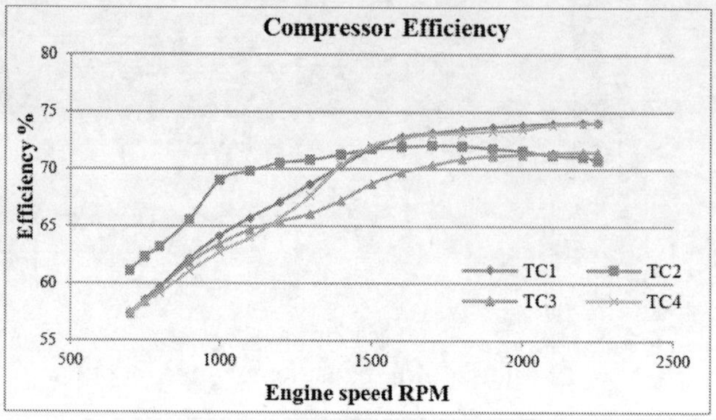

Figure 17 Compressor efficiency

The figure 17 above represents average efficiency against ideal condition. Theoretically speaking it is a ratio of ideal (Isentropic) power of the compressor to the actual power that it can possess. Thus, lower this value higher the engine torque

The performance of an actual turbo machine is always inferior to that of a frictionless and loss free ideal machine. Generally losses occur in compressors principally due to mechanical friction at bearings, windage losses, friction between blade and moving fluid etc.

The actual engine performance on turbocharged version was run with turbocharger 1and data was recorded on a modern and efficient engine test bed facility, equipped with high speed data recording devices. In addition to pressures and temperatures at different locations, the captured data also included typical engine performance describing data such as power, torque, turbocharger critical data, evaluation of lubrication & cooling systems etc. The recorded performance was compared with simulation results. Refer figure 18.

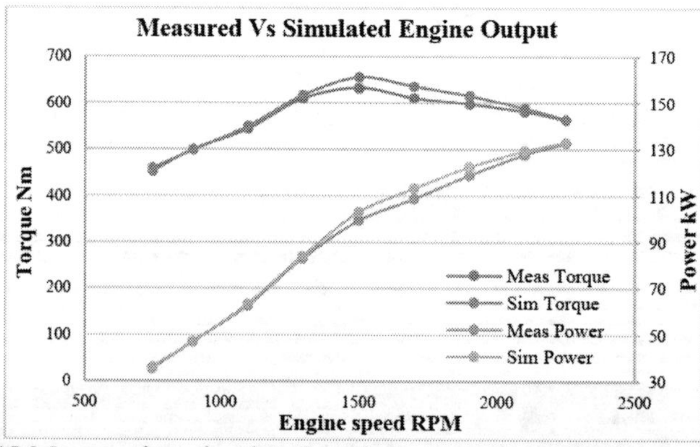

Figure 18 Measured vs simulated engine output for turbocharger sample 1

From the table 4, it can be seen that there is good representation between simulated data and engine test bed performance parameters, acknowledging accuracy of virtual engine model and measured variables. Thus as per the virtual model pressures and other simulated engine output, based on crank angle domain prove the true representation of model and same can be used for evaluating different means that affect the low speed torque.

Table 4. Test data and simulation output comparison

Parameters	Test Data	Sim. data	% Diff
1500 RPM @ Max Power			
Torque	651.11	655.34	-0.65
Power	102.22	102.89	-0.65
BMEP	14.41	14.51	-0.69

Since these match closely, the simulated model adequately represents the turbocharger sample 1 and hence the simulations of other turbocharger and subsequently the supercharger can be taken as applicable.

6. Supercharging

Supercharging refers to use of a compressor that is mechanically powered by crankshaft using an appropriate means such as gear or belt drive etc. This compressor draws air from atmosphere through an air filter and delivers it to intake manifold via an intercooler. As the compressor is mechanically coupled, the boosted charge is available almost instantly without any lag, but at the same time, as the energy is drawn from the crankshaft, it takes away a significant part of useful engine output, ultimately affecting fuel consumption.

As the compressor draws power from engine mechanically, the mass flow output of compressor does not diminish with respect to reduced load or air flow. As spark ignited engines are throttled with help of throttle valve, obviously air flow to engine reduces at part loads, when the throttle valve is partially closed. To overcome this situation where the compressor mass air flow does not reduce whereas engine fair flow demand reduces, a typical re-circulating valve is introduced that re-circulates or by-passes air from high side of the compressor to the low side, while maintaining the desired air flow to engine. This also helps to limit the maximum boost pressure at each speed and load as the valve's position changes as per driver demand. On the positive note, this behaviour of compressor can assist to have higher engine output at low engine speeds.

Virtual model of supercharger used in our study can be seen in figure 19. Essential characteristics are captured in required detail to enable setup to be as close as possible to realistic scenario. A typical representation is made for re-circulating valve to bypass the compressed air at part load running of engine. Similarly throttle body is introduced as a primary control of air entering into the engine.

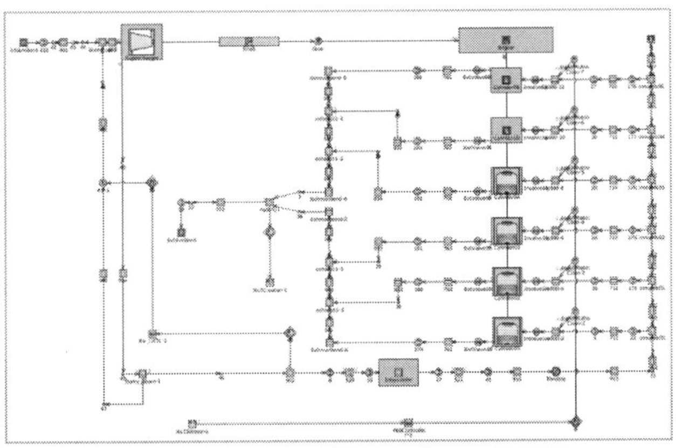

Figure 19 Simulated supercharged engine

The critical parameters from simulation of supercharged engine are compared against those from simulation of turbocharged versions. To have a realistic representation the simulated supercharged model is programmed to deliver same output as turbocharged version of engine so that the comparison can be more realistic.

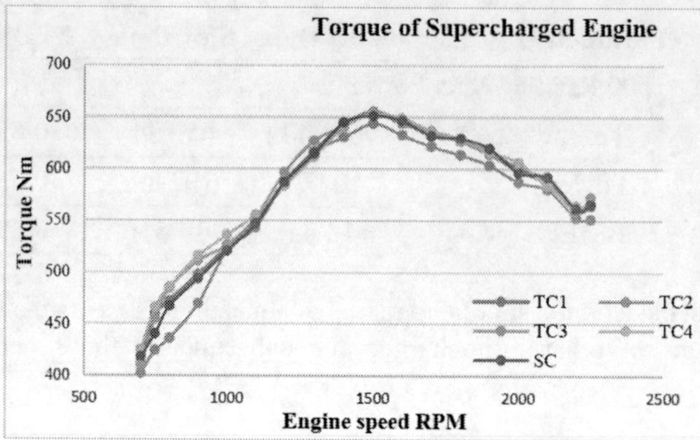

Figure 20 Turbocharged & Supercharged engine torque

Figure 20 represents torque output from 5 different configurations that are simulated one after the other on same base naturally aspirated engine, under consideration. Throttle valve position is varied to produce the desired torque. The intent is to compare other parameters such as specific fuel consumption etc. that are to be compared to decide better of two options when it comes to enhance low speed torque of engine.

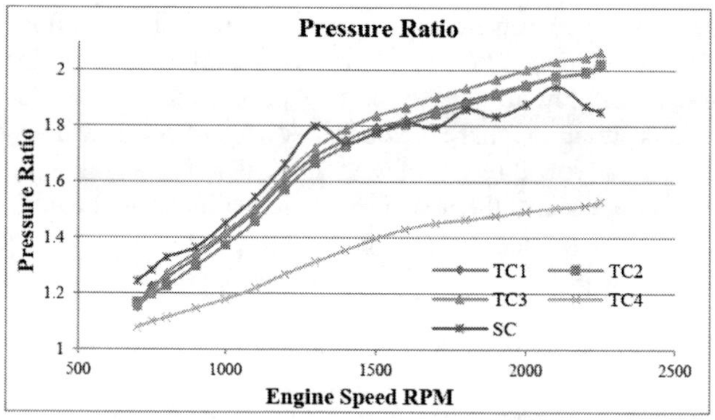

Figure 21 Turbocharged & Supercharged engine pressure ratio

The figure 21 above represents pressure ratio for compressor for supercharger. It is to be noted that boost pressure requirements to deliver same torque in turbocharger as well as supercharger remains more or less the same, across the engine speed. However, to deliver these levels of boost while maintaining the similar air flow rate, the supercharger draws considerable power from engine. In the figure 22, magnitude of supercharger power consumption can be observed.

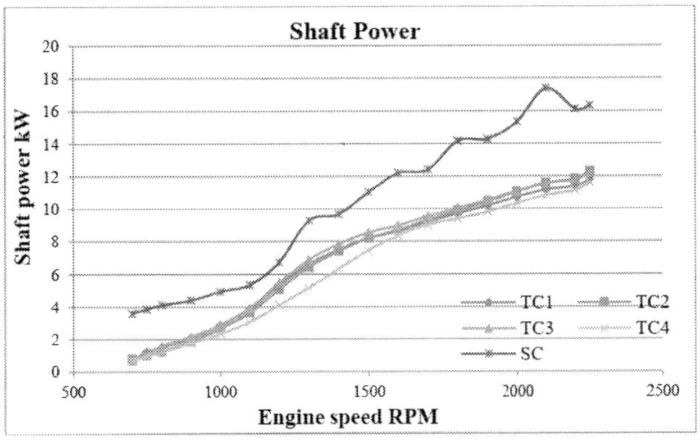

Figure 22 Turbocharger & Supercharger shaft power

The power consumed to run the supercharger is drawn from crankshaft and is a part of usable power that could have been made available to propel the vehicle. In case of turbochargers, this power is usually extracted from exhaust gases, thus enhancing specific fuel consumption. Superchargers thus suffer from poor fuel efficiency. Refer figure 23.

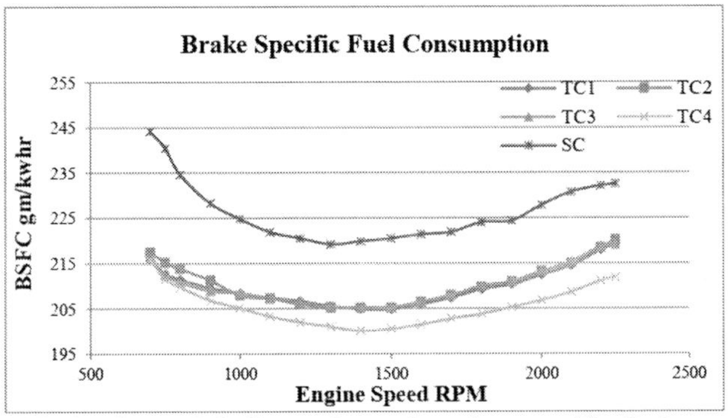

Figure 23 Turbocharged & Supercharged engine specific fuel consumption

The figure 24 represents efficiency of supercharger modelled to the base engine under consideration. In the area of interest i.e. the low engine speed region, it can be observed that the efficiency of supercharger as a compressor drops away rapidly. The compressor runs close to surge limit while operating in this zone. This is also a reason for poor efficiency of supercharger in desired zone of engine running.

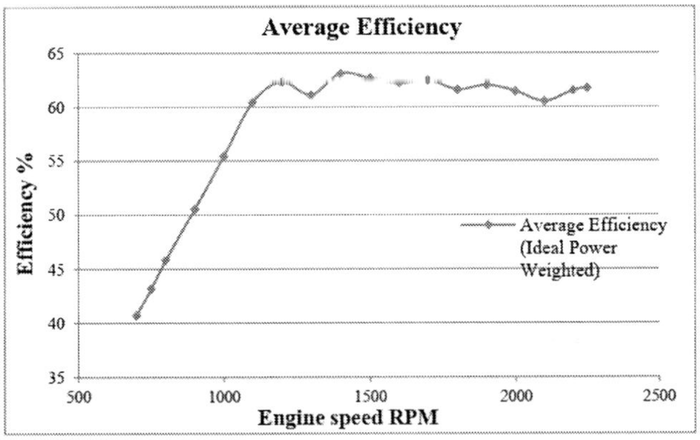

Figure 24 Efficiency of Supercharger

IOP Conf. Series: Materials Science and Engineering **810** (2020) 012050 doi:10.1088/1757-899X/810/1/012050

7. Conclusion

It is seen that both superchargers as well as turbochargers are capable of delivering the desired torque and provide low engine speed boost. However, it is to be noted that the supercharger does so at expense of useful work being derived from crankshaft, which ultimately lowers the overall fuel efficiency. Essential aspect of this study is to enhance output at low engine speeds. Investigation shows effect turbocharger has on overall impact on natural gas engine performance. It is further seen that there is a scope of improvement for in near future. Important conclusion that can be drawn are as follows: -

It is observed that the characteristics of turbocharger assembly considerably raises the availability of air for combustion, based on pressure ratio and mass flow rate. This presents opportunity for combusting more fuel and leads to higher engine outputs. In many cases, the turbocharger selection is done taking into consideration the maximum engine power and the maximum engine torque. It so happens that the availability of air and other critical turbocharger performance parameters specific to low engine speeds are neglected.

If adequate boost is made available at low speed zone, the turbocharger may prove to be smaller for higher engine outputs and can unnecessarily raise the inlet turbine pressures, causing difficulties to engine to expel the exhaust gases. This leads to waste of useful work. The benefits of turbocharger are thus not evident in such cases. Thus an ideal balance is to be maintained so that one achieves best combination of low engine speed torque as well as desired maximum engine power. A turbocharger aimed to enhance output at low engine speeds needs to bypass a lot of exhaust gases at high engine speeds, else the boost is high and engine tends to knock

Turbocharging presents its own set of challenges. Firstly, most of the turbochargers for commercial vehicles are made considering diesel engine as end applications. These usually operate at lower exhaust temperatures and high boost pressures. Diesel engines are quality governed and large lambda fluctuations are normal in its usual operations. Natural gas engines in the other hand need precise control of air as well as fuel. The stoichiometric air fuel ratio is to be maintained at all times, as the engine is quantity governed, a throttle valve controls the amount of charger entering the engine as per load demand. Thus there is a big difference in minimum and maximum air flow of a natural gas engine. The high exhaust temperatures associated with typical natural gas engines are beyond safe working temperature limits of most common turbine and turbine housing materials used for diesel engine. At times it is required to have coolant circulated through bearing housing of main shaft to keep bearings from overheating, thus making it a complicated design to manufacture.

Apart from availability, the matching of turbochargers for natural gas engines is critical. Usually the boost pressures associated with natural gas engines are low than diesel counterpart. Beyond this, the matching of turbochargers to provide desired boost with help of wastegate is a concern. The wastegate for natural gas engines are softer than diesel applications. The spring is soft and questions of consistency of performance are always raised. Spring hysteresis makes matter more difficult. Since the turbochargers for natural gas engines run hotter, the pneumatic hose leading from compressor to wastegate is to be specially designed to withstand high temperatures. For these reasons as well, use of electronic wastegate is a problem with natural gas engines.

On the other hand, using supercharger to enhance low speed torque seems to be a relatively simple option considering the fact the high exhaust temperature material requirement is no longer significant. However, just like turbochargers, superchargers are not designed specifically for commercial vehicles. Thus, the drive arrangement of superchargers has to be carefully evaluated as most of the superchargers are designed for engine that runs at relatively higher speeds than commercial vehicles. This means that drive ratio has to be larger than normally used for cars, thus raising additional design problems. It is also to be noted that the use of supercharger shall significantly lower the fuel efficiency as all the power is drawn from crankshaft.

Secondly, a lot of compressed air is to be re-circulated at part loads. This wastes a lot of useful work done as most of the compressed air re-introduced into low pressure side. Considering tighter emission norms, a close loop control is to be introduced in addition to electronic throttle for the engine, making the engine management more complicated.

The actual application of means of boosting the engine to enhance torque output at low engine speeds involves careful evaluation of many parameters. However, these means provide significant enhancement as compared to naturally aspirated version of base engine. These, however are expensive and demand a lot of piping, controls, etc. that should be fitted in tight spaces, under the engine hood. The elevated maximum pressures of cylinders, mostly requiring careful re-consideration of materials strength and design as well. High exhaust temperatures associated with high power engines is yet another concern. However, considering the flexibility of different application that can be driven by a high torque engine such as air conditioner, pneumatic suspensions etc. use of these means to boost output at low engine speeds is worth considering.

References

[1] B S Reddy "Transportation, Energy and Environment: A Case Study of Bangalore", Economic and Political Weekly: Vol. 30(3), 161-170, Jan. 1995.
[2] R S Parker and C E Pettijohn "The Use of Alternative Fuels in the Private Trucking Industry: Is There a Viable Target Market?" Journal of Marketing Theory and Practice: Vol. 5(4), 88-93, Fall 1997.
[3] Frost & Sullivan, "HD Transit Bus Market—Global Analysis", presentation in market engineering, NF63-18, Mar. 2016.
[4] Frost & Sullivan, "Heavy Duty Truck Engine Platform Strategies", presentation in market engineering, NECE-18, Mar. 2016.
[5] A T Khan, V Kakde, V G Ramkrishnan "Strategic Analysis of the Light and Small Commercial", presentation in market engineering, Frost & Sullivan, P61F-18. Dec. 2012.
[6] Frost & Sullivan, "Global Powertrain Outlook 2015", presentation in market engineering, NF22-18, Mar. 2015.
[7] J Jacob, Y Abhimanyu, A H Kumar and S Singh "Strategic Analysis of Compressed Natural Gas (CNG) Passenger Cars Market in Europe", presentation in market engineering, MA30-18, Sep. 2014, Frost & Sullivan
[8] M I Khan, T Yasmeen, A Shakoor, B N Khan et. al., "Exploring the Potential of Compressed Natural Gas as a Viable Fuel Option to Sustainable Transport: A Bibliography (2001-2015)", Journal of Natural Gas Science and Engineering, Vol. 31(2016), doi:10.1016/j.jngse.2016.03.025
[9] S Sudhakar, B Lakshminarsimhan, S Kar and S Singh "Strategic Analysis of the Medium- to Heavy-duty Natural Gas Commercial Vehicle Market in Europe", presentation in market engineering, M8D2-18, Feb. 2013, Frost & Sullivan
[10] Frost & Sullivan, "Natural Gas Engines & Bio-Methane as Transportation Fuel", presentation in Energy & Power Systems, TechVision, D960 Oct. 2015.
[11] C Kailasam, V Shankar, F Leveque and L Whalen "Strategic Analysis of the Global City Truck Market", presentation in strategic insignt, ND32-18, Oct. 2014, Frost & Sullivan
[12] S Kumar and V Kakde "Impact of Regulatory Trends on Commercial Vehicles Industry in India", presentation in market engineering, P4ED-18, Jun. 2012, Frost & Sullivan
[13] N Kulkarni, M Sharma, V Kakde and V G Ramkrishnan "Indian Vehicle Technologies Evolution Road Map", presentation in market engineering, P62C-18, Jan. 2013, Frost & Sullivan
[14] E Pipitone, F Cagnes and A Beccari "Performance Prevision of a Turbocharged Natural Gas Fuelled S.I. Engine", SAE Technical Paper 2008-36-0058, series E, doi:10.4271/2008-36-0058
[15] E G Giakoumis "Review of Some Methods for Improving Transient Response in Automotive Diesel Engines through Various Turbocharging Configurations", Front. Mech. Eng. 2:4, May. 2016, doi:10.3389/fmech.2016.00004
[16] K T Garrett, K Newton and W Steeds "Turbocharging and supercharging" Chapter 16, The Motor Vehicle, Sep. 2007, ISBN 9780080537016

[17] U Kesgin, "Effect of turbocharging system on the performance of a natural gas engine", Energy Conversion and Management 46 (2005) 11–32, doi:10.1016/j.enconman.2004.02.006
[18] S Marelli, G Marmorato and M Capobianco "Evaluation of heat transfer effects in small turbochargers by theoretical model and its experimental validation", Energy 112 (2016), 264-272, doi:10.1016/j.energy.2016.06.067
[19] A Gharehghani, M Koochak, M Mirsalim and T Yusaf "Experimental investigation of thermal balance of a turbocharged SI engine operating on natural gas", Applied Thermal Engineering, 60 (2013) 200-207, doi:10.1016/j.applthermaleng.2013.06.029
[20] E Codan and C Christen "Further development of two-stage turbocharging systems for large engines", 11th International Conference on Turbochargers and Turbocharging, 2014, Pages 189–203, Woodhead Publishing, doi:10.1533/978081000342.189
[21] L Eriksson, L Nielsen, J Brugard, J Bergstrom et. al., "Modelling Of A Turbo Charged SI Engine", IFAC Proceedings Volumes, (34) 1, Mar. 2001, doi:10.1016/S1474-6670(17)34425-7
[22] S Thipse, A Dsouza, S Sonawane, S Rairikar et al., "Development of Multi Cylinder Turbocharged Natural Gas Engine for Heavy Duty Application," SAE Int. J. Engines 10(1):2017, doi:10.4271/2017-26-0065.
[23] A Boretti and S Jiang "Development of a two stroke direct injection jet ignition compressed natural gas engine", Journal of Power Technologies, 94 (3) (2014) 145–152.
[24] M A Kalam and H H Masjuki "An experimental investigation of high performance natural gas engine with direct injection", Energy 36 (2011) 3563-3571, doi:10.1016/j.energy.2011.03.066
[25] D Sankesh, J Edsell, S Mazlan, and P Lappas "Comparative study between early and late injection in a natural-gas fuelled spark-ignited direct-injection engine", Energy Procedia 110 (2017) 275 – 280, doi:10.1016/j.egypro.2017.03.139
[26] J Song, M Choi and S Park "Comparisons of the Volumetric Efficiency and Combustion Characteristics between CNG-DI and CNG-PFI engines", Applied Thermal Engineering (2017), doi:10.1016/j.applthermaleng.2017.04.110
[27] M Costa, F Catapano, P Sementa, U Sorge et. al., "Mixture preparation and combustion in a GDI engine under stoichiometric or lean charge: an experimental and numerical study on an optically accessible engine", Applied Energy 180 (2016) 86–103, doi:10.1016/j.apenergy.2016.07.089
[28] T Wang, X Zhang, J Zhang and X Hou "Numerical analysis of the influence of the fuel injection timing and ignition position in a direct-injection natural gas engine", Energy Conversion and Management (2017), doi:10.1016/j.enconman.2017.03.004
[29] J Sevik, M Pamminger, T Wallner, R Scarcelli et al., "Performance, Efficiency and Emissions Assessment of Natural Gas Direct Injection compared to Gasoline and Natural Gas Port-Fuel Injection in an Automotive Engine," SAE Int. J. Engines 9(2):1130-1142, 2016, doi:10.4271/2016-01-0806.
[30] P Zoldak and J Naber "Spark Ignited Direct Injection Natural Gas Combustion in a Heavy Duty Single Cylinder Test Engine - AFR and EGR Dilution Effects," SAE Technical Paper 2015-01-2808, 2015, doi:10.4271/2015-01-2808.
[31] P Zoldak and J Naber "Spark Ignited Direct Injection Natural Gas Combustion in a Heavy Duty Single Cylinder Test Engine - Start of Injection and Spark Timing Effects," SAE Technical Paper 2015-01-2813, 2015, doi:10.4271/2015-01-2813.
[32] P Zoldak and J Naber "Spark Ignited Direct Injection Natural Gas Combustion in a Heavy Duty Single Cylinder Test Engine - Nozzle Included Angle Effects," SAE Technical Paper 2017-01-0781, 2017, doi:10.4271/2017-01-0781.
[33] Y T Anbese, A A Rashid and Z A Karim "Flame Development Study at Variable Swirl Level Flows in a Stratified CNG DI Combustion Engine using image Processing Technique", Journal of Applied Sciences, 11(10): 1698-1706, 2011 ISSN 1812-5654, doi:10.3923/jas.2011.1698.1706
[34] J Wheeler, J Stein and G Hunter "Effects of Charge Motion, Compression Ratio, and Dilution on a Medium Duty Natural Gas Single Cylinder Research Engine," SAE Int. J. Engines 7(4):1650-1664, 2014, doi:10.4271/2014-01-2363

[35] Semin, A Ismail and T Nugroho "Experimental and Computational of Engine Cylinder Pressure Investigation on Port injection Dedicated CNG Engine Development", Journal of Applied Sciences 10(2):107-115
[36] R Tilagone and S Venturi "Development of natural gas demonstrator based on an urban vehicle with a down-sized turbocharged engine", Journal of Oil and Gas Science and Technology, 59 (6) 581-591, Nov. 2004, doi:10.2516/ogst:2004042.
[37] A Kaleli, M Ceviz and K Erenturk "Controlling spark timing for consecutive cycles to reduce the cyclic variations of SI engines", Applied Thermal Engineering 87 (2015) 624-632, doi:10.1016/j.applthermaleng.2015.05.042
[38] I Semin, A Ismail and A Bakar "Investigation of Torque Performance Effect of development of Sequential Injection CNG Engine", Journal of Applied Sciences 9(13): 2416-2423, 2009.
[39] D R Johnson, R Heltzel, A C Nix, N Clark, et.al., "Greenhouse gas emissions and fuel efficiency of in-use high horsepower diesel, dual fuel, and natural gas engines for unconventional well development", Applied Energy 206 (2017) 739–750, doi:10.1016/j.apenergy.2017.08.234
[40] Q Tang, J Fu, J Liu, F Zhou, et al. "Performance improvement of liquefied natural gas (LNG) engine through intake air supply", Applied Thermal Engineering 103 (2016) 1351–1361, doi:10.1016/j.applthermaleng.2016.05.031
[41] B Yan, H Wang, Z Zheng, Y Qin, et.al., "The effect of combustion chamber geometry on in-cylinder flow and combustion process in a stoichiometric operation natural gas engine with EGR", Applied Thermal Engineering (2017), doi:10.1016/j.applthermaleng.2017.09.067
[42] B Yadollahi and M Boroomand "The effect of combustion chamber geometry on injection and mixture preparation in a CNG direct injection SI engine", Fuel 107 (2013) 52–62, doi:10.1016/j.fuel.2013.01.004
[43] C Wu, K Deng and Z Wang "The effect of combustion chamber shape on cylinder flow and lean combustion process in a large bore spark-ignition CNG engine", Journal of the Energy Institute xxx (2015) 1-8, doi:10.1016/j.joei.2015.01.023
[44] Q Zhang, M Li, G Li, S Shao and P Li "Transient emission characteristics of a heavy-duty natural gas engine at stoichiometric operation with EGR and TWC", Energy (2017), doi:10.1016/j.energy.2017.05.039
[45] A A Sabaruddin, S Wiriadidjaja, A S MohdRafie, F I Romli et. al., "Engine Optimization By Using Variable Valve Timing System At Low Engine Revolution", Journal of Engineering and Applied Sciences, 10 (20), Nov. 2015, ISSN 1819-6608
[46] A-F M Mahrous, A Potrzebowski, M L Wyszynski, H M Xu et.al., "A modeling study into the effects of variable valve timing on the gas exchange process and performance of a 4-valve DI homogeneous charge compression ignition (HCCI) engine", Energy Conversion and Management 50 (2009) 393–398, doi:10.1016/j.enconman.2008.09.018
[47] D Ramasamy, Z A Zainal, K Kadirgama, H W Briggs, "Effect of dissimilar valve lift on a bi-fuel CNG engine operation", Energy 112 (2016) 509-519, doi:10.1016/j.energy.2016.06.116
[48] M Baratta, D Misul, J Xu, A Fuerhapter et al., "Development of a High Performance Natural Gas Engine with Direct Gas Injection and Variable Valve Actuation," SAE Int. J. Engines 10(5):2017, doi:10.4271/2017-24-0152
[49] L Romani, G Vichi, A Bianchini, L Ferrari et. al., "Optimization of the Performance of a Formula SAE Engine by means of a Wastegate Valve Electronically Actuated", presented at 71st Conference of the Italian Thermal Machines Engineering Association, ATI2016, 14-16, Sep. 2016, Energy Procedia 101 (2016) 654 – 661.
[50] S H Khudhur, A M Saleh and M T Chaichan "The Effect of Variable Valve Timing on SIE Performance and Emissions", International Journal of Scientific & Engineering Research, 6(8), Aug. 2015, ISSN 2229-5518
[51] J Zhao, F Ma, X Xiong, J Deng et. al., "Effects of compression ratio on the combustion and emission of a hydrogen enriched natural gas engine under different excess air ratio", Energy 59 (2013) 658-665, doi:10.1016/j.energy.2013.07.033
[52] H B Mathur and L M Das "Performance Characteristics Of A Hydrogen Fuelled S.I. Using Timed Manifold Injection", Int. J. Hydrogen Energy, Vol. 16, No. 2, pp. 115-127, 1991.

IOP Conf. Series: Materials Science and Engineering **810** (2020) 012051 doi:10.1088/1757-899X/810/1/012051

Metamodel-based parametric study of composite laminates

Vishal Fegade, Shannay Rawal, Ramachandran M

MPSTME, SVKM's NMIMS, Shirpur, Dhule, Maharashtra 425405, India

vishalfegade@gmail.com

Abstract. Composite laminates are an integral part of structural engineering. To exploit the full potential of such laminated structures, it is important to understand the effect of various material and geometric parameters on its static and dynamic behaviour. Conventionally high-fidelity numerical procedures like finite element methods have been used to carry out such analysis. Despite their overwhelming accuracy, such methods are computational time intensive making them unsuitable to carry out exhaustive parametric studies. To counter this, a polynomial regression-based metamodeling strategy and its subsequent use as a quick parametric analysis tool is depicted in this study. The metamodels are meticulously built and rigorously validated on the training and testing data using suitable statistical measures like〚 R〛^2,〖〖 R〗^2〗_adj and 〖Q^2〗_F3.

1 Introduction

The use of composite materials is at an all-time high. Every day more and more of the traditional structural elements are being replaced by the composite materials. This is due to the superior qualities of composite materials like high strength-weight and high stiffness-weight ratios [1]. The use of composite laminates in structural engineering has ushered the need to investigate their behaviour under static and dynamic loading conditions.
Conventionally numerical structural analysis of composite plates has relied on used of high-fidelity but time-intensive approaches like finite element method. Owing to notable advances in various machine learning and allied approaches, a major shift in development and use of local approximate methods has been seen since the last decade. These 'local approximations' or metamodels are special purpose, computationally cheap models often developed to provide a quick but rough estimation of the modelled process or phenomenon. Among these, the artificial neural network [2], radial basis functions [3] and response surface methodology (RSM) [4] [5] [6] are most commonly used.
Response surface methodology starts by choosing a polynomial form to which it tries to fit the training data. In past, RSM has been used by various researchers to approximate the free vibration behaviour of composite plates [7] [8] and shells [9]. Ganguli [10] used a central composite design based RSM methodology to model a helicopter rotor. He fitted a second-order RSM model for 3 design variables to conduct an aeroelastic analysis. Ju et al. [11] used a similar CCD RSM approach to carry out a structural optimization of trusses. Dey et al. [12] used an RSM D-optimal design to model composite shells.
In this paper, we develop a set of RSM metamodels to conduct a parametric study on the free vibration behaviour of composite plates. The built metamodels are validated using independent test datasets.

2 Problem Definition

The objective is to develop appropriate RSM metamodels to study the effect of various material and geometric parameters on the fundamental frequency of composite plates. The various material properties considered are— orthotropy ratio$\left(\frac{E_1}{E_2}\right)$, major shear modulus to Young's modulus ratio$\left(\frac{G_{12}}{E_2}\right)$, minor shear modulus to Young's modulus ratio$\left(\frac{G_{23}}{E_2}\right)$ and Poisson's ratio(ϑ_{12}). The geometric parameters considered in the study are— composite plate height-to-width ratio$\left(\frac{b}{a}\right)$, plate thickness to width ratio$\left(\frac{h}{a}\right)$, skew angle of plate(α) and number of plies(n).

The range of the various parameters used in the study are- $\frac{E_1}{E_2} \in \mathbb{R}[20,60], \frac{G_{12}}{E_2} \in \mathbb{R}[0.4,0.8], \frac{G_{23}}{E_2} \in \mathbb{R}[0.3,0.7], \vartheta_{12} \in \mathbb{R}[0.19,0.31], \frac{b}{a} \in \mathbb{R}[1,5], \frac{h}{a} \in \mathbb{R}[0.01,0.1], \alpha \in \mathbb{R}[0,60]$ $and\ n \in \mathbb{Z}[2,10]$. In total four different cases are considered for the parametric study.

Case 1: An all side simply-supported, 8-layer symmetric ([45/-45/45/-45]$_s$), square composite plate with h/a of 0.01 is chosen as the structure of interest.

Case 2: Same as case 1, except instead of square, a 30° rhombic composite plate is considered.

Case 3: For studying the effect of geometric parameters, an all side simply-supported symmetric composite laminate is considered. The material properties are chosen as $\frac{E_1}{E_2} = 40, \frac{G_{12}}{E_2} = 0.6, \frac{G_{23}}{E_2} = 0.5\ and\ \vartheta_{12} = 0.25$.

Case 4: Same as case 3, except instead of all side simply-supported, an all side clamped symmetric composite laminate is considered.

3 Methodology

A response surface methodology based polynomial regression approach is adopted for metamodeling. A second-order model is selected as the polynomial basis function.

$$y = \beta_0 + \sum_{i=1}^{n} \beta_i\, x_i + \sum_{i=1}^{n} \sum_{j=i+1}^{n} \beta_{ij}\, x_i\, x_j + \sum_{i=1}^{n} \beta_{ii}\, x_i^2 + \varepsilon$$

One disadvantage of using a second-order polynomial basis function is that it cannot capture the very high non-linearity and as such would fail miserably on non-linear hard-to-model processes. To counter this, a Box-Cox transformation approach is adopted [13] [14].

$$y' = \begin{cases} (y + C)^{\lambda} & (\lambda \neq 0), \\ \ln(y + C) & (\lambda = 0); \end{cases}$$

Where C is a constant and λ is the power of transformation. The commonly used transformations are,

$\lambda = 1.0$, no transformation

$\lambda = 0.5$, square root

$\lambda = 0$, natural log

$\lambda = -0.5$, inverse square root.

$\lambda = -1.0$, inverse

In the current research, a D-optimal design is used as per the recommendations of Kalita et al. [15] to model the training dataset. The training data needed to fit the above equation for the 4 selected cases is generated by using Kalita et al.'s special purpose finite element program [16]. More details regarding the finite element formulation can be obtained at [17] [18].

The difference between the actual design points (finite element) $(y_i)$and the metamodel predicted design points $(\hat{y}_i)$ is called as residue.

$$\varepsilon_i = y_i - \hat{y}_i$$

The β_i estimates in the second-order RSM equations are selected such that the sum of squares of the residuals is minimized. Sum of residuals is given as,

$$SS_R = \sum_{i=1}^{n} (y_i - \hat{y}_i)^2$$

Similarly, the total sum of squares is given as,

$$SS_T = \sum_{i=1}^{n} (y_i - \bar{y})^2$$

Where, $\bar{y}$ represents mean of the actual training dataset.

To measure the accuracy of the built RSM metamodels criteria like R^2 and $R^2{}_{adj}$ are used.

$$R^2 = 1 - \frac{\sum_{i=1}^{n} (y_i - \hat{y}_i)^2}{\sum_{i=1}^{n} (y_i - \bar{y})^2}$$

$$R^2{}_{adj} = 1 - \frac{n-1}{n-k-1}(1 - R^2)$$

Since R^2 and $R^2{}_{adj}$ provide only a measure of goodness of fit of the training data, it is important to rely on external validation metrics like $Q^2{}_{F3}$. $Q^2{}_{F3}$ proposed by Consonni et al. [19] is given as,

$$Q^2{}_{F3} = 1 - \frac{\sum_{i=1}^{n_{test}} (\hat{y}_i - y_i)^2 / n_{test}}{\sum_{i=1}^{n_{train}} (y_i - \bar{y}_{train})^2 / n_{train}}$$

To calculate the $Q^2{}_{F3}$ additional 100 sample points are generated which constitute the testing data.

4 Results and Discussion

4.1 RSM metamodels

Using the training data generated by the finite element program, four RSM metamodels are built for the 4 cases discussed in section 2.

Case 1:

$$\frac{1}{(\lambda_1)^2} = 0.0005 - 1.23 \times 10^{-5}\left(\frac{E_1}{E_2}\right) - 3.66 * 10^{-6}\left(\frac{G_{12}}{E_2}\right) - 2.93 * 10^{-6}\left(\frac{G_{23}}{E_2}\right) - 9.93 * 10^{-8}(\upsilon_{12}) + 1.16 * 10^{-8}\left(\frac{E_1}{E_2}\right)\left(\frac{G_{12}}{E_2}\right) + 1.84 \times 10^{-6}\left(\frac{G_{12}}{E_2}\right)\left(\frac{G_{23}}{E_2}\right) + 1.02 * 10^{-7}\left(\frac{E_1}{E_2}\right)^2 + 9.2 * 10^{-7}\left(\frac{G_{12}}{E_2}\right)^2 + 8.85 * 10^{-7}\left(\frac{G_{23}}{E_2}\right)^2$$

Case 2:

$$(\lambda_1)^2 = 219.8785 + 213.5809\left(\frac{E_1}{E_2}\right) + 439.0622\left(\frac{G_{12}}{E_2}\right) + 279.9152\left(\frac{G_{23}}{E_2}\right) + 250.7809(\upsilon_{12}) + 4.9276\left(\frac{E_1}{E_2}\right)\left(\frac{G_{12}}{E_2}\right) + 4.5119\left(\frac{E_1}{E_2}\right)\left(\frac{G_{23}}{E_2}\right) - 246.4965\left(\frac{G_{12}}{E_2}\right)\left(\frac{G_{23}}{E_2}\right) - 252.7662\left(\frac{G_{23}}{E_2}\right)(\upsilon_{12}) - 0.0551\left(\frac{E_1}{E_2}\right)^2 - 152.6717\left(\frac{G_{12}}{E_2}\right)^2 - 147.8168\left(\frac{G_{23}}{E_2}\right)^2$$

Case 3:

$$\frac{1}{\sqrt{\lambda_1}} = 0.1079 + 0.0246\left(\frac{b}{a}\right) - 0.3217\left(\frac{h}{a}\right) - 0.0005(\alpha) + 0.0021(n) + 0.01171\left(\frac{b}{a}\right)\left(\frac{h}{a}\right) - 0.0011\left(\frac{b}{a}\right)(n) + 0.0089\left(\frac{h}{a}\right)(\alpha) - 0.00004(\alpha)(n) - 0.0017\left(\frac{b}{a}\right)^2 - 0.000007(\alpha)^2$$

Case 4:

$$\ln(\lambda_1) = 4.4282 - 0.3614\left(\frac{b}{a}\right) - 1.3347\left(\frac{h}{a}\right) - 1.2572(\alpha) + 0.1514(n) + 8.6191\left(\frac{b}{a}\right)(\alpha) - 0.0742\left(\frac{h}{a}\right)(\alpha) - 0.0391\left(\frac{h}{a}\right)(n) - 0.0002(\alpha)(n) + 0.0428\left(\frac{b}{a}\right)^2 + 0.0004(\alpha)^2 - 0.0071(n)^2$$

4.2 Metamodel Validation

The metamodels are validated using R^2 and $R^2{}_{adj}$ on the training data and $Q^2{}_{F3}$ on the testing data. From Figure 1, it is clear that all the built models are highly accurate. In fact, the metamodels built for case 1 and case 2 have ideal R^2, $R^2{}_{adj}$ and $Q^2{}_{F3}$ values.

To further evaluate the predictive power of the metamodels, scatter plots between the actual finite element models and the predictive response surface methodology based metamodels are constructed. The closer the data points are to the diagonal line better is the accuracy of the metamodels [20]. If the data points lie above the line it means the metamodel is over-predicting while any data point below the diagonal signifies an underprediction by the metamodel. The scatter plots of actual model versus the metamodel for the four cases are reported in Figure 2 and Figure 3.

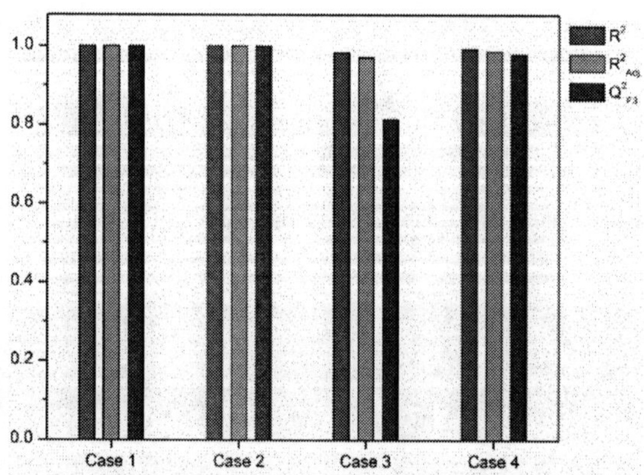

Figure 1. R^2, $R^2{}_{adj}$ *and* $Q^2{}_{F3}$ of all the four metamodels.

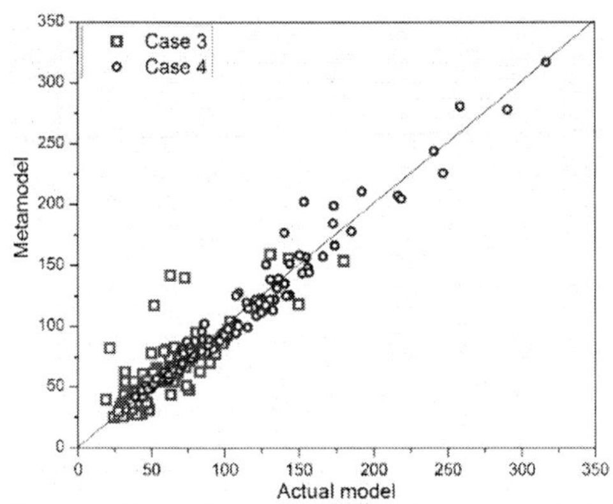

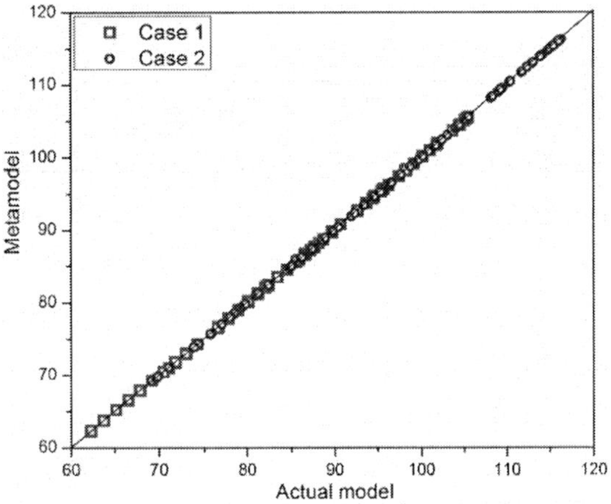

Figure 2. Predictive power of RSM metamodels in Case 1 and Case 2. Figure 3 Predictive power of RSM metamodels in Case 3 and Case 4.

4.3 Parametric study

The metamodels built in section 4.1 are used for conducting a parametric study on free vibration behaviour of composite plates. The effect of orthotropy ratio$\left(\frac{E_1}{E_2}\right)$ on the fundamental frequency is depicted in Figure 4. It is seen that the fundamental frequency monotonically increases as the orthotropy ratio increases. This is because the increasing orthotropy would increase the stiffness of the composite plates thereby increasing the frequency parameter. Also, it is clear that Case 2 i.e. rhombic composite plate has a higher frequency parameter as compared to Case 1 i.e. Square plate. The effect of major shear modulus to Young's modulus ratio$\left(\frac{G_{12}}{E_2}\right)$ on fundamental frequency is depicted in Figure 5. Fundamental frequency increases with increase in$\frac{G_{12}}{E_2}$. However, the increase is prominent only in case of rhombic plates. Similarly in Figure 6, increase in $\frac{G_{23}}{E_2}$ nominally increases the fundamental frequency. The effect of Poisson's ratio(υ_{12}) on fundamental frequency is depicted in Figure 7. The Poisson's ratio has a negligible effect on the frequency parameter of the composite plates.

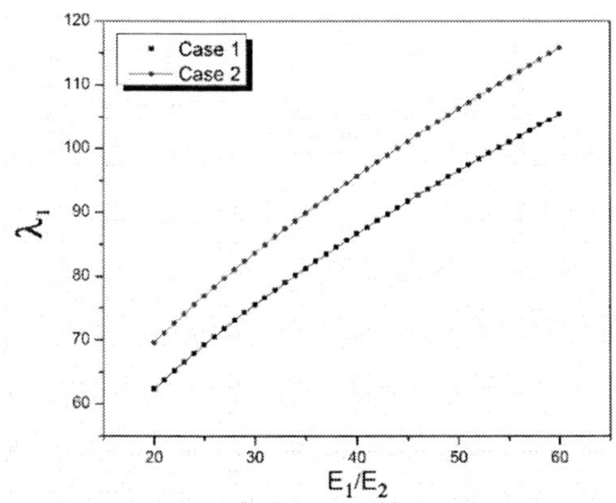

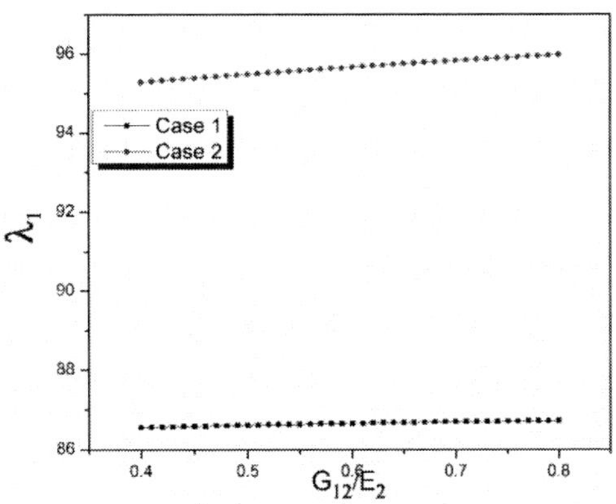

Figure 4. Effect of $\frac{E_1}{E_2}$ on fundamental frequency. Figure 5. Effect of $\frac{G_{12}}{E_2}$ on fundamental frequency.

The effect of composite plate height-to-width ratio$\left(\frac{b}{a}\right)$ on fundamental frequency is illustrated in Figure 8. In general, the increase in $\frac{b}{a}$ causes decrease in fundamental frequency. However, the boundary condition also plays a dominat role in the behaviour of the composite plate.

The influence of thickness ratio, $\frac{h}{a}$ on fundamental frequency is illustrated in Figure 9. It is clear that increase in $\frac{h}{a}$ causes a monotic decrease in fundamental frequency. This is due to the increase in mass of the composite plate. The drop in fundamental frequency with increase in thickness is more prominent in all sides clamped plate as compared to the all sides simply supported one.

Figure 10 demonstrates the influence of skew angle on the fundamental frequency. Increase in skew angle causes an increase in stiffness of the composite laminates, thereby increasing the fundamental frequency. In all sides simply supported laminates the increase is gradual whereas in clamped plates there is a rapid increase in fundamental frequency between 30° to 60° skew angle.
In general, increase in the number of layers for the same thickness causes a slight increase in fundamental frequency as shown in Figure 11. This is because as more as more plies are added, the rigidity of the composite plate increase which causes the fundamental frequency to increase.

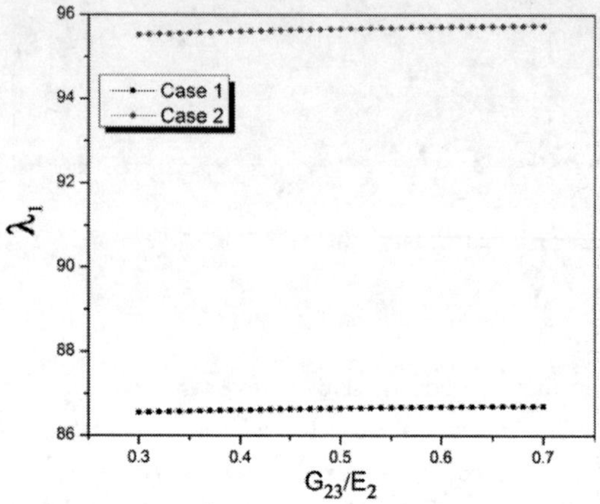

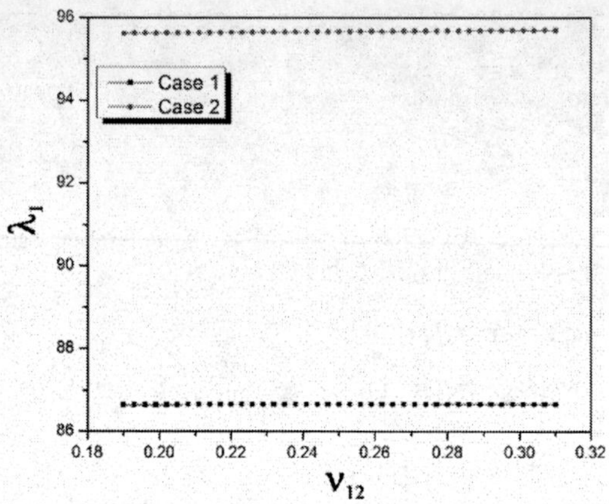

Figure 6. Effect of $\frac{G_{23}}{E_2}$ on fundamental frequency. Figure 7. Effect of υ_{12} on fundamental frequency.

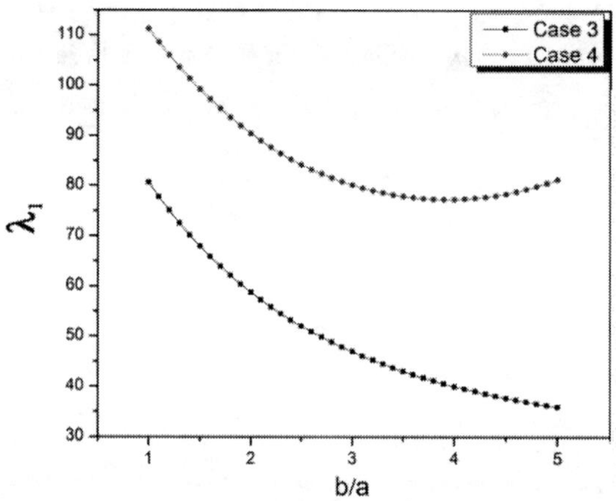

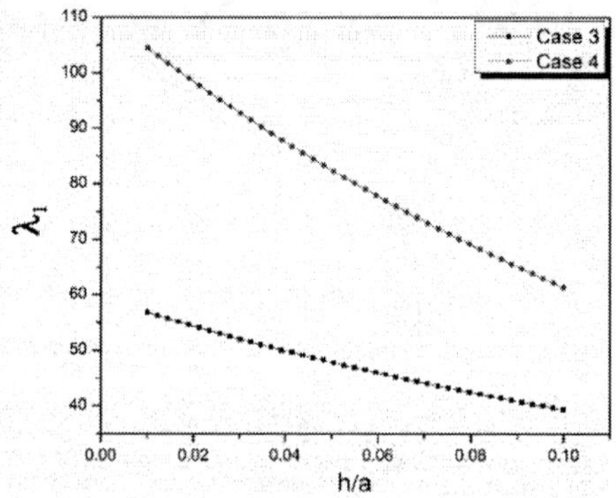

Figure 8. Effect of $\frac{b}{a}$ on fundamental frequency. Figure 9. Effect of $\frac{h}{a}$ on fundamental frequency.
It is also worth mentioning here that for all the parametric studies reported in Figure 4–Figure 11, effect of one factor is studied while keeping the other paramters in the metamodel at their mean level.

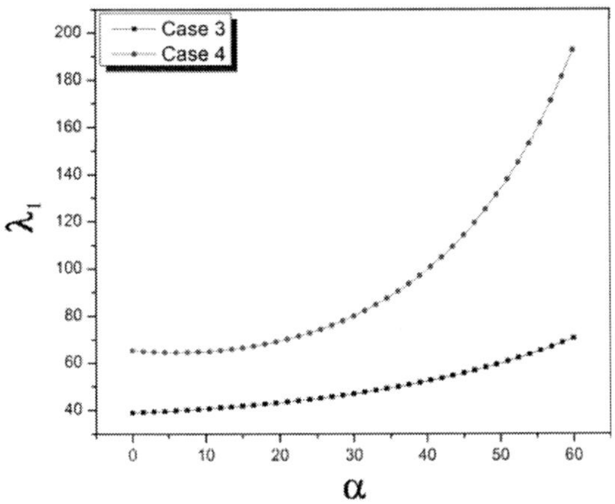

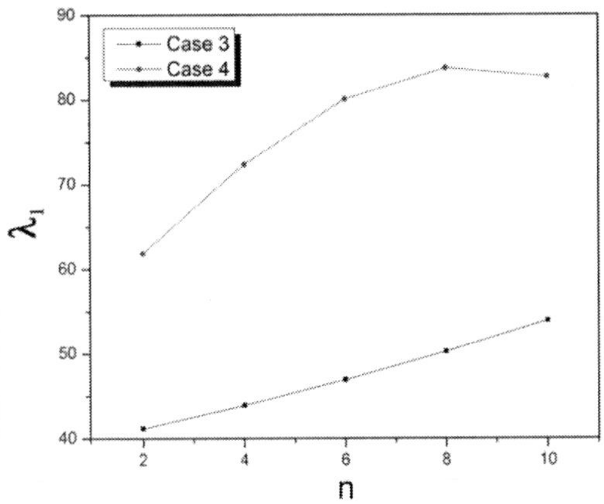

Figure 10. Effect of skew angle, α on fundamental frequency. Figure 11. Effect of number of layers on fundamental frequency.

5 Conclusions

In this work, a numerical study is conducted on the dynamic behaviour of composite laminates. Instead of using time-intensive finite element models, computationally cheap and easy to implement local approximate models (metamodels) are used. The metamodels require only a few sample points to train itself and replace the finite element model. These sample points or the training data may be obtained by conducting finite element simulations or experiments. Similarly, it is well advised to check the performance of the metamodels on independent testing data sets thereby making sure that the metamodels are not 'over-fitted' on the training data. Use of statistical metrics like R^2, $R^2{}_{adj}$ and $Q^2{}_{F3}$ provide a quick and reliable knowledge on the usability of the metamodels. Alternately, scatter plots between the actual model and the metamodel provides a quick visual assessment on the predictive power of the metamodel. Any under-prediction or over-prediction by the metamodel can be quickly identified. In the final part of the manuscript, the efficacy of the metamodels in carrying out a quick but approximate parametric analysis of composite laminates is illustrated. Thus, appropriately built robust response surface methodology based metamodels can be a useful alternative to finite element models.

References

[1] K. Kalita, P. Dey and S. Haldar, "Robust Genetically-Optimized Skew Laminates," *Proceedings of the Institution of Mechanical Engineers, Part C: Journal of Mechanical Engineering Science,* vol. DOI: 10.1177/0954406218756943, 2018.

[2] R. R. Behera, R. K. Ghadai, K. Kalita and S. Banerjee, "Simultaneous prediction of delamination and surface roughness in drilling GFRP composite using ANN," *International Journal of Plastics Technology,* vol. 20, pp. 424-450, 2016.

[3] R. L. Hardy, "Multiquadric equations of topography and other irregular surfaces," *Journal of geophysical research,* vol. 76, no. 8, pp. 1905-1915, 1971.

[4] G. E. P. Box and K. B. Wilson, "On the experimental attainment of optimum conditions," in *Breakthroughs in Statistics,* Springer, 1992, pp. 270-310.

[5] R. Ghadai, K. Kalita, S. C. Mondal and B. P. Swain, "Genetically optimized diamond-like carbon thin film coatings," *Materials and Manufacturing Processes,* vol. Doi: 10.1080/10426914.2019.1594273, pp. 1-12, 2019.

[6] K. Kalita, P. K. Mallick, A. K. Bhoi and K. R. Ghadai, "Optimizing Drilling Induced Delamination in GFRP Composites using Genetic Algorithm& Particle Swarm Optimisation," *Advanced Composites Letters,* vol. 27, p. 096369351802700101, 2018.

[7] A. Y. Abu-Odeh and H. L. Jones, "Optimum design of composite plates using response surface method," *Composite structures,* vol. 43, no. 3, pp. 233-242, 1998.

[8] K. Kalita, P. Nasre, P. Dey and S. Haldar, "Metamodel based multi-objective design optimization of laminated composite plates," *Structural Engineering and Mechanics,* vol. 67, pp. 301-310, 2018.

[9] S. Dey, T. Mukhopadhyay, H. H. Khodaparast and S. Adhikari, "Stochastic natural frequency of composite conical shells," *Acta Mechanica,* vol. 226, no. 8, p. 2537, 2015.

[10] R. Ganguli, "Optimum design of a helicopter rotor for low vibration using aeroelastic analysis and response surface methods," *Journal of Sound and Vibration,* vol. 258, no. 2, pp. 327-344, 2002.

[11] S. Ju, R. A. Shenoi, D. Jiang and A. J. Sobey, "Multi-parameter optimization of lightweight composite triangular truss structure based on response surface methodology," *Composite Structures,* vol. 97, pp. 107-116, 2013.

[12] S. Dey, T. Mukhopadhyay, H. H. Khodaparast and S. Adhikari, "A response surface modelling approach for resonance driven reliability based optimization of composite shells," *Periodica Polytechnica. Civil Engineering,* vol. 60, no. 1, p. 103, 2016.

[13] K. Kalita, I. Shivakoti and R. K. Ghadai, "Optimizing Process Parameters for Laser Beam Micro-Marking Using a Genetic Algorithm and Particle Swarm Optimization," *Materials and Manufacturing Processes,* vol. 32, no. 10, pp. 1101-1108, 2017.

[14] R. K. Ghadai, K. Kalita, S. C. Mondal and B. P. Swain, "PECVD process parameter optimization: towards increased hardness of diamond-like carbon thin films," *Materials and Manufacturing Processes,* vol. 33, no. 16, pp. 1905-1913, 2018.

[15] K. Kalita, T. Mukhopadhyay, P. Dey and S. Haldar, "Genetic programming-assisted multi-scale optimization for multi-objective dynamic performance of laminated composites: the advantage of more elementary-level analyses," *Neural Computing and Applications,* pp. 1-25, 2019.

[16] K. Kalita, U. Ragavendran, M. Ramachandran and A. K. Bhoi, "Weighted sum multi-objective optimization of skew composite laminates," *Structural Engineering and Mechanics,* vol. 69, pp. 21-31, 2019.

[17] K. Kalita and S. Haldar, "Eigenfrequencies of Simply Supported Taper Plates with cut-outs," *Structural Engineering and Mechanics,* vol. 63, no. 1, pp. 103-113, 2017.

[18] K. Kalita, M. Ramachandran, P. Raichurkar, S. D. Mokal and S. Haldar, "Free Vibration Analysis of Laminated Composites by a nine node isoparametric plate bending element," *Advanced Composites Letters,* vol. 25, no. 5, p. 108, 2016.

[19] V. Consonni, D. Ballabio and R. Todeschini, "Evaluation of model predictive ability by external validation techniques," *Journal of chemometrics,* vol. 24, pp. 194-201, 2010.

[20] U. Ragavendran, R. K. Ghadai, A. Bhoi, M. Ramachandran and K. Kalita, "Sensitivity Analysis and Optimization of EDM Process Parameters," *Transactions of the Canadian Society for Mechanical Engineering,* 2018.

IOP Conf. Series: Materials Science and Engineering **810** (2020) 012052 doi:10.1088/1757-899X/810/1/012052

Finite Element Analysis for Time Varying Mesh Stiffness behavior of different shapes of spalling

Akhilesh Lodwal, Dr. Ashesh Tiwari

Mechanical Engg. Department, IET DAVV, Indore, India
alodwal@ietdavv.edu.in, atiwari@ietdavv.edu.in

Abstract :Time Varying gear mesh stiffness is the variation of stiffness for the one contact period of a gear pair. Spalling on gear tooth is one of the most common defects in gear transmission. The loss of surface materials due to tooth spall reduces the Time-Varying Mesh Stiffness (TVMS) of the gear pair. The time varying mesh stiffness is the change in mesh stiffness for a gear pair in one contact period. The evaluation of the TVMS of the pair of gear tooth under gear tooth spalling conditions may be useful in finding its dynamic behavior or instant change in TVMS behavior may be the signal of fault developed. The potential energy method is used for analytical calculation of time varying mesh stiffness of the gear. The tooth spall may be of any shape and in this paper we have assumed the spall to be circular, triangular and elliptical and conducted analytical calculation to find TVMS and compare with healthy gear pair. The same gear pair with three shape of spall and modelled and verified through finite element analysis. The data may be used for further analysis. Also a number of shape may be modelled in future.
Keywords: TVMS, Potential Energy, Spalling,FEA.

I. Introduction

Gear is a important mechanical element which is used to transmit power from one shaft to another shaft. Gear is used in many industrial applications such as automobile, marine, aviation etc. Spalling occur on the gear face of the tooth profile, due to the insufficient lubrications, high service load, bad operating conditions, high contact speed and hence transmission of power is not uniform. The loss of surface materials due to tooth spall reduces the Time-Varying Mesh Stiffness (TVMS) of the gear pair, and thus modifies the vibration response of the gear transmission. The evaluation of the TVMS of the gear tooth pair under gear tooth spalling conditions plays an important role in gear dynamic Simulation and the corresponding fault feature analysis[1]. Based on the probability distribution a new analytical model for tooth pitting was developed and mesh stiffness was investigated [1]. Xihui Liang, et al. study a circular shape spall having specific location, radius and depth was considered. With the help of potential energy equations mesh stiffness of gears with single and multiple pits are evaluated[2]. Yang Luo, et al Gear tooth spall was studied by curved bottom shape method that better represents the geometries of spall. In this study different shape of spall like circular and ellipsoid are taken and mesh stiffness result were compared with FEA [3]. Ankur Saxena et. al. study the effect of spalling defect on the gear an analytical formulation was developed for the calculations of the TVMS.. The TVMS was calculated for rectangular, circular, ellipsoid shape of spall and observed TVMS reduction when spall in in mesh. H Ma, et al calculated error of TVMS under double tooth engagement by improved analytical method and finite element method l. [5] Yan Ding, F. Reiger In this study spalling is differentiated spalling from pitting. Experimental study in a test rig was done at the contact surfaces about the thickness of 10 microns and spalling appears as a deeper cavityat contact surface about the thickness of 20 to 100 microns. [6] Fakher Chari, Wallid Baccar l given modified the potential energy method for finding the spall and tooth breakage effect on vibratio response[7] Z. Chen, Y. Shao [8] In this study six degree of freedom dynamic spur gear pinion model is taken to examine the gear defects such as tooth root crack along the width and depth. Time varying mesh stiffness is calculated analytically and validated by finite element method. So many research was done to find the spalling defect analytically and its comparison either by FEA or dynamic response. In this paper along with circular and elliptical spall a new triangular error is used for TVMS change response analytically and its FEA was done further verification.

IOP Conf. Series: Materials Science and Engineering **810** (2020) 012052 doi:10.1088/1757-899X/810/1/012052

Methodology

For the study of time varying mesh stiffness a tooth is considered cantilever and spalling is assumed symmetrical about pitch line. . The spalling of triangular, Circular and elliptical are taken for study. The shapes are shown in figure 1.

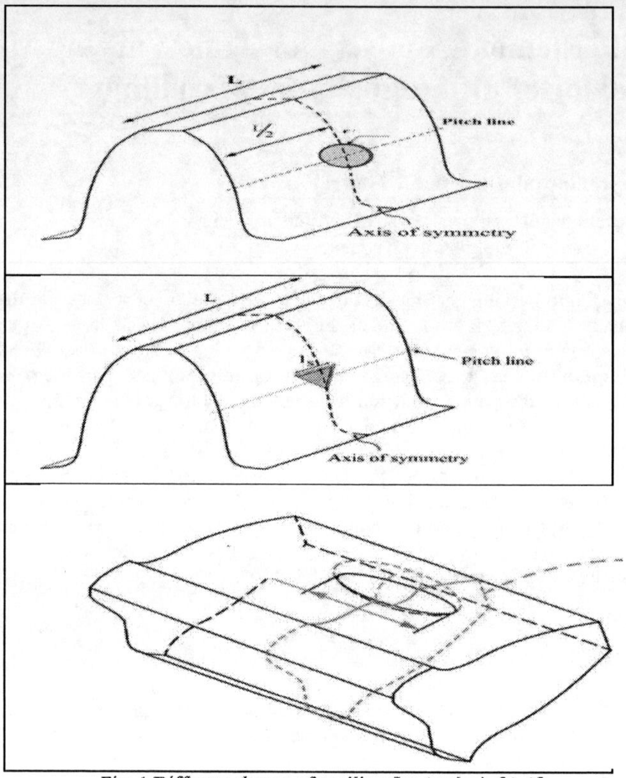

.Fig. 1 Different shapes of spalling for Analysis [4, 3]

The gear tooth geometry is as shown in fig. 2 spalling is located near the pitch circle.

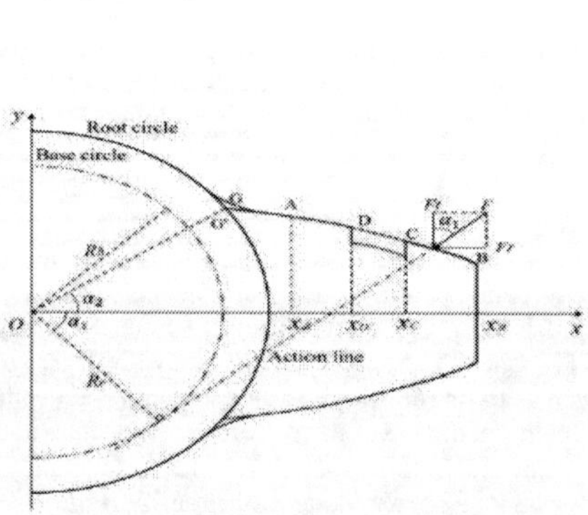

Fig. 2 The cross section of gear tooth [6]

In this case the gear fault is assumed symmetrical about pitch point so for the analysis the gear may be divided as figure no.2.

The length of DC will depend on the one of the dimension of the fault. The Values of x_A, x_B, x_C, x_D are the values of distance of points A, B, C and D which has been calculated from reference point A. and y values are useful for finding area and moment that is necessary in further calculations.

For the analysis purpose the teeth may be divided into a number of strips, the area of strip without spall is given by

$$A=-2yl \qquad \text{-------------(1)}$$

While for the spalled region it is calculated by

$$As=(y+ys)\,l \qquad \text{-------------(2)}$$

Where y, ys, represent the vertical values of the gear tooth contour line with respect to axis of gear teeth in healthy and spalled region.

During the meshing, when teeth contact force F is transmitted from one teeth to another teeth along the line of action as shown in fig4. This force F is having two one is in radial direction and another in tangential direction given by

$$F_r = F\sin\alpha_1 \qquad \text{----------(3)}$$

$$F_t = F\cos\alpha_1 \qquad \text{-----------(4)}$$

The bending moment of the teeth is also an important data that is necessary for Bending energy calculations. The value of bending moment for the healthy part of the tooth is given by

$$M= F_t(x_\beta - x_A) - F_r y \qquad \text{------------------(5)}$$

For the spalled region it will be

$$M_s= F_t(x_\beta - x_A) - F_r ys \qquad \text{-------------------(6)}$$

In the similar way the moment of inertial for healthy and spalled region of the teeth are given by equation 7 and 8.

$$I = \frac{2}{3}y^3 l \qquad \text{------------------------(7)}$$

$$Is= \frac{2}{3}ys^3 l \qquad \text{-----------------------(8)}$$

Analytical Calculation:

The potential energy method is used for all the calculations. The Bending stiffness, Shear Stiffness, Axial Stiffness and Contact stiffness were calculates by using the outcome of the following equations. [4,9].

$$\frac{1}{K_a}=\int_{-\alpha_1}^{\alpha_2}\frac{R_b\sin^2\alpha_1(\alpha_2-\alpha)\cos\alpha}{EA_s}d\alpha \qquad \text{-------------(9)}$$

$$\frac{1}{K_s}=\int_{-\alpha_1}^{\alpha_2}\frac{2.4R_b(1+\mu)\cos^2\alpha_1(\alpha_2-\alpha)\cos\alpha}{EA_s}d\alpha \qquad \text{-------------(10)}$$

$$\frac{1}{K_b}=\int_{-\alpha_1}^{\alpha_2}\frac{R_b{}^3[1+\cos\alpha\{(\alpha_2-\alpha)\sin\alpha-\cos\alpha\}]^3(\alpha_2-\alpha)\cos\alpha}{EI_s}d\alpha \qquad \text{----------(11)}$$

Contact stiffness- The Hertzian contact stiffness is a constant value along the line of action it is independent of contact position and expressed for healthy gear as-

$$K_h = \frac{\pi El}{4(1-\mu^2)} \qquad \text{------------------(12)}$$

The Hertzian Contact stiffness for faulty gear can be expressed as –

$$K_h = \frac{\pi E(l-l_s)}{4(1-\mu^2)} \qquad \text{----------------(13)}$$

Where E, L, μ represents young's modulus, tooth contact width and Poisson's ratio respectively.

The total equivalent mesh stiffness of single tooth pair in meshing can be expressed as-

$$K_{total} = \frac{1}{\frac{1}{K_h}+\frac{1}{K_{b_1}}+\frac{1}{K_{s_1}}+\frac{1}{K_{a_1}}+\frac{1}{K_{b_2}}+\frac{1}{K_{s_2}}+\frac{1}{K_{a_2}}} \qquad \text{-------------(14)}$$

Where subscripts 1and 2 denotes the driving gear and driven gear. The total equivalent mesh stiffness of double teeth pair in meshing can be expressed as

$$K_{total} = \sum_{i=1}^{2}\frac{1}{\frac{1}{K_{hi}}+\frac{1}{K_{b_{1i}}}+\frac{1}{K_{s_1}}+\frac{1}{K_{a_{1i}}}+\frac{1}{K_{b_{2i}}}+\frac{1}{K_{s_{2i}}}+\frac{1}{K_{a_{2i}}}} \qquad \text{---------------(15)}$$

Modelling for different shape of spall: In this paper the analytical calculation are done by assuming the spall to be symmetrical about pitch line. Three types of spall were modelled and the time varying mesh stiffness calculated.

1 Circular spall- A circular spall of radius r =1.5 mm and depth 1 mm is created at about pitch line .Using the equation of circle

x2+y2=r2

Where, r = radius of the circular spall and y varies from −r to+r. The length of spall ls=2x

2. Triangular Spall: A triangular spall of size 3mm x 3mm x 3mm and depth 1 mm is created on the gear surface at pitch line. The height of the triangle represent the length of spall and one of the side that is parallel with the width represents the maximum width of spall. For other region of the triangle the width of spall may by calculated manually marking a number of strips.

3. Elliptical Spall: The parametric function of an ellipsoid is given by

$$\frac{X^2}{a^2}+\frac{y^2}{b^2}=1$$

If a=b then it will be a circle

After modelling of the gear the minor axis will represent the length of spall and the major axis with depth will cause the change in width as width of contact is reduced by width of spall. In this case the major axis were taken as 5mm and the minor axis was 3 mm.

.

Gear Modeling: For the analytical as well as the finite element analysis the gear-pinion arrangement was modeled in software. Gear parameters used in this study are given below in table 1 [9].

Table 1: Parameters for gear modeling

Parameters	Gear	Pinion
Teeth numbers	30	25
Mass (Kg)	0.588	0.46
Tooth width (mm)	20	20
Modulus of elasticity (Gpa)	210	210
Poisson's ratio	0.3	0.3
Shear modulus(Gpa)	78	78
Shear factor	2	2
Module	2	2
Pressure angle	20°	20°

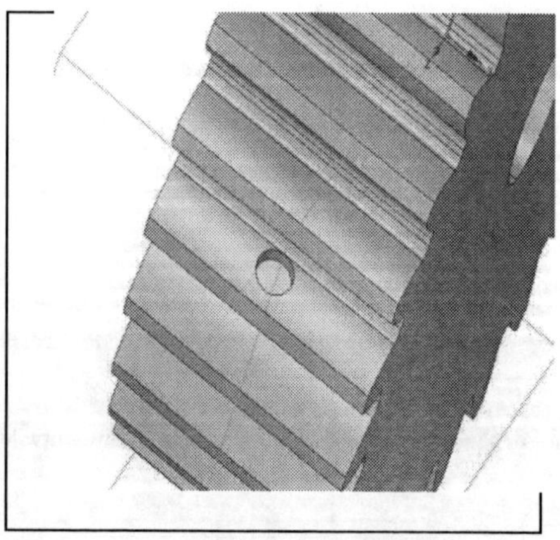

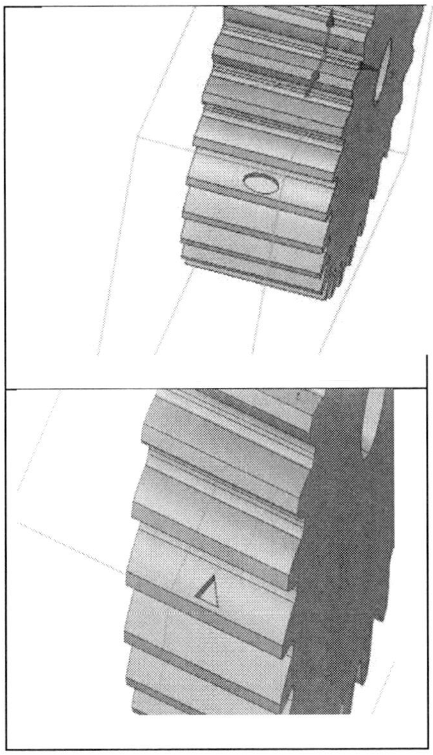

Fig-3 : Model used for the finfite element analysis.

Finite element analysis:

After analytical calculation model was tested in the ansys software for finding its stiffness. The model is imported in preprocessing, meshed and applied by fixing one and moment applied by other. A force of 1000 N is supposed to apply on the gear teeth of pinion by teeth of gear. The deflection of gear is used for its stiffness value. The number of load position used and data were compared with the analytical data. The Figure 4 is showing the deflection in the teeth when the load is applied.

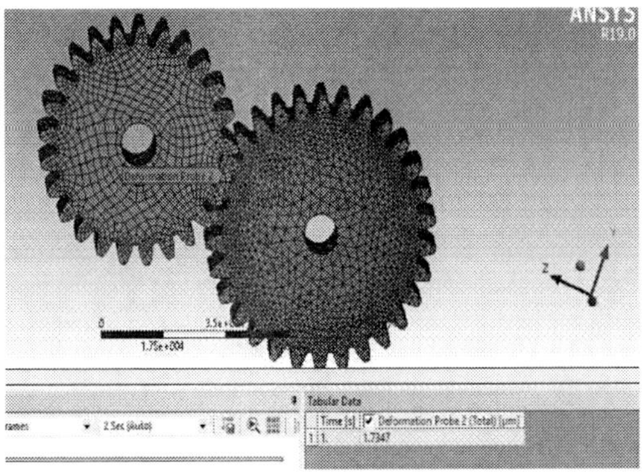

Figure 3: Finite Element Analysis on gear pair.

Results

The analytical result shown in the form of time varying mesh stiffness for different shape of spall like circular, triangular and elliptical are given in figure below.

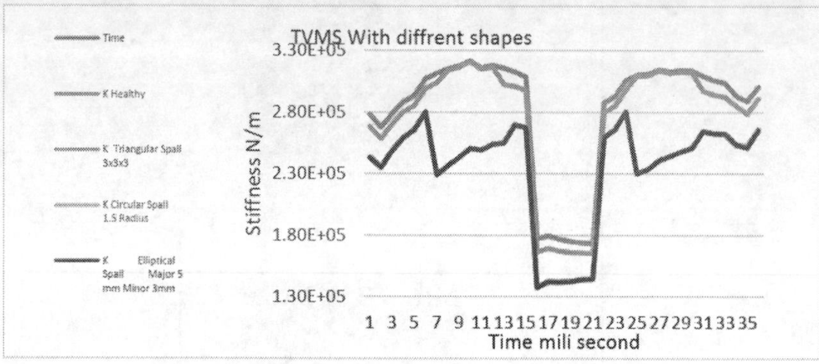

Comparison of mesh stiffness of healthy gear in analytical calculations with healthy gear in FEA:

Gear Condition	Analytical Stiffness (N/m)	FEA Stiffness (N/m)	Variation (%)
Healthy	312920	317918	1.57

Time varying mesh stiffness is calculated through analytical calculations and FEA. And maximum percentage variation at a point where double tooth contact occurs is evaluated. Maximum Variation in TVMS is 1.57%

Comparison of mesh stiffness when spall shape is circular with Analytical Result –

Gear Condition	Analytical Stiffness (N/m)	FEA Stiffness (N/m)	Variation (%)
Circular Spalled	258591	277937	6.96

Time varying mesh stiffness is calculated through analytical calculations and FEA. And percentage variation in faulty gear having circular spall is evaluated. Maximum Variation in TVMS is 7.98%

Comparison of mesh stiffness when spall shape is elliptical with Analytical Result –

Gear Condition	Analytical Stiffness (N/m)	FEA Stiffness (N/m)	Variation (%)
Elliptical Spalled	255693	271846	6.31

Time varying mesh stiffness is calculated through analytical calculations and FEA. And percentage variation in faulty gear having elliptical spall is evaluated. Maximum Variation in TVMS is 6.31%

Comparison of mesh stiffness when spall shape is Triangular with Analytical Result –

Gear Condition	Analytical Stiffness (N/m)	FEA Stiffness (N/m)	Variation (%)
Triangular Spalled	305642	297235	2.69

Time varying mesh stiffness is calculated through analytical calculations and FEA. And percentage variation in faulty gear having Triangular l spall is evaluated. Maximum Variation in TVMS is 2.69%

Conclusion

Finite element conducted on the analytical results shows a good agreement. So the method is used for different shape of spall. The data may also be used for dynamic analysis and vibration behavior of the system.

References

[1]. Lei, Y., Liu, Z., Wang, D., Yang, X., Liu, H., & Lin, J. (2018). A probability distribution model of tooth pits for evaluating

time-varying mesh stiffness of pitting gears. Mechanical Systems and Signal Processing,106(2018) 355-366 Elsevier Publication.
[2]. Liang, X., Zhang, H., Liu, L., & Zuo, M. J. (2016). The influence of tooth pitting on the mesh stiffness of a pair of external spur gears. Mechanism and Machine Theory,106(2016) 1-15. Elsevier Publication.
[3]. Luo, Y., Baddour, N., Han, G., Jiang, F., & Liang, M. (2018). Evaluation of the time-varying mesh stiffness for gears with tooth spalls with curved-bottom features. Engineering Failure Analysis, Elsevier Publication 92(2018) 430-442.
[4]. Saxena, A., Parey, A., & Chouksey, M. (2016). Time varying mesh stiffness calculation of spur gear pair considering sliding friction and spalling defects. Engineering Failure Analysis,70 (2016) 200-211, Elsevier Publication.
[5]. Ma, H., Pang, X., Feng, R., Zeng, J., & Wen, B. (2015). Improved time-varying mesh stiffness model of cracked spur gears. Engineering Failure Analysis, 55, 271-287.
[6]. Ding, Y., & Rieger, N. F. (2003). Spalling formation mechanism for gears. Wear,254 (2003), 1307-1317, Elsevier Publication.
[7]. Chaari, F., Baccar, W., Abbes, M. S., & Haddar, M. (2008). Effect of spalling or tooth breakage on gearmesh stiffness and dynamic response of a one-stage spur gear transmission. European Journal of Mechanics - A/Solids,27 (2008), 691-705, Elsevier Publication.
[8]. Chen, Z., & Shao, Y. (2011). Dynamic simulation of spur gear with tooth root crack propagating along tooth width and crack depth. Engineering Failure Analysis, 18 (8), 2149-2164, Elsevier Publication.
[9] Ankur Saxena, Manoj Chouksey, Anand Parey (2017) Effect of Mesh Stiffness of healthy and cracked gear tooth on modal and frequency response characteristics of geared rotor system. Mechanism and Machine Theory,107 (2017), 261-273, Elsevier Publication.

IOP Conf. Series: Materials Science and Engineering **810** (2020) 012053 doi:10.1088/1757-899X/810/1/012053

Effect of Copper Nano Powder on Kevlar Fiber Reinforced Epoxy Resin Composites

Rachel Reuben, Parth Joshi, Ramachandran Manickam

MPSTME NMIMS Shirpur Campus.
rachelreubenn44@gmail.com, pparthjoshii@gmail.com,

Abstract. With developing technologies, more durable and efficient manufacturing materials is needed. One of such materials include "Fiber reinforced composites" which are robust and strong. They have a major demand in the industries of engineering and manufacturing. Carbon, glass, basalt and aramid fibers are mainly used for manufacturing of fiber reinforced polymer composites. The property which makes kevlar unique is that it is very strong compared to most of the metals while also being light in weight. It possesses excellent heat and shock resistance and is stable at very high temperatures. It has a wide range of uses in military applications such as bulletproof jackets, fireproof bodysuits and also in automobile industry in vehicles, aircrafts and spacecrafts. The tensile, flexural and impact strength of Kevlar 29 (K-29) fiber reinforced polymer composite is examined in this paper. The composite samples were fabricated by hand lay-up method. All Material testings were ensured to comply with the ASTM standards. According to the test results the developed Kevlar-29 fiber reinforced polymer composite showed better flexural, tensile and impact strength. Kevlar possesses superior and unique mechanical properties which makes it a prominent material for industrial applications.

1. Introduction.

Composites are combination of two or more materials with different properties which consists of two phases. The primary phase is the reinforcing phase which can be fibers, sheets or particles which is integrated with the matrix phase. The lattice shields the strands from natural and outer harm and moves the heap between the filaments. The strands, thus, give quality and firmness to fortify the network—and assist it with opposing breaks and cracks. The combined phases don't blend completely or lose their individual characters; they consolidate and contribute their most helpful attributes to improve the result or last item. Composites are ordinarily planned in light of a specific use, for example, increased strength, decreased weight, low density, low cost and durability. Composite has the upside of integrating various properties not typically found together in a solitary material. Specifically, it consolidates high quality and low weight, while simultaneously it is non-destructive and has thermal and electrical protection properties. Another advantage of composites is that it is highly flexible while designing. Composites have powered the development of new applications in business like transportation, aviation, consumer products, appliances etc.

2. Materials and Method

Kevlar Para-Aramid is an aromatic polyamide that is characterized by long rigid crystalline polymer chains. It is characterized by high tensile strength, high tensile strength to weight ratio, ability to absorb impacts and high chemical stability at high temperature. Kevlar is used independently or mixed with a composite material to give added strength. It has been widely used in bulletproof vests and body armour, and other applications such as reinforcement in car tires, brakes, and in car, boat, and aircraft bodies. It can be also used in buildings and structures, but not as the primary structural material because it's less flexible and compressible. Polyetherimide, an amorphous engineering thermoplastic offers an impressive collection of attributes such as high stiffness, heat resistance, transparency, impact strength, high mechanical strength, good electrical properties, high flame resistance, and broad chemical resistance. It is considered as an ideal material of choice in automotive, medical, electrical and other industrial applications.

Figure 1. Processed Kevlar fibre; Figure 2. Copper nanopowder.

ARALDITE AY105, a medium viscosity unmodified epoxy resin based on biphenyl-A, is used in this study. Araldite, which was a registered trademark of Huntsman LLC, is used to bind the Kevlar fabric layers together. The chemical properties of the epoxy resin used in this study are presented in TABLE 2. Epoxy resin is treated with the addition of a curing agent commonly called a hardener. The hardener used in this research is Low viscosity cycloaliphatic polyamine - Hardener HY 2962. The addition of curing agent is usually a higher ratio of resin to hardener. The ratio of resin to hardener used in this study is 5:1. Araldite LY556 is known for excellent mechanical, dynamic, thermal properties and chemical resistance to acids at temperatures below 80 °C. Specific gravity is between 1.15-1.2 at 25°C and vapour pressure is found to be 0.01Pa at 20°C. The decomposition temperature and the boiling point is greater than 200°C. Hardener (HY 951) is selected to improve the strength and adhesion to the composite. Hardener HY951 is colourless in nature. It is found in liquid state and has odor like ammonia with its pH value 13. The boiling point and the thermal decomposition temperature must be greater than 200oC. Its vapour pressure is 0.3 Pa at 20°C and its density is 1g/cm3 at 20°C and flash point is 110°C. Nano powders and nanoparticles are widely used in engineering and manufacturing field. Properties of materials change when they are converted into Nano powders. Their behaviour changes with change in their shape, size and when exposed to other chemicals. Copper possess high ductility and very high thermal and electrical conductivity. Copper Nano powder appears circular particles, and is a brown to black powder .Copper being ductile it is mixed with other metals to form alloys like - brass. Brass is a copper-zinc alloy. Copper Nano powder is graded as highly flammable solid. And it is highly toxic. Compression Molding is a method of molding is used. In this method the mold is preheated in heated mold cavity which is open. First the composite is pre- formed in the shape of the mold and inserted into the heated mold which is then compressed under very high pressure. The composites were made of Kevlar, epoxy and copper Nano powder. The 4 samples differed in the content of copper Nano powder. All the samples weighed 200 grams.

Table 1. Composition of Composite material in weight %.

Specimen	Kevlar weight %	Epoxy weight %	Copper Nano powder weight %
1	15	85	0
2	15	84	1
3	15	83	2
4	15	82	3

Now after measure the materials to its most possible accuracy and mix epoxy resin and hardener completely for nearly 4 to 5 minutes. Spread the mixture properly on the mold after this properly adjust the Kevlar sheet in proper shape. Repeat this until all the material is completely used. Fix the upper lid carefully, so as it can be get fixed in the mold. Then the mold is placed in Compression molding machine. The Compression molding machine then starts. The temperature of plunger reaches approximately to 180°C. Once the required temperature is obtained the plunger is brought down till it touches the mold then the pressure of 100 kg/cm² is applied for nearly 5 minutes, and then the pressure is released. In order to remove any accumulated vapours inside the mixture it is left in the machine for almost a day. In the consecutive day mold is removed from the machine and dismantled. The composite is removed by some hammering. The above procedure is repeated for all the remaining samples. Only the proportions of hardener i.e. copper Nano powder with mixing epoxy resin change.1%, 2% and 3% of copper Nano powder to sample 2, sample 3 and sample 4 respectively. The first sample consists of only epoxy and Kevlar fiber with the proportion of 85:15 respectively.

Figure 3. Sample 1 and sample 2.

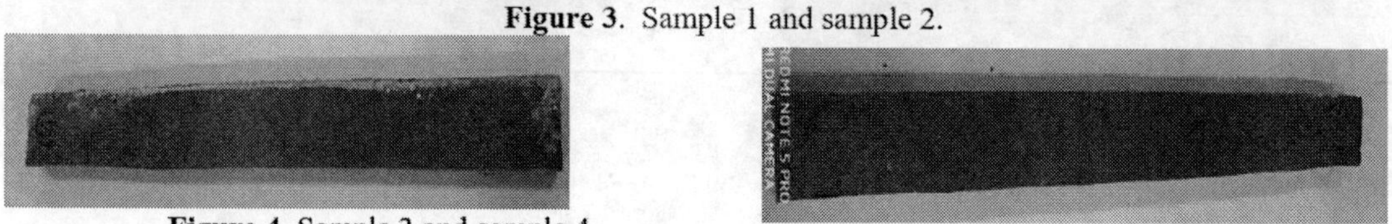

Figure 4. Sample 3 and sample 4.

3. Impact test

Impact test measures toughness of composite and energy absorbed during fracture of specimen. To determine the impact strength of Kevlar fiber reinforced composite samples Charpy impact test was performed. Specimens were prepared according to ASTM D-256 standards. The dimension of specimen are 55 mm ×10mm × 10mm.While performing the test, the force was applied perpendicular of fiber orientation. The arrangement of fabricated samples for Charpy test is shown in Fig. 5. The fiber type, orientation, fiber-matrix interface and matrix nature contribute to Mechanical properties of the composites. Tests were performed on 3 samples of KFRPC laminates. The average value was taken as strength of prepared composite. The table mentioned below shows average values of tensile strength, flexural strength and impact strength obtained from all three samples. Izod test measures the resistance to impact and toughness of materials. The strength depends on the dimension of specimen being tested even small cracks can act as acts as stress raisers and reduce the fracture toughness of the laminate. Izod impact tests were performed on various specimens produced as per ASTM D-256 standard. The dimension of the specimens were 75 mm × 10 mm × 10 mm and radius of the notch was 0.25R having notch length of 2mm.

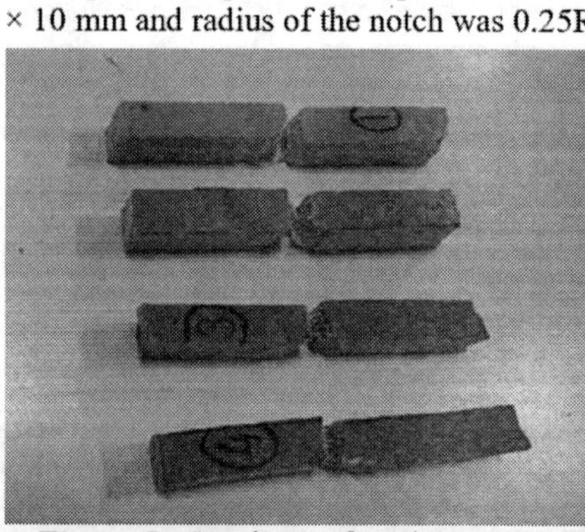

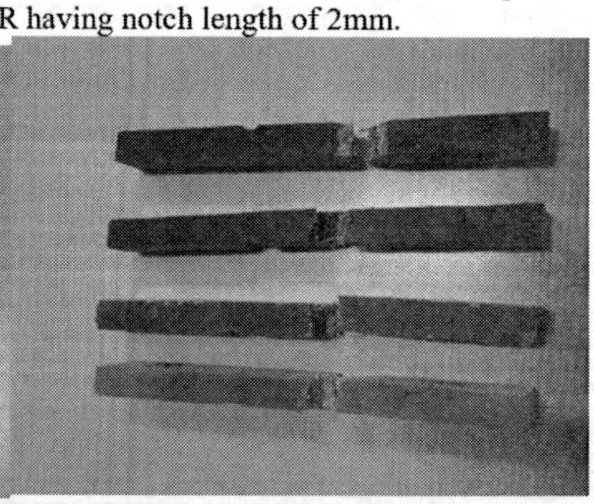

Figure 5. Specimen after charpy test; **Figure 6.** Specimen after izod test.

Table 2. Results for Izod impact test.

Sr. no.	Sample(Kevlar/Epoxy/CNP)	Izod test (Joule)	Charpy test(Joules)
1.	15/85/0	10	4
2.	15/84/1	8	4

3.	15/83/2	8	4
4.	15/82/3	8	8

4. Brinell hardness test.
The Brinell hardness test was used to measure the hardness of the material complying to the American society for testing materials (ASTM). This type of hardness test is selected to test materials that is too coarse have rough surface and is large. A test load (predetermined) is subjected to the test material for few seconds and indented with a 10mm diameter indenter. The resulting impression is measured with a microscope, horizontally and vertically for more accuracy. The two diameters is averaged to use in the calculation. The Brinell formula is used to calculate the brinell hardness number as shown figure .

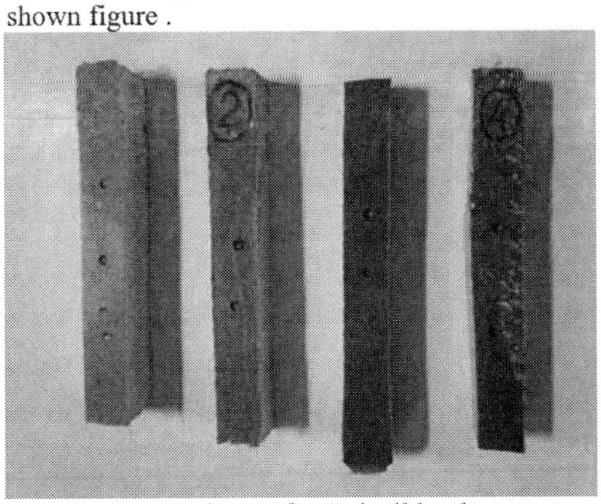

Figure 7. Specimen after Brinell hardness test.

Table 3. Results for Brinell hardness test.

Sr. no.	Sample(Kevlar/Epoxy/CNP)	Brinell hardness number
1.	15/85/0	65.275
2.	15/84/1	71.587
3.	15/83/2	78.655
4.	15/82/3	100.377

5. Conclusion
Effect of Copper Nano Powder on Kevlar Fiber Reinforced Epoxy Resin Composites was studied in this work. The composites were made using compression molding machine. Tests performed on four samples of Kevlar 29 were charpy, Izod and Brinell hardness test. Sample with 15% Kevlar, 82% Epoxy and 3% CNP showed the highest results for charpy test (8 Joules). Sample with 15% Kevlar, 85% Epoxy and 0% CNP showed the highest results for Izod test (10 Joules). Sample with 15% Kevlar, 82% Epoxy and 3% CNP showed the highest results for Brinell hardness test (BHN 100.377).

References
1. Karuppuchamy, S. "Thermo-mechanical properties of coconut shell powder reinforced plastic composites." International Journal of ChemTech Research Vol.8, No.6 (2015): 852-857.
2. Kanak Kalita, U Ragavendran, M. Ramachandran, Akash Bhoi, Weighted sum multi-objective optimization of skew composite laminates, Structural Engineering and Mechanics, Vol. 69, No. 1 (2019) 21-31.
3. Sefiu Adekunle Bello, Isiaka Ayobi Raheem, Nasir Kolawole Raji, Study of tensile properties, fractography and morphology of aluminium (1xxx)/coconut shell micro particle composites, Journal of King Saud University - Engineering Sciences, Vol29, Issue 3, July 2017, Pages 269-277

4. Liyanage, Chinthani D., and Mevan Pieris. "A Physico-Chemical Analysis of Coconut Shell Powder." Procedia Chemistry 16 (2015): 222-228.
5. Agarwal, Rakshit, M. Ramachandran, and Stanly Jones Retnam. "Tensile Properties of Reinforced Plastic Material Composites with Natural Fiber and Filler Material." ARPN Journal of Engineering and Applied Sciences 10, no. 5 (2015): 2217-2220.
6. Siti Nurbazilah Ab Jabal, Yew Been Seok, Wee Fwen Hoon, carbon composition, surface porosities and dielectric properties of coconut shell powder and coconut shell activated carbon composites, ARPN Journal of Engineering and Applied Sciences, VOL. 11, NO. 6, MARCH 2016, 3832- 3838.
7. Shyamkumar Shah, Akshaykumar Patil, M. Ramachandran, Kanak Kalita, Effect of coal ash as a filler on mechanical properties of glass fiber reinforced material, International Journal of Applied Engineering Research, 9(22) (2014) pp. 14269-14277.
8. Salleh, Z., M. M. Islam, M. Y. M. Yusop, and MA Mun'aim M. Idrus. "Mechanical properties of activated carbon (AC) coconut shell reinforced polypropylene composites encapsulated with epoxy resin." APCBEE Procedia 9 (2014): 92-96.
9. D. Bino prince raja, B. Stanly Jones Retnam, M. Ramachandran, Analysis of mechanical properties of glass and carbon fiber reinforced polymer material, International Journal of Applied Engineering Research. ISSN 0973-4562 Volume 10, Number 11 (2015) pp. 10387-10391.
10. Gojny, F. H., M. H. G. Wichmann, U. Köpke, B. Fiedler, and K. Schulte. "Carbon nanotube-reinforced epoxy-composites: enhanced stiffness and fracture toughness at low nanotube content." Composites science and technology 64, no. 15 (2004): 2363-2371.
11. Alex. S, Stanly Johns Retnam, M. Ramachandran, A review on Biodegradability of Hybrid Bamboo/Glass fiber polymer composites, International Journal of Applied Engineering Research. ISSN 0973-4562 Volume 10, Number 11 (2015) pp. 10565-10569.
12. Kim, Jin Ah, Dong GiSeong, Tae Jin Kang, and Jae RyounYoun. "Effects of surface modification on rheological and mechanical properties of CNT/epoxy composites." Carbon 44, no. 10 (2006): 1898-1905.
13. Bisanda, E. T. N., and Martin P. Ansell. "The effect of silane treatment on the mechanical and physical properties of sisal-epoxy composites." Composites Science and Technology 41, no. 2 (1991): 165-178.
14. Joshi, Satish V., L. T. Drzal, A. K. Mohanty, and S. Arora. Are natural fiber composites environmentally superior to glass fiber reinforced composites. Composites Part A: Applied science and manufacturing 35, no. 3 (2004): 371-376.
15. Sgriccia, N., M. C. Hawley, and M. Misra. "Characterization of natural fiber surfaces and natural fiber composites." Composites Part A: Applied Science and Manufacturing 39, no. 10 (2008): 1632-1637.
16. M. Ramachandran, , Sahas Bansal, Pramod Raichurkar, Scrutiny of Jute Fiber Poly-Lactic Acid (PLA) Resin Reinforced Polymeric Composite, Journal of the Textile Association, Volume 76, Issue 6, 2016, pp. 372-375.
17. Herrera-Franco, PJp, and A. Valadez-Gonzalez. "A study of the mechanical properties of short natural-fiber reinforced composites." Composites Part B: Engineering 36, no. 8 (2005): 597-608.
18. Ramanan. G, Rajesh Prabha. N, Diju Samuel. G, Jai Aultrin. K. S, M Ramachandran, Prediction of Machining Characteristics of Hybrid Composites Using Response Surface Methodology Approach, International Journal of Engineering & Technology, 2018, 7(3.1), 162-165.
19. Bachtiar, Dandi, S. M. Sapuan, and M. M. Hamdan. "The effect of alkaline treatment on tensile properties of sugar palm fibre reinforced epoxy composites." Materials & Design 29, no. 7 (2008): 1285-1290.
20. Sahas Bansal, M. Ramachandran, Pramod Raichurkar, Comparative Analysis of Bamboo using Jute and Coir Fibre Reinforced Polymeric Composites, Materials Today: Proceedings,4(2), 2017, 3182–3187.
21. Yousif, B. F., A. Shalwan, C. W. Chin, and K. C. Ming. "Flexural properties of treated and untreated kenaf/epoxy composites." Materials & Design 40 (2012): 378-385.
22. Mohanty, A. K., ManjusriMisra, and Lawrence T. Drzal. "Surface modifications of natural fibers and performance of the resulting biocomposites: an overview." Composite Interfaces 8, no. 5 (2001): 313-343.
23. Gassan, Jochen. "A study of fibre and interface parameters affecting the fatigue behaviour of natural fibre composites." *Composites part A: applied science and manufacturing* 33, no. 3 (2002): 369-374.
24. P. M. Bhagwat, M. Ramachandran, Pramod Raichurkar, Mechanical Properties of Hybrid Glass/Carbon Fiber Reinforced Epoxy Composites, Materials Today: Proceedings, 4(8), 2017:1788–1793.

A Review on New Technology in Internal Combustion Engine-HCCI engine

Ashish Mogra, Krishna Kumar Gupta
Department of Mechanical Engineering, SVKM's, NMIMS, MPSTME, Shirpur, Maharashtra, India
ashish.mogra@nmims.edu

Abstract. In present study a concise literature review on Homogeneous Charge Compression Ignition (HCCI) engine have been performed by taking the many practical problem which are hurdle for implementation of this concept. HCCI combustion is also termed as Low Temperature Combustion (LTC) technology because of its prospective to attain high efficiency and low Nox and particulate matter emission. The typical quality of HCCI technique is the need of preparation of homogeneous mixture early at combustion. Combustion processes in HCCI engine were controlled by chemical kinetics and proper mixing of fuel. Numerous researchers are working towards the challenges and solution for commercialization of this engine. Present review covers the elementary feature for the evaluation of HCCI engine, its working principle and challenges for practical implementation. This study elaborates the prospective solution and future growth in HCCI engine. The most critical challenges for HCCI engine is controlling method for direct ignition of combustion. Therefore in HCCI combustion engine implementation of this problem is way for practical availability of this concept.
Keywords: HCCI, Injection timing, low temperature combustion, EGR, VCR.

1. Introduction

The requirements for the crude oil are continuing increasing and it is anticipated to increase about 280 million ton by the upcoming days. It is shown by India Energy Outlook –IEA 2017 in their report that almost three-fourth of Indian energy demand is met by fossil fuels. According to the world energy outlook 2011 it is reported that, India and China will be reach in very high growth rate in consumption of primary energy requirements by the year 2035, which will be more energy demand from the rest of the world. The interest for higher mileage and lower fumes gas outflows of pressure start motor can't be accomplished by simply utilizing mechanically administered fuel infusion frameworks. The extreme need of an automobile industry is new technologies which can minimize the gas emissions in effective manner. They should be clean and more efficient, increase the atmosphere air condition and also support in energy protection. It is well known that Internal Combustion (IC) engine is old decade technique used in automotive sector. Due to the strict emission norm provided by the government such as Europe and Bharat Stages, it is required to change in combustion technologies as well as in design concept. In today's scenario automotive industry are facing twin challenges. The challenges are improve the fuel economy simultaneously reducing the emission. Researchers are moving to find out the solution for the problem.

IOP Conf. Series: Materials Science and Engineering **810** (2020) 012054 doi:10.1088/1757-899X/810/1/012054

Many peoples are working on development of highly effective and more friendly environment combustion technique. Experimental studies are carried out for developing the system which is capable of applying the alternative fuel. Low Temperature Combustion (LTC) has shown the future prospectus which meets the environmental emission standard. Researchers has focused their attention on new LTC technique which shows its prospective for minimizing the emission and attain the higher efficiency as compared to normal combustion modes. Low temperature Combustion (LTC) is classified in different techniques. All techniques are different only with respect to their function but the fundamental principal remains the same. The present paper review the fundamental understanding for the development of HCCI engine, its research area with various challenges and different method which can commercialize this new technique.

2. Principal of HCCI engine

HCCI was suggested as a new combustion concept at the end of 1970s. The reason for the proposed HCCI was difficulties facing by the peoples in operation of conventional SI and CI engines. The credit for the most acknowledgeable and established original work towards the development of HCCI engine goes to Onishi [1] who developed the concept of Active Thermo Atmospheric Combustion (ATAC) and Noguchi et al.[2] who introduced the concept of Toyota-Soken (TS) combustion on two stroke engines. They was motivated to achieve controlled irregular combustion which caused by auto-ignition of charge present in cylinder. The objective behind is to get stable lean combustion in traditional two stroke gasoline engine on part load. There work was to development of HCCI principal on two stroke engine. To gain an additional understanding and the physic behind the HCCI engine, Najt and Foster was the first investigator to apply this principal on four stroke engine. Performance evaluation had done on HCCI engine which was operated with full blended gasoline four stroke engines [3, 4]. The concept which distinguishes between conventional engine and HCCI engine is its mixing phenomenon in combustion chamber before ignition. The air-fuel mixture is prepared in such a way that it auto-ignite and combustion starts. It is observed that concept of Homogeneous Charge Compression Ignition (HCCI) is somehow alike to Spark Ignition (SI) by the fact that both required the premixed charge and it is also close to Compression Ignition (CI) engine as it is required auto-ignition temperature of fuel for staring the combustion. The HCCI engines are more effective as compared to SI engine in terms of thermal efficiency which is nearer to Compression Ignition engine. A homogeneous mixed air fuel charge is introduced in the combustion chamber of the cylinder during intake stroke. Piston will start move toward to Top Dead Centre (TDC) in compression stroke while the inlet valve close. In compression stroke, temperature and pressure of the charge increased. When the piston reaches the TDC, charge gains the auto-ignition point. The Figure 1 describes the HCCI combustion phenomena. In HCCI engine the start of combustion can be direct by various parameters like compression ratio (CR), intake charge pressure and temperature, ignition timing, swirl, and charge preparation. It has been found that during the compression stroke when the fuel reached its auto ignition temperature it get oxidized and instantaneously released its chemical energy. There are many locations in HCCI engine where auto ignition occurs simultaneously called as hot spots. As compared to traditional engines the pressure increase significantly in minimum time period because of quick heat release rate; on the other hand the maximum temperature in the cylinder keep relatively low. During, combustion process the charge temperature and pressure increase further. The work done by the piston in the expansion stroke due to expanding the hot burn gases will be available at the crank shaft. The cycle competed when piston reaches the TDC after performing the exhaust stroke, in which the burn hot gases thrown out from the cylinder. In brief the processes are summarized in following steps:

i. Exhaust Gas Recirculation (EGR) approach used for preparing the To control the combustion and heat removal rate highly dilute fuel –air mixture was prepared using the exhaust gas recirculation (EGR).

ii. The charge reaches the auto-ignition temperature at the end of the compression stroke, which lead the simultaneous instant ignition at several location of the engine.
iii. Exact control of heat removal rate to attains trade-off between combustion efficiency and exhaust emissions.

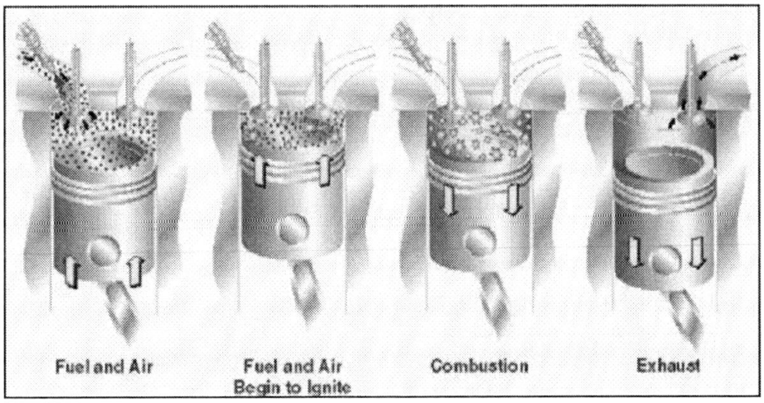

Figure.1 HCCI process [5]

3. Advantages of HCCI

In consideration to strict engine emission norm, researches are doing effort to make HCCI combustion as an alternate to traditional combustion concept. Homogeneous Charge Compression Ignition provides the multiple advantages over the traditional engine combustion concept.

i. HCCI has almost similar heat addition like Otto cycle. In which combustion takes place at constant volume and attains the higher relatively thermal efficiency due to its higher compression ratio and shorter combustion timing.
ii. It has the lower radiation as well throttling losses due to its low temperature combustion.
iii. It has potential for lower emission compared to compression ignition (CI) and Spark Ignition (SI) engine. Simultaneously it decreases the NOx and Particulate Matter (PM).
iv. HCCI has its main advantages of fuel flexibility. It has the ability to operate on gasoline as well as mineral fuel like diesel. It can also use the combination of mineral fuel with alternative fuels such as biodiesel.
v. HCCI engines can be joined with traditional engine to get higher efficiency i.e. they can utilized in due mode also. This engine is suitable for replacement of traditional engine such as CI and SI engine.

4. Challenges of HCCI

The concept of HCCI engine has been known over the past three decade but because of number of challenges coupled with it; this technology has not been commercialized today. HCCI engine can work only for those engines which having highly equipped with modern electronic system. Researchers are working on several challenges which are hurdles for replacement of this engine in traditional engine. Some of the challenges are listed below:

a) Injection timing
b) Maximum levels of Noise, UHC and CO emissions
c) Cold start capability
d) Operation range
e) Homogeneous mixture preparation
f) Intake charge temperature

4.1 Injection Timing

Unlike traditional engines, HCCI engines do not have direct method of controlling start of combustion. Start of combustion in HCCI engine occurs due to chemical kinetics. So homogeneous charge is required for combustion in HCCI. Initially the port fuel injection would appear to produce the homogeneous mixture by providing sufficient time for mixing. Because of the high boiling property of mineral diesel fuel poor vaporization take place at intake manifold. The delivery of the diesel fuel is not feasible via port injection [6].

4.2 High levels of emissions and Noise

In continuation with the injection timing exhaust emission and high level of noise are one of the practical hurdles for the implementation of HCCI engine over conventional engine [7]. In homogeneous combustion technology, some part of fuel has been already kept in cylinder as the same time of the compression stroke. When unburned fuel re-enters in the cylinder at the time of expansion stroke, they comes in contact with too low burned gas temperature. That process significantly increases both CO and HC emission level as compared to traditional combustion. The combustion efficiency degenerates suddenly due to the too low temperature (less than 1400K) of burned gas. This completes the CO to CO_2 oxidation at low loads [8].

The effectiveness of the HCCI combustion at particular higher load was up to the limit due to the difficulties faced in loss of combustion efficiency. The engine noise increases significantly and it becomes too large enough at higher load due to increase in pressure rise. The effectiveness of HCCI combustion limits at lighter load found difficulties due to loss of combustion efficiency [9]. After treatment techniques which can control the HC and CO emission at higher engine loads. The CO emission can be reduced at low engine loads by applying the direct in-cylinder method [10]. It is found that most of the Un-burn Hydro Carbons present during the last stage of exhaust process in cylinder. It indicates that origin of the un-burn hydro carbons from the wall of the cylinder and on the top of the piton layer [11].

For identifying the source of CO and HC emission they used two dimensional planer laser-induced fluorescence (PLIF). The circulation of carbon oxides and hydrocarbon emissions which was affected by concentration of oxygen and injection timing was evaluated [12]. A comparison was further been made in between the hydrocarbon and carbon oxides emission for the thermal efficiency of dilution controlled of HCCI engine [13]. For removal of HC and CO emission catalyst convertor technique is well understood. It was found that hydrocarbons and carbon oxide emission decreased in the range of 90-95% over the prototype catalyst on the other hand maximum NOx reduced in the range of 35-55% under the lean engine operating condition [14].

Engine-operating conditions appear to exert a strong influence on the total mass emissions of carbonyls measured before catalyst [15]. It was found that a twin way catalytic convertor under HCCI combustion conditions showed minimum hydrocarbon conversion performance. It was found that conversion efficiency for homogeneous charge compression ignition varied in the range of 53 to 61%, while for SI combustion it was around 85% [37].

4.3 Cold Start capability

One major difficulty which HCCI engine facing is its firing in cold start operation. Heat loss from the compressed charge is too high and temperature is very low in combustion chamber. To control this difficulty, the engine should have start in traditional mode, then redirect to HCCI mode. The temperature needed for achieving the auto ignition changes with respect to properties of test fuel and operating condition of engine.

For enhancement of HCCI combustion during the cold start, there are lots of technique on which peoples are working they include fuel additives, increasing the compression ratio and variable valve timing [12]. Although the dual mode method is known as most realistic, during this approach engine have to go two way combustion. In first phase engine has to run on SI or CI mode and in second phase it will switch to HCCI mode [48]. Hence, it is a tough challenge to keep real homogeneous combustion after cold start. This is the area of HCCI operation where more development efforts are required.

4.4 Operation range

Incorporation with other obstacle, one more basic hurdle in HCCI development is it operating range. To get maximum benefit from HCCI engine it should be operated on high load. Increasing the operation range is very important parameter, as auto ignition depends on it. Insufficient thermal energy at the end of compression stroke which required for auto ignition of charge is also one limitation. Studies rivals that when an engine is operating at higher load, combustion rate is increasing, which result in increasing in high rate of pressure leads to pressure oscillation. Due to inadequate noise engine can be damage and NOx emission increase. Due to that power decreased and increased the basic load requirement on HCCI engine [48]. Therefore the better understanding towards to reduce the chemical kinetics of homogeneous charge new techniques should be developed. So it will help to minimize the higher noise and damage of engine when duration of combustion increased particularly at higher load. To expand the operating range of HCCI engine there are different techniques such as supercharging [16,17] and turbocharging [18-20] have been used. Operating range can also be increased by using the dual fuel mode like SI-HCCI [21, 22] and CI-HCCI [23, 24].

4.5 Preparation of Homogeneous charge

Efficient preparation of homogeneous charge and keep away the interaction between fuel and wall is critical for attaining the maximum thermal efficiency, minimizing the hydrocarbons and particulate matter emission and intercepting the oil dilution [16]. The homogeneous mixture preparation had an effect on auto-ignition reactions, which can control the combustion phasing in HCCI engine [17]. Outstanding proof has been found that with less degree of mixture inhomogeneity low Nox emission can be produced within the combustion chamber.

It is found that preparation of homogeneous mixture is most difficult for such types of fuel which has minimum volatility like diesel. For preparation of homogeneous air fuel mixture, injection of fuel had been done at a maximum turbulent port flow in case of gaseous fuel and highly volatile in case of liquid fuel [45,46].Possible Solution of HCCI engine. Researchers presented various techniques or methodology by using them one can go for practical implementation of HCCI technology. Various parameters can affect the performance of HCCI engine. In present paper some of the parameters and methodology has been discussed.

4.6 Intake charge temperature

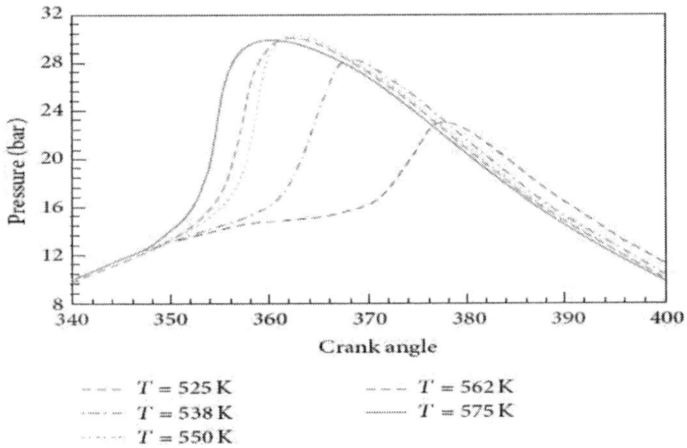

Figure 2 At various inlet temperature graph between Cylinder pressure and crank angel degree (CAD) [60].

Many researchers have been reported the effect of inlet mixture temperature on working of HCCI combustion. It was showed in 1983, that in spark ignition engine which has a lower compression ratio, a lean mixture for HCCI engine can be achieved with inlet temperature in the range of 300° to 500°C [19]. Intake charge temperature had vigorous effect on combustion timing of HCCI engine. Intake charge temperature significantly affects the Indicating Mean Effective Pressure (IMEP) which is produced under HCCI combustion mode [66]. Figure 2 shows the relation between cylinder pressure and crank angle variation for different intake temperature. It is observed that by increasing the inlet charge temperature HCCI combustion timing affects significantly. At temperature of 575 K, advanced ignition occur although the combustion is not so effective but by decreasing the intake temperature ignition would retard [99].

The influence of intake temperature on spark assisted and unassisted HCCI combustion for particular load and negative valve overlap condition, experiments have been performed within the limit of 15°C and 50°C [61]. It has been found that ignition of richer mixture is possible in HCCI combustion mode at low inlet temperature [62]. Experiment results revealed that nevertheless of the EGR, by reducing the intake charge temperature, IMEP increased which in turn increases the in-cylinder charge mass. It is shown that while decreasing the temperature from105°C to 30°C, the IMEP increased from 2.7 bar to 3.7 bar [63].

4.7 Injection pressure

To encourage better mixing of the charge in cylinder higher fuel injector pressure plays a vital role. This is especially applicable when it is used for combination of small nozzle orifice [63]. For better combustion and desirable spray structure a high fuel injection pressure as well as speed is required [64]. Moreover, researcher's shows that by increasing the fuel injection pressure PCI combustion are produced [65].

4.8 Variable Compression Ratio (VCR)

Research showed that by using higher compression ratio and lean mixture one can achieve high efficiency and low NOx emission [25]. To achieve HCCI combustion control compression ratio is an efficient parameter. They investigated carefully and shows that compression ratio in the range of 9.6 to 22.5 had vigorous impact on ignition timing and decreasing in inlet temperature [38, 39]. The compression ratio in the range of 12 to 18.6 has been demonstrated and present a significantly effect on combustion rate. Their researches reveal that homogeneous charge compression ignition engine showed a higher compression ratio permit for lower inlet mixture temperature and higher inlet density which gives the higher power output. Moreover higher compression ratio provides the higher thermal efficiency [40]. It has been found that the change in compression ratio had a strong influence on homogeneous charge compression ignition combustion [41]. Researchers also studied the effect of higher compression ratio on natural gas fuelled in HCCI engine. They used Variable Compression Ratio (VCR) method for changing the compression ratio. They observed the compression ratio in the range of (21:1, 20:1, 17:1) according to their operating condition [42]. Variable compression ration technique was used in multi cylinder engine in closed loop combustion control for investigating the HCCI combustion phasing. The study showed that acceptable performance was given by fast variable compression ratio [43].

4.9 Exhaust Gar Recirculation (EGR)

Exhaust gas recirculation is the very efficient way to moderate the rise in pressure and extend the homogeneous charge compression ignition operating range to maximum load. EGR is necessary to attain the less soot as well as NOx emission form HCCI engine without using high rate of fuel consumption. Exhaust Gas Recirculation is the generally used method to regulate the temperature within the cylinder which in turn starts of combustion. Early, Thring [44] succeed the HCCI combustion by changing the intake charge temperature and fraction of EGR over a wide range of equivalence ratio. Some of the gas species of combustion like CO2, water, nitrogen, CO, PM, oxygen, HC, NOx which control the reaction constitute high heat capacity of EGR [45-47]. These elements of species describe mainly four effects on emission and combustion. The first effect happen when the hot EGR was mixed with air fuel ratio due to that the intake charge temperature increased this is called preheating effect. Second effect is called as dilution effect, where reduction in the oxygen concentration taken place by introducing the EGR. Next is called heat capacity effect during that all the heat carrying capacity of charge of EGR and fuel would be maximum for higher heat capacity of CO_2 and water vapor [48]. Last is the chemical effect, where for completing the reaction unburnt product of EGR would take part.

Considering all these effects and by regulating the EGR quantity overall homogeneous charge compression ignition combustion can be managed. EGR also helps in start of combustion. Consequently, EGR makes it possible to extinguish excessively advanced start of combustion at low temperature in HCCI engine [49]. This quick and unnecessarily combustion causes knocking, in turn which minimize the performing limit of the homogeneous charge compression ignition combustion [51, 53].

Ishikawa et al. [52] studied the influence of efficiency of heat exchanger on an exhaust gas recirculation cooler on combustion. The increases the rejection of cooler heat in the range of 3.1 kW to 8.5 KW. Increasing the charge density and concentration of oxygen, intake temperature decreased from 110°C to 36°C. They observed the minimization in soot emission is because of reduction in equivalence zone, although the hydrocarbon emission increased little bit. Boyarski and Reitz [54] found that exhaust gas recirculation technique is highly effective as controlling parameters for start of combustion. They used up to 68% of EGR. They also concluded that to control the rate of pressure rise high EGR can be used. Approximately 40% of EGR can be used to control the NOx emission [55]. Idicheria and Pickett [56] investigated that due exhaust gas recirculation early development of cool flames taken place after fuel injection in constant volume chamber. The timing of soot emission was largely dependent on EGR, with increasing the EGR delay time for soot emission increases. Hardy et al. [57] studied the influence of maximum equivalence ratio on mixing time used in heavy duty engine. They used high EGR. They showed that by applying the high EGR, high equivalence ratio can be obtain, which in turn slow down the PM and NOx emission.

4.10 Dual Fuel

Two fuels with different auto-ignition properties are used in dual fuel mode. In that system primary fuel with high octane number and secondary fuel with low octane number is used [58]. Wilhelmsson et al. [59] did expanded study on HCCI; they used natural gas (NG) and n-heptane as a dual fuel. The influence of various primary fuel blends on homogeneous charge compassion ignition performance limit, start of combustion and fuel consumption was studied by Atkins and Koch, they concluded that by varying the octane number of fuel performance limit of HCCI engine can be developed [68]. It has been investigated in many studies that use of commercial fuel or mixture of single -component fuel and primary reference fuel are advised for HCCI engine [69]. One consider the examples of these types of dual fuel are mixture of ethanol and n- heptane, combination of n-heptane and mineral diesel, mixture of gasoline and mineral diesel, mixture of gasoline and natural gas [76].

5. Conclusion

Currently research happening related to HCCI engine affect the advancement of new combustion technology. Due to certain technical challenges such as ignition timing, preparation of mixture and its operating limitation to lower load, HCCI combustion concept is facing difficulties for its commercialization. It represents a way to replace the conventional combustion with advanced combustion technology. Although for its practical existence for higher load and other application purpose huge research effort is required. Present literature review found some identical gaps toward the developments of HCCI engine:

(i) Enlargement of controlling methods for attain the ignition timing is a major challenge of this technique, which need to be solved.

(ii) Expansion of control system for emission and planning to resolve the challenge of constant level of NOx emission and particulate matter.

(iii) Development of new methodology to reduce the heat removal rate in HCCI engines is another challenge, which needs attention.

(iv)Multi mode engines capability seems to be worthy for adaption in HCCI combustion technique. Research and developments need more effort to analysis the shortcoming in these strategies for easily transition in HCCI to SI or CI mode.

(v) Multi fuel mode is another aspect in HCCI engine. It has capability to operate on two fuels with different octane and cetane number. This technique can also work on biofuels.

In all descendent of HCCI, controlling of combustion is almost a complicated issue, which can be solved using different technique such as variable compression ratio, exhaust gas recirculation, multi-fuel injection, variable valve timing etc. Although, these methods have certain advantages as well as challenges. It was found from published data, many peoples proposed different techniques or methodology for feasibility of this concept. It is observed that dual mode combustion system shows an efficient solution for commercializing the HCCI engine. In this method conventional engines are coupled with HCCI, although complexity and cost will remain a key issue. Exhaust gas recirculation reduce the NOx and particulate emission. Achieving the homogeneous preparation of mixture minimize the soot emission. Extending the operating range of HCCI engine in turns increase the break thermal efficiency. Multiple injections provide the homogeneous mixture preparation but start of combustion is quite difficult. Hence, for continuing the progress of HCCI engine adjustable fuel injection system, exhaust gas recirculation for betterment of mixture preparation and to extend the higher load are few key aspects, which required extensive investigation then only a HCCI engine may be developed and made commercialized.

Reference

[1]. Onishi S, Jo S, Shoda K, Jo P, Kato S. Active thermo atmosphere combustion (ATAC)-A new combustion process for internal combustion engines. SAE Technical Paper 1979; 790501; 1979.

[2]. Noguchi M, Tanaka Y, Tanaka T, Takeuchi Y. A study on gasoline engine com bustion by observation of intermediate reactive products during combustion. SAE Technical Paper 1979; 790840; 1979.

[3]. P.M.Najt andD. E. Foster, "Compression-ignited homogeneous charge combustion," SAE

[4]. Paper 830264, 1983.

[5]. R. H. Thring, "Homogeneous-charge compression-ignition (HCCI) engines," SAE Paper 892068, 1989.

[6]. Gan S, Ng KH, Pang KM. Homogeneous Charge Compression Ignition (HCCI) combustion: Implementation and effects on pollutants in direct injection diesel engines. Applied Energy 88 (2011) 559–567.

[7]. Takeda Y, Keiichi N, Keiichi N. Emission characteristics of premixed lean diesel combustion with extremely early staged fuel injection. SAE paper no. 961163;1996.

[8]. Dec JE.A computational study of the effects of low fuel loading. SAE paper 2002;01-1309; 2002.

[9]. Eng JA. Characterization of pressure waves in HCCI combustion. SAE paper 2002;01-2859; 2002.

[10]. [10]US Department of Energy. A report to the U.S. Congress: homogeneous charge compression ignition (HCCI) technology, 2001.

[11]. Colban WF, Miles PC, Oh S. On the cyclic variability and sources of unburned hydrocarbon emissions in low temperature diesel combustion systems. Society of Automotive Engineers; 2007. SAE 2007-01-1837.

[12]. Kim D, Ekoto I, Colban WF, Miles PC. In-cylinder CO and UHC imaging in a light-duty diesel engine during PPCI low-temperature combustion. Society of Automotive Engineers; 2008. SAE 2008-01-1602.

[13]. Colban WF, Miles PC, Oh S. Effect of intake pressure on performance and emissions in an automotive diesel engine operating in low temperature combustion regimes. Society of Automotive Engineers; 2007. SAE 2007-01- 4063.

[14]. Hasan AO, Abu-Jrai A, Alaa H. Al-Muhtaseb, Athanasios Tsolakis, Hongming Xu, HC, CO and NOx emissions reduction efficiency of a prototype catalyst in gasoline bi-mode SI/HCCI engine, Journal of Environmental Chemical Engineering 4 (2016) 2410–2416

[15]. Hasan AO, Abu-jrai A, Alaa H. Al-Muhtaseb , Athanasios Tsolakis , Hongming Xu, Formaldehyde, acetaldehyde and other aldehyde emissions from HCCI/SI gasoline engine equipped with prototype catalyst, Fuel 175 (2016) 249–256.

[16]. Iwabuchi Y,Kawai K, Shoji T, Takeda Y .Trial of new concept diesel combustion system premixed compression-ignition combustion.SAETechnicalPaper;1999-01-0185;1999. SunR,ThomasR,GrayCL.An HCCI Engine : Power plant for a hybrid vehicle ,SAE Technical Paper ;2004-01-0933 ; 2004.

[17]. [18]Olsson JO, Tunestal P, Johansson B. Boosting for high load HCCI .SAE Technical Paper 2004 ;2004 - 01 -0940 ;2004.

[18]. Olsson JO ,Tunestal P, Haraldsson G, Johansson B. A Turbocharged dual fuel HCCI engine. SAE Technical Paper 2000 ;2001 -01 -1896 ; 2000.

[19]. Koopmans L, Strom H ,Lundgren S, Backlund O, Denbratt I. Demonstrating a SI - HCCI-SI mode change on a Volvo 5-cylinder electronic valve control engine. SAE Technical Paper 2003 ;2003 -01 -0753 ;2003.

[20]. Santoso H, Matthews J, Cheng WK. Managing SI-HCCI dual mode engine operation .SAE Technical Paper 2005; 2005-01-0162; 2005.

[21]. Guezennec Y, Musardo C, Staccia B,Midlam Mohler S ,Calo E, PisuP. Supervisory control for NOx reduction of an HEV with a mixed mode HCCI / DI engine. SAE Technical Paper; 2004-05-0123; 2004.

[22]. Midlam- Mohler S, Haas S ,Guezennec Y, Bargende M, Rizzoni G. Mixed mode diesel HCCI/DI with external mixture preparation. SAE Technical Paper2004;2004-05-0446;2004. Hou JX, Qiao XQ. Characterization of knocking combustionin HCCI DME engine using wavelet packet transform. Applied Energy 2010;87:1239–46.

[23]. J.O. Olsson, P. Tunest°al, and B. Johansson, "Closed-loop control of an hcci engine," SAE Paper 2001-01-1031, 2001.

[24]. D. S. Kim and C. S. Lee, "Improved emission characteristics of HCCI engine by various premixed fuels and cooled EGR," Fuel, vol. 85, no. 5-6, pp. 695–704, 2006.

[25]. Jacek Hunicz , Alejandro Medina, Experimental study on detailed emissions speciation of an HCCI engine equipped with a three-way catalytic converter, Energy 117 (2016) 388-397.

[26]. M. Christensen, A. Hultqvist, and B. Johansson, "Demonstrating the multi fuel capability of a homogeneous charge compression ignition engine with variable compression ratio," SAE Paper1999-01-3679, 1999.

[27]. M. Christensen, B. Johansson, and P. Einewall, "Homogeneous charge compression ignition (HCCI) using isooctane, ethanol and natural gas—a comparison with spark ignition operation," SAE Paper 972874, 1997.

[28]. K. Hiraya, K. Hasegawa, T. Urushihara, A. Iiyama, and T. Itoh, A Study on gasoline fueled compression ignition engine—a trial of operation region expansion. SAE Paper 2002-01-0416, 2002.

[29]. N. Iida and T. Igarashi, "Auto-ignition and combustion of n-butane and DME/air mixtures in a homogeneous charge compression ignition engine," SAE Paper 2000-01-1832, 2000. J.O.Olsson, P. Tunestal, B. J. Johansson, S. Fiveland, R.Agama, and M. Willi, "Compression ratio influence on maximum load of a natural gas fueled HCCI engine," SAE Paper 2002-01-0111, 2002.

[30]. G. Haraldsson, P. Tunest°al, and B. Johansson, "HCCI combustion phasing with closed-loop combustion control using variable compression ratio in a multi cylinder engine," SAE Paper 2003-01-1830, 2003.

[31]. Thring RH. Homogeneous charge compression ignition (HCCI) engines. SAE Technical Paper 1989; 892068; 1989

[32]. Maurya RK, Agarwal AK. Experimental investigation of close loop control of HCCI engine using dual fuel approach. SAE Technical Paper;2013-01-1675;2013.

[33]. Strandh P,Bengtsson J,Johansson R,Tunest lP,Johansson B.Variable valve actuation for timing control of a homogeneous charge compression ignition engine. SAE Technical Paper; 2005-01-0147;2005.

[34]. Larimore J, Hellstrom E, Sterniak J, Jiang L, Stefanopoulou AG. Experiments and analysis of high cyclic variability at the operational limits of spark assisted HCCI combustion. In: American control conference 2012, ACC2012 ; 27June2012.

[35]. A.K Agrwal, A.P. Singh, R.K. Maurya, Evolution, challenges and path forward for low temperature Progress in Energy and Combustion Science 61 (2017) 1-56. 2017.

[36]. Agrwal AK, Singh AP, Maurya RK. Evolution, challenges and path forward for low temperature combustion engines. Progress in energy and combustion science 61, 2017 ,1-56.

[37]. Nishi M, Kanehara M, Iida N. Assessment for innovative combustion on HCCI engine by controlling EGR ratio and engine speed. Appl Therm Eng. 2016;99:42–60.

[38]. Shi L , Cui Y , Deng K , Peng H , ChenY .Study of low emission homogeneous charge compression ignition (HCCI) engine using combined internal and external exhaust gas recirculation (EGR). Energy2006;31:2665–76.

[39]. Ishikawa N, Ohkubo Y, Kudou K. Study on the effects of EGR cooler performance on combustion properties of the pre mixed compression ignition combustion by multi-cylinder DI diesel engine. SAE Technical Paper D1090XX; 2007-01-1881;2007.

[40]. Kanda T, Hakozaki T, Uchimoto T, Hatano J , Kitayama N , Sono H. PCCI operation with early injection of conventional diesel fuel. SAE Technical Paper ;2005-01-3837;2005.

[41]. Boyarski NJ, Reitz RD. Premixed Compression Ignition (PCI) combustion with modeling-generated piston bowl geometry in a diesel engine. SAE Technical Paper; 2006-01-0198;2006.

[42]. Mathivanan K, Mallikarjuna JM, Ramesh A. Influence of multiple fuel injection strategies on performance and combustion characteristics of a diesel fuelled HCCI engine-An experimental investigation .Exp Therm Fluid Sci 2016 ;77:337–46.

[43]. Idicheria C, Pickett LM. Ignition, sootformation, and end of combustion transients in diesel combustion under high EGR conditions. Int J Engine Res 2011; 12:376. doi: 10.1177/ 1468087411399505.

[44]. Hardy WL ,Reitz RD . A study on the Effects of high EGR, high equivalence ratio ,and mixing time on emissions levels in a heavy duty diesel engine for PCCI combustion. SAE Technical Paper ;2006-01-0026;2006.

[45]. OlssonJ,JohanssonB.ClosedloopcontrolofanHCCIengine.SAETechnicalPaper2001;2001-01-1031;2001.

[46]. C. Wilhelmsson, P. Tunest, and B. Johansson, "Operation strategy of a dual fuel HCCI engine with VGT," SAEPaper 2007- 01-1855, 2007.

[47]. M. IzadiNajafabadi, A. A.Nuraini,A.NorMariah, and L.Abdul Mutalib. Effects of intake temperature and equivalence ratio on HCCI ignition timing and emissions of a 2-stroke engine Applied Mechanics and Materials Journal, vol. 315, pp. 498–502, 2013.

[48]. H. Persson,M. Agrell, J. O. Olsson, B. Johansson, and H. Strom. The effect of intake temperature on HCCI operation using negative valve overlap. SAE Paper 2004-01-0944, 2004.

[49]. R. K. Maurya and A. K. Agarwal, Experimental investigation on the effect of intake air temperature and air-fuel ratio on cycle-to-cycle variations of HCCI combustion and performance parameters. Applied Energy, vol. 88, no. 4, pp. 1153–1163, 2011.

[50]. Okude K, Mori K, Shiino S, Moriya T. Premixed Compression Ignition (PCI) combustion for simultaneous reduction of NOx and soot in diesel engine. SAE paper no. 2004-01-1907; 2004.

[51]. Suzuki T, Kakegawa T, Hikino K, Obata A. Development of diesel combustion for commercial vehicles. SAE paper no. 972685; 1997.

[52]. Wahlin F, Cronhjort A. Fuel sprays for premixed compression ignited combustion - characteristics of impinging sprays. SAE paper no. 2004-01-1776; 2004.

[53]. Peng Z, Zhao H, Tom, Ladommatos N. Characteristics of homogeneous charge compression ignition (HCCI) combustion and emissions of n-heptane. Comb Sci Tech 2005;177(11):2113–50.

[54]. Harada A, Shimazaki N, Sator, S, Miyamoto T, Akagawa H, Tsujimura K. The effects of mixture formation on premixed lean diesel combustion. SAE paper 980533; 1998.

[55]. M. J. Atkins and C. R. Koch, "The effect of fuel octane and dilutent on homogeneous charge compression ignition combustion," Journal of Automobile Engineering, vol. 219, no. 5, pp. 665–675, 2005.

[56]. Aldawood A, Mosbach S, Kraft M. HCCI Combustion control using dual fuel approach: experimental and modeling investigations, SAE Technical Paper ; 2012-01-1117;2012.

[57]. Xingcai L, Yuchun H, Linlin Z, Zhen H. Experimental study on the auto ignition and combustion characteristics in the homogeneous charge compression ignition (HCCI) combustion operation with ethanol /n-heptane blend fuels by port injection. Fuel 2006; 85:2622–31.

[58]. Ma J, Lu X, Ji L ,Huang Z. An experimental study of HCCI DI combustion and emissions in a diesel engine with dual fuel. Int J Therm Sc i2008 ;47:1235–42.

[59]. Kokjohn S, Hanson R, Splitter D, Reitz R. Experiments and modeling of dual fuel HCCI and PCCI combustion using in-cylinder fuel blending. SAE Technical Paper ;2009-01-2647;2009.

[60]. Singh AP, Agarwal AK. Diesoline, diesohol and diesosene fuelled HCCI engine development. J Energy Res Technol 2016:15–1090.

[61]. Genchi G, Pipitone E. Preliminary experimental study on double fuel HCCI combustion. Energy Procedia 2015 ;81:784–93.
[62]. Hasegawa R, Yanagihara H. HCCI combustion in DI diesel engine. SAE paper no. 2003-01-0745; 2003.

Analysis of Curve Fitting for Case Studies for Indian Power Sector

Rajkumar P Narkhede[1], Prabha Rastogi[1]

[1]Dept. Of Mathematics, J. J. T. University , Churu , Rajasthan, India

Rajkumar.p.narkhede@gmail.com

Abstract. The requirements for Indian electrical power sector is always a challenging task in terms of predication of units of generation of electrical units, load factor (central, state and private) for the power plants, peak demand, peak demand meet and peak demand shortage. The historical data from 1997 to 2018 are used for the analysis. The straight line function and polynomial functions used for the analysis of the data to predict the future. The paper outlines appropriates of the both model in context with the Indian power sector. Further, provide the opportunity to understand the regression coefficient as test condition for the model. The regression coefficients are evaluated for load factor such as 0.9448, 0.9659 and 0.9566 for central, state and private. While regression coefficients for the peak demand and peak demand meet are fairly closer to one. However, the model data values are evaluated and the minimum error in the peak demand is 44% and for peak demand meet is 274 % while maximum deviation in peak demand is 107 and fr peak demand meet is 743%.

Key Words Regression Coefficient curve fitting straight line method linear regression

1. Introduction

In statistical modelling, regression analysis is a set of statistical processes for estimating the relationships between a dependent variable and one or more independent variables. The most common form of regression analysis is linear regression, in which a researcher finds the line or a curve that most closely fits the data according to a specific mathematical criterion. For example, the method of ordinary least squares compute the unique line that minimizes the sum of squared distances between the true data and that line. To examine certain examples of the above method used by various researchers in various forms to develop models for physical phenomenon. The models can further predict the future values for certain independent variables provided certain dependent variables are known.

A major part of statistical modelling goes into predicting and optimizing current processes to increase profits. In one such study by Zhilin Lu [1] which discusses optimizing wind and photovoltaic energy. The author uses stochastic collocation method for optimizing the wind and photovoltaic energy through IEEE 39 bus as well as 1009 bus systems. The authors conclude that by changing the method of optimization from stochastic economic dispatch (SED) to nested sparse grid-based allocation method (NS-SCM) there is an increase in the economic dispatch of power. This is because the NS-SCM models the process better than SED model. A similar study is carried by Debalina Sahaa et. al. [2], on restructuring the power sector in West Bengal into public sector and private sector utilities is estimated. Various parameters are evaluated and their impact on the overall impact on welfare of the people is evaluated. The authors have taken into account various factors like rationalization of tariffs, cross-subsidy, generation, transmission, and distribution losses, etc. By using statistical methods authors have evaluated impact of each change in each parameter to its impact. A study is carried by Jungwoo Leea using levelized cost of electricity (LCOE) for predicting the future investment. Another study is estimating electricity consumption to match generation. This is done to ensure smooth working of the grid system. A study by Yang Zhoua [3] has created a model using 2,66,000 firms in Shanghai to predict their requirements to ensure continuous electricity supply. The algorithm used here is simple linear regression. A comparison of linear regression is done with LASSO, PCA and Ridge regression and it is realized that linear regression is best suited for dimension reduction compared to other algorithms. A study by Anastasia Ioannoua et. al. [4] has found ways in which impact of technological, political, social, environmental factors, etc. is evaluated before the commencement of a renewable energy project. The method assists the stakeholders to take decisions to make power generation sustainable. Semi-quantitaive methods like scenario analysis and MCDA are used for addressing non-statistical parameters such as social factors and emergence of competitive technologies. Paraskevas [5] suggests a bottom-up approach for long-term energy planning of the power supply sector integrating mainland and insular electric systems. The model which is developed has a composite cost objective function expressing the NPV of total electricity generation costs over a specific period, comprising annualized investment costs, fixed and variable O&M expenses, fuel costs, GHG emission rights which are all related to power plants plus the cost of electricity imports and the annualized capital costs for the islands-to-mainland interconnections. In conclusion use of such a multi-objective optimization model is not only economical but also better for the environment. Fenghao Cui, et. al.[6] have used cubic smoothing spline, polynomial curve-fitting and nonlinear least squares for fitting models to estimate river water quality. The challenge here is that there are various unknown parameters which are very difficult to model and hence various techniques need to be tried before implementing a specific model. The models were tested by changing the sensitivity index. The variation of model outputs showed a slight difference at a sensitivity index of less than 10% and a significant difference at a sensitivity index of more than 50%. The one-way analysis of variance showed a large p-value of 0.8431, indicating that differences between model data and measured data means are not significant. Thus, the comprehensive application of the numerical curve-fitting techniques of cubic smoothing spline, polynomial curve-fitting, and nonlinear least squares provided a robust estimation procedure to support data analysis and parameter calibration when estimating the unknown parameters of the river water quality management (RWQM). An application of the modelling process is to be able to predict future independent parameters by using historical data. This is done by Farhad Mehmanpazir, et. al. [7] for evaluating supply and demand for future by analysing historical data of past 60 months. A similar research takes place in modelling stock market data to predict which stocks will raise in the future. The critical part here is the choice of various dependent parameters on the values of independent parameters and their weighted influence on the same. Such curve fits work only for a short period of time as the influence of various parameters is not linear. The accurate modelling of the charging characteristics of electric vehicles (EVs) is the basis for the load forecasting, infrastructure planning, and orderly charging management required for proper use of electric vehicles. A study by Zhong Chen et al [8], have done an investigation in Nanjing, China based on operational data available from an EV charging company. On this basis, they carried out stochastic simulation for the load curve of disordered charging and analyzed the potential of the EV charging load participating in the orderly management and its coordination with the output of power generation using renewable energy. Statistical analysis is widely used to predict non- linear phenomenon in physics experiments to derive new laws or to predict certain properties of materials under certain conditions. For example, bending of Aluminium alloys is inelastic in nature and hence it cannot be predicted by natural laws. In a study by Cheng Ming et al [9], Aluminium alloy bending is observed and a model is prepared to estimate the bending in aluminium alloys. Semi-empirical formulas are created which can be used to design Box beams and I beam sections. In evaluating properties of Non-

newtonian fluids (Mohammadhossein Irani [10]), curve for shear stress Vs shear rate is plotted and indexes for temperature and solid volume fraction are evaluated for curve fitting. Thus the model created with two variables, temperature and volume fraction, to evaluate the shear stress/shear strain for other environmental conditions. In GPS systems, there are two important aspects in optimally processing the acoustic measurements: the establishment of the functional model and the refinement of the associated stochastic model. By improving the stochastic model the accuracy of positioning underwater at 3000m depth was increased from 0.426m to 0.306m. (Zhenjie Wang et al [11]). In study, Raman spectra of carbonaceous matter are evaluated automatically by using curve-fitting. This is done as Raman spectra are biased by subjectivity which reduces overall comparability of Raman spectra of carbonaceous material (RSCM). It is concluded that RSCM parameter reflects the metamorphic conditions [12].
The present paper discusses the appropriateness of the curve fitting tool used for the Indian electrical power generation data with the regression analysis. The history data for the 1997 to 2018 were taken for the analysis and the future is predicated with the different curve fitting functions. The typical errors are evaluated for the various functions used for the data analysis.

2. Results and Discussion
The actual electrical power generation data from year 1997 to 2018 is plotted. Initially, straight line functions for the data analysis. The regression coefficient obtained for the data point is 0.8895 which suggest the better fitting ability. Once the constants obtained for the straight line the model is the used to predict the electrical power generation for each year. It is observed that model and actual electrical power generation is not deviates much. The error in the actual and model electrical power generation capacity is limited within 16.42%. Thus, it is appropriate to use the straight line model to predict the future electrical power generation in Indian context.

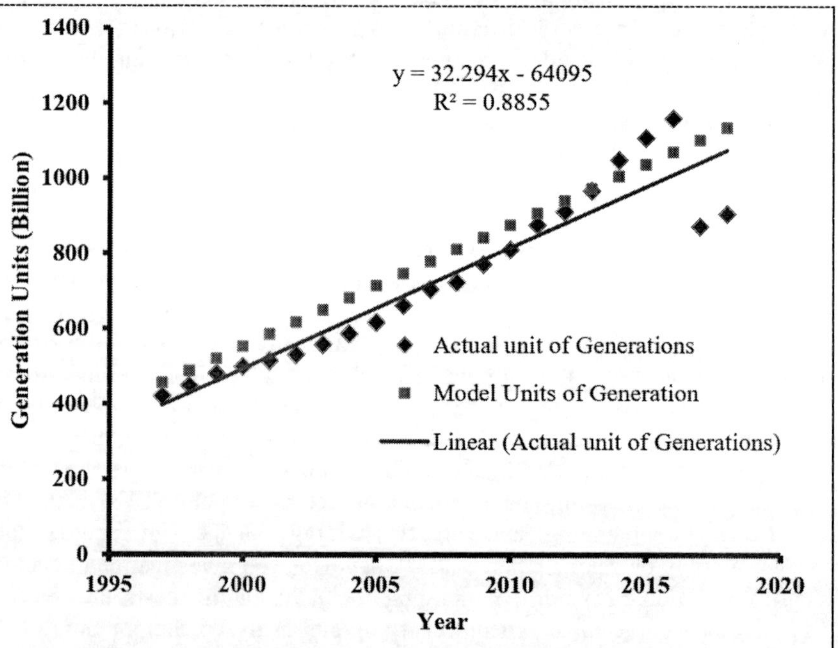

Figure 1. Generation of unit per year from 1997 to 2018

Table 1. Load factor for the Indian power plants operated by central, state and private sector.

Year	Actual Load Factor				Model Polynomial Load Factor			Model Straight line fit		
	Central	State	Private	Total	Central	State	Private	Private	State	Central
1997	64.7	70.4	60.9	65	3705911002	-6761185015	-4214739774	79	72	75
1998	64.6	64.6	60.7	63	3707941778	-6766122230	-4217825678	78	71	75
1999	67.3	67.3	63.7	66	3709971444	-6771060558	-4220912420	77	70	76
2000	74.3	65.6	73.1	71	3712000000	-6776000000	-4224000000	77	70	76
2001	77	67	74.7	73	3714027444	-6780940558	-4227088420	76	69	76
2002	77.1	68.7	78.9	75	3716053774	-6785882234	-4230177682	75	69	76
2003	78.7	68.4	80.5	76	3718078990	-6790825028	-4233267786	75	68	76
2004	81.7	69.6	85.1	79	3720103090	-6795768943	-4236358733	74	67	76
2005	82.1	67.1	85.4	78	3722126073	-6800713979	-4239450525	73	67	76
2006	84.8	70.6	86.3	81	3724147938	-6805660138	-4242543163	73	66	77
2007	86.7	71.9	90.8	83	3726168683	-6810607421	-4245636647	72	65	77
2008	84.3	71.2	91	82	3728188308	-6815555828	-4248730978	71	65	77
2009	85.5	70.9	82.4	80	3730206811	-6820505363	-4251826158	71	64	77
2010	85.1	66.7	76.7	76	3732224192	-6825456024	-4254922186	70	63	77
2011	82.1	68	76.2	75	3734240449	-6830407814	-4258019064	69	63	77
2012	79.2	65.6	64.1	70	3736255581	-6835360733	-4261116791	69	62	78
2013	76.1	59.1	62.1	66	3738269588	-6840314783	-4264215369	68	61	78
2014	74	59.8	60.6	65	3740282468	-6845269964	-4267314797	67	61	78
2015	72.5	55.4	60.5	63	3742294220	-6850226277	-4270415077	67	60	78
2016	72	54.3	55.7	61	3744304844	-6855183724	-4273516208	66	59	78
2017	70.4	53.9	55.4	60	3746314339	-6860142304	-4276618191	66	59	78
2018	70.9	54.5	55.8	60	3748322704	-6865102019	-4279721025	65	58	78

Table 1 shows the load factor for the power plants operated by the central, state and private power plants. The historical data used for the analysis is from 1997 to 2018. The load factor helps us the aggravator that how much electrical power that could be available for grid planning and distribution. The load factor data is then plotted with the straight line function for all three power production companies. The constants are evaluated and regression coefficient for central, state and private sector obtained are 0.0205, 0.4925 and 0.1241, respectively (refer figure 2). The model developed and the data evaluated for the same it is observed that the variation in the load factor is closer for the central power plant, whereas for state and private power plants model load factor is deviation a large extent within 30 %. Thus, the straight line function may be appropriate to some extent. Same data is used to plot with the forth order polynomial. The regression coefficients are evaluated such as 0.9448, 0.9659 and 0.9566 for central, state and private. The regression coefficients are closer to unity, thereby suggests that the polynomial model works closer with the actual data points. The constants obtained are then used for the model evaluation. It is observed that the model data points deviated in the exuberant manner and the polynomial fit is no longer the appropriate for the data analysis (refer Table -1).

IOP Conf. Series: Materials Science and Engineering **810** (2020) 012055 doi:10.1088/1757-899X/810/1/012055

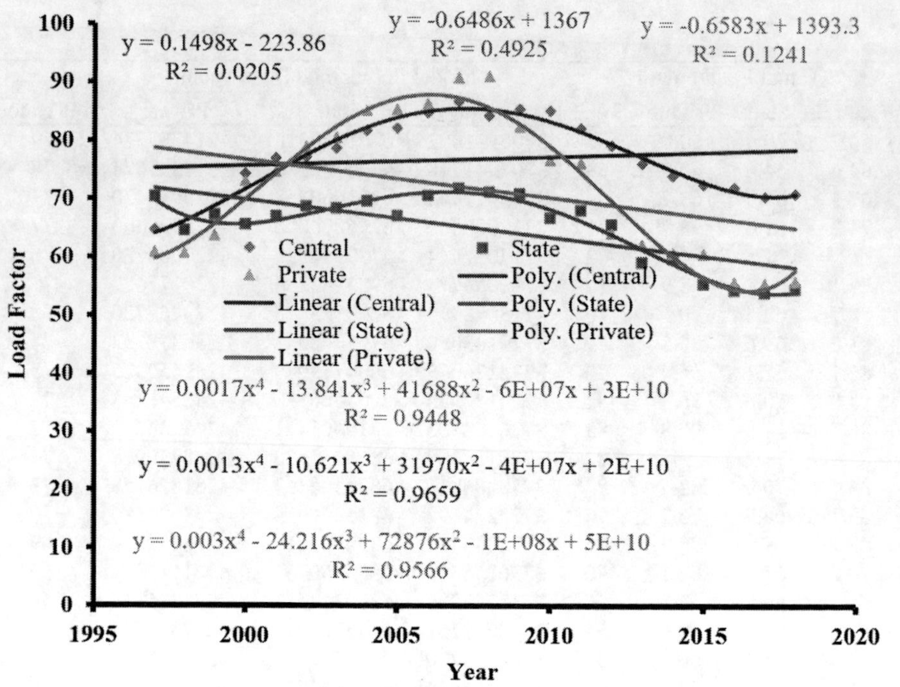

Figure 2. Load factor for the central, private and state power plants from year 1997 to 2018

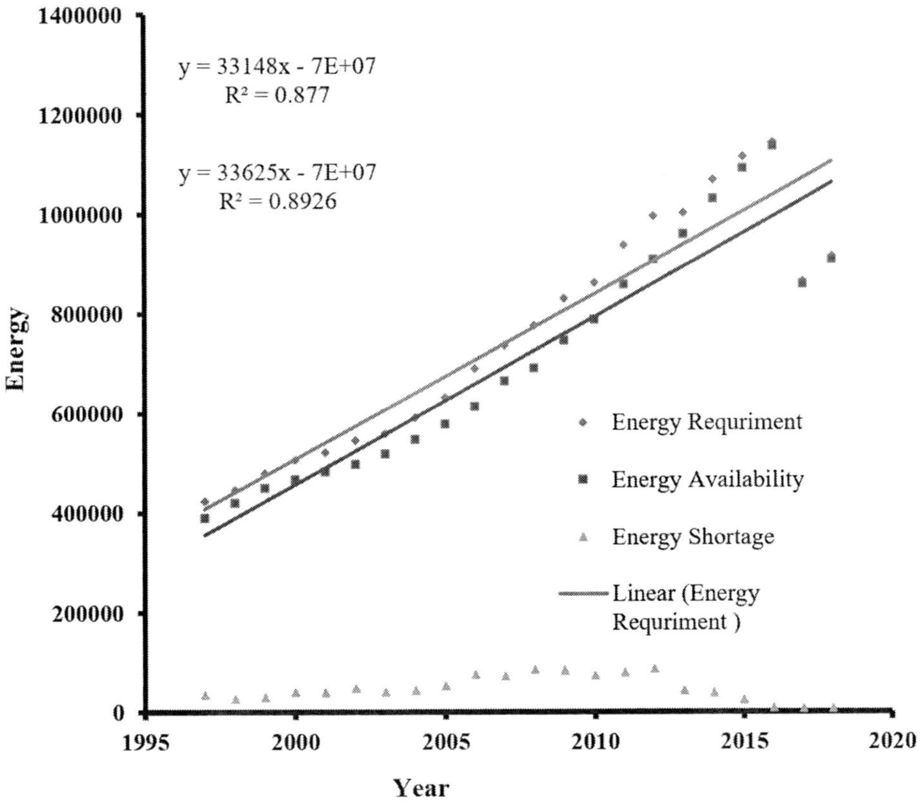

Figure 3. Energy required and demand meet by the Indian Power sector from 1997 to 2018

Figure 3 shows the energy availability, demand and shortage for Indian context from year 1997 to 2018. Similar way data analyses are carried out for the all three energy indices with the straight line fit. The constant for the all energy indices are evaluated and the model is fitted for the data. It is observed that the predicted model data points are fairly close to the actual data points. Thereby it is appropriate to use the straight line fit the energy availability, demand and shortage.

IOP Conf. Series: Materials Science and Engineering **810** (2020) 012055 doi:10.1088/1757-899X/810/1/012055

Table 2. Peak demand and peak demand meet for the Indian power sector.

Year	Peak Demand (MW)	Peak Demand Meet (MW)	Model Peak Demand (MW)	Model Peak Demand Meet (MW)	% Model Peak Demand (MW)	% Model Peak Demand Meet (MW)
1997	65435	58042	135374	487645	107	740
1998	67905	58445	140449	492897	107	743
1999	72669	63691	145525	498148	100	682
2000	78037	67880	150600	503400	93	642
2001	78441	69189	155675	508652	98	635
2002	81492	71547	160751	513903	97	618
2003	84574	75066	165826	519155	96	592
2004	87906	77652	170901	524407	94	575
2005	93255	81792	175977	529659	89	548
2006	100715	86818	181052	534910	80	516
2007	108866	90793	186127	540162	71	495
2008	109809	96785	191202	545414	74	464
2009	119166	104009	196278	550665	65	429
2010	122287	110256	201353	555917	65	404
2011	130006	116191	206428	561169	59	383
2012	135453	123294	211504	566420	56	359
2013	135918	129815	216579	571672	59	340
2014	148166	141160	221654	576924	50	309
2015	153366	148463	226730	582176	48	292
2016	159542	156934	231805	587427	45	274
2017	164066	160752	236880	592679	44	269

Table 2 shows the peak demand and peak demand meet from 1997 to 2018 for the Indian context. In the earlier discussion it is observed that the straight line function provides better model fit as against the polynomial fit. The straight line fit is used for the analysis of the both power indices. The regression coefficients for the both cases are fairly closer to one (i.e. 0.9845 and 0.9609). The actual data points fairly lie closer to the straight line fit (refer figure 4). However, the model data values are evaluated and the minimum error in the peak demand is 44% and for peak demand meet is 274 % while maximum deviation in peak demand is 107 and for peak demand meet is 743% (refer Table 2). Thus, it is observed that always straight line fit for the data under consideration will not meet. The appropriateness for the straight line or polynomial fit is always a matter of the detail investigation and only regression coefficient cannot be a tool provide the information with confidence that the model will work.

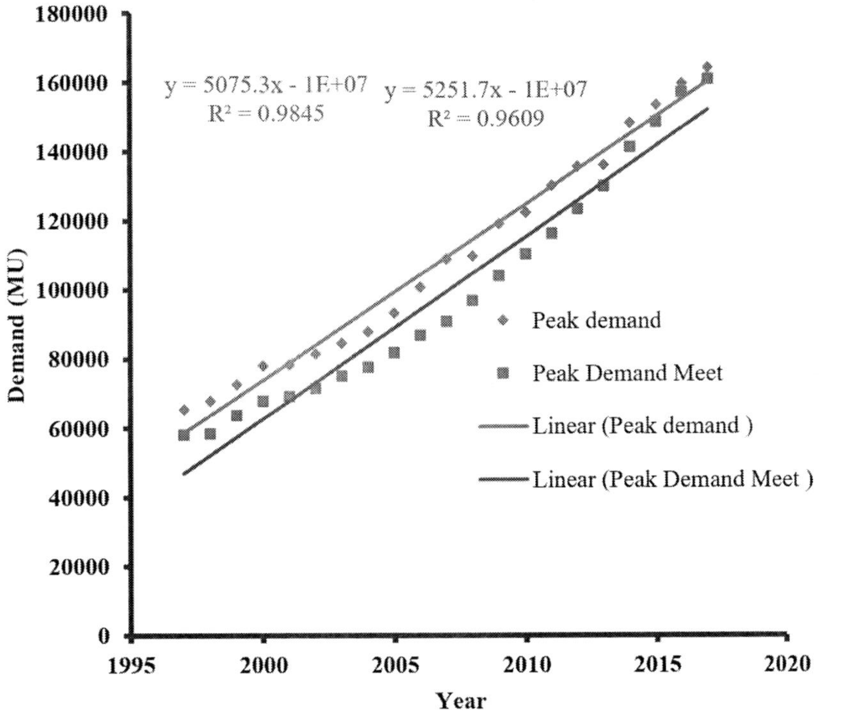

Figure 4. Peak Demand and peak demand meet for Indian context from 1997 to 2018

3. Conclusions.

The two prominent curve fitting functions used for the electrical power generation indices for the Indian context are analysed. It is observed that the polynomial fit do not yield the model parameters appropriate in the context of the problem under consideration. The straight line functions on the contrary work better for the few data sets whereas for few data sets it deviates to greater extent. Thereby, for straight line function further mathematical better precise conditional parameter is required and only regression coefficient do not provide the appropriateness of the straight line model. In the present case the straight model provides the better and appropriate for fit for the units of electricity generated, while fairly closer to load factor for central own power plants and deviates to greater extent in all other cases.

4. References

[1]Zhilin Lu, Mingbo Liu, Wentian Lu, and Zhuoming Deng 2019 *IEEE-Access* 7 91827
[2]Debalina Saha and Rabindra N. Bhattacharya 2019 *Utilities Policy* **5** 62
[3]Yang Zhou, Shuaishuai Zhang, Libo Wu and Yingjie Tian 2019 *Applied Energy* **255** 113790
[4]Anastasia Ioannou, Andrew Angusc and Feargal Brennan 2017 *Renewable and Sustainable Energy Reviews* **74** 602
[5]Paraskevas N. Georgiou 2016 *Computer and Operation Research* **66** 292

IOP Conf. Series: Materials Science and Engineering **810** (2020) 012055 doi:10.1088/1757-899X/810/1/012055

[6]Fenghao Cui, Chul Park and Moonil Kim 2019 *Journal of Environmental Management* **249** 109375
[7]Farhad Mehmanpazir, Kaveh Khalili-Damghani and Ashkan Hafezalkotob 2019 *Resources Policy* **63** 101409
[8]Zhong Chen, Ziqi Zhang, Jiaqing Zhao, Bowen Wu, and Xueliang Huang 2018 *IEEE Access* **6** 24475
[9]Cheng Ming , Shi Yongjiu and Wang Yuanqing 2006 *Tsinghua Science and Technology* **11** 648
[10]Mohammadhossein Irani, Masoud Afrand and Babak Mehmandoust 2019 *Physica A* **531** 120837
[11]Zhenjie Wang, Shuang Zhao, Shengyue Ji, Kaifei He, Zhixi Nie, Huimin Liu and Rui Shan 2019 *Applied Acoustics* **150** 36
[12]N.K. Lünsdorf and J.O. Lünsdorf 2016 *Journal of Coal Geology* **160–161** 51

Modal and frequency response characteristics of vehicle suspension system using full car model

Sanjay Sharma[1], **Manoj Chouksey**[2], **Vinod Pare**[3], **Pranay Jain**[4]

[1,2,3]Department of Mechanical Engineering, SGSITS, 23, Park Road, Indore, 452003, M.P., India

[4]VECV, Pithampur, Dhar, M.P., India

[1]sanjaysharma1124@gmail.com, [2]machouksey@gmail.com, [3]v_pare@yahoo.co.in, [4]pjain10@vecv.in

Abstract. This paper involves simulated studies of a vehicle suspension system for modal and frequency response characteristics. Ride comfort and car handling are two important considerations for finalizing the design of the vehicle suspension. However, both of these are conflicting to each other. Ride comfort and car handling depend on dynamic characteristics of the vehicle (i.e. natural frequencies, mode shapes, frequency response functions etc.) too among other parameters. Hence it is very important to carry out modal examination of the vehicle suspension system to find out the natural characteristics of the system. The same has been attempted in this work using full car model having seven degree of freedoms. The degrees of freedom include pitching, bouncing and rolling motion along with motion of the un-sprung masses. The software tools namely ANSYS Workbench, MATLAB, and SIMULINK environment has been employed to discover the system's modal and frequency response features. The vehicle's natural frequencies and mode shapes were evaluated and bode plots are drawn to explore the system's frequency response features. The results produced in different software tools have also been compared and discovered to be in close agreement with the results present in the literature.
Keywords. Natural Frequency, Modal Analysis, Frequency Response, Ride Comfort.

1. Introduction

A Suspension system is a complicated system of vibration, which can be conveniently modelled with various degrees of freedom.

Suspension system is intended to separate the car body from the inputs of the highway. Suspension sys tem serves a dual objective - to contribute to the road holding and handling of the vehicle.

It is important from the point of view of vehicle stability during braking, driving pleasure and comfort, and isolation from road bumps, vibration and noise, etc.

Today, there are many challenges to automobile companies in the cutting edge competition to survive and excel in the market. The competition among automobile companies has forced them to seek better alternative strategies in suspension systems through research studies. One of the performance demands is sophisticated suspension systems aimed at improving ride comfort and handling while improving riding skills and smooth driving performance. The suspension system aims at ensuring a smooth ride and maintaining car control over rough roads or in the event of sudden brakes. Larger suspension travel and lower damping in wheel bounce mode result in improved riding convenience [1].

Suspensions systems may be classified on various basis. On the basis of control, it is characterized into three types, namely (i) Passive suspension systems, (ii) Semi active suspension system, and (iii) Active suspension system. Most of the cars are hang in passive way, where the suspension system includes basically springs and dampers. In passive system, the component features, i.e. the spring constant values and the damping coefficients, are fixed. The disadvantage of the passive suspensions system is that their performance remains effective only over a certain frequency range. In the early 1970s, semi-active suspensions were suggested and can be almost as efficient as full-active suspension to enhance the quality of car behavior. A semi-active system is a combination of passive spring element and a controllable damper element. It enables automatic adjustment/control of the damping coefficient based on control strategy. The controllable damper works using external energy by means of an integrated controller and a set of sensors. The semi-active suspension may continue to function under a passive condition if the control system fails. An active suspension system enables variation/adjustment of damper element as well as spring element. Passive devices that can only dissipate energy, while the force actuator in active systems can add and dissipate energy from the system. Unlike the passive suspension, the active suspension can enhance the efficiency of the suspension mechanism over a wide spectrum of frequencies. Aly and Salem [2] addressed the use of smart technology to regulate a continually variable suspension system for automotive damping. An active suspension system was suggested in their research to enhance riding convenience. Many studies have been carried out on suspensions system models to make it function efficiently by optimizing its parameters.

For many years, active and semi-active suspensions were used to enhance the stability and handling of all vehicle kinds. Bose, Daimler Chrysler, Land Rover, and Delphi have effectively used these kinds of devices as factory or aftermarket application. Controlling or helping to stabilize the roll, pitch, and yaw of a car through the suspension would be particularly useful for high-gravity vehicles such as SUVs, trucks, and ATV.

Griffin [3] has discussed the human response to vibration excitation discussing about both whole body vibration and hand-transmitted vibration using analytical as well as experimental studies while considering various aspects like psychology, physiology etc. The author reports that the human whole-body fundamental Natural frequency ranges from 3 Hz to 7 Hz.

Jazar RN [4] has categorized passenger cars in his book ' Vehicle Dynamics Theory and Application, ' a vehicle intended to carry ten or fewer people depending on their size and weight. The author claims that for 205/50R15 passenger car tire the value of $k = 200000$ N / m is a good approximation, and for X31580R22.5 truck tire $k = 1200000$ N / m is a good approximation.

Gillespie [5] in his book 'fundamental of vehicle dynamics' explained the stiffness ratio, r_k (ratio of tire stiffness to suspension stiffness) which is the main criterion for deciding the car suspension stiffness. The stiffest suspension, r_k =5, would be indicative of sports vehicles and performance vehicles whereas the softest, r_k =20, would be relate to luxury vehicles with air suspensions. Figure 1 shows the stiffness ratios for most commercial vehicles. However, in practice it is not possible to use the entire range of performance shown in the figure 1 shaded region. The low damping concentrations lead to suspension strokes that are beyond the range available for most passenger vehicles, and low damping is also inadequate to regulate wheel hop oscillations that compromise the behaviour of road holding.

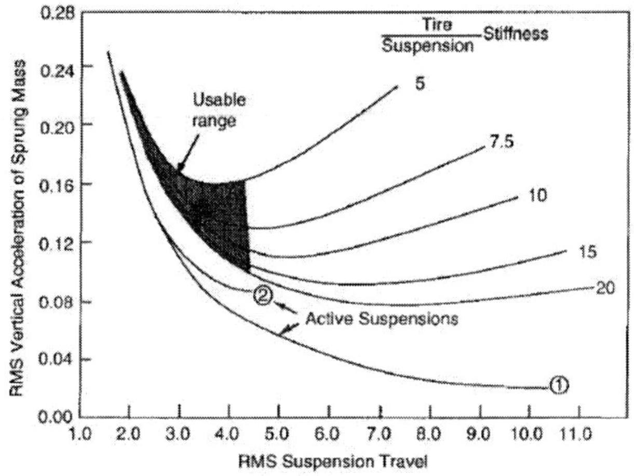

Figure 1. The practical range of performance for passive Suspensions (shaded area) [5].

Gillespie also described that a soft suspension helps in ride isolation as the sprung mass acceleration was found to be minimum for lower values of natural frequencies. In the case considered, the sprung mass acceleration was found to be minimum at 1 Hz natural frequency. However, other practical considerations restricts the natural frequencies in the range from 1 to 1.5 Hz for most vehicles. The natural frequency in performance cars, where vehicle handling is more important than ride comfort, is between 2 to 2.5 Hz.

Correct estimation of such frequencies is very important. Hence, it becomes necessary to model the vehicle suspension system accurately. This paper, therefore, discusses modelling of the vehicle suspension system through developing own code (MATLAB in this work) and using ready software tools (in this work ANSYS workbench environment). Bode plots are also drawn with the SIMULINK model to study frequency response of the system. The results produced in this way have also been compared and found to be in close agreement with the result present in the literature.

2. Modelling of the suspension system

Figure 2 shows the complete car model considered in this work. Vehicle wheels and Suspension System are regarded as spring-Damper element. M_{U1}, M_{U2}, M_{U3}, M_{U4} denotes the car's un-sprung masses while M_S denotes the car's sprung mass. Z_{U1}, Z_{U2}, Z_{U3}, and Z_{U4} denote the un-sprung masses vertical displacement while Z_S denotes the sprung mass vertical displacement. Θ and $\emptyset$ denote the degree of freedom corresponding to pitching and rolling motion respectively. K_{S1}, K_{S2}, K_{S3}, K_{S4} and C_{S1}, C_{S2}, C_{S3}, C_{S4} indicates the Suspension System Spring and damping variables, Similarly K_{W1}, K_{W2}, K_{W3}, K_{W4} and C_{W1}, C_{W2}, C_{W3}, C_{W4} are the Spring and damping variables of four tires of car. a and b indicates distance of front and back tires from sprung mass center of gravity (CG), whereas c and d indicates distance of right and left tire from the Sprung mass CG. The seven coordinates which describe the vibrating system are Z_S, Θ, $\emptyset$, Z_{U1}, Z_{U2}, Z_{U3}, and Z_{U4}.

In this work the un-damped natural frequencies of full car model are analysed. Hence, the damping parameters of front suspension (C_{S1}, C_{S2}), rear suspension (C_{S3}, C_{S4}), and four wheels (C_{W1}, C_{W2}, C_{W3}, C_{W4}) of a vehicle are taken as zero.

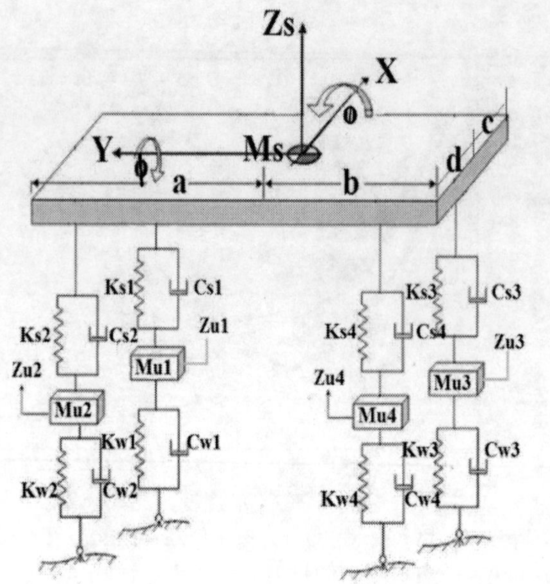

Figure 2. 7-dof Full Car Model.

The values of various vehicle parameters for numerical simulations are considered, as shown in Table 1 which are close to the practical passenger car.

Table 1. Vehicle Parameters.

S. No	Model Parameter	Value	Unit
1	M_S	1243.4	kg
2	M_{U1},M_{U2},M_{U3},M_{U4}	62.8	kg
3	K_{S1},K_{S2}	17000	N/m
4	K_{S3},K_{S4}	20000	N/m
5	C_{S1},C_{S2},C_{S3},C_{S4}	0	N-s/m
6	K_{W1},K_{W2},K_{W3},K_{W4}	200000	N/m
7	C_{W1},C_{W2},C_{W3},C_{W4}	0	N-s/m
8	a , b	1.21 , 1.59	m
9	c = d	0.8	m
10	I_{XX}	940.85	Kg-m^2
11	I_{YY}	265.57	Kg-m^2

2.1. Equations of motion

The motion equations for a full car model of seven degrees of freedom, as shown in Figure 2, were extracted and written in the following matrix as follows:
Where M is mass Matrix and K is stiffness matrix.

$$M\ddot{Z} + KZ = 0 \tag{1}$$

$$M = \begin{pmatrix} M_S & 0 & 0 & 0 & 0 & 0 & 0 \\ 0 & I_{XX} & 0 & 0 & 0 & 0 & 0 \\ 0 & 0 & I_{YY} & 0 & 0 & 0 & 0 \\ 0 & 0 & 0 & M_{U1} & 0 & 0 & 0 \\ 0 & 0 & 0 & 0 & M_{U2} & 0 & 0 \\ 0 & 0 & 0 & 0 & 0 & M_{U3} & 0 \\ 0 & 0 & 0 & 0 & 0 & 0 & M_{U4} \end{pmatrix} \quad (2)$$

$$\ddot{Z} = \begin{pmatrix} \ddot{Z}_S \\ \ddot{\theta} \\ \ddot{\varphi} \\ \ddot{Z}_{U1} \\ \ddot{Z}_{U2} \\ \ddot{Z}_{U3} \\ \ddot{Z}_{U4} \end{pmatrix} \quad (2) \qquad Z = \begin{pmatrix} Z_S \\ \theta \\ \varphi \\ Z_{U1} \\ Z_{U2} \\ Z_{U3} \\ Z_{U4} \end{pmatrix} \quad (3)$$

$$K = \begin{pmatrix} (K_{S1}+K_{S2}+K_{S3}+K_{S4}) & (-aK_{S1}-aK_{S2}+bK_{S3}+bK_{S4}) & (cK_{S1}-dK_{S2}+cK_{S3}-dK_{S4}) & -K_{S1} & -K_{S2} & -K_{S3} & -K_{S4} \\ (-aK_{S1}-aK_{S2}+bK_{S3}+bK_{S4}) & (a^2K_{S1}+a^2K_{S2}+b^2K_{S3}+b^2K_{S4}) & (-acK_{S1}+adK_{S2}+bcK_{S3}-bdK_{S4}) & aK_{S1} & aK_{S2} & -bK_{S3} & -bK_{S4} \\ (cK_{S1}-dK_{S2}+cK_{S3}-dK_{S4}) & (-acK_{S1}+adK_{S2}+bcK_{S3}-bdK_{S4}) & (c^2K_{S1}+d^2K_{S2}+c^2K_{S3}+d^2K_{S4}) & -cK_{S1} & dK_{S2} & -cK_{S3} & dK_{S4} \\ -K_{S1} & aK_{S1} & -cK_{S1} & (K_{S1}+K_{W1}) & 0 & 0 & 0 \\ -K_{S2} & aK_{S2} & dK_{S2} & 0 & (K_{S2}+K_{W2}) & 0 & 0 \\ -K_{S3} & -bK_{S3} & -cK_{S3} & 0 & 0 & (K_{S3}+K_{W3}) & 0 \\ -K_{S4} & -bK_{S4} & dK_{S4} & 0 & 0 & 0 & (K_{S4}+K_{W4}) \end{pmatrix} \quad (4)$$

2.2. Ansys Workbench model

The block diagram based approach in "Ansys Workbench" software have been used for modal analysis of the Suspension System to extract natural characteristics. The use of line body elements (spring) enables accurate modelling of the system with lesser mobility as compare to the use solid elements. The sprung mass has three mobility: first, vertical translation in Z direction (bouncing): Second, rotation about X direction (pitching): third, rotation about Y direction (rolling). Whereas, each un-sprung mass is provided with one mobility i.e. the translation in Z direction. It may be noted that the degrees of freedom, which are not considered, have been constrained in the software tool. The fully developed Ansys workbench model is shown in figure 3. Similar analysis has also been carried out by developing a code in the MATLAB environment and the results have been listed in the next section.

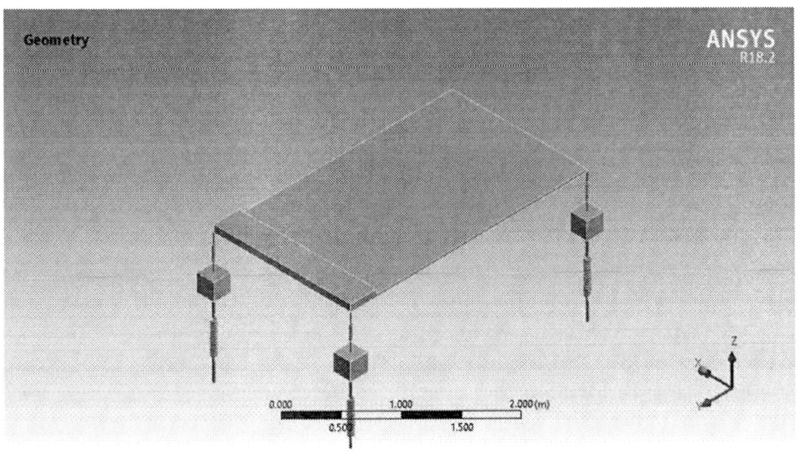

Figure 3. Ansys Workbench Model with seven degrees of freedom.

IOP Conf. Series: Materials Science and Engineering **810** (2020) 012056 doi:10.1088/1757-899X/810/1/012056

2.3. MATLAB code for modal analysis

Modal assessment is used to evaluate the natural frequencies, damping variable and mode shapes of the system [6, 7]. A MATLAB code has also been created for the seven degrees of freedom model. The code can be used to compute the Eigen-values and Eigen-vectors of the suspension systems. Eigenvalues give information about natural frequencies and damping in various modes, whereas eigenvectors provide information about the mode shapes. The "eig" command has been used for this purpose [8].

[V, D] = eig (K, M) generates a diagonal matrix D of Eigenvalues and a complete matrix V with the corresponding Eigen vectors in columns such that "K*V = M*V*D".

The results as obtained through ANSYS Workbench and MATLAB are also correlated and validated in the next section.

3. Result and discussion

3.1. Natural Frequencies

The natural frequencies obtained in ANSYS Workbench and using MATLAB code are shown in Table 2 along with percentage difference. It may be noted that the results as computed through the two different approaches are in close agreement with each other as well as the results presented in the literature and their percentage error is almost negligible. This also establishes the validation of the work. Hence, any of the two approaches can be reliable employed for the modal analysis of the system.

Table 2. Natural Frequencies

Mode	Frequency (Hz)		Difference (%)
	Ansys workbench	Matlab	
	1.1349	1.1349	0.0
1	1.9448	1.9449	0.005
2	2.0278	2.0283	0.024
3	9.3631	9.3631	0.0
4	9.3639	9.3639	0.0
5	9.4340	9.4340	0.0
6	9.4382	9.4382	0.0

3.2. Mode shapes

The analysis of the mode shapes figure 4, shows the relative displacement of the seven vibrating coordinates (Z_S, ϴ, Ø, Z_{U1}, Z_{U2}, Z_{U3}, and Z_{U4}) at different natural frequencies, it can be clearly observed that first mode corresponds to bouncing motion, second mode corresponds to pitching motion and third mode corresponds to rolling motion of the car. The bounce center, pitch center, and the roll center of vehicle affects vehicle handling and ride comfort. Further, the mode shapes (eigen-vectors) may also be used to compute the pitch, bounce and roll centers of the car. It can be seen that the un-sprung masses M_{U1} and M_{U2} move in opposite direction in the fourth mode whereas these mass move in same direction in the fifth mode. Similarly, M_{U3} and M_{U4} move in opposite direction in the sixth mode and in the same direction in the seventh mode.

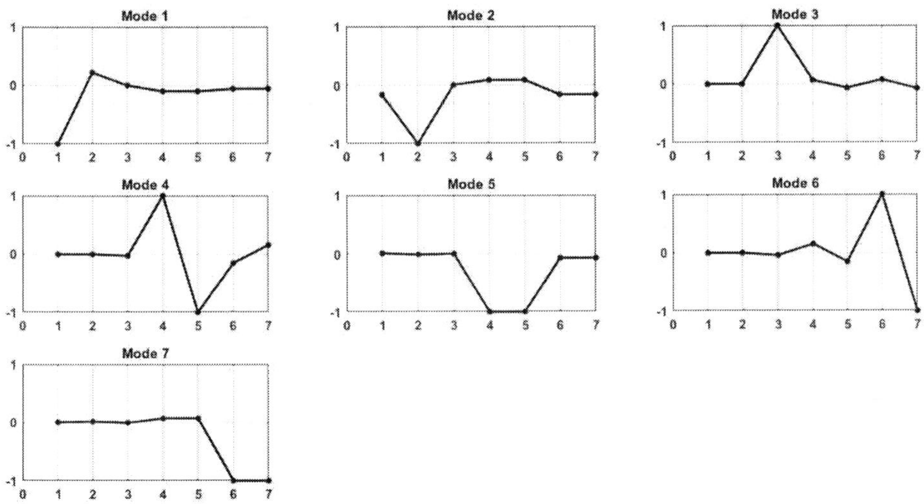

Figure 4. Mode Shapes

3.3. Frequency response (Bode diagram)

Figure 5 shows the SIMULINK model for the complete vehicle suspension system of 7-dof. Bode plots are used to determine system response at different frequencies. The Simulink model is employed to study the Bode plots for the system. Figure 6 shows the Bode plot for the system. The peaks, at the natural frequencies as listed in Table 2, are clearly visible in the plot. The acceleration values at these peaks is also mentioned in the plot. The plot shows distinct peaks at the first three natural frequencies corresponding to bouncing, pitching and rolling motion. Next four modes are close modes and are not clearly distinguishable in the bode plot. These are the modes corresponding to the un-sprung masses.

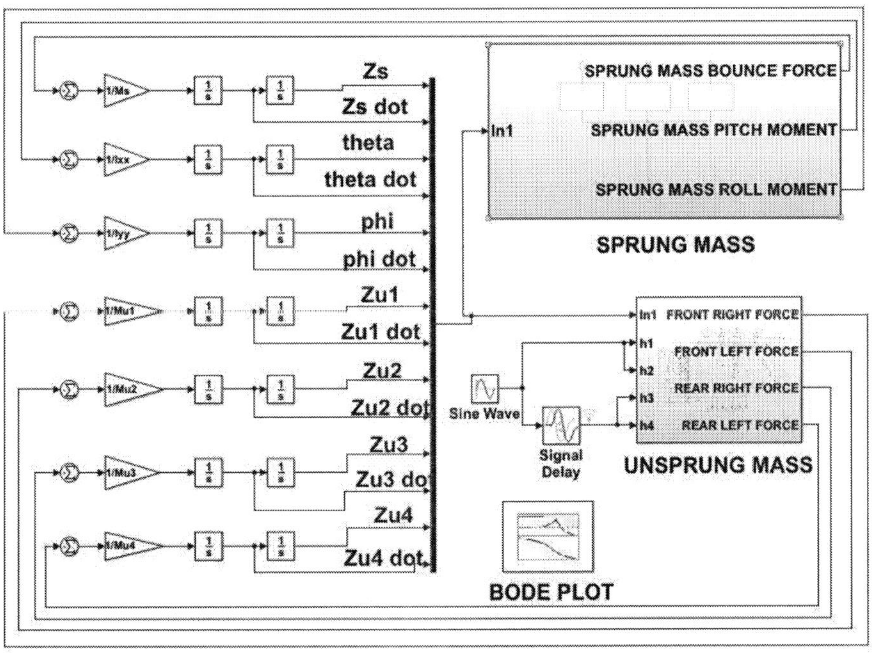

Figure 5. SIMULINK diagram for seven degree of freedom Full Car model.

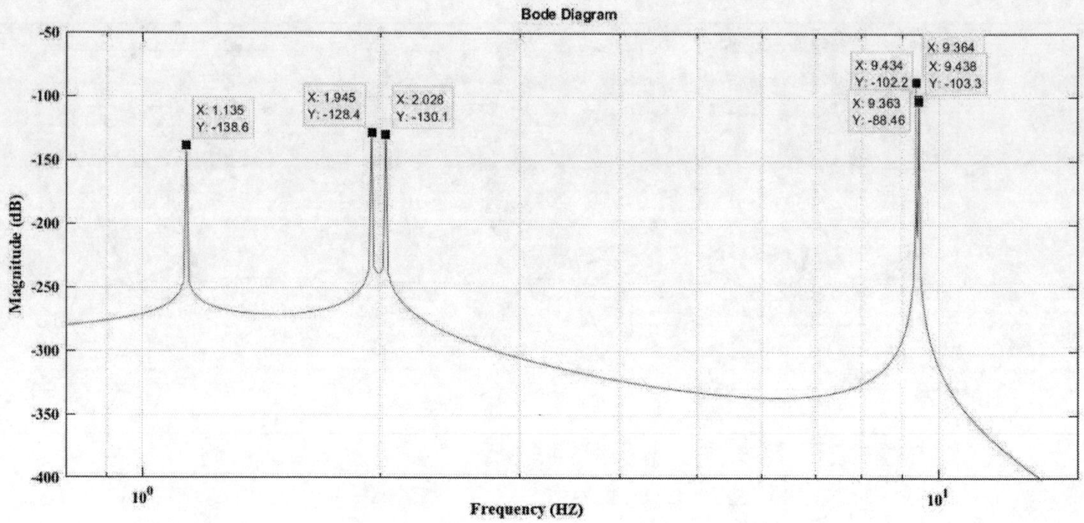

Figure 6. Bode plot for the system.

4. Conclusion

The work presented studies in modal and frequency response features of a vehicle using seven degrees of freedom full car model. The results have been computed using two different approaches, i.e. first writing a code and second using a software tool. The results obtained by the two approaches are found to be in close agreement. Hence, it is concluded that any of the two approaches can be reliably employed for the modal analysis of the system. System's Mode shapes are also studied in the seven modes. Such mode shapes may be helpful in locating the pitching, bouncing and rolling centers of the vehicle. Bode plot has been studied for the study to find out the force vibration characteristics of the system. The peaks are observed at the natural frequencies of the vehicle in the plot.

5. References

[1] Sharma SK, Pare V, Chouksey M, Rawal BR. Numerical studies using full car model for combined primary and cabin suspension. Procedia Technology. 2016 Jan 1;23:171-8.

[2] Aly AA, Salem FA. Vehicle suspension systems control: a review. International journal of control, automation and systems. 2013 Jul;2(2):46-54.

[3] Griffin MJ. Handbook of human vibration. Academic press; 2012 Dec 2.

[4] Jazar RN. Vehicle Roll Dynamics. InAdvanced Vehicle Dynamics 2017 Springer, Cham.

[5] Gillespie TD. Fundamental of Vehicle Dynamics (SAE International). ISBN 878-1-56091-199-9; 1992.

[6] Chouksey M, Dutt JK, Modak SV. Model updating of rotors supported on journal bearings. Mechanism and Machine Theory. 2014 Jan 1;71:52-63.

[7] Chouksey M, Modak SV, Dutt JK. Influence of rotor-shaft material damping on modal and directional frequency response characteristics. Proceedings of ISMA-2010. 2010:1543-57.

[8] MathWorks, Inc. MATLAB: The Language of Technical Computing. Getting started with MATLAB, version 7. MathWorks, Incorporated; 2005.

IOP Conf. Series: Materials Science and Engineering **810** (2020) 012057 doi:10.1088/1757-899X/810/1/012057

Monte Carlo simulation based prediction model for ultimate tensile strength of Friction Stir Welded AA6063-ETP copper joint

[1]N J Panaskar, [2]R P Terkar

[1,2]Mukesh Patel School of Technology Management and Engineering, NMIMS University, Vile Parle, Mumbai- 400056, Maharashtra, India

njpanaskar@gmail.com

Abstract. In the present study, Al–6063 and ETP Cu were lap welded using friction stir welding wherein the aluminum alloy sheet was placed on top of the Copper sheet. The process parameters, namely, tool rotational speed, tool traverse speed, and thickness of zinc inter-filler material, were optimized using Taguchi L9 orthogonal design of experiments. The optimum process parameters were determined with respect to the ultimate tensile strength of the weld. The predicted optimal value of ultimate tensile strength was confirmed through confirmation run using the optimum parameters. The analysis of variance showed that tool traverse speed and the zinc foil thickness were the most dominant factors contributing to the weld strength. The most critical parameters were found by using the full factorial analysis. Using full factorial analysis, tool traverse speed, zinc foil thickness, and their interactive effect, were found to be the most critical parameters. The Anderson-Darling test indicates that the ultimate tensile strength tends to follow a normal distribution.

1. Introduction

Aluminium and copper joining have applications in industries such as automotive, HVAC, and refrigeration [1], [2]. These industries prefer substituting the comparatively costly copper parts with aluminium, either partially or entirely [3], [4]. Entire substitution of copper with aluminium is impractical as it unfavourably affects the system efficiency. Hence, it is crucial to join copper to aluminium. The usual methods used to join copper to aluminium are friction welding [5], [6], ultrasonic welding [7], and laser welding [8]. The dissimilar physical properties of aluminium and copper make them difficult to lap weld. Intermetallic compounds (IMCs) are formed during the welding process. Thick layers of intermetallic layers comprises microcracks, which decrease the strength of the joint. It is hard to control the thickness of the intermetallic compound layers [9], [10]. Friction Stir Welding (FSW) is newly used technique to join aluminium and copper feasibly.

FSW was developed by W.M. Thomas et al. at The Welding Institute (TWI) in Cambridge, UK, in 1991 [11]. This process has been widely used for joining aluminium and aluminium alloys [12]. In some studies, an intermediate layer such as zinc is used, which is compatible with aluminium as well as copper [13], [14]. From the present literature [15]–[18], the FSW process parameters, namely tool traverse speed and tool rotational speed, have significantly effect on the quality of the weld. Further, few studies pertaining to dissimilar metal welding, an inter-filler material was found to significantly affect the weld quality [13], [19].

Taguchi method is a technique for robust design and performance improvement of product, process, and system with significantly reduced time and cost of experimentation, which helps in the production of a low-cost good quality product. It provides a systematic approach to optimize the design for performance, cost, and quality. Taguchi has established Orthogonal arrays (OA) to depict a large number of experimental situations. However, it seems that the optimization of dissimilar (Al-Cu) FSW process parameters when an inter-filler layer is employed has not been reported using Taguchi technique.

2. Materials And Methodology

AA6063 and ETP copper sheets, both with 3 mm thickness, 130 mm length, and 90 mm width, were chosen for producing lap joints. 0.2 mm and 0.4 mm thick foils of zinc were used as an intermediate layer. Heat-treated H13 steel tools were used. Four distinct tools with pin lengths of 4.4 mm, 4.6 mm, 4.8 mm, and 4.94 mm were used to achieve 50% penetration in the bottom copper material. The tool pin profile was inverse conical with top and bottom diameters 5.4 mm and 6 mm, respectively. A flat shouldered tool of 24 mm diameter was used, and the shoulder penetration of 0.1 mm was used for producing adequate frictional heat for welding. The work pieces were placed on a mild steel base plate and firmly clamped.

Preliminary experiments were performed to set the operational range of process parameters, namely, tool rotational and traverse speeds. Accordingly, the tool rotational speed used was 1000, 1200, 1400 rpm; and the tool traverse speed used was 5, 10, and 15 mm/min; and the thickness of zinc inter-filler (ZFT) used was 0.2 mm and 0.4 mm. Taguchi L9 OA is used in the present study. The level-wise process parameters are shown in Table 1. The chemical composition and mechanical properties of the base materials are presented in Table 2. and Table 3., respectively. For testing the mechanical properties, tensile shear tests were conducted. The tensile shear test specimens 20 mm wide and 130 mm long were cut from the weld portion.

Post Taguchi analysis, further analysis was carried out using the full factorial DOE to find the most critical parameters from selected parameters. Monte Carlo simulation is used to predict the ultimate tensile strength. The normality of the response variable 'ultimate tensile strength (UTS)' is assessed by the Anderson-Darling test.

Table 1. Process parameter values and their levels

Level	A Rotational speed (rpm)	B Traverse speed (mm/min)	C Thickness of zinc foil (mm)
Level 1	1000	5	0
Level 2	1200	10	0.2
Level 3	1400	15	0.4

Table 2. Main chemical compositions of the base materials

Sheet Metal	Al	Cu	Mg	Mn	Zn
Al 6063	Base	0.08	4.8	0.8	0.1
ETP Cu	0.02	Base	–	–	4.7

Table 3. Mechanical properties of the base materials

Sheet Metal	Ultimate Tensile strength (MPa)	Microhardness (HV)
Al6063	160	80-85
ETP Cu	263	85–90

3. Results And Discussion

3.1. Taguchi Method

3.1.1 Signal to noise ratio

Taguchi technique uses the S/N ratio to assess the deviation of the quality characteristic from the desired value. The S/ N ratio characteristics can be separated into three modes: the nominal is best, the smaller is better, and the larger is better. In the present work, the objective is to maximize the ultimate tensile strength through optimum FSW process parameters, and larger is better characteristic is used. The SN ratio is calculated using the following formula :

$$\frac{S}{N} = -10\ log_{10}\frac{1}{N}\sum_{i=1}^{n}\frac{1}{{y_i}^2} \quad (1)$$

where y_i is the value of ultimate tensile strength for the ith test, n is the number of tests, and N is the total number of data points.

The ultimate tensile strength of the welded joints is analyzed to understand the effect of the FSW process parameters. The analysis is done using the MINITAB 18 software. Table 4. shows the mean S/N ratio for each level of the welding parameters. The higher S/N ratio corresponds to better quality characteristics. On the basis of S/N ratio values, the optimal level setting was achieved at the rotational speed of 1000 rpm (A_1), traverse speed of 10 mm/min (B_2), and zinc foil thickness of 0.2 mm (C_2). The response table for mean effect and S/N ratio are given in Table 5. and Table 6., respectively. The main effects plot for means and S/N ratios are given in figure 1. and figure 2., respectively.

Table 4. S/N response for the weld strength

Sr. No.	A Rotational Speed (rpm)	B Traverse Speed (mm/min)	C Zinc foil Thickness (mm)	Mean Ultimate Tensile Strength (MPa)	Signal to noise Ratio (S/N)
1	1000	5	0	65.65	36.3447
2	1000	10	0.2	106.1	40.5143
3	1000	15	0.4	43.5	32.7698
4	1200	5	0.2	79.67	38.0259
5	1200	10	0.4	61.89	35.8324
6	1200	15	0	48.97	33.7986
7	1400	5	0.4	43.99	32.8671
8	1400	10	0	69.8	36.8771
9	1400	15	0.2	63.14	36.0061

Table 5. Response Table for Means

Level	A	B	C
1	71.75	63.1	61.47
2	63.51	79.26	82.97
3	58.98	51.87	49.79
Delta	12.77	27.39	33.18
Rank	3	2	1

Table 6. Response Table for Signal to Noise Ratios : Larger is better

Level	A	B	C
1	36.54	35.75	35.67
2	35.89	37.74	38.18
3	35.25	34.19	33.82
Delta	1.29	3.55	4.36
Rank	3	2	1

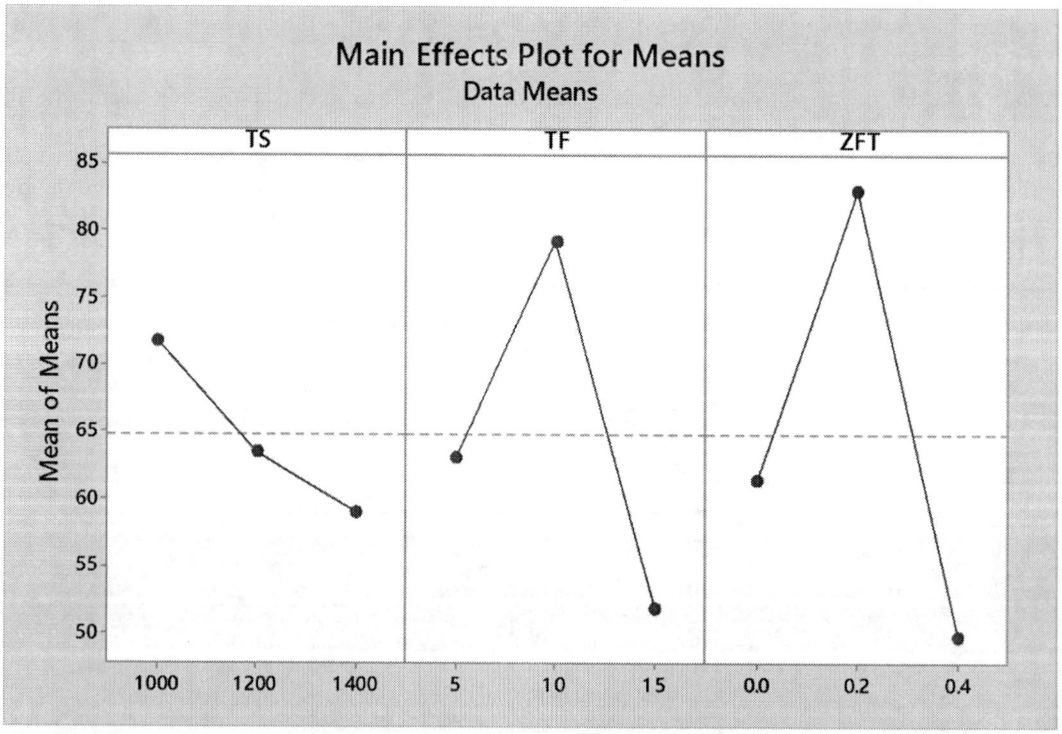

Figure 1. Main effects plot for Means

IOP Conf. Series: Materials Science and Engineering **810** (2020) 012057 doi:10.1088/1757-899X/810/1/012057

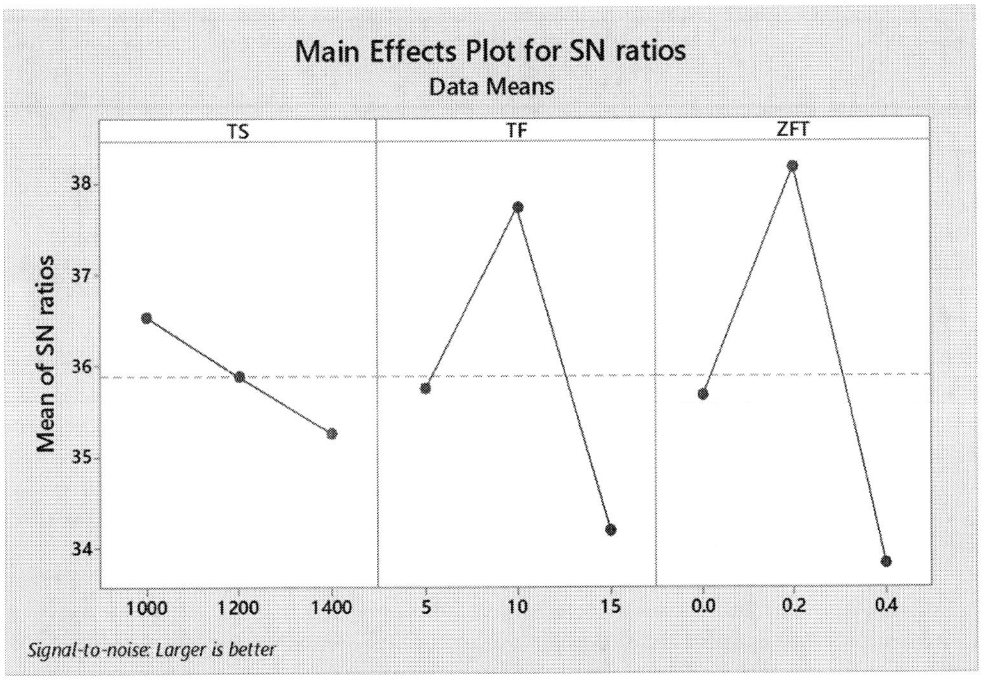

Figure 2. Main effects plot for SN ratios

3.1.2 Signal to noise ratio

The analysis of variance is used to evaluate the effect of process parameters on ultimate tensile strength. The results from ANOVA of means and ANOVA of signal to noise ratios are shown in Table 7. and Table 8., respectively. In this investigation, the tool traverse speed and the zinc foil thickness were found to be the most significant factors contributing to the weld strength.

Table 7. Analysis of Variance for Means

Source	DF	Seq SS	Adj MS	F	P	% Contribution
A	2	251.61	125.803	20.12	0.047	8.114
B	2	1137.73	568.864	90.97	0.011	36.688
C	2	1699.22	849.61	135.87	0.007	54.795
Residual Error	2	12.51	6.253			0.4
Total	8	3101.06				100

Table 8. Analysis of Variance for signal to noise ratios

Source	DF	Seq SS	Adj MS	F	P	% Contribution
A	2	2.5074	1.2537	14.95	0.063	4.976
B	2	18.999	9.4993	113.25	0.009	37.702
C	2	28.718	14.359	171.19	0.006	56.989
Residual Error	2	0.1678	0.0839			0.333
Total	8	50.392				100

[a] Dof – degrees of freedom, Seq SS – sequential sum of squares, Adj MS – adjusted mean square, F – fisher ratio

3.1.3 Determination of the maximum ultimate tensile shear strength

From the experiments, the optimum level setting is $A_1B_2C_2$. The average values of the factors at their levels are taken from Table 5. and the predicted value of the ultimate tensile strength is given below

$$Ultimate\ tensile\ strength = A_1 + B_2 + C_2 - 2T \quad (2)$$

= 71.75 + 79.26 + 82.97 - 2 X 64.74 = 104.5 MPa

where A_1 is the mean value of tool rotational speed at 1st level, B_2 is mean value of tool traverse speed at 2nd level, and C_2 is mean value of thickness of zinc inter-filler material at 2nd level..

3.1.4 Confirmation test

The improvement of the quality characteristic is verified using the optimal level of the design parameters. The tool rotational speed, tool traverse speed, zinc foil thickness were set at 1000 rpm, 10 mm/min, and 0.2 mm, respectively. The average ultimate tensile strength value of the friction stir welded AA6063 and ETP copper joint was 106.1 MPa.

3.2. Full factorial analysis

Further analysis was carried out using the full factorial DOE to find the most critical parameter from selected parameters. The full factorial DOE suggested eight combinations for the three parameters. The ultimate tensile strength for the different combination is computed and analyzed using a normal plot of effect, main effect plot, Pareto chart, and interaction plot. The normal effect plot shows that the tool traverse speed, the zinc foil thickness, and their interaction have a significant effect on ultimate tensile strength at 95 % confidence interval, which is also evident from the Pareto chart. The normal effect plot, the main plot of the effects, and the Pareto chart of the effects are shown in figure 3., figure 4., and figure 5., respectively. The main effect plot identified that a high ultimate tensile strength is obtained at higher values of tool traverse speed and zinc foil thickness. Variation in tool rotational speed had an insignificant effect on ultimate tensile strength.

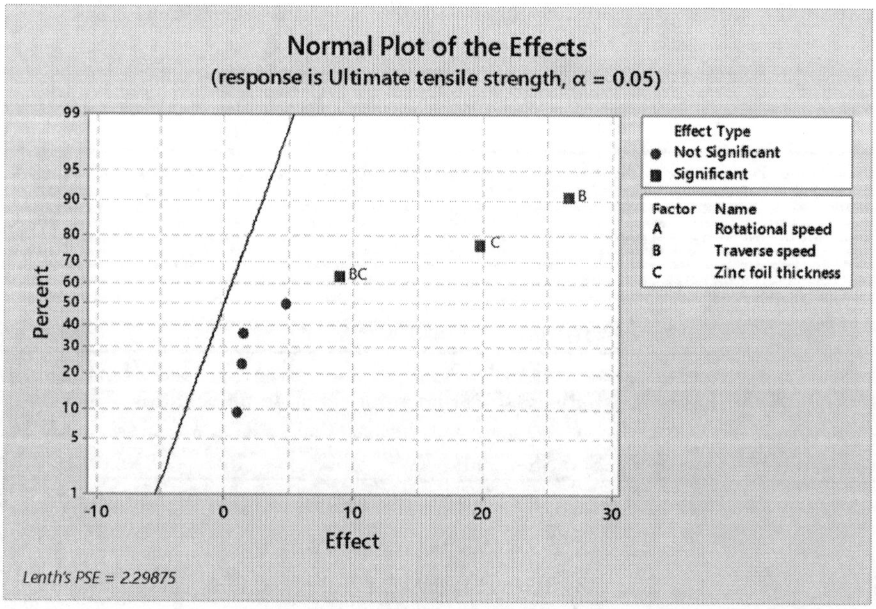

Figure 3. Normal plot of the effects

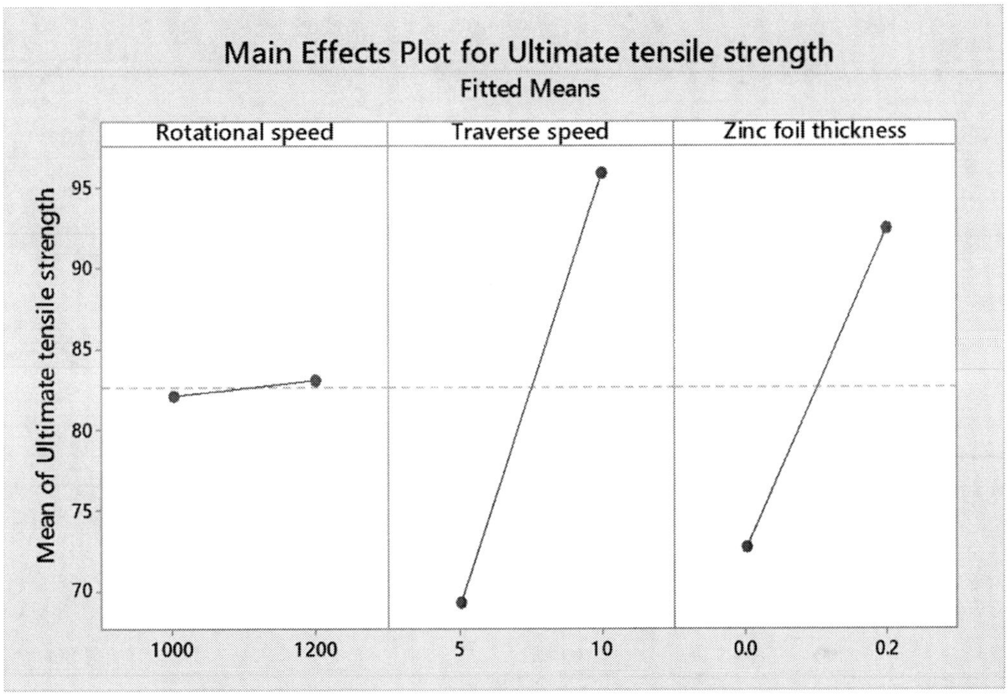

Figure 4. Main plot of the effects

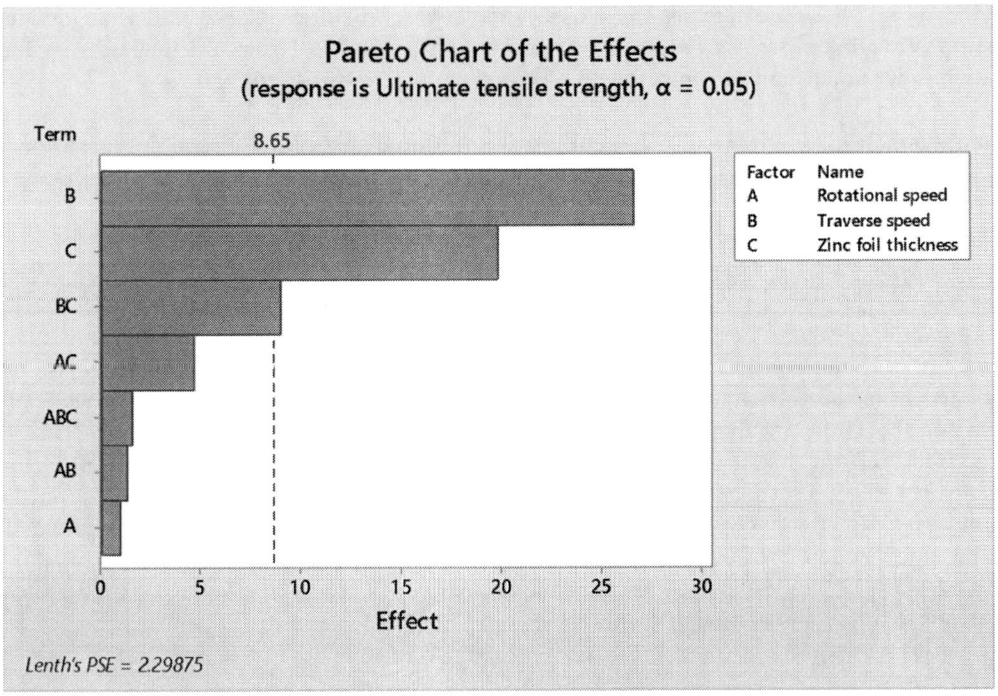

Figure 5. Pareto chart of the effects

IOP Conf. Series: Materials Science and Engineering **810** (2020) 012057 doi:10.1088/1757-899X/810/1/012057

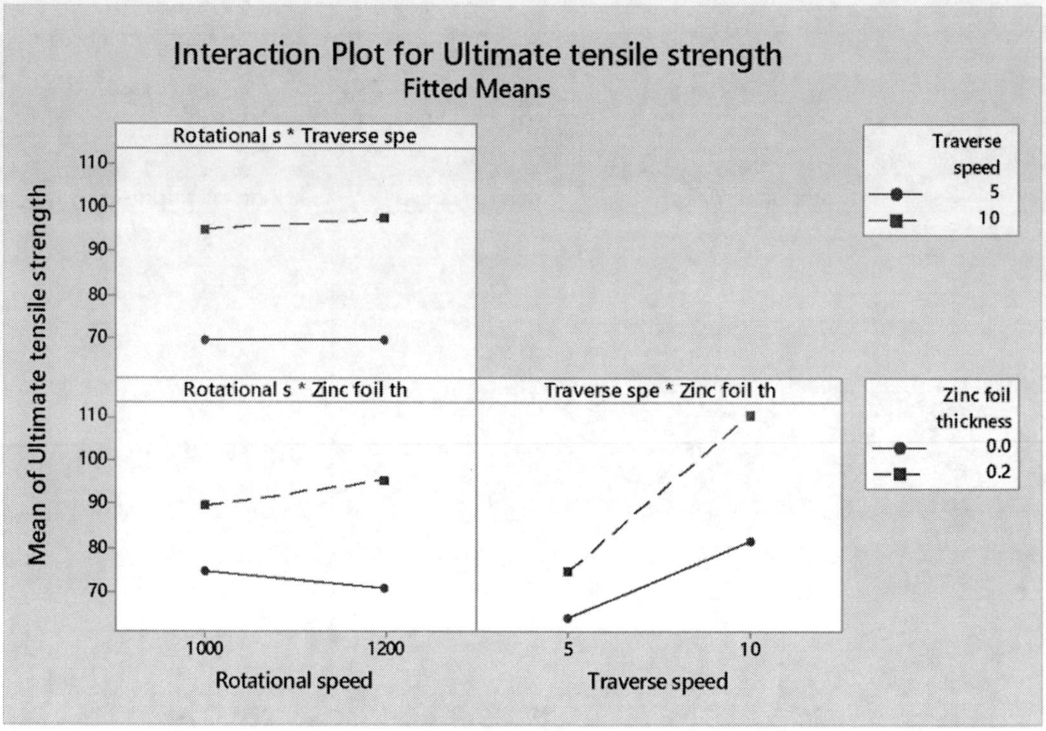

Figure 6. Interaction plot of the effects

Interaction plot, as shown in figure 6., shows the interaction between any two selected parameters with respect to ultimate tensile strength. Negligible interaction is observed between rotational speed and traverse speed. Moderate interaction is observed between rotational speed and zinc foil thickness. Strong positive interaction is observed between traverse speed and zinc foil thickness. A functional equation is developed and shown below to predict the ultimate tensile strength.

$$\text{Ultimate tensile strength} = 63.07 - 0.0154 \text{ x 'Rotational speed'} + 4.026 \text{ x 'Traverse speed'} - 41.1 \text{ x 'Zinc foil thickness'} - 0.00044 \text{ x 'Rotational speed' x 'Traverse speed'} + 0.00525 \text{ x 'Rotational speed' x 'Zinc foil thickness'} - 15.78 \text{ x 'Traverse speed' x 'Zinc foil thickness'} + 0.03065 \text{ x 'Rotational speed'} * \text{'Traverse speed' x 'Zinc foil thickness'} \quad (3)$$

3.3. Monte Carlo simulation

Monte Carlo simulation is applied to predict the ultimate tensile strength by using the equation developed from the full factorial analysis. The random data was generated with about 10,000 data points for the three parameters using the normal distribution. The simulation was performed to compute the performance using the equation for ultimate tensile strength. The mean ultimate tensile strength is 119.65, based on about 10,000 samples. The minimum and maximum ultimate tensile strength, are 112.54 and 127.26, respectively, with a standard deviation of 1.92. The probability plot and the summary report of ultimate tensile strength are shown in figure 7. and figure 8, respectively. The Anderson-Darling test is used to check the normality of UTS. The p-value of 0.073 indicates that UTS tends to follow the normal distribution.

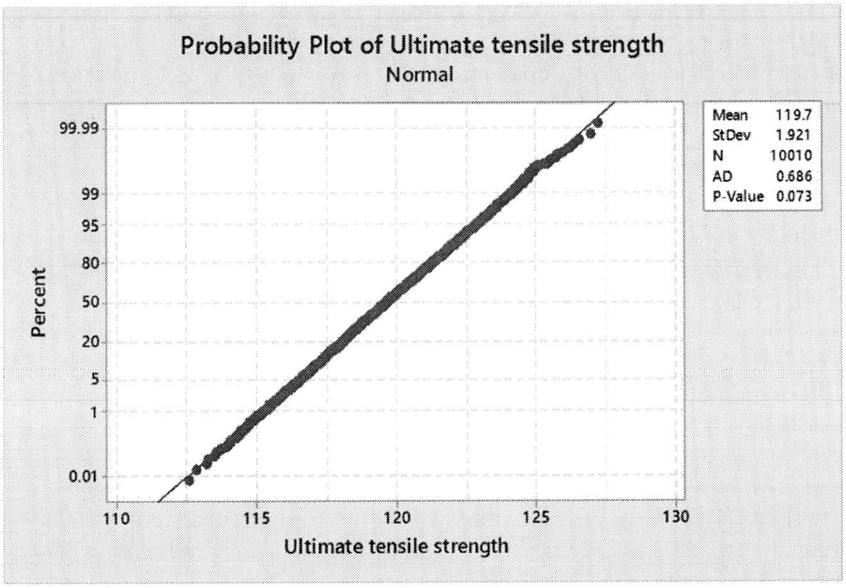

Figure 7. Probability plot of Ultimate tensile strength

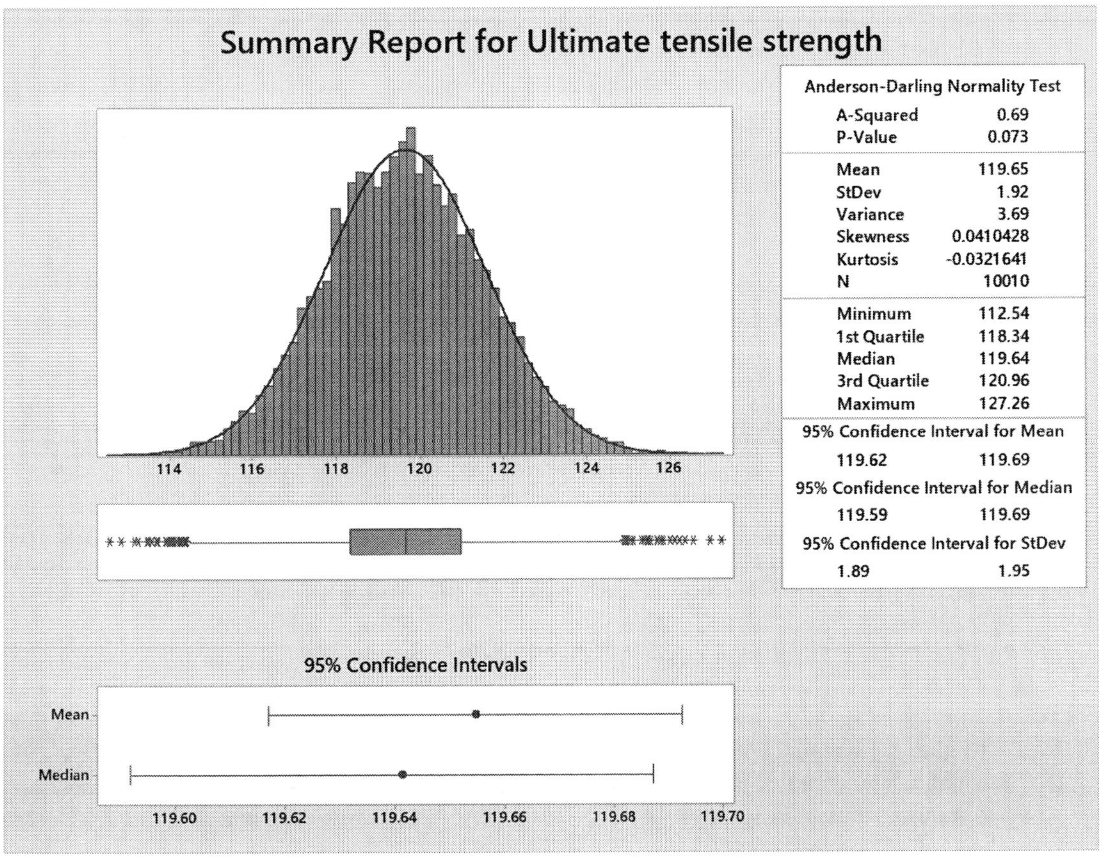

Figure 8. Probability plot of Ultimate tensile strength

4. Conclusion

Friction-stir welding experiments were conducted to lap weld the AA6063 and ETP copper sheets. The following conclusions were reached:

1. The optimum levels of tool rotational speed, tool traverse speed, and thickness of zinc inter-filler material are 1000 rpm, 10 mm/min, and 0.2 mm, respectively.
2. It is found that the tool rotational speed, tool traverse speed, and thickness of zinc inter-filler contribute 8.11%, 36.69%, and 54.80%, respectively, to the ultimate tensile strength of welded joints.
3. The error between the predicted and the experimental value of the ultimate tensile strength was 1.53%.
4. The factorial analysis shows that tool traverse speed, zinc foil thickness, and their interaction have the most dominant effect on ultimate tensile strength.
5. By Monte Carlo simulation, the range of ultimate tensile strength observed varies from 112.54 MPa to 127.26 MPa, with a mean value of 119.65 MPa.
6. The ultimate tensile strength tends to follow the normal distribution.

5. References

[1] Xue P, Xiao B L, Wang D and Ma Z Y 2011 Achieving high property friction stir welded aluminium/copper lap joint at low heat input *Science and Technology of Welding and Joining* **16** 657–61

[2] Mehta K P and Badheka V J 2016 A Review on Dissimilar Friction Stir Welding of Copper to Aluminum: Process, Properties, and Variants *Materials and Manufacturing Processes* **31** 233–54

[3] Technische Universitaet Muenchen Press Release 2011 Aluminum to replace copper as a conductor in on-board power systems *ScienceDaily*

[4] Onstad, E., Obayashi, Y. and Shamseddine, R. 2016 Auto, power firms save millions swapping copper for aluminum *Reuters*

[5] Yilbaş B S, Şahin A Z, Kahraman N and Al-Garni A Z 1995 Friction welding of St-Al and Al-Cu materials *Journal of Materials Processing Technology* **49** 431–43

[6] Sahin M 2010 Joining of aluminium and copper materials with friction welding *Int J Adv Manuf Technol* **49** 527–34

[7] Matsuoka S and Imai H 2009 Direct welding of different metals used ultrasonic vibration *Journal of Materials Processing Technology* **209** 954–60

[8] Jones M G 1980 Laser welding aluminum to copper

[9] Esmaeili A, Rajani H Z, Sharbati M, Givi M B and Shamanian M 2011 The role of rotation speed on intermetallic compounds formation and mechanical behavior of friction stir welded brass/aluminum 1050 couple *Intermetallics* **19** 1711–1719

[10] Mishra R S and Ma Z Y 2005 Friction stir welding and processing *Materials Science and Engineering: R: Reports* **50** 1–78

[11] Thomas W M, Nicholas E D, Needham J C, Murch M G, Temple-Smith P and Dawes C J 1995 Friction welding

[12] Thomas W M and Nicholas E D 1997 Friction stir welding for the transportation industries *Materials & Design* **18** 269–73

[13] Elrefaey A, Takahashi M and Ikeuchi K 2005 Preliminary investigation of friction stir welding aluminium/copper lap joints *Weld. World* **49** 93–101

[14] Elrefaey A, Takahashi M and Ikeuchi K 2004 Microstructure of aluminum/copper lap joint by friction stir welding and its performance *Journal of High Temperature Society* **30** 286–292

[15] Devaiah D, Kishore K and Laxminarayana P 2018 Optimal FSW process parameters for dissimilar aluminium alloys (AA5083 and AA6061) Using Taguchi Technique *Materials Today: Proceedings* **5** 4607–14

[16] Muthu Krishnan M, Maniraj J, Deepak R and Anganan K 2018 Prediction of optimum welding parameters for FSW of aluminium alloys AA6063 and A319 using RSM and ANN *Materials Today: Proceedings* **5** 716–23

[17] Ugrasen G, Bharath G, Kumar G K, Sagar R, Shivu P R and Keshavamurthy R 2018 Optimization of Process Parameters for Al6061-Al7075 alloys in Friction Stir Welding using Taguchi's Technique *Materials Today: Proceedings* **5** 3027–35

[18] Shojaeefard M H, Khalkhali A, Akbari M and Tahani M 2013 Application of Taguchi optimization technique in determining aluminum to brass friction stir welding parameters *Materials & Design (1980-2015)* **52** 587–92

[19] Kuang B, Shen Y, Chen W, Yao X, Xu H, Gao J and Zhang J 2015 The dissimilar friction stir lap welding of 1A99 Al to pure Cu using Zn as filler metal with "pinless" tool configuration *Materials & Design* **68** 54–62

Mapping & Analyzing Mechanical Properties of Dental Restorations with Product Composition

[1]**Savita Ranba Sonwane**, [2]**M. Ramachandran**
[1]Asst. Professor at Sinhgad college of Engineering, Vadgaon, Pune
[2]MPSTME, SVKM'S NMIMS University, Dhule 425405, Maharashtra, India
savita.sonwane26@gmail.com, sweetestchandran@gmail.com

Abstract. This study compares the compressive strength, flexural strength and micro-hardness of three types of dental composite materials {Nano-filled, Nano-hybrid, Micro-hybrid} by respective filler fraction–composition. SEM, EDS technique is used for analysis of obtained results. These days, dental restorations play a very vital role considering the dental health. Esthetically, reliably composite resins are much more superior to ancient tooth filling materials. Hence it is necessity of today's research is to study and map the mechanical properties of different brands of composite resins and hence provide fruitful results to doctors utilizing these dental composites for satisfactory results. By this, research will be beneficiary to the society. The findings of this study indicates that for Compressive strength, Brilliant NG < Tetric N Ceram < Polofil NHT <Z 350 < Charisma. Flexural Strength, Brilliant NG < Tetric N Ceram < Charisma < Z 350 < Polofil NHT. Micro-hardness, Charisma < Tetric N Ceram < < Z 350 < Brilliant NG < Polofil NHT. In other words, Micro-hybrid- Have highest compressive strength than Nano-filled and Nano-hybrid. Nano-Hybrid- Highest flexural strength & micro-hardness than Nano-filled and Micro-hybrid.
Keywords: Compressive strength; Flexural strength; Micro-hardness; Micro-hybrid; Nano-hybrid; Nano-filled.

1. Introduction
Depending on the intended use, composites with different mechanical properties may be required. In some situations, the materials must be hard and strong, while in other situations the flexibility of the material is more important and strength is not an important factor; in contrast, stiffness or resilience may be the properties of interest. Hence, the physical and mechanical properties of resin composites depend on many variables which clinicians should be aware.

2. Material and Methods
Materials: In this experimental in-vitro study following five groups containing three different types of composite resins is used.

Group No.	Material	Syringes	Specimen
Group 1	Filtek Z350 (3M)	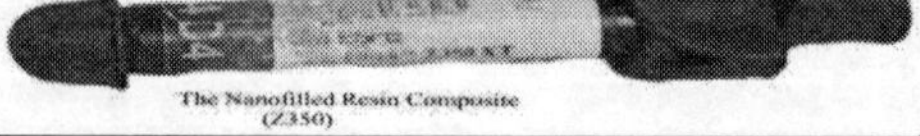 The Nanofilled Resin Composite (Z350)	

IOP Conf. Series: Materials Science and Engineering **810** (2020) 012058 doi:10.1088/1757-899X/810/1/012058

Group 2	TetricNCeram (3M ESPE)	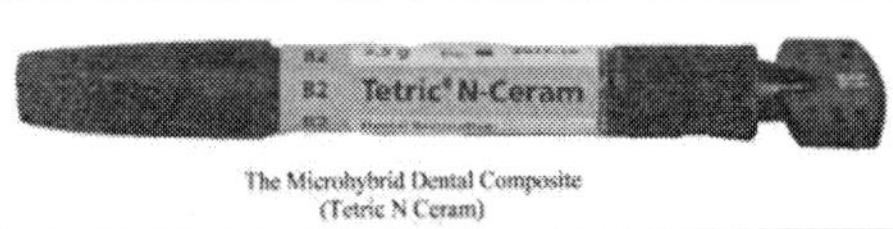	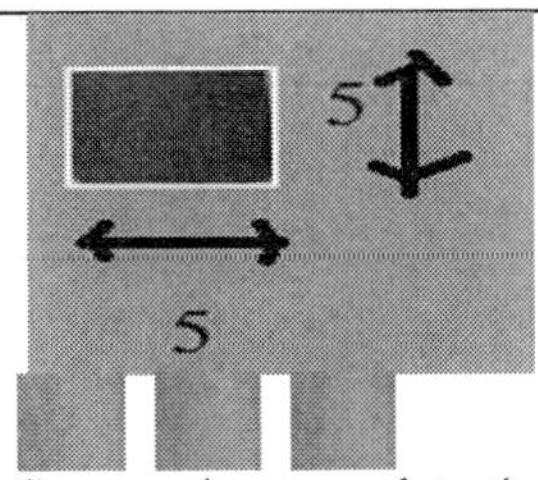 Compression strength testing
Group 3	Charisma		
Group 4	Briliant N G	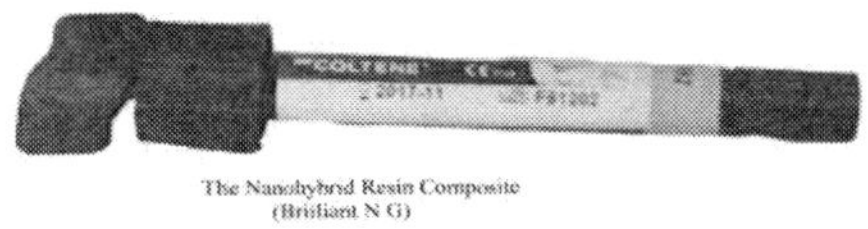	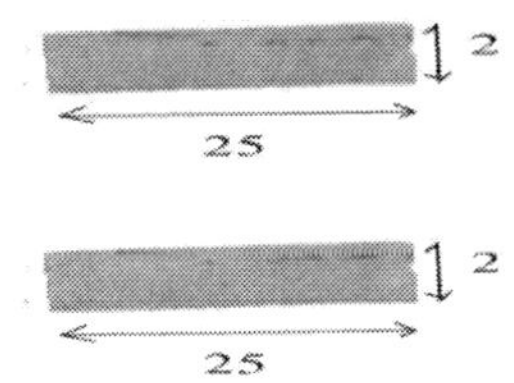 Flexural strength & Micro hardness testing
Group 5	Polofil NHT	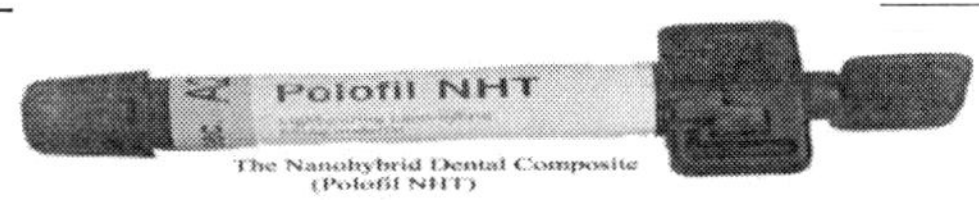	

Total forty numbers of dental composites specimens were used in this study. Cylindrical specimens (n=3) for compressive strength & rectangular shaped specimens (n=5) each for flexural strength and micro-hardness testing were made according to manufacturer's recommendations. Five different types of dental composites comprising three kinds are utilized in this study.

Table 2: Product Composition of the Five Different Types of Light Activated Nano- hybrid Composite Resins Provided by manufacturers'.

Sr. No.	Composite Name & Shade	Manufacturer	Filler Fraction (Wt. %, Vol. %)	Particle Size (µm)	Filler Composition
Group 1	Filtek TM Z350XT (C4 Dentin)	Haerious Kulzera	78/58	0.02-2µm	Combination of agglomerated - Nonagglomerated Zr /Si cluster fillers.
Group 2	Tetric N Ceram (B2)	3MESPEa (USA)	78.5/59.5	0.6-10µm	BA glass, ytterbium trifluoride, mixed oxide, Sio2.
Group 3	Charisma Smile (A3)	Ivoclar Vivadenta	80.5/56	0.004-3µm	BA-Al-B-F-Si, glass, pyrogenic Sio2
Group 4	Brilliant TM NG (A2/B2)	Coltene Whaldenta	80/65	0.01-2.5µm	Dental glass, amorphous silica
Group 5	Polofil NHT (A2)	Voco	83/68	0.01-0.1 µm	Nano scaled particles with glass ceramic fillers.

3. Methods

The mechanical strengths (compression and flexural strength) testing of prepared samples is done on Universal Testing Machine (UTM) as shown in fig.1&2. For micro-hardness strength testing, Vickers hardness tester is used. Mould on which hardness testing done is as shown in fig.3. The value of micro hardness is calculated by measuring the diagonals as shown in fig 4.

Figure 1: Compression Strength;

Figure 2: flexural Strength;

Figure 3: Micro-hardness Mould

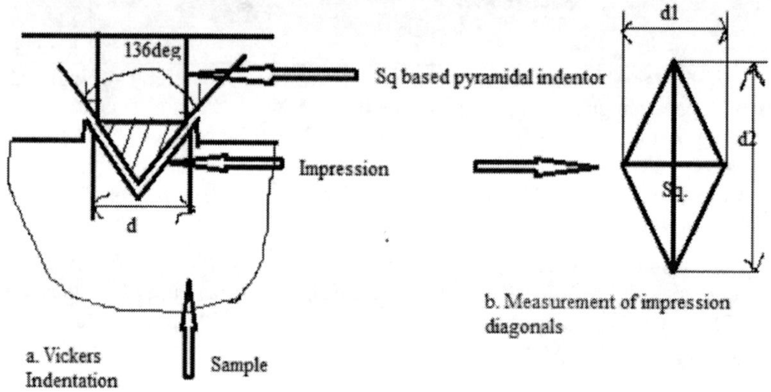

Figure 4: Vickers Micro-hardness Testing

4. Results & Discussion:

Table 2 Mean values of mechanical strengths of different resin composites.

Brand Name	Flexural Strength (MPa)	Compressive Strength (MPa)	Micro-hardness (hv)
Z350	102.52	166.35	67.73
Tetric N Ceram	88.63	137.65	56.21
Charisma	88.63	**176.45**	52.43
Brilliant NG	92.77	54.20	79.28
Polofil NHT	**171.34**	143.57	**93.40**

From above table, it is coming to know that the flexural strength and Micro hardness of group 5 (Polofil NHT) is higher among all other four. And, Compressive strength of group 3 (Charisma) is higher than other four groups. To know the reasons, we came across the tests like Scanning Electron Microscope & Elemental Deposition Structure. Following are the SEM images showing the structure at different sizes.

Figure 5: SEM Image of Z350 reinforced composite; Figure 6: SEM Image of Tetric N reinforced composite

Figure 7: SEM Image of Charisma reinforced composite; Figure 8: SEM Image of Brilliant NG reinforced composite

Figure 9: SEM Image of Polofil NHT reinforced composite

For this discussion going back to table I of composites composition and came to following conclusion of mapping product composition with their mechanical strengths. The figure 5, 6, 7, 8 and 9 shows the uniform distribution of filler particles in the resin.

5. Conclusion

From results, it is concluded that, the compressive strength of dental restoration strongly depends upon the type of filler particles rather than the filler weight % in the composition. Similarly, the flexural strength and micro hardness are strongly affected by the filler weight-volume % as well as the size of filler particles i.e. Micro, Macro Nano particle. The SEM images of all composites having different fillers shows uniform distribution of fillers in the composite

References

1. Karuppuchamy, S. "Thermo-mechanical properties of coconut shell powder reinforced plastic composites." International Journal of ChemTech Research Vol.8, No.6 (2015): 852-857.
2. Yew, B. S., Siti Nurbazilah, F. H. Wee, S. B. B. Mohamed, Ahmad Humaizi, and W. N. F. Mohamad. "The effect of surface texture and carbonaceous material composition on the dielectric properties measurement of coconut shell-polymer (CSP) composites." ARPN Journal of Engineering and Applied Sciences 2016b 11, no. 6 (2006): 3822-31.
3. Agarwal, Rakshit, M. Ramachandran, and Stanly Jones Retnam. "Tensile Properties of Reinforced Plastic Material Composites with Natural Fiber and Filler Material." ARPN Journal of Engineering and Applied Sciences 10, no. 5 (2015): 2217-2220.
4. Sefiu Adekunle Bello, Isiaka Ayobi Raheem, Nasir Kolawole Raji, Study of tensile properties, fractography and morphology of aluminium (1xxx)/coconut shell micro particle composites, Journal of King Saud University - Engineering Sciences, Vol29, Issue 3, July 2017, Pages 269-277
5. Liyanage, Chinthani D., and Mevan Pieris. "A Physico-Chemical Analysis of Coconut Shell Powder." Procedia Chemistry 16 (2015): 222-228.
6. Udhayasankar, R., and B. Karthikeyan. "A Review on Coconut Shell Reinforced Composites." International Journal of ChemTech Research 8, no. 11 (2015): 624-637.
7. Siti Nurbazilah Ab Jabal, Yew Been Seok, Wee Fwen Hoon, carbon composition, surface porosities and dielectric properties of coconut shell powder and coconut shell activated carbon composites, ARPN Journal of Engineering and Applied Sciences, VOL. 11, NO. 6, MARCH 2016, 3832- 3838.
8. Sahas Bansal, M. Ramachandran, Pramod Raichurkar, Comparative Analysis of Bamboo using Jute and Coir Fibre Reinforced Polymeric Composites, Materials Today: Proceedings,4(2), 2017, 3182–3187.
9. Arena, Noemi, Jacquetta Lee, and Roland Clift. "Life Cycle Assessment of activated carbon production from coconut shells." Journal of Cleaner Production 125 (2016): 68-77.
10. Salleh, Z., M. M. Islam, M. Y. M. Yusop, and MA Mun'aim M. Idrus. "Mechanical properties of activated carbon (AC) coconut shell reinforced polypropylene composites encapsulated with epoxy resin." APCBEE Procedia 9 (2014): 92-96.
11. Khraisheh, Majeda, Jongkyu Kim, Luiza Campos, Ala'A. H. Al-Muhtaseb, Gavin M. Walker, and Mohammad AlGhouti. "Removal of Carbamazepine from Water by a Novel TiO2–Coconut Shell Powder/UV Process: Composite Preparation and Photocatalytic Activity." Environmental engineering science 30, no. 9 (2013): 515-526.
12. Nishant Godbole, Shajit Yadav, M. Ramachandran , Sateesh Belemkar, A Review on Surface Treatment of Stainless Steel Orthopedic Implants, Int. J. Pharm. Sci. Rev. Res., 36(1), 2016; Article No. 33, Pages: 190-194.
13. Gojny, F. H., M. H. G. Wichmann, U. Köpke, B. Fiedler, and K. Schulte. "Carbon nanotube-reinforced epoxy-composites: enhanced stiffness and fracture toughness at low nanotube content." Composites science and technology 64, no. 15 (2004): 2363-2371.
14. Gojny, Florian H., Malte HG Wichmann, Bodo Fiedler, and Karl Schulte. "Influence of different carbon nanotubes on the mechanical properties of epoxy matrix composites–a comparative study." Composites Science and Technology 65, no. 15 (2005): 2300-2313.
15. Rong, Min Zhi, Ming Qiu Zhang, Yuan Liu, Gui Cheng Yang, and Han Min Zeng. "The effect of fiber treatment on the mechanical properties of unidirectional sisal-reinforced epoxy composites." Composites Science and technology 61, no. 10 (2001): 1437-1447.
16. Kim, Jin Ah, Dong GiSeong, Tae Jin Kang, and Jae RyounYoun. "Effects of surface modification on rheological and mechanical properties of CNT/epoxy composites." Carbon 44, no. 10 (2006): 1898-1905.
17. Bisanda, E. T. N., and Martin P. Ansell. "The effect of silane treatment on the mechanical and physical properties of sisal-epoxy composites." Composites Science and Technology 41, no. 2 (1991): 165-178.
18. Joshi, Satish V., L. T. Drzal, A. K. Mohanty, and S. Arora. Are natural fiber composites environmentally superior to glass fiber reinforced composites. Composites Part A: Applied science and manufacturing 35, no. 3 (2004): 371-376.

Investigation of tribological behaviour of 88WC-12Co coating

S Rakhecha[1], V Pare[2], S Tiwari[3] and G Yadav[4]

[1,2,3,4]Mechanical Engineering Department, Shri G S Institute of Technology & Science Indore- 452003 India

[2]v_pare@yahoo.co.in, [3]sudhir2609@gmail.com

Abstract. High velocity oxy-fuel (HVOF) is a coating technique of thermal spray process usually employed for protecting the substrate, where high velocity is used instead of high temperature for coating the substrate. In this paper, the wear behaviour of tungsten carbide-cobalt (88WC-12Co) coating was studied experimentally. The coating of 0.3 mm was deposited by HVOF technique on SS202 substrate. The dry abrasive wear experiments were carried out using pin-on-disc test rig at fixed sliding velocity 0.603 m/s, and 20, 40, 60 N load for 1000 m distance. The weight loss was measured at each 250 m distance. For the preparation of abrasive surface 80 grit SiC (silicon carbide) emery paper was used. Results show that at initial stage wear rate of coated substrate was high which reduces with increase in sliding distance and linear increment in coefficient of friction with increase in load.

1. Introduction

In many engineering areas, the mechanical components are subjected to extreme operating conditions like force, speed, temperature and chemical environment. The surface of the components for such application needed to be modified to prevent them from various types of degradation. For achieving a goal of high production rate, low break down time and low maintenance cost in industries, it is necessary to reduce wear in machine. There are several methods suggested for controlling wear and applying protective coating is widely used technique. Coating is a layer of protection, which isolate substrate from effect of high temperature, corrosion and wear [1-4].

Tungsten carbide (WC) is used for components subjected to wear and corrosion due to its excellent hardness properties. This can be added with alloying metals such as cobalt which will improve its toughness and overall performance. The high velocity oxy-fuel (HVOF) spraying process is used for tungsten carbide cermet powder spraying. by using high velocity oxy-fuel (HVOF) spraying processes. The coating obtained by this method have high hardness and small porosity and good adhesion to substrate. Literature showed that coating from HVOF has better properties compared to other thermal spraying methods. Components in relative motion, coating with high hardness compared to counter surface is required to minimize abrasive wear [2, 4-6].

Cemented carbides are widely used in industrial application compared to other products obtained from powder metallurgy due to excellent wear resistance. It is used in application like mining tools, agriculture machineries, metalworking accessories and application where high wear resistant required. Its application includes tool bits and drill bits, gas, oil and mining drilling, earthmoving and heavy construction machinery, metal working and agriculture machinery [1]. Pin-on-disk test is broadly used method to characterize the tribological properties of material. It is preferred due to simplicity, ease to control operating parameters and cost effectiveness.

Many studies are available for characterization of thermal spray coatings. Umale et al. [7] studied abrasive wear behaviour of two reinforced copper matrix composites. The silicon carbide (SiC, 12%) and silica particles (SiO_2, 9%) were used as reinforced particles and prepared by powder metallurgy. The two composites were analyzed for metallography, abrasive wear experiment, surface imaging and surface hardness. They reported that copper-SiC (12%) composites showed high abrasive wear resistance. The SEM analysis demonstrated that the wear mechanism is a function of test conditions. Khan et al. [8] investigated aluminium (Al6082-T6) and Al-6082T6 coated by plasma spray technique. The coating of Alumina (Al_2O_3), Titania (TiO_2) and Alumina-Titania (Al_2O_3+40% TiO_2) ceramic with different coating thicknesses of 100 µm, 150 µm and 200 µm were deposited on Al6082-T6 substrate. These coating were studied for tribological and microstructural properties. The results showed excellent wear resistance by alumina coating compare to other coatings (Titania and Alumina-Titania coating). The imaging by SEM of the wear out surfaces of coated specimen shows better properties than that of bare substrate Al 6082-T6. Guilemany et al. [9] work was focused on WC-Co cermets. They studied two coatings of conventional and nanostructured cermets, sprayed by high-velocity oxy-fuel (HVOF). The coating deposited by nanostructured cermets showed better properties between two, however under dry abrasion test wear results are more or less same. Tavoosi et al. [10] worked on experimental and numerical simulation of wear in nanostructured NiAl coating. The coating were put on substrate of mild steel through HVOF. Mechanical characterization was carried out by nano-indentation test. The test results indicate high wear-rate initially, which reduces gradually. The results of experiment and simulation shows good agreement of wear rate (volume losses rate) obtained from simulation and experimental results.

2. Material and experiment

2.1. High velocity oxy-fuel (HVOF) coating

The SS 202 circular pins of diameter 10 m and length 30 mm were chosen as substrate material for coating. WC-Co powder is used as coating materials, percentage of coating material are shown in table 1. In this investigation 0.3 mm thick WC-Co coating was developed on the top surface of the substrate by High Velocity Oxy Fuel (HVOF) coating technique. The Coating parameter are given in table 2.

Table 1. Composition (%) of of coating material

Coating Material	Percentage
Tungsten Carbide (WC)	88
Cobalt	12

Table 2. Coating parameters

Parameters	Value
Fuel gas flow rate	~60 lpm
Compressed air flow rate	~650 lpm
Spray distance	~7 inches
Spray rate	~38 gm/min
Coating thickness	~0.3 mm

2.2. Sliding wear test apparatus

Pin on disc (POD) apparatus used for characterizing the wear behavior of coating material in present work. In POD test, pin is arranged perpendicular to a disc. The disc is circular and flat and rotated through a motor in machine at fixed speed. The pin specimen is stationary with respect to disc and pressed at constant load against the disc. The loading is applied by a lever arm as shown in Fig. The data of wear loss of specimen are measure as volume loss in cubic mm. The wear results were measure at regular interval of sliding distance at selected value of load and speed. Figure 1 shows the Pin on Disc setup used for conducting the sliding wear test available in the laboratory.

Figure 1. Pin on disc experimental set-up

2.3. Abrasive wear test

The abrasive wear tests were conducted on pin on disc tribometer working at constant sliding velocity and fixed wear path diameter and the specific wear rate (k) is calculated, the specific wear rate helps in determining wear resistance offered by the metal under running conditions. By fixing various parameters of test, wear rate of 88WC-12Co was calculated. Coating of WC-Co is performed by HVOF (High Velocity Oxy-Fuel) process and the thickness of the coating was 0.3 mm, SS202 was substrate material for coating. For creating abrasive conditions in pin on disc wear test a waterproof silicon carbide (SiC) emery paper was used, which was properly glued with the surface of disc. Working parameters used present work wear experiment are shown in table 3

Table 3. Wear test parameters

Parameter	Value
Sliding Distance	1000 m (in interval of 250 m)
Sliding Velocity	0.6 m/sec
Load	20, 40, 60 N
Wear Track Diameter	80 mm

In this tribological testing, coated WC-Co pin run for 1000 m at each load condition, and weight loss of each pin was measured at interval of 250 m. Wear track diameter, sliding velocity and time was fixed for each experiment. Tests were repeat for different load conditions [11]. Specific wear rate was calculated using expression.

$$K = \frac{\text{Volume of Wear}}{\text{(Applied Load)x(Sliding Distance)}} \; mm^3/_{N.m} \tag{1}$$

3. Results and discussion

In this work dry abrasive wear test was performed for different load of 20 N, 40 N and 60 N, constant sliding distance of 1000 m by using pin-on-disc. The results incurred from the above experimental work are plotted in figure 2, 3, 4 and 5 given below.

- Volume loss of any specimen can be calculated by dividing mass loss during test by density of softer material. Mathematically it is expressed as

$$\text{Volume Loss} = \frac{\text{Initial weight} - \text{Final weight}}{\text{Density of material}}$$

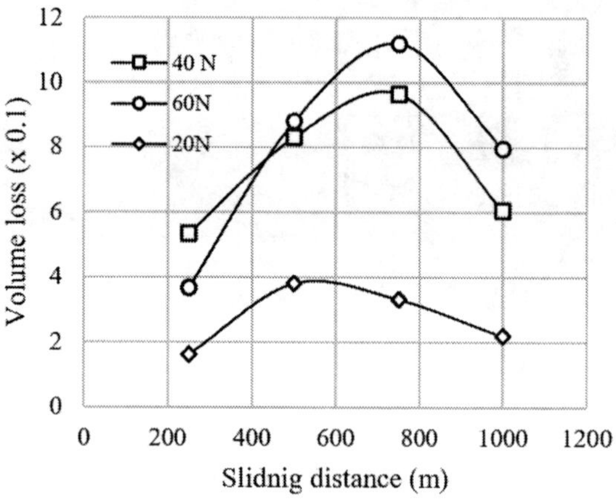

Figure 2. Loss vs. sliding distance at different Load

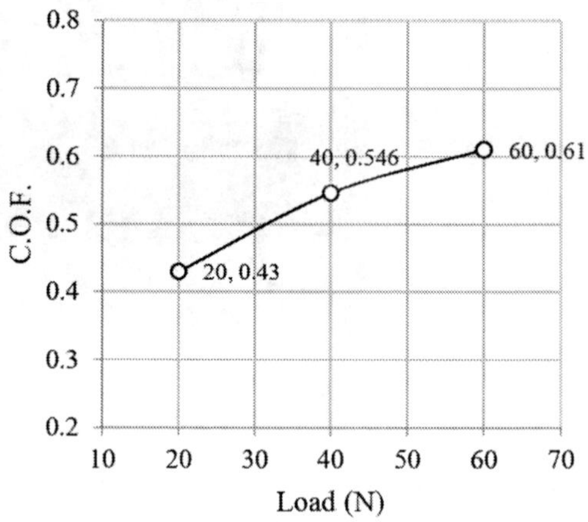

Figure 3. Coefficient of Friction vs. Load

Graph in figure 2 is plotted in between of volume loss and various sliding distance. Graph shows that volume loss showing same pattern in each load condition (i.e. 20, 40, 60 N). With increase in sliding distance firstly it is increasing and there after decrement in volume loss is taking place.

- A coefficient of friction is a value that shows the relationship between the force of friction between two objects and the normal force between the objects. Graph is plotted in between coefficient of friction (C.O.F) and load; change in value of coefficient of friction relatively to load is shown in figure 3. Coefficient of friction is showing linear increment with respect to load.
- Weight loss is loss in material of wearing surface, which can be measured by cleaning and weighing sample before and after wear test. Weight loss i.e reduction in weight is a difference between initial weight of sample and final weight of sample. Change in weight loss with time at different different load condition is shown in figure 4. Graph is showing that at the starting

of experiment weight loss is increasing but after some time it is showing decrement in weight loss.

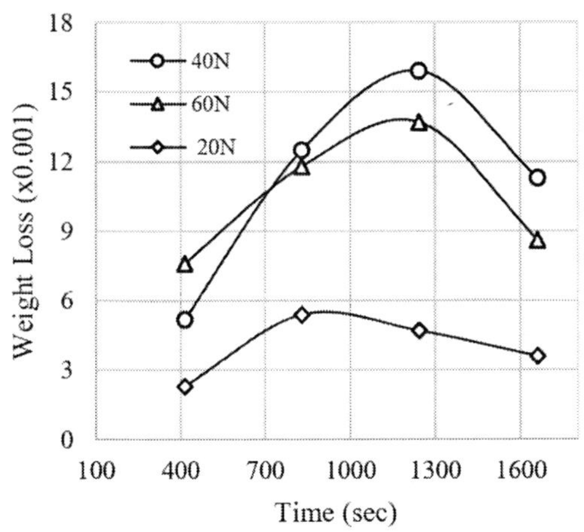

Figure 4. Graph between weight loss and time

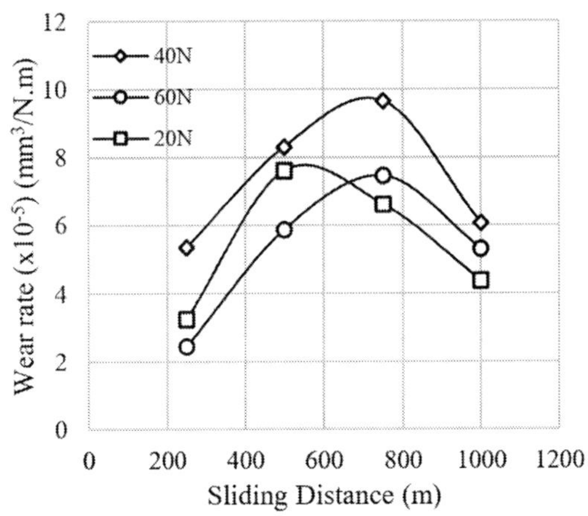

Figure 5. Graph Between wear rate and sliding distance

- The wear rate is defined as the wear volume of the pin specimen corresponding to the unit sliding distance under the application of a unit vertical load. The weight loss is divided by the density and converted into the wear volume. The graph is showing same behaviour as other above in this graph wear rate at 40N load condition is high then 60N load conditions and it is also incremental at the start of test and decreasing after a certain sliding distance as shown in figure 5.

  In each graph wear volume, wear rate and weight loss corresponding to time and sliding distance initially high but with the increase of time and distance it gradually reduces. The high wear rate at beginning due to low area of contact between disc and coated pin due to surface roughness of coated surface, however with sliding it decrease with smoothening of surface [11-13].

4. Conclusion

In this investigation, coating of cemented carbide (88WC-12Co) with thickness 0.3mm deposited on SS202 substrate through High Velocity Oxy Fuel (HVOF) technique. To identify the wear characteristics of the cemented carbide (88WC-12Co) coating, pin-on-disc dry abrasive wear test was performed. For the preparation of abrasive surface 80 grit SiC (silicon carbide) emery paper was used. Test was performed by applying 20, 40 and 60 N normal loads on pin and the tests were continued for a distance of 1000 m. The reduction of mass at different sliding distances (i.e. 250 m, 500 m, 750 m and 1000 m) was determined. The results indicated an increased wear-rate at beginning, which slowly decreases [11]. At beginning of the experiment, surface roughness of newly coated pin results into small actual area contact with disc which increases with increase in sliding distance.

5. References

[1] Barbezat G 2005 Advanced thermal spray technology and coating for lightweight engine blocks for the automotive industry *Surface and Coatings Technology* **200** 1990-3

[2] Meng R, Deng J, Liu Y, Duan R and Zhang G 2018 Improving tribological performance of cemented carbides by combining laser surface texturing and WSC solid lubricant coating *International Journal of Refractory Metals and Hard Materials* **72** 163-71

[3] Domínguez-Meister S, Rojas T C, Brizuela M and Sánchez-López J C 2017 Solid lubricant behavior of MoS2 and WSe2-based nanocomposite coatings *Science and Technology of advanced MaTerialS* **18** 122-33
[4] de Villiers Lovelock H 1998 Powder/processing/structure relationships in WC-Co thermal spray coatings: A review of the published literature *Journal of thermal spray technology* **7** 357-73
[5] de Villiers Lovelock H, Richter P, Benson J and Young P 1998 Parameter study of HP/HVOF deposited WC-Co coatings *Journal of thermal spray technology* **7** 97-107
[6] Yao Z, Stiglich J J and Sudarshan T 1999 Nano-grained tungsten carbide–cobalt (WC/Co) *Mater. Modif* 1-27
[7] Umale T, Singh A, Reddy Y, Khatitrkar R and Sapate S *International Journal of Modern Physics: Conference Series,2013),* vol. Series 22*)*: World Scientific) pp 416-23
[8] Kumar A and Suresh P 2014 Tribological Behaviour of Plasma Sprayed Al2O3-TiO2 Coating on Al-6082T6 Substrate *behaviour* **3**
[9] Guilemany J, Dosta S, Nin J and Miguel J 2005 Study of the properties of WC-Co nanostructured coatings sprayed by high-velocity oxyfuel *Journal of thermal spray technology* **14** 405-13
[10] Tavoosi H, Ziaei-Rad S, Karimzadeh F and Akbarzadeh S 2015 Experimental and finite element simulation of wear in nanostructured NiAl coating *Journal of Tribology* **137** 041601
[11] Sharma A, Singh N and Rohatgi P 2013 Study of Wear Pattern Behavior of Aluminum and Mild Steel Discs Using Pin on Disc Tribometer *European Journal of Applied Engineering and Scientific Research* **2** 37-43
[12] Sahraoui T, Fenineche N-E, Montavon G and Coddet C 2003 Structure and wear behaviour of HVOF sprayed Cr3C2–NiCr and WC–Co coatings *Materials & design* **24** 309-13
[13] Hajare A S and Gogte C 2018 Comparative study of wear behaviour of Thermal Spray HVOF coating on 304 SS *Materials Today: Proceedings* **5** 6924-33

IOP Conf. Series: Materials Science and Engineering **810** (2020) 012060 doi:10.1088/1757-899X/810/1/012060

Multistage forming analysis of spoke resonator end-wall

N K Sharma[1], P Karma[2], V Pare[2#], A Chaturvedi[1], N Nigam[1], G V Kane[1] and M Chouksey[2]

[1]Raja Ramanna Centre for Advanced Technology, Indore 452013, India;
[2]Mechanical Engineering, Shri G. S Institute of Technology and Science Indore 452003, India;

Email: [#]vinod.pare@gmail.com

Abstract. Spoke resonator components are typically formed by sheet metal forming operations such as deep drawing and bending. End wall, Spoke and Shell are the main components of spoke resonators. A sheet metal blank of niobium material is used to form the End wall and Spoke for a spoke resonator. This paper presents an elasto-plastic finite element analysis for single stage and multi stage forming of spoke resonator End wall. Numerical simulations were performed to investigate the effect of forming process parameters on localized thinning, residual stresses and study of failure such as tearing or wrinkling. Forming tools like die, punch and binder were modeled by UGNX 8.0, a modeling software and finite element analysis of forming has been simulated by Altair HYPERFORM software. The results of the analysis were used to assess the various forming process. The analysis showed that the multistage forming leads to better uniformity in thickness of component. This data will help in selection of appropriate forming process.

1. Introduction

Raja Ramanna Centre for Advanced Technology (RRCAT), Indore is involved in the development of a spallation based neutron research facility for future projects. The facility will consist of a 1GeV high power proton linac, an accumulator ring, spallation target and various beam lines. The neutrons will be generated by bombarding a high power proton beam on heavy atom targets through spallation process. The spallation neutrons will be used for research in the areas of condensed matter physics, materials science, biology, chemistry etc. The 1 GeV proton linac will consist of an ion source and different normal and superconducting Radio Frequency (RF) cavities to accelerate ions to the final energy. These cavities are classified based on β-value ($\beta = v_0/c$), which is defined as the ratio of velocity of particle, v_o, to the velocity of light, c. Spoke resonator is a low velocity structure with typical range of β varies from 0.1 to 0.6 and used for acceleration of proton and heavy atoms [1]. Physics design of three families of spoke resonator cavities with beta values of 0.11, 0.22 and 0.42 respectively has been carried out by Accelerator & Beam Physics Section, RRCAT. These cavities will accelerate ion beam from 3 MeV to 160 MeV [2]. The cavities will be fabricated using high purity niobium material, which becomes superconducting at very low temperature (~ 9.2 K) [3,4]. After spoke resonators, super conducting elliptical cavities would be employed for beam acceleration up to 1 GeV.

IOP Conf. Series: Materials Science and Engineering **810** (2020) 012060 doi:10.1088/1757-899X/810/1/012060

A spoke resonator consists of mainly three components such as End-wall, Spoke and Shell. The end-wall of spoke resonator consists of protruding nose. It is made by either welding two components, nose and outer end wall or stamping/ deep drawing operation from single sheet [5]. The second method is preferable, though it requires use of stamping setup such as die and punch. The sheet metal deep drawing or stamping is a typical seamless forming process to get desired shape of component by plastic deformation. In this process a blank is placed on a die cavity, held in position using a holding plate and pressed against the die cavity using solid punch to get the desired shape. The resonator surface is chemically polished to remove any macroscopic defects developed during forming of component and soft nature of niobium material and then heat treated [6].

Deep drawing is one of the important and widely used metal forming processes to form symmetric and asymmetric components of sheet metals in industries like automobile, packaging etc. Forming operations can be performed in single stage and multiple stages depending upon the complexity of the geometry. From deep drawing, uniform thickness distribution of sheet and highest drawing ratio expected, however residual stresses induced and localized thinning of the component occurs based on contour and shape of die [7]. In multistage forming, rapid increase in drawing force is observed during subsequent stages after first stage forming due to strain hardening [8]. Higher strain hardening coefficient delayed the necking phenomenon so subsequent drawing is possible and formability of the material increases, and springback effect decreases. The multistage forming facilitates the forming of complex and asymmetrical parts, where more than one forming operations are involved. With the advancement of technology, the multi-stage forming has become feasible concurrently with the help of progressive dies. However the multistage forming incurred high cost due to expensive setup. Use of software based on finite element analysis is becoming more important to simulate the multistage forming of metal sheets as well as to design the complex tools for forming. Simulation reduced the time and cost involve in design and development of prototype and helped in optimizing the process parameters [9].

In the present work, finite element analysis of forming processes of spoke resonator endwall performed to see and compare the effect of single stage and multistage forming processes on formability of the material. The various forming tools like die, punch and binder were modeled by UGNX 8.0, modeling software. The finite element (FE) analysis of forming was conducted by Altair HyperForm software [10]. It is a FE based software for incremental sheet metal forming simulation and used for formability analysis. The study of effect of single stage and multistage forming on distribution of thickness and equivalent stress can help to develop the product without expense of material with favorable properties. The efficient and accurate finite element simulation will help in selecting optimal forming process and parameters of tools for manufacturing of spoke resonator endwall with better quality [11].

2. Material and modeling

A niobium material is used for forming of spoke resonator components due to favorable properties. Niobium ore processed by electron beam melting and forged into ingots, which is further annealed and rolled into sheets. Annealing under controlled temperature and time is carried out for recrystallization and to obtain the desired grain size [3, 4]. Elasto-plastic material model with mechanical properties of niobium given in Table 1were used in the present simulation work.

Table 1. Mechanical properties of Niobium

Modulus of elasticity	106 GPa
Yield strength	38 MPa
Ultimate tensile strength	183 MPa
Poisson's Ratio	0.3
Strain Hardening Exponent	0.38
Density of material	8570 Kg/m^3

A CAD model of spoke resonator end-wall and forming tool for single and multistage analysis has been prepared in UGNX modeling software. The components were meshed in HYPERFORM software. Forming tools like Punch, Die and binder were considered to be rigid and meshed by Rigid R-Mesh and blank was meshed by B-Mesh. Coulomb friction model is used for interacting surfaces. After meshing various inputs like material properties, blank holding force, clearance between die and punch, punch velocity, symmetricity and drawing direction were specified for the analysis as per requirement. Boundary conditions were specified with respect to real process.

The Altair Hyperform, a finite element software was used to solve the model [10]. During multistage forming analysis data file from previous stage imported to next stage for subsequent level analysis. The output of the finite element simulation were analyzed to obtain results of material thickness, percentage thinning, von Mises stresses, and forming limit diagram.

3. Deformation mechanism

Sheet metals are usually processed through cold rolling and their properties differs in rolling and transverse directions. Therefore, the sheet metal forming generally deals with non-linear plastic deformation of anisotropic material [12]. For an isotropic ductile material, von Mises and Tresca's criteria are used, while for anisotropic material Hill's anisotropic criteria is applied. The Hill's criterion is a generalization of von Mises criteria [13] and given by $F(\sigma_y - \sigma_z)^2 + G(\sigma_z - \sigma_x)^2 + H(\sigma_x - \sigma_y)^2 + 2L\tau_{yz}^2 + 2M\tau_{zx}^2 + 2N\tau_{xy}^2 = 2f(\sigma_{ij})^2$, where *F, G, H, L, M* and *N* are constants. There values determined by anisotropy property of material. The equation will reduce von Mises criteria for isotropic material i.e. *F, G, H = 1* and *L, M, N = 3*.

In deep drawing of cup, the metal is subjected to three different kinds of deformations. To understand the deformation mechanism of sheet metal forming, an example of deep drawing of a cylindrical cup is shown in figure 1 from [14]. The Figure shows stresses and deformation in pie-shaped segment of circular blank.

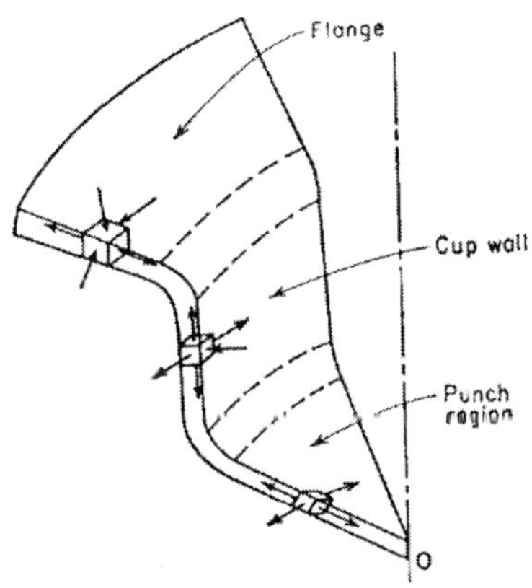

Figure 1. Schematic of stresses and deformation in a section from a drawn cup [14].

During forming operation middle part of the metal sheet comes in contact under the punch tool. It is wrapped around the profile of the punch. The sheet metal in this region is subjected to 2-D stresses (plane) due to the action of the punch result into reduction in thickness. The surrounding part of metal sheet of the blank is drawn radial outward around the punch in the neck region of die. As the punch progressed the outer circumference subjected to compressive strain and tensile strain in tangential and radial direction respectively. The two principal strains leads to thickening of the component as the metal moves inward. When the metal passes over the die shoulder radius during forming, first bending and then straightening occurs. Due to this plastic bending under tension a considerable thinning occurs [14].

4. Results and analysis

4.1. Single stage forming

The figure 2 shows the geometric details of the spoke resonator endwall, which is modeled and simulated under single stage and multi-stage forming. Figure 3 shows the model of the endwall forming tool kinematics. The forming tools consist of blank, binder, punch and die. The initial blank size of 680 mm diameter and 4 mm thickness for endwall, estimated by Radios One Step module in Altair Hyperform. The blank size of 690 mm with machining tolerance of 10 mm has been taken for forming analysis. The die is provided with die shoulder radius of 12 mm and clearance between punch and die is 0.8 mm. These parameters were optimized by finite element analysis. Blank holding force of 10 kN applied by pressure plate or binder to hold the blank during forming.

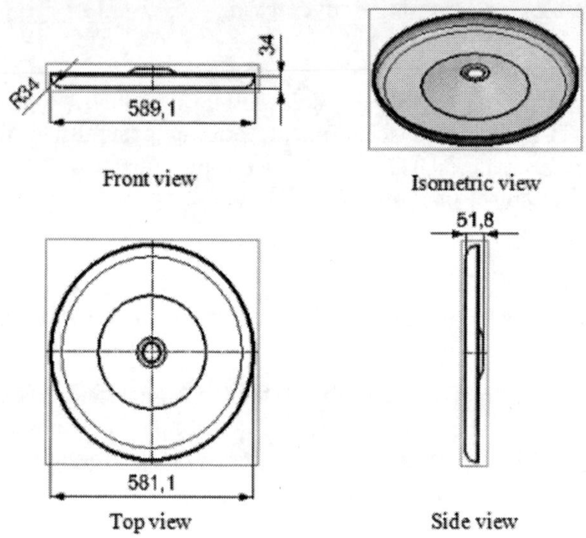

Figure 2. Geometry of formed component being analyzed

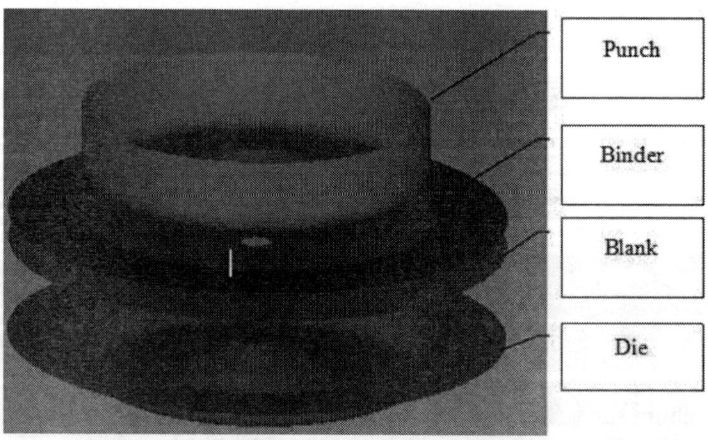

Figure 3. Modeling of Single stage forming Process

A forming limit diagram (FLD), which is the plot of plastic strain history, is used as a tool to establish limits of forming process such as wrinkling, necking and failure during the sheet metal forming process. The forming limit curve (FLC) is plotted by combinations of maximum strain that a component can undergo in various forming operations, without failure [15]. Circular grid method proposed by [16] used for determine sheet metal formability. Figure 4 shows forming limit diagram of endwall obtained from simulation under different forming conditions. Red line indicates the fracture limit, yellow line indicates marginal limit and green line indicates limit for wrinkle formation of the formed component. If FLC lies beyond the red line then component gets fractured, whereas if FLC lies between marginal line and green line the component is said to be safe. From the FLD, it has been clearly shown that the component is completely safe during forming.

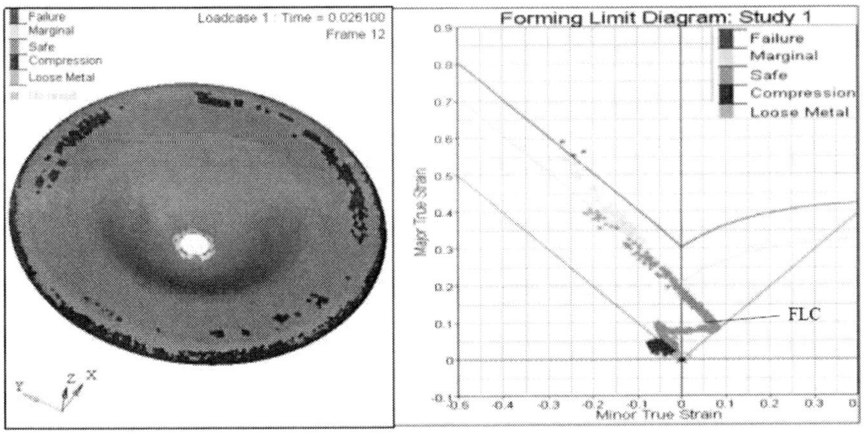

Figure 4. Forming limit diagram (FLD)

A quarter section of geometry has been picked for the forming analysis because of circular symmetry of the endwall. This saved the simulation time. The thickness analysis of single stage forming of endwall showed that maximum percentage thinning i.e. 9.1% occurred at center nose region of the endwall marked by red contour (Figure 5). The increase in friction coefficient limits the flow of material result into change in thickness. The equivalent stress diagram after forming in single stage is shown in figure 6. The maximum equivalent stress was noted as ~148 MPa in bending region of the sheet around contour. The equivalent stress at the nose section were also high ~135 MPa. However it is smaller than the outer bending zone.

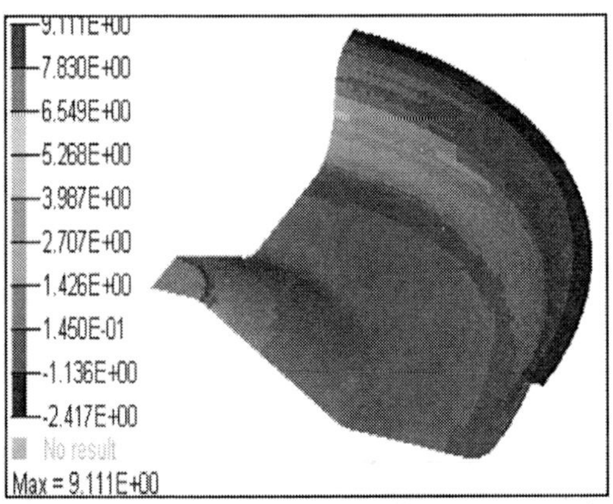

Figure 5. Percentage thinning contour in endwall

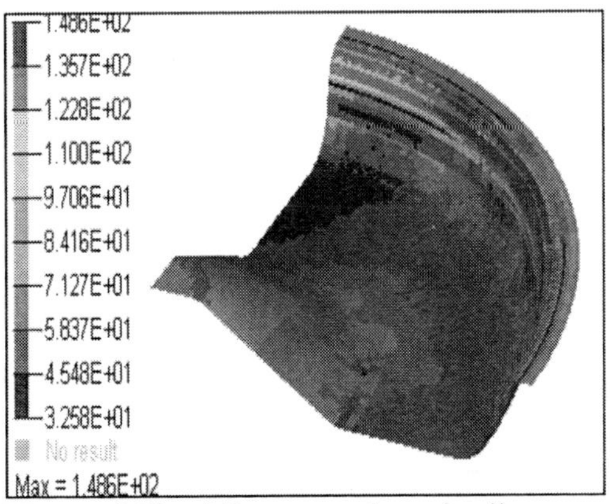

Figure 6. Von Mises stress in endwall

4.2. Multi-stage forming

Simulation of multistage forming has been carried out for two-stages and three-stages. In two-stage forming, the component was formed by two sheet metal forming operations namely (a) deep drawing and (b) bending. In first stage niobium sheet was drawn into conical shape of height 47.80 mm and which is followed by bending in 30 mm radius of flange section in the second stage. The kinematics of the forming steps for two-stage forming process is shown in figure 7.

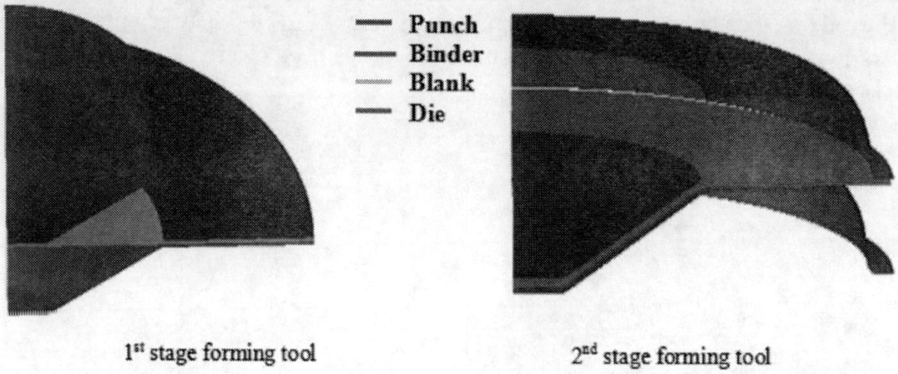

Figure 7. Stages of the modeling of multistage (two-stages) forming process

In two-stage forming of endwall, it has been observed that the percentage thinning improved and equivalent stress decreases as compared to single stage forming as in single stage the complete blank stretched simultaneously, which causes excessive thinning and stresses in the blank. The thinning percentage contour is presented in figure 8. The maximum percentage thinning was observed in nose region similar to single stage forming process however it was limited to 7.5 %. The maximum equivalent stress is noted as ~78 MPa in outer ring portion which was subjected to bending as has been shown in figure 9.

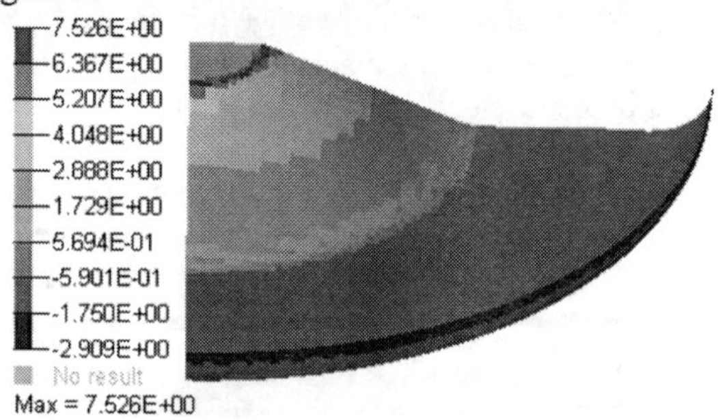

Figure 8. Percentage thinning contour in 2-stage forming

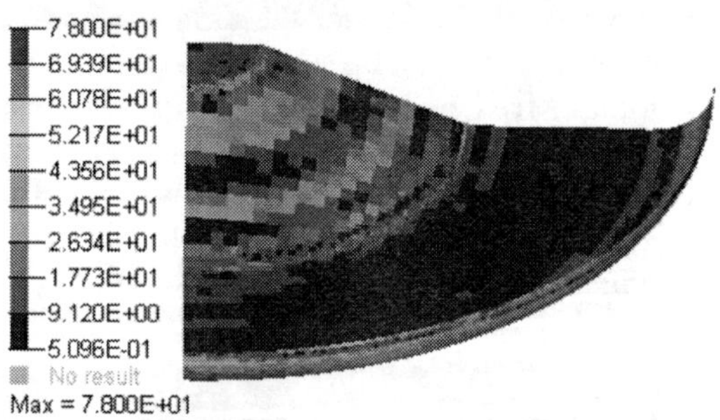

Figure 9. Von Mises stress in 2-stage forming

In three-stage forming a curved profile punch was used for first stage drawn, which enables the maximum amount of material flow into the conical cavity. In the second stage complete deep drawing of component obtained from first stage was done and in the third stage bending of flange part was performed to form the final shape of endwall. After final stage, in three-stage forming, the percentage thinning noted as 6.4% as shown in figure 10, which is better than single stage and two stage results. The improvement in percentage thinning is attributed to use of profile punch during first stage of three-stage forming [17]. Figure shows that the maximum percentage reduction of thickness took place at center part i.e. nose of the endwall. The equivalent stresses were also decreased as compared to the single stage forming and observed as ~93 MPa as shown in figure 11. However there is also middle region where stress are also high ~82 MPa compared to single stage forming. The increase in residual stresses in three stage forming were more compared to two-stage forming and due to strain hardening

IOP Conf. Series: Materials Science and Engineering **810** (2020) 012060 doi:10.1088/1757-899X/810/1/012060

of the niobium material in the previous stages. Hence, it can be concluded that the percentage thinning decreases as the forming conducted in multiple stages and uniformity in thickness increases.

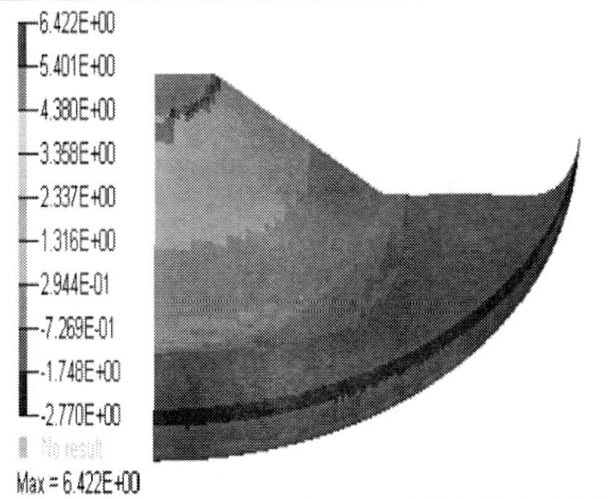

Figure 10. Percentage thinning contour in 3-stage forming

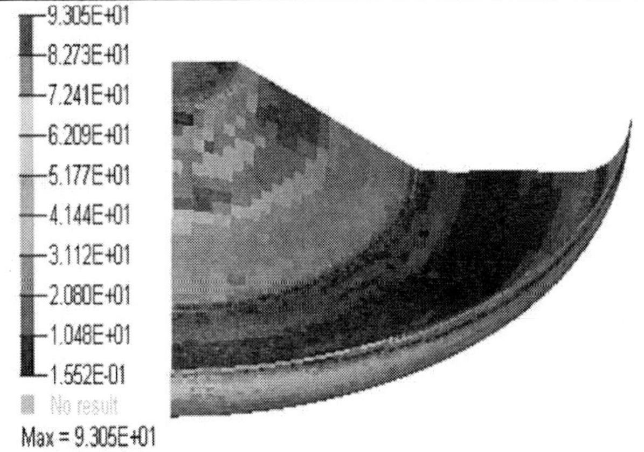

Figure 11. Von Mises stress in 3-stage forming

5. Conclusion

In this work the finite element analysis of forming operating under single stage and multi stage processes were carried out for spoke resonator endwall. Following points were noted from analysis:

- The comparison of reduction in thickness of contour showed that the maximum thinning i.e. 9.1% occurred at center/ nose part of the endwall in single stage forming. While multi-stage forming analysis showed that the percentage thinning improved to 7.5% and 6.4% under two-stage and three-stage forming operation respectively for the same geometry. The improvement in percentage thinning helped in uniformity of thickness of the spoke resonator endwall by employing proper design and selection of forming tool and multistage forming.
- High equivalent stresses were observed in endwall i.e. ~148 MPa in single stage forming. In multistage forming (three-stage) maximum von Mises stress is recoded at edge of flange of magnitude ~93 MPa. This is lower than the stresses induced in single stage forming however more than stresses noted in two-stage forming and due to strain hardening while plastic deformation.

6. References

[1] Hopper C S and Delayen J R 2013 Superconducting spoke cavities for high-velocity applications *Phys. Rev. ST Accel. Beams* **16** 102001

[2] Pal M K, Gaur R and Kumar V, *RRCAT/ABPL/SSR/01 Internal Report*, RRCAT Indore, 2015.

[3] Antoine C, Foley M and Dhanaraj N *Physical properties of niobium and specifications for fabrication of superconducting cavities* (No. FERMILAB-TM-2503-TD). Fermi National Accelerator Lab. (FNAL), Batavia, IL (US), 2011

[4] Myneni G R 2007 Physical and Mechanical Properties of Niobium for SRF Science and Technology *AIP Conference Proceedings* Single Crystal - Large Grain Niobium Technology: International Niobium Workshop vol 927 (Araxa (Brazil): AIP) pp 41–7

[5] Lanfranco G, Apollinari G, Gonin I V, Khabiboulline T N, Romanov G, Wagner R and Bosotti A 2007 Production of 325 MHZ single spoke resonators at FNAL *IEEE Particle Accelerator Conference (PAC)* 2007 IEEE Particle Accelerator Conference (PAC) pp 2262–2264

[6] Delayen J R, Kennedy W L and Roche C T 1992 Design and Test of a Superconducting Structure for High-Velocity Ions *Proceedings of the Linear Accelerator Conference*, p. 695, Ontario, Canada.
[7] Leu D-K 1997 Prediction of the limiting drawing ratio and the maximum drawing load in cup-drawing *International Journal of Machine Tools and Manufacture* **37** 201–13
[8] Paćko M, Dukat M, Śleboda T and Hojny M 2010 The Analysis of Multistage Deep Drawing of AA5754 Aluminum Alloy *Archives of Metallurgy and Materials* **55** 1173–84
[9] Önder E and Tekkaya A E 2008 Numerical simulation of various cross sectional workpieces using conventional deep drawing and hydroforming technologies *International Journal of Machine Tools and Manufacture* **48** 532–42
[10] *Altair HyperForm user manual*, v6.0, 2003, Altair Engineering Inc. Troy MI
[11] Kobayashi S, Oh S, Altan T and Chaudhary A 1990 Metal forming and the finite-element method *J. Mater. Shaping Technol.* **8** 65–65
[12] An Y G, Vegter H, Melzer S and Romano Triguero P 2013 Evolution of the plastic anisotropy with straining and its implication on formability for sheet metals *Journal of Materials Processing Technology* **213** 1419–25
[13] Hosford W F and Caddell R M 2011 *Metal Forming: mechanics and metallurgy* (Cambridge ; New York: Cambridge University Press)
[14] Dieter G E *Mechanical Metallurgy*. 3rd ed., Mc Graw-Hill Book Co., New York 1998
[15] Geiger M and Merklein M 2003 Determination of forming limit diagrams – a new analysis method for characterization of materials' formability *CIRP Annals* **52** 213–216
[16] Keeler S P 1968 Circular Grid System - A Valuable Aid for Evaluating Sheet Metal Formability *SAE Transactions*, 371-379
[17] Zein H, El-Sherbiny M, Abd-Rabou M and El Shazly M 2013 Effect of Die Design Parameters on Thinning of Sheet Metal in the Deep Drawing Process *AJME* **1** 20–29

IOP Conf. Series: Materials Science and Engineering **810** (2020) 012061 doi:10.1088/1757-899X/810/1/012061

Numerical analysis of microwave heating: Fundamentals and applications

P K Loharkar[1] and A Ingle[2]

[1]Department of Mechanical Engineering
SVKM's NMIMS MPSTME Shirpur Campus
Dhule, India
[2]Department of Mechanical Engineering
SVKM's NMIMS MPSTME Mumbai, India

Email ID:ploharkar@gmail.com

Abstract. In recent years, microwave-assisted materials processing applications are on the rise in terms of novel processes being developed and increased coverage of materials that can be processed. In general, the studies have been of experimental nature. Since microwave heating has extensive application in food processing at an industrial scale, literature on the research involving numerical studies is available in this domain. It is primarily focussed on determination of the distribution of temperature and the rate of heating of a food sample. The evolution in materials processing applications using microwaves has paved the way for exploring numerical analysis to predict and control the response parameters required for efficient material processing. This paper presents the theoretical background of electromagnetic wave propagation along with the heat transfer mechanism required in the numerical studies of microwave heating. The steps for this multiphysics-based numerical analysis are also discussed. An account of the state of microwave heating simulation research is presented as well. The unpredictability of microwave heating mechanisms coupled with fast-developing novel applications makes microwave heating simulation quite challenging. In addition, the merits of such numerical analysis present a case for extensive future research.

Symbols

E intensity of the electric field, Vm^{-1}

E_{rms} Root mean square (rms) value of Electric field intensity, Vm^{-1}

H Intensity of magnetic field, Am^{-1}

B Magnetic flux density, Wm^{-2}

D Electric flux density, Cm^{-2}

J Electric current density, Am^{-2}

ρ Volume charge density, Cm^{-3}

Q_{em} Power dissipated or Heat generated per unit volume, Wm^{-3}

ε' Dielectric constant of the material, dimensionless

ε_0 Permittivity of free space, 8.852×10^{-12} Fm^{-1}

ε'' Dielectric loss factor, dimensionless

ω Angular frequency of radiations, $rads^{-1}$

σ Electrical conductivity, Sm^{-1}

ρ Density, kgm^{-3}

C_p Specific heat capacity at constant stress, J $kg^{-1}K^{-1}$

t Heating time, s

T Temperature, deg C

λ Free space wavelength, cm

1. Introduction

Microwaves, an important constituent of electromagnetic (EM) spectrum has numerous engineering applications. Many such applications as based on its heating ability. Microwave heating has significant potential in the field of materials processing [1–4]. With the advent of technologies and processes to maximize the transformation of the energy possessed by electromagnetic radiation into thermal energy, extensive developments have taken place in materials processing applications involving thermal energy. These applications have demonstrated the merits of microwave heating over the conventional ways of heating the materials. These merits are quantified in the form of time and energy savings [5–7]. Still, industrial heating using microwaves is not carried out on a large scale partly due to limitations of versatility on the part of developed technologies and challenges owing to uncontrolled heating of materials which is a function of dielectric and magnetic properties. In addition, limited knowledge on mechanisms driving the rate of temperature rise in a material specimen also attributes to a lack of large scale commercialization. Numerical techniques such as finite element analysis can serve to enhance the predictability of the heating process and provide a platform for effective experimentation [8]. In the case of numerical analysis of microwave heating, a multiphysics approach is required that involves the coupling of the equations governing microwave propagation and the equations governing the conversion of microwave energy into heat and its transfer [9]. The subsequent sections present the fundamentals of wave propagation and methodology involved in numerical analysis. An overview of related studies is also presented.

1.1. Fundamentals of EM wave propagation

The microwave propagation is mathematically represented by Maxwell's equations which are basically a description of physical laws [10]. These equations relate to the change of magnetic field and electric field with respect to space and time. The equations are as follows [11]:

$$\nabla \times \mathrm{E} = -\frac{\partial \mathrm{B}}{\partial \mathrm{t}} \tag{1}$$

$$\nabla \times \mathrm{H} = \mathrm{J} + \frac{\partial \mathrm{D}}{\partial \mathrm{t}} \tag{2}$$

$$\nabla . D = \rho \tag{3}$$

$$\nabla . B = 0 \tag{4}$$

From equations (1)-(4), it can be inferred that "electric field" and "magnetic field" are interdependent and time-varying. These fields in combination contribute in the subsequent generation of each other thereby leading to continuous wave propagation [10, 11]. For instance, the variation of the electric field in space as represented by equation (1) is a function of the variation in the magnetic field with respect to time. The same is true for the generation of the magnetic fields, according to equation (2) with an addition of current density term. The presence of partial derivatives in the expressions also points toward the fact that these variations (in the electric and magnetic field) co-exist. These equations enable us to predict the intensity of the electromagnetic field inside the microwave cavity. Detailed discussion on these equations is available in the ref. [10].

1.2. Fundamentals of power dissipation and heat transfer

Microwave radiation's heating ability is owing to the conversion of microwave energy into thermal energy. The rate of heat generated per unit volume due to microwave energy in dielectric materials is given by equation (5) and in conducting materials is given by equation (6) [12].

$$Q_{em}(x,y,z,t) = \omega\varepsilon_0\varepsilon''\mathrm{E}_{rms}^2 \quad (5)$$

$$Q_{em}\,(x,y,z,t) = \sigma\mathrm{E}_{rms}^2 \quad (6)$$

Thus, equations (5) and (6) express the loss in power per unit volume owing to the transformation of electromagnetic energy into thermal energy, in terms of root mean square (RMS) value of the electric field intensity in the local region with coordinates x, y, z and at time t.

The distribution of the temperature in the load (material being heated) is governed by the "heat transfer" equation based on the thermo-physical properties of the load being heated [13, 14]. Equation (7) expresses the rate of heat transfer.

$$\rho\mathrm{C_P}\frac{\partial T}{\partial t} = \nabla(k\nabla T) + Q_{em} \quad (7)$$

The combined solution of equation (5), equation (6) and equation (7) along with the equations governing electromagnetic wave propagation yields temperature distribution and distribution of electromagnetic field in and across the material being processed. The analysis of these equations therefore results in multiphysics simulation.

2. General Steps of Simulation

A general flow-chart for numerical simulation of microwave heating is presented in figure 1. The first step is to create a model, either 2D or 3D that represents the actual physical geometry of the whole set-up. It includes the oven cavity, material to be processed and the waveguide and the port and other components of the heating set-up. The use of 2D is generally preferred and applicable only in such cases where the load (material to be processed) is symmetric. However, the 3D model is the best alternative to ensure better accuracy of the desired solution. The subsequent steps are discussed in the following sections.

2.1. Material properties

The accuracy of the analysis depends significantly on the correctness of the material property values given as an input to the general-purpose programs such as COMSOL, QUICKWAVE, HFSS, etc. or custom codes. The important properties required at the pre-processing stage are dielectric properties of the materials, thermal properties such as conductivity and heat capacity, electrical conductivity, and density. In the case of metal processing using microwaves, susceptors play a critical role to achieve the desired temperatures. In fact, until it was successfully demonstrated by Roy et al. [15] that metals can also be processed using microwaves by using microwave hybrid heating concept (MHH). In MHH, a material with superior interaction with microwaves which is quantified on the basis of its dielectric property which is dielectric loss factor to be specific is used. Once it gets heated volumetrically, it transfers the heat to the surrounding region. Few of these susceptor materials used in various microwave processing applications are silicon carbide (SiC), charcoal, graphite, pure alumina (Al_2O_3), zirconia, etc. [16].

The key challenge in pre-processing step is to accurately determine the required properties. Loharkar et al. [16] have also summarized the various methods used for characterizing the dielectric properties which are critical in numerical analysis. In addition, these properties are temperature-dependent and therefore, the analysis gets quite non-linear, which can also have an effect on the accuracy of the temperature distribution and field distribution profiles represented by the solution.

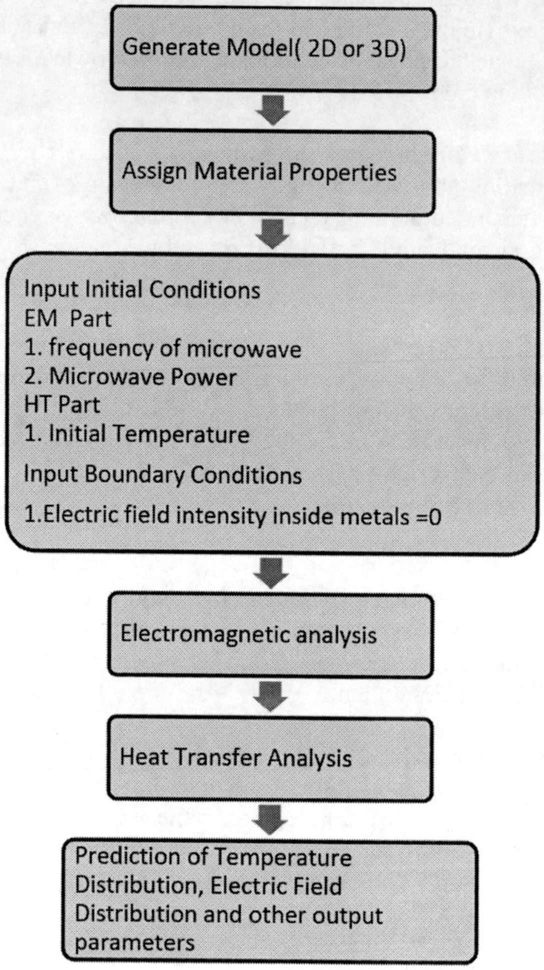

Figure 1. General steps in microwave heating simulation.

2.2. Initial and boundary conditions

Microwave heating is a combination of initial value and boundary value problems. Typically, the oven walls and the waveguide are assigned the boundary condition of perfect metallic conductors which is expressed mathematically as follows (refer to equation (8)) [17]:

$$E_{tangential} = 0 \tag{8}$$

The frequency of microwave irradiation is fixed at 2450 MHz. The initial condition in heat transfer modeling is primarily the temperature of the load to be heated before it is subjected to microwave irradiation.

2.3. Meshing

Mesh size and quality has a significant impact on the accuracy of the final solution along with the time required for processing. It has been established that the mesh size in dielectric material has a relationship with the wavelength of radiation in the free space and dielectric constant of material (refer to equation (9)) [14]. It results in minimizing the computational time along with the error in the prediction of temperature at 2450 MHz.

$$h_{es} = \frac{\lambda}{6\sqrt{\varepsilon'}} \tag{9}$$

The element quality, another important criterion used in COMSOL multiphysics commercial package varies from 0 to unity for faster and robust convergence. An acceptable value of minimum mesh quality is 0.1 [17]. In addition, memory usage can also be reduced by using half models instead of full models.

3. Recent applications of microwave heating simulation

Pioneering work in simulation of microwave heating was performed in the field of food engineering and technology. Several researchers have used the numerical simulation tools to predict the rate of temperature rise in the food sample under consideration, possible locations of hot spot creation and effect of cavity size and geometry. One of the most popular numerical techniques used in the analysis of microwave propagation is a finite-difference time-domain (FDTD) method [18]. Krishnamoorthy [19] developed a mathematical model based on the solutions obtained by combining the electromagnetics and heat transfer phenomenon using Quickwave (v 7.5) software. The combined equations were processed using the FDTD method. Optimization of simulation parameters such as mesh size, electric field strength, heating time step and frequency was carried out to predict the temperature. This numerical model was validated using experimental results. Yarahmadi et al. [18] also used FDTD for determining the distribution of the electromagnetic field and to determine temperature distribution based on this information. In continuation of his work, Pitchai [20] used the finite element method (FEM) to estimate surface temperature patterns on various food samples. FEM is another extensively used numerical technique that offers the ability to analyze irregular geometries and complex boundary conditions [14].

COMSOL multiphysics software package is based on the finite element method [21]. Quite a few of the numerical simulation-based studies have been carried out using this software owing to its ability to model and analyze problems involving multiple scientific principles. For instance, it has been used in the food technology domain to study the drying of food grains. ElGamal et al. [21] developed models to predict the moisture and temperature profile across the bed of rice grains. Similarly, Pitchai [20] obtained the solution for temperature distribution using a commercial package based on FEM. The slight discrepancy in the results of the experiment and simulation studies was observed owing to unaccounted phenomena such as chemical changes in the sample, mass transfer, evaporation, etc. during the pre-processing stage.

Due to the inherent benefits of numerical simulation and evolving research in materials engineering domain, several simulation studies have been reported in materials processing as well [22–31]. Few of these materials are coal [22, 30], biomass [25], iron powder [29], silicon carbide [32]. As discussed earlier, the properties of the specimen to be heated are of utmost importance and govern the accuracy of the solution.

A detailed parametric study to quantify the effect of microwave heating of biomass has been presented by Halim and Swithenbank [25]. These parameters were the height of the specimen from the cavity bed, waveguide position, properties of the specimen to be heated, etc. This study has laid down the basis of further investigation, especially the optimization of microwave-based heating applications which is still an open field for research. A similar study on microwave heating of asphalt mixture has been carried out by Wang et al. [26]. The properties of asphalt mixture were determined using laboratory tests and the samples were heated in a commercial microwave oven to validate the simulation results. The model was based on fairly linear characteristics and therefore, the authors had proposed the inclusion of non-linear effects to enhance the applicability of the model for extensive analysis.

Lovas et al. [27] determined the dielectric properties experimentally and modeled the microwave heating of minerals such as andesite and other minerals at a frequency of 2.216 GHz using COMSOL multiphysics. The model was used to predict the temperature distribution in the mineral samples under the effect of irradiation.

To summarize, quite a few microwave-assisted heating applications have been simulated using numerical techniques. Commercial software packages are available that are based on a multiphysics approach and enable parametric investigation of microwave heating in a quick time without being dependent on physical experimentation.

4. Conclusion

Numerical techniques such as the finite element method have made a major contribution to enhancing the rate of advancements in the field of scientific and engineering analysis. The advent of general-purpose analysis programs and fast computational media have fostered a reduction in research and development cycle times. The discussion on the steps of the simulation process and recent applications presents an overview of modeling and analysis of microwave heating. Since, the domain of microwave-assisted materials processing is relatively in its earlier stages of development as far as its commercialization is concerned. The use of numerical simulation would contribute immensely to this task.

It has also been inferred that the accuracy of these simulations would require a thorough characterization of the materials being processed. Consequently, the numerical simulations will aid in achieving enhanced control and process predictability, which is still a major challenge in microwave-assisted processing of materials.

5. References

[1] Thostenson E T and Chou T-W 1999 Microwave processing: fundamentals and applications *Compos. Part A Appl. Sci. Manuf.* **30** 1055–71

[2] Agrawal D 2010 Latest global developments in microwave materials processing *Mater. Res. Innov.* **14** 3–8

[3] Singh S, Gupta D, Jain V and Sharma A K 2015 Microwave processing of materials and applications in manufacturing industries: A Review *Mater. Manuf. Process.* **30** 1–29

[4] Kim J, Mun S C, Ko H U, Kim K B, Khondoker M A H and Zhai L 2012 Review of microwave assisted manufacturing technologies *Int. J. Precis. Eng. Manuf.* **13** 2263–72

[5] Mishra R R and Sharma A K 2016 A Review of Research Trends in Microwave Processing of Metal-Based Materials and Opportunities in Microwave Metal Casting *Crit. Rev. Solid State Mater. Sci.* **41** 217–55

[6] Buttress A, Jones A and Kingman S 2015 Microwave processing of cement and concrete materials - Towards an industrial reality? *Cem. Concr. Res.* **68** 112–23

[7] El Khaled D, Novas N, Gazquez J A and Manzano-Agugliaro F 2018 Microwave dielectric heating: Applications on metals processing *Renew. Sustain. Energy Rev.* **82** 2880–92

[8] Srinath M S, Murthy P S, Sharma A K, Kumar P and Kartikeyan M V 2012 Simulation and analysis of microwave heating while joining bulk copper *Int. J. Eng. Sci. Technol.* **4** 152–8

[9] Li H, Shi S, Lin B, Lu J, Lu Y, Ye Q, Wang Z, Hong Y and Zhu X 2019 A fully coupled electromagnetic, heat transfer and multiphase porous media model for microwave heating of coal *Fuel Process. Technol.* **189** 49–61

[10] Collin R E 2007 *Foundations for Microwave Engineering* (John Wiley and Sons)

[11] Rao N N 2007 *Fundamentals of Electromagnetics for Electrical and Computer Engineering in 108 Slides : A Tutorial Nannapaneni Narayana Rao*

[12] Lanz J E 1998 A Numerical Model of Thermal Effects in a Microwave Irradiated Catalyst Bed 113

[13] Curet S, Rouaud O and Boillereaux L 2008 Microwave tempering and heating in a single-mode cavity: Numerical and experimental investigations *Chem. Eng. Process. Process Intensif.* **47** 1656–65

[14] Liu S, Fukuoka M and Sakai N 2013 A finite element model for simulating temperature distributions in rotating food during microwave heating *J. Food Eng.* **115** 49–62

[15] Roy R, Agrawal D, Cheng J and Gedevanlshvili S 1999 Full sintering of powdered-metal bodies

in a microwave field *Nature* **399** 668–70
[16] Loharkar P K, Ingle A and Jhavar S 2019 Parametric review of microwave-based materials processing and its applications *J. Mater. Res. Technol.* **8** 3306–26
[17] COMSOL 2013 Comsol Multiphysics Reference Manual (Version 4.3b)
[18] Saremi-Yarahmadi S, Whittow W and Vaidhyanathan B 2013 Electromagnetic simulation studies of microwave assisted heating for the processing of nanostructured iron oxide for solar driven water splitting *Appl. Surf. Sci.* **275** 65–70
[19] Krishnamoorthy P 2011 Electromagnetic and Heat Transfer Modeling of Microwave Heating in Domestic Ovens 194
[20] Pitchai K 2015 A Finite Element Method Based Microwave Heat Transfer Modeling of Frozen Multi-Component Foods *Thesis*
[21] ElGamal R, Ronsse F and Pieters J G 2013 Modeling Deep-Bed Grain drying Using COMSOL Multiphysics *2013 COMSOL Conf.* 5
[22] Hong Y Du, Lin B Q, Li H, Dai H M, Zhu C J and Yao H 2016 Three-dimensional simulation of microwave heating coal sample with varying parameters *Appl. Therm. Eng.* **93** 1145–54
[23] Mimoso R M C, Albuquerque D M S, Pereira J M C and Pereira J C F 2017 Simulation and control of continuous glass melting by microwave heating in a single-mode cavity with energy efficiency optimization *Int. J. Therm. Sci.* **111** 175–87
[24] Ratanadecho P 2002 A numerical and experimental investigation of the modeling of microwave heating for liquid layers using a rectangular wave guide (effects of natural convection and dielectric properties) **26** 449–72
[25] Halim S A and Swithenbank J 2018 Simulation study of parameters influencing microwave heating of biomass *J. Energy Inst.* 1–22
[26] Wang H, Zhang Y, Zhang Y, Feng S, Lu G and Cao L Laboratory and Numerical Investigation of Microwave Heating Properties of Asphalt Mixture
[27] Lovás M, Kováčová M, Dimitrakis G, Čuvanová S, Znamenáčková I and Jakabský Š 2010 Modeling of microwave heating of andesite and minerals *Int. J. Heat Mass Transf.* **53** 3387–93
[28] Tamang S and Aravindan S 2019 3D numerical modelling of microwave heating of SiC susceptor *Appl. Therm. Eng.* **162** 114250
[29] Devi J, Akhtar M J and Datta P 2017 Multiphysics Model of Iron Powder Compacts for Efficient Microwave Processing *Prog. Electromagn. Res. M* **56** 11–23
[30] Huang J, Xu G, Hu G, Kizil M and Chen Z 2018 A coupled electromagnetic irradiation, heat and mass transfer model for microwave heating and its numerical simulation on coal *Fuel Process. Technol.* **177** 237–45
[31] Suzuki M, Ignatenko M, Yamashiro M, Tanaka M and Sato M 2008 Numerical Study of Microwave Heating of Micrometer Size Metal Particles *ISIJ Int.* **48** 681–4
[32] Bansal A and Sharma A K 2014 3D Electromagnetic Field Simulation of Silicon Carbide and Graphite Plate in Microwave Oven *Int. J. Mech. Eng. Robot. Res.* **1** 7–12

IOP Conf. Series: Materials Science and Engineering **810** (2020) 012062 doi:10.1088/1757-899X/810/1/012062

Numerical evaluation of Knuckle Joint with stainless steel material for total deformation and Von Mises analysis

[1]Tejas Pawar, [1]Yogeshwar Chaudhari, [1]Pradip Jamadar, [2]M. Ramachandran
[1]RC Patel Institute of technology, Shirpur, Maharashtra 425405
[2]MPSTME, SVKM's NMIMS, Shirpur, Dhule, Maharashtra 425405, India
tejas1935@gmail.com

Abstract. A knuckle joint is used to connect the two rods which are under the tensile load, when there is requirement of small amount of flexibility or angular moment is necessary. There is always axial or linear line of action of load. So modelling and analysis of knuckle joint under a certain condition is carried out. Modelling and analysis of a knuckle joint was performed by using 3D software CATIA & Finite Element Analysis (FEA) respectively. For analysis we use Stainless Steel materials. The maximum elastic strain intensity is 0.00052959 m/m and the minimum elastic strain intensity is 5.1094e-10 m/m. The maximum equivalent elastic strain is 0.00038217 m/m and the minimum equivalent elastic strain is 4.4217e-10 m/m. The maximum equivalent (Von Mises) stress is 7.6421e7 Pa and the minimum equivalent (Von Mises) stress is 72.422 Pa. The maximum total deformation is 7.0468e-6 m and the minimum total deformation is 0 m.

1. Introduction

A knuckle joint is used to connect the two rods which are under the tensile load, when there is requirement of small amount of flexibility or angular moment is necessary. There is always axial or linear line of action of load. The knuckle joint assembly consist of following major components: 1.Single eye. 2. Double eye or fork and 3.Knuckle pin. At one end of the rod the single eye is formed and double eye is formed at the other end of the rod. Both, single and double eye are connected by a pin inserted through eye. The pin has a head at one end and at other end there is a taper pin or split pin. For gripping purpose the ends of the rod are of octagonal forms. Now, when the two eyes are pulled apart, the pin holds them together .The solid rod portion of the joint in this case is much stronger than the portion through which the pin passes. The modes of failure are: Shear failure of pin (single shear), Crushing of pin against rod, Tensile failure of flat end bar. A Knuckle joint joint permits angular misalignment of the rods and may take compressive load if it is guided. These joints are used for different types of connections i.e. tie rods, tension links in bridge structure. In this, one of the rods as an eye at the rod end and other end is forked with eyes at the both the legs. If failure of knuckle joint occurs then possibilities of accident. So it is necessary that design and analysis of knuckle joint should be proper enough to withstand in working condition without failure.

2. Design of knuckle joint

At the first instance we have prepared the CAD model of knuckle using CATIA software as shown in Figure. While preparing the CAD model, we have assigned the proper dimensions and the material properties of the above mention parts. Then after completing design CATIA, We store it in IGES format. This IGES format file then use in CAE Software i.e. Ansys. At first the Knuckle is designed in CATIA software and then the file is saved as IGES format and imported in the ANSYS 18.2 workbench software. The mechanical properties of Stainless Steel are Density with value 7.7 g/cc g/cc, Tensile Strength with value 450 Mpa, Yield Strength with value 480 Mpa and Poisson's Ratio with value 0.275.

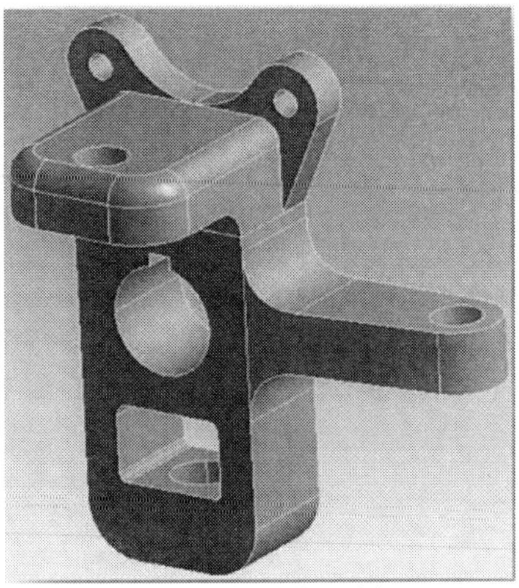

Figure 1: Design of knuckle joint with CATIA software

3. Numerical Analysis of knuckle joint

For doing the numerical Analysis of knuckle joint, meshing is most important part. The meshing of knuckle joint is done with 56447 nodes and with 16621 elements. In this analysis hex dominant mesh type is used and adaptive meshing is used for analysis. Each element is around 3mm size after meshing. For analysis 2060.1 N force is used.

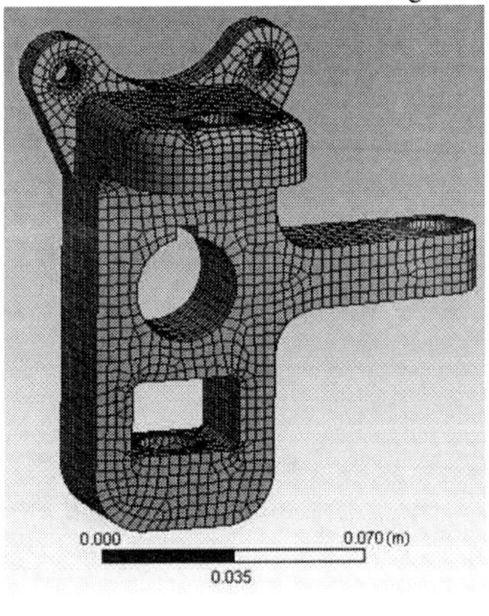

Figure 2: Meshing of knuckle joint

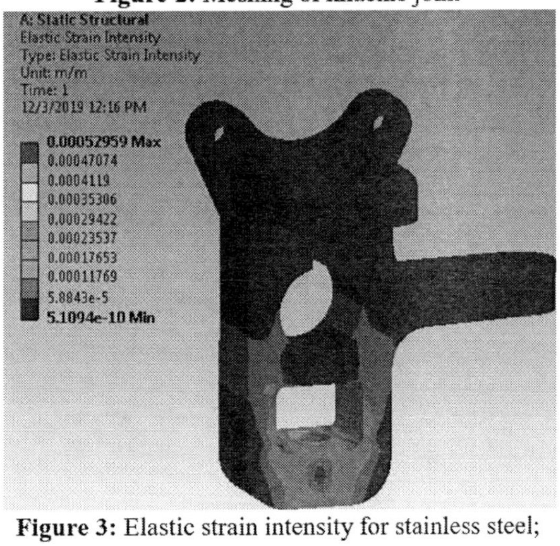

Figure 3: Elastic strain intensity for stainless steel;

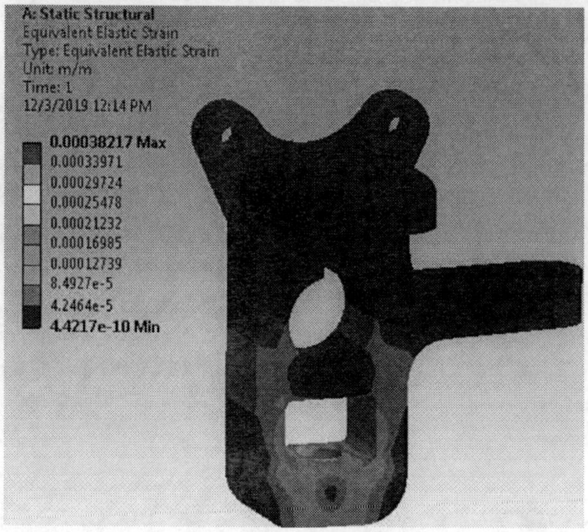

Figure 4: Equivalent elastic strain for stainless steel

From the Figure 3 which shows Elastic strain intensity of knuckle joint with stainless steel material, the maximum elastic strain intensity is 0.00052959 m/m and the minimum elastic strain intensity is 5.1094e-10 m/m. From the Figure 4 which shows equivalent elastic strain for stainless steel of knuckle joint, the maximum equivalent elastic strain is 0.00038217 m/m and the minimum equivalent elastic strain is 4.4217e-10 m/m.

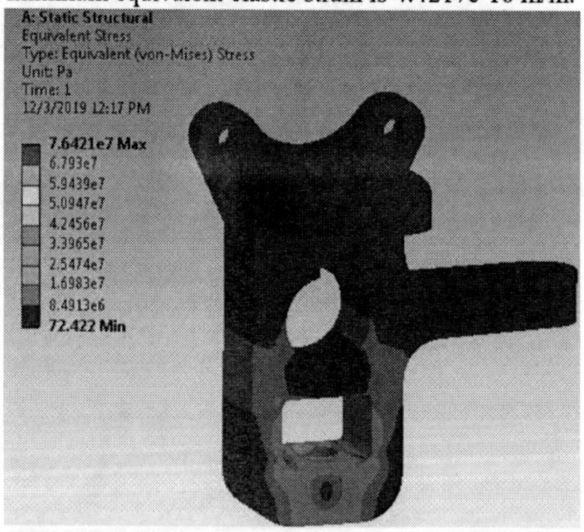

Figure 5: Equivalent (Von Mises) stress for stainless steel

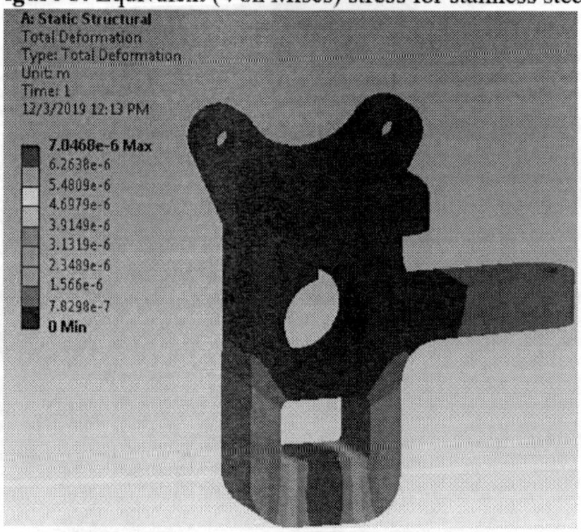

Figure **6:** Total deformation for stainless steel

From the Figure 5 which shows equivalent (Von Mises) stress of knuckle joint with stainless steel material, the maximum equivalent (Von Mises) stress is 7.6421e7 Pa and the minimum equivalent (Von Mises) stress is 72.422 Pa. From the Figure 6 which shows total deformation for stainless steel of knuckle joint, the maximum total deformation is 7.0468e-6 m and the minimum total deformation is 0 m.

4. Conclusion

At the first instance we have prepared the CAD model of knuckle using CATIA software as shown in Figure. While preparing the CAD model, we have assigned the proper dimensions and the material properties of the above mention parts. Then after completing design CATIA, We store it in IGES format. This IGES format file then use in CAE Software i.e. Ansys. At first the Knuckle is designed in CATIA software and then the file is saved as IGES format and imported in the ANSYS 18.2 workbench software. The mechanical properties of Stainless Steel are Density with value 7.7 g/cc g/cc, Tensile Strength with value 450 Mpa, Yield Strength with value 480 Mpa and Poisson's Ratio with value 0.275. For doing the numerical Analysis of knuckle joint, meshing is most important part. The meshing of knuckle joint is done with 56447 nodes and with 16621 elements. In this analysis hex dominant mesh type is used and adaptive meshing is used for analysis. Each element is around 3mm size after meshing. For analysis 2060.1 N force is used. The maximum elastic strain intensity is 0.00052959 m/m and the minimum elastic strain intensity is 5.1094e-10 m/m. The maximum equivalent elastic strain is 0.00038217 m/m and the minimum equivalent elastic strain is 4.4217e-10 m/m. The maximum equivalent (Von Mises) stress is 7.6421e7 Pa and the minimum equivalent (Von Mises) stress is 72.422 Pa. The maximum total deformation is 7.0468e-6 m and the minimum total deformation is 0 m.

Reference

1. Das, Sourav, Vishvendra Bartaria, and Prashant Pandey. "Analysis of Knuckle Joint of 30C8 Steel for Automobile Application." *International Journal of Engineering* 3, no. 1 (2014).
2. T. Sunder selwyn, R Kesavan, M. Ramachandran, FMECA Analysis of Wind Turbine Using Severity and Occurrence at High Uncertain Wind in India, International Journal of Applied Engineering Research. ISSN 0973-4562 Volume 10, Number 11 (2015) pp. 10250-10253.
3. Beisel, Winfried, and Ingo Oberste-Dommes. "Precision embossing Alternative method of forming flat parts." *Journal of materials processing technology* 71, no. 1 (1997): 18-24.
4. Vishal Fegade, Gajanan Jadhav, M. Ramachandran, Design, Modelling and Analysis of Tilted Human Powered Vehicle, IOP Conf. Series: Materials Science and Engineering 377 (2018) 012215.
5. Wujun, Chen, Luo Yaozhi, Fu Gongyi, Gong Jinghai, and Dong Shilin. "Design Conception and Deployment Simulation for a Highly Synchronized Extendable/Retractable Space Mast." *International Journal of Space Structures* 16, no. 4 (2001): 261-269.
6. Seering, W., K. Armbruster, C. Vesely, and D. Wormley. "Experimental and analytical study of pantograph dynamics." *Journal of dynamic systems, measurement, and control* 113, no. 2 (1991): 242-247.
7. Yossifon, S., and R. Shivpuri. "Optimization of a double knuckle linkage drive with contrast mechanical advantage for mechanical presses." *International Journal of Machine Tools and Manufacture* 33, no. 2 (1993): 193-208.
8. Singh, Pradeep K., Satish C. Jain, and Pramod K. Jain. "Advanced optimal tolerance design of mechanical assemblies with interrelated dimension chains and process precision limits." *Computers in Industry* 56, no. 2 (2005): 179-194.
9. Mavroidis, Constantinos, Kathryn J. DeLaurentis, Jey Won, and Munshi Alam. "Fabrication of non-assembly mechanisms and robotic systems using rapid prototyping." *Journal of mechanical design* 123, no. 4 (2001): 516-524.
10. Halicioglu, Recep, L. Canan Dulger, and A. Tolga Bozdana. "Mechanisms, classifications, and applications of servo presses: A review with comparisons." *Proceedings of the Institution of Mechanical Engineers, Part B: Journal of Engineering Manufacture* 230, no. 7 (2016): 1177-1194.
11. Yossifon, S., and R. Shivpuri. "Analysis and comparison of selected rotary linkage drives for mechanical presses." *International Journal of Machine Tools and Manufacture* 33, no. 2 (1993): 175-192.
12. Yan, Hong-Sen, and Wei-Ren Chen. "A variable input speed approach for improving the output motion characteristics of Watt-type presses." *International Journal of Machine Tools and Manufacture* 40, no. 5 (2000): 675-690.

IOP Conf. Series: Materials Science and Engineering **810** (2020) 012062 doi:10.1088/1757-899X/810/1/012062

13. Ashish Mogra, M. Ramachandran and Hemant Kumar Mehta, Experimental Investigation on Droplet Formation through a Minichannel Y- Junction, IOP Conf. Series: Materials Science and Engineering 377 (2018) 012189.
14. Yossifon, S., and R. Shivpuri. "Optimization of a double knuckle linkage drive with contrast mechanical advantage for mechanical presses." *International Journal of Machine Tools and Manufacture* 33, no. 2 (1993): 193-208.
15. Kanak Kalita, M. Ramachandran, Pramod Raichurkar, S. Haldar, Free Vibration Analysis Of Laminated Composites By A Nine Node Isoparametric Plate Bending Element, Advanced Composites Letters, 25(4), 2016.
16. Pantazopoulos, G., A. Sampani, and E. Tsagaridis. "Torsional failure of a knuckle joint of a universal steel coupling system during operation–A case study." *Engineering Failure Analysis*14, no. 1 (2007): 73-84.
17. Kumar, Nipun, Gian Bhushan, and Pankaj Chandna. "Analysis of Knuckle Joint of Various Materials using CAE Tools." *International Journal of Engineering, Management and Applied Sciences* 5 (2017): 45-52.
18. Shajit Yadav, M.Ramachandran, Analysis of Optimized Wind Turbine Failure Using Computational Fluid Dynamics, Materials Today: Proceedings, 4(2), 2017:1788–1793.
19. Tso, P. L., and K. C. Liang. "A nine-bar linkage for mechanical forming presses." *International Journal of Machine Tools and Manufacture* 42, no. 1 (2002): 139-145.
20. Hegazy, S., H. Rahnejat, and K. Hussain. "Multi-body dynamics in full-vehicle handling analysis under transient manoeuvre." *Vehicle System Dynamics* 34, no. 1 (2000): 1-24.
21. Kim, H. W., D. S. Bae, and K. K. Choi. "Configuration design sensitivity analysis of dynamics for constrained mechanical systems." *Computer methods in applied mechanics and engineering* 190, no. 40-41 (2001): 5271-5282.
22. M. Ramachandran, Benildus, Failure Analysis of Turbine Blade Using Computational Fluid Dynamics, International Journal of Applied Engineering Research. ISSN 0973-4562 Volume 10, Number 11 (2015) pp. 10230-10233.
23. Tang, Liang, Wen-Bin Shangguan, and Lin Dai. "A calculation method of joint forces for a suspension considering nonlinear elasticity of bushings." *Proceedings of the Institution of Mechanical Engineers, Part K: Journal of Multi-body Dynamics*226, no. 4 (2012): 281-297.
24. Saxena, Nishant Vibhav, and Rohit Rajvaidya. "Study & Analysis of Knuckle Joint With the Replacement of Material By Using Teflon." *Int. Journal of Engineering Research and Application* 5 (2015): 67-72.
25. Bowonder, B. "Competitive and technology management strategy: a case study of TELCO." *International Journal of Technology Management* 15, no. 6-7 (1998): 646-680.*

IOP Conf. Series: Materials Science and Engineering **810** (2020) 012063 doi:10.1088/1757-899X/810/1/012063

Optimisation of Flexural strength in Zirconia nanoclusters of the Bis-GMA & TEGDMA based dental composites

Mohd. Saif Mohd sadiq Siddiqui, Dr.Umesh Hambire

Department of Mechanical Engineering, Government College of Engineering, Aurangabad (MS), Maharashtra, India

saifsiddiqui59@gmail.com

Abstract. Chewing food is a dynamic and continuous process. tooth is an important component which can not be neglected. Accordingly to WHO(world health organization) report nearly 100% people have decay once in life. Decay can be filled with gold, silver amalgam. Composites have advantages by providing natural colour of tooth, less tooth removal required to fill cavity and natural tooth structure can be maintained. The aim of this work is to present an analysis of the composites and furnish dentists with a basis that can provide criteria for choosing one or another to suit their treatment. The mechanical properties of composite resins are importantly depends upon filler content. Dental composites are using silica, quartz and alumina glass as filler for long time. Zirconia is the latest addition as filler. Chewing of food is a dynamic situation in which many forces are acting in all possible direction, the most important being the flexural force. Our aim is to optimise the flexural strength in Zirconia nanoclusters of the Bis-GMA & TEGDMA based dental composites. The aim of work is to find out best combination of composite material.

1. Introduction

The purpose of this paper to obtain higher flexural strength of dental composite by using zirconia and hydroxyapatite as a filler. Proper composition of Zirconia, hydroxyapatite and silica is used to have favourable outcome by gaining optimum flexural strength now a days people have many problems such as dental decay, cavities, and root canal infections mainly cause due to bacterial infections. These problem hampers the tooth structure [1]. For treatment of the affected tooth, dentists recommend the removal of the dental caries and filling of the cavities with appropriate materials. Human health get affected during filling of cavities with dental amalgam. So, now in recent times, polymer composites have showed remarkable advantages over those traditional ones. Excellent properties such as biocompatibility, antibacterial and nontoxic ness makes them use highly as a dental restorative material. More than that its advantages over aesthetic makes it more popular. Also, Physical, mechanical, thermal and tribological properties of polymer composites is good [2]. Bisphenol A-glycidyl methacrylate (Bis-GMA) polymer composites are the favourable and most commonly practice in dentistry for anterior and posterior tooth filling. It has low polymerization shrinkage, low volatility, and high viscosity. Urethane dimethacrylate has lower viscosity but higher toughness than Bis-GMA. It can be used a substitute for Bis-GMA [15].Recently, reinforcements based on Nano fillers with saline treatment are being widely used in dental polymer composites [3].

Dental adhesives are applied to restorative materials to tooth structure [1]. Adhesion to the enamel is much durable than the dentin due to the dynamic properties of the wet dentin [2,3].Dentin is a hydrated biological dynamic compositwhich consists of30% water and organic materials and 7

IOP Conf. Series: Materials Science and Engineering **810** (2020) 012063 doi:10.1088/1757-899X/810/1/012063

0% minerals [11;12].A weak bonding between the restorative materials and dentin results in problems such as secondary caries, sensitivity after restoration, colour change and micro leakage Several mechanisms are involved in the adhesion between dental adhesives and tooth structure including physical adsorption, micro mechanical interlocking, chemical and ionic bonding and acid base interactions[7;8]. In the developments of dental adhesives attempts have been made to improve the mechanisms or to impart more mechanisms in the adhesion providing a stronger and more reliable bond for adhesive dentistry.

2. Materials and Methods

The monomer system can be viewed as the pillar of the composite resin system. Bis-GMA (*bis*phenol glycidyl methacrylate) and TEGDMA (triethylene glycol dimethacrylate) continue to be the most used monomers for manufacturing present-day composites [7]. Recent studies reveal introduction of zirconia as filler for improvement in mechanical properties.
The resin matrix was a visible light- cured system composed of Bis-GMA (Bisphenol A Glycidyl Methacrylate) and TEGDMA (Triethylene glycol dimehtacrylate) as matrix and Nano filler amounts of percentage volume shown in table 1 the resin specimens were polymerized by means of the ICI light. The light output was monitored before and after use by means of a light meter designed to measure the output in the critical wave length range. Minimal acceptable output was established at 1200 W/m^2. The nano hybrid fillers were: silica, hydroxyapatite and Zirconia with varying volume percentage as mentioned in table 1

Process parameters	Level 1	Level 2	Level 3
Hydroxyapatite%	5.8	11	15.4
Zirconia%	4.7	9.6	14.3
Silica%	4.1	7.8	12.7

Table 1 Percentage of filler volume at each level.

Experim ent number	Filler volume levels			flexural strength	Signal to noise Ratio
	Hydroxyap atite	Zirco nia	Silic a		
1.	1	1	1	75.91	37.606 0
2.	1	2	2	41.92	32.448 4
3.	1	3	3	30.55	29.700 2
4.	2	1	2	44.83	33.031 4
5.	2	2	3	54.98	34.804 1
6.	2	3	1	32.05	30.116 6
7.	3	1	3	34.26	30.695 7
8.	3	2	1	37.57	31.496 8
9.	3	3	2	19.28	25.702 1

Table 2 Taguchi L9 orthogonal array for flexural strength.

One more filler which was responsible for improvement in mechanical properties was Zirconia. Presently few commercially available dental composite is using Zirconia Nano clusters. So in addition to silica and hydroxyapatite we have added Zirconia nanoparticles as filler.
Matrix used his combination of the following components.
1) bisphenol A glycidyl methacrylate (**Bis-GMA**)

IOP Conf. Series: Materials Science and Engineering **810** (2020) 012063 doi:10.1088/1757-899X/810/1/012063

2) triethylene glycol dimethacrylate (**TEGDMA**)
We have varied the volume fraction of the three fillers namely. 1) Hydroxyapatite. 2) Silica. 3) Zirconia

2.1 Preparation of test specimen for flexural strength:
The test specimens for flexural strength experimental composites were made by the following procedure. A hydroxyapatite rectangle of 25×3×3mm used as a mold for the preparation of a rectangular test specimen for flexural strength. Each mold was positioned upon polyester film of 0.1 mm thick. The mold was slightly overfilled with the test composite and all air bubbles were excluded. A second piece of polyester film was placed onto the material in the mold and covered with a hydroxyapatite plate, then pressure was applied gently, thus exuding excess material from the mold Each of the 3 molds filled with the composite were prepared for flexural strength and placed in a visible light irradiation apparatus & irradiated for 15min, then turned upside down and irradiated for another 15 min.The light-cured specimen was removed from its mold and stored in distilled water at 37°c for 24hr prior to testing. The rectangular specimen was tested in a universal testing machine (Make-Autograph) & utilizing a crosshead speed of 1 mm/min. Three specimens of each experimental group were tested. The mean value of the three was accepted as an observed strength of the tested composite resin. The purpose of the study was to study the contribution of different filler particles of Dental composites in the market to investigate their mechanical properties and their filler contents.

2.2 Procedure adopted for testing Flexural strength:
The ASTM Standard adopted was ASTM D7264 [5]. We measured the width and thickness of the specimen on the performance of dental composite in general and flexural strength in particular. Experimentation was done and testing of all the leading commercially available brands to the nearest 0.01 mm (0.001 in.) at several points along its length. Calculate and record the minimum value of the cross-sectional area was calculated. We measured the length of the specimen and recorded the value. Test specimen was placed flexural testing machine. We move the crosshead of the testing machine until it just contacts the top of the tool plunger was adjusted. The test machine is equipped with an automatic recording device, record the complete load-deformation curve. After the yield point has been reached it may be desirable to increase the speed from 5 to 6 mm/min (0.20 to 0.25 in./min) and allow the machine to run at this speed until the specimen breaks.

2.3 Preparation of test specimen for flexural strength
We have taken three levels of filler volume percentages, Low, medium and high. These three levels are shown in table 1. We have varied the three levels of the volume percentage of silica, hydroxyapatite and Zirconia. By L9 (Taguchi's orthogonal array) we have nine experimental composites. For each trial three readings are taken and the average flexural strength is calculated. The graphs for each filler and flexural strength is plotted keeping the other volume percentage constant, to study the contribution of each filler in the flexural strength. Then we have calculated Regression equation for each mechanical property and Analysis of variance (Table No 2) and Sensitivity analysis (Table No 3) for each mechanical property,

Figure 1 experimental setup for flexural strength

2.4 Taguchi experiment: Design and analysis

Traditional experimental design procedures are too complicated and not easy to use. When the number of process parameters are increases, at large number of experimental works have to be carried out. Dr Genichi Taguchi has designed a method which reduces the number of experiments by the use of an orthogonal array. The information the behaviour of the process is obtained from a plan of experiments with the objective of acquiring data in a controlled way. This design of high-quality system called Taguchi's design method which reduces efforts of conducting a number of experiments, thus saving time and cost and enabling discovery significant factors quickly. But the main advantage of Taguchi's method is by conducting less experiments, we get a reliable result and also the higher the value of sum of square of an independent variable, the more it has influence on the performance parameter. One can also calculate the ratio of individual of square of a particular

Independent variable to the total sum of squares of all the variables. This ratio gives the percent contribution of the independent variable on the performance parameter. In addition to above, one could also find the optimal solution to the problem. A statistical analysis of variance (ANOVA) and signal to noise ratio (S/N ratio) can be employed to study the influence of different filler volume percentages on flexural strength values. Following are the steps followed in Taguchi's optimization study (Taguchi 1990; Ross 1996). (1) Select noise and control factors. (2) Select Taguchi orthogonal array. (3) Conduct experiments and measure the flexural strength for each of the experimental composites. (4)Analyse results (signal-to-noise ratio). (5) Predict optimum .(6) Confirmation experiment.

2.5 Planning of experiments:

Here Taguchi method is used, which includes the experiment design theory and the quality loss function concept. The degrees of freedom for three parameters in ach of three levels were calculated as follows: degree of freedom (DOF) = number of levels –1. For each factor, DOF is as follows: for hydroxyapatite DOF= 3 – 1 = 2; for zirconia, DOF = 3 – 1 = 2; for silica, DOF = 3 – 1 = 2. Taguchi L9 orthogonal array was used, which has three columns at three levels and nine rows corresponding to the number of tests [11]. Nine experiments were conducted at different parameters in this study. L9 orthogonal array has eight DOFs, in which six were assigned to three factors (each one 2 DOFs) and two DOF were assigned to the error. Table 4 shows the degree of influence of the process parameters, i.e. the filler content (volumetric percentage), three factors each at three levels, are taken into account [11]. The flexural strength was tested for each experimental composite with filler volume percentage corresponding to each level. The flexural strength values corresponding to each experimental composite are shown in table2.

3. Results and discussion

3.1 Regression analysis:

A mathematical model was developed with the volumetric percentages of three fillers in the composite, namely, hydroxyapatite, zirconia and silica and its effect on the flexural strength. The correlation between factors (hydroxyapatite, zirconia, silica and flexural strength was obtained by multiple linear regressions. To derive the models, the standard commercial statistical software package 'MINITAB' is used

$$CS = 31.733 + 1.518\ A\_1 + 0.917\ A\_2 - 2.435\ A\_3 + 2.044\ B\_1 + 1.183\ B\_2 - 3.227\ B\_3 + 1.340\ C\_1 - 1.340\ C\_2 - 0.000\ C\_3$$

3.1a Analysis of the signal to noise ratio: For the minimization of quality characteristic variation due to uncontrollable parameters the response variation is studied using signal-to-noise ratio (S/N ratio) as enumerated in Taguchi method. In the dental composites, larger flexural strength will be better, so the flexural strength was considered as the quality characteristic and the concept of 'the larger-the-better' [11] was used. The S/N ratio used for this type response is taken as 'larger-the-better' and given by the relation

$$S/N = -\ log_{10}\left(\frac{1}{n}\Sigma\frac{1}{y^2}\right) \quad (1)$$

Where y is the measured value in a run/row and n the number of measurements in a trial/row (n= 1 in this case). Equation (1) was used to calculate the S/N ratio. The S/N ratios and the corresponding values of flexural strength are shown in table 2. For the volumetric percentages of hydroxyapatite zirconia and silica concentrations, the flexural strength response table and the rank are shown in table 3. A greater S/N ratio value will give better performance, irrespective of the performance characteristic. The greater S/N ratio value will give the optimum level filler volume percentage. The optimal flexural strength was obtained at hydroxyapatite% = 5.8 (level 1), zirconia= 4.7 (level 1), silica% = 4.1 (level 1), based on the analysis of the S/N ratio

Larger is better

Level	A	B	C
1	33.25	33.78	33.07
2	32.65	32.92	30.39
3	29.30	28.51	31.73
Delta	3.95	5.27	2.68
Rank	2	1	3

Table No.3 Response Table for Signal to Noise Ratios

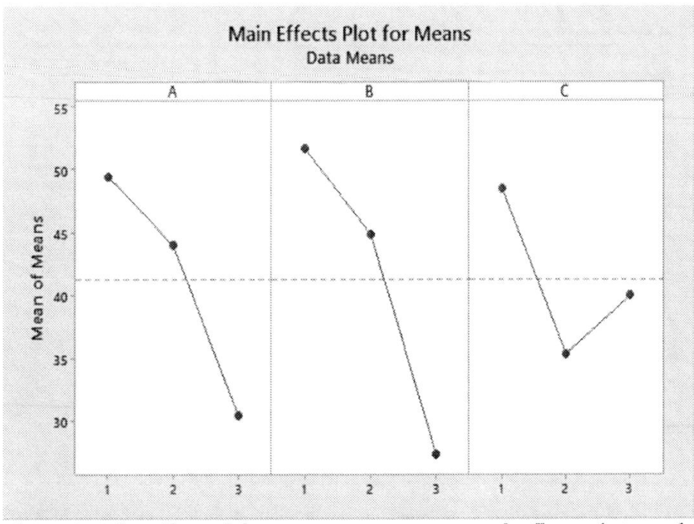

Figure 2 : The effect of the process parameters on the flexural strength.

The effect of the process parameters on the flexural strength is shown in figure 1, which is evaluated using statistical software package MINITAB.

3.2 Analysis of variance (ANOVA)

For finding out the differences in the average performance of the groups of items tested, a statistical tool named ANOVA, which is an objective decision-making tool is used. At a specific confidence level, the mean square is compared with an estimate of experimental error in ANOVA The interaction and significance of all the main factors are tested. The following equation is used to find out the total sum of squared deviations *SST* from the total

mean S/N ratio, n_m

$$SS_T = \sum_{i=1}^{n}(n_i - n_m)^2 \qquad (2)$$

where n_i is the mean S/N ratio for the *i*th experiment and *n* the number of experiments in the orthogonal array. For calculation of *P*, i.e. percentage contribution

$$P = \frac{ss_d}{ss_t} \qquad (3)$$

where is SS_d the sum of the squared deviations Table 4 shows the results of ANOVA. To find out the significance of a factor, a ratio of the mean square error to the residual error is taken and called as *F*-ratio, named after Fisher. Fisher test is used to study the significant effect on quality characteristic by the design parameters. To study the effect of the filler volume percentage on the flexural strength we use the percentage (%) shown in table 4. From table 4, it can be seen that nano filler hydroxyapatite has 26.97%, zirconia has 44.15% and silica has 12.48%, which influences on the flexural strength, respectively.

IOP Conf. Series: Materials Science and Engineering **810** (2020) 012063 doi:10.1088/1757-899X/810/1/012063

Source	DF	Adj SS	Adj MS	F-Value	P-Value	Total %
A	2	579.3	289.6	1.65	0.378	26.97
B	2	948.2	474.1	2.69	0.271	44.15
C	2	268.0	134.0	0.76	0.568	12.48
Residual	2	351.8	175.9			16.38
Total	8	2147.3				100

Table 4. Analysis of variance (ANOVA) results for FLEXURAL strength of dental composite.

Experimental dental composite no.	**Filler volume percentages**		
	Hydroxyapatite	Zirconia	Silica
1	5.8	4.7	4.1
2	5.8	9.6	7.8
3	5.8	14.3	12.7
4	11	4.7	7.8
5	11	9.6	12.7
6	11	14.3	4.1
7	15.4	4.7	12.7
8	15.4	9.6	4.1
9	15.4	14.3	7.8

Table 5. Fillers used in the experimental dental composite.

3.3 Confirmation test:

Taguchi strongly recommended a confirmation test for verifying the results [11;14]. The experiments are conducted with the optimum conditions and the strength is found. This flexural strength is compared with that calculated value using the relation enumerated above. We had obtained optimum filler volume as A1B1C1 and the flexural strength found experimentally was 71 MPa.

4. Conclusions

(1)In our study,we have prepared a experimental dental composite. Contemporary dental composite use hydroxyapatite and silica individually or in combination as fillers. Zirconia was introduced as nano filler. We have used three nanofillers, namely, hydroxyapatite, zirconia and silica in combination (conglomerate).

(2) It is observed that the influence of zirconia as a filler material on experimental dental composite is very significant and highest in our study, effect of zirconia on flexural strength being 44.15%. Statistical results (at a 95% confidence level) show that the zirconia, hydroxyapatite, silica affect the flexural strength by 44.15%, 26.97% and 12.48% of the experimental dental composites,

(3) Statistically designed experiments based on Taguchi methods were performed using L9 orthogonal arrays to analyse the flexural strength as a response variable. Conceptual S/N ratio and ANOVA approaches for data analysis drew similar conclusions.

(4) In this study, the analysis of the confirmation experiment for flexural strength has shown that Taguchi parameter design can successfully verify the optimum filler volume percentages ($A1B1C1$), which are hydroxyapatite = 5.8% ($A1$), zirconia = 11% ($B1$) and silica = 15.4% ($C1$).

(5) The confirmatory experimental tests results are matching with the optima value of flexural strength calculated.`

References

1. Wu YR, Chang CW, Ko CL, Wu HY, Chen WC. The morphological effect of calcium phosphates as reinforcement in methacrylate-based dental composite resins on mechanical strength through thermal cycling. Ceram Int 2017;43: 14389e94.
2. Wille S, Holken I, Haidarschin G, Adelung R, Kern M. Biaxial € flexural strength of new Bis-GMA/TEGDMA based composites with different fillers for dental applications. Dent Mater 2016;32:1073e8.
3. Ai M, Du Z, Zhu S, Geng H, Zhang X, Cai Q, et al. Composite resin reinforced with silver nanoparticleseladen hydroxyapatite nanowires for dental application. Dent Mater 2017;33:12e22
4. RamkumarYadav,MukeshKumar.Dental restorative composite materials: A review Journal of Oral Biosciences 61 (2019) 78e83
5. ASTM D 695-08, Standard test method for compressive properties of rigid plastics, 2008
6. Atai M, Nekoomanesh M, Hashemi S A and Amani S 2004 Dent. Mater. **20** 663
7. Bowen, R L 1964 J. Am. Dent. Assoc. **69** 482
8. Eliseu A M et al 2012 J. Contemp. Dent. Practice **13** 299
9. Manhart J, Kunzelmann K H, Chen H Y and Hickey R Dent. Mater. 2000 **16** 33
10. Mitra S B, Dong W U and Holmes B N 2003 J. Am. Dent. Assoc. **134** 1382
11. Ross P J 1996 Taguchi techniques for quality engineering (Singapore: McGraw-Hill International Editions)
12. Ruddell D E, Maloney M M and Thompson J Y 2002 Dent. Mater. **18** 72
13. Soh S, Sellinger M, Alan U J and Adrian Y 2006 Curr. Nanosci. **2** 373
14. Taguchi G 1990 Introduction to quality engineering (Tokyo: Asian Productivity Organisation)
15. Taira Miyasaka 1996 Dent. Mater. J. **15** 98

IOP Conf. Series: Materials Science and Engineering **810** (2020) 012064 doi:10.1088/1757-899X/810/1/012064

Optimization of Turning Parameters Using Taguchi's Method and Artificial Bee Colony Algorithm

Yadnyesh Bhor, Nikhil S Kakuste, Shreyash Gaynar

U.G. Student, Savitribai Phule Pune University, Pune, Nashik, India

Email Address; shreyashag7@gmail.com

Abstract Manufacturing industries are in need of various methods for obtaining the specified output parameters of a product. To fulfil such parameters in accordance to the specification, it is essential to quantitatively analyse the effect of each and every input functions on the output characteristics of the product. With ever increasing customer requirement in functionality and service of a product economically with predetermined quantity CNC machines are now finding application in majority of the industries to get further and better output on demand. In CNC machine, spindle speed, feed and depth of cut are most significant controllable parameters which majorly define the output characteristics of product like surface finish. Industries are now more interested in trends of optimization and automation which ease the process of overcoming constraints. This paper presents experimental study of such parameters of CNC turning centre to optimize the effect of cutting parameters on surface finish of mild steel material. The paper also illustrates how Artificial Bee Colony technique of optimization was used to achieve the results. The experiment is designed and performed by using concept of Design of Experiments (DOE).

Keywords: Artificial Bee Colony technique, Design of Experiments (DOE), Taguchi's Method

IOP Conf. Series: Materials Science and Engineering **810** (2020) 012064 doi:10.1088/1757-899X/810/1/012064

1. INTRODUCTION

A good understanding of machining process and its parameters is essential for evaluating the desired output parameters. While machining any workpiece the input parameters need to be selected in such a manner that their combination as a cumulative effort will result in the best obtainable output. The development of manufacturing technologies to improve machining and obtain high productivity has become a very important goal in modern industry. The challenge of modern machining industries is mainly focused on the achievement of high quality, in terms of work piece dimensional accuracy, surface finish, high production rate, less wear of the cutting tools, economy of machining in terms of cost saving and increase the performance of the product with reduced environmental impact. Quality in the form of surface finish is the most important criteria in many fields and it plays great importance in evaluating machining parameters While inspecting the finished surface of the product some major considerations are regarding the tool marks that are left behind, the maximum distance between the peak and valley along with the surface texture.In case of turning operation, the workpiece is rotated by the spindle at a given speed and the turning tool is advanced to take a depth of cut. Once the desired depth of cut is achieved, feed is given to the tool parallel to the spindle axis. The combination of speed of rotation, feed and depth of cut act as input parameters for the machine while performing turning operation. In order to establish an adequate functional relationship between the responses (such as surface roughness, cutting force, tool life/wear) and the cutting parameters (cutting speed, feed, depth of cut, nose radius, cutting time, etc.), a large number of tests are needed, requiring a separate set of tests for each and every combination of cutting tool and work piece material.

The paper deals with obtaining a functional relationship amongst input parameters that are speed, fed and depth of cut and output parameter surface roughness using linear and non-linear modes of regression analysis along with utilization of Taguchi's Method. The regression analysis will provide the ability to fit a curve for set of equations obtained by regression analysis. The fitness of the curve with respect to the equations are determined by co-relation coefficient, the higher the value of co-relation coefficient more is its suitability of fit. The equation of curve will act as an objective function for surface roughness. As the ultimate objective is to maximize the surface finish, minimization of surface roughness will be performed using Design of Experiments (D.O.E) method and Evolutionary optimization algorithm that are gradient free in nature to avoid local minima.

There are two objectives of this seminar study, to obtain optimal process parameters for a turning mild steel work piece on DX200-4B CNC turning machine by using a carbide cutting tool type TNMG 160408. The first is to apply Taguchi's method for D.O.E. to find out the best possible combination for turning a Mild Steel component. The second is to demonstrate the use of Artificial Bee Colony is optimization method in order to find out the optimum surface roughness with a particular combination of cutting parameters in a turning operation by multimodal type of objective optimization

1.1 Introduction to CNC machine: -

The introduction of CNC machine radically changed the manufacturing industry to produce complex curved profile as easy to cut as straight lines and also easy to produce intricate structures. With improved automation techniques of manufacturing processes and reduction on dependency of human operators led to development of CNC machines. These machines resulted into improvement over machine control, machine capacity and machine capability as well as the quality of the finished product. In CNC turning surface finish is the important factor that is significantly influenced by machining performance. CNC machines are capable of achieving reasonable accuracy, surface finish and very low processing time as compared to the conventional machine. Figure 1 shows the block diagram of the working principle of CNC machine.

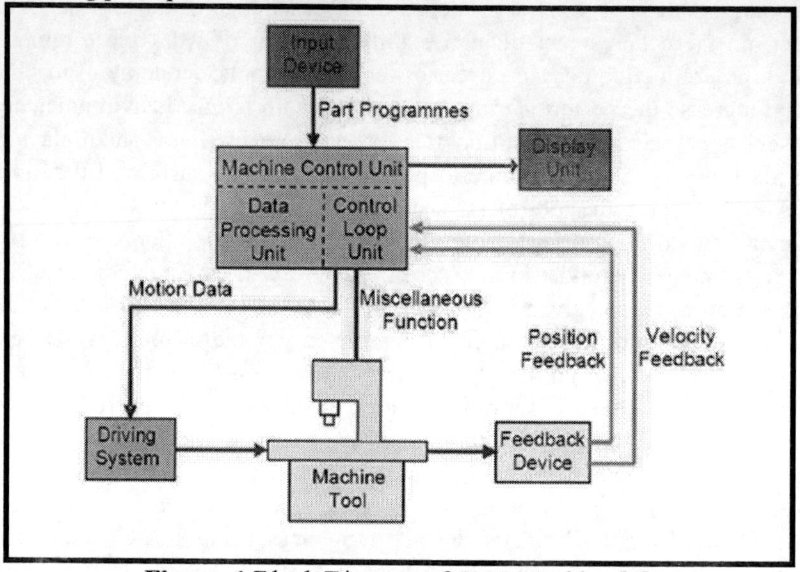

Figure. 1 Block Diagram of CNC Machine [7]

1.2 Introduction to Turning: -

Turning is a machining operation wherein material is removed from the outer periphery of the workpiece, resulting in reduction of diameter of the workpiece. Generally, turning operation is most widely performed on Lathe machines and hence they are also termed as Turning Machine. This study has considered three machine tool parameters: Spindle Speed for 4 levels and Feed, Depth of Cut for 3 levels.

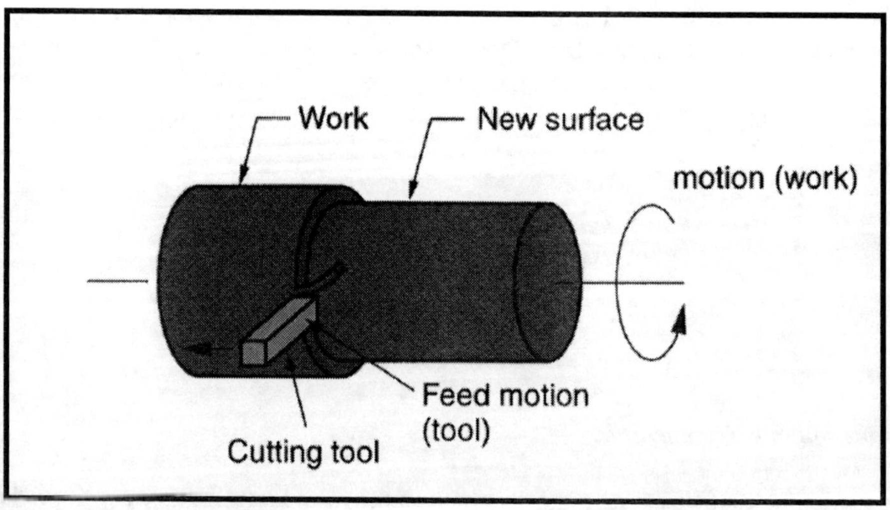

Figure. 2. Turning operation [4]

2.LITERATURE REVIEW

Table1: Literature Review

Sr. No.	Year	Author	Material	Input Parameter	Output Parameter	Method	Significant Parameter
[1]	2013	Mussa I. Mgwatu	Low Carbon Steel	Cutting Speed, Feed, DOC	Cycle Time, Production Cost	Lingo Non-Linear Software Package	Feed
[2]	2014	SayakMukherje, Anurag Kamal, Kaushik Kumar	SAE 1020	Speed, Feed, DOC	Material Removal Rate	Taguchi Technique of Orthogonal Array	DOC, Feed
[3]	2011	WeifengGao,SanyangLiu,Lingling Huang	-	No. of bees, search zone, objective function, constraint	Optimized solution	Artificial Bee Colony	Operation characteristics of bees are mimicked
[4]	2014	Jitendra Thakkar	SS 410	Cutting Speed, Feed, DOC	Surface Roughness and Material Removal Rate	ANOVA Method	Feed, DOC
[5]	2015	KrupalPawar and R. D. Palhade	High Speed Steel (M2)	Insert Nose Radius, Cutting Speed, Feed, DOC	Surface Roughness	Taguchi Technique and ANOVA Analysis	Feed Rate and Nose Radius

[6]	2014	G. Harinath Gowda, M. VenugopalGoud, K. DivyaTheja, M. Gunasekhar Reddy	EN-31	Cutting Speed, Spindle Speed, Feed, DOC	Force Fx, Temperature	MATLAB (ANN Module)	Speed, DOC
[7]	2010	M. Kaladhar, K. VenkataSubbaiah, Ch. Srinivasa Rao and K. Narayana Rao	AISI 202 (Austenitic Stainless Steel)	Cutting Speed, Spindle Speed, Feed, DOC	Surface Roughness	ANOVA Method	Feed

Mussa I. Mgwatu in his paper on "Machining Optimization and Operation allocation for NC lathe machines in a job shop manufacturing system" explains the economics of a manufacturing process in relation to the various types of times that are encountered during the machining operation. If the cycle time is minimized with the production time equal to cycle time, then slack times among machines can totally be eliminated. The study has also confirmed that if the minimized cycle time is treated as the constraint in the production cost model, then the production cost can further be reduced.The objective is solved for effective assignment of cutting operations for each machine and optimal cutting speed, feed rate, depth of cut for all rough and finish operations for different part features. [1]

SayakMukherje, Anurag Kamal, Kaushik Kumar in their paper on" Optimization of Material Removal Rate During Turning of SAE 1020 Material in CNC Lathe using Taguchi Technique" elaborate the Taguchi technique as a powerful tool for identification of affect of various process parameters based on orthogonal array (OA) experiments which provides much reduced variance for the experiments with an optimum setting of process control parameters.[2]

WeifengGao,SanyangLiu,Lingling Huang in their journal paper" A global best artificial bee colony algorithm for global optimization" explain the algorithm wherein a candidate source position that is produced and evaluated by the artificial bee, its performance is compared with that of the old one. If the new food source has a better quality of nectar amount the bees gather along it revising its source position. The number of cycles for which the bees will explore and exploit will depend on a predetermined number also called as the abandonment of bees from further exploration. [3]

Jitendra Thakkar, Mitesh Patel in their research article" A Review on Optimization of Process Parameters for Surface Roughness and Material Removal Rate for SS 410 Material During Turning Operation" investigated the effects of process parameters feed, speed, depth of cut and nose radius on surface finish and material removal rate(MRR) to obtain the optimal setting of these process parameters. The aim of experimental investigation is to evaluate the effects of the process parameters on AISI 304 austenitic stainless-steel work piece surface roughness and material removal rate by employing Taguchi's orthogonal array design and Analysis of Variance (ANOVA) using PVD coated Cermets tool on CNC lathe under dry environment. They conclude that most effected parameters to cutting condition are cutting speed, feed rate and depth of cut and they are easily controlled by operator at the machine at same time or simultaneously. [4]

KrupalPawar, R. D. Palhade in their journal paper" Multi-objective Optimization of CNC Turning Process Parameters for High Speed Steel (M2) Using Taguchi and ANOVA Method" determine the best design required for the use of a designed experiment. According to them Taguchi converts the objective function values to Signal-to-Noise ratio (S/N ratio) to measure the performance characteristics of the levels of control factors that affect the entire parametric space with a limited number of experiments. The four machining parameters considered in this study are cutting speed, feed rate, depth of cut, and insert nose radius. [5]

G.Harinath Gowda, M. VenugopalGoud , K. DivyaTheja, M. Gunasekhar Reddy I their journal paper " Optimal Selection Of Machining Parameters In CNC Turning Process Of EN-31 Using Intelligent Hybrid Decision Making Tools", explain the three main variables that affect the formation of the chips in cutting are the tool geometry, work and tool materials. The interaction between the tool and work material is also significant: this is often mentioned as the fourth main variable. All the above variables are tested for different network configurations, as per the value of performance error obtained, best model is identified and selected. The models are evaluated by calculating the percentage deviation using predicted values and actual values.[6]

M. Kaladhar, K. VenkataSubbaiah, C. Srinivasa Rao, K. Narayana Rao in their journal paper" Optimization of process parameters in Turning of AISI202 austenitic stainless steel" have observed a slight inclination of significance in case of interaction between speed and nose radius to influence surface roughness when compared with other interactions. These results are analyzed using ANOVA for the purpose of identifying the significant factors, which affects the surface roughness. [7]

3. PROBLEM STATEMENT & OBJECTIVES OF STUDY: -

Optimization of turning parameters is usually a complex task where the aspects such as domain knowledge of machining and the specification of machine tool capabilities is mandatory. The level of selection of input parameters is calculated on the basis of threshold value that signify the major difference of output response amongst the consecutive levels and determine the range of optimization. In a turning operation, it is important task to select the best combination of parameter levels for achieving high cutting performance i.e. surface roughness. The main aim of study is to find the optimal values of input machine parameters to get minimum surface roughness. Following are the objectives of the study:

1. To study the optimization and optimization technique.
2. To study the working of CNC machine parameters and their effects on output parameter.
3. To study regression modelling of experimental data in the form of linear or nonlinear regression form.
4. To establish the relationship equation between Surface Roughness and process parameters.
5. To obtain the optimum value of machine parameters to get global optimum result of output parameter i.e. Surface Roughness.

3.1 PARAMETER SELECTION: -

3.1.1 SELECTION OF MACHINE PARAMETERS: -

1. Speed: -

   Speed always refers to the spindle and the work piece. When it is stated in revolutions per minute (rpm) it tells their rotating speed.
2. Feed: -

   Feed always refers to the cutting tool, and it is the rate at which the tool advances along its cutting path. On most power-fed lathes, the feed rate is directly related to the spindle speed and is expressed in mm (of tool advance) per revolution (of the spindle), or mm/rev.
3. Depth of Cut: -

   Depth of cut is the thickness of the layer being removed (in a single pass) from the work piece or the distance from the uncut surface of the work to the cut surface, expressed in mm. It is important to note that the diameter of the work piece is reduced by two times the depth of cut because this layer is being removed from both sides of the work.
   Depth of cut = D_0-D_1/2 mm
   Here, D_0 and D_1 represent initial and final diameter (in mm) of the job respectively.

3.1.2 RESPONSE PARAMETER: -

- Surface Roughness: -
  Roughness is a measure of the texture of a surface. It is quantified by the vertical deviations of a real surface from its ideal form. If these deviations are large, the surface is rough; if they are small the surface is smooth. Roughness is typically considered to be the high frequency, short wavelength component of a measured surface. Surface roughness is an important measure of product quality since it greatly influences the performance of mechanical parts as well as production cost. Surface roughness has an impact on the mechanical properties like fatigue behavior, corrosion resistance, creep life, etc. surface roughness tester to measure surface roughness of work piece

3.2 EFFECT ON OUTPUT PARAMETER: -.

3.2.1. EFFECT OF INDIVIDUAL PARAMETER ON SURFACE ROUGHNESS (Ra): -

1. Spindle Speed: - Spindle speed is a significant parameter in the turning operation of CNC machine. It has a sizeable influence on the surface roughness of turning operation. As the spindle speed increases surface roughness also increases.

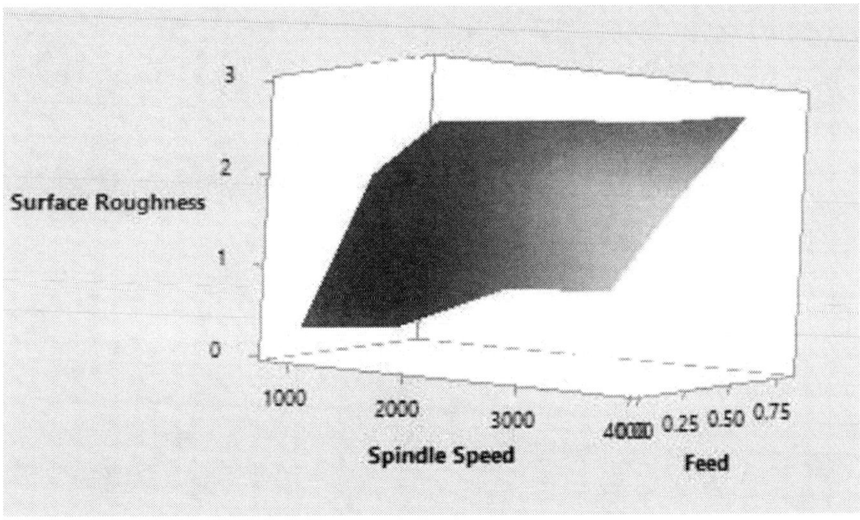

Figure. 3 Surface Roughness vs. Spindle Speed and Feed [10]

2. Feed: -_Feed has the maximum influence on the surface roughness. As the feed increases, the surface roughness also increases.

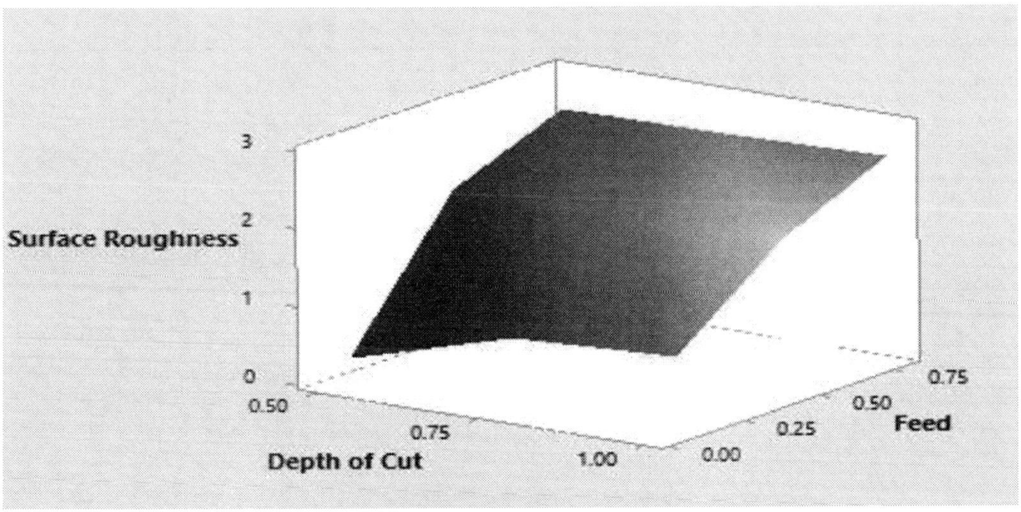

Figure. 4 Surface Roughness vs. Feed and Depth of Cut [10]

3. Depth of Cut (DOC): -

Depth of Cut has the second most influence on Surface Roughness after feed. As the Depth of Cut increases surface roughness also increases.

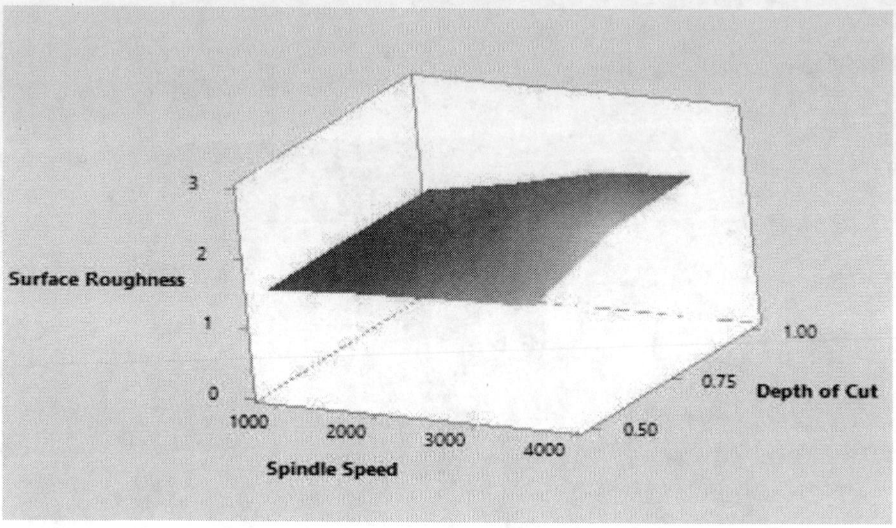

Figure. 5 Surface Roughness vs. Depth of Cut and Spindle Speed [10]

3.3 SELECTION OF LEVELS OF PARAMETERS:

Table 2: Levels of Parameters [10]

Cutting parameters	Designation	Units	Operational range	Level 1	Level 2	Level 3	Level 4
Spindle speed	**N**	RPM	900-3600	900	1800	2700	3600
Feed	**F**	mm/rev	0.1-0.8	0.1	0.45	0.8	-
Depth of cut	**D**	mm	0.5-1.0	0.5	0.75	1.0	-

Table 3: Range of Parameters [10]

CUTTING PARAMETERS	RANGE
Spindle Speed	50-4000 RPM
Feed	0.1-0.8 mm/rev
Depth of cut	0.1-1.5mm

IOP Conf. Series: Materials Science and Engineering **810** (2020) 012064 doi:10.1088/1757-899X/810/1/012064

4.CASE STUDY:

4.1 EXPERIMENTAL DATA: -

This experimental data is obtained by performing full factorial Design of Experiment (DOE).

Table 4: Experimental Data after S/N Processing

Sr. No.	Expt. No.	Spindle Speed	Feed	Depth of Cut	Surface Roughness	S/N Ratio
		x_1 (RPM)	x_2 (mm/rev)	x_3 (mm)	Y_1 (μm)	
1.	**E-1**	900	0.1	1	0.282	-10.995
2.	**E-2**	900	0.1	0.75	0.261	-11.6672
3.	**E-3**	900	0.1	0.5	0.33	-9.62972
4.	**E-4**	900	0.45	1	1.855	5.366878
5.	**E-5**	900	0.45	0.75	1.93	5.711146
6.	**E-6**	900	0.45	0.5	1.827	5.234771
7.	**E-7**	900	0.8	1	2.404	7.618689
8.	**E-8**	900	0.8	0.75	2.428	7.704974
9.	**E-9**	900	0.8	0.5	2.332	7.317244
10.	**E-10**	1800	0.1	1	0.542	-5.32001
11.	**E-11**	1800	0.1	0.75	0.555	-5.11414
12.	**E-12**	1800	0.1	0.5	0.189	-14.4708
13.	**E-13**	1800	0.45	1	2.051	6.239313
14.	**E-14**	1800	0.45	0.75	2.001	6.024942
15.	**E-15**	1800	0.45	0.5	2.011	6.068241
16.	**E-16**	1800	0.8	1	2.628	8.569175
17.	**E-17**	1800	0.8	0.75	2.433	7.722842
18.	**E-18**	1800	0.8	0.5	2.469	7.850422
19.	**E-19**	2700	0.1	1	1.39	2.860296

20.	**E-20**	2700	0.1	0.75	1.233	1.819262
21.	**E-21**	2700	0.1	0.5	0.199	-14.0229
22.	**E-22**	2700	0.45	1	1.998	6.01191
23.	**E-23**	2700	0.45	0.75	1.969	5.844914
24.	**E-24**	2700	0.45	0.5	2.034	6.167019
25.	**E-25**	2700	0.8	1	2.596	8.286094
26.	**E-26**	2700	0.8	0.75	2.546	8.117168
27.	**E-27**	2700	0.8	0.5	2.562	8.171583
28.	**E-28**	3600	0.1	1	1.498	3.510236
29.	**E-29**	3600	0.1	0.75	1.322	2.424629
30.	**E-30**	3600	0.1	0.5	0.214	-13.3917
31.	**E-31**	3600	0.45	1	1.981	5.93769
32.	**E-32**	3600	0.45	0.75	1.924	5.684101
33.	**E-33**	3600	0.45	0.5	1.928	5.702141
34.	**E-34**	3600	0.8	1	2.744	8.767682
35.	**E-35**	3600	0.8	0.75	2.782	8.887143
36.	**E-36**	3600	0.8	0.5	2.729	8.720071

4.2 METHODOLOGY: -

4.2.1 Taguchi's Method for D.O.E.

The method states that the amount of experiments required to perform in accordance to full factorial design method are generally not feasible due to its large population. To overcome this only a certain amount of experiments is needed so as to imitate the same effect as that obtained from full factorial design method. The limited number of experiments to be performed in a preset combination to deliver the same flavor as Full Factorial Design method are evaluated as per Taguchi's Orthogonal Array. Although as the full factorial design of experiments is already available for a combination 4 levels of speed, and 3 levels of fed and depth of cut as variables from [14] the number of experiments that have been performed are 36. Therefore, proceeding further calculations in accordance to Taguchi's method for D.O.E.

Post the conduction of experiments, a Signal to Noise ratio is calculated the S/N ratio estimates the effect of Desirable controllable parameters with respect to Undesirable non-controllable parameters. In case of the experiment the desirable controllable parameters are speed, feed and depth of cut respectively, whereas the undesirable and non-controllable parameters are stiffness of spindle, degree of vibration, etc.

The objective function of the study is Surface Roughness (Ra value), it is essential to minimize the objective function so as to obtain the maximum surface finish. This is done by substituting the value of S/N ratio as:

$$\frac{S}{N} = 10 * (\log_{10}[\![Ra^2)]\!]$$

The greater the S/N ratio the greater is the controllability of input machine parameters. For each combination of each level of variables the average of S/N ratio is obtained individually indicated in Table 5, Table 6 and Table 7 for speed, feed, and depth of cut respectively. Once the average of S/N ratio is obtained a relationship matrix is to be developed indicating the average S/N ratio for each level of input parameter i.e. speed, feed and depth of cut, Refer Table 8.

Table 5: Speed data rearranged in accordance to S/N ratio of individual level

No. of combinations	**s/n1**	**s/n2**	**s/n3**	**s/n4**
Speed level (R.P.M)	900	1800	2700	3600
1	-10.995	-5.32001	2.860296	3.510236
2	-11.6672	-5.11414	1.819262	2.424629
3	-9.62972	-14.4708	-14.0229	-13.3917
4	5.366878	6.239313	6.01191	5.93769
5	5.711146	6.024942	5.884914	5.684101
6	5.234771	6.068241	6.167019	5.702141
7	7.618689	8.569175	8.286094	8.767682
8	7.704974	7.722842	8.117168	8.887143
9	7.317244	7.850422	8.171583	8.720071
Sum	6.661774	17.57002	33.29531	36.24197
Mean	0.740197	1.952224	3.699478	4.026885

Table 6: Feed data rearranged in accordance to S/N ratio of individual level

No. of combinations	**s/n1**	**s/n2**	**s/n3**
Feed level(mm/rev)	0.1	0.45	0.8
1	7.618689	5.366878	-10.995
2	7.704974	5.711146	-11.6672
3	7.317244	5.234771	-9.62972
4	8.569175	6.239313	-5.32001
5	7.722842	6.024942	-5.11414
6	7.850422	6.068241	-14.4708
7	8.286094	6.01191	2.860296
8	8.117168	5.884914	1.819262
9	8.171583	6.167019	-14.0229
10	8.767682	5.93769	3.510236
11	8.887143	5.684101	2.424629
12	8.720071	5.702141	-13.3917
Sum	97.73309	70.03307	-73.9971

Mean	8.144424	5.836089	-6.16642

Table 7: D.O.C data rearranged in accordance to S/N ratio of individual level

No. of combinations	s/n1	s/n2	s/n3
DOC level(mm)	1	0.75	0.5
1	-10.995	-11.6672	-9.62972
2	5.366878	5.711146	5.234771
3	7.618689	7.704974	7.317244
4	-5.32001	-5.11414	-14.4708
5	6.239313	6.024942	6.068241
6	8.569175	7.722842	7.850422
7	2.860296	1.819262	-14.0229
8	6.01191	5.884914	6.167019
9	8.286094	8.117168	8.171583
10	3.510236	2.424629	-13.3917
11	5.93769	5.684101	5.702141
12	8.767682	8.887143	8.720071
Sum	46.85293	43.19979	3.716343
Mean	3.904411	3.599983	0.309695

The relationship matrix represents the average S/N ratio of each level of speed, feed and depth of cut. The levels of parameters speed, feed and depth of cut are as indicated in Table 2. For individual parameter level, the difference amongst the greatest and least value for that parameter itself is calculated, indicated in Table 8. The magnitude of difference indicates the significance of that parameter on the objective function. The significance of each parameter on objective function is ranked in accordance to their magnitude of difference amongst the maximum and minimum average S/N ratio, indicated in Table 10.
The result indicates that the effect of feed on surface roughness is maximum followed by depth of cut and speed. The effect of feed on surface roughness is almost five times that of the remaining parameters.

Table 8: Relationship matrix amongst parameters to evaluate rank of significance

Parameter	level1	level2	level3	level4	difference	Rank of significance
Speed (R.P.M)	0.74019708	1.952224	3.699478	4.026885	3.286688	3
Feed (mm/rev)	-6.166424	5.836089	8.144424		14.31085	1
DOC (mm)	3.90441096	3.599983	0.309695		3.594716	2

Each level of a parameter is ranked in terms of its magnitude of average S/N ratio in Table 8. The overall results obtained indicate that the best level of parameters for machining the workpiece are indicated with red color marking. The cumulative of all the best levels of parameters lead to the best possible combination for turning the workpiece. The best combination is shown on Table 10, along with the resultant Surface Finish.

Table 9: Rank matrix

	level1	level2	level3	level4
Speed	4	3	2	1
Feed	3	2	1	
Doc	1	2	3	

Table 10: Result table for D.O.E.

BEST COMBINATION	
Speed (R.P.M)	3600
Feed (mm/rev)	0.8
DOC (mm)	1
RESULTANT Ra= 2.744 micron	

4.2.2 Regression Modelling

The experimental matrix that was adopted in this study in the form of actual values is shown in the Table 2. To study the effect of process parameters, i.e. Spindle Speed, Feed, Depth of Cut, on response parameters, i.e. Surface Roughness (R_a), a first order and second-order polynomial response is fitted into the following equation.

Let, N=x1,
F=x2 &
D=x3

4.2.2.1 Linear Regression

From experimental data, following values are collected:

n=36, $\sum x_1$ =81000, $\sum x_2$ =16.2, $\sum x_3$ =27, $\sum y$ =62.2, $\sum x_1x_2$ =36450, $\sum x_1x_3$ =60750, $\sum x_2x_3$ =12.15, $\sum x_1^2$ =218700000, $\sum x_2^2$ =10.23, $\sum x_3^2$ =21.75, $\sum x_1y$ =145328.4, $\sum x_2y$ =35.90295, $\sum x_3y$ =47.419

A B = C

$$\begin{bmatrix} n & \sum x_1 & \sum x_2 & \sum x_3 \\ \sum x_1 & \sum x_1^2 & \sum x_1x_2 & \sum x_1x_3 \\ \sum x_2 & \sum x_1x_2 & \sum x_2^2 & \sum x_2x_3 \\ \sum x_3 & \sum x_1x_3 & \sum x_2x_3 & \sum x_3^2 \end{bmatrix} \begin{bmatrix} a_0 \\ a_1 \\ a_2 \\ a_3 \end{bmatrix} = \begin{bmatrix} \sum y \\ \sum x_1y \\ \sum x_2y \\ \sum x_3y \end{bmatrix}$$

The matrices were solved on Scilab Software

```
--> a=[36, 81000, 16.2, 27; 81000, 218700000, 36450, 60750;16.2, 36450, 10.23, 12.15; 27, 60750, 12.15, 21.75;]
 a  =

   36.      81000.      16.2     27.
   81000.   2.187D+08   36450.   60750.
   16.2     36450.      10.23    12.15
   27.      60750.      12.15    21.75


--> c=[62.2; 145328.4; 35.90295; 47.419]
 c  =

   62.2
   145328.4
   35.90295
   47.419

--> a\c
 ans  =

  -0.199888
   0.0001476
   2.6914796
   0.5126667
```

R^2=87.90%

a_0= -0.199888, a_1=0.0001476, a_2=2.69147965, a_3=0.5126667

The equation obtained by linear regression is:

$$Y = (-0.199888) + (0.0001476 * x1) + (2.6914796 * x2) + (0.5126667 * x3)$$

4.2.2.2 Non-Linear Regression

From experimental data, following values are collected:

A B = C

$$\begin{bmatrix} n & \Sigma x_1 & \Sigma x_2 & \Sigma x_3 & \Sigma x_4 & \Sigma x_5 & \Sigma x_6 & \Sigma x_7 & \Sigma x_8 & \Sigma x_9 \\ \Sigma x_1 & \Sigma x_1^2 & \Sigma x_1x_2 & \Sigma x_1x_3 & \Sigma x_1x_4 & \Sigma x_1x_5 & \Sigma x_1x_6 & \Sigma x_1x_7 & \Sigma x_1x_8 & \Sigma x_1x_9 \\ \Sigma x_2 & \Sigma x_1x_2 & \Sigma x_2^2 & \Sigma x_2x_3 & \Sigma x_2x_4 & \Sigma x_2x_5 & \Sigma x_2x_6 & \Sigma x_2x_7 & \Sigma x_2x_8 & \Sigma x_2x_9 \\ \Sigma x_3 & \Sigma x_1x_3 & \Sigma x_2x_3 & \Sigma x_3^2 & \Sigma x_3x_4 & \Sigma x_3x_5 & \Sigma x_3x_6 & \Sigma x_3x_7 & \Sigma x_3x_8 & \Sigma x_3x_9 \\ \Sigma x_4 & \Sigma x_1x_4 & \Sigma x_2x_4 & \Sigma x_3x_4 & \Sigma x_4^2 & \Sigma x_4x_5 & \Sigma x_4x_6 & \Sigma x_4x_7 & \Sigma x_4x_8 & \Sigma x_4x_9 \\ \Sigma x_5 & \Sigma x_1x_5 & \Sigma x_2x_5 & \Sigma x_3x_5 & \Sigma x_4x_5 & \Sigma x_5^2 & \Sigma x_5x_6 & \Sigma x_5x_7 & \Sigma x_5x_8 & \Sigma x_5x_9 \\ \Sigma x_6 & \Sigma x_1x_6 & \Sigma x_2x_6 & \Sigma x_3x_6 & \Sigma x_4x_6 & \Sigma x_5x_6 & \Sigma x_6^2 & \Sigma x_6x_7 & \Sigma x_6x_8 & \Sigma x_6x_9 \\ \Sigma x_7 & \Sigma x_1x_7 & \Sigma x_2x_7 & \Sigma x_3x_7 & \Sigma x_4x_7 & \Sigma x_5x_7 & \Sigma x_6x_7 & \Sigma x_7^2 & \Sigma x_7x_8 & \Sigma x_7x_9 \\ \Sigma x_8 & \Sigma x_1x_8 & \Sigma x_2x_8 & \Sigma x_3x_8 & \Sigma x_4x_8 & \Sigma x_5x_8 & \Sigma x_6x_8 & \Sigma x_7x_8 & \Sigma x_8^2 & \Sigma x_8x_9 \\ \Sigma x_9 & \Sigma x_1x_9 & \Sigma x_2x_9 & \Sigma x_3x_9 & \Sigma x_4x_9 & \Sigma x_5x_9 & \Sigma x_6x_9 & \Sigma x_7x_9 & \Sigma x_8x_9 & \Sigma x_9^2 \end{bmatrix} \begin{bmatrix} a_0 \\ a_1 \\ a_2 \\ a_3 \\ a_4 \\ a_5 \\ a_6 \\ a_7 \\ a_8 \\ a_9 \end{bmatrix} = \begin{bmatrix} \Sigma y \\ \Sigma x_1y \\ \Sigma x_2y \\ \Sigma x_3y \\ \Sigma x_4y \\ \Sigma x_5y \\ \Sigma x_6y \\ \Sigma x_7y \\ \Sigma x_8y \\ \Sigma x_9y \end{bmatrix}$$

Similarly, by solving on Scilab, we get:

$$A=\begin{bmatrix} 36 & 81000 & 16.2 & 27 & 36450 & 12.15 & 60750 & 218700000 & 10.23 & 21.75 \\ 81000 & 218700000 & 36450 & 60750 & 98415000 & 27337.5 & 164025000 & 6.561E+11 & 23017.5 & 48937.5 \\ 16.2 & 36450 & 10.23 & 12.15 & 23018.5 & 7.6725 & 27337.5 & 98415000 & 7.2495 & 9.7875 \\ 27 & 60750 & 12.15 & 21.75 & 27337.5 & 9.7875 & 48937.5 & 164025000 & 7.6725 & 18.5625 \\ 36450 & 98415000 & 23018.5 & 27337.5 & 62147250 & 17623.125 & 73811250 & 2.95E+11 & 16311.375 & 22021.875 \\ 12.15 & 27337.5 & 7.6725 & 9.7875 & 17623.125 & 6.180625 & 22021.875 & 73811250 & 5.437125 & 8.353125 \\ 60750 & 164025000 & 27337.5 & 48937.5 & 73811250 & 22021.875 & 132131250 & 4.921E+11 & 17263.125 & 41765.625 \\ 218700000 & 6.561E+11 & 98415000 & 164025000 & 2.95E+11 & 73811250 & 4.921E+11 & 2.09E+15 & 62147250 & 132131250 \\ 10.23 & 23017.5 & 7.2495 & 7.6725 & 16311.375 & 5.437125 & 17263.125 & 62147250 & 5.40848 & 6.1806 \\ 21.75 & 48937.5 & 9.7875 & 18.5625 & 22021.875 & 8.353125 & 41765.625 & 132131250 & 6.1806 & 16.547 \end{bmatrix}$$

$$B=\begin{bmatrix} a_0 \\ a_1 \\ a_2 \\ a_3 \\ a_4 \\ a_5 \\ a_6 \\ a_7 \\ a_8 \\ a_9 \end{bmatrix}, C=\begin{bmatrix} 62.177 \\ 145328.4 \\ 35.903 \\ 47.42 \\ 82505 \\ 27.062 \\ 111276 \\ 401646600 \\ 24.4586 \\ 38.7 \end{bmatrix}$$

R^2=94.80%

$\boldsymbol{a_0}$ = *-1.612,* $\boldsymbol{a_1}$ = *0.0001,* $\boldsymbol{a_2}$ = *7.135,* $\boldsymbol{a_3}$ = *2.547,* $\boldsymbol{a_4}$ = *-0.0002,* $\boldsymbol{a_5}$ = *-1.786,* $\boldsymbol{a_6}$ = *0.00033,* $\boldsymbol{a_7}$ = *-2.17E-08,* $\boldsymbol{a_8}$ = *-2.840,* $\boldsymbol{a_9}$ = *-1.317*

$$Y = (-1.612) + (0.0001 * x1) + (7.135 * x2) + (2.547 * x3) + (-0.0002 * x1 * x2) + (-1.786 * x2 * x3) + (0.00033 * x1 * x3) + (-0.0000000217 * x1 * x1) + (-2.84 * x2 * x2) + (-1.317 * x3 * x3)$$

As the second order polynomial response for the data has a greater value of co-relation co-efficient i.e. R^2. Therefore, selecting the second order polynomial as response equation to input machine parameters.

$$y = a_0 + \sum_{i=1}^{p} a_i x_i + \sum_{i=1}^{p} a_{ii} x_i^2 + \sum_{j>1}^{p} a_{ij} x_i x_j$$

Where y is the response and x_i are experimental levels of p quantitative variables. The coefficients a_0 is the free term, the coefficients a_i are the linear terms, the coefficients $a_{ij}$are the quadratic terms and the coefficients $a_{ij}$are the interaction terms. Matrices (A) and (B) are then derived by determining the value of the coefficients using the Cramer's Rule for the observations collected as shown in Table 2, for Surface Roughness (R_a).

4.2.3 Artificial Bee Colony:

In ABC, the colony of artificial bees contains three groups of bees: employed bees associated with specific food sources, onlooker bees watching the dance of employed bees within the hive to choose a food source, and scout bees searching for food sources randomly. Both onlookers and scouts are also called unemployed bees. Initially, all food source positions are discovered by scout bees. Thereafter, the nectar of food sources is exploited by employed bees and onlooker bees, and this continual exploitation will ultimately cause them to become exhausted. Then, the employed bee which was exploiting the exhausted food source becomes a scout bee in search of further food sources once again. In other words, the employed bee whose food source has been exhausted becomes a scout bee. In ABC, the position of a food source represents a possible solution to the problem and the nectar amount of a food source corresponds to the quality (fitness) of the associated solution. The number of employed bees is equal to the number of food sources (solutions) since each employed bee is associated with one and only one food source.
In summary, the ABC algorithm,

i. is inspired by the foraging behaviour of honeybees,

ii. is a global optimization algorithm,

iii. has been initially proposed for numerical optimization,

iv. can be also used for combinatorial optimization problems,

v. can be used for unconstrained and constrained optimization problems,

vi. employs only three control parameters (population size, maximum cycle number and limit) that are to be predetermined by the user,

vii. is quite simple, flexible and robust

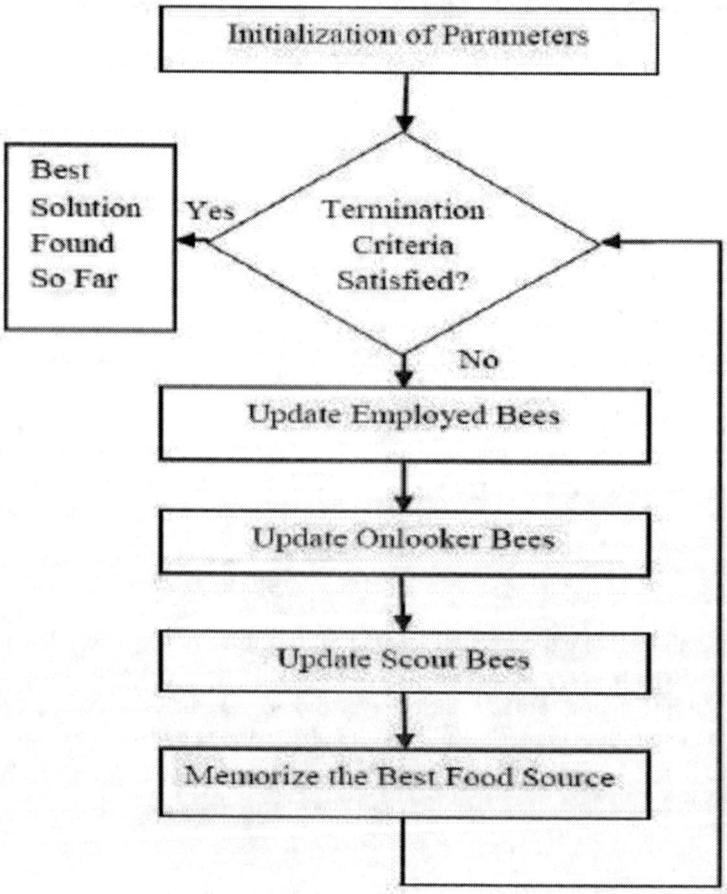

Figure. 6 Flowchart of A.B.C.

Figure. 6 Flowchart of A.B.C.

4.3 Results:

The result obtained after using artificial bee colony method, determine that the optimum combination of input parameters Spindle speed, Feed and Depth of cut to be selected are depicted in Table 11. and 12. Table 11 indicates the results obtained from linear type of regression. Table 12 indicates the results obtained using non-linear regression. The optimum combination of all the input parameters would result in a best output surface roughness value of 0.050001 micron.

Table 11: Result Table: Linear Regression

BEST COMBINATION	
Speed (R.P.M)	50
Feed (mm/rev)	0.1
DOC (mm)	0.1
RESULTANT Ra= 0.127907 micron	

Table 12: Result Table: Non-Linear Regression

BEST COMBINATION	
Speed (R.P.M)	2112.225
Feed (mm/rev)	0.208031
DOC (mm)	0.1
RESULTANT Ra= **0.050001** micron	

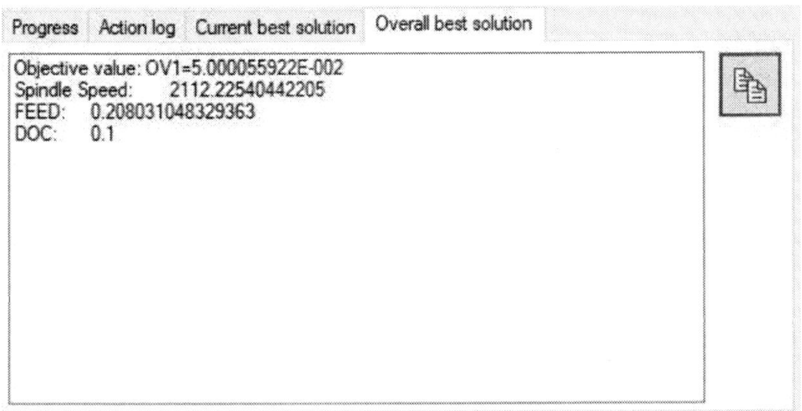

7 Output window showing final result

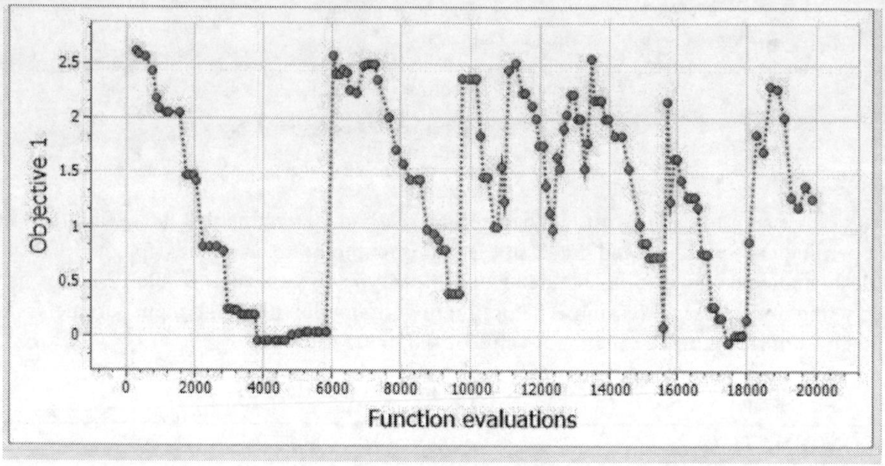

Figure. 8 Surface Roughness value (Ra) vs. No. of iterations

5. CONCLUSION:

1. According to Taguchi's method, the result obtained is best possible combination of desirable input parameters that can be set to obtain minimum surface roughness.
2. The result obtained is viable only when the input parameters operate in the predefined levels. Better surface roughness value is evident in Table 4, although the roughness value indicated has a S/N ratio below par compared to other values.
3. The result of surface roughness(Ra) obtained by Artificial Bee Colony can be defined as the global minima, wherein the best possible parameters combination of Spindle speed, feed and depth of cut required is indicated in Table12.
4. If a comparison is to be made between the results obtained by Taguchi's Method and that obtained by Artificial Bee Colony method, the improvement in results in percentage is 98.17%. The significant improvement in result can be attributed to the difference of local and global minima that have been obtained while following the numerical methods.
5. This study shows that for the linear regression modelled, the optimum conditions of variable parameters are the combination of lower boundary condition. But for the Non-linear regression it should vary as per initial guess assumed.
6. The study shows that the FEED has the maximum effect on the variable parameter i.e. surface roughness followed by Depth of Cut and Spindle Speed.

References

[1.] M.I. Mgwatu, "Machining Optimization and Operation Allocation for CNC Lathe Machines in a Job Shop Manufacturing System," in *Journal of Engineering Science and Technology,* vol. 8, no. 4 (2013) 482 – 492

[2.] S. Mukherjee, A. Kamal, K. Kumar, "Optimization of Material Removal Rate during Turning of SAE 1020 Material in CNC Lathe using Taguchi Technique," in *Procedia Engineering 97,* no.1 (2014) 29 – 35

[3.] W. Gao, S. Liu, L. Huang,"A global best artificial bee colony algorithm for global optimization," in *Journal of Computational and Applied Mathematics*, no. 236(2012) 2741-2753

[4.].J.Thakkar, "A Review on Optimization of Process Parameters for Surface Roughness and Material Removal Rate for SS 410 material during Turning Operation," in*Journal of* *engineering research and applications*, ISSN : 2248-9622, Vol. 4, Issue 2(version 1), February 2014, pp.235-242

[5.]. K. Pawar, R. Palhade, "Multi-Objective Optimization of CNC Turning Process Parameters for High Speed Steel (M2) using Taguchi and ANOVA method," in *International Journal of Hybrid Information Technology*, vol.8, no.4 (2015), pp.67-80

[6.]. G.H. Gowda, M.V. Goud, K.D. Theja, M.G.Reddy, "Optimal Selection of Machining Parameters in CNC Turning Process of EN-31 using Intelligent Hybrid Decision Making Tools," in *Procedia Engineering 97 (2014) 125 – 133*

[7.]. M. Kaladhar, K.V. Subbaiah, C.S. Rao and K.N. Rao, "Optimization of Process Parameters in Turning of AISI202 Austenitic Stainless Steel," in *ARPN Journal of Engineering and Applied Science,* vol. 5, no. 9, September 2010 ISSN 1819-6608

[8.]. K. Deb, "Optimization for Engineering Design: Algorithms and Examples," in *1st Edition, PHI publication*, ISBN-978-81-203-4678-9.

[9.].S. K. Saini, S. K. Pradhan, "Soft Computing Techniques for the Optimization of Machining Parameter in CNC Turning Operation," in *IJETAE,* ISSN 2250-2459, ISO 9001:2008 certified journal, volume 4, issue 3 march 2014

[10.]. N. Sharma, "Optimization of CNC Machine Parameters using Simulated Annealing Optimization Method," a Seminar Report submitted to *S.P.P.U. Pune* (2016-17), Page 11,12, Table 4.

IOP Conf. Series: Materials Science and Engineering **810** (2020) 012065 doi:10.1088/1757-899X/810/1/012065

Parametric review on Fuel Cells and their Applications

Sumeet Shanbhag[a], Girish Joshi[b]
[a] Undergraduate Student, Mechanical Engineering department, Mukesh Patel School of Technology Management and Engineering, Mumbai, India.

[b] Assistant Professor, Mechanical Engineering department, Mukesh Patel School of Technology Management and Engineering, Mumbai, India.

E-mail address: sumeetdineshshanbhag@gmail.com, girish.joshi@nmims.edu

Abstract. This study aims to review the issues affecting the long term performance and the life span of the fuel cell in accordance with the various surveys of the currently available. According to current research, parameters such as temperature, pressure along with other issues such as fuel and oxidant starvation (stoichiometric effect), corrosion, poor water management, humidity, and uncontrolled chemical reactions are some of the reasons leading to poor performance in the fuel cell. Poor water management can either lead to flooding or dehydration, both of which are extremely detrimental for the longevity of the fuel cell since the former facilitates corrosion of electrodes, membrane and catalyst layers whereas the latter leads to shrinkage of the membrane. Also, contamination of fuel cell membranes due to corrosion products or any impurities from outside leads to the poisoning of the cell. The construction of fuel cells in the future taking into consideration all these issues and mechanisms can lead to a performance-enhanced and long-lasting fuel cell.

1. Introduction

The evolution pertaining to fuel cells date back to the 1800's with Sir William Grove being recognized as the first one to develop the model. Various attempts were being made by scientists to develop various cells with different electrolytes and other variations in the parameters throughout the remainder of the century. Further developments also took place in the early 1900s which laid the foundation for the advancements in fuel cells. In 1959, Francis T. Bacon revealed the first fully operational fuel cell. NASA used proton exchange membrane fuel cells (PEMFC) as part of their Gemini space programs in 1960 for a few missions. At that time, hydrogen and oxygen were used as the two main reactant gases and were not a commercially viable option as the production of these gases was an extremely expensive process. Due to the energy crisis in 1973, there was a boost in the development of the fuel cell with NASA taking a particularly keen interest in researching it. The research has continued and many varieties of fuel cells have been successfully developed and are used in an extensive number of applications.

The fuel cell working is based on converting chemical energy into electrical energy. Due to the absence of any moving parts in this device, the efficiency of this device is much more than the conventional engines such as the spark ignition and combustion ignition/Diesel engines. In spite of high system efficiencies and the benefits to the environment that come along with it due to its clean emission, it is very difficult to develop fuel cells which can be used as industrial products. They are not commercially viable yet and are still not ready to be used on mass scale. These problems arise due to its failure to compete with the existing technology due to the absence of appropriate materials and the manufacturing processes presently existing, thus increasing the cost of electricity per kWh [1].

Working of Fuel Cell:-
A fuel cell comprises of four basic components, two electrodes namely the cathode and anode, an electrolyte as shown in figure 1. There is an external circuit which connects the anode and the cathode. Hydrogen is used as fuel for most of the fuel cells and sufficient oxygen must also be provided for the oxidation processes to occur. The hydrogen enters the fuel cell from anode side and at the cathode, it is divided into electron and proton, where the former are forced to travel through the external circuit, thus generating electricity on its way while the latter travels through the electrolyte towards the cathode. The external source may be connected to a bulb or any other load that requires electricity to operate. The electrons then recombine with the protons and oxygen to form water at the cathode. An attractive feature of fuel cells is its ability to produce less harmful products and generation of electricity with very less pollution and the formation of water as a byproduct.

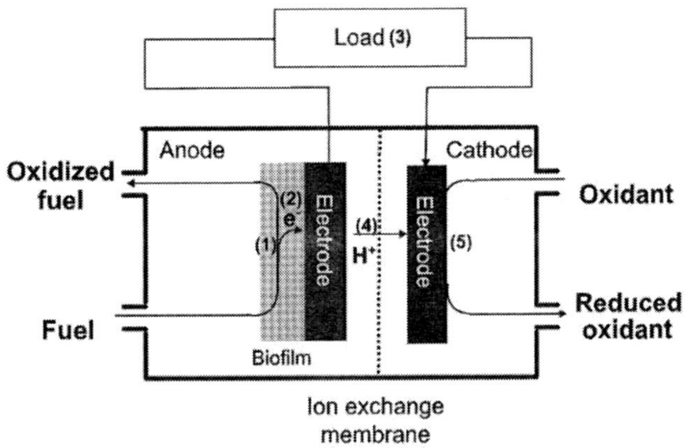

Figure 1. Schematic Diagram of a Fuel Cell

Generally, a fuel cell stack is used because a single fuel cell produces very tiny amount of DC current by itself. The main intention of a fuel cell is to yield an electric current which can be used outside the fuel cell for doing any kind of work pertaining to electricity, such as illumination of a bulb or for providing power to an electric motor. This electricity eventually returns back to the fuel cell completing an electric circuit. The chemical reactions taking place at the cathode, anode and the overall chemical reactions are listed below.
At anode: $H_2 \rightarrow 2H^+ + 2e^-$
At cathode: $(½)O_2 + 2H^+ + 2e^- \rightarrow H_2O$
Overall reaction: $H_2 + (½)O_2 \rightarrow H_2O$
Chemical energy is converted into electrical energy in galvanic cells. A fuel cell can also be considered as a galvanic cell, like a battery. The consumption rate of oxygen and hydrogen can be determined by the electric current load. But in an actuality, a variety of electrical loads are applied to the fuel cell [1].
There are various types of fuel cells available in market and they are mainly classified into two categories based on the temperature. The fuel cells operating on low temperature are proton exchange membrane, direct methanol, alkaline, phosphoric acid and biofuel cells. Molten carbonate fuel cells and solid oxide fuel cells have high operating temperatures. A regenerative fuel cell is a new type of fuel cell and still under development and also operates at high temperatures.

There are several parameters that affect the performance of the fuel cell and even the slightest change in these parameters lead to noticeable changes in the performance of the fuel cell. Parameters such as temperature, pressure, humidity among other parameters play a crucial role in the overall functioning of fuel cell. Also providing the fuel cells with suitable catalysts speeds up the reaction.

2. Types of Fuel cells

Please follow these instructions as carefully as possible so all articles within a conference have the same style to the title page. This paragraph follows a section title so it should not be indented.

2.1. Proton Exchange Membrane Fuel Cell:-

It is also called as Polymer electrolyte membrane as in figure 2, these fuel cell provides high power density and the advantages such as low weight and volume as compared to other types of fuel cells. These fuel cells also employ porous carbon electrodes containing a platinum catalyst and a solid polymer as the electrolyte. Hydrogen, water and oxygen from air are required in order for the polymer electrolyte membrane fuel cells to operate. Polymer electrolyte membrane fuel cells are usually fueled with pure hydrogen provided from onboard reformers or storage tanks. Polymer electrolyte membrane fuel cells are mainly used for transportation applications such as cars and buses and they operate at relatively low temperatures [2].

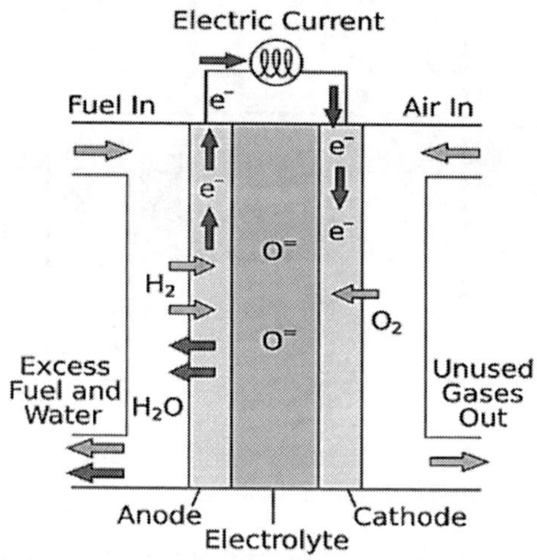

Figure 2. Schematic Diagram of Proton Exchange Membrane Fuel Cell

2.2. Direct Methanol Fuel Cell:-

These fuel cells are powered by pure methanol. Since methanol has higher energy density than hydrogen but a lower density than fuels such as gasoline and diesel; methanol being similar liquid to gasoline, is also easier for transportation and stream to the public using existing infrastructure. The methanol used in this fuel cell is mixed with steam which is then fed directly to the fuel cell anode. Direct methanol fuel cell technology is comparatively novel as compared to other hydrogen fuel cells [3].

2.3. Alkaline Fuel Cell:-

These fuel cells employ a large number of non-precious metals as a catalyst at the cathode and anode and a solution of potassium hydroxide in water as the electrolyte, thus they are also economical since they don't need to use costly platinum catalysts. The reason for high performance of these fuel cells is due to the faster rate at which the chemical reactions take place inside the cell. However, they are found to be easily poisoned by carbon dioxide; a small amount of carbon dioxide in the air can affect the operation of the fuel cell [4].

2.4. Phosphoric Acid Fuel Cell:-
This fuel cell comprises of porous carbon electrodes containing a platinum catalyst and a liquid phosphoric acid, which is stored in a Teflon-bonded silicon carbide matrix as an electrolyte. These fuel cells have an efficiency up to 85 percent if they are used for the cogeneration. However, they are less efficient if they are used only for the generation of electricity. Since these fuel cells are very large and heavy, they are not as powerful as the other fuel cells. Phosphoric acid fuel cells are also more expensive since the fuel cells require an expensive platinum catalyst [5].

2.5. Molten Carbonate Fuel Cell:-
These fuel cells use a molten carbonate salt mixture electrolyte which is suspended in a lithium aluminum oxide ceramic matrix which is porous and also chemically inert as shown in figure 3. These fuel cells operate at high temperatures of greater than 600°Cfuel cells that offer significant cost reductions over PAFC. They also offer significantly higher efficiency of about 60% when it is used to generate electricity than compared to phosphoric acid fuel cells. The major drawback of current molten carbonate fuel cell technology is its low durability and lifespan [6].

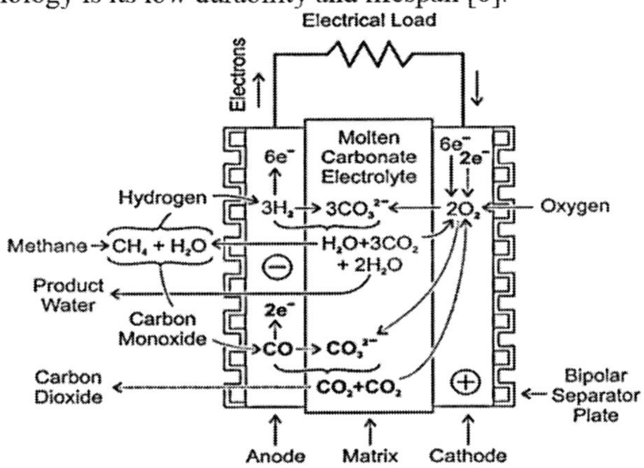

Figure 3. Schematic Diagram of Molten Carbonate Fuel Cell [6]

2.6. Solid oxide Fuel Cell:-
These fuel cells use a hard and non-porous ceramic compound used as the electrolyte. An efficiency of about 50-60% can be expected when solid oxide fuel cells are used to convert fuel into electricity, but can be much greater if used for cogeneration. Since solid oxide fuel cells operate at very high temperatures of about 1000°C, there is no need for using expensive catalysts made out of precious metals and as a result, the cost of these fuel cells is reduced [7,8].

2.7. Regenerative Fuel Cell:-
Similar to other fuel cells, regenerative fuel cells produce electricity from oxygen and hydrogen and heat and water as by-products. Regenerative fuel cells can also be used solar power or other sources of electricity to split water produced as by-product into hydrogen and oxygen in a process referred to as electrolysis. These are comparatively new type of fuel cell technology which is currently under development by organizations such as NASA [9].

2.8. Biofuel Cell:-
A biofuel cell uses living organisms to produce electricity. Cells of this type use biological moieties such as enzymes of living cells to directly generate power from the chemical energy contained within

various organic or inorganic species. In this fuel cell, two electrodes which are separated by a semi permeable membrane are kept in a solution as displayed in figure 4. Any biological species such as microbes or enzymes can be placed in the solution. Once a suitable fuel is introduced, usually hydrogen, then the partial or complete oxidation process takes place at the anode and the electrons released by this process are used to reduce oxygen at the cathode. There are two types of biofuel cells, an Enzyme based and a microbial fuel cell. An Enzyme based biofuel cell is a specific type of fuel cell that uses enzyme as a catalyst to oxidise its fuel, rather than precious metals. They are a popular focus for research due to the high turnover rates associated with the enzymes leading to a high catalysis rate. One of the main issues associated with this type of fuel cell is that although the biological moieties will readily produce a supply of electrons, they cannot be used unless they can be transferred to the cathode. A microbial fuel cell is a bio-electrochemical system that drives the electric current by using bacteria and mimicking bacterial interactions found in nature. Microbes offer some major advantage over enzymes as they are able to catalyse a more thorough oxidation of many biofuels and can be less susceptible to poisoning, which makes them a better option for use in biofuel cells [10].

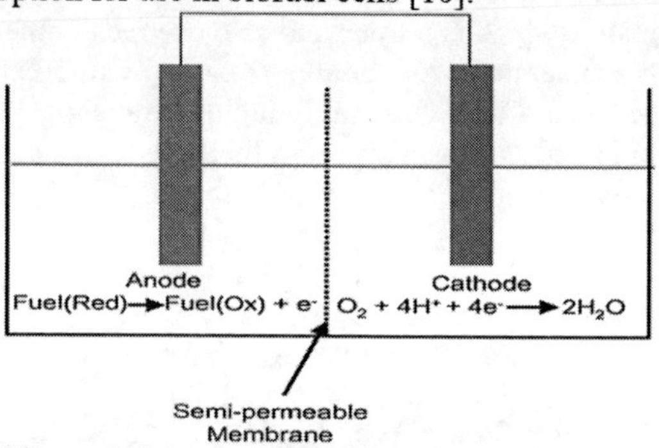

Figure 4. Schematic Diagram of a Biofuel Cell

3. Parameters affecting fuel cell performance

3.1. Fuel Cell System Efficiency:-

It tells us about the overall performance of a system/power plant installed with fuel cells. The fuel cell stack functions properly only if it is provided with properly pressurized air and hydrogen and then cooled by a coolant. It is observed that additional equipments for regulation of gas/fluid streams, managing the output electricity, providing lubrication, operating other auxiliary equipment and control the process in actual fuel cell systems. Some system comprises of reformers for processing fuel. All of these multiple equipments bring about losses and collectively reduce the total efficiency of the system thus deviating from its ideal. While drawing comparisons between fuel cells and other systems generating power, it is crucial that each of the system must be defined in a similar way. While comparisons are made between an IC engine and a fuel cell power plant for an automotive application, they both can be defined in a similar way as device that delivers mechanical output power to the shaft by taking in fuel and air. In both cases, the fuel is stored in a tank after refining/reforming and other processing and drawn from the tank in either a gaseous or liquid form.A similarity between the two is the compression action; while the IC engine uses the piston action to do so, the fuel cell power plant accomplishes this by using an external compressor. The fuel cell power plant uses an inverter and electric motor to deliver power to the driveshaft while the IC engine delivers the mechanical power directly to the driveshaft. Heat is rejected to the surroundings in a similar way in both systems, i.e. using a radiator, a coolant pump and other heat management equipment. The overall efficiency of an IC engine is between 15 to 25%. The output obtained at the output of the flywheel ranges between 30 to 35% and are even higher for Diesel engines. A fuel cell power plant that operates on hydrogen, the efficiency breakdown at flywheel output is as follows:

Fuel cell efficiency: 40 to 50%
Air compression: 85% (uses 15% of gross power)
Inverter efficiency: 95%
Electric motor efficiency: 97%
An overall efficiency of 30 to 39%is yielded when all these values are multiplied. The overall efficiency is further reduced if a reformer is used and the overall efficiency drops to about 20 to 29%. Also, fuel system systems are heavier than IC engines of comparable power and range, thereby using more power on an ongoing basis [11].

3.2. Polarization curves:-

A theoretical operating voltage of 1.2 V is realizedat all operating currents ideally. But in practical use, fuel cells reach their uppermost output voltage in an open circuit and the voltage drops with increase in current. This is known as Polarization and is represented by Polarization curve as shown in figure 5.
Batteries and fuel cells have similar polarization curves. Both of them have extremely good performances at partial loads since the voltage increases as the load decreases. The IC engines on the other hand perform excellently at full load conditions but fail to do so at partial load conditions. Polarization is mainly caused due to various chemical and physical factors related with different elements of the fuel cell [11-12].

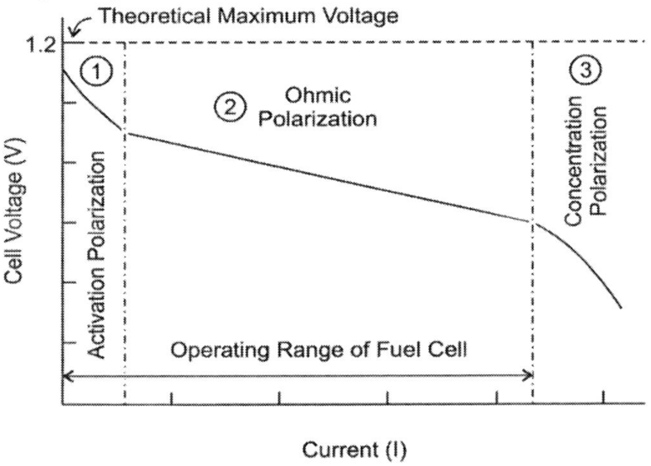

Figure 5. Polarization curves denoting active, ohmic and activation regions

Activation Polarization: - The energy barrier that must be overcome to initiate a chemical reaction between reactants is called Activation Polarization. The electron transfer rate is slow at low current draw and a portion of the electrode voltage is lost in order to reimburse for the lack of electro-catalytic activity.
Ohmic Polarization: - Ohmic Polarization occurs due to resistive losses in the cell. These resistive losses occur within the electrolyte (ionic), in the electrodes (electronic and ionic), and in the terminal connections in the cell (electronic). Since the stack plates and electrolyte obey Ohm's law (V=IR), the amount of voltage lost in order to force conduction varies linearly throughout this region.
Concentration Polarization: - A Concentration polarization is usually encountered when the electrode reactions are hindered by mass transfer effects. In this region, the reactants are utilized at greater rates than they can be supplied while the product gathers at a greater rate than it can be removed. Eventually these effects hinder further reaction altogether and the cell voltage drops to zero.

3.3. Power characteristics:-

The product of voltage and current is the electric power (P=VI). A fuel cell's polarization curve denotes a relationship between current and voltage at any point of the curve. Hence this can be used to derive a corresponding power curve. The instantaneous power is represented graphically at any point as the rectangular area that just touches the curve.As seen from the figure.6, the maximum power occurs when the voltage ranges from 0.5 to 0.6 V which corresponds to a high current. At this peak value, the electrical resistance of the external circuit is equal to the internal resistance of the cell. There is a tradeoff between high power and high efficiency since the efficiency drops with increasing voltage. The desired operating range must be selected based on whether the efficiency or the power is of more importance. Care should be taken to operate in the range before which the curve drops off [11].

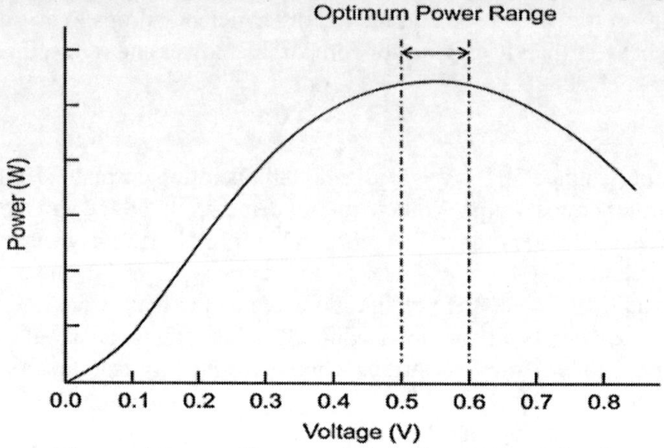

Figure. 6 Power Characteristics of a Fuel Cell [11]

3.4. Pressure and temperature:-

Pressure and temperature of the stack play a crucial role in determining the shape of the polarization curves. To characterize the stack performance over the entire operating pressure and temperature, a family of polarization curves is drawn. Thus we can conclude that in general, any parameter that leads to the polarization curves increasing or going up is beneficial, since it results in higher electrochemical efficiency and greater power. The converse is also true.

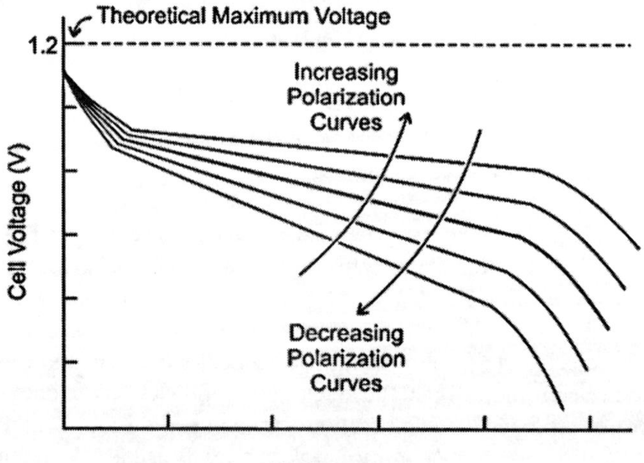

Fig.7 Effect on polarization curves due to pressure and temperature [11]

With increasing operating pressure, the fuel cell polarization curves typically increase. Conversely, with decreasing operating pressure the polarization curves decrease as well. This is due to the rate of the chemical reaction being proportional to the partial pressures of the hydrogen and the oxygen. Higher pressures help to power the hydrogen and oxygen to come in contact with the electrolyte. The pressure sensitivity is greater at high currents. In spite of promoting electrochemical reactions, the increase in

IOP Conf. Series: Materials Science and Engineering **810** (2020) 012065 doi:10.1088/1757-899X/810/1/012065

pressure causes other problems. Fuel cell stack flow field plates exhibit smaller flow-induced pressure losses at low pressure and thus should be operated at low pressures. For higher pressure, a higher compression ratio is also required, which absorbs more gross power. The redesigning of a few fuel cell components has to be done which is not economical while also increasing the size. The fuel stack and overall system efficiency have negligible changes when operating pressure is increased. Owing to these reasons, the PEM fuel cells are operated near to the atmospheric pressure range. The fuel cell polarization curves typically increase, with increasing operating temperature. Conversely, the polarization curves decrease with decreasing operating temperature well. This happens because at higher temperatures, mass transfer within the fuel cells is improved and results in a net decrease in cell resistance. These effects improve the reaction rate altogether.
The operating temperatures are limited to below 100°C due to the buildup of product water within the oxidant stream. The water boils and the resulting steam rigorously decreases the partial pressure of the oxygen at this temperature. The performance is then dramatically reduced due to lack of oxygen. This can potentially harm the fuel cells lifespan. Higher temperatures can be attained by operating at higher pressures since this increases the water boiling point accordingly but only to some extent. The fuel cell voltage increases with increasing temperature until it approaches the boiling point of water after which the voltage declines with increasing temperature. At 80°C, an optimum temperature is obtained where the two effects balance each other out. Typically operating temperatures are between 70 to 90°C [13].

3.5. Humidity effects:-
An appropriate amount of humidification is important for PEM fuel cell operation since water molecules move with the hydrogen ions during the ion exchange reaction. Insufficient humidification water can lead to dehydration of the membrane and cracks or holes in the membrane. This collectively results in a chemical short circuit, local gas mixing, hot spots, and also the possibility of fire. But it has also been observed that excess humidification water leads to condensation and flooding within the flow field plates.
A phenomenon known as cell reversal occurs due to this where the affected cells produce a zero or negative voltage and if a large enough negative voltage occurs, the affected fuel cells start to act like an electrolyser. Excessive heat is produced which can potentially lead to the destruction of the cell. Cell reversal is detected by cell monitoring systems before cell damage occurs.
Humidity is usually measured by using the term "relative humidity"; the word is used since relative since it depends on the pressure and temperature of the gas. A gas is said to be saturated and has a relatively humidity of 100% when it has absorbed as much water as it is physically able to at a given pressure and temperature. The relative humidity drops if the saturated gas becomes hotter. But if the gas cools, the gas remains saturated at the new temperature due to the condensation of water [14].
Fuel cells are generally operated at or near saturated conditions at the fuel cell operating temperature so that the relative humidity is 100% or close to it. This provides the maximum amount of water possible while also preventing flooding due to excess water. Thus the use of water to provide humidification effectively limits fuel operating and storage temperature between 0 and 100°C. Outside of these limits, the water freezes and boils respectively [14].

3.6. Stoichiometric Effects:-
With increasing reactant gas stoichiometry, the fuel cell polarization curves increase and the polarization curves decreases with decreasing reactant gas stoichiometry. This effect occurs because higher stoichiometry increases the chance of hydrogen and oxygen molecules to sufficiently interact with the electrolyte. Insufficient stoichiometry starves the fuel cell stack of sufficient reactants for reaction and thus can cause permanent damage to the fuel cell. The fuel cell performance degrades and the cell voltage drops when starved from fuel or oxygen. Stoichiometry is defined as the ratio of the amount of gas present relative to the amount of that gas that is needed to complete the reaction. A stoichiometric ratio of 1.0 provides exactly the correct number of gas molecules to theoretically complete the reaction.

IOP Conf. Series: Materials Science and Engineering **810** (2020) 012065 doi:10.1088/1757-899X/810/1/012065

Stoichiometric ratios greater than 1.0 provide excess gas and ratios less than 1.0 provide insufficient gas. Thus it is always desirable to keep the stoichiometric ratio above 1.0 for the reactions to take place sufficiently [15].

3.7. Fuel Cell Flooding:-

Flooding is the buildup of excess water and it takes place at both the electrodes of the fuel cell membrane. As discussed in humidity effects, it is essential to keep the membrane humidified for high proton conductivity, because the membrane's conductivity is directly related to its water content but it should be noted that accumulation of too much water also impacts performance and lifespan of the fuel cell. Reactant starvation occurs due to excess water blocking the flow channels and the pores of the gas diffusion layer (GDL), thus instantly leading to reactant starvation. The transport rate of the reactants to the electro catalyst sites is significantly reduced i.e. flooding leads to an instant increase in the mass transport losses, particularly at the cathode. Water blocks the pores of the GDL which creates a sterical(caused by the spatial arrangement of atoms in a molecule) hindrance which prevents the reactants to reach the catalysts eventually leading to gas starvation and an immediate drop in cell potential .The voltage can be recovered relatively fast by purification of the cathode and anode. The pore size may be reduced due to water layer present on the GDL surface[2,16].

3.8. Contamination of the Cell:-

PEM fuel cells contamination can also have adverse effects on the performance and lifespan. Contamination of the fuel cell is the process when impurities pollute and penetrate into cell components slowing down the actual reactions that take place in a cell by initiating a chemical attack. These contamination products originate from components placed inside the cell and are transported into the cell via the reactants. Metal, alkaline metal and ammonium ions, silicon and catalyst particles along with gases such as carbon monoxide (CO), nitrogen oxides (NOx) or sulfur dioxide (SO2) can be present in the cell. Trace amount of impurities can also lead to considerable degradation of performance [2].

Table No. 1 Summary of different fuel cells

Parameters → Types	Efficiency	*Electrolyte used*	Ions in the electrolyte	*Operation Temperature*	*Fuel*	*Catalyst used*
PEM	Up to 40%	Nafion membrane	H^+	60 - 120°C	Hydrogen	Platinum or platinum/ Ruthenium
DMFC	Around 10%	Nafion membrane	H^+	60 - 120°C	Methanol	Platinum or platinum/ Ruthenium
AFC	60 – 70%	Potassium hydroxide	OH^-	> 100°C	Hydrogen	Nickel and Silver
PAFC	40 – 50%	Phosphoric Acid	H^+	160 – 200°C	Hydrogen	Platinum or platinum/ Ruthenium
MCFC	Up to 60%	Carbonate salts of sodium & potassium	CO_3^{2-}	600 – 800°C	Hydrogen as a fuel + CO2 to replenish	Nickel chromium

					CO32- ions	
SOFC	Up to 60%	Yttria stabilized zicronia	O^{2-}	800 – 1000°C	Methane / Propane	Nickel – YSZ

4. Applications of Fuel Cells

Various applications of fuel cells along with their parameters have been discussed in the above sections. One of the main reasons for using fuel cells instead of the conventional systems is due to its ability to produce clean power with highly reduced emissions and also due to its higher efficiency and overall performance as compared to other systems [13].

4.1. PEMFC:-

PEMFC are capable of producing power up to hundreds of kW and thus can be used for a wide variety of applications where electricity generation is a requirement. They have wide range of applications such as vehicles, bicycles, aerospace and defence applications which may comprise of space shuttles and submarines. They are also utilized for distributed power generation inside individual homes, buildings or communities due to their modularity and excellent flexibility in power supply

4.2. AFC:-

Due to the high reactivity of OH- ions with carbon and its compounds, even the slightest quantity of carbon dioxide affects the efficiency considerably. Due to the lack of any carbon and its compounds in space this fuel cell can easily be used in space environments and they are still the main technology used in related applications. Also due to the production of extremely clean byproducts which in turn reduces pollution in the space environment these fuel cells are nearly irreplaceable with the current available technology.

4.3. MCFC & SOFC:-

They demand very high operating temperature above 600 degrees and 1000 degrees Celsius respectively and mostly, all of the applications regarding these kinds of fuel cells are restricted to large and emission free power plants. The heat from these fuel cells can be used for cogeneration or can be used for other applications such as industrial processes or steam turbines in order to generate more electricity. Developments have been made for using MCFC assubstitute power supply for ships. This will be cleaner and avoid the pollution of the marine Diesel engines. SOFC is used for tri-generation, which is a simultaneous process of cooling, heating and power generation from only one fuel point.

4.4. Biofuel Cell:-

They can be used for power production which require only low power, but where replacing batteries may be unfeasible, such as wireless sensor networks. They can function as biosensors and microbial cells can measure the solute concentration of waste water, as the current generated from a microbial cell is directly proportional to the energy content of waste water used as fuel. A device constructed in 2010 was capable of producing electricity and reducing Cu2+ ions to copper metal. A first self-powered and autonomous biochemical oxygen demand (BOD) biosensor has been developed and allows to detect organic contaminants in fresh water.

5. Conclusion

In this review article, development of the fuel cell with improved performance and various impelling parameters has been discussed. Overall efficiency of the fuel cell is affected by the interactions between different parameters. Also, it has adverse impact on fuel cell durability. Hence, to avoid any losses concerning to working parameters of the fuel cell, essential remedial actions should be incorporated to improve performance as well as life of fuel cells. It has been revealed that working performance and the life span of the fuel cell is most influenced by polarization, working temperature, operating pressure and humidity. Major conclusions are as follows

- With increase in temperature and pressure, polarization curve also increases which affects rate of reaction resulting into oxygen starvation. This in turn hampers durability of fuel cell.
- In order to achieve higher efficiency of the fuel cell, ideal stoichiometric ratio is found to be 1.4 for fuel (hydrogen) and 2.0 for air.
- Poor water management leads to humidity imbalance. This results into massive reduction of the electrochemical active surface area (ECSA) and affects gas diffusion layer (GDL).
- Corrosion of the electrode and membrane strongly influenced performance in long run and major difficulty in commercialization of fuel cell.
- Biofuel cell is one of the most dynamic areas towards power generating sources for prosthetic devices within humans as it has numerous benefits over present technologies.

6. References

[1] O'hayre, Ryan, Suk-Won Cha, Whitney Colella, and Fritz B. Prinz. Fuel cell fundamentals. John Wiley & Sons, 2016.

[2] Schmittinger, Wolfgang, and Ardalan Vahidi. "A review of the main parameters influencing long-term performance and durability of PEM fuel cells." Journal of power sources 180, no. 1 (2008): 1-14.

[3] Surampudi, S., S. R. Narayanan, E. Vamos, H. Frank, G. Halpert, A. N. T. H. O. N. Y. B. LaConti, J. Kosek, GK Surya Prakash, and G. A. Olah. "Advances in direct oxidation methanol fuel cells." In Across Conventional Lines: Selected Papers of George A Olah Volume 2, pp. 1226-1234. 2003.

[4] Coutanceau, C., L. Demarconnay, C. Lamy, and J-M. Léger. "Development of electrocatalysts for solid alkaline fuel cell (SAFC)." Journal of Power Sources 156, no. 1 (2006): 14-19.

[5] Sammes, Nigel, Roberto Bove, and Knut Stahl. "Phosphoric acid fuel cells: Fundamentals and applications." Current opinion in solid state and materials science 8, no. 5 (2004): 372-378.

[6] Bove, Roberto, Angelo Moreno, and Stephen McPhail. "International status of molten carbonate fuel cell (MCFC) technology." JRC Scientific and Technical Report (2008).

[7] Boudghene Stambouli, A.; Traversa, E. Solid oxide fuel cells (SOFCs): a review of an environmentally clean and efficient source of energy. Renew. Sustain. Energy Rev., 2002, 6 (5), 433-455.

[8] A.M. Abdalla, et al., "Nanomaterials for solid oxide fuel cells: A review", Renewable and Sustainable Energy Reviews 82 (2018) 353-368.

[9] McElroy, James, Matthias Gottmann, John Finn, and Fred Mitlitsky. "Solid oxide regenerative fuel cell." U.S. Patent 6,821,663, issued November 23, 2004.

[10] F. Davis and S.P.J. Hisgon, “Biofuel Cells – Recent Advances and Applications”, Biosensors and Bioelectronics 22 (2007) 1224–1235.
[11] Blomen, Leo JMJ, and Michael N. Mugerwa, eds. Fuel cell systems. Springer Science & Business Media, 2013.
[12] Zhao, Feng, and Anil V. Virkar. "Dependence of polarization in anode-supported solid oxide fuel cells on various cell parameters." Journal of power sources 141, no. 1 (2005): 79-95.
[13] L. Giorgi and F. Leccese, “Fuel Cells: Technologies and Applications”, The Open Fuel Cells Journal, July 2013, 6, 1-20
[14] P. Sridhar et al. , “Humidification studies on polymer electrolyte membrane fuel cell”, Journal of Power Sources 101 ,2001) 72-78.
[15] Qu, Shuguo, Xiaojin Li, Ming Hou, Zhigang Shao, and Baolian Yi. "The effect of air stoichiometry change on the dynamic behavior of a proton exchange membrane fuel cell." Journal of Power Sources 185, no. 1 (2008): 302-310.
[16] Z. Qi, A. Kaufman, “Improvement of water management by a microporous sublayer for PEM fuel cells, Journal of Power Sources 109 (2002) 38-46.

Quantification of the phase fraction in multiphase steel and 2D design using mini CNC plotter

Bikram Sarkar[1], Asif Iqbal Khan[1], Amit Kumar Rana[1], Sanjib Kundu[1,*]

[1] Department of Mechanical Engineering, Academy of Technology, Hooghly-712121, India

* Corresponding author: sanjib.kundu@aot.edu.in

Abstract. This present experimental setup is about designing a low cost three axis Mini CNC plotter using stepper motor, Arduino microcontroller and motor control software. This present experimental plotter deals with a low cost CNC plotter which can be further modified into small - medium size or an open structure as per the requirement. The demand for Computer Numerical Control (CNC) plotter machines in educational institutions and laboratories are rapidly rising with the advancement of technology. Current investigation will present an affordable model of a CNC plotter machine which is able to draw different microstructural phase fraction identification layout for further FE analysis of multiphase steel, automotive layout and any other solid surface using simple algorithm. First of all, image file or text file is converted into G code using Inkscape software and then feed it to the machine using processing software. The microcontroller converts G-code into a set of machine language instruction to be sent to the motor driver of the CNC plotter. L293D Motor Shield accordingly instructs the stepper motor how to move, combine, and synchronize motions of stepper motor ends in a good image.

Keywords: Computer Numerical Control, Inkscape software, microstructural phase fraction, CNC plotter

1. Introduction

Mini CNC 2D plotter is based on Arduino Nano and L293D motor shield. G codes are the functions generally used in computer numerical control machine. According to the movement of machine the G-code retrieves. Instruction to the design tool in X, Y, Z directions is provided through G-codes. Anything can be used in place of a design tool such as pen, pencils, sketch pen and all as a tool of CNC. The aim of this investigation is to make a mini design tool plotter machine which can draw any design on the writable surface. Two stepper motors with timing belts and pulley in Cartesian coordinate X and Y directions are used for functioning with a greater accuracy and a servo motor for Z axis has been used [1]. CNC is commonly used for cut, etch, mill, engrave, turn and can perform manufacturing operations on various materials. Typically, a CNC machine can move a cutting or printing head in three axes, hence it can position the tool head at a precise point on the material to create the operation desired at that point [2]. The 2D plotter is mainly designed for to plot the drawings and images. Park et al., [3] discussed the mechanism of dual-drive servo and develop an XY framework model consisting of two motors for Y control with another motor sliding on the framework in X direction.

The main aim of the current work is to make a low cost mini CNC plotter machine and quantification of the phase fraction in steel using mini CNC plotter. Another concern is the accuracy level of the machine. Few literatures have been discussed on plotters in term of their adjustment methods on the accuracy and movement of plotter [4]. Open source microcontroller platform Arduino is used to control of the motors, and it is used for executing the G code for the machining applications. The Arduino does not need a separate piece of hardware (called a programmer) in order to load new code onto the board, an USB cable can be simply used. Under CNC machining, machine tools function through numerical control [5].

2. Model Description

Microcontrollers are interactive devices that can sense and control both physically and digitally. Arduino microcontroller is used as a brain in this current work. Any image file or text file is converted into G code using Inkscape software and then input it to the machine using processing software. Here control device is used Arduino Nano with an ATmega328P microcontroller. The complete mechanical system was designed in the metallic CD drive cover. The three main sections of mini CNC 2D Plotter are hardware, software and industrial design. Electronic hardware consists of interconnected electronic components which perform analog or logic operations on received and locally stored information to produce as output resulting new information to provide control for output actuator mechanisms. Electronic hardware can range from individual chips/circuits to distributed information processing systems. The XY-plotter consists of two axes operating orthogonally to each other. The axes are introduced by a CD drive and a third axis is introduced by a stepper motor. The software system of the current work consists of Inkscape (version 0.48.5), Arduino IDE and Processing 3.0.2. Stepper motors of CD drives are used so that the movement should be very accurate. Y-axis: carries design tool that move from front to back. X-axis: carries design tool that move from left to right. CD drives containing stepper motor are used for X-Y direction co-ordinate movements as shown in the Figure 1. For design tool setup (Z axis) high-density bar board (HDF) is used. It is a type of bar board, which is petroleum by product. It is of light weight. Servomotor is adjusted inside the HDF to get the up and movement required to plot the object. Z-axis: carries design tool part that move up to down. The outer metallic cover of the CD drive is used for holding CD drive in the X and Y axis. All the sections are integrated together to get the complete setup of the plotter. The Figure 2(a) and 2(b) shows the design of the design tool stand and also the view angle of that is side view, Figure 2(a) and also front view, Figure 2(b).

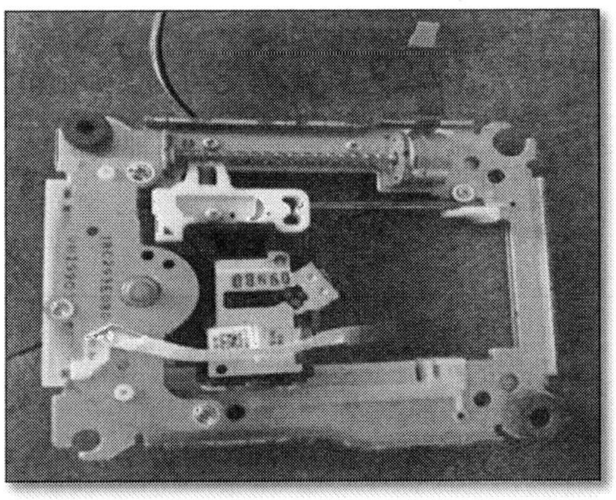

Figure1. Lens Frame in CD Drive (Containing stepper motor)

IOP Conf. Series: Materials Science and Engineering **810** (2020) 012066 doi:10.1088/1757-899X/810/1/012066

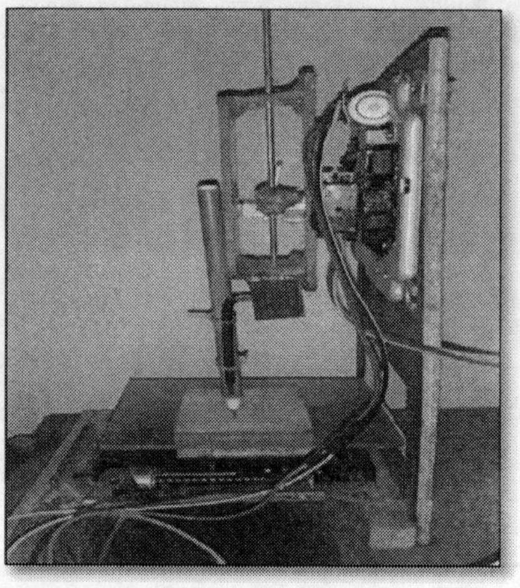

a) side view

b) front view

Figure 2. Design of tool stand (a) side view (b) front view

The following steps show the building stages of a low cost mini CNC plotter. For X and Y axis, the stepper motors from CD drive is used. Servo motor is used for Z axis. Inkscape, Processing and Arduino IDE provide the command from the computer as G code to the arduino board to get the plotted output. Figure 3 shows the flow chart of low cost CNC plotter in which the G code of a scanned image enters into the Arduino Nano and Arduino Nano instructs the L293D motor shield which further controls the movement of the stepper and servo motors.

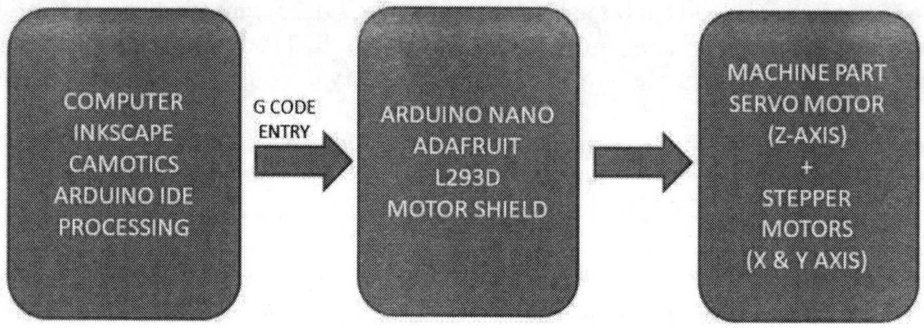

Figure 3. Main Block Diagram

The Arduino Nano is a small, complete, and bread board-friendly board based on the ATmega328 (Arduino Nano 3.x) or ATmega168 (Arduino Nano 2.x). It has more or less the same functionality of the Arduino Duemilanove, but in a different package. It lacks only a DC power jack, and works with a Mini-B USB cable instead of a standard one. The Arduino Nano can be powered via the Mini-B USB connection, 6-20V unregulated external power supply (pin 30), or 5V regulated external power supply (pin 27). The power source is automatically selected to the highest voltage source. Figure 4 shows the Arduino microcontroller with pin layouts.

Figure 4. Arduino NANO

Table 1: description of arduino microcontroller

Pin No.	Name	Type	Description
1-2, 5-16	D0-D13	I/O	Digital input/output port 0 to 13
3, 28	RESET	Input	Reset (active low)
4, 29	GND	PWR	Supply ground
17	3V3	Output	+3.3V output (from FTDI)
18	AREF	Input	ADC reference
19-26	A7-A0	Input	Analog input channel 0 to 7
27	+5V	Output or	+5V output (from on-board regulator) or
		Input	+5V (input from external power supply)
30	VIN	PWR	Supply voltage

Arduino is an open-source electronics platform based on easy-to-use hardware and software. Arduino boards are able to read inputs - light on a sensor, a finger on a button, or a twitter message - and turn it into an output - activating a motor, turning on an LED, publishing something online. One can direct the board what to do by sending a set of instructions to the microcontroller on the board. To do so you use the Arduino programming language (based on wiring), and the Arduino software (IDE), based on processing. Over the years Arduino has been the brain of thousands of current works, from everyday objects to complex scientific instruments. A world wide community of makers-students, hobbyists, artists, programmers, and professionals has gathered around this open source platform. There are total 30 pins in the Arduino nano, among them 14 are digital i/o pins and 8 analog pins, Reset, Gnd, 3V3, Aref, +5v and a Vin pin.Table 1 shows the description of arduino microcontroller.

The motor driver is a module for motors that allows you to control the working speed and direction of two motors simultaneously. This motor driver is designed and developed based on L293D IC. L293D is a 16 Pin Motor Driver IC. The device is a monolithic integrated high voltage, high current four channel driver designed to accept standard DTL or TTL logic levels and drive inductive loads (such as relays solenoids, DC and stepping motors) and switching power transistors. To simplify use as two bridges each pair of channels is equipped with an enable input. A separate supply input is provided for the logic, allowing operation at a lower voltage.This device is suitable for use in switching applications at frequencies up to 5 kHz. Figure 5 shows a L293D IC which instructs the stepper motors.

Figure 5. L293D Motor Drive IC

The servo motor is most commonly used for high technology devices in the industrial application like automation technology. It is an electrical device that rotates parts of a machine with high efficiency and great precision. The output shaft of this motor can be moved to a particular angle. Servo motors are mainly used in home electronics, toys, cars, airplanes, etc. Servo motors are classified into different types based on their application, such as AC, DC, brushless DC servo motor and linear servo motor etc. RC servo motor is the most common type of servo motor used in hobby applications, robotics due to their simplicity, affordability and reliability of control by microprocessors. Figure 6 shows a servo motor which is used for Z axis movement of the designing tool. Mini CNC Plotter machine is worked on input as G codes of design and converting it via use of software further transferring to Arduino, CNC Shield, stepper motors, servo motor. The set up has been worked on to maintain lowest possible cost. A simple construction of the 2D plotter has been designed. This is easier way to use stepper motor with pulleys and belts, CNC shield, stepper drivers, Arduino Board, etc. The Setup of machine is flexible that's why it will be easily transported and Maintenance time is short. The basic flow diagram of CNC Plotter machine is shown in the above Figure 7.

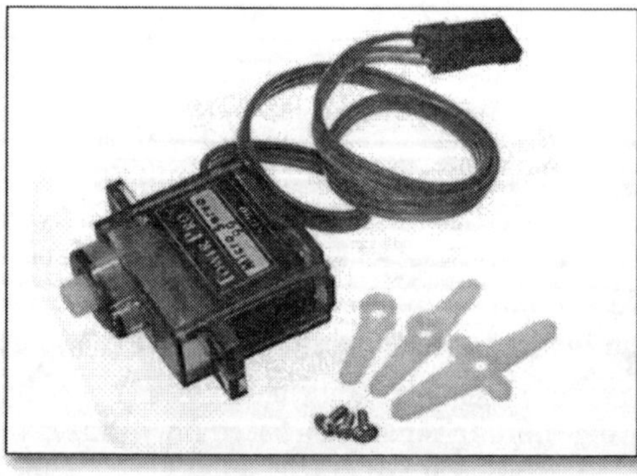

Figure 6. Servo Motor

IOP Conf. Series: Materials Science and Engineering **810** (2020) 012066 doi:10.1088/1757-899X/810/1/012066

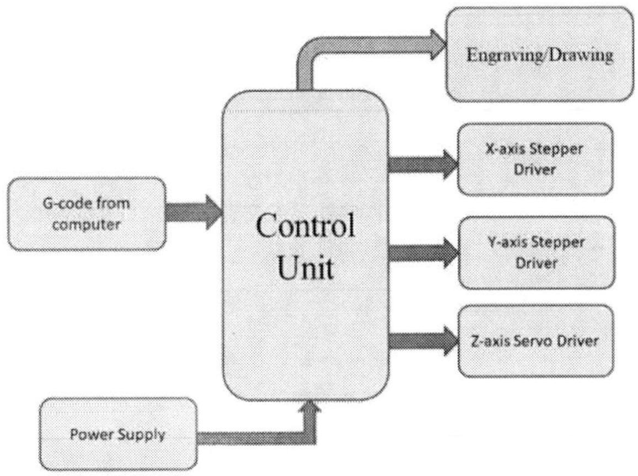

Figure 7. Flow diagram of design tool plotter

3. Software implementation and Coding

To complete the task of the entire current work two software's are used Inkscape and Processing. Inkscape is used to design the plotted image or text in this current work by using this software G-code file of a selected image or text is created. G-code is a commonly used numerical control programming language which includes X, Y, Z coordinates. G-code is a language in which people tell computerized machine tools how to make desire path. The G-code instructions provided to a machine controller (industrial computer) that tells the motors where to move, how fast to move, and what path to follow. The two most common situations are that, within a machine tool such as a lathe or mill, a cutting tool is moved according to these instructions through a tool path cutting away material to leave only the finished work piece and/or, an unfinished work piece is precisely positioned in any of up to three axis around the dimensions relative to a tool path and, either or both can move relative to each other. The same concept also extends to non cutting tools such as forming or burnishing tools, photo-plotting, additive methods such as 3D printing, and measuring instruments. The CNC plotter in this current work will work within 5cm×5cm (Width × Height). Initially the axes have been kept at the nearest end of the motors which is considered as origin to easily modify the design. Figure 8 depicts that an image of car is feed into the Inkscape software then it is converted into G-Code file. Then the G-Code is feed into the Arduino controller which instructs the motor shield and the synchronous movement of the two stepper motors and the servo motor the outer layout of the same car is plotted by the mini CNC plotter.

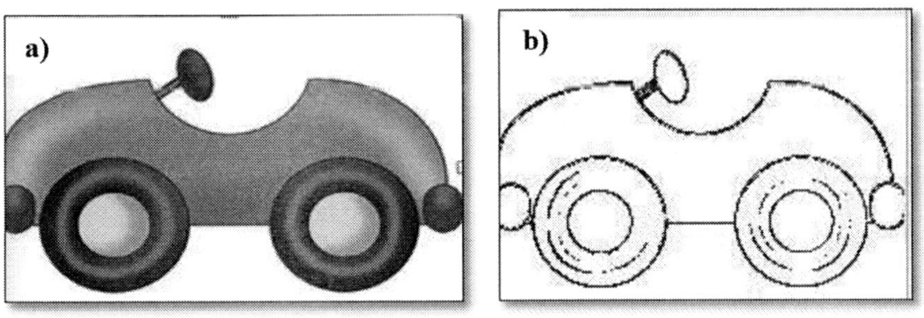

Figure 8. a) Input and b) plotted output image of a car

4. Results and Discussions

By using Mini CNC 2D plotter outer layout of any specific 2D image of different solid surfaces can be drawn easily.

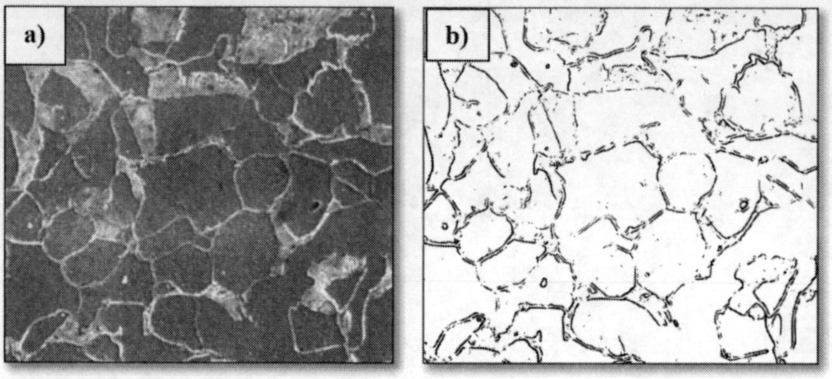

Figure 9. a) micrograph image b) plotted microstructural dual phase layout

This paper present an affordable low cost model of a CNC plotter machine which is able to draw different microstructural phase identification layout, automotive layout or any other solid surface using simple algorithm and available components. Here an outer layout of given scanned image will be printed, along with the image will also saved in different formats like .iges, .igs, .sat, .stp.

Figure 10. Plotted Layout of automotive design

Figure 9 is shown a microstructure of multiphase steel is provided into the Inkscape software. It converts the following image into G codes and feed it into the Arduino Nano microcontroller. Arduino instructs the stepper motor via L293D motor shield to move accordingly as per the provided G codes of the image. Finally due to the synchronize movement of two stepper motors and a servo motor a microstructural dual phase layout is plotted in the paper which can be further saved in different image format like .iges, .igs, .sat etc. and can be used for microstructural phase identification. Similarly different structural layout of automotive designs can be plotted effortlessly using this design tool plotter as shown in Figure 10. Figure 11 shows the Mini CNC plotter during working condition while sketching the architecture of the input image of a car. Following are the highlights of this work:

- It is aimed at students of multidiscipline engineering in professional degree and focuses on the hand on experience with the principles operations and practices of training professional, technical and multidiscipline engineers and those in related fields.

- This setup is the designing a low cost three axis Mini CNC Plotter using stepper motor, Arduino microcontroller and motor control software.
- The demand for Computer Numerical Control (CNC) plotter machines in Educational Institutions and Laboratories is rapidly rising with the advancement of automatic control system.
- Current investigation is presented an affordable model of a CNC plotter machine which is able to draw different circuit layouts on PCB, different microstructural phase identification layout for further FE analysis for multiphase steel, automotive layout and any other solid surface using simple algorithm and available components as a lab demonstration.

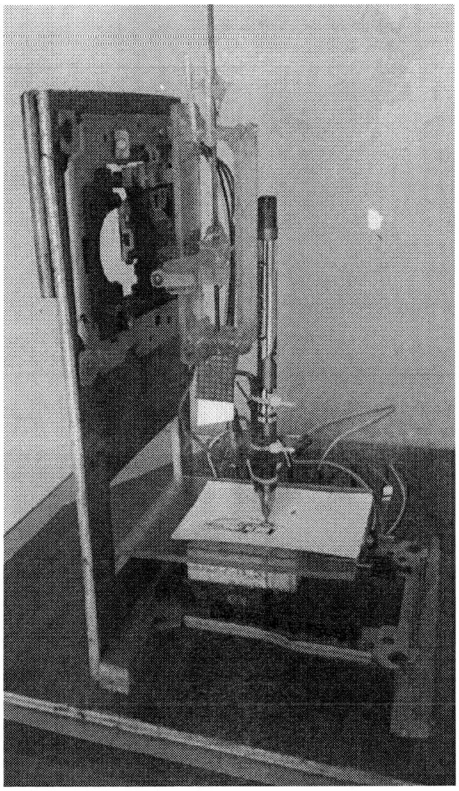

Figure 11. During operation of mini CNC plotter

5. Conclusion

This work is about building a mechanical prototype of a low cost CNC plotter machine which is able to draw a micro structural phase diagram layout, automotive designs, and any 2D image on a given solid surface. It consumes low power and works with significant accuracy due to precise controlling of stepper motors. This is a low cost set up as compared to other CNC product. It is made with easily available components and spare parts. It is designed for private manufacturing and small scale designing applications in educational institutes. The machine is designed with a very simple construction scheme and can be carried anywhere without much effort. The algorithm used is simple and can be easily modified as per the user requirement. The design tool can be replaced with a pinhead or laser head or any other cutting tool for different purpose of use. Software that has been used is open source, user-friendly and can be modified further.

IOP Conf. Series: Materials Science and Engineering **810** (2020) 012066 doi:10.1088/1757-899X/810/1/012066

References

[1] Joshi H, Khole K and Kulkarni N, 2018 *IOSR-JMCE* **6**,44
[2] Pabolu VK and Srinivas KNH 2010 Int. J. of Comp. Sci. and Eng. **2(8)** 2567
[3] Park HK., Kim SS,. Park JM, Cho T. Y and Hong D.2001 *In ISIE. IEEE Int. Symp. on Ind Electron Proc (Cat. No. 01TH8570)* **3** 1996
[4] Madekar KJ, Nanaware KR, Phadtare PR and Mane VS 2016 *Int. Res. J. Eng. Techn.(IRJET)* **3(2)** 1107
[5] Jeon JW, and Ha YY 2000 *IEEE Trans. Ind. Electron* **47(1)** 133

Recent research status on modern friction materials-an Overview

Udaykumar V Saindane[1], Sandeep Soni[2], Jyoti V Menghani[3]
[1]Research Scholar, Mechanical Engineering Department, S V National Institute of Technology, Surat, Gujarat, India
[2]Assistant Professor, Mechanical Engineering Department, S V National Institute of Technology, Surat, Gujarat, India
[3]Associate Professor, Mechanical Engineering Department, S V National Institute of Technology, Surat, Gujarat, India

udaysaindane@gmail.com

Abstract. Modern friction materials have application in automotive, aerospace and industrial brake systems. It should also possess various resistance to water and heat, low wear rate and high thermal stability, but It is quite impossible to achieve all this in practical conditions. To meet these requirements, composite friction materials have been designed often based on experience or a trial and error method to make a new formulations. Selection of Binders, fibers, fillers, and modifiers plays key role in deciding new formulation of friction composite. Listing out these ingredients from literature give rise to idea for selection of suitable ingredient of new formulation which can be acceptable worldwide. Inferences and discussions are drawn from fact revealed of modern friction materials. In this Paper, A through literature survey of work carried out so far in field of manufacturing and characterization of friction materials is presented. It reports various methods of evaluation of friction performance, thermal stability of test samples. Types of mechanical, physical, thermal, morphological properties of test samples are reported and optimum formulation among other test samples is predicted. This gives idea about friction materials which fulfills the need of future.

1. Introduction

Discovery and application of friction materials dates back to approximate 117 years ago. Due to advantages such as light weight and easy processing conventional materials are being replaced by modern composite materials in many fields [1, 2, 3, 4]. Asbestos reinforced composites was used in the first generation of composites which are used in brake friction materials [27]. Asbestos fiber reinforced polymeric composites are used in brake couplings, brake pads and brake linings; however, asbestos has been found to be hazardous to the environment and to human health [11, 14, 15, 16]. Characteristics of friction materials, classification and factors affecting the friction and wear behavior of friction materials are presented. A Modern friction material is a multi-component polymer matrix composite. In general more than 10 ingredients are used for obtaining high friction materials Researchers have been working on improving the properties of existing friction materials varying the composition of their Ingredients. Types and proportions of various ingredients used by researchers are listed and studied in this paper. In a nutshell, Manufacturing methods, types and proportions of ingredients used in new formulation, characterization methods of friction materials are discussed in detailed and an approach towards developing eco-friendly organic friction composite is investigated.

2. Present status on research work done related to natural fiber reinforced polymer composite

A wide range of published work relating to the development of new eco-friendly friction materials and their characterization is critically survey and summarized. Present status discusses some of the important findings of literature under headings listed as preparation method of composites, types of characterization methods used, types of matrix used, types of reinforcement used, types of fillers used, types of friction modifiers used to give exact direction of working for future researchers.

2.1. Preparation method of composites

Manufacturing methods of composites depend on types of ingredients involve in the preparation of friction composites. 70 % of Investigators have used Powder metallurgy technique to fabricate the friction samples. 20% researchers have used Compression molding methods for fabrication wherein only 10% researchers have preferred the casting method. Homogeneity and compactness of various ingredients is assured by powder metallurgy method. With the help of Powder metallurgy technique, samples can be made with a great variety of compositions. It is therefore very easy to have samples of desired mechanical and physical properties like hardness, density, stiffness, toughness, etc compression molding method is used to shape the raw material with the help of mould. Figure 1 shows manufacturing methods of friction composites.

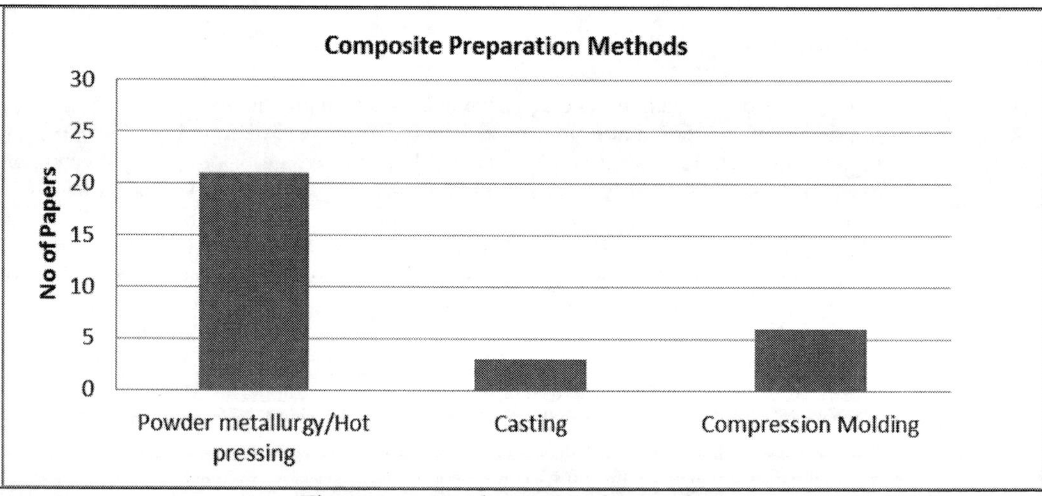

Figure 1 composite preparation methods.

2.2. Types of characterization methods used

In order to evaluate the performance of friction material testing methods are employed and test are conducted using different Test equipment and Test rigs. Generally Pin on disc test rig is used as to check the coefficient of friction and wear rate , Brake Dynamometer, Friction assessment and screening Test machine (FAST), Krauss Machine, CHASE machine, Tribometer are the optional equipments found in literature to check friction and wear performance of friction material following different standard procedures Mechanical Properties such as Compression test, Hardness test carried out using Universal testing machine, Rockwell and Brinell hardness Tester. Density of samples is measured by taking weight of sample and Archimedes principle is applied. Thermal stability of samples is checked with help of thermogravimetric analysis. Infra red Thermometers are used to check and control the temperature of samples. Morphology and microstructure of sample is checked by scanning electron microscopy and elemental composition is checked with the help of Energy dispersive X ray spectroscopy method. Figure 2 shows different characterization methods used by researchers in the literature.

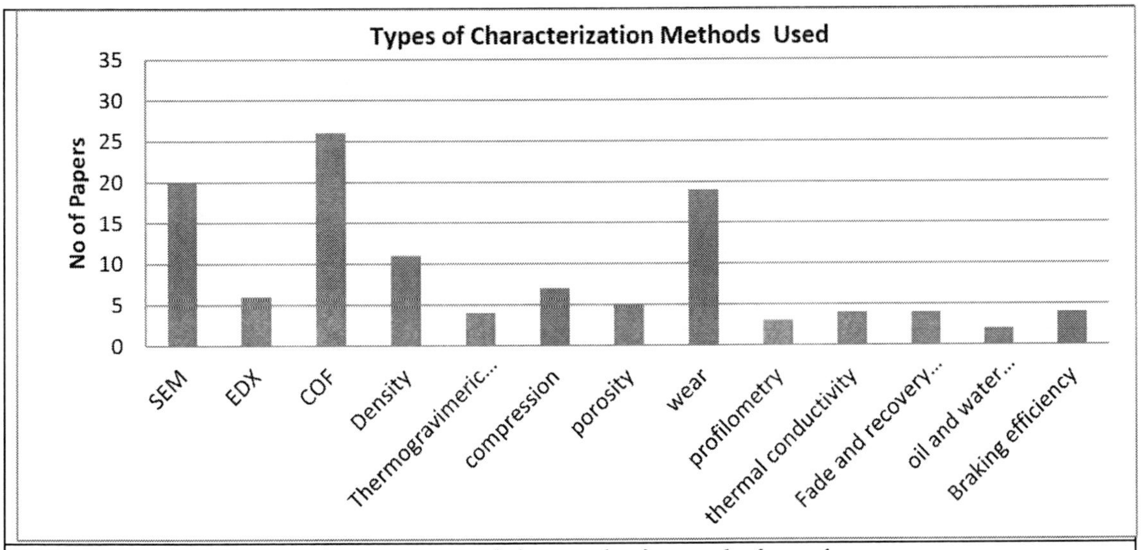

Figure 2 Types of characterization methods used.

2.3. Types of matrix used

Binders are mainly responsible for holding all ingredients of composite together. It should possess resistance to high temperature [6, 10]. Investigators have used different matrices based on adhesion and compatibility with other ingredients. 60% researchers have given their preference to Phenolic resin/Phenol Formaldehyde [26, 28, 30]. Figure 3 shows different matrices used in literature.

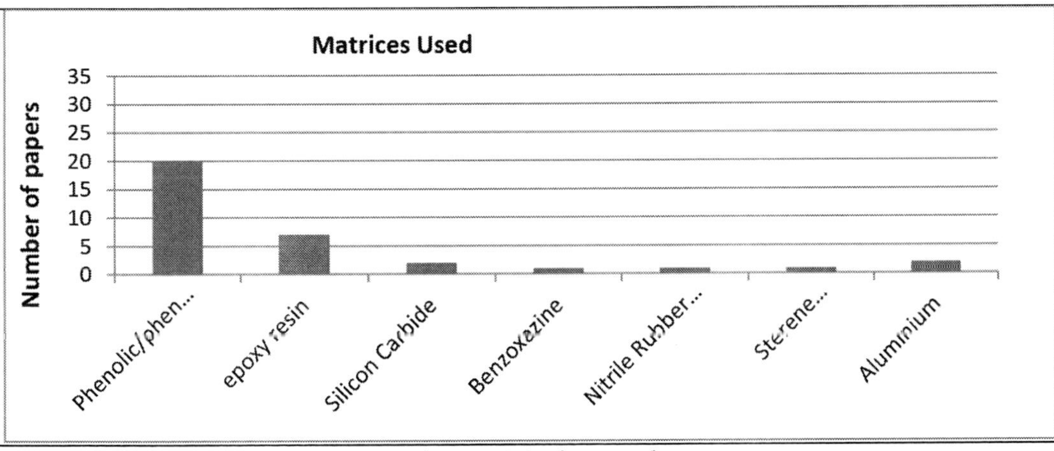

Figure 3 Matrices used.

2.4. Types of Reinforcement used

Reinforcement is generally provided in the form of fibers to mechanically strengthen the composite. Fig. 4 shows types of reinforcing fibers used in literature. 23% researchers have used Steel fiber as reinforcement fiber. 13% of investigators have used natural fiber such as Coconut powder [9, 12, 13, 17, 21, 22]. Baryte is also used as reinforcing fiber by 13% of the researchers mentioned in the literature. Remaining investigators have used other reinforcing fibers shown in below figure 4. Selection of reinforcing fiber mainly depends on strength requirement of friction composite [30, 31].

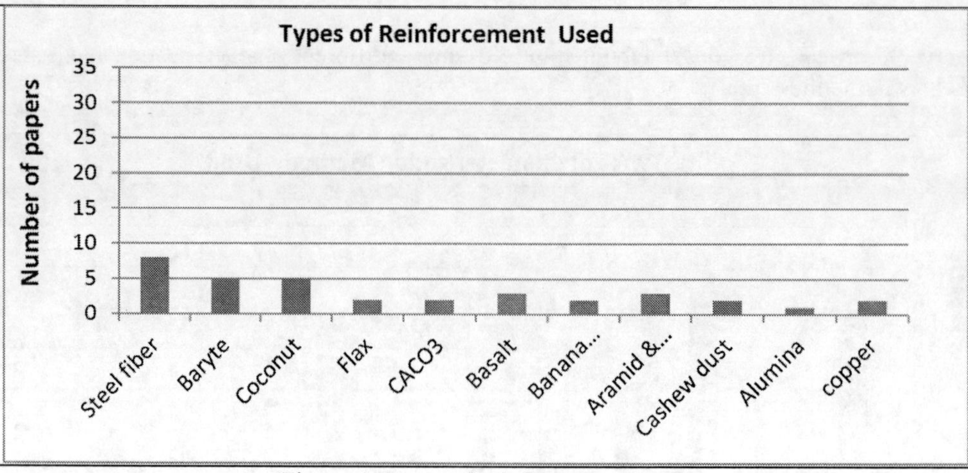

Figure 4 Types of reinforcement used.

2.5. *Types of Fillers used*

The role of fillers is to increase the ease of manufacturing and reduce the cost of manufacturing [18, 20, 25]. Several types of fillers such as Barytes, Vermiculites, Graphite, Coconut, Cashew, Coca Bean Shell, Latex rubber are used as Fillers in given literature. 20% of Investigators have used Vermiculite as filler material whereas Baryte is used as Filler by 16% of researchers. Five of the researchers have not used any type of Filler in their formulations. Figure 5 shows all types of fillers used by number of Researchers in their research.

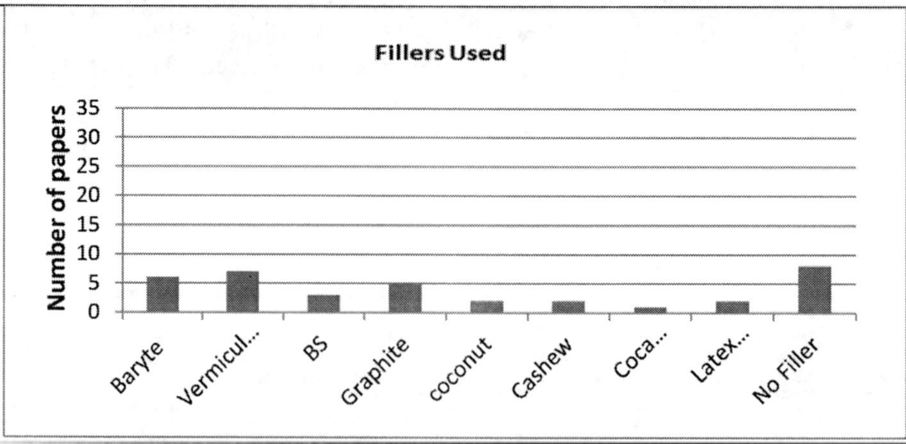

Figure 5 Types of fillers used.

2.6. *Types of Friction modifiers*

Friction modifiers raises the friction and react with oxygen to control interfacial films. Friction additives like abrasives which increases friction and controls the build up of friction films, included in the friction modifiers. Lubricants are also come under friction modifiers. Friction modifiers such as Carbon, Aluminum oxide, Alumina, Brass, Zirconium silicate/ oxide, Silicon, Titanium, Magnesium, Lead, Copper, Antimony sulphide have used by Investigators as shown in fig. 6. 43% of researchers have used Carbon as Modifier whereas Aluminum oxide have been used by 26% of researchers.

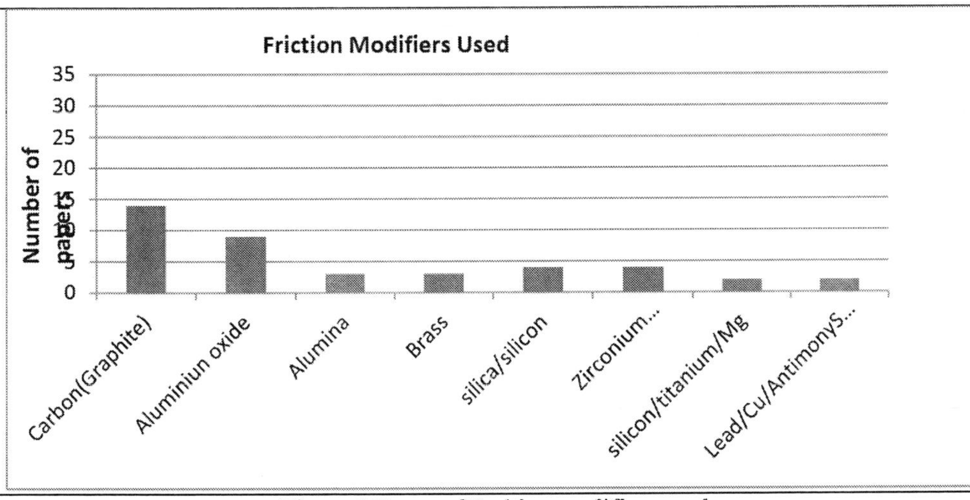

Figure 6 Types of Frcition modifiers used

3. Inferences drawn from research overview

From the research overview of recent years, Approximate 32 papers on Development of modern friction composites are taken under consideration and inferences are drawn, based on it this section discusses some of the important findings in the area of fabrication and characterization of friction materials and possible future research directions as mentioned below.

3.1. It is found that most of the researchers are working on developing asbestos free organic friction material used in Brake pad of automobiles. Powder metallurgy technique is used by most of the researchers. Powder metallurgy technique to fabricate composite friction material includes mixing, hot pressing, sintering and curing of mixture. This technique has its own advantages, namely it is easy to mix various composite elements in powder form, homogeneous mixing is achieved. So Powder metallurgy method could be used to develop friction material in future research.

3.2. Many investigators have focused on using Phenolic resin as a matrix to bind and hold together other elements in composite formulation. Phenolic resin is used as binder which gives good bonding action to the various constituents. The disadvantage of phenolic resin is brittle, Low impact resistance, highly toxic and decomposes at relatively low temperatures. Phenolic resin is selected as Matrix for future research due to brittleness property since brake pad is made up of Brittle material.

3.3. Due to carcinogenic nature of Asbestos fiber current trend in fabrication of friction material is to use non asbestos organic fibers in manufacturing of friction materials. Steel fibers have been used by maximum researchers as a reinforcement in the literature. However the aim of future research would be to use natural fibers as reinforcement due to their biodegradable and eco friendly nature [5, 7]. It is shown in literature that many researchers have been using Coconut shell powder, Flax like natural fibers as substitute natural fiber for asbestos. These materials may find their application in future research work.

3.4. Several Testing methods and equipment have been used by investigators to check performance of newly formulated friction materials. Coefficient of friction and wear is measured under different loading, temperature, pressure and environmental condition in the literature. Various types of characterization methods are available to check quality and properties of friction material [9, 19, 23, 24, 29, 32]. According to our proposed research application Pin on Disc, XRD. EDX, Microhardness, microstructure and compression test can be perform by considering testing standard of friction brake pad material.

4. Conclusion

IOP Conf. Series: Materials Science and Engineering **810** (2020) 012067 doi:10.1088/1757-899X/810/1/012067

For each and every literature introduced and employed in development of friction material and characterization, the objectives are critically studied and based on same some important conclusions are derived and summarized as given below.

4.1. The brake friction material selected for the design and manufacturing should not release any toxic substances in to the atmosphere.

4.2. It is required to develop a friction material by selecting a proper reinforcing fiber and Chemical treatment method on fiber to use it more effectively as reinforcement in polymer matrix composites.

4.3. There is also need to develop a friction material by selecting a proper resin which will bind all ingredients of composites firmly and should have excellent heat resistance all the time.

4.4. There is also need to concentrate on types of filler materials and friction modifiers which can lead to fabrication of high performance friction materials.

4.5. The interfacial adhesion, and bonding strength between fiber and matrix for a friction material to achieve better mechanical and tribological properties needs to be considered.

5. References

[1] Afiqah, O., Fauziana, L., Rasid, O., and Wong S.V. (2013). Elemental Composition Study of Commercial Brake Pads for a Passenger Vehicle: A Case Study. Materials and Design. 177-184.

[2] Aranganathan, N., Bijwe, J. (2016). Development of copper-free eco-friendly brake-friction material using novel ingredients. Wear, 352-353, 79-91.

[3] Ashok Kumar, I., Vijayaraghavan, S. (2016). Wear behavior of asbestos-free eco-friendly composites for automobile brake materials. Friction,4(2),144–152.

[4] Asvath, R., Suresh, M. (2016). Experimental Investigation on Behavior of Friction Materials in Brake Liners. Journal of Material Sciences & Engineering, 5(2), Retrieved from http://dx.doi.org/10.4172/2169-0022.1000228

[5] Bashar, D. A., Madakson, P. B., Manji, J. (2012). Material Selection and Production of a Cold-Worked Composite Brake Pad. World J of Engineering and Pure and Applied Sci. 2(3).Retrieved from http://www.rrpjournals.com/.

[6] Bernard, S. S., Jayakumari, L. S. (2014). Effect of the Properties of Natural Resin Binder in a High Friction Composite Material. Polímeros, 24(2) ,149-152.

[7] Chandgude, S. B., Ganiger, S. G. (2014). Development of Composite Material for Automobile Disc Brake Pad.International Engineering Research Journal,144-149.

[8] Craciun, A. L., Heput, T., and Bretotean, C. P. (2017). Formulation of materials with natural fiber for brake system components. Annals of Faculty Engineering Hunedoara– International Journal of Engineering, 3, 17-20.ISSN: 1584-2665 (online).

[9] Darius, G.S., Berhan, M. N., David, N.V., Shahrul, A.A., and Zaki, M. B.(2005). Characterization of brake pad friction materials: Computational Methods and Experiments in Material Characterization II. WIT Transactions on Engineering Sciences, 51,43-50.

[10] Gurunath, P.V., Bijwe, J.(2007). Friction and wear studies on brake-pad materials based on newly developed resin. Wear ,263 ,1212–1219

[11] Idris, U. D., Aigbodion, V.S., Abubakar, I. J., and Nwoye, C. I. (2015). Eco-friendly asbestos free brake-pad: Using banana peels. Journal of King Saud University – Engineering Sciences,27,185-192.

[12] Keskin, A. (2011). Investigation of using natural zeolite in brake pad. Scientific Research and Essays ,6(23), 4893-4904.http:// doi: 10.5897/sre10.1072.

[13] Kumar, D., Boopathy, R., and Sangeetha, D. (2016). Investigation on tribological properties of horn fibre reinforced phenol formaldehyde composites. International Journal of Advanced Engineering Technology,7(2),599-607.

[14] Kumar, N., K. et al. (2017). Review of brake friction materials for future development. Journal of Mechanical and Mechanics Engineering,3(2),1-29.

[15] Lagel, M.C., Hai, L., Pizzi, A., Basso, M.C., Delmotte, L., Abdallac, S., Zahedc, A., and Marzouki, F.M.A. (2016). Automotive brake pads made with a bioresin matrix. Industrial Crops and Products, 85,372-381. Elsevier Publication.

[16] Liew, K.W., Nirmal, U. (2013). Frictional performance evaluation of newly designed brake pad materials. Materials and Design, 48,25-33.

[17] Maleque, M.A., Atiqah, A., Talib, R.J., and Zahurin, H.(2012). New natural fibre reinforced aluminium composite for automotive brake pad. International Journal of Mechanical and Materials Engineering (IJMME),7(2), 166-170.
[18] Mutlu, I., Sugozu, I., & Keskin A. (2015). The effects of porosity in friction performance of brake pad using waste tire dust.Polímeros, 25(5), 440-446.
[19] Nagesh, S. N., Siddaraju, C., Prakash, S. V., and Ramesh, M. R. (2014), Characterization of Brake pads by variation in composition of friction materials presented at International conference on Advances in manufacturing and Materials Engineering,AMME2014. Procedia Materials Science, 5,295-302.
[20] Olabisi, A., Ademoh,N., Boye, T.(2016). Development of Asbestos-Free Automotive Brake Pad Using Ternary Agro-Waste Fillers. Journal of Multidisciplinary Engineering Science and Technology,3(7),5307-5323. ISSN: 2458-9403.
[21] Olele, P.C., Nkwocha, A.C., Ekeke, I.C., Ileagu, M.O., and Okeke, E.O., (2016). Assessment of Palm Kernel Shell as Friction Material for Brake Pad Production. International Journal of Engineering and Management Research, 6,(1),281-284.ISSN :2250-0758 (Online).
[22] Rajmohan, B., Arunachalam, K., Sundarapandian, G. (2017). Predict the tribological properties on brake pad using coconut shell/sugarcane/sic powder hybrid composites. International Journal of Engineering and Innovative Technology,7(3),43-49.ISSN:2277-3754.
[23] Rongping, Yun., Filip, P., and Yafei, L. (2010). Performance and evaluation of eco-friendly brake friction materials. Tribology International,43,2010–2019.
[24] Roubicek, V., Raclavska, H., Juchelkova, D., Filip, P. (2008). Wear and environmental aspects of composite materials for automotive braking industry. Wear, 265, 167–175.
[25] Rudramurthy, Chandrashekara, K., Ravishankar, R., and Abhinandan, S. (2014). Evaluation of the Properties of Eco-friendly Brake Pad Using Coconut Shell Powder as an Filler Materials. International Journal of Research in Mechanical Engineering & Technology,4, (2),98-106.ISSN : 2249-5762 (Online).
[26] Singh, T., Patnaik, A., (2014). Performance assessment of lapinus–aramid based brake pad hybrid phenolic composites in friction braking. .Archives of civil and mechanical engineering 15, 151-161.
[27] Sloan, M. R. (2008). Eco-efficient friction materials. Ph.D. Thesis. University of Exeter.
[28] Subagia, I.D.G., Atmika, K.T., Setiadi, W.N., and Parwata, M.D. (2016). Wear Behavior of Basalt Powder Reinforced Phenolic Resin Matrix Composites Brake Lining Pads presented at International Conference on Mechanics, Materials and Structural Engineering (ICMMSE 2016), published by Atlantis Press. 20-25.
[29] Yafei, L. (2006). A combinatorial approach for automotive friction materials: Effects of ingredients on friction performance. Composites Science and Technology 66, 591–598.
[30] Yuvaraj, L., Jeyanthi, S. (2015). An investigation on chemical treatment of phenol formaldehyde with natural fibers for brake pads presented at National Conference On Recent Trends And Developments In Sustainable Green Technologies. Journal of Chemical and Pharmaceutical Sciences,419-421.ISSN: 0974-2115.
[31] Zhezhen, F., Baoting, S., Rongping, Y., Yimei, L., Wang, H., Shicheng, Q, Shengling, J., Yafei, L., and Matejka, V. (2012). Development of eco-friendly brake friction composites containing flax fibers. Journal of Reinforced Plastics and Composites ,31(10) ,681–689.
[32] Zula, H., Ghetiya, N.D., and Pandya, D. (2017) .Development of friction pad and study of its wear characteristics. International Journal of Mechanical And Production Engineering, 5(2), 5-8. ISSN: 2320-2092.
[33] Bongarde, U S, Shinde V D, (2014).Review on Natural fiber Reinforcement polymer composite, International journal of Engineering science and Innovative technology"vol 3,issue2
[34] Vijay choudhary, Pramendra kumar Bajpai and Sachin Maheshwari (2017). An Investigation on Wear and Dynamic Mechanical Behaviour of Jute/Hemp/Flax Reinforced composites and its Hybrid for tribological Applications, Journal of Fibers and polymers 2018,vol19,No2,403-415.
[35] Jorda-Vilaplana , A. Carbonell -Verdú , M.D. Samper A.Pop , D.Garcia-Sanoguera (2017) Development and characterization of a new natural fiber reinforced thermoplastic (NFRP) with Cortaderia selloana (Pampa grass) short Fibers, Elsevier journal Composites Science and Technology 145, 1-9.

Remanufacturing Benefits: An Analysis of Recent Trends in Indian Context

Vinay Ahuja, Dr. Ravi Terkar

Mukesh Patel School of Technology Management & Engineering, Mumbai, India

vinayahuja.nmims@gmail.com , ravi.terkar@nmims.edu

Abstract. In the past years, remanufacturing processes have become a popular topic in the industrial area and has become one of the most effective and potential strategies for the "end of life" project management. Material restoration from end of life (EOL) items is the key to most of the remanufacturing processes that have discovered as of now. This paper talks about two case studies – 1. Indian Cartridge Remanufacturers. 2. Remanufacturing at Timken. Remanufactured products are high in demand and its relevance is growing currently due to various factors. This high demand in the market is due to its low price and remarkable functional quality. Remanufacturing is recognized only 10 and a half years back in India and hence hasn't captured a very high market share compared to scenarios worldwide. Remanufacturing is a $35 Billion industry today and it justifies it cost just as it significantly increases the potential life of components by repair. In this research paper, the research work has studied the remanufacturing operation at Timken India Ltd. As well as their customers like Lafarge Cement, Tinplate Company of India and Usha Martin and has highlighted the advantages of remanufacturing.
Keywords: Remanufacturing Process, Product Cannibalization, Original Equipment Manufacturers, End of life, Indian Cartridge Remanufacturers, Remanufacturing

1. Introduction

Material restoration from end of life (EOL) items or product recovery is a key strategy resulting products being returned to the market in new condition or even better, becoming one of the key issues in green growth and sustainable product development area of research. A product is said to be remanufactured when an EOL item is collected from any customer to restore its value and engineer it to yield a ready to use product again. This has done by carrying out a series of remanufacturing tasks so that the product can be reused or recycled and brought back to the market as good as a new product [1, 2]. Customers are pulled in towards remanufacturing because of attractive quality dimension and low cost when compared with new counterpart products. Amid the most recent couple of decades, awareness about environment protection studied in large extent. At present, the industry part that has the most involvement in the remanufacturing zone is the car business. Be that as it may, the idea of remanufacturing has turned out to be prevalent in the ongoing decades because of tremendous interest of remanufactured items like PCs, cartridges, family unit apparatuses, hardware and cell phones. Perfect Launching of a remanufactured item is vital for determining most extreme offer in the commercial center. Demand of remanufactured products relies on the EOL of a current item. Foreseeing the interest of a remanufactured item and its ideal propelling isn't a simple errand for an organization. Investigation of disposal rate of existing item is likewise another urgent issue for remanufacturing ventures. It is also necessary to study the problem of product cannibalization due to admission of remanufactured products in the market [3]. Remanufacturing ventures do have tremendous turnover in United States, the turnover of the business being assessed at $40.5 billion out of 2003. In the United Kingdom as well, the turnover of the remanufacturing business was evaluated to be £5 billion of every 2004.

2. Problems of Remanufacturing in India

As indicated by Associated Chambers of Commerce and Industry of India (ASSOCHAM), Original Equipment Manufacturers (OEMs) should bolster the remanufacturing businesses in India. Some OMEs do have a dread that the remanufactured product will eat the market of new product. ASSOCHAM in a report to the Ministry of Commerce and Industry have recommended with the expectation of complimentary exchanging of remanufacturing items in India. The Automotive Parts Remanufacturers Association (APRA) feels that the Indian Government should move quickly to change their situation on exchange remanufactured items and support remanufacturing sector for green growth and product sustainable development. In fact, considering all the benefits, they suggest the OEMs to start their own venture in remanufacturing products. Many OEMs fear product cannibalization due to remanufacturing but this cannibalization can also help them increase their overall profit. As OEMs normally utilize remanufactured items, under contractual understanding for repairs and overhaul of its own products under warranty already in the market place eliminating fears of cannibalization of other products in the market place [4-6].

3. Extent of Remanufacturing in Indian Market

Remanufacturing is a sizzling business prospect worldwide and the matter of e-squander is the most advantageous and accordingly effective result. India is the second most populated nation on the planet and the need of remanufacturing is recognized 10 years and a half back in India. As indicated by driving Indian makers, Indian businesses, however have made a moderate begin, are yet to deal with it and consequently the business is still in its initial life. A lot of study suggests that remanufacturing helps in going g green as well has its own economic advantages and this has highly influenced firms in the cartridge manufacturing business. Today remanufacturing is a $35 billion industry, which speak to about 30 percent of the $115 billion worth printer cartridges advertise all around. However, the Indian remanufacturing industry keeps on assuming an optional job and even after over a time of presence, it is yet to classify itself. The optional pretended by the Indian remanufacturing industry is primarily because of absence of affiliations and negative end client suppositions. As indicated by Indian Cartridge Remanufacturers and Recyclers Association (ICRRA), just fewer than 10 percent remanufactured and refilled inkjet cartridges and laser toners were sold and among these, just 20 percent cartridges were remanufactured in India. Cartridge making firms in India has to a great extent been into refilling than remanufacturing. The expense is a key integral factor, as individuals are reluctant to pay 75 percent of the cost of the Original item. Aggressive strategies for remanufacturing in India have also been framed by Static Control Components (SCC). The organization creates exactly 13,000 unique parts including drums, ink, housings and chips for 1,200 distinct toners and cartridges. As per International Director Sales of India SCC, the Indian market has extremely intense conditions and is immature yet has tremendous potential and this market will be vital later on with firms contributing a great deal of time, vitality and cash into it. Timken is an international leader in friction management technologies and provides one of the most complete offerings of antifriction bearings in the industry. They also offer remanufacturing options for bearings and related equipment's. They recommend and suggest that remanufacture and repair of qualified components is frequently an economical alternative to replacing it with new ones.

4. Timken Repair Section

Timken industries offers a perfect repair solution to all its clients who have signed a contract with them. In this section, Timken offers its customers various reconditioning and remanufacturing facilities with aid of services like rectifications, reclaim, rebuild and repair. The clients are given the flexibility to choose their new specifications as per their needs.

5. Recertify, Reconditioning & Remanufacturing

This involves the renewing of the certification of a bearing, which is valid for all products with an outdated shelf life. Timken offers this service usually for its unused products. This service allows Timken to restore good condition of its bearings by polishing, horning, tumbling and other processes that involve removal of minor surface defects. Bearings being made of metals often get corroded and rusted which greatly decreases the performance as well as the life of bearings. These processes help increase these parameters. This service provided by Timken is in high demand due to its capability to bring back most of the defective bearings back to the market by converting them to bearings with life and quality of new bearings. The processes involved in this include machining, grinding, turning and many other operations that aid in fixing aggressive surface damage. In this process they also make sure to replace all the non-repairable parts with new components.

6. Potential Life Due to Repair

Life span of a bearing decreases due to corrosion, rusting and contamination. Any bearing remaining life is always inversely proportional to time in service. Regular maintenance of a bearing is very necessary to extend the life of it. Overseeing the defects and not fixing them will drastically reduce its life. Therefore it is stated that regular repair increases potential life of bearing.

IOP Conf. Series: Materials Science and Engineering **810** (2020) 012068 doi:10.1088/1757-899X/810/1/012068

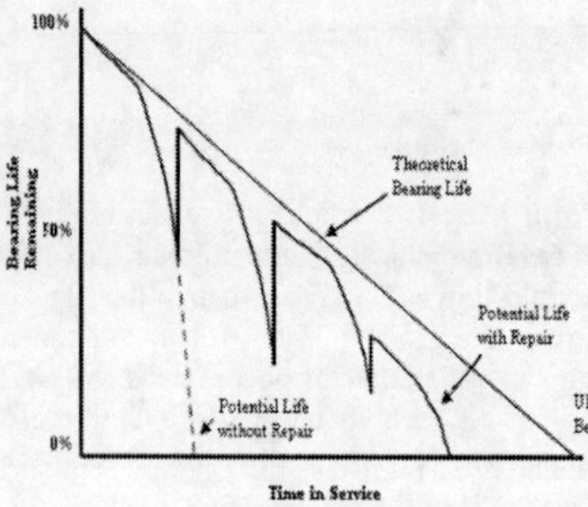

Figure 1: Potential Life of Bearing with Repair

7. Remanufacturing at Timken

Timken provides its clients with the state-of-the-art bearing remanufacturing facilities. It repairs and remanufactures all types of bearings from any bearing manufacturer. They ensure all their clients know that remanufacturing can help them save about 60% than purchasing a new bearing as well as shorter lead times. Providing a wide range of refurbishing/remanufacturing and maintenance services, they bring the bearings to almost its new condition. By this, the life of the bearing is considerably increased, providing economic sustainability as well as functionality. Timken manufactures various kinds of bearings involving tapered roller bearings, spherical roller bearings, cylindrical roller bearings, ball bearings, thrust bearings and many more. As per Timken, a few signs always help indicate if a bearing must be remanufactured. One of it suggests that when a bearing goes for inspection and its life is found to be near to the suggested life expectancy, if the operating temperature exceeds 200F due to vibration then the most suitable option is remanufacturing.

Figure 2: Remanufacturing of Bearings at Timken

8. Element of Customer Expectation

Under its repair section, Timken offers recertify reclaim, reconditioning & remanufacturing services. Customer expectation generally refers to the needs and wants of any individual investing in a certain product. Meeting the parameters of customer expectations helps gain customer satisfaction. Information regarding customer expectations also helps evaluate the value and quality of existing product. Timken meets most of its clients expectation by remanufacturing of bearings due to the benefit of increased potential life. users and clients expect the following from Timken. Most of the users are interested in remanufacturing of bearings due to its improved potential life.

Users expect the following elements from Timken.

A. Lower down Capital Cost

Operation of plant and machinery with lowest cycle cost of component.

B. Boosting of Productivity

The effort is to minimize the equipment delays and unproductive time with planned maintenance.

C. Enhancement of Product Quality

User expects the enhancement of quality of product by up gradation & preventive maintenance.

D. Minimum Inventory of Factor

User always expects to retain the optimum inventory level to control capital investment at a minimum level.

9. Remanufacturing Stages

Many EOL products are returned to the repair centre, but not all need to be remanufactured. Also it is not always advantageous to remanufacture a product so the company selects the bearings that should be sent for remanufacturing operations. Once End-of-Life (EOL) product is returned to repair centre, company selects the bearings for remanufacturing operations.

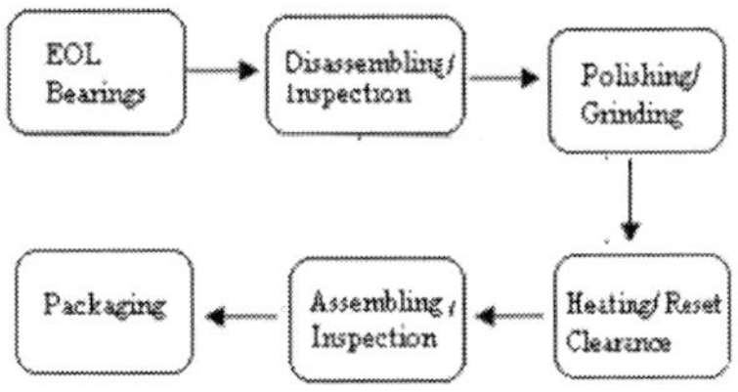

Figure 3: Remanufacturing Operation

Any product brought in is first sent over to the information recording center where data regarding the actual internal clearances is recorded. This is the data necessary for disassembly and inspection of a product. The products then disassembled and parts are tagged with unique identifiers. The entire comprehensive inspection of a product involves two stages. The initial inspection covers locating major damages like fractures and bluing due to thermal damage. This inspection declares if a bearing needs basic repairs or needs to be refabricated which is the second stage of inspection. Any damaged part found to be irreparable is replaced with a new one. The others are sent over for grinding and polishing. Resetting the metallurgical and clearance aspects up to the mark is done by heating, and these are the challenges usually faced during remanufacturing. Finally the remanufactured bearing parts are tested individually and assembled. The entire bearing is then tested for quality.

10. Remanufacturing Benefits

Timken **sells differing kinds** of bearings **within the** market as per the customers' **need**. The Company **additionally** provides repair services to **clients** creating a **large** secondary **marketplace for** business. **Several main firms in the** world purchase **top quality** bearings from the company with a maintenance agreement [9-11]. At the EOL of Bearings, Timken Company offers remanufactured bearings repair service. Quality of remanufactured bearing is **almost of the standard of recent** bearing and remanufacturing operation saves material, energy and capital investment of company. **Sustainable product development is also a really necessary** key **issue to protect** environmental pollution **connected issues. Altogether**, remanufactured **products are useful** to the OEMs and purchasers.

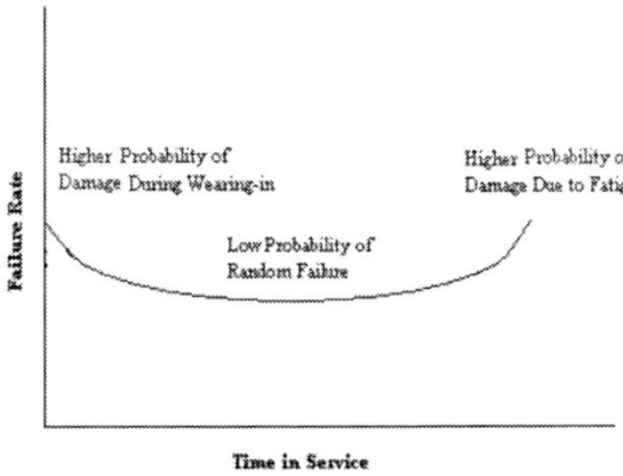

Figure 4: Likelihood of Premature Bearing Damage

Fig. 4 shows the failure rate of remanufactured bearings. The initial probability of bearing being damaged during the wearing-in period is high, however after this initial stage, bearings tend to work smoothly. Any remanufactured bearings also have a high initial failure rate, which then dips in the intermittent period but due to high fatigue, the failure rate rises again as a bearing reaches its EOL. This shows that the reliability of a remanufactured bearing is same as that of any new bearing. Here there is also a brief description put together using data collected by investigating some case studies of top Indian client companies that opt for remanufacturing as a solution. This will help us understand remanufacturing benefits.

A. Remanufacturing Benefit: Essar Steel Hot Strip Mill& Lafarge Cement

Essar Steel Hot Strip Mill (ESHSM) is one of the top Indian clients of Timken Industry. Essar Steel purchased the bearing part no. – M272749Assy902B3 & M268749 Assy90118 which were mounted on Hot Coil Box Mandrel. As soon as these bearings were approaching their EOL period, the company decided to invest in new bearings worth value of $ 27,500.

Table1: Remanufactured Bearing ESHSM & Lafarge Cement

Remanufacturing Service Offered	Timken India Ltd.	Timken India Ltd.
Client	Essar Steel Hot Strip Mill	Lafarge Cement
Bearings Part No.	M272749Assy902B3 and M268749 Assy90118 1	FAG – 517680A
Price of New Bearings	$ 27,500	$ 102,360
Price of Remanufactured Bearings offered by Timken	$ 13,475	$ 61462
Direct Cost Saving due to Remanufactured Bearings	$ 14,075	$ 40,898
Percentage of Cost Saving due to Remanufactured Bearings	51 %	40 %

IOP Conf. Series: Materials Science and Engineering **810** (2020) 012068 doi:10.1088/1757-899X/810/1/012068

Lead Time for new bearings	12 Months	12 Months
Lead Time of Remanufactured Bearings	04 Months	06 Months
Percentage of Lead Time Saving due to Remanufactured Bearings	67 %	50 %

The process of bearing removal from the costly mandrel would be done by gas cutting operation damaging the mandrel. This is why Timken advised ESHSM to opt for remanufacturing as a repair solution. Timken's design engineers came up with the best fixture solution, dismantling all the bearings without damaging the mandrel. After carrying out rigorous manufacturing operations for 16 weeks, Timken successfully remanufactured the EOL bearings. As mentioned in Table 1, remanufacturing of theses bearings helped save $14,075, which is higher than $13,475, the total money invested in remanufacturing saving 51% of the amount for ESHSM. Not only did ESHSM benefit from this, Timken also did, by earning more profits in remanufacturing operations. This is because remanufacturing drastically reduced the lead time, drastically increasing the productivity. Another example of the same is the Lafarge cement case, a company that entered India in 1999 through the cement market. The Company uses the bearing part no: – FAG – 517680A for KHD Roller Press. According to Table 1 the company saved $40,898 by selecting remanufacturing as a solution .The company did start by investing $102,360 in new bearings, bust as soon as theses bearings reached their EOL, Timken suggested they remanufacture the old bearings just for $61,462 , saving 40% of cost invested in new bearings. Not only this, Lead time for new bearings is 12 months but remanufacturing was possible in 6 months, reducing the lead time by 50%.

B. Chock Remanufacturing

The Tinplate Company of India (TCI) is a venture started by the Tata Group and is the largest producer of tin plate since 1922. Table 2 explains the advantages of remanufacturing operation with the help on TCI's case study. The company uses 4 Hi Skin Pass Mill Chock whose life is 12 years. At the EOL of the Product, the Company decided it's best to replace the part by a new mil chock. This is when Timken India Ltd offered them a remanufacturing solution which proposed that it would save 67% of TCI's revenues. The price of a new mill choke is around $60,000 and Timken India completed the operation in $ 14,000 without making any compromises on the quality of the product. Not only saving capital, remanufacturing also saved material and lead time. This resulted in sustainable product development by the environment protection view. Usha Martin which is a wire rope manufacturing company at Ranchi, India also preferred remanufacturing service from Timken as the cost benefit is around 60% as shown in Table 2.

Table 2: Remanufactured Bearings Chock for Tinplate Company of India

Remanufacturing Service Offered	Timken India Ltd.	Timken India Ltd.
Client	Tinplate Company of India	Usha Martin
Bearings Chock	4 Hi Skin Pass Mill Chock	Bar Mill Chock
Price of New Chock	$ 60,000	$ 1,25,000
Price of Remanufactured Bearings Chock offered by Timken	$ 14,000	$ 50,000
Direct Cost Saving due to Remanufactured Bearings Chock	$ 46,000	$ 75,000
Percentage of Cost Saving due to Remanufactured Bearings Chock	77 %	60%
Lead Time for new bearings	12 Months	NA
Lead Time of Remanufactured Bearings Chock	04 Months	NA
Percentage of Lead Time Saving due to Remanufactured Bearings Chock	67 %	NA

C. Comparison of Price/ Cost Savings

Fig. 5 Portrays the benefits of remanufacturing operation in three well-known companies of India. Timken India Bearings offered remanufacturing services toss many industries who are ready to accept the benefits of these services.

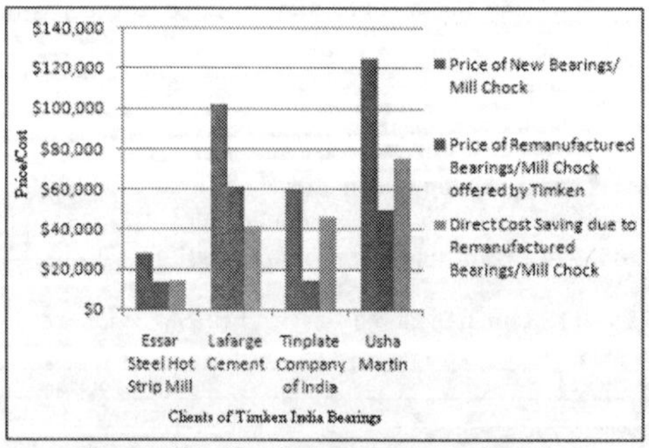

Figure 5: Price/Cost Saving due to Remanufacturing

TCI got 77% benefits due remanufacturing operation. The amount of cost saving due to remanufacturing operation is really huge. ESHSM and Lafarge Cement also saved 51% and 40% cost due to remanufacturing operation. Companies save on the cost as well as lead time due to remanufacturing. Large amount of lead time is saved and productivity and overall profitability of organisation is increased.

IOP Conf. Series: Materials Science and Engineering **810** (2020) 012068 doi:10.1088/1757-899X/810/1/012068

11. Price Saving Due to Remanufacturing

Analysis of data derived by four companies is used for calculation of maximum cost saving due to remanufacturing,. Companies such as ESHSM, Lafarge Cement, TCI and Usha Martin benefited due to remanufacturing operation and is as shown in Table 1&2.

Table 3: Expected Remanufactured Bearings Cost Saving

<table>
<tr><td>Average cost Saving due to Remanufacturing ($\bar{x}$)</td><td>Expected cost Saving due to Remanufacturing (μ)</td><td>Maximum cost Saving due to Remanufacturing (μ1)</td></tr>
<tr><td>57%</td><td>62.66%</td><td>68.5%</td></tr>
<tr><td colspan="3">Case I, Null Hypothesis, μ1=68.66 %,
Alternative Hypothesis, μ1 <68.5%,
t-test, Level of Significance (α) =0.05, μ1= 57 %, Population Distribution (N) =4, Degree of freedom=3, Standard Deviation (S) =9.11%, t = -2.5246 is less than -2.353 so Null Hypothesis (μ1) = 68.66 % is rejected and alternative hypothesis μ1 <68.5%, has been accepted</td></tr>
<tr><td colspan="3">Case II, Null Hypothesis, μ=62.66 %,
Alternative Hypothesis <62.66%,
t-test, Level of Significance (α) =0.05, μ= 57 %, Population Distribution (N) =4, Degree of freedom=3, t = -1.24 is greater than -2.353 so Null Hypothesis (μ) = 62.66 % accepted</td></tr>
</table>

The maximum cost saving due to remanufacturing is calculated by taking an average of the top two values which is about 68.7%, with the average cost saving to be about 57%. It is interesting to find the nearby value of cost saving due to remanufacturing. On available data as shown in Table 3, t-test is the most feasible analysis to find out nearby value of cost saving due to remanufacturing. Null hypothesis & alternative hypothesis have been shown in the table 3. In Case I, value of t-test is less than -2.353 and hence null hypothesis is rejected and alternative hypothesis $\mu 1<68.5\%$ has been accepted. Since the value of $\mu 1<68.5\%$, we conducted case II, as it is necessary to find nearby value of cost saving percentage due to remanufacturing. From case II, it is seen that the expected cost saving due to remanufacturing is 62.66% i.e. average of top three values of cost saving percentage. In the t-test, it has been proved that null hypothesis is accepted and μ=62.66 % (see the table 3). Due to the remanufacturing initiative taken by Timken, bearings in the Indian market saved cost up to around 62.66% and this is really a significant cost saving.

12. Cannibalization Issues in Indian Market

Product cannibalization is the term used for drastic reduction in sales volume, sales revenue or market share of any existing product as a result of introduction of a new product by the same or any competitor manufacturer. Many companies carefully plan a strategy to use the negative impacts of product cannibalization to their bone. Product cannibalization due to entry of a new product is a very common occurring in the market, but here in this paper we are discussing cannibalization due to the entry of remanufactured products, and how it has become a big issue for the OEMs [7-8]. The entry of a remanufactured product completely starts a new product life cycle which is same as the product life cycle of an existing product. The above analysis of cases have proven that remanufactured products are always profitable for the OEMs and customers. However many Indian OEMs have the fear of cannibalization of their new products due to the entry of remanufactured products. What they do not seeyet is that OEMs can also profit by starting remanufacturing/Refurbishing business along with new product development, without fearing cannibalization by deciding which process is profitable when. Timken is the best example that suggests the same. The important questions to be asked sre why are industries like Timken offering remanufacturing options to its clients? And is the profit percentage of remanufactured products more than the profit percentage of new products?. The cost associated to the remanufactured products is comparatively lower than that of new products, and this can be proved by the bearings/chocks cases addressed by Timken. The only reason of this cost precent difference being reduction of new material, reduction of lead times and therefore reduction of hours input by labour, saving up on primary and secondary expenses a company has to make for product development. Due to these factors, any company offers remanufactured products at a lower rate, which is also beneficial for the clients. A top manufacturer like Timken is positively grasping both the new product as well as their manufacturing market, as not all items can be profitably remanufactured. It is important to gain the clients trust by giving the correct advice that is not only beneficial to one's own business but also the customer, making the entire situation a win-win one. Timken has grabbed the remanufacturing market as a secondary source of income which has proven to be highly beneficial. An OEM needs to well understand the fact that if due to their fear of new product cannibalization they do not enter the remanufacturing market; other competitors will take lead into this sector cannibalizing their product. Now that it has been proven that entering the remanufacturing sector increases the overall profitability, Indian OEMs need to change their perception about cannibalization concerns and need to enter the remanufacturing market. There is a lot of scope of Indian manufacturers and OEMs to shine in the remanufacturing sector , which is in turn expected to boost the economy as well as create thousands of job opportunities here , however the Indian OEMs oppose free globe trading of remanufactured products. The Indian government also does not permit the trading of remanufactured items within the country. The Indian government needs to have a wider spectrum regarding the same and see how it is going to help the Indian economy reach the sky. All that is needed is a little consideration and encouragement.

13. Conclusion

The fear of product cannibalization of new products has reached to the Original Equipment Manufacturers up to such an extent that they are always posing free globe trading of remanufactured products in India. However it is found that the Indian market has huge potential to rise in the remanufacturing industry. This is proven as the cartridge industries have shown remarkable impact of remanufacturing over the last decade. The remanufacturing industry is still at a very naïve position in the Indian market and in the upcoming years it is expected to leave a very positive impact on the market. For boosting the economy of the country, the government of India must permit free globe trading of remanufactured products in the country. In this research paper, the effects of remanufacturing services in the market are depicted and understood by taking help of the case study of an OEM, Timken. Timken is seen providing remanufacturing solutions to its Indian clients like Essar Steel Hot Mill, Lafarge Cement, Tinplate Company of India and Usha Martin. Remanufacturing of EOL bearings and mill chock on an average resulted in 57% of cost saving as compared to new price. A T-test, which is a test to calculate the nearby value of cost saving due to remanufacturing, was conducted on observed values were noted as shown in Table 3. Analysis shows that 62.66% of the cost is saved due to remanufactured products. Remanufacturing facility not only helps in cost cutting but a also reduces lead time and this helps in improving productivity. It is evidently noticed that Timken has earned healthy ptofits as the outcome of its remanufacturing services. Instead of cannibalizing the new product, remanufacturing increased the overall profit of the OEM. Not only Timken but many other industries are also helping their customers by providing remanufacturing solutions, and the Indiam market must accept these products with open arms. This will not only help save the Indian economy but also help boost awareness about these products and processes in the market. Afterall it is said that poverty is notonly measured in the terms of money, but also knowledge.

14.References

1. W.M. Hauser and R.T. Lund, (May 2008). "Remanufacturing: Operating Practices and Strategies," Boston University.
2. M. Ferguson, (2009). "Strategic Issues in Closed-Loop Supply Chains with Remanufacturing, "College of Management, Georgia Tech, Atlanta, GA 30332.
3. H. Vasudevan, V. Kalamkar. And R. Terkar, (November 10-12, 2011). " Product Cannibalization due to Remanufacturing and its Decisive Strategies," *International Conference on Sustainable Manufacturing: Issues, Trends and Practices*, Bits, Pilani, pp. 13-18.
4. Johan Östlin, Erik Sundin and Mats Björkman, (2009). "Product Lifecycle Implications for Remanufacturing Strategies," *Journal of Cleaner Production*, (17), 11, pp. 999-1009.
5. H. Hermansson and E. Sundin, (2005). "Managing the Remanufacturing Organization for an Optimal PLC,"*1-4244-0081-3/05/$20.OO ©) IEEE.*
6. V.D.R. Guide, (2000). "Production planning and control for remanufacturing: Industry practice and research needs," *Journal of Operations Management* 18_2000.467–483.
7. V.D.R. Guide and Li, J, (2010). "The Potential for Cannibalization of New Products Sales by Remanufactured Products," *Decision Sciences Journal*, Volume 41 Number 3.
8. Atasu, V.D.R. Guide and L.N.V.Wassenhove, (2010). "So what if remanufacturing cannibalizes my new product sales?" *California Management Review*, Vol. 52, No. 2, CMR.Berkeley. Edu.
9. L. Debo, L.Toktay and L. Wassenhove, (2005). "Life Cycle Dynamics for Portfolios with Remanufactured Products," Carnegie Bosch Institute.
10. M. Matsumoto and Y. Umeda, (2011). "An analysis of remanufacturing practices in Japan,", Journal of Remanufacturing, Springer open journal.
11. Y. Ogush and M. Kandlikar, (2005). "The Impact of End-of-Life Vehicle Recycling Law on Automobile Recovery in Japan," *1-4244-0081-3/06/$20.00 @2005 IEEE.*
12. R.Terkar, H. Vasudevan, (November 17-19, 2011). " Need for Product Cannibalization in Small Scale Manufacturing Industries through Product Life Cycle Management," *International Conference on Modern Trends in Industrial Engineering*, November 17-19, 2011 S.V. National Institute
13. of Technology, Surat – 395 007, Gujarat, India.
14. R. Terkar, H. Vasudevan, V. Sunnapwar and V. Kalamkar, (2011). "Perfect Product Launching Strategies in the Context of Survival of Small Scale Consumer Product industries," *ICSTM 2011,* CCIS 145, pp.321-326.
15. R. Terkar, H. Vasudevan and V. Sunnapwar, (2011). "Importance of Innovative Product Launching and Product Life Cycle Management in Small Scale Consumer industries," *International Journal of Computer Application*, Number-3, Article-2.
16. Wi Jun-yu, Y. and Wei, ZHAOtty, (2010). "Points worth Re-Consideration Concerning Product Life Cycle Management, *IEEE*".
17. Sarvary and M.L. Wassenhove, (2009). "Remanufacturing as a Marketing Strategy," *Management Science* Vol. 54 (10), pp. 1731-1747.
18. Peizhi, Y. Jukun and Z. Sheng, (2010). "Information-based Remanufacturing Upgrade Study," *ICRM2010-Green Manufacturing,* Ningbo, China.
19. Sasikumar, G Kannan, (2008). " Issues in reverse supply chains, part I: end-of-life product recovery and inventory management - an overview," *International Journal of Sustainable Engineering*, Volume: 1, Issue: 3,Publisher: Taylor & Francis, Pages: 154-172.
20. Kim, H. W., and D. H. Lee., (2017). "An Optimal Algorithm for Selective Disassembly Sequencing with Sequence-Dependent Set-Ups in Parallel Disassembly Environment," *International Journal of Production Research* 1–17.
21. Yan, W., Chai, J., Qian, Z., Tsai, S.B., Chen, H., Xiong, Y., (2018). "Operational Decisions on Remanufacturing Outsourcing Involved with Corporate Environmental and Social Responsibility—A Sustainable Perspective," *Sustainability* **2018**, *10*, 1132.
22. Rajeev, A.; Pati, R.K.; Padhi, S.S.; Govindan, K., (2017). "Evolution of sustainability in supply chain management: A literature review," *J. Clean. Prod.* **2017**, *162*, 299–314.
23. Vasudevan, Hari, Vilas Kalamkar, and Ravi Terkar. "Remanufacturing for Sustainable Development: KeyChallenges, Elements, and Benefits." International Journal of Innovation, Management and Technology 3, no. 1 (2012): 84.

IOP Conf. Series: Materials Science and Engineering **810** (2020) 012069 doi:10.1088/1757-899X/810/1/012069

Review of Micro cantilever sensors for detecting diseases: Fabrication Methodologies and sensing techniques

Sanket S Devkare [1] **and Kiran S Bhole** [2]

[1,2] Department of Mechanical Engineering, Sardar Patel College of Engineering, Mumbai, India.

sanketdevkare95@gmail.com [1], kiran_bhole@spce.ac.in [2]

Abstract. Recently Lab on chip devices are widely used for detecting various diseases. Cantilever being used as commonly among variety of sensors used on such chips. This paper presents review of various fabrication methodology and analysis techniques used in sensing diseases using micro-cantilever sensors. The principal aim in detecting diseases using blood sample is to know concentration of antigens or bacteria in sample. This also provides stage of disease in the patient. Various techniques such as ELISA (Enzyme Linked Immunosorbent Assay), Radio-Immuno Assay, X-ray are used for detecting the stages of disease. However, time required in these techniques is very large. To reduce the time in detection, simple micro cantilever Bio-sensor is used widely. For the purpose of detection, cantilevers are coated with antibody to attract the antigens from blood sample which is dropped on cantilever beam. Antigens that get adhered to the surface of cantilever thus develop a surface stress causes bending of cantilever. This redirection of the cantilever can be detected by various strategies and thus we can predict about the concentration of the antibodies in a particular blood sample. This paper exhaustively presents the fabrication methodology and disease detection technique using micro-cantilever.

1. Introduction

This Micro Cantilever Bio-sensor usually uses an array of cantilever beams whose sizes are in range of microns and have an antibody immobilized on the surface of the cantilever beam. For the purpose of sensing the antibody is immobilized on the surface of the cantilever beam. The antigen present in the blood sample dropped on the beam gets attracted or repelled from the antibodies immobilized on the beam. The concentration of antigens to antibody thus leads to the deflection of cantilever beam developing surface stress.[1-2]

Antibodies are used to attract the antigens from the blood sample so that they get adhered to the cantilever beam. As different diseases have different antigens, different antibodies for different diseases have to be used. Antibody varies from disease to disease. A counter acting agent (antibody) is fundamentally a 'Y' formed protein that is used to identify foreign objects like bacteria and viruses. Each tip of 'Y' has a paratope that is specific for one particular epitope or an antigen.[2-4]

Cantilever sensor consists of simple structure fixed at one end and free at the other end (see figure For increase in sensitivity the structure has different variants (see figure 2, 3, 4). Here the dimensions width, length and thickness are varied to enhance the sensitivity (deflection) for the same loading conditions. Although the variants increase the sensitivity of the structure, it is difficult to develop with present micro-fabrication techniques. Bulk lithography [5] has presented the capability to produce cantilever with varying thickness. However, the minimum feature size of structure, formed are in order of 100 microns and surface roughness is in order of 10 microns. This restricts its utility as sensor. A Micro Cantilever Bio-sensor has its dimensions usually in the order of few microns ranging from 1 to 100 microns.

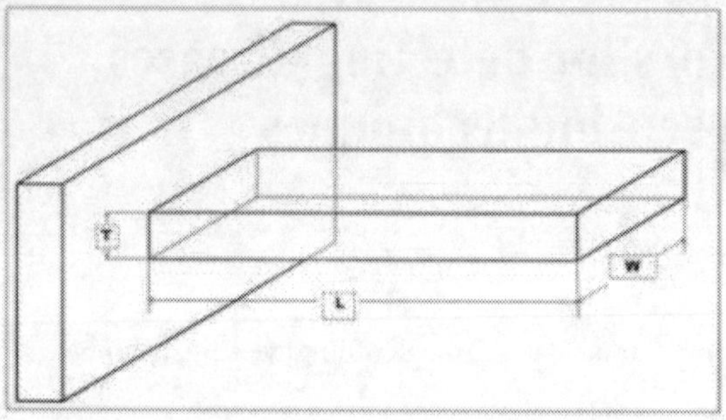

Figure 1. Dimensions of Micro-cantilever

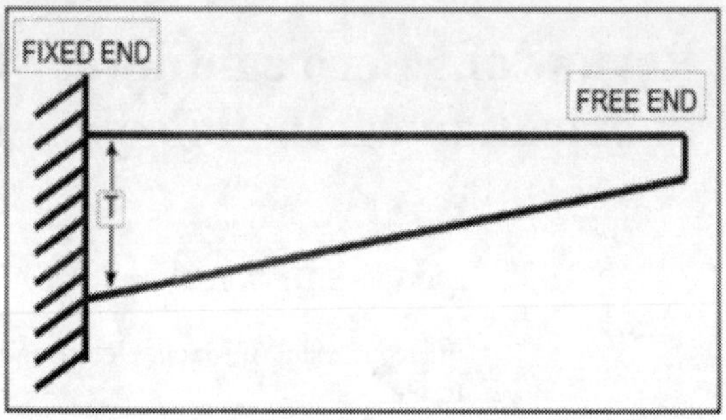

Figure 2. Front view of Micro Cantilever showing decrease in Thickness

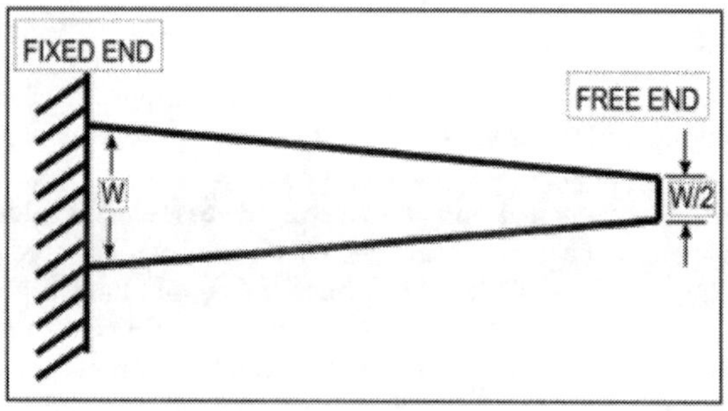

Figure 3. Top view of Micro Cantilever showing decrease in Width

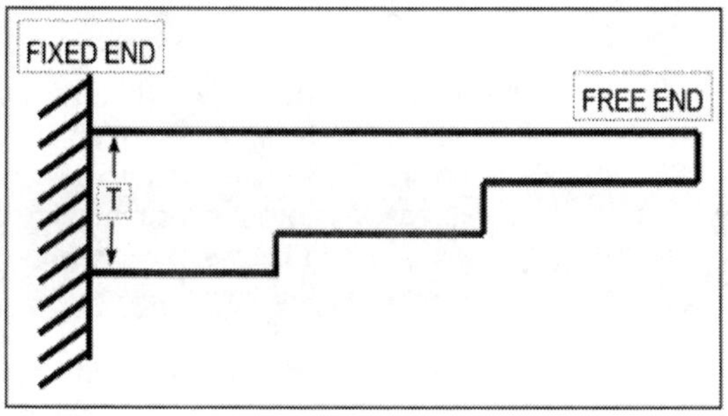

Figure 4. Front view of Stepped Cantilever

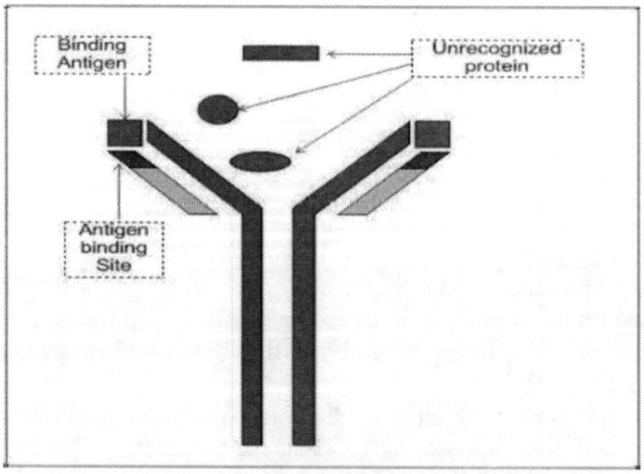

Figure 5. Y-shaped protein.

Such a Micro Cantilever beam has a series of different processes that are to be carried out in order to fabricate it. Fabrication process mainly includes the different sub processes like Thermal Oxidation,

IOP Conf. Series: Materials Science and Engineering **810** (2020) 012069 doi:10.1088/1757-899X/810/1/012069

Hot Wire Chemical Vapour Deposition (HWCVD) and also Inductively Coupled Plasma Chemical Vapour Deposition (ICPCVD). Once the deflection is obtained on the Micro Cantilever the only task remains is to sense the deflection of the beam. Many different methods have also been developed to measure the deflection of cantilever beam. These methods are mainly Optical Detection Method, Piezo Resistive Method and also Capacitive Method. [6]. This paper reviews these fabrication processes [7-11] and present their merit and demerits. Further paper also presents the different deflection measurement techniques which are used as one of the sensing parameter for determination of disease.

2. Fabrication.

The Fabrication of Micro Cantilever involves different steps of layering as follows:-

2.1 Step i. Selection of Silicon Wafer.

A p-type Silicon Wafer is used as the substrate on which the complete setup of Micro Cantilever Bio sensor is mounted. Very first the selected substrate material is cleaned by a standard process called as RCA (Radio Corporation of America) cleaning. This is done to evacuate organic as well as metallic contaminants. This is done by dipping it in different chemicals like NH_4OH , DI water, HCL HF which is followed by subsequent heating and cooling of substrate.[6]

Figure 6. RCA Cleaned Silicon Wafer

2.2 Step ii. Deposition of Silicon di-oxide (SiO_2) layer.

After RCA cleaning Silicon di-oxide (SiO_2) is deposited by a method called as Thermal oxidation. A layer of 250 nm of SiO_2 is to be deposited by thermal oxidation as shown in the setup diagram as shown below in figure 7. This is done to provide isolation between the upper layers to be deposited and the silicon substrate.[6]

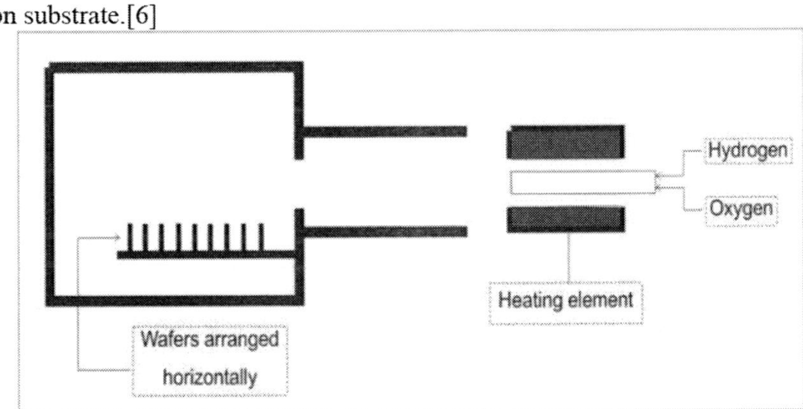

Figure 7. Thermal Oxidation of silicon wafer.

It consists of wafers arranged horizontally in the heating furnace which contains water vapor. Hydrogen and oxygen gas get heated when it is passed through the heating element. The temperature of silicon wafer is raised to temperature of about 1100°C. The Oxygen reacts with the silicon surface to form Silicon-di-Oxide (SiO_2). Thus the layer of Silicon-di-oxide is formed just by oxidizing its upper layer[5]. SiO_2 layer formed is shown as in figure 8.

IOP Conf. Series: Materials Science and Engineering **810** (2020) 012069 doi:10.1088/1757-899X/810/1/012069

Figure 8. Deposition of 250 nm of Silicon Di-oxide layer by Thermal Oxidation.

2.3 Step iii. Deposition of Poly Silicon layer:-

After deposition of SiO_2 the next step is deposition of Polysilicon and it is done by a method called HWCVD (Hot Wire Chemical Vapour Deposition). This is an important layer as it forms a structural part of Micro cantilever. The setup diagram is shown in figure 9. Subsequent to setting the substrate temperature is raised up to 300°C. Here proportion 1:5:10 is set in gas stream controller for SiH_4:B_2H_6:H_2.

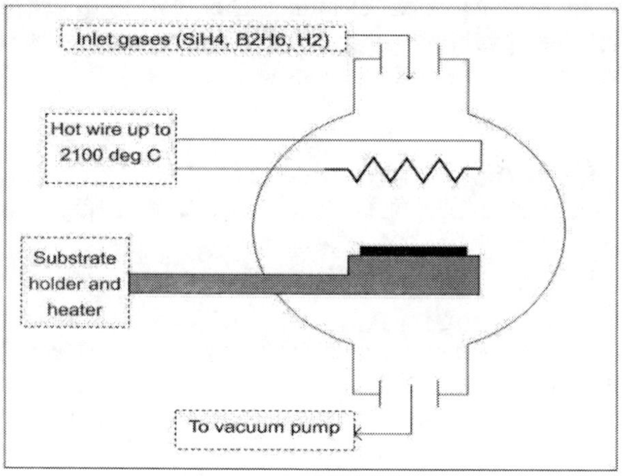

Figure 9. Hot Wire Chemical Vapour Deposition.

When these gases pass from the filament that is heated to the temperature of 2100°C, Silane gas gets decomposed to one Si atom and four H atoms. Thus a layer of Polysilicon is formed on the substrate. And when it is continued for further 30 minutes a complete layer of 1 micron is formed as shown in Figure 10. This method is specially used to have small layer deposition which is usually in the order of microns. [12]

Figure 10. Deposition of Polysilicon layer of thickness 1 micron

2.4 Step iv. Patterning of Polysilicon layer.

The subsequent stage after testimony of Polysilicon layer incorporates patterning of Polysilicon. The patterning is finished by technique called RIE (Reactive Ion Etching). It is a sort of a Dry Etching

Technique where a material is evacuated, commonly a masked material by presenting it to the barrage or bombardment of ions and is appeared in figure 11.

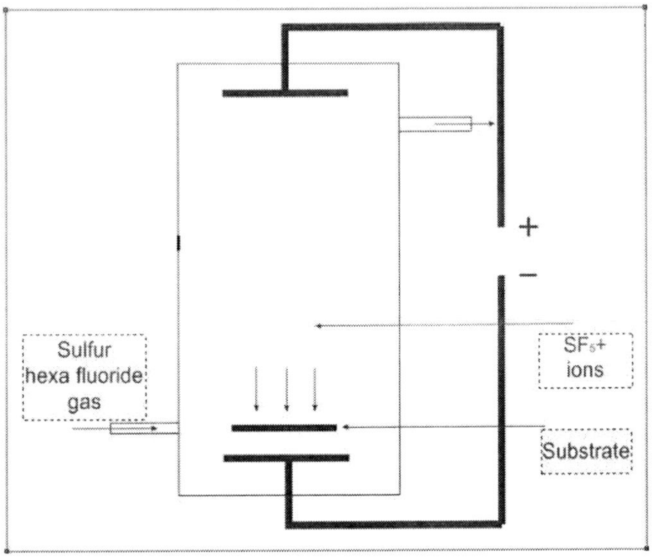

Figure 11. Reactive Ion Etching.

In this procedure for Polysilicon scratching we utilize the gas SF_6 (Sulfur Hexa-fluoride). The test arrangement comprises of the encased chamber with substrate holder with negative terminal beneath and positive terminal over the substrate. Hexafluoride gas is presented inside the chamber and at the same time a solid RF electromagnetic field of 1500W is connected. This causes the ionization of SF_6 particles making SF_5^+ particles. Because of enormous potential contrast connected over the posts SF_5^+ particles will in general float towards the Silicon substrate. These ions slam into the samples to be etched [13]. As a result the removal of the entire layer takes place and we get the electrode formed from the previous Polysilicon layer as shown in figure 12.

Figure 12. Patterning of Polysilicon.

2.5 Step v. Deposition of Sacrificial layer.

This layer is stored by the strategy for ICPCVD (Inductively Coupled Plasma Chemical Vapor Deposition). It utilizes the gases like SiH_4 and N_20 to store the layer of SiO_2. In the exploratory arrangement of ICPCVD as appeared in figure 13 the encased chamber has a channel opening that is associated with the gas chamber that contains the gases like SiH_4 and N_2O. The substrate is put on the substrate holder and is raised to the temperature of 400 °C. At the point when gases enter through the opening RF intensity of around 30 W is connected which causes a chemical reaction between SiH4

IOP Conf. Series: Materials Science and Engineering **810** (2020) 012069 doi:10.1088/1757-899X/810/1/012069

and N_2O which results into the arrangement of SiO_2. This Silicon-di-oxide layer that is created acts just as a conciliatory layer. The designing that is required as appeared in figure 14 is finished by RIE.[14]

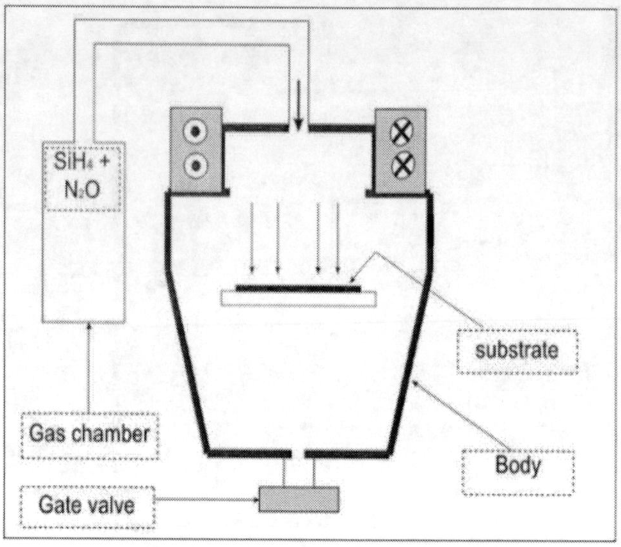

Figure 13. Inductively Coupled Plasma Chemical Vapor Deposition

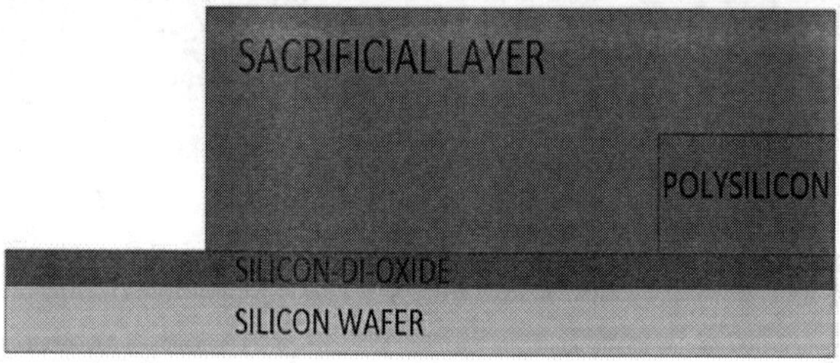

Figure 14. Deposition and Patterning of Sacrificial layer.

2.6 Step vi. Deposition of Polysilicon layer.

In the wake of patterning of Sacrificial layer, again the Polysilicon layer of thickness 1 micron must be stored as appeared in figure 15. Same HWCVD technique is utilized to deposit Polysilicon.

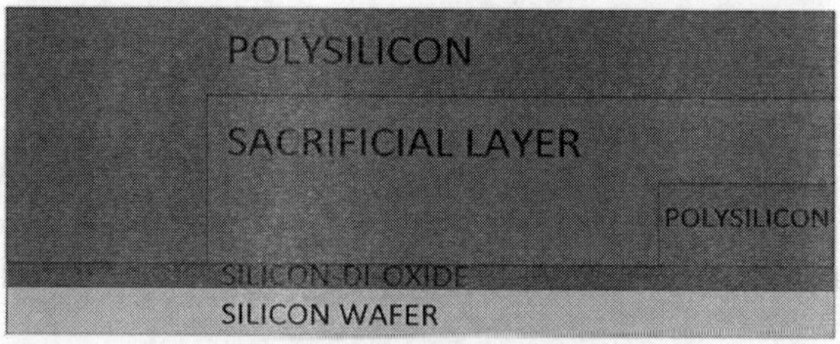

Figure 15. Deposition of Polysilicon layer.

2.8 Step vii. Patterning of Polysilicon and Removal of Sacrificial layer.

The last advance comprises of patterning upper layer and evacuation of Sacrificial layer as appeared in figure 16. Patterning is finished by same RIE strategy while Removal of Sacrificial layer is finished by dunking it in Acetone $(CH_3)_2CO$ and evaporating. This method is utilized to accomplish total evacuation of Sacrificial layer including all the base corners.

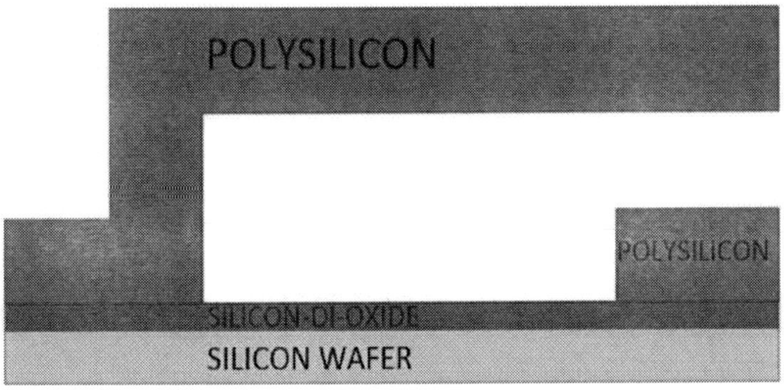

Figure 16. Patterning and removal of sacrificial layer.

3. Measuring the deflection of Micro-cantilever.

The capacity to delicately distinguish the physical changes coming about because of stresses connected is a basic part of any micro scale mechanical optical location sensor. By knowing the deflection of Micro Cantilever we can guess the amount of antigens that are immobilized on the cantilever beam from the blood sample. The methods to detect the deflection of Micro Cantilever Bio sensor are :-

a) Optical detection method.
b) Piezo Resistive Transducer.
c) Capacitive Transducer.

3.1. Optical Detection Method.

Optical Detection method utilized for estimating the diversion of miniaturized scale cantilever essentially utilizes the low power laser and Photo Sensitive Detector (PSD). At the point when light is episode on the cantilever bar by the laser diode, it gets reflected toward PSD due to gold covering present on the outside of the cantilever as appeared in figure 17. This situation of light on PSD is noted down.[15]

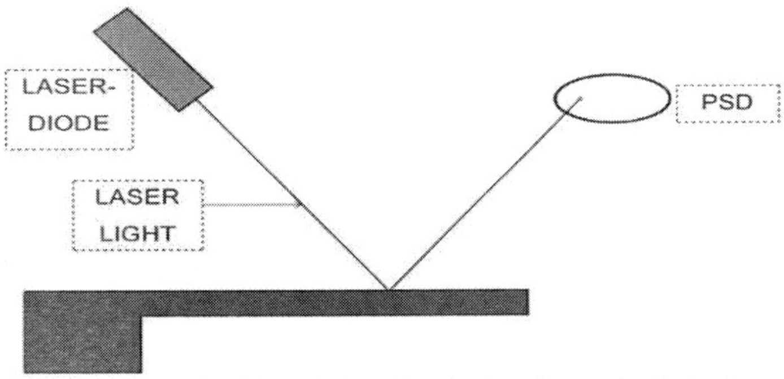

Figure 17. Reflection of light when Cantilever is undeflected.

At the point when the cantilever is deflected because of the surface burdens built up the situation of light on PSD changes which is appeared in figure 18. The distinction between two positions is only the deflection of Micro Cantilever. This sort of diversion location gadget utilizes Laser diode as well as Photo Sensitive Detector, it makes the framework progressively cumbersome and furthermore cost of the entire arrangement of framework is more. In this manner, this estimating method has turned out to be out of date.[15]

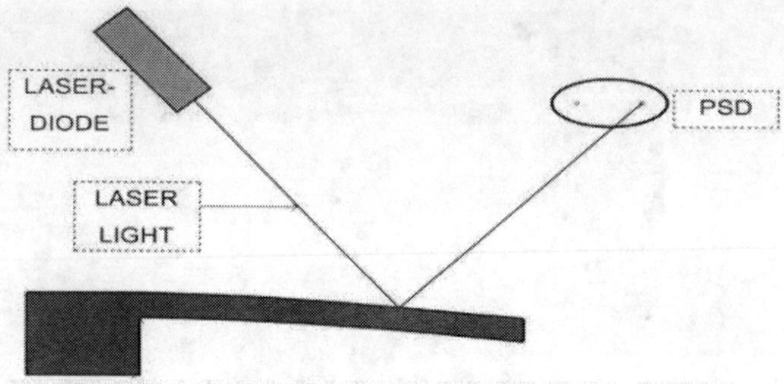

Figure 18. Reflection of light when Cantilever is deflected.

3.2. Piezo Resistive Transducer.

Creation of the Micro Cantilever including Piezoresistive transducer is somewhat unique in relation to that of standard Micro Cantilevers. The Micro Cantilever comprises of a Structural layer, Piezoresistive layer and an Antibody layer as appeared in figure 19.

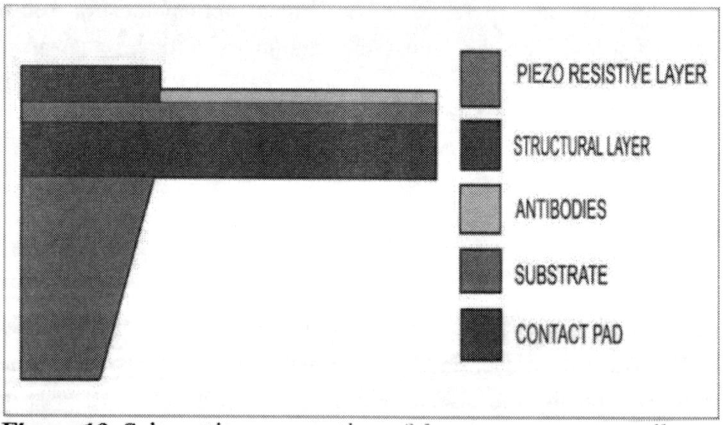

Figure 19. Schematic cross-section of the measurement cantilever.

Alongside the customary basic layer of SU8 polymer it likewise comprises of Piezoresistive layer as appeared in above figure. A twin-cantilever structure in a single chomp, the residue was seen as where one is the reference and other is the measurement cantilever as showed up in figure 20. Simply the measurement cantilever will be secured with bio-identifying particles. Subsequently just measurement cantilever will be diverted because of stress created because of antigens. This redirection is contrasted and the undeflected reference Micro Cantilever and the difference in deflection is determined. Measurement of diversion utilizing Piezoresistive transducer includes the electrical circuit as appeared in figure 21. The two resistors indicated are only the reference and measurement cantilevers separately.[16]

IOP Conf. Series: Materials Science and Engineering **810** (2020) 012069 doi:10.1088/1757-899X/810/1/012069

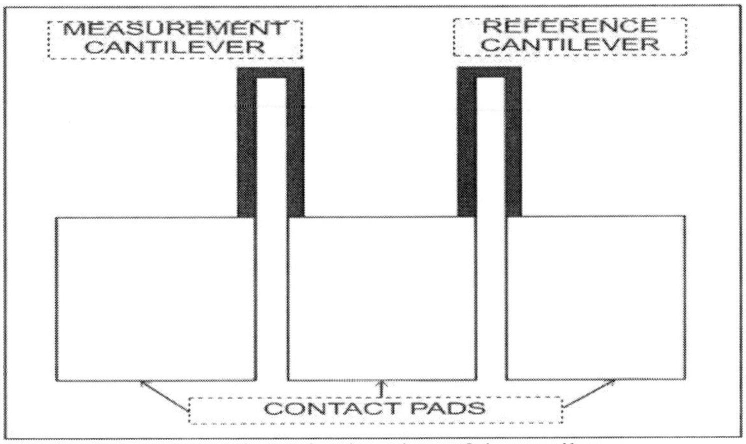

Figure 20. Schematic plan view of the cantilever

Two Piezoresistors are related to equal expansion between centers X and Y, and are continued with sine waves drivers, that are 180° stage isolated for each Piezoresistor. The voltage divider center point A sets at 0mV if the resistors are undefined. Due to (delta R) changes in the estimation cantilever in diverted conditions, a little AC voltage appears at the center point A. This signal is intensified and associated with the signal contribution of a lock in Amplifier (LIA), in light of the fact that the significance of the sign at A degrees from nanovolts up to two or three microvolts and thus it ought to be enhanced. The yield of LIA is a DC voltage which is relative to the block assortment (del R) which is with respect to avoidance in the measurement cantilever.[16]

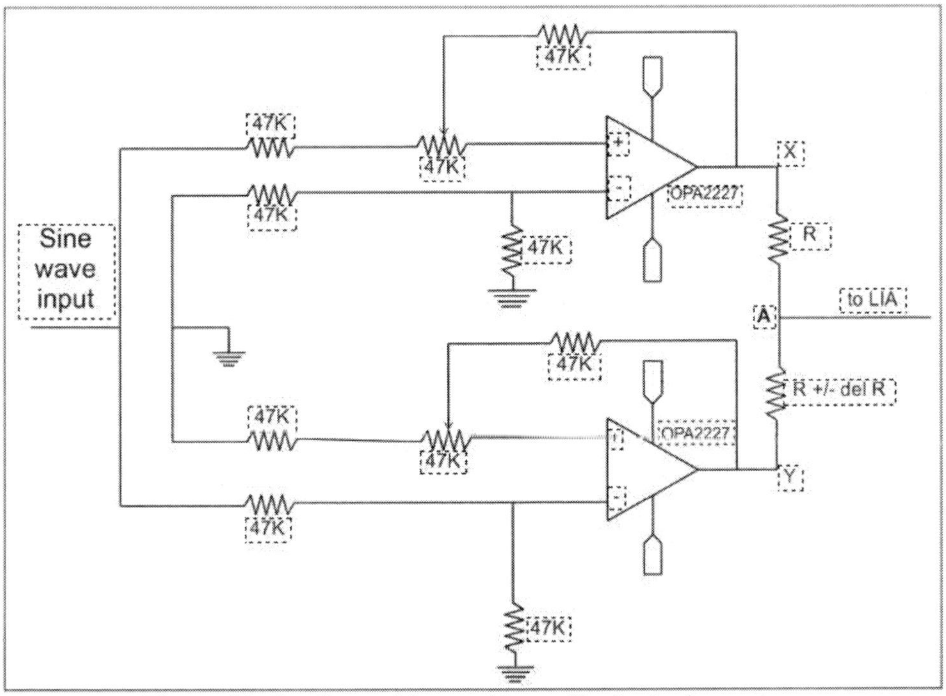

Figure 21. Cantilever drivers with amplitude altering system.

3.3. Capacitive Transducer.

The Capacitive technique depends on the rule that when the cantilever diversion happens because of the adsorption of the analyte, the Capacitance of a plane Capacitor is changed. The figure 22 proposes the plan of Micro-cantilever and fixed shaft. The contraption showed in the figure works for a couple of stages, and the amount of portable and fixed electrodes may be extended for further stages. The capacitance between the versatile and fixed electrodes change as the column diverts on the usage of bio molecule. The adjustment in capacitance is subsequently transuded as the proportion of deflection. [17-19]

Figure 22. Bio-MEMS cantilever device

4. Conclusion.

Micro Cantilevers have been demonstrated helpful to determine the illness to have a less time when contrasted with different procedures. Its creation is an impressive huge procedure that includes distinctive sub procedures like RCA cleaning, warm oxidation, HW-CVD, ICP-CVD and Reactive Ion Etching. Out of the considerable number of strategies for estimating the deflection of cantilever Piezo Resistive technique is for the mostly utilized because of its precision.

5. References.

[1] P Sangeetha, Dr. A Vimala Juliet 2014 *Simulation and Analysis of Micro Cantilever Sensor for Enhanced Bio sensing of Disease causing Pathogens, IJEDR Conference Proceeding (NCETSE-2014)* 131.

[2] Tzuen-Rong J Tzeng, Yunyan R Cheng 2011 *Adhesin-Specific Nano mechanical Cantilever Biosensors for Detection of Microorganisms, ASME* 1.

[3] Sneha Dhakane and W V Patil 2013 *Piezo-resistive method for tuberculosis detection using Micro-cantilever biosensor, Micro system Technologies* 457.

[4] P.Sangeetha and Dr. A Vimala Juliet 2013 *Biosensor for Tuberculosis detection using MEMS device, 3rd International Conference on Electronics, Biomedical Engineering and its Applications* 52.

[5] Deep Kishore Parsediya 2016 *Deflection and Stresses of Effective Micro-cantilever Beam designs under low mass loading, International Conference on Electrical Power and Energy Systems (ICEPES)* 1111.

[6] Dr. Rajendra Patrikar , Mrs. Jayu Kalambe 2013 *Fabrication of Micro-cantilever Based Sensor, Research Project at Indian Institute of Technology Bombay under Indian Nano electronics User Program (INUP) sponsored by DIT, MCIT Government of India* 1.

[7] Prasanna Gandhi and Kiran Bhole 2011 *USA 3D Microfabrication using Bulk Lithography, ASME International Mechanical Engineering Congress and Expositio* 393.

[8] Kiran Bhole, Sunil Ekshinge, and Prasanna Gandhi 2014 *USA Fabrication of Continuously Varying Thickness Micro-cantilever using Bulk Lithography Process, ASME International Manufacturing Science and Engineering Conference* V001T01A009.

[9] Kiran Bhole, Prasanna Gandhi, and T Kundu,*2014 On the Evolution of Cured Voxel in Bulk Photopolymerization upon Gaussian Laser Exposure, Journal of Applied Physics*, 116.

[10] Prasanna Gandhi and Kiran Bhole 2013 *Characterization of 'Bulk Lithography' Process for Fabrication of 3D Micro-structures, ASME Journal on Nano and Micro Manufacturing* 1.

[11] Prasanna Gandhi, Suhas Deshmukh, Rahul Ramtekkar, Kiran Bhole, and Alem Baraki 2013 *On-axis Linear Focused Spot Scanning Microstereolithography System: Optomechatronic Design, Analysis and Development, Journal of Advanced Manufacturing Systems* **12(1)** 43.

[12] J Cifre, J Bertomeu, J Puigdollers, M C Polo, J Andreu, A Lloret 1994 *Polycrystalline silicon films obtained by Hot Wire Chemical Vapour Deposition, Applied Physics* **A** 645.

[13] M Avaram, A Avaram, F Comanescu, A M Popescu, C Voitincu 2009 *Romania Reactive Ion Etching for patterning high aspect ratio and Nano scale features, National Institute for Research and Development in Micro technologies* 253.

[14] Zewen Zuo, Yu Wang, Yu Xin, Jin Lu, Junzhuan Wang, Lin Pu, Yi Shi 2010 *Structural evolution of the incubation layer in microcrystalline silicon films deposited by Jet- ICPCVD* 1.

[15] Sandeep Kumar Vashist 2007 *A Review of Micro cantilevers for Sensing Applications, AZojono , A journal on nanotechnology online* 1.

[16] Sudip Nag, Nitin S Kale, V Ramgopal Rao, Dinesh K Sharma 2009 *An Ultra-sensitive R/R Measurement System for Biochemical Sensors using Piezoresistive Micro-Cantilevers, 31st Annual International Conference of the IEEE EMBS* 3794.

[17] M Ahsan Saeed , Shahbaz M Khan, Nisar Ahmed, M Umair Khan *2016 Design and Analysis of Capacitance based Bio MEMS Cantilever Sensor for Tuberculosis Detection , IEEE* 1.

[18] E A Wachter, T Thundat, P I Oden, and R J Warmack 1996 *Remote optical detection using micro cantilevers, American Institute of Physics* 3434.

[19] Christiane Ziegler 2004 *Cantilever-based biosensors, Analytical and Bio analytical Chemistry* 946.

IOP Conf. Series: Materials Science and Engineering **810** (2020) 012070 doi:10.1088/1757-899X/810/1/012070

Review on Mechanical, Thermal and Morphological Characterization of Hemp Fiber Composite

Divyansh Sankhla, Mihir Sancheti, M. Ramachandran
MPSTME, SVKM'S NMIMS University, Dhule 425405, Maharashtra, India
sweetestchandran@gmail.com

Abstract. Hemp, (Cannabis sativa), also known as industrial hemp, a family plant of Cannabaceae cultivated for its fiber or edible seeds. Hemp is sometimes confused with cannabis plants that serve as sources of prescription weeds and preparation of hashish products. Although all three products— hemp, marijuana, and hashish — contain tetrahydrocannabinol (THC), a compound that induces psychoactive effects in humans, the cannabis variety grown for hemp has only small amounts of THC compared to that grown for marijuana or hashish growth. Bast fiber plants are distinguished by long slender primary fibers on the outer portion of the stem. Hemp grows in countries like Canada, USA, France, Hungary, Belgium, Holland, Thailand, Austria, Italy, China, and Philippine island, Russia, Mexico, Germany, West Indies and India. In this paper we analyze all these test on hemp Fibre- differential scanning Calorimetry test, scanning electron microscope test, tensile test, thermo gravimetric analysis and flexural test.

1. Introduction

The surface morphologies of untreated strands, silane coupling operator treated filaments, NaoH treated filaments were analyzed, utilizing a filtering electron magnifying lens and the break surfaces of the tractable composite examples were inspected and covered with gold. When soluble base treatment the cracked surfaces of tractable tried composite examples were examined with the assistance of hemp strands. The treated and untreated fiber surfaces were graphically spoken to by utilizing DSC (Differential examining Calorimetry) bends. Utilizing a DSC Q2000 from TA Instruments Differential checking Calorimetry. (DSC) was performed under helium stream. The DSC bends demonstrated that filaments had less measure of level of dampness in light of the fact that before the treatment they have been dried. Hemp filaments proclaimed as noil because of their short length and there treatment was finished with coupling specialist. At the point when a coupling specialist was included there was an improvement found in the rigidity of the composites. The elasticity is emphatically impacted by the (Hemp/Starch) proportion. To distinguish the warm corruption procedure of the untreated and treated fiber Thermo gravimetric Analysis (TGA) was utilized and unadulterated PLA and hemp-PLA composites were recognized. ASTM D 790 were led utilizing a steam ductile testing machine three-point bowing flexural tests. 130 ×15 × 5 mm measurement was utilized to cut the Flexural examples. A thickness of 1.19 g/ cm^3 and hardness of 95 (shore A) was utilized from Thermoplastic polyurethane from Epaflex (Epamould 295 A10). Test Methods for Reinforced Plastics, Flexural Properties of Unreinforced, and Electrical Insulating Materials utilizing an INSTRON SATEC APEX T5000 testing outline with a crosshead speed of 5 mm/min was done.

2. Differential Scanning Calorimetry

In the terms of warmth stream the DSC bends for untreated and surface-treated strands were expressed. The liquefying point (M.P) of cellulose has been accounted for to be close to 300°C, yet this made a contention because a temperature of 30°C and 160°C was acquired for the glass change .The peak on the diagram was gotten because of the weight reduction in light of the water vanishing which was found in tests of all-fiber, and in hemp filaments it was gone from 10°C to 130°C, contingent on the level of hydrogen holding collaboration. This change moved towards the right after surface treatment of the hemp fiber. Since they had been dried before any sort of treatment so the filaments had less dampness rate. This guaranteed the inalienable dampness of the filaments, which is 12–14% of the all-out fiber weight, was not ready to connect with the surface medications in any capacity. The differential checking calorimeter (DSC) assessment was performed utilizing a Perkin Elmer DSC-6 chipping away at a Pyris stage. DSC was adjusted utilizing indium and cyclohexane before the tests. In a nitrogen atmosphere The Perkin Elmer film of 5 mg was analyzed. It was chilled off to 30 °C after the filtering pace of the DSC tests was set to be 10 °C min−1. Tests were warmed to 200 °C, tempered for 20 min. The cycle was then rehashed. Use of a TA Instruments DSC Q2000 under helium stream (100 mL / min) Calorimetry differential analysis (DSC) has been performed. From room temperature to 200 ° C, isothermal for 3 min, cooled at that point to 50 ° C and warmed again to 190 ° C at a steady warming / cooling rate of 10 ° C / min, examples weighing 10 and 12 mg were heated up. HF was treated with composites by the DSC dissolving bends (second sweep) 40 wt percent. For silane treated filaments compared to PP for untreated HF, temperature expansion was observed for composites, higher for PP / HF–MPS, approximately 58 ° C, lower for PP / HF–GPS (1.7 ° C) and moderate for PP / HF–APS (38 ° C) was observed.

3. Scanning Electron Microscope

The significant purpose behind their improved weariness quality, properties, and affectability, as a result of these micrographs improved interfacial attachment of fiber or lattice contrasted and non-alkalized fiber composites. For correlation, the Scanning electron magnifying instrument (SEM) micrograph of the "crack" surface of 5% alkalized test that didn't bomb after 106 cycles of weariness testing at an elastic worry of 34 Mpa (practically 70% of static quality). The example that were browsed among all the alkalized fiber tests weakness tried, these example showed the most noteworthy continuance limit. Use of a pinnacle weariness worry of 70% of the static quality didn't debilitate the interfacial holding and the material was sufficiently able to stop the proliferation of subsurface splits radiating in the material. Distinctive surface treatment focuses were examined utilizing SEM micrographs of the hemp or PP composites. It was seen that the interfacial holding between the untreated hemp fiber and the PP pitch isn't solid, as showed by the hole between them. Subsequently, the elasticity of this composite is low. NaoH treatment can expel the slight layer from the outside of the filaments, bringing about improved grip with the PP sap. The difference in fiber morphology is affirmed by SEM perceptions on cleaned segments of RPP or hemp from various cycles that were watched. These perceptions have been done at the center of the example, the photos feature the nearness of groups and long strands after the primary infusion cycle. These packs have almost vanished after numerous cycles; just short and well-isolated filaments remain. SEM micrographs of the ductile break surface of tests containing 30% and 40 % of filaments. Fiber haul out is seen in certain spots and the crack surfaces are not one way and have rough ways, this demonstrated bendable disappointments. The grid can't completely wet the fiber surfaces as fiber content increment to 40%, and keeping in mind that experiences malleable pressure, fiber destroy out happens because of fiber and lattice diverse pressure strain practices. Fiber fibrillation in which micro fibrils are isolated and turn out from the cell divider is watched. Composite crack surfaces upheld improved interfacial holding impacting quality. Break surfaces for more fragile composites, for example, F13 for the most part indicated de-holding among fiber and network, while it was hard to perceive the genuine interface between the fiber and framework for more grounded composites, for a consolidated soluble base and D2B treatment. The SEM images for a superior association of cryo-cracked PP composite surfaces with untreated and handled HF at a comparable amplification. The composite with untreated strands revealed holes in the pullout filaments and interstices at the fiber or frame interface, indicating weak interfacial bonding. Just the high soft stream speed of PP used as a lattice without treatment and the hardness of the surface of HF will cover PP filaments. Due to the weak grip between the APS treated HF and, in particular, due to GPS treated filaments and PP, the SEM images of the HF–APS and HF–

IOP Conf. Series: Materials Science and Engineering **810** (2020) 012070 doi:10.1088/1757-899X/810/1/012070

GPS composites display broken strands and at the fiber or grid interface often haul out filaments and gaps. It suggests stronger HF or PP interface partiality due to MAPP's compacting operation. Great interface connection was also observed for HF–MPS and HF–KP composites. The expectations of SEM are in great competition with the variety of versatility modules which is the pressure of the filaments with regard to their strain.

4. Tensile Strength

Hemp filaments having short length was announced as noil and treated with NaoH, and it was sandwiched with (BA-a) benzoxazine to frame composites. At an estimation of 48.87 ± 1.24 Mpa the THF composite fizzled at a mean worry when contrasted with 23.28 ± 0.76 Mpa of local poly (BA-a) tar, this brought about an expansion of 109 % in tractable pressure when 25% THF was consolidated in the composite. The abatement in strain esteems was seen of THF by joining, 1.2% on 25 qualities came to THF fuse. 1.6% was the recorded base an incentive for poly (BA-a) framework. At the point when 10% of THF is treated with poly (BA-an) it was fizzled at mean pressure estimation of 36 Mpa, when 15% of THF is treated with poly (BA-an) it was fizzled at mean pressure estimation of 41.26 Mpa. At the point when 20% of the is treated with poly (BA-a) faile at mean pressure estimation of 46.82 Mpa. The elasticity of the composites improved when a coupling specialist was included. The malleable worry of pp was seen as 31.1MPa. The rigidity of 10% of hemp without including any coupling specialist was discovered to be 26.7 Mpa. The elasticity of 20% of hemp without including any coupling specialist was discovered to be 28.9 Mpa. The elasticity of 30% of hemp without including any coupling specialist was discovered to be 28 Mpa. The rigidity of 40% of hemp without including any coupling specialist was discovered to be 28.1 Mpa. The rigidity of half of hemp without including any coupling operator was discovered to be 29.3 Mpa. The rigidity of 60% of hemp without including any coupling operator was discovered to be 24.8 Mpa. This delineates the effect of MAPOE and MAPP on ductile composites with 40 wt. % and 30 wt% noil hemp fiber. Almost no difference was identified between 2.5 wt. percent fuse and 5 wt. % PP-MAP is 39.7MPa, 39.3MPa, 36.5 Mpa, and 40.8 Mpa individually. While the expansion of 5 wt. percent PP-MAP resulted in the best improvement among the composites for the composite with 40 wt. percent hemp fiber. The H/S proportion for size S2 is taken out to be 6, the limit was discovered to be at 0.084 Mpa. The H / S proportion for size S2 is taken out to be 6 and the maximum was found to be 0.084 Mpa. For 8, 10, 12, 14, 0.076MPa, 0.06 Mpa, 0.051MPa and 0.036 Mpa are discovered separately. The standard stiffness ranges from 0.13 Mpa to 0.08 Mpa for S1 and from 0.08 Mpa to 0.03 Mpa for S2 when the H / S ratio varies from 6 to 14. The results show that when the H / S ratio increases, the stiffness decreases. The hardness of the S1 size composite materials for a given H / S ratio is greater than that of the S2 size composite materials. Those values are not exactly those obtained on the concrete of 0.7e 1.2 Mpa lime-hemp. The elasticity and module of untreated hemp strands based on bio composite were 45 percent higher and 325 percent higher than that of flawless gum, respectively. The elasticity of fiber-based bio composite treated with a soluble base was 34% higher than that of untreated hemp-based composite, whereas that of silane-based bio composite was 48% higher than that of untreated hemp composite. The toughness of the composite E-glass–UPE was 130% higher. The half breed UPE composite E-glass–hemp tangle had an extra stiffness of 76 percent. The durability of untreated and handled base hemp fiber is illustrated as the hemp fiber ductile was found to be 600 Mpa at 0 % antacid. The hemp fiber is discovered to be 780 Mpa with 5 percent salt NaoH.

5. Thermo Gravimetric Analysis

The thermo gravimetric study shows the TGA bends for the fiber and virgin treated and untreated and reused HDPE. The fiber had two-pronged forms of corruption: the primary degradation was in the range of 225–275 ° C due to hemicellulose degradation, and the subsequent was in the range of 325–360 ° C due to lignin degradation. The warm degradation of 99wt percent polymer began for virgin HDPE at 402.84 ° C and for r-HDPE at 420 ° C. The deterioration was completed by a 2-3 percent increase for v-HDPE at 486.52 ° C and r-HDPE at 486.69 ° C. Temperatures of underlying degradation of woven hemp (HH), woven PET (PP) and half-and - a-half (HP / PH) composites were 92.5 ° C, 100 ° C, 98.8 ° C. The last temperatures of abuse were 470 composites of woven hemp (HH), woven PET (PP) and intertwined cross breed (HP / PH). The deterioration in the example speaks to the corruption of gelatin, hemi-cellulose, and lignin past 170 °C, and the distinction between these bends can be clarified by the way that the fume fixation started emphatically from the center of the example. In this way, the center of the example was debased at 190°C. From that point forward, the corruption of the cellulose began when the temperature expanded to more than 200-230 °C. This conduct is generally observed in lignocellulose materials. TGA was utilized to distinguish the warm debasement procedure of the untreated and treated fiber, unadulterated PLA and hemp-PLA composites. At first, the impact of fiber stacking on warm decay was contemplated and dependent on these outcomes it was inferred that 30% fiber volume division is the ideal stacking of hemp fiber. In this manner, the initiation vitality of warm debasement for the composites was researched with a 30% fiber volume division (both with untreated and treated hemp-PLA composites). The warming rate ran from 5 to 30 °C/min in a nitrogen environment.]. A important capacity for the production of 23 fiber / polymer composites is the warm solidity of the filaments. The thermoplastic composites have a temperature reduction of around 150–220 ° C, whereas normal strands have a higher rate of contamination during this period, which limits the support of characteristic filaments in composites to 25. The TGA and DTG bends from 50 to 800 ° C in nitrogen air for NHF, 4 THF, immaculate poly(BA-an) and composites are depicted and T5 percent, T10 percent, and Yc.

6. Flexural Strength

Flexural properties of TPU-treated hemp fiber composite at various fiber substance. At 0% fiber substance, the flexural quality was 7.55((±0.05) Mpa and the flexural modulus was 156.78(±13.30) Mpa. At the point when the fiber content was 20% the flexural quality was 10.73(±0.85) Mpa and the flexural modulus was 217.93(±28.04) Mpa. At the point when the fiber content was 30% the flexural quality was 20.40(±2.03) Mpa and the flexural modulus was 554.85(±138.90) Mpa. At the point when the fiber content was 40%the flexural quality was 22.14(±0.48) Mpa and the flexural modulus was 586.83(±7.03) Mpa. Polypropylene was utilized. At the point when the fiber was slick PP flexural quality was between 56(±2) Mpa. At the point when the fiber was untreated with NaoH the flexural quality was 54(±1) Mpa. At the point when the fiber was treated with 2% NaoH the flexural quality was 53(±1) Mpa. At the point when the fiber was treated with 4% NaoH the flexural quality was 60(±2) Mpa. At the point when the fiber was treated with 6%NaOH the flexural quality was 55MPa. At the point when the fiber was treated with 2%silane the flexural quality was 55MPa and when the fiber was treated with 4%silane and 6% silane the flexural quality was 55 Mpa for both separately. The fiber utilized was flax PP composites. At the point when the fiber was treated with 22% NaoH for 60 min untreated and temperature at 4C, 10C, 20C the qualities for flexural quality were 148,226,228,219 individually and the estimation of flexural modulus was 5919,11995,11726,12351 separately. Medications to the Flexural Strength of Unidirectional Flax-PP-Composites, when untreated 77N/mm2, when mercerized 115N/mm2, MAH PP coupling specialist 127N/mm2 and when both mercerized and treated with MAH PP coupling operator 149N/mm2 [19]. The fiber utilized was short length hemp fiber proclaimed as noil treated with NaoH. At the point when the THF content was 0%,5%,10%,15%,20% and 25% the flexural quality were 122MPa, 124 Mpa, 125MPa, 126.5MPa, 127 Mpa and 123 Mpa separately and the flexural modulus was 3.5Gpa, 4.75Gpa , 5GPa , 5.5Gpa, 5.65Gpa and 5.3Gpa individually.

7. Conclusion

Hemp fibres contains very less amount of THC content. Hemp fibres have properties that make it a suitable material to replace glass fibres as reinforcements in composite material. The composite made from hemp fibres shows promising mechanical properties which in some cases, surpass even those of glass fibres composite. For silane treated filaments compared to PP for untreated HF, temperature expansion was observed for composites, higher for PP / HF–MPS, approximately 58 ° C, lower for PP / HF–GPS (1.7 ° C) and moderate for PP / HF–APS (38 ° C) was observed. The H / S proportion for size S2 is taken out to be 6 and the maximum was found to be 0.084 Mpa. As the H / S ratio increases, the stiffness decreases. The TGA and DTG bends from 50 to 800 ° C in nitrogen air for NHF, 4 THF, immaculate poly(BA-an) and composites are depicted and T5 percent, T10 percent, and Yc. At the point when the THF content was 20% we get the highest value of flexural quality [127 Mpa] and the flexural modulus was 5.65Gpa. At 0% alkali the tensile of hemp fibres was found out to be 600 MPa. The interfacial holding between the untreated hemp fiber and the PP pitch isn't solid without any treatment.

References

[1]. Haghighatnia, Tina, Ali Abbasian, and Jalil Morshedian. "Hemp fiber fortified thermoplastic polyurethane composite: An examination in mechanical properties." Industrial Crops and Products 108 (2017): 853-863.
[2]. Pickering, Kim L., et al. "Interfacial adjustment of hemp fiber-fortified composites utilizing contagious and soluble base treatment." Journal of Biobased Materials and Bioenergy 1.1 (2007): 109-117.
[3]. P. M. Bhagwat, M. Ramachandran, Pramod Raichurkar, Mechanical Properties of Hybrid Glass/Carbon Fiber Reinforced Epoxy Composites, Materials Today: Proceedings, 4(8), 2017:1788–1793.
[4]. Panaitescu, Denis Mihaela, et al. "Impact of hemp filaments with the adjusted surface on polypropylene composites." Journal of Industrial and Engineering Chemistry 37 (2016): 137-146.
[5]. Shahzad, Asim. "Impacts of alkalization on elastic, effect, and weakness properties of hemp fiber composites." Polymer Composites 33.7 (2012): 1129-1140.
[6]. Mehta, Geeta, et al. "Impact of fiber surface treatment on the properties of biocomposites from nonwoven modern hemp fiber mats and unsaturated polyester pitch." Journal of applied polymer science 99.3 (2006): 1055-1068.
[7]. Sahas Bansal, M. Ramachandran, Pramod Raichurkar, Comparative Analysis of Bamboo using Jute and Coir Fibre Reinforced Polymeric Composites, Materials Today: Proceedings,4(2), 2017, 3182–3187.
[8]. Dayo, Abdul Qadeer, et al. "Normal hemp fiber strengthened polybenzoxazine composites: Curing conduct, mechanical and warm properties." Composites Science and Technology144 (2017): 114-124.
[9]. Etaati, Amir, et al. "The investigation of fiber/framework bond quality in short hemp polypropylene composites from the dynamic mechanical examination." Composites Part B: Engineering 62 (2014): 19-28.
[10]. Le, A. T., et al. "Trial examination on the mechanical presentation of starch–hemp composite materials." Construction and Building Materials 61 (2014): 106-113.
[11]. Ramanan. G, Rajesh Prabha. N, Diju Samuel. G, Jai Aultrin. K. S, M Ramachandran, Prediction of Machining Characteristics of Hybrid Composites Using Response Surface Methodology Approach, International Journal of Engineering & Technology, 2018, 7(3.1), 162-165.
[12]. Le, A. T., et al. "Impact of different starch/hemp blends on mechanical and acoustical conduct of starch-hemp composite materials." Composites Part B: Engineering 75 (2015): 201-211.
[13]. Sair, S., et al. "Impact of surface adjustment on morphological, mechanical and warm conductivity of hemp fiber: Characterization of the interface of hemp–Polyurethane composite." Case thinks about in warm designing 10 (2017): 550-559.
[14]. M. Ramachandran, , Sahas Bansal, Pramod Raichurkar, Scrutiny of Jute Fiber Poly-Lactic Acid (PLA) Resin Reinforced Polymeric Composite, Journal of the Textile Association, Volume 76, Issue 6, 2016, pp. 372-375.
[15]. Oza, Shubhashini, Ruoyang Wang, and Na Lu. "Warm and mechanical properties of reused high-thickness polyethylene/hemp fiber composites." International Journal of Applied Science and Technology 1.5 (2011).
[16]. D. Bino prince raja, B. Stanly Jones Retnam, M. Ramachandran, Analysis of mechanical properties of glass and carbon fiber reinforced polymer material, International Journal of Applied Engineering Research. ISSN 0973-4562 Volume 10, Number 11 (2015) pp. 10387-10391.
[17]. Ahmad, M. A. An., et al. "Dynamic mechanical examination and impacts of dampness on mechanical properties of intertwined hemp/polyethylene terephthalate (PET) half breed composites." Construction and Building Materials 179 (2018): 265-276.
[18]. Almusawi, An., et al. "Proposition of assembling and portrayal trial of binderless hemp shive composite." International Biodeterioration and Biodegradation 115 (2016): 302-307.
[19]. Alex. S, Stanly Johns Retnam, M. Ramachandran, A review on Biodegradability of Hybrid Bamboo/Glass fiber polymer composites, International Journal of Applied Engineering Research. ISSN 0973-4562 Volume 10, Number 11 (2015) pp. 10565-10569.
[20]. Oza, Shubhashini, et al. "Impact of surface treatment on the warm strength of the hemp-PLA composites: Correlation of enactment vitality with warm debasement." Composites Part B: Engineering 67 (2014): 227-232.
[21]. Haghighatnia, Tina, Ali Abbasian, and Jalil Morshedian. "Hemp fiber fortified thermoplastic polyurethane composite: An examination in mechanical properties." Industrial Crops and Products 108 (2017): 853-863.
[22]. Wu, Yingji, et al. "Water-safe hemp fiber-fortified composites: In-situ surface assurance by polyethylene film." Industrial Crops and Products 112 (2018): 210-216.
[23]. Suardana, N. P. G., Yingjun Piao, and Jae Kyoo Lim. "Mechanical properties of hemp filaments and hemp/pp composites: impacts of synthetic surface treatment." Materials material science and mechanics 11.1 (2011): 1-8.
[24]. Bourmaud, Alain, Antoine Le Duigou, and Christophe Baley. "What is the specialized and natural enthusiasm for reusing a reused polypropylene–hemp fiber composite?." Polymer Degradation and Stability 96.10 (2011): 1732-1739.

[25]. Kanak Kalita, U Ragavendran, M. Ramachandran, Akash Bhoi, Weighted sum multi-objective optimization of skew composite laminates, Structural Engineering and Mechanics, Vol. 69, No. 1 (2019) 21-31.
[26]. Sefiu Adekunle Bello, Isiaka Ayobi Raheem, Nasir Kolawole Raji, Study of tensile properties, fractography and morphology of aluminium (1xxx)/coconut shell micro particle composites, Journal of King Saud University - Engineering Sciences, Vol29, Issue 3, July 2017, Pages 269-277
[27]. Liyanage, Chinthani D., and Mevan Pieris. "A Physico-Chemical Analysis of Coconut Shell Powder." Procedia Chemistry 16 (2015): 222-228.
[28]. Agarwal, Rakshit, M. Ramachandran, and Stanly Jones Retnam. "Tensile Properties of Reinforced Plastic Material Composites with Natural Fiber and Filler Material." ARPN Journal of Engineering and Applied Sciences 10, no. 5 (2015): 2217-2220.
[29]. Siti Nurbazilah Ab Jabal, Yew Been Seok, Wee Fwen Hoon, carbon composition, surface porosities and dielectric properties of coconut shell powder and coconut shell activated carbon composites, ARPN Journal of Engineering and Applied Sciences, VOL. 11, NO. 6, MARCH 2016, 3832- 3838.
[30]. Shyamkumar Shah, Akshaykumar Patil, M. Ramachandran, Kanak Kalita, Effect of coal ash as a filler on mechanical properties of glass fiber reinforced material, International Journal of Applied Engineering Research, 9(22) (2014) pp. 14269-14277.

IOP Conf. Series: Materials Science and Engineering **810** (2020) 012071 doi:10.1088/1757-899X/810/1/012071

SonReb Method for Evaluation of Compressive Strength of Concrete

Narayan R Chandak, Hemraj R Kumavat

Professor and Head, Civil Engineering Department, MPSTME, NMIMS Shirpur

Research Scholar, Civil Engineering Department, MPSTME, NMIMS Shirpur

narayan.chandak@nmims.edu, kumavathr1981@gmail.com

Abstract. Ultrasonic Pulse Velocity (UPV) and Rebound Hammer (RH) are continually adopted methods by the researchers to assess the characteristics and to estimate the compressive strength of the concrete. As matrix of concrete is very complex, it's obvious that the strength of concrete can be influenced due to even minor change in any of the factor. The factors that influenced the strength of concrete largely are the type and size of aggregates, cement content, physical and mechanical factors. To evaluate the compressive strength considering these factors with higher precision, UPV and RH methods are combined henceforth called SonReb. The results of testing of on-site casted cube samples and cores samples taken from the existing buildings in Pune, India are presented here. The application of the regression model to observe destructive and non-destructive testing were used through SonReb method to evaluate the reliability of concrete. Equations are obtained using multiple regression statistical analysis (MRSA) to access concrete compressive strength and the accuracy of the strength prediction technique is discussed in this paper. The regression curve of proposed model strongly consistent with the models suggested by Meynink, Ramyar et al. and Khedar et.al. It is also concluded that the results obtained from core are underestimate, which reflects the age factor when compared with the samples tested in laboratory. It is also seen that the regression coefficient of both the formulation are closely matching, which indicates no effect of age of samples tested. SonReb results seems to be more accurate and reliable than any individual test performed on the same samples.

1. Introduction

These It is of great interest to evaluate concrete properties, whether to detect enhanced regions or to control the quality of concrete and to estimate its compressive strength. Typically the most searched over attribute is the concrete compressive strength. This allows the combination of UPV and RH results [1]. Major benefit of non-destructive test approach is to prevent concrete disruption to the reliability of structural elements of building. Concrete testing in structures involves the use of less expensive equipment in which the cores cannot be drilled. [2]. All obtainable techniques for in situ concrete assessment has constrained, their efficiency is often called into question mark, as well as the blend of two or more techniques arises as a reaction to all these problems. [3].

The application of certain non-destructive testing techniques is very often applied scientifically, incorporating two most commonly used methods to improve the accuracy of the estimation of concrete compressive strength; the theory is focused on comparisons between the measurements obtained and the expected value. [4]. The most frequently adopted standard combination method is the SONREB technique provided by RILEM [5], which was first developed in Romania and implemented in Europe and Australia.

The formulation suggested by different authors varies with the location and majorly influenced by locally available materials. That is why the consistency of results is depends upon the formulation and applicable only to its own use. This paper intended to bring the formulation using SonReb method as per the locally available material used in casting of samples tested in laboratory and in-situ testing condition.

1.1. Objective of Research

The main objective of this paper is to assess in-situ compressive strength of the concrete. The development of correlation between the independent variable (rebound number, ultrasonic pulse velocity) and dependent variable (compressive concrete strength) is used to evaluate the strength of standard laboratory samples and core samples taken from existing buildings. To present comparative analysis between proposed and suggested models by different researchers using multiple regression statistical analysis (MRSA).

2. Literature Review

In mixing with those acquired by ultrasonic pulses, even Schmidt's use of rebound hammer dates back to the mid-sixties, although Facaoaru should be assigned the first significant scientific article. [6].

Researcher's recommends techniques for assessing concrete strength predicated on certain correction factors depending on the type and dosage of the cement, and also the aggregate type and size. Afterward, throughout 1970s, Samarin [7] started using the combined SonReb method, in which the only recognized factors prior to the test were the type of aggregate and age of concrete. Many research papers have been published since the end of the seventies and based on formulations that were calibrated informal to determine the mechanical strength of the concrete using the SonReb process. [8, 9]. In 1993, RILEM NDT4 [10] also suggested using this SonReb technique and supplied ISO based strength relation in terms of curves where both the rebound index and the ultrasonic velocity are known to evaluate the compressive strength of concrete. Application of both techniques simultaneously, certain improvement in the effect of several factors that affect the RH and UPV test could be accomplished in part. A contemporary example of this implementation is the SonReb process, produced by RILEM technical committees 7 NDT and TC-43 CND in particular [5]. Improving the accuracy of the Facaoaru strength prediction [6] had accomplished by using correction factors that take into account the effect of cement type, cement content, fine aggregate fraction and total aggregate volume.

The reliability of the compressive strength will not reflect the overall performance of the concrete building material when it is located, but it is altered to take into consideration the special circumstances which may impact the strength of the sample [17], such as the slenderness and diameter of the sample, the presence of steel bars and the disruption caused by the mode of extraction.

Table 1. Formulation suggested by different authors

Year	Author	Equation used
1979	Bellander [8]	(Polynomial) $fc = -25.568 + 0.000635 \times R^3 + 8.397V$
1979	Meynink et al [9]	(Polynomial) $fc = -24.668 + 1.427 \times R + 0.0294V^4$
1991	Arioglu et al. [11]	(Power) $fc = 18.6E^{0.019 \times R} + 0515V$
1993	RILEM [10]	(Power) $fc = 9.27 \times 10^{-11.11} \times R^{1.4} \times V^{2.6}$
1994	Di Leo, Pascale [12]	(Power) $fc = 1.2X\ 10^{-9} \times V^{2:446} \times R^{1.058}$
1996	Ramyar et al. [13]	(Linear) $fc = -39.570 + 1.532 \times R + 5.0.614 \times V$
1996	Arioglu et al. [11]	(Power) $fc = 0.00153 \times (R^3 \times V^4)0.611$
1998	Khedar et al. [14]	(Power) $fc = 0.0158 \times R^{1.1171} \times V^{0.4254}$

2004	Menditto et al. [15]	(Power) fc= $0.00004\times R^{1.88148}\times V^{0.80840}$
2009	Faella et al. [16]	(Power) fc= $2.6199\times10^{-8}\times R^{0.5341}\times V^{2.2878}$

Table 1 shows the correlation between rebound number, UPV and compressive strength suggested by different authors in terms of equation given in power and polynomial function. The formulation given in table is the results obtained from laboratory tested cube samples. Where R is rebound number, V is ultrasonic pulse velocity, fc is compressive strength of cube sample.

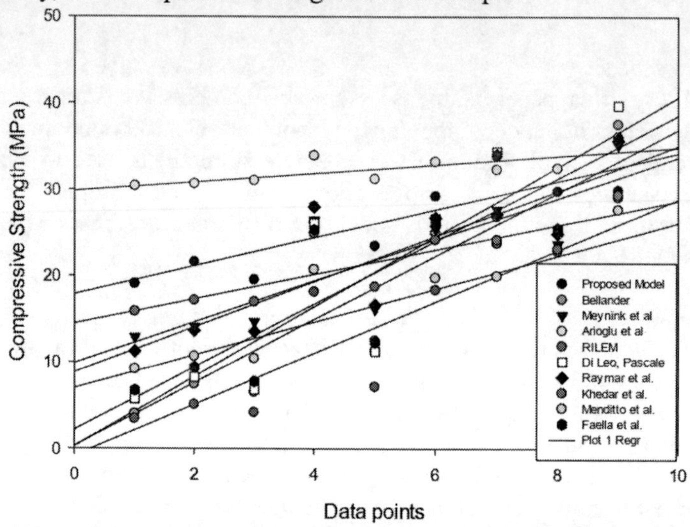

Figure 1. Regression analysis of proposed and suggested models of various authors (RH & UPV)

These equations are formulated and calibrated based on the experimental results obtained in laboratory. It is observed that several equations greatly underestimates the type of sample used by researchers, while others tend to substantially overestimate the cubic compressive strength. Compared to the model proposed, the Arioðlu model in Fig.1 showing a much higher estimate of strength than the experimental values. Also the laboratory-prepared samples of Meynink and Ramyar et.al. and Khedar et.al. Models provides accurate forecast, with numerical values closely matching to those of the experimental values. The Bellander, Menditto et. al., DI Leo, Pascle, RILEM, and Fallea models underestimating the strength than experimental values and is unsatisfactory. The proposed models showing very similar strength predictions when compared with experimental values.

3. Experimental Program

For conducting this research work, huge data were expected and the same were obtained from testing laboratory of Pune. The data obtained from laboratory is consist of non-destructive test (UPV and RH) and destructive test conducted on core samples taken from the buildings having age of 20 year and 30 year. Samples prepared in laboratory considering same parameters and tested at various ages. Results are interpreted by combining correlations between non-destructive techniques and mechanical test techniques i.e. SonReb. The materials properties are estimated in laboratory and used for making of concrete are shown in Table 2. Casting of 27 cubes of grade M20, size 150x150x150 mm is done in laboratory strictly as per IS: 10262-1982 and IS: 456-2002. The curing is done for a period of 7, 14 and 28 days under standard temperature conditions. Relationships between 7, 14 and 28 day compressive strength specimens prepared in laboratory and measurements of non-destructive test for RH and UPV is done. Before extracting cores from buildings, NDT tests has been carried out using UPV with direct transmission. The mean diameter of core is in the range of 70mm -75 mm. The core cylinder strength is converted into equivalent cube strength applying uniform loading at the rate of 5MPa/s.

Table 2. Material Properties

Ingredients	Cement	Coarse Aggregate	Fine Aggregate
Grade	OPC- 53 Grade	Max. nominal size: 20 mm	Max. nominal size: 4.75 mm
Specific gravity	2.76	2.65 (Zone II of IS: 383)	2.74
Water absorption	-	4.71%	1.52%
w/c ratio	0.5		

4. Statistical Analysis

Research has shown that the various factors have a significant impact on RH and UPV values. It is important to calibrate the measurements from the two tests in order to resolve inconsistent outcomes and enhance the accuracy of results. Regression analysis tends to be the response to investigate experimental data utilizing mechanical test results on specimens and cores and is essential to calibrate these tests. It is observed that, the better correlation has been identified between the experimental outcome from core test data and non-destructive test (UPV and RH). The results are plotted in graphical form and for each regression line, SIGMAplot is used to obtain the coefficients of the curves (Regression line) and (R2) determination. The concrete specimen from a structural framework can be considered as part of a population of infinite dimensions whose mechanical properties are the subject matter. Since an unlimited number of experimental observations cannot be conducted.

5. Result and discussion

The proposed model was built by conducting a non-linear regression analysis, which varies from each other for its mathematical structure. The application of regression enables a mathematical relationship to be established between a dependent variable and one or more independent variables. The independent variables are assumed to be the velocity of the ultrasonic pulse and the average number of the rebound, while the compressive strength was considered a dependent variable are shown in Table 3. The power expression was calibrated for the entire sample of experimental data by reducing the error between numerical prediction and experimental data.

Table 3. Summary of results obtained from laboratory testing and in-situ testing.

Rebound value	UPV	Compressive strength (MPa)	Age	Rebound value	UPV	Compressive strength (MPa)	Age
25.82	2220	19.8	7	40.02	3598	23.36	20
26.28	2560	20.9	7	39.01	3074	41.25	20
26.95	2340	19.4	7	38.08	2834	25.10	20
31.62	3780	25.4	14	39.93	2060	12.47	20
27.25	2860	25.5	14	39.77	3401	29.55	20
30.52	3870	23.1	14	15.96	2777	14.63	30
29.07	4388	37.2	28	12.44	2516	9.37	30
29.33	3856	30.6	28	13.55	2655	11.30	30
34.90	4300	34.5	28	12.37	2277	12.48	30

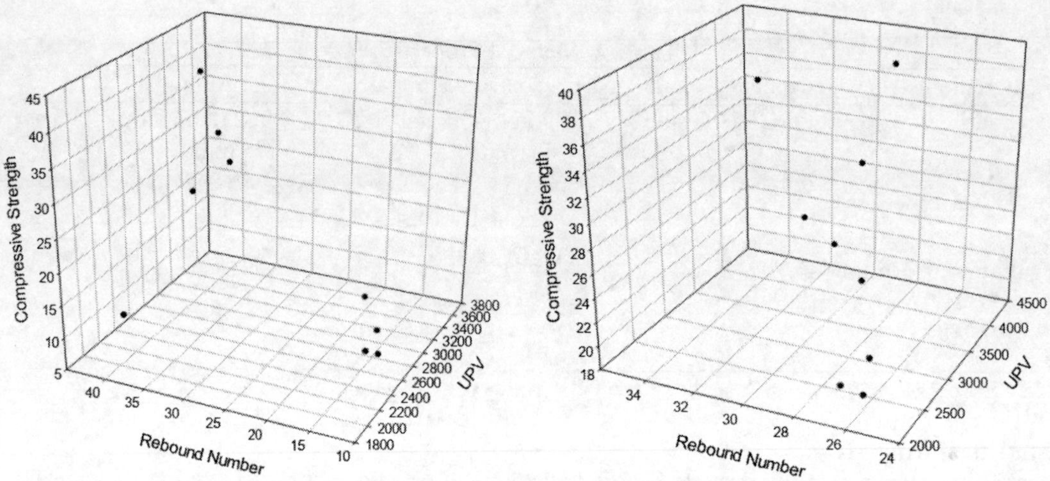

Figure 2. Compressive strength vs. Rebound number and UPV (a. Laboratory test, b. Core Test)

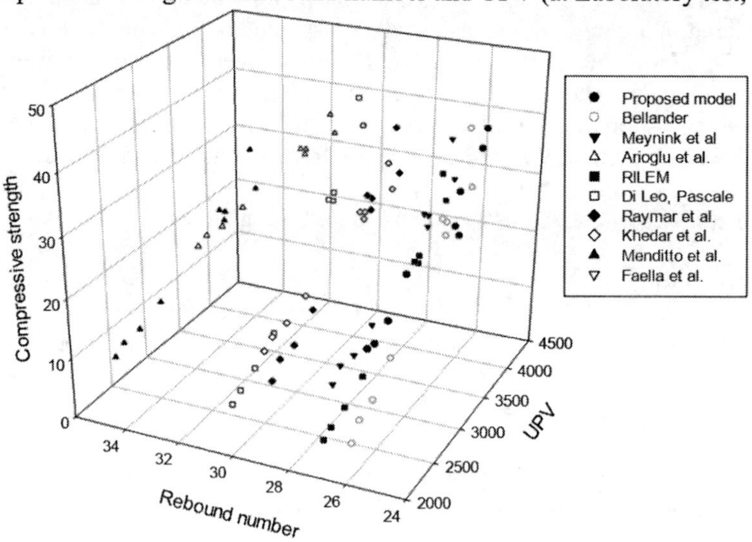

Figure 3. Comparison of proposed model with predicated model results

Table 4. Proposed model equation

Plot equation	R^2	Condition
fc= 0.0841 x $R^{-0.572}$ x $V^{0.945}$	0.769	Laboratory testing
fc= 0.517 x $R^{-0.546}$ x $V^{0.997}$	0.757	Core sample testing

The relationships are categorized as linear, polynomial, power, exponential and logarithmic as a further significant parameter for classifying the forecast relationships between concrete strength and independent variables. The rebound values and UPV values obtained from the lab are always rational prediction for the proposed model of the researcher. Due to greater variation in results as shown in Fig. 2 and Fig. 3, results obtained from cores are not significantly matched with the suggested model. The SonReb method is adopted and the results of testing done on laboratory and core samples are checked with proposed equations. The proposed equations and results are shown in Table 4 for two different testing conditions. It is observed that, the proportion of the variance for a dependent variable are closely matched for both the conditions with different age samples. In case of laboratory testing done by RH, rebound values showing no difference in spite of age criteria. Whereas, in case of UPV the result reflects the change in compressive strength with age. This indicates that only consideration of the results obtained by RH are inadequate for strength estimation.

6. Conclusion

Based on the review of models suggested by different researchers and proposed model on experimental observations, following conclusions can be drawn:

1. The regression curve of proposed model strongly consistent with the models suggested by Meynink, Ramyar et al. and Khedar et.al.
2. It is also seen that the regression coefficient of both the formulation are closely matching, which indicates no effect of age of samples tested.
3. Only consideration of the results obtained by RH are inadequate for strength estimation as compared to UPV in case of laboratory testing in spite of same age criteria, this supports use of SonReb methodology.
4. SonReb results seems to be more accurate and reliable than any individual test performed on the same samples. It can help to assess concrete performance more accurately and it will be easier to track concrete behaviour after long period of service.

7. References

[1] D. Breysse, "Quality of NDT Measurements and Accuracy of Physical Properties," Concrete NDTCE'09, Nantes, 30 June-3 July, 2009.

[2] Hobbs and K. Tchoketch, "Nondestructive Testing Techniques for the Forensic Engineering Investigation of Reinforced Concrete Buildings," Forensic Science International, Vol. 167, No. 2-3, 2007, pp. 167-172.

[3] J. H. Bungey and S. G. Millard, "Testing of Concrete in Structures," 3rd Edition, Blackie Academic & Professional, London, 1996.

[4] Breysse, "Combining Information, Reliability and Maintenance," Diagnosis of NEC Works, 2008

[5] Rilem Report TC43-CND, "Draft Recommendation for in Situ Concrete Strength Determination by Combined Non- Destructive Methods," 1983.

[6] Facaoaru, "Non-Destructive Testing of Concrete in Romania, Symposium on NDT of Concrete and Timber," Institute of Civil Engineers, London, 1970, pp. 39-49.

[7] Samarin, G. Smorchevsky, The non-destructive testing of concrete, Central Research Laboratory, International Tech. Rep. 54 (1973).

[8] U. Bellander, NDT testing methods for estimating compressive strength in finished structures – evaluation of accuracy and testing system, in: RILEM Symp. Proc. On Quality Control of Concrete Structures, Session 2.1 vol. 1, Swedish Concrete Research Institute, Stockholm, Sweden, 1979, pp. 37–45.

[9] P. Meynink, A. Samarin, Assessment of compressive strength of concrete by cylinders, cores, and non-destructive tests, in: RILEM Symp. Proc. On Quality Control of Concrete Structures, Session 2.1, Swedish Concrete Research Institute Stockholm, Sweden, 1979, pp. 127–134.

[10] RILEM NDT 4, Recommendation for in Situ Concrete Strength Determination by Combined Non-destructive Methods, Compendium of RILEM Technical Recommendations, E&FN Spon, U.K., London, 1993.

[11] Arioglu, O. K€oylüoglu, Discussion of prediction of concrete strength by destructive and non-destructive methods by Ramyar and Kol, Cem. Concr. World 3, (1996) 33–34.

[12] Di Leo, G. Pascale, "Prove non distruttive sulle costruzioni in cemento armato", Convegno Sistema Qualita e Prove non Distruttive per l'affidabilita e la sicurezza delle strutture civili, 1994. Bologna SAIE (in Italian).

[13] K. Ramyar, P. Kol, Destructive and non-destructive test methods for estimating the strength of

concrete, Cement Concr. World 2 (1996) 46–54 (in Turkish).
[14] G.F. Kheder, A two stage procedure for assessment of in-situ concrete strength using combined non-destructive testing, Mater. Struct. 32 (1999) 410–417.
[15] 1G. Menditto, S. Bufarini, V. D'Aria, M. Massaccesi, Nuove curve di correlazione per lo sclerometro, ultrasuoni e metodo combinato, in: Concreto, vol. 57, Maggioli Editore, Italia, 2004, pp. 84–89 (in Italian).
[16] G. Faella, M. Guadagnuolo, A. Donadio, L. Ferri, Calibrazione sperimentale del metodo SonReb per costruzioni della Provincia di Caserta degli anni in: atti del XIV Convegno Nazionale L'Ingegneria Sismica in Italia, 18-22 settembre 2011, 2011. Bari, Italia (in Italian).
[17] M.T. Cristofaro, A. D'Ambrisi, M. De Stefano, M. Tanganelli, "Concrete compressive strength extracted from existing buildings", in: L. Dezi, G. Moriconi, R. Realfonzo (Eds.), The New Boundaries of Structural Concrete Session D: Concrete Quality Control on Site, ACI Italian chapter, 2011.

Root cause analysis of newsprint waste using pareto analysis and cause and effect matrix

[1]**Jatin H Varma**, [2]**Dr. Sachin Lal**

Mechanical Engineering, Sir Padampat Singhania University, Udaipur, India

[1]jatin.varma@spsu.ac.in, [2]sachin.lal@spsu.ac.in

Abstract. The dynamic business environment has changed the mindset of printing industry, which enhanced the focus towards waste management as its critical issue in printing industry. Printing waste is also leading to financial loss and developing environmental issues. This scenario motivates to perform root cause analysis of newsprint waste, which can help to reduce waste and to improve the environmental conditions. In this paper, the root cause of newsprint waste is identified using Pareto analysis and cause and effect matrix. The result shows that web break between operation and the wrong roller setting are the root causes of newsprint waste. At the end of this study, few improvement measures are suggested and implemented for validating the improvement.

Keywords: Pareto Analysis, Newsprint waste, Cause & effect matrix.

1. Introduction

In the newspaper printing industry, high-speed web machines are used; this high speed causes enormous wastage. The waste leads financial losses for printing industry as well as develops environmental issues. The competitive business culture motivates printing industry to focus on environmental issues, which change the approach of the printing industry towards waste management.

2. Methodology

Root Cause Analysis (RCA) is a problem-solving process, which helps in conducting an investigation into an identified incident and problem. RCA practice tries to solve problems by identifying the root causes of events and proposes some improvement actions to reduce the intensity of root cause. Root cause analysis (RCA) requires the investigators to look beyond the solution to the immediate problem and understand the fundamental or underlying causes of the situation and put them right, thereby preventing re-occurrence of the same issue. This may involve the identification and management of processes, procedures, activities, inactivity, behaviors or conditions. The benefits of comprehensive root cause analysis are identification of permanent solutions, prevention of recurring failures, and introduction of a logical problem-solving process applicable to issues and non-conformities of all sizes. The key to effective problem prevention is detecting the causes of a problem that has occurred. Root cause analysis involved five steps, which are shown in Figure 1.

IOP Conf. Series: Materials Science and Engineering **810** (2020) 012072 doi:10.1088/1757-899X/810/1/012072

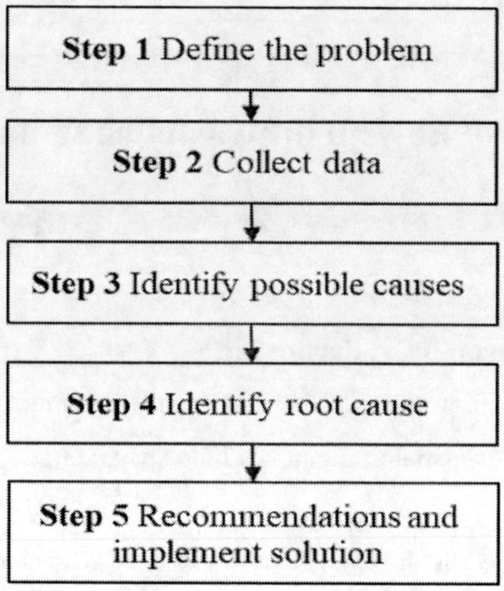

Figure 1. Root cause analysis flowchart.

2.1. Define the problem

In this step of RCA, the problem is identified based on the literature survey and available historical data. The historical data is collected and analyzed, which shows that printing waste is enormous, and its effects on the efficiency of an organization. This scenario motives for developing the problem statement as root cause analysis of newsprint waste using Pareto analysis and cause and effect matrix.

2.2 Collect data

In this step of RCA, needs to validate the developed problem statement. The newsprint waste is a combination of printed waste, sweeping waste, tear of waste, reel end waste, and others. The data was collected for one month based on identified categories and plotted the Pareto chart for identifying the most contributing factor for newsprint waste. Pareto chart (figure 2) concludes that printed waste is contributing 53.5 %, and hence it is the critical factor for newsprint waste. The Pareto chart also shows that there is huge newsprint waste, i.e., 2516 Kg/month, and the organization needs to focus on reducing the newsprint waste; hence, the developed problem statement is valid and shows the need for RCA for newsprint waste.

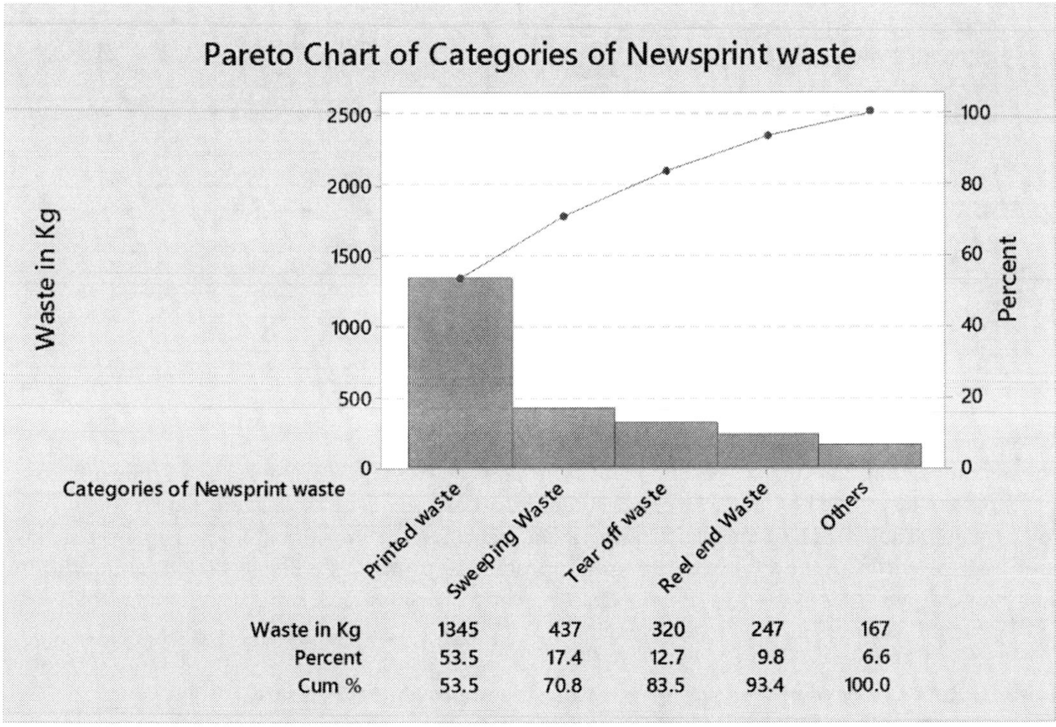

Figure 2. Pareto chart of Newsprint waste.

2.3. Identify possible causes

In this stage, identify as many causal factors or causes as possible for selected problems or issues. The literature survey and discussion with experts have identified few causes for printed waste, such as web breaks between operation, auto pasting of copies, incorrect pre-setting, inappropriate machine setting, unbalance ink-water combination, unsuitable dosing of a fountain, wrong roller setting, variation in cut off, and irregular inking. A total nine causes were identified for printed waste issue.

2.4. Identify root cause

This step helps to investigate the root cause of a problem or issue. There are different methods available for identifying the root cause of problems such as fishbone diagram, failure mode effect analysis, Why-why analysis, and cause and effect matrix. In this study, the cause and effect matrix has been used to investigate the root cause of printed waste, which comprises potential causes and evaluation criteria. The experts are selected in the category of managers, team leaders, and maintenance engineers to identify the evaluation criteria, who have the experience of more than five years. A total 13 experts are involved for identifying the evaluation criteria and experts have suggested the evaluation criteria such as frequency of incidence, effect on waste, cost-saving possibility and achievability of solution. The matrix was developed and shown in Table 1. Further, experts are also involved in investigating the root cause of printed waste, for that experts have asked to share their opinion based on rating scale of 1-10, which is varying from 1-no effect to 10-very strong effect. The experts have suggested for considering the equal importance for all evaluation criteria, i.e. 0.25. Table 1 shows the scale for which experts have given maximum response. It also shows that the web breaks between operation and wrong roller setting are root causes of printed waste.

Table 1. Cause and effect matrix for printed waste.

Sr. No	Potential causes	Cost-saving possibility	Frequency of incidence	Effect on waste	Achievability of solution	Total
1	Web breaks between operation	6	8	9	6	7.25
2	Auto pasting of copies	7	9	4	4	6
3	Incorrect presetting	5	7	6	3	5.25
4	Inappropriate machine setting	4	5	4	7	5
5	Unbalance ink-water combination	6	5	4	7	5.5
6	Unsuitable dosing of fountain	7	7	5	4	5.75
7	Wrong roller setting	7	5	7	8	6.75
8	Variation in cut off	5	5	5	9	6
9	Irregular inking	7	6	6	4	5.75

2.5 Recommendations and implement solutions

The recommendations are suggested to avoid the root cause of printed waste. The improvement measures are developed by considering cost and effort aspects. The suggested improvement measures are shown in Table 2. The proposed improvement measures are implemented, and again data has been collected for ensuring the improvement in the process. The data is collected for one month after the improvement and represented by Pareto chart as shown in Figure 3, which shows the total waste as 853 Kg/month. The result shows that the quantity of printed waste is reduced significantly.

Table 2. Root causes of printed waste and suggested improvement measures.

Sr. No.	Root Causes of printed waste	Improvement Measures
1	Web breaks between operation	Check the quality of the reel before mounting on a machine for printing.
		Develop standard operating procedures (SOP) for reel handling and stocking.
		Regularly monitor the speed of the roller.
		Develop SOP for daily maintenance, cleaning, and electrical maintenance.
2	Wrong roller setting	Develop SOP for roller setting.
		Regularly provide training to the operator related to machine handling and especially related roller setting.
		Develop the mechanism for monitoring the roller setting.

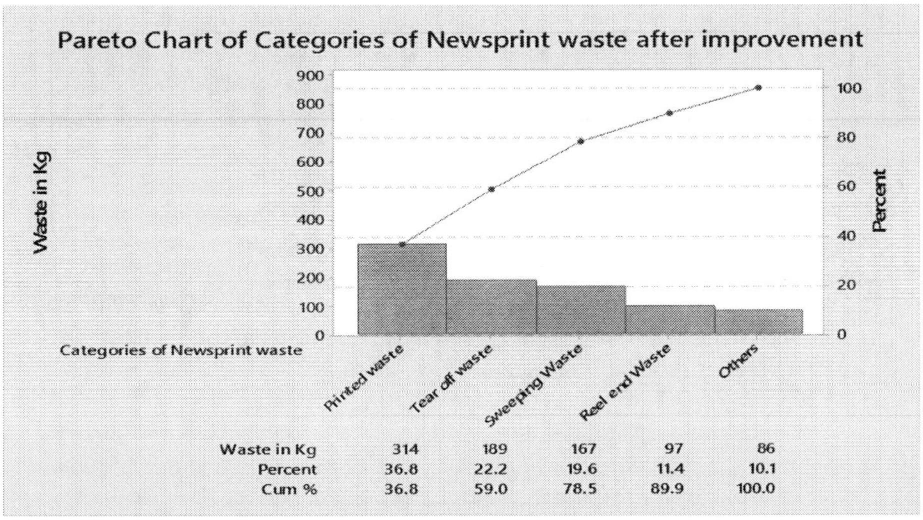

Figure 3. Pareto chart of newsprint waste after improvement.

3. Conclusions

The identification of the root cause of newsprint waste is more straightforward and systematically by using Cause effect matrix and Pareto analysis. In this study, various causes of newsprint waste have been identified and investigated root causes for newsprint waste. The suggested improvement measures have reduced the waste by 1303 Kg/month i.e. it helps to reduce the waste by 60.43%. The reduction in waste improves efficiency and reduces environmental issues. It can be concluded that training and operation awareness to operator with a practical waste management approach can help printing industry to maintain the newsprint waste at minimum level, which can further help to achieve the waste target of 3.5% or less can be obtained easily.

4. Acknowledgement

I would like to thank my guide, Dr. Sachin Lal, for giving me valuable input from time to time and making available all the resources. I would like to thank Sir Padampat University of Singhania, Udaipur for providing all the facilities required.
Finally, heartfelt thanks to all members of my family for their constant encouragement and support.

5. References

[1] An Analysis: Productivity Improvement in Printing Industry, International Journal of Innovative Technology and Exploring Engineering (IJITEE) ISSN: 2278-3075, Volume-8 Issue-9S3, July 2019.
[2] Ambrish Pandey, Sabharwal Sandeep, Wastage Minimization of Printing Substrate in Newspaper Organization, International Journal of Science, Engineering and Computer Technology, Vol 2, No 2 (2012), p. 47-49.
[3] Helmut Kipphan (ed.), Handbook of Print Media, Springerverlag Berlin Heidelberg 2001.
[4] Newsprint Waste Management, SFN Report No. 2, 2014, www.wan-ifra.org/rmtc.
[5] An Analysis: Productivity Improvement in Printing Industry, International Journal of Innovative Technology and Exploring Engineering (IJITEE) ISSN: 2278-3075, Volume-8 Issue-9S3, July 2019.
[6] H. S. Elsayad & S. El-sherbiny., A Study into the Influence of Paper Coatings on Paper Properties and Print Quality of Dye Sublimation Thermal Prints. Polymer Plastics Technology and Engineering, 47(2), 122-136 (2008a).
[7] M. Ataeefard., Influence of paper surface characteristics on digital printing quality. Surface

Engineering, Institute of Materials, Minerals and Mining, 30(7), 529-534(2014a).
[8] Nicholas Kingsley Graham Isaac Owusu., Achieving quality in printing: some determining factors in the printing industry in Ghana. The TQM Journal, 27, 6(2015).
[9] Dong Shang Chang, Yi Chun Kuo & Tz Yu Chen., Productivity measurement of the manufacturing process for outsourcing decisions: the case of a Taiwanese printed circuit board manufacturer. International Journal of Production Research, 46(24), 6981-6995(2008b).
[10] Nazim Emre Islamoglua, Kwangyeol Ryub & Ilkyeong Moon., Labour productivity in modular assembly: a study of automotive module suppliers. International Journal of Production Research, 1- 17 (2014b).
[11] Pardeep Gupta & Sachit Vardhan., Optimizing OEE, productivity and production cost for improving sales volume in an automobile industry through TPM: a case study. International Journal of Production Research, 1-13(2016a).
[12] Anupama Prashar., Process improvement in farm equipment sector (FES): a case on Six Sigma adoption. International Journal of Lean Six Sigma, 5(1), 62-88(2014c).
[13] Siddra Qayyum Siddiqui Fahim Ullah Muhammad Jamaluddin Thaheem Hamza Farooq Gabriel., Six Sigma in construction: a review of critical success factors. International Journal of Lean Six Sigma, 7(2), 1-38(2016b).
[14] Ramakrushna Padhy, Six Sigma project selections: a critical review, International Journal of Lean Six Sigma, Vol. 8 Issue: 2, pp.244-258(2016).
[15] Rajesh, N. and Klas, S., Strategic outsourcing through specifications. Omega, 28, 525–540(2000a).
[16] Tomas, F.E.R. and Victor, P.R., A resource-based view of outsourcing and its implications for organizational performance in the hotel sector. Tourism Mgmt., 26, 707–721(2005a).
[17] Tsai, C.Y., Chiu, C.C. and Chen, J.S., A case-based reasoning system for PCB defect prediction. Expert Syst. Appl., 28, 813–822(2005b).
[18] Batool, A. and Batool, B.., Effects of employees training on the organizational competitive advantage: Empirical study of Private Sector of Islamabad, Pakistan, Far East. Journal of Psychology and Business, 6(1), 59-72(2012a).
[19] Beheshti, H. M. and Lollar, J. G.., An empirical study of US SMEs using TQM. Journal of TQM and Business Excellence, 14(8), 839-84 (2003a).
[20] Curkovic, S., Melnyk, S., Calantone, R. and Handfield, R.., Validating the Malcolm Baldrige National quality award framework through structural equation modelling. International Journal of Production Research, 38(4), 765-791(2000b).

Static structural and thermal analysis of disc brake pad model

Yashvardhan singh Chouhan[1], Sudhir Tiwari[2]

[1,2]Department of Mechanical Engineering, SGSITS, 23, Park Road, Indore, 452003, M.P., India

[1]ysinghchouhan810@gmail.com, [2]sudhir2609@gmail.com

Abstract. This paper attempts to present a study of the automotive disk brake's thermo-mechanical behaviour during the braking period using the Static Structural and Transient Thermal analysis for vehicle disk brake pad system using finite elemental analysis. In this work Thermal and static analysis was performed on solid and ventilated disc for two configurations of pads using ANSYS 15. Two pad Profiles with pure and single slot along with both discs have been studied. The analysis is conducted on the model that lacks thermal properties first. It is predicted for structural performance such as deformation and von-mises stress. After that, with the introduction of heat flux, convection and adiabetic components, thermo-mechanical analysis is performed on the same model. The estimate results of temperature distribution, total deformation and von-mises stress are presented. The software tools CATIA and ANSYS Workbench have been selected for modelling and analysis. CATIA is a widely used 3D modelling technology in the design process. ANSYS is a software package for general purpose finite element analysis (FEA). The outcome of this investigation assists brake engineers in choosing appropriate research to determine the disk brake assembly in structural and contact behaviour.

Keywords. Ventilated Disc, Brake Pads, Structural Analysis, Thermo-mechanical Coupling, Catia, Ansys.

1. Introduction

A vehicles used in a transportation sector are being developed to be able to transport a passengers and products at higher speeds and higher capacities to achieve a desired high performance. This requires absorbing more kinetic energy during stopping or slowing down of vehicles. Thus, brake system is required in a vehicle to stop & adjust a speed in altering conditions of street and rush-hour traffic. A brake is functioned with a help of caliper through which frictional resistance occurs on wheel in order to slow down or break a motion of a vehicle. In general we find two types of a frictional brake namely disc brake & drum brakes. Disc brake in comparison with drum brake cools sooner because of larger cleared range & comparatively high acquaintance of air stream & also shows self-cleaning skill because of centrifugal force.

A hydraulic braking system was invented over 100 years ago (Newcomb & Spur 1989) and has been universally used in passenger cars for over 60 years. A hydraulic braking system relies upon a 'muscular' energy of a driver, which may be amplified by a suitable 'booster'. A basic principle is that incompressible 'brake fluid' is pressurised by a 'master cylinder' piston connected to a brake pedal & a pressure generated creates an actuation force from a 'slave cylinder' piston, which actuates a brake shoe or pad. In the disc brake, a rotating disc is pressed against by pads and heat is generated because

IOP Conf. Series: Materials Science and Engineering **810** (2020) 012073 doi:10.1088/1757-899X/810/1/012073

of friction at the interface of disc and pad. The frictional heat is transfers directly to environment and disc becomes cool. A disc brake assembly is shown in figure 1.

Figure 1. A simplified disc brake assembly

Brake disc is placed secure to axle, so that it exchanges with an equivalent speed as a wheel. Brake disc normally are of two types.
1. Solid brake disc
2. Ventilated brake disc

The simplest structure & involving of single strong plate is known as Solid disc. While in ventilated disc vanes separate two partitions of disc and stretches way for a mid-air to pass or stream. It also performs faster cooling & subsequently, outcomes low outside temperature. Disc brakes now become critical safety components whose functionality depends on the working members. Rashid [1] presents a review paper which depicts various geometries of parts & materials used in disc handing systems. Despite an extensive number of improvements, there are until now many working issues for different systems which need to be understood. Vehicle safety is becoming more concerning subject now days in automobile industry. Using a finite element approach to define and predict structural quality of the disk brake is becoming very popular in the research community. Belhocine et. al [2] presents research of a thermal and mechanical manner of a car disc brake while braking stage is being introduced Likewise, a basic execution between a two breaks down (mechanical & thermo mechanical) was additionally analyzed. Talati and Jalalifer [3] presented the heat equation for pad and disc in the form of transient heat equations.

In the work performed by Kumar et.al [4] a selective material for brake pads & rotor is proposed. Two unique materials, S2 glass fiber & carbon fiber, are picked as cushion materials and aluminium alloy & grey cast iron are picked for a rotor material. Warm investigation performed in solid & ventilated disc for different materials of pad utilizing ANSYS 15. Diverse pad profiles along with solid, single slot & two slots have been examined by fluctuating material properties of pad. Parab et.al [5] depicts a designing of brake disc using CATIA. Basic & thermal investigation is performed on disc brakes utilizing 3 different materials. Reddy et.al [6] carried out research in a solid and ventilated disc using Pro-E and ANSYS. He performed a couple analyses done on both disc to predict strength of disc brake. Comparisons can be made among three materials to get best material for FSAE car. Babukant and Teja [7] performed transient thermal analysis using ansys for distribution of temperature and heat flux on each friction surface of contacting bodies. Hwang and Wu [8] investigate the thermal stress and temperature in ventilated disc pad model. The work is done using thermo-mechanical multi body technique and it is validated with experimental results. Yildiz and Duzgun [9] presents three diverse ventilated brake disc and braking power examinations were researched tentatively together with a solid disc. In the work performed by Kumar and Bijwe [10] copper as filler material is investigated for a brake pad.

The finite element modelling is the main research method used in this project. The model is prepared in the CATIA software based on the assumptions raised in the literature survey then; the model is implemented to analysis software in ANSYS where it is being introduced with basic theories of structural and thermo-mechanical analysis. The basic equations such as heat transfer equations, frictional heat flux and convection were implemented in thermo-mechanical analysis.

2. Modelling of disc brake-pad model

CATIA is multi stage 3D programming created by Dassault Systems, including CAD, CAM and CAE. CATIA is a solid modeling tool displaying pictures that joins 3D parametric highlights with 2D devices & furthermore addresses each structure to-assembling process. CATIA likewise gives creating orthographic, projections, isometric or point by point 2D drawing perspectives. It also produces model measurements & makes reference measurements in a drawing. Table 1 shows dimension of parts and figure 2 (a-d) shows catia model of parts.

Table 1. Dimension of disc-pad model

Parameter Name	Disc(mm)	Pad(mm)
Outer diameter	221	220
Inner diameter	133	137
Thickness	10	12

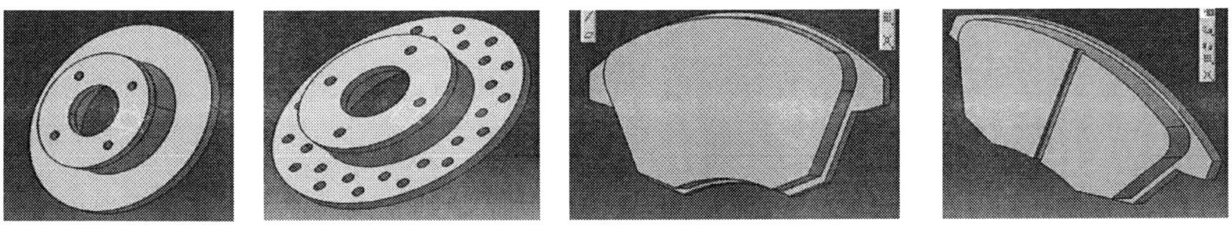

(a) (b) (c) (d)

Figure 2. Different parts of disc and pad

3. Introduction of FEA

A FEA is a simulation technique of physical problem through numerical method which is also known as FEM. Experts tried to decrease an amount of hard models , optimize by redesign parts in their structure stage to enhance effectively things, quickly . FEA is a numerical procedure used for prediction of how an area or assembly together acts under given conditions. FEA method has transformed into a favoured system in performing thermal assessment on a brake assembly together recently. It is communicated that thermal examination is generally performed using a finite part method, as this procedure has transformed into an outstanding resource for a numerical game plans of a wide extent of structure problems.

3.1. Ansys

Dr. John Swanson built up ANSYS in 1970 with an idea to advertise a possibility of computer mirrored structure, setting up himself as one of pioneers of Finite Element Analysis (FEA). ANSYS is all around valuable finite element assessment (FEA) programming group. ANSYS 15 is totally used to reproduce assistant deformation, temperature, stresses and pressure distribution in brake applications. In the present work, a 3-D FE model includes a solid disc & two pads are utilized. The material of a disc is grey cast iron & brake pads have an isotropic behavior whose mechanical features of two portions are given in table 2.

Table 2. Properties of disc and pad

Properties	Grey Cast Iron (Disc)	Isotropic Material (Pad)
Density, ρ	7000 kg/m^3	2.31 g/cc
Young Modulus (E)	100Gpa	2.30 Gpa
Thermal Conductivity, k	54W/m.k	2.41 W/m.k
Specific heat, Cp	586J/kg.k	1076 J/kg.k
Poisson ratio, μ	0.28	0.25

3.2. Meshed model

Cross section includes division of a whole model into little pieces called elements. Elements utilized for a coincided model are tetrahedral 3D components. A limited component model of a rotor was completed to such an extent that a subsequent components came to 10325 elements with an aggregate of 19843 nodes. Cross section of disc pad model is appeared in figure 3 and meshed properties in table 3. Working force to the brake pad is 2785.24 N. The pressure applied externally between the disc and pad is 1 N/mm^2. The rotational speed of the disc is 123.44 rad/s.

Table 3. Finite elements meshed properties

Description	Nodes	Elements
Disc	12402	6836
Pad 1	1803	884
Pad 2	1840	913

Figure 3. Meshed model of disc-pad assembly

3.3. Boundary condition and loading

In the present model, boundary conditions are shown in figure 4 for pressure applied on single side of pad. The disk is fixed tightly in all directions at the bolt holes except in its direction of rotation and pads are fixed at the edges in all degree of freedom excluding in the normal direction. So the pad moves up and down and makes contact with the disc surface.

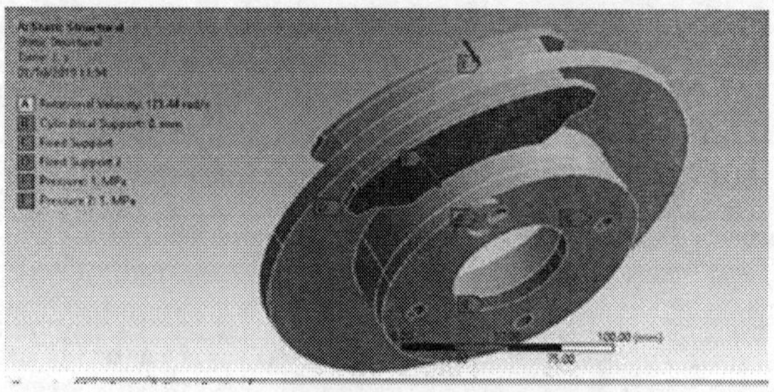

Figure 4. Boundary condition

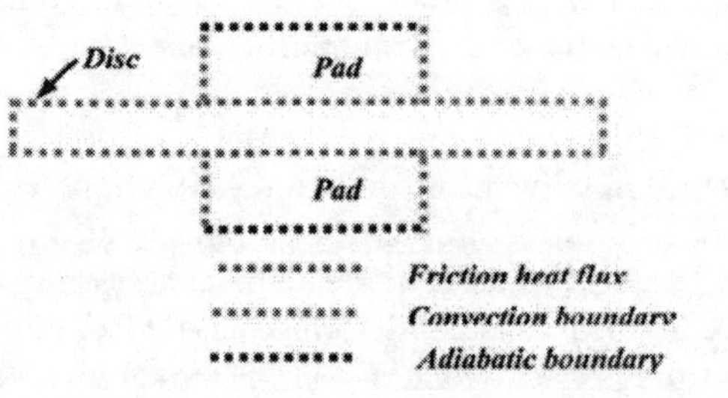

Figure 5. Thermal boundary condition

IOP Conf. Series: Materials Science and Engineering **810** (2020) 012073 doi:10.1088/1757-899X/810/1/012073

3.4. Thermal boundary condition
Thermal boundary conditions as appeared in figure 5 are applied in the model for expressing heat transfer. At the boundary between the disc & brake pad, heat is created because of friction which is appeared in blue lines. For uncovered region of disc & brake pad, it is accepted that heat is traded with surrounding over convection. In this way, convection limit condition is functional there. At outside of a back plate, adiabatic boundary condition is utilized.

4. Result and discussion

4.1. Disc-pad deformation without thermal effects

4.1.1 Total deformation
Figure 6 (a-d), presents the disc deformations in contrast to braking time. From the figures, it is seen that large deformations mostly found at discs outer radius that are in contact with the pads. Ventilated disc deform more compare to the solid disc through both configurations of pads. The highest deformation is 29μm for ventilated disc with single slot pad. As for the pads, deformations occur at the outer radius of the pads. Table 4 shows total deformation in a model.

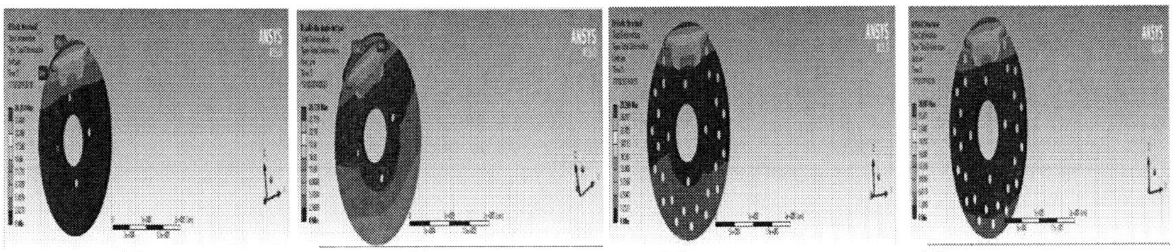

(a) (b) (c) (d)

Figure 6. Total deformation in both discs along with pads

Table 4. Total deformation

Configurations	Total deformation(μm)
Solid disc-solid pad (a)	26.351
Solid disc-single slot (a)	26.729
Ventilated disc-solid pad (c)	28.881
Ventilated disc-single slot (d)	29.269

4.1.2 Stress distribution
Figure 7 (a-d) shows the von-mises stress distribution of the disc-pad model. The maximum stress generates in the single slot-solid disc model of the order of 25Mpa. The stresses are maximum at the bowl of the disc. Table 5 shows stress values in different configuration of model.

Table 5. Von-mises stress

Configurations	Von-mises stress (Mpa)
Solid disc-solid pad (a)	16.331
Solid disc-single slot pad (b)	25.082
Ventilated disc-solid pad (c)	14.384
Ventilated disc-single slot pad(d)	21.136

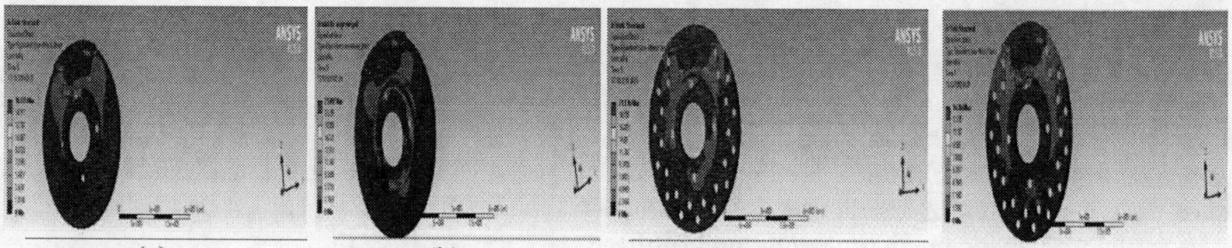

(a) (b) (c) (d)

Figure 7. Stress distribution in both discs with pads

4.2. Disc-pad deformation with thermal effects

4.3. Solid disc with solid pad

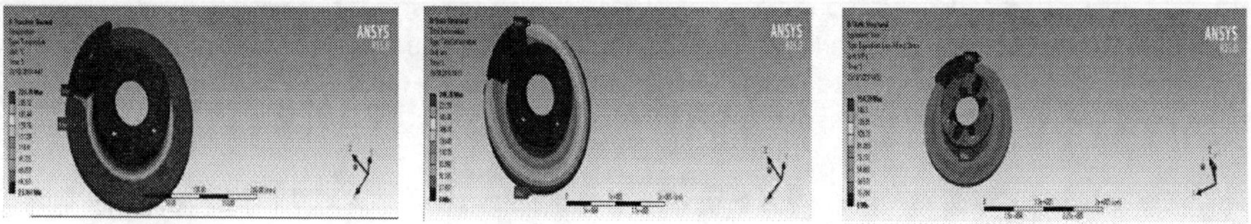

(a) (b) (c)

Figure 8. Thermo-mechanical analysis of solid disc with pad

Figure 8 (a-c) shows temparature distribution, total deformation and von-mises stress in a solid disc model based on thermo-mechanical examination. It is seen that the highest temparature it attains is equals to 228°c while total deformation is shaped like a umbrella with the highest percentage value of 249μm. It is also seen from the analysis that stress is generated more on outer radius of the disc.Based on this coupling model stress is also predicted which equals to 249Mpa.The mamimum stress value is geneted at the bowl of the disc and minimum occurs at the pad. Figure 9 (a-c) shows temparature distribution, total deformation and von-mises stress in a ventilated disc model based on thermo-mechanical examination.

(a) (b) (c)

Figure 9. Thermo-mechanical analysis of ventilated disc with pad

Based on the table 6, it shows that total deformation is maximum in solid disc but stress generation is maximum in ventilated disc. On the basis of temperature distribution ventilated disc is suited best for temperature distribution.

Table 6. Thermo-mechanical results of solid and ventilated disc

	Solid disc		Ventilated disc	
	Maximum	Minimum	Maximum	Minimum
Total deformation, µm (a)	249.28	0		
Von-mises stress, Mpa (b)	164.29	0	191.34	$2.7555e^{-9}$
Temperature distribution, °c (c)	227.79	23.707	226.55	11.436

5. Conclusion

The Work attempted studies in static structural and transient thermal analysis of disc brake pad model using mechanical and thermo-mechanical approaches. Deformation, von mises stresses and temperature distribution of disc-pad model has been carries out in order for improving the braking effectiveness and provides good strength to the vehicle in the design stage. From the results it can deducted that:

- The outer radius of the disc deforms more.
- When dealing with without thermal effects, large stresses generated at the bowl of the disc. Based on total deformation solid disc-solid pad is best and based on von-mises stress ventilated disc with pure profile pad is taken as the best.
- Von-mises stress and total deformation of disc pad model increase in noticeable way in thermo-mechanical analysis.
- Large deformations are found in solid disc compare to ventilated disc but stresses are reduced in solid disc.
- Ventilated disc provide good temperature distribution effect compare to solid disc to dissipate heat to the surroundings

6. References

[1] Rashid A, Overview of Disc Brakes and Related Phenomenon – a review, 2014, *Intl. J. of Vehicle Noise and Vibration*, 10(4), pp. 257-301.

[2] Belhocine A, Bouchetara M, Ghazali N M and Abdullah O I, Structural and Contact Analysis of a 3-Dimensional Disc-Pad Model With and Without Thermal Effects, 2014, *Turkish J. of Engg. Science and Tech.*, pp. 188-202.

[3] Talati F and Jalalifer S, Analysis of Heat Conduction in a Disc Brake System, 2008, *J. of Applied Sciences*, 8(20), pp. 3583-3552.

[4] Kumar K N, Raju B N and Roy M R, Static and Thermal Analysis of Disc Brake With Pad, 2015, *Intl. J of Engg. Trends and Tech.*, 23(8), pp. 411-420.

[5] Parab V, Naik K, Dhale A D, Structural and Thermal Analysis of Brake Disc, 2014, *Intl. J. of Engg Develp. and Research*, 2(2), pp. 1398-1403.

[6] Reddy V C, Reddy M G, Gowd G H, Modeling and Analysis of FSAE Car Disc Brake Using FEM, 2013, *Intl. J. of Emerging Tech. and Advan. Engg*,3(9), pp. 383-389.

[7] Babukanth G and Teja M V, Transient Analysis of Disc Brake By Using Ansys Software, 2012, *Intl. J of Mech. and Industrial Engg.*,2(1), pp. 21-25.

[8] Hwang P and Wu X, Investigation of Temperature and Thermal Stress in Ventilated Disc Brake Based on 3D Thermo-mechanical Coupling Model, 2010, *J. of Mech. Science and Tech.*, vol. 24, pp. 81-84.

[9] Yildiz Y and Duzgun M, Stress Analysis of Ventilated Brake Discs Using Finite Element Method, 2010, *Intl. J. of Automotive Tech,* 11(1), pp. 133-138.

[10] Kumar M and Bijwe J, Non-Asbestos Organic (NAO) Friction Composites: Role of Copper; It's Shape and Amount, 2011, *Elsevier*, pp. 269-280.

IOP Conf. Series: Materials Science and Engineering **810** (2020) 012074 doi:10.1088/1757-899X/810/1/012074

Review on Mechanical, Thermal and Morphological Characterization of Sisal Fibre Composite

Sahil Jain, Ritika Das, Ramachandran M

MPSTME, SVKM's NMIMS, Shirpur, Dhule, Maharashtra 425405, India
sweetestchandran@gmail.com*

Abstract. Growing worldwide ecological risks and energy crisis, fibers from different plants reinforced into composites made up of polymer have initiated extraordinary research interest leading to their quality of acting as an alternate substitute for synthetic composites made up of fiber. Comparing it with commonly used carbon fibers or glass fibers, fibers from plants have numerous advantages like environmental kindly, renewable, less in weight, minimal cost, and good specific mechanical performance. Seeking to the existing situation the globe is hunting for sustainable and substitutable resources, therefore one of such natural fiber which is ordinarily being used across the world is sisal fiber.
Keywords. Sisal Fibre; mechanical properties; morphological properties; Thermal properties.

1. Introduction

Sisal fibers can be gathered from the plant leaves which belongs to the agave family of plants. The fiber from sisal plant is scratched out from leaves extracted from stronger and stiffer plants, with a little bit of elasticity. Sisal fiber have several applications and multiple traditional uses like; ropes, twine, yarn and string, and also can be woven into mats, carpets and numerous handicrafts. Also sisal cloth has proved its use in polishing few materials. Moreover, sisal has a great prospective as its reinforcement in polymer composites because of its comparatively low density. Major benefit of the composites is that their strength to their weight ratio, which is usually tends to be high and providing good strength. Moreover sisal fibers are also used in multiple industries such as in automotive for making components and also in making of furniture. Furthermore the left-out waste after the making of composite from sisal and its extraction it is generally being used for producing ingredients in pharmaceutical, biogas and material manufactured for building. Sisal biomass consist of good proportion of cellulose, and its mash is further useful as an alternative for fibers from wood and being used in the paper manufacturing industry. Tremendous research is being conducted for validating it to the cementitious and polymeric grounds as reinforcement. For studding further in detail, a wide range of datum of test results of various different composites made up of sisal fiber were explored to reach out a certain conclusion.

2. Tensile Strength

To achieve continuous and unidirectional sisal reinforced composites. Characteristics such as dependence of tensile property and flexural properties, sisal volume fraction, impact strength mechanical properties, and tensile strength were assessed and compared. Relation between reinforced natural fiber composites and tensile strength was observed to rise along with the fiber composite to a level of optimization and then begins to decrease.

Paper number	Composition	Tensile strength (MPa)
3	Short sisal tensile properties: 95% Benzylated and 5% sisal (vol %)	20

	30% Benzylated and 70% sisal	20
3	Tensile properties of unidirectional sisal: 11% Benzylated and 89% sisal 40% Benzylated and 60% sisal	 30 69
7	Mechanical and physical properties: (B-banana fiber, G-glass, S-sisal)(vol %) 29% S 21% S, 9% G 15% S, 13% G 16% B, 15% S 12% B, 11% S, 8% G	 23 52 93 25 64
14	Tensile strength of PE, PEox, PE/sisal and PEox/sisal composites PE/S10 PEox/S5 PE/sac5 PEox/sac10	 21.1 18.5 19 18.2
16	Relation between sisal fiber content on tensile results of PP and MAPP matrix composites (tensile strength) Fiber content (wt%) with PP 10% 20% Fiber content (wt%) with MAPP 10% 20%	 30 32 35 45
25	The relation between the tensile strength along with some glass fibers in the composite made up of sisal–glass hybrid. Glass fiber content (vol %) for 5% volume 25 100 Glass fiber content (vol %) for 8% volume 25 100	 21 34 23 40
29	The consequences of exposing untreated sisal fiber to boiling water in terms of tensile strength, CTDIC-treated composites of sisal fiber. At zero hours Untreated Treated	 12.5 16.9
30	Change in tensile properties of sisal fiber to its test length (Ø of fiber: 200μm). Test Length (in mm) 15 65	 Observations (MNm^2) 793.80 620.81
30	Change in tensile properties of sisal fiber to its speed of testing (Ø of fiber: 200μm; test length: 50mm). Testing speed (mm/min) 1 500	 Observations (MNm^2) 481.00 441.60

The linear correlation between tensile strength and fiber loading was observed as far as 60 vol % in sisal. Also, unidirectional fibers show long tails after the major damage that was concluded by the brittle failure theory. This improves the safety of use as it must be related to the interfacial failure characteristics and post-failure crack propagation resistance. Which means, the generalized shear deformation of the ductile material is brought due to the reinforcing fiber, which causes the phenomenon. So, with an increase of interaction of solid in the present laminates would result in a higher elongation breaking point.

3. Impact Test

Impact test determines qualities of the material such as brittleness or ductility. In order to do that the net energy consumed by a material during fracture has to be calculated. This energy depicts material's toughness and helps to evaluate temperature-dependent brittle-ductile transition. The capability of the fiber to tolerate sudden load is termed as its impact strength. Izod impact test rig was used here in this case to conclude the impact strength of the laminates. The purpose of the test was to measure the kinetic energy that is needed to initiate the fracture and also to break the specimen completely. The standard dimension for the Izod test is ASTM: D4812. The specimen was held vertically upwards with the help of jaws and the pendulum swung from one side and struck the fiber with kinetic energy. Energy being consumed by the material prior to its fracture point was verified on the scale which concluded the ductility and toughness of the material. The table below shows some Impact test on different composites made from sisal fiber;

Paper Number	Composition	Impact Test
3	 100% Benzylated fir sawdust 70% Benzylated fir sawdust 30% sisal	charpy (kJ/m^2) 5.8 4.4
7	Mechanical and physical properties: (B-banana fiber, G-glass, S-sisal) (vol %) 16% B, 14% G 29% S 15% S, 13% G 7% B, 6% S, 14%G	Izod impact test - Impact energy (J) 12.5 8.4 13.3 12.8
14	PE- polyethylene; PP- polypropylene; PEox- oxidized polyethylene matrix; S- sisal fibers; 10- 10 wt%; Sac - acetylated sisal fibers as reinforcement. PE/S10 PEox/S5 PE/Sac10 PEox/Sac10	Izod impact strength J/m 63 55 64 56
11	Compositions of the SF (Sisal fiber) /MA-g-PP (maleic-anhydride-grafted polypropylene) /PP (polypropylene) were studied composites. Sisal fiber (wt%) MA-g-PP (wt%) PP (wt%) 10SF 10 9 81 20SF 20 18 62 30SF 30 27 43	Izod impact strength (kJ/m2) at no immersion in water 2.25 4.5 6 or 5.9
16	Change of sisal fiber content on the Charpy impact strength of PP(Polypropylene) and MAPP (maleic-anhydride -polypropylene) composites with change in percent of Sisal Fiber Sisal Fiber content (wt%) 0% 10% 20%	Charpy impact strength (kJ/m^3) for; PP matrix MAPP matrix 3.25 3.5 3.60 2.4 4.7 3.2
19	Composites with different patterns of layers, with the Ratio of Volume Fraction of Banana (B) and Sisal (S) 4 : 1 B/S S/B/S B/S/B	Izod impact (kJ/m2) 34 30 33

23	Polypropylene grafted with maleic anhydride (PP-g-MA) PP,Sisal fiber, Glass fiber, PP-g-MA (wt%) (wt%) (wt%) PP/sisal 70 30 – 3 PP/sisal/PP-g-MA 70 30 – 3 PP/10sisal/20GF/PP-g-MA 70 10 20 3	Izod impact (kJ/m2) 12.78 14.42 20.01
24	Magnesium hydroxide (Mg(OH)2), zinc borate (Zb), ammonium polyphosphate (APP), and Zb in sisal and combination of APP with Mg(OH)2fiber/polypropylene (PP) composites PP/30sisal PP/30sisal/40APP PP/30sisal/40Mg PP/30sisal/40Zb PP/30sisal/30APP/10Mg PP/30sisal/30APP/10Zb	Izod impact Impact strength (kJ/m2) 14.02+- 0.52 10.49+- 0.31 11.22+- 0.54 12.96+- 0.20 10.49+- 0.31 10.72+- 0.37
54	 Glass fiber + sisal fiber composite Glass fiber + jute fiber + sisal fiber composite	Charpy impact test 18 12
37	Fiber content banana / sisal 100/0 0/100	Impact strength kJ/m2 13.25 22.54

After evaluating the impact ability and losses of energy in different laminates with the help of Izod impact test. Impact results for the glass fibers are much better than the natural fibers. The specimen starts to break with a crack due to impact which further propagates because of loss of adhesion among fibers and matrix which eventually results in initiating the fiber breakage and pullout. Therefore, it is observed that there is a steep rise in the impact characteristics with increase in glass fibers content along with the natural fibers. Impact results are improved by 34%, 39% and 53% for the two-layer glass ply addition as extreme ply to the banana (G/B/G), sisal (G/S/G) and banana–sisal (G/BS/G) laminates. Best results for impact test are observed for three-layer glass fiber samples with sisal fiber (G/S/G/S/G) which is 3.1% and 6.4% greater than the glass hybrid banana (G/B/G/B/G) and banana–sisal fiber (G/BS/G/BS/G) laminates.

4. Differential scanning calorimetry

Differential scanning calorimetry (DSC) was used to investigate the behavior of sisal fiber that undergo solid-liquid phase transitions. It is performed to conclude the declination in crystallinity and breaking down of natural plant cellulose. The outcome of this test is found by analyzing first exothermic peak (decomposition) observed in DSC which associates with the data collected with the help of second endothermic peak which is stronger than the first peak, generally tends to be around 350°C to measure the thermal degradation of crystallites in plant fibers.

Paper number	composition	DSC
36	DSC thermo grams of untreated sisal Temperature (°C) 100 500	Heat Flow (W/g) -0.4 0.3
36	DSC thermo grams of alkalized sisal Temp. (°C) 100 500	 Heat Flow (W/g) -0.9 -0.4

36	Temp. (°C) 347.5 350	Soda Concentration (%) 5 35
15	DSC curves of unmodified and modified sisal TEMP. (°C) for Unmodified sisal (mW) 100 400 TEMP. (°C) for Esterified sisal (mW) 100 400 TEMP. (°C) for Ionized air-treated Sisal (mW) 100 400 TEMP. (°C) for Mercerized sisal (mW) 100 400	 00 58 00 21 00 30 00 02

The results demonstrated that the degradation procedure relies upon the atmospheric conditions. In dormant climate, the DSC curves presented definite peaks for sisal fiber and its constituents, though in normal room conditions the thermal curves exhibited two exothermic peaks for each of the constituents. The degradation of cellulose in dormant conditions happened by endothermic procedures though, in atmospheric conditions, it happened by exothermic procedures.

5. Hardness Test

The material hardness was tested with the use of a shore-A hardness tester at room temperature, with the help of test method (Durometer) ASTM D-2240-81.

Page number	Composite	Hardness
37	SBR-1502" (Styrene-butadiene rubber) 100 A- sisal -, length -0mm H- sisal -20, length -6mm	 43 71
34	MMA-methyl methacrylate Untreated %v/v MMA-Length 10 40 0.025%v/vMMA-Length 10 40 0.050%v/vMMA-Length 10 40 0.075%v/vMMA-Length 10 40 0.100%v/vMMA-Length 10 40	 71.4 71.8 71.6 72.1 71.75 72.25 72.4 72.8 72.4 72.8

The observation indicates that the hardness properties is directly proportional to the fiber length and composites made up from ad micellar-treated sisal filaments additionally had higher hardness and impact strength. The mechanical characteristics of short sisal-fiber-strengthened composite of SBR have been evaluated as length of fiber, direction, stacking, and a holding specialist.

6. Flexural test

Flexural test was conducted for sisal fibers to measure the force required to bend the sisal fiber. Tensile stress and compressive stress were subjected to the sisal fibers respectively which in result created shear stress for bending the material. Three point bending and tensile strength tests took place using an LWK-5 universal tester at 1 and 5 millimeters per minute crosshead speeds.

Paper number	composite	Flexural test (MPa)
3	Benzylated fir sawdust composites short sisal considered as a function of sisal content. Sisal vol % 0 14	 35 55
7	ASTM: D790 standards. Where; G- glass fiber, S-Sisal fiber, B-banana fiber, BS-banana sisal fiber. Laminates GBGBG GSGSG GBSG GBSGBSG	 175 185 190 165
9	ASTM D-790; Sisal fiber-reinforced rubber seed oil-based polyurethane Content of sisal fiber (wt.%) 0 30	 53.2 (+-13.9) 96.1 (+-7.9)
20	Isopthalic polyester resin HSR 8131. Volume fraction of fiber sisal 0.2 0.5 Bs1:1 0.2 0.5 Bs1:3 0.2 0.5 Bs3:1 0.2 0.5	 45.5 52.0 47.5 56.0 46.0 53.0 47.0 57.0
23	PP- Polypropylene; MAPP - Maleated polypropylene; Polypropylene-PP PP NBPP/sisal PP/10sisal/20GF/PP-g-MA	 55.40 60.80 68.84
26	cobalt naphthenate and Methyl ethyl ketone peroxide; Volume of sisial fibers; Sisal rest is glass fiber (5% total fiber content) 100 0 Sisal 8% 100 0	 32.06 80.80 50.21 110.03
32	Sisal content in %vol in Benzylated fir sawdust	

	0 40	30 115
34	UTSF - Untreated sisal fiber; ATSF- Alkali treated sisal fiber(2% NaOH) **vol %(UTSF)** 26 75 **vol %(ATSF)** 30 70	 160 285 185 315

Sisal composite fibers displayed maximum results in tensile and flexural properties when a concentration of up to 30 weight percentage of sisal fiber was used. The results drastically improved when tested unidirectionally.

7. Thermal gravimetric analysis

Thermal gravimetric analysis (TGA) usually is used for the observation of the decomposition attribute and also the constituent phase of the composites in high temperature. TGA being done with the help of Perkin–Elmer TGA-7 thermo gravimetric analyser. Wherein samples were ranging between 5 to 10mg being heated at 25 up till 600°C and also at 20°C for 1 minute in flowing nitrogen.

Paper number	Composition	TGA
27	TGA curves of untreated and Benzylated sisal fiber. TEMP. (°C) for Untreated 100 500 TEMP. (°C) for Benzylated 100 500	 90g 05g 94g 00g
24	TGA curves of sisal fiber TEMP. (°C) 100 600	 94 18
16	Temp. (°C) for He 100 400 Temp. (°C) for He + O_2 100 400	 91g 19g 95g 18g

Observing various datum the thermal decomposition of every specimen happens in the fixed temperature varying between 50–600°C. Talking about sisal fiber, weight lost between the range of 60 and 100°C matches to vaporization taking place of the water molecules present in the specimen. Another loss of weight which takes place about 325 °C in because of thermal polymerization of the hemicellulose and in addition to the glucosidal cleavage of cellulose links. For EVA, the decomposition is being observed in two steps. The initial step consist of VA group degradation, step two involves degradation of the main chain. From the datum we can analyses that the fiber starts to degrade before the EAV matrix, plus the composites tend to be stable that EAV and other fiber. The strength of the composite do not rely much on the quantity of sisal in it, this is because in maximum cases the decomposition of sisal takes place first and then EVA decomposes.

8. Conclusion

Application of the fiber sisal as a good reinforcing material in compounds based on the polymers, after studying, analyzing various aspects and also desiring a future with a natural fibers in common, properties and construction of fiber sisal, surface alteration, mechanical and physical qualities of sisal fiber grounded on their composites made up of polymer. Sisal fiber got an impressive potential as polymer reinforced composites. Having a good specific characteristics of sisal fiber and low in density, the composites made using sisal fibers tend to have a considerable implications in multiple sectors like transportation and automotive industries. Furthermore, reduced apparatus, scrape and successive reduction of re-tooling costs which makes composites made up of sisal fibers prone to become a better choice. Raw material in plastic industries being replaced by sisal fibers as an alternative source does not only makes it a sustainable resource, but also is able to produce a non-traditional path towards economic growth and development for sectors like agriculture for the farmers.

References

[1]. Joseph Kuruvilla, Tolêdo Filho Romildo Dias, James Beena, Thomas Sabu, de Carvalho Laura Hecker. A review of sisal fiber reinforced polymer composites. Revista Brasileira de Engenharia Agrícola Ambiental 1999; 3(3):367–79.

[2]. Kanak Kalita, U Ragavendran, M. Ramachandran, Akash Bhoi, Weighted sum multi-objective optimization of skew composite laminates, Structural Engineering and Mechanics, Vol. 69, No. 1 (2019) 21-31.

[3]. Gradwell SE, Renneckar S, Esker AR, Heinze T, Gatenholm P, Vaca-Garcia C, et al. Surface modification of cellulose fibers: towards wood composites by biomimetics. C R Biol 2004; 327:945–53.

[4]. Agarwal, Rakshit, M. Ramachandran, and Stanly Jones Retnam. "Tensile Properties of Reinforced Plastic Material Composites with Natural Fiber and Filler Material." ARPN Journal of Engineering and Applied Sciences 10, no. 5 (2015): 2217-2220.

[5]. Hepworth DG, Hobson RN, Bruce DM, Farrent JW. The use of unretted hemp fibre in composite manufacture. Composites: Part A 2000; 31:1279–83.

[6]. Cyras VP, Iannace S, Kenny JM, Va´zquez A. Relationship between processing conditions and properties of a biodegradable composites based on PCL/starch and sisal fibers. Polym Compos 2001; 22(1): 104–10.

[7]. Paul SA, Boudenne A, Ibos L, Candau Y, Joseph K, Thomas S. Effect of fiber loading and chemical treatments on thermo physical properties of banana fiber/polypropylene commingled composite materials. Composites Part A 2008; 39:1582–8.

[8]. Shyam kumar Shah, Akshay kumar Patil, M. Ramachandran, Kanak Kalita, Effect of coal ash as a filler on mechanical properties of glass fiber reinforced material, International Journal of Applied Engineering Research, 9(22) (2014) pp. 14269-14277.

[9]. Firestone D. Official methods and recommended practices of the American Oil Chemists Society. 5th ed. Chapaign Illinois: AOCS; 1998.

[10]. J. I. Kroschwitz, Polymers: Fibers and Textiles, Wiley, New York, 1990

[11]. Joseph PV, Joseph K, Thomas S. Effect of processing variables on the mechanical properties of sisal-fiber-reinforced polypropylene composites. Compos Sci Techol 1999; 59:1625–40.

[12]. Brandrup J, Immergut EH. Polymer handbook. 3rd ed. New York (USA): Wiley-Interscience; 1989.

[13]. Boynard CA, d'Almeida JRM. Morphological characterization and mechanical behavior of sponge gourd (luffa cylindrica) — polyester composite materials. Polym Plast Technol Eng 2000; 39(3):489–99.

[14]. Bourmaud A., Baley C.: Rigidity analysis of polypropylene/ vegetal fibre composites after recycling. Polymer Degradation and Stability, 94, 297–305 (2009).

[15]. J. R. Perkins, "Some General Remarks on Corona Discharges", in: Engineering Dielectrics. Corona Measurements and Interpretation,R. Bartnikas, E. J. McMahon, Eds., n Vol. 1, ASTM STP 669 (American Society for Testing and Materials), Baltimore 1984, p. 3.

[16]. Ramanan. G, Rajesh Prabha. N, Diju Samuel. G, Jai Aultrin. K. S, M Ramachandran, Prediction of Machining Characteristics of Hybrid Composites Using Response Surface Methodology Approach, International Journal of Engineering & Technology, 2018, 7(3.1), 162-165.

[17]. K. G. Satyanarayana, P. C. Pai, K. Sukumaran and S.G. K. Pfllai, Handbook of Ceramics and Composites. Vol. 1. N. P. Cheremisinoff, ed. Marcel Dekker, New York (1990)

[18]. Alex. S, Stanly Johns Retnam, M. Ramachandran, A review on Biodegradability of Hybrid Bamboo/Glass fiber polymer composites, International Journal of Applied Engineering Research. ISSN 0973-4562 Volume 10, Number 11 (2015) pp. 10565-10569.

[19]. Philips, L. N.; Murphy, D. J. Carbon Fibre Reinforced Plastics— The New Technology; RAE technical memorandum, MAT, 1977; p. 273.

[20]. Jacob, M., Thomas, S. and Varughese, K. T. (2004). Natural Rubber Composites Reinforced with Sisal/oil Palm Hybrid Fibres: Tensile and Cure Characteristics, J. Appl. Polym. Sci., 93(5): 2305–2312.

[21]. M. Ramachandran, , Sahas Bansal, Pramod Raichurkar, Scrutiny of Jute Fiber Poly-Lactic Acid (PLA) Resin Reinforced Polymeric Composite, Journal of the Textile Association, Volume 76, Issue 6, 2016, pp. 372-375.

[22]. L.E. Nielson, Mechanical Properties of Polymers and Composites, Marcel Dekker, New York (1974).

[23]. Joseph PV, Joseph K, Thomas S. Effect of processing variables on the mechanical properties of sisal-fiber-reinforced polypropylene composites. Compos Sci Technol 1999;59:1625–40.

[24]. Sahas Bansal, M. Ramachandran, Pramod Raichurkar, Comparative Analysis of Bamboo using Jute and Coir Fibre Reinforced Polymeric Composites, Materials Today: Proceedings,4(2), 2017, 3182–3187.

[25]. Park, S.J. and Kim, T.J. (2001). Studies on Surface Energies of Glass Fabrics in an Unsaturated Polyester Matrix System: Effect of Sizing Treatment on Glass Fabrics, J. Appl. Polym. Sci., 80: 1439–1445.

[26]. D. Bino prince raja, B. Stanly Jones Retnam, M. Ramachandran, Analysis of mechanical properties of glass and carbon fiber reinforced polymer material, International Journal of Applied Engineering Research. ISSN 0973-4562 Volume 10, Number 11 (2015) pp. 10387-10391.

[27]. Lightsey, G. R., Organic fillers for thermoplastics, Polym. Sci. Technology, Vol. 17. Plenum Press, New York, 1983.

[28]. PAUL, A.; THOMAS, S. Electrical properties of natural fibre reinforced low density polyethylene composites: A comparison with carbon black and glass fibre filled low density polyethylene composites. Journal of Applied Polymer Science, New York, v.63, p.247-266, 1997 .

[29]. H. Morawetz, Macromolecules in Solution, 2nd ed., Interscience, New York, 1975.

[30]. C. Klason, J. Kubat, and P. Gatenholm, "Wood fiber reinforced composites," in Viscoelastdy of BwmateriaIs, p. 82, W. G. Glasser and H. Hatakeyama, eds., American Chemical Society, Washington, D.C. (1992).

[31]. Joseph, K.; Varghese, S.; Kalaprasad, G.; Thomas, S.; Prasannakumari, L.; Koshy, P.; Pavithran, C. Eur Polym Mater 1996, 32, 1243.

[32]. Robson, J.; Hague, J.; Newman, G.; Jeronimidis, G.; Ansell, M. P. Report No. EC/431/92 to DTI LINK Structural Composites Committee, January 1993.

[33]. E. Baer and A. Moet (Eds.) , High Performance Polymers, Hanser, New York, 1991, p. 206.

[34]. John K, Venkata Naidu S. Sisal fiber/glass fiber hybrid composites: impact and compressive properties. J Reinf Plas Compos 2004;23(12):1253–8.

[35]. Joseph K, Varghese S, Kalaprasad G, Thomas S, Prasannakumar L, Koshy P et al. Influence of interfacial adhesion on the mechanical properties and fracture behavior of short sisal fiber reinforced polymer composites. Eur Polym J 1996;32(10):1243–50.

[36]. Bismarck A, Mohanty AK, Aranberri-Askargorta I, Czapla S, Misra M, Hinrichsen G, et al. Surface characterization of natural fibres: surface properties and the water up-take behavior of modified sisal and coir fibres. Green Chem 2001;3(2):100–7.

[37]. P. M. Bhagwat, M. Ramachandran, Pramod Raichurkar, Mechanical Properties of Hybrid Glass/Carbon Fiber Reinforced Epoxy Composites, Materials Today: Proceedings, 4(8), 2017:1788–1793.

Study and analysis of WO-CO Turning Tool Using Finite Element Method

Nishant Dhengre[1], Dr M. K.Pradhan[2], Dr. Rajnish Katarne[3]

[1,3] Assistant Professor, NMIMS, MPSTME, Shirpur
[2] Assistant Professor, MANIT , Bhopal
1.Nishant.Dhengre@nmims.edu, 2.mohanrkl@gmail.com, 3.Rajnish.Katarne@nmims.edu

Abstract. Machining is the most vital operation, widely used in the manufacturing industry to remove materials and to achieve the desired geometry. As with most engineering problems we want to get the uppermost return, with the least investment. In this case we want to minimize costs, while increasing cutting speeds. But the major challenge faced is about understanding the mechanics of machining and significant influence of tool geometry, cutting velocity, depth and feed for cut that can be iterated to bring the best in the process. In this study, emphasizes will be made on range of speed causing stress on turning tool during metal cutting. A model has been developed using ANSYS software that can help in understanding the machining process. A different approach is proposed which will lead to the minimization of the damage to work material as well as tool material and to increase the profit or minimizing manufacturing cost per unit. In cutting operation, the region near the tool tip is the most important region and conditions at the tool tip is examined with precision so that improvements in tool performance can be achieved.

In the present work, study of tool wear caused by the change in cutting depth, feeding rate and speed of cutting of single point cutting tool for a turning operation to predict the tool life in orthogonal cutting using Finite Element Method (FEM) is carried out. A simulation model was developed using ANSYS software in order to estimate the magnitude of the von-misses and residual stresses evolved during the machining process.

The result shows the influence of depth during cut in the machining and influence of induced stresses on carbide tool. The result obtained have been verified by the available results from the literature for the variation of x-y stress, von-misses stress with the temperature, on carbide tool and was found to be in good agreement.

1. Introduction

Metal cutting test has been practiced for about past 150 years using various cutting parameters in immensely huge volume *Astakhov, V. P. (2006)*. F. W. Taylor is viewed as a pronounced historic figure in the field of cutting especially metal, whose efforts has been broadcasted till the ending of the 19th century and had a excessive influence on the manufacturing industries. In the metal cutting industry, the usual method to predict the optimize technique for metal cutting, perform numerous investigational tests with changing cutting parameter and examine its result on process outputs. Experimental study of machining process is equally

IOP Conf. Series: Materials Science and Engineering **810** (2020) 012075 doi:10.1088/1757-899X/810/1/012075

essential as the theoretical analysis. Unfortunately, to optimize through an experimental approach, a large number of experiments in search of desired accuracy are required and this approach is very time consuming and expensive. Therefore, modeling is used because it has the capacity to assist the solution of many important and fundamental problems in manufacturing engineering. Currently, the practice of non-conventional processes of metal cutting are being extended in the advance for industries dealing with manufacturing. The fact is that requirement of high tool life as it effect quality of surface and force required for cutting. *Lotfi M et.al (2017).* The application of digital computers has brought revolutionary changes in the domain of various engineering and scientific disciplines, and out of them first was the area of mechanics of solids *Pavel Kovac,2011*. Modified processor enabled computer can significantly improve the relationship between price systems and their performance, resulting in an increased number of users in industry who uses FEM systems integrated into the process of construction and development. This allows one, to maintain the intricacy of the exercise on one hand, and the preferred precision on the other. In recent years, finite element analysis has become a major method in simulation of metal cutting. Recently developed software systems are based on the method of finite elements that are exclusively designed for cutting process simulation. FEM facilitate coalesced simulation of thermal and plastic activity, it is capable of operate umpteen reliance of physical constraints with each other. FEM of chip removal process is being considered as a favorable approach to study the cutting process, permits reduce in experimental cost. It provide process related some essential aspects which calculate like stress, energy or temperature which help in understanding of chip formation process *P.J.Arrazola,2005.*

2. Literature Review

L.J.Xie et.al.(2004) used an arrangement of formation of chips in three steps, with early chip formation, chip development and steady state chip , to analyse continuous chip formation processes during turning process. Jijun Zhang et. al. (2011) made a 3D model of turning the tool on ANSYS workbench. He made different calculations for both static and thermal models. Pavel Kovac et. al. (2011) analyzed the effects of cutting force in turning was realized using the ANSYS Workbench 11 software system and its module for structural analysis. He used Mechanical modeling and thermal load of cutting tool. Pietro Stolf et.al(2019) made simulation using collected information (data) to predict and state the relationship between chip formation, wear mechanism tool life, cutting speed, coolant presser and coolant temperature

3. Finite Element Modeling

FEM is a technique for approximately answering the so-called space problem. It is a method of computer analysis of problems in mechanics, fluid mechanics, thermodynamics, etc. In industries, mostly in the metal cutting process, the finite element technique has proved to be an important tool while solving research problems and it helps in the advancement of industrial applications. Typical work out of the method involves dividing the continuum (space) of the problem into a collection of elements or parts, with each particular element represented by a set of elemental equations to the original problem *Shubhabrata D. 2016*. FEM is numerically stable, it means that if an error occurred in the input data or in-between calculations does not show the error or react but causes the resulting output to be of no use.

The benefits of using FEM rather than analytical and empirical approach in machining process can be stated as Variables namely, flunk, stress, region of stress, temperature, strain, strain rate, etc which are difficult to obtain, can be attained quantitatively, in supplement to chip geometry and cutting forces *E. Bagci,2011*. Boundries of Non-linear geometries of free surface can be signified and worked. Properties of Material can be operated with function of temperature, strain and strain rate. FEM aids extremely in chip generating of the tool and also in minimizing cost, material and weight.

Properties of material can be controlled as they are dependent on stress, strain and temperature. FEM extremely helps in analysis of chip formation and strength visualizations of the tool and also in reducing weight, material, and costs. The resistance offered between tool and the chip can be designed in FEA as primary deformation zone called as sticking zone and secondary deformation zone called as sliding conditions. Simple FEM calculations in the design process are imperative for further reducing the time of development.

Thus, Analysis based on FEA delivers described qualitative and quantitative information in to the chip generation process that is very much essential for thoughtful understanding of the influence of machining parameters. FEM leads to more responsive and better refine knowledge in the field of metal cutting. So, in this research attempt has been made to take the advantage of the method to understand the machining process. The various parameters are analyzed to model the process.

4. Software

During turning, for stress analysis ANSYS 2013 software package was used. ANSYS is a general-purpose software package designed for analysis using finite elements. ANSYS provides engineering imitation explanation, Predict and acceptable solution that requires in design processes. Manufacturing Companies widely use ANSYS software. Software modeling provide a computer-generated artefact through a laborious testing procedure before it becomes a physical object. ANSYS Contains equations that govern the behaviour of finite elements, fixes them and gives a comprehensive explanation of the functioning of the system as a whole. These results can be presented in tabular or graphical form. The steps in solving problems of ANSYS software package are Build Geometry, Define Material Properties, Generate Mesh, Present the Results

5. Cutting Mechanism

The basic idea of orthogonal cutting is to analysis the 3-dimensional cutting in 2-dimensional plane. Here in this below fig. Shows how a 2D turning is analysis. Rake angle and clearance angle are the most significant for all the cutting tools. Rake angle is the most important parameter in tool geometry. It affects almost every phenomenon in cutting tool. Rake angle are of three types positive, negative and zero. Each one of them has its own importance.

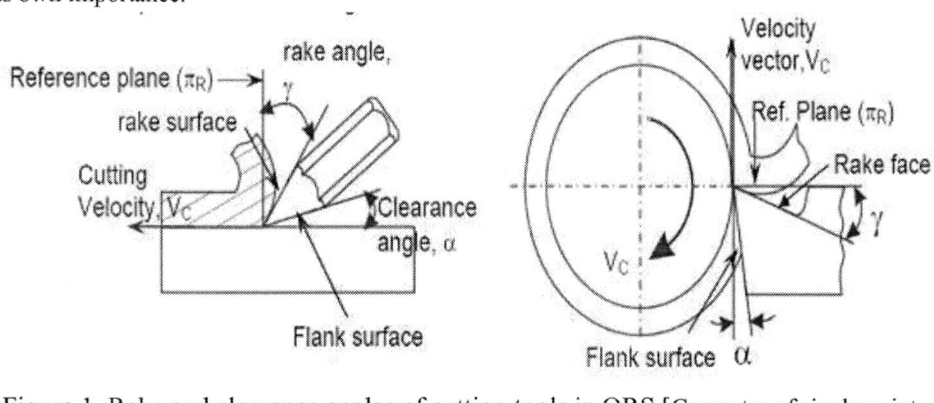

Figure 1. Rake and clearance angles of cutting tools in ORS [Geometry of single point cutting tool. ME IIT, Kharagpur.]

6. Forces Involved in Cutting

Study of force throughout turning process includes the amount of force required for peeling and their components and also the region of the action of those forces as well as the nature of the forces, statically or dynamically *L.N. Lakale,2000*
Fc = Tangential cutting force working in line of action of cutting velocity.
X_f = Force acting in line of action of feed.
Y_f = Radial force acting in the direction perpendicular to the field force.
T_f = Thrust force

7. Zone of Deformation

The actual deformation in this machining process occurs in two regions close tothe cutting tool edge i.e.

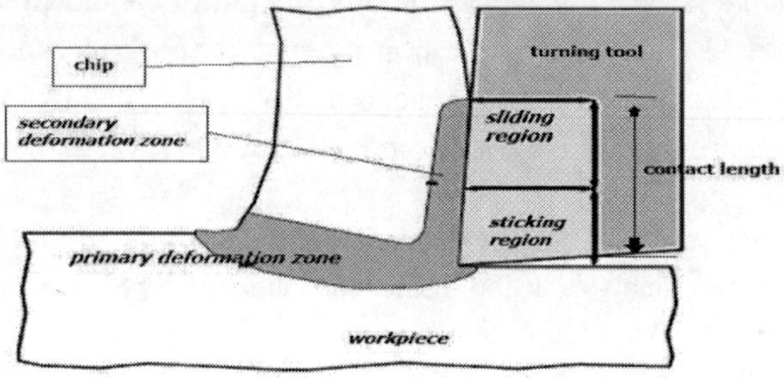

Primary deformation zone and Secondary deformation zone. The primary deformation region begins from the tip of cutting tool and extend to the junction of unreformed material and deformed chip. During deformation, the work piece is subject to high strain rate in this region. The secondary region is divided into two regions [8,9]. The sticking region and sliding region. In sticking region, the workpiece material adheres to the tool and shear occurs on the chip. The friction force is high so it also causes heat generation. The heating is due to energy dissipation from the plastic deformation. Sticking region is also affected by temperature, tool/work material, surrounding surface of tool material. Coefficient of friction is not constant

Figure 2. Locations of the primary and secondary deformation zones and the sliding and sticking regions[*K.Paul,1991*]

8. Modeling

In designing a model, ANSYS 2013 was used. In 2D analysis we consider two faces, rake angle, relief angle and curved face known as tool radius.
Rake face length = 3cm
Flank length = 2cm
Clearance angle = 6°
Rake angle = 6°
For designing the 2D line diagram, 8 nodes wear selected at definite distance. Each node has given its special coordinate in the x-y system. After drawing point, the lines are made connecting two distinct points completing the whole drawing. After drawing the line diagram, area is made by selecting all lines. If the lines don't form closed loop or can say that my f line is not connected between two nodes, software will not generate the area. Here in this 2-d line diagram sticking contact length and sliding contact length are made considering two different lines. In next, mesh is generated mesh. Material property is the behaviour of materials under different parameters such as physical property like strength, plasticity, modulus of elasticity, thermal expansion,

IOP Conf. Series: Materials Science and Engineering **810** (2020) 012075 doi:10.1088/1757-899X/810/1/012075

poisons ratio, heat capacity, temperature etc. In this study WO-CO carbide is used as a tool material. Its physical and thermal properties shown in the table.

Temperature [°c]	20	200	400	600
Ultimate strength R_M [MPa]	2600	2550	2350	2000
Young's modulus E[GP a]	800	570	560	540
Poisson's ratio	0.22	0.22	0.22	0.22
Density[K g/m^3]	14500			
Thermal expansion	5.4	5.3	5.4	5.6
Heat capacity c_p	220			

Table 1 : Defining material property

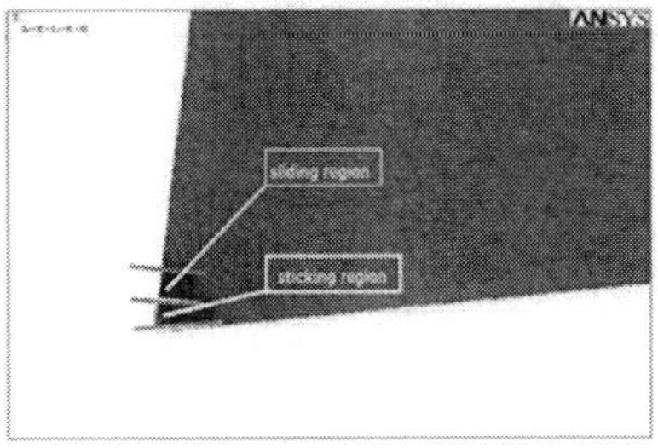

Figure 3 : Fine mesh region

9. Result and discussion

The results obtained on application of load on designed tool shows, its deformation maximum von-mises stress, XY stress and its behaviour on variation of stress. The table 2 below shows at a particular speed, feed, depth of cut what, where, the stress evolved in designed tool. The Figure shows the deformation in the tool geometry due to force applied. The Figure shows the trend of change in tangential and cutting force.

Exp.	Speed	Feed	depth	F_C	F_T	F_n	F_s	T	Result	Von m stress	xy stress
	[mm/rev]	[mm/min]	[mm]	N	N	N	N	°C	N/m^2	N/m^2	N/m^2
1	120	0.05	3	422	390	431	377	26	366	378	168
2	120	0.15	3	1005	729	830	923	32	139	205	51
3	180	0.05	3	404	360	400	365	30	657	501	79
4	180	0.15	3	890	584	824	674	36	146	136	128
5	240	0.05	3	391	357	396	351	32	734	492	55
6	240	0.15	3	857	520	606	798	38	525	534	150

Figure 4. shows a graph of the trend variation in von-misis stress and its experimental result (from literature). The result derived from a computer simulated for different forces on designing tools are shown in Figure below. Figure 5 shows un-deformed tool and deformation of the tool under applied load

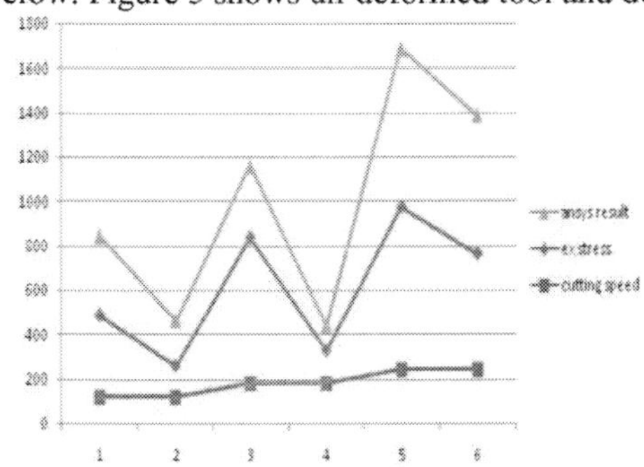

Figure 4 : Graph of result showing the same trend of variation in Experimental result

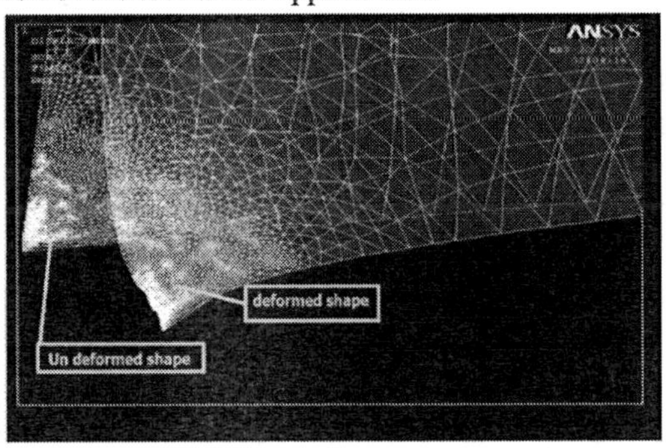

Figure 5: Deformed shape of tool under load applied

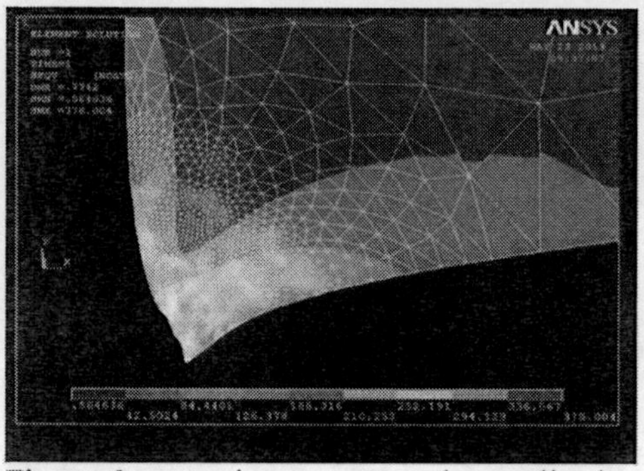

Figure 6: von-misses stress under application of first loaded condition

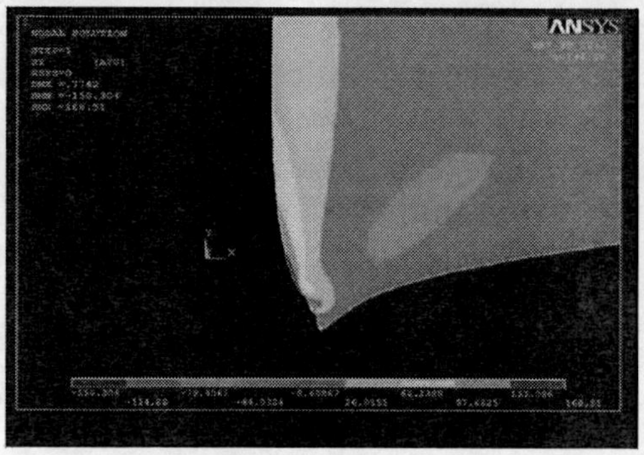

Figure 7: von-misses stress under the application of second loaded condition

10. Conclusion

In present study, Finite Element Model is used in investigation of cutting toll. For the purpose, finite element analysis of AISI 1045 insert had investigated. Cutting tool was analysis using Von-mises equivalent stress criteria. Following conclusions were drawn from the analysis: While calculation, it was observed that at a particular feed, the normal force is greater than friction force and when its feed is increased friction force becomes greater than normal force. When, at a particular speed feed is increased, maximum von-misses stress decreases as well as is the variation in XY stress. Von-misses give the information about max stress a material bear in that particular condition, and XY stress shows stress affecting zone in x-y direction. In x-y stress, it was observed that maximum region of stress is near the sliding region where crater wear is most likely to be observed in previous (literature) studies. Von-misses maximum region of stress is observed near the flank face of the tool wear micro facture can cause flank wear. The results provided in this work indicates that the prediction model implemented to obtain thermal information can be used to give useful data about cutting temperature occurring in other tool geometry. The existence of such model allows us to develop analyses concerning optimization issues for turning tool process. As this study concern about WO-CO wearing within speed range. The same may be extended to different type of material for different speed. This can also further lead to coating the tool face with different material with redesigning until design and optimizing criteria will be satisfied.

References

[1]. Arrazola, P. J., Ugarte, D., Montoya, J., Villar, A., & Marya, S. (2005). Finite element modeling of chip formation process with Abaqus/Explicit 6.3. VII International Conference on Computational Plasticity, Barcelona.
[2]. Astakhov, V. P. (2006). Tribology of metal cutting (Vol. 52). Access Online via Elsevier.
[3]. Bagci, E. (2011). 3-D numerical analysis of orthogonal cutting process via mesh-free method.Int.
[4]. J. the Physical Sciences, 6(6), 1267-1282.
[5]. Geometry of single point cutting tool. ME IIT, Kharagpur. Kohnke,
P. (Ed.). (1999). ANSYS theory reference. Ansys.
[6]. L.N. Lakale, A. Guttierrez, I.J. Llorente, A.J. Sanchez, and J.Abnoiga. Using high pressure coolant in the drilling and turning of low Machinabilityn alloys. International Journal of Adv. Tech, 16:85–91, 2000.
[7]. Pavel Kovač, Borislav Savković, Branislav Serdar, Milenko Sekulić (2011). Modeling Mechanical and Thermal Load of Gutting tool, Acta Technica Corviniensis– Bulletin ofEngineering University Politehnica Timisoara, Faculty of Engineering Hunedoara, 5, Revolutiei, Romania
[8]. Wright, P. K., & Trent, E. M. (2000). Metal Cutting. Butterworth-Heinemann.

[9]. Xie L.-J., Schmidt, Schmidt C., Biesinger F. 2004, 2D FEM estimate of tool wear in turning operation, Wear, 258-10, p. 1479-1490.

[10]. Zhang, J., Jia, X., Liu, X., & He, G. (2011, August). Static and thermo-mechanical coupled analysis of Turing tools based on ANSYS. In Electronic and Mechanical Engineering and Information Technology (EMEIT), 2011 International Conference on (Vol. 7, pp. 3836-3839). IEEE.

[11]. Zorev, N. N. (1966). Metal Cutting Mechanics.Pergamon, Headington hill hall, Oxford, England. 1966, 526 p.

[12]. Lotfi, M., & Amini, S. (2018). FE simulation of linear and elliptical ultrasonic vibrations in turning of Inconel 718. Proceedings of the Institution of Mechanical Engineers, Part E: Journal of Process Mechanical Engineering, 232(4), 438-448.

[13]. Datta, S., & Davim, J. P. (Eds.). (2016). Computational Approaches to Materials Design: Theoretical and Practical Aspects: Theoretical and Practical Aspects. IGI Global.

IOP Conf. Series: Materials Science and Engineering **810** (2020) 012076 doi:10.1088/1757-899X/810/1/012076

Study on the Development of Banana Fibre Reinforced Polymer Composites for Industrial and Tribological Applications: A Review

Upendra S. Gupta, Dr.Sudhir Tiwari

Asst. Prof. Mechanical Engineering Department, SVVV Indore, India ,
Professor Department of Mechanical Engineering, SGSITS Indore, India

upendrasharangupta@gmail.com,sudhir2609@gmail.com

Abstract. Fibres form one of the basic tenets of the human society. Fibres are used as the building block of human existence and thrive. The fibre technology is mostly comprised of artificial fibres in the present day due to their longevity, stability and inertness. However, these fibres destroy the integrity of nature as they are non biodegradable and pollutant in nature. This has led to an increase in the usage of natural fibre based composites. These composites add to the effectiveness of the artificial composites. One such fibre is banana fibre. Banana fibre is obtained from the trunk of the banana tree (Musa acuminata) which is a plant native to Southeast Asia. Effectively considered as a waste material after the harvest of the fruit, banana fibre has shown remarkable properties as an additive in the preexisting fibres. The addition of banana fibres in the materials have shown to produce a remarkable improvement in the chemical, mechanical and physical attributes of the fibres. This paper gives the overall view in the field of banana fibre reinforced polymer composites and its physical, mechanical and tribological properties of the composites. Emphasis is laid in evaluating all the factors of the advantage of banana fibre. Banana fibre reinforced composites, in turn, enable the development of industrial and tribological components having excellent wear, friction and lubrication properties which finds numerous applications

1. Introduction

Composites are widely used in our day to day life. Due to their low weight and ability to be moulded for specific usage they have gained a considerable fame in performance applications, such as aerospace and automotive industry. Nowadays the industry is being by natural fibre reinforced composites. The reinforced composites are on a rise due to the properties of biodegradability and revamped properties of the base fibre. The reinforcing property of the natural fibers is due to their lignistic content and their crystallinity. In recent years, natural fibers are gaining considerable attention due to their properties of low cost and varied applications. The recent developments in the area of fiber technology has led to the realization of the potential of the composites in the near future[1]. One of the most important developments has been in the discovery of uses of banana fibers. Banana fibres obtained from the trunk of the trees of the Musacae species of plant have started to become as a renowned interest for scientist all over the world. Cellulose has been found out to be the main constituent of plant fibres followed by lignin, hemicellulose and pectin. Cellulose serves as the reinforcement for all other constituents. Higher the aspect ratio of the fibers, higher is the tensile strength aiding reinforcement. The variability of these properties enables plant fibers to exhibit characteristics of a composite material.As examine by Satyanarayana K.G. et.al. [2] Several plant fibers such as sisal, coir and banana find resourceful usage in industry. Palanikumar et.al. [4] Reviewed the inherent properties of fibers along with their extraction techniques employed as resins in the composite. Justiz Smith and Satyanarayana K.G. et.al. [2,3,5] have also shown the mechanical and physical properties of some naturally occurring fiber as shown in Table1 &Table 2

Table 1 Physical and Mechanical Properties of Natural Fibres Courtesy of Satyanarayana, K. G (1984) [2]

Fibers	Width or Diameter(μm)	Density (kg/m^3)	Cell l/d ratio	Microfibrillar angle (degrees)	Initial Modulas(Gpa)	Ultimate tensile Strength (Mpa)	Elongation (%)
Coir	100-450	1150	35	30-40	4-6	106-175	17-47
Banana	80-250	1350	150	10±1	7.7-20.0	54-754	10.35
Sisal	50-200	1450	100	10-22	9.4-15.8	568-640	3-7
Pineapple Leaf	20-80	1440	450	8-14	34.5-82.5	413-1627	0.8-1
Palmyra	70-1300	1090	43	29-32	4.4-6.1	180-215	7-15

Table 2 Physical Properties of Different Cellulosic Fibers, Courtesy of Justiz-Smith(2008) [5]

Fibers	Moisture Content (wt %)	Ash Content (wt %)	Carbon Content (wt %)	Water absorption (wt %)	Tensile Strength (Mpa)
Banana	85.6	8.3	50.9	40	142.9
Coconut	27.1	5.1	51.5	169	138.7
Bagasse	52.2	4.5	53	235	29.6

Laly et.al. [6] has investigated on banana fibers to deduce the optimum quantity in composites to be about 40%. Justiz smithet.al. [5] have examined the longitudinal sections of banana fiber reinforced composites to calculate the presence of metal ions in the following percentages as shown in Table 3 .

Table 3 Metal elements present as ions in Cellulosic Fibers Courtesy of Justiz Smith (2008) [5]

Fibers	Al3+	Ca+	Mg+	Na3+	Si+4
Banana	0.14	5.72	1.77	0.28	1.41
Coconut	0.03	2.44	0.76	2.53	2.56
Bagasse	3.89	3.87	1.32	0.97	27

Reddy.N.et.al. [7] gave the various properties of cellulose based fibers as shown in. Table 4 and Table 5

Table 4 Chemical Properties of Natural cellulosic Fibers Courtesy of Reddy N.(2003) [7]

Fibers	Lignin (wt %)	Cellulose (wt %)	Hemicellulose (wt %)
Banana	9	43.46	38.54
Coconut	59.4	32.65	7.95
Bagasse	13	30.27	56.73

Table 5 Structure and Properties of Bio Fibers Courtesy of Geethama, V. G.(1995) [8]

Fiber	Fiber Structure Cell Dimensions			Fiber properties		
	Length (mm)	Width (μm)	Crystanilty	Tenacity g/den[a]	Elongation %	Moisture %
Coir	0.3-1.0	100-450	27-33	2	17-47	10-12
Bagasse	0.8-2.8	10-34	-	0.9-1.25	5.5-11.8	8.8
Banana	0.9-4.0	80-250	45	2.4-3.7	1.0-3.5	10-15
Palf	3-9	20-80	44-60	0.7-3.8	0.8-1.6	10-13
Wheat Straw	0.4-3.2	8-34	45	2.4-3.7	1.0-3.5	10-15
Rice Straw	0.4-3.4	4-16	40	-	-	6.5

Corbiere Nicollier et.al.[9] has investigated mechanical and physical properties of banana cement fiber composites. Natural fibers are called as reinforced materials and the cellulose content and microfibril angle determine the physical properties. High value of cellulose content and low microfibril angle for banana fibers indicate its apparent sustainability as a reinforcing material [10]. Hybridization of two different fibers' has proved to be an effective method in the design of materials [10,11,12]. The aging studies of reinforced composites show that reduction in mechanical properties of hybrid composite is nearly half of hybridized composite [10,11,12]. Thomas et.al. [13] has proved that banana fibers' and sisal fibers' gives better performance and can be used as an effective reinforcement agent in polyester matrix. Laly et al [13,16] and Kalaprasad G et.al. banana fibres were hybridized with glass fibres to get better mechanical performance .Mohan Rao et.al. [17] showed that cross-sectional area of banana fibre taken for analysis is 0.3596 mm2. The cross-sectional dimensions are measured along different orientations. It is noticed that, magnitude of laser beam diffractions is dependent on the curvature of the cross-section. Hence we can conclude that the test gives us relative structures instead of actual structure. Laly A. Pothan et.al. [18] have compared the cellulose content and fibrillar angle to strengthen the notion of use of banana fibres as a reinforcing material. An increase in the quality of the fibres has been observed wherever banana fibre has been used as a reinforcing material. All the classified properties show an improvement since the reinforcing causes change at molecular level. The various aspects can be evenly controlled by varying the percentage of different contents. Kulkarni et.al.[19] investigated the structure of fibre using optical microscopy. Banana fibres constitute of xylem, phloem, sclerenchyma and parenchyma. Also the number of various cells and helix angle for different diameter of the fibres are given in Table 6. Also the variation in mechanical properties with reference to diameter of fibre is also shown in Table 7.

Table 6 Number of Various cells and helix angle of banana Fibers for Various Diameter Courtesy of Kulkarni, A. G (1982) [19]

Diameter of fiber(μm)	Average no. of xylem cells	Average no. of Phloem cells	Average no. of Sclerenchyma cells	Total No. of cells	Fraction of Sclerenchyma cells to the total no. of cells (Strength donating)	Helix or Microfibrillar angle (θ)
100	3	6.25	53	62.25	0.85	12±1°
150	3	8	70	81	0.875	11±2°
200	4	7.75	92	103.25	0.886	11±1°

Table 7 Mechanical Properties of Banana fibers of different diameters. Guage length =50 x 10-3 Courtesy of Kulkarni, A. G(1982) [19]

Sample Number	Diameter of fiber (μm)	Initial Youngs Modulas (GN/m^2)	SD Initial Youngs Modulas (GN/m^2)	Breaking strength (MN/m^2)	SD Breaking Strength (MN/m^2)	% Strain	SD % Strain
1	50	32.703	8.19	779.078	209.3	2.75	0.957
2	100	30.463	4.689	711.661	239.614	2.469	0.798
3	150	29.748	8.561	773.002	297.104	3.583	1.114
4	200	27.698	7.083	789.289	128.558	3.34	0.688
5	250	29.904	4.059	766.605	165.515	3.244	1.284

Laly A. Pothan and. Hoyur et.al. [20,21] have discussed the various technique for the production of the fibre and the strength of the bio-composites. Banana fibres obtained from the dried stalk of banana tree, offers possibilities for engineering applications. Even though considered as a waste by-product, it has good specific strength properties comparable to conventional materials [23,24,25]. The increasing population has led to the necessity of waste control and use of biodegradable composites lends a helping hand in this area. The further addition in all properties has just led to remarkable development being made in the field of bio composite production. Therefore, composites of high-strength pseudo-stem banana reinforced composite can be used in a broad range of applications.

2. Methodology

To begin with the Mechanical, Tribological, Morphological and Thermo gravimetric tests, The specimen should undergo steps for the preparation of woven banana fiber reinforced epoxy composites [22]. These steps are illustrated below

i. Abstracting the Natural Continuous Banana Fibre
ii. Thinning the Dried Banana Fibre (B.F.)
iii. Weaving the Continuous fibre
iv. Mould Preparation
v. Mixing the Epoxy and Hardener with a ratio
vi. Preparation of the Specimen
vii. Mechanical ,Tribological, Morphological & Thermo Gravimetric Tests
viii. Results and Analysis

3. Results and Discussions

Bio-composites have served as an important factor in the production of economical, lightweight and superior composite structures. The mechanical properties of these bio-composites are superior to their base counterparts. The bio-composites developed vary in properties depending on the additive used.

Sapuan et.al.[22] studied the property of woven banana fibre experimenting with the tensile and flexural strengths combined with different materials.Joseph et.al.[25] compared the property of phenol formaldehyde resin when mixed with banana and glass fibre. In banana fibre composites, a small decrease in impact strength compared to neat resin occurs at lower fibre loading and a good increase in impact strength is observed with an increase in fibre loading .Thwe et.al [26] examined the effect of fibre content, fibre length, bamboo to glass fibre ratio, and coupling agent (maleic anhydride polypropylene) on Tensile and flexural properties of bio-composites. As observed by Murali M. Rao et.al.[27] natural fibres used as fillers in a polymeric matrix enable the production of economical and lightweight composites for load carrying structures. Geethama et.al. [28] estimated the tenacity and elastic modulus of banana fibre, and are in the range of 529.759 MPa, and 8.20 GPa, respectively. The percentage elongation at break of banana fibre varies between 1.0 and 3.5. The diameter of banana fibre will vary between 0.08 and 0.25 mm.The composites are well capable to be used in wider sectors. Zhu at.al. [29] Investigated the air-cured banana fibre reinforced cement composites and found that at a fibre loading of14% by mass, flexural strength is about 25MPa and the fracture toughness value is1.74 kJm2. They suggested that these mechanical properties with the water absorption of less than 25% are suitable for the use of building material. Mukherjee et.al.[30] showed in their investigation of banana fibre with polyester composites without weathering test that the specific modulus of the composites is 2.39 which are of the same order as that of glass fibre plastics. Maries Idicula et.al. [31] showed in their investigation of Banana fibres have remarkable addition to the matrix structure of the pre-existing fibres leading to an increase in strength of the fibres. The effect of having varying percentage of banana fibres is to increase tensile, reinforcing and elongation properties. Tensile stress-strain behaviour of neat polyester and banana composite having different fibre loading by incorporating 0.20 and 0.40 Vf of fibre, the modulus is increased by 30% and 54%, respectively, as compared to neat polyester. As the fibre loading is increased, the elongation at break is increased. The addition of cellulose fibers makes the matrix ductile. K V P P Chandu et.al[32] had shown in his research that direct relationship between the sliding distance and load. Coefficient of friction almost remains constant by evaluated the Banana-Pineapple fiber reinforced hybrid composite with defined constitutes of fiber and found the wear rate proportionate due to increase in the normal applied. Hybrid composites 30B-70P shows lower weight loss, frictional force, wear and coefficient of friction with different loads at 1m/s sliding velocity when compared with pure composites and other hybrid composites due to proper bonding between matrix and reinforcements .A Shadrach Jeyasekarann et al [33] has investigated to find the tensile characteristics of hybrid composite made by intruding unidirectional banana and glass fibers into epoxy resin mixture. The hand lay-up method of fabrication was used for preparing the composites of unidirectional glass fiber (UGF) and unidirectional banana/glass fiber (U B/G F). Tensile properties of the composites are verified on ANSYS software. It was observed that the numerical analysis results were higher than experimental analysis. Hybridization of banana fiber has better tensile properties. The specimen composite material was produced with the banana (stem) particulate as reinforcement using compression moulding. Results showed that density and elastic Modulus of the composite decreases and increases respectively with increasing weight fraction of the particulate reinforcement by B. Dan asabe et.al [34]. he shown that The tensile strength increased to a maximum of 42 MPa and then decreased steadily. The composition with optimum mechanical property (42 MPa) was determined at 8, 62 and 30 % formulation of banana stem particulates (reinforcement), PVC (matrix) and Kankara clay (filler) respectively with corresponding percentage water absorption of 0.79 %, Young's Modulus of 1.3 GPa, flexural strength of 92 MPa and density of 1.24 g/cm3. Comparison with conventional piping materials showed the composite offered a price savings per meter length of 84 % and 25 % when compared with carbon steel and PVC material .M. A. Maleque et.al[35] conclude that the tensile strength on the pseudo-stem banana woven fabric reinforced epoxy composite is increased by 90% compared to virgin epoxy and with epoxy material banana woven fabric flexural strength increased also he showed that approximately 40% increment in the impact strength of the virgin epoxy material, and due to higher impact strength value it has higher toughness properties. Claudia Merlini et.al [36] In his context, the author used plant fibers and polyurethane derived from castor oil in polymeric composites can be a good alternative. The aim of study was to evaluate the influence of the fiber volume fraction, fiber length and alkaline treatment on the mechanical and thermal properties of short random banana fiber reinforced polyurethane derived from castor oil. The banana fibers were chemically modified through contact with 10 wt.% sodium hydroxide solutions for 1 h and characterized through Fourier transformed infrared spectrometry (FTIR), scanning electron microscopy (SEM), tensile strength and density were measured through the FTIR spectra and SEM micrographs, modifications in the chemical structure and morphology of the treated fibers were observed compared with untreated fibers. The composites were analyzed by SEM, dynamic mechanical analysis (DMA), FTIR spectrometry, tensile strength and pull-out tests. The tensile strength and Young's modulus increased with increasing fiber volume fraction and length for the untreated and treated banana fiber polyurethane composites. On the other hand his result shown that, the treated banana fiber composites displayed higher tensile strength and Young's modulus values than the untreated fiber composites, due to the stronger interfacial interactions between the treated fibers and the polyurethane matrix. These effects occurred due to the morphological and chemical changes in the treated fiber surfaces which promote better adhesion between the fibers and the polyurethane matrix. The effects of Banana fibre length on the physical and mechanical properties of banana fibre/epoxy composite were investigated by M.Sumaila1et.al.[37] Five different samples were produced by varying the length of the fibre between 5mm and 25mm at 30% wt. fibre loading using the hand lay-up moulding technique. The mean density, percent moisture absorption, void content, tensile strength, tensile modulus, % elongation, compressive strength, impact energy, flexural strength and modulus of the composite were analysed. The results showed that the percent moisture absorption, void content and the compressive strength increased with increase fibre length while a decrease in density was observed. Result showed that at 15mm fibre length suggesting critical fibre length for effective and maximum stress transfer. Carolyn Palma et.al.[38] also conclude that use of banana peel as promising material for the development of a global bioremediation strategy for wastewater containing hazardous compounds, such as dyes. Its high adsorption capacity of Acid Black 1 (250 mg g-1) as well as the high decolourizing efficiency of the ligninolytic enzymes produced by Inonotus sp SP2 (97%) and Stereum hirsutum RU 104 (82%) allow to postulate the use of BP to implement a hybrid strategy of remediation based on the combination of physic chemical biological treatments. Laly A. Pothana et.al [39] The dynamic mechanical analysis of banana fiber reinforced polyester composites was carried out At lower temperatures (in the glassy region), the E' values are maximum for the neat polyester whereas at temperatures above Tg, the E' values was found to be maximum for composites with 40% fiber loading, indicating that the incorporation of banana fiber in polyester matrix induces reinforcing effects appreciably at higher temperatures. The loss modulus and damping peaks were found to be lowered by the incorporation of fiber. The height of the damping peaks depended on the fiber content. When higher fiber content of 40% was used, an additional peak in the tan δ curve, pointing to micro mechanical transitions due to the immobilized polymer layer was evident. The glass transition temperature associated with the damping peak was lowered up to a fiber content of 30%. The Tg values were increased with higher fiber content. To understand the phase behaviour of the composite samples .Cole- Cole analysis was made and he constructed a master curve based on time temperature super position principle, which allows the prediction of long-term effects. Satish Pujari et.al[40]. In this research, biodegradable and eco-friendly composites with various volume fractions of fenugreek banana composite materials were successfully developed and their thermal properties were investigated as a function of volume fraction. From obtained results, the following conclusions were drawn. The thermal conductivity of composites decreased with increase in volume fraction of fenugreek due to non-conducting nature of the natural fibers. At the maximum volume fraction of fenugreek, the thermal conductivity of the composite material is 0.259 W/m K. The results of this study indicate that the fenugreek banana composites are economical and possess good thermal insulating properties. Hence, the newly developed composite material can be used for different applications like steam pipes, automobile interior parts, electronic packages and building construction materials.T.P. Mohan et.al.[41] Aim of this research work was to study the tribological properties of natural fiber based composites using nanotechnology. The naturally available banana plant fibers were treated with nano clay particles and these treated fibers were then reinforced in an epoxy polymer to form composites. The friction and wear properties of nanoclay treated banana fiber (NC BF) reinforced composites were compared with untreated banana fiber (UT BF) reinforced composites. Short NC BF and UT BF reinforced composites with fiber concentration ranging from 20 wt.% to 60 wt.% was prepared by vacuum resin infusion processing method. The result indicates that the NC BF reinforced composites has shown improved friction and wear properties. Microscopy examination revealed that NC BF reinforced composites were able to form transfer layer between wear test specimens, resulting in improved wear properties. The nanoclay particles were also added to increased hardness and friction to the composites and improving the braking properties. Jinguo Wang et.al [42] showed in their investigation that, coir fiber board was reinforced with banana stem fiber to improve the mechanical properties of the board. Different ratios of banana stem fiber and coir fiber were mixed to make composite boards. These ratios had different effects on the physical (TS) and mechanical (MOR, MOE, and IB) properties of the particleboard composites. Banana fiber improved the MOE and MOR of the hybrid coir fiber boards.. SEM and FTIR, and SSA were used to analyze the fiber micro-structures, and bonding mechanisms of the particleboard were examined. These results confirmed that a novel hybrid particleboard composite could be produced. They suggested that coir fiber/banana stem fiber Particleboard composites were suitable for non-load bearing partitions and outdoor building materials. S Vigneshkumar et. al [43] has carried out an Experiment with fiber reinforced polymer composite on Pin on disc equipment. And investigated the changes of wear and friction performance of the sisal and rice husk composite weight .The carbon, sisal fiber and rice husk composite wear and frictional performance was improved with the increased load an orthogonal array method. with the increase in the load and sliding velocity the sisal, banana and kenaf fiber composite weight loss was more and Sisal and rice husk composite wear and frictional force was more with respect to increasing the speed and time .A. Alavudeenet.al [45] showed in their investigation that a high wear resistant material using banana fibre has been used to develop a hybrid clutch plate composite matrix with strength and rigidity. The composite material also improves the life-cycle of the composite Haneefa et al. [47] concluded that the hybridization of banana fiber and glass fiber increases the tensile strength and Young's modulus of the composite with increase in volume fraction of glass fiber because of greater compatibility of glass fiber with polystyrene matrix .Zainuddin et al. [48] on the work found that the decomposition temperature of banana pseudo stem fiber/unplasticized polyvinyl chloride decreases from 279°C to 256°C on the addition of banana filler from 10-40%. They also predicted that the thermal stability of acrylic modified composites was found to be more stable than unmodified banana pseudo stem fiber/unplasticized polyvinyl chloride composite. Kumar et.al. [49] have carried out research on preparing green composites using banana fibre and soy protein as matrix. The results show that at 0.3 volume fraction tensile strength and modulus of alkali treated fibre reinforced with soy protein composites increased to 82% and 96.3%respectively compared with soy protein film without fibres. Biodegradability test shows the composites are 100% bio degradable Due to low density, higher tensile strength, higher tensile modulus, and lower elongation at break of banana fibers, composites based on banana fibres have potential use in the various sectors like construction, automobiles, machinery, etc., Thus the use of bio-additives to existing fibre technology can lead to an improvement in the overall mechanical properties of the fibre.

Table 8: Summary of Banana Reinforced polymer Composites and its outcomes

Authors/Title	**Materials Used**	**Methodology**	**Outcomes**
K V P P Chandu et.al "Experimental Analysis on Wear Behaviour of Banana - Pineapple,Hybrid Natural Fiber Composites"(2018) [32]	-Banana Pineapple fiber reinforced hybrid composite by using hand layup technique -Epoxy resin -Hardener	-Wear test by Pin on disc type Friction and Wear monitor (DUCOM TL20) with data acquisition system	-Evaluation using composites with defined constitutes of fibre and matrix the wear rate proportionate due to increase in the normal applied load. -Coefficient of friction almost remains constant. - Direct relationship between the sliding distance and wear rate as the sliding distance increases .proportionately with respect to wear Rate. -Hybrid composites 30B-70P shows lower weight loss, frictional force, wear and coefficient of friction with different loads at 1m/s sliding velocity when compared to pure composites and other hybrid composites due to proper bonding between matrix and reinforcements.
A Shadrach Jeyasekarann et.al."Numerical and experimental analysis on tensile properties of banana"(206) [33]	-Unidirectional glass fiber epoxy laminate (UGF), and unidirectional banana/glass fiber - Epoxy laminate (U B/G F) by using hand lay up technique - Epoxy resin LY556 -Hardener HY951 as binder	-Tensile test by according to the ASTM 3039 standard -Tensile properties of composites by using ANSYS software - Surface Morphology through scanning electron microscope (SEM).	-UGF composite possesses higher ultimate tensile strength among the samples and can withstand the strength of up to 567 Mpa at a load of 60 kN - Experimental stress value was 567.7 Mpa and ANSYS Von misses stress value was predicted to be 616.69 Mpa for UGF composite. - The average load displacement was found to be 10.8 mm for the UGF composite and 11 mm for UB/G F composite. - It can be used in applications which require moderate strength, especially in house hold articles such as partition boards. - Based on experimental and numerical analysis using ANSYS software , the deviation was mainly due to the assumptions in numerical analysis such as increase of interfacial bonding, between fibre and matrix, alignment of

			fibre, non porous, uniform diameter of fibres and these assumptions made the predicted tensile value to be higher for UGF and U B/U F composites - In experimental analysis fabrication defects in composites such as voids, cracks, matrix and fibre fracture, poor bonding, improper matrix distribution and nonalignment of fibres were the probable causes for obtaining lower tensile value for UGF and U B/G F composites when compared with the numerical analysis - By chemically treating the natural fibres can enhance the tensile strength of the composites
B.Dan-asabeet.al. "Mechanical, Spectroscopic and Micro structural Characterization of Banana Particulate Reinforced PVC Composite as Piping Material"(2016) [34]	Kankara kaolin clay (200 g) - Polyvinyl chloride (PVC, 500 g) -Banana stem (200 g) -Sodium hydroxide (NaOH) - distilled water (5 litres)	-Carver-3851 compression machine. characterization tests Fourier Transform Infrared (FTIR) analysis	-Density and Elastic Modulus of the composite decreases and increases respectively with increasing weight fraction of the particulate reinforcement. -Tensile strength increased to a maximum of 42 MPa and then decreased steadily -The composition with optimum mechanical property (42 MPa) was determined at 8, 62 and 30 % formulation of banana stem particulates (reinforcement), PVC (matrix) and Kankara clay (filler) respectively with corresponding percentage water absorption of 0.79 %, Young's Modulus of 1.3 GPa, flexural strength of 92 MPa and density of 1.24 g/cm³. -Fourier Transform Infrared (FTIR) analysis of the constituents showed identical bands within the range 4000–1000 cm. - Scanning Electron Microscopy (SEM) result showed fairly uniform distribution of constituents' phases. - X Ray Fluorescence (XRF) confirms the Xray diffraction (XRD) result of the presence of minerals of kaolinite, quartz, rutile and illite in the kaolin clay. - Comparison with conventional piping materials showed the composite offered a price savings per meter length of 84 % and 25 % when compared with carbon steel and PVC mater
M. A. Maleque et.al."Mechanical properties study of pseudo stem banana fiber reinforced epoxy composite"(2005) [35]	-Pseudo-stem banana woven fabric reinforced epoxy composite by the hand layup method - Epoxy resin grade 3554A	-Tensile test according to the ASTM D638 standard - flexural test according to BS EN ISO 14125 - weight impact test as per	- The tensile strength on the pseudo stem banana woven fabric reinforced epoxy composite was increased by 90% compared to virgin epoxy. - The flexural strength increased when

	-Hardener grade 3554B	BS 2782 - Fracture Surface Study by SEM	banana woven fabric was used with epoxy material. - Impact strength test showed that the pseudo stem banana fibre improved the impact strength properties of the virgin epoxy material by approximately 40%. Higher impact strength value leads to higher toughness properties of the material. - The banana fibre composite exhibits a ductile appearance with minimum plastic deformation.
Sevgi Hoyur et.al."Production of banana / glass fiber biocomposite profile and its bending strength"(2012) [21]	-Banana fibers -Polyester resin (CE 92 N8) -Multi faceted glass fibers (MAT8) -Hardener (MEK) -Cobalt naftanat as a catalyst	-Bending tests were carried out on a SHIMADZU AG-IS unit with a capacity of 50 kN.	-The highest and lowest bending strengths for a single layer specimen were found to be 13.085 N/mm^2 and 8.957 N/mm^2. -The highest and lowest bending Strength for the double layer specimens was found to be 18.196 N/mm^2 and16.834 N/mm^2. -Produced banana / glass fibre bio composites can be used for indoors and outdoors applications.
Claudia Merlini et.al."Influence of fiber surface treatment and length on physicochemical polyurethane composites"(2011)[36]	-Banana fibers -Sodium hydroxide -Polyol -Prepolymer -Diphenylmethane diisocyanate (MDI) -Resin (IMPERVEGUG 132 A)	-Attenuated Total Reflectance-Fourier Transform Infrared spectroscopy (ATR-FTIR) -Scanning electron microscopy (SEM), Jeol model JSM- 6390LV -EMIC testing machine (model DL2000) following ASTM C1557, D638 and D3039 standard methods. -Pull out test by Universal Testing Machine -Dynamic mechanical analyzer (DMA-983 interfaced to a TA 2000).	-To evaluate the influence of the fibre volume fraction, fibre length and alkaline treatment on the mechanical and thermal properties of short random banana fibre reinforced polyurethane derived from castor oil. -The banana fibres were chemically modified through contact with 10 wt.% sodium hydroxide solutions for 1 h and characterized through Fourier transformed infrared spectrometry (FTIR), scanning electron microscopy (SEM), tensile strength and density measurements. -The tensile strength and Young's modulus increased with increasing fibre volume fraction and length for the untreated and treated banana fibre polyurethane composites. -The treated banana fibre composites displayed higher tensile strength and Young's modulus values than the untreated fibre composites, due to the stronger interfacial interactions between the treated fibres and the polyurethane matrix. -The morphological and chemical changes in the treated fibre surfaces which promote better adhesion between the fibres and the polyurethane matrix.
M.Sumaila1et.al."Effect of fiber length on the physical and Mechanical properties of	-Mature banana pseudo-stem -5% sodium hydroxide	-Moisture Absorption Test(Mettler Toledo balance, Model XP603S).	-The percentage moisture absorption and void content increases with increase fibre length at constant fibre

random oriented non woven short(musa balbisiana)Fibre/epoxy composite Banana"(2013) [37]	(NaOH) solution -Epoxy resin (LY556) -Hardener (HY591) -Petroleum jelly (debonding agent)	-Density Measurement (Mettler Toledo balance, Model XP603S with a precision of ±0.001g) -Tensile Test according to ASTM D638 standard -Compressive Test According to ASTM D 695-02a standard -Izod impact test machine according to ASTM D 256 standard -Flexural Test according to ASTM D790M standard	loading of 30%wt while there was a decline in composite density. -The tensile strength, tensile modulus and percentage elongation of the composite attained a maximum in composite fabricated from 15mm fibre length. - The compressive strength increases whereas the impact energy decreased with increasing fibre length. -The mean flexural strength and modulus of the composite increased with increasing fibre length up to 25mm.
Maries Idiculae et.al."A Study of the Mechanical Properties of Randomly Oriented Short Banana and Sisal Hybrid Fiber Reinforced Polyester Composites"(2004) [31]	-Banana fiber and sisal fiber -Unsaturated polyester resin HSR 8131 -Curing agent, methyl ethyl ketone peroxide -Catalyst cobalt napthenate	-Mechanical testing FIE universal testing machine of 500 KN capacities. -fiber diameter was measured using a Leica (DMLP) Polarizing light microscope. -load displacement curves -tensile as well as flexural strength and modulus were calculated. -Izod impact test (25 Joules pendulum impact testingmachine) - Fracture Surface Study by SEM	- Mechanical properties, such as tensile, flexural, and impact properties, of short banana/sisal hybrid fibre reinforced polyester composites were evaluated by varying the relative volume fraction of banana and sisal at a total constant fibre loading of 0.40 volume fractions. - The tensile strength was found to be increased in banana/sisal hybrid fibre reinforced polyester composites when the volume fraction of banana was increased. - Higher compatibility was obtained in hybrid composites, which leads to better stress transfer ability in composites. When the Vf of sisal is 0.26, the maximum modulus is obtained - A positive hybrid effect can also be observed for flexural strength and flexural modulus. When the Vf of banana is 0.19, the flexural modulus is maximum - It can be concluded that the reinforcement of banana and sisal fibre in polyester results in a positive hybrid effect for tensile and flexural properties - Value added and cost-effective composites having high tensile and flexural properties could be well developed by the judicious selection of banana and sisal fibre
Carolyn Palma et.al."Eco friendly Technologies Based on Banana Peel Use for the Decolourization of the Dyeing Process Wastewater(2010)"[38]	- Banana peel (Mussa ssp) - Microorganisms Inonotus sp SP2 and Stereum hirsutum RU 104 - Dyes of analytical grade from Sigma–Aldrich Co	- Particle size Distribution by laser diffraction using the Mastersizer X equipment (MALVERN INSTRUMENTS, MSX1) -Specific Surface by by nitrogen gas adsorption according to Brunauer, Emmett and Teller (BET)	-Banana peels as promising material for the development of a global bioremediation strategy for wastewater containing hazardous compounds. - High decolourizing efficiency of the ligninolytic enzymes produced by Inonotus sp SP2 (97%) and Stereum hirsutum RU 104 (82%) allow to postulate the use of BP to implement a

		isotherm method. - Point Zero Charge (pHPZC) - Bioadsorption of Dyes In Banana Peel(UV–V is spectrophotometer Spectronic Helios Gamma model)	hybrid strategy of remediation based on the combination of physicochemical biological treatments.
Laly A. Pothana et.al."Dynamic mechanical analysis of banana fiber reinforced polyester composites"(2001)[6]	- Banana fiber - Unsaturated polyester HSR 8131 - Methyl ethyl ketone peroxide -Cobalt naphthenate	- A dynamic mechanical thermal analyzer M K I I - scanning electron microscope (SEM) model Jeol JSM-35 C and Cambridge 250 MK3 stereo scan	- The dynamic modulus shown a decrease with incorporation of fibre below the glass transition temperature and had a positive effect on the modulus at temperatures above Tg. -The maximum improvement in properties was observed for composites with 40% fibre loading, which is chosen as the critical fibre loading. - Increase of frequency shifts the Tg to higher temperatures supporting the good fibre/matrix interaction. - An additional peak occurred at high fibre loading in the tan δ curves, due to the interlayer effect. -Addition of fibre lowers the tan δ peak height, which again points to the improved fibre/matrix adhesion. - Cole Cole plots shown an imperfect semicircle showing the heterogeneity of the system as well as the good interfacial. Adhesion at high fibre loading . - Use of master curve and the relationship of the shift factors were useful in predicting the long-term behaviour of the composite
Laly A. Pothana et.al."The Static and Dynamic Mechanical Properties of Banana and Glass Fiber Woven Fabric Reinforced Polyester Composite" (2005) [39]	- Banana fiber - Unsaturated polyester HSR 8131 - Methyl ethyl ketone peroxide -Cobalt naphthenate	-The tensile test (ASTM D 638-76) -Charpy impact tester Instron Wolpert PW5, ASTM D256 -Fractography optical microscopy using 'Citoval' stereo microscope Carl Zeiss, Jena -Dynamic mechanical analysis DuPont Instrument DMA 983	-Woven fabric composites of banana and glass fibre had proved to be an effective reinforcement in imparting high strength to polyester. -The tensile properties for composites with two layers of the fabric with closed weave architecture were found to be max. -The flexural strength, showed the highest value for composites with three layers of the fabric. -The highest impact strength was obtained for composites with four layers of the fabric with closed weave Architecture. -The damping peaks were found to be lowered by the incorporation of higher number of layers of the fabric indicating an improved fibre/matrix interaction. -The mechanical properties of

			composites made out of woven fabrics of natural and synthetic fibres were comparable with those of synthetic fibre composites whereas from an economic point of view, these composites have an edge over the other conventional materials.
S.M. Sapuan et al. "Mechanical properties of woven banana fibre reinforced epoxy composites"(2004) [22]	-Banana fibre -Epoxy 3554A -Hardener 3554B	- Tensile and flexure tests Galdabini 1890 - Statistical analysis analysis of variance-one way (ANOVA)	-The maximum value of stress in x direction was 14.14 MN/m^2 and the maximum value of stress in y direction was 3.398 MN/m^2. -The Young's modulus, the value of 0.976 GN/m^2 in x direction and 0.863 GN/m^2 in y direction were obtained -Three point bending (flexural), the maximum load applied was 36.25 N to get the deflection of woven banana fibre specimen beam of 0.5 mm. -Gave a good opportunity to use woven banana fibre especially in producing household utilities.
Satish Pujari et.al."Experimental Investigations on Thermal Conductivity of Fenugreek and Banana Composites" (2016)[40]	-The banana stem layers -Fenugreek	-Thermal Conductivity Measurementas per ASTM E 1225-87 -	-The thermal conductivity of composites decreased with increase in volume fraction of fenugreek due to non-conducting nature of the natural fibres. -The maximum volume fraction of fenugreek, the thermal conductivity of the composite material was 0.259 W/m K. -The fenugreek banana composites were economical and possess good thermal insulating properties. -The newly developed composite material can be used for different applications like steam pipes, automobile interior parts, electronic packages and building construction materials.
T.P. Mohan et.al."Tribological properties of nanoclay infused banana fiber (NC-BF) reinforced epoxy composites" (2018) [41]	-Banana fibers -Untreated raw Na+ montmorillonite (MMT) clay -Diglycidyl ether of bisphenol A (DGEBA) based epoxy resin and cyclic aliphatic amine based epoxy hardener -Nanoparticle infusion were NaOH, acetic acid. -Steel material of surface roughness	-Scanning electron microscopy (SEM) technique(Zeiss Environmental Scanning Electron Microscopy ESEM ESEM: model EVO HD 15) -wear test(pin-on-disc tribometer, ASTM G99-05 2010)	- The NCBF reinforced composites had shown improved friction and wear properties. - Microscopy examination revealed that NCBF reinforced composites were able to form transfer layer between wear test specimen wear surface and counter face, resulting in improved wear properties. - The nanoclay particles also induced increased hardness and friction to the composites and improving the braking properties.
Jinguo Wang et.al."Novel Particleboard Composites Made from Coir Fiberand	-Raw coir fibers -Waste banana stem fibers -Type 14L962 phenol	-Mechanical Testing(modulus of rupture (MOR), modulus	-Banana stem fibre had a significant impact on the mechanical performance (MOR,MOE, IB, and 24-hTS) of CB

Waste Banana Stem Fiber" (2016) [42]	formaldehyde (PF) adhesive -Paraffin wax as a waterproofing agent	of elasticity (MOE), internal bond strength (IB), and thickness expansion) -Best fit curves -Regression equations -Specific Surface Area (SSA) of Fibers Analysis -SEM Analysis -FTIR Analysis	particleboard. -The mechanical properties of the CB particle board had linear correlations with banana fibre content. -The properties of microstructure and high surface area for banana fibre brought about the increase in the PF resin content and 24-hTS for the CB particleboard. -The chemical composition of banana fibre influenced the performance of the CB particleboard, which was ascribed to two characteristics of banana fibre: higher cellulose and hemicelluloses contents and lower lignin content.
S Vigneshkumar et. al. "Experimental analysis on tribological behavior of fiber reinforced Experimental analysis on tribological behavior of fiber reinforced composites" (2018) [43]	-Sisal, banana, kenaf and carbon fiber filler material is rice husk -unsaturated polyester resin -1.5% accelerator (Cobalt Octate) -1.5% catalyst (Methyl ethyl ketone peroxide).	-wear with ASTM G99. -Wear and Friction monitor- TR 201,(Pin on disc) -Scanning Electron Microscopy of FEI Quanta 200 FEG (FEI Company)	-Experiment was carried out with fibre reinforced polymer composite on Pin on disc equipment. The carbon, sisal fibre and rice husk composite wear and frictional performance were improved with the increased load an orthogonal array method. - The sisal, banana and kenaf fibre composite weight loss was more with the increase in the load and sliding velocity. - Sisal and rice husk composite wear and frictional force was more with respect to increasing the speed and time.
T. Hariprasadet.al. "Study of mechanical properties of banana coir hybrid composite using experimental and FEM techniques" (2013) [44]	- Pseudo stem banana woven fiber - Coir - Epoxy or polyepoxide -Polyamine "hardener". - Hardener (EH758)	- Tensile Test(ASTM Standard D638) - UTM Lloyd LR100K Testing Machine - flexural test (ASTM D 790) - Izod and Charpy test. - Finite-Element Analysis (FEA)	- An alkali treated banana coir epoxy hybrid composite had greater tensile strength and impact strength than an untreated banana coir epoxy hybrid composite. - The alkali treated banana coir epoxy hybrid composite had less flexural strength than the untreated banana coir epoxy hybrid composite. - This analysis was useful for realizing the advantages of hybrid fiber reinforced composites in structural applications and for identifying where the stresses were critical and damage the interface under varying loading conditions. - The simulated stress distribution for the mechanical properties of the treated and untreated banana-coir hybrid samples, under different loading conditions, was obtained. The maximum stresses are averages of the values of the center elements used in the FEA.
A. Alavudeen et.al. "Clutch plate using woven	- Banana (Musa sepentium)/kenaf (Hibiscus	- The tensile test (Shimadzu AG- 50 kN as	-The tensile strength test of the woven hybrid reinforced polyester composites

hybrid composite materials" (2011) [45]	cannabinus) - unsaturated polyester matrix - 10%NaOH and 10% sodium lauryl sulphate (SLS) - cobalt naphthenate(catalyst) -methyl ethyl ketone (accelerator)	per ASTM D638 standard) - Wear tests per ASTM G-99 standard - Thermal deformation test	showed the ultimate stress of 137.221 MPa. - A high wear resistant material created using banana/kenaf woven hybrid composite matrix could successfully develop a beneficial clutch plate composite, particularly in terms of strong and rigidity. - This composite material would increase the life of the clutch plate and reduce the energy loss during power transmission and also exhibits high machinability. - Banana and kenaf fibre was a feasible reinforcement to form a reinforced composite as an alternative and new material for clutch plate applications at low cost.
Lina Herrera-Estrada et.al."Banana fiber composites for automotive and transportation applications" (2015) [46]	- Banana fibers - 6% NaOH solution - SC-15 epoxy resin - ENVIREZ 1807 polyester - catalyst (alkyl polyamine and benzoyl peroxide/ tert-butyl perbenzoate mixture, respectively). - Polytetrafluoroethylene (PTFE) release agent -Teflon Cloth -	- Three point bending test method as per ASTM D 790-03 - compressive strength as per ASTM D 695- 96. - universal testing machine SATEC T-5000.	-The flexural strength of banana fibre/eco polyester composites was 40.16 MPa, which is 14.78% higher than the strength of banana fibre/epoxy due to fibre/matrix interaction. -The highest compressive strength of 122.11 MPa of the banana fibre/epoxy composite was attained after fibre pre-treatment and was 38.35% higher than the observed strength without the treatment - The highest compressive strength in banana fibre/eco-polyester composites was 122.88MPa and was achieved without fibre pre-treatment. - It was observed that environmental resistance is higher in banana fibre/epoxy composites with alkaline pre-treatment, followed for banana fibre/polyester composites without any treatment. Because of improvement in fibre/matrix interaction with the fibre chemical pre treatment in epoxy composites and to deterioration of the inter phase in polyester composites.
W. H. Zhu et.al."Air-Cured Banana Fibre Reinforced Cement Composites"(1994)[29]	- Banana stalk - 9% active alkali (as Na_2O) - 25% sulphidity - 7:1 liquor to sample ratio - Ordinary Portland cement	- Modulus of rupture (MOR) as perASTMD790 - Modulus of elasticity in flexure (MOE) as per ASTM-D790 - Program for routine flexural testing of cellulosic fibre-cement -Fracture Energy - Water absorption /density tests As perASTM-C220-75.	- Pulped banana fibre (Cavendish) was a satisfactory fibre for incorporation into a cement matrix - At a loading in the range of 8-16% by mass, flexural strengths in excess of 20 MPa (as high as those found in some synthetic fibres reinforced cements, and up to 90% that of pulped softwood fibre cement materials). - At a fibre loading of 14% by mass, the flexural strength is about 25 MPa and the fracture toughness value is 1-74 kJm^2.

H. Savastano Jr et. al "Brazilian waste fibres as reinforcement for cement based composites"(2000) [50]	- Sisal (Agave sisalana) - Banana (Musa cavendishii) pseudo-stem fibres - Waste Eucalyptus grandis pulp. - Active alkali (as Na2O) - Sulphidity (as Na2O) - Plain cement paste and Pinus radiata kraft reinforced cement composites	- Modulus of rupture (MOR) - Modulus of elasticity - flexural testing of cellulosic fibre-cement -Fracture Energy - Water absorption /density tests As per ASTMC948-81	-Mechanical testing showed that optimum performance of the various waste fibre reinforced composites was obtained at fibre content of around 12% by mass, with flexural strength values of about 20 MPa and fracture toughness values in the range of 1.0±1.5 kJ m^2 -E. grandees was the preferred reinforcement for low cost fibre cement
M. Jannah et.al."Effect of Chemical Surface Modifications on the Properties of Woven Banana Reinforced Unsaturated Polyester Composites"(2009) [51]	-Banana fibers -Unsaturated polyester P9509 -Methyl ethyl ketone peroxide (55%) as hardener.	-Tensile test as per ASTM D3822-01 -Fourier-transform infrared spectroscopy (FTIR) model Perkin Elmer System 2000 -flexural test as per (ASTM D790-98) -Izod impact tests as per(ASTM D4812) -Scanning electron microscopy (FESEM) model ZEISS SUPRA 35VP.	-The flexural strength and modulus of woven banana fibre composites increased with the increase in fibre content to optimum fibre content of 10 volume %. -Treated composite systems showed improvements in flexural properties. - The impact strength increased with the fibre content up to 15 volume %, with an obvious trend shown by acrylic acid-treated banana reinforced polyester composites -Treated fibre composites reduced the water absorption compared to the untreated composite system
Anshida Haneefa et .al." Studies on Tensile and Flexural Properties of Short Banana/Glass Hybrid Fiber Reinforced Polystyrene Composites"(2008) [47]	-Banana fiber -Glass fiber (E-Glass) -Polystyrene (PS) crystal grade -Polystyrene maleic anhydride as a surface modifier	-Tensile test (Universal testing machine) -Flexural tests as per ASTM D 790 -Scanning Electron Microscopy (SEM) JEOL JSM 35C model SEM.	-The tensile strength and Young's modulus of the composite increased with respect to increase in volume fraction of glass fibre due to the greater compatibility of glass fibre than banana fibre with polystyrene. -The influence of fibre content, fibre loading, and hybrid effect on the mechanical properties such as tensile strength, Young's modulus, elongation at break, and flexural properties of the composites, was evaluated. - The volume fraction of glass fibre based on total fibre content increased all the mechanical properties, except elongation at break. -The tensile and flexural properties of composites were observed to had improved as the fibre loading (volume %) increases. - Modification of the banana fibre improved the optimum fibre-matrix properties. - Hybrid effect was calculated using additive rule of hybrid mixtures.

E.S. Zainudin et.al." "Thermal degradation of banana pseudo stem filled unplasticized polyvinyl chloride (UPVC) composites"(2008) [48]	-Banana pseudo-stem (BPS) -Unplastisized polyvinyl chloride (UPVC) -BPS filler -Acrylic FM50	-Thermogravimetric analysis (TGA)	-The thermal stability of the composites was found to be higher than that of BPS fibre and the UPVC matrix. - The incorporation of BPS filler decreased the thermal stability of BPS/UPVC composites for the case of non acrylic. -The thermal stability of BPS/UPVC composite with acrylic was found to be higher compared to BPS/UPVC composite without acrylic.
Vijay Sitaram Shivankar et.al "Some studies on 100% banana parallellaid and 60:40% banana: polypropylenecross laid non-woven fabrics"(2019) [52]	-Three varieties of banana fibres	-The tensile Test Instron (model 4301) machine -Non woven fabric ASTM D 5729-97 standard -GSM of non-woven fabric ASTM D 3776-96 standard. -Stiffness of non-woven fabric ASTM D 5732-95 standard. -Air permeability FX 3300 Air permeability tester ASTM D 737-96 standard. -Bursting strength Eureka bursting strength tester ASTM D 3786-87 standard	-Graint Naine banana nonwoven fabric had shown higher elongation% for machine cross direction. -Increased air permeability of C.L. (60:40) banana/polypropylene blend non-woven fabric observed than P.L. 100% banana non woven fabric. -Bursting strength of P.L. nonwoven fabric was higher, both in the machine as well as in cross direction than C.L. nonwoven fabric for three varieties of banana fibres -Bending length of the C.L. banana nonwoven fabric was higher than the P.L. nonwoven fabric.

4. Conclusions

Therefore we can conclude that banana fibre due to its varying properties and differing usages is going to serve an important part in the development of the future of fibre industry. The results of different tests conducted on the fibre indicate that it can be used in varying applications across different fields. The fibre can be used in combination with other fibres as an additive and enhance the properties of the base fibres.

I. Value intensive and cost effective composites having high tensile and flexural properties can be well-developed by judicial usage of banana fibres.
II. A high wear resistant material using banana fibre has been used to develop a hybrid clutch plate composite matrix with strength and rigidity. The composite material also improves the life cycle of the composite.
III. High variability along the length and between fibers is displayed by the banana fibers, This is a characteristic of natural fibers to completely illustrate the banana fibers and provide suitable applications in natural fiber reinforced composites
IV. Banana peel shows potential of being the feasible material for the development of a global bioremediation plan for wastewater containing hazardous compounds.
V. Biodegradable and eco-friendly composites with volume fractions of fenugreek banana composite materials were successfully achieved. Moreover, thermal properties were examined as a function of volume fraction. Therefore, it can be seen recently made composite material can be used for various applications like steam pipes, automobile interior parts, electronic packages and building construction materials.

VI. Banana natural reinforced polymer composites have balance of structure and property, due to which these composites can be manufactured at a lower cost.

VII. Banana natural reinforced polymer composites can be infused with the nanoclay for the application in wear and braking works.

VIII. To improve tensile properties of banana natural reinforced polymer chemical modification (alkali treatment, benzoyl chloride treatment and PSMA) can be added. The addition of chemical displayed decreased hydrophilicity of banana fibre, and improvement in the fibre/matrix compatibility through mechanical anchoring, physical and chemical bonding.

IX. User friendly and cost effective banana natural reinforced polymer with sufficient stiffness and damping behaviour can be developed by hybridization of banana and sisal fibre.

As the current situation of environment is depleting while the demand for the alternative energy is increasing. To provide a solution banana natural reinforced polymer can be best suited for energy applications and diminish its consumption

5. References

[1], Van Norstrand Reinhold Bodig, J and Jayne, BA (1982): Characteristics Of Wood Composites, Mechanics Of Wood and Wood Composites Company Inc., New York, **1-47**

[2] Satyanarayana, K. G., Sukumaran, K., Mukherjee, P. S., Pavithran, C. and Pillai, S. G. K. (1990). Natural Fiber-polymer Composites, Journal of Cement and Concrete Composites, 12: **117-136.**

[3] Satyanarayana, K. G., Sukumaran, K., Kulkarni, A. G., Pillai, S. G. K. and Rohatgi, P. K. (1984).Performance of Banana Fabric - Polyester Composites, In: Marshall, I. H. (ed.), Proceedings of Second International Conference on Composite Structure, 13-16 September, 1983, London, pp. **535-537**

[4] Palani Kumar K and Shadrach Jeya Sekaran A 2014 Some natural fibers used in polymer composites and their extraction processes: A review. J. Reinforced Plastics Compos. 33(20): **1879–1892**

[5] Justiz-Smith, N. G., Virgo, G. J. and Buchanan, V. E. (2008). Potential of Jamaican Banana, Coir, Bagasse Fiber as Composite Materials, Journal of Material Characterization, 59: **1273-1278**

[6] Laly A. Pothana, Zachariah Oommenb, and Sabu Thomas, "Dynamic Mechanical Analysis of Banana Fiber Reinforced Polyester Composites", Composites Science and Technology, 63(2)(2003), pp. **283–293**.

[7] Reddy, N. and Yang, Y. (2003). Bio-fibers from Agricultural By-products for Industrial Application, Trends in Biotechnology, 23**: 22-27**

[8] Geethama, V. G., Reethama, J. and Thomas, S. (1995). Short Coir Fiber Reinforced Natural Rubber Composites: Effect of Fiber Length, Orientation and Alkali Treatment, Journal of Applied Polymer Science, 55: **583-594.**

[9] T. Corbière-Nicollier, B. G. Laban, L. Lundquist, Y. Leterrier, J. -A. E. Månson, and O. Jolliet, "Life Cycle Assessment of Biofibers Replacing Glass Fibers as Reinforcement in Plastics", Resources, Conservation and Recycling, 33(4)(2001), pp. **267–287**

[10] Fu S, Lauke B, Mader E, Yue C, Hu X, Mai YW. J Mater Sci 2001;36:**1243.**

[11] Thwe MM, Liao K. Composites Part A 2002;**33:43.**

[12] Thwe MM, Liao K. J Mater Sci 2003;35(2):**363.**

[13] Pothan LA, Thomas S, Neelakantan NR. J Reinf Plast Comp 1997;16**:744.**

[14] Joseph K, Varghese S, Kalaprasad G, Thomas S, Prasannakumari L, Koshy P, et al. Eur Polym J 1996;3210:1243.

[15] Pothan LA, Thomas S. In: Proceedings of ICCM 12-Paris paper 1999, **1267**

[16] Kalaprasad G, Joseph K, Thomas S. J Compos Mater 1997;31(5):**509.**

[17] Murali Mohan Rao, K. and Mohana Rao, K. (2007). Extraction and Tensile Properties of Natural Fibers: Vakka, Date and Bamboo, Journal of Composite structures, 77: **288-295**

[18] Laly A. Pothan, Sabu Thomas, N. R. Neelakantan; Short Banana fibre reinforced polyester composite: Mechanical Failure and Aging Characteristics.

[19] Kulkarni, A. G., Satyanarayana, K. G., Rohatgi, P. K. and Vijayan, K. (1982). Mechanical Properties of Banana Fibers, Journal of Material Science, 18: **2290-2296**

[20] Pothan, L.A., Thomas, S. and Neelakantan, N.R. (1997). J. Reinf. Plast. and Comp., 16: 744 Aziz S. H., Ansell M. P. The effect of alkalinization and fibre alignment on the mechanical and thermal properties of kenaf and hemp bast fibre composites: Part 1-polyester resin matrix.Composites Science and Technology, 64: 1219-**1230 (2004).**

[21] Sevgi Hoyur, Kerim Çetinkaya; Production of banana / glass fibre bio–composite profile and its bending strength. Usak University Journal of Material Sciences 1 **(2012) 43 – 49.**

[22] S.M. Sapuan, A. Leenie, M. Harimi, Y.K. Beng; Mechanical properties of woven banana fibre reinforced epoxy composites.Materials and Design 27 **(2006) 689–693**

[23]Satyanarayana, K.G.,A.G. Kulkarni and P.K. Rohtagi.1981. J. scient.Ind. Res., **40:222**

[24] Idicula M., Maholtra S.K., Joseph K., Thomas S.. Effect of layering pattern on dynamic mechanical properties of randomly oriented short banana/sisal hybrid fiber reinforced composites. Journal of Applied Polymer Science, 97: **2168-2174 (2005).**

[25] Seena Joseph, M.S. Sreekala, Z. Oommen, P. Koshy, Sabu Thomas; A comparison of the mechanical properties of phenol formaldehydecompositesreinforced with banana fibres and glass fibres . Composites Science and Technology 62 **(2002) 1857–1868**

[26] Thwe MM, Liao K.; Effects of environmental aging on themechanical properties of bamboo-glass fibre reinforced polymermatrix hybrid composites. Composites Part A 2002;**33:43–52.**

[27] K. Murali Mohan Rao, K. Mohana Rao; Extraction and tensile properties of natural fibres:Vakka, date and bamboo. Composite Structures 77 (2007) **288–295.**

[28] Geethama, V. G., Mathew, K. T., Lakshminarayana, R. and Thomas, S.(1998). Composite of Short Coir Fibres and Natural Rubber: Effect of Chemical Modification, Loading and Orientation of Fibre, Journal ofPolymer Science, 65: **1483-1491.**

[29] Zhu, W. H., Tobias, B. C., Coutts, R. S. P. and Langfors, G. (1994). Air-cured Banana-fibre-reinforcedCement Composites, Journal of Cement and Concrete Composites, **16: 3-8**

[30] Mukherjee, K. G. and Satyanarayana, K. G. (1984). Structure and properties of some Vegetable Fibres,Journal of Material Sciences, 19: **3925-3934**

[31] Maries Idicula, N. R. Neelakantan, Zachariah Oommen, Kuruvilla Joseph,Sabu Thomas; A Study of the Mechanical Properties of RandomlyOriented Short Banana and Sisal Hybrid Fibre ReinforcedPolyester Composites. DOI 10.1002/app**.21636**.

[32] K V P P Chandu, S.V.Gopala Krishna, B V Subrahmanyam,K Venkateswar Rao; Experimental Analysis on Wear Behaviour of Banana - Pineapple,Hybrid Natural Fiber Composites, International Journal of Engineering Science Invention (IJESI), ISSN (Online): 2319 – 6734, ISSN (Print): 2319 – 6726, www.ijesi.org / PP. **14- 20.**

[33] A Shadrach Jeyasekarann .K Palanikumar and S Rajarajan;Numerical and experimental analysis on tensile properties of banana, Sadhana,Indian Academy of Sciences.

[34] B. Dan-asabe , A.S. Yaro , D.S. Yawasa, S.Y. Aku , I.A. Samotua, U. Abubakar , D.O. Obada ;Mechanical, Spectroscopic and Micro-structural Characterization of Banana Particulate Reinforced PVC Composite as Piping Material, Tribology in Industry Vol. 38, No. 2 (2016) **255-266**

[35] M. A. Maleque, F. Y. Belal; Mechanical properties study of pseudo-stem banana fiber reinforced epoxy composite, The Arabian Journal for Science and Engineering, Volume 32, Number **2B**, October **2007**

[36] Claudia Merlini ,Valdir Soldi ,Guilherme M.O Barra .;Influence of fiber surface treatment and length on physico-chemical polyurethane composites, Polymer Testing **30 (2011) 833–840** ,doi:10.1016/j.polymertesting.2011.08.008

[37] M.Sumaila1,I. Amber, M. Bawa,Effect of fiber length on the physical and Mechanical properties of random oriented non woven short(musa balbisiana)Fibre/epoxy composite Banan, Vol.

2. No. 1. March 2013 , ISSN: 2186-8476

[38] Carolyn Palma, Elsa Contreras , Johana Urra ,Marıa Jesus Martınez; Eco friendly Technologies Based on Banana Peel Use for the Decolourization of the Dyeing Process Wastewater, Waste Biomass Valor **(2011) 2:77–86,**

[39] Pothan,L.APetra Potschke,RudigerHabler and Sabu Thomas;The Static and DynamicMechanicalProperties of Banana and Glass FiberWoven Fabric-ReinforcedPolyester Composite, Journal of Composite Materials,Journal of Composite Materials **2005 39: 1007**, DOI: 10.1177/0021998305048737,

[40] Satish Pujari, Talari Venkatesh . Hepsiba Seeli."Experimental Investigations on Thermal Conductivityof Fenugreek and Banana Composites; J. Inst. Eng. India Ser. D, DOI **10.1007/s40033-017-0146-z.**

[41] T.P. Mohan, K. Kanny ;Tribological properties of nanoclay infused banana fiber (NC-BF) reinforced epoxy composites, Journal of Tribology. Received February 26, 2018; February **12, 2019. doi:10.1115/1.4042873**

[42] Jinguo Wang , Yingcheng Hu ;Novel Particleboard Composites Made from Coir Fiberand Waste Banana Stem Fiber, Waste Biomass Valor, DOI **10.1007/s12649-016-9523-3**

[43] S Vigneshkumar T Rajasekaran. ;Experimental analysis on tribological behavior of fiber reinforcedExperimental analysis on tribological behavior of fiber reinforced composites, 2nd International conference on Advances in Mechanical Engineering (ICAME 2018), IOP Conf. Series: Materials Science and Engineering **,402 (2018) 012198 .**

[44] T. Hariprasad,G. Dharmalingam2 and P. Praveen Raj.;Study of mechanical properties of banana-coir hybrid composite using experimental and FEM techniques, Journal of Mechanical Engineering and Sciences (JMES) , ISSN (Print): **2289- 4659**; e-ISSN: 2231-8380; Volume 4, pp. 518-531, June 2013,

[45] A. Alavudeen, M. Thiruchitrambalam and A. Athijayamani.;Clutch plate using woven hybrid composite materials, Materials Research Innovations 2011,Vol 15 no 4, **229, 30** November 2010,.

[46] Lina Herrera-Estrada; Selvum Pillay and Uday Vaidya ;Banana fiber composites for automotive and transportation applications, https://www.researchgate.net/publication/**265891790,** 16 February 2015

[47] Haneefa, A., Bindu, P., Arvind, I. and Thomas, S. (2008). Studies of Tensile and Flexural Properties of Short Banana/Glass Hybrid Fiber Reinforced Polystyrene Composites, Journal of Composite Materials, 42:**1471-1489**

[48] Zainuddin, E. S., Sapuan, S. M., Abdan, K. and Mohamad, M. T. M. (2008). Thermal Degradation of Banana Pseudo-stem Filled Unplasticized Polyvinyl Chloride Composites, Journal of Materials and Design,30(3): **557-562**

[49] Kumar, R., Choudary, V., Mishra, S. and Varma, I. K. (2008). Banana Fibre Reinforced Biodegradable SoyProtein Composites, Frontiers of Chemistry in China**, 3(3): 243-250.**

[50] H. Savastano Jr, P.G. Warden, R.S.P.Coutts; Brazilian waste fibres as reinforcement for cement-based composites, Cement & Concrete Composites 22 (2000) **379±384**, 6 June 2000,

[51] Jinnah, M., Mariatti, M. and Abubacker, A. (2008). Effect of Chemical Surface Modification on the Properties of Woven Banana Reinforced Unsaturated Polyester Composites, Journal of Reinforced Plastics and Composites, 41(9): **2371-2380.**

[52] Vijay Sitaram Shivankar, Samrat Mukhopadhyay ;Some studies on 100% banana parallellaid and 60:40% banana: Polypropylene cross laid non-woven fabrics, https://doi.org/10.1186/s40691-018-0157-x, **(2019) 6:7**

To Evaluate the Optimum Parameters in Vehicle Chassis Model Vibrations

Nilesh Ahirrao[1] and Dr. Santosh Bhosle[2]

[1] Research Scholar, Maharashtra Institute of Technology, Aurangabad,India.
[2] Maharashtra Institute of Technology, Aurangabad, India.

nileshahirrao15@gmail.com

Abstract. New designs and ideas are always evolving in automobile industries. Along with these, there are many aspects and complexities faced in controlling noise and vibrations in vehicles. Due to uneven roads, vibrations are generated which in turn degrades not only vehicle performance but also passenger comfort. Accidents are undoubtedly the most dangerous which may cause damage to human lives. Proper material selection of vehicle chassis is essential for reducing vibrations and increasing the safety. Wheelbase and the track dimension also influencing the performance of the vehicle while running on the roads. Road roughness is the key factor which may lead to the unnecessary vibrations that may harm the riders or passengers comfort and vehicle accuracy as well. Different techniques are commonly used to reduce such undesired vibrations. This paper describes a study of chassis transient vibrations and response with different materials for different damping properties. The behaviour of vehicle chassis under vibrations is observed through modelling in these conditions with different parameters. The aim of the presented paper is to show the effect of different computational techniques on designing a vehicle along with Taguchi's method for optimizing the parameters.

Keywords: chassis, damping, materials, optimization, transient, vibrations

1. Introduction

Design of vehicles focuses on lessening the vibrations in order to give comfort for the riders. There are so many developments in the automobile field which results in great improvement in the vehicle performance. We are able to evaluate the performance of the chassis vibrations with various parameters. Walter V. Wedig [1] states that the vibrations are generated by uneven roads are vertical vibrations. The vertical disturbance of vehicle body governs the nature of the driving force required to maintain a constant speed. This is a major reason for investigations of many researchers about vehicles in motion. The vibration measurement and analysis is also the key issue of investigations for many researches. Automobile vibrations are of multiple frequencies, magnitude and orientations. Lifu Wanh et al. [2] states that, active suspensions aims on major motion modes. Their major task is the use of neural networks to identify motion modes with considerable accuracy. Vehicle dynamics is the prediction of responses under extreme conditions and thereby study the effects of these responses on the vehicle itself. Nowadays, the negative side of automobile industries and its growth results in increased traffic, accidents and other fatalities. Temporal instability is always present due to many basic behavioral causes. Research by Fred Mannering [3] is done by the findings and analyzing many latest accident data available for this study. Roads are never perfect and they provide irregularities. Erdem Y et al. [4] deals with interior noise inside the passenger cabin of automobiles. They investigated a very important source of noise coming from structure. This structure borne noise is an output of vibrations in panels. Panels are enclosing the complete vehicle, and this study greatly contributes in such investigations of noise and vibrations. Ride Dynamics is always critical to investigate. Verros et al. [5] investigated the large vehicles by modeling and simplifying the equations of motions. They also considered the tire and shock absorber effects along with the separation of vehicle from ground for small amount of time and integrated equations are presented for transient road conditions. Newly constructed roads have some amount of irregularities. These irregularities directly interact with wheels of vehicles. Forces are obviously getting induced in wheels due to these irregularities. Even a bump in the road is responsible for the vertical movement of wheels. This movement is perpendicular to the road surface. The magnitude of movement and forces obviously depends on size of a bump. Watts et al. [6] investigated ground borne vibrations by vehicle crossing road humps. Wheels of vehicles always experiences acceleration in vertical direction as they passes over any kind of imperfections. De Moerlooze et al. [7] discovered a latest model for transient motion with rolling and friction effects. They applied it to different dynamic systems. The approach is to simulate the rolling friction effects on the dynamics of any mechanical parametric system. They discovered very interesting results which shows that the behavior of systems depends on the amplitude and the frequency of oscillation. A two level optimization method has been presented by Torsten Butz et al. [8], which categorically make thrust on evaluating unknown chassis model parameters. Their research mainly focuses on parametric estimation process. It leads to improvement in parametric selection done by excitation strategies, adopted by solving the optimal design problem. Karpel et al. [9] presented a paper about steps in analysis of complicated structures. An approach of fictitious mass component mode synthesis is adopted by them. They made the rigid connections in terms of displacement equations. Very importantly, the soft connections are made through stiffness and damping connecting matrices. The effort is made without altering the degrees of the freedom. They also proved the betterment in using fictitious mass that each substructure is presented as a single set of free interfacing modes. Shock absorbers are supposed to be the main nonlinearities responsible for structure related noise in vehicles. The prediction of this noise at the design stage is always a challenge to automotive industries. Marouane Benaziz et al. [10] proposed a unique method in predicating this structure borne noise. They compared the simulation results with the experimental results which identify undesired transitory phenomena. Moment of inertia and vertical oscillations of vehicles are correlated strongly and are directly proportional to each other. A two terminal mass based absorber which provides a variable moment of inertia is proposed by Tongyi Xu et al. [11]. Performance of this newly developed absorber has been analyzed through simulation. In comparison to constant moment of inertia, this method of changing moment of inertia provides early response, very good comfort and a much needed safety. The added advantage is about the reduced suspension deflections. Vehicles are subjected to random excitation by the irregularities and bumps in a road. Werner Schiehlen et al. [12] provided a study on suspension model for vehicles. They showed the uncertainties in shock absorber and tire springs, as these two elements are subjected to wear. For a better ride comfort and driving safety suspensions are very important in designing a vehicle. Heydinger et al. [13] provides an outlook on development of a vehicle dynamic simulation model. Their fundamental study aspects on models used to present a chassis and suspension systems along with tires. Taguchi method for robust design and optimization of control parameters is significant in the vehicle vibrations. Mitra et al. [14] highlights on the effective use of Taguchi method for optimized selection of control parameters.

2 Vibration response of vehicle on road bumps

Steel and aluminium are two common materials for chassis. Now a days European cars consist of around 135-140 kg of aluminium parts in the body structure, mainly chassis and suspensions. For

reducing the vibration levels, general concern is to achieve considerable damping in terms of a whole system response.
In case of vehicle travel on road bumps, jump or bounce of wheels occurs. This is a transient condition that occurs with high speed vehicles. With jump, the vehicle and road surface separate owing to high amount of unbalanced forces. In this situation all of the wheel energy is transferred to the chassis

2.1 Chassis model for transient vibrations

Chassis materials should be easy to weld and form so that repairs and modifications can be easily done. Also chassis materials should get easily painted in wide variety of colours. A chassis model as shown in fig.1 is built to perform the transient analysis through ANSYS for simulating the vehicle movement on road bumps like speed breakers. Table1 shows the significant properties and dimensions of the chassis model.

Table 1.Significant dimensions and properties of chassis model

Material Properties	Values for Structural Steel	Values for Aluminium
Young's modulus	200x10^3 N/mm^2	69 x10^3 N/mm^2
Yield stress	250 N/mm^2	290 N/mm^2
Density	7.85e-006 kg/ mm^3	2.86e-006 kg/ mm^3
Major dimensions of chassis model in three directions	X=807 mm; Y= 500 mm; Z= 233.2 mm	

Fig 1. Vehicle chassis model

3. Taguchi's Method

Many different factors, inputs, or variables need to be considered when making a product.
However, there is a better way to design an experiment to find out the best combination of variables to make your product. The Taguchi method is a standardized approach for determining the best

combination of inputs to produce a product or service. This is accomplished through design of experiments (DOE). The Taguchi method provides:

- A basis for determining the functional relationship between controllable product or service design factors and the outcomes of a process.
- A method for adjusting the mean of a process by optimizing controllable variables.

3.1 Controlling factors

Control factors are those variables in a process that management can manipulate. Optimal levels are the targets or measurements for performance. The goal is to fine the most efficient process and service design. These parameters can be determines through experimentation. Table 2 shows the variable selected for this research with L16 analysis.

Table 2. Controlling factors for L16 analysis

Sr. No	Stiffness N/m	Speed rpm	Mass kg	Damping Ratio
1	180	300	20	0.21
2	180	300	50	0.22
3	180	400	20	0.23
4	180	400	50	0.24
5	260	300	20	0.25
6	260	300	50	0.26
7	260	400	20	0.27
8	260	400	50	0.28
9	180	300	20	0.29
10	180	300	50	0.30
11	180	400	20	0.32
12	180	400	50	0.34
13	260	300	20	0.39
14	260	300	50	0.42
15	260	400	20	0.45
16	260	400	50	0.50

4. Results

The results for chassis transient vibrations through ANSYS are important to observe in deciding the most significant parameter.

4.1 Transient vibrations and stress-strain values for steel and aluminium chassis

For a time period of 6 sec as an interval period for moving a chassis model over bouncing surface revels following results are found:

- Structural Steel shows induced stress as 10.565 N/mm^2.
- Aluminum shows induced stress as 10.589 N/mm^2.
- The maximum deflection for steel is 7.5289e-002 mm.
- The maximum deflection for aluminum is 0.21849 mm.
- Strain Energy for Steel chassis is 12.367 mJ.
- Strain Energy for aluminum chassis is 35.94 mJ.

IOP Conf. Series: Materials Science and Engineering **810** (2020) 012077 doi:10.1088/1757-899X/810/1/012077

4.2 Taguchi results for parameter optimization
It is important to evaluate the signal to noise ratio on the basis of larger is better. It is shown in table 3 and the response for means is shown in table 4.

Table 3. Signal to noise ratio levels in Taguchi's analysis

Level	Damping Ratio	Stiffness N/m	Speed rpm	Mass kg
1	-12.255	-11.543	-10.922	-10.687
2	-8.643	-9.355	-9.977	-10.211
Delta	3.612	2.188	0.945	0.476
Rank	1	2	3	4

Table 4. Response for means through Taguchi's method

Level	Damping Ratio	Stiffness N/m	Speed rpm	Mass kg
1	0.2450	0.2688	0.2925	0.3013
2	0.3763	0.3525	0.3288	0.3200
Delta	0.1312	0.0838	0.0363	0.0187
Rank	1	2	3	4

4.3 Taguchi's results for main effects
Taguchi's experimental analysis reveals the main effect plots as shown in figure 2.The graphs for all four factors, mass, stiffness, damping and speed are available through this technique and are important in robust design of chassis.

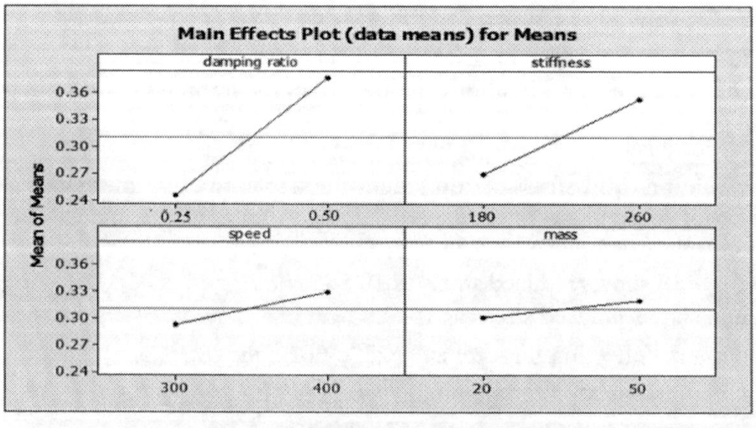

Fig 2. Main effect plot for means

5. Conclusion

Strain Energy is nothing but the internal work done in deforming the body by the action of external forces. Aluminium is seen better than steel in this regard. On comparing with the amount of energy dissipated for steel, the amount of energy dissipated for aluminium is more. This feature provides good damping to the structural vibrations. Weight saving is an additional benefit with aluminium chassis, as compared with conventional steel chassis. Benefits in terms of improving driving comfort along with safety, due to the reduction of the mass is a key feature of using aluminium as a chassis material. Also, aluminium provides good manufacturing feasibilities like formability and weld ability. Good corrosion resistance of aluminium is an important feature in wet conditions, which is desirable for chassis design. Table 3 and table 4 shows through signal to noise ratio and response for means, the most critical component is damping factor in the vehicle vibrations and comfort . Fig2 shows the similar effect through main effect plots. Analysis also shows that the damping factor should be selected in the range of 0.2 to 0.5 for the effective vibration damping and comfortable riding of vehicles.

6. References

[1] Wedig W V 2016 Velocity Jumps in road-vehicle dynamics *Procedia Engineering, 12th International Conference on Vibration Problems* **144** 1076-1085
[2] Wang L, Zhang N and Haiping D 2015 Real-time identification of vehicle motion-modes using neural networks *Mechanical Systems and Signal Processing* **50 632**-645
[3] Mannering F 2018 Temporal instability and the analysis of highway accident data *Analytic Methods in Accident Research* **17** 1-13
[4] Yuksel E, Kamci G and Basdogan I 2012 Vibro- Acoustic design optimization study to improve the sound pressure level inside the passenger cabin *Journal of Vibration and Acoustics* **134** 061017 /1-9
[5] Verros G and Natsiavas S 2002 Ride dynamics of non-linear vehicle models using component mode synthesis *Journal of Vibration & Acoustics* **124** 427-432
[6] Moerlooze K De, Al-bender F and Brussel H V 2011 Modeling of the dynamic behavior of systems with the rolling elements *International Journal of Nonlinear Mechanics* 46 222-233
[7] Watts G R and Krylov V V 2000 Ground-borne vibration generated by vehicles crossing road humps and speed control cushions *Applied Acoustics* **59** 221-236
[8] Butz T, Simeon B and Stadler M 2010 Optimal design of experiments for estimating parameters of vehicles dynamics simulation model *Journal of Computational and Nonlinear Dynamics* **5** 031011/1-7
[9] Karpel M, Moulin B and Feldgun V 2007 Component mode synthesis of a vehicle system model using the factitious mass methods *Journal of Vibration and Acoustics* **129** 73-83
[10] Benaziz M, Nacivet S and Thouverez F 2015 A shock absorber model or structure borne noise Analyses *Journal of Sound and Vibration* **349** 177-194
[11] Xu T, Liang M, Chuan Li and Yang S 2015 Design and analysis of a shock absorber with variable moment of inertia for passive vehicle suspensions *Journal of Sound and Vibration* **355** 66-85
[12] Werner S and Igor I 2015 Uncertainties in road vehicle suspensions *Procedia IUTAM Symposium on Dynamical Analysis of Multibody Systems with Design Uncertainties* **13** 151-159
[13] Heydinger G J, Salaani M K, Garrott W R and Grygier P A 2002 Vehicle dynamics modelling for the national advanced driving simulator *Proceedings of the Institution of Mechanical Engineers,Part D Journal of Automobile Engineering* **216** 307-317
[14] Mitra A C, Jawarkar M, Soni T and Kiranchand G R 2016 Implementation of Taguchi method for robust suspension design *Procedia Engineering, 12th International Conference on Vibration Problem* **144** 77-84

IOP Conf. Series: Materials Science and Engineering **810** (2020) 012078 doi:10.1088/1757-899X/810/1/012078

Survey of the emerging bio-inspired Unmanned Aerial Underwater Vehicles

A. Narayanan, P. Rajeshirke, A. Sharma, K. Pestonjamasp
MPSTME, SVKM'S NMIMS University, Maharashtra, India

ajaykrishna.nmims@gmail.com

Abstract. Unmanned Aerial-Underwater Vehicles have been envisioned as a new kind of hybrid unmanned system capable of performing equally well in multiple mediums and seamlessly transitioning between them. Currently, bio inspired vehicles are leading the way in the development of such a vehicle, that can fly like an eagle and swim like a fish. This path-breaking amphibious aerial underwater vehicle is sought after because of its wide range of applications in military, commercial and civil environments. From aerial surveillance, deep water exploration to underwater non-destructive testing, this vehicle aims to do it all. However, the development of a fully functional version is hindered by the significantly different properties of air and water, considering the air water transition, different propulsion systems and stability underwater. We evaluate these available architectures that aim to accomplish the goal of creating a flawless multi-medium unmanned vehicle which has its limitations as one of the modes is dominant leaving the other to be at a lesser than expected measure. This paper presents a survey discussing the evolution of these systems by first looking at their parent systems and then evaluating the various existing types of UAUVs by understanding their working mechanisms. The paper aims to provide a comprehensive overview of the work done in the field so far which can provide direction to the further development of such vehicles.

1. Introduction

In this world of growing domination of technology in almost every field, robots have been at the center of this revolution. Many mechanical systems have been automated partially or completely by making robots for specific applications. Drones or unmanned vehicles are a class of robots that have been used in air or in water for a variety of applications from underwater exploration to door step package delivery. The application base for these robots is increasing with every passing day and this expands the scope of its usage. Another approach to evolution of these systems is integrating the aerial and underwater abilities into one to perform tasks with ease across mediums. Mapping and exploration of inhospitable regions where any means of human access is either incapable or extremely dangerous to life, an unmanned hybrid vehicle is needed. It is desirable to launch the unmanned vehicle from the control station or ship to the target position. The interest has further grown as unmanned vehicles have seen a dramatic increase in their use for civil and military applications. This has incentivized the interest in a platform capable of air and underwater operation.

The variety of applications and possible future scope makes it an exciting proposition to look forward to. It is important to understand the systems that this new concept is derived from. It is a bio-inspired design with an aim to emulate the flying ability of a bird and the swimming capabilities of a fish. UAVs are the parent that gives the new technology its flying ability and its underwater operation is derived from UUVs and USVs. This makes UAVs, UUVs and USVs germane to the discussion. Since aerial

IOP Conf. Series: Materials Science and Engineering **810** (2020) 012078 doi:10.1088/1757-899X/810/1/012078

underwater vehicles are still a relatively new phenomenon, its novelty demands us to look into the evolution of this idea to understand the system better and help us build a product that is the best techno-economic solution to the problem. So, the paper first look at aerial remotely operated vehicles and underwater remotely operated vehicles or autonomous vehicles which forms the skeleton of this discussion. The focus then shifts to other multi-media vehicles previously developed. This gives an understanding of what went wrong or what desirable characteristics these machines were able to incorporate. A thorough understanding of the path gives direction to the efforts of the future and by objectively evaluating current systems in conjunction with past and their predecessors can suggest the best way forward for UAUVs.

2. Overview of the types of Unmanned Vehicles

2.1. Unmanned Aerial Vehicles (UAVs)

Unmanned aerial vehicles are a class of aerial vehicles which can be operated remotely without having any onboard presence, with the help of a communication system between the vehicle and the operator. These can be fixed wing or rotary wing UAVs. Both these types have further subdivisions, and the type of UAV used depends on the specific purpose the UAV is designed for. Fixed winged aircrafts have a simple aerodynamic structure, using airflow over their wings to provide lift. Thus, they need to be in constant motion, while rotor wing aircrafts can be stationary mid-air due to generating lift from constant rotary motion of rotor blades. Fixed wing UAVs inherent design promotes gliding without power and thus these UAVs have higher range as compared to rotor wing UAVs. Rotor wing aircrafts have vertical take-off and landing capabilities, and are often seen in the form of quadcopters, which are sought after due to their small size, easy control and high manoeuvrability [1].

Recent years have seen a surge in the evolution of these technologies, with unmanned aerial vehicles being developed for applications ranging from military applications to drone enabled package delivery. Search and rescue operations also have great potential for integration of drones, which can provide better success rates and safety in search and rescue operations [2]. Use of small drones for collection of traffic control data has also been theorized as a field of use, to counter the costs and restrictedness of current methods using human counters and video recording devices [3]. One the most widely seen civil applications of drones is in photography and videography. In all the above applications drones prove to be more efficient and cost effective than the traditional methods, and thus have been proposed to be the new norm instead of many existing technologies.

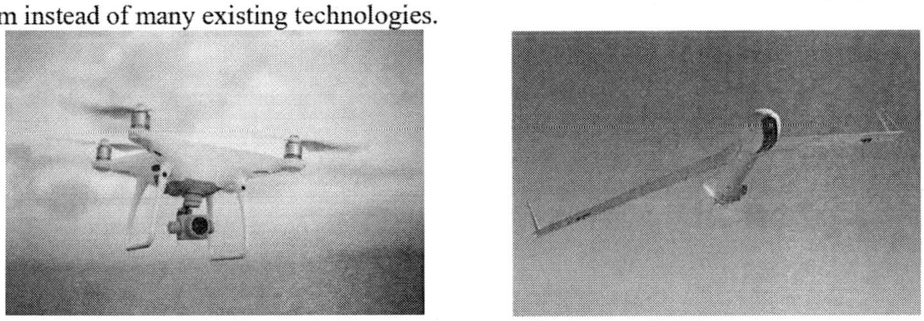

Figure 1. Rotary wing and fixed wing UAV

2.2. Unmanned Surface Vehicles

Unmanned surface vehicles are vessels which operate on the surface of water, and can be remotely operated or autonomous. USVs have been developed from as early as the second world war, although only from the 1990s has this technology seen a surge in development [4]. USVs can be made in many different forms but the mains components remain the same, which include propulsion system, GNC systems, communication systems, data collection equipment and ground stations. USVs prove to be advantageous as they are able to operate in conditions where human safety would be at risk, and are compact vehicles with low maintenance costs [5]. Small coastline teams can ensure quick deployment of USVs, without establishing any prominent infrastructure [6]. Despite these advantages, the widespread popularity of UAVs is not reciprocated in USVs.

USVs were originally developed for naval applications like surveillance and reconnaissance, but have now moved into the civil domain with applications such as environmental monitoring and assessment. Onboard camera enabled USVs can be used for surveillance in conditions where substantial human risk is involved. Developing applications also include search and rescue, autonomous shipping and offshore surveying in oil and gas industries. Future developments in USVs aim to provide lower operational costs, improved range and autonomy, to push these devices towards mainstream usage [5].

Figure 2. Unmanned Surface Vehicle [4]

2.3. Unmanned Underwater Vehicles
Underwater vehicles (UUVs), are robots that operate under the surface of water with minimal or no human operator intervention. They include both Remotely Operated Vehicle (ROV) and Autonomous Underwater Vehicle (AUV).

2.3.1 Remotely operated vehicles (ROVs)
These tele-operated vehicles used for installations underwater and various tasks of repairing. These vehicles are majorly used by offshore industries because of characteristics like high safety, reaching great depths, having a longer endurance and their less demand for support equipment. Information is sent to the ROV through a tethered cable or acoustic link by an operator onboard a surface ship.
2.3.2 Autonomous Underwater Vehicles (AUVs)
These vehicles require no constant monitoring or supervision from a human operator. AUVs are not limited in the depths they can dive up to due to the absence of an umbilical cable as in the case of ROV's. AUVs can be categorised as: a) Safe distance operation from the sea bed, including the use of Sonar for observation of the sea floor, water composition examination, sampling of floating creatures; b) Close proximity inspection of the sea floor and structures made by man, such as hydrothermal activity inspection, creatures on the seafloor and underwater structures; c) Interactions with the sea floor and man-made structures

One of the main difficulties faced by AUVs is the inherent nonlinearity of the underwater vehicle dynamics. Due to many other uncertainties a prediction or calculation of hydrodynamic coefficients need to be performed. Some other challenges faced are limited operational underwater navigation sensors, low visibility during the use of vision sensors and underwater external disturbances.

AUVs operate within a highly unstructured environment which pose difficulties in navigation via information received from satellites. AUV navigation takes place by INS, and acoustic systems, but due to an increase in the endurance of AUVs, their utilization has been more restrictive in terms of affordability and range. Current involvements include improvement in battery technology, underwater communication, fuel cells, sensor fusion and propulsion systems.

Figure 3. USV

2.4 Need and Evolution
UAVs, UUVs and UAUVs in their respective fields are able to perform tasks with extremely high levels of precision and are constantly evolving and bettering its previous iterations. This increased interest in unmanned vehicles has led to the rise of various applications never thought of before. Underwater ship hull testing and delivery of relief aid in times of disaster are all possible in today's world where technology is making human life easier with each passing day. The gradual improvement of these systems is key to further develop and make use of these systems in more testing real world environments.

Another tangent of development is the integration of two or more of these systems to achieve synergistic interaction of these mechanisms to perform multi-media operations. This ability allows the vehicle to transition with a relative degree of ease between different mediums and that opens up new avenues in the applications of drones. By integrating the best of flying abilities and swimming abilities of UAVs and UUVs, a new line of devices is created that are equally able in both air and water. Military grade reconnaissance to recreational photography or exploration are some of the possible applications of this newly emerging technology.

2.5 Unmanned Aerial Underwater Vehicles (UAUVs)
There are biological species like the squid, the flying fish or puffins that are capable of operating in water and air. However, they have a mode of operation they are better at and the other mode is an additional underdeveloped ability. Taking inspiration from this concept, UAUVs are vehicles that are capable of operating in both air and water. They may have a dominant characteristic being more functional in air or water or be equally able in both media. Development of these systems has been tries and various such machines have been successfully implemented. The table below from a survey by Beihang University [11] shows the different types of air water systems that have existed.

Table 1. Different air water systems developed in the past

Name	Institute	Morphing structure design	Transition	
			Water-Air	Air-Water
Gull Series	Warrior (Aero-Marine Limited)	None	Running	Floating
Flying Fish	University of Michigan	None	Running	Floating

Switchblade	AeroVironet Inc.	Variable sweptback angle wing	Catapulting	Plunge diving
Flying Fish prototype	Beihang University	Variable sweptback angle wing	Running	Floating
Gannet	MIT Lincoln Laboratory	Variable sweptback angle wing	Hand launched	Plunge diving
Flying Squid prototype	Imperial College, London	Variable sweptback angle wing	Water jet propulsion	Plunge diving
Cormorant	Lockheed Martin Company	Folding wing	Rocket propelled	Splashing down

These systems have had limitations in one of their two core functionalities as one of them becomes a dominant trait and the other is merely an added feature. Systems today emphasize on equally able operations in air and water, along with seamless transition between mediums and the ability to float or glide on water. The goal is to create a system that can swim like a fish and fly like a bird. There are a few mechanisms that have attempted to achieve this and the main systems are discussed in the following section of this article.

3. Types of UAUVs

3.1 Ballast System

A hybrid unmanned aerial-underwater vehicle can use a ballast tank for active buoyancy control and seamless air-water and water-air transition. The depth control system in such a vehicle will consist of the following components: 1) Water-proof hull 2) Water pump 3) Two-way valve 4) Water cylinder 5) Water pump controller. UUVs' can dive in one of two ways, Static dive or Dynamic dive. 4 concepts are comprised within these two methods: 1) Air compressor-based ballast system 2) Hydraulic pump-based ballast system 3) Piston tank ballast system 4) Direct thrust system

Dynamic diving is achieved by direct thrust system, whereas static diving is accomplished by the first three concepts. The system reviewed in this survey uses a combination of hydraulic pump and piston tank-based ballast system in order to reduce power load on rotors. As compared to dynamic diving, static diving allows depth control of vehicle underwater and 90-degree flip with minimal power consumption.

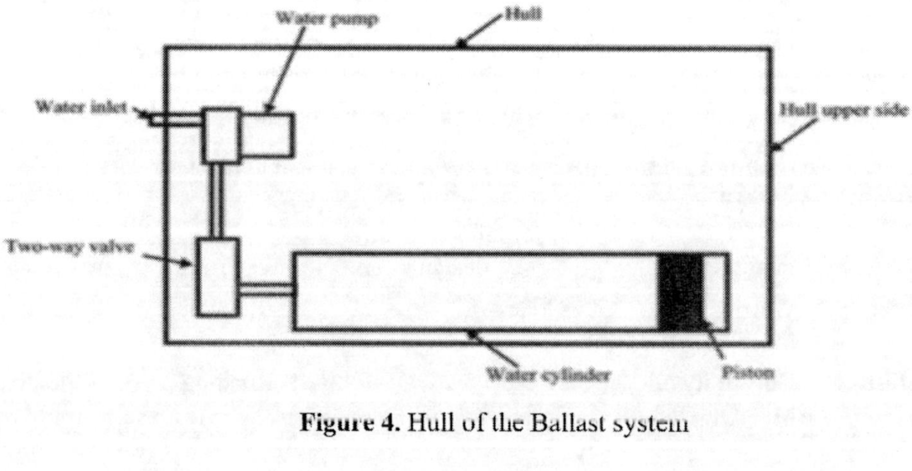

Figure 4. Hull of the Ballast system

Figure (4) shows the components of a ballast system as constructed in the Loon Copter. It consists of a hull, a water pump, a two-way valve, a piston and a water cylinder. All the components are mounted in the hull and it also holds the compressed air. The vehicle's stability is improved as during diving the movable piston holds the water in the water cylinder. The capacity of the water cylinder is 150 ml, out of which, 100 ml is used for diving process and the remaining capacity is used to control the depth of the vehicle. When underwater, the compressed air will attempt to push the water out of the water cylinder, this action is countered by deploying a two-way valve.

When the vehicle is on the water surface, the cylinder and the hull will have air at ambient pressure. As water is pumped into the water cylinder, the transition begins. The air in the hull starts to get pressurised as water is pumped into the cylinder. The process of tilting and submerging occurs simultaneously when the cylinder is filled with water. In a similar manner, the vehicle achieves water-air transition when the water in the cylinder is pumped out. Figure (5) shows a visualisation of the air-water transition.

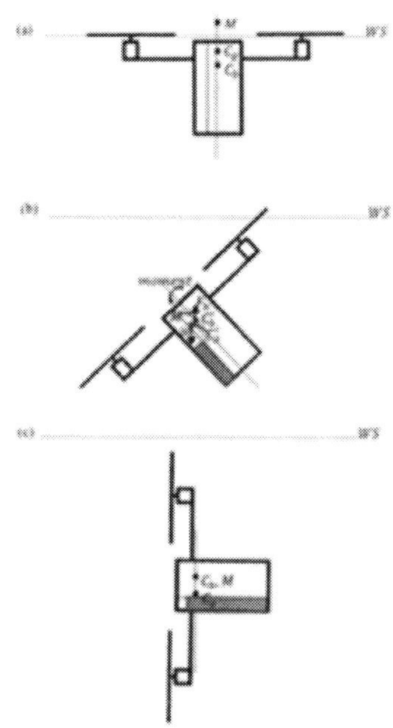

Figure 5. Air-water Transition in Ballast system

3.2 AquaMAV

AquaMAV is another unmanned aerial underwater vehicle that is different in concept than its counterparts. It was ideated with the sole application of water sampling in mind and it provides one of the most efficient, safe and cost-effective systems that can achieve collection of water samples and go back to its launch site where the recorded data can be evaluated to then perform necessary action. This is another bio-inspired mechanism that tries to emulate the flying or gliding ability of squids. These animals are known to generate a high impulse gliding leap by discharging a jet of water that gives them enough momentum to fly out of water. The researchers at Imperial College, London took this very mechanism and tried to implement it on a tiny fixed wing aircraft that could perform a similar task.

The AquaMAV works on the principle same as a jet thruster. The thrust in this case is obtained from the expulsion of pressurized gas that imparts high velocities to the vehicle which facilitates it's launch out of water. 5ml of 57 bar CO2 gas is stored inside a little tank in the vehicle which allows this mechanism of air-water transition.

IOP Conf. Series: Materials Science and Engineering **810** (2020) 012078 doi:10.1088/1757-899X/810/1/012078

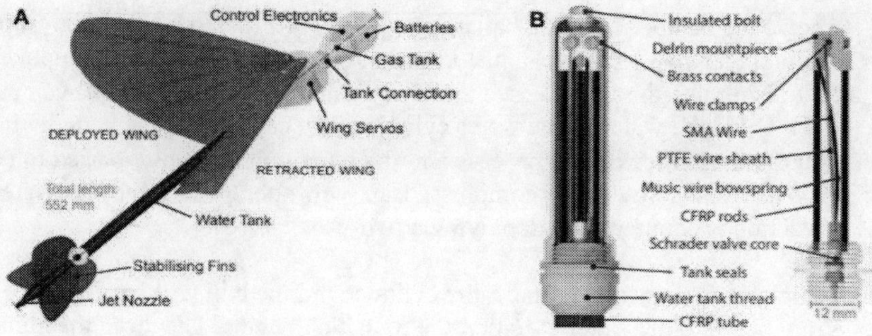

A: Components of AquaMAV B: Gas Release Mechanism

Figure 6. Details of the AquaMAV

The robot has a few important components that contribute to the success of the mechanism. The wing is divided into six segments with the leading edge acting as the free end and the root segment hinged to the main body. It is regulated in way such that when the leading edge is regulated, all segments of the wing follow. This is needed because while entering and exiting or plunging out of water the robot has its wings retracted and opens only when in flight. In this mechanism, the robot dives into water at a steep angle and collects the water samples it needs and then plunges out of water when the plunging mechanism actuated which release the pressurized gas. This provides the robot with the thrust required by the vehicle to fly out of water. Since it jumps out of water at a considerably high speed, it already has the takeoff velocity required for it to fly in the air phase. This unique way of achieving seamless air water transition is extremely effective as shown in calculations in (citation).

The design is based on basic fluid mechanics by using the unsteady Bernoulli's equation. The objective function to be maximized in this mechanism is impulse. All calculations are based on creating maximum impulse from minimum space and the properties of gas is decided according to this consideration.
Dual air-water operations always come with their own challenges. The other mechanisms where a quad or octa rotor system is used, is also capable of performing underwater tasks using the same propellers that generate thrust in air but operation in water demands a much lesser rotor speed. However, this would mean that the rotors spin at speeds lesser than their designed range. But these multi rotor systems can be used for multiple missions back to back, while that is something the AquaMAV is not capable of. This robot is more suitable for more frequent applications where it can be used in short intervals for a lesser period of time.

As discussed earlier, the AquaMAV is primarily made for testing of water samples in different civil, military and disaster relief applications. It's ability to accurately reach points to test is something that the plunging mechanism provides for. The same is very difficult to achieve in multi-rotor systems as the change on rotor speed affects its positioning. An additional advantage of this mechanism is that it is relatively cheaper. A more cost-effective and robust design makes this the most suitable device for this particular application. It can be used by government environmental bodies and researchers to test the health of an ecosystem. This is done periodically and by using this vehicle collecting and testing samples becomes faster. In times of disaster, when locations are rendered inaccessible to personnel, say after a flood or in case of an earthquake, the robot can reach the isolated location with relative ease and bring back essential data like the quality of water or the presence of toxic materials on the water.

The AquaMAV in this way takes an alternative path to solve a particular problem and brings the most optimum solution to the table. Moreover, the AquaMAV is important to understand this evolution of multi-media unmanned vehicles and analyse its future scope. This revolutionary take on doing things is sure to encourage researchers to look outside the confines of multi-rotor systems to build an unmanned vehicle that can fly and swim with equally able mechanisms.

3.3 Octa-rotor Mechanism

Single or multi rotor vehicles are an alternative to fixed-wing aircrafts. The major advantage offered by single/multi rotor vehicles is their ability to take off and land vertically. Octa-rotor system redefines multi rotor VTOL operations in order to achieve smooth air-water and water-air transition.
As shown in figure (3), each vehicle arm is equipped with dual motors. There is a column gap between the two motors that facilitates the transition.

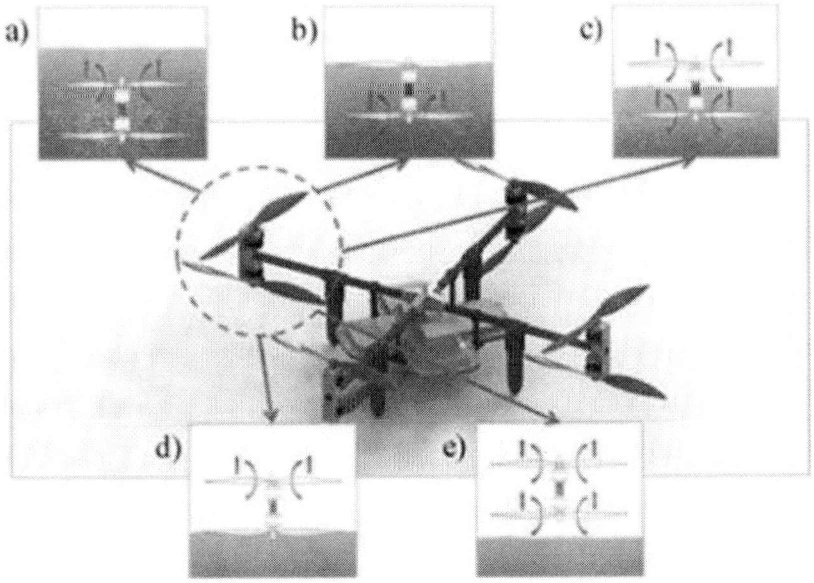

Figure 7. Octa-rotor Mechanism with air-water transition stages

The octa-rotor system is completely reversible, which allows for an air-water as well as water-air transition. Both the propellers on an arm are generating lift as the vehicle approaches the water-air interface. The top propellers are slowed while going through the interface in order to achieve a smooth transition. They accelerate again as soon as they are completely above the surface of water to generate lift. The bottom propellers follow a similar pattern where they slow down as they approach the water-air interface and accelerate again as they exit the water surface to generate lift.
The water-air transition involves complex interaction of the propellers with the surface of water, so in order to avoid a spike in drag, the method of slowing or stopping of propellers is adopted. The transitions demonstrated in figure are performed under two seconds, which realises the seamless transition.
The capability of propellers to generate lift and thrust is fully exploited in this method. Underwater vehicles need to be almost neutrally buoyant and generate thrust at the same time. The propellers are used to generate the required thrust underwater.

4. Applications and Future scope

Civil: Utilisation of drone technology for civil applications has been blooming in recent times.
Drones are used in disaster management and relief supply operations. Offshore technology and ecosystem monitoring, including testing of water samples, near water sampling, flora and fauna monitoring can be carried out with the help of drones.

Military: Out of the multiple applications of drones utilised by the military sector, drones can be majorly used as single combat weapons for air or above sea weapon detection, underwater submarine and torpedo detection along with assisting in information acquisition, communication relay.

Development of drone technology and applications of drones in multiple fields is increasing at an unprecedented rate. Future drones can make their way into fields of geoscience, environmental monitoring, agriculture, atmospheric sciences, metrology including atmospheric sciences. The addition of hybrid aerial underwater vehicles to the fleet of unmanned will provide a completely new paradigm to the applications of drone. It gives an additional angle to military reconnaissance and this will remain an area of interest and hence an area of development. Further work is needed in defining specific application-based robots that specialize in tasks that are well suited to this kind of system. In the civil or recreational domain, further development will be focussed on making the drone more accessible and easier to control in a way that it appeals to the audience.

5. Conclusion

With every emerging technology, it is important to understand its course of evolution to understand its present and future scope. As this paper surveys aerial underwater vehicles, it glances through its potential parents' UAVs, UUVs and USVs. In this survey, we have looked at a few relatively new technologies, capable of multi-medium operation, focusing on the transition between the two mediums, namely unmanned aerial underwater vehicles (UAUVs). A brief analysis of bioinspired vehicles suggests that for similar performance in both mediums of air and water, conditions which produce lift and thrust for flight, and conditions which provide near neutral buoyancy and thrust for swimming, overlap. The transition between the 2 mediums is seen to be the biggest challenge in the development of such technology being able to perform both air and underwater missions. The paper discusses two systems which can achieve the desired results, ballast tank with active buoyancy control and a dual propeller system, analysing the advantages and limitations of each system. While both use rigid body dynamics to tilt themselves underwater, the mechanisms deployed for the transition of medium are significantly different. It also explains the AquaMAV, which works on an aquatic jet thruster mechanism. These mechanisms provide the basis for future development by incorporating them in focussed applications that can be catered to. The wide range of possibilities makes this technology an exciting prospect.

6. References

[1] Norouzi Ghazbi, Somayeh & Aghli, Yasaman & Alimohammadi, M. & Akbari, Ali Akbar. (2016). Quadrotors unmanned aerial vehicles: A review. International Journal on Smart Sensing and Intelligent Systems. 9. 309-333. 10.21307/ijssis-2017-872.

[2]: Burke, C.; McWhirter, P.R.; Veitch-Michaelis, J.; McAree, O.; Pointon, H.A.; Wich, S.; Longmore, S. Requirements and Limitations of Thermal Drones for Effective Search and Rescue in Marine and Coastal Areas. Drones 2019, 3, 78

[3] Khan, Muhammad Arsalan & Ectors, Wim & Bellemans, Tom & Ruichek, Yassine & Yasar, Ansar-Ul-Haque & Janssens, Davy & Wets, Geert. (2018). Unmanned Aerial Vehicle-based Traffic Analysis: A Case Study to Analyze Traffic Streams at Urban Roundabouts. Procedia Computer Science. 130. 636-643. 10.1016/j.procs.2018.04.114.

[4] Bertram, Volker. (2008). Unmanned Surface Vehicles–A Survey.

[5] Liu, Zhixiang & Zhang, Youmin & Yu, Xiang & Yuan, Chi. (2016). Unmanned surface vehicles: An overview of developments and challenges. Annual Reviews in Control. 41. 10.1016/j.arcontrol.2016.04.018.

[6] Man Zhu and Yuan-Qiao Wen, "Design and Analysis of Collaborative Unmanned Surface-Aerial Vehicle Cruise Systems," Journal of Advanced Transportation, vol. 2019, Article ID 1323105, 2019

[7] Pogmohone [CC BY-SA 4.0 (https://creativecommons.org/licenses/by-sa/4.0)]

[8] Agus Budiyono ,Advances in unmanned underwater vehicles technologies: Modeling, control and guidance perspectives, CSIR, IJMS Vol.38(3) Sep-2009
[9] Alzubi, Hamzeh & Mansour, Iyad & Rawashdeh, Osamah. (2018). Loon Copter: Implementation of a hybrid unmanned aquatic-aerial quadcopter with active buoyancy control. Journal of Field Robotics. 35. 10.1002/rob.21777
[10] Siddall, Robert, and Mirko Kovac. "Fast aquatic escape with a jet thruster." IEEE/ASME Transactions on Mechatronics 22.1 (2016): 217-226.
[11] Yang, Xingbang, Tianmiao Wang, Jianhong Liang, Guocai Yao, and Miao Liu. "Survey on the novel hybrid aquatic–aerial amphibious aircraft: Aquatic unmanned aerial vehicle (AquaUAV)." Progress in Aerospace Sciences 74 (2015): 131-151.
[12] Marx, Andrew, Yu-Hsi Chou, Kevin Mercy, and Richard Windisch. "A Lightweight, Robust Exploitation System for Temporal Stacks of UAS Data: Use Case for Forward-Deployed Military or Emergency Responders." Drones 3, no. 1 (2019): 29.
[13] Verfuss, Ursula K., Ana Sofia Aniceto, Danielle V. Harris, Douglas Gillespie, Sophie Fielding, Guillermo Jiménez, Phil Johnston et al. "A review of unmanned vehicles for the detection and monitoring of marine fauna." Marine pollution bulletin 140 (2019): 17-29
[14] Sharma, Vishal. "Advances in Drone Communications, State-of-the-Art and Architectures." (2019): 21.
[15] Chabot, Dominique. "Trends in drone research and applications as the Journal of Unmanned Vehicle Systems turns five." Journal of Unmanned Vehicle Systems 6, no. 1 (2018): vi-xv.
[16] Maia, Marco M., Parth Soni, and Francisco J. Diez. "Demonstration of an aerial and submersible vehicle capable of flight and underwater navigation with seamless air-water transition." arXiv preprint arXiv:1507.01932 (2015).

IOP Conf. Series: Materials Science and Engineering **810** (2020) 012079 doi:10.1088/1757-899X/810/1/012079

Optimization of wear characteristics of PLA material reinforced with HAP nanoparticle by using taguchi technique

Sumit R Agrawal and Saili R Kulkarni

Department of Mechanical Engineering, GEC, Aurangabad, India

Email. sumit.agrwl992@gmail.com

Abstract. In the field of biomedical application, the polymer composite has been played a very challenging role and its used has been increasing since it possess good tribological properties when reinforce with appropriate material. In this regard, PLA (poly lactic acid) and it's composite with HAP (Hydroxyapatite) was studied for its tribological behaviour which may be used as an implant material. PLA with varying percentage of HAP nanoparticles was examined on linear reciprocating tribometer under dry sliding test for specific wear rate. Design of experiment was based on Taguchi technique for optimizing specific wear rate parameter. In this study, effect of reinforcement and operating parameters that is load and frequency on tribological performance of PLA Nano composite was studied. ANOVA (Analysis of Variance) has been used to determine the influence of each parameter on wear characteristics. Regression analysis was used to establish a regression equation between all affecting parameters. The results shows that reinforcement plays a major role while load and frequency plays an average role on wear characteristic of PLA Nano composite. It was observed that PLA reinforced with HAP nanoparticle significantly improve wear characteristic up to 5% of reinforcement while beyond it, there is reverse nature of composite. The optimum parameter for wear rate were 5% reinforcement of HAP, 80 N of load, and 7 Hz frequency.
Keywords. PLA, HAP, Taguchi technique, Wear, ANOVA, nanoparticles

1. Introduction

In past decades, it has been seen that the researchers has shift from metal to its alloy and then ceramics and their composite, to achieve combine effect of both material into single. With the evolution of composite material, it is possible to achieve light weight with high strength of material which improves the efficiency as well as reduces the cost incurred in product life cycle. Now a days, the composite is one of the focused area of most of the researchers trying to find more and more reliable material.

Now the recent trend is on polymer and it's composite. Basically polymer is lighter in weight as compare to other material and also possess good strength which can be enhanced to greater extent by reinforcing other element like ceramics and fibrous material. Polymer also proven its wide area of application in biomedical field, automobile, aircraft etc. replacing other material due to its good compatibility and ease of manufacturing. Mainly for biomedical application of polymer, lot of study is going on, since it has been seen that some previously used implant has some drawbacks as shown in detail in following table no. 1 which lowers the life of implant.

Various polymer like HDPE (High density polyethylene), UHMWPE (ultrahigh molecular weight polyethylene), PEEK (Poly ether ether ketone), PTFE (Poly tetra fluro ethylene) etc. shows improvement in the wear property when reinforced with ceramics like Al_2O_3, TiO_2, Carbon fibre(CF), Kevlar fibre, SiO_2 etc. The reinforcement of HDPE with Kevlar fibre at 10% shows optimum results in term of wear rate and biocompatibility also enhanced with increase in percentage of Carbon fibre [1]. The UHWPE when reinforced with biocompatible material like Zirconia also shows improvement in wear resistance [4] but composite may not act as biocompatible material. PEEK reinforced with the oxidized CF shows reduction in COF by 60 % at 15 % reinforced as compare to unfilled PEEK which is also better than unoxidised state and in case of wear rate, it is found 85 % reduction as compared to unfilled PEEK at 15 % CF reinforcement [5] but it is said that PEEK is not viable as biocompatible material. PTFE also shows reduced in wear up to 162 time and COF (Coefficient of friction) up to 38% when reinforced with 2 wt. % graphite nano particles [16]. UHMWPE and carbon nanotubes composite shows improved mechanical characteristic and wear property but in vitro and vivo test results shows adverse effect [15]. The reinforcement of HAP (Hydroxyapatite) into polymer has shown significant improvement in wear property of matrix material. The reinforcement of HAP into UHMWPE up to 20% has reduction of wear rate up to 60% than pristine form of UHMWPE and also COF of 20 % reinforcement also shows 34 % reduction as compare to its pristine form [2][3].

Table 1. Implant material Merit and Demerit

Implant material	Merit	Demerit	Reference No.
Stainless steel	Good strength and modulus	Corrosive, large wear debris form, aseptic loosening of bone	[6],[7]
Cobalt based alloy	Corrosion resistance and sufficient strength.	Releasing harmful ion like Ni, Cr, etc.	[8],[9],[10]
Titanium based alloy	Lighter in weight, high strength and resistance to corrosion	Prolong use may cause Alzheimer disease	[9],[11]
HDPE composite and UHMWPE composite other polymer composite.	Lighter in weight, good strength and corrosion resistance	Lack in biocompatible and biodegradable	[2],[12], [13]

The PLA (poly lactic acid) which is thermoplastic and not only biocompatible but also a biodegradable which may be feasible for implant material if it has enough wear resistance. The pristine form of PLA has poor tribolgical properties and it can be improve with the reinforcing material as various other polymer material has proven this phenomenon as mention above. PLA when reinforced with TiO_2 grafted flax fibre show enhanced physical and mechanical properties [20]. It is seen that property of PLA can be enhanced with appropriate reinforcement. The PLA/HAP nanocomposite also shows enhanced thermal stability which overcome one of its demerit [14].

In this study, PLA/HAP nanocomposite with different composition as shown in below table no. 2 was studied at parameter of load and frequency as mentioned in section 2.2. The Tribological behaviour of composite was studied for determining percentage feasibility of composite as an implant. The various researchers has found that PLA reinforce with HAP may be suitable for implant or scaffold or tissue engineering application [17] [18] [19].

Table 2. Material Composition of PLA nanocomposite

Sample Id	PLA (wt. %)	HAP (wt. %)
A	100	0
B	95	5
C	90	10

2. Experimental details

2.1 Material and Composite preparation

In experimentation, PLA material was chosen as matrix material while Hydroxyapatite (HAP) as a reinforcing material. PLA material was purchased in the form of pallets from additive manufacturing pvt. Ltd., Hyderabad and HAP was obtain in the form of nanoparticle from Nano research lab, Jharkhand. The properties of HAP nanoparticles are average particle size 20-80 nm with 99.5% purity, molecular formulae as $Ca_5(OH)(PO_4)_3$, density : 1.038 g/cm^3 and melting point: 1100^0 C. To prepare PLA nanocomposite, HAP with different weight percentage (i.e. 5, 10 and 15) was added into matrix material. To make proper mixing, the mix material was preheated at 50^0 to 60^0 C under stirring mechanism and allowing it to cool at room temperature. Due to this preheating, nanoparticles get adhere to PLA pallets which approximately ensuring proper mixing between both materials.

2.2 Friction and wear test

The preheated composite material was processed through vertical injection moulding machine for preparation of sample. The semi liquid state material was injected into mould cavity and sample with size of 10×10×15 (all dimension are in mm). The semi liquid state was attain approximately at 70^0 to 80^0 C for pristine PLA while 90^0 to 100^0 for composite, as reinforcement of HAP increases thermal stability of PLA[14]. The samples of wear was tested on linear reciprocating tribometer (DUCOM) (fig.1) under dry sliding condition and parameter of test was as shown in below table no.3. On basis of pre experimentation, it is found that pristine form of PLA shows significant wear rate in range of load (i.e.70 N to 95 N) and frequency (i.e. 6 Hz to 10 Hz). So, the range of experimentation was selected accordingly to study the effect of reinforcement. The sample for wear was tested against EN31 plate of 40×40×5 (all dimension are in mm) having high surface finish. Each sample of wear and EN31 was cleaned with acetone before and after test. The specific wear rate was calculated with the (eq.1) as shown below and to measure mass loss, before and after mass was measure by using electronic weighing machine. Each sample was measure for mass three time for improving accuracy. For wear, sample were tested two times and their average reading are taken into count.

$$Wear\ Rate = \frac{\Delta m \times 10^3}{\rho L F} \tag{1}$$

where,

Wear rate is in mm^3/Nm .Total mass loss (Δm) is in 'g'.

Density of Composite material (ρ) in 'g/cc'. Load (F) is in Newton.

Sliding distance (L)= $2 \times \text{freq.} \times t \times l \times 10^{-3}$.

Sliding distance is in 'm'. Frequency (freq.) is in 'Hz'.

Time (t) is in seconds. Stroke length (l) is in 'mm'.

2.3 Optimization Technique

Design of experiment was planned by using Taguchi technique as it optimises the number of experiment and gives a systematic approach to experimentation. This technique is widely used because it not only reduces the number of test to be performed but also cost and time incurred in doing experimentation. As per the level of design and number of factors that to be examine, it suggest some set of orthogonal array that to be chosen for experiment. In the following experiment, the number of level of design was '3' since there are two composite material and one pristine material while number of factors are three that is reinforcement, load and frequency. Based on above details, L_9 was selected as orthogonal array for experimentation in which number of runs are sixteen. After performing all runs, experimental results were analysed by using ANOVA (Analysis of variance) to find influence of each parameters on output. On basis of pre experimentation, it is found that pristine form of PLA shows significant wear rate in range of load (i.e.70N to 95 N) and frequency (i.e. 6 Hz to 10 Hz). So, the range of experimentation was selected accordingly to study the effect of reinforcement.

IOP Conf. Series: Materials Science and Engineering **810** (2020) 012079 doi:10.1088/1757-899X/810/1/012079

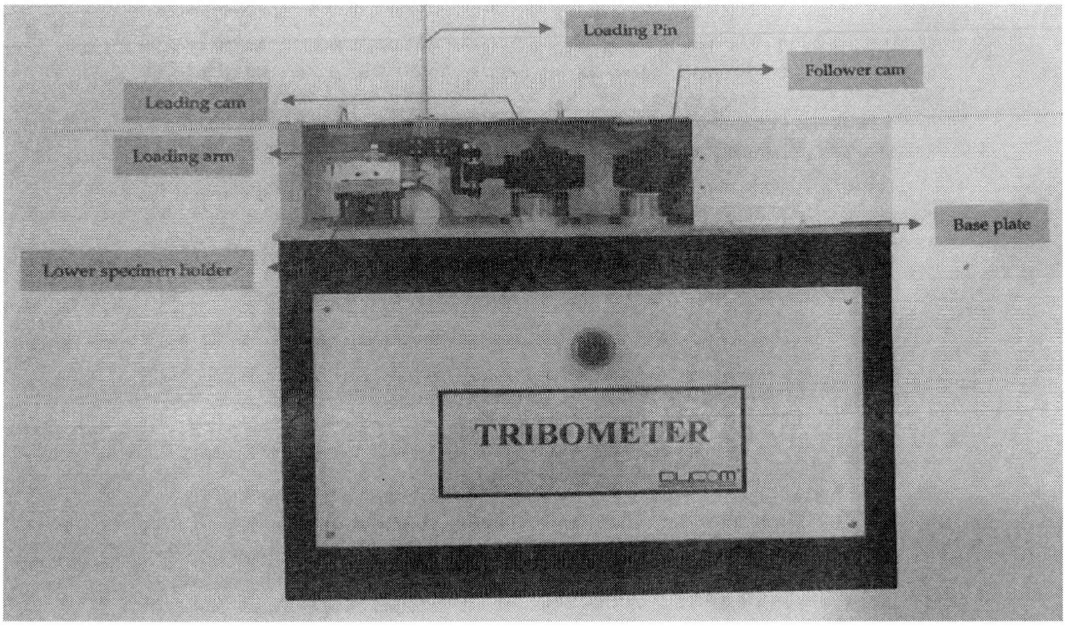

Figure 1. Linear reciprocating tribometer

Table 3. Experimental Parameters

Parameters	Value
Reinforcement of HAP nanoparticle (%)	0, 5, 10
Load(N)	80, 85, 90
Frequency(Hz)	7,8,9
Stroke(mm)	10 (constant)
Time (hr.)	1 (constant)
Temperature (^{0}C)	37 (constant)

3. Result and Discussion

With the help of mass of samples before and after experimentation, the total mass was calculated and by using equation (1), the corresponding wear rate for defined parameter was calculated. The calculated reading was use for optimisation by using Taguchi method. The percentage composition was selected up to 10 % because in experimentation it is found that wear rate for 15 wt. % composite increases instead of decreasing. Because of this reverse nature of composite, the 15 wt. % composite was omitted to reduce calculation and error causing in experimentation.

3.1 S/N ratio plot

The Signal to noise ratio is plotted by using Minitab software which minimizes the effect of noise that affecting the results. In this experiment, smaller is better S/N ratio is selected since wear should be small and the equation for smaller is better is mentioned below in equation no. 2. The following table no. 4 shows the corresponding results for all experiments with S/N ratio by using Taguchi technique.

Smaller is better:

$$(S/N)LB = -10\log(\left(\frac{1}{n}\right)\sum_{j=1}^{n} y_j^2) \quad (2)$$

Table 4. L_{09} Orthogonal Array with experimental results and S/N ratio

Reinforcement (%)	Load(N)	Frequency (Hz)	Wear rate $(mm^3/Nm)\times10^{-4}$	SNRA1	MEAN1
0	80	7	4.0197	-12.0839	4.0197
0	85	8	4.6595	-13.3668	4.6595
0	90	9	5.591	-14.9498	5.591
5	80	8	2.941	-9.3699	2.941
5	85	9	3.451	-10.7589	3.451
5	90	7	3.33	-10.4489	3.33
10	80	9	3.891	-11.8012	3.891
10	85	7	3.251	-10.2403	3.251
10	90	8	4.2015	-12.4681	4.2015

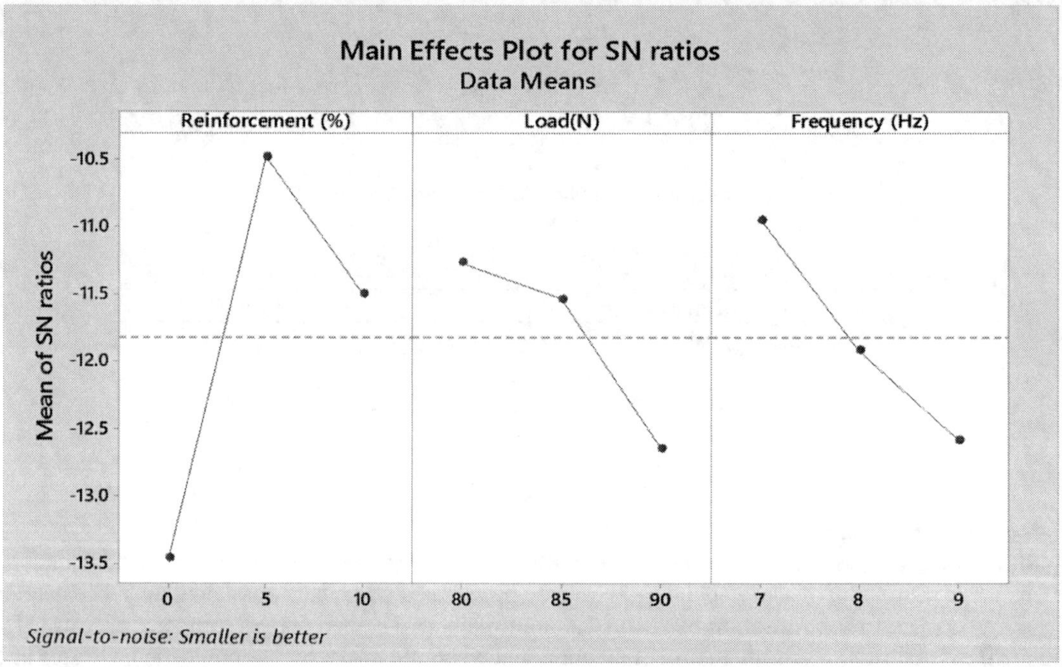

Graph 1. S/N Plot of experimental results

The graph (1) shows the main effect plot for S/N ratio for PLA/HAP Nano composite and from the graph the level of factor having highest S/N ration is selected as optimum respond measured. The optimum parameter for wear rate were 5% reinforcement of HAP (level 2), 80 N of load (level 1), and 7 Hz frequency (level 1).

3.2 Analysis of Variance

It is standard statistical technique to interpret the experimental results mainly used to study the relative influences of multiple variables and their significance. It is a statistical tool for testing null hypothesis for designed experimentation, where a number of different variables are being studied simultaneously. This analysis was carried out for a level of significance of 5% i.e. the level of confidence 95%. The results of ANOVA analysis is shown in below table 5, it is suggested that P value less than 0.05 in all parameter will influence the output parameter. Therefore from the results it is clear that reinforcement, load and frequency has influence on wear characteristic of composite material. Below pie diagram (graph 2) shows the percentage contribution of each parameter on wear rate and it suggest that reinforcement has largest influence on wear property of composite material while both load and frequency comparatively less.

Table 5. ANOVA for S/N ratio

Source	DF	Seq. SS	Adj. SS	Adj. MS	F	P	% Contribution
Reinforcement (%)	2	13.5919	13.8019	6.90093	231.16	0.004	64.34
Load(N)	2	3.1869	3.1869	1.59343	53.37	0.018	15.085
Frequency (Hz)	2	4.0772	4.0772	2.03859	68.29	0.014	19.29
Residual Error	2	0.2697	0.0597	0.02985			1.277
Total	8	21.1256					100

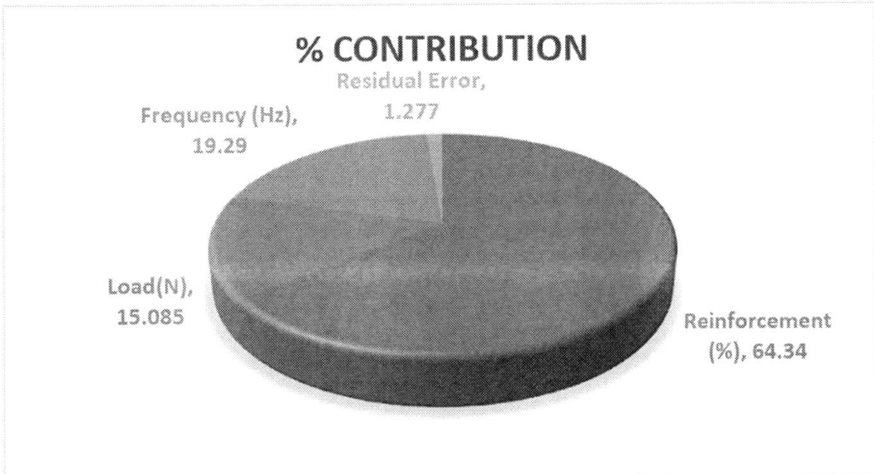

Graph 2. Percentage contribution of factors to wear

3.3 Regression Analysis

The relationship between input and output parameter for PLA/HAP composite was set with the help of linear regression model and equation 3 shows the relation between them. By substituting the values of input variable, we can estimate the predicted value for specific wear rate. The following graph 3 show the graphical comparison between the predicted and experimental specific wear rate.

$$\text{Wear Rate} = -4.76 - 0.0982\ \text{Reinforcement (\%)} + 0.0701\ \text{Load(N)} + 0.401\ \text{Frequency (Hz)} \quad (3)$$

3.4 Confirmation Test

The confirmation was carried out at optimum parameter i.e. for wear rate were 5% reinforcement of HAP (level 2), 80 N of load (level 1), and 7 Hz frequency (level 1). And table 6 shows the comparison between predicted by mathematical model (i.e. eq. 3) and experimentally evaluated specific wear rate.

Table 6. Confirmation test results

Parameter	Predicted value	Experimental value	Error %
Wear rate	3.164	2.935	7.21%

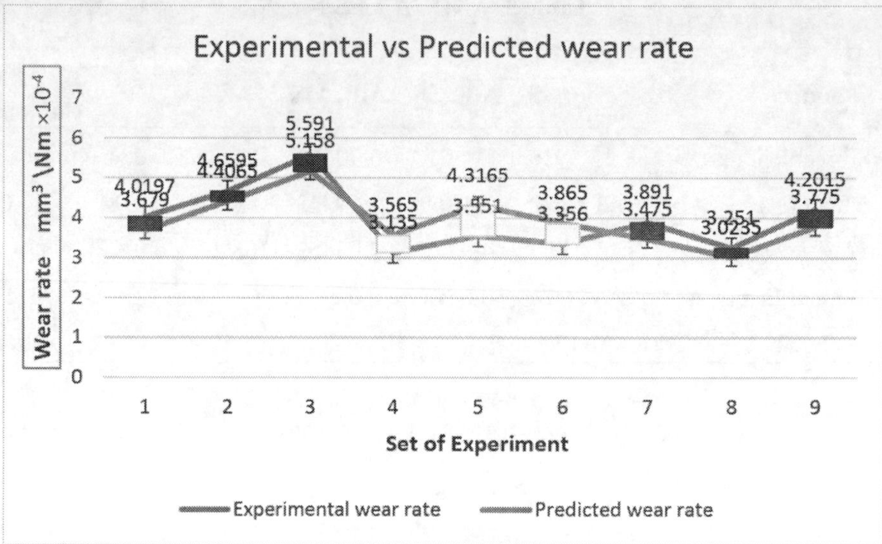

Graph 3. Experimental vs Predicted wear rate of PLA/HAP nanocomposite

4. Conclusion

The Taguchi technique for optimizing the parameter was significantly analyzed the experimental data and its gives the optimum results for PLA/HAP nano composite for its wear characteristics under dry sliding environment.

- The main effect of S/N ratio plot shows that at optimum percentage of reinforcement is 5%, while load and frequency are optimum at 80 N and 7 Hz.
- It is also seen that beyond 5% of reinforcement the wear behavior of material gets changed that is wear rate increases.
- The experiment also reveals that the PLA/HAP nanocomposite may offer better wear resistance in the range 0 to 5 or 5 to 10 wt. %.
- ANOVA results shows that reinforcement has greater influence while Load and Frequency has less for considered range.
- PLA/HAP nanocomposite exhibits lower wear rate at all its level of experiment as compare to other i.e. pristine form of PLA and PLA with 10 wt. % of HAP nanoparticle.

Acknowledgements

Authors express their gratitude to Government College of Engineering, Aurangabad and Central Institute of Plastic and Engineering, Aurangabad.

References

[1] S .K. Roy Chowdhury, A. Mishra, B. Pradhan, D. Saha, " Wear Characteristic And Biocompatibility of some Polymer Composite Acetabular Cups", Wear 256 (2004) 1026-1036.

[2] Liu Jin-Long, Zhu Yuan-Yuan, "Biotribological Behavior of Ultra High Molecular Weight Polyethylene Composites Containing Bovine Bone Hydroxyapatite", J China Univ Mining & Technol 18 (2008) 0606-0612.

[3] Qinglang Wang, Jinlong Liu, Shirong Ge, "Study of Biotribological Behavior of the Combined Joint Of Cocrmo And Uhmwpe/Bha Composite in a Hip Joint Simulator", Journal of Bionic Engineering 6 (2009) 378-386.

[4] Kevin Plumlee, Christain J. Schwartz, "Improved Waer Resistance of Orthopedic Uhmwpe By Reinforcement Of Airconium Particles", Wear (2009) 710-717.

[5] Xiaoduao Zhao, Dansheng Xiong, Xinxin Wu, "Effects of Surface Qxidation Treatment Of Carbon Fibers Onbiotribological Properties of Cf/Peek Materials", Jornal Of Bionic Engineerng 14 (2017) 640-647.

[6] P. Christel, A. Meunier, A.J.C. Lee (Eds.), "Biological and Biomechanical Performance of Biomaterials," Elsevier, Amsterdam, The Netherlands, 1997, pp. 81–86.

[7] Kalpana S. Katti, "Biomaterials in total joint replacement", Colloids and Surface B: Biointerfaces 39 (2004) 133-142.

[8] Ramsden, J. J, Allen D. M, Stephenson D. J, Alcock J. R, Fuller, G. N., Goch, G. D. "The Design and Manufacture of Biomedical Surfaces," CIRP Annals-Manufacturing Technology, 2007. 56(2):687-711.

[9] Sachin G. Ghalme, Ankush Mankar, and Yogesh Bhalerao, " Biomaterials in Hip Joint replacement," International Jouranal of Materials Science and Engineering, volume 6, June 2016.

[10] Ozturk, O., Turkan, U., and Eroglu, A. E., "Metal ion release from nitrogen ion implanted CoCrMo orthopaedic implant material," Surface & Coatings Technology, 2006. 200(20-21):5687-5697.

[11] Eisenbarth, E., Velten, D., Muller, M., Thull, R., Breme, J., "Biocompatibility of a stabilizing elements of titanium alloys," Biomaterials, 2004. 25(26):5705-5713.

[12] Roy S, Pal S., "Characterization of silane coated hollow sphere alumina-reinforced ultra high molecular weight polyethylenec omposite as a possible bone substitute material". *Bull Mater Sci*, 2002, 25(7): 609–612.

[13] Hashimoto M, Takadama H, Mizuno M, et al., "Titanium dioxide/ultra high molecular weight polyethylene composite for bone-repairing applications: preparation and biocompatibility." *Key Engineering Materials*, 2003(240/242): 415–418.

[14] Ming Gong, Qian Zhao, Liming Dai, Yingying, Tingshun Jiang, "Fabrication pf Polylactic acid/ Hydroxyapatite /graphene oxide/ composite and teir thermal stability, hyfdrophobic and mechanical properties," Journal of asian Ceramic Societies 5, (2017) 160-168.

[15] Kanagaraj, S., Mathew, M. T., Fonseca, A., Oliveira, M. S., Simoes, J. A., Rocha, L. A.. "Tribological characterisation of carbon nanotubes/ultrahigh molecular weight polyethylene composites: the effect of sliding distance." *Int J Surf Sci Eng*. 2010, 4(4- 6):305-21.

[16] R.K.Goyal, M.Yadav, " The wear and friction behavior of Novel Polytetrfluroethylene/ expanded Graphite Nanocomposite for tribological application", Joi\urnal of tribology, April 2014, Vol. 136/ 021601-1.

[17] Siriporn Tanodekaew, Somruethai Channasanon, Pakkanun Kaewkong, Paweena Uppanan, "PLA-HA scaffolds: preparation and bioactivity", Procedia Engineering 59 (2013) 144 – 149.

[18] Maria Persson, Gabriela S. Lorite, Hana E. Kokkonen, Sung Woo Cho, " Effect of bioactive extruded PLA/HA composite film on focal adhesion formation of preosteoblastic cells," Colloids and Surface B: Biointerfaces 121 (2014) 409-416.

[19] Ming Gong, Qian Zhao, Liming Dai, Yingying Li, Tingshun Jiang, "Fabrication of polylactic acid/hydroxyapatite/graphene oxidecomposite and their thermal stability, hydrophobic and mechanicalproperties," Journal of Asian Ceramic Societies 5 (2017) 160–168.

[20] MReza Foruzanmehr, Pascal Y. Vuillaume, Saïd Elkoun, Mathieu Robert, "Physical and mechanical properties of PLA composites reinforced by TiO2 grafted flax fibers", Materials and Design 106 (2016) 295–304

IOP Conf. Series: Materials Science and Engineering **810** (2020) 012080 doi:10.1088/1757-899X/810/1/012080

A review on straightening of bars and application of probabilistic approach on Moment-curvature relationship

Sanjib Roy[1]
Narsee Monjee Institute of Management Studies, India
sanjibbhel@gmail.com

Arun Kiran Pal[2]
Jadavpur University, India
arunkiranpal@yahoo.co.in

Shannay Rawal[3]
Narsee Monjee Institute of Management Studies, India
shannay.nmims@gmail.com

Abstract. Bar straightening is a significant process in the value stream of bar manufacturing processes. Any improvement in the process has a substantial impact on quality and productivity as a whole. This paper presents the outline of the fundamentals of the straightening process from a material science perspective and discusses the state of art in terms of process analysis and development. It has been inferred that the major aspect of the process is to understand the moment-curvature relationship in bars subjected to re-rolling. The process is governed by several parameters such as bar speed, the angle between osculating plane and loading plane, bar size or radius of bar, applied reverse bending moment, initial and final curvature, etc. It has also been observed that the research in this domain has seldom applied a statistical approach to enhance process capability. A brief account of the potential use of statistical process control has been presented. Use of statistics shall enable a better understanding of various process parameters. This, in turn, will help in deciding whether the process is in control and therein to assist in deciding the acceptance or rejection of lot.

Nomenclature:

c = a constant
E = Elastic modulus
E_p = Plastic modulus of the bar.
σ = Stress
ϵ, e = strain
Y = Yield Point, Yield Stress
M = Bending moment for producing change in curvature, κ
M_o = Moment needed to produce zero curvature
κ = curvature change along length of bar.
κ_r = residual curvature along bar length
κ_f = final residual curvature along bar length
η = curvature function
σ_a , σ_b , σ_c = stresses at point a, b and c
h = depth of the section of bar
ϵ_y = yield strain

IOP Conf. Series: Materials Science and Engineering **810** (2020) 012080 doi:10.1088/1757-899X/810/1/012080

ϵ_r = residual strain
ϵ_{rl} = residual strain at extreme fiber
ϵ_1 = change in strain at ½ depth of bar section or at *h/2* in fiber
λ = a function dependent on section parameter, line segment
I_z = Moment of inertia about z-axis
v_x = Throughput speed of bar
θ = Angle between osculating plane and loading plane
r = radius of bar
P = Probability of occurrence
$f_\kappa(\kappa_f)$ = probability density function of final curvature
μ_K = mean curvature
σ_K = standard deviation of curvature

1. Introduction:

Bar Straightening is a process of straightening of bars of various sections in order to improve degree of straightness [1]. Straight Bars of various sections are a general requirement in industries. Straight Bars are required as raw material for the purpose of production of various parts, machine components etc. in industries. There are many large number of parts that are formed from cut pieces of straight bars. Hence, degree of straightness [1] is always a general concern in any industry where straight bars are required as raw materials where straightness of rollers, rolling shafts, idlers are significant machine components. In printing machineries, rolling shafts, idlers are used in significant numbers from start of feed to end products. In all such cases use of straight bars have significant importance for smooth operation of machines and quality output.

In re-rolling industries degree of straightness is of prime consideration. Rolling mills provide light sections and bars in slightly bent and distorted conditions, which need to be straightened before end use. Commercial straightness of metallic bars and sections usually ranges from 1 in 750 to 1 in 5000 depending on their application.[1]

Cross-roll straightening machines are widely used to straighten circular bars and tubes. The product axis is used as a reference about which the bar is rotated as it moves along the axis through a series of staggered rolls.[2] The principle of operation is used as a primary classifier for commercially straightening machines. The main groups are section straighteners, cross-roll straighteners and stretch straighteners.

The product bar passes through a series of staggered rolls in section straighteners and cross-roll straighteners. Plastic strains are redistributed in the bars due to the presence of reverse kinematic loading. In the process it secures the desired straightening [2]. Several work have been done on bar straightening, mostly in the theoretical development with general analysis of straightening process[2] including numerical modelling approach, experimental simulation[22], optimization, etc., of straightening process.

In the re-rolling industries, statistical considerations are of much importance since many quality related decisions are usually based on statistical considerations. Rejections of products are also due to process based on control limits or specification limits set by quality standards. It is always intended to keep the process close to estimated value or mean value of the desired parameters. Rejection of products if takes place do necessarily have an impact on overall cost of production and ultimately on the price of final product.

In the subsequent sections, the background of the straightening process is reviewed. In addition, an account of introducing statistical consideration on controlling parameters in the straightening process is presented.

IOP Conf. Series: Materials Science and Engineering **810** (2020) 012080 doi:10.1088/1757-899X/810/1/012080

2. Mechanics of Bar Straightening

Bar straightening involves applications of stresses by the rolls during the axial travel of the bars. To understand the mechanism, an analysis was carried out by "Das Talukder, N.K. et al." This analysis was based on several assumptions as listed below.

i. Elastic or linear work hardening is assumed to be present in the material of the straightened bar. [12] [Figure 1a]
ii. Under the condition of pure bending, the bar has a circular cross section and the neutral axis passes through the centre of the bar end axes. [12]
iii. The cross section remains on a plane and always make right angles to the axis of the material.[9]
iv. At any assumed point in the bar section, the strain varies proportionally with the distance from the perpendicular point on the neutral axis. [12]
v. The material is perfectly elastic-plastic, and its stress-strain relations in uniaxial tension and compression are the same. It has the same modulus of elasticity and yield point in tension and compression.[9]
vi. Every stress component other than the stress in longitudinal direction is zero.[9]

The basics of stress-strain relationship have been stated for enabling understanding how the analysis has actually proceeded. We are familiar the stress-strain relationship of any isotropic material having elastic modulus, E up to elastic limit and plastic modulus, E_p beyond E. A simplified diagram on the same is presented below [Figure 1a]. Straightening of bar essentially involves moment-curvature relationship as reverse bending due to applied moment results in reduction of curvature. This is quite a usual practice in industry and also experimentally shown by [Biju, B et.al. IRJET, 2016] [23].

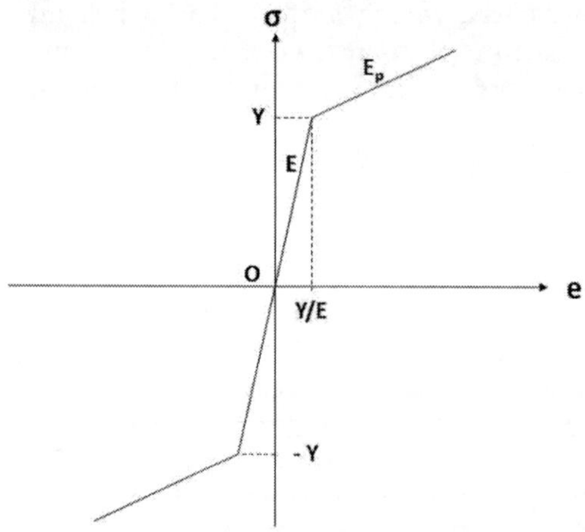

Figure 1a: The Stress-strain relationship of the material

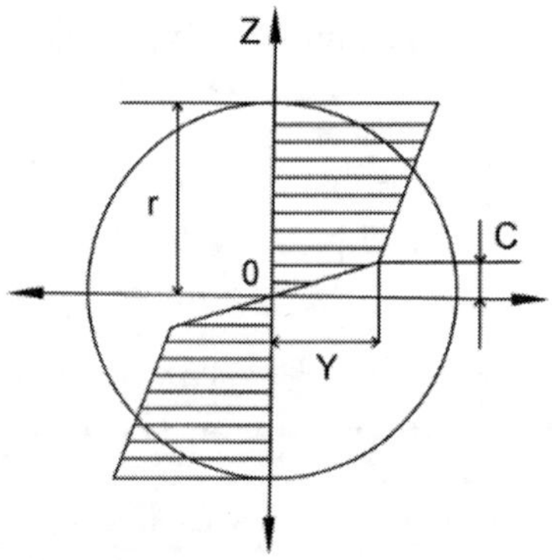

Figure 1b: The Stress distribution through the section of a circular bar when subjected to pure elastic-plastic bending

where E = Elastic modulus and E_p is plastic modulus of the bar.

This phenomenon is well explained by Bauschinger effect.

The distribution of stresses in the bar cross section when subjected to elastic-plastic pure bending which can be observed in Figure 1b. The stress-strain relationship is customarily expressed as below:

$$\sigma = E\varepsilon \qquad for \qquad z < c$$
$$= Y + E_p\,[\varepsilon - \tfrac{Y}{E}] = (1-\mu)\,Y + E_p\,\varepsilon \qquad for \qquad z \geq c$$

where, $\mu = \frac{Ep}{E}$, E is elastic modulus and E_p is plastic modulus and Y is yield point.

2.1 Bauschinger effect [18]

Previous research findings have found that repeated plastic deformation introduces three major effects in pipes and bars: strain hardening, Bauschinger effect and strain aging. However, point of interest is much restricted to mainly on Bauschinger effect. This can be attributed to the cyclic loading and unloading of bars or pipes in reverse directions in addition to the unidirectional loading beyond σ_b when the initial yield point is at σ_a. [Figure 1c].

Bauschinger effect is a metallurgical phenomenon wherein the stress/strain characteristics of the studied material change due to the influence of microscopic stress distribution in the material.[18] The effect is notably observed in real metals, whenever there is a reversal of stress. In the process of re-rolling, consequently, there is a change in yield point in reverse loading.

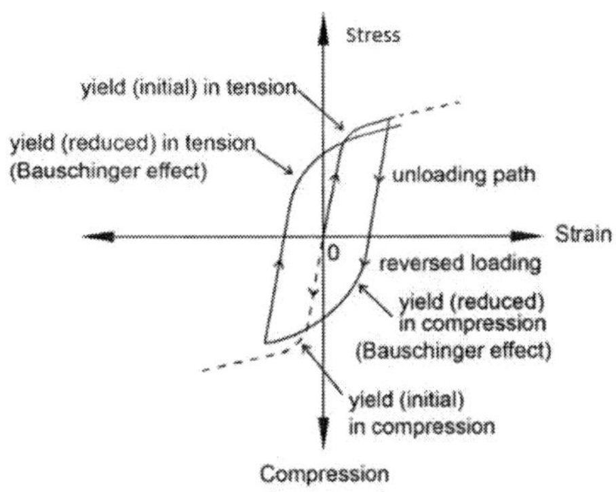

Figure 1c

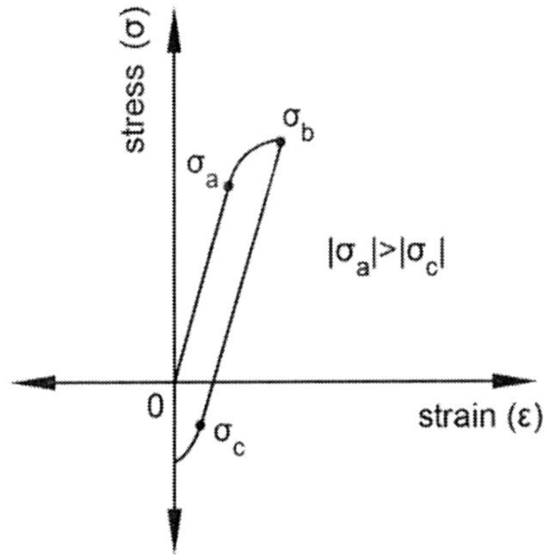

Figure 1d

Figure 1c & 1d: Bauschinger effect

The above-mentioned loadings are widely regarded as the cause for occurrence of Bauschinger effect. This is due to the close association of the result where after initial plastic deformation, reverse yield strength σ_c is reduced in comparison to the initial value [Figure 1d].

There are three different known residual stresses, viz. Type-1, Type-2 and Type-3 residual stresses. Commercially manufactured components possess all three residual stresses in varying degrees that are generated during the process of manufacturing. Type-2 and Type-3 residual stresses are primary causal agents of Bauschinger effect.[3] [4]
The physical presence of these stresses could be imperceptible or in the form of minute cracks. Cast and welded components can fail catastrophically even without loading at room temperature due to residual stresses. They are a proven agent in reducing the working lifespan of metal components.[5]

As further discussion is mainly on reverse loading of bars, our consideration will be made on such analysis where discussion is centred around reverse loading and its impact on final curvature, consequently to degree of straightness and statistical consideration on the same.

3. Analysis of Bar Straightening

Detailed analysis of the involved mechanics has been mainly done by Das Talukder, Johnson and Yu in the processes of straightening of bars.[2][6][12] In the analysis, it has been assumed that bars behave both anisotropically as well as non-homogeneously, along with the degree of straightness of the finished component.[6] More thorough analyses of the processes have also been done for different types of machines.[2]

When analysing the process of general straightening under the action of reverse kinematic loading, prime considerations are of curvature change and resisting moments. Assumption is made that stress-strain curve on the basis of the findings of Davis et. al. are indicative curves showcasing the results for ductile nature of materials.[4] The stress-strain relationship, between the reverse loading stress, σ and the change in longitudinal strain, ε [Figure 2(a)] is collectively demonstrated in Figure 2(b).

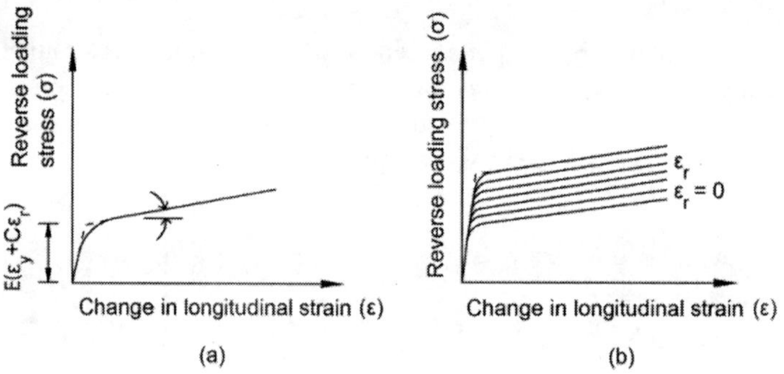

Figure 2. Stress-strain change curves for various different initial residual strains.

Based on the results as reported[6][7], it shows that reverse loading does not have a clearly defined yield point and instead curves show sharp changes in the yield region. In addition to assumptions made earlier, it is also necessary that following assumptions are made:

(a) Yield point stress is the same in tension and in compression for an initially unstrained bar.

(b) During reverse loading, stress-strain relation is bilineal. This relationship is very close to the results reported by Davis, E. A, et al (1950) [7]. The bilineal relationship as assumed can be expressed as

$$\sigma = E\varepsilon \qquad \varepsilon < (\varepsilon_y + c\varepsilon_r)$$
$$\sigma = E\,(\varepsilon_y + c\varepsilon_r) + E_p\,[\varepsilon - (\varepsilon_y + c\varepsilon_r)] \qquad \varepsilon \geq (\varepsilon_y + c\varepsilon_r) \qquad (1)$$

Where, E is elastic modulus and E_p is plastic modulus,

ε_y is the yield strain during reverse loading without initial residual strain for an element,

c is a constant and $c \ll 1$. [18]

3.1 The Moment ~ Curvature (M ~ κ) relationship for Circular Bars

Considering a bar with the following characteristics:

i. Sectional symmetry about two axes in the y and z directions such that they are mutual perpendicular.

ii. initial residual curvature κ_r in one plane of symmetry which will be different at different distances since strains in fibre at different distances will be different from z-axis.

IOP Conf. Series: Materials Science and Engineering **810** (2020) 012080 doi:10.1088/1757-899X/810/1/012080

Figure 2(b) shows the variation in strain change-stress curves as a result of an induced strain change. According to the fibre in consideration, the curve of the initial residual strain is noted to calculate the stress.

If residual strain in the extreme fibre is ε_{r1} and the residual strain in the fibre at a distance y from the z-axis is ε_r then assuming h to be the depth of the studied section, ε_r to be the residual strain at a distance of y units from the z-axis and the residual strain in the furthest fibre to be ε_{r1} then

$$\frac{\varepsilon_{r1}}{h/2} = \frac{\varepsilon_r}{y} = \kappa_r \tag{2}$$

If κ is the change in curvature along the length of the bar as a result of the straightening, then from Figure 3:

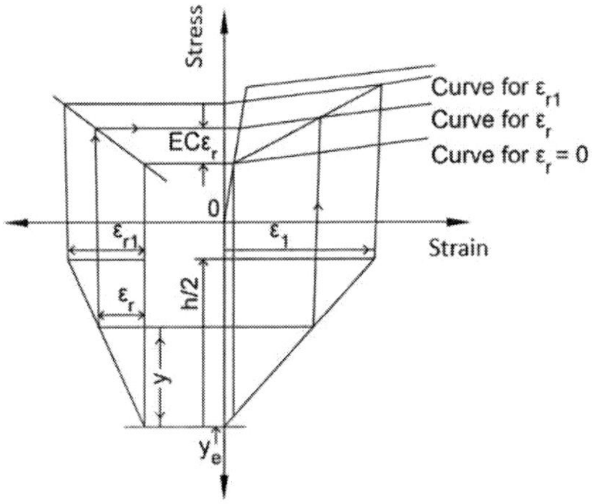

Figure 3: Variation of stress across the depth in bending

$$\kappa = \frac{2\varepsilon_1}{h}$$
$$\text{and } \varepsilon = \frac{2y\varepsilon_1}{h} \tag{3}$$

where the change in strain in the fiber at distances y and $h/2$ are given by ε and ε_1 respectively. If the value of y at elastic-plastic boundary is given by y_e,[6] then from equation (3)

$$\begin{aligned} \sigma &= \kappa E y & -y_e \le y \le y_e \\ &= (E - E_p)\,\varepsilon_y + \xi_y & -y_e \le |y| \le \frac{h}{2} \\ &\text{where } \xi = (E - E_p)\, c\kappa_r + E_p\kappa \end{aligned} \tag{4}$$

Thus, the moment required to produce this change, M, can be found as follows [6]:

$$\begin{aligned} M &= [\lambda E_\kappa + (1-\lambda)\,\xi]\, I_z & \kappa \ge (\kappa_y + c\kappa_r) \\ & & 0 \le \lambda \le 1 \\ &= \kappa E I_z & \kappa \le (\kappa_y + c\kappa_r) \end{aligned} \tag{5}$$

where λ is a function dependent on the section parameters, y_e and change in curvature.

We know, $$y_e = \frac{\varepsilon_y}{h} \quad (6)$$

If the moment required to produce the same curvature change κ in a bar length, having zero initial curvature is represented by $M_o(\kappa)$, then from equations (4) and (5),[6]

$$M_o = [E + (1 - \lambda)\, E_p]\, \kappa I_z \quad (7)$$

from equations (5) and (7)

$$\Delta M = M - M_o = (1 - \lambda)(E - \;E_p\,]\, c\kappa_r\, I_z \quad (8)$$

where M is the extra moment required to produce a particular curvature, due to initial curvature in the bar. It has been seen that the curvature falls down from roller to roller (Groover, 2010)[24] when a round bar is subjected to reverse bending according to the absolute values.[23]

4. Statistical approach to the Bar Straightening

It has been seen experimentally that deflection reduces after initially bent bar passes through straightening process by reverse bending. Three specimens have been used by Biju et. al. and experimentally observed that at given points deflections reduced in each specimen.[23] Considering the data published by Biju et. al. for three different specimens, it has been observed that changes in deviation from central axis are much dominant in the first pass and subsequent changes in deviation are much lesser in comparison to initial change. The Figures 4(a), 4(b) and 4(c) indicate the above-mentioned observations.

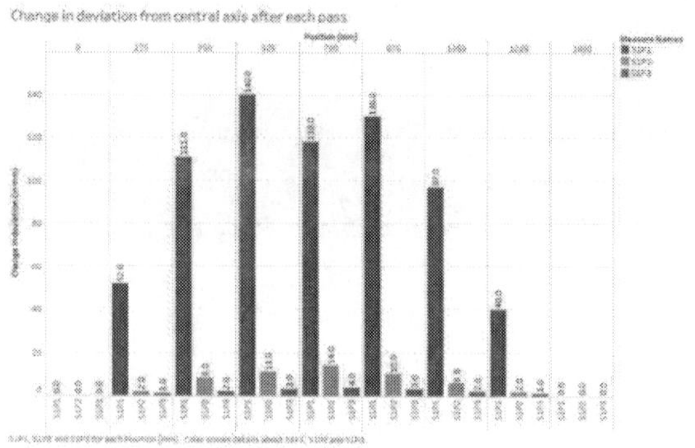

Figure 4(a)

Figure 4(a): Change in deviation from central axis after three passes for specimen-1

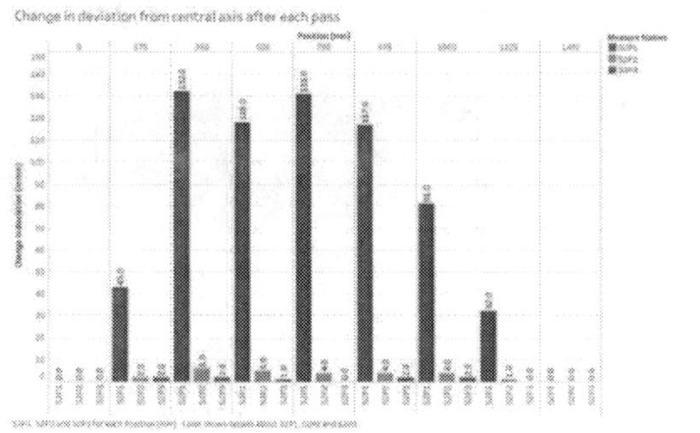

Figure 4(b)

Figure 4(b): Change in deviation from central axis after three passes for specimen-2

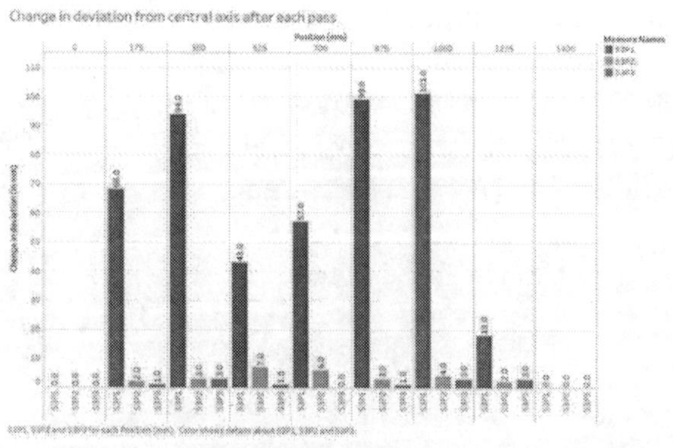

Figure 4(c)

Figure 4(c): Change in deviation from central axis after three passes for specimen-3

IOP Conf. Series: Materials Science and Engineering **810** (2020) 012080 doi:10.1088/1757-899X/810/1/012080

Although the data available is meagre in nature in the experimentation part, but it can be ascertained from above that with reduction of deflections, curvatures are actually reduced and degree of straightness is improved significantly with initial pass of reverse bending and also to some extent at subsequent passes. This is also in line with the very purpose of manufacturing of bar straightening machines. As each specimen enters the roller arrangement for reverse bending, bending moments thus faced by specimen causes improvement in reduction of curvature. However, due to insufficient experimental data, the frequency distribution followed by the reduction of deflection cannot be accurately ascertained.

It can be seen that moment and curvature have relationship in reverse kinematic loading for straightening of bar viz. bending moment M, initial and final curvatures κ including residual curvatures κ_r are therefore random variables in the event of straightening process which is actually a continuous one. As for continuous variables, probability density function in any arbitrary interval (a,b) or (λ_a, λ_b) of length segments of bar under reverse kinematic loading is applied.

However, values of interval (λ_a, λ_b) shall be close to the spacing between set of rolls producing bending moment and not exceeding the length between two rolls. Even if, rolls are somewhat evenly spaced, then too depending on throughput speed and residual stresses and residual curvatures, applied bending moments will be random in nature. This is more so, as bar under kinematic loading is also rotating about axis making an angle between loading plane and osculating plane affecting the process [6].

When a bent bar with initial residual curvature k_r is subjected to a loading moment in reverse direction (which may be assumed positive), an additional residual curvature k_r in the reverse direction causes it to become straightened, i.e. the resultant curvature tends to reduce to zero. However, in reality, it will be indeed probabilistic only, that meeting various parameters under consideration, i.e. speed of bar, angle between loading and osculating plane [Figure 4], applied load through cross rolls, the effect of actual reverse loading for curvature in opposite direction amounting same value in reverse direction to enable final residual curvature which shall tend to be zero at every point along the axis from beginning of process to end. However, there will indeed be various values of curvature at every point along the bar which will be random in nature.

Several variables are involved in the process of bar straightening. These are viz. bending moment (M) arising due to reverse bending, angle (θ) between loading plane and osculating plane, throughput speed (v_x) of the bar in progress, section parameter of bar like radius of the bar (r), elastic and plastic moduli [E and E_p] of the bar, initial curvature (κ_r)and final curvature (κ_f).

An important addition to make here is that the zero probability of a random variable taking any selected value 'x' does not imply impossibility for the random variable taking on the value of x. Similarly, in the case of continuous assessment, zero probability does not correlate to logical impossibility. Additionally, owing to the practical limitations of our ability to measure and observe the entirety of the generated results, the study of this field is largely academic by nature and we ideally work within probabilities dealt in defined intervals and not for isolated points per se.

In the continuous case, probabilities associated with individual points are always zero, and therefore if we deal in the probability associated with the interval present between a and b, the inclusion of the endpoint becomes irrelevant. In the case of continuous intervals, the importance of specifying interval probabilities drops drastically.

As we know, P(X=x) =0 if X is continuous random variable. [9]

Symbolically,

$$P(a<X<b) = P(\lambda_a<X<\lambda_b) = P(\lambda_a<X\leq\lambda_b) = P(\lambda_a\leq X<\lambda_b) = P(\lambda_a\leq X\leq\lambda_b) \text{ [18]} \quad (9)$$

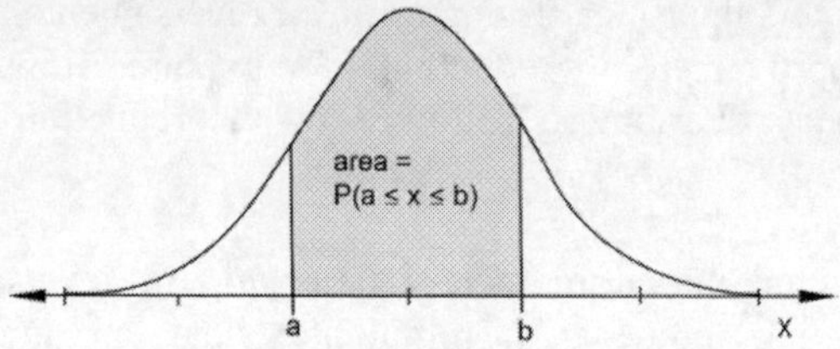

Figure 5: Probability as area under *f(x)*

Since, this is a continuous case of re-rolling, hence all the parameters figuring in the process will also be continuous random variables. The probability between any intervals, more specifically between the rolls of reverse bending shall be in consideration.

However, certain assumptions can be made which is quite logical and also to make our present case a bit simple. Since all rollers are running by standard commercial motors, revolution of these motors will be nearly constant or with insignificant variation which may be arising out of voltage fluctuation, if any. However, on stabilized power supply, the possibility of variation will actually be insignificant. Hence, throughput speed (v_x) of bars can be treated as constant excepting start and end points. Similarly, average radius of bar can also be treated as constant since dimensional variations will be insignificant and for all practical purpose considering variation is actually not much meaningful. Now, the angle between loading plane and osculating plane (θ) will be indeed varying. Roller positions are by and large fixed and rolling about their fixed axis. Variability aspect of θ can be considered suitably. Hence, at present, we can actually deal primarily with two main variables, i.e. change of bending moment ΔM and final curvature, κ_f.

Therefore, using equation (8), we can say that probability of occurrence of actual loading towards of change in bending moment under straightening process may be defined as

$$P(\lambda_a \leq \Delta M \leq \lambda_b) = P(\lambda_a \leq \kappa_f \leq \lambda_b) = \int_{\lambda_a}^{\lambda_b} f_\kappa(\kappa_f)\, d\kappa \quad (10)$$

where, $f_\kappa(\kappa_f)$ is the probability density function of final curvature due to change in bending moment on actual loading event during straightening process in the cross-roll arrangement.

Now as the value of μ_κ is available for a process, then by using the equation (10) it is possible to ascertain the value of final curvature κ_f at any interval region between a set of rolls causing reverse bending. The probability of any such value for κ_f can also be found out using equations (9) and (10).

Therefore, the probability or likelihood of falling values of final curvature κ_f within the range of application of reverse bending in the range of any set of rolls (λ_a , λ_b) *or (a, b)* will be as below:

$$P(\lambda_a \leq \kappa_f \leq \lambda_b) = \int_{\lambda_a}^{\lambda_b} f_\kappa(\kappa_f)\, d\kappa \quad (11)$$

5. Discussion and Conclusion:

Probability density function can be evaluated when sufficient amount of experimental data is available. The probability distribution function of the data set will play significant role in ascertaining the probability of occurrence of required straightness between set of rolls under reverse bending. Mean curvature and standard deviation of the bar can also be estimated accordingly.

The use of bar straighteners as a valuable preparatory process can be established by comparison between the conventional bar machining process and the use of bar straightening as a preliminary process before the machining stage. This can directly translate into improvements in multiple aspects such as productivity, energy requirements and monetary benefits.

Concept of statistical consideration has been applied to bar straightening process for estimation of final curvature of line segment of a given bar. The probability of occurrence of final curvature at a point within length segment can be expressed if probability density function is expressed between any two rollers in case of equidistant rollers. For any other arrangement of rollers, limit values of length segment of bar λ_a and λ_b have to be suitably changed and equation shall be changed accordingly. As the mean value of curvature and standard deviation of the bar after straightening process is within acceptable limits, same can be used for commercial purposes. The viability of using this for preparing raw material to produce various components increases as compared to current process.

Acknowledgements
The authors are grateful to Dr. N.K. Das Talukder for providing his valuable initial guidance in making this paper. Several changes has been brought on the final manuscript based on guidance received time to time. The authors are deeply indebted and sincerely acknowledge significant advices and discussions with Dr. (Prof.) K.K.Gupta, Prof. Praveen Loharkar, and Dr. Milan Joshi, NMIMS, Shirpur Campus.

References:
[1] Kemshall, G E 1974 Light Bar and Section Mill Technology, *Steel Times* **13** 243
[2] Das Talukder, N K, Singh A N 1991 *ASME J. Eng. Ind.* **113** 232.
[3] Meiwes Karl Christoph, Erdelen-Peppler, Marion and Brauer, Holger 2014 Canada. *Proc. of the 10th Int. Pipeline Conf. IPC2014*
[4] Wu B J, Chan L C, Lee T C, Ao L W 2000 *J. of Materials Proc. Technology*, **99** 202.
[5] Liscic B, Tensi H M and Luty W (Eds.), 1992 *New York, Springer-Verlag*
[6] Das Talukder N K, Singh A N 1991 *ASME J. of Engg. for Industry* **113** 224
[7] Davis E A, Rose B A, and Nadai A 1950 *Theory of Flow and Fracture of Solids*, 2nd *Ed. McGraw-Hill* **I** 20
[8] Nastran M and Kuzman K 2002 *J. of Materials Proc. Technology* **125-126** 711
[9] Tokunaga H 1961 *Bulletin of the J.S.M.E.* **4 15** 605
[10] Das Talukder N K, Johnson W, 1979 Varanasi. *Research Monograph No. 79121,* Mech. Engg. Dept., Banaras Hindu University
[11] Das Talukder, N K and Johnson W 1981 *Int. J. Mech. Sci.,* **23** 213
[12] Yu T X, Johnson W 1981 *Proc. 22 M.T.D.R. Con., Macmillan Press*
[13] Tselikov A I, Smirnoff V V 1975 Oxford. *Straightening and Levelling Machines, Rolling Mills, Pergamon Press* **5** 254,
[14] White J S, and Briggs G C, 1951 *Progress Report on Work on an Experimental Roller Leveler, Iron and Steel Research Association* Rep. No. MW/E/59/51,
[15] Ang A H S and Tang W H, *Probability Concepts in Engg. Planning and Design Vol-1, Basic Principle*
[16] Ang A H S and Tang W H, *Probability Concepts in Engg. Planning and Design Vol-2,* Decision, Risk and Reliability
[17] Mitra Amitava Third Edition "Fundamentals of Quality Control and Improvement", Wiley India Pvt. Ltd. 613

[18] Chakrabarty J , 2014 *Theory of Plasticity*, Third Edition, Elsevier, Butterworth-Heinemann.
[19] Jain, RK, 2011 *Engineering Metrolog*, 20th Edition, Kahnna Publishers
[20] Johnson, Richard A, 2011 New Delhi, *Miller & Freund's Probability and Statistics for Engineers, PHI Learning Pvt. Ltd ,Eighth Edition* 121
[21] Casella George, Berger, Roger L, 2000 *Statistical Inference* 2nd Ed, Clenage Learning, 35
[22] Macura P , Petruska, J , 1996 *J. of Matl. Proc. Tech.,* **60**, 55
[23] Biju B, Dijin J S, Anujith C, Augustine, Arun, Mohammad, Anas P, 2016 *IRJET* **3** 379
[24] Groover Mikell P, 2010 John Wiley & Sons, Inc., *Fundamentals of Modern Manufacturing*, 472,

IOP Conf. Series: Materials Science and Engineering **810** (2020) 012081 doi:10.1088/1757-899X/810/1/012081

Review of effect of meso scale combustor wall material, Catalytic material, & Porous media on flame Stabilization

Gajanan Z. Jadhav[1], Dr. K K Gupta[2]

SVKM'S, NMIMS, MPSTME, Shirpur, Maharashtra, India 425405.
gajanan.jadhav@nmims.edu

Abstract. In recent trends demand of energy for portable devices like notebook, microsatellite, laptops, UAV are increasing at high rate, a high energy density hydrocarbon fuel creates opportunity to alternate chemical Li-ion batteries with micro combustion based power generation system. Mainly, this review focus on experimental, mathematical modelling techniques, and numerical simulation performed on heat recirculating combustor to find the effect of combustor wall material, catalytic material and porous media. It also summarizes the electricity generation possibilities with micro combustor turbine, thermo electric material & thermo-photovolatic combustion system. The effect of combustor wall material, catalytic material, and porous media on flame stabilization are enlisted with its summary.
Keywords: - Combustion, Wall conductivity, Catalysis, Thermoelectric

1. Introduction

The recent development in manufacturing technology, MEMS (micro electro mechanical system), & rapid prototyping, minimization of portable devices have accelerated. Because of this rapid growth of portable devices like cell phone, notebook, video cameras are dramatically increased. With this development demand of energy are increased currently chemical batteries are used in this devices, but this battery having low energy density about 0.2KWh/Kg [1].

This chemical battery with low energy density as compared to hydrocarbon fuel require more time recharge and it continues for few hours. It also had limited number of recharging cycle limitations. As well as its disposal creates environmental pollution which is not desirable. Moreover, new devices like 4G cellular phone, laptops require high energy and minimum recharging time, with this view in mind increase in the operational life time of power devices, the energy density fuel & the energy conversion system need to be re-examined.Fig.1 displays judgement of energy densities of different fuels. It is seen that most progressive Li-ion batteries having one sixteenth energy density of hydrocarbon fuels. Several researchers took the advantages of this high energy density of hydrocarbon fuel to generate power at small scale.

At macro scale conventional power plant were more efficient but at micro scale it difficult to fabricate, still at MIT micro gas turbine, Wrankle Engine at [FernadezPello], HCCI by [Aichlmajr et al], IC Engine by Dahm et al] were developed and tested for power generation. Table-1 shows list of micro engine developed at micro scale with its capacity. As it had moving parts which leads to friction subsequent causes less efficiency as well as manufacturing cost at microscale increases therefore this research leads to the development of non-moving part combustion system for power generation and propulsion system.

Several approaches were used to developed combustor based power generation system at micro scale level. Firstly, Thermoelectric, Thermophotovolatic, Thermoelectromagnetic material to convert heat energy into the electric energy. But at micro and meso level the flame stabilization problem were found. To tackle this problem researcher used excess enthalpy concept with number of techniques likewise change of combustor wall material, use of catalyst, use of porous media and its effect are revisited in this paper.

Table 1. Micro Engine with its capacity

Researcher	Engine Type	Capacity	Fuel
FernadezPello	Wankel Engine	4W	Hydrogen/ Air
Aichlmajr et al	HCCI	10W	n Heptane/ Air
Dahm et al	Micro I C Engine Swing	20W	Butane Fuel

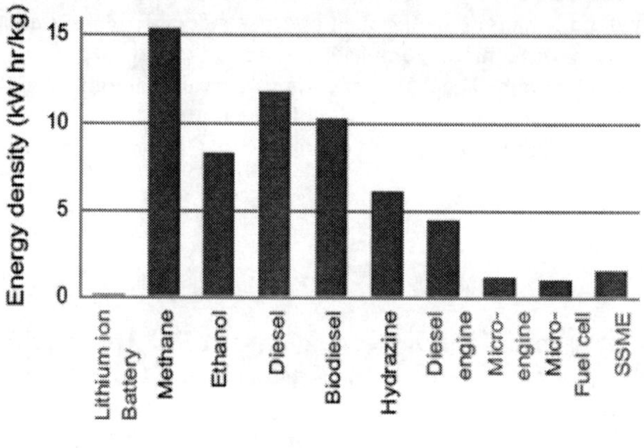

Figure 1. "Comparison of specific energy densities of lithium ion batteries with hydrocarbon and oxygenated hydrocarbon fuels as well as different engines" [1].

1.1 Decision parameters for sizing of micro/meso combustor:

During last decades' development of micro devices like engines, thrusters & reactors uses different reference scale. It leads to confusion regarding micro scale &meso scale size. Table -2 shows three different lengths have been used in previous studies one length are physical dimensioning of combustor. When the size of the combustor is less than 1 mm is referred as microscale combustor whereas size of the combustor is in between 1 mm to 1 cm is referred as meso scale combustor. This definition mostly used in micro engines [2].

Secondly length of flame is used to define size of the combustor if combustor size is smaller than quenching diameter. Thirdly relative length scale is used. In this the size of the combustor are compared with conventional large size combustor for example size of combustor for micro satellite would be (1 to 10 Kg) compared to large satellite of 1000 Kg and above. Micro combustion process depends on chemical kinetics & physical process such as gas phase & surface reaction, radiation, transport, convection, mass diffusion.

Table 2 Sizing of micro- meso scale combustion using different length scales [5]

Sizing based on	Combustion regime	Length scale	Applications

Physiacal Length	Mesoscale Microscale	1-10mm 1-1000μm	MEMS Power Thruster Power generation
Flame quenching diameter	Mesoscale Microscale	Quenching diameter Quenching diameter	Energy conversion
Device Scale	MIcroscale	Smaller than conventional engine size	Micro-satellites Micro-air planes

2. Experimental and Numerical Studies: -

Several experimental numerical studies conducted over meso and micro scale combustor. It was observed that flame instability in pre-mixed, non-pre-mixed, non-equilibrium, catalyst combustion system. But when combustion occurs at small scale surface area to volume ratio increases and thermal characteristic inertia time also reduced which subsequently leads to the heat loss from combustor wall, flameless combustion, and flame instabilities. To reduce this flame instability different techniques were reported, likewise use of excess enthalpy concept by using heat recirculation. With heat recirculation combustor wall material conductivity affects flame stability. These effect of Combustor wall material are revisited in this study. Also this study focuses on use of catalyst, porous media on flame stabilization.

2.1 Effect of use of catalytic material:

The large surface area to volume ratio has concern with higher heat loss tend itself use of catalytic material for low temperature combustion. The use of catalytic allows combustion of fuel at low temperature. (~55°C minimum temperature observed [12] and this also enables low level power production without any waste. Ronney et al [5] demonstrated 3W of thermal energy release with near complete combustion fuel (propane). This low temperature combustion allows use of plastics and polymers combustion that can manufacture in different geometries with rapid prototyping and CNC machining.

Lee & Kin, [12] studied effect of catalytic material on combustion of methane. Platinum were good catalytic material for combustion. The oxygen: methane feed ratio, particle size, support nature, loading of precious metal affects the catalytic combustion.

P D Ronney [6] performed experimental study on Swiss roll burner to evaluate extinction limits of flame with use of catalyst. Effect of catalyst with Re number were evaluated. With use of catalyst combustion of Propane-air at lean stoichiometric ratio were observed. Gas phase combustion occurred in "Flameless mode", with high Re, Stoichiometric combustion didn't require heat recirculation but it occur out of the centre, to centre it catalyst were used. It was concluded by the researcher that with use of catalyst in heat recirculating burner with low Re greatly benefited by selecting proper mixture. Figure -2 shows Extinction limit map for gas phase combustion on Swiss Roll Combustor.
Kim et al [7] in a flat type counter current combustor and Pt type catalysis arise largely same conclusion i.e. flame rich extinction regime increases & flame lean extinction regime lone marginally.

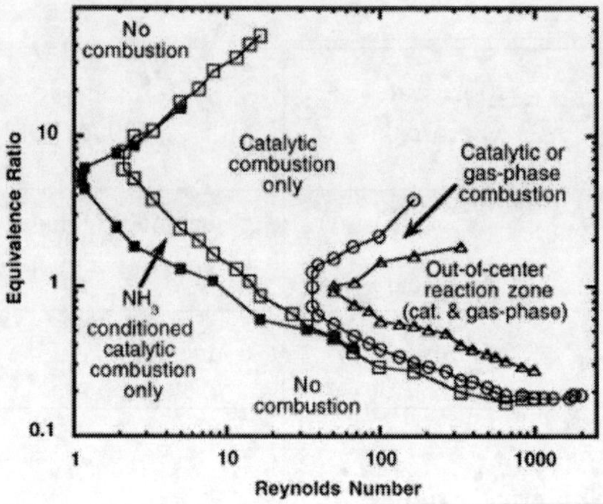

Figure 2. "Extinction limit map for catalytic and gas phase combustion in the Inconel Swiss-Roll" [6].

Teresa A. Wierzbicki et al [11] compared the performance of meso scale heat recirculator combustor with and without Platinum & Rhodium catalysing material. The result showed that addition of catalyst greatly increased stability and extinction limits. It also observed that Propane with Rh had rich stoichiometric ratio, whereas propane with Pt had leaner stoichiometric ratio.
Junwei et al reported experiments results on N-Heptane fuel used in micro combustor. This experiment results shows the effect of inlet reynold number and mass flow rate on flame shape, combustor wall material temperature and outlet temperature. Whereas other studies which [14] evaluates energy efficiency of micro combustor by comparing heat recuperation micro combustor with non-heat recuperation micro combustor. They performed numerical simulation with SIMPLE algorithm. It had been observing that average wall temperature, emitter efficiency notably improved with the incorporation of heat recuperator. Junjie Chen et al performed CFD analysis to observe combustor wall material conductivity, feed composition, and flow rate effect on thermal uniformity during exothermic catalyst reaction. With H_2-air, equivalence ratio of 0.9, 25.8%improvement of total radiation power and 30.6% emitter efficiency were observed.
Yei Chinchao et al [3] performed theoretical, experimental analysis as well as numerical simulation which indicates as the size of tube decreases, which results in shift of flame from low velocity region to high velocity region.

Summary:-Different investigator performance Numerical and experimental evaluation of catalyst with heat recirculation techniques in combustor with varying inlet velocity, mass flow rate, Equivalence ratio on flame stabilization. It had been observed that use of catalyst likewise Platinum, Rhodium had positive effect over flame stabilization as compared to without catalyst combustion. Catalytic extended the rich extinction intensely and lean extinction only slightly. It is applicable only when gas phase combustion cannot be achieved.

2.2 Effect of thermal conductivity of combustor wall material:
At small / meso scale level combustion affects because of large surface to volume ratio, the researchers examine the effect of combustor wall material with different thermal conductivity on flame stabilization. It was observed that material with high thermal conductivity with heat recirculation geometry supports to flame stabilization V Vijayan performed experimental and numerical simulation on heat recirculating combustor, which shows thermal conductivity of wall material plays significant roll to stabilize the flame. With low thermal conductivity material, it was found that amount of heat recirculation increased while heat loss decreases. Also investigate the lowest thermal conductivity material.

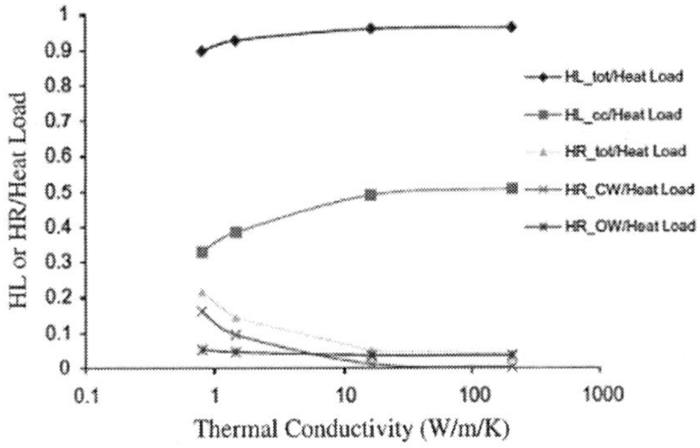

Figure 3. "Effect of thermal conductivity on heat transfer rates for a heat load of 29W." [13]

Aiwu Fan et.al. [10] Perform numerical investigation on Swiss roll combustor made of Steel, Silicon Carbide, and Quartz with bluff body. Heat loss play important role in flame blow off limits for this newly designed combustor with bluff body. This study summarizes the effect of heat loss, heat recirculation and flow recirculation zone on SS, SiC, Quartz combustor, which causes largest blow off limit in SS, Smallest blow off limit in SiC, and Medium blow off limit in Quartz.

Junijie et al [15] performed CFD and experimental analysis which shows with low thermal conductivity wall material shows flame stability as well as for the heat recirculation inlet, outlet stream temperature equal to room temperature and exhibits high thermal and combustor efficiency.

Upendra Taywade et al [16] find the effect of mixture velocity, equivalence ratio and heat recirculation on the wall temperature. It had been observed that for high velocity >4m/s wall temperature is lower. It also reported that for rich equivalence ratio high wall temperature obtained.

Summary: - Several researchers conducted experimental and numerical study on material of combustor wall like SS, Sic, and Quartz etc. to found its effect on flame stabilization with heat recirculation techniques. It had been observed from above discussion that material with thermal conductivity helps to stabilize the flame inside combustor. Fig.-3 notable indicates Effect of combustor wall material thermal conductivity on heat transfer rates for a heat load of 29W.

2.3 Flame Dynamics and instability:

Junjie Chen et al [15] compare the flame stability of single pass combustor with respect to heat loss coefficient and total power loss in between CFD and single channel combustor.

It had been observed that heat recirculation affect blows out limits but had minimal effect over extinction.

Upendra W taywade [16] et al made numerical and experimental investigation three step combustor to find effect of heat recirculation on flame stability. It was observed by the researcher that heat recirculation is function of cup dimension, material of cup and flow velocity and equivalence ratio for the cylindrical shape heat recirculating combustor. Thermal outputs of 2.2W for 0.5 equivalence ratio were observed with stable flame. This study was also reported the variation in position of flame with respect to mixture inlet flow velocity.

V Vijayan et al [17] performed experimentation, acoustic emission incorporation with high speed flame imaging diagnostic. It had been observed that stable self-sustained combustion at high mean velocity in meso scale combustor. The reactant preheat temperature of 700K-1000K range were observed. Tradition FREI was absent. Uttam rana et al [18] performed thermodynamic model to evaluate exergy transfer. It were reported that flame stabilization by heat recirculation were justified from energy quality preservation perspective.

IOP Conf. Series: Materials Science and Engineering **810** (2020) 012081 doi:10.1088/1757-899X/810/1/012081

Chien Hua et al [19] performed numerical and experimental study on Swiss roll burner. It were reported that thermal management of 3D swiss roll combustor were satisfied double spiral burner. Effect of third dimension heat loss on extinction compared numerically as well as experimentally.
Aiwu et al [10] carried out investigation for flame blow out limits on swiss roll heat recirculating combustor with bluff body. It were concluded with numerical simulation and experimentation that bluff body helps to stabilize the flame and the position of bluff body affects inlet and outlet velocity.
Summary: -The extinction of flame greatly depending on the thermal characteristic of combustion geometry. Increase in heat recirculation channel length of combustor allows combustion of lean mixture. As researcher tried to vary the heat recirculation geometry which causes pre-heating of reactant leads to significant flame dynamics under the correct conditions. It notably indicates that combustor of high thermal conductivity material had positive response as compared to material with low thermal conductivity.

2.4 Porous Media Material Effect: -
Daoguan Wing et al [20] mesotubes of 4, 5,6mm in diameter Y shaped combustor were developed. The effects of adding fibrous porous media on flame stabilization, ignition temperature were systematically investigated. Jonwei Li et al [21] performed experimentation on n-heptane air combustion with porous media in meso scale burner.

3. Micro-power Generation

It were easy to convert heat energy into electric energy or kinetic energy at macro scale but at micro scale level it's challenging task. Researcher developed combustor at meso /micro level to generate power and thrust force. This section discussion about the power generation with combustor.
Some of the researcher tried miniaturization of macro scale power generation system like wise Fernadez Pello developed "Wankle" engine, Aichlmayr et al. designed HCCI engine with a power of 10W. But because of moving parts its fabrication are difficult and costly as well as friction between moving parts leads to loss of efficiency. Therefore, most researchers focused on micro level power generation with solid conversion system instead of moving parts likewise conventional macro level power plant. Most of the researcher preferred thermoelectric and thermophotovolatic material based conversion system over the other.

3.1 Electricity Generation
The conversion of heat energy into electrical energy has two scenarios. Firstly, conversion of shaft power into electric energy through dynamo. Secondly conversion of electrical energy through Seeback & Peltier Effect as well as thermophotovolatic conversion (TPV) Thermoelectric system consist of major four components:
i) Combustor
ii) Thermal Spreader
iii) Thermoelectric Material
iv) Heat sink
The thermoelectric material are of P type and N type semiconductor. As because of heat electrons tends to flow from n type of semiconductor material toward cooling region, whereas holes from p type semiconductor flows toward power load. It is reported that power output of the TE material are proportional to square of temperature gradient, but it has to operate at 300 ^{0}c. This limiting the application of combustor with catalyst which operates comparatively at lower temperature. Fig-4 describes working of conversion of heat energy into electrical energy by using thermoelectric materials.

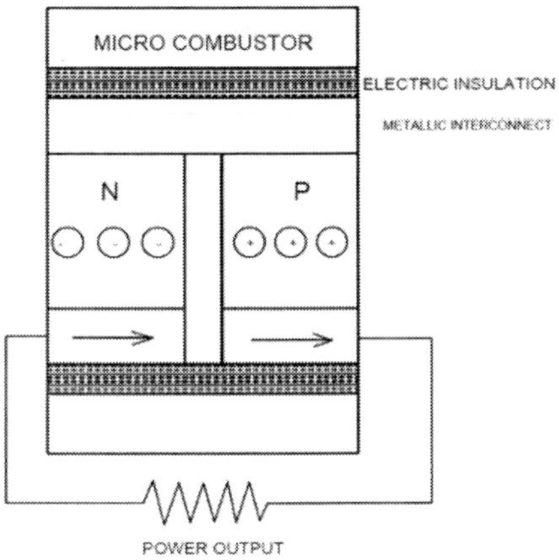

Figure 4. "Scematic of typical Micro Thermoelctric Power conversion system." [13]

TanujSingh[24] et al developed noval combustor and optimise it to obtained stable enclosed premixed flame. This study found by using environmental control chamber a relation between ambient temperature and thermoelectric power generation. Jianlong et al [25] reported length effect of a plate flame holder on blow off limit of flame in preheated channel of combustor. It had been observed that firstly blow limit decreases with increase in plate length. It were found that largest blow out limit at Lb=1.0mm. Another Solid power conversion system i.e. use of thermophotovolatic material to absorb thermal radiation from the heat source and convert into electrical energy were also investigated by different researchers. Fig-5 shows working of TPV based power conversion system.

Roman Fursenko, et al [26] performed numerical as well as experimental in multichannel system for flame penetration and its propogation. With varying mixture flow rate, equivalence ratio, and channel size different combustion regime were observed. Y.-C. Chao, [27] reported results of experimental and numerical analysis of reverse tube TPV combustor with a porous media. It had been observed that wall with emitter material with swirling combustor improve flame structure and uniformity in illumination by the flame.

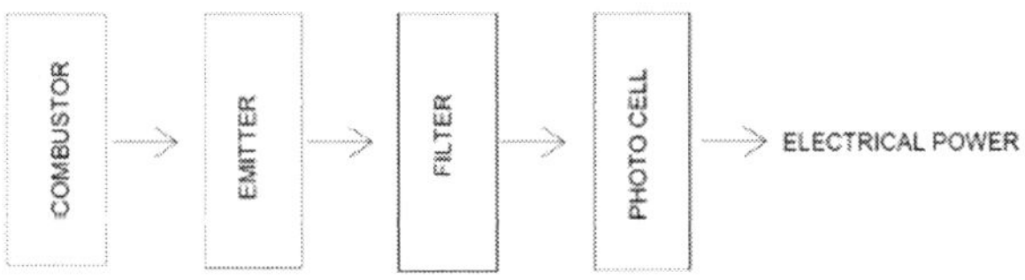

Figure 5. "The schematic layout of the thermophotovoltaic (TPV) power system" [24].

4. Conclusion

Effect of combustor wall material interpreted by most researcher that wall material with high thermal conductivity had better effect on flame stabilization.

Similarly use of catalyst and porous media also helps to stabilize the flame inside the meso scale combustor. Thermal conversion of electric energy through shaft work, thermoelectric or thermophotovolatic is still in its infancy if modest efficiency is reported.

IOP Conf. Series: Materials Science and Engineering **810** (2020) 012081 doi:10.1088/1757-899X/810/1/012081

5. References

[1] A. C. Fernandez-Pello, Micropower generation using combustion: Issues and approaches, Proceedings of the Combustion Institute 29 (1) (2002) 883–899.

[2] L. C. Chia, B. Feng, The development of a micropower (micro-thermophotovoltaic) device, Journal of Power Sources 165 (1) (2007) 455–480.

[3] Yei Chinchao et al, Operational characteristic of catalytic combustion in a platinum microtube, Combustion sci. & Tech., 2004

[4] Y. Ju, K. Maruta, Microscale combustion: Technology development and fundamental research, Progress in Energy and Combustion Science 37 (6) (2011) 669–715.

[5] K. Maruta, Micro and mesoscale combustion, Proceedings of the Combustion Institute 33 (1) (2011) 125–150.

[6] P D Ronney, Gas-phase and catalytic combustion in heat-recirculating burners, Proceedings of the Combustion Institute 30 (2005) 2463–2472

[7] A. Catalano, Thermophotovoltaics: A New Paradigm for Power Generation National Renewable Energy Laboratory, 1996, pp. 495–499.

[8] S. Raimondeau, D. Norton, D. Vlachos, R. Masel, Modeling of high-temperature microburners, Proceedings of the Combustion Institute 29 (1) (2002) 901–907.

[9] Ju Y, Maruta K, Microscale combustion: Technology development and fundamental research, Progress in Energy and Combustion Science (2011), doi: 10.1016/j.pecs.2011.03.001.

[10] Aiwu Fan, He Zhang, Jianlong Wan Numerical investigation on flame blow-off limit of a novel microscale Swiss-roll combustor with a bluff-body, Energy 123 (2017) 252e259

[11] Teresa A. Wierzbicki, Ivan C. Lee, Ashwani K. Gupta, Combustion of propane with Pt and Rh catalysts in a meso-scale heatrecirculating combustor, Applied Energy 130 (2014) 350–356

[12] Joo.Lee DavidL.Trimm, Catalytic combustion of methane, Volume 42, Issues 2–3, April 1995, Pages 339-359

[13] V Vijayan, Combustion and Heat Transfer in Meso-Scale Heat Recirculating Combustors, 2010

[14] Junwei Experimental study on flame stability and thermal performance of an N Heptane fuelled microscale combustor, Combustion science & Technology

[15] Junijie Chen, Flame stability of Propane air premixed combustion in heat recirculating Micro Combustors, Frontiers in Heat & Mass transfer.

[16] Upendra W Taywade Thermal performance of micro combustor with heat recirculation, Fuel Processing Technology, 109(2013), 179-188

[17] V Vijayan, Combustion and heat transfer at meso scale with thermal recuperation, Applied Energy 87 (2010)2628-2639

[18] Uttam Rana, Thermodynamics of premixed combustion in a heat recirculating micro combustor, Energy 68 (2014)510-518.

[19] Chien Hua Chen et al, Numerical and Experimental Study in Swiss Roll Heat-Recirculating Burner.

[20] Daoguan Wing et al, Experimental investigation on non-premixed methane/air combustion in Y-shaped meso-scale combustors with/without fibrous porous media, Energy Conversion and Management 138 (2017) 22–29

[21] Junwei Li et al, Experimental study of n-heptane/air combustion in meso-scale burners with porous media, Experimental Thermal and Fluid Science 52 (2014) 47–58

[22] N. S. Kaisare, D. G. Vlachos, A review on microcombustion: Fundamentals, devices and applications, Progress in Energy and Combustion Science 38 (3) (2012) 321–359.

[23] Y. B. Zeldovich, G. Barenblatt, V. Librovich, G. Makhviladze, the Mathematical Theory of Combustion and Explosions. 1985, Consultants Bureau, New York, 1985.

[24] Tanuj Singh, Development and investigation of a non-catalytic self-aspirating meso-scale premixed burner integrated thermoelectric power generator, Energy Conversion and Management 117 (2016) 431–441

[25] Jianlong Wan, Effect of the length of a plate flame holder on flame blowout limit in a micro-combustor with preheating channels, Combustion and Flame 170 (2016) 53–62

[26] Roman Fursenko, et al, Experimental and numerical study of premixed flame penetration and propagation in multichannel system, ISSN: 0010-2202 (Print) 1563-521X
[27] Y.-C. Chao, Numerical and Experimental Studies of Mixing Enhancement and Flame Stabilization in a Meso-Scale TPV Combustor with a Porous-Medium Injector and a Heat-Regeneration Reverse Tube, DOI: 10.1080/01457632.2013.810968

IOP Conf. Series: Materials Science and Engineering **810** (2020) 012082 doi:10.1088/1757-899X/810/1/012082

Stress Concentration Factor for Jute Fibre Reinforced PLA Composite Plate with Circular Cut-out and its Effect on Coconut Shell Powder

Ramachandran M [a,b] and B. A. Modi[a,*]
[a]Institute of Technology, Nirma University, Ahmedabad, Gujarat 382481, India
[b]MPSTME, SVKM's NMIMS, Shirpur, Dhule, Maharashtra 425405, India
bharat.modi@nirmauni.ac.in

Abstract. In strictly engineering terms, the design of machine elements is focused reasonably on the regions where stress concentration occurs. Plates having holes are a practical requirement in various aeronautical, marine, naval, mechanical and civil arrangements. In many aerospace and naval machine components these kinds of holes are provided to reduce the overall weight of the assembly and to lay electrical and fuel lines. These two parameters were then used in the modified ANSYS, virtual crack closure technique to obtain the stress intensity factors. Very good Pressure vessel. Agreement between the finite element stresses and the theoretical stresses is seen. The intend of this work was to analyse stress concentration factor of Jute fibre reinforced PLA composite plate with circular cut-out and compare it with addition of coconut shell powder. The Ansys 2019 R2 Finite Element Analysis software was used to get the stress analysis of the composite plate. The shell 281 elements were used in meshing the models and this was due to their regular agreement around the circular cut-out that enabled the nodal displacements to be obtained at a constant distance from the circular cut-out. From the Ansys analysis, it is clearly evidenced that addition of CSP decreases the stress concentration factor

1. Introduction

In strictly engineering terms, the design of machine elements is focused reasonably on the regions where stress concentration occurs. Plates having holes are a practical requirement in various aeronautical, marine, naval, mechanical and civil arrangements. In many aerospace and naval machine components these kinds of holes are provided to reduce the overall weight of the assembly and to lay electrical and fuel lines. Many a times such cut-outs are provided for fulfilling some maintenance function like entry and examination of internal sections in aircrafts and in bridges and aero structures having plated structures such as box girders, bridge deck. These cut-outs reduce the strength of the structure and make the structure prone to failure. Simha and Mahapatra analytically studied the stress concentration caused by irregular holes [1]. Tafreshi used FEM (Finite Element Method) and BEM (Boundary Element Method) for stress analysis of thick flat plates having oblique cut-outs under uniaxial tension and out-of-plane [2]. Any disruption in structure causes change in stress flow patterns and it reduces the strength of the structure. Stress concentration always occurs near the discontinuity in structure. It is seen that the analytical equations for calculating SCF around elliptical holes can be used to calculate SCF around circular holes with sufficient accuracy [3]. For composite Laminates, SCF depends on material properties. It increases with increase in Ey/Ex ratio and decreases with increase in µxy.SCF in Y direction is are much more as compared to the SCF for normal stresses in X direction and shear stress under biaxial load [4].The stress concentration factor is the ratio of the greatest stress at the crack tip to the corresponding nominal stress that existed at the location of the crack tip before the formation of the crack. That is;

$$K_f = \frac{\sigma_{max}}{\sigma_o}$$

The severity of the stress concentration depends on the geometry of the crack. Designers should therefore always try to reduce stress concentrations as much as possible in order to avoid fatigue problems.

2. Design of rectangular plate with and without circular cut-out

The design of composite plate was made with the help of solid works software with 2019 version. The dimensions of the plates are 200mm x 100mm x 10mm (Lx B x T) with volume 1.9688e+005 mm^3. For circular cut-out, we taken 10mm diameter. Figure 1 shows the design of rectangular plate without circular cut-out and Figure 2 shows the design of rectangular plate with circular cut-out

Figure 1: Design of rectangular plate without circular cut-out

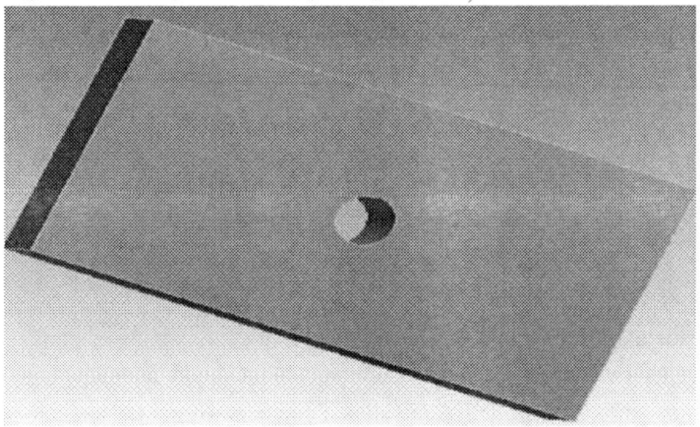

Figure 2: Design of rectangular plate with circular cut-out

The mechanical properties of jute reinforced poly lactic acid polymeric composite without coconut shell powder addition which is considered for analysis are Yield Strength with value 45 Mpa, Tensile Strength with value 56 Mpa, Young's Modulus with value 1800 Mpa, Poisson's Ratio with value 0.27, Bulk Modulus with value 1304.3 Mpa and Shear Modulus with value 708.66 Mpa. The mechanical properties of jute reinforced poly lactic acid polymeric composite with coconut shell powder addition which is considered for analysis are Yield Strength with value 52 Mpa, Tensile Strength with value 62 Mpa, Young's Modulus with value 2450 Mpa, Poisson's Ratio with value 0.42, Bulk Modulus with value 5104.2 Mpa and Shear Modulus with value 862.68 Mpa. It was then meshed using shell 281 elements which is present in Ansys library. These elements were chosen because they could be arranged in a regular manner around the circular cut out and also because they allowed for sufficient mesh refinement around the circular cut out. During the meshing, the model was first freely meshed. The meshing of the whole model was then refined. Finally mesh refinement was done at the circular cut out. The next stage of the analysis involved the application of boundary conditions. The displacement boundary conditions were applied to prevent the cylinder from rotating about its own axis. The precondition conjugate solver was chosen to carry out the final analysis since it is fast and accurate. Figure 5 shows the stress distribution along the crack tip. It is clear from this figure that the stress along the crack tip is higher than the stress at any other point on the cylinder. After the solution, the deformed and the unreformed model were observed. The maximum stress was also checked so as to verify that it was

IOP Conf. Series: Materials Science and Engineering **810** (2020) 012082 doi:10.1088/1757-899X/810/1/012082

below the model yield stress. The variation of the radial, hoop and axial stresses was also examined along the crack, i.e. the variation of the stresses from the point of crack initiation on the inner surface of the cylinder up to the crack tip.

3. Numerical analysis of Composite plate for stress intensity factor

For doing the numerical Analysis of stress intensity factor, meshing is most important part. The meshing of rectangular composite plate with and without circular cut-out is done with 9481 nodes and with 1622 elements. Figure 3 shows the mesh model of rectangular plate without circular cut-out and Figure 4 shows the mesh model of rectangular plate circular cut-out. The Ansys 2019 R2 Finite Element Analysis software was used in this study.

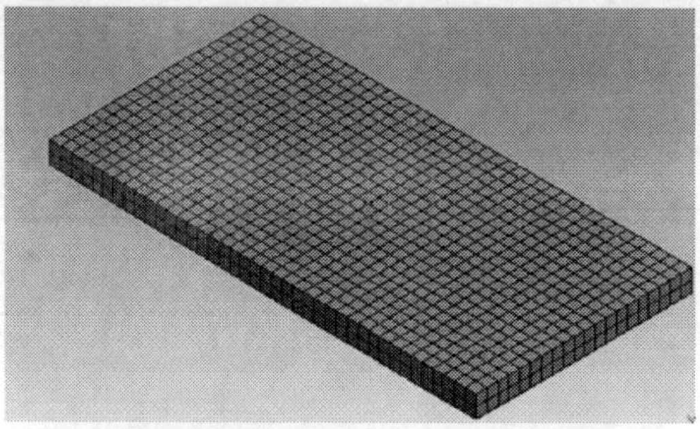

Figure 3: Mesh model of rectangular plate without circular cutout

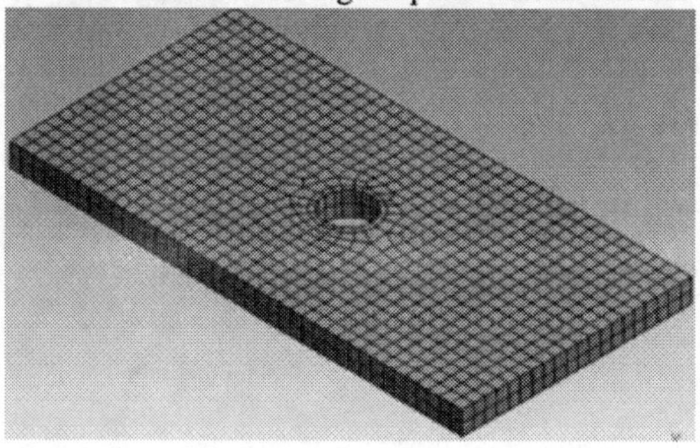

Figure 4: Mesh model of rectangular plate without circular cutout

In this analysis hex dominant mesh type is used and adaptive meshing is used for analysis. Each element is around 3mm size after meshing. For analysis 500N (ramped) magnitude of force is used.

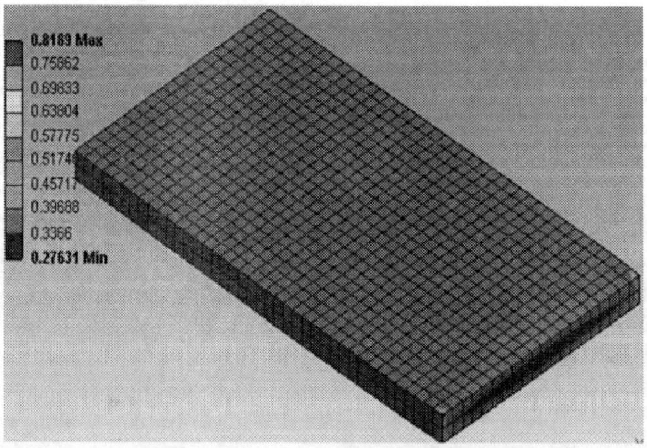

Figure 5: Equivalent stress for PLA Jute Composite plate without central cut-out

IOP Conf. Series: Materials Science and Engineering **810** (2020) 012082 doi:10.1088/1757-899X/810/1/012082

From the Figure 5 which shows equivalent stress for PLA Jute Composite plate without central cut-out, the maximum equivalent stress for PLA Jute Composite plate without central cut-out is 0.8189 Mpa and the minimum equivalent stress for PLA Jute Composite plate without central cut-out is 0.27631 Mpa.

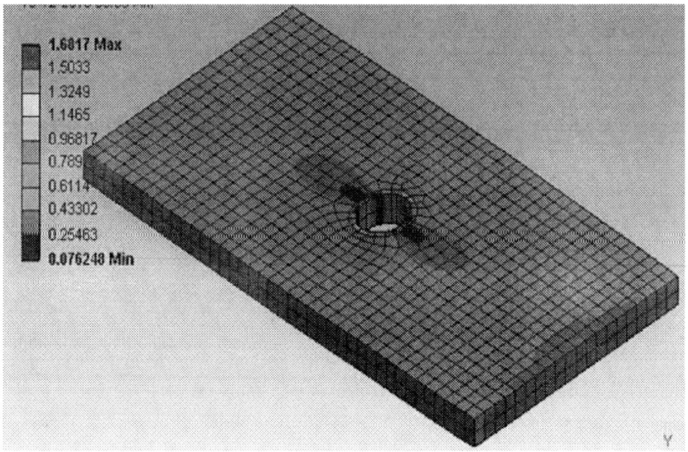

Figure 6: Maximum Equivalent stress PLA Jute Composite plate with central cut-out

From the Figure 6 which shows Equivalent stress PLA Jute Composite plate with central cut-out, the maximum equivalent stress PLA Jute Composite plate with central cut-out is 1.6817 Mpa and the minimum equivalent stress PLA Jute Composite plate with central cut-out is 0.876248 Mpa.

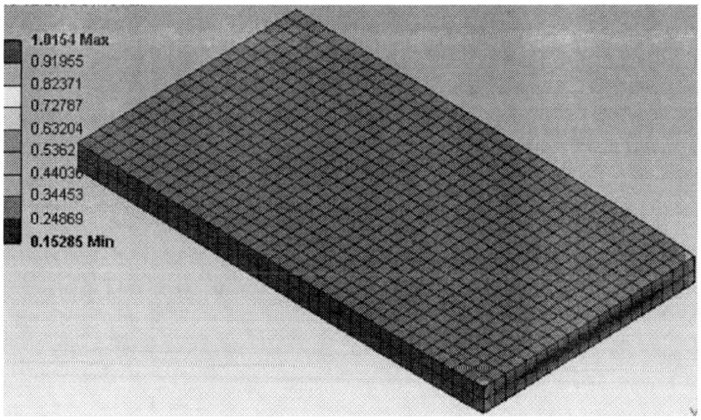

Figure 7: Maximum Equivalent stress PLA Jute with coconut shell powder Composite plate without central cut-out

From the Figure 7 which shows Equivalent stress PLA Jute with coconut shell powder Composite plate without central cut-out, the maximum stress PLA Jute with coconut shell powder Composite plate without central cut-out is 1.0154 Mpa and the minimum equivalent stress PLA Jute with coconut shell powder Composite plate without central cut-out is 0.15285 Mpa.

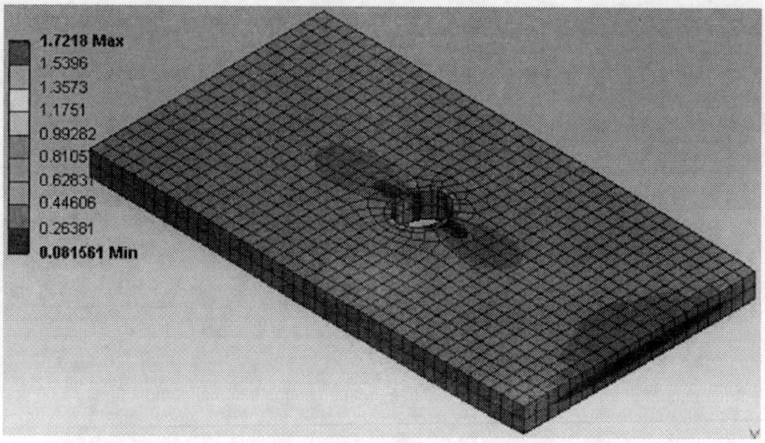

Figure 8: Maximum Equivalent stress PLA Jute with coconut shell powder Composite plate with central cut-out

From the Figure 8 which shows Equivalent stress PLA Jute with coconut shell powder Composite plate without central cut-out, the maximum stress PLA Jute with coconut shell powder Composite plate without central cut-out is 1.7218 Mpa and the minimum equivalent stress PLA Jute with coconut shell powder Composite plate without central cut-out is 0.081561 Mpa.

Table 1: Stress concentration factor for composite plate with and without circular cut-out

Materials	**Maximum Equivalent stress without circular cut-out (MPa)**	**Maximum Equivalent stress with circular cut-out(MPa)**	**Stress concentration factor**
PLA + Jute	.8189	1.6817	2.05
PLA+ Jute+ CSP	1.0154	1.7218	1.69

Table 1 shows the Stress concentration factor for composite plate with and without circular cut-out. From the Ansys analysis, it is clearly evidenced that addition of CSP decreases the stress concentration factor

4. Conclusions

The intend of this work was to analyse stress concentration factor of Jute fibre reinforced PLA composite plate with circular cut-out and compare it with addition of coconut shell powder. The Ansys 2019 R2 Finite Element Analysis software was used to get the stress analysis of the composite plate. The shell 281 elements were used in meshing the models and this was due to their regular agreement around the circular cut-out that enabled the nodal displacements to be obtained at a constant distance from the circular cut-out. From the Ansys analysis, it is clearly evidenced that addition of CSP decreases the stress concentration factor

References

[1]. K. R. Y. Simha, S. S. Mohapatra. 1998. Stress concentration around irregular holes using complex variable. Sadhana. 23, Part 4: 393-412.

[2]. Tafreshi. 1995. Numerical analysis of stresses at oblique holes in plates subjected to tension and bending. Journal of strain analysis. 30(4): 317-323.

[3]. Thirumump, Mohit, KanakKalita, M. Ramachandran, and Ranjan Ghadai. "A numerical study of SCF convergence using ANSYS." ARPN Journal of Engineering and Applied Sciences 10, no. 5 (2015).

[4]. Kumar, Aditya, Akshay Agrawal, RanjanGhadai, and Kanak Kalita. "Analysis of stress concentration in orthotropic laminates." Procedia Technology 23 (2016): 156-162.

[5]. Das, Sourav, Vishvendra Bartaria, and Prashant Pandey. "Analysis of Knuckle Joint of 30C8 Steel for Automobile Application." *International Journal of Engineering* 3, no. 1 (2014).

[6]. T. Sunder selwyn, R Kesavan, M. Ramachandran, FMECA Analysis of Wind Turbine Using Severity and Occurrence at High Uncertain Wind in India, International Journal of Applied Engineering Research. ISSN 0973-4562 Volume 10, Number 11 (2015) pp. 10250-10253.

[7]. Beisel, Winfried, and Ingo Oberste-Dommes. "Precision embossing Alternative method of forming flat parts." *Journal of materials processing technology* 71, no. 1 (1997): 18-24.

[8]. Vishal Fegade, Gajanan Jadhav, M. Ramachandran, Design, Modelling and Analysis of Tilted Human Powered Vehicle, IOP Conf. Series: Materials Science and Engineering 377 (2018) 012215.

[9]. Yossifon, S., and R. Shivpuri. "Optimization of a double knuckle linkage drive with contrast mechanical advantage for mechanical presses." *International Journal of Machine Tools and Manufacture* 33, no. 2 (1993): 193-208.

[10]. Singh, Pradeep K., Satish C. Jain, and Pramod K. Jain. "Advanced optimal tolerance design of mechanical assemblies with interrelated dimension chains and process precision limits." *Computers in Industry* 56, no. 2 (2005): 179-194.

[11]. Yossifon, S., and R. Shivpuri. "Analysis and comparison of selected rotary linkage drives for mechanical presses." *International Journal of Machine Tools and Manufacture* 33, no. 2 (1993): 175-192.

[12]. Yan, Hong-Sen, and Wei-Ren Chen. "A variable input speed approach for improving the output motion characteristics of Watt-type presses." *International Journal of Machine Tools and Manufacture* 40, no. 5 (2000): 675-690.

[13]. Ashish Mogra, M. Ramachandran and Hemant Kumar Mehta, Experimental Investigation on Droplet Formation through a Minichannel Y- Junction, IOP Conf. Series: Materials Science and Engineering 377 (2018) 012189.

[14]. Kanak Kalita, M. Ramachandran, Pramod Raichurkar, S. Haldar, Free Vibration Analysis Of Laminated Composites By A Nine Node Isoparametric Plate Bending Element, Advanced Composites Letters, 25(4), 2016.

[15]. Pantazopoulos, G., A. Sampani, and E. Tsagaridis. "Torsional failure of a knuckle joint of a universal steel coupling system during operation–A case study." *Engineering Failure Analysis*14, no. 1 (2007): 73-84.

[16]. Kumar, Nipun, Gian Bhushan, and Pankaj Chandna. "Analysis of Knuckle Joint of Various Materials using CAE Tools." *International Journal of Engineering, Management and Applied Sciences* 5 (2017): 45-52.

[17]. Shajit Yadav, M.Ramachandran, Analysis of Optimized Wind Turbine Failure Using Computational Fluid Dynamics, Materials Today: Proceedings, 4(2), 2017:1788–1793.

[18]. Kim, H. W., D. S. Bae, and K. K. Choi. "Configuration design sensitivity analysis of dynamics for constrained mechanical systems." *Computer methods in applied mechanics and engineering* 190, no. 40-41 (2001): 5271-5282.

[19]. M. Ramachandran, Benildus, Failure Analysis of Turbine Blade Using Computational Fluid Dynamics, International Journal of Applied Engineering Research. ISSN 0973-4562 Volume 10, Number 11 (2015) pp. 10230-10233.

[20]. Bowonder, B. "Competitive and technology management strategy: a case study of TELCO." *International Journal of Technology Management* 15, no. 6-7 (1998): 646-680.

Topological Data Analysis Based Feature Selection for Predicting Fatigue Strength of Steel Using Machine Learning

M A Joshi[1] and D L Joshi[2]
[1]Department of Applied Science and Humanities, MPSTME NMIMS, Shirpur, India
[2]Department of Computer Engineering, MPSTME NMIMS, Shirpur, India
mlnjsh@gmail.com

Abstract. Routine Machine learning methodologies are based on Statistics and traditional data analysis methods, which may account for loss of some important information from the data during analytical process. So we need a new set of tools and algorithms that can effectively make most of such information. This paper investigates the implementation of new and growing data analytics technology known as topological data analysis (TDA). We use Mapper algorithm from TDA on data taken from public domain database of National Institute for Material Science (NIMS), for selecting key process parameters or features that impacts fatigue strength of steel. Mapper algorithm outputs a network graph called topological network that captures the important connectivity information and complex interactions between the clusters present in the data where traditional methods fail to capture such interactions. We analyse the shape of topological network obtained from mapper for selecting important features that impact the fatigue strength of steel. We then use XG-Boost prediction model with tuned hyper parameters using various tuning algorithms, to evaluate the impact of the selected features. This approach of data analysis and modelling helps to monitor and control the process in a more cost-effective manner. The experimental results have successfully proved the efficacy of TDA and machine learning (ML) and hence developed the predictive model for the same.
Keywords: TDA, Mapper, Feature Selection, Material Informatics, Machine Learning, XG-Boost, Regression, Steel, Fatigue, Processing-Property Linkages

1. Introduction

Fatigue strength plays major role for design and failure analysis of mechanical components, and serves as a basic data required for the same. Reports shows that fatigue is responsible for over 90% of all mechanical failures of structural components [1]. Enfeebling outcomes of fatigue failures with high cost and time for fatigue testing, automating prediction of fatigue life is of utmost importance not only for field of materials science but also for mechanical engineering i.e. it is important to have a prediction model that predicts fatigue strength of steels. In this work we use large number of heat treatment process and composition parameters for prediction. Most of the conventional methods failed to capture the complex interaction between the various input variables, so we give an attempt to use TDA which is proven to be powerful way to capture connectivity information between the input parameters [2, 3]. This study serves as an introductory guide for potential researchers in this field. The scope of this paper includes application of ML and TDA methods together to predict fatigue strength of steels using composition and processing parameters. In TDA we use mapper algorithm that gives network graph which is a powerful and compressive representation of a high dimensional complex dataset [4, 5].

Plotting the network provides a direct and insightful visualization of the underlying 'data shape', which can help to explore hidden patterns of the data. Nodes in the network correspond to groups of similar samples that have been clustered together. An edge between two nodes means that there are shared/common samples between the two nodes. Both large scale global structure and local group of nodes can be revealed in the network. Furthermore, mapping a specified target variable onto the network, using a color scheme, can show how the target variable change along with the network. Therefore, plotting a TDA network and coloring the network with interested target variables is the first step to understand the data. Once a TDA network has been constructed, network-based enrichment analysis can be performed for target variables, or properties of the studied data samples. We use the spatial analysis of functional enrichment (SAFE) for network enrichment analysis. This algorithm is originally developed as a systematic method for annotating biological networks and examining their functional associations. [6]. Once SAFE scores are calculated for target variable, we selected and rank the features that contribute for prediction of fatigue strength, on newly selected variable we apply XG-Boost which is modification of Gradient Boosting algorithm. The details of Gradient Boosting algorithm can be found in [7]. We use XG-Boost with tuned hyper parameters using various algorithms like Greed Search, Random Search, Genetic algorithm, differential Evolution. The details of XG-Boost can be found in [8].We use RMSE, MAE and R-square to evaluate the performance of our model on test data.

2. Data And Methods

2.1 Data

Data is the most important part any research, correct and accurate data endeavour fruitful results. In this research, Fatigue dataset for steel is used for predicting fatigue strength. It is accessed from National Institute of Material Science (NIMS) MatNavi [9]. The following table shows the data in detail.

Contents of data	Input parameters
Chemical composition	- %C, %Si, %Mn, %P, %S, %Ni, %Cr, %Cu, %Mo (all in wt. %)
Upstream processing details	- ingot size, reduction ratio, non-metallic inclusions
Heat treatment conditions	- temperature, time and other process conditions for normalizing, through-hardening, carburizing-quenching and tempering processes
Mechanical properties	- YS, UTS, %EL, %RA, hardness, Charpy impact value (J/cm2), fatigue strength.
The 437 data instances include 371 carbon and low alloy steels, 48 carburizing steels, and 18 spring steels. This data pertains to various heats of each grade of steel and different processing conditions.	

Table 1: Dataset Description

There are 25 input features and 1 target feature (Fatigue Strength) which pertain to rotating bending fatigue tests at room temperature conditions,. Total observations are 437. In this study fatigue strength of steel is to be predicted using given input parameters.

2.2 Methods

Figure 1 below illustrate the overall approach. The first step is to pre-process the raw data for consistency using domain knowledge [1]. We start with data and pre-processed data obtained from [9, 10]. The pre-processed data is then input to Mapper algorithm which outputs a network called topological network. Based on shape of network and colouring of network with respect to target variable Fatigue strength SAFE scores are calculated for each node, based on SAFE score the features are selected as they are ranked. The new set of features are then selected for final study, we apply XG-Boost algorithm for prediction, finally effectiveness of model is tested on test set using performance measures like RMSE, MAE and R^2. Details of each of the 8 stages is represented below.

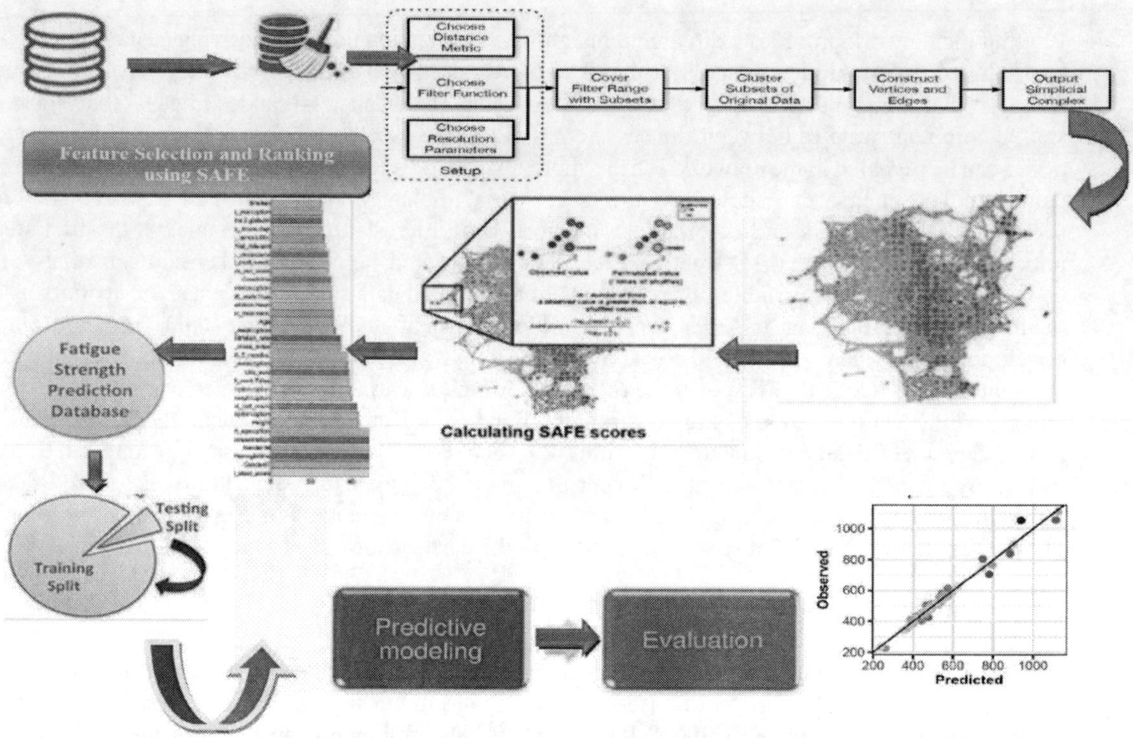

Figure 1. Pipe line for analysis and modelling using TDA and Machine Learning

3. TDA and MAPPER

TDA is branch of applied mathematics. TDA uses techniques from algebraic topology. Topology is very old branch of mathematics which studies shapes. TDA revolves around the principle that "Data has shape and shape has meaning". Natural data offers hard challenges. Extracting information from noisy, incomplete and high-dimensional datasets is generally challenging. To analyse such data, TDA provides magical framework in a manner that is it does not depend of choice of particular metric and it also provides dimensionality reduction and it is immune to noise.

In Mapper, we construct a network graph or in general (Simplicial complex which is generalized version of graph) from data in a way that captures topological features of the data. Mapper is visual representation of data in the form of cluster like that estimates various complex connections between the input parameters and those network clusters can colour according to various target variables, in our case target variable is Fatigue strength of steel. Such visual exploration of the data that can often reveal new insights of the data, it is an unsupervised method. Other methods cannot reveal such new insights. Informally, the Mapper algorithm works by performing a local clustering based on given clustering method guided by a projection/filter function based on given filter function. The steps are as follows:

1. When we get a real world dataset, first we need to choose a distance metric to describe the distance between each pair of points. For example, a three-dimensional Euclidean distance metric. It is not necessary to use the Euclidean metric Predefined metric can also be used. The defined metric will be used by a filter function for data projection (or dimension reduction).
2. Next, *Mapper* projects the original data cloud to a low dimensional space, by using a filter function, which can be *PCA, MDS, t-SNE*, or any other dimension reduction method. In our case, the filter function is PCA with two components, in which the data points are coloured by the filter function. Different filters can also be combined to form *lens*, or a low dimensional space. Usually, *Mapper* uses a two-dimensional space of projected data.

3. After filtering, the projected dataset is binned into groups (*hypercube*) of data points, by a cover, which is a collection of intervals along the filters. A cover can be specified via two resolution parameters (resolution: number of intervals, and overlap: percentage of overlap between adjacent intervals).
4. The final step of *Mapper* is clustering and network construction. Clustering is performed on each bin/hypercube of data points to identify clusters of data points in the original data space, rather than in the projected data space. Network construction is based on the clustering results: each cluster forms a node in the network, and edges are made between two nodes if they have common samples shared by their corresponding clusters. Therefore, the resulted TDA network, as shown in figure 2, is a compressive network representation of the underlying 'data shape', which is an abstract shape of a circle in this case.

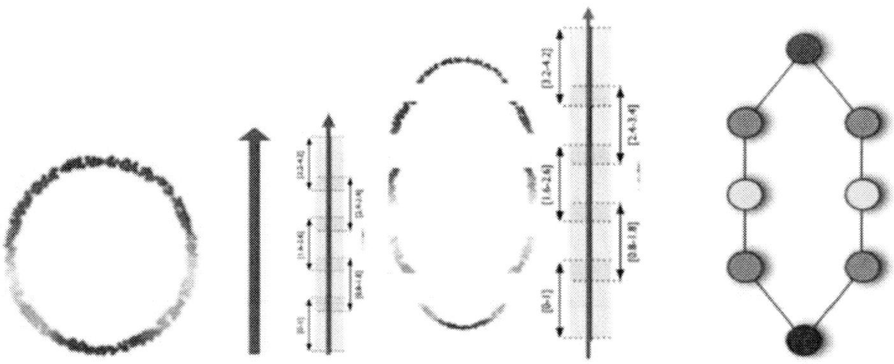

(a) (b) (c) (d)

Figure 2: Mapper Implement of circle (a) Circle (b) Each data point on circle projected on y-axis (Lenses/Filter) (c) Filter is divided into Bins or number of overlapping intervals and then coloured according to some colour function (d) Taking Inverse –Image of coloured bins forming clusters€ each cluster is treated as node of graph connected with edges capturing connectivity information showing there by compressed representation of circle.

Our Approach A.

1. Bin the Fatigue strength in to three bins (224.035, 546.667], (546.667,868.333],(868.333,1190.0]
2. Label the bins 0, 1, 2 respectively.
3. Split the whole data into input features and target features
4. Apply mapper algorithm to our dataset with target variable as fatigue strength labelled as 0,1,2
5. Construct mapper graph by trying various parameters of interest we have selected the best as per in table 2
6. Try colouring nodes of graphs using our target variable Fatigue strength
7. Graphs are also coloured using single proportion levels

The details of the approach is shown in the following diagrams. We use Kepler-mapper python implementation/Scikit-tda library. The details can be found in [13, 14].

IOP Conf. Series: Materials Science and Engineering **810** (2020) 012083 doi:10.1088/1757-899X/810/1/012083

Parameter	**Value**
Projection 1	Isolation forest (That detects the anomaly in the input data)
Projection 2	L2 norm
Projection	[Projection1,Projection2]
Clustering algorithm	K-means (with 5 clusters)
Colour function	Proportion of y-class labels
Percent overlap	70
Number of bins	15

Table 2: Mapper Parameters

Topological Network Representing the Steel Fatigue Dataset:

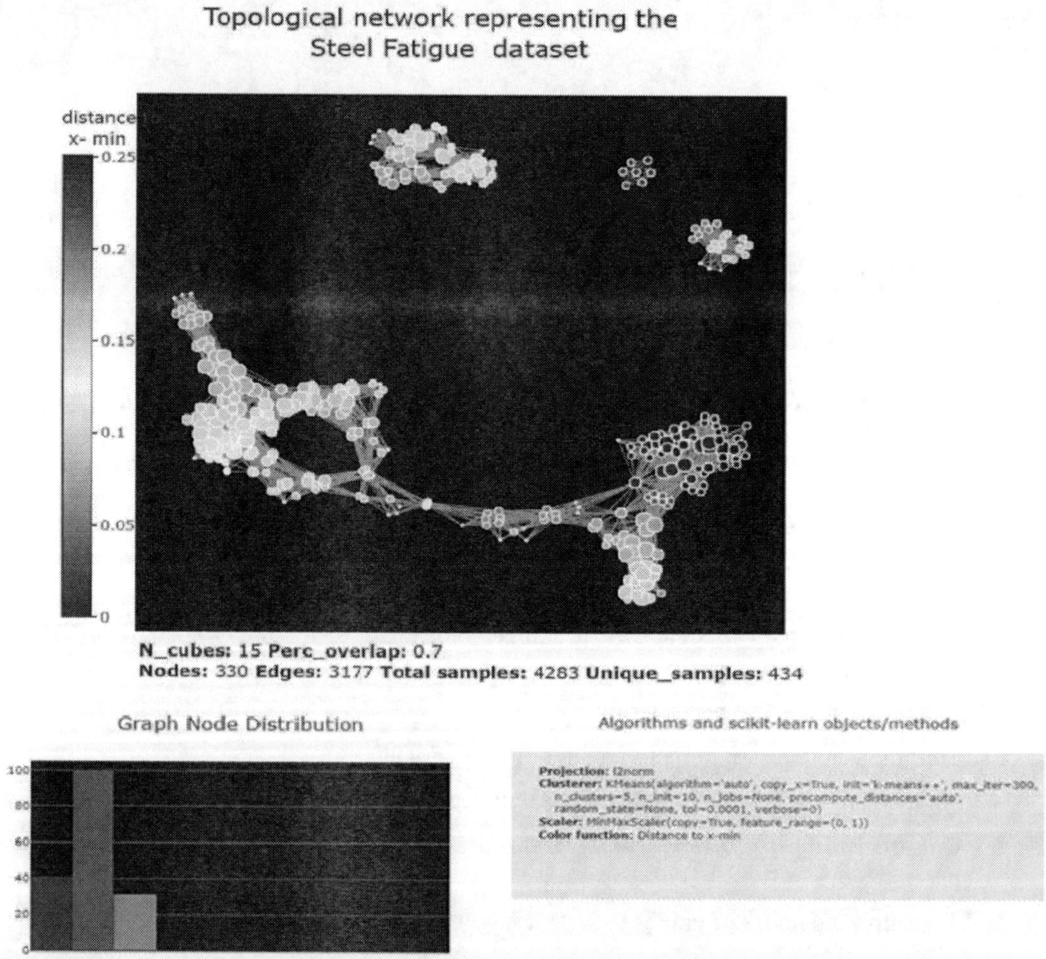

Figure 3: Topological Network Representing dataset Fatigue strength of steel with Graph Node distribution and information of network.

Following Diagram shows the same network with different clusters encircled, we also see interesting flares shown in blue circles, which is major finding of this work. Deep analysis of these flares and study of deep interaction between the circled parts of this network is beyond the scope of this paper and can be incorporated in the future research.

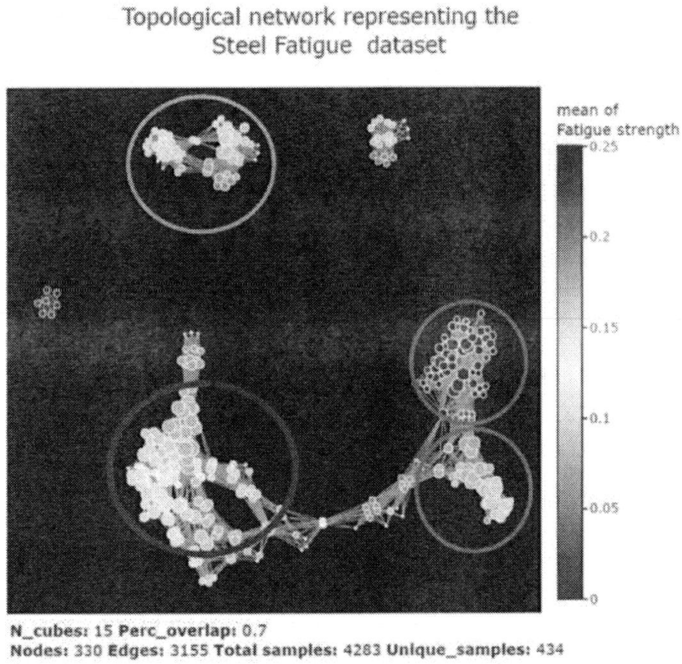

Figure 4: Topological Network Representing dataset Fatigue strength of steel with clusters& Flares

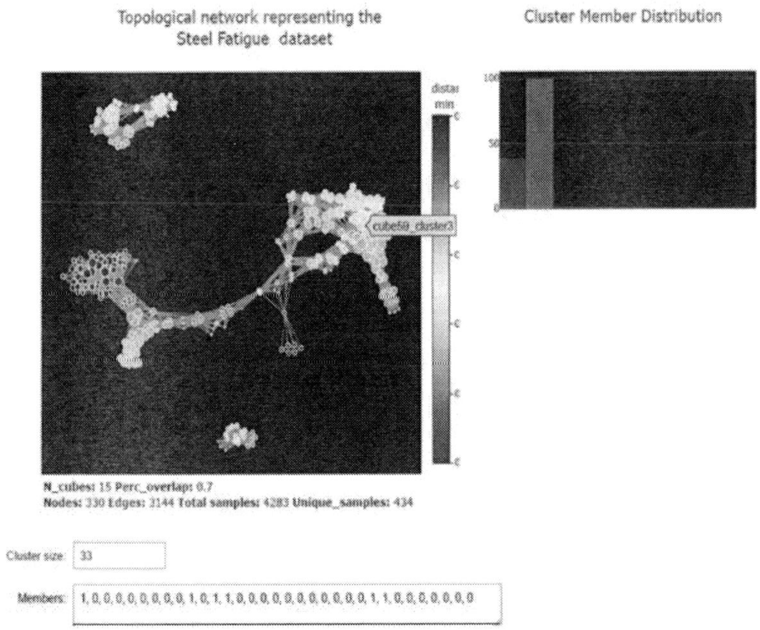

Figure 5: Topological Network Representing dataset Fatigue strength of steel with target variable distribution

IOP Conf. Series: Materials Science and Engineering **810** (2020) 012083 doi:10.1088/1757-899X/810/1/012083

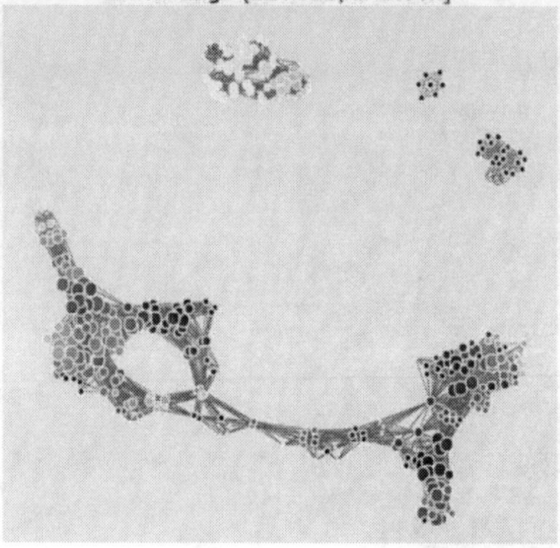

(a)

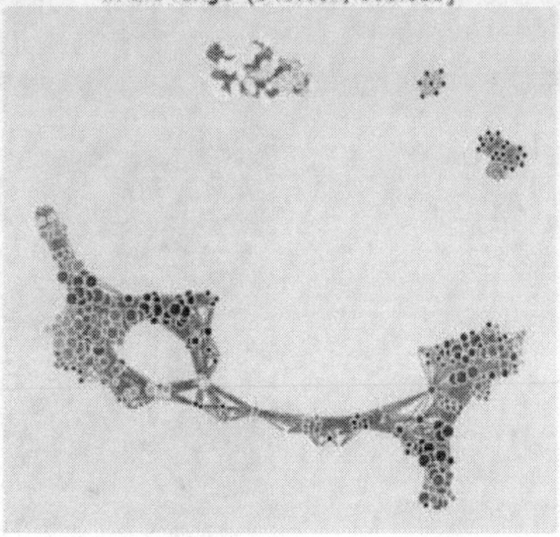

(b)

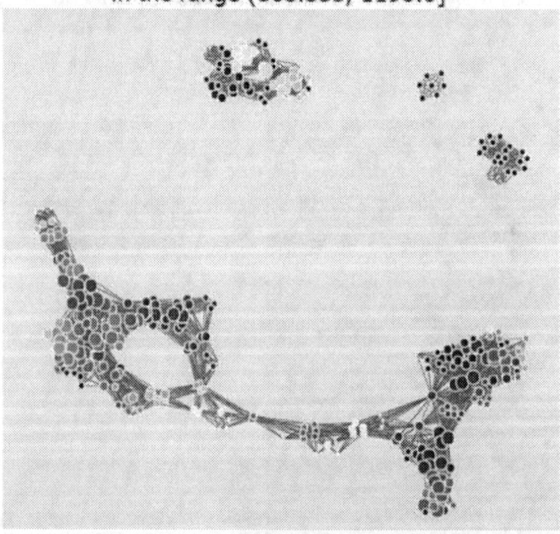

(c)

Figure 6: Nodes are coloured according to the production of fatigue strength

IOP Conf. Series: Materials Science and Engineering **810** (2020) 012083 doi:10.1088/1757-899X/810/1/012083

4. SAFE algorithm and Feature Selection

Once a TDA network has been constructed, network-based enrichment analysis can be performed for target variables, or properties of the studied data samples. We use SAFE for network enrichment analysis. This algorithm is originally developed as a systematic method for annotating biological networks and examining their functional associations.

The **SAFE** algorithm takes the following steps to calculate an enrichment score (SAFE score) of each node in a network for a given target variable:

1. Input TDA network graph and Target variable
2. For each node u in the network graph **SAFE** defines a local neighbourhood of u by identifying any other nodes that are closer than or equal to a maximum distance threshold d to u. Node distance is measured via shortest path length between nodes. By default, the maximum distance threshold d is set to be equal to the 0.5th percentile of all pairwise node distances in the network.
3. For each neighbourhood, SAFE sums the values of neighbour nodes for a target variable as an observed neighbourhood score S. The score is then compared with the distribution of permuted neighbourhood scores obtained by randomly shuffling the target variable among nodes in the network. The significance of enrichment P is determined as the probability that a random observation from the distribution will fall into the interval between the observed neighbourhood score S and the largest value of the permuted scores. The neighbourhood significance of enrichment p is converted into a neighbourhood enrichment score o, named SAFE score, as below (normalized to a range from 0 to 1):

$$P_n = \frac{m}{I}$$

Where m is the number of times a observed value is greater than or equal to shuffled values; I is the number of shuffles;

$$O_n = \frac{-log_{10}(max(P_n, \frac{1}{I+1}))}{-log_{10}\frac{1}{I+1}}$$

Where I is the number of shuffles; P is the neighbourhood significance of enrichment of node n; and O is the neighbourhood enrichment score **(SAFE score)** of node n. Random shuffle is performed independently for each target variable when there are more than one.

A node is considered to be significantly enriched under a p-value threshold of 0.05, which can be translated to a threshold of SAFE score by:

$$O_n \geq \frac{-\log_{10} 0.05}{-\log_{10} \frac{1}{I+1}}$$

Filter and rank target variables using number of significant nodes or sum of SAFE score of significant nodes. We use tmap package implemented in python to calculate safe score, the details can be found in [15]

IOP Conf. Series: Materials Science and Engineering **810** (2020) 012083 doi:10.1088/1757-899X/810/1/012083

Our Approach B.

1. Once the mapper graph is constructed we use SAFE score to select important variables
2. According to SAFE scores We see that in the following diagram S, THQCr, Ni, dA ,P, RedRatio, DT, SL.No, CU, Tt, Cr, NT, CT, Mo,. Mn, TT, DT DC,DB are important features

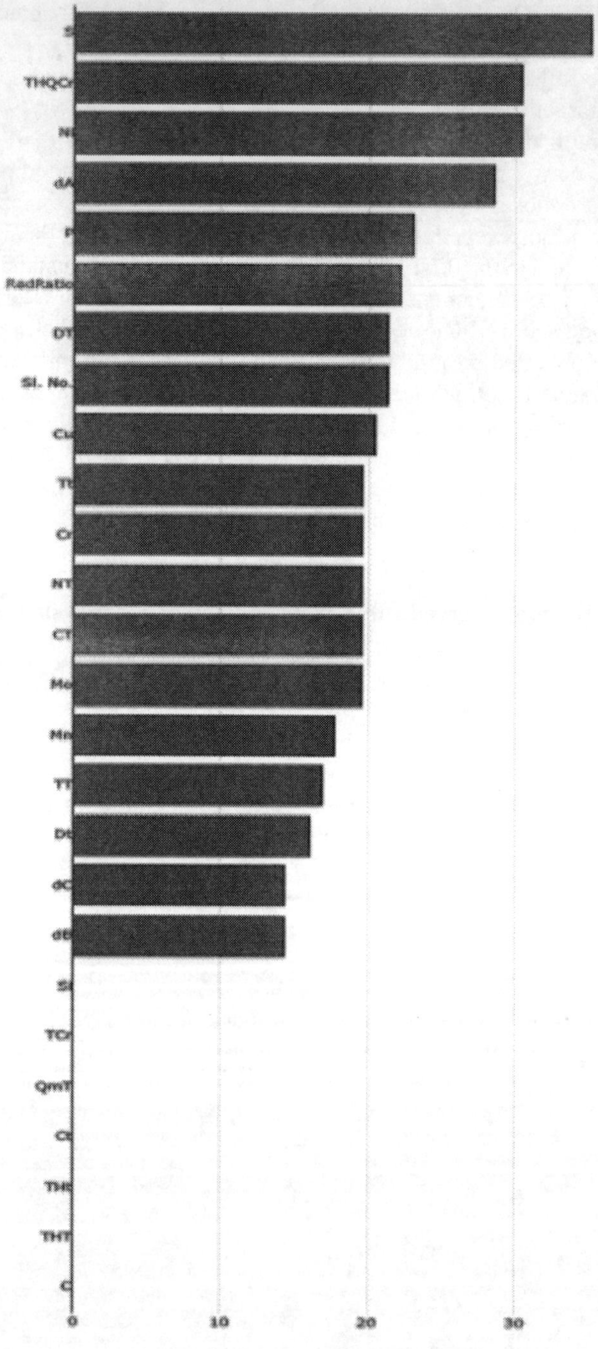

Figure 7: feature selection and ranking using SAFE Scores based on topological network with respect to target variable as steel fatigue strength

5. **Gradient Boosting (GB) Algorithm**

GB is a powerful machine learning algorithm it can do regression classification and ranking. Gradient Boosting has two steps Gradient Descent and Boosting. GB produces a prediction model by combining collection of weak prediction models also known as ensemble of weak learners, basically boosting Models comprises ensemble of base learners with Low variance and High Bias. The base learners are basically shallow decision trees. The core idea of Boosting is, given training data $\mathfrak{D}_{train} = \{x_i, y_i\}_{i=1}^{n}$ we train our First model say $\boldsymbol{M_0}$ on whole of our training data. $\boldsymbol{M_0}$ needs to have high bias and low variance ($\boldsymbol{M_0}$ could be shallow decision tree or any other model). High bias means high graining error. Let $\boldsymbol{M_0}: \boldsymbol{y} = \boldsymbol{h_0(x)}$ .Now we define an error at $\boldsymbol{i^{th}}$ data point to be simply $\boldsymbol{Error_i{=}y_i - h_0(x_i)}$. Now we have $\{x_i, y_i, error_i\}_{i=1}^{n}$ for each data point $\boldsymbol{i}$ for which $\leq \boldsymbol{i} \leq \boldsymbol{n}$, Now we will construct another model $\boldsymbol{M_1} : \boldsymbol{y} = \boldsymbol{h_1(x)}$ with new training data $\{x_i, error_i\}_{i=1}^{n}$, now our final model will become $\boldsymbol{F_1(x) = \alpha_0 h_0(x) + \alpha_1 h_1(x)}$. We may stop or we can go ahead as $\boldsymbol{F_2(x) = F_1(x) + \nu\, \alpha_2 h_2(x)}$ etc. Here we are building models in a stage wise fashion with optimizing the arbitrary differentiable loss function. Implementing gradient boosting in python or in R takes lot of time and space, so we need scalable/parallelizable, quick to execute tool to work. eXtreme Gradient Boosting or XG-Boost is a library of gradient boosting algorithms optimized for modern data science problems and tools. XGBoost does Parallel Computing, Regularization, Enabled Cross Validation, Missing Values, Flexibility, Availability, hese problems and typically out performs other algorithm. More about XG-Boost can be found in [8]. We give the pseudo code for gradient boosting s follows

1. Initialize $f_0(x) = \arg\min_\gamma \sum_{i=1}^{N} L(y_i, \gamma)$
2. For $m = 1$ to M:

   (a) For $i = 1, 2, \ldots, N$ compute the residuals

   $$r_{im} = -\left[\frac{\partial L(y_i, f(x_i))}{\partial f(x_i)}\right]_{f=f_{m-1}}$$

   (b) Fit a regression tree to the target r_{im} giving terminal regions R_{jm}, $j = 1, 2, \ldots, J_m$.

   (c) For $j = 1, 2, \ldots, J_m$ compute the prediction in the region R_{jm}

   $$\gamma_{jm} = \arg\min_{\gamma} \sum_{x_i \in R_{jm}} L(y_i, f_{m-1}(x_i) + \gamma).$$

   (d) Update $f_m(x) = f_{m-1}(x) + \sum_{j=1}^{J_m} \gamma_{jm} I(x \in R_{jm})$, where $I(x \in R_{jm}) = 1$ if $x \in R_{jm}$, otherwise $I(x \in R_{jm}) = 0$.

3. Output $\hat{f} = f_M(x)$.

Our Approach C.

1. Once important features are selected using SAFE scores, we use modified version of Gradient Boosting called XG-Boost
2. Hyper-parameters of XG-Boost are listed in the table 3
3. We optimize set of hyper-parameters of XG-boost using following optimization algorithms.
4. Grid search, Random search, genetic algorithm, differential evolution.
5. We describe details of Grid search in this paper in table 4
6. Grid search does the best job, it can be seen in table 5 and 6 reporting RMSE 23.6631, MAE 15.1262 and R^2 0.9888
7. Finally, we have plotted the predicted vs. actual

Parameter	**Description**
Nrounds (M)	Total number of trees
Eta($\mathbf{0 < \eta \leq 1}$)	Learning rate
Max_depth	Maximum depth of tree
Min_child_weight	Minimum number of samples in each node. The larger the value , the more conservative algorithm will be
Subsample ($\mathbf{0 < \omega \leq 1}$)	Proportion of samples used in each round, ω=1 indicates using all samples in the training set
Colsmple_bytreee ($\mathbf{0 < \nu \leq 1}$)	Proportion of variables used in each round, ν =1 indicates using all variables
Lambda ($\boldsymbol{\lambda \geq 0}$)	Ridge regularization (L2) term on weight
Alpha ($\boldsymbol{\alpha \geq 0}$)	Lasso regularization (L1) term on weight

Table 3: Hyper-Parameters in XG-Boost

Prameters	Description
Number of trees (nrounds)	[min = 50,max = 500,step=50]
Learning rate(eta)	[min = 0.01,max = 1,step=0.5]
Lambda # L2 Regularization (Ridge Regression)	[min = 0.1,max = 1,step=0.5]
Alpha # L1 Regularization (Lasso Regression)	[min = 0.1,max = 1,step=0.5]

Table 4: Grid search of best hyper-parameters for XG-Boost

5.1 Prediction and Results Using XG-Boost

Training Set Statistics and Hyperparameter Values of XGBoost Models.

Method	RMSE	MAE	Rsquared	nrounds	eta	lambda	alpha	Processing Time
Grid Search	29.76842	19.69660	0.97498	50	1.00000	0.10000	1.00000	3 mins
Random Search	29.75865	19.53965	0.97491	48	0.32730	0.09358	0.04431	1 mins
Genetic Algorithm	30.30175	19.95909	0.97416	887	0.00054	0.09803	0.08894	12 mins
Differential Evolution	31.23318	20.76231	0.97271	71	9.07471	0.04417	0.03711	31 mins

Note:
Summary statistics obtained using 5-fold cross validation repeated 3 times.
Grid and Random Search performed with adaptive resampling.

Table 5: XG-Boost performance with hyper-parameter tuning algorithms on train set

Test Set Statistics of XGBoost Models.

Method	RMSE	MAE	R-squared
Grid Search	23.6631	15.1262	0.9888
Random Search	24.3655	15.4806	0.9881
Genetic Algorithm	24.7095	15.8165	0.9879
Differential Evolution	23.8552	15.5625	0.9886

Table 6: XG-Boost performance with hyper-parameter tuning algorithms on test set

IOP Conf. Series: Materials Science and Engineering **810** (2020) 012083 doi:10.1088/1757-899X/810/1/012083

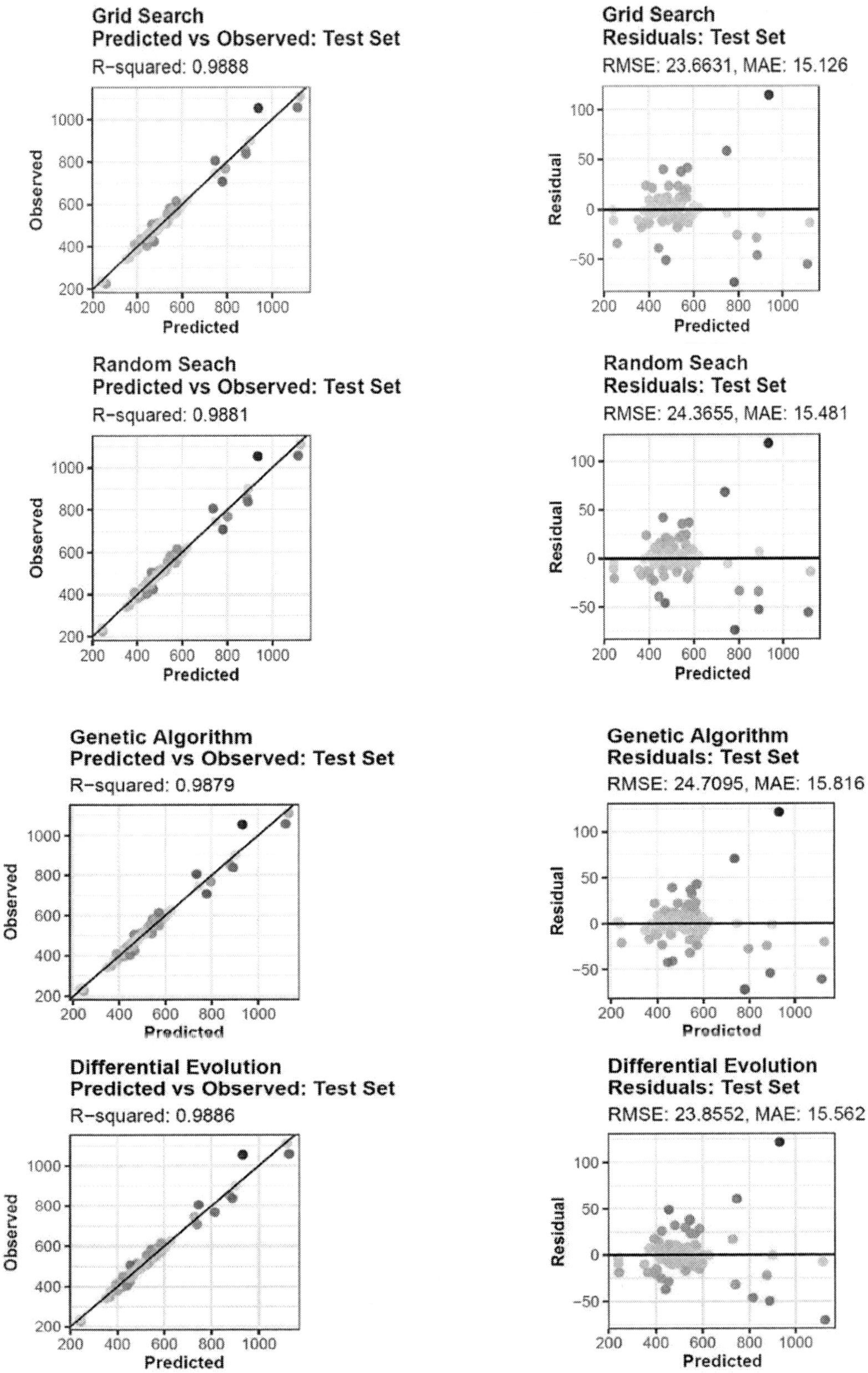

Figure 8. Plotting predicted vs. actual for all optimization algorithms

Conclusion: In this paper we have successfully applied topological data analysis method called mapper to select key process variables that affects the fatigue strength of steel using SAFE scored method which is used for enrichment analysis in biological network. Further we have built prediction model XG-Boost as regression model based on selected parameters, which shows pretty good accuracy. So we have effectively utilize TDA and Machine learning to build prediction model. Further the network graph can be analysed into more details, for example in this paper we did not talk about the clusters and the flares, which can be explored in a great detail to get more insight of data as an extension of this research in future.

References

[1] Dieter GE: Mechanical Metallurgy. Mc Graw-Hill Book Co. 3rd edition, ISBN: 0–07–016893–8 26. Deshpande PD, Gautham BP, Cecen A, Kalidindi S, Agrawal A, Choudhary (2013) Application of Statistical and Machine Learning Techniques for Correlating Properties to Composition and Manufacturing Processes of Steels. In *2nd World Congress on Integrated* Computational Materials Engineering. John Wiley & Sons, Inc.; 1986:155–160. ISBN: 9781118767061 ISBN: 9781118767061

[2] G. Carlsson, "Topology and data", Bull. Amer. Math. Soc., vol. 46, no. 2, pp. 255-308, 2009

[3] Edelsbrunner H, Harer J. Computational topology: an introduction [M]. American mathematical Soc., 2010

[4] Lum P.Y, Singh G, Lehman A , et al. Extracting insights from the shape of complex data using topology{j} Scientific reports , 2013, 3:serp01236

[5] Gurjeet Singh, Facundo Mémoli and Gunnar Carlsson, *Topological Methods for the Analysis of High Dimensional Data Sets and 3D Object Recognition*, Eurographics Symposium on Point Based Graphics, European Association for Computer Graphics, 2007, pp. 91–100

[6] Baryshnikova A. Systematic functional annotation and visualization of biological networks [J]. Cell systems, 2016, 2(6): 412-421.

[7] Friedman, J. H. Greedy function approximation: a gradient boosting machine. *Ann. Stat.* **29**, 1189–1232 (2001).

[8] Chen, Tianqi and Guestrin, Carlos A Scalable Tree Boosting System Proceedings of the 22nd ACM SIGKDD International Conference on Knowledge Discovery and Data Mining.

[9] URL: National Institute of Materials Science. http://smds.nims.go.jp/fatigue/index_en.html .

[10] Agrawal, A., Deshpande, P.D., Cecen, A. et al. Exploration of data science techniques to predict fatigue strength of steel from composition and processing parameters Integrating Materials (2014) 3: 90.

[11] W. Guo and A. G. Banerjee, "Toward automated prediction of manufacturing productivity based on feature selection using topological data analysis," *2016 IEEE International Symposium on Assembly and Manufacturing (ISAM)*, Fort Worth, TX, 2016, pp. 31-36.

[12] Rajan K, Suh C, Mendez P: Principal component analysis and dimensional analysis as materials informatics tools to reduce dimensionality in materials science and engineering. Stat Anal Data Min 2009, 1: 361–371. 10.1002/sam.10031

[13] van Veen et al., (2019). Kepler Mapper: A flexible Python implementation of the Mapper algorithm. Journal of Open Source Software, 4(42), 1315

[14] Hendrik Jacob van Veen, Nathaniel Saul, Eargle, David, & Sam W. Mangham. (2019, October 14). Kepler Mapper: A flexible Python implementation of the Mapper algorithm (Version 1.3.3). Zenodo. http://doi.org/10.5281/zenodo.3485851

[15] Tianhua Liao, Yuchen Wei, Mingjing Luo, Guoping Zhao, Haokui Zhou *tmap*: topological analysis of population-scale microbiome data bioRxiv 396960;

Understanding of Defect in Photonic Crystal using Solitons and Spectral Theory

Vivek S. Sharma

Department of Applied Science and Humanities, MPSTME NMIMS, Shirpur, India
Email id: vivek.sharma@nmims.edu

Abstract. The concept of mechanical deformation is generally understood in terms of propagation of crystal defect. The deformability of the crystals such as uranium can be studied if its local vibrations are known. These local vibrations also give insight into the nature of the defect if it's dynamics and bifurcation are known. One such local vibration is soliton or topological soliton. Solitons also form solutions to the differential equations modelling thermophoretic motion in superconductors such as graphene and traveling wave type solutions to elastic rods with deformations. Also spectral theory is widely used to study defects in structures. In this paper, soliton theory and spectral theory approaches are combined and analytical and numerical analysis is carried out for a 10*7 um dielectric material with 1.5 refractive index. The defect is introduced by changing the refractive index, changing the dimensions of one of the dielectric rod in the crystal and removing one of the dielectric rods using FDTD simulations. The results obtained are new and have not been reported in literature.

Keywords: Defect Analysis, Solitons, PDEs, Sine-Gordon Equations, Maxwell's Equations, Resonance Photonic Crystals, Photonic Band Gap Crystals.

1. Introduction

The solitons are exceptionally useful structures that widely occur in nature. The unique characteristic of solitons is their shape preserving nature. Because of this reason, they have applications in numerous domains of science and engineering such as fiber optics, plasma flows, fluid dynamics, high-energy physics, chemical kinematics, chemical physics and geochemistry. Crystal defect modeling for non-linear periodic systems is a recent branch of engineering studied using nonlinear partial differential equations (PDEs) having solitonic solutions. These problems are of utmost importance from both theoretical and experimental point of view. Three most widely used PDEs for studying such systems are the Maxwell's equations, the nonlinear Schrodinger equations and the non-linear Sine-Gordan equations.

The solitonic solutions to Sine-Gordan equations particularly explains diverse range of physical phenomenon. The Resonance photonic crystal (RPhC) are one of such systems where light propagation is described by Sine-Gordan Equations [1], [2]. The non-linearity has its own importance as it detunes the band gap locally as energy can propagate in the form of solitary waves [3]. Igor et al. [4] have shown for the first time that Gap Solitons exists in resonant photonic crystals. Gap solitons are stable in the deep of gap and near the band edge they feature specific oscillatory instabilities similar to those observed in fiber Bragg gratings [6], [7]. They have used Schrodinger equations [6] and Laplace equations [7] to explain the propagation of light. Photonic crystals with resonant nonlinearities have applications in manufacturing of periodic bragg semiconductor Nano-structures.

Whereas, Schrodinger's equations are basic quantum mechanical equations used to model wave-particle nature of photons. Soliton solutions to these equations are extensively studied in literature. These equations along with Maxwell's equations are widely used in the study of photonic band gap crystals (PBG). The Photonic Bandgap (PBG) materials are dielectric structures which are periodic in nature. They form the energy structure of photons allowing or forbidding the propagation of electromagnetic waves having very specified ranges of frequency. The working of PBGs crystals is similar to that of periodic potential for electrons in atomic crystals as discussed by Maka et al. [5]. In general, a defect is any interruption of the translation symmetry [8]. The defects can be either point defects or extended defects including rows or planes of atoms.

This paper is arranged as follows: In section 1, the resonance photonic crystals (RPhc) and photonic band gap crystals [PBG] are briefly discussed. In section 2, the Maxwell equation modelling defect are discussed. Also the design specifications of the simulated environment using Optiwave software are discussed. In section 3, FDTD analysis of the defect is carried out and the results are presented. In section 4, Graphs and Plots of the observer point and plane with and without defect are plotted and specifications discussed. In section 5, the simulations of the photonic crystal with and without defects are compared. In section 6, conclusions and inferences are drawn and future directions using solitons are discussed.

2. THE MAXWELL'S WAVE EQUATION IN BRIEF

One of the most important theoretical analogy between photonic and solid state crystals is the band of photons. It enables one to create two widely applicable features of photonic band gap and localization of light [9]. In solid state physics one of the most fundamental equations (Maxwell's wave equation) is changed into an eigenvalue equation having a periodic boundary condition and is solved using computations. Also the photonic Band-Gap Crystals contains periodic array of optical materials called Photonic Band Gap (PBG) fibers. Particle physics at smaller scale can be studied using higher accelerating gradients. These accelerating gradients can be created by small structures with large field strengths [3], [21]. Therefore, in a photonic crystal, light cannot enter the crystal and photons cannot be emitted by it. Implying that the doping of an optical item with impurity or defect creates the opposite situation of the strong localization of the resonant photons [9]. The term band gap means a range of energies and the direction in which the electrons are not allowed to travel [11]. As a photonic bandgap [PBG] crystal behaves like an insulator of light, flexible and dense photonic circuit such as an electronic LSI [Large Scale Integrated circuit] can be created. The light velocity in PBG crystal is changed from vacuum velocity **c** to zero (stopping condition) so that the light matter interaction can be arbitrarily controlled. Point defects may be defined within PBG materials to build embedded optical cavities. Such cavities which contains an emitting material may be used to inhibit spontaneous emission [10]. One of the exiting domains where photonic crystals has wide applications is optical biosensors [12]. In photonic crystals, if the refractive indices of dielectrics differ greatly, the modulation of the light waves would be sufficient, and so a photonic bandgap may form, in which optical models are absent [16]. The variation of refractive indices is the most common way to introduce defect in a dielectric photonic crystal. The other ways that can be used to introduce a point defect are **1.** Removal of a dielectric rod from the structure and **2.** Changing the dimensions of a dielectric rod. In this section, firstly, a brief introduction to Maxwell's equations which forms basis of the Finite Domain Time Difference (FDTD) method applied here to analyze the defects in the photonic bandgap crystal is given. Further, the design specifications of a photonic bandgap crystal using Optiwave software are given.

The Maxwell's Equations

The Maxwell's equations are four equations which together form a complete description of the production and interrelation of electric and magnetic fields. If the free space propagation model is assumed, then these equations become

$$\nabla.\boldsymbol{E} = \frac{\rho}{\epsilon_0} = 4\pi k\rho$$

$$\nabla.\boldsymbol{B} = 0 \qquad (1)$$

$$\frac{\partial \mathbf{E}}{\partial t} = \frac{1}{\epsilon_0} \nabla \times \boldsymbol{H} \quad and$$

$$\frac{\partial \mathbf{H}}{\partial t} = -\frac{1}{\mu_0} \nabla \times \boldsymbol{E} \tag{2}$$

Where, **E** = Electric field, **B** = Magnetic Field, **H** = Magnetic Field Strength, μ_0 = Permeability of the medium and ϵ_0 = Permittivity of free space. Where equation (1) specifically is Gauss law of Electricity and Magnetism and equation (2) is Faraday's Law of Induction. The constant $k = \frac{1}{4\pi\epsilon_0}$ = Coulomb's Constant. With $\nabla.\boldsymbol{E}$ and $\nabla \times \boldsymbol{E}$ and similarly $\nabla.\boldsymbol{B}$ and $\nabla \times \boldsymbol{H}$ representing the vector operations of divergence and curl respectively.

Of the above two equations, the equation (2) is discretized using lee's scheme [17] following leap-frog algorithm for implementation of Finite Domain Time Difference [FDTD] Scheme. The details of this scheme can be found in [17].

3. Design Specifications of the Photonic Band Gap (PBG) Crystal

The layout of the Photonic bandgap crystal is designed using waveguides. An anisotropic dielectric material with wafer dimensions of 10*7 um is Consider. The refractive index of the 2-dimensional wafer dielectric material used being 1.5 and with PBG atom have unit width and thickness. The PBG lattice atom is of 2-dimensional hexagonal type with elliptic waveguide. Two input planes are considered with the photonic source emitting a sine-modulated Gaussian pulse with wavelength of 1.55 um. Three observation points and an observation area is selected on the design. As seen in Figure 1(a), 1(b) and 1(c), the 6 equally spaced dielectric rods are used in the design. The design has three observers and an observation area A.

IOP Conf. Series: Materials Science and Engineering **810** (2020) 012084 doi:10.1088/1757-899X/810/1/012084

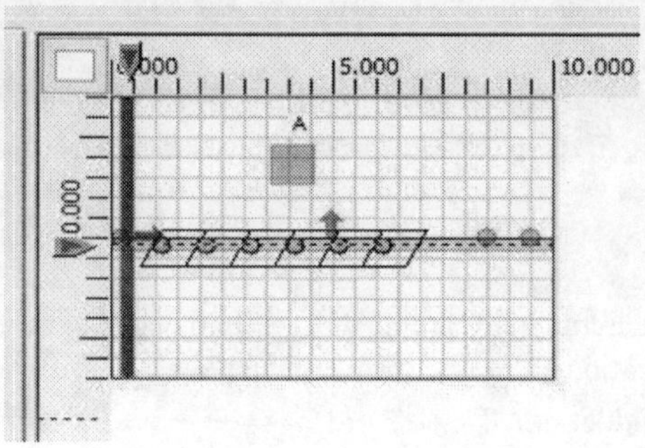

Figure 1(a). Layout of Designed Photonic Bandgap Crystal

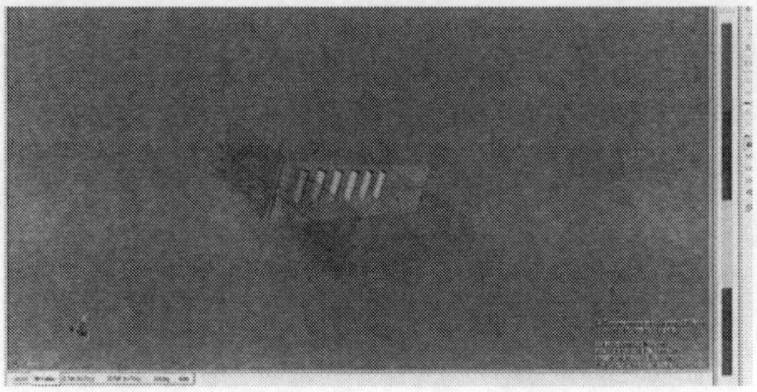

Figure 1(b). 3-Dimensional view of the designed PBG crystal

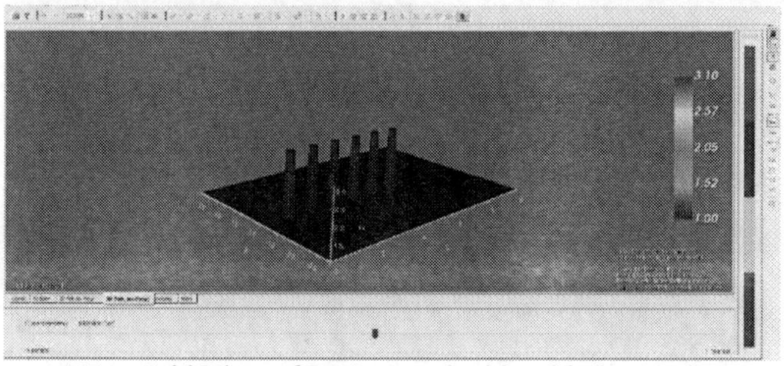

Figure 1(c). Grid View of PBG crystal with grid size 101*71.

Simulation Specifications and detector analysis of the Photonic Band Gap (PBG) Crystal

The design environment is 2-D TE(s-polarized) simulated with mesh size of 0.05*0.05 and for 12000 time steps and the simulation images are obtained and the detector analysis of PBG crystal is carried out at the three observation points and the observation area. In detector analysis all observation points at ordinates (0.00, 0.00, 0.25), (0.00, 0.00, 8.5) and (0, 0, 9.5) are selected and spectral analysis is done using Discrete Fourier Transform (DFT). These normalized frequency Discrete Fourier Transform graphs are plotted with wavelength range from 1.5 um to 2.3 um and at 1000 sample points Fig 2 (a). The observation area integral surface is obtained for DFT with wavelength 1.9 um and center position being (1.78, 0.00, 4.11) as shown in Fig 2(b) and Fig 2(c).

IOP Conf. Series: Materials Science and Engineering **810** (2020) 012084 doi:10.1088/1757-899X/810/1/012084

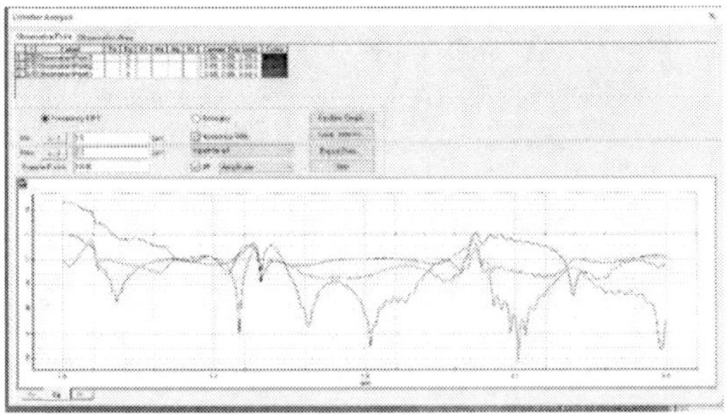

Figure 2(a). Discrete Fourier plot at observation points

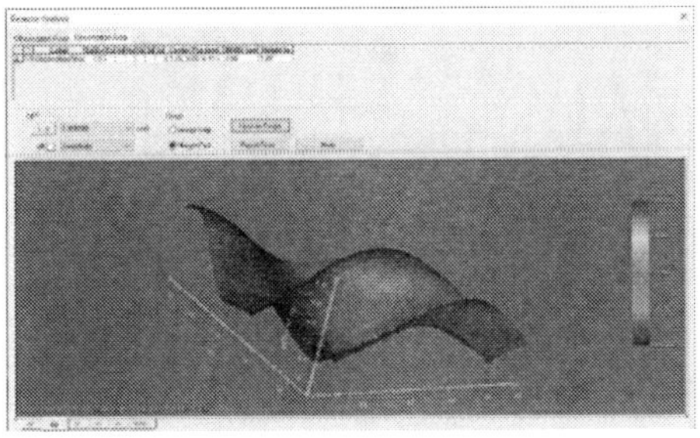

Figure 2(b). Integral Surface plot for the observation area

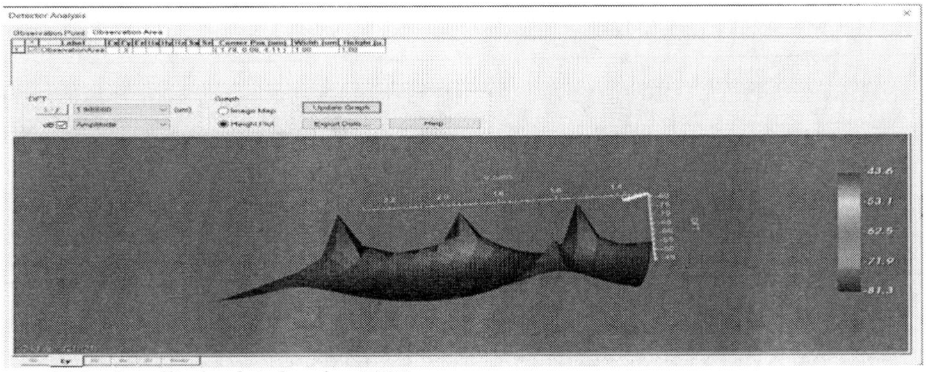

Figure 2(c). Amplitude plot showing DFT

The DFT plots show that spectral plots in form of Discrete Fourier Transform (DFT) give an insight into the output of the FDTD scheme for dielectric crystal considered. This crystal is without any defect in it. In the next section, the PBG crystal is designed with point defects

i By changing the major and minor axis radius of a cell for the dielectric crystal

ii. By removing a dielectric rod from the crystal lattice

and the environment is simulated. Further detector analysis is carried out similar to crystal without defect. Graphs and Plots of the observer points and plane are plotted with defects.

4. Defect Study of the Photonic Bandgap Crystal

The defect in the photonic bandgap crystal can be introduced by any of the following three ways:

i. By changing the major and minor axis radius of a cell for the dielectric crystal

ii. By removing a dielectric rod from the crystal lattice.

iii. By changing it's refractive index.

Of the above three ways, the defects are introduced by using first two approaches.

Study of defect by changing radius of major and minor axis

The major radius and minor radius of the cell for dielectric rods with ordinates (0, 0, 2) are changed from both 0.2 to new major radius of 0.3 and new minor radius of 0.25. The design and simulations thus obtained are noted in Fig 3(a, b, c).

Fig 3(a, b, c) depict Design and Simulation of 2D Photonic Bandgap Crystal with defect introduced by changing major and minor radius in cell (0,0,2).

Further, the normalized Discrete Fourier Transform (DFT) graphs are plotted for the detector points and area A. These DFT values are plotted for all the three observers and the observer plane. The integral surface and amplitude plots are plotted for the DFT in Figure 4(a, b, c).

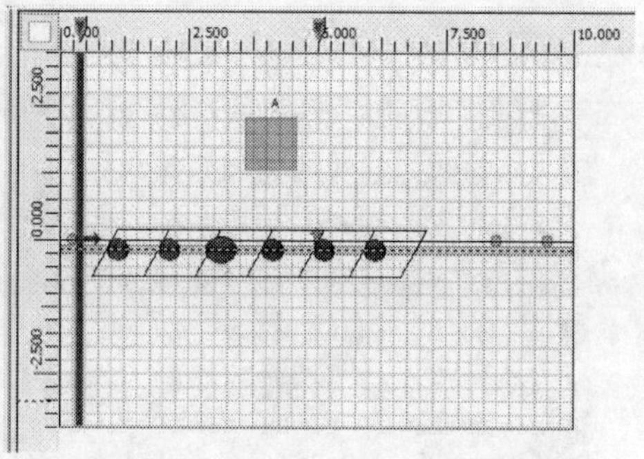

Figure 3(a). Design of defective PBG crystal by removing dielectric rod at (0,0,3)

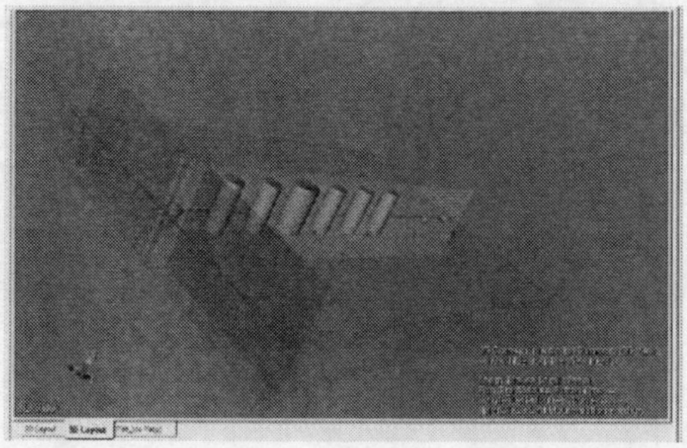

Figure 3(b). 3-Dim. Layout of defective PBG crystal by by removing dielectric rod at (0,0,3)

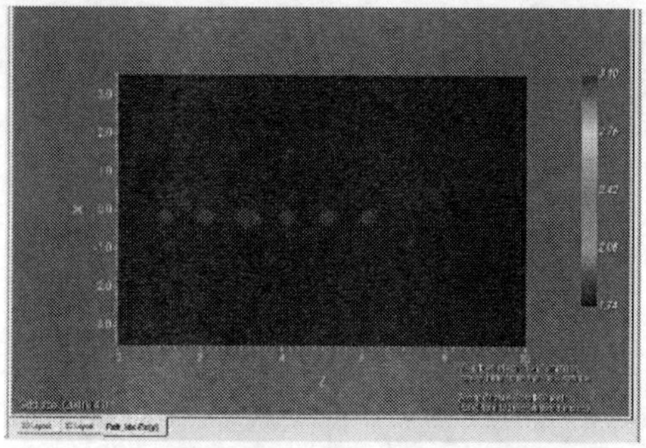

Figure 3(c). Simulation of 2D PBG Crystal with defect introduced by removing a dielectric rod at positon cell (0,0,3)

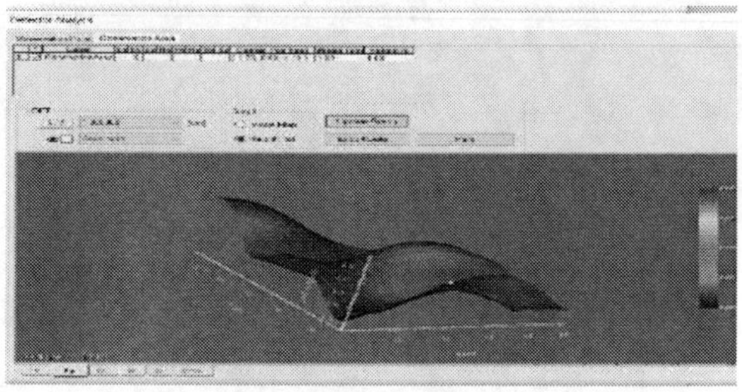

Figure 4(a). The integral surface induced with defect by changing major and minor radius in cell (0, 0, 2)

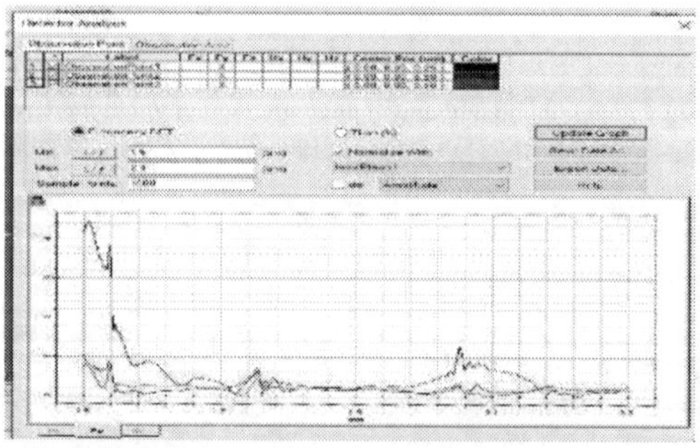

Figure 4(b). The DFT plot for the PBC induced with defect by changing major and minor radius in cell (0, 0, 2)

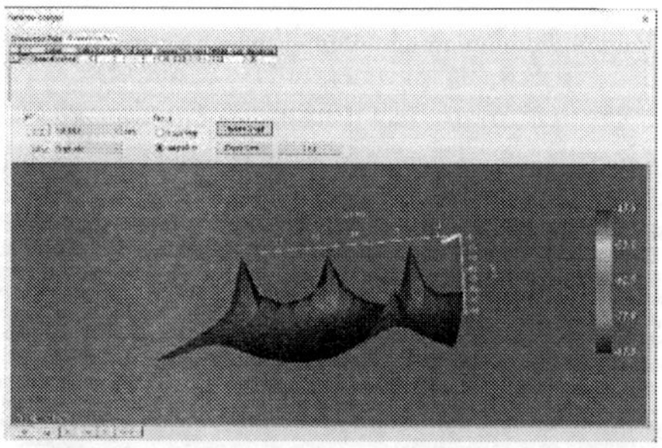

Figure 4(c). The Amplitude plot for the PBC induced with defect by changing major and minor radius in cell (0, 0, 2)

5. **Study of defect by removing one of the dielectric rod in the PBC lattice**

The major radius and minor radius of the cell for dielectric rods with ordinates (0, 0, 2) are not changed and they remain 0.2 but one of the dielectric rod is removed from the position (0, 0, 3) . The design and simulations thus obtained are noted in Fig 5(a, b, c).

IOP Conf. Series: Materials Science and Engineering **810** (2020) 012084 doi:10.1088/1757-899X/810/1/012084

Further, the normalized Discrete Fourier Transform (DFT) graphs are plotted for the detector points and area A. These DFT values are plotted for all the three observers and the observer plane. The integral surface and amplitude plots are plotted for the DFT in Figure 6(a,b,c).

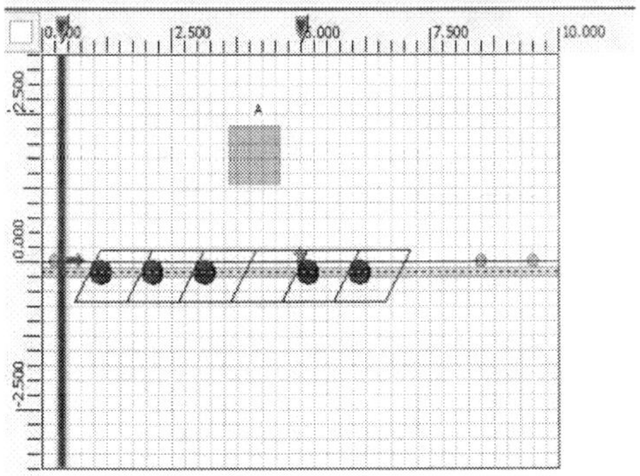

Figure 5(a). Design of defective PBG crystal by by removing dielectric rod at (0,0,3)

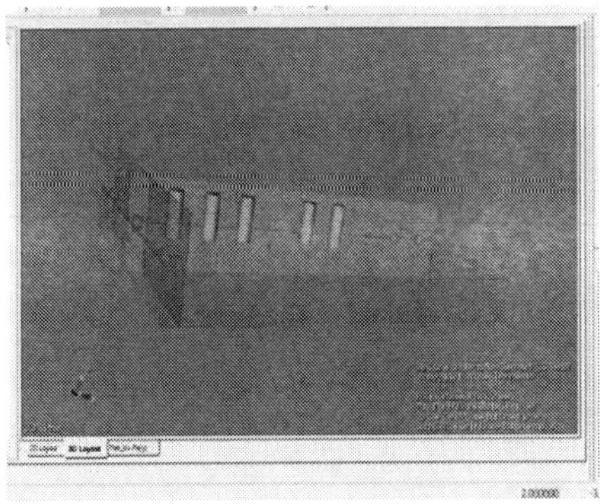

Figure 5(b). 3-Dim. Layout of defective PBG crystal by by removing dielectric rod at (0,0,3)

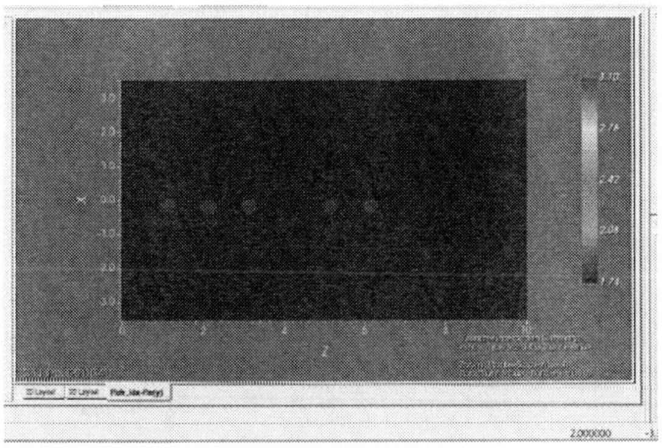

Figure 5(c). Simulation of 2D Photonic Bandgap Crystal with defect introduced by removing a dielectric rod at positon cell (0,0,3)

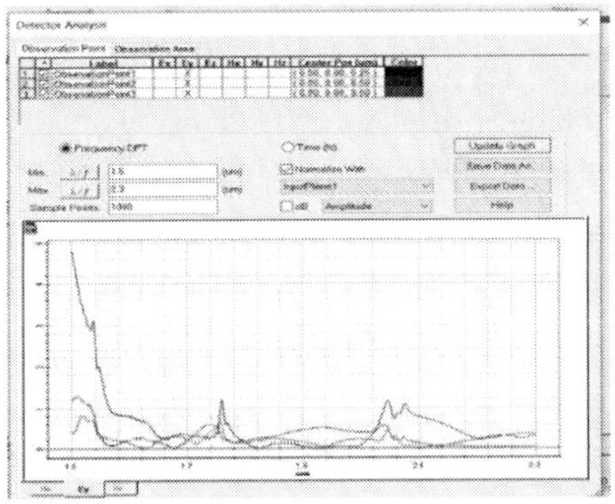

Figure 6(a). The DFT plot for PBC with defect at (0,0,3)

IOP Conf. Series: Materials Science and Engineering **810** (2020) 012084 doi:10.1088/1757-899X/810/1/012084

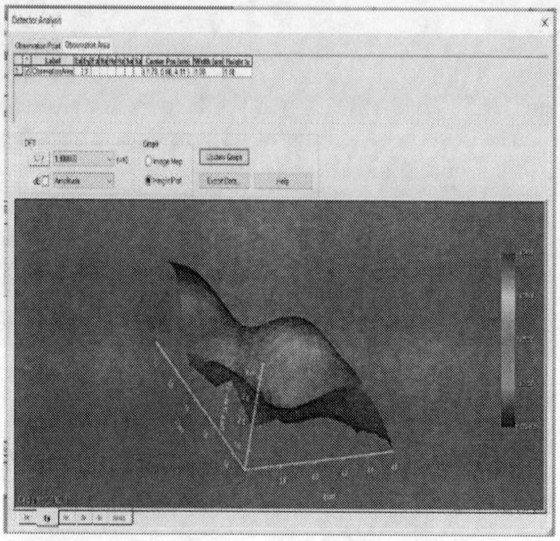

Figure 6(b). The Integral Surface plot for PBC with defect at (0,0,3)

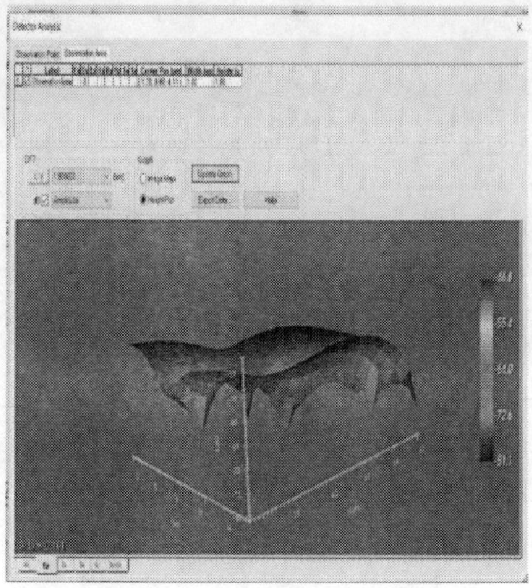

Figure 6(c). The amplitude plot of DFT for the PBC induced with defect by removing the dielectric rod at cell (0,0,3)

6. Remark

Spatial Kerr Soliton solutions exist for Maxwell equation discussed in equations (1) and (2) in the form of both in bright and dark version, as TM [19]. These solutions are linearly polarized, exactly integrable one-dimensional solitons. In two-dimensional solids, these solutions are obtained as azimuthally polarized and circularly symmetric dark solitons. In the current simulations of Photonic Band Gap Crystal (PBG) and its defects it is observed the integral surfaces obtained seem to have dark soliton type solutions which can be inferred by their DFTs. It also seems to be in agreement with the analytical results obtained by Ciatonni et al. [19] and Snyder et al. [20].

7. Conclusion and Future Research

The design and simulation of crystal defects in Photonic Band Gap crystal is carried out in this work. The defects considered are introduced by increasing the radius of major and minor axis of the cell of dielectric crystal and by removing a dielectric rod from the PBG crystal. The simulations are carried out using optiwave software assuming three observers and one observation plane. These simulations are new and have not been reported in literature.

8. References

[1] I.V.Mel'nikov , A. Knigavko, J.S. Aitchison and C.A.Merchent, "Generation of slow intense optical solitons in a resonance photonic crystal," .Phys. J. Special Topics, **160**, pp 301-308 (2008). doi: 10.1140/epjst/e2007-00733-8.

[2] I.V.Mel'nikov and J. Stewart Aitchison, "Gap soliton memory in a resonalnt photonic crystal", Applied Physics Letters **87** 201111(2005) : doi: 10/1063/1.2132070.

[3] Elliott C.Johnson, "Introducing Defects in Photonic Band-Gap (PBG) Crystals", Office of Science, Science undergraduate laboratory Internship (SULI), North Dakota State University, Stanford Linear Accelerator Center, Stanford, CA. August 2007.

[4] Igor V.Mel'nikov, J.Stewart Aitchson, "Gap soliton gating, dissociation, and retrival via defect mode exitation in a resonant photonic crystal", Nonlinear Guided waves and their applications. Dersden Germany 6-9 September 2005 ISBN: 1-55752-791-1.

[5] T. Maka, D.N.Chigrin, S.G.Romanov and C.M. Sotomayor Torres , "Three dimensional photonic crystals in the visible regime", Progress in Electromagnetics Research, PIER 41, pp 307-335, 2003.

[6] Yaroslav V.Kartashov, Victor A.Vysloukh and Lluis Torner, "Soliton, Shape and Mobility Control in Optical Lattices", Progress in Optics, Volume 52, 2009,63-148 https://doi.org/10.1016/S0079-6638(08)00004-8.

[7] Shuai Sun, Mario Miscuglio, Engin Kayraklioglu, Chen Shen, Jeffery Anderson, Tarek El-Ghazawi, Bahram Jalali, Volker J. Sorger, "Analog Photonic Computing Engine as Approximate Partial Differential Equation Solver", arXiv:1911.00975v1, 3 November 2019.

[8] Vorgelegt Von, "Defect Analysis in Aluminium Nitride", Ph.D. Thesis, Berlin, 2010.

[9] Toshihiko Baba, "Photonic Crystals-An Introduction", *https://www.jsps.go.jp › j-bilat › fos_ja › data › jishi_06 › abst08*

[10] Paul V.Braun, Stephanie A. Rinne and Florencio Gracia-Santamaria, "Introducing Defects in 3D Photonic Crystals: State of the Art", Advanced Materials, Wiley InterScience, 2006, 18,2665-2678.

[11] Guihua Chen, Shaoqiang Zhang and Muying Wu, "Optical Solitons in a Trinal Channel inverted nonlinear photonic crystal", Journal of Nonlinear Optical Physics & Materials Vol. 22, No.1 (2013) 1350012. https://doi.org/10.1142/S0218863513500124.

[12] Giampaolo Pitruzzello and Thomas F. Krauss, "Photonic crystal resonances for sensing and imaging", J.Opt. 20 (2018) 073004 (23pp)

[13] Yufeng Song, Xujie Shi, Chengfa Wu, Dingyuan Tang and Han Zhang "Recent progress of study on optical solitons in fiber lasers", Appl. Phy. Rev. 6, 021313 (2019). https://doi.org/10.1063/1.5091811.

[14] H. Jaeger and J.V. Sanders, "Surface Structure of defects in crystals". J. Res. In st. Catalysis, Hokkaido Unizo., Vol. 16, No.1, pp. 287 to 808 (1968)

[15] F.N. van de Vosse, P.D. Minev, "Spectral element methods: theory and applications", (EUT report W, Department of Mechanical Engineering, Vol 96-W-001) Eindhoven, Eindhoven University of Technology, ISSN 0167-9708

[16] Biao Wang, Guang Tao Fei, Min Wang, Ming Guang Kong and Li De Zhang, "Preparation of photonic crystals made of air pores in anodic alumina", IOP Publishing, Nanotechnology 18 (2007) 365601 (4pp).

[17] Archambeault B., Ramahi O.M., Brench C., "The Finite-Difference Time Domain Method", In: EMI/EMC Computational Modeling Handbook. Springer, Boston, MA (1998).

[18] S. Chakravarty, "Soliton solutions of coupled Maxwell-Bloch equations", Physics Letters A, Elsvier, 2015. https://doi.org/10.1016/j.physleta.2015.10.031.

[19] Alessandro Ciattoni, Bruno Crosignani, Paolo Di Porto, and Amnon Yariv, "Perfect Optical Solitons: spatial kerr solitons as exact solutions of Maxwell's equations", Journal of the Optical Society of America B, Vol. 22, Issue 7, pp 1384-1394 (2005).

[20] Allan W. Snyder, D. John Mitchell, and Yijiang Chen, "Spatial Solitons of Maxwell's equations", Optics Letters, Vol. 19, Issue 8 pp 524-526 (1994). https://doi.org/10.1364/OL.19.000524

[21] Igor V.Mel'nikov, Anton N.Knigavko, J. Stewart Aitchson and Clark A. Merchant "Slow light generation in a resonance photonic crystal", Proceedings Volume 6904, Advances in Slow and Fast Light: 69040A (2008) https://doi.org/10.1117/12.763068.

AUTHOR INDEX